AF251441

Applied Computational Aerodynamics

Edited by
P. A. Henne
Douglas Aircraft Company
McDonnell Douglas Corporation
Long Beach, California

Volume 125
PROGRESS IN
ASTRONAUTICS AND AERONAUTICS

A. Richard Seebass, Editor-in-Chief
University of Colorado at Boulder
Boulder, Colorado

Published by the American Institute of Aeronautics and Astronautics, Inc.
370 L'Enfant Promenade, SW, Washington, DC 20024-2518.

American Institute of Aeronautics and Astronautics, Inc.
Washington, DC

Library of Congress Cataloging in Publication Data

Applied computational aerodynamics/edited by P. A. Henne.
  p.cm. – (Progress in astronautics and aeronautics;v.125)
  1. Aerodynamics.   I. Henne, P. A.   II. Series.
TL507.P75   vol. 125   90-78321
[TL570]
629.1 s – dc20
[629.132'3]
ISBN 0-930403-69-X

Copyright © 1990 by the American Institute of Aeronautics and Astronautics, Inc. All rights reserved. Reproduction or translation of any part of this work beyond that permitted by Sections 107 and 108 of the U.S. Copyright Law without the permission of the copyright owner is unlawful. The code following this statement indicates the copyright owner's consent that copies of articles in this volume may be made for personal or internal use, on condition that the copier pay the per-copy fee ($2.00) plus the per-page fee ($0.50) through the Copyright Clearance Center, Inc., 21 Congress Street, Salem, Mass. 01970. This consent does not extend to other kinds of copying, for which permission requests should be addressed to the publisher. Users should employ the following code when reporting copying from this volume to the Copyright Clearance Center:

0-930403-69-X/90 $2.00 + .50

Progress in Astronautics and Aeronautics

Editor-in-Chief
A. Richard Seebass
University of Colorado at Boulder

Editorial Board

Richard G. Bradley
General Dynamics

John R. Casani
California Institute of Technology
Jet Propulsion Laboratory

Allen E. Fuhs
Carmel, California

George J. Gleghorn
TRW Space
and Technology Group

Dale B. Henderson
Los Alamos National Laboratory

Carolyn L. Huntoon
NASA Johnson Space Center

Reid R. June
Boeing Military Airplane Company

John L. Junkins
Texas A&M University

John E. Keigler
General Electric Company
Astro-Space Division

Daniel P. Raymer
Lockheed Aeronautical Systems
Company

Joseph F. Shea
Massachusetts Institute
of Technology

Martin Summerfield
Princeton Combustion Research
Laboratories, Inc.

Charles E. Treanor
Arvin/Calspan
Advanced Technology Center

Norma J. Brennan
Director, Editorial Department
AIAA

Jeanne Godette
Series Managing Editor
AIAA

Table of Contents

Part 1.   History

Chapter 1.   The Panel Method: Its Original Development 3
A. M. O. Smith, *Douglas Aircraft Company, McDonnell Douglas Corporation, Long Beach, California*

Part 2.   Computational Aerodynamic Schemes

Chapter 2.   Linear Potential Schemes 21
J. L. Hess, *Douglas Aircraft Company, McDonnell Douglas Corporation, Long Beach, California*

Part 3.   Airfoils, Wings, and Wing Bodies

Chapter 6.   Elements of Airfoil Design Methodology 167

Mark Drela, *Massachusetts Institute of Technology, Cambridge, Massachusetts*

Chapter 7.   Inverse Airfoil Design: A Classical Approach Updated for Transonic Applications 191

G. Volpe, *Grumman Corporate Research Center, Bethpage, New York*

Part 4.   High-Lift Systems

**Chapter 11.   Development of High-Lift Wing Modifications
for an Advanced Capability EA-6B Aircraft 435**
Edgar G. Waggoner, *NASA Langley Research Center, Hampton, Virginia*

Part 5. Propulsion Systems

**Chapter 12. Navier-Stokes Methods for Internal and Integrated
Flow Prediction. 461**
Raymond R. Cosner, *McDonnell Douglas Corporation, St. Louis, Missouri*

Chapter 13.   Computational Analysis of Rotor-Stator Interaction in Turbomachinery Using Zonal Techniques 481

Nateri K. Madavan, *Sterling Federal Systems, Palo Alto, California,* and Man Mohan Rai, *NASA Ames Research Center, Moffett Field, California*

Part 6.   Rotors

Chapter 14.   Euler/Navier-Stokes Computations of the Flowfield of a Helicopter Rotor in Hover and Forward Flight 533

Ramesh K. Agarwal and Jerry E. Deese, *McDonnell Douglas Corporation, St. Louis, Missouri*

Part 7. Complex Configurations

Chapter 15. CFD Applications to Complex Configurations: A Survey ... 559

Edward N. Tinoco, *The Boeing Company, Seattle, Washington*

Chapter 16. Computational Aerodynamics Applied to General Aviation/Business Aircraft ... 617

Neal J. Pfeiffer, *Beech Aircraft Corporation, Wichita, Kansas*

**Chapter 19. Toward the Routine Aerodynamic Analysis
of Complex Configurations 701**
Michael D. Madson and Larry L. Erickson, *NASA Ames Research Center,
Moffett Field, California*

Chapter 20.   Computational Aerodynamic Simulation Experience....753
E. Bonner, C. J. Woan, and G. J. Sova, *Rockwell International Corporation,
North American Aircraft, Los Angeles, California*

Chapter 24.   Flow Computations for the Space Shuttle in Ascent Mode Using Thin-Layer Navier-Stokes Equations 863

F. W. Martin Jr., *NASA Lyndon B. Johnson Space Center, Houston, Texas,* and J. P. Slotnick, *Lockheed Engineering and Sciences Company, Houston, Texas*

Part 8.   Forecast

Preface

The focus of this volume is the application of computational aerodynamics methods to aircraft design problems. Since application is stressed as opposed to algorithm development, the volume is aimed at the group of engineers and scientists that is application-oriented rather than code development-oriented. The material in the volume largely emphasizes recent studies and design developments in which computational aerodynamics played a key or enabling role. An in-depth review of such applications under one cover identifies the current state of the art and establishes an excellent reference for future design efforts. Additionally, direction for future advancements and extensions of the computational technology itself is easily found.

The field of computational aerodynamics, as a subset of computational fluid dynamics (CFD), has undergone an exceptional growth in the last 30 years. In the late 1950s the computation of simple academic flows using linear equations and a few hundred unknowns was the state of the art. The state of the art has now evolved to the ability to evaluate the flow about complex, complete aircraft configurations using nonlinear equations and several million unknowns. Furthermore, the rate of change of this state does not display any evidence of maturing or slowing. A key factor in this growth has been the substantial increase in computer speed and memory capabilities. Advancing computer capabilities have enabled ever-increasing sophistication and complexity in the governing equations and boundary conditions that can be effectively and practically utilized in real design studies.

In an attempt to reflect this contemporary situation and to limit the scope of this monograph to a tractable level, several guidelines were established for contributions to this volume:

1) Aerodynamic problems considered are of the steady type. Problems associated with unsteady or periodic types of boundary conditions are considered to be outside the scope of this volume. The exceptions to this restriction are the periodic nature of rotor-type flows and the current widespread use of a time-dependent means of iterative convergence to a steady solution.

2) Computational aerodynamics has progressed to the point where exact boundary conditions are the practical and preferred approach for most problems. Current computational power and algorithm development have reached a state that the use of a "thin" boundary condition assumption is no longer a practical necessity. Hence, computational schemes making the thin assumption for body geometry, such as lifting surface theory and small-disturbance theory, have not been emphasized.

3) Viscous effects are necessarily reviewed in the introductory material covering the governing equations and numerical schemes. Boundary-layer theory is mentioned in conjunction with potential and Euler schemes. Turbulence modeling is introduced with the discussion of the Navier-Stokes equations. The treatment of these topics is meant to be of an overview nature and is not meant to dominate the overall discussion of the numerical schemes.

4) Application type has also been limited. Areas such as missile aerodynamics, wind-tunnel simulations, and hypersonic problems have not been emphasized. This lack of emphasis certainly does not reflect a lack of computational aerodynamics applications in these areas. Rather, interest in producing a volume of reasonable scope dictated such a restriction.

The volume is divided into eight parts. The first two parts cover a historical review and a summary of the contemporary computational aerodynamics schemes. The historical review (Smith) describes one of the earliest developments in computational aerodynamics and presents some precious insights on the importance of application in promoting real advancements. The summary of numerical schemes includes a discussion of the various equation sets (Hess, Jameson) as well as grid generation (Steinbrenner and Anderson). These three chapters, although very much an overview nature, represent the essence of the fundamental algorithm work that has been accomplished by many researchers and reflect the basis for most current computational aerodynamics applications today.

The next five parts contain the bulk of the volume. These five parts are organized by problem geometric type and include 20 chapter contributions. The geometric types of problems range from the classical airfoil problem (Liebeck, Drela, Volpe) to the complexity of complete aircraft such as the 737 and 7J7 (Tinoco), the F-26 (Madson and Erickson, Holst et al.), and the Space Shuttle launch configuration (Martin and Slotnick). Applications to various components such as wings (Henne, Hicks et al.), high-lift systems (Brune and McMasters, Waggoner), propulsion system components (Cosner, Madavan and Rai), and rotors (Agarwal and Deese) indicate component design and analysis continues to be an important aspect in contemporary aerodynamics organizations. However, it is clear that complex configuration application effort permeates the industry (Pfeiffer, Narramore, Raj, Bonner et al., Welge, and Verhoff).

The final part is devoted to identifying future directions in computational aerodynamics (Bradley). This final chapter represents a global look at the state of computational aerodynamics and identifies areas where future efforts are needed to further advance the technology.

Chapters have been contributed by representatives of NASA, private aerospace companies, and academic institutions. Consequently, the various results presented reflect a broad spectrum of applications. The applications also display varying degrees of success and levels of sophistication. In some applications various codes have been treated like black boxes and have failed without the aid of a coaxing expert developer. In other cases it would appear that a particular code is quite robust and performs well hands-off. Additionally, comparison of different chapters can provide evidence of conflicting conclusions or results. This difference of opinion can often be related to which equation set is appropriate or practical to use. Conflicting opinion is to be expected in a rapidly advancing subject such as computational aerodynamics. No attempt has been made to resolve differences of opinion between the various authors. Rather, the views of each author are provided for the reader's review and consideration.

P. A. Henne

Douglas Aircraft Company
McDonnell Douglas Corporation
Long Beach, California

April 1990

Part 1.   History

Chapter 1

The Panel Method: Its Original Development

A. M. O. Smith*
*Douglas Aircraft Company, McDonnell Douglas Corporation,
Long Beach, California*

Introduction

THE panel method is an extremely general method for solving Laplace's equation that governs low-speed inviscid flow. The flow being solved for may be about a body of any shape or past any boundary. Furthermore, the solution can be subject to nearly any kind of boundary condition, not just those imposed by the flow of a fluid. If the body is two dimensional or axisymmetric, the profile is approximated by a many-sided inscribed polygon. If it is three dimensional, it is approximated by flat quadrilateral elements. The name "panel method" comes from these treatments of the body shape. We at Douglas did not invent this rather appropriate name; instead, we called the calculation method either the Neumann method or Neumann program because it was solving the classical second boundary-value problem, that is, the Neumann problem. The method as generally programmed solves mainly for the kind of kinematic boundary values determined by a fluid flow. The method has also been adapted for purely arbitrary boundary conditions or ones determined by other kinds of physical situations, such as heat flow. About the only restriction is that basic existence proofs for solution do not exist when there is a discontinuity in boundary conditions, as exists for sharp corners on a body immersed in fluid flow. But in practice it has been found that any kind of convex corner can be handled with a high degree of accuracy. There was some trouble with concave sharp corners such as the hinge line on the bottom side of a lowered plain flap. However, ways were found to minimize this problem.

Conventional computational fluid dynamics (CFD) methods require calculation for the entire three-dimensional field about the body. While the panel method also can calculate the entire three-dimensional field, it requires only calculation over the surface of the body, that is; a two-dimensional calculation. Therefore, it inherently requires much less calculation than CFD, especially if flow values only on the surface of a body are sought.

Copyright © 1989 by the American Institute of Aeronautics and Astronautics, Inc. All rights reserved.
*Formerly Chief Aerodynamics Engineer, Research (retired).

Getting into the Problem

About 1952 Richard Whitcomb had just introduced the Area Rule. All the design aerodynamicists were then converting their airplanes into equivalent bodies of revolution. But converting an airplane to an equivalent body of revolution leads to a rather bumpy type of body. Rather naturally design aerodynamicists wished to calculate the flow; that is, the velocity distribution about these bumpy equivalent bodies. But the methods available seemed unable to handle the more obnoxious shapes like that shown in Fig. 3.

One day late in 1952, I was approached by my superior, K. E. Van Every, Chief of the Aerodynamics section at the old El Segundo plant of the Douglas Aircraft Company. He asked me if I would look at the methods currently available for calculating the flows about bodies of revolution and recommend the best, if possible. This was my motivation; the primary concern originally was bodies of revolution, not airfoils or other two-dimensional bodies. Because bodies of revolution do not have lift, at least inviscidly, I was not particularly interested in vortex flow. That is a major reason why our original panel method (i.e., Neumann) was so strongly source biased.

Preliminary Work

I was certainly no expert on the various methods of calculating flows about bodies of revolution, but I probably was more knowledgable than others in the aerodynamics section. For instance, in connection with the inlet design problem on the D-558-1 research airplane first flown in Aug. 1947, a helper, Sue Hart, and I had worked up a large plot of the flowfield due to a ring source. Our plot was on 1-m wide and 2-m long millimeter roll graph paper. We went both axially and radially to five-ring radii. With this size plot plus the accuracy we maintained in our calculations, our results could be read to three decimal places. This work was done during the early design of the airplane in late 1944 or early 1945. At the end of World War II, while on a U.S. Navy mission in Europe, I met Dr. Dietrich Kuchemann in Germany. In 1940 he had completed a paper on the flow due to both ring sources and ring vortices.[1] But an interesting fact is that his tables also were to three decimal places and extended the same distance as my graph—five radii in both directions. This work was one bit of background; the large graph was a substantial project.

Later, in 1946, I found that I could analytically integrate an axial line source that had a quartic longitudinal variation in strength. A quartic variation can generate many kinds of Fuhrmann-type bodies, from ones with cusped noses and tails to ones with very blunt noses and tails. Therefore, we worked up a full report on nose and tail shapes with varying degrees of bluntness and cuspiness. Everything was solved to four-place accuracy. A report was issued, most of the work being done by G. Brazier,[2] presenting 11 nose and tail shapes plus 1 complete body, which was used for our first test case of the panel method several years later.

Still later, about 1948, I believe, I wondered if anything could be learned about supersonic delta wings by looking at the flowfield of a flat triangular shaped source, whose shape was approximately that of the delta wing. Nothing came of this, but I did find that I could indeed integrate such a three-dimensional distribution, and this knowledge came in handy later when we undertook to solve the three-dimensional Neumann problem since it, too, basically uses triangular elements.

The Aerodynamics Section was currently using the British method of Young and Owen[3] for analyzing bodies of revolution. I probably looked at this report first, but I also looked at the methods of von Kármán,[4] von Wijngaarden,[5] R. H. Smith,[6] and two methods of Kaplan.[7,8] There also were articles about disk sources such as the one by van Tuyl[9] that might lead to blunter bodies generated by the inverse (Fuhrmann) method. Other geometric arrangements of sources were also apparent, such as point sources off the axis. In fact, for awhile I wondered about point sources or ring sources that were located slightly below the surface of a body. But still nothing looked very promising. Every method seemed inadequate, so I kept looking and reading. I am sure I passed on my opinions and findings periodically to Van Every, but since I regarded his question as a general assignment that could last a long time, I kept looking and thinking.

Getting on to the Method

The last report about a method for the direct calculation of flow about an arbitrary body of revolution that I studied was one by Prof. L. Landweber.[10] I read it through slowly. It looked rather good, but at one point I came to a relation I could not understand, try as I might. I decided I did not know enough about potential flow and theory and decided to take some time off from looking at specific methods and instead read Kellogg's *Foundations of Potential Theory*[11] to learn more about the theory. (As I slowly read through Landweber's report I forgot about its title and took the method to be exact. A year or two later I happened to glance at his report and noticed the word "elongated" in the title. This explains why I could not understand a certain relation. But the report was very useful in an indirect sort of way because it caused me to read Kellogg.)

About two-thirds of the way through the book I came to the following equation, but rewritten here in the notation we were to use:

$$\frac{\partial \phi}{\partial n^{-}}(p) = 2\pi\sigma(p) + \iint_{S} \sigma(q)\,\frac{\partial}{\partial n}\left(\frac{1}{r}\right) dS, \qquad \text{region } R$$

$$\frac{\partial \phi}{\partial n^{+}}(p) = -2\pi\sigma(p) + \iint_{S} \sigma(q)\,\frac{\partial}{\partial n}\left(\frac{1}{r}\right) dS, \qquad \text{region } R'$$

Figure 1 will be used to show some notation and to help explain the equation. It is taken from the original Smith and Pierce report (1958), where it is identified as Fig. 13. Now consider the figure. The term ϕ is, as usual, the potential. One equation applies to the region R' external to the body, the other to region R, internal to it. The $+$ and $-$ subscripts to $\partial\phi/\partial n$ indicate the direction of the normal derivative. The term $\partial\phi/\partial n$ is the component of the freestream or onset flow that is normal to the boundary. It is the boundary condition. The term σ denotes the area density of a sheet of sources covering the complete surface of the shape being analyzed. It is the unknown to be found. The term p represents a point where a flow is being calculated, and q is a variable point. The term r is the distance between points p and q as shown on the figure, and n represents the normal to the surface at any point. S represents surface area, and the two equations tell us that the indicated operations are done over the entire surface of the body. The figure will be referred to periodically. Although this kind of solution can be found more or less buried in Lamb's *Hydrodynamics*, this was the first time I became aware that a pure surface source-sink method had possibilities of solving for the kinds of flow we were interested in. Before this discovery I had not known that sources and sinks could be right on the surface.

Furthermore, the equation looked tractable, being a Fredholm integral equation of the second kind, since high-speed computers were becoming available. Upon further study I could see no special restrictions. The surface shape could be entirely arbitrary, and the $(\partial/\partial n)(1/r)$ term required nothing more than directional differentiation and was just a function of the geometry of the body. The pair of equations shows that an internal problem is really no different from an external one except for the change in sign of the $\sigma(p)$ term. Of course, conservation of mass is a requirement for the internal flow.

Besides looking solvable and quite general, Kellogg makes the following statements on pages 311 and 314, respectively: "II. The Neumann problem is solvable for the infinite region R' for any continuous values of the normal derivative on the boundary." "V. The Neumann problem is solvable for a single one of the bounded regions R: under the essential condition that the integral over the bounding surface of the values assigned to the normal derivative vanishes."

Although not immediately obvious, it soon became apparent that for external flows there could be more than one body, thus solving interference problems and the like. As I looked more and more into this kind of solution, it became clearer and clearer that the two statements made by Kellogg really said a mouthful. Seldom in aerodynamics are to be found such unqualified statements accompanied by equations that look as if they can be solved. (We aerodynamicists indeed have other essentially unqualified statements, like the Navier-Stokes equation, but it cannot be solved in general.)

There are essentially no restrictions on the kinds of derivatives $(\partial\phi/\partial n)p$ that can exist. Low body slopes or linearization of any kind are not even mentioned in the two statements given by Kellogg. The quantity $\partial\phi/\partial n$ can apply for all kinds of conditions in fluid flow, from rectilinear flow, porous flow, vortex flow, rotating bodies, onset flows caused by other bodies, etc. If Laplace's equation applies to different physical situations, $\partial\phi/\partial n$ can be determined for that situation too, for example, temperature distribution in a solid. Moreover, the equation applies equally to two-dimensional, axisymmetric, and three-dimensional shapes. I can think of very few situations where a governing equation has essentially no restrictions,† yet seemed fairly readily solvable using machine calculation. In actual application it was found that even discontinuous boundary conditions which are excepted in Kellogg's statement II could be calculated accurately as noted in the introduction.

Trying to Solve the Equations

Serious work to solve the equations began in the summer of 1953. I told Van Every that it looked as if we might be on to something, and he supported us by authorizing the necessary budget. There are, of course, many approaches to solving the two equations, but most have some kind of limitation. Many methods will handle only very regular solutions, not our bumpy body problem. I thought about the solution and got more familiar with the equation for several months, including considerable time looking in the literature for methods of solution. In particular, I remember spending one entire day in the California Institute of Technology physics and mathematical libraries trying to find leads to a method of solution. The first and second boundary-value problems are famous problems, and there is extensive literature about them. But I got no help. All I found were proofs of the existence and uniqueness of the solution, in line with Kellogg. I found nothing on practical methods for solving the problem. I came away with a mild irritation at the mathematicians. Of course, knowing that a solution exists and is unique is important because one would not find himself following a blind alley. But I already knew that from Kellogg.

The Solution

Although there are numerous ways of evaluating the integration required by the equation (e.g., Martenson[12] and Jacob[13]), after a few months of looking around and thinking about the problem I rather quickly settled on a method. Because the calculations would be lengthy, only one method was worked out at the time, although theoretical refinements were seen. The method chosen was one I thought was the simplest possible that would converge to the exact answer for the case of an infinite number of panels. It had the following main features:

1) The body would be divided into a series of sections, later called panels (see Fig. 1).

2) Continuous source density distribution over each of these sections would be used; i.e., no point or line sources would be used.

3) These sections would be flat in profile, i.e., panels.

4) The panels might each have a different source density σ, but the density would be constant for each panel.

5) The indicated integration would be replaced by a simple summation.

Now we will discuss these assumptions in order and the thinking behind them.

---

†That is, where Laplace's equation applies. More generally there are indeed restrictions; the flow cannot be compressible or viscous.

1) Because our interest was bumpy, i.e., nonfair bodies, no kind of continuous or analytic treatment seemed likely to work, yet Kellogg's statements were truly applicable to nonfair bodies. Hence, as in numerical quadrature, it seemed that the best course was to divide up the body into a number of sections as in Fig. 1.

2) In dealing with a continuous surface distribution, when the variable point q approaches the control point p (see Fig. 1), the normal flow $\partial\phi/\partial n$ due to the sheet of sources takes on the very finite value $2\pi\sigma$. If the source material were concentrated in lines or points, then when q approaches p, i.e., $r \rightarrow 0$ in the figure, the velocity becomes infinite. Therefore, in order to avoid all kinds of singularities, it seemed to me to be imperative to use a continuous surface source distribution.

3) A flat element in profile seemed much the easiest for integration of the $(\partial/\partial n)(1/r)$ term applicable to each panel. In fact, for two-dimensional flow the integration across a strip is analytic. It also is analytic for flat quadrilateral elements in three-dimensional flows that were solved later. Only in the axisymmetric case where each element is a frustum of a cone (see Fig. 1) is it nonanalytic, but even here accurate results can be obtained by a combination of numerical integration and series expansion. Using flat elements essentially amounts to solving for the flow about a polygon inscribed inside the true shape. By using flat elements and assuming constant σ on each, it became possible to perform the first integration of the indicated double integration analytically or at least in advance.

4) Having constant source density σ on each panel made integration simpler and helped lead to a simple system of linear algebraic equations.

5) It is probably apparent by now what will be the system for solving this Fredholm integral equation of the second kind. Referring to Fig. 1 and the equation, it is seen that the double integral is approximated by breaking the body into a sum of values for strips, each one having an unknown value of σ. The $\sigma(q)$ is the unknown, but the $(\partial/\partial n)(1/r)$ term is directly calculable from the geometry of the body. Then a system of linear equations can be formed. On the other side of the equal sign is the $\partial\phi/\partial n$ term. This quantity is a given by the boundary conditions. If the flow being analyzed is a rectilinear kind, $\partial\phi/\partial n$ is just proportional to the sine of the local slope of the body.

What is being done is really simple and straightforward. At any point p we are writing an expression for the contribution of all the ring sources for the entire body (see Fig. 1). It seemed to me at the time that the $(\partial/\partial n)(1/r)$ value at p for an inscribed frustum would differ exceedingly little from the value that would exist for the exact shape, thus further justifying the use of flat elements. On the other side of the equation we have $(\partial\phi/\partial n)p$. This is evaluated at the middle of each element; hence, it is a central difference calculation and should have good accuracy. Note that the equation is represented by a simple sum of terms; there were no quadrature formulas, etc.

Approximating a body shape by an inscribed polygon was a rather bold step. At the center of a two-dimensional strip where σ values are calculated, there is indeed no self-induced tangential flow, but there is some for cone frustums. But at the edge of any strip there is a logarithmic infinity that exists because 1) the next strip has a different σ, and 2) there is usually an angle between any two adjacent strips. However, we felt that by working in the middle of elements (strips) this unruly trait of the elements would not cause a problem. Yet the only way to know for sure was to program the method and run some cases.

Programming and First Results

All of the formulas or algorithms were worked out around the middle of 1954. Programming using the IBM/701 calculator was begun in the fall. The programming was going along smoothly enough but in the middle of it I had an assignment that took me to Europe from January 2 to March 10, 1955. During this trip I often wondered how the programming was coming along. At this early stage the programming was all in machine language; no FORTRAN, etc., was used. When I returned I found the program was working, and as far as I knew no special bugs had been encountered.

The original program was written for 24-point solutions, and the first case to be run was a Fuhrmann-type body taken from Ref. 2 for which we knew the accuracy to four places. The accuracy of our new 24-point solution turned out very good, and this very first case is shown as Fig. 29 in the long Hess-Smith paper, but is shown here for convenience as Fig. 2 taken from the Smith and Pierce report. Various cases were then run. We made extensive use of Milne-Thomson's book *Theoretical Hydrodynamics* to find theoretical solutions of various kinds to serve as test cases.[6] Also, two special cases come to mind. One was Fig. 2 of the Smith and Pierce report, shown here as Fig. 3. It was a very severe test; in fact, at first I contemplated making a wind-tunnel model of the same shape to help check our method. But the calculation turned out to be so accurate that I dropped the idea. The dashed line in Fig. 3 identified as the conventional method was calculated by use of Ref. 3, which was the method being used by the Aerodynamics Section at that time. As you can see, the improvement in accuracy was substantial, to say the least. The conventional method completely ignored the bump.

Another special body was in Figs. 1 and 27 of the Smith and Pierce report. This body was chosen 1) because it had a flat nose that up to then could not be calculated by any existing method and 2) because there were test data on the shape. The shape and velocity distribution are shown by Figs. 4 and 5. More will be mentioned about Fig. 4 later.

As can be seen, the original program was for bodies of revolution, but after we found that it was working we began programming the two-dimensional variation. By December 7, 1955, we had run a 150-point solution for the flow about a circular cylinder. Extensions like this generally came easy. Of course, I enjoyed seeing good results occur with each new try or extension, but it was almost more enjoyable to see the pleasure and enthusiasm that Jesse Pierce showed with each successful new case. More will be said about Pierce a bit later.

Our first report was the Smith and Pierce Report, in April 1958. In June 1958 I gave a shortened version of the paper at Brown University. The method was first revealed at this meeting. At the meeting Prof. Irmgard Flugge-Lotz attended my lecture. Afterward, she got up and commented that she had used the surface method in some work she had done on airships at yaw. This was the first I knew that the basic idea had been used before. Later, I learned that Dr. W. Prager had published a paper[14] in 1928 also using surface sources. Flugge-Lotz's paper[15] helped Hess in his crossflow work, but Prager's paper was too primitive to be of any assistance, except that it called attention to the fact that a surface source treatment was a possibility.

Up to 1958 all of this work was financed by Douglas, but we could see much more work to be done, and extensions were proposed to the Office of Naval Research (ONR). An interesting coincidence developed. The very first illustration in the Smith and Pierce report was that of the flat-nosed body, Fig. 4, mentioned earlier. The man in charge of the Fluid Mechanics Branch at ONR at the time was Phillip Eisenberg. He had tested the flat-nosed model himself in a water tunnel at the David Taylor Model Basin in 1947. In contacting ONR we had sent in advance a copy of the Smith and Pierce report; thus, when he looked at the figure it immediately caught his eye. He arranged for a meeting with representatives of the Model Basin, the principal one being Dr. Avis Borden. We told her and the others what we thought we could do, and shortly thereafter they chose to give us a contract to extend the method to three-dimensional nonlifting flows, which would be directly applicable to ships. The work would also be directly useful to Douglas, since free surface effects were not being considered under this contract.

Other Early Extensions

After the original body of revolution program including inlets, the first extension was to the nonlifting two-dimensional problem. The next extension of significance as far as I remember was Hess' extension of the body of revolution problem to crossflow. This extension allowed bodies of revolution to be analyzed at angles of attack just like airfoils. At low angles of attack this method is quite useful, only losing its validity at higher angles of attack when separation develops. The next development was the three-dimensional nonlifting problem already mentioned.

At about the same time we began working on incorporating lift into two-dimensional flows. A lifting airfoil has circulation; hence, at first we simply put a vortex inside the airfoil and calculated for still another kind of onset flow. Algebraically it makes no difference where the vortex is located inside the airfoil, but computationally it made a great difference. A point vortex generally gave the correct gross pressure distribution and lift coefficient but created two bumps in the velocity distribution near wherever we put the vortex. After a number of tries with various vortex treatments, it was found that a vortex onset flow created by turning all the source strips into vortex strips worked very well and eliminated the irregularities in the pressure distribution. J. P. Giesing did most of the airfoil work. But since then further refinements made in years subsequent to this chronology have made the results with lift even better. Our method of solving most of the lifting problem with sources may seem peculiar to many, but the source method was working so well that it seemed natural to try to extend it. Also, of course by the very generality of the method of solution, the extensions to cascades, multielement airfoils, and ground effect were very simple. Still another extension was to find added mass or inertia coefficients for arbitrary bodies of revolution.

Another early extension was to solve the hydrofoil problem subject to a linearized free surface condition. Except for complications caused by this linearized surface condition, a hydrofoil could be just as arbitrary in shape as an airfoil. I became interested in this problem because of our continuing relation with the David Taylor Model Basin. A very complicated elementary singularity that automatically satisfies the linearized free-surface condition is called the Havelock source. We hoped to apply it to the three-dimensional problem but were unable to perform one integration. However, we could and did do it for the two-dimensional case—the hydrofoil case. Although I started this line of action, Giesing did most of the work; thus, he is the first author in the Giesing-Smith paper. Hydrofoil and other free-surface work had a lower priority because Douglas was an airplane company, not a ship company.

There were other extensions, such as suction over part of a body, rotating bodies, bodies in shear flow, effects of compressibility by use of the Goethert rule, nonlinear unsteady airfoil theory, and applications in physics to problems involving superconductivity. Most of this work is written up in the long Hess-Smith paper. The early work is considered to end with this publication. Much more work has been done by now; for instance, the three-dimensional problem with lift, exact propellor calculation (except that the location of the wake is not precisely known), higher-order solutions using curved instead of flat elements, more use of vortex elements to eliminate some final problems near the trailing edge of airfoils, and, finally, good inverse solutions.

Four Key Helpers

My first helper was a retired mathematics professor, Dr. Jesse Pierce. His son, E. W. Pierce, was assistant chief of the Aerodynamics Section where I worked. Jesse Pierce was 67 years old when he was transferred to my group. He had been head of the mathematics department at a small school, Heidelberg College, in Tiffin, Ohio. He was still interested in working and got a job in the computing section at Douglas, in El Segundo, California. After awhile, about the middle of 1954, I believe, the manager of computing felt that he would be more useful in my group, especially since I needed help. Accordingly, he was transferred to work mainly but not exclusively on the Neumann Program. He advised the actual programmers, mainly George W. Timpson and William E. Moorman, worked up details of test cases, and performed various other duties.

Most of the theory was all worked out before Dr. Pierce joined our group, but he did contribute one important observation. He noticed that the flow of interest is always on the left as one traverses a shape. Thus, on a closed body of revolution, if the coordinates of the body are written for the top side from nose to tail, the flow will be the conventional external flow. If the coordinates for the bottom side are used instead, then in loading in the deck the flow calculated will be one internal to the body, if that is possible. Jesse Pierce worked in my group from 1954 to 1958. He helped a great deal in preparing the Smith and Pierce report, writing about half of it. About the time it was finished he decided to quit work for good, since he was 71 years old by then.

John Hess came into my group in July 1956. He had fulfilled all of the requirements for a PhD in Applied Math at the Massachusetts Institute of Technology, except for the research requirements. I hired him mainly because I needed more help on a variety of projects. At first he did indeed perform a variety of duties, but showed an interest in the potential flow work that Pierce and I were doing. In fact, he was soon giving proofs and demonstrations of relations we thought were true, but where questions existed.

Then when Jesse Pierce quit it seemed only natural for Hess to take over. In retrospect, it was a good decision, as evidenced by all of the papers and extensions of the method that he has made. My decision was confirmed very early by his extension of the body of revolution problem to the crossflow case. I had practically nothing to do with this problem. Then when we collaborated on the ONR three-dimensional flow contract, besides doing things like developing the quadrilateral element input system, he worked out a simplified method for evaluating the flowfield of a quadrilateral source element. He applied a multipole expansion frequently used in electrostatic problems. If the p point in Fig. 1 is at some relative distance from the q point, the lengthy exact formula could be bypassed, thus saving much computing time. Obviously, if p is sufficiently far from q, a quadrilateral source could be treated as a point source. I only mention these two points because Hess' more recent contributions are well known. Since he was handling the potential flow work very well, perhaps better than I could, I gradually eased out of it, turning most further development over to him.

Joe Giesing had been in our group earlier but went back to school to get his Master's degree in Aeronautics at the California Institute of Technology on a Douglas scholarship. He returned to us in June 1962 and immediately began working on the airfoil and cascade problem, studying various distributions and locations of vortices to get lift. I am not sure who gets credit for the final quite successful method, converting all of the source elements into vortex elements to give lift. It rather grew out of discussions among Hess, Giesing, and myself, but in any case Giesing wrote the final report. Later he did most of the work on the hydrofoil problem. In May 1968 he was offered a better position in the structures group at Douglas, and he left my group.

Sue Schimke, née Faulkner, was also instrumental in the early development of the method. Sue was a University of California, Los Angeles, graduate in meteorology but never practiced it. She came to Douglas in

1957 and transferred to our group in October 1958. She made herself very useful; being able to find things and knowing where everthing was, was one of her strong points. Although there was nothing formal about it, she gradually became John Hess' assistant. She soon became very adept at running any sort of Neumann problem case. All that was necessary was to tell her the general problem, and often she would work out the formulas for it if needed. Then she would compute the necessary body coordinates and finally run the case on the IBM/704, 7090, 7094, 360, or whatever. It could be said that after awhile Hess did the basic theory and Sue would turn it into practice.

References

[1]Küchemann, D., "Tafeln für die Strömfunktion und die Geschwindigkeits componenten von Quellring und Wirbelring," *Jahrbuch der Deutschen Luftfahrforschung*, Rufus Oldenburg, Munich and Berlin, 1940, pp. 1547–1564.

[2]Brazier, J. G., and Smith, A. M. O., "Development of Nose and Tail Shapes in Incompressible, Irrotational Flow," Douglas Aircraft Co., Long Beach, CA, Rept. E.S. 20875, June 1947.

[3]Young, A. D., and Owen, P. R., "A Simplified Theory for Streamline Bodies of Revolution, and its Application to the Development of High-Speed Low-Drag Shapes," Aeronautical Research Council, London, R & M 2071, 1943.

[4]von Kármán, T., "Calculation of Pressure Distribution on Airship Hulls," NACA TM-574, 1930.

[5]von Wijngaarden, A., "Écoulment Potential Autour d'un Corps de Revolution," *Colloques Internationaux du Centre National de la Recherche Scientifique, XIV. Methods de Calcul dans des Problèms de Mécanique*, Paris, 1948.

[6]Smith, R. H., "Longitudinal Potential Flow about Arbitrary Body of Revolution with Application to Airship 'Akron'," *Journal of the Aeronautical Sciences*, Vol. 3, No. 1, Sept. 1935, pp. 26–31.

[7]Kaplan, C., "Potential Flow about Elongated Bodies of Revolution," NACA Rept. 516, 1935.

[8]Kaplan, C., "On a New Method for Calculating the Potential Flow Past a Body of Revolution," NACA Rept. 752, 1943.

[9]van Tuyl, A., "On the Axially Symmetric Flow around a New Family of Half Bodies," *Quarterly of Applied Mathematics*, Vol. 7, No. 4, 1950, pp. 399–409.

[10]Landweber, L., "The Axially Symmetric Potential Flow about Elongated Bodies of Revolution," David W. Taylor Model Basin, U.S. Navy, Washington, DC, Rept. 761, 1951.

[11]Kellogg, O. D., *Foundations of Potential Theory*, Ungar, New York, 1929; also Dover, New York.

[12]Martenson, E., "Berechnung der Druckverteilung an Gitterprofilen in Ebner Potentialströmung mit einer Fredholmschen Integralgleichung," *Archives of Rational Mechanics and Analysis*, Vol. 3, No. 3, 1959, pp. 235–270.

[13]Jacob, K. W., "Some Programs for Incompressible Aerodynamic Flow Calculations," California Inst. of Technology Computing Center, TR-122, Feb. 1964.

[14]Prager, W., "Die Druckverteilung an Körpern in Ebener Potentialströmung," *Physikalische Zeitschrift*, Vol. XXIX, 1928, pp. 865–869.

[15]Flügge-Lotz, I., "Calculation of Potential Flow Past Airship Bodies in Yaw," NACA TM-675, July 1932.

Bibliography

Geising, J. P., and Smith, A. M. O., "Potential Flow about Two-Dimensional Hydrofoils," *Journal of Fluid Mechanics*, Vol. 28, Pt. 1, April 1967, pp. 113–129.

Hess, J. L., and Smith, A. M. O., *Calculation of Potential Flow about Arbitrary Bodies*, Vol. 8, Progress in Aeronautical Sciences, Pergamon, Oxford, UK, 1966, pp. 1–138.

Smith, A. M. O., and Pierce, J. "Exact Solution of the Neumann Problem. Calculation of Non-Circulatory Plane and Axially-Symmetric Flows About or Within Arbitrary Boundaries," Douglas Aircraft Co., Long Beach, CA, Rept. E.S. 26988, April 1958 (this is the first report about the method); also available as Armed Services Technical Intelligence Agency, Rept. AD 161445. (A shortened version of the paper but with the same title can be found in *The Proceedings of the III U.S. National Congress of Applied Mechanics*, 1958, p. 807.

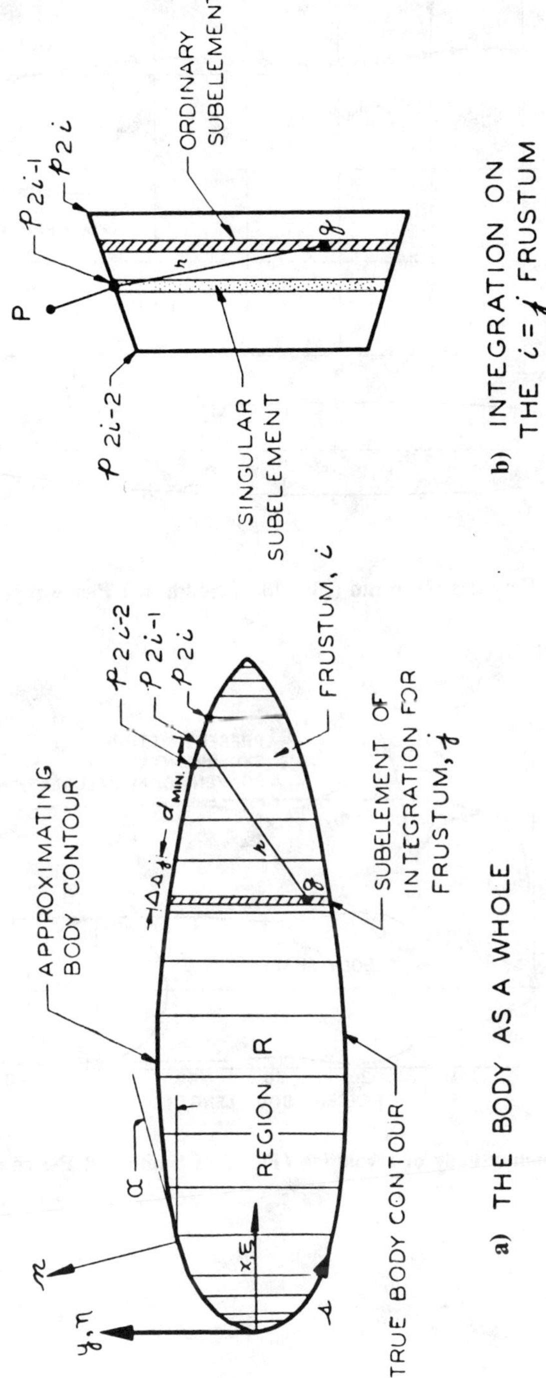

Fig. 1   Notation and method of approximating a body of revolution (Fig. 13 of Smith and Pierce report).

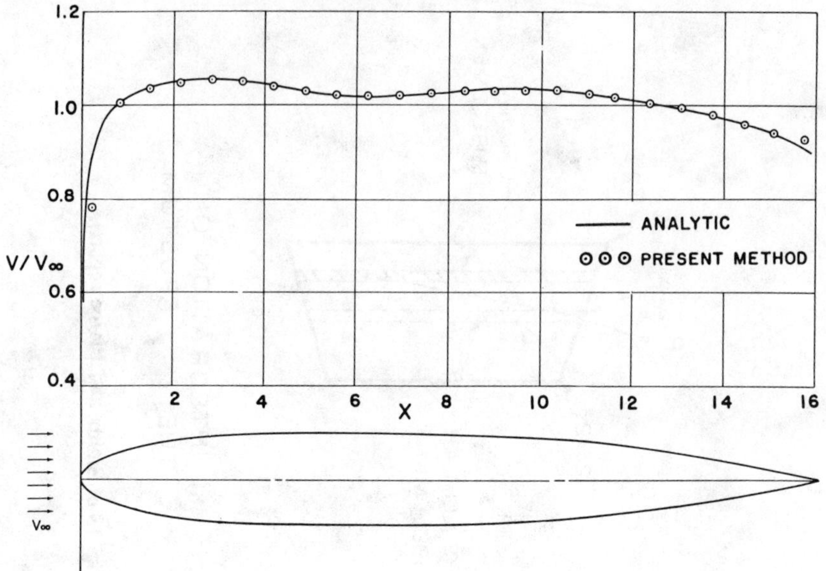

Fig. 2   Very first case run (Fig. 28 of Smith and Pierce report).

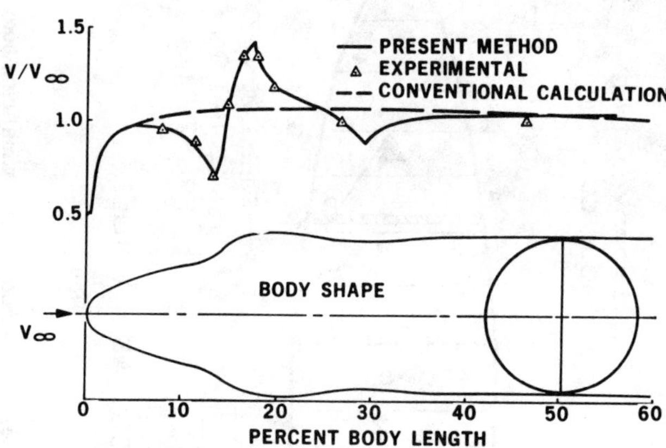

Fig. 3   A bumpy body of revolution (Fig. 2 of Smith and Pierce report).

Fig. 4   A flat-nosed body of revolution (Fig. 1 of Smith and Pierce report).

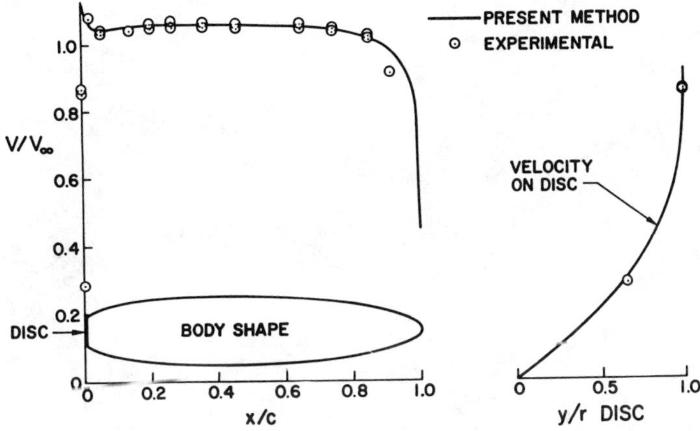

Fig. 5   Further data on the flat-nosed body of revolution (Fig. 27 of Smith and Pierce report).

Part 2.   Computational Aerodynamic Schemes

Chapter 2

Linear Potential Schemes

J. L. Hess*
*Douglas Aircraft Company, McDonnell Douglas Corporation,
Long Beach, California*

Introduction

LINEAR potential flow, that is, incompressible, inviscid, irrotational flow, has been extensively studied for well over a century. Originally, it attracted investigators because almost all of the analytic solutions to flow problems are such flows. At first, its importance for design applications went unrecognized. Fluid dynamicists were divided into two classes: hydraulic engineers who observed many things they could not explain and theoretical hydrodynamicists who explained many things they could not observe. Textbooks featured linear potential flow prominently because of its suitability for classroom exercises. Occasionally, comparisons were presented between theoretical and experimental surface pressures for a circular cylinder or a sphere, and the lack of agreement was suitably lamented. Indeed, if airplane wing sections were circles or fuselages spheres, there would be little practical reason for pursuing linear potential flow. However, for the streamline shapes actually used in applications, theory and experiment agree very adequately in many cases of interest. It is almost as if what makes a shape good aerodynamically is that the flow about it closely approximates potential flow.

Efforts to calculate linear potential flow over realistic configurations have been pursued seriously for many years. It was recognized very early that, although analytic solutions can provide guidance, such a method would have to be numerical. Because of the linearity of the governing equations, the principle of superposition is applicable, and, as described in Chapter 1, the technique that has proved most universally successful is based on superposition of certain singular solutions. This method, which is applicable to both two- and three-dimensional problems, is the main topic of the present section. However, no discussion of linear potential-flow methods could neglect a very efficient and powerful technique that is limited to two-dimensional flow, namely, solution by conformal mapping.

Copyright © 1989 by the American Institute of Aeronautics and Astronautics, Inc. All rights reserved.
*McDonnell Douglas Research Fellow, CFD Development.

Mapping Solutions

Mapping solutions are based on the intimate connection between any problem governed by the two-dimensional Laplace equation and the theory of functions of a complex variable. Suppose there are two planes: an x,y plane, where the actual body is defined, and a ξ,η plane. Define the complex variables $z = x + iy$ and $\zeta = \xi + i\eta$ and let a coordinate transformation between the two planes be defined by the real and imaginary parts of an analytic mapping function $z = z(\zeta)$. Then the Laplacians of any function ϕ at corresponding points of the two planes are related by

$$\nabla^2\phi(z) = \frac{\nabla^2\phi(\zeta)}{|dz/d\zeta|^2} \tag{1}$$

Thus, if ϕ satisfies Laplace's equation in one plane, it also satisfies it in the other. The gradients are related by

$$\nabla\phi(z) = \frac{\nabla\phi(\zeta)(dz/d\zeta)}{|dz/d\zeta|^2} \tag{2}$$

Moreover, angles are preserved by the mapping, so that, if two curves intersect at right angles in one plane, they do in the other also. Now suppose a problem of linear potential flow is defined in the z plane with normal velocity zero on the boundary and with velocity approaching freestream V_∞ at infinity. If $z = z(\zeta)$ is any conformal mapping with

$$\lim_{|z| \to \infty} \left(\frac{dz}{d\zeta}\right) = 1 \tag{3}$$

then the same problem obtains in the ζ plane, such that the potential ϕ is equal at corresponding points in the two planes and the velocities are related by Eq. (2).

Classically, the preceding procedure was applied in an indirect fashion to produce analytic solutions. Starting with a circle in the ζ plane, for which the solution is well known, a closed-form mapping is used to yield the solution for another body in the z plane. For example, the well-known Joukowski transformation gives a certain class of airfoils having zero trailing-edge angles, and the Kármán-Trefftz transformation gives a class with finite trailing-edge angles. For practical single-airfoil problems this process must be reversed. The starting point is a given airfoil in the z plane, and it is required to determine a mapping that takes this airfoil to a circle in the ζ plane and thus yields the flow solution. For completely general airfoils the mapping must be obtained numerically, and a number of methods for accomplishing this have been formulated, first by Theodorsen[1] in 1932. A problem of great interest in airplane design is the flow about multielement airfoils which represent high-lift configurations. In order to solve such problems by mapping, the image boundary in the ζ plane, which consists of the same number of separate bodies as the multielement airfoil in the z plane, must have a solution that can be obtained more easily than that for the original body. Garrick[2] and Ives[3] solved this problem for a two-element airfoil by mapping to a circular annulus. However, a practical numerical procedure for multielement airfoils of more than two elements was first constructed by Halsey,[4] who mapped the multielement airfoil to a set of disjoint circles. His solution, which has been put on a production basis, is fast and reliable, and includes all previous methods as special cases. The discussion of this subsection follows his article.[4] One result of his work is to reduce greatly the importance of two-dimensional panel methods.

The mapping of a multielement airfoil into a set of disjoint circles consists of two main stages. The first may be called the corner-removing mapping, which uses the Kármán-Trefftz mapping in the reverse or "inverse" direction, that is, from the z plane to the ζ plane. If such a mapping is applied to a single Kármán-Trefftz airfoil, it yields an exact circle in the ζ plane. Applying it to a different single airfoil produces a smooth near-circular shape without corners. If it is applied to one element of a multielement airfoil, it gives a smooth image for the element to which it is applied and distorted shapes still having corners at their trailing edges for the other elements. The corner-removing mapping applies the inverse Kármán-Trefftz mapping successively to each element of a multielement airfoil. The result is a set of disjoint smooth near-circles in the ζ plane.

The second stage of the mapping procedure is based on a numerical procedure for mapping a single body to a circle. It is a variant of a technique developed by James,[5] which uses the fast Fourier transform. When this procedure is applied to one of the smooth shapes obtained by the corner-removing mapping, the image consists of a circle for that shape and modified smooth shapes for the others in the ensemble. The numerical single-body mapping is applied to each of the smooth shapes successively, and the process is repeated until all elements have been reduced to circles. Convergence is very rapid. Figure 1 (taken from Halsey[4]) shows three phases of the process for an example. The first is the actual four-element airfoil, and the second consists of the four smooth shapes produced by the corner-removing mapping. Applying the single-body mapping only once per body is sufficient to produce circles to graphical accuracy, the third portion of Fig. 1, and two additional mappings per body yield circles to five significant figures.

The solution to the problem of flow about the disjoint circles is obtained iteratively and is based on the fact that the flow about a circle in the presence of external doublet and vortex singularities may be effected by putting image doublets and vortices inside the circle; i.e., if this is done properly, the circle remains a streamline.[6] Flow about a single circle in a uniform stream may be generated by placing a doublet and a vortex at the center, where the doublet strength is proportional to the square of the circle's radius, and the vortex strength serves to place the rear stagnation point at a prescribed location, the Kutta condition. If another doublet is placed in the flow exterior to the circle, the circle will remain a streamline if a related doublet is placed at the image point inside the circle. (The product of distances of the doublet and its image from the center of the circle equals the radius of the circle.) Similarly, the circle remains a streamline in the presence of an exterior vortex by placing inside a pair of equal counter-rotating vortices, one at the center of the circle and one at the image point. The flow about the set of circles proceeds as follows. First, each circle is replaced by a doublet and vortex at its center with strengths appropriate to the case where that circle is alone in the uniform stream. Next, within each circle are placed image systems of the doublet/vortices of all other circles. Then within each circle images are placed of all the doublet/vortices added at the previous step, and the process is iterated until convergence is obtained. Although the number of singularities grows rapidly with the number of iterations, their strengths fall rapidly. Halsey[4] introduces certain series expansions as efficiency measures, and his method obtains solutions in much less computation time than a panel method. Results for the four-element airfoil are shown in Fig. 2 (from Halsey[4]), which verified that the mapping method obtains the same solution as a panel method.

Panel Method Singularity Distribution

The linear nature of the potential-flow problem considered here allows the principle of superposition to be employed in order to construct solutions to problems of interest by summing simpler solutions. Flows due to sources, doublets, and vortices are the elementary building blocks out of which general solutions are constructed. Rankine first used this technique to obtain flow about the ovals that bear his name, and the technique was considerably generalized by others. However, these efforts all addressed what may be called the indirect problem, where the strengths and locations of the singularities are specified initially, and the body about which flow is calculated comes out as one of the results. In fact, the calculation of the body shape is the most difficult portion of this procedure. The direct problem, where the body shape is specified and the singularities and their resulting flowfields are calculated, is much more difficult. Indeed, methods for complicated bodies had to await the availability of computers. A good portion of the early history of development of such methods is presented in Chapter 1, which carries the narrative into the mid 1960s. Singularity techniques have continued to be developed in the intervening years and have obtained not only great generality but also extreme reliability, so that they represent one of the two or three "workhorse"-type flow-calculation methods in major aerospace companies. Since the vast majority of applications are three dimensional, and most of them involve lift, it is that problem with which the discussions of this section are mainly concerned. It is known from Lamb[7] that under fairly general circumstances the disturbance potential due to a nonlifting body maybe expressed as an integral over the body surface of a source distribution σ and a doublet distribution μ, i.e.,

$$\phi = \int_{\text{body}} \left[\frac{1}{r}\sigma + \frac{\partial}{\partial n}\left(\frac{1}{r}\right)\mu \right] \mathrm{d}S \qquad (4)$$

where $1/r$ is the point source potential, and its normal derivative is the point doublet potential. Since ϕ as given by Eq. (4) satisfies Laplace's equation and approaches zero at infinity for arbitrary functions σ and μ, these are available for satisfying the boundary conditions of the problem— mainly the zero normal-velocity condition on the body surface. The system is underdetermined, since there are two functions available to satisfy one principal boundary condition. The indeterminacy is greater than stated because arbitrary singularities may be introduced anywhere inside the body surface. This nonuniqueness applies only to the body singularity distributions, which are simply devices for effecting the solution. Once a flow solution that satisfies all boundary conditions is obtained by whatever means, the resulting velocity field is unique. In three-dimensional lifting flows there must necessarily be a trailing vortex wake that is a stream surface of the flow across which there is no pressure discontinuity. Since the representation of the wake is not simply a device for calculating the flow but is an approximation to physical reality, there is no conceptual nonuniqueness, but the numerics may be formulated at different levels of accuracy. Lift implies satisfaction of a Kutta condition along all of the trailing edges from which vortex wakes leave the body, but this does not change the earlier discussion of nonuniqueness in any essential way.

When the zero normal-velocity boundary condition is applied to Eq. (4), the rest is an integral equation for the singularity strengths. To effect a solution, two separate discretizations are required: that of the body surface and that of the singularity distributions. Of the several methods that have been constructed, every one without exception has discretized the body by means of small quadrilateral "panels," which are either plain or nearly so (Fig. 3). This universal choice has caused these approaches to be designated panel methods. Probably the chief factor dictating this choice is the ease of organizing the geometrical input data. Once a method has been constructed and reduced to production status, however difficult this may be, attention shifts to the recurring difficulty of inputting cases to the method. Input costs can dominate overall costs; thus, reducing them is always a leading consideration in code development. The second discretization, that of the singularity distributions, in most methods is accomplished by assuming a simple polynomial variation over each quadrilateral panel, e.g., constant[8,9] or variations up to quadratic.[10-12] This matter is discussed further in later sections.

One point of each panel is selected where the normal-velocity boundary condition is to be applied. It is customary to call this the control point. One of the two major computational tasks in a panel method is to obtain the matrices of panel influences at each others' control points. The key formulas are obtained by integrating over a panel the point source and doublet formulas weighted by the proper polynomial variations. Some formulations require both potential and velocity influences of a panel, and some require only one or the other. These integrations, which express the panel influences at a general field point in space, may be performed analytically over a plane panel to obtain exact closed-form expressions, which, however, are rather complicated. For efficiency approximate expressions are employed if the point in question is far from the panel. Some methods use different approximate expressions for various ranges of distance. All methods replace the panel by a point source and doublet if the point is sufficiently far away. In large practical cases well over 90% of the panel influences are calculated from the point singularity formulas.

The relevant integral equation, as discretized by means of panels and their mutual interferences, is solved by collocation. First, the internal logic of the program reduces the number of unknown parameters defining the basic source and doublet distributions to equal the number of panel control points plus the number of points where the Kutta condition is applied. This yields a system with the same number of equations as unknowns. Accomplishing this requires some assumption to remove the nonuniqueness from the representation of Eq. (4) plus analytic procedures to express the various coefficients of the polynomial representation over each panel in terms of a single unknown, e.g., the value of the unknown singularity at the control point. For example, if one singularity in Eq. (4) is assumed known, and the other is constant over each panel, this process is trivial. Otherwise, some further analysis must be done, but it is relatively straightforward. Before discretization the Kutta condition must be applied all along the wing trailing edge. In the discretized problem it is applied at each trailing-edge segment (Fig. 3), which represents the trailing edge of a lifting strip of panels, usually at a particular spanwise location. For a fuselage or other portion of the configuration without a well-defined trailing edge, no Kutta condition is imposed and no trailing vortex wake is assumed. It should be emphasized that the wing-fuselage shown in Fig. 3 is intended simply to be representative of more general configurations, which may have several lifting portions with their associated trailing edges and wakes and several nonlifting portions without trailing edges.

Removal of the nonuniqueness of Eq. (4) by selection of the "mix" of source and doublet singularity to be employed usually receives chief emphasis in descriptive articles with various advantages claimed for one or other choice based mainly on theoretical considerations. When and if these methods are brought to production status for use by engineers in design problems, many of the predicted advantages turn out to be unimportant, and attention shifts to details of the numerics, such as sensitivity to panel aspect ratio, mismatch at panel edges, or irregular panel spacing. Generally, such problems can be solved by suitably modifying the code, with the result that methods that have actually been used for a period of years in a variety of applications have attained generality, but perhaps not in the manner originally envisaged. Other presumed advantages turn out in practice to be more closely related to expediency. An example is the use of interior singularities, particularly the "lift carryover doublicity" often placed inside, say, a fuselage. Some authors make a point that their approach need not use interior singularities. Strictly speaking, no method has to use them. This is guaranteed by the uniqueness theorems for Laplace's equation. The important consideration is whether their use leads to more accurate solutions and/or reduced panel numbers.

Several selections of singularity "mix" have been tried by various investigators. However, after several years of experimentation only two choices have survived, and both have been so extensively exercised as to merit the designation of well-proven design tools. The first type, which is designated a source method, uses one independent value of source density per panel plus a doublet distribution over the wing (Fig. 3) that depends on a number of adjustable parameters equal to the number of trailing-edge segments where the Kutta condition is applied. The second approach, the doublet or Green's identity method, assumes the source strength on each panel to be proportional to the inclination of the freestream to the panel normal vector, and solves for the doublet strengths in such a way that the panel boundary condition and the Kutta condition are satisfied. Both first-order[8-10] and higher-order[11,12] versions of both techniques have been constructed.

Higher-Order Singularities

Whether the simplest discretization of the singularity distributions suffices for good accuracy or whether a more elaborate representation is necessary depends on the method being used, but it depends even more strongly on the particular flow situation being considered. There is considerable disagreement between investigators as to what constitutes a higher-order method. Many investigators claim a method is higher order if account is taken of the variation of singularity over a panel even if the panel remains flat. However, analysis[11] indicates that it is mathematically inconsistent to include effects of singularity variation without also accounting for the variation over the panel of the direction of its normal vector. Put in other terms, the effects of singularity derivative and panel curvature have the same mathematical order.

Doublet method developers disagree on the requirement for a higher-order approach. Most doublet methods[8,9] use a first-order implementation consisting of constant source and doublet distributions on flat panels and apparently obtain reliable results for most cases. Other investigators[12] stress the importance of singularity continuity across panel edges. This is certainly necessary in linearized supersonic flow where infinities occur, not only at singularity discontinuities, but all over the Mach cones issuing from them. It is not clear why continuity is an important requirement in subsonic flow, especially in view of the contrary experience of other investigators. Continuity is enforced[12] by dividing the quadrilateral into five subpanels: a central quadrilateral and four corner triangles whose planes may be inclined to it. This permits edges of adjacent panels to be coincident, but does not represent local curvature. If the original quadrilateral panels have coincident adjacent edges, as, for example, those that occur on bodies of revolution, the original flat panels are used. To obtain singularity continuity, separate polynomial variations are assumed for each of the five subpanels, and matching conditions are applied across all edges.

The so-called first-order source method[10] uses a constant source distribution on a flat panel as its principal singularity for satisfying the panel boundary condition. However, in constructing the auxiliary doublet variable for satisfying the Kutta condition, the logic calls for a quadratic distribution over a panel. The first-order source method is satisfactory for the great majority of external flows. However, for internal flows, flows with strong lift interaction, and, to a lesser degree, flow in highly concave regions, a higher-order formulation[11] is required. This uses a linear source and quadratic doublet distributions and accounts for curvature effects as equivalent source distributions by projecting the curved panel to a tangent flat panel. A particularly severe test case[13] is shown in Fig. 4. It consists of symmetric NASA 10% thick airfoil section, rotated about an axis parallel to its chord and displaced from it by one airfoil thickness. The resulting "ring-wing" is at zero angle of attack; hence, the flow is axisymmetric. Thus, this represents an interior flow of area ratio four and a case of extreme lift interaction, because of the large chord-to-diameter ratio. Figure 4 shows four calculated internal surface pressure distributions, one obtained by a highly accurate axisymmetric method,[14] which was refined until numerical convergence was obtained, and three distributions obtained from three-dimensional panel methods all using the same paneling: first-order source[10] (Hess Prog.), first-order doublet[9] (QUADPAN), and higher-order source[11] (H-O Hess). The failure of the first-order source method is quite dramatic. The higher-order source method gives results that are substantially exact and a significant improvement over those of the doublet method.

Boundary Conditions

Boundary conditions for a panel method consist of three types: 1) the zero normal-velocity boundary condition applied to one control point for each panel on the body surface; 2) the Kutta condition applied at or near the trailing-edge segments of lifting portions of the configuration (Fig. 3); and 3) the zero normal-velocity and zero pressure-jump conditions to be applied along the trailing vortex wake.

In the source method the zero normal-velocity boundary condition is applied in a straightforward manner. The relevant panel influences at the control points are velocities; thus, the matrix of aerodynamic influence is a matrix of vectors. Taking normal components gives the scalar coefficient matrix of the linear equations for singularity strengths. The right sides of these equations are the negative of the normal components of the onset flow. When these equations have been solved together with the Kutta conditions at the trailing-edge segments (see below), velocities at the control points are obtained directly as the product of the vector matrix of panel influences with the singularity strengths. Values of the potential are not required.

Doublet methods have an alternative means of applying the zero normal-velocity boundary condition. This possibility is most easily explained for a pure doublet method, i.e., one with zero source density. Since the normal velocity is continuous through a doublet sheet, the doublet distribution that makes the normal velocity zero on the exterior surface of the body also makes it zero on the interior surface. The interior flow satisfies Laplace's equation subject to a zero normal-velocity boundary condition: thus, the interior velocity field is identically zero, and the interior potential is a constant, which may be taken as zero. If now instead of zero source density the source density implied by Green's identity is distributed over the body, i.e., one proportional to the inclination of the local normal vector to the freestream, it turns out that the interior flow equals freestream. Thus, the perturbation potential is zero in the interior, and this may be used as an alternative to the normal-velocity boundary condition. The relevant panel influences at the control points are potentials, and the matrix of aerodynamic influences is a scalar matrix, which is the coefficient matrix of the linear equations for the doublet strengths. This results in a large reduction in storage requirements. However, the doublet strengths, which are equal to the exterior perturbation potentials at the control points, must be differentiated numerically to obtain surface velocities. Recently published experience[15] with a doublet method indicates that, for portions of the configuration where flow conditions are extreme, such as wing flaps and slats, *both* types of boundary condition should be applied using both source and doublet as unknowns. Thus, both scalar and vector matrices are required for these portions of the configuration, and the order of the system of equations is increased since two equations per panel must be satisfied.

The Kutta condition, which must be applied along trailing edges of lifting portions of the configuration (Fig. 3), determines the distribution of bound vorticity and of course the overall lift. This condition is more important than any other boundary condition, not only because total lift is the quantity of greatest interest in most cases, but because the circulation distribution to a great extent determines the entire flowfield. In theoretical treatments the Kutta condition consists simply of avoiding infinite velocity at a sharp trailing edge. However, in a numerical scheme some other related condition must be used. Several are available. All are theoretically equivalent, and any of them may be used with any form of the panel method. It is in the choice of Kutta condition that the panel methods exhibit their greatest variety. Any method that has been extensively applied to design problems must have developed a satisfactory form of the Kutta condition. Popular types of Kutta condition include specifying doublet continuity at the trailing edge, a highly nonphysical consideration, or requiring the wake to leave the trailing edge in a certain direction, which in practice may be difficult to determine. A desirable Kutta condition should be as physically motivated as possible, and on this basis the choice would appear to be the equality of pressure at the upper and lower surfaces of the trailing edge. This form is not widely used, probably because it is quadratic in the singularity strengths. Thus, a small number of nonlinear equations must be appended to the linear equations for the zero normal-velocity boundary condition. The solution of such a system has been perceived to be more difficult than that of an entirely linear system, which is obtained using the other forms of Kutta condition. However, physically impossible pressure distributions may result. Figure 5 shows three calculated chordwise pressure distributions on a wing,[16] one using an equal-pressure Kutta condition[10] (Hess) and two using other linear forms[8,9] (QUADPAN, VSAERO). The latter two forms exhibit a pressure mismatch at the trailing edge of about half of freestream dynamic pressure.

The third type of boundary condition, that on the wake, is evidently less important than the other two in the vast majority of cases. Variation of doublet, or equivalent vorticity, strength in the wake is determined by conditions at the trailing edge downstream of which the vortex lines must be tangent to the flow direction. Since the flow depends on the wake location and the wake location on the flow, this introduces a nonlinearity into the problem, and the location must be determined by iteration. Unless the wake happens to pass near some downstream portion of the configuration, surface pressures are quite insensitive to its location, which accordingly is simply input in most cases. Cases where its location is important use various wake-iteration routines, which for the steady problem tend to be quite problem-specific, to minimize the number of iterations. Alternatively, the problem may be formulated as an unsteady one and solved step by step in time.[17] Such a procedure is more general, but the number of steps is much greater.

Matrix Solutions

The set of linear equations expressing the zero normal-velocity boundary condition is very favorable for computation. Although not diagonally dominant in the strict mathematical sense, it is very nearly so. It can be shown that to within an amount equal to the truncation error of the numerics the sum of the diagonal terms equals the sum of the off-diagonal terms. Thus, for the average row the diagonal term equals the sum of all the others, although in any particular row it may be somewhat greater or somewhat less. If these equations were to be solved by direct Gaussian elimination, there would be no need for pivoting, and round off would not increase excessively. The equations corresponding to a linear form of the Kutta condition (previous section) do not possess so favorable a diagonal, but in any case the number of such equations is relatively small, and the total system of equations can be solved by direct elimination without difficulty. As is well known, the computational effort for this solution procedure increases as the cube of the order, and its use is limited to relatively small panel numbers. Some investigators[12] use a direct solution for panel numbers of 2000 and even beyond, but most are unwilling to use it beyond 1000 panels, which in practice means that they use the alternate of an iterative solution for all meaningful design cases.

Although most panel methods[8,9,11] solve the equations by an iterative technique, for which computing time varies as the square of the order, the relevant publications do not give details of the solution procedure. The exception is the method of Clark,[18] which is somewhat nontypical because it has been applied mainly to a panel method[11] using the nonlinear equal pressure form of the Kutta condition. This procedure may be categorized as block Gauss-Seidel with acceleration based on the previous iterates. On lifting portions of the configuration each block consists of a single lifting strip of panels (Fig. 3). Within the block, which is solved directly, the unknowns are the source strengths on the panels plus the doublet strength associated with that strip, and the equations are those for the zero normal-velocity boundary condition at the panel control points plus the Kutta condition at the trailing-edge segment. On nonlifting portions panels are associated arbitrarily into blocks. The nonlinear equations for the Kutta condition are linearized at each iteration. These latter equations may be thought of as linear equations whose coefficients change for each iteration. Using the convergence procedure,[18] solutions are usually obtained within 20 iterations. Use of an iterative solution makes feasible greatly increased panel numbers. Currently, 5000–10,000 panels each side of the symmetry plane are quite routine. What this means to the flow solution is illustrated in Figs. 6 and 7. Figure 6 shows a transport aircraft represented by 1000 panels, and the inadequacy of the paneling is evident on the vertical tail and even more evident on the fuselage and aft nacelle. In contrast, a transport aircraft represented by 7000 panels (Fig. 7) has adequate paneling for all components even in the high-lift (flapped) configuration shown.

References

[1]Theodorsen, T., "Theory of Wing Sections of Arbitrary Shape," NACA Rept. 411, 1932.

[2]Garrick, T. E., "Potential Flow about Arbitrary Biplane Wing Section," NACA Rept. 542, 1936.

[3]Ives, D. C. "A Modern Look at Conformal Mapping Including Multiply Connected Regions," *AIAA Journal*, Vol. 14, Aug. 1976, pp. 1006–1011.

[4]Halsey, N. D., "Potential Flow Analysis of Multi-element Airfoils Using Conformal Mapping," *AIAA Journal*, Vol. 17, Dec. 1979, pp. 1281–1288.

[5]James, R. M., "A New Look at Two-Dimensional Incompressible Airfoil Theory," McDonnell Douglas Rept. MDC J0918/1, 1971.

[6]Milne-Thomson, L. M., *Theoretical Hydrodynamics*, Macmillan, New York, 1967.

[7]Lamb, H., *Hydrodynamics*, Cambridge Univ. Press, London, 1932.

[8]Maskew, B., "Prediction of Subsonic Aerodynamic Characteristics: A Case for Low-Order Panel Methods, *Journal of Aircraft*, Vol. 19, Feb. 1982, pp. 157–163.

[9]Coopersmith, R. M., Youngren, H. H., and Bouchard, E. E., "Quadrilateral Element Panel Method (QUADPAN)," Lockheed, Burbank CA, LR 29671, 1981.

[10]Hess, J. L., "The Problem of Three-Dimensional Lifting Flow and Its Solution by Means of Surface Singularity Distribution," *Computer Methods in Applied Mechanics and Engineering*, Vol. 4, 1974, pp. 283–319.

[11]Hess, J. L., Friedman, D. M., and Clark, R. W., "Calculation of Compressible Flow about Three-Dimensional Inlets with Auxiliary Inlets, Slats, and Vanes by Means of a Panel Method," NASA CR-174975, 1985.

[12]Johnson, F. T., "A General Panel Method for the Analysis and Design of Arbitrary Configurations in Incompressible Flows," NASA CR-3079, 1980.

[13]Miranda, L. R., "Reply by Author to J. L. Hess," *Journal of Aircraft*, Vol. 22, April 1985, p. 352.

[14]Hess, J. L., "Improved Solutions for Potential Flow about Arbitrary, Axisymmetric Bodies by Use of a Higher-Order Surface Source Method," *Computer Methods in Applied Mechanics and Engineering*, Vol. 5, 1975, pp. 297–308.

[15]Tinoco, E. N., Ball, D. N., and Rice, F. A., II, "PAN AIR Analysis of a Transport High-Lift Configuration," *Journal of Aircraft*, Vol. 24, March 1987, pp. 181–187.

[16]Margason, R. J., Kielgaard, S. O., Sellers, W. L., Morris, C. E., Walkey, K. B., and Shields, E. W., "Subsonic Panel Methods—A Comparison of Several Production Codes," AIAA Paper 85-0280, 1985.

[17]Katz, J., and Maskew, B., "Unsteady Low-Speed Aerodynamic Model for Complete Aircraft Configurations," *Journal of Aircraft*, Vol. 25, April 1988, pp. 302–310.

[18]Clark, R. W., "A New Iterative Matrix Solution Procedure for Three-Dimensional Panel Methods," AIAA Paper 85-0176, 1985.

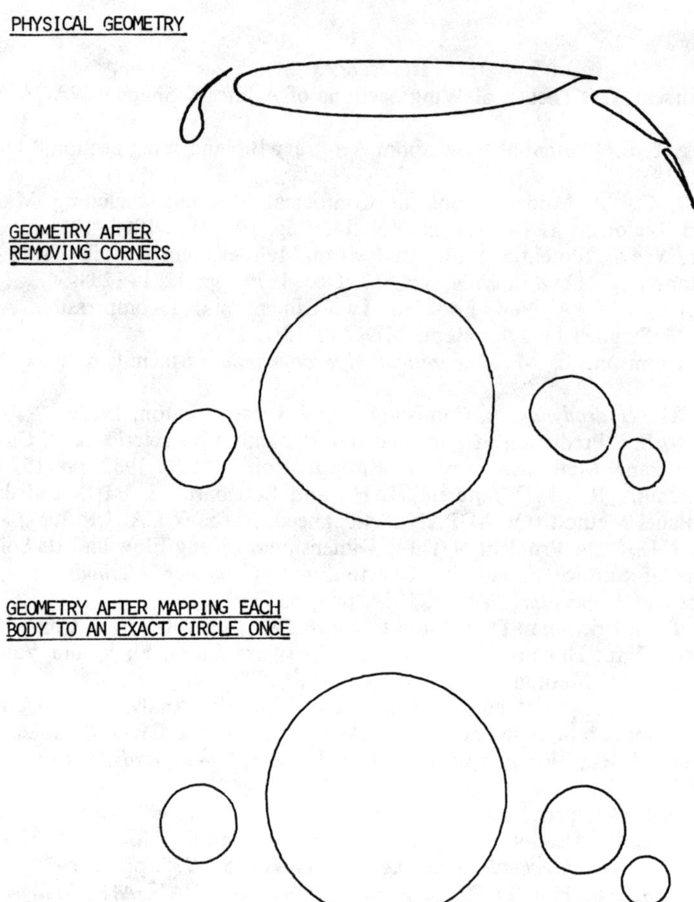

Fig. 1   Transformation of a four-element airfoil into four circles.

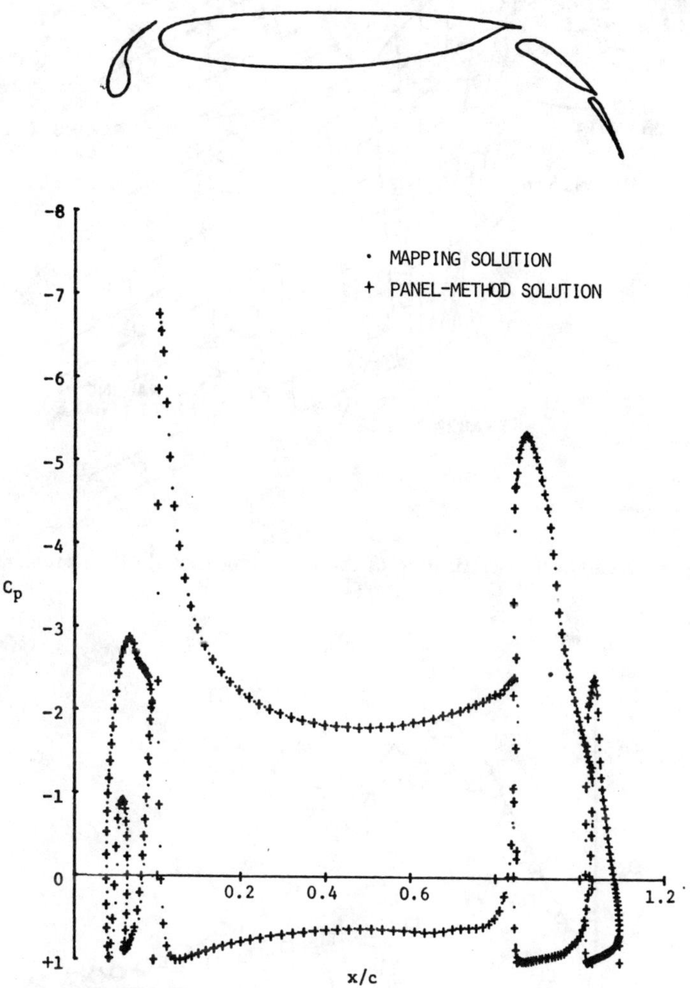

Fig. 2   Comparison of results of a panel method and the conformal mapping method (four-element airfoil case).

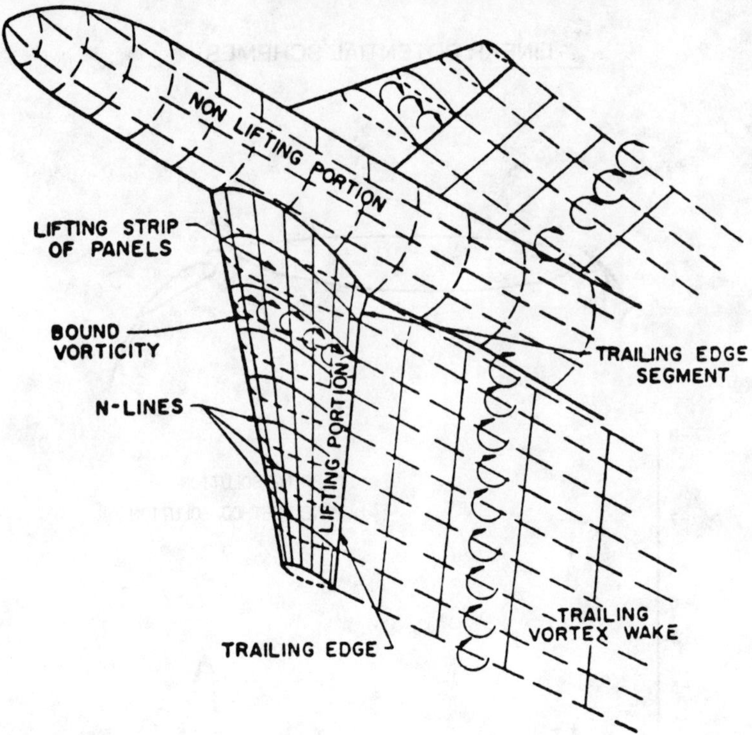

Fig. 3   Numerical discretization of the three-dimensional lifting problem.

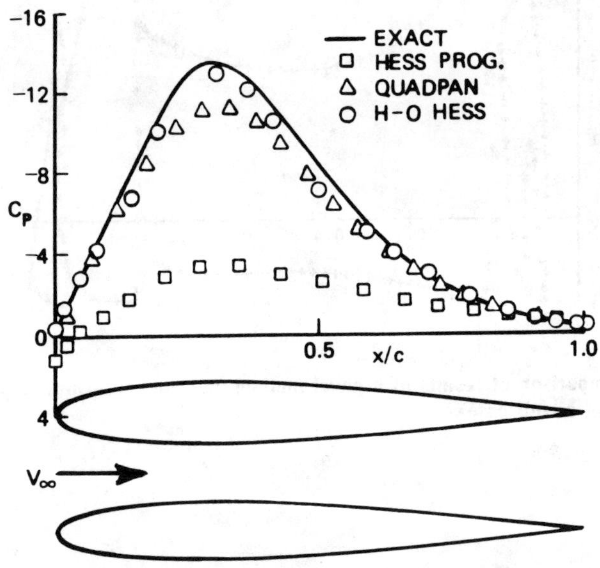

Fig. 4   Pressure distributions on the internal surface of a ring wing in axisymmetric flow.

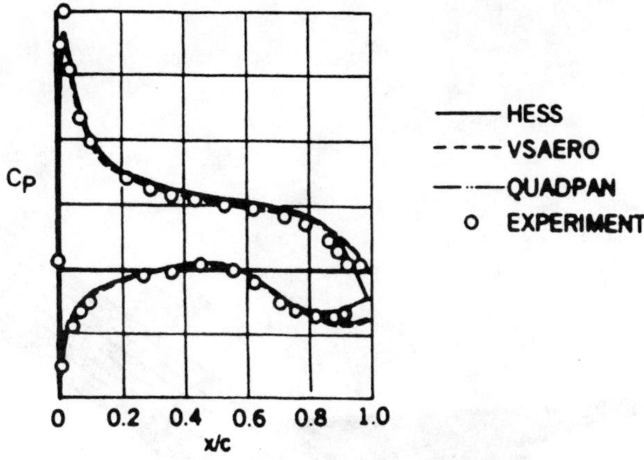

Fig. 5   Mismatch in trailing-edge pressures due to alternate Kutta conditions.

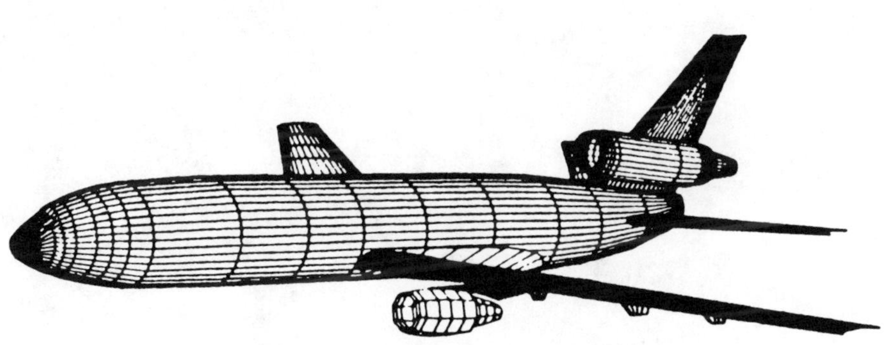

Fig. 6   A 1000-panel representation of a transport aircraft.

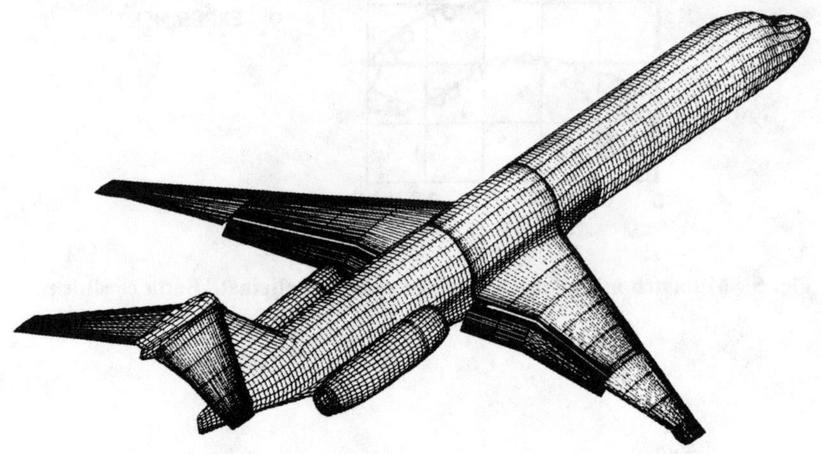

Fig. 7   A 7000-panel representation of a transport aircraft.

Chapter 3

Full-Potential, Euler, and Navier-Stokes Schemes

Antony Jameson*
Princeton University, Princeton, New Jersey

Introduction

THE purpose of this chapter is to survey some of the highlights of computational fluid dynamics (CFD) schemes for solving the full-potential, Euler, and Navier-Stokes equations. Prior to the advent of the computer, there was a rather comprehensive mathematical formulation of fluid mechanics already in place. This formulation had been developed by elegant mathematical analysis, frequently guided by brilliant insights. Well-known examples include the airfoil theory of Kutta and Joukowski, Prandtl's wing and boundary-layer theories, von Kármán's analysis of the vortex street, and, more recently, Jones' slender wing theory[1] and Hayes' theory of linearized supersonic flow.[2] These methods require simplifying assumptions of various kinds and cannot be used to make quantitative predictions of complex flows dominated by nonlinear effects. The computer opens up new possibilities for attacking these problems by direct calculation of solutions to more complete mathematical models.

The main uses of CFD in aeronautical science fall into two broad categories. First, there is the objective of providing reliable aerodynamic predictions, which will enable designers to produce better airplanes. Second, there is the possibility of using CFD for purely scientific investigations. It seems possible that numerical simulation of complex flows not readily accessible to experimental measurements can provide new insights into the underlying physical processes. In particular, computational methods offer a new tool for the study of structures in turbulent flow and the mechanisms of transition from laminar to turbulent flow.

Most of this chapter is devoted to the use of computational methods for aerodynamic prediction. This is a comparatively recent development. Prior to 1965 computational methods were hardly used in aerodynamic analysis, although they were already widely used for structural analysis. The primary tool for the development of aerodynamic configurations was the wind tunnel. Experimental aerodynamicists could arrive at efficient shapes

---

Copyright © 1989 by Antony Jameson. Published by the American Institute of Aeronautics and Astronautics, Inc. with permission.
*James S. McDonnell Professor, Department of Aerospace Engineering.

through testing guided by good physical insight. Notable examples of the power of this method include Whitcomb's discovery of the area rule for transonic flow and his subsequent development of aft-loaded supercritical airfoils.[3,4] By the 1960s it began to be recognized that computers had become powerful enough to make it worthwhile to attempt calculations of aerodynamic properties of at least isolated components of an aircraft. It was also apparent that, depending on the intended application, useful simulations might be achieved with a range of mathematical models of varying complexity. Commercial aircraft fly largely with attached flows, in which the viscous effects are confined to the boundary layer. Consequently, they have a relatively small effect on the global flow pattern, other than their role in establishing circulatory flows through the shedding of start-up vortices off the trailing edges of lifting surfaces. Inviscid flow predictions then serve a useful role and can take advantage of irrotationality to simplify the equations through the introduction of a velocity potential. This reduction led to the first major advance, the introduction of panel methods to solve the linearized potential-flow equation. The initial demonstration of this approach by Hess and Smith[5] was soon followed by its extension to lifting flows[6] and to linearized supersonic flow.[7]

The 1970s saw widespread efforts to develop methods of predicting transonic flows with shock waves, which required the use of a nonlinear mathematical model. The first major breakthrough was the scheme of Murman and Cole[8,9] for treating the transonic small-disturbance equation. This was the catalyst for widespread development of methods for calculating transonic potential flows in two and three dimensions using either the small-disturbance equation or the full-potential-flow equation.

In parallel, efforts were underway to devise efficient algorithms for solving the Euler and Navier-Stokes equations. Following the pioneering efforts of Magnus and Yoshihara,[10] MacCormack introduced his famous explicit difference scheme in 1970.[11] Efforts to improve efficiency led to the implicit scheme of Beam and Warming,[12] which was extended to general curvilinear coordinates by Steger.[13] The need to find a better shock-capturing method was also apparent and stimulated the introduction of flux splitting.[14] By 1979, however, Euler methods remained very expensive and had not attained levels of accuracy that justified their routine use for engineering design. The GAMM Workshop of 1979 served to highlight the deficiencies of the methods then available.[15] Nevertheless, it was already evident that advances in the available computing power would soon make it entirely feasible to solve the three-dimensional Euler equations, and the 1980s have seen widespread efforts to realize this objective. The alternating-direction method has been systematically developed into an effective tool, and the current state of the art is represented by ARC2D and ARC3D.[16] Implicit schemes using LU decomposition[17] and relaxation have also proved successful. A parallel path of development that has also led to efficient programs has been the use of multistage explicit time-stepping schemes.[18] The author's FLO52 and FLO57 programs using this concept have been widely used. Stemming from the mathematical theory of shock waves, procedures have also been developed for the design of effective shock-capturing schemes. There have been intensive efforts to find more rapidly convergent methods to find steady-state solutions. In particular, the use of multiple grids, first introduced by Fedorenko[19] and subsequently developed by Brandt,[20] has been extended to the treatment of hyperbolic systems[21-23] and has proved to be extremely effective.

We are now at a point where a variety of efficient algorithms for the solution of the Euler and Navier-Stokes equations have been developed, and the principles underlying their construction are quite well understood. Their application to date has largely been limited to relatively simple configurations because of the difficulty of generating meshes around complex shapes. Viscous effects in attached flows can be fairly well predicted by making boundary-layer corrections. Military aircraft frequently fly in conditions of separated flow. The appropriate mathematical model is then the Navier-Stokes equations. At Reynolds numbers typical of full-scale flight, however, the flow becomes turbulent, and the disparity of scales in a turbulent flow is so large that direct simulation is probably not feasible without radical developments in computer technology. Therefore, it becomes necessary to resort to Reynolds averaging, and the equations must be closed by a turbulence model. Progress in simulating separated viscous flows may now be more dependent on improvement in turbulence modeling than it is on algorithm development.

Computational aerodynamics has reached a point of maturity where it may be worthwhile to take stock of the present situation and to consider which directions of future efforts are likely to be most profitable. In this chapter some of the algorithmic concepts believed to be a foundation for future developments will be identified. It seems useful for this purpose first to consider the objectives of computational aerodynamics. Three levels of desirable performance can be identified:

1) the capability of predicting the flow past airplanes in different flight regimes (takeoff, cruise at transonic speed, flutter);

2) the interactive calculations to allow immediate improvement of the design; and

3) the integration of the predictive capability into an automatic design method using computer optimization and artificial intelligence.

To date not even the first level has been fully realized for all regimes of flight. Some methods are fast enough that the second level is already feasible, say, for airfoil evaluation. Some pioneering attempts have been made at the third level, and it is clear that advances in computational power and algorithmic efficiency will make this feasible for useful applications within the coming decade.

It is also important to understand what kind of information the designer may be seeking. For the final design he may need accurate quantitative predictions of design parameters such as the lift and drag coefficients. In the early stages he may be more interested in acquiring a qualitative understanding of the nature of the flowfield and the impact of design changes on the onset of separation, for example, or the location of the regions of separated flow.

The requirements to be met by an effective method include the following:

1) the capability of simulating the main features of the flow, such as shock waves and vortex sheets;

2) the prediction of viscous effects;
3) the ability to handle geometrically complex configurations; and
4) the efficiency in both computational and human effort.

In any case it is clear that the value of the information provided must be measured against the cost of producing it. In the application of computer simulations to engineering design, we can therefore anticipate that simplified mathematical models will continue to be useful for preliminary estimations and tradeoff studies for which full details of the flowfield are not essential. On the other hand, there is a pervasive need to predict flows over exceedingly complex configurations, and future computational methods must be designed to address this requirement.

The remaining sections review some of the main algorithmic developments of the past two decades in this context. The next section reviews the mathematical models. The section on algorithms for potential flow covers potential-flow methods and the section on algorithms for Euler equations and the section on viscous flow calculations cover methods for the full inviscid and viscous equations. In the conclusion, I try to identify what I believe to be the principal remaining problems, including algorithmic issues such as the construction of schemes with a higher order of accuracy, convergence acceleration, and shock-capturing or front-tracking schemes, and also computer science issues such as concurrent calculation on vector, pipelined, or parallel processors; optimization and design techniques; and expert systems.

Mathematical Models of Fluid Flow

The equations for flow of a gas in thermodynamic equilibrium are the Navier-Stokes equations. Let ρ, u, v, E, and p be the density, Cartesian velocity components, total energy, and pressure, respectively, and let x and y be Cartesian coordinates. Then, for a two-dimensional flow these equations can be written as

$$\frac{\partial w}{\partial t} + \frac{\partial f}{\partial x} + \frac{\partial g}{\partial y} = \frac{\partial R}{\partial x} + \frac{\partial S}{\partial y} \tag{1}$$

where w is the vector of dependent variables, and f and g are the convective flux vectors:

$$w = \begin{bmatrix} \rho \\ \rho u \\ \rho \\ \rho E \end{bmatrix}, \qquad f = \begin{bmatrix} \rho u \\ \rho u^2 + p \\ \rho uv \\ \rho uH \end{bmatrix}, \qquad g = \begin{bmatrix} \rho v \\ \rho vu \\ \rho v^2 + p \\ \rho vH \end{bmatrix} \tag{2}$$

Here H is the enthalpy,

$$H = E + \frac{p}{\rho}$$

and the pressure is obtained from the equation of state

$$p = (\gamma - 1)\rho[E - \tfrac{1}{2}(u^2 + v^2)] \tag{3}$$

The flux vectors for the viscous terms are

$$\boldsymbol{R} = \begin{bmatrix} 0 \\ \tau_{xx} \\ \tau_{xy} \\ u\tau_{xx} + v\tau_{xy} \end{bmatrix}, \quad \boldsymbol{S} = \begin{bmatrix} 0 \\ \tau_{xy} \\ \tau_{yy} \\ u\tau_{xy} + v\tau_{yy} \end{bmatrix} \tag{4}$$

where the viscous stresses are

$$\tau_{xx} = 2\mu u_x - \frac{2\mu}{3}(u_x + v_y)$$

$$\tau_{yy} = 2\mu v_y - \frac{2\mu}{3}(u_x + v_y)$$

$$\tau_{xy} = \mu(u_y + v_x)$$

and μ is the coefficient of viscosity. The computational requirements for the simulation of turbulent flow have been estimated by Chapman.[24] They are clearly beyond the reach of current computers.

The first level of approximation is to resort to time averaging of rapidly fluctuating components. This yields the Reynolds equations, which require a turbulence model for closure. Since a universally satisfactory turbulence model has yet to be found, current turbulence models have to be tailored to the particular flow. The Reynolds equations can be solved with computers of the class of the Cray 1 or Cyber 205, at least for two-dimensional flows, such as flows over airfoils.

The next level of approximation is to eliminate viscosity. Equation (1) then reduces to the Euler equation

$$\frac{\partial w}{\partial t} + \frac{\partial f}{\partial x} + \frac{\partial g}{\partial y} = 0 \tag{5}$$

It is quite feasible to solve complex three-dimensional flows with this model, as will be discussed.

If we assume the flow to be irrotational, we can introduce a velocity potential ϕ and set

$$u = \phi_x, \qquad v = \phi_y$$

The Euler equation [Eq. (5)] now reduces to the potential-flow equation

$$\frac{\partial}{\partial x}(\rho\phi_x) + \frac{\partial}{\partial y}(\rho\phi_x) = 0 \tag{6}$$

or, in quasilinear form,

$$(c^2 - u^2)\phi_{xx} - 2uv\phi_{xy} + (c^2 - v^2)\phi_{yy} = 0 \tag{7}$$

where c is the speed of sound. This is given by

$$c^2 = \frac{\gamma p}{\rho}$$

where γ is the ratio of specific heats. According to Crocco's theorem, vorticity in a steady flow is associated with entropy production through the relation

$$q \times \zeta + T \cdot \nabla S = 0$$

where q and ζ are the velocity and vorticity vectors, respectively, T is the temperature, and S is the entropy. Thus, the introduction of a potential is consistent with the assumption of isentropic flow. Then, if M_∞ is the freestream Mach number, the units may be normalized so that

$$p = \frac{\rho^\gamma}{\gamma M_\infty^2}, \qquad \rho = (M_\infty^2 c^2)^{1/\gamma - 1} \tag{8}$$

while the local speed of sound can be determined from the energy equation, written as

$$\frac{c^2}{\gamma - 1} + \frac{q^2}{2} = \frac{1}{(\gamma - 1)M_\infty^2} + \frac{1}{2} \tag{9}$$

Because shock waves generate entropy, they cannot be exactly modeled by the potential-flow equation. However, weak solutions admitting isentropic jumps that conserve mass but not momentum are a good approximation to shock waves, as long as the shock waves are quite weak (with a Mach number <1.3 for the normal velocity component upstream of the shockwave). Stronger shock waves tend to separate the flow, with the result that the inviscid approximation is no longer adequate. Thus, this model is well balanced and has proved extremely useful for estimating the cruising performance of transport aircraft.

If one assumes small disturbances and a Mach number close to unity, the potential equation can be reduced to the transonic small-disturbance equation. A typical form is

$$[1 - M_\infty^2 - (\gamma + 1)M_\infty^2 \phi_x]\phi_{xx} + \phi_{yy} = 0 \tag{10}$$

Finally, if the freestream Mach number is not close to unity, the potential-flow equation can be linearized as

$$(1 - M_\infty^2)\phi_{xx} + \phi_{yy} = 0 \tag{11}$$

Algorithms for Potential Flow

Overview

Although the Euler- and Reynolds-averaged Navier-Stokes equations can now be solved with quite moderate computational costs, algorithms for potential flow remain useful because they can provide extremely inexpensive quick estimates. Also, certain ideas for shock-capturing and convergence acceleration that were first developed for potential-flow calculations have proved transferable to more complex models such as the Euler equations.

Upwind Differencing

When the potential-flow equation [Eq. (6)] is used to predict transonic flows, the difficulty arises that the solution is invariant under a reversal of the velocity vector ($u = -\phi_x$, $v = -\phi_y$). Consider a transonic flow past an ellipse with a compression shock wave. Then there is a corresponding solution with an expansion shock wave (see Figs. 1a and 1b). In fact, a central-difference scheme would preserve fore-and-aft symmetry, leading to a solution of the type illustrated in Fig. 1c. In 1970 the landmark paper of Murman and Cole[8] appeared. This demonstrated a simple way to obtain physically relevant solutions of the transonic small-disturbance equation [Eq. (10)]. Writing this equation as

$$A\phi_{xx} + \phi_{yy} = 0 \qquad (12)$$

where A is the nonlinear coefficient in Eq. (10), they proposed the use of central differencing if $A > 0$ (subsonic flow), but upwind differencing for ϕ_{xx} if $A < 0$ (supersonic flow), as illustrated in Fig. 2. The equations were then solved by a line-relaxation scheme, in which the unknowns were determined simultaneously on each successive vertical line, marching in the streamwise direction. The scheme amounted to a combination of a relaxation method for the subsonic zone, in which the equation is elliptic, with an implicit scheme for the wave equation in the supersonic zone.

This work was extremely important both because it pointed the way to reasonably inexpensive simulations of transonic flows and also because it demonstrated for the first time the possibility of an effective shock-capturing scheme with a sharp, nonoscillatory discrete shock structure. Within the next few years the concept of Murman and Cole was generalized to the full transonic potential-flow equation and applied to a wide variety of flow simulations.

In order to treat the full quasilinear potential-flow equation [Eq. (7)], one may rewrite it in a coordinate system locally aligned with the flow. Equation (7) then becomes

$$(c^2 - q^2)\phi_{ss} + c^2\phi_{nn} = 0$$

where

$$\phi_{ss} = \frac{u^2}{q^2}\,\phi_{xx} + \frac{2uv}{q^2}\,\phi_{xy} + \frac{v^2}{q^2}\,\phi_{yy}$$

and

$$\phi_{nn} = \frac{v^2}{q^2}\,\phi_{xx} - \frac{2uv}{q^2}\,\phi_{xy} + \frac{u^2}{q^2}\,\phi_{yy}$$

Upwind differencing is now used for all second derivatives contributing to ϕ_{ss} whenever $q > c$. This leads to Jameson's rotated difference scheme[25] (see also Albone[26]). A convergent iterative scheme can be derived by regarding the iterations as time steps in an artificial time coordinate. The principal part of the equivalent time-dependent equation has the form

$$(M^2 - 1)\phi_{ss} - \phi_{nn} + 2\alpha\phi_{st} + 2\beta\phi_{nt} = 0$$

Introducing a new time coordinate

$$T = t - \frac{\alpha s}{M^2 - 1} + \beta n$$

this becomes

$$(M^2 - 1)\phi_{ss} - \phi_{nn} - \left(\frac{\alpha^2}{M^2 - 1} - \beta^2\right)\phi_{TT} = 0$$

If the flow is locally supersonic, T is spacelike and either s or n is timelike. Since s is the timelike direction in the steady-state problem, and the time-dependent problem is compatible with the steady-state problem only if

$$\alpha^2 > \beta^2(M^2 - 1)$$

This generally requires the explicit addition of a term in ϕ_{st}.

In his paper of 1973, Murman recognized that the switch in the difference scheme could violate the conservative form of the equations, leading to shock jumps that violated the conservation of mass.[9] This difficulty can be corrected by reformulating the switch to upwind differencing by the introduction of artificial viscosity. The dominant discretization error in the upwind difference formula for ϕ_{xx} is $-\Delta x\phi_{xxx}$, and terms of this nature can be added explicity in conservation form, leading to special transition operators across the sonic line. An appropriate form of artificial viscosity for the potential flow equation [Eq. (6)] is a difference approximation to

$$\Delta x\,\frac{\partial}{\partial x}\,\mu|u|\rho_x + \Delta y\,\frac{\partial}{\partial y}\,\mu|v|\rho_y$$

where Δx and Δy are the mesh widths, and μ is a switch function

$$\mu = \max\left\{0,\, 1 - \frac{1}{M^2}\right\}$$

which cuts off the viscosity in the subsonic zone.[27] It was realized by several authors that a term of this kind can be added simply by biasing the density in an upwind direction.[28-30] This has facilitated the development of discretizations on arbitrary subdivisions of the domain into hexahedrons or tetrahedrons.

Convergence Acceleration

Transonic flow calculations by relaxation methods generally require a very large number of iterations to converge (on the order of 500–2000). This inhibited the more widespread use of these methods, particularly for three-dimensional calculations, and stimulated numerous efforts to find more rapidly convergent methods. The two most effective approaches have been approximate factorization of the difference operator and acceleration by the use of multiple grids.

Let the difference equations be written as

$$L\phi = 0 \tag{13}$$

where L is a nonlinear difference operator, and ϕ is the solution vector. Then a typical iterative scheme can be written as

$$N\delta\phi + L\phi^n = 0 \tag{14}$$

where $\delta\phi$ is the correction, and N is a linear operator that can be inverted relatively cheaply and should approximate L (in the linear case the error is reduced at each cycle by $I - N^{-1}L$). In an approximate factorization method, N is formed as a product

$$N = N_1 N_2 \ldots N_q$$

of easily invertible operators. Ballhaus et al.[31] found that a good choice for the small-disturbance equation [Eq. (12)] is

$$(\alpha - AD_x^{\sim})(\alpha D_x^- - D_y^+)\delta\phi + \alpha L\phi^-$$

where D_x^+ and D_x^- are forward- and backward-difference operators, and

$$D_x^{\sim} = \begin{cases} D_x^+ & \text{if} \quad A > 0 \\ D_x^- & \text{if} \quad A < 0 \end{cases}$$

Very efficient schemes of this type have been developed for the transonic potential-flow equation by Holst.[32]

The multigrid method was first proposed by Fedorenko,[19] and some promising results for the small-disturbance equation were obtained by Brandt and South.[33] The idea is to use corrections calculated on a sequence of successively coarser grids to improve the solution on a fine grid. Consider a linear problem and let

$$L_h \phi_h = 0 \qquad (15)$$

be the discrete equations for a mesh with a spacing proportional to h. Let u_h be an estimate of ϕ_h, and let v_h be a correction that should reduce $L_h(u_h + v_h)$ to zero. Then, instead, one can write an equation for v on a mesh with twice as large a spacing:

$$L_{2h}v_{2h} + Q_{2h}^h L_h u_h = 0 \qquad (16)$$

where Q_{2h} is a collection operator that forms a weighted average of the residuals on the fine grid in the neighborhood of each mesh point of the coarse grid. The correction is finally interpolated back to the fine grid:

$$u_h^{\text{new}} = u_h + P_h^{2h} v_{2h} \qquad (17)$$

where P_h^{2h} is an interpolation operator. Corrections to the solution of Eq. (16) can in turn be calculated on a still coarser grid, and so on. The same basic iterative scheme can be used on all of the grids in the sequence. It has been proved that solutions to elliptic problems with N unknowns can be obtained in $\mathcal{O}(N)$ operations by the use of multiple grids.[34] A condition for the successful use of multiple grids is that, before passing to a coarser grid, the high-frequency error modes should be reduced to the point that the remaining error can be properly resolved on the coarser grid.

The method can be reformulated for a nonlinear problem by explicitly introducing the solution vector u_{2h} on the coarse grid. An updated solution vector $\bar{u}_{2h}$ is then calculated from the equation

$$L_{2h}\bar{u}_{2h} + Q_{2h}^h L_h u_h - L_{2h}u_h = 0$$

where the difference between the collected residuals from neighboring points on the fine grid and the residual calculated on the coarse grid appears as a forcing function. The correction $\bar{u}_{2h} - u_{2h}$ is then interpolated back to the fine grid.

Figure 3 shows the result of a calculation in which a generalized alternating-direction (ADI) method was used to drive the multigrid iteration.[35] The ADI scheme differs from the standard ADI scheme in replacing the scalar parameter by a difference operator (which also operates on the residuals). The purpose of this is to retain a well-posed problem in the supersonic zone. An efficient strategy is to use a simple V cycle in which one ADI iteration is performed on each grid until the coarsest grid is reached, and then one ADI iteration on each grid on the way back up to the fine grid. A solution on a 192×32 grid accurate to four figures was obtained by 3 V cycles on a 48×8 grid, followed by three V cycles on a 96×16 grid, and 3 V cycles on the 192×32 grid. The total calculation is equivalent to 4 V cycles on the 192×32 grid. It seems likely that this must be close to the lower bound for the number of operations required to solve 6144 simultaneous nonlinear equations.

Treatment of Complex Geometry

An effective approach to the treatment of two-dimensional flows over complex profiles is to map the exterior domain conformally onto the unit disk.[25] Equation (6) is then written in polar coordinates as

$$\frac{\partial}{\partial \theta}\left(\frac{\rho}{r}\phi_\theta\right) + \frac{\partial}{\partial r}(r\rho\phi_r) = 0$$

where the modulus h of the mapping function enters only in the calculation of the density from the velocity

$$q = \frac{\nabla\phi}{h}$$

This procedure is very accurate.

Applications to complex three-dimensional configurations require a more flexible method of discretization, such as that provided by the finite-element method. Jameson and Caughey[36] proposed a scheme using isoparametric bilinear or trilinear elements. The discrete equations can most conveniently be derived from the Bateman variational principle. This states that the integral

$$I = \int\int p \, \mathrm{d}x \, \mathrm{d}y$$

is stationary in two-dimensional potential flow. It follows from Eqs. (8) and (9) that

$$\frac{\partial p}{\partial u} = -\rho u, \qquad \frac{\partial p}{\partial v} = -\rho v$$

whence, in potential flow,

$$\delta I = -\int\int (\rho u \delta\phi_x + \rho v \delta\phi_y) \, \mathrm{d}x \, \mathrm{d}y$$

and Eq. (6) is recovered on integrating by parts and allowing arbitrary variation $\delta\phi$. In the scheme of Jameson and Caughey I is approximated as

$$I = \Sigma \, P_k V_k$$

where p_k is the pressure at the center of the kth cell, and V_k is its area (or volume), and the discrete equations are obtained by setting the derivative of I with respect to the nodal values of potential to zero. Artificial viscosity is added to give an upwind bias in the supersonic zone, and an iterative scheme is derived by embedding the steady-state equation in an artificial, time-dependent equation. Several widely used codes (FLO27, FLO28, FLO30) have been developed using this scheme.

An alternative approach to the treatment of complex configurations has been developed by Bristeau et al.[37] Their method uses a least-squares formulation of the problem, together with an iterative scheme derived with the aid of optimal control theory. The method could be used in conjunction with a subdivision into either quadrilaterals or triangles, but in practice triangulations have been used. The least-squares method in its basic form allows expansion shocks. In early formulations these were eliminated by penalty functions. It was subsequently found to be best to use upwind biasing of the density. The method has been extended at Avions Marcel Dassault to the treatment of extremely complex three-dimensional configurations, using a subdivision of the domain into tetrahedrons. A striking success was achieved in 1982 with the first simulation of transonic flow past a complete aircraft by solution of the full quasilinear potential-flow equation, as illustrated in Fig. 4.

Algorithms for the Euler Equations

Overview: Time-Dependent Formulation

In parallel with the development of effective algorithms for potential flow, there were ongoing efforts to derive fast, accurate, and reliable methods for solving the Euler equations. Steady-state solutions are typically needed for design applications. The introduction of a space discretization procedure then reduces the problem to the solution of a large number of coupled nonlinear equations. These equations might be solved by a variety of iterative methods. Two possibilities in particular are the least-squares method[37] and the Newton iteration.[38] However, it has generally been found expedient to use the time-dependent equations as a vehicle for reaching the steady state. Some advantages of this strategy include the following:

1) The strategy is simple.

2) It is possible to use the same computer program to calculate steady and unsteady flows.

3) The time-dependent problem provides a natural framework for the design of nonoscillatory shock-capturing schemes that reflect the physics of wave propagation.

4) Algorithms can be devised for concurrent computation on vector, pipelined, or parallel processors either through the use of an explicit time-stepping scheme or through the use of an iterative procedure at each time step of an implicit scheme.

It has also been found that satisfactory schemes should be designed to conform to some general guidelines. Some of these include the following:

1) The conservation laws of gasdynamics should be satisfied in discrete form by the numerical approximation.

2) Shock waves and contact discontinuities should be automatically captured by the difference scheme.

3) In steady flow calculations the final steady state should be independent of the time-stepping scheme.

4) Invariant quantities in the flowfield, such as entropy upstream of a shock wave or total enthalpy in a steady flow, should also be invariant in the numerical solution.

5) Uniform flow should be an exact solution of the difference equations on an arbitrary mesh.

An alternative to guideline 2 is automatic detection of shock waves in conjunction with front tracking. In this case guideline 1, which is needed to ensure the satisfaction of correct jump conditions by a shock-capturing scheme,[39] is no longer strictly necesssary, but it remains desirable since it ensures global conservation of mass, momentum, and energy.

The early standard for time-stepping methods was set by the two-stage scheme of MacCormack,[11] which has been very widely used. To solve the one-dimensional system

$$\frac{\partial w}{\partial f} + \frac{\partial}{\partial x} f(w) = 0 \tag{18}$$

the scheme advances from time level n to time level $n + 1$ by setting

$$\tilde{w} = w^n - \Delta t D_x^+ f(w^n)$$

and

$$w^{n+1} = w^n - \frac{\Delta t}{2} D_x^+ [f(w^n) + D_x^- f(\tilde{w})] \tag{19}$$

where the superscripts denote the time level, and D_x^+ and D_x^- are forward- and backward-difference operators approximating $\partial/\partial x$:

$$D_x^+ f_i = \frac{f_{i+1} - f_i}{\Delta x}, \qquad D_x^- f_i = \frac{f_i - f_{i-1}}{\Delta y}$$

The value at the end of the time step is first predicted using forward differences, and then the predicted value is used in the calculation of the final corrected value w^{n+1} by a formula that is centered about the middle of the time step.

This is the simplest known two-level scheme that is both stable and second-order-accurate. Additional dissipative terms have to be introduced to eliminate oscillations in the vicinity of shock waves. The scheme also does not satisfy principle 3, since it yields a steady state that depends on the time step Δt. Nor is the enthalpy constant in discrete steady solutions. However, the algorithm performs well in the absence of discontinuities in the flow.

A convenient way to meet requirement 3 is to separate the space-marching procedure entirely from the time-marching procedure by applying first a semidiscretization. This has the advantage of allowing the problems of spatial discretization error, artificial dissipation, and shock modeling to be studied independently of the problems of time-marching stability and convergence acceleration.

Space Discretization of the Euler Equations

Following the lead of MacCormack and Paullay,[40] the space discretization of the Euler equation [Eq. (5)] can be derived in a very natural way from the integral form

$$\frac{\partial}{\partial t} \int_S w \, dS + \int_{\partial S} [f(w) \, dy - g(w) \, dx] = 0 \tag{20}$$

for a domain S with boundary ∂S.

If we divide the domain into a large number of small subdomains, we can use Eq. (20) to estimate the average rate of change of w in each subdomain. This is an effective method to obtain discrete approximations to Eq. (5), which preserve its conservation form. In general, the subdomains could be arbitrary, but it is convenient to use either quadrilateral or triangular cells. Correspondingly, it is convenient to use either distorted cubic or tetrahedral cells in three-dimensional calculations. Alternative discretizations may be obtained by storing sample values of the flow variables at either the cell centers or the cell corners. These variations are illustrated in Fig. 5 for a two-dimensional case.

Figures 5a and 5b show cell-centered schemes on rectilinear and triangular meshes.[18,41] In either case Eq. (20) is written for the cell labeled 0 as

$$\frac{d}{dt}(Sw) + Q = 0 \tag{21}$$

where S is the cell area, and Q is the net flux out of the cell. This can be approximated as

$$Q = \sum_k (\bar{f}_{0k} \, \Delta y_{0k} - \bar{g}_{0k} \, \Delta x_{0k}) \tag{22}$$

where the sum is over the edges of cell 0, Δx_{0k} and Δy_{0k} are measured along the edge separating cell 0 from cell k, and the flux vectors $\bar{f}_{0k}$ and $\bar{g}_{0k}$ are evaluated by taking the average of their values in cell 0 and cell k:

$$\bar{f}_{0k} = \tfrac{1}{2}(f_0 + f_k), \qquad \bar{g}_{0k} = \tfrac{1}{2}(g_0 + g_k) \tag{23}$$

An alternative averaging procedure is to multiply the average value of the convected quantity, ρ_{0k}, in the case of the continuity equation, for example, by the rate of transport

$$Q_{0k} = \tfrac{1}{2}(u_0 + u_k)\,\Delta y_{0k} - \tfrac{1}{2}(v_0 + v_k)\,\Delta x_{0k} \qquad (24)$$

obtained by taking the inner product of the mean of the velocity vector $\boldsymbol{q}$ with the unit normal multiplied by the edge length.

Figures 5c and 5d show corresponding schemes on rectilinear and triangular meshes in which the flow variables are stored at the vertices.[42] We can now form a control volume for each vertex by taking the union of the cells meeting at that vertex. Equation (21) then takes the form

$$\frac{\mathrm{d}}{\mathrm{d}t}\left(\sum_k V_k\right)\boldsymbol{w} + \sum_k Q_k = 0 \qquad (25)$$

where V_k and Q_k are the area and flux balance, respectively, for the kth cell in the control volume. The flux balance for a given cell is now approximated as

$$Q = \sum_\ell (\bar{f}_\ell\,\Delta y_\ell - \bar{f}_\ell\,\Delta x_\ell) \qquad (26)$$

where Δx_ℓ and Δy_ℓ are measured along the ℓth edge, and f_ℓ and g_ℓ are estimates of the mean flux vectors across that edge. Fluxes across internal edges cancel when the sum $\Sigma_k Q_k$ is taken in Eq. (25), so that only the external edges of the control volume contribute to its flux balance. The mean flux vector across an edge can be conveniently approximated as the average of the values at its two endpoints,

$$\bar{f}_{12} = \tfrac{1}{2}(f_1 + f_2), \qquad \bar{g}_{12} = \tfrac{1}{2}(g_1 + g_2)$$

in Fig. 5c or 5d, for example. The sum $\Sigma\,Q_k$ in Eq. (25), which then amounts to a trapezoidal integration rule around the boundary of the control area, should remain fairly accurate even when the mesh is irregular.

Dissipation, Upwinding, and Total Variation Diminishing Schemes

Equations (21) and (22) represent nondissipative approximations to the Euler equations. Dissipative terms may be needed for two reasons. First, there is the possibility of undamped oscillatory modes. For example, when either a cell-centered or a vertex formulation is used to represent a conservation law on a rectilinear mesh, a mode with values ± 1 alternately at odd and even points leads to a numerically evaluated flux balance of zero in every interior control volume. Although the boundary conditions may suppress such a mode in the steady-state solution, the absence of damping at interior points may have an adverse effect on the rate of convergence to the steady state.

The second reason for introducing dissipative terms is to allow the clean capture of shock waves and contact discontinuities without undesirable oscillations. Following the pioneering work of Godunov,[43] a variety of dissipative and upwind schemes designed to have good shock-capturing properties have been developed during the past decade.[44–53] The one-dimensional scalar conservation law

$$\frac{\partial u}{\Delta t} + \frac{\partial}{\partial x} f(u) = 0 \tag{27}$$

provides a useful model for the analysis of these schemes. The total variation

$$\text{TV} = \int_{-\infty}^{\infty} \left| \frac{\partial u}{\partial x} \right| \, dx$$

of a solution of Eq. (27) does not increase, provided that any discontinuity appearing in the solution satisfies an entropy condition.[55] The concept of total variation diminishing (TVD) difference schemes, introduced by Harten,[49] provides a unifying framework for the study of shock-capturing methods. These are schemes with the property that the total variation of the discrete solution

$$\text{TV} = \sum_{-\infty}^{\infty} \left| v_j - v_{j-1} \right|$$

cannot increase. The general conditions for a multipoint one-dimensional scheme to be TVD have been stated and proved by Jameson and Lax.[56]

TVD schemes preserve the monotonicity of an initially monotone profile, because the total variation would increase if the profile ceased to be monotone. Consequently, they prevent the formation of spurious oscillations. In this simple form, however, they are at best first-order-accurate. Harten devised a second-order-accurate TVD scheme by introducing anti-diffusive terms, and flux limiters to improve shock resolution can be traced to the work of Boris and Book.[44] The concept of the flux limiting was independently advanced by Van Leer.[45] A particularly simple way to introduce a second-order-accurate TVD scheme is to introduce flux limiters directly into a higher-order dissipative term.[53]

There are difficulties in extending these ideas to systems of equations and also to equations in more than one space dimension. First, the total variation of the solution of a system of hyperbolic equations may increase. Second, it has been shown by Goodman and Leveque that a TVD scheme in two space dimensions is no better than first-order-accurate.[57]

If one wishes to use one-sided differencing, one must allow for the fact that the general one-dimensional system defined by Eq. (18) produces signals traveling in both directions. One way of generalizing one-sided differencing to a system of equations is the flux-vector-splitting method proposed by Steger and Warming.[14]

Another approach to the discretization of hyperbolic systems was originally proposed by Godunov.[43] Suppose that Eq. (18) is approximated by

$$w_i^{n+1} = w_i^n - \frac{\Delta t}{\Delta x}(F_{i+1/2} - F_{i-1/2}) \tag{28}$$

where the numerical flux function $F_{i+1/2} = F(w_i, w_{i+1})$ is an approximation to the flux across the cell boundary $x_{i+1/2}$. This function must satisfy the consistency condition $F(w, w) = F(w)$. In the Godunov scheme $F_{i+\frac{1}{2}}$ is taken to be the flux value arising at $x_{i+1/2}$ in the exact solution of the initial value problem defined by piecewise constant data between each cell boundary. This simulates the motion of both shocks and expansion fans, but it is expensive.

Various simpler schemes designed to distinguish between the influence of forward- and backward-moving waves have recently been developed, based on the concept of flux difference splitting introduced by Roe.[47]

The use of flux splitting allows precise matching of the dissipative terms to introduce the minimum amount of dissipation needed to prevent oscillations. This in turn reduces the thickness of the numerical shock layer to the minimum attainable, one or two cells for a normal shock. In practice, however, it turns out that shock waves can be quite cleanly captured without flux splitting by using adaptive coefficients. The dissipation then has a low background level that is increased in the neighborhood of shock waves to a peak value proportional to the maximum local wave speed. The second difference of the pressure has been found to be an effective measure for this purpose. The dissipative terms are constructed in a similar manner for each dependent variable by introducing dissipative fluxes that preserve the conservation form.

For a two-dimensional rectilinear mesh the added terms have the form

$$d_{i+\frac{1}{2},j} - d_{i-\frac{1}{2},j} + d_{i,j+\frac{1}{2}} - d_{i,j-\frac{1}{2}} \tag{29}$$

These fluxes are constructed by blending first and third differences of the dependent variables. For example, the dissipative flux in the i direction for the mass equation is

$$d_{i+\frac{1}{2},j} = R(\epsilon^{(2)} - \epsilon^{(4)}\delta_x^2)(\rho_{i+1,j} - \rho_{i,j}) \tag{30}$$

where δ_x^2 is the second difference operator, $\epsilon^{(2)}$ and $\epsilon^{(4)}$ are the adaptive coefficients, and R is a scaling factor proportional to an estimate of the maximum local wave speed normal to the cell boundary. The coefficient $\epsilon^{(4)}$ provides the background dissipation in smooth parts of the flow and can be used to improve the capability of the scheme to damp high-frequency modes. Shock capturing is controlled by the coefficient $\epsilon^{(2)}$, which is made proportional to the normalized second difference of the pressure

$$v_{i,j} = \frac{|p_{i+1,j} - 2p_{i,j} + p_{i-1,j}|}{|p_{i+1,j} + 2p_{i,j} + p_{i-1,j}|}$$

in the adjacent cells.

Schemes constructed along these lines combine the advantages of simplicity and economy of computation, at the expense of an increase in thickness of the numerical shock layer to three or four cells. They have also proved robust in calculations over a wide range of Mach numbers (extending up to 20 in recent studies[58]). They can also be quite easily modified for calculations on triangular or tetrahedral meshes.[42]

Time-Stepping Schemes

The discretization procedures of the section on mathematical models of fluid flow lead to a set of coupled ordinary differential equations, which can be written in the form

$$\frac{dw}{dt} + R(w) = 0 \tag{31}$$

where w is the vector of the flow variables at the mesh points, and $R(w)$ is the vector of the residuals, consisting of the flux balances defined by Eqs. (21) or (25), together with the added dissipative terms. These are to be integrated to a steady state. Since the objective is simply to reach the steady state, and details of the transient solution are immaterial, the time-stepping scheme may be designed solely to maximize the rate of convergence without having to meet any constraints imposed by the need to achieve a specified level of accuracy, provided that it does not interfere with the definition of the residual $R(w)$. Figure 6 indicates some of the principal time-stepping schemes that might be considered. The first major choice is whether to use an explicit or an implicit scheme.

Explicit schemes that might be considered include linear multistep methods such as the leap frog and Adams-Bashforth schemes and one step multistage methods such as the classical Runge-Kutta schemes. The one-step multistage schemes have the advantages that they require no special start-up procedure, and they can readily be tailored to give a desired stability region. They have proved extremely effective in practice as a method of solving the Euler equations.

Let w^n be the result after n steps. The general form of an m-stage scheme is

$$w^{(0)} = w^n$$

$$w^{(1)} = w^{(0)} - \alpha_1 \, \Delta t R^{(0)}$$

$$\cdots$$

$$w^{(m-1)} = w^{(0)} - \alpha_{m-1} \, \Delta t R^{(m-2)}$$

$$w^{(m)} = w^{(0)} - \Delta t R^{(m-1)}$$

$$w^{n+1} = w^{(m)} \tag{32}$$

The residual in the $(q + 1)$th stage is evaluated as

$$R^{(q)} = \sum_{r=0}^{q} \beta_{qr} R(w^{(r)}) \tag{33}$$

where

$$\sum_{r=0}^{q} \beta_{qr} = 1$$

In the simplest case,

$$R^{(q)} = R(w^{(q)})$$

It is then known how to choose the coefficients α_q to maximize the stability interval along the imaginary axis and consequently the time step.[59] Since only the steady-state solution is needed, it pays to separate the residual $R(w)$ into its convective and dissipative parts $Q(w)$ and $D(w)$, respectively. The residual in the $(q + 1)$th stage is now evaluated as

$$R^{(q)} = \sum_{r=0}^{q} \{\beta_{qr} Q(w^{(r)}) - \gamma_{qr} D(w^{(r)})\} \tag{34}$$

where

$$\sum_{r=0}^{q} \beta_{qr} = 1, \qquad \sum_{r=0}^{q} \gamma_{qr} = 1$$

Blended multistage schemes of this type, which have been analyzed in Ref. 60, can be tailored to give large stability intervals along both the imaginary and negative real axes.

The properties of multistage schemes can be further enhanced by residual averaging.[60] Here the residual at a mesh point is replaced by a weighted average of neighboring residuals. The average is calcualted implicitly. In a one-dimensional case $R(w)$ is replaced by $\bar{R}(w)$, where at the jth mesh point,

$$-\epsilon \bar{R}_{j-1} + (1 + 2\epsilon)\bar{R}_j - \epsilon \bar{R}_{j+1} = R_j$$

It can easily be shown that the scheme can be stabilized for an arbitrarily large time step by choosing a sufficiently large value for ϵ. In a nondissipative one-dimensional case one needs

$$\epsilon \geqslant \frac{1}{4}\left[\left(\frac{\Delta t}{\Delta t^*}\right)^2 - 1\right]$$

where Δt^* is the maximum stable time step of the basic scheme, and Δt is the actual time step. The method can be extended to three dimensions by using smoothing in product form

$$(1 - \epsilon_x \delta_x^2)(1 - \epsilon_y \delta_y^2)(1 - \epsilon_z \delta_z^2)\bar{R} = R \tag{35}$$

where δ_x^2, δ_y^2, and δ_z^2 are second difference operators in the coordinate directions, and ϵ_x, ϵ_y, and ϵ_z are the corresponding smoothing coefficients. Residual averaging can also be used on triangular meshes.[41] The implicit equations are then solved by a Jacobian iteration.

One can anticipate that implicit schemes will yield convergence in a smaller number of time steps, since the time step is no longer constrained by a stability limit. However, this will only pay if the decrease in the number of time steps outweighs the increase in the computational effort per time step consequent upon the need to solve coupled equations. The prototype implicit scheme can be formulated by estimating $\partial w / \partial t$ at $t + \mu \Delta t$ as a linear combination of $R(w^n)$ and $R(w^{n+1})$. The resulting equation

$$w^{n+1} = w^n - \Delta t\{(1 - \mu)R(w^n) - \mu R(w^{n+1})\} \tag{36}$$

can be linearized as

$$\left(I + \mu \Delta t \frac{\partial R}{\partial w}\right)\delta w + \Delta t R(w^n) = 0 \tag{37}$$

Equation (37) reduces to the Newton iteration if one sets $\mu = 1$ and lets $\Delta r \to \infty$. In a three-dimensional case with an $N \times N \times N$ mesh, its bandwidth is of order N^2. Direct inversion requires a number of operations proportional to the number of unknowns multiplied by the square of the bandwidth, that is, $\mathcal{O}(N^7)$. This is prohibitive and forces the recourse to either an approximate factorization method or an iterative solution method.

The main possibilities for approximate factorization are the alternating-direction method and the LU decomposition method. The alternating-direction method, which may be traced back to the work of Gourlay and Mitchell,[61] was given an elegant formulation for nonlinear problems by Beam and Warming.[12] In a two-dimensional case Eq. (37) is replaced by

$$(I + \mu \Delta t D_x A)(I + \mu \Delta t D_y B)\delta w + \Delta t R(w) = 0 \tag{38}$$

where D_x and D_y are difference operators approximating $\partial / \partial x$ and $\partial / \partial y$, and A and B are the Jacobian matrices,

$$A = \frac{\partial f}{\partial w}, \qquad B = \frac{\partial g}{\partial w}$$

This may be solved in two steps.
   Step 1:

$$(I + \mu \Delta t D_x A)\delta w^* = - \Delta t R(w)$$

Step 2:

$$(I + \mu \Delta t D_y B)\delta w = \delta w^*$$

Each step requires block tridiagonal inversions and may be performed in $\mathcal{O}(N^2)$ operations on an $N \times N$ mesh. The algorithm is amenable to vectorization by simultaneous solution of the tridiagonal equations along parallel coordinate lines. The method has been refined to a high level of efficiency by Pulliam and Steger,[16] and Yee has extended it to incorporate a TVD scheme.[54] Its main disadvantage is that its extension to three dimensions is inherently unstable according a Von Neumann analysis.

The idea of the LU decomposition method[17] is to replace the operator in Eq. (20) by the product of lower and upper block triangular factors L and U,

$$LU\delta w + \Delta t R(w) = 0 \tag{39}$$

Two factors are used independently of the number of dimensions, and the inversion of each can be accomplished by inversion of its diagonal blocks. The method can be conveniently illustrated by considering a one-dimensional example. Let the Jacobian matrix $A = \partial f / \partial w$ be split as

$$A = A^+ + A^-$$

where the eigenvalues of A^+ and A^- are positive and negative, respectively. Then we can take

$$L \equiv I + \mu \Delta t D_x^- A^+, \qquad U \equiv I + \mu \Delta t D_x^+ A^- \tag{40}$$

where D_x^+ and D_x^- denote forward- and backward-difference operators, respectively, approximating $\partial / \partial x$. The reason for splitting A is to ensure the diagonal dominance of L and U independently of Δt. Otherwise, stable inversion of both factors will only be possible for a limited range of Δt. A crude choice is

$$A^\pm = \tfrac{1}{2}(A \pm \rho I)$$

where ρ is at least equal to the spectral radius of A. If flux splitting is used in the calculation of the residual, it is natural to use the corresponding splitting for L and U. An interesting variation is to combine an alternating-direction scheme with LU decomposition in the different coordinate directions.[62,63]

If one chooses to adopt the iterative solution technique, the principal alternatives are variants of the Gauss-Seidel and Jacobian methods. These may be applied to either the nonlinear equation [Eq. (36)] or the linearized equation [Eq. (37)]. A Jacobian method of solving Eq. (36) can be formulated by regarding it as an equation.

$$w - w^{(0)} + \mu \Delta t R(w) + (1 - \mu)\, \Delta t R(w^{(0)}) = 0$$

to be solved for w. Here $w^{(0)}$ is a fixed value obtained as the result of the previous time step. Such a procedure is a variant of the multistage time-stepping scheme described by Eqs. (32) and (33). It has the advantage of permitting simultaneous or overlapped calculation of the corrections at every mesh point and is readily amenable to parallel and vector processing.

A symmetric Gauss-Seidel scheme has been successfully employed in several recent works.[64] Consider the case of a flux split scheme in one dimension, for which

$$R(w) = D_x^+ f^-(w) + D_x^- f^+(w)$$

where the flux is split so that the Jacobian matrices

$$A^+ = \frac{\partial f^+}{\partial w} \quad \text{and} \quad A^- = \frac{\partial f^-}{\partial w}$$

have positive and negative eigenvalues, respectively. Now Eq. (37) becomes

$$\{I + \mu\Delta t(D_x^+ A^- + D_x^- A^+)\}\delta w + \Delta t R(w) = 0$$

At the jth mesh point this is

$$\{I + \alpha(A_j^+ - A_j^-)\}\delta w_j + \alpha A_{j+1}^- Dw_{j+1} - \alpha A_{j-1}^+ \delta w_{j-1} + \Delta t R_j = 0$$

where

$$\alpha = \mu \frac{\Delta t}{\Delta x}$$

Set $\delta w_j^{(0)} = 0$. A two-sweep-symmetric Gauss-Seidel scheme is then
Sweep 1:

$$\{I + \alpha(A_j^+ - A_j^-)\}\delta w_j^{(1)} - \alpha A_{j-1}^+ \delta w_{j-1}^{(1)} + \Delta t R_j = 0$$

Sweep 2:

$$\{I + \alpha(A_j^+ - A_j^-)\}\delta w_j^{(2)} + \alpha A_{j+1}^- \delta w_{j+1}^{(2)} - \alpha A_{j-1}^+ \delta w_{j-1}^{(1)} + \Delta t R_j = 0$$

Subtracting sweep 1 from sweep 2 we find that

$$\{I + \alpha(A_j^+ - A_j^-)\}\delta w_j^{(2)} + \alpha A_{j+1}^- \delta w_{j+1}^{(2)} = \{I + \alpha(A_j^+ - A_j^-)\}\delta w_j^{(1)}$$

Define the lower triangular, upper triangular, and diagonal operators L, U, and D as

$$L \equiv I - \alpha A^- + \mu\Delta t D_x^- A^+$$

$$U \equiv I + \alpha A^+ + \mu\Delta t D_x^+ A^-$$

$$D \equiv I + \alpha(A^+ - A^-)$$

It follows that the scheme can be written as

$$LD^{-1}U\delta w = -\Delta t R(w)$$

The iteration is usually terminated after one double sweep. The scheme is then a variation of an LU implicit scheme.

Some of these interconnections are illustrated in Fig. 6. Schemes in three main classes appear to be the most appealing:

1) variations of multistage time stepping, including the application of Jacobian iterative method to the implicit scheme (indicated by a single asterisk);

2) variations of LU decomposition, including the application of a Gauss-Seidel iterative method to the implicit scheme (indicated by a double asterisk); and

3) alternating-direction schemes, including schemes in which an LU decomposition is separately used in each coordinate direction (indicated by a triple asterisk).

The optimal choice may finally depend on the computer architecture. One might anticipate that the Guass-Seidel method of iteration could yield a faster rate of convergence than a Jacobian method, and it appears to be a particularly natural choice in conjunction with a flux split scheme that yields diagonal dominance. However, this class of schemes restricts the use of vector or parallel processing. Multistage time stepping, or Jacobian iteration of the implicit scheme allows maximal use of vector or parallel processing. The alternating-direction formulation removes any restriction on the time step (at least in the two-dimensional case), while permitting vectorization along coordinate lines. The ADI-LU scheme is an interesting compromise.

Acceleration Methods: Multigrid Technique

Clearly one can anticipate more rapid convergence to a steady state as the time step is increased. Accordingly, the rate of convergence of an explicit scheme can generally be substantially improved by using a variable time step close to the local stability limit throughout the flowfield. Assuming that the mesh cells are clustered near the body and expand as one moves away from the body, this effectively increases the rate at which disturbances are propagated through the outer part of the mesh. A similar strategy also pays with implicit schemes. In this case the terms in Δt^2 or Δt^3 resulting from factorization become dominant if Δt is too large, and the optimum rate of convergence is typically realized with a time step corresponding to a Courant number on the order of 10.

Further radical improvements in the convergence rate can be realized by the multigrid time-stepping technique, which extends the multigrid concept to the treatment of hyperbolic systems. Whereas relaxation methods for elliptic equations typically force the solution towards equilibrium by repeated smoothing, the transient behavior of hyperbolic systems is generally dominated by wave propagation. Accordingly, it seems that it should be possible to accelerate the evolution of the system to a steady state by using large time steps on coarse grids, so that disturbances are more rapidly expelled through the outer boundaries. This is a quite different mechanism for convergence from smoothing. However, the interpolation of corrections back to the fine mesh will introduce errors that cannot be rapidly expelled from the fine mesh and should be locally damped if a fast rate of convergence is to be attained. Thus, it remains important that the driving scheme should have the property of rapidly damping out high-frequency modes. A relatively simple way to analyze the behavior of multigrid time-stepping schemes is proposed in Ref. 23.

A novel multigrid time-stepping scheme was proposed by Ni[21] in 1981. In his scheme the flow variables are stored at the mesh nodes, and the rates of change of mass, momentum, and energy in each mesh cell are estimated from the flux integral appearing in Eq. (20). The corresponding change δw_0 associated with the cell is then distributed unequally between the nodes at its four corners by the rule

$$\delta w_0 = \tfrac{1}{4}(\delta w_0 \pm A\delta w_0 \pm B\delta w_0)$$

where δw_0 is the correction at a corner, and A and B are the Jacobian matrices. The signs are varied in such a way that the accumulated corrections at each node correspond to the first two terms of a Taylor series in time, like a Lax-Wendroff scheme. When several grid levels are used, the distribution rule is applied once on each level down to the coarsest grid, and the corrections are then interpolated back to the fine grid. Distributed correction schemes of this type have been further developed by Hall,[65] with very good results. They have also been extended to the Navier-Stokes equations by Chima and Johnson.[66]

An alternative formulation of multigrid time-stepping schemes was proposed by the present author.[22] This formulation, which can be combined with a variety of time-stepping schemes, corresponds to the full approximation scheme of Brandt.[20] It is most easily described by using subscripts to indicate the grid level. Several transfer operations need to be defined. First, the solution vector on grid k must be initialized as

$$w_k^{(0)} = T_{k,k-1} w_{k-1}$$

where w_{k-1} is the current value on grid $k-1$, and $T_{k,k-1}$ is a transfer operator. Next, it is necessary to transfer a residual forcing function such that the solution on grid k is driven by the residuals calculated on grid $k-1$. This can be accomplished by setting

$$P_k = Q_{k,k-1} R_{k-1}(w_{k-1}) - R_k(w_k^{(0)})$$

where $Q_{k,k-1}$ is another transfer operator. Then $R_k(w_k)$ is replaced by $R_k(w_k) + P_k$ in the time-stepping scheme. For example, the multistage scheme defined by Eq. (32) is reformulated as

$$w_k^{(1)} = w_k^{(0)} - \alpha_1 \, \Delta t_k (R_k^{(0)} + P_k)$$

$$\ldots$$

$$w_k^{(q+1)} = w_k^{(0)} - \alpha_{q+1} \, \Delta t_k (R_k^{(q)} + P_k)$$

$$\ldots$$

The result $w_k^{(m)}$ then provides the initial data for grid $k + 1$. Finally, the accumulated correction on grid k has to be transferred back to grid $k - 1$. Let w_k^+ be the final value of w_k resulting from both the correction calculated in the time step on grid k and the correction transferred from grid $k + 1$. Then one sets

$$w_{k-1}^+ = w_{k-1} + I_{k-1,k}(w_k^+ - w_k^{(0)})$$

where w_{k-1} is the solution on grid $k - 1$ after the time step on grid $k - 1$ and before the transfer to grid k, and $I_{k-1,k}$ is an interpolation operator. A W cycle of the type illustrated in Fig. 7 proves to be a particularly effective strategy for managing the work split between the meshes.

Both cell-centered and vertex-based schemes can be devised along these lines,[22,23,67] and they seem to work about equally well. With properly optimized coefficients the multistage time-stepping scheme is a very efficient driver of the multigrid process. Some results are presented in Figs. 8 and 9. Figure 8 shows a result for the RAE 2822 airfoil computed on an O mesh with 160 cells around the profile and 32 cells in the normal direction. This was obtained with a five-stage time-stepping scheme in which the dissipative terms were evaluated three times in each step. A cell-centered formulation was used for the space discretization, with adaptive dissipation of the type defined by Eqs. (29) and (30). The average residual measured by the rate of change of the density was reduced from 0.124 to 0.219×10^{-10} in 100-W cycles. This corresponds to an average reduction of 0.797 per cycle. The solution after 10 cycles is also displayed, and it can be seen that the solution is virtually identical. The lift coefficient is 1.1258 after 10 cycles and 1.1256 after 100 cycles. Figure 9 shows a three-dimensional calculation for a swept wing using a vertex scheme on a $192 \times 32 \times 48$ mesh. In this case the mean convergence rate over 100 cycles is 0.8222, and a fully converged result is obtained in 25 cycles. Computer times for these calculations are small enough their use in an that interactive design method could be contemplated. A two-dimensional calculation with 10 cycles on a 160×32 mesh can be performed on a Cray in several seconds. A three-dimensional calculation with 15 cycles on a $96 \times 16 \times 16$ mesh requires about 25 s using one processor of a Cray XMP.

Alternating-direction and LU implicit time-stepping schemes, as well as symmetric relaxation schemes, have been explored as alternatives to the multistage time-stepping procedure as a driver of the multigrid scheme.[68-71]

They are also effective. Very good results have been obtained by Anderson et al.,[72] who used an *ADI* scheme with Van Leer flux splitting, and by Hemker and Spekreijse,[73] who used relaxation with Osher flux splitting. Multigrid methods have also been extended to unstructured triangular meshes.[74,75]

Grid Generation and Complex Geometry

If computational methods are to be really useful to airplane designers, they must be able to treat extremely complex configurations, ultimately extending up to a complete aircraft. A major pacing item in the effort to attain this goal has been the problem of mesh generation. For simple wing-body combinations it is possible to generate rectilinear meshes without too much difficulty.[76] For more complicated configurations containing, for example, pylon-mounted engines, it becomes increasingly difficult to produce a structured mesh that is aligned with the body surface.

A wide variety of grid-generation techniques have been explored by numerous investigators. Algebraic transformations can be used to generate grids for quite complex shapes.[77,78] A popular alternative, pioneered by Thompson et al.,[79] is to generate grid surfaces as solutions of elliptic equations. Hyperbolic marching methods have also proved successful in some applications.[80]

The algebraic and elliptic methods can be extended to treat more complex configurations by dividing the flowfield into subdomains and generating the mesh in separate blocks. The mesh blocks may be required to match at the interfaces,[81] or they may be allowed to overlap each other.[82] A striking example of what can be achieved by these methods is exhibited in the work of Sawada and Takanashi,[83] who have calculated the flow over a four-engined short takeoff aircraft with overwing nacelles, using a flux-difference-split upwind discretization of the Euler equations.

An alternative procedure is to use tetrahedral cells in an unstructured mesh that can be adapted to conform to the complex surface of an aircraft. References 84 and 85 present a method based on such an approach. Separate overlapping meshes are generated around the individual components to create a cluster of points surrounding the whole aircraft. The swarm of mesh points is then connected together to form tetrahedral cells that provide the basis for a single finite-element approximation for the entire domain. This use of triangulation to unify separately generated meshes bypasses the need to devise interpolation procedures for transferring information between overlapping meshes. The triangulation of a set of points is in general nonunique. The method adopted in this work is to generate the Delaunay triangulation,[86] which is dual to the Voronoi diagram[87] that results from a division of the domain into polyhedral neighborhoods, each consisting of the subdomain of points nearer to a given mesh point than to any other mesh point. The Euler equations are discretized by establishing conservation of mass, momentum, and energy in polyhedral control volumes with a three-dimensional generalization of Eq. (25) and are solved by a multistage time-stepping scheme.

Figures 10a–10c and Plates 1–3 (see the color section) illustrate a transonic flow solution about a McDonnell Douglas MD-11 commercial transport aircraft at cruise conditions. The surface grid and flow solution are displayed with closeup views that emphasize the details of the engine regions. In the solution figures, high-pressure regions and low-pressure areas are shaded black, whereas near-freestream areas are very light. In the plates blue indicates regions of low velocity, and red indicates regions of high velocity. Flow is allowed through the nacelles and core-cowls (which are modeled as open tubes) without specification of mass flux. Discretization of the flowfield is accomplished with an unstructured mesh that contains 334,595 nodes and 1,953,286 packed tetrahedrons. The aircraft's surface is described with 34,521 triangles. A solution of this type currently requires about .4 Cray 2 CPU hours for the grid generation and nearly 4 Cray 2 CPU hours for a 300-iteration flow solution that reduces the residuals 3.5 orders of magnitude. Both the grid generation and flow solution use 50 megawords of core for this case.

Viscous Flow Calculations

Boundary-Layer Corrections

Although it is true that the viscous effects are relatively unimportant outside the boundary layer, the presence of the boundary layer can have a drastic influence on the pattern of the global flow. This will be the case, for example, in the event that the flow separates. The boundary layer can also cause global changes in a lifting flow by changing the circulation. These effects are particularly pronounced in transonic flows. The presence of a boundary layer can cause the location of the shock wave on the upper surface of the wing to shift 20% of the chord.

Although we must generally account for the presence of the boundary layer, the accuracy attainable in solutions of the Navier-Stokes equations for complete flowfields is severely limited by the extreme disparity between the length scales of the viscous effects and those of the gross patterns of the global flow. This has encouraged the use of methods in which the equations of viscous flow are solved only in the boundary layer, and the external flow is treated as inviscid. These zonal methods can give very accurate results in many cases of practical concern to the aircraft designer. The underlying ideas have been comprehensively reviewed in papers by Lock and Firmin,[88] Le Balleur,[89] and Melnik.[90]

In the outer region the real viscous flow is approximated by an equivalent inviscid flow, which has to be matched to the inner viscous flow by an appropriate selection of boundary conditions. In most of the boundary layer the viscous flow equations may consistently be approximated by the boundary-layer equations. This is sufficient in regions of weak interaction, in which the viscous effect on the pressure is small. However, there are regions of strong interaction in which the classical boundary-layer formulation fails, because of the appearance of strong normal pressure gradients across the boundary layer. Coupling conditions for the interaction between the inner viscous flow and the outer inviscid flow can be derived from an asymptotic analysis in which the Reynolds number is assumed to become very large. The coupled viscous and inviscid equations are solved iteratively. Semi-inverse methods in which transpiration boundary conditions are prescribed for both the inviscid flow calculation and the boundary-layer analysis have allowed these methods to be extended to treat flows with separated regions.[91,92]

The method of Bauer et al.[93] was the first to incorporate boundary-layer corrections into the calculation of transonic potential flow. This method only accounted for displacement effects on the airfoil and modeled the wake as a parallel semi-infinite strip. Nevertheless, this simple model substantially improved the agreement with experimental data. Several more complete theoretical models, including effects due to the wake thickness and curvature, have been developed.[94-96]

The simulation of attached flows by zonal methods now rests on a firm theoretical foundation and has reached a high level of sophistication in practice. The treatment of three-dimensional flows is presently limited by a lack of available boundary-layer codes for general configurations. Zonal methods have the disadvantage that extensions to more general configurations require a separate asymptotic analysis of each component region, such as the corner between a wing and a nacelle pylon, with the result that they can become unmanageable as the complexity of the configuration is increased.

Reynolds-Averaged Navier-Stokes Equations

Advances in algorithms and also in the speed and memory of currently available computers have brought us to a point where solutions of the Reynolds-averaged Navier-Stokes equations are entirely feasible for both two- and three-dimensional flows. The hope is that it will be possible to develop a fairly universal method that will be able to predict separated flows where present zonal methods fail. The principal requirements for a satisfactory solution of the Reynolds-averaged Navier-Stokes equations include the following:

1) the reductions of the discretization errors to a level such that any numerically introduced dissipative terms are much smaller than the real viscous terms; and

2) the closure of the equations by a turbulence model that accurately represents the turbulent stresses.

The development of the necessary numerical methods is already quite advanced. The methods described in the previous section can generally be carried over to the Navier-Stokes equations. The viscous terms can be discretized by standard techniques for numerical approximation.

The principal difficulty in producing an adequate numerical approximation is the need to use a mesh with very fine spacing in the direction normal to the wall to resolve the extreme gradients in the boundary layer. Typically it has been found that there should be of the order of 32 intervals inside the boundary layer and another 32 intervals between the boundary layer and the far field. Meshes of this type generally contain cells with a very high aspect ratio, on the order of 1000, adjacent to the wall and in the wake region. When the aspect ratio of the cells becomes so large, discretization schemes are prone to suffer both loss of accuracy and attrition of their rate of convergence to a steady state. These difficulties can be remedied by very careful control of the numerical dissipation introduced by the discretization and improvements in the iterative scheme.

The recent Viscous Transonic Airfoil Workshop at the 25th AIAA Aerospace Sciences Meeting provided an opportunity to assess the current state of the art. Results were presented for a variety of different numerical methods and turbulence models. Among the more highly developed methods were those of Coakley,[97] who used an upwind flux-split scheme with a variety of turbulence models, Rumsey et al.,[98] who used an upwind scheme with Van Leer splitting and a Baldwin-Lomax turbulence model, and Maksymiuk and Pulliam,[99] who used the ARC2D program with central differencing and a Baldwin-Lomax turbulence model. All three of these methods use alternating-direction time-stepping schemes. King showed the results of substituting alternative turbulence models in ARC2D, including the recently developed Johnson and King model.[100] Results obtained by the rational Runge-Kutta method with a Baldwin-Lomax turbulence model were presented by Morinishi and Satofuka.[101] A comparison of these results indicates that simulations using quite different numerical methods were in excellent agreement with each other as long as they used the same turbulence model, but that a change in the turbulence model could produce a drastic change in the solution, particularly in the case of a strong shock-induced separation. Predictions using the Baldwin-Lomax model agree quite well with experimental data when the flow is attached or only slightly separated, but an examination of the velocity profile in the boundary layer indicates that the model does not correctly represent the shock-wave–boundary-layer interaction. The two-equation models tested by Coakley showed no substantial improvement. The new Johnson and King model produced a better simulation of strongly separated flows, but seemed to be less accurate in the regions of attached flow.

An extension of the multigrid multistage scheme to treat the Navier-Stokes equations was presented in Ref. 103. This method is the subject of ongoing research to improve its accuracy and efficiency,[104] and some recently obtained results are presented in Figs. 11–14. The first two figures show a verification of the ability of the method to produce accurate Euler solutions on meshes with very-high-aspect-ratio cells, designed to resolve the boundary layer in Navier-Stokes calculations. Figure 11 shows a shock-free Euler solution for the Korn airfoil[93] on a 320×64 Navier-Stokes mesh, obtained in 50 multigrid cycles. Figure 12 shows an Euler solution for the RAE 2822 airfoil on a 512×64 Navier-Stokes mesh. This is case 9 from Ref. 102, and the experimental data are also displayed. Figure 13 shows the prediction obtained for the same case using the Baldwin-Lomax turbulence model. This result also agrees well with the result obtained by Coakley for this case using the Baldwin-Lomax model. Figure 14 shows the same calculation with the lower-order artificial dissipation defined by $\epsilon^{(2)}$ in Eq. (38) deleted. It can be seen that essentially the same result is obtained. Apparently the high-order dissipative terms are sufficient for numerical stability, and the shock wave is captured without oscillations with the aid of the eddy viscosity. These calculations exhibit a less rapid rate of convergence to a steady state than Euler calculations on less highly bunched meshes. Nevertheless, 100–200 cycles have consistently proved sufficient for convergence in numerous calculations.

Usefully accurate three-dimensional Navier-Stokes simulations are now also within the range of existing supercomputers. This has been demonstrated by the calculations of Shang and Scherr[105] and Fujii and Obayashi.[106] Chandler and MacCormack[107] have recently developed an effective relaxation method for the Navier-Stokes equations.

Conclusion

Computational aerodynamics has come of age during the last two decades, and in several of its branches it is now a mature discipline. Some of the more striking successes, which have been referred to in this necessarily incomplete survey point the way to its future acceptance as a primary tool for aerodynamic analysis and design. Basic numerical algorithms for the treatment of viscous and compressible flows with shock waves are now in hand, and rapidly convergent solution methods are well established. The concepts of total variation diminishing difference schemes and multigrid acceleration methods, in particular, provide examples that mathematical elegance can be just as important in the development of computational methods as it has been for analytical methods.

Some areas of likely concentration for future research can be identified. These include the following

1) The quest for numerical approximation schemes with a higher order of accuracy: The power of the spectral method[108,109] for numerical approximation of smooth solutions beckons efforts to extend it to treat discontinuous solutions. It is already used as an effective tool in simulations of turbulence.[110] The recently proposed concept of essentially nonoscillatory (ENO) schemes[111] is another interesting direction of research.

2) Front tracking: Ultimately, it should be possible to attain improved representation of shock waves and contact discontinuities by treating them as internal boundaries.[112] The use of triangular meshes offers new opportunities to represent more complex features such as triple points.

3) Adaptive mesh redistribution and refinement: It is clear that greatly improved accuracy can be attained for a given computational cost by adapting the mesh to the solution as it develops during the calculation. This can be a key to obtaining adequate resolutions of all the important features of really complex flows, and it may be an effective alternative to front tracking for the representation of discontinuities. Widespread research on the realization of adaptive methods is in progress.[113–117] Repeated local subdivision ultimately destroys any coherence of the structure of the mesh, and unstructured triangular meshes provide a natural basis for the use of such a procedure.[74,75,116,117]

4) Concurrent computation: The best opportunity for further increases in computing speed lies in the use of concurrent computation, which may be realized through the introduction of vector, pipelined, and parallel

arithmetic processors. Algorithms of the future must be designed to take full advantage of these architectures.

5) Optimization and design: An aerodynamic analysis may warn a designer that his proposed configuration is unsatisfactory, but it is not very helpful to him if it provides no indication of how to make an improvement. Ultimately, computational methods for aerodynamic analysis ought to be incorporated in automatic design procedures, which will use computer optimization methods, and perhaps expert systems to refine and improve the configuration. Some early efforts have already demonstrated the feasibility of automatic design.[118,119]

6) Turbulence modeling: Improved simulations of separated viscous flows will require advances in turbulence modeling. More complex multi-equation formulations may be necessary for the realization of more universally applicable models. Renormalization group theory offers another avenue towards the construction of a rational turbulence model.[120]

Once these various challenges have been surmounted, the simulation of both external and internal flows will become a routine practice. Methods of the future must be capable of treating arbitrary configurations, including complex aircraft, and flows in complex propulsive systems. Simulations for hypersonic aircraft will require the inclusion of real gas effects and chemical reactions at high temperatures. New ideas will surely be brought to bear on some of these problems.

References

[1]Jones, R. T., "Properties of Low Aspect Ratio Pointed Wings at Speeds Below and Above the Speed of Sound," NACA Rept. TR-835, 1946.

[2]Hayes, W. D., "Linearized Supersonic Flow," North American Aviation Rept. AL-222, 1947.

[3]Whitcomb, R. T., "A Study of the Zero Lift Drag-Rise Characteristics of Wing-Body Combinations Near the Speed of Sound," NACA TR-1273, 1956.

[4]Whitcomb, R. T., "Reviews of NASA Supercritical Airfoils," ICAS Paper 74-10, 1974.

[5]Hess, J. L., and Smith, A. M. O., "Calculation of Non-Lifting Potential Flow About Arbitrary Three-Dimensional Bodies," Douglas Aircraft Rept. ES 40622, 1962.

[6]Rubbert, P. E., and Saaris, G. R., "A General Three-Dimensional Potential Flow Method Applied to V/STOL Aerodynamics," Society of Automotive Engineers Paper 680304, 1968.

[7]Woodward, F. A., "An Improved Method for the Aerodynamic Analysis of Wing-Body-Tail Configurations in Subsonic and Supersonic Flow, Part 1—Theory and Application," NASA CR-2228, May 1973.

[8]Murman, E. M., and Cole, J. D., "Calculation of Plane Steady Transonic Flows," *AIAA Journal*, Vol. 9, No. 1, 1971, pp. 114–121.

[9]Murman, E. M., "Analysis of Embedded Shock Waves Calculated by Relaxation Methods," *AIAA Journal*, Vol. 12, No. 5, 1974, pp. 626–633.

[10]Magnus, R., and Yoshihara, H., "Inviscid Transonic Flow Over Airfoils," *AIAA Journal*, Vol. 8, No. 12, 1970, pp. 2157–2162.

[11]MacCormack, R. W., "The Effect of Viscosity in Hyper-Velocity Impact Cratering," AIAA Paper 69-354, 1969.

[12]Beam, R. W., and Warming, R. F., "An Implicit Finite Difference Algorithm for Hyperbolic System in Conservation Form," *Journal of Computational Physics*, Vol. 23, 1976, pp. 87–110.

[13]Steger, J. L., "Implicit Finite Difference Simulation of Flow About Arbitrary Two Dimensional Geometries," *Journal of Computational Physics*, Vol. 16, 1978, pp. 679–686.

[14]Steger, J. L., and Warming, R. F., "Flux Vector Splitting of the Inviscid Gas Dynamics Equations with Applications to Finite Difference Methods," *Journal of Computational Physics*, Vol. 40, 1981, pp. 263–293.

[15]Rizzi, A., and Viviand, H., "Numerical Methods for the Computation of Inviscid Transonic Flows with Shock Waves," *Proceedings of the GAMM Workshop*, Vieweg, 1981.

[16]Pulliam, T. H., and Steger, J. L., "Implicit Finite Difference Simulations of Three-Dimensional Compressible Flow," *AIAA Journal*, Vol. 18, No. 2, 1980, pp. 159–167.

[17]Jameson, A., and Turkel, E., "Implicit Schemes and LU Decompositions," *Math. Comp.*, Vol. 37, 1981, pp. 385–397.

[18]Jameson, A., Schmidt, W., and Turkel, E., "Numerical Solution of the Euler Equations by Finite Volume Methods Using Runge-Kutta Time Stepping Schemes," AIAA Paper 81-1259, 1981.

[19]Fedorenko, R. P., "The Speed of Convergence of One Iterative Process," *USSR Computational Mathematics and Mathematical Physics*, Vol. 4, 1964, pp. 227–235.

[20]Brandt, A., "Multi-Level Adaptive Solutions to Boundary Value Problems," *Math. Comp.*, Vol. 31, 1977, pp. 333–390.

[21]Ni, R. H., "A Multiple Grid Scheme for Solving the Euler Equations,' *AIAA Journal*, Vol. 20, No. 11, 1982, pp. 1565–1571.

[22]Jameson, A., "Solution of the Euler Equations by a Multigrid Method," *Applied Math. and Computation*, Vol. 13, 1983, pp. 327–356.

[23]Jameson, A., "Multigrid Algorithms for Compressible Flow Calculations," *Lecture Notes in Mathematics*, Vol. 1228, edited by W. Hackbusch and V. Trottenberg, Springer-Verlag, 1986, pp. 166–201.

[24]Chapman, D. R., "Computational Aerodynamics Development and Outlook," *AIAA Journal*, Vol. 17, No. 12, 1979, pp. 1293–1313.

[25]Jameson, Antony, "Iterative Solution of Transonic Flows Over Airfoils and Wings, Including Flows at Mach 1," *Comm. Pure. Appl. Math.*, Vol. 27, 1974, pp. 283–309.

[26]Albone, C. M., "A Finite Difference Scheme for Computing Supercritical Flows in Arbitrary Coordinate Systems," Royal Aircraft Establishment Rept. 74090, 1974.

[27]Jameson, A., "Transonic Potential Flow Calculations in Conservation Form," *Proceedings of AIAA 2nd Computational Fluid Dynamics Conference*, AIAA, New York, 1975, pp. 148–161.

[28]Eberle, A., "A Finite Volume Method for Calculating Transonic Potential Flow Around Wings from the Minimum Pressure Integral," NASA TM-75324, translated from MBB UFE 1407(0), 1978.

[29]Hafez, M., South, J. C., and Murman, E. M., "Artificial Compressibility Method for Numerical Solutions of Transonic Full Potential Equation," *AIAA Journal*, Vol. 17, No. 8, 1979, pp. 838–844.

[30]Holst, T. L., and Ballhaus, W. F., "Fast Conservative Schemes for the Full Potential Equation Applied to Transonic Flows," *AIAA Journal*, Vol. 17, No. 2, 1979, pp. 145–152.

[31]Ballhaus, W. F., Jameson, A., and Albert, J., "Implicit Approximate Factorization Schemes for Steady Transonic Flow Problems," *AIAA Journal*, Vol. 16, No. 6, 1978, pp. 573–579.

[32]Holst, T. L., and Thomas, S. D., "Numerical Solution of Transonic Wing Flow Fields," AIAA Paper 82-0105, 1982.

[33]South, J. C., and Brandt, A., "Application of a Multi-Level Grid Method to Transonic Flow Calculations," *Proceedings of Workshop on Transonic Flow Problems in Turbomachinery, Monterey, 1976,* edited by T. C. Adamson and M. F. Platzer, Hemisphere, Washington, DC, 1977, pp. 180–206.

[34]Hackbusch, W., "On the Multi-Grid Method Applied to Difference Equations," *Computing,* Vol. 20, 1978, pp. 291–306.

[35]Jameson, A., "Acceleration of Transonic Potential Flow Calculations on Arbitrary Meshes by the Multiple Grid Method," *Proceedings of AIAA 4th Computational Fluid Dynamics Conference,* AIAA, New York, 1979, pp. 122–146.

[36]Jameson, A., and Caughey, D. A., "A Finite Volume Method for Transonic Potential Flow Calculations," *Proceedings of AIAA 3rd Computational Fluid Dynamics Conference,* AIAA, New York, 1977, pp. 35–54.

[37]Bristeau, M. O., Glowinski, R., Periaux, J., Perrier, P., Pirronneau, O., and Poirier, G., "On the Numerical Solution of Nonlinear Problems in Fluid Dynamics by Least Squares and Finite Element Methods (II), Application to Transonic Flow Simulations," *Computer Methods in Applied Mechanics and Engineering,* Vol. 51, 1985, pp. 363–394.

[38]Giles, M., Drela, M., and Thompkins, W. T., "Newton Solution of Direct and Inverse Transonic Euler Equations," AIAA Paper 85-1530, 1985.

[39]Lax, P. D., and Wendroff, B., "Systems of Conservation Laws," *Comm. Pure Appl. Math.,* Vol. 13, 1960, pp. 217–237.

[40]MacCormack, R. W., and Paullay, A. J., "Computational Efficiency Achieved by Time Splitting of Finite Difference Operators," AIAA Paper 72-154, 1972.

[41]Jameson, A., and Mavriplis, D., "Finite Volume Solution of the Two-Dimensional Euler Equations on a Regular Triangular Mesh," AIAA Paper 85-0435, Jan. 1985.

[42]Jameson, A., "Current Status of Future Directions of Computational Transonics," *Computational Mechanics—Advances and Trends,* edited by A. K. Noor, American Society of Mechanical Engineers, New York, Publ. AMD 75, 1986.

[43]Gudonov, S. K., "A Difference Method for the Numerical Calculation of Discontinuous Solutions of Hydrodynamic Equations," *Mathematicheskii Sbornik,* Vol. 47, 1959, pp. 271–306; translated as JPRS 7225 by U.S. Dept. of Commerce, 1960.

[44]Boris, J. P., and Book, D. L., "Flux Corrected Transport, 1 SHASTA, A Fluid Transport Algorithm that Works," *Journal of Computational Physics,* Vol. 11, 1973, pp. 38–69.

[45]Van Leer, B., "Towards the Ultimate Conservative Difference Scheme. II Monotonicity and Conservation Combined in a Second Order Scheme," *Journal of Computational Physics,* Vol. 14, 1974, pp. 361–370.

[46]Van Leer, B., "Flux Vector Splitting for the Euler Equations," *Proceedings of the 8th International Conference on Numerical Methods in Fluid Dynamics, Aachen, 1982,* edited by E. Krause, Springer-Verlag, Berlin, 1982, pp. 507–512.

[47]Roe, P. L., "Approximate Riemann Solvers, Parameter Vectors, and Difference Schemes," *Journal of Computational Physcs,* Vol. 43, 1981, pp. 357–372.

[48]Osher, S., and Solomon, F., "Upwind Difference Schemes for Hyperbolic Systems of Conservation Laws," *Math. Comp.,* Vol. 38, 1982, pp. 339–374.

[49]Harten, A., "High Resolution Schemes for Hyperbolic Conservation Laws," *Journal of Computational Physics,* Vol. 49, 1983, pp. 357–393.

[50]Osher, S., and Chakravarthy, S., "High Resolution Schemes and the Entropy Condition," *Journal of Numerical Analysis,* Vol. 21, 1984, pp. 955–984.

[51]Sweby, P. K., "High Resolution Schemes Using Flux Limiters for Hyperbolic Conservation Laws," *SIAM Journal of Numerical Analysis*, Vol. 21, 1984, pp. 995–1011.

[52]Anderson, B. K., Thomas, J. L., and Van Leer, B., "A Comparison of Flux Vector Splittings for the Euler Equations," AIAA Paper 85-0122, Jan. 1985.

[53]Jameson, A., "Nonoscillatory Shock Capturing Scheme Using Flux Limited Dissipation," *Large Scale Computations in Fluid Mechanics*, Vol. 22, Pt. 1, edited by B. E. Engquist, S. Osher, and R. C. J. Sommerville, American Mathematical Society, Providence, RI, 1985, pp. 345–370 (Lectures in Applied Mathematics).

[54]Yee, H. C., "On Symmetric and Upwind TVD Schemes," *Proceedings of the 6th GAMM Conference on Numerical Methods in Fluid Mechanics, Göttingen, FRG, Sept. 1985*.

[55]Lax, P. D., "Hyperbolic Systems of Conservation Laws and the Mathematical Theory of Shock Waves," *SIAM Regional Series on Applied Mathematics*, Vol. 11, 1973.

[56]Jameson, A., and Lax, P. D., "Conditions for the Construction of Multi-point Total Variation Diminishing Schemes," *Applied Numerical Mathematics*, Vol. 2, 1986, pp. 335–345.

[57]Goodman, J. B., and Leveque, R. J., "On the Accuracy of Stable Schemes for 2-D Scalar Conservation Laws," *Math. Comp.*, Vol. 45, 1985, pp. 15–21.

[58]Yoon, S., and Jameson, A., "An LU Implicit Scheme for High Speed Inlet Analysis," AIAA Paper 86-1520, June 1986.

[59]Kinmark, I. P. E., "One Step Integration Methods with Large Stability Limits for Hyperbolic Partial Differential Equations," *Advances in Computer Methods for Partial Differential Equations*, Vol. V, edited by R. Vichnevetsky and R. S. Stepleman, IMACS, 1984, pp. 345–349.

[60]Jameson, A., "Transonic Flow Calculations for Aircraft," *Numerical Methods in Fluid Dynamics*, Vol. 1127, edited by F. Brezzi, Springer-Verlag, Berlin, 1985, pp. 156–242 (Lecture Notes in Mathematics).

[61]Gourlay, A. R., and Mitchell, A. R., "A Stable Implicit Difference Method for Hyperbolic Systems in Two Space Variables," *Numer. Math.*, Vol. 8, 1966, pp. 367–375.

[62]Obayashi, S., and Kuwakara, K., "LU Factorization of an Implicit Scheme for the Compressible Navier Stokes Equations," AIAA Paper 84-1670, June 1984.

[63]Obayashi, S., Matsukima, K., Fujii, K., and Kuwakara, K., "Improvements in Efficiency and Reliability for Navier-Stokes Computations Using the LU-ADI Factorization Algorithm," AIAA Paper 86-0338, Jan. 1986.

[64]Chakravarthy, S. R., "Relaxation Methods for Unfactored Implicit Upwind Schemes," AIAA Paper 84-0165, Jan. 1984.

[65]Hall, M. G., "Cell Vertex Multigrid Schemes for Solution of the Euler Equations," IMA Conference on Numerical Methods for Fluid Dynamics, Reading, MA, April 1985.

[66]Chima, R. V., and Johnson, G. M., "Efficient Solution of the Euler and Navier Stokes Equations with a Vectorized Multiple-Grid Algorithm," AIAA Paper 83-1893, 1983.

[67]Jameson, A., "A Vertex Based Multigrid Algorithm for Three Dimensional Compressible Flow Calculations," *Numerical Methods for Compressible Flow — Finite Difference, Element and Volume Techniques*, edited by T. E. Tezduar and T. J. R. Hughes, Americal Society of Mechanical Engineers, New York, Publ. AMD 78, 1986.

[68]Jameson, A., and Yoon, S., "Multigrid Solution of the Euler Equations Using Implicit Schemes," AIAA Paper 85-0293, Jan. 1985.

[69]Jameson, A., and Yoon, S., "LU Implicit Schemes with Multiple Grids for the Euler Equations," AIAA Paper 86-0105, Jan. 1986.

[70]Yoon, S., Choo, Y. K., and Jameson, A., "An LU-SSOR Scheme for the Euler and Navier Stokes Equations," AIAA Paper 87-0599, Jan. 1987.

[71]Caughey, D. A., "A Diagonal Implicit Multigrid Algorithm for the Euler Equations," AIAA Paper 87-453, Jan. 1987.

[72]Anderson, W. K., Thomas, J. L., and Whitfield, D. L., "Multigrid Acceleration of the Flux Split Euler Equations," AIAA Paper 86-0274, Jan. 1986.

[73]Hemker, P. W., and Spekreijse, S. P., "Multigrid Solution of the Steady Euler Equations," Proc. Öberwolfach Meeting on Multigrid Methods, FRG, Dec. 1984.

[74]Jameson, A., and Mavriplis, D., "Multigrid Solution of the Euler Equations on Unstructured and Adaptive Meshes," Third Copper Mountain Conference on Multigrid Methods, Copper Mountain, April 1985.

[75]Lallemand, M. H., and Dervieux, A., "A Multigrid Finite Element Method for Solving the Two Dimensional Euler Equations," Third Copper Mountain Conference on Multigrid Methods, Copper Mountain, April 1988.

[76]Baker, T. J., "Mesh Generation by a Sequence of Transformations," *Applied Numerical Mathematics*, Vol. 2, 1986, pp. 515–528.

[77]Eiseman, P. R., "A Multi-Surface Method of Coordinate Generation," *Journal of Computational Physics*, Vol. 33, 1979, pp. 118–150.

[78]Eriksson, L. E., "Generation of Boundary-Conforming Grids around Wing-Body Configurations using Transfinite Interpolation," *AIAA Journal*, Vol. 20, No. 10, 1982, pp. 1313–1320.

[79]Thomspon, J. F., Thames, F. C., and Mastin, C. W., "Automatic Numerical Generation of Body-Fitted Curvilinear Coordinate System for Field Containing any Number of Arbitrary Two-Dimensional Bodies," *Journal of Computational Physics*, Vol. 15, 1974, pp. 299–319.

[80]Steger, J. L., and Chaussee, D. S., "Generation of Body-Fitted Coordinates Using Hyperbolic Partial Differential Equations," *SIAM Journal of Scientific and Statistical Computing*, Vol. 1, 1980, pp. 431–437.

[81]Weatherill, N. P., and Forsey, C. R., "Grid Generation and Flow Calculations for Aircraft Geometries," *Journal of Aircraft*, Vol. 22, No. 10, 1985, pp. 855–860.

[82]Benek, J. A., Buning, P. G., and Steger, J. L., "A 3-D Chimera Grid Embedding Technique," AIAA Paper 85-1523, June 1985.

[83]Sawada, K., and Takanashi, S., "A Numerical Investigation on Wing/Nacelle Interferences of USB Configurations," AIAA Paper 87-0455, Jan. 1987.

[84]Jameson, A., Baker, T. J., and Weatherhill, N. P., "Calculation of Inviscid Transonic Flow over a Complete Aircraft," AIAA Paper 86-0103, Jan. 1986.

[85]Jameson, A., and Baker, T. J., "Improvements to the Aircraft Euler Method," AIAA Paper 87-0452, Jan. 1987.

[86]Delaunay, B., "Sur la Sphere Vide," *Bulletin of the Academy of Sciences of the USSR VII: Class Scil. Mat. Nat.*, 1934, pp. 793–800.

[87]Voronoi, G., "Nouvelles Applications des Parametres Continus a la Theorie des Formes Quadratiques. Deuxieme Memoire: Recherches Sur les Parallell oedres Primitifs," *Journal Reine Angew. Math.*, Vol. 134, 1908, pp. 198–287.

[88]Lock, R. C., and Firmin, M. C. P., "Survey of Techniques for Estimating Viscous Effects in External Aerodynamics," *Proceedings of the IMA Conference on Numerical Methods in Aeronautical Fluid Dynamics, Reading, 1980*, edited by P. L. Roe, Academic, New York, 1982, pp. 337–430.

[89]Le Balleur, J. C., "Numerical Viscid-Inviscid Interaction in Steady and Unsteady Flows," Proceedings of the 2nd Symposium on Numerical and Physical Aspects of Aerodynamic Flows, Long Beach, CA, 1983.

[90]Melnik, R. E., "Turbulent Interactions on Airfoils at Transonic Speeds—Recent Developments," AGARD CP-291, Paper 10, 1980.

[91]Carter, J. E., "A New Boundary Layer Inviscid Iteration Technique for Separated Flow," *Proceedings of AIAA 4th Computational Fluid Dynamics Conference*, AIAA, New York, 1979, pp. 45–55.

[92]Le Balleur, J. C., "Strong Matching Methods of Computing Transonic Viscous Flow Including Wakes and Separations—Lifting Airfoils, "*La Recherche Aerospátiále*, Vol. 3, 1981.

[93]Bauer, F., Garabedian, P., Korn, D., and Jameson, A., *Supercritical Wing Sections*, Vol. II, Springer-Verlag, New York, 1975.

[94]Le Balleur, J. C., Peyret, R., and Viviand, H., "Numerical Studies in High Reynolds Number Aerodynamics," *Computers and Fluids*, Vol. 8, 1980, pp. 1–30.

[95]Collyer, M. R., and Lock, R. C., "Prediction of Viscous Effects in Steady Transonic Flow Past an Aerofoil," *Aeronautical Quarterly*, Vol. 30, 1979, pp. 485–505.

[96]Melnik, R. E., Chow, R. R., Mead, H. R., and Jameson, A., "A Multigrid Method for the Computation of Viscid/Inviscid Interaction on Airfoils," AIAA Paper 83-0234, 1983.

[97]Coakley, T. J., "Numerical Simulation of Viscous Transonic Airfoil Flows," AIAA Paper 87-0416, Jan. 1987.

[98]Rumsey, C. L., Taylor, S. L., Thomas, J. L., and Anderson, W. K., "Application of an Upwind Navier-Stokes Code to Two-Dimensional Transonic Airfoil Flow," AIAA Paper 87-0413, Jan. 1987.

[99]Maksymiuk, C. M., and Pulliam, T. H., "Viscous Transonic Airfoil Workshop Results Using ARC2D," AIAA Paper 87-0415, Jan. 1987.

[100]King, L. S., "A Comparison of Turbulence Closure Models for Transonic Flows About Airfoils," AIAA Paper 87-0418, Jan. 1987.

[101]Morinishi, K., and Satofuka, N., "A Numerical Study of Viscous Transonic Flows Using RRK Scheme," AIAA Paper 87-0426, Jan. 1987.

[102]Cook, P. H., McDonald, M. A., and Firmin, M. C. P., "Aerofoil RAE 2822 Pressure Distributions, and Boundary Layer and Wake Measurements," AGARD Advisory Rept. 138, 1979.

[103]Martinelli, L., Jameson, A., and Grasso, F., "A Multigrid Method for the Navier Stokes Equations," AIAA Paper 86-0208, Jan. 1986.

[104]Martinelli, L., "Calculations of Viscous Flows with a Multigrid Method," MS Thesis, Princeton Univ., Princeton, NJ, May 1987.

[105]Shang, J. S., and Scherr, S. J., "Navier Stokes Solution of the Flow Field Around a Complete Aircraft," AIAA Paper 85-1509, July 1985.

[106]Fujii, K., and Obayashi, S., "Navier Stokes Simulation of Transonic Flow Over Wing-Fuselage Combinations," AIAA Paper 86-1831, June 1986.

[107]Chandler, G. V., and MacCormack, R. W., "Hypersonic Flow Past 3-D Configurations," AIAA Paper 87-0480, Jan. 1987.

[108]Orszag, S., and Gottlieb, D., "Numerical Analysis of Spectral Methods," *SIAM Regional Series on Applied Mathematics*, Vol. 26, 1977.

[109]Canuto, C., Hussaini, M. Y., Quarteroni, A., and Zang, D. A., "Spectral Methods in Fluid Dynamics" (to be published).

[110]Hussaini, M. Y., "Stability, Transition and Turbulence," Proceedings of the NASA Symposium on Supercomputing in Aerospace, NASA CP-2454, March 1987.

[111]Chakravarthy, S. R., Harten, A., and Osher, S., "Essentially Non-Oscillatory Shock Capturing Schemes of Uniformly Very High Accuracy," AIAA Paper 86-0339, Jan. 1986.

[112]Moretti, G., "An Efficient Euler Solver, with Many Applications," AIAA Paper 87-0352, Jan. 1987.

[113]Dwyer, H. A., Smooke, M. O., and Kee, R. J., "Adaptive Grid Method for Problems in Fluid Mechanics and Heat Transfer," *AIAA Journal*, Vol. 18, No. 10, 1980, pp. 1205–1212.

[114]Berger, M., and Jameson, A., "Automatic Adaptive Grid Refinement for the Euler Equations," *AIAA Journal*, Vol. 23, No. 4, 1985, pp. 561–568.

[115]Dannenhoffer, J. F., and Baron, J. R., "Robust Grid Adaptation for Complex Transonic Flows," AIAA Paper 86-0495, Jan. 1986.

[116]Lohner, R., Morgan, K., and Peraire, J., "Improved Adaptive Refinement Strategies for the Finite Element Aerodynamic Configurations," AIAA Paper 86-0499, Jan. 1986.

[117]Holmes, D. G., and Lamson, S. H., "Adaptive Triangular Meshes for Compressible Flow Solutions," *Proceedings of the International Conference on Numerical Grid Generation in Computational Fluid Dynamics, Landshut, July 1986*, edited by J. Hauser and C. Taylor, Pineridge, Swansea, UK, 1986, pp. 413–424.

[118]Hicks, R. M., and Henne, P. A., "Wing Design by Numerical Optimization," AIAA Paper 79-0080, 1979.

[119]Gregg, R. D., and Misegades, K., "Transonic Wing Optimization Using Evolution Theory," AIAA Paper 87-0520, Jan. 1987.

[120]Yakhot, V., and Orszag, S., "Renormalization Group Analysis of Turbulence. 1. Basic Theory," *Journal of Scientific Computing*, Vol. 1, 1986, pp. 3–51.

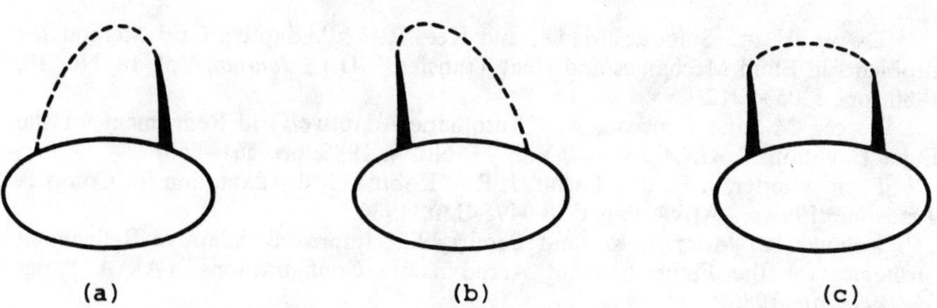

Fig. 1 Alternative solutions for an ellipse: a) compression shock; b) expansion shock; c) symmetric shock.

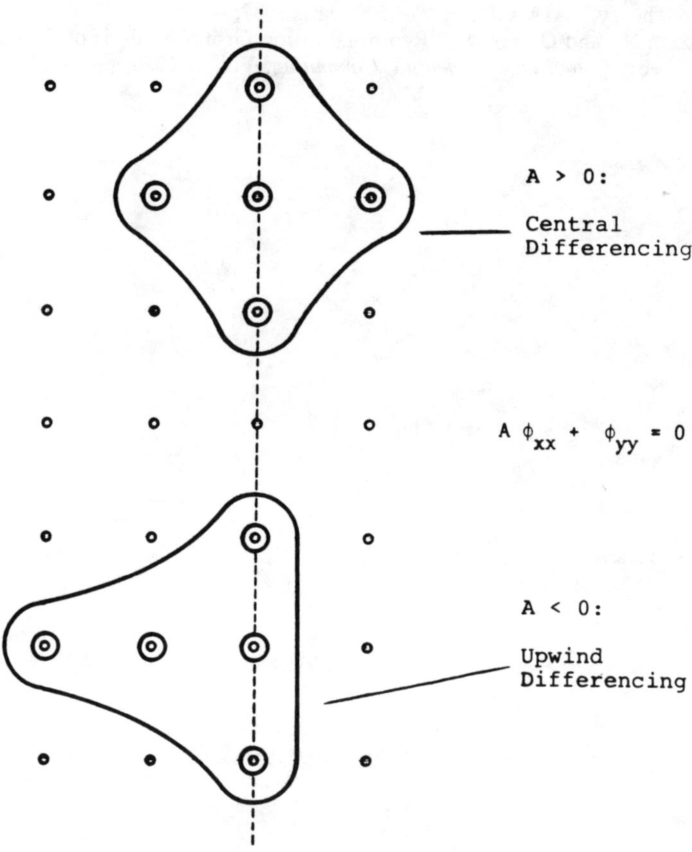

Fig. 2 Murman-Cole difference scheme.

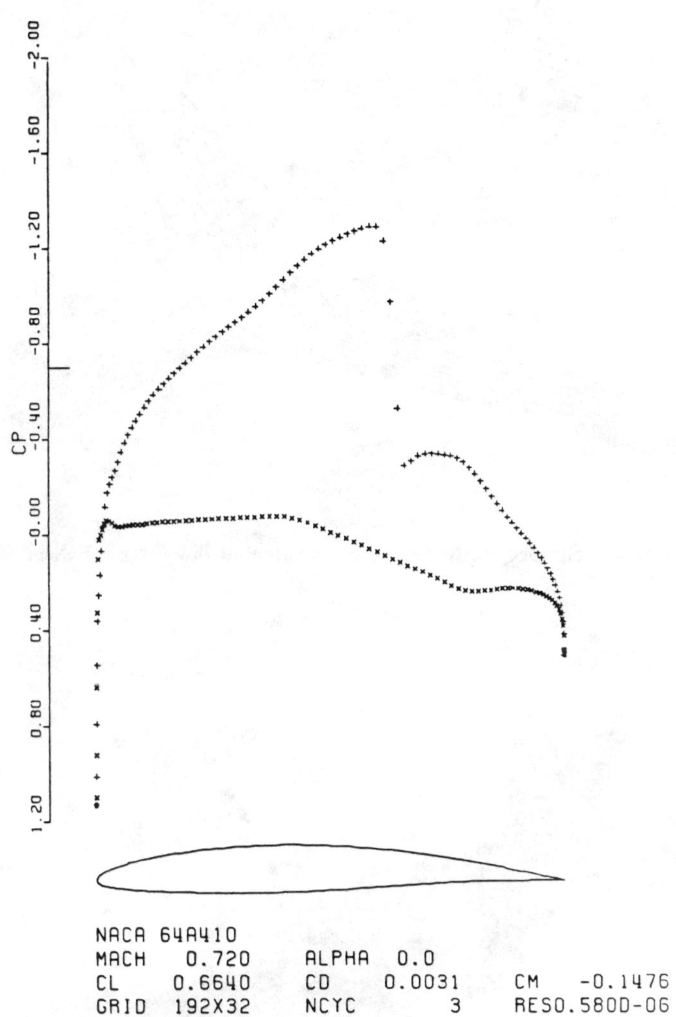

Fig. 3   Transonic potential-flow solution calculated with three multigrid V cycles.

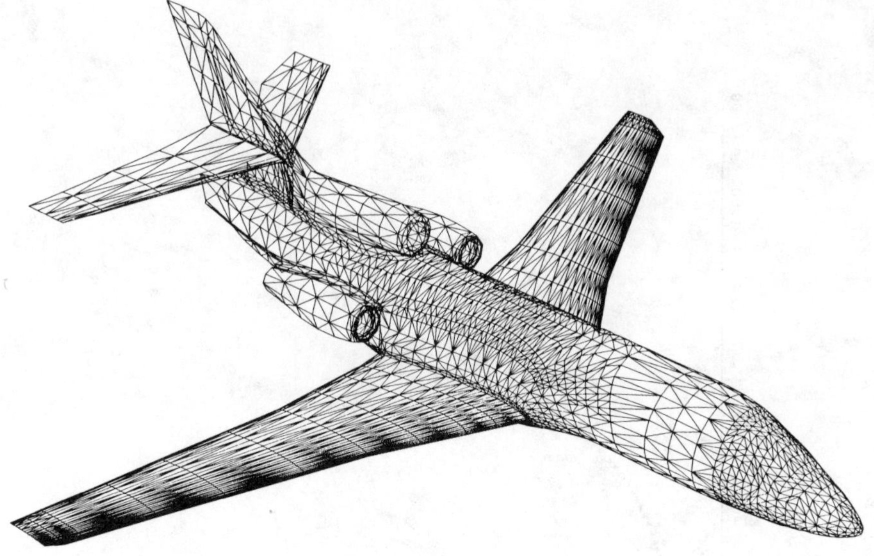

Fig. 4a   Surface mesh for transonic potential flow over a Falcon 50.

Fig. 4b   Surface Mach contours for transonic potential flow over a Falcon 50.

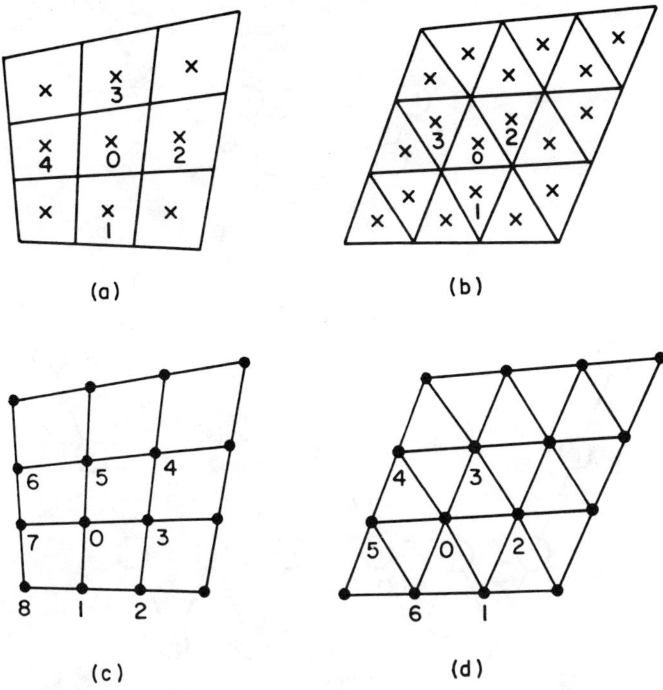

Fig. 5 Alternative discretization schemes: a) cell-centered rectilinear; b) cell-centered triangular; c) vertex rectilinear; d) vertex triangular.

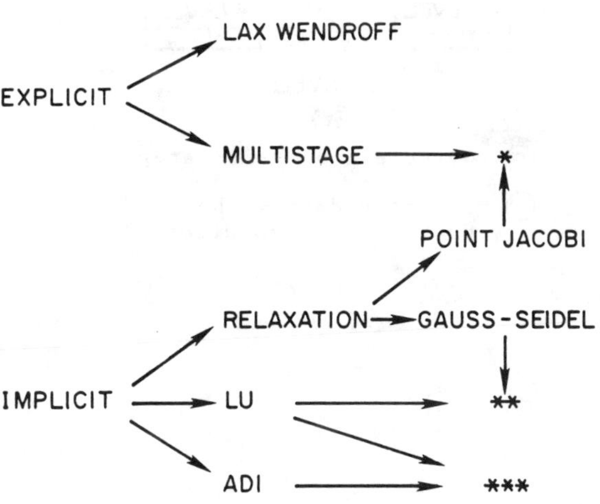

Fig. 6 Time-stepping schemes.

3   LEVELS
(a)
4 LEVELS
(b)
E
4 LEVEL
CYCLE
4 LEVEL
CYCLE
5 LEVELS
(c)
Calculate one time step
Transfer data without
updating the solution

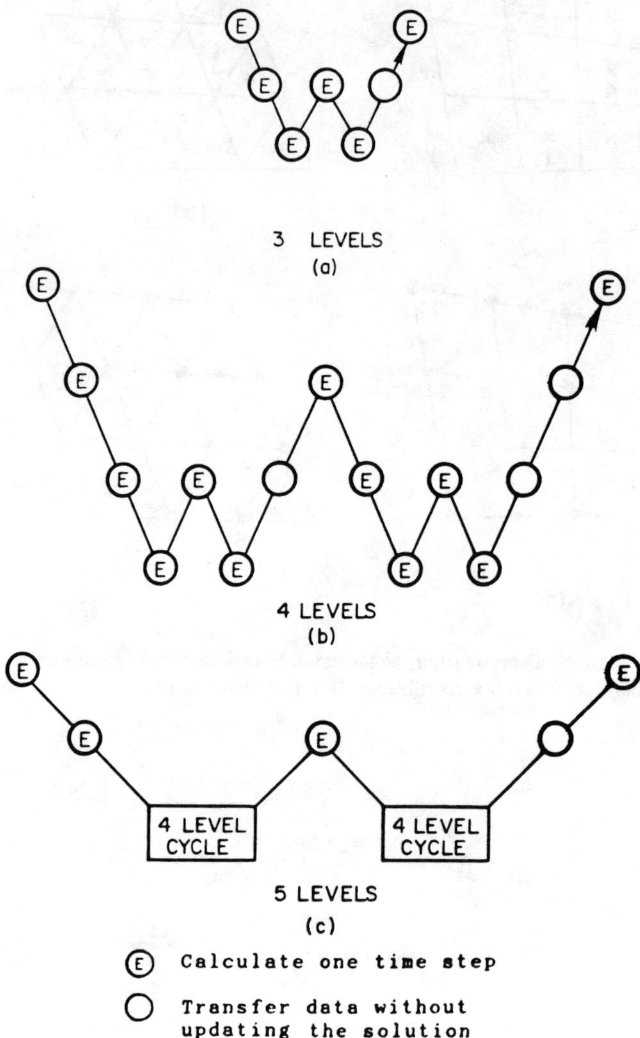

Fig. 7   *W* cycle.

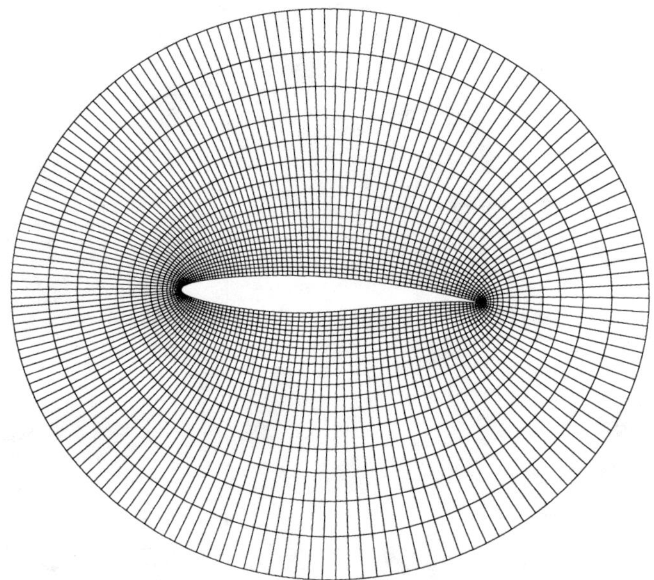

Fig. 8a   Inner grid for Euler solution on RAE 2822 airfoil.

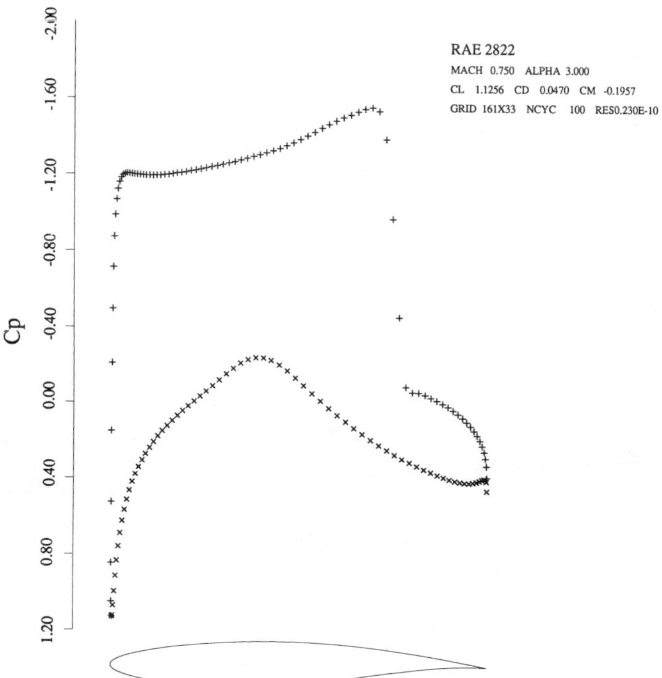

Fig. 8b   Euler solution on RAE 2822 airfoil after 100 cycles.

 A. JAMESON

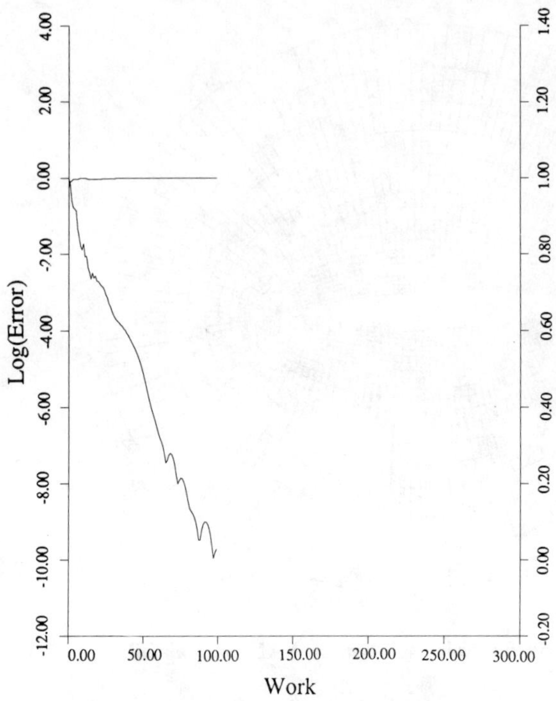

Fig. 8c   Convergence history for Euler solution on RAE 2822 airfoil.

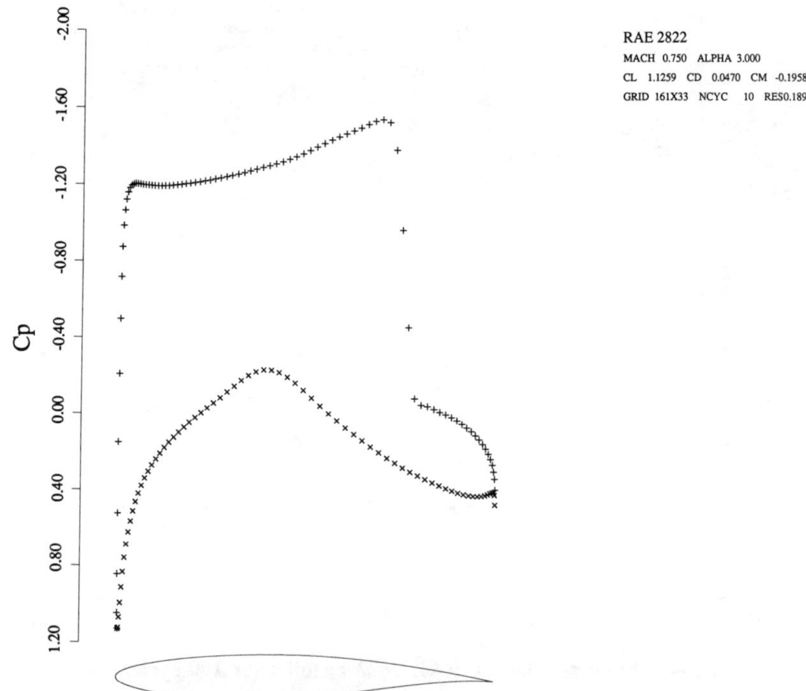

Fig. 8d   Euler solution on RAE 2822 airfoil after 10 cycles.

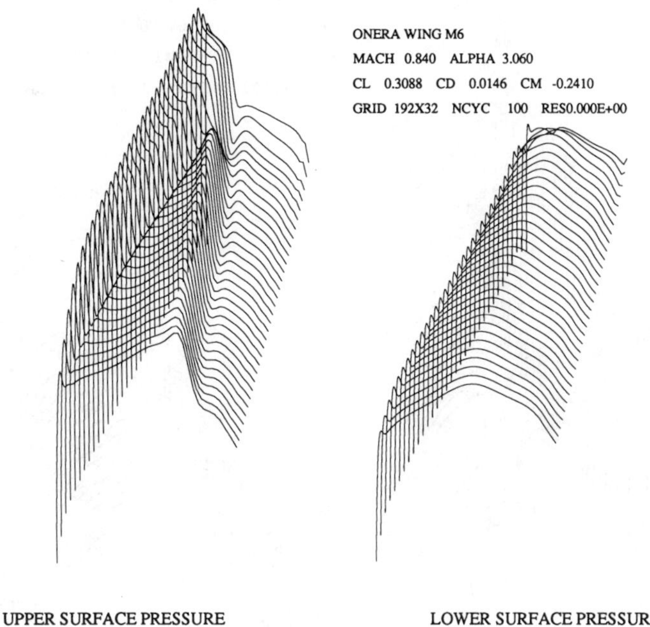

Fig. 9a   Euler solution on ONERA M6 wing after 100 cycles.

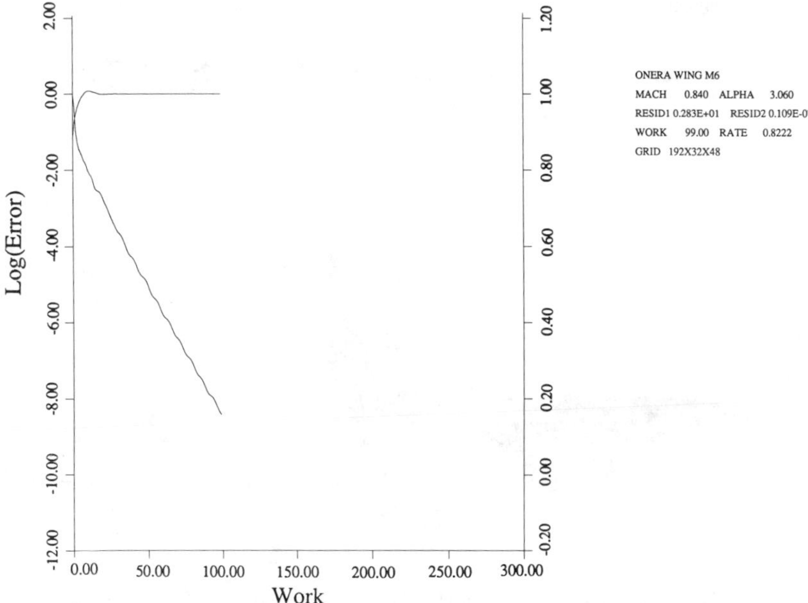

Fig. 9b   Convergence history for Euler solution on ONERA M6 wing.

A. JAMESON

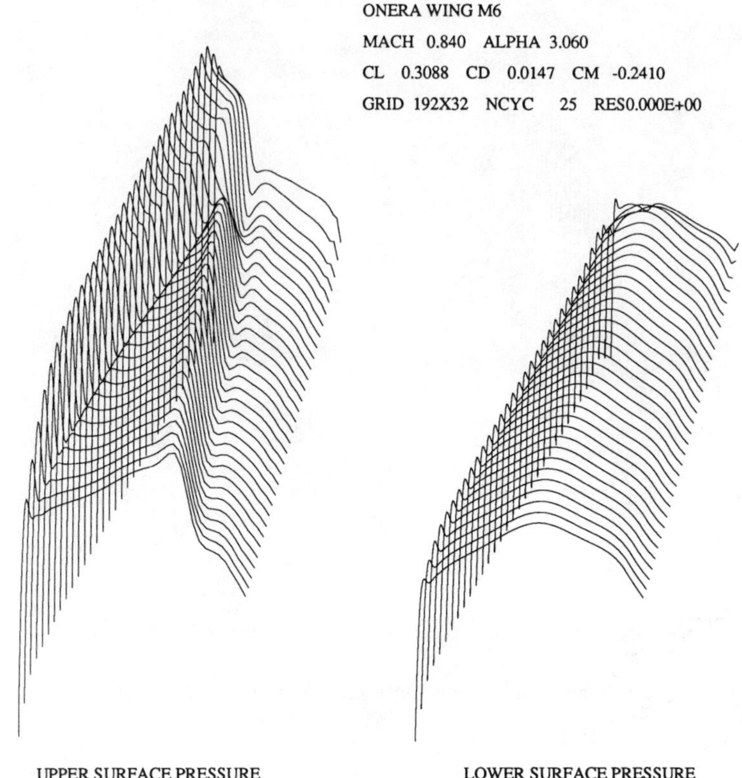

Fig. 9c   Euler solution on ONERA M6 wing after 25 cycles.

Fig. 10a   McDonnell Douglas MD-11 transport shaded surfaces.

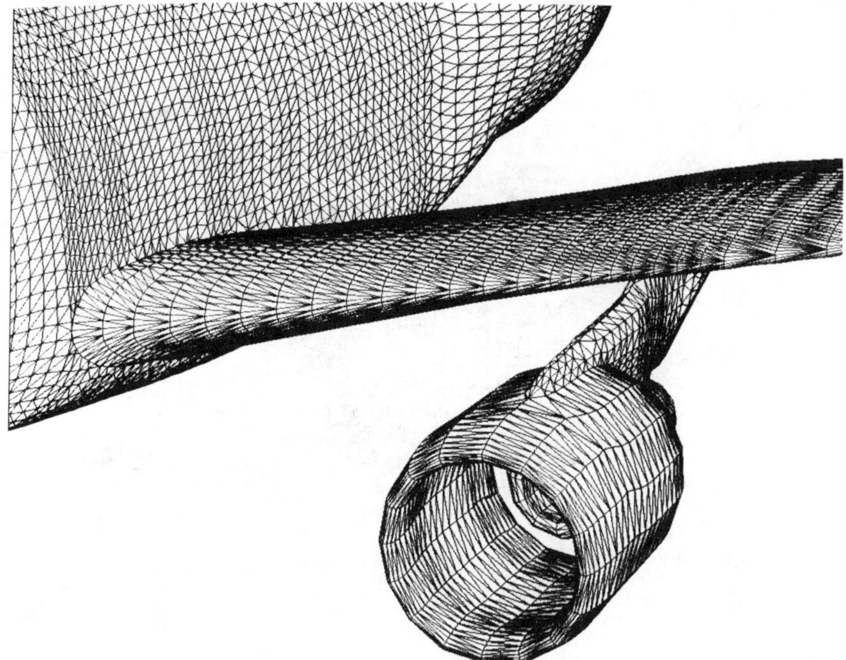

Fig. 10b   Detailed grid for McDonnell Douglas MD-11 transport fuselage/wing pylon nacelle.

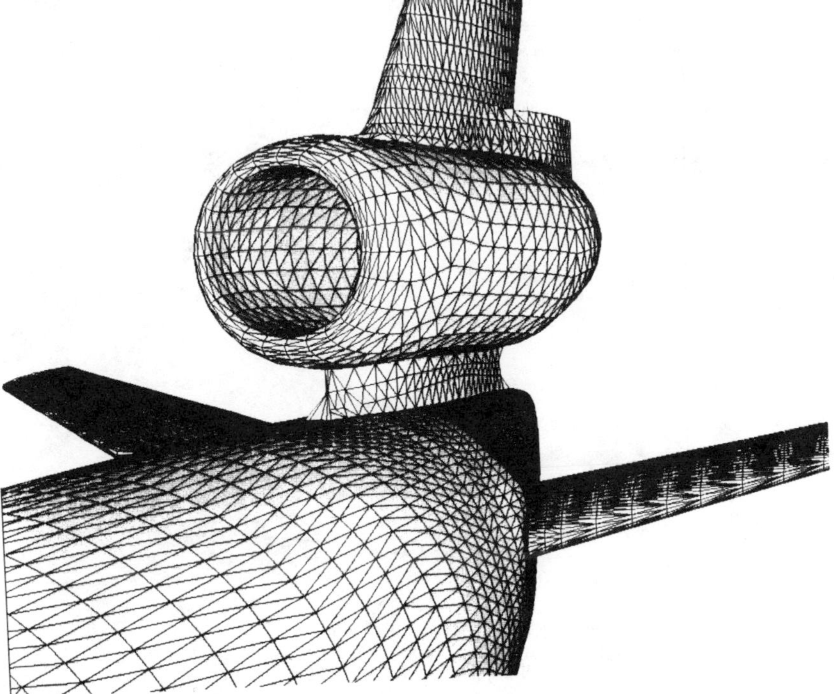

Fig. 10c   Detailed grid for McDonnell Douglas MD-11 transport empennage/aft nacelle.

A. JAMESON

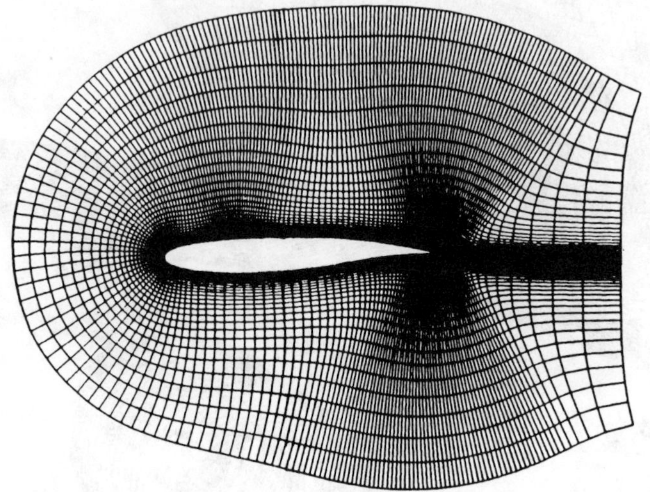

Fig. 11a   Navier-Stokes mesh for Euler solution on Korn Airfoil.

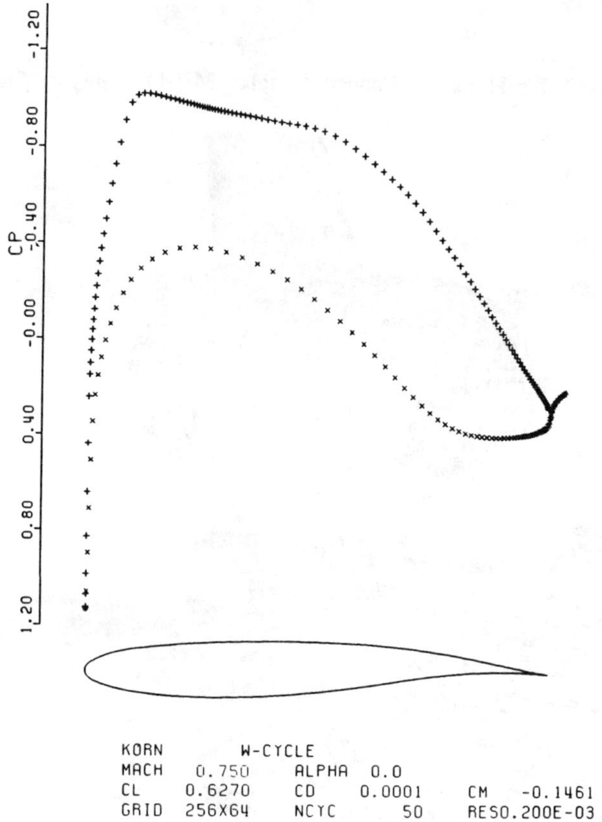

Fig. 11b   Euler solution for Korn airfoil on Navier-Stokes mesh.

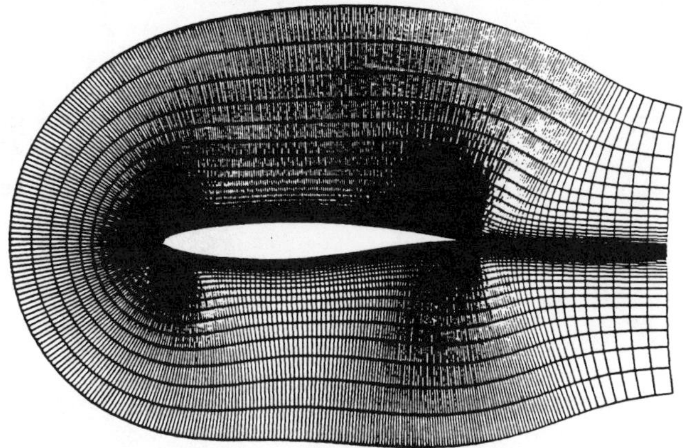

Fig. 12a   512 × 64 Navier-Stokes mesh for Euler solution on RAE 2822.

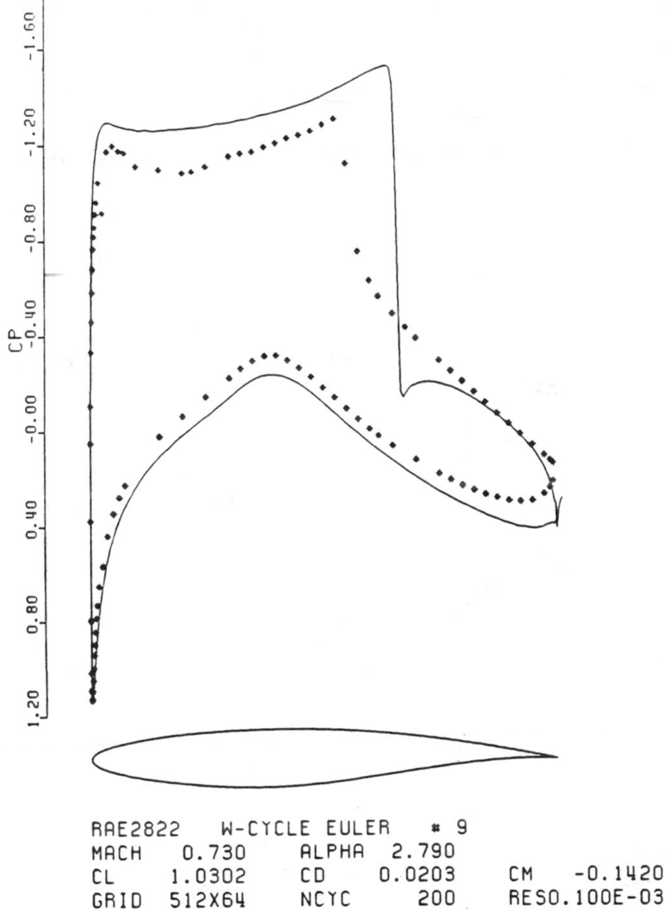

Fig. 12b   Euler solution for RAE 2822 on Navier-Stokes mesh.

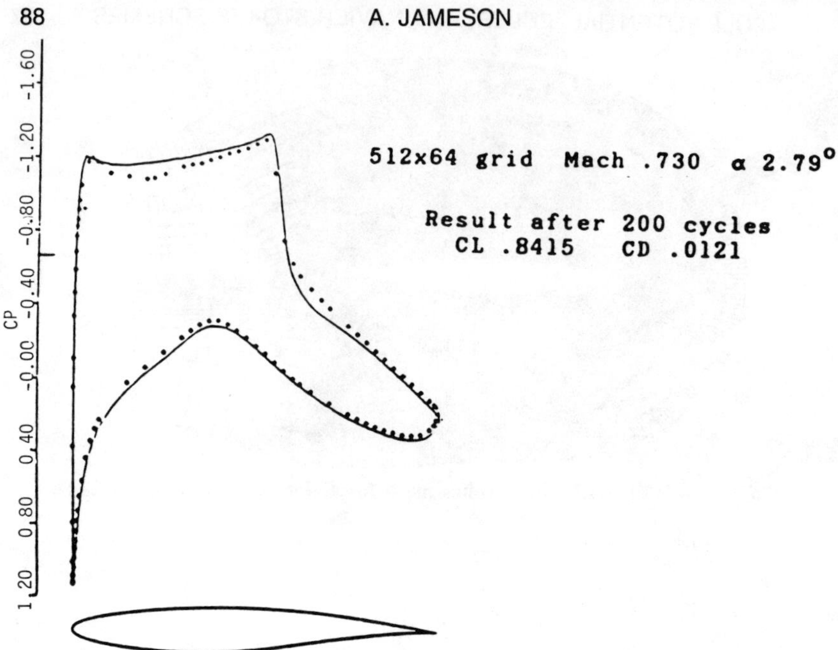

Fig. 13   Navier-Stokes solution for RAE 2822-Baldwin-Lomax turbulence model.

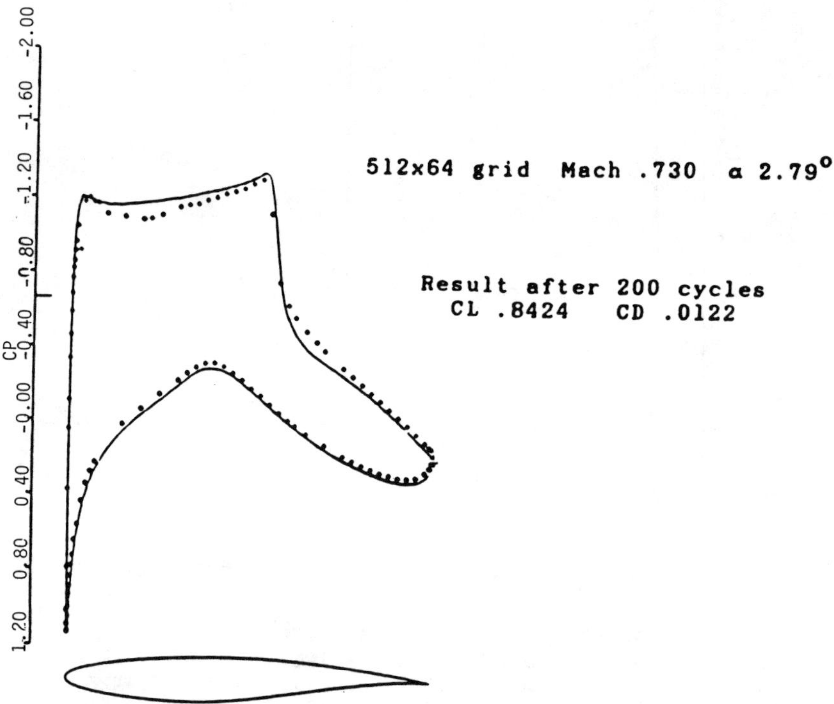

Fig. 14   Navier-Stokes solution for RAE 2822-Baldwin-Lomax turbulence model, artificial dssipation from fourth difference only.

Chapter 4

Grid-Generation Methodology in Applied Aerodynamics

John P. Steinbrenner*
General Dynamics, Fort Worth, Texas
and
Dale A. Anderson†
University of Texas at Arlington, Arlington, Texas

Introduction

THE growth of applied computational aerodynamics over the past two decades has largely been attributed to advances made in numerical flow algorithms and the massive increase in available computational power. Another important yet underappreciated factor has been the advances made in methods for discretizing flowfield domains, i.e., grid generation. Once considered a trivial step in the numerical process, the constant push toward higher orders of flowfield accuracy and increasing geometric complexity has forged a sophisticated grid-generation technology, one that can now stand on its own as a technical discipline.

In this chapter, we examine the existing grid-generation methodology as applied to computational aerodynamics, the driving force behind its development. The review begins with basic considerations in flowfield domain discretization, including structured and unstructured grid concepts and single- and multiple-block ideas. Next, a thorough but necessarily brief overview of the major mathematical classifications of grid generation is offered. This section presents several methods of grid generation in a simple form, allowing the casually familiar reader to appreciate the mathematical basis for the relative strengths and weaknesses of each. Finally, current aerodynamic applications of these grid-generation methods are reviewed.

Because of the limited scope of this review, it is necessary to ignore a number of important grid-generation topics. For example, methods of surface grid generation for panel codes are not included, and neither are multigrid methods, which are more closely related to solution convergence than grid generation. Also, we limit the discussion to steady-state flows.

Copyright © 1989 by the General Dynamics Corporation. Published by the American Institute of Aeronautics and Astronautics, Inc., with permission.
*Senior Engineer, Fort Worth Division.
†Professor, Department of Aerospace Engineering.

Hence, methods of grid generation for geometries with relative movement are not included. Furthermore, adaptive gridding, an interesting idea that is still under development for time-accurate problems, is considered only from a static standpoint. By static adaption we refer to the use of an approximately converged flow solution to modify the original grid in order to enhance the overall acuracy of the steady-state flow solution.

Figures have been used sparingly herein and are included only where they are necessary to the understanding of the problem at hand. Fortunately, a number of conferences and publications have been devoted to grid generation over the past few years and are excellent sources of state-of-the-art grids and grid methods. The reader is encouraged to peruse these volumes, listed in Refs. 1–5.

Domain Discretization Ideas

Before generating a numerical solution to the partial differential equations (pde's) that govern fluid flow, the physical flow domain must be discretized. This discretization may be based on structured or unstructured concepts. In a structured grid, points are arranged so that their relative positioning in physical space is preserved in their computational storage; i.e., points adjacent to a given point in physical space are also adjacent in computational space (Fig. 1a). On the other hand, there is not necessarily any correspondence between a point's physical and computational neighbors in unstructured grids. As such, separate arrays are needed to define the manner in which points connect. Also, to avoid cell area problems, unstructured grids are usually composed of triangular elements in two dimension and tetrahedrons three dimension, compared to a structured mesh's quadrilateral and octahedral elements (see Fig. 1b). Unstructured grids are discussed in more detail later.

For structured grids, the physical domain is discretely represented in three dimension by three independent families of intersecting coordinate curves. Associated with these families are three computational indices, ξ, η, and ζ, which generally take integer values equivalent to their storage locations in three-dimensional array. Physical coordinates are related to computational coordinates by a generalized transformation given by $r = r(\xi,\eta,\zeta)$ where r is a vector with components x, y, and z. This transformation is illustrated in Fig. 2. In a Cartesian-type grid, each physical coordinate is a constant linear function of its corresponding (ξ,η,ζ) variable, and lines of constant ξ, η, and ζ are aligned with the physical coordinate axes (Fig. 1b). Such a grid, which is trivial to generate, allows the governing flow equations to be solved directly in physical rather than transformed space, thereby increasing derivative computational accuracy and efficiency. A serious detriment to Cartesian systems, however, is that domains with curved boundaries will pierce the Cartesian grid at irregular locations, requiring complicated interpolation schemes and problem-specific coding. The compromised accuracy inherent in Cartesian grid systems makes their application to higher-order fluid flow problems (Euler, Navier-Stokes) a rarity today.

The problems associated with curved boundaries in Cartesian grids is avoided if body-fitted coordinates are used. This approach can be likened to deforming, rotating, and translating a Cartesian mesh to fit within the entire flow domain, so that the boundaries of the physical domain coincide with the computational boundaries (Fig. 2). In this type of grid, the transformation defined by $r = r(\xi,\eta,\zeta)$ is much more general, and the governing flow equations are transformed so that the derivatives are expressed in terms of computational coordinates. This decreases the efficiency of the flow calculation, but permits very accurate representation of curvilinear boundaries. With a body-fitted coordinate system it is also possible to generate a fairly general flow code, since boundary conditions will always be applied on the outer surface of the computational cubes. Unlike Cartesian grids, however, body-fitted structured grids are not generated trivially but require application of the involved methods described in the next section. Despite the difficulty of generation, body-fitted coordinates had emerged as the preferred choice for fluid problems by the mid-1980s. Only recently has this position been challenged by the rapidly advancing unstructured methods. We note here that, since Cartesian grids are of secondary interest in this chapter, we will use the terms "structured" and "body-fitted" interchangeably.

The use of a single-structured grid on a given geometry requires the existence of a transformation that will map the entire flow domain into a computational parallelepiped, or block. For simple configurations, a considerable amount of ingenuity may be needed to envision such a transformation. For more complex shapes, however, it may be practically impossible to map the entire domain into a single block, especially if any degree of grid smoothness or orthogonality is required. This problem is alleviated in some cases by the incorporation of interior boundary conditions. In the simplest case, part of the flow domain boundary would be represented by computational points interior to the block, rather than on the block's outer surface. For more complex configurations, whole regions of the computational block could be zeroed out. Interior boundary conditions are used in the code described in Ref. 6.

Since the difficulties associated with stretching a single block to conform to a complex domain are often insurmountable, it is logical to divide the domain into a number of smaller, more manageable blocks. This multiple-block idea, investigated initially by Lee and Rubbert,[7] is now a widely used general structured grid form. In this technique, the entire flow domain is represented by the union of a number of subdomains (blocks), each of which has six computational faces in three dimension, and four edges in two dimension. Information is communicated between blocks in the flow solver via a connectivity table containing boundary interface data. A few of the block interface types in use today are described in the following paragraphs, arranged in order of increasing generality.

The most accurate (in a flow-solver sense) block interface type is one that will maintain complete grid continuity across block boundaries. Each boundary point on an abutting surface will have an identical companion point on another block, and the slope of the grid transverse to the abutting surface will remain continuous, as illustrated in Fig. 3a. Such a scheme requires no special boundary point treatment in the flow solver, but it can be difficult to generate the grid. Elliptic pde schemes described in the next section are often used to enforce continuity. One approach is to load in an outer "ghost layer" of grid points, which is obtained from points residing one layer inside of the abutting blocks, and then to calculate boundary points in the same manner as interior points are calculated. Boundary point locations will then evolve in the iterative process, as explained in Refs. 8 and 9. Another idea is to keep boundary points fixed, but to drive interior points toward orthogonality at block surfaces. This approach is adopted in Ref. 10.

A more easily implemented interface condition is point continuity alone (Fig. 3b). Here, each block may be generated independently, as long as interface grid points are common. This relaxed requirement reduces the complexity of the grid solver, but can also introduce solution errors due to slope discontinuities at the interface points. An even less restrictive interface condition is one where two grids of varying point density abut, so that one face is an ordered subset of another (Fig. 3c).[11] If no point continuity of abutting surfaces is maintained, then the scenario in Fig. 3d results. This type of block interface is supported in the flow solver of Ref. 11, and conservative interpolation methods for such interfaces are examined in Ref. 12.

Probably the most general interface type in use today is the grid embedding or chimera technique.[13,14] Individual blocks are generated for each geometric component, with component blocks overlapping to fill the entire domain (Fig. 3e). Naturally, this general interface requires the most work to implement into a flow solver. The major difficulty lies in the automated determination of block overlap and in the development of interpolation routines to transfer flow data across component blocks in a conservative manner.[15] The major benefit is that the overall grid is easily generated even for complex configurations, since the component grids have no external constraints due to overlapping. This technique is also well-suited to geometries with relative movement and has recently been applied to the problem of underwing store separation.[16]

A final multiple-block interfacing technique is to combine structured blocks with unstructured blocks, as recently investigated by Weatherill.[17] Two philosophies to this approach may be taken. First, nonoverlapping structured grids can be placed around the geometric components of the domain, with an unstructured mesh used to fill in the gaps and the far-field domain. Second, unstructured grids may be used to discretize the geometrically complex regions of the domain, with structured grids used outside of the geometric intricacies. Both of these approaches combine the numerical efficiency of structured schemes with the ease of generation associated with unstructured schemes. With this in mind, it is logical to expect this type of hybrid scheme to become an area of active research.

A variety of grid-generation codes have been written that utilize combinations of the discretization methods of this section and the grid-generation techniques of the next section, and several are referenced throughout the paper. Each espouses the ideas of the flow solvers on which it was intended to be used. Since most flow solvers are applied to a wide range of geometric domains, it is has been necessary to develop these grid-generation codes to handle equal generality. Disappearing for the most part are the grid-generation codes that are hard-wired to a specific geometric shape. The added generality increases the length of the code significantly, and it also increases the number of execution options available to the user. Hence, a number of these codes have been developed for interactive operation. Many of these have also been written with a large number of graphic commands, providing necessary visual feedback to the user. The sophistication of the existing codes vary, but it is clear that the trend is toward graphic, interactive codes. This is true even for unstructured grid methods, many of which still rely on user intervention to generate surface grids. Although the generality of the code considerably increases grid-generation time (compared to hard-wired codes), the overall turnaround time for complex geometries is now measured in days (unstructured grids) and weeks (structured grids), rather than the weeks and months required just a few years ago. This rapid increase in speed can be attributed to many factors, including the effective use of graphics and user interactivity, improved and newly developed grid algorithms, and advances in workstation technology.

Basic Grid-Generation Methods

Having discussed the different ways of discretizing domains, we are now in a position to investigate the methods employed to generate the grids. For structured grids, three major classes of techniques prevail: algebraic methods, pde methods, and conformal mapping methods. Although in a mathematical sense conformal mapping methods have distinct ties to both algebraic and pde methods, the techniques employed differ enough from (and predate) the others to warrant separate treatment. In fact, each class has developed its own terminology and conventions, and each is well-represented in the literature. Unstructured grid-generation methods, on the other hand, have been in use for some time, but their general application to fluid mechanics is relatively new. The three major classes of unstructured grid techniques will be discussed in a single section.

Algebraic Methods

By algebraic grid generation, we refer to methods of grid construction that use interpolative techniques to distribute grid points from a discrete set of data, consisting of grid boundary points and/or interior control points. A feature common to all algebraic methods is computational efficiency, since the grids are generally calculated directly from functional relations, rather than interatively as in pde methods. Thus, algebraic methods occupy an important volume in the library of grid-generation methods. The importance of efficiency continues to grow as methods using algebraic techniques increase in generality.

The techniques used in algebraic grid generation are thoroughly explained in a number of excellent references.[18,19] Since the intricacies of these methods are beyond the scope of this paper, we present here only the basic ideas behind the more successful algebraic methods currently in use.

Unidirectional Interpolation

Given two points in physical space and a specified number of points to interpolate between them (ξ_{max}), intuition might lead one to suggest the following distribution formula:

$$r(\xi) = r(1) + \bar{\xi}[r(\xi_{max}) - r(1)] \tag{1}$$

where

$$\bar{\xi} = \frac{(\xi - 1)}{(\xi_{max} - 1)}$$

where r corresponds to the physical coordinate (x, y, or z), and ξ is the computational index, equal to 1 at one of the specified points and equal to ξ_{max} at the other. Equation (1) represents a linear fit between the two endpoints, but any continuous functions that satisfy the two boundary conditions are equally acceptable. In general, then,

$$r(\xi) = a_1(\xi)r(1) + a_2(\xi)r(\xi_{max}) \tag{2a}$$

where

$$a_1(1) = 1, \qquad a_1(\xi_{max}) = 0 \tag{2b}$$

$$a_2(1) = 0, \qquad a_2(\xi_{max}) = 1 \tag{2c}$$

Because this interpolation proceeds in only one computational direction, it is referred to as unidirectional interpolation. Furthermore, if $a_i(\xi)$ is a polynomial, this equation represents Lagrange interpolation, and if the polynomial is linear [as in Eq. (1)], it is known as a shearing transformation. Equation (2) can be extended to satisfy an exact specification of derivatives at the endpoints as well. Specifically,

$$r(\xi) = [(1 + 2\bar{\xi})(1 - \bar{\xi})^2]r(1) + [(3 - 2\bar{\xi})\bar{\xi}^2]r(\xi_{max})$$

$$+ [(1 - \bar{\xi})^2\bar{\xi}]r'(1) + [(\bar{\xi} - 1)\bar{\xi}^2]r'(\xi_{max}) \tag{3}$$

can easily be shown to satisfy conditions for r and r' at both endpoints. This case is an example of Hermite interpolation, in that cubic polynomials are used to satisfy four boundary conditions. Likewise, the order of the polynomials can be increased by two again to satisfy second derivatives at the endpoints.

Now moving on to two-dimensions, replace the $r(1)$ and $r(\xi_{max})$ endpoints in Eqs. (1–3) with two curves already discretized in the transverse (η) computational index. Equation (2) becomes

$$r(\xi,\eta) = a_1(\xi)r(1,\eta) + a_2(\xi)r(\xi_{max},\eta) \tag{4}$$

Equation (4) is in effect interpolating an entire $\xi = $ const line between the $\xi = 1$ and $\xi = \xi_{max}$ curves, resulting in a two-dimensional grid from only the bounding curves. Smith and Everton[20] refer to the two-dimensional version of Eq. (3) as the two-boundary technique, which they have since exploited in several applications.[20,21] Notice that the two-boundary technique provides no control of the two remaining boundaries in the resulting grid $[r(\xi,1),r(\xi,\eta_{max})]$. Rather, they are a result of the application, dependent entirely on the choice of the interpolating functions and the corner points.

Univariate interpolation as described earlier is easy to apply, but does not always afford the user a sufficient degree of grid point control, particularly at interior points. In response to this problem, Eiseman[22] developed a general method known as the multisurface transformation. In this technique, a number of intermediate surfaces (curves) are defined to influence the path of grid points in the interpolated (η) direction, but points in the generated grid do not necessarily pass through the intermediate surfaces. Rather, the intermediate surfaces are used to construct a field of tangent vectors in the ξ direction at each η surface, which are then interpolated smoothly between η surfaces and ultimately used to determine grid point locations. B-splines[23] and Bezier splines[24] are two other examples of unidirectional interpolation schemes that rely on interior control points and surfaces for grid point placement.

Transfinite Interpolation

Since many applications of grid generation insist on a specific domain boundary, the uncontrollable boundary phenomenon associated with unidirectional interpolation methods is often unsuitable. Higher-order interpolative schemes may be constructed, but the one of most interest satisfies the grid boundary point conditions on all four edges of the domain (two dimensional). Such a scheme, referred to as transfinite interpolation, was surprisingly not developed until a relatively short time ago.[25] The concept in its most general form can be derived quickly and neatly using the projector idea of Boolean algebra[18]; however, the idea can be derived from a more physical standpoint using the argument given later, borrowed from Ref. 18. Start with a two-dimensional grid and apply a unidirectional interpolation of grid points in the, say, η direction. Exact boundary conditions are of course satisfied on the two $\eta = $ const edges, but not on the two $\xi = $ const edges. An error distribution thus exists on the ξ edges, equal to the grid boundary value minus the η-directed interpolated value. Next, apply a second unidirectional interpolation, this time in the ξ direction, and using the error distribution as the interpolated value. Since the error values on the four corners are zero, they will vanish on the $\eta = $ const edges as well (by virtue of the ξ-directed interpolation), meaning that the summation of the two unidirectional interpolations will yield an expression that satisfies boundary conditions on all four edges of the grid.

When simple linear interpolations are used, this results in

$$\bar{r}(\xi,\eta) = (1 - \bar{\xi})r(1 - \bar{\eta}) + (\bar{\xi})r(\xi_{\max},\eta)$$

$$+ (1 - \bar{\eta})r(\xi,1) + (\bar{\eta})r(\xi,\eta_{\max})$$

$$- (1 - \bar{\xi})(1 - \bar{\eta})r(1,1) - (1 - \bar{\xi})(\bar{\eta})r(1,\eta_{\max})$$

$$- (\bar{\xi})(1 - \bar{\eta})r(\xi_{\max},1) - (\bar{\xi})(\bar{\eta})r(\xi_{\max},\eta_{\max}) \tag{5a}$$

where

$$\bar{\xi} = \left(\frac{\xi - 1}{\xi_{\max} - 1}\right), \qquad \bar{\eta} = \left(\frac{\eta - 1}{\eta_{\max} - 1}\right) \tag{5b}$$

The three-dimensional equivalent of Eq. (5) may be derived from an analogous argument, although the Boolean algebra method is easier to apply. More complex interpolant types may also be used. Ericksson, for example, incorporates an arbitrary number of derivative boundary conditions into his generalized transfinite interpolation formulation.[26] As long as the interpolants are direct functions of the computational indices, however, there is a tendency for transfinite methods to exhibit grid line crossing when a large amount of stretching is imposed on grid boundaries. This symptom is by and large removed when interpolants based on normalized edge arc lengths are used. One such application is given in Ref. 27.

Finally, one of the more general algebraic grid-generation methods proposed to date is the control point form.[19] This method is the extension of the multiple-surface technique to several variables, with the control surfaces replaced with a sparse array of control points. With this technique, a sparse grid obtained by any means can be used to control the distribution of points on the resulting grid's interior. This method, which is best suited to an interactive environment, has been applied to several complex configurations.

Partial Differential Equation Methods

The speed of generation associated with algebraic grid-generation methods is well-known, but the smoothness of the resulting grids is usually not a certainty. Since the numerical accuracy of flow solutions obtained on body-fitted coordinate grids is dependent on grid smoothness, it is natural to look toward a generating mechanism that has smoothness implicit in the driving equations. Various classes of pde's exhibit this feature, many of which have been used with success to generate grids. Since almost all of the resulting grid-generating pde's are nonlinear (the exception being simple conformal mapping methods), there is little chance for closed-form analytical solutions, and numerical approximations (usually finite differences) are made. In doing so, the resulting grid is either obtained iteratively, or is obtained in a single sweep with a prespecification of some grid variables (Jacobians, point locations, etc.). In either case, it is usually necessary to calculate a provisional grid before the pde solution proceeds. For this it is convenient to rely on the algebraic methods described earlier, due to their speed and quality. From this viewpoint it is possible to consider pde methods as refinements to algebraic grids, although the algebraic grids themselves are often entirely satisfactory.

Equidistribution
If we begin our discussion in one dimension, then the simplest meaningful differential equation that can be written to govern the spacing on a grid is

$$\xi_x = C \tag{6}$$

Since in $1-d$ $\xi = \xi(x)$, $d\xi/dx = 1/(dx/d\xi)$, and hence $dx/d\xi = \text{const}$ as well. If instead C is a functional variable, then we have the expression

$$\xi_x = Cw(\xi,x) \tag{7a}$$

or

$$x_\xi w(\xi,x) = C \tag{7b}$$

This latter form is the statement of equidistribution, which will turn out to be a basis for several more complex methods. Here, w can be thought of as a user-definable weighting function chosen so that w values are large where the required grid clustering is also high and vice versa. This equation may be differentiated into a second order ordinary differential equation:

$$x_{\xi\xi} + x_\xi\left(\frac{w_\xi}{w}\right) = 0 \tag{8}$$

The concept of equidistribution can be applied toward generation of grids in higher dimensions by replacing the x variable in Eqs. (7) with an arc length S, so that equidistribution proceeds along a given set of coordinate curves. In two dimensions, this would result in

$$S_\xi w_1 = \text{const} \tag{9a}$$

$$S_\eta w_2 = \text{const} \tag{9b}$$

The solution to these equations may be obtained either by integration (cf. Ref. 28) or by differentiation.[29] The equations are generally applied in a curve-by-curve manner, either only in a single set of coordinate curves, or in multiple directions, in alternating iterative sweeps. Eiseman attempted to place the curve-by-curve method on firmer mathematical ground in Ref. 30. The solution of Eqs. (9) requires an a priori specification of grid points (to calculate S values), and the differentiated form of this equation results in a decoupling of the independent (ξ,η) variables with a coupling of the physical x, y, and z variables. This results in a rather stiff set of equations that usually require numerical fixes, such as derivative filtering[30] and orthogonality control.[31]

Despite the inherent problems, the method of equidistribution in multiple directions has been used in many adaptive grid applications. The reason for this is that the weighting function's influence on grid point spacing is direct; thus, grid point control is affected straightforwardly through the weight function. Conventional (nonadaptive) grids could equivalently be generated using equidistribution, but this is usually not done because more robust grid-generation techniques are available (these techniques will be explained later). Among the adaptive applications of equidistribution (in various forms) are those of Dwyer,[28] Fritz et al.,[32] Gnoffo,[33] and Nakahashi and Deiwert.[31] These last two methods employ ideas of potential energy systems. Gnoffo uses a spring analogy to adapt grid points in one computational direction. Nakahashi and Deiwert use an elaborate tension and torsion spring model to provide adaption (tension spring) and orthogonality (torsion spring) in multiple computational directions. Both of these methods take first principles from equidistribution, although their divergence from strict equidistribution is significant, resulting in a great deal more applicability.

Elliptic Methods

A desirable feature of any body-conforming grid is smoothness of coordinate lines, and a way to guarantee this feature is to distribute grid points according to a smoothing operator, viz., Laplace's equation:

$$\nabla^2 \xi = 0 \tag{10a}$$

$$\nabla^2 \eta = 0 \tag{10b}$$

This idea was first explored by Winslow[34] more than 20 years ago. These equations do indeed exhibit solution smoothness, but grid point clustering and orthogonality are usually compromised. Recognizing this, Thompson et al.[35] inserted a degree of user control by solving instead the Poisson equations,

$$\nabla^2 \xi = P = (\nabla \xi \cdot \nabla \xi)\Phi \tag{11a}$$

$$\nabla^2 \eta = Q = (\nabla \eta \cdot \nabla \eta)\Psi \tag{11b}$$

Both P,Q and Φ,Ψ are viable representations of control functions, although the latter form appears to be more prevalent today. Using the Φ,Ψ form, Eqs. (11) can be recast in the Thomas and Middlecoff[36] form of the Thompson equations:

$$\alpha(r_{\xi\eta} + \Phi r_\xi) - 2\beta r_{\xi\eta} + \gamma(r_{\eta\eta} + \Psi r_\eta) = 0 \tag{12a}$$

$$\alpha = r_\eta \cdot r_\eta, \quad \beta = r_\xi \cdot r_\eta, \quad \gamma = r_\xi \cdot r_\xi, \quad r = [x\,y]^T \tag{12b}$$

These equations are usually solved iteratively, linearizing the a, b, and c terms by computing their values from known values. For reasonable Φ and Ψ values, iterative determination of the grid points proceeds with much greater stability than equidistribution methods based on Eqs. (9). The inherent smoothness in the governing equations also make elliptic methods a popular choice of grid generation. The equations as written earlier are valid only for strict two-dimensional grids, although the three-dimensional equations are similar and are identical when each is written in tensor notation.[18] Halfway between the purely two-dimensional and fully three-dimensional cases is the surface grid case, which has two computational variables (say, ξ and η) that span three physical variables x, y, and z. These applicable surface grid equations can be derived from the three-dimensional equations, among other ways, by assuming that grid lines transverse to a surface in a three-dimensional volume intersect the surface orthogonally and with zero curvature.[37] The resulting equation is also similar to Eqs. (12), although a nonhomogeneous term arises that accounts for the curvature of the surface. Notice that the one-dimensional analog of Eqs. (12) is simply

$$x_{\xi\xi} + \Phi x_{\xi} = 0 \tag{13}$$

which is equivalent to the equidistribution law of Eq. (8), provided that

$$\Phi = w_{\xi}/w \tag{14}$$

Thus, in the one-dimensional case we have a relationship between weighting functions of equidistribution methods and control functions of elliptic methods.

Selection of suitable control functions is not always an easy task, and in fact the extremum principle applicable to Laplace's equation (i.e., the guarantee of no grid line crossing) will not hold for any but carefully developed functions. Thomas and Middlecoff[36] derived one such suitable function form, calculating Φ and Ψ from boundary data and then interpolating these values into the interior. Expressions for the control functions were backed out of the governing equations by assuming again that transverse grid lines intersect bounding surfaces orthogonally and with zero curvature. Sorenson[38] developed another control function form by calculating the first derivative values at boundaries required to provide orthogonality and a specified cell height normal to the bounding surface. Second derivative terms in the governing equations (and hence the control functions) are updated iteratively in this method. Both of these basic methods have been modified and rederived from different viewpoints since their inception,[8,39] but they remain two of the more commonly used control function types of today.

Although the connection now seems almost trivial, the idea of using the weighting function-control function relationship [Eq. (14)] in elliptic solvers for the purpose of adaption was not developed until a few years ago.[40] Using the same forms of w as in Eqs. (9), expressions for the control functions in two dimensions are immediately written as

$$\Phi = w_{1\xi}/w_1 \tag{15a}$$

$$\Psi = w_{2\eta}/w_2 \tag{15b}$$

Again, for adaption, w_i is any function obtained from the flow solution that assumes large values where grid clustering is needed. The resulting grid from this control function form applied to Thompson's equations [Eqs. (12)] will not satisfy exactly an equidistribution law [Eqs. (7)] in any computational direction, but virtually all of the skewness and instabilities associated with a strict equidistribution application are eliminated. The simplicity of the method makes this elliptic adaptive approach an easy and robust means of obtaining adaptive grids in multiple dimensions.[41]

Parabolic Methods

The major pitfall to elliptic methods of grid generation is the significant computational effort needed for large, multidimensional problems, due to the iterative nature of the solution procedure. Nakamura[42] first explored the idea of solving elliptic equations in a noniterative fashion by experimenting with two parabolic heat equationlike relations:

$$\frac{\partial x}{\partial \eta} = A \frac{\partial^2 x}{\partial \xi^2} + S_x \tag{16a}$$

$$\frac{\partial y}{\partial \eta} = A \frac{\partial^2 y}{\partial \xi^2} + S_x \tag{16b}$$

He found that, with a certain formulation of the source (S_x, S_y) terms, the discretized form of Eqs. (19) would be identical to the discretized Laplace form of Eqs. (12), with derivatives in the ξ direction replaced with central differences and with derivatives in the η direction replaced with known values estimated from previously calculated grid points and prespecified outer boundary data. With this formulation, the elliptic equations were effectively "parabolized" in the η direction, in that calculation of the grid could proceed by marching outward in the η direction in a single iterative pass. The manner in which the derivatives in the marching direction are calculated is the driver of the grid, but Nakamura demonstrated that a degree of orthogonality can be achieved with careful approximation of the η derivatives. This method was later extended to multiple dimensions[43,44] and to the control function (Poisson) form of Eqs. (12).[45,46] Reference 46 used the adaptive control function form of Eqs. (15) to effect solution adaptation in the grid.

Today, parabolic grid generation is generally considered to be a single-pass approximation to elliptic grid generation; hence, one cannot expect the same degree of grid line smoothness, orthogonality, or clustering control with parabolic methods that may be obtained with elliptic methods. Still, the significantly reduced computational requirement makes it a method worth considering. Furthermore, the fact that all boundaries may be specified before calculation of the grid makes it a viable scheme for use with multiple-block techniques, as explained later.

Hyperbolic Methods

In computational fluid dynamics (CFD), the determination of physical flow domain boundaries can be a nebulous task, especially in external flow problems, which analytically are bounded on at least one side by infinity. Nevertheless, the CFD practitioner must choose an "infinity" (i.e., the far-field boundary), which he deems to be sufficiently far from the geometric surface. At a certain distance, the exact boundary location and shape will no longer affect the flowfield solution, but will, however, continue to affect the effort needed to generate the grid. Thus, it would be advantageous to have a grid-generating scheme that could determine the outer boundary directly from the governing grid equations. In other words, why not generate a grid by marching outward from the solid surface boundary, until the grid has either reached an acceptable distance from the inner surface, expended a maximum number of grid points in marching, or, more likely, both? This whole idea suggests the use of hyperbolic pde's for generation of the grid, somewhat analogous to the method of characteristics for linearized supersonic (hyperbolic) flow.

In two dimensions, a hyperbolic set of equations has been derived by Steger and Chaussee,[47] who linearized the following equations about a known surface ($\eta = $ const):

$$x_\xi x_\eta + y_\xi y_\eta = 0 \tag{17a}$$

$$x_\xi y_\eta - y_\xi x_\eta = J(\xi,\eta) \tag{17b}$$

The first of these equations is a condition of grid point orthogonality, and the second is an expression for the cell area at a given point [$J(\xi,\eta)$]. Hence, the linearized form of Eqs. (17) is marched outward from the inner surface ($\eta = 1$), resulting in a grid that maintains orthogonality and a predetermined cell area distribution throughout. The outer, or infinity boundary is calculated as the last marching step.

This type of technique is extremely efficient, since the entire grid is generated in a single sweep, and excellent results may be obtained for reasonably smooth bodies. However, because the governing equations are hyperbolic, discontinuities on the inner surface may be propagated into the interior. Also, it is possible to find grid overlap, particularly near regions of boundary concavity. This latter problem makes it necessary to specify the area distribution either from physical reasoning or from an initial grid. The hyperbolic equations given earlier [Eqs. (17)] are easily extended to three dimensions by adding another condition of orthogonality, resulting in three governing equations,[48] with marching still proceeding in a single computational direction. Note that orthogonality may not be imposed within surfaces normal to the marching direction, since a fourth equation overconstrains the problem. Finally, the use of hyperbolic grid generation has recently been extended to the solution adaptive realm by Klopfer.[49] Here, instead of linearizing Eqs. (17), a second-order smoothing term is added, along with two additional terms similar in concept to the tension-torsion spring idea described earlier.[31] The last two terms provide adaption to a prescribed solution without excessive deviation from orthogonality.

Optimization Methods

Most of the procedures for grid generation mentioned to this point are based on somewhat ad hoc approaches, in that everyday applications are the major justification of their viability. An additional approach based on more concrete mathematical reasoning would be to use optimization methods to distribute points within a grid. In this method we first construct an integral that we wish to minimize within the problem domain. The integrand is a function whose local value is a measure of departure from what is considered to constitute a good grid, and the integral is then a measure over the entire domain. Since orthogonality and grid line smoothness are desired features, two possible integrals to minimize (two dimensional) are as follows:

$$I_O = \int\int (\nabla\xi \cdot \nabla\eta) J_a^3 \, dx \, dy, \qquad J_a = \frac{\partial(x,y)}{\partial(\xi,\eta)} \tag{18a}$$

$$I_S = \int\int (\nabla\xi \cdot \nabla\xi) + (\nabla\eta \cdot \nabla\eta) \, dx \, dy \tag{18b}$$

These two equations are minimized when global smoothness and orthogonality, respectively, are maximized. Other forms are obviously acceptable, but the two preceding have been used successfully in Ref. 50. A third integral can be used for solution adaptivity, again with w specified to be large where tight grid clustering is needed. This form is also used in Ref. 50:

$$I_W = \int\int W^2(x,y) J_a \, dx \, dy \tag{18c}$$

Generally, two or more of the preceding integrals are minimized simultaneously so that more than one desirable feature will be present in the resulting grid.

There are basically two ways to minimize the integrals of Eqs. (18). The first way, through the use of variational calculus, involves developing the Euler-Lagrange equations that govern the minimum solution of the functional. In two dimensions, this results in two pde's for each integral which are solved by iterative means. This technique was pioneered by Brackbill[51] and Saltzman,[50] who created a single global integral consisting of dimensionally correct combinations of Eqs. (18). These authors demonstrated that the Euler-Lagrange equations corresponding to the smoothness integral above [Eq. (18a)] were equivalent to the Laplace system of equations given by Eqs. (10). Furthermore, Saltzman[51] showed that the one-dimensional equivalent Euler-Lagrange equation of the volume integral [Eq.

(18c)] was equivalent to the one-dimensional statement of equidistribution given by Eq. (8). The actual form of the resulting Euler-Lagrange equations in two dimensions is considerably complex and requires the numerical solution of many derivative terms. The form in three dimensions is even more complex, and the subsequent evaluation of the component terms makes the entire iterative process significantly slower than even the elliptic solver methods described earlier. However, the grids that may be obtained with this method can exhibit a striking blend of clustering, smoothness, and orthogonality, and for this reason, variational methods are still under investigation.[41,52]

The other common method of solving optimization integrals can best be described as a direct method. Here, the integral equations are first discretized and replaced by summations. The resulting summation becomes an algebraic functional, which is dependent on a vector of variables (grid point coordinates) of length three (x, y, and z) times the number of points in the grid. The solution vector (the grid) is then obtained by any standard numerical optimization technique, such as the conjugate gradient method.[53] A method such as this requires a large number of functional evaluations, and it is necessary to make the functional form as simple as possible, from computational considerations. In application, the integrands of Eqs. (18) are modified for ease of calculation. Examples of this type of technique may be found in Refs. 54–56.

Optimization as a method of grid generation is not as routinely used as the other methods described earlier. However, the firm mathematical basis of the ideas behind the method can easily justify the amount of research underway today. Progress continues, as optimization methods have recently been applied to three-dimensional examples of reasonable complexity.[55]

Other Methods

A number of pde methods of grid generation have not yet been treated, and for brevity, only a couple of the more proven forms are mentioned here. Conformal mapping methods, which contain elements of differential equations and algebra, are treated separately in the next section.

The elliptic method of surface grid generation, of which Eqs. (12) are a subset, may be rederived from a mathematically more formal standpoint if one uses concepts borrowed from differential geometry. The resulting equations are presented in Ref. 18 and are derived by Warsi in Ref. 57. The whole idea of solving pde's on prescribed surfaces requires a definition of the surface beforehand and a means of transferring the information from the surface to the governing equations. Two approaches to this problem are suggested in Refs. 58 and 59.

Higher-order equations may also be used in grid generation. For example, the biharmonic equation is frequently solved with the computational indices used as the independent coordinates, resulting in $\nabla^4 r = 0$. Since this equation form does not exhibit a maximum principle,[60] applications are limited to geometric domains that do not have excessive skewing or sharp corners. By virtue of the higher order of the equation, we are afforded an additional boundary condition (b.c.) at each boundary point. This b.c. is implemented as strict boundary orthogonality in Refs. 60 and 61. Alternatively, the biharmonic equation may be reduced to two sets of Poisson-type equations, namely, $\nabla^2 r = P$ and $\nabla^2 p = 0$.[62] In this case, the P variables [similar in function to the control functions used in the second-order system of Eqs. (11)] are used as the additional boundary condition. To enforce boundary orthogonality and clustering via the P variable, Schwartz suggests using a sixth-order equation.[62] As a final higher-order example, Thompson et al.[18] indicate that the method of transfinite interpolation explained earlier is actually the solution to a homogeneous fourth-order pde.

Conformal Mapping Methods

By definition, a conformal transformation is an analytically defined complex mapping whose derivative never vanishes.[63] An important feature of such a mapping is that the local magnification of the elements via the transform is not dependent on direction, meaning that a single parameter will properly scale elements locally between physical and computational space. More important, however, is the fact that relative angles in a plane are preserved after transformation. This means that an orthogonal computational space (as is generally used) will yield an orthogonal physical space as well. Since complex variables are being used, we are restricted to a two-dimensional transformation which Ives[64] represents as follows:

$$\left\{\begin{matrix} \partial x \\ \partial y \end{matrix}\right\} = H \begin{bmatrix} \cos\theta & -\sin\theta \\ \sin\theta & \sin\theta \end{bmatrix} \left\{\begin{matrix} dx \\ dy \end{matrix}\right\} \tag{19}$$

where H and θ are the relative scaling factor and rotation angle, respectively. Since the 2×2 matrix contains the metric terms defining the mapping, it is immediately apparent that

$$x_\xi = y_\eta \tag{20a}$$

$$y_\xi = -x_\eta \tag{20b}$$

These are one form of the famous Cauchy-Riemann equations that govern conformal mappings. Thus, conformal mappings are equivalently described by a set of two pde's. What separates this method from pde methods is that analytical solutions to these equations are attainable for a number of different boundary condition sets. The determination of the analytic mapping from computational to physical space is indeed the entire objective of conformal techniques.

Another important point to mention here is the fact that a large number of terms involved in a generalized coordinate transformation of the fluid mechanics equations vanish if the transformation is conformal—more terms than the simplification afforded by strict grid point orthogonality alone. This results in a computational savings in the flow code if the vanishing terms are not included in the code. In generalized flow codes, the increased efficiency is minimal, although the benefit of grid orthogonality will enhance solution accuracy.

Although conformal mapping methods continue to be used widely, several detrimental factors prohibit their global mainstream usage. First, such methods are not extendable to three dimensions, although three-dimensional surface grids may be generated. For three-dimensional applications, conformal mappings are invariably used in combination with other methods, resulting in grids that are conformal in at most two coordinate directions. Second, for effective application of the methods, it is necessary to possess a considerable understanding of functions of complex variables.[65] Even experts may find extension of existing techniques to differing configurations to be a formidable task.

What follows are several of the more important conformal mapping methods used today in grid generation. As in other sections, the intent here is only to glean the technology, and the reader is encouraged to consult Refs. 18, 64, and 66 for more thorough explanations. Finite-difference applications of conformal mappings, explained in Ref. 18, are not covered herein.

The most basic type of conformal mapping method is one that describes the mapping as a simple function of the complex computational variable. For example, the well-known Joukowski transform maps a circle into an airfoil-like shape:

$$Z = \zeta + 1/\zeta \tag{21}$$

where Z is the complex physical location given by $x + iy$, and ζ is the complex computational location given by $\xi + i\eta$. The value of i is, of course, $\sqrt{(-1)}$. Another example is the equation given as

$$Z = \zeta + e^{\zeta} \tag{22}$$

which is often used for the generation of grids around inlets. These are but a few examples, and an extensive catalog of conformal mappings is given in Ref. 67.

Unfortunately, the applicability of analytic mappings to generally shaped geometries is limited, and in grid generation, we address the inverse problem: What transformation will map the given geometry into a simple shape? Using the inverse Joukowski transform, we could hope that a non-Joukowski airfoil would be mapped into a shape that is nearly circular. When this is the case, the near-circle can then be mapped into a circle following the method of Theodorsen and Garrick.[68] However, for many airfoil shapes, an inverse application of the Joukowski (or more generally the von Kármán-Trefftz) transform will result in a very poor near-circle. In this case, a new transformation must be employed, which is far from a trivial task.

Another problem associated with conformal mapping methods is the multivaluedness of many of the transformations. It is often possible that a number of values of ζ exist for a given physical Z plane, and Ives[64] presents a method to ensure the correct root for several analytic function types. Because of the root selection problem, many have found it easier to apply a complex transformation as a series of simpler conformal mappings. Since the roots of each simpler transform are more likely to be understood, the entire problem proceeds with greater ease. This type of procedure was embraced both by Ives and Liutermoza[69] and by Frith[70] for turbomachinery cascade problems.

A technique that is useful for a wider range of geometries is known as the Schwarz-Christoffel transformation. With this function, canonical shapes such as circular disks and half-planes are mapped into straight-sided polygons. This transform has been used successfully to generate mappings about slender re-entry shapes in Ref. 71. In a significant extension of the Schwarz-Christoffel mapping, Davis[72] later modified the governing function to allow for curved-sided polygons. This extended technique has made it possible to use conformal mapping methods on a variety of complex geometries.[73,74]

As the complexity of the geometry increases, and the push toward three dimensions grows, the need to use auxilliary methods with conformal mappings becomes more apparent. Mixing algebraic mappings with conformal mappings usually presents no problem as long as the user is aware that strict conformality will no longer be present. In fact, mixing of conformal and algebraic methods is standard practice today. For example, Thompson et al.[18] suggest the addition of an algebraic function to map the near-circle domain from an inverse Joukowski transform into a true circle. In another study Ives and Menor[75] used conformal transforms in tandem with algebraic shearing transformations to generate a grid around a three-dimensional engine nacelle. This type of hybrid scheme is commonplace today.

Unstructured Methods

In recent years increasing interest has appeared in applying unstructured grids to problems in computational aerodynamics. This interest has resulted from the development of efficient finite-volume schemes for solving the equations of fluid dynamics. Since these methods can be applied to arbitrarily shaped volume elements, the use of unstructured grids in solving fluid flow problems for irregularly shaped domains is a logical development.

There are a number of advantages in using an unstructured mesh. The biggest advantage is that an unstructured mesh always exists for an arbitrary domain. A nodal point distribution can always be selected to cover the physical domain for any topology. In addition, grid point densities can be increased or decreased in zones where resolution requirements differ. Thus, mesh refinement can be achieved without regard for the spacing or mapping requirements that plague structured grids. This feature allows unstructured grids to be generated in a fraction of the time required by their structured counterparts. Since grid point densities are easily modified on a local basis, adaptive refinement is easily achieved, either by the addition of new grid points or by the redistribution of existing grid points.

The largest disadvantage of unstructured grids is that the nodal point information must include explicit connectivity data. In a structured mesh, the identification of points common to a volume element or cell is automatically carried through the storage array indices; i.e., in the structured mesh, the $(i + 1)$st point lies adjacent to the ith point. In an unstructured mesh, information defining the nodes that are neighbors or nodes common to that designated cell must also be stored. With the storage of explicit connectivity information, neighboring nodes in physical space appear in storage at non-neighboring locations and nonconstant distances. Consequently, implementation of parallelism and vectorization is not straightforward.

Most unstructured grid methods in use today employ one or more of the following three methods: subdivision schemes, Voronoi-based methods, and advancing front techniques. The basics behind each of these methods are explained in the following paragraphs.

Subdivision Methods

Subdivision schemes are based on recursively subdividing the physical domain until a termination criterion is reached. Typical of these schemes are the quadtree and octree methods used by Shephard et al.,[76] among others. In two dimensions, an initial rectangle is drawn around the domain. From here, the rectangle is divided into four cells (hence, quadtree). Elements that are determined to lie completely within or outside of the domain are left alone, but those that pierce a domain boundary are further subdivided into another four cells. This procedure is continued until a minimum cell size is approached, and the result is a tree structure, with cells successively branching into four subcells as necessary. The final step is to create triangular elements from the quadrilateral cells, which is straightforward for interior or exterior cells, but is logically intensive for cells that pierce the problem domain. This type of generation scheme is well-suited to solution adaptivity, since given cells in regions of low solution accuracy may be further divided into any number of branches of subcells. Examples of this type of unstructured grid generation may be found in Refs. 77 and 78.

Voronoi-Based Methods

Methods based on the idea of creating Voronoi cells have probably received more interest than other techniques. When a number of points arranged in two-dimensional space are considered, the Voronoi cell is the total area that is closer to that point than any other point. The shape of the Voronoi cell is a straight-sided closed polygon when the point is interior to the domain and is an open polygon with one or more sides of infinite length when the point is on the domain boundary. The process of determining the Voronoi cells for a given set of points is often referred to as Dirichlet tesselation. From the Voronoi cell determination, the Delauney triangulation of the grid is obtained by joining together (with a straight-line segment) any two points whose Voronoi cells share a common edge. The Delauney triangulation of the grid points is attractive because such triangular grids are nearest to equilateral in an average sense.[79] Consequently, the resulting mesh will have quite regularly shaped triangles with a small probability of high-aspect-ratio cells appearing. The manner in which the triangulation is calculated is very important, since large grids are expensive to triangulate. One efficient technique is known as Bowyer's algorithm,[80] in which grid points are added successively to the existing triangulation, and the triangulation is modified to account for the new point. Baker[81] has employed an octtree-type search algorithm to find all elements disturbed by the introduction of new grid points and has reported a computational speed proportional to $(n) \log(n)$, where n is the number of points in the grid.

Prior to triangulation, one must distribute the intended number of points throughout the domain. Various approaches have been used, including using overlapping structured meshes to patch the entire domain[82] and using a quadtree-like (actually, tritree) branching algorithm.[83] One of the more promising techniques, however, is one proposed in two dimensions by Holmes and Snyder.[84] Here, after boundary points are generated, an initial triangulation proceeds using only the boundary nodes. New grid points are then inserted at the circumcenter of targeted triangles. Cells are targeted in the first sweep if their total area exceeds a given value and in the second sweep if their aspect ratio is too large. In this manner, the entire grid generation proceeds automatically after the boundary points are defined, and the resulting triangulation will satisfy a predetermined level of quality.

Advancing Front Methods

The advancing front methods are based on starting with a given point distribution on the domain boundary and moving away from the initial front or boundary into the domain interior using various criteria to establish the next row of interior point locations. These interior point locations are formed one at a time and are connected to the previous front to establish triangles, tetrahedrons, or other geometric shapes. Introduced by Thomasset,[85] this scheme has an advantage that it will allow for element stretching in selected directions, based on information stored in a very sparse background unstructured mesh. This method has recently been extended and applied by Lohner et al.,[86,87] who have produced extraordinary results for a variety of complex configurations.

Applications

Viewed by many as pure academic exercise in its early stages, the field of CFD was established as a viable analysis tool by the early 1980s. Much of this change in public perception was due to the rapidly advancing numerical algorithms, which were being applied to aircraft component shapes of ever-increasing complexity. One near-term goal in the CFD community at this time was to calculate the flowfield around a complete aircraft configuration, since many believed that this milestone would be indicative of the field's arrival into the realm of real-world applications. By the time the flow solver methods had matured to a level that would permit such a calculation, it became apparent that grid-generation methods had lagged behind in development. Consequently, many of the techniques of grid-generation described earlier were refined, extended to three dimensions, and incorporated into generalized codes so that grids about realistic geometries could be created.

In this section we examine the grid-generation methodology applied today to aerodynamic configurations. The discussion begins with aerodynamic components and continues through entire aircraft configurations. Literature is very sparsely cited in this section, but the references should serve as representative examples of the various geometric shapes encountered.

Aircraft Components

Until the CFD analysis of an entire aircraft engine becomes feasible, computational engine analysis will continue to proceed on a component basis. Physical flow domains for engine components vary widely between the inlets, ducts, and turbomachinery, and so do, of course, the grids generated about them. Structured single-block grid generation around simple three-dimensional ducts is usually straightforward, in that one of the three computational coordinates traverses the grid nearly parallel to the duct centerline. The transverse plane, however, containing two computational coordinates, may be generated in either of two ways. If the duct is approximately square, then the four boundaries of the transverse plane lie along the four sides of the duct (cf. Ref. 88). If the duct is more circular, then a polar coordinate system is used. This latter choice works well for duct flows with centerbodies, so that one of the computational boundaries will wrap around the centerbody, the other around the duct. This approach is taken in Ref. 89, which employs differing algebraic functions in each computational direction to produce the grid. When there is no centerbody, the flow domain will contain a singular line, which must be accounted for properly in the flow code. For ducts that vary between square and round, neither of these topologies will be entirely satisfactory. A square topology will make it difficult to cluster grid points near the duct walls in the round portion, and a polar topology will have similar problems in the square region. Elliptic pde methods will tend to pull points from corners, and algebraic methods will tend to propagate slope discontinuities from the corners well into the grid interior. The user must take into consideration the anticipated flowfield to determine which topology is best for his application.

If the duct opens forward into the inlet of an engine, then the flow domain contains both internal and external flows. Here, the domain is no longer regularly shaped, but is a complex region containing multiple-length scales. The natural grid-generation type for such a flowfield is multiple block (or unstructured or single block with interior boundary conditions), with component blocks used to represent either the internal or external portions of the flowfield. This approach has been taken in Refs. 90–92. When single-block grids are necessary, conformal mapping methods may be used for near-axisymmetric inlets, since they automatically control the grid line skewness. As described earlier and shown in Fig. 4, Ives and

Menor[75] used a polar-type topology on an inlet-centerbody combination, employing conformal mapping in circumferential planes followed by algebraic stretching. Halsey uses three separate mappings to generate a nacelle-pylon-aft fuselage combination grid in Refs. 93 and 94. In this example, conformal mappings are used to transform the complicated region into simpler shapes, and then algebraic methods are used to create the grid conforming to the simplified geometry.

Turbomachinery applications of CFD usually have a cyclic geometric pattern that allows the geometries to be discretized without a sizable effort. The grid for a two-dimensional cascade of airfoils, for example, can be generated by dividing the flowfield into repeating geometric patterns of equal area and shape, by creating a grid in one region, and then by translating the resulting grid to the other regions. The grid in each region may be either an H, C, or O topology, named for the letter it resembles, as illustrated in Fig. 5. Both O- and C-type topologies allow for enhanced resolution about the airfoil surface, particularly near the leading edge. O-type grids for cascade flows are used in conjunction with pde methods in Ref. 38, with algebraic methods in Ref. 95, and with conformal mapping methods in Ref. 64. This indicates that there is really no preferred grid-generation method for turbomachinery applications. The same general considerations exist for three-dimensional turbomachinery applications, although the grid is usually not so readily generated. Hilgenstock et al., employed an elliptic pde solver[96] to enforce orthogonality near surfaces of three-dimensional turbine stator and rotors. Neury[97] instead used transfinite interpolation with orthogonality control to produce similar results on similar geometries. Finally, Ozell and Camarero[88] developed a code specifically for the generation of turbomachinery grid components.

Even though the entire flowfield around an aircraft can be computed today, the aerodynamic analysis of the aircraft on an individual component basis remains an effective design tool. Airfoil design, for example, utilizes CFD methodology extensively. Generation of grids for conventional airfoils is simple, but is considerably tougher for multiple-element airfoils. If the airfoil components lie along a line, then trailing and leading edges of adjacent components may be connected with branch cuts, with a C- or O-type grid wrapped around the connected component-branch cut combination. This topology was adopted by Halsey[98] in a conformal mapping study. If components overlap in the chordwise direction, however, branch cuts are not practical, and either a multiple-block or an unstructured approach must be taken. The multiple-block approach has been applied by a number of authors, including Sorenson[99] and Schuster and Birckelbaw.[100] C-type grids were used around airfoil components in this second study, as illustrated in Fig. 6, but H grids could equivalently have been used. In fact, a number of acceptable blocking schemes will always exist for any geometric shape. Andrews[101] has examined the airfoil blocking problem closely and has written a code for the automatic blocking of general two-dimensional domains. The blocking proceeds from a set of rules developed from user mesh generation experience and flowfield considerations, resulting in a block structure with adjacent but discontinuous interfaces. Andrews found that block structures developed automatically and manually tended to differ, although each was acceptable. This suggests that a conclusive measure of blocking quality has not yet been determined.

Blocking is, of course, no concern on multielement airfoil problems if the grid is unstructured. Representative unstructured examples are given in Refs. 83 and 102. Figure 7 is taken from Ref. 83. Also, the hybrid structured-unstructured method of Ref. 17 is validated on a three-component airfoil.

Grid generation for wings is sometimes no more complicated than that for airfoils, since the additional computational direction (from two to three dimensional) can be made to traverse the wing in the spanwise direction. However, single-block grid generation about wing-body configurations, the next level of geometric complexity, does require a differing strategy. One solution is to treat the fuselage as an extension of the wing, so that the fuselage and the entire wing map into a single wall of the computational block. Alternatively, the fuselage could be mapped into a wall adjacent to the wing wall, as in Ref. 103. This latter approach generally provides more orthogonally shaped grid cells near the wing-fuselage junction. Yet another approach is to map the wing into one wall, the plane of symmetry into the wall on the opposite side of the block,[104] and the fuselage on the face between these two walls. Countless other approaches for generating wing-body grids fill the literature, but rather than discuss them and approaches for geometries of slightly more complexity (e.g., wing-body-tail), we instead consider the most general type of aerodynamic configuration analyzed computationally today—the complete aircraft.

Full Aircraft Configurations

Single-Block Grids

Most of the early structured grid systems for full aircraft configurations utilized single-block topologies mapping the entire flow domain into one computational cube. Some of these techniques were automated, so that the grid could be generated in a batch environment with a minimum number of inputs. Automated generation techniques for particular topologies required a fair amount of ingenuity. For example, Caughey and Jameson[105] used a blend of algebraic and conformal mapping techniques to generate a grid around a wing-fuselage combination. Specifically, a Joukowski transformation was applied to reduce the fuselage to a nearly planar surface, and then a complex parabolic mapping was used to unwrap the wing. Shearing transformations were then used to distribute points outward from the surfaces. Conformal mapping techniques continue to be applied today, although more user control is usually available. Takanashi et al.[106] have extended these ideas in generating the wing-body-tail combination shown in Fig. 8.

Full aircraft grids have also been generated quickly by purely algebraic means. Pagendarm et al.[107] have developed an algebraic algorithm to define grid lines between two surfaces: the wing-fuselage-wake surface and the far-field surface. In this method, grid lines emanating from the two surfaces do so orthogonally. This idea is similar to the "two-boundary technique" proposed by Smith,[108] who has since advocated the use of transfinite interpolation for blending a grid in all three computational directions[20] and has developed a dual-block methodology for wing-body configurations.

Since algebraic methods tend to propagate surface discontinuities far into the field, pde methods have often been applied to full aircraft configurations. Single-block applications include elliptic methods developed by Sorenson[109] and by Eberle and Schwarz.[110] Sorenson's method is the three-dimensional analog of his two-dimensional method, solving Poisson's equation with control functions providing cell height control and orthogonality. Eberle and Schwarz solve the fourth-order biharmonic equation in physical space, resulting in a linear set of pde's, rather than the nonlinear set of equations associated with the Poisson solver of Thompson et al.[35] In their biharmonic approach, one large computational cube is generated, with portions of the cube blocked off that fall inside the body. This approach limits the topology to an H-type around the wing, but otherwise produces acceptable results.

Elliptic methods, generally solved in an iterative manner, require a large computational effort for solution. Hyperbolic and parabolic pde's, however, are marching problems that are computed with a single sweep through the domain. Steger and Rizk[48] have exploited this feature in generating a hyperbolic grid around a Shuttle-like configuration. In their work, grid points exterior to the surface are obtained by marching outward from the body. Two pde's are generated by enforcing orthogonality of grid points in the two directions transverse to the surface normal (marching direction), and closure to the system is obtained by prespecifying a Jacobian distribution in the entire domain. Estimation of the Jacobian field appears to be the most arbitrary aspect of the method, but Steger and Rizk adopt a logical approach that results in an outstanding grid around the Shuttle-like geometry. The ease of computation benefit is somewhat offset by the applicability of the scheme, however, as problems have been reported for body surfaces with discontinuities and for surface grids with excessively irregular cell sizes. In addition, this technique often requires the addition of an "artificial viscosity" term for stability. Finally, since the scheme is hyperbolic, the outer boundary cannot be specified a priori, but instead evolves with the solution.

As mentioned earlier, this last restriction is eliminated when parabolic pde's are used for grid generation. Using an extension of Nakamura's original parabolic scheme,[42] Edwards[43] marches in two computational directions using a modified difference form of the standard elliptic (Laplace) equations. The three equations are in effect parabolized in two directions, using differences that take into account the previous point and the far boundary in each marching direction. Edwards has applied this idea to a simplified F-16 wing-fuselage geometry in Ref. 43. The whole idea of a parabolic scheme is to provide much of the grid quality offerred by elliptic methods, but with significantly reduced computational effort.

Multiple-Block Grids

The examples cited earlier indicate that there are a number of viable means of generating single-block grids about complete albeit simplified aircraft configurations. Despite the success, the single-block idea has likely reached its geometric limit of complexity, since more complex applications depend entirely on the ingenuity of the developer. This fact has opened the door for multiple-block methods, which have surged in popularity in recent years. By reducing a flow domain into smaller subdomains, complex geometries may be modeled more easily and with greater numerical accuracy. Rather than having to decide how to map a domain into a single cube, the user may place artificial boundaries into the domain at convenient locations, using these artifices as boundaries of connecting blocks. Shown in Fig. 9 are the bounding lines of a 27-block structure around a pylon-store configuration, taken from Ref. 8.

Naturally, the added geometric complexity afforded by the multiple-block flowfield concept does not come without a price. When more than one block is used to represent the domain, connectivity between the blocks must be specified. This typically entails the creation of tabular data, which is needed to transfer information across blocks. The connectivity data is accessed throughout the analysis during the generation of the grid and during the flowfield solution. A true multiple-block grid-generation scheme is one that treats the collection of blocks as a single domain, so that complete continuity of grid points (point, slope, etc.) is maintained across abutting faces. This requirement is not always necessary, but does provide the ultimate in generality.

As described earlier and illustrated in Fig. 3, the flow solver determines the type of permissible interblock connections in a grid system, which may be point continuous, discontinuous, or even overlapping. In addition, these connections may either be implemented on the point level or on the full face level. Point level implementation means that a face of a particular block may contain several types of b.c.'s and connections, as long as each point on that face has either a b.c. or an equivalent point on another block. Full-face implementation of course means that only 1 b.c. type or connection is applied on the entire face. Interblock connections are usually determined (at least conceptually) before the grid is generated, meaning that provisional grid values are required for all six boundary faces of each block. Hence, methods that determine the boundaries of a block during execution, such as hyperbolic pde methods, are generally not used in multiple-block schemes. In fact, since the shapes and sizes of component blocks may vary widely in a given block system, only block grid-generation methods that exhibit considerable robustness are acceptable. For the most part, this eliminates conformal mapping techniques, which are not easily applied to all conceivable block shapes. Exceptions to these two rules do exist, however, most notably the multiple-block grid system around an X-24C lifting body developed by Scherr and Shang[111] which uses a hyperbolic grid on the interior of the domain, with an algebraic grid outside.

If hyperbolic pde and conformal mapping methods are removed from the field of candidate multiple-block schemes, we are left only with algebraic techniques and elliptic and parabolic pde techniques. These are the grid-generation methods incorporated into the multiple-block methods described later.

Creation of connectivity tables is usually not a difficult task for configurations represented by a moderate number of blocks, and several researchers have used this kind of grid system for complete aircraft. Sorenson[10] used a multiple-block version of the elliptic code described in Ref. 109 to construct a multiple-block model about an F-16 fuselage-wing-inlet combination. Yu et al.[112] used Thompson's multiple-block elliptic EAGLE code[8] to wrap a three-block grid around a wing-fuselage combination with multiple winglets. Using an algebraic statement of the equidistribution Eqs. (7), Abolhassani and Smith[113] generated an adaptive grid around a two-block fighter configuration. Their formulation managed complete grid line slope continuity and spacing continuity at the block interface.

In another application, Fritz et al.,[32] generated multiple-block grids for fighter and transport aircraft, using both algebraic and elliptic methods to distribute grid points within the blocks. The approach taken here was to use as few blocks as possible during generation, dividing the large blocks as necessary after generation. This ensures slope continuity without having to develop the logic associated with true multiple-block solvers. The authors have since advocated the use of repetitive mapping procedures[114] in lieu of the algebraic and elliptic methods of distribution previously and still popularly used.

As the number of blocks in a grid system increases, the complexity of the interblock connections increases accordingly. This problem has prompted several researchers to develop tools for the decomposition of the domain. For example, Jacobs et al.[115] use interactive graphics codes to draw suitable blocking schemes and then inspect the resulting structure carefully to aid in determining the data input to the connectivity tables. Their full aircraft examples include a 48-block wing-nacelle-propeller combination and a 66-block wing-fuselage combination. In an earlier study, Karman et al.[116] created a 20-block system about an F-16 model, complete with inlet, airbrake, ventral fins, and ECS inlet, but without wing-tip missiles or external stores. Surface grids encompassing seven of the upper surface blocks are shown in Fig. 10. The methodology used to create this grid system has since been enhanced with graphics codes for interactive decomposition of the domain, as described in Ref. 117. In this method, interblock connections and flow boundary conditions set graphically are transferred automatically to the connectivity files.

The methods described in Refs. 8, 109, 112, 116, and 117 all allow the interblock connections to be set on the point, rather than full-face level. This feature is helpful in keeping the number of blocks small for complex configurations. If full-face connections are required, it is usually necessary to significantly increase the number of blocks in the system significantly for complex geometries. For example, Shaw et al.[118] built a 468-block system around a wing-tail-canard-inlet configuration, and Allwright[119] generated a 1697-block system around a fuselage-wing-tail-nacelle business jet configuration. Both of these methods employ elliptic solvers for interior grid point distribution, but what separates these two examples from the rest of the field are the techniques they adopt for domain decomposition. Naturally, the effort needed to design a multiblock system with hundreds of blocks would be excruciating, and it would probably be more prudent to allow for boundary conditions to be applied point by point to eliminate a large number of blocks. In both of these examples, the domain is divided into blocks using semiautomated methods, thus eliminating much of the confusion and the interactive determination of the interblock connections.

Shaw et al.[118] have extended their evolving methodology of grid generation with a code that allows geometries to be represented schematically as planar regions in a Cartesian coordinate system. Once input by the user, the code will divide the Cartesian domain into a number of subdomains, subject to the condition of a single b.c. type per block face. By defining the surface in parametric coordinates, they have found it possible to formulate a set of rules that will allow the Cartesian representation to be divided automatically.

Allwright[119] uses an independent methodology of a similar nature. Here, the geometry is again defined schematically, but as component blocks rather than planes. The grid structure local to each component (wing, tail, etc.) is initially considered to be a "hypercube," with the component placed at the center block and with a separate block abutting each face of the component. From here, some of the abutting blocks may be collapsed to allow for different types of topological structures, such as C-H, O-H, or C-O grids. Regions of the domain not blocked automatically are completed by the user in a code displaying the domain schematically. At this point the user specifies the distribution of grid points along control lines in the domain, and when completed, the multiple-block system is generated automatically with elliptic methods. The idea of spending most of the time generating the grid schematically allows slight geometric (but not topological) changes in a system to be implemented very quickly. For example, a slight change in a wing shape or location would not require a total remake of the block, since the schematic representation would not change.

In both of these cases, the large number of blocks used does not necessitate the need for automated methods; rather, it is the automated methods that create such a complex multiple-block structure. However, the fact that in each method the user does not need a detailed understanding of the block structure allows the overall multiple-block process to proceed at a rapid pace. The reader is encouraged to consult both of these references for a better understanding of automated decomposition techniques. It is likely that automated methods of domain decomposition such as these two will increase in importance in the next few years.

Unstructured Grids

Unstructured grid-generation methods were well-established by the mid-1980s, but their representation in the applied aerodynamic literature was at best spotty. Much of the attention focused on grid generation at this time was on the developing multiple-block structured methods, which were finally being applied to full aircraft configurations. This changed abruptly in early 1986 when unstructured methods were used to generate a grid about a complete 747 aircraft.[82] In this work, overlapping structured blocks were generated about individual components of the aircraft, and then the resulting collection of points was meshed using the Delauney triangulation scheme. The significant feature of this work by Jameson et al.[82] was that it raised the level of public perception of unstructured methods, in that grid systems could finally be generated for complex geometries using either of two equally convincing philosophies.

Development of unstructured grid-generation methods since that time have secured this discipline's importance in CFD. Today unstructured grids are generated for complete aircraft with relative regularity. The main reason for this is because methods exist to generate interior tetrahedrons in three dimensions virtually automatically. For example, Baker developed a generalized Delauney triangulation algorithm that he applied to three separate complex aircraft: the Boeing 747, the Lockheed S3A, and the McDonnell Douglas F-15.[120] Lohner et al., on the other hand, used the advancing front technique to generate grids around the 747 geometry[86] and the F-18 geometry[87] shown in Fig. 11. In both of these cases, surface grids were generated first, and the interior tetrahedrons were then generated automatically.

The generation of surface grids poses the single biggest difficulty in unstructured grid applications. Like structured grids, unstructured surface grids are best generated in a graphic, interactive environment. Lohner et al.[86] have developed such a code, which uses numerical geometry description files, generally obtained from CAD/CAM systems, as input. The user interactively divides the given geometric data into a number of patches and then proceeds to triangulate each patch. The distribution of points on each patch is controlled by a sparse background mesh defined interactively by the user.

Some of the ideas in Lohner's code resemble the methodology incorporated into today's interactive structured grid-generation codes (cf. Refs. 121–124). When codes of similar sophistication are compared, the unstructured surface grid code will always be faster than the structured surface grid code. This is a logical statement if one considers that surface mesh points may be placed in any geometrically convenient arrangement on an unstructured mesh, but must be arranged according to blocking requirements in structured grids. However, the difference in surface generation times for each method is minor compared with the disparity between overall grid-generation times (volume and surface points). It is believed that this gap could reduce if the automated domain decomposition methods of structured methods (see earlier discussion) continue to show promise.

Summary

This chapter has attempted to review many of the prevalent grid-generation methods in use today and to survey how these methods are applied to computational aerodynamics. It has been seen that grid-generation methods developed for specific geometry types are falling out of favor now that very general multiple-block and unstructured methods are being developed. A chronic problem with the overall grid-generation process is the time needed for construction of complex meshes, particularly with structured multiple-block systems. Fortunately, a number of efforts are underway to reduce generation time, and grids are already generated with twice the speed available a few years back. Most of these efforts incorporate an interactive graphic framework into the codes. Some of the more general codes allow virtually any aerodynamic configuration to be treated, so that, from an absolute sense, it is no longer the grid-generation aspect that is limiting the state of the art of computational aerodynamics. Until faster generation methods are developed, however, it will continue to be the grid-generation aspect that limits routine aerodynamic analysis of complex configurations.

References

[1]Ghia, K. N., and Ghia, U. (eds.), *Advances in Grid Generation*, Vol. 5, American Society of Mechanical Engineers FED, New York, 1983.

[2]Hauser, J., and Taylor, C. (eds.), *Numerical Grid Generation in Computational Fluid Dynamics*, Pineridge, Swansea, UK, 1986.

[3]Sengupta, S., Häuser, J., Eiseman, P. R., and Thompson, J. F. (eds.), *Numerical Grid Generation in Computational Fluid Mechanics '88*, Pineridge, Swansea, UK, 1988.

[4]Thompson, J. F., and Steger, J. L., "Three Dimensional Grid Generation for Complex Configurations—Recent Progress," AGARD-AG-309, 1988.

[5]Thompson, J. F. (ed.), *Numerical Grid Generation*, North-Holland, Amsterdam, 1982.

[6]Cooper, G. K., "The PARC Code: Theory and Usage," Arnold Engineering Development Center, Tullahoma, TN, TR-87-24, Oct. 1987.

[7]Lee, K. D., and Rubbert, P. E., "Transonic Flow Computations Using Grid Systems with Block Structure," *7th International Conference on Numerical Methods in Fluid Dynamics*, Springer-Verlag, New York, 1980, pp. 266–271.

[8]Thompson, J. F., "A Composite Grid Generation Code for General 3-D Regions," AIAA Paper 87-0275, Jan. 1987.

[9]Weatherill, N. P., and Forsey, C. R., "Grid Generation and Flow Calculations for Complex Aircraft Geometries Using a Multi-Block Scheme," AIAA Paper 84-1665, June 1984.

[10]Sorenson, R. L., "Three-Dimensional Zonal Grids About Arbitrary Shapes by Poisson's Equation," *Numerical Grid Generation in Computational Fluid Mechanics '88*, edited by S. Sengupta, J. Häuser, P. R. Eiseman, and J. F. Thompson, Pineridge, Swansea, UK, 1988.

[11]Raj, R., Brennan, J. E., Keen, J. M., Mani, K. K., Olling, C. R. Sikora, J. S., and Singer, S. W., "Three-Dimensional Euler Aerodynamic Method (TEAM) Volume 1: Computational Method," Air Force Wright Aeronautical Lab., Wright-Patterson AFB, OH, TR-87-3074, Dec. 1987.

[12]Rai, M. M., "A Conservative Treatment of Zonal Boundaries for Euler Equation Calculations," AIAA Paper 84-0164, Jan. 1984.

[13]Steger, J. L., Dougherty, F. C., and Benek, J. A., "A Chimera Grid Scheme," *Advances in Grid Generation*, Vol. 5, edited by K. N. Ghia and U. Ghia, American Society of Mechanical Engineers FED, June 1985.

[14]Benek, J. A., Buning, P. G., and Steger, J. L., "A 3-D Chimera Grid Embedding Technique,' AIAA Paper 85-1523, July 1985.

[15]Berger, M. J., "On Conservation at Grid Interfaces," Institute for Computer Applications in Science & Engineering Rept. 84-43, 1984.

[16]Dougherty, F. C., and Kuan, J. -H., "Transonic Store Separation Using a Three-Dimensional Chimera Grid Scheme," AIAA Paper 89-0637, Jan. 1989.

[17]Weatherill, N. P., "On The Combination of Structured-Unstructured Meshes," *Numerical Grid Generation in Computational Fluid Mechanics '88*, Pineridge, Swansea, UK, 1988, pp. 729–739.

[18]Thompson, J. F., Warsi, Z. U. A., and Mastin, C. W., *Numerical Grid Generation Foundation and Applications*, North-Holland, Amsterdam, 1985.

[19]Eiseman, P. R., and Smith, R. E., "Applications of Algebraic Grid Generation," AGARD Fluid Dynamics Panel Specialists' Meeting Applications of Mesh Generation to Complex 3-D Configurations, Loen, Norway, May 1989.

[20]Smith, R. E., and Everton, E. L., "Interactive Grid Generation for Fighter Aircraft Geometries," *Numerical Grid Generation in Computational Fluid Mechanics '88*, edited by S. Sengupta, J. Häuser, P. R. Eiseman, and J. F. Thompson, Pineridge, Swansea, UK, 1988.

[21]Smith, R. E., and Weigel, B. L., "Analytic and Approximate Boundary-Fitted Coordinate Systems for Fluid Flow Simulation," AIAA Paper 80-0192, Jan. 1980.

[22]Eiseman, R. R., "Orthogonal Grid Generation," *Numerical Grid Generation*, edited by J. F. Thompson, North-Holland, Amsterdam, 1982.

[23]Roberts, A., "Automatic Topology Generation and Generalized BB Spline Mapping," *Numerical Grid Generation*, edited by J. F. Thompson, North-Holland, Amsterdam, 1982.

[24]Bezier, P., *Numerical Control: Mathematics and Applications*, Wiley, New York, 1970.

[25]Gordon, W. J., "Blending Function Methods of Bivariate and Multvariate Interpolation," *SIAM Journal of Numerical Analysis*, Vol. 8, 1971, pp. 158–177.

[26]Eriksson, L. E., "Generation of Boundary-Conforming Grids Around Wing-Body Configurations Using Transfinite Interpolation," *AIAA Journal*, Vol. 20, Oct. 1982, pp. 1313–1320.

[27]Soni, B. K., "Two- and Three-Dimensional Grid Generation for Internal Flow Applications of Computational Fluid Dynamics," AIAA Paper 85-1526, July 1985.

[28]Dwyer, H. A., "Grid Adaption for Problems with Separation, Cell Reynolds Number, Shock-Boundary Layer Interaction, and Accuracy," AIAA Paper 83-0449, Jan. 1983.

[29]Anderson, D., "Adaptive Grid Methods for Partial Differential Equations," ASME Applied Mechanics, Bioengineering, and Fluids Engineering Conference, Houston, TX, 1983.

[30]Eiseman, P. R., "Alternating Direction Adaptive Grid Generation," AIAA Paper 83-1937, July 1983.

[31]Nakahashi, K., and Deiwert, G. S., "A Three-Dimensional Adaptive Grid Method," AIAA Paper 85-0486, Jan. 1985.

[32]Fritz, W., Haase, W., and Seibert, W., "Mesh Generation for Industrial Application of Euler and Navier Stokes," AGARD-AG-309, 1988.

[33]Gnoffo, P. A., "A Vectorized, Finite-Volume, Adaptive Grid Algorithm Applied to Planetary Entry Problems," AIAA Paper 82-1018, June 1982.

[34]Winslow, A., "Equipotential Zoning of Two-Dimensional Meshes," *Journal of Computational Physics*, Vol. 149, pp. 153–172, 1966.

[35]Thompson, J. F., Thames, F. C., and Mastin, C. W., "Boundary-Fitted Curvilinear Coordinate Systems for Solution of Partial Differential Equations on Fields Containing Any Number of Arbitrary Two-Dimensional Bodies," NASA CR-2729, July 1977.

[36]Thomas, P. D., and Middlecoff, J. F., "Direct Control of the Grid Point Distribution in Meshes Generated by Elliptic Equations," *AIAA Journal*, Vol. 18, June 1980, pp. 652–656.

[37]Thomas, P. D., "Composite Three-Dimensional Grids Generated by Elliptic Systems," *AIAA Journal*, Vol. 20, Sept. 1982, pp. 1195–1202.

[38]Sorenson, R. L., "A Computer Program to Generate Two-Dimensional Grids About Airfoils and Other Shapes by the Use of Poisson's Equation," NASA TM-81198, May 1980.

[39]Hilgenstock, A., "A Fast Method for the Elliptic Generation of Three-Dimensional Grids with Full Boundary Control," *Numerical Grid Generation in Computational Fluid Mechanics '88*, edited by S. Sengupta, J. Häuser, P. R. Eiseman, and J. F. Thompson, Pineridge, Swansea, UK, 1988.

[40]Anderson, D. A., and Steinbrenner, J., "Generating Adaptive Grids with a Conventional Grid Scheme,' AIAA Paper 86-0427, Jan. 1986.

[41]Kim, H. J., and Thompson, J. F., "Three-Dimensional Adaptive Grid Generation on a Composite Block Grid," AIAA Paper 88-0311, Jan. 1988.

[42]Nakamura, S., "Marching Grid Generation Using Parabolic Partial Differential Equations," *Numerical Grid Generation*, edited by J. F. Thompson, Elsevier, Amsterdam, 1982.

[43]Edwards, T. A., "Noniterative Three-Dimensional Grid Generation Using Parabolic Partial Differential Equations," AIAA Paper 85-0485, Jan. 1985.

[44]Nakamura, S., "Noniterative Grid Generation Using Parabolic Partial Differential Equations for Fuselage-Wing Flow Calculations," *Lecture Notes in Physics*, Vol. 170, Springer-Verlag, New York, 1982.

[45]Hodge, J. K., Leone, S. A., and McCarty, R. L., "Noniterative Parabolic Grid Generation for Parabolized Equations," *AIAA Journal*, Vol. 25, April 1987, pp. 542–549.

[46]Noack, R. W., and Anderson, D. A., "Solution Adaptive Grid Generation Using Parabolic Partial Differential Equations," AIAA Paper 88-0315, Jan. 1988.

[47]Steger, J. L., and Chaussee, D. S., "Generation of Body-Fitted Coordinates Using Hyperbolic Partial Differential Equations," *SIAM Journal of Scientific Statistical Computing*, Vol. 1, March, 1980, pp. 431–437.

[48]Steger, J. L., and Rizk, Y. M., "Generation of Three-Dimensional Body-Fitted Coordinates Using Hyperbolic Partial Differential Equations," NASA TM-86753, June 1985.

[49]Klopfer, G. H., "Solution Adaptive Meshes with a Hyperbolic Grid Generator," *Numerical Grid Generation in Computational Fluid Mechanics '88*, edited by S. Sengupta, J. Häuser, P. R. Eiseman, and J. F. Thompson, Pineridge, Swansea, UK, 1988.

[50]Brackbill, J. U., "Coordinate System Control: Adaptive Meshes," *Numerical Grid Generation*, edited by J. F. Thompson, North-Holland, Amsterdam, 1982.

[51]Saltzman, J., "A Variational Method for Generating Multidimensional Adaptive Grids," Courant Mathematics and Computing Lab., New York Univ., New York, DOE/ER/03077-174, Feb. 1982.

[52]Steinberg, S., and Roache, P. J., "Variational Grid Generation," *Numerical Methods for Partial Differential Equations*, Vol. 2, 1986, pp. 71–96.

[53]Fletcher, R., and Reeves, C. M., "Function Minimization by Conjugate Gradients," *The Computer Journal*, Vol. 7, No. 2, July 1964, pp. 148–154.

[54]Castillo, J. E., "A Direct Variational Grid Generation Method: Orthogonality Control," *Numerical Grid Generation in Computational Fluid Mechanics '88*, edited by S. Sengupta, J. Häuser, P. R. Eiseman, and J. F. Thompson, Pineridge, Swansea, UK, 1988.

[55]Jacquotte, O. P., and Cabello, J., "A Variational Methods for the Optimization and Adaption of Grids in Computational Fluid Dynamics," *Numerical Grid Generation in Computational Fluid Mechanics '88*, edited by S. Sengupta, J. Häuser, P. R. Eiseman, and J. F. Thompson, Pineridge, Swansea, UK, 1988.

[56]Kennon, S. R., and Dulikravich, G. S., "A Posteriori Optimization of Computational Grids," AIAA Paper 85-0483, Jan. 1985.

[57]Warsi, Z. U. A., "A Note on the Mathematical Formulation of the Problem of Numerical Coordinate Generation," *Quarterly of Applied Mathematics*, Vol. 41, No. 221, July, 1983, pp. 221–236.

[58]Lee, K. D., and Loellbach, J. M., "Geometry-Adaptive Surface Grid Generation Using a Parametric Projection," AIAA Paper 88-0522, Jan. 1988.

[59]Woan, C. J., "An Integrated Surface Patch/Surface Elliptic Grid Generator," AIAA Paper 88-0521, Jan. 1988.

[60]Shubin, G. R., Stephens, A. B., and Bell, J. B., "Three-Dimensional Grid Generation Using Biharmonics," *Numerical Grid Generation*, edited by J. F. Thompson, North-Holland, Amsterdam, 1982.

[61]DeHeer, D. C., Sottos, N. R., and Guceri, S. I., "Application of Biharmonic Grid Generation to Thermal Stress Analysis," *Numerical Grid Generation in Computational Fluid Mechanics '88*, edited by S. Sengupta, J. Häuser, P. R. Eiseman, and J. F. Thompson, Pineridge, Swansea, UK, 1988.

[62]Schwartz, A. L., and Connett, W. C., "Evaluating Algebraic Adaptive Grid Strategies," *Numerical Grid Generation in Computational Fluid Dynamics*, edited by J. Häuser and C. Taylor, Pineridge, Swansea, UK, 1986.

[63]Churchill, R. V., Brown, J. W., and Verhey, R. F., *Complex Variables and Applications*, McGraw-Hill, New York, 1948.

[64]Ives, D. C., "Conformal Grid Generation," *Numerical Grid Generation*, edited by J. F. Thompson, North-Holland, Amsterdam, 1982.

[65]Thompson J. F., "Grid Generation Techniques in Computational Fluid Dynamics," *AIAA Journal*, Vol. 22, Nov. 1984, pp. 1505–1523.

[66]Moretti, G., "Conformal Mappings for Computations of Steady, Three-Dimensional, Supersonic Flows," *Numerical/Laboratory Computer Methods in Fluid Mechanics*, American Society of Mechanical Engineers, New York, 1976.

[67]Kober, H., *Dictionary of Conformal Representations*, Dover, New York, 1957.

[68]Theodorsen, T., and Garrick, I. E., "General Potential Theory of Arbitrary Wing Sections," NACA TR-452, 1933.

[69]Ives, D. C., and Liutermoza, J. F., "Analysis of Transonic Cascade Flow Using Conformal Mapping and Relaxation Techniques," *AIAA Journal*, Vol. 15, May 1977, pp. 647–652.

[70]Frith, D. A., "Inviscid Flow Through a Cascade of Thick, Cambered Airfoils, Part 1; Incompressible Flow," American Society of Mechanical Engineers, New York, ASME Paper 73-GT-84, April 1973.

[71]Skulsky, R. S., "A Conformal Mapping Method to Predict Low-Speed Aerodynamic Characteristics of Arbitrary Slender Re-Entry Shapes," *Journal of Spacecraft and Rockets*, Vol. 3, Dec. 1967, pp. 247–253.

[72]Davis, R. T., "Numerical Methods for Coordinate Generation Based on Schwarz-Christoffel Transformation," AIAA Paper 79-1463, July 1979.

[73]Anderson, O. L., Davis, R. T., Hankins, G. B., and Edwards, D. E., "Solution of Viscous Internal Flows on Curvilinear Grids Generated by the Schwarz-Christoffel Transformation," *Numerical Grid Generation*, edited by J. F. Thompson, North-Holland, Amsterdam, 1982.

[74]Grossman, B., "Numerical Procedure for the Computation of Irrotational Conical Flows," *AIAA Journal*, Vol. 17, Aug. 1979, pp. 828–837.

[75]Ives, D. C., and Menor, W. A., "Grid Generation for Inlet and Inlet-Centerbody Configurations Using Conformal Mapping and Stretching," AIAA Paper 81-0997, June 1981.

[76]Shephard, M. S., "Finite Element Modeling Within an Integrated Geometric Modeling Environment: Part 1—Mesh Generation," *Engineering with Computers*, Vol. 1, Jan. 1985, pp. 61–71.

[77]Baehmann, P. L., Shephard, M. S., and Flaherty, J. E., *Adaptive Analysis for Automated Finite Element Modeling*, Vol. 6, edited by J. Whiteman, Academic, New York, 1987.

[78]Cheng, J. H., Finnigan, P. M., Hathaway, A. F., Kela, A., and Schroeder, W. J., "Quadtree/Octree Meshing with Adaptive Analysis," *Numerical Grid Generation in Computational Fluid Mechanics '88*, edited by S. Sengupta, J. Häuser, P. R. Eiseman, and J. F. Thompson, Pineridge, Swansea, UK, 1988.

[79]Dukowicz, J., "Langrangian Fluid Dynamics Using the Voronoi-Delaunay Mesh," *Numerical Methods for Coupled Problems*, Pineridge, Swansea, UK, 1981.

[80]Bowyer, A., "Computing Dirichlet Tessellations," *The Computer Journal*, Vol. 24, No. 2, Feb. 1981, pp. 162–166.

[81]Baker, T. J., "Generation of Tetrahedral Meshes Around Complete Aircraft," *Numerical Grid Generation in Computational Fluid Mechanics '88*, edited by S. Sengupta, J. Häuser, P. R. Eiseman, and J. F. Thompson, Pineridge, Swansea, UK, 1988.

[82]Jameson, A., Baker, T. J., and Weatherill, N. P., "Calculation of Inviscid Transonic Flow over a Complete Aircraft," AIAA Paper 86-0103, Jan. 1986.

[83]Fang, J., and Kennon, S. R., "Unstructured Grid Generation for Non-Convex Domains," AIAA Paper 89-1983-CP, June 1989.

[84]Holmes, D. G., and Snyder, D. D., "The Generation of Unstructured Triangular Meshes Using Delaunay Triangulation," *Numerical Grid Generation in Computational Fluid Mechanics '88*, edited by S. Sengupta, J. Häuser, P. R. Eiseman, and J. F. Thompson, Pineridge, Swansea, UK, 1988.

[85]Thomasset, F., "Appendix to Navier-Stokes Problems," *Navier-Stokes Equations: Theory and Numerical Analysis*, edited by R. Teman, North-Holland, Amsterdam, 1977.

[86]Lohner, R., Parikh, P., and Gumbert, C., "Interactive Generation of Unstructured Grids for Three-Dimensional Problems," *Numerical Grid Generation in Computational Fluid Mechanics '88*, edited by S. Sengupta, J. Häuser, P. R. Eiseman, and J. F. Thompson, Pineridge, Swansea, UK, 1988.

[87]Lohner, R., and Parikh, P., "Generation of Three-Dimensional Unstructured Grids by the Advancing-Front Method," AIAA Paper 88-0515, Jan. 1988.

[88]Ozell, B., and Camarero, B., "CAGD in Turbomachinery," *Numerical Grid Generation in Computational Fluid Mechanics '88*, Pineridge, Swansea, UK, 1988, pp. 865–874.

[89]Marcum, D. L., and Hoffman, J. D., "Calculation of Three-Dimensional Inviscid Flowfields in Propulsive Nozzles with Centerbodies," AIAA Paper 86-0449, Jan. 1986.

[90]Buers, H., and Leicher, S., "Numerical and Experimental Investigations of Engine Inlet Flow with the Dornier EM2 Supersonic Inlet Model," AGARD-CP-437, 1988.

[91]Cosner, R., "Propulsion System Flowfield Analysis," Test and Design Applications of CFD—A Short Course, Huntsville, AL, May 1988.

[92]Woan, C. J., "Three-Dimensional Elliptic Grid Generations Using a Multi-Block Method," AIAA Paper 87-0278, Jan. 1987.

[93]Halsey, N. D., "Grid Generation for an Aft-Fuselage-Mounted Nacelle/Pylon Configuration," *Numerical Grid Generation in Computational Fluid Mechanics '88*, Pineridge, Swansea, UK, 1988, pp. 775–784.

[94]Halsey, N. D., "Conformal Mapping as an Aid in Grid Generation for Complex Three-Dimensional Configurations," AIAA Paper 86-0497, Jan. 1986.

[95]Choo, Y. K., Eiseman, P. R., and Reno, C., "Interactive Grid Generation for Turbomachinery Flow Field Simulations," *Numerical Grid Generation in Computational Fluid Mechanics '88*, Pineridge, Swansea, UK, 1988, pp. 895–904.

[96]Hilgenstock, A., Kursawe, M., Pfost, H., and von Lavante, E., "Elliptic Generation of Three-Dimensional Grids for Internal Flow Calculations," *Numerical Grid Generation in Computational Fluid Dynamics*, edited by J. Häuser and C. Taylor, Pineridge, Swansea, UK, 1986.

[97]Neury, C., "3-D Mesh Generation for Calculating Flow Through Radial-Axial Turbines," *Numerical Grid Generation in Computational Fluid Dynamics '88*, edited by J. Häuser and C. Taylor, Pineridge, Swansea, UK, 1986.

[98]Halsey, N. D., "Conformal Grid Generation for Multielement Airfoils," *Numerical Grid Generation*, edited by J. F. Thompson, North-Holland, Amsterdam, 1982.

[99]Sorenson, R. L., "Grid Generation by Elliptic Partial Differential Equations for a Tri-Element Augmentor-Wing Airfoil," *Numerical Grid Generation*, edited by J. F. Thompson, North-Holland, Amsterdam, 1982.

[100]Schuster, D. M., and Birckelbaw, L. D., "Numerical Computation of Viscous Flowfields about Multiple Component Airfoils," AIAA Paper 85-0167, Jan. 1985.

[101]Andrews, A. E., "Knowledge-Based Flow Field Zoning," *Numerical Grid Generation in Computational Fluid Mechanics '88*, Pineridge, Swansea, UK, 1988. pp. 13–22.

[102]Mavriplis, D. J., "Adaptive Mesh Generation for Viscous Flows Using Delauney Triangulation," *Numerical Grid Generation in Computational Fluid Mechanics '88*, edited by S. Sengupta, J. Häuser, P. R. Eiseman, and J. F. Thompson, Pineridge, Swansea, UK, 1988, pp. 611–620.

[103]Yu, N. J., "Grid Generation and Transonic Flow Calculations for Three-Dimensional Configurations," AIAA Paper 80-1391, July 1980.

[104]Smith, R. E., Jr., and Kudlinski, R. A., "Algebraic Grid Generation for Wing-Fuselage Bodies," AIAA Paper 84-0002, Jan. 1984.

[105]Caughey, D. A., and Jameson, A., "Progress in Finite-Volume Calculations for Wing-Fuselage Combinations," *AIAA Journal*, Vol. 18, Nov. 1980, pp. 1281–1288.

[106]Takanashi, S., Obayashi, S., Matsushima, K., and Fujii, K., "Numerical Simulation of Compressible Viscous Flows Around Practical Aircraft Configurations," AIAA Paper 87-2410, Aug. 1987.

[107]Pagendarm, H. G., Laurien, E., and Sobieczky, H., "Interactive Geometry Definition and Grid Generation for Applied Aerodynamics," AIAA Paper 88-2515, June 1988.

[108]Smith, R. E., "Two-Boundary Grid Generation for the Solution of the Three-Dimensional Navier-Stokes Equations," NASA TM-83123, May 1981.

[109]Sorenson, R. L., "Three-Dimensional Elliptic Grid Generation for an F-16," AGARD-AG-309, 1988.

[110]Eberle, A., and Schwarz, W., "Grid Generation for an Advanced Fighter Aircraft," AGARD-AG-309, 1988.

[111]Scherr, S. J., and Shang, J. S., "Three-Dimensional Body-Fitted Grid System for a Complete Aircraft," AIAA Paper 86-0428, Jan. 1986.

[112]Yu, N. J., Chen, H. C., Chen, A. W., and Wittenberg, K. R., "Grid Generation and Flow Analyses for Wing/Body/Winglet Configurations," AIAA Paper 88-2548-CP, June 1988.

[113]Abolhassani, J. S., and Smith, R. E., "Multiple-Block Grid Adaption for an Airplane Geometry," *Numerical Grid Generation in Computational Fluid Mechanics '88*, edited by S. Sengupta et al., Pineridge, Swansea, UK, 1988.

[114]Seibert, W., "A Graphic-Iterative Program-System to Generate Composite Grids for General Configurations," *Numerical Grid Generation in Computational Fluid Mechanics '88*, edited by S. Sengupta et al., Pineridge, Swansea, UK, 1988.

[115]Jacobs, J. M. J. W., Kassies, A., Boerstoel, and Buijsen, F., "Numerical Interactive Grid Generation for 3-D-Flow Calculations," *Numerical Grid Generation in Computational Fluid Mechanics '88*, edited by S. Sengupta et al., Pineridge, Swansea, UK, 1988.

[116]Karman, S. L., Jr., Steinbrenner, J. P., and Kisielewski, K. M., "Analysis of the F-16 Flow Field by a Block Grid Euler Approach," AGARD-CP-412, France, 1986.

[117]Steinbrenner, J. P., Chawner, J. R., and Fouts, C. L., "A Structured Approach to Interactive Multiple Block Grid Generation," AGARD, Fluid Dynamics Panel Specialists' Meeting on Mesh Generation for Complex Three-Dimensional Configurations, Loen, Norway, 1989.

[118]Shaw, J. A., Georgala, J. M., and Weatherill, N. P., "The Construction of Component-Adaptive Grids for Aerodynamic Geometries," *Numerical Grid Generation in Computational Fluid Mechanics '88*, edited by S. Sengupta, J. Häuser, P. R. Eiseman, and J. F. Thompson, Pineridge, Swansea, UK, 1988.

[119]Allwright, S. E., "Techniques in Multiblock Domain Decomposition and Surface Grid Generation," *Numerical Grid Generation in Computational Fluid Mechanics '88*, edited by S. Sengupta, J. Häuser, P. R. Eiseman, and J. F. Thompson, Pineridge, Swansea, UK, 1988.

[120]Baker, T. J., "Unstructured Mesh Generation by a Generalized Delauney Algorithm," AGARD Fluid Dynamics Panel Specialists' Meeting Applications of Mesh Generation to Complex 3-D Configurations, Loen, Norway, May 1989.

[121]Amdahl, D. J., "Interactive Multi-Block Grid Generation," *Numerical Grid Generation in Computational Fluid Mechanics '88*, Pineridge, Swansea, UK, 1988, pp. 579–588.

[122]Coleman, R. M., "INMESH: An Interactive Program for Numerical Grid Generation," David W. Taylor Naval Ship Research and Development Center, Bethesda, MD, DTN-SRDC-85-054, 1985.

[123]Sonar, T., and Radespiel, R., "Geometric Modelling of Complex Aerodynamic Surfaces and Three-Dimensional Grid Generation," *Numerical Grid Generation in Computational Fluid Mechanics '88*, Pineridge, Swansea, UK, 1988, pp. 795–804.

[124]Soni, B. K., "GENIE: GENeration of Computational Geometry-Grids for Internal-External Flow Configurations," *Numerical Grid Generation in Computational Fluid Mechanics '88*, Pineridge, Swansea, UK, 1988, pp. 915–924.

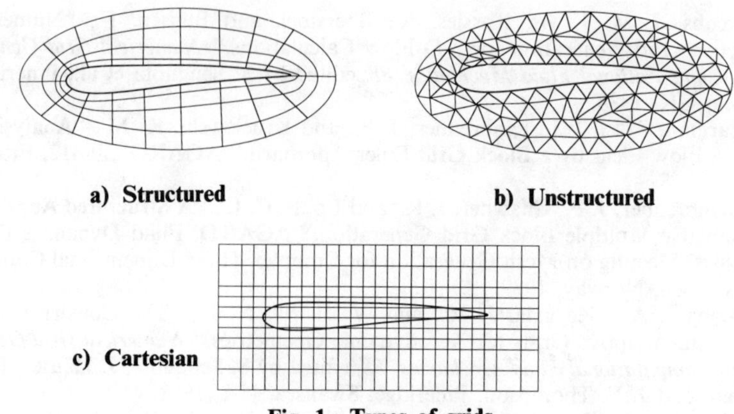

Fig. 1   Types of grids.

Fig. 2   Transformation between coordinates.

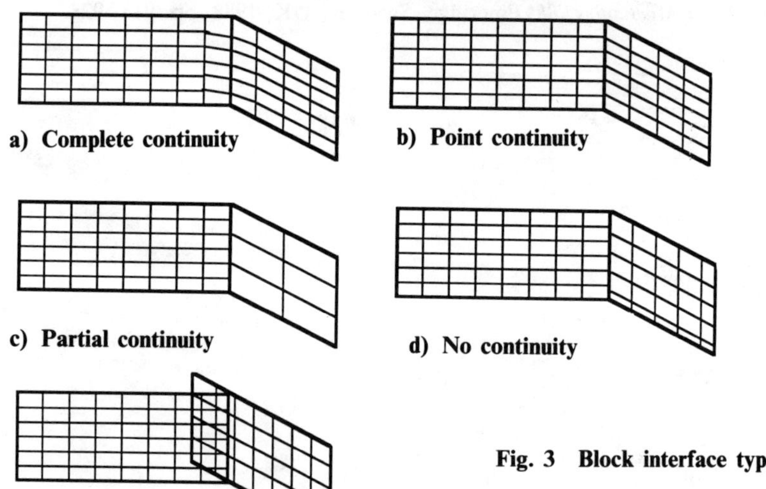

Fig. 3   Block interface types.

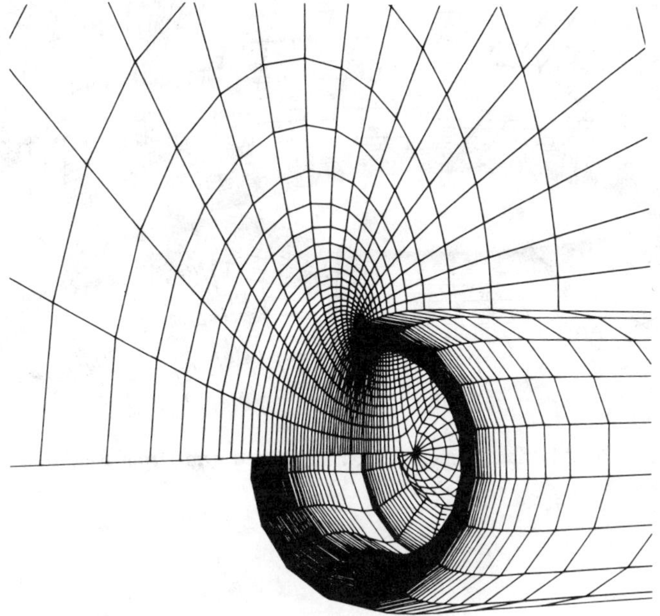

Fig. 4   Conformally mapped inlet grid (from Ref. 75).

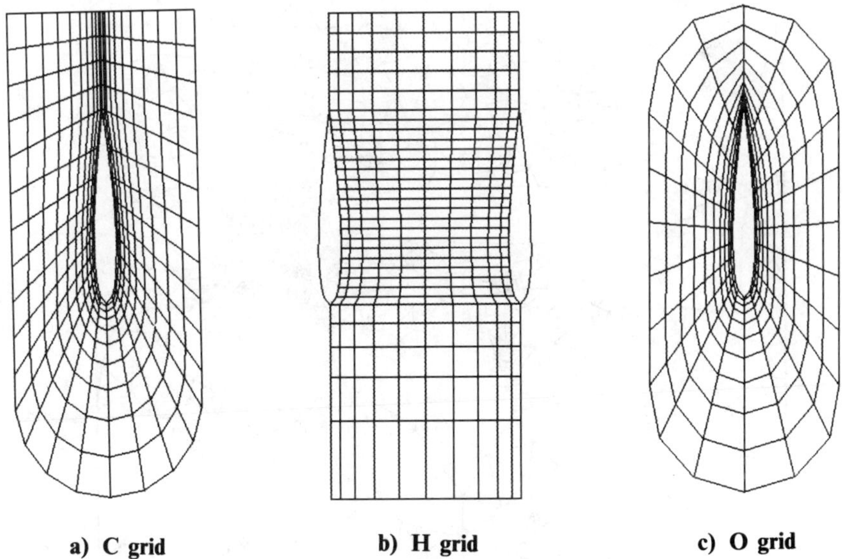

a) C grid   b) H grid   c) O grid

Fig. 5   C-, H-, and O-type topologies.

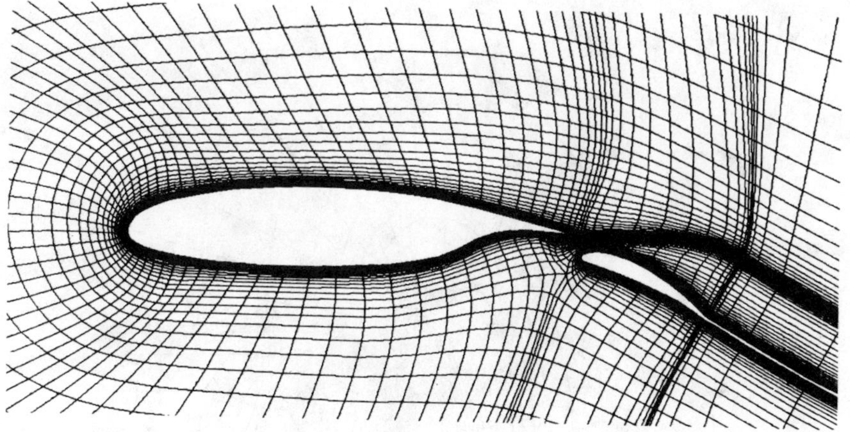

Fig. 6   Multiple-element airfoil structured grid (from Ref. 100).

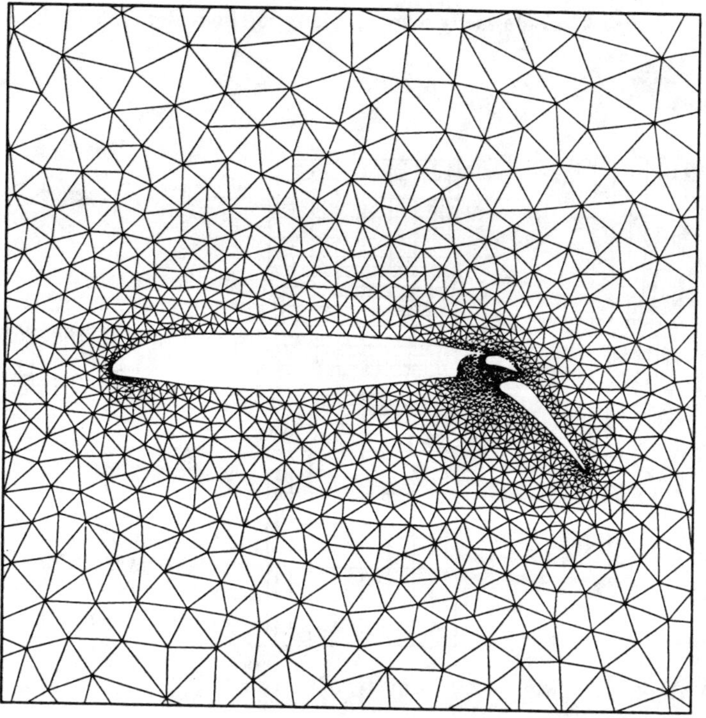

Fig. 7   Multiple-element airfoil unstructured grid (from Ref. 83).

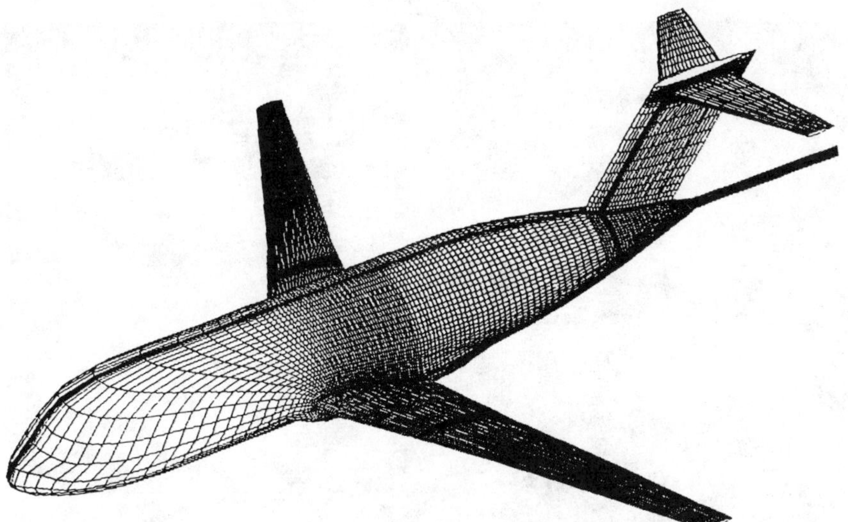

Fig. 8 Single-block grid for wing-body-tail configuration (from Ref. 106).

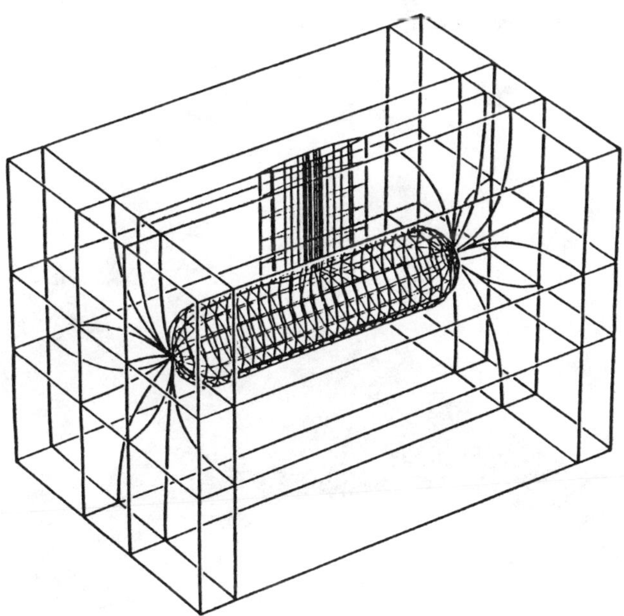

Fig. 9 Edges of 27-block grid for pylon-store configuration (from Ref. 8).

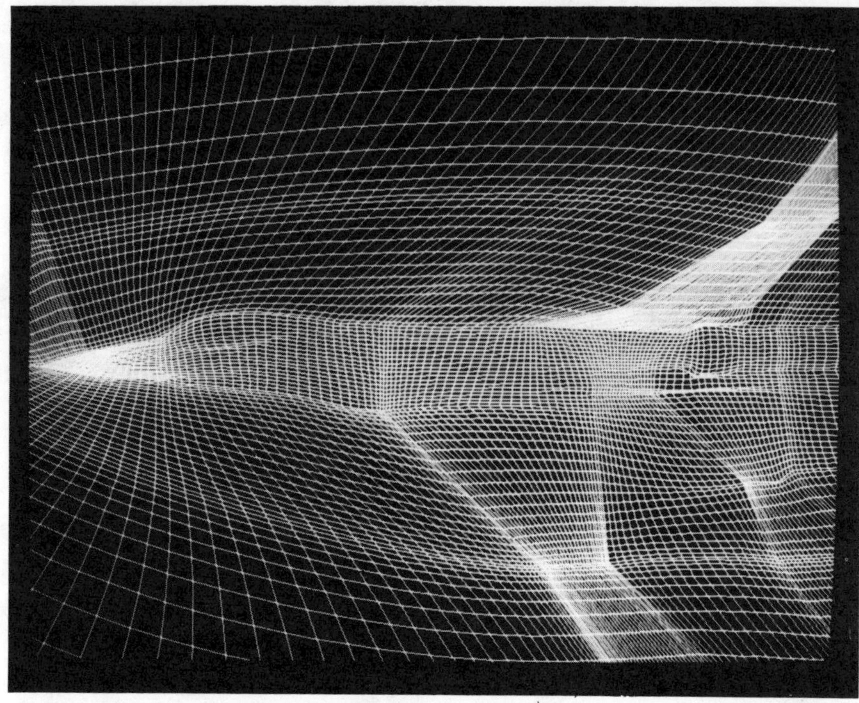

Fig. 10   Surface grids around 20-block F-16 configuration (from Ref. 116).

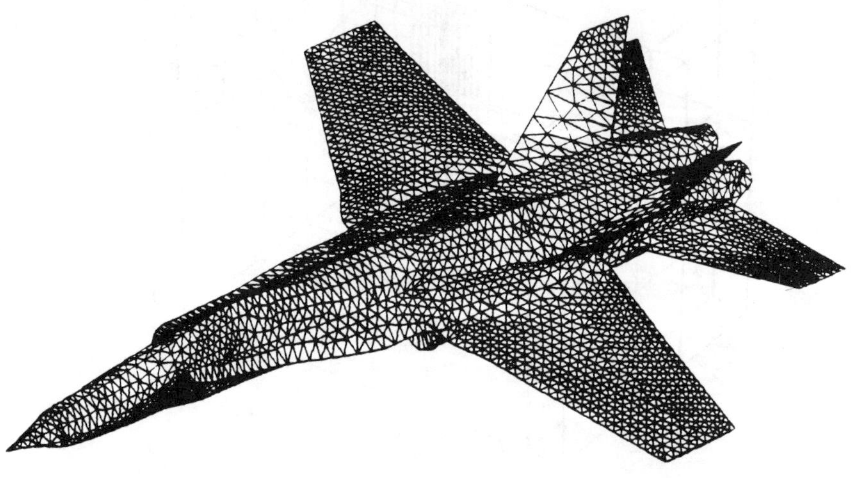

Fig. 11   Unstructured surface grid around F-18 configuration (from Ref. 87).

Part 3.   Airfoils, Wings, and Wing Bodies

Chapter 5

Subsonic Airfoil Design

Robert H. Liebeck*
*Douglas Aircraft Company, McDonnell Douglas Corporation,
Long Beach, California*

Introduction

THE airfoil design problem has existed since the time man first expressed the desire to fly. Initially, airfoils were copied from the sections of birds' wings, then various modifications were made to adapt to man's airplane application and his fabrication capability. This basic process has continued to the present day. With few exceptions, a "new" airfoil design is in fact a derivative or modification of an existing airfoil.

Typically, an airfoil design problem begins with a set of aerodynamic performance requirements that include lift, drag, and pitching moment in a specified Mach/Reynolds number flow regime plus geometric constraints on thickness. There are two basic approaches to solution of this problem: direct and inverse. For the direct method, one begins with an airfoil geometry (usually based on the geometry of an existing airfoil whose performance is already known) and calculates and/or tests to obtain the aerodynamic performance. The geometry is then modified and retested until the desired results are achieved. Alternatively, the inverse approach begins with an airfoil pressure (or velocity) distribution that provides the desired performance. An "inverse" calculation procedure is then used to obtain the corresponding airfoil shape. Although any given airfoil geometry has a corresponding pressure distribution (at least in terms of potential flow), an arbitrarily prescribed pressure distribution cannot be guaranteed a priori to provide a closed and non-re-entrant airfoil. Consequently, the prescribed pressure distribution must be adjusted until an acceptable airfoil shape is obtained. Thus, both the direct and inverse approaches to airfoil design are iterative.

Methods of Solution: A Brief Historical Perspective

Airfoil design began using the direct approach where a candidate geometry was evaluated in the wind tunnel (if not in actual flight) by the pioneer

---

Copyright © 1989 by the American Institute of Aeronautics and Astronautics, Inc. All rights reserved.
*McDonnell Douglas Fellow.

researchers such as the Wright brothers. Anderson[1] identifies Englishman Horatio F. Phillips as the "grandparent of the modern airfoil," based on Phillips' 1884 patent of a series of double-surface (albeit thin), cambered airfoils. Early airfoil development depended primarily on measurement of the macro parameters: lift, drag, angle-of-attack range, and maximum lift. Again, Anderson cites a quote from NACA Rept. 18 (1917):

> Mathematical theory has not, as yet, been applied to the discontinuous motion past a cambered surface. For this reason, we are able to design aerofoils only by consideration of those forms which have been successful, by applying general rules learned by experience, and by then testing the airfoils in a reliable wind tunnel.

At this time, airfoil design was predicated solely on the direct approach: define a candidate geometry and then evaluate its aerodynamic performance.

A unified approach to airfoil design and analysis appeared in 1932 with a publication by Jacobs et al.[2] Airfoil geometry was separated by the superposition of camber and thickness distributions, and the attributes/liabilities of various forms of these distributions evolved. Measurement of chordwise pressure distributions, together with the rapidly developing boundary-layer theory, began to offer a systematic understanding of airfoil performance and capability. Several airfoil families based on particular camber and thickness distributions were developed and tested in a wind tunnel. The resulting array of classical NACA airfoils have been successfully applied to airplane designs from 1933 to the present.

Airfoil theory was also progressing during this period. Because of the extremely limited computational capability (no computers), linearized procedures were the most popular. An excellent example is the thin airfoil theory method given by Allen[3] in 1948. Both the direct and inverse problems are considered. A modification to an existing airfoil geometry or its corresponding pressure distribution is prescribed, and the method calculates the resulting pressure distribution or airfoil geometry, respectively. This method was amenable (albeit tedious) to application via desktop calculating machines. Weber[4] developed a second-order linearized procedure that eliminated the requirement for beginning with an existing airfoil geometry/pressure distribution. Her method was effective for application to airfoils with low to moderate camber and thickness and was used by this author in his initial studies[5] of the high-lift airfoil problem. Unfortunately, the extreme camber implied by the prescribed high-lift pressure distributions caused the solutions from the Weber method to be at best approximate, and the main utility of the early results was the demonstration that the high-lift problem could be solved. An exact inverse method was essential.

For incompressible flow, the exact solution for the direct problem is fundamentally straightforward using complex potential theory and conformal mapping techniques. Although the theory itself was well understood before the turn of the century, application to the airfoil problem was slowed by the lack of computational capability. Around 1932, Theodorsen[6] published an exact solution for the direct problem. Use of the method was tedious due to the requirement of forming numerical derivatives and a cumbersome iterative process, particularly without a computer. Goldstein[7] provided an approximation to Theodorsen's theory that substantially improved the utility of the method with a minimal decrease in accuracy.

Practical solution of the inverse problem lagged due to concern over the existence and uniqueness of solutions. In 1945, Lighthill[8] offered what may be considered the first exact solution to the airfoil design problem. Existence of a solution was provided by the requirement that the prescribed velocity distribution meets three integral constraints: one for uniform freestream flow at infinity and two for closure of the resulting airfoil geometry. An awkward feature of the Lighthill method is that it requires the desired velocity distribution to be specified as a function of the angular coordinate of the circle into which the resulting airfoil is mapped. However, this method does provide the exact solution of the airfoil design problem with no restriction on thickness and camber. Most modern incompressible design methods are variants of Lighthill's basic approach.

Subsonic airfoil design capability at the Douglas Aircraft Company was significantly enhanced in 1970 by Malcolm James with his "New Look at Two-Dimensional Incompressible Airfoil Theory,"[9] which provides exact solutions to the direct and inverse problems. For the inverse problem the method uses as input the desired airfoil velocity distribution prescribed as a function of arc length along the airfoil surface itself, as shown in Fig. 1. As discussed earlier, an arbitrarily prescribed velocity distribution will not necessarily conform to a closed airfoil with the proper flow conditions at infinity. Some compromise to the input distribution is inevitable, and the James method returns as an output solution an airfoil shape and a velocity distribution that are an exact pair. The output velocity distribution has been changed to meet the closure and infinity conditions while preserving the input details and modifying certain overall characteristics. Since the basic theory of the method depends on the implicit mapping of the airfoil to the unit circle domain, the source of this compromise is through modification of a Fourier transform associated with the input velocity. The first two terms of this transform are modified by the closure and infinity conditions, and in the sense that the Fourier transform is a truncation, the output exact solution pair could be said to represent the closest fit to the input conditions in a least-squares sense, but in the circle domain. By comparing the input and resulting modified (output) velocity distributions, the input distribution is easily adjusted so that agreement between the input and output distributions is obtained. An example solution is shown in Fig. 2, where the disagreement has been intentionally exaggerated for the purpose of illustration.

Leading- and trailing-edge singularities are treated exactly by the James method, and this implies some special consideration at the trailing edge. It is necessary to specify the desired trailing-edge angle as an input, and this in principle amounts to an overspecified problem. The input velocity distribution itself implicitly defines the trailing-edge angle. In reality it is not practical to expect that an input velocity distribution will contain sufficient detail in the last 2% or so of chord to provide an accurate definition of the trailing-edge angle. Since the input velocity distribution will be modified as described earlier, the minor adjustment of the distribution near the trailing edge, which is required in order to obtain the specified angle, becomes insignificant.

The James method is robust and has been used by the author for the past 20 years for the design of subsonic airfoils for a wide variety of applications. No intermediate judgments or decisions are required by the designer during the calculation procedure. Input and output velocity distributions are simply compared, and the new input distribution is adjusted as required. Specification of the velocity distribution in the physical plane (as opposed to, for example, the circle plane) eliminates ambiguity in the choice of the input distribution. For the inverse problem, the only point of the airfoil geometry that can be considered as known when the design is begun is the point at the trailing edge. Thus, specifying the velocity distribution as a function of arc length along the airfoil surface for an airfoil of unit perimeter is the logical format for the input. Surface distance is also the natural independent variable for boundary-layer theory that is used in the specification of the velocity distribution.

Viscosity and the Airfoil Design Problem

Specification of the input pressure distribution for the airfoil design problem is typically predicated on obtaining some desired behavior of the boundary layer, e.g., no separation, minimum profile drag, extensive laminar flow, a well-balanced laminar separation bubble, or whatever. Thus, an "inverse" boundary-layer theory is required to obtain the pressure distribution, which in turn is used with the inverse airfoil theory to obtain the airfoil geometry. Most boundary-layer theory, however, is set in the direct mode of predicting the behavior of the boundary layer for some given pressure distribution. Relatively qualitative techniques, such as favorable gradients ensuring laminar flow (at low and moderate Reynolds numbers) and transition occurring coincident with the change in sign of the pressure gradient, have been used to guide the design of airfoils for the past several decades (e.g., the NACA 60 series laminar flow sections[2]). Formal inverse boundary-layer methods are not common. A notable exception is the method of Stratford.[10]

Stratford has developed an analytical method that provides a pressure recovery distribution that continuously avoids separation of the turbulent boundary layer by a specified margin. This form of pressure distribution in principle recovers a given pressure rise in the shortest possible distance, or it can be interpreted as recovery of the maximum pressure rise in a given distance. For a given Reynolds number and trailing-edge pressure, an infinite family of flat rooftop plus Stratford pressure recovery distributions such as those shown in Fig. 3 can be generated. These pressure distributions are limiting cases for rooftop length vs rooftop level, and they can be used as the foundation for the specification of an inverse airfoil design problem.

High-Lift Airfoil Development

The author's introduction to the airfoil design problem came in 1967 when A. M. O. Smith asked the following question: "What is the maximum lift that can be obtained from an airfoil, and what is the shape of that airfoil?" A linearized solution was obtained in Ref. 5, where lift was expressed as $\int C_p \, \mathrm{d}x$ with the freestream aligned with the x axis. Utilizing the Stratford pressure recovery and the calculus of variations, a family of maximum lift airfoil pressure distributions were defined. Weber's second-order inverse airfoil method[4] was used to calculate the corresponding airfoil shapes, as shown in Fig. 4. The highly-cambered airfoils indicated that a linearized inverse method was inadequate for this problem; however, the results of this early study demonstrated that an optimization problem for airfoil design could be formulated and solved in a relatively closed form.

Enter the James method.[9] The high-lift problem was reformulated with lift coefficient (c_l) expressed as the integral around the airfoil of the velocity on the surface of the airfoil as shown in Fig. 1. Since $v(s) < 0$ everywhere on the lower surface, c_l will be maximized by keeping $v(s)$ as close to stagnation as possible. On the upper surface, it is desired to maximize the area under the $v(s)$ vs s curve, subject to the constraint that the boundary layer does not separate. As in the linearized problem, this leads to the selection of the Stratford recovery, and applying the calculus of variations yields the solution that a flat rooftop preceding said recovery maximizes the contribution of the upper surface to the airfoil lift. For a given Reynolds number the resulting optimum velocity distribution for the case of a laminar boundary layer on the rooftop will have a longer and higher rooftop than that for a turbulent boundary layer on the rooftop. Of course, the boundary layer must be turbulent in all cases in the recovery region. A detailed development of this high-lift design problem is given in Ref. 11.

The preceding analysis says that c_l will be maximized by a velocity distribution of the form shown by the solid line in Fig. 5. However, it will not provide an airfoil shape. The discontinuities implied at the leading and trailing edges and the fact that stagnation can only occur at a single point prevent a meaningful potential-flow solution. Therefore, the velocity distribution has been modified as shown by the broken line in Fig. 5. The slope of the $v(s)$ curve through the leading-edge stagnation point affects the leading-edge radius and thickness of the resulting airfoil shape, and the accelerating portion of the upper-surface rooftop region is rounded over to allow for operation at angles of attack above the design value. A boundary-layer transition ramp has been located at the rooftop peak for those cases where the rooftop is laminar and also to ease a turbulent boundary-layer's introduction to the severe initial Stratford gradient. For the maximum lift problem, the lower-surface distribution is modified according to two general constraints: the velocity remains as low as possible to maximize lift, and the flow continuously accelerates to minimize drag. Near the leading edge, the lower-surface distribution is also rounded to provide off-design performance at lower angles of attack.

The trailing-edge velocity ratio v_{te}/V is a very important parameter. For example, increasing v_{te}/V by 10% may increase the upper-surface lift by

15–20%. However, v_{te}/V is severely limited by the requirement of obtaining a proper trailing-edge geometry. Potential-flow theory limits v_{te}/V to values of <1 for cusped trailing edges. It is possible to have $v_{te}/V > 1$ immediately upstream of the trailing edge, but this typically implies a large trailing-edge angle with an increase in drag and a loss in lift due to a modified Kutta condition. The case of aft-cambered airfoils where v_{te}/V is quite large near the trailing edge will be discussed later. For the present, v_{te}/V should be considered to take on values between 0.80 and 0.95.

At this point, the airfoil velocity distribution has been optimized for maximum lift while satisfying boundary-layer theory for no separation and then modified and adjusted to allow potential-flow theory to provide an airfoil shape. The free parameters include the location of the leading-edge stagnation point $s = sp$, the slope $v'(sp)/V$, and the shape of the upper- and lower-surface acceleration regions. In a purely mathematical sense, the modified velocity distribution can no longer be called optimum. The class of velocity distributions defined in Fig. 5 has been extended to a wide variety of airfoil design problems that are well beyond the original high-lift problem, and some examples will be presented later. The next discussion presents two early solutions to the original problem.

Stated in a simple form, the minimum requirements for an airfoil geometry are that it is non-re-entrant, has a rounded leading edge, and has a pointed trailing edge. According to the variational analysis, it is desirable to obtain an airfoil velocity distribution that is as close as possible to the solid line in Fig. 5. By strictly following this approach, the airfoil shape and corresponding velocity distribution shown in Fig. 6 was obtained. This says that the maximum lift coefficient of an airfoil at a Reynolds number of 5×10^6 is about 3 with a lift-to-drag ratio of 600! From a practical standpoint this airfoil may not be very useful. Its relatively sharp leading edge will limit it to a narrow angle-of-attack range, and it is probably too thin for most applications. By simply changing the velocity distribution on the lower surface and the slope $v'(s)$ through the leading-edge stagnation point, airfoil L1002 (Fig. 7) was created. Airfoil L1002 has a more practical thickness, but its design C_l is reduced compared to airfoil L1001 by more than 20%.

The first wind-tunnel test for this class of airfoils was conducted in 1972 at the McDonnell Douglas low-speed wind tunnel in St. Louis. Airfoil L1003 (Fig. 8), which assumes (and in principle requires) a laminar rooftop, and airfoil L1004 (Fig. 9), which assumes a turbulent rooftop, were tested at Reynolds numbers of 1.0×10^6 and 3.0×10^6, respectively, and their resulting performance curves are given in Figs. 10 and 11. Both airfoils exceeded their design lift coefficients by a significant margin and offered a wide c_l range over which the drag remained low. Figure 12 shows a comparison of the experimental pressure distributions and those predicted by potential flow, without a correction for boundary-layer thickness. These results, together with the observation of tufts on the pressure recovery region, indicated that the flow remained firmly attached all the way to the trailing edge. At stall the entire recovery region separated instantaneously. Reducing the angle of attack less than one-half of a degree resulted in an instantaneous and complete reattachment, which indicated an almost total lack of hysteresis effect on stall recovery.

Generalized Airfoil Design Capability

The high-lift airfoil design problem and its solution described in the preceding section has led to a procedure that is applicable to a wide variety of airfoil design problems. It is based, quite simply, on the James potential-flow inverse solution and the Stratford boundary-layer inverse solution. The capability of the James method has been described in a previous section. Stratford's imminent separation pressure recovery distribution has been verified experimentally and shown to exhibit "a good margin of stability," to quote Stratford himself.[12] Moreover, the theoretical and experimental results from the high-lift airfoil problem just discussed confirmed the viability of the application of the Stratford theory to airfoil design.

The maximum lift flat rooftop plus Stratford recovery pressure distribution discussed earlier is only one member of an infinite family of such pressure distributions for a given Reynolds number and trailing-edge pressure. Figure 3 shows an example family of this class, including the member that provides maximum upper-surface lift. (Here, upper-surface lift is defined as $\int v \, ds$, where the integral extends from the leading-edge stagnation point to the trailing edge.) By definition, each of these pressure distributions carries its given rooftop pressure level as far aft as possible before entering the Stratford recovery to the specified trailing-edge pressure. It should be noted that upper-surface lift varies slowly with rooftop level, which makes the lower rooftop distributions appealing for cases where high lift is desired but the pressure level is limited by compressibility or cavitation.

Instead of viewing the Stratford distribution as imminently separating everywhere, it is more properly interpreted as a recovery distribution that avoids separation by an effectively constant margin along its entire length. The boundary layer is no more ready to separate at the trailing edge than it is near the beginning of the recovery region. In principle, a Stratford recovery extends to downstream infinity with a continuously decreasing gradient.

An important feature of the flat rooftop plus Stratford recovery distribution is that it is defined analytically and requires only three parameters for its specification: Reynolds number, trailing-edge pressure, and rooftop level. In addition, the state of the boundary layer on the rooftop region must be given: laminar, turbulent, or laminar plus turbulent with transition occurring somewhere on the rooftop region. As expected, for a given set of the preceding three parameters, a laminar rooftop will be longer than a turbulent one.

The airfoil velocity distribution sketched in Fig. 5 can be generalized to include arbitrary rooftop levels on the upper surface, and it is of course not essential that the flow be continuously accelerating on the lower surface. A sinusoidal variation of $v(s)$ through the leading-edge stagnation point is exact for the singularity that exists there, and the remaining portions of the velocity distribution are readily defined by second- and third-order curves. Thus, an entire input airfoil velocity distribution can be defined analytically by the specification of a relatively small set of parameters. They include Reynolds number, rooftop level, trailing-edge pressure, location of the leading-edge stagnation point and slope $v'(s)$ there, and the various parameters needed to define the lower-surface distribution and roundover into the rooftop region (typically five or so).

Parametric representation of the input velocity distribution for the James inverse airfoil design code provides a powerful capability for the airfoil designer. Since an arbitrarily specified input distribution is unlikely to be acceptable for defining a corresponding airfoil shape, this input distribution must be modified so that it will define an airfoil. As described earlier, the James method returns an acceptable distribution (from a potential-flow point of view) that does in fact define an airfoil. However, this acceptable distribution is not likely to satisfy boundary-layer constraints. This distribution can be compared to the originally desired input distribution, and the input is then modified accordingly until agreement between the input and output velocity distributions is achieved. The capability to simply adjust parameters such as trailing-edge velocity or stagnation-point location and have the entire input distribution redefined accordingly makes convergence to a final airfoil design a straightforward process. Boundary-layer constraints are, by definition, always satisfied by the parametric input velocity distribution.

The combination of the James code and the parametric input velocity distribution offers an accurate and efficient method that is adaptable to an interactive procedure. Airfoil design is a subjective process, and the ability to immediately observe the results for an input velocity distribution and change/adjust the distribution based on the observed results becomes a powerful tool. The designer's thoughts and reactions to a particular solution can be immediately translated into a new input distribution, which in turn yields a corresponding new solution. There are no interruptions due to time required to generate the new input, and consequently the engineer can concentrate on the design problem itself. This capability is not, in principle, unique to the James plus Stratford combination; however, it is believed to be the first example put into practice.

Example Airfoil Designs

The airfoil design procedure described above has been applied to a wide variety of problems with good results. Some representative examples are offered in the following subsections.

Low-Reynolds-Number Airfoils

One of the more popular applications of the method has been in the development of low-Reynolds-number airfoils ($RN < 10^6$). Typically, these airfoils have been developed for use on high-altitude long-endurance airplanes, and this requires design lift coefficients in excess of 1.0 and thickness-to-chord ratios on the order of 15%.

The basic method described in this chapter has been applied to the low-Reynolds-number design problem with one significant distinction. A fundamental requirement of the Stratford pressure recovery distribution is that the boundary layer be established as fully turbulent at the onset of the severe initial pressure rise. For Reynolds numbers above 10^6, this is accomplished via a small transition ramp such as that sketched in Fig. 5. Here the laminar boundary layer on the rooftop region has become sufficiently unstable by the time it reaches the transition ramp, and the mild adverse gradient of the ramp causes almost instantaneous transition to a turbulent boundary layer, particularly for $RN > 5 \times 10^6$. When the Reynolds number is less than 10^6, the transition mechanism is typically a laminar separation bubble. In this case, the laminar boundary layer in an adverse pressure gradient separates, and transition occurs in the free shear layer at the edge of the bubble, followed by the reattachment of the turbulent boundary layer. Bubble length varies and increases with decreasing Reynolds number. A stable bubble length may be on the order of 2% to as much as 10% of the airfoil chord. Beyond this length, the bubble is likely to burst, and the flow separates downstream over the entire airfoil. At higher Reynolds numbers, although a bubble may still exist, it is so short that transition may be considered as instantaneous.

A primary factor controlling the behavior of the laminar separation bubble is the local Reynolds number RN_P at the rooftop peak (i.e., RN based on the velocity at the rooftop peak and the length of the rooftop). As a general rule, if $RN_P < 0.10 \times 10^6$, the bubble is likely to burst. As described earlier, for a given freestream Reynolds number, there exists an infinite family of flat rooftop plus Stratford recovery pressure distributions such as the example given in Fig. 3. As the rooftop length increases, the rooftop velocity decreases; however, the value of RN_P increases as shown in Fig. 3. Consequently, long rooftops are regarded as desirable for low-Reynolds-number airfoils.

Specification of the boundary-layer transition region of the pressure distribution represents one of the more difficult areas of low-Reynolds-number airfoil design. A laminar boundary layer is very sensitive to any adverse pressure gradient, and the goal is to set the gradient such that it causes instability and subsequent transition without creating a large separation bubble. When applied to the class of airfoils considered here, this has resulted in a substantial rounding of the corner at the rooftop peak. (Note: airfoil L1003 performed well without a rounded corner due to the high turbulence level of the St. Louis wind tunnel. This would not be the case in the Douglas Long Beach low-speed wind tunnel where the turbulence is very low—less than 1%. All of the experimental data given later were taken in the Long Beach wind tunnel.) A more detailed discussion of the development of low-Reynolds-number airfoils at the Douglas Aircraft Company is given in Ref. 13. Some representative designs are now described.

Airfoils LA5104E (Fig. 13) and LA203A (Fig. 14) are example high-lift designs for the Reynolds number range of $0.25–0.50 \times 10^6$. Their primary distinction is that airfoil LA203A is aft loaded, and airfoil LA5104E is not.

Aft loading permits the specification of a lower value of the trailing-edge pressure coefficient that allows a longer rooftop for a given rooftop level, which in turn increases the value RN_P. The penalty for aft loading is the increased value of negative pitching moment. Although this does not affect an airfoil's ability to perform in terms of lift and drag, it can have a measurable effect on the trimmed lift and drag of the complete airplane.

A comparison of the performance of airfoils LA5104E and LA203A and the popular Wortmann FX63-137 airfoil (Fig. 15) is given in Fig. 16. The Wortmann airfoil has the most aft loading and consequently the most negative pitching moment. Drag polars of all three airfoils are somewhat equivalent; however, both of the LA series airfoils are thicker and have lower drag above $c_l = 1.0$ and slightly higher $c_{l_{max}}$. Note that three distinct airfoils developed by two designers have very similar overall performance.

If the pitching moment is constrained to near-zero or positive values, an airfoil such as LA2573A (Fig. 17) results. Here it is necessary to front load the upper-surface pressure distribution with a short, high-level rooftop. As discussed earlier, this reduces the value of RN_P which increases the potential for difficulty with a laminar separation bubble. In addition, it was necessary to add reflex to the lower-surface pressure distribution near the trailing edge in order to achieve the desired positive pitching moment. This causes a reduction in $c_{l_{max}}$, since this reflex is in effect unloading the aft portion of the airfoil. The conflict between positive pitching moment and $c_{l_{max}}$ is apparent and is intensified by the low-Reynolds-number laminar separation bubble threat. Wind-tunnel performance is given in Fig. 18. Note that there is a significant increase in drag in the mid c_l range at $RN = 0.25 \times 10^6$. This is due to a laminar separation bubble on the order of 10% chord in length. (Some details of this will be discussed shortly.)

Airfoils LA2573A, LA5104E, and LA203A offer a spectrum of pitching moments from $+0.05$ to -0.17, which could be considered as their primary distinction. (All three airfoils were designed for high lift in the Reynolds number range of $0.25-0.75 \times 10^6$.) The pitching moment of LA203A has already been criticized for its potential for high airplane trim drag. Airfoil LA2573A was designed for a tailless configuration and could be said to have no trim drag. However, the pressure distribution of Fig. 17 indicates that the last 10% of the airfoil chord is in fact lifting downward. In addition, the lower $c_{l_{max}}$ of LA2573A implies an increase in wing area over LA203A for the same stall speed, and this will result in an increase in profile drag that must be traded with the implied trim drag of LA203A. There is no "free lunch." Design trade studies such as this are conveniently carried out using airfoils based on the rooftop plus Stratford recovery distributions. Since these are regarded as limiting forms of the flow's capability without separation, trades based on this approach are quite valid.

As a final example of low-Reynolds-number airfoil design, compressibility effects are included. Airfoil LC111A (Fig. 19) was developed for the Air Force for use in their design studies of high-altitude long-endurance airplanes. For this application the requirement was for high lift in the Reynolds number range of $0.5-5.0 \times 10^6$ at Mach numbers as high as 0.4.

This relatively moderate Mach number produces significant compressibility effects when design lift coefficients in excess of 1.5 are desired. The recently developed MIT ISES airfoil analysis code[14] has been used to theoretically predict the performance of airfoil LC111A given in Fig. 20. At this writing, the airfoil has not been tested in a wind tunnel. Figure 21 shows the compressibility drag rise predictions, of the ISES code which indicate that the airfoil can be operated at a c_l of 1.8 at a Mach number of 0.4.

Chordwise pressure distributions at $c_l = 1.4$ for airfoil LC111A as calculated by the ISES code are shown in Fig. 22. It appears that the shock at $M = 0.47$ and 0.50 is coincident with the laminar separation bubble at the end of the rooftop. (The bubble is evidenced by the short chordwise plateau in the pressure distributions of Fig. 22, which is not in the design pressure distribution of Fig. 19.) The ISES code does not account for the interaction between the shock and the bubble, and consequently the drag rise performance awaits experimental verification. (This is not a criticism of the ISES code, and a brief discussion of the capability of ISES will follow shortly.)

It is interesting to note that both low-Reynolds-number performance and compressibility induced drag rise are enhanced by extending the length of the rooftop of the upper-surface pressure distribution, for mutually exclusive fluid mechanical arguments: higher local Reynolds number at transition and weaker shock due to lower rooftop level.

Before the discussion of low-Reynolds-number airfoil design is closed, some comments on the ISES code are in order. The appeal of this code is that it is uniquely capable of predicting airfoil performance at low Reynolds numbers where laminar bubble formation is virtually inevitable. Such bubbles can cause an increase in drag and a reduction in maximum lift. The ISES code not only predicts the existence of these bubbles, it also accurately predicts the drag and maximum lift with a bubble present. Figure 23 gives a comparison of the theoretical predictions of the Douglas MADAAM airfoil analysis code[15] and the ISES code with wind-tunnel results for the drag polars of airfoil LA2573A. At a Reynolds number of 0.25×10^6 this airfoil suffers from a significant drag penalty in the middle C_l range due to a laminar separation bubble. The extent of the penalty is shown by the difference between the experimental drag and that predicted by the MADAAM code, which does not account for the bubble. Alternatively, the ISES code gives a very good prediction of the drag polar and maximum lift. Figure 24 compares the ISES prediction of the airfoil's chordwise pressure distribution with the wind-tunnel data, and the location and extent of the bubble are accurately reproduced. This remarkable capability of the ISES code will serve to enhance the development of low-Reynolds-number airfoils well beyond the present state of the art.

Windmill Blade Sections

This method has been applied to the design of airfoils for windmill blades where the goal was low drag at $c_l = 1.0$, $c_{l_{max}} = 1.5$, and $t/c = 15\%$ at a Reynolds number range of $0.5-1.5 \times 10^6$. Airfoil LW101B (Fig. 25) was designed for this requirement, and it has the additional feature of a flat lower surface. This is useful in the manufacturing of windmill blades in that the radial twist distribution can be easily defined and monitored.

Airfoil LW101B was used as the blade section for the windmill shown in Fig. 26. This machine was designed and developed by Industrial Design Laboratories of Culver City, CA, and the prototype has performed beyond the theoretical predictions for start up and power at low wind speeds. At these conditions, low drag at high lift becomes a significant requirement for the blade section.

Airfoil LW108A (Fig. 27) is a second generation design, and its intended application is for larger machines where the design Reynolds number is on the order of 3.0×10^6. This airfoil maintains a $c_{l_{max}}$ of 1.5, and its main distinction from LW101B is its increased thickness (19% vs 15%). Most of the thickness comes from the increased design Reynolds number. The increased thickness offers a significant reduction in blade weight and cost.

Aerobatic Airplane Airfoils

In 1986 the author was approached by the late Harold Chappell and Henry Haigh to design an airfoil for a "world class aerobatic airplane." The Russians had just won the World Aerobatic Contest in England with a Sukhoi 26M airplane that had an airfoil section somewhat similar to Douglas symmetric designs based on the method in the present article.

The design process was begun by first evaluating the airfoils used on existing aerobatic airplanes. This was surprisingly straightforward, since almost all American airplanes use the classic NACA 0015 at the root and 0012 at the tip (or minor variants of these thicknesses). These airfoils offer very good performance for this application, and it was not particularly clear as to what should be "improved."

For a given thickness, improvement in performance of a symmetric airfoil is limited to lower drag and higher $c_{l_{max}}$. Airfoil LAB121A15 (Fig. 28) was designed with both of these improvements as goals. Comparison of the theoretically predicted performance indicated a moderate reduction in profile drag at lift coefficients above 0.8 and an increase in $c_{l_{max}}$ of about 0.3 compared to the NACA 0015 section.

Since airfoil LAB121A15 is a symmetric section, using a rooftop plus Stratford distribution might be regarded as improper. In principle, this limiting distribution should imply that the airfoil will stall at any angle of attack much different from zero. However, by using a Reynolds number lower than the flight value to specify the recovery distribution, additional conservatism is achieved. Also, the forward portion of the rooftop is shaped with a generous acceleration region that simply fills in to form a flat rooftop as the angle of attack is increased.

Airfoil LAB121A15 and its thinner companion LAB121A12 were used to define the wing of Henry Haigh's new aerobatic airplane, the Ratsrepus 360 (Fig. 29), which was completed in December 1988. Initial flight-test results are encouraging. Haigh reports that it is the first aerobatic airplane he has found capable of an 8-g turn. Observed stall speed suggests that the trimmed airplane $C_{l_{max}}$ is greater than predicted, and poststall behavior is benign and symmetric. The airplane is easy to snap roll, which is often not the case when conventional stall behavior is proper. Before taking too much credit for the airfoil's contribution to the performance of the Ratsrepus 360, it is noted that Henry Haigh won the 1988 World Aerobatic Championship in his "old" airplane.

Sailboat Rudder Section

The 1987 America's Cup competition embraced aerospace technology on all possible fronts. Douglas was enlisted to develop rudder sections for the St. Francis Yacht Club, San Francisco entry. A unique underwater configuration was comprised of two large rudders, one forward and one aft, with an axisymmetric streamlined bobweight on a strut at midship. The rudders had a span greater than a conventional keel, and a chord of about one-fourth of a conventional keel. Side loads were predicted to be of the same order as those on a keel; consequently, there was concern over the strength of the rudders.

A popular and successful section for sailboat rudders has been the NACA 63a–010 symmetric airfoil. It was desired to design a new section with the same thickness-to-chord ratio, but with increased volume for added strength. In addition, the new section was to have equal if not greater $c_{l_{max}}$ and the same or lower profile drag. Airfoil LSB119A (Fig. 30) was the result, and its geometry is compared with that of the NACA 63a-010 airfoil in Fig. 31.

The strut to support the central bobweight was a different design problem. Maximum thickness and volume were desired, and airfoil LSB117A (Fig. 32) was prepared as a response to these requirements. This airfoil has a thickness-to-chord ratio of 25%, and its low-drag angle-of-attack range is ± 6 deg.

In the competition, the St. Francis boat was second among seven challenger candidates. Difficulties with "stick force gradients" of the complex dual rudder system, which had to allow for both co- and counter-steering modes, were one of the main problem areas of this radical new design.

Wings for Racing Cars

As an airfoil design problem, the common goal of a race car wing is to provide as much lift (actually downforce) as possible within the physical constraints of the allowable dimensions of the wing. Because of the high lift coefficients, the primary source of drag is the induced contribution, and profile drag is of concern only in that separation is avoided.

When considered in a general form, the problem of maximizing lift may be viewed as attempting to fill a "box" in the C_p vs x plane. As shown in Fig. 33, the box is bounded below by $C_p = 1.0$ and above by $C_p = C_{p_{crit}}$ as implied by the value of the critical Mach number. Figure 33 gives an example comparison of a single- and two-element air-foil with respect to their ability to fill the box. For the two-element airfoil, the flap induces a high velocity field about the trailing-edge region of the main element. This allows the main element to be designed with a Stratford recovery distribution based on a trailing-edge c_p, which is substantially more negative than that of a single-element airfoil, and consequently the main element can carry much more lift. The sketch of Fig. 33 is representative of the relative capabilities of single- and two-element airfoils.

Figure 34 gives the geometry of airfoil L175 that was used on the race car (Fig. 35) that won the 1975 Indianapolis 500. Both elements are based on Stratford pressure recovery distributions. As an aside, the demonstrated ability of this theoretically optimized airfoil to operate successfully in the hostile environment of automobile racing was of significant value in establishing the viability of the entire airfoil design approach being discussed in the present article.

Pterosaur Airfoil

In 1984, AeroVironment, Inc., of Morovia, CA, was commissioned by the Smithsonian National Air and Space Museum to develop a flying replica of the giant pterosaur (flying reptile) that lived during the Mesozoic era (64–200 million years ago). An artist's rendering of this creature is shown in Fig. 36. Douglas was asked to provide an airfoil that satisified two basic requirements: it should be representative of what the original geometry was thought to be, and it should have structural and aerodynamic characteristics that would allow the replica to be constructed and flown.

Original wing construction is thought to have been a thin-walled hollow bone near the leading edge with a membrane forming most of the wing surface. Thus, the replica airfoil was to be as thin as possible with maximum thickness far forward. Aerodynamic requirements included a design c_l of 1.2, $c_{l_{max}}$ of 1.6, and pitching moment of 0.0 at a Reynolds number of 0.25×10^6. Forward thickness and low pitching moment are compatible constraints. Airfoil LPT102B (Fig. 37) was the result. The full-scale replica of the largest pterosaur made several successful flights in 1987 and now resides in the Smithsonian National Air and Space Museum in Washington DC.

Closing Remarks

This Chapter presented the development of and results from an airfoil design method that was initiated at the Douglas Aircraft Company 22 years ago. Various aspects have been modified and refined over the years; however, the basic computational fluid dynamics model comprised of the James inverse potential-flow method and the Stratford inverse boundary-layer method has remained as the core. Success of the method is a consequence of its accuracy and the capability to operate it interactively with no abstractions.

Acknowledgment

Many individuals have made important contributions to the work described in this article. At the risk of offending those who might be inadvertently omitted, acknowledgement of the following aerodynamicists is given: R. M. James, P. B. S. Lissaman, A. I. Ormsbee, and A. M. O. Smith.

References

[1]Anderson, J. D., Jr., *Introduction to Flight*, McGraw-Hill, New York, 1978.

[2]Jacobs, E. N., Ward, K. E., and Pinkerton, R. M., "The Characteristics of 78 Related Airfoil Sections from Tests in the Variable-Density Wind Tunnel," NACA Rept. 460, 1933.

[3]Allen, H. J. "General Theory of Airfoil Sections Having Arbitrary Shape or Pressure Distribution," NACA TR-833, 1948.

[4]Weber, J., "The Calculation of the Pressure Distribution on the Surface of Thick Cambered Wings and the Design of Wings with Given Pressure Distribution," Aeronautical Research Council, London, R&M 3026, June 1955.

[5]Liebeck, R. H., "Optimization of Airfoils for Maximum Lift," Ph.D. Thesis, Univ. of Illinois, Urbana, IL, 1968.

[6]Theodorsen, T., "Theory of Wing Sections of Arbitrary Shape," NACA TR-411, 1932.

[7]Goldstein, S., "Approximate Two-Dimensional Airfoil Theory," Pts. I–VI, *Current Papers of the Aeronautical Research Council*, London, 1952.

[8]Lighthill, M. J., "A New Method of Two-Dimensional Airfoil Design," Aeronautical Research Council, London, R&M 2112, April 1945.

[9]James, R. M., "A New Look at Two-Dimensional Incompressible Airfoil Theory," Douglas Aircraft Co., Rept. MDC-JO918, May 1971.

[10]Stratford, B. S., "The Prediction of Separation of the Turbulent Boundary Layer," *Journal of Fluid Mechanics*, Vol. 5, 1959, pp.

[11]Liebeck, R. H., "A Class of Airfoils Designed for High Lift in Incompressible Flow," *Journal of Aircraft*, Vol. 10, Oct. 1973, pp.

[12]Stratford, B. S., "An Experimental Flow with Zero Skin Friction Throughout Its Region of Pressure Rise," *Journal of Fluid Mechanics*, Vol. 5, 1959.

[13]Liebeck, R. H., "Low Reynolds Number Airfoil Design at the Douglas Aircraft Company," Royal Aeronautical Society Conference on Aerodynamics at Low Reynolds Numbers, London, Oct. 1986.

[14]Drela, M., and Giles, M. B., "Viscous Inviscid Analysis of Transonic and Low Reynolds Number Airfoils," AIAA Paper 86-1786, June 1986.

[15]Callaghan, J. G., and Beatty, T. D., "A Theoretical Method for the Design and Analysis of Multi-Element Airfoils," *Journal of Aircraft*, Vol. 9, Dec. 1972, pp.

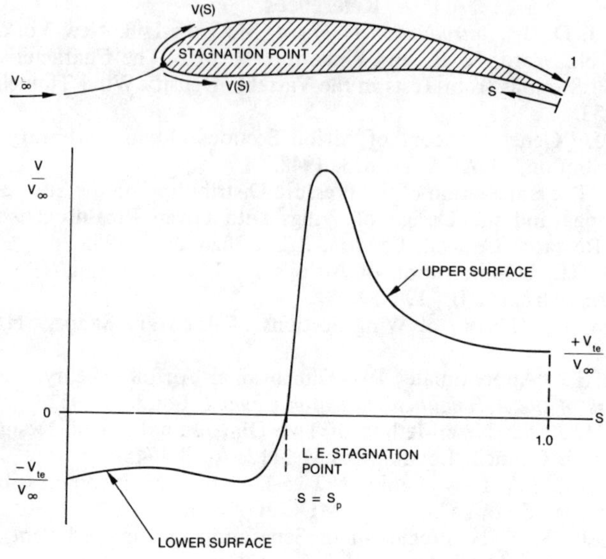

Fig. 1 Definition of airfoil velocity distribution.

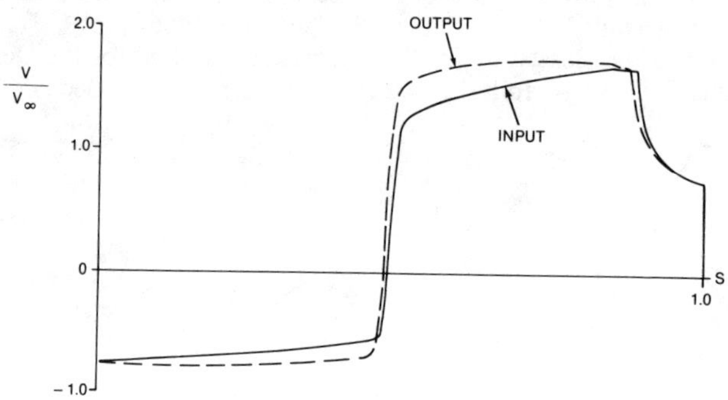

Fig. 2 Input and output velocity distributions (James airfoil design method).

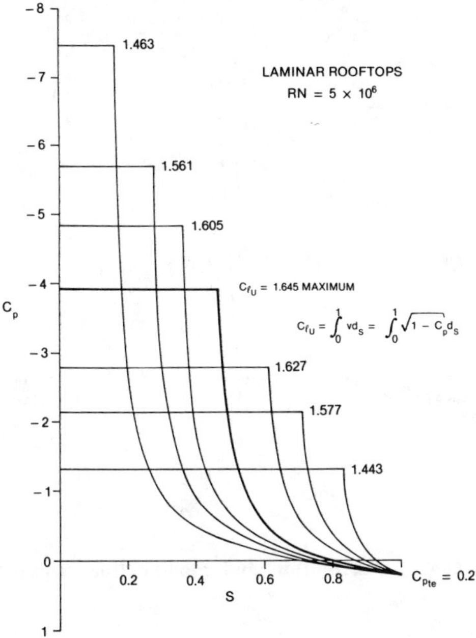

$$C_{l_U} = \int_0^1 v\,d_s = \int_0^1 \sqrt{1 - C_p}\,d_s$$

Fig. 3   Family of flat rooftop plus Stratford pressure recovery distributions.

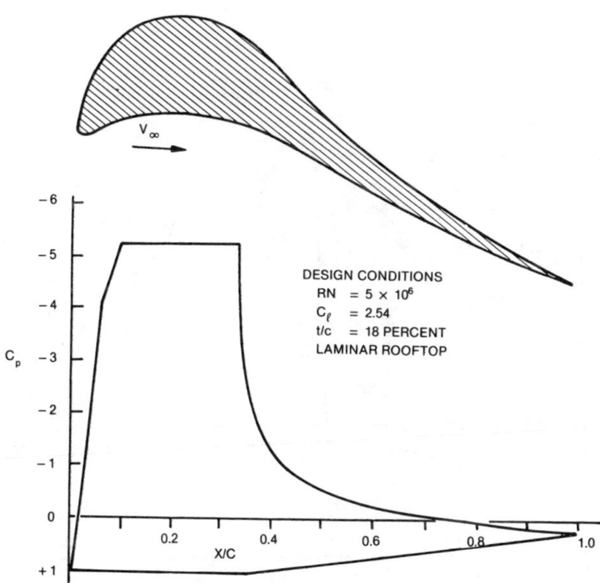

Fig. 4   Early high-lift airfoil solution.

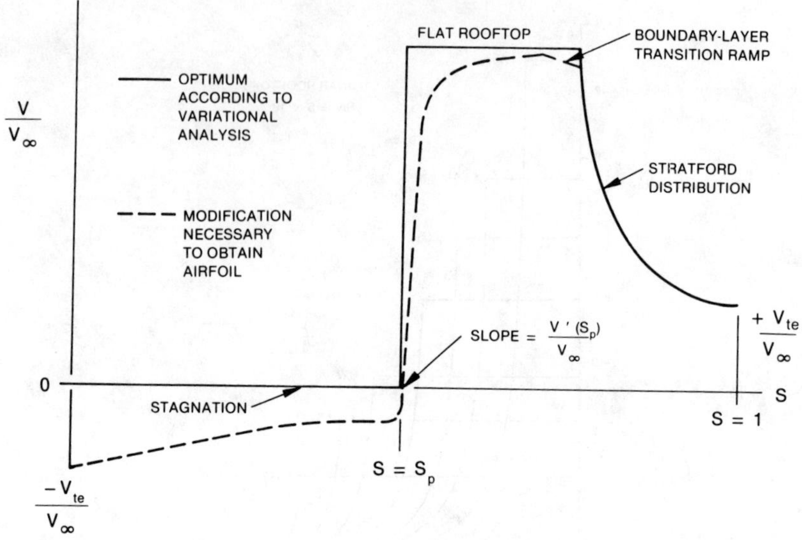

Fig. 5   Optimum velocity distribution and modification required to obtain airfoil shape.

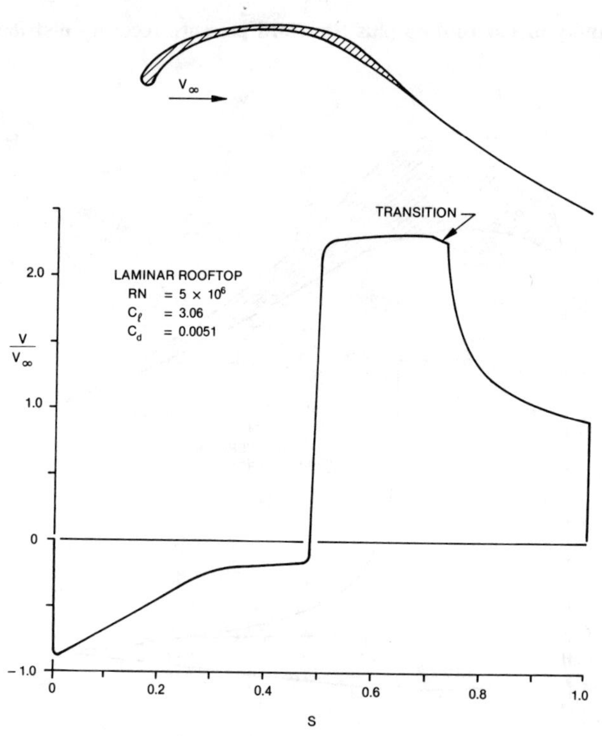

Fig. 6   Maximum lift airfoil and design velocity distribution.

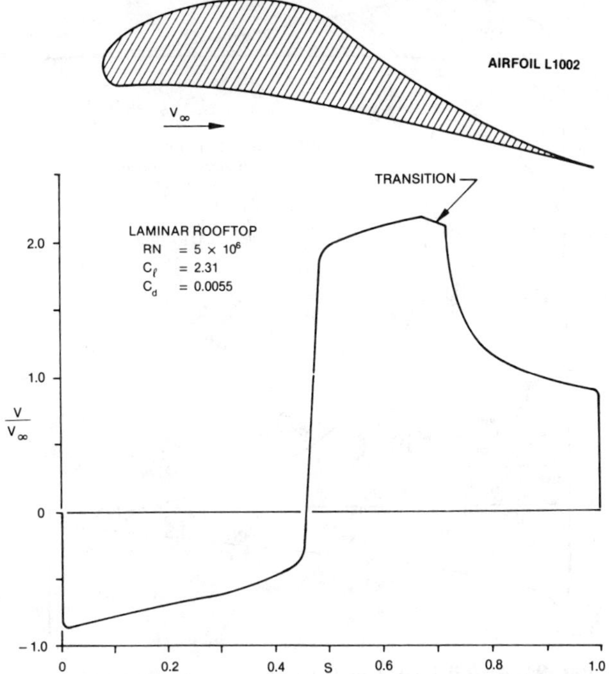

Fig. 7   Airfoil of Fig. 6 with a practical thickness distribution.

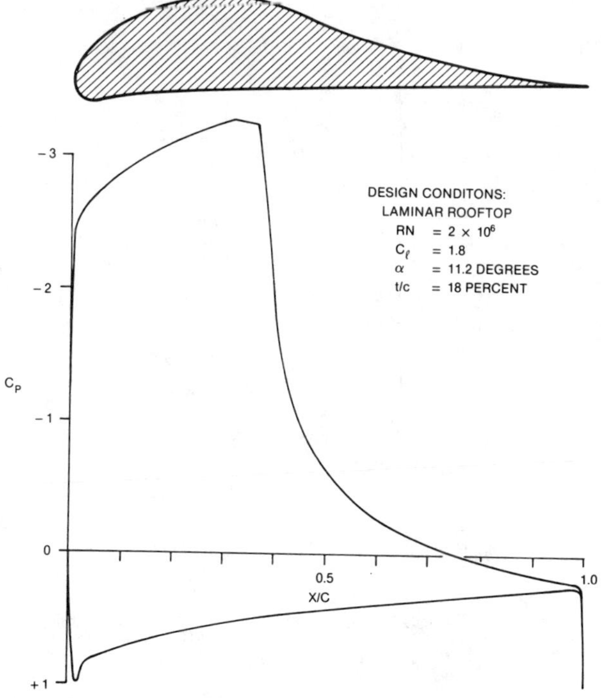

Fig. 8   Airfoil L1003 and design pressure distribution.

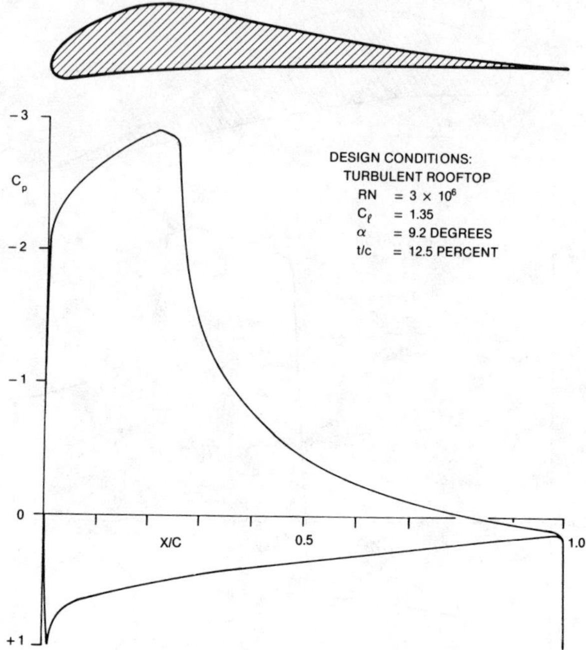

Fig. 9   Airfoil L1004 and design pressure distribution.

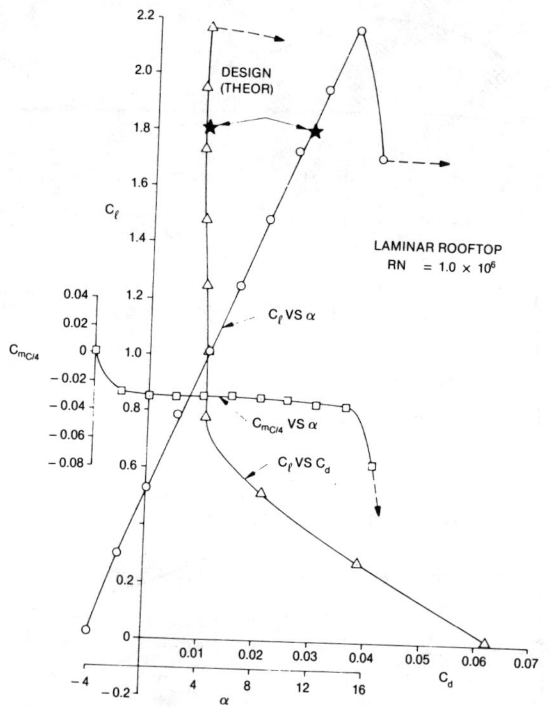

Fig. 10   Experimental performance curves for airfoil L1003.

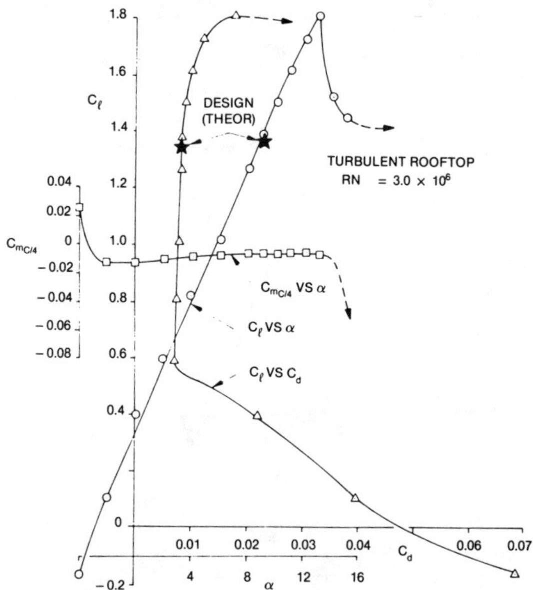

Fig. 11 Experimental performance curves for airfoil L1004.

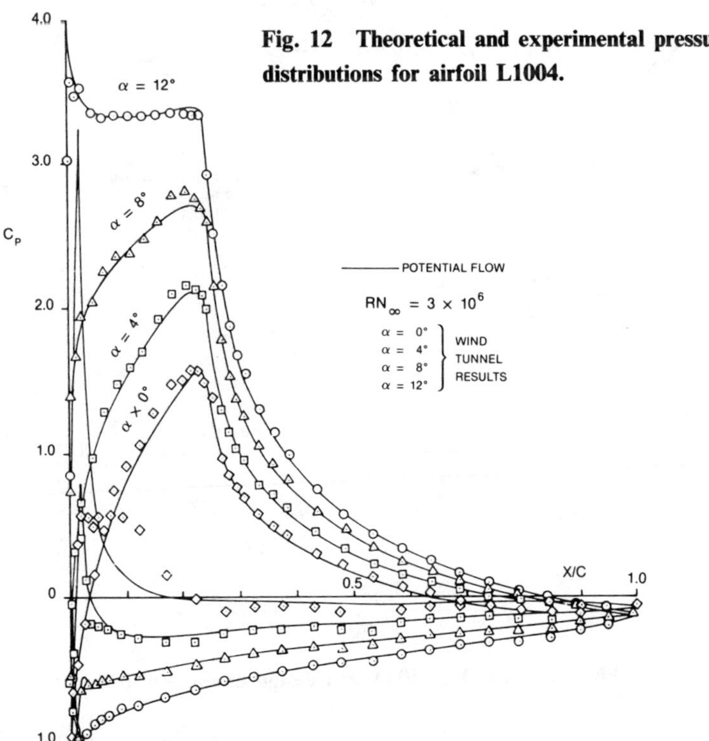

Fig. 12 Theoretical and experimental pressure distributions for airfoil L1004.

R. H. LIEBECK

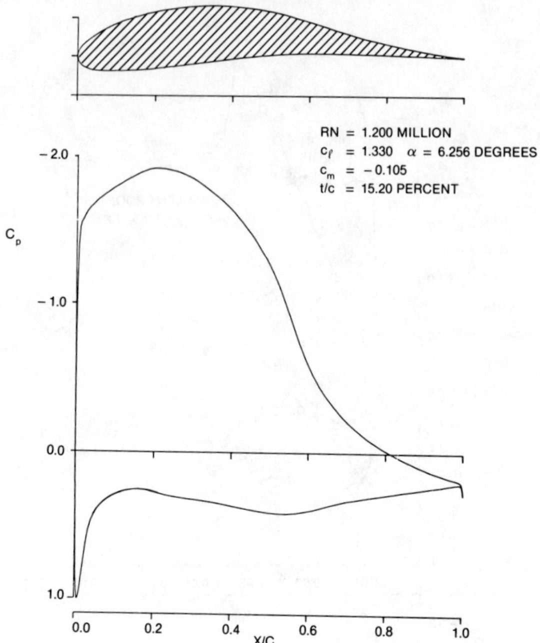

Fig. 13   Airfoil LA5104E and design pressure distribution.

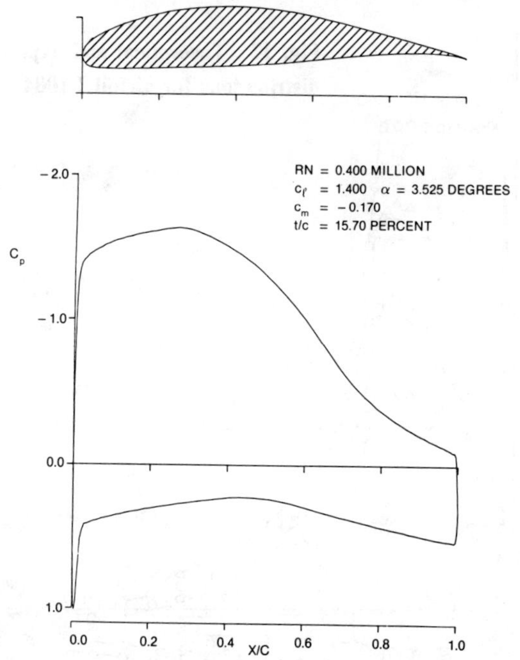

Fig. 14   Airfoil LA203A and design pressure distribution.

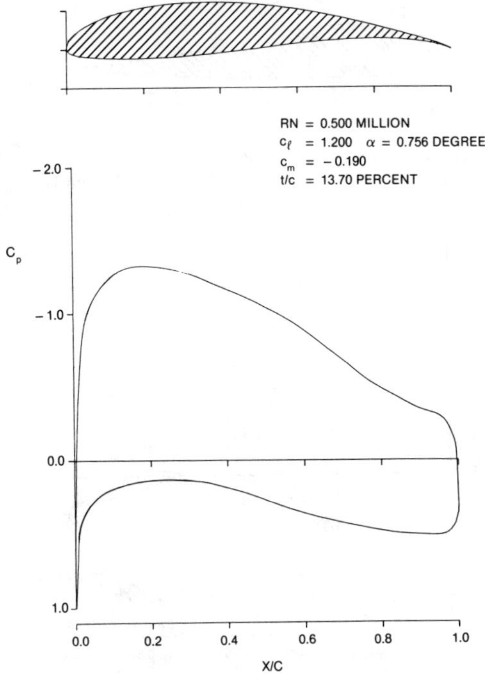

Fig. 15   Wortmann FX63-137 airfoil and pressure distribution.

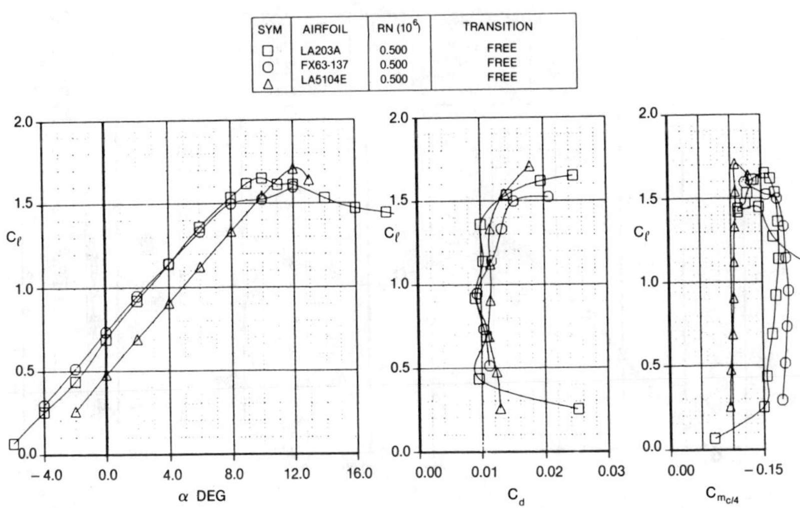

Fig. 16   Comparison of the experimental performance of airfoils LA5104E and LA203A with the Wortmann FX63-137 airfoil.

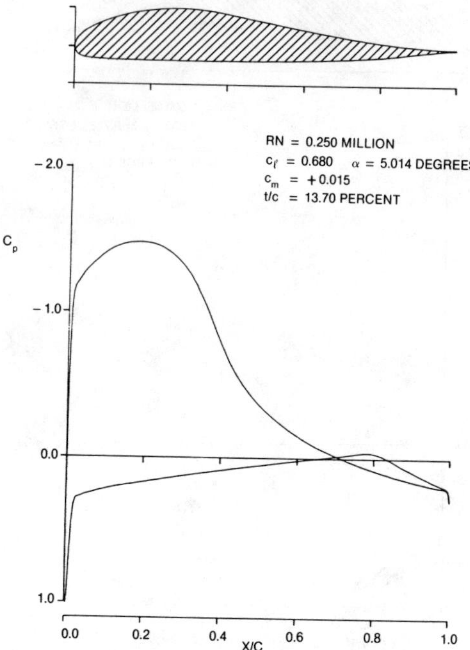

Fig. 17   Airfoil LS2573A and design pressure distribution.

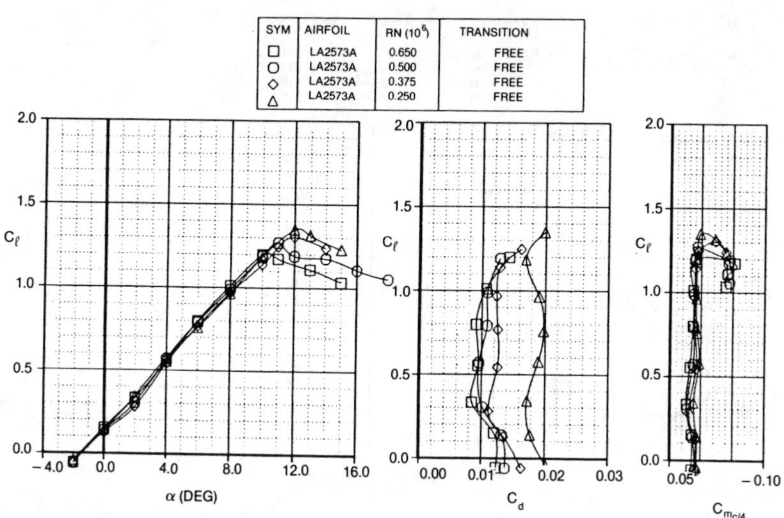

Fig. 18   Experimental performance curves for airfoil LA2573A.

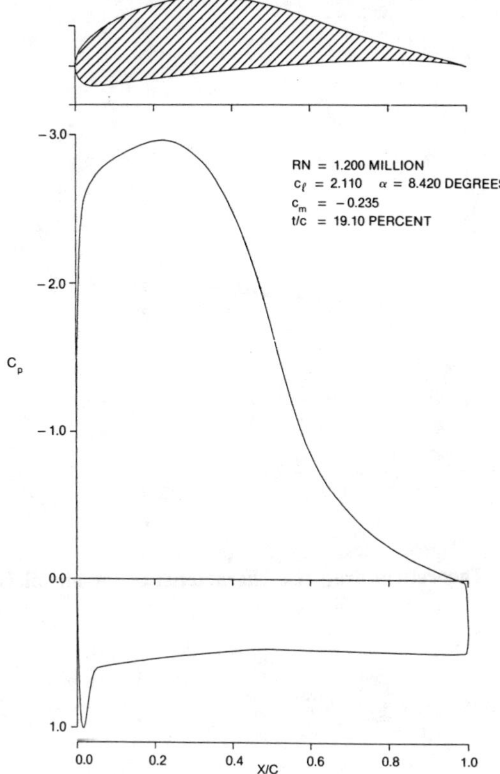

Fig. 19   Airfoil LC111A and design pressure distribution.

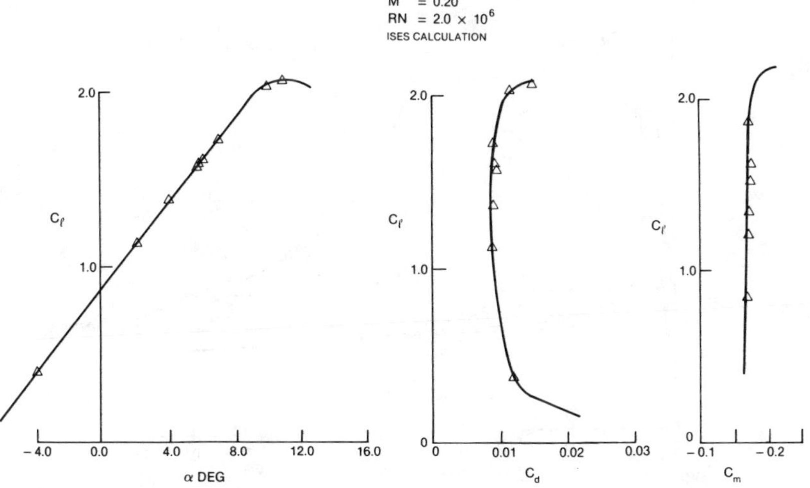

Fig. 20   Theoretical performance curves for airfoil LC111A.

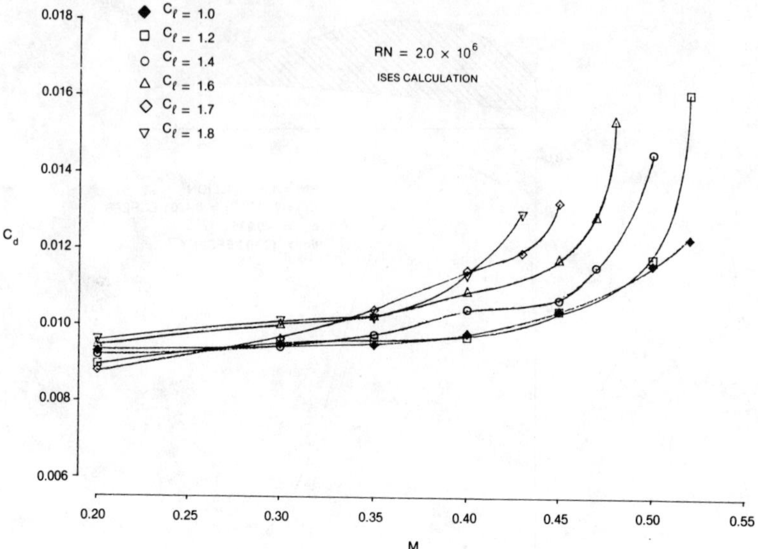

Fig. 21   Theoretical drag rise characteristics for airfoil LC111A.

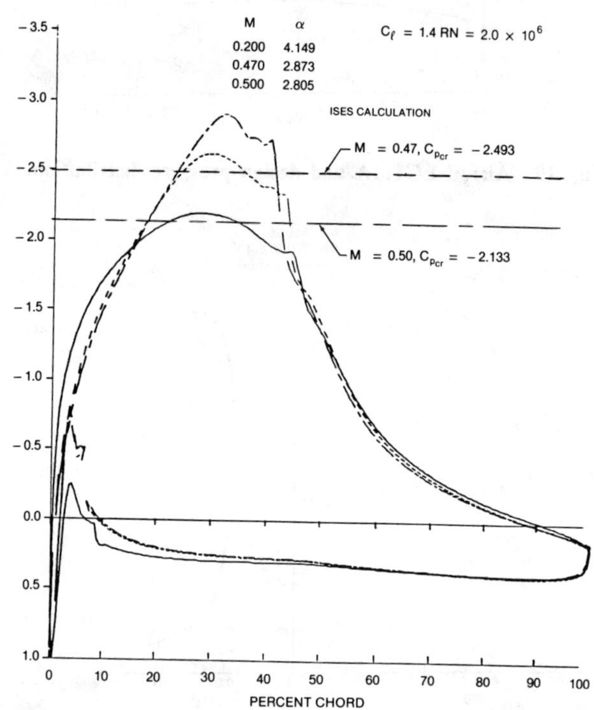

Fig. 22   Theoretical pressure distribution for airfoil LC111A at $C_\ell = 1.4$ for varying Mach number.

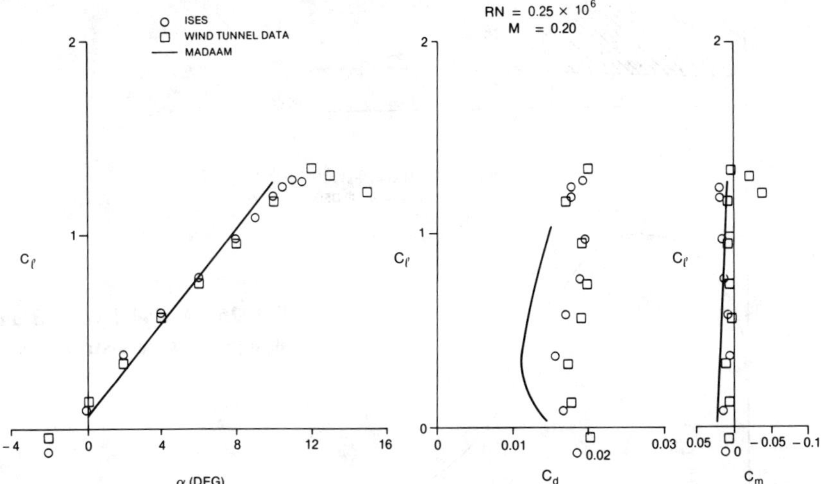

Fig. 23 Comparison of theoretical and experimental drag polars for airfoil LA2573A.

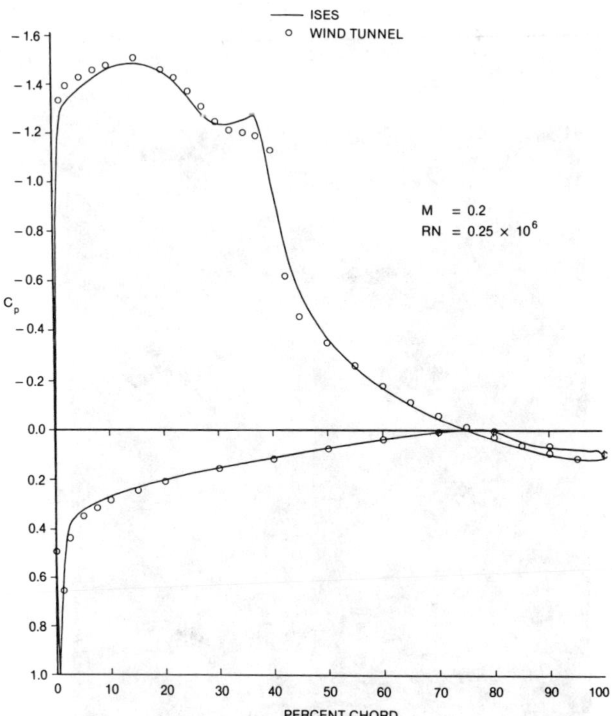

Fig. 24 Comparison of theoretical and experimental pressure distributions for airfoil LA2573A.

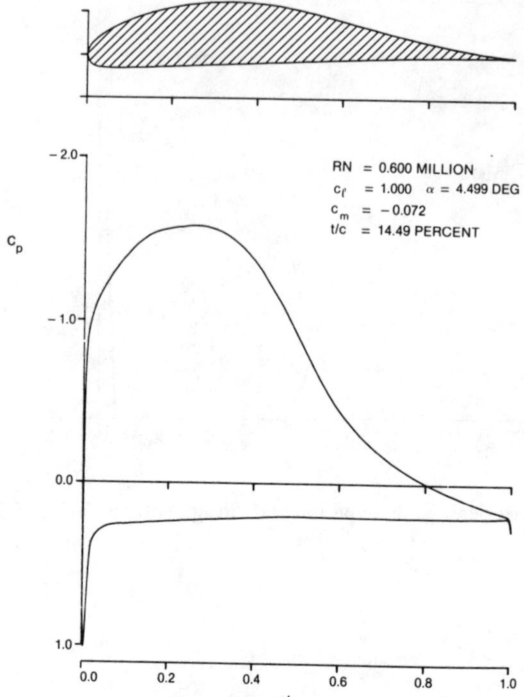

Fig. 25   Airfoil LW101B and design pressure distribution.

Fig. 26   Prototype windmill developed by Industrial Design Laboratories.

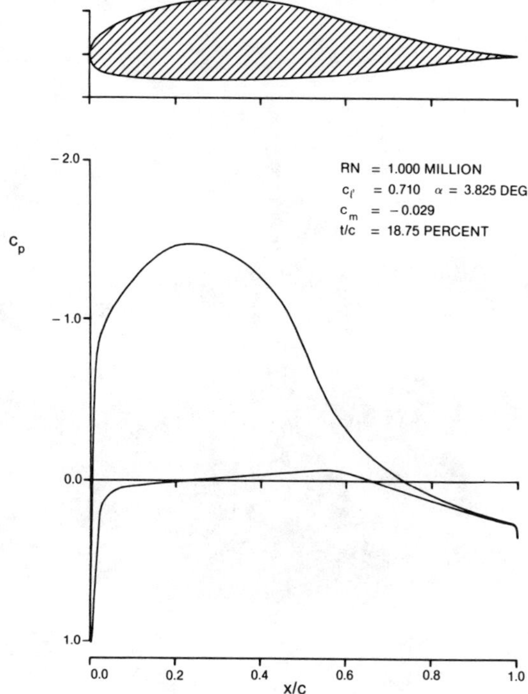

Fig. 27   Airfoil LW108A and design pressure distribution.

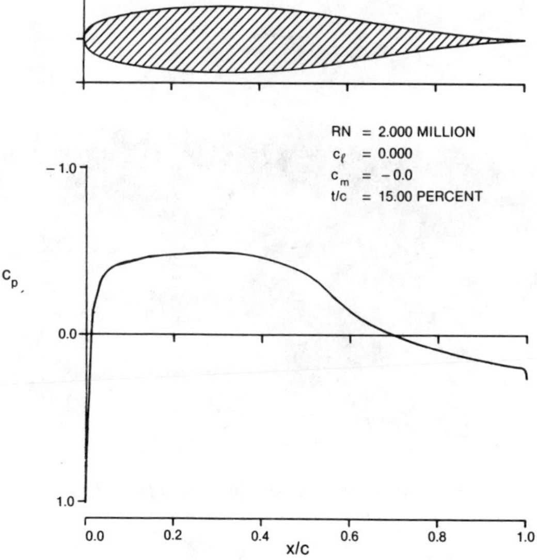

Fig. 28   Airfoil LAB121A15 and design pressure distribution.

Fig. 29   Ratsrepus 360 aerobatic airplane designed by Henry Haigh.

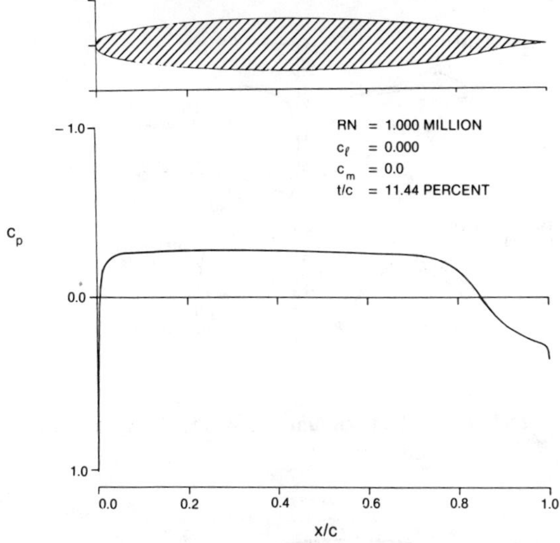

Fig. 30   Airfoil LSB119A and design pressure distribution.

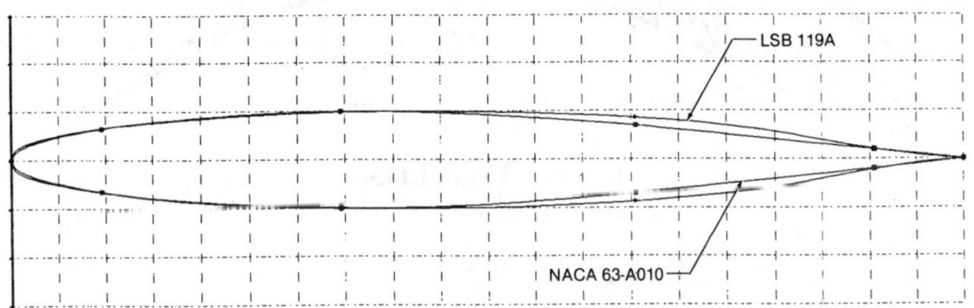

Fig. 31   Geometry comparison of the LSB119A and NACA 63a-010 airfoils.

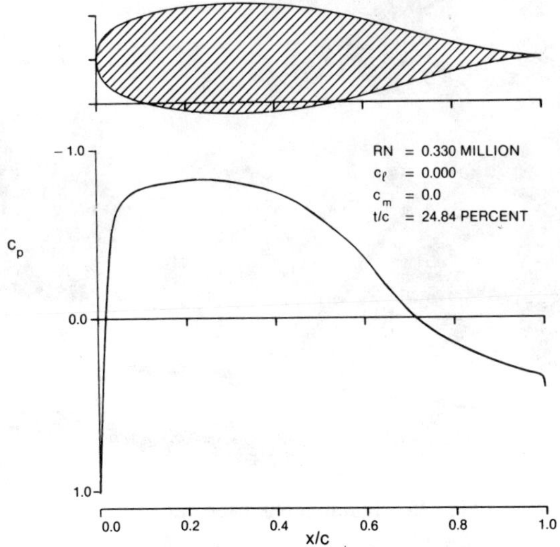

Fig. 32   Airfoil LSB117A and design pressure distribution.

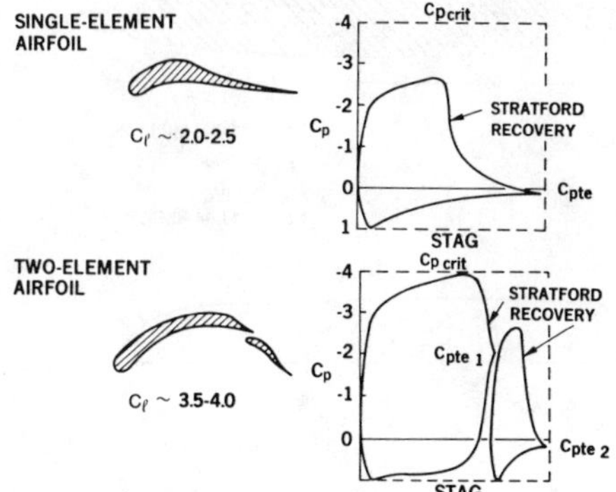

Fig. 33   Definition of maximum lift in the C_p vs x plane.

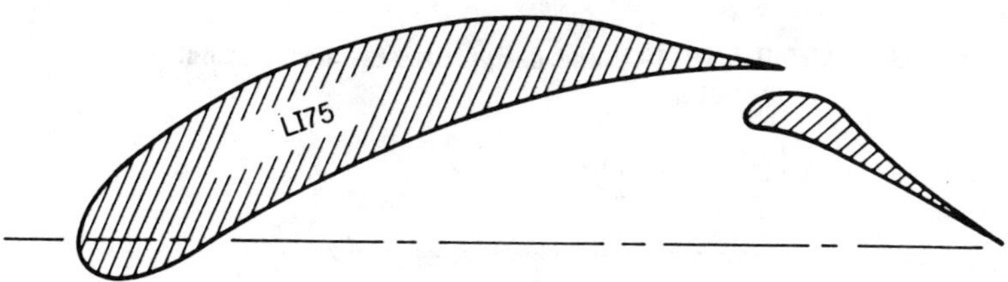

Fig. 34   Airfoil L175.

Fig. 35   Indianapolis race car.

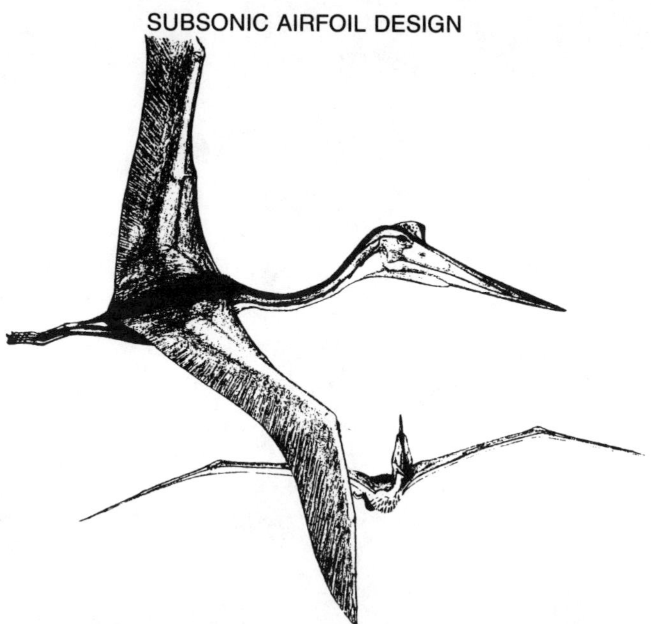

Fig. 36 Rendering of giant pterosaur (Gregory Paul, artist).

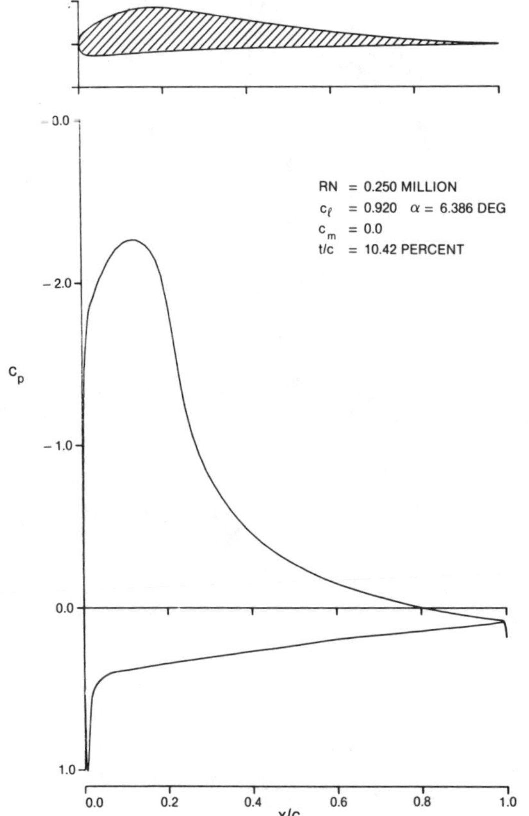

Fig. 37 Airfoil LPT102B and design pressure distribution.

Chapter 6

Elements of Airfoil Design Methodology

Mark Drela*
Massachusetts Institute of Technology, Cambridge, Massachusetts

Introduction

RECENT advances in direct and inverse aerodynamic computational technology have had a strong impact on the overall airfoil design process. An unprecedented level of aerodynamic design can now be performed by numerical methods, with wind-tunnel testing frequently being used only as a final check on the design. Old-fashioned iterative shape guessing and subsequent inviscid analysis (or tunnel testing) has been replaced by efficient inverse algorithms and remarkably accurate viscous analysis methods. Despite these computational advances, however, airfoil design remains to a great extent a cut-and-try procedure, albeit a very sophisticated one. Fully automating the design process has simply not proved to be possible. The fundamental reason is that airfoil design is a multidisciplinary field, involving aerodynamics, structural mechanics, stability and control, and manufacturing and maintenance considerations. Conflicting airfoil requirements that invariably arise between these disciplines must be resolved, but this is an extremely complex task that is very problem-specific and must be done by an experienced designer. The main obstacle to automating such design decisions is that most constraints and performance-limiting flow phenomena become known *during the design process* and cannot be coded or even quantified a priori. This is true even when relatively modest changes are made to an already existing airfoil. Also, many important constraints are not absolute and can usually be mildly violated if sufficient advantages are produced in some other way. The designer in the design loop is essential to identify and resolve requirement conflicts and put the appropriate amount of emphasis on constraints that arise as the design matures.

The amount of detailed design performed on modern airfoils is made possible by viscous numerical simulation codes for performance analysis; using only a wind tunnel in a cut-and-try mode to achieve the same effect is impossible if real cost and time constraints are present. In addition to an analysis code, the designer's numerical "toolbox" must also include an inverse code that permits generation of airfoil geometry from specified surface pressures. This allows the designer to tailor and specify aerody-

---

Copyright © 1989 by the American Institute of Aeronautics and Astronautics, Inc. All rights reserved.
*Carl Richard Soderberg Assistant Professor, Department of Aeronautics and Astronautics.

namic characteristics directly. Since geometric constraints invariably arise in airfoil design, the designer's productivity can be further enhanced by software that permits direct manipulation of airfoil geometry without resorting to an inverse solver.

The scope of this chapter will be limited to two-dimensional airfoil design. It is recognized that, in the context of three-dimensional wing design, two-dimensional airfoil design is an essential starting point. As pointed out by Rubbert and Goldhammer,[1] resolving wing profile design conflicts is far more productive when done at the two-dimensional level. The far greater speed, accuracy, and simplicity of two-dimensional analysis methods give the designer a clearer picture of the tradeoffs between key airfoil design parameters and allows a substantial narrowing of the range of feasible designs for the three-dimensional problem.

This chapter will review the current state of the art in the elements that constitute a two-dimensional computational airfoil design system. It represents a somewhat personal point of view accumulated from airfoil design experience and contact with other designers in the field. Specific examples will be used to illustrate the use of current computational methods to refine airfoil designs and the role these methods play in developing airfoil design philosophies.

Airfoil Design

The central point in airfoil design is the identification and control of important performance characteristics. Some of the more common characteristics addressed in practice include the following:

1) low drag coefficient (C_D) over design lift coefficient (C_L) range;
2) turbulent drag (for laminar airfoils);
3) Mach number sensitivity (transonic drag rise);
4) Reynolds number sensitivity;
5) simple flap and aileron deflection effects; and
6) surface waviness tolerance.

Also important are characteristics that do not directly reflect the airfoil's aerodynamic performance, but strongly contribute to the performance and cost of the entire aircraft in an operational setting. Typical examples include the following:

7) structural merit (e.g., thickness, enclosed area);
8) stability and control merit (e.g., pitching moment, stall severity);
9) high-lift system compatibility; and
10) manufacturing method compatibility.

More often than not, a substantial number of these characteristics will need to be considered, making airfoil design a complex and severely over-constrained problem. This section will discuss a few of the approaches that are effective at dealing with the airfoil design problem and illustrate these with two examples. A brief review of the current viscous analysis and inverse calculation methods suitable for airfoil design work will be given in the following sections.

Design Methodology

On the most basic procedural level, airfoil design consists of simple geometry manipulation. However, user-level procedures with which the geometry is altered are key to the effectiveness of the overall design process. Geometry alteration can take place directly through explicit geometry changes (camber, thickness, trailing-edge angle), or indirectly through specification of the surface pressure distribution. Typical design parameters (geometric and aerodynamic) that can be varied by the designer consist of the following:

1) surface speed (or C_P) distribution;
2) maximum thickness;
3) camber;
4) leading-edge radius;
5) trailing-edge thickness;
6) local surface curvatures;
7) flap hinge position (if any); and
8) boundary-layer trip position (if any).

Naturally, having all of the preceding parameters adjustable is redundant to a large extent. Indeed, the geometry can be indirectly prescribed with complete generality through the surface speed distribution. However, this is a very inconvenient way to enforce a required airfoil thickness or to specify a flap deflection! For such cases, direct geometry control is usually much more effective. Other more general airfoil geometry manipulation is frequently overlooked as a very useful tool in the overall design process. This is especially true for subcritical airfoils where geometric changes have a fairly predictable effect on airfoil characteristics, at least on a qualitative level. The airfoil designer should have at his disposal software that permits at least the following changes in a straightforward and predictable manner:

change of camber line shape with fixed thickness distribution;
change of thickness distribution with fixed camber line;
change of leading-edge radius;
change of blunt trailing-edge thickness;
conventional flap deflection;
explicit coordinate input (via splines); and
smoothing of airfoil contour.

Specified geometric shape changes (of the camber line or actual surface) are easily defined by splining a small number of points input by the user with a screen cursor, as long as the result can be smoothed. For subcritical flow, it is also useful to specify the camber shape change via a specified change in the shape of the loading $\delta(\Delta C_P) \equiv \delta(C_{P\text{lower}} - C_{P\text{upper}})$. The change in the camber line corresponding to any specified change in ΔC_P is easily derived from Glauert's thin airfoil theory.[2] Such camber line manipulation is one of the most effective ways of controlling a subcritical airfoil's pitching moment, especially during preliminary design stages. On the other hand, it is less suitable for transonic airfoils, where each surface is best designed more or less independently.

In addition to the geometric manipulation utilities, an inverse code is also essential for effective airfoil design. Specified surface pressures are best specified in terms of surface arc length on either the as yet unknown airfoil shape, or some sufficiently close "seed" airfoil. The inverse code must make automatic adjustments to the specified surface pressures, since airfoil surface pressures can never be completely arbitrarily prescribed (this will be discussed in detail later). For maximum flexibility in the design process, it is useful to have two types of inverse code available: 1) a full-inverse code, which determines the overall airfoil geometry from the overall surface pressure distribution: and 2) a mixed-inverse code, which determines part of the airfoil contour while holding the rest unchanged. The former is most useful for preliminary airfoil design, whereas the latter is most useful for redesign of existing airfoils.

From this author's point of view, the most effective overall design environment is a toolbox, which includes most if not all of the calculation methods and utilities described earlier. Such a toolbox should allow the designer to modify the airfoil with either geometric and/or aerodynamic specifications in a natural and intuitive manner and allow him to rapidly analyze the resulting performance.

Design Examples

Airfoil design philosophy strongly depends on the particular type of airfoil and application targeted. For example, design principles that lead to efficient sailplane airfoils are totally inappropriate for transonic transport airfoils. For established classes of airfoils, the relevant measures of performance are well known and often are quite specific. For example, Liebeck[3] has defined the requirements and features of single-element airfoils designed for maximum lift, resulting in profiles having exceptional performance. Wortmann[4] presents a design rationale for obtaining low drag at relatively low chord Reynolds numbers (below 1/2 million), which has been applied to airfoil design for human-powered aircraft with structural and manufacturing constraints being considered.[5] In general, the performance goals for any airfoil design problem are clear. The particular design approach taken to achieve these goals is rarely obvious, however, and is often formulated as the design evolves. In fact, discovering and applying effective approaches to achieving performance while eliminating shortcomings are the primary activities that take place in the course of the design of an airfoil. Two modest examples will be given to illustrate this process: a flapped laminar sailplane airfoil and an all-turbulent transonic transport airfoil. The geometry manipulations, inverse calculations, and viscous analysis calculations that will be presented for illustration were performed with the author's XFOIL[6] and ISES[7,8] codes.

Sailplane Airfoil

The two basic requirements of a sailplane flown in competition are to achieve a high speed between thermals while maintaining an acceptably flat glide and to maximize climb rate while circling in a thermal. Also important is a large maximum lift-to-drag ratio (L/D), which increases the pilot's ability to roam in search of thermals.

The high-speed/low-sink requirements cause a sailplane airfoil to be basically a two-point design that requires a very low C_D at low C_L and a reasonably large endurance parameter $C_L^{3/2}/C_D$ at high C_L. Modern composite construction causes structural and manufacturing constraints having a relatively small impact on the airfoil, leaving the designer more or less free to concentrate on aerodynamic performance alone.

Because of the large difference in C_L required between the high-speed and low-sink flight modes, a sailplane airfoil benefits substantially from a flap that effectively increases the width of the low-drag bucket. Figure 1 shows calculated polars for a typical sailplane airfoil with a number of flap settings. The flap hinge is at $x/c = 0.8$ near the bottom surface. In these polars the Reynolds number is automatically scaled as $Re \sim 1/\sqrt{(C_L)}$ as it would be in actual flight with a fixed total wing lift. A striking feature in Fig. 1 is the relatively high drag in the middle of the "bucket" of each polar curve. Examination of the pressure and boundary-layer variable distributions in Fig. 2 for the zero flap deflection reveals very substantial separation bubbles that have a high form drag associated with them. Most of this additional drag can be eliminated by causing transition on each surface to occur at the optimum location nearer the separation point, either by forced turbulation (mechanical or pneumatic)[9,10] or by applying a weak inviscid adverse pressure gradient before and at the bubble.[4,5] The latter method is simpler, and better in the sense that the transition location moves favorably with the optimum location as the angle of attack is varied, whereas a mechanical turbulator works best at a narrow angle of attack only, and pneumatic turbulators are complicated and can be costly to manufacture. Figure 3 shows the C_p distribution and geometry modification made to reduce the bubble drag. Figure 4 shows three of the polars for the modified airfoil compared with the original. The drag in the middle of each bucket is now substantially reduced, especially at higher lift coefficients where the Reynolds number is lower [since $Re \sim 1/\sqrt{(C_L)}$]. No significant disadvantage appears with the modification; thus, in practice it would certainly be retained.

Transonic Transport Airfoil

Compared to the laminar sailplane airfoil, a transonic transport airfoil has a much different and somewhat wider set of requirements and constraints. It must have a large range parameter $M\,C_L/C_D$ at the cruise condition along with adequate performance over a wide range of Reynolds and Mach numbers and is strongly constrained by structural thickness. An intuitive approach to producing good aerodynamic performance in such an airfoil might involve specifying the upper surface C_p distribution with an inverse solver so that the shock wave is quite weak—almost a smooth recompression. This eliminates much of the wave drag and achieves the desired increase in the range parameter $M\,C_L/C_D$. Figure 5 shows an airfoil designed with this rationale at its viscous design point. The suction side shape parameter distribution indicates that the boundary layer is strongly stressed, but has a reasonable margin from separation.

The point design airfoil must be analyzed over a range of Mach numbers and lift coefficients to ensure that the separation margin and the off-design performance are still adequate. Figure 6 shows the calculated pressure distributions at the design Mach number $M = 0.74$ at a sequence of lift coefficients, with the design lift coefficient $C_L \simeq 0.79$ included. Such calculations reveal that a stronger shock wave appears at lift coefficients above and below the design point. This results in the drag advantage of the weakly shocked flow being limited to a quite narrow C_L range about the design point, which might limit the aircraft to an unreasonably narrow C_L range for good cruise economy. Figure 7 shows entire drag polars for a sequence of Mach numbers. Such a calculation clearly reveals the very narrow extent of the weakly shocked design condition. Note that all of the advantage is lost if the Mach number is decreased from 0.74 to 0.73 and that drag divergence occurs at $M = 0.75$ and above.

A more extended Mach sweep may also be necessary to verify that the airfoil is performing satisfactorily at lower Mach numbers and higher lift coefficients, such as might occur during climbout. Figure 8 shows that the drag is somewhat higher than might be expected at $M = 0.65$ for $C_L > 0.8$, which could limit the climb performance of the aircraft. Examination of the C_p distributions in Fig. 9 reveals that the cause is a rather strong shock that forms far forward on the airfoil.

The calculations shown here indicate that the airfoil under consideration has some shortcomings that it may be possible to eliminate. To reduce the excessive drag rise for higher lift coefficients at $M = 0.65$, the forward shock is weakened with an inverse solver, and a new geometry is obtained, as shown in Fig. 10. Polars for this new airfoil, shown in Fig. 11, indicate that the break in the drag polar is delayed to substantially higher lift coefficients at $M = 0.65$. Unfortunately, the break now occurs at lower lift coefficients at the higher $M = 0.74$. Whether the advantage at $M = 0.65$ outweighs the $M = 0.74$ disadvantage depends on the operating envelope of the aircraft for which the airfoil is intended. The new airfoil is likely to be superior in climb performance, which favors a high C_L at lower speeds, but inferior at the cruise condition, which favors a high speed. Clearly, evaluating the merits of the seemingly simple redesign may in fact require a very complicated decision involving the economic factors that drive the design of the aircraft under consideration. The scenario of a solution to one problem causing or aggravating another problem is all too typical, and the airfoil designer's judgment (or that of another expert in the appropriate field) is needed to advance the design.

Role of "Optimal" Solutions

The examples in the preceding subsection serve to illustrate the use of flexible airfoil design/analysis software to investigate and possibly improve airfoil technology. It is clear that the designer in the loop is essential to identify problems as they arise and to try out possible solutions. A stand alone optimization approach is probably unable to deal with the complexity of actual design tradeoffs that occur in practice. Nevertheless, a potential application of optimization techniques may be valuable as an additional tool to be exercised at the discretion of the designer. In the preceding examples, it may be useful to know the characteristics of an "optimal airfoil" defined and constrained in necessarily simple ways. This airfoil could then serve as a *guide* to designing an airfoil with more realistic, perhaps unquantifiable constraints. For example, one could in principle determine the sailplane airfoil or transonic transport airfoil with the lowest possible drag under the constraints of a specified C_L and a specified maximum thickness or enclosed area. This would indicate to the designer what features produce low drag and perhaps save much trial and error. Such a tool would naturally be quite computationally expensive, since a substantial number of geometric degrees of freedom are required to describe an airfoil with adequate generality, as discussed by Vanderplaats.[11] To this author's knowledge, such restrained use of optimization techniques has not yet been employed in "production" airfoil design work and may be a fruitful area for investigation. Whether it will truly be accepted as another effective design tool remains to be seen.

Viscous Analysis Methods

As shown in the previous examples, a viscous analysis method is crucial in modern airfoil design techniques, being relied on as the main discriminator between candidate designs. Wind-tunnel simulation is used primarily as a final check on the finished design and to predict performance in regimes where the accuracy of the analysis method may be suspect, such as past stall.

Method Classification

All current viscous analysis methods can be grouped in one of three categories: 1) Navier-Stokes methods, 2) interacted zonal viscous/inviscid methods, and 3) noninteracted inviscid plus boundary-layer methods. The latter methods, such as that of Eppler and Somers,[12] are very fast, but do not account for the very important viscous displacement effect. This plays a key role in the determination of form drag and is also essential to the prediction of lift in the presence of separation. The noninteracted methods are therefore unsuitable for transonic, low-Reynolds-number, or near-stall airfoil flows. The prediction of these flows requires either Navier-Stokes or interacted zonal methods, both of which accurately represent the physics of the viscous displacement mechanism.

In interacted zonal methods, the displacement effect on the outer inviscid flow is approximated by the displacement body or wall transpiration model. Either model gives the correct displacement effect to first order in the limit of a thin viscous layer,[13] and both models are normally quite accurate in practice. In contrast, the Navier-Stokes methods model the displacement effect exactly no matter how thick the viscous layer may be. Unfortunately, current Navier-Stokes methods[14-16] require between two and three orders of magnitude more computational effort than zonal methods and hence are simply too slow to be considered as practical design tools. The Mach-sensitivity polars presented for the transonic airfoil example shown earlier altogether consisted of about 100 individual point solutions and required a total of 20 h of CPU time with the ISES code on a MicroVAX II (0.1 Mflop) machine. On a 100-Mflop supercomputer, only a few minutes would be required for the zonal ISES code, whereas a Navier-Stokes solver would consume many hours or even days of CPU time for the same task. Furthermore, many airfoil design problems require an even greater computational effort than that required for the Mach number sweep. If the Reynolds number sensitivity and flap deflections were also to be investigated, the CPU requirement could easily increase tenfold. This would still allow the zonal method to perform the task on the supercomputer, whereas using the Navier-Stokes solver would be totally unfeasible.

The Viscous Transonic Airfoil Workshop organized by Holst[17] has served the very useful purpose of giving direct comparisons between most of the currently available airfoil analysis methods. A significant finding was that, despite the speed difference of two or three orders of magnitude, the accuracy of the best Navier-Stokes methods was no better than the accuracy of the best zonal methods. This difference in accuracy is perhaps not too surprising, given that the common factor that limits accuracy in any viscous aerodynamic solver is the turbulence model (usually in the form of an empirical eddy viscosity model or integral closure relation). The approximate model for the displacement mechanism and the additional assumption of negligible normal pressure gradients in the boundary-layer formulations of zonal methods have less effect than any uncertainty of the turbulence model, even in flows with shocks or substantial separation regions. Almost invariably, whenever the displacement model or zero normal pressure gradient assumptions become suspect, so does the turbulence model.

Because of their relatively low cost and no clear accuracy disadvantage, the zonal approaches will continue to represent the method of choice for airfoil analysis computations. Only when very substantial improvements in turbulence models appear and the computation time of Navier-Stokes methods drops by two orders of magnitude will they become attractive for airfoil design work.

One possible application where Navier-Stokes methods may be required is the design of high-lift multielement airfoil systems. These configurations experience substantial pressure gradients across the viscous layers, which cast doubt on the thin shear-layer assumptions inherent in zonal methods. Also, zonal methods require that each viscous layer have a distinct identity, and this becomes uncertain if boundary layer/wake merging occurs—not an uncommon occurrence. However, the currently best turbulence models used in Navier-Stokes methods, such as the Johnson-King model,[18] also require the identification of shear layers; hence, this may be a moot point. These issues are the topics of current research, and the outcome is not yet clear.

Current Zonal Methods

Numerous zonal methods have been developed to date to solve the viscous airfoil analysis problem. Lock and Williams[19] give a very extensive review of the basic theory behind these methods and give limited comparisons of accuracy between several current methods. For brevity, only the codes used most commonly in present industrial and research applications will be described here.

The GBK code[20] solves the nonconservative potential equation with a method using complex characteristics and a fast Poisson solver. The Nash-MacDonald[21] boundary-layer method is solved in the turbulent regions with the inviscid solution edge velocity as input. The resulting displacement thickness is then added to the airfoil contour to obtain a new effective airfoil shape, and the solution process is repeated. Unless the displacement thickness is heavily smoothed every iteration, this simple technique is quite unstable as shown by Wigton and Holt.[22] It also invariably must fail if any separation occurs and hence is unreliable near maximum lift. Nevertheless, the overall code is quite fast and has been employed in the industry to a great extent. Also, recent improvements have been incorporated, resulting in the VGK code.[23] The original Nash-Mac-Donald boundary-layer method has been replaced by a modified Green's Lag-Entrainment method,[24] and the direct coupling method has been replaced by the semi-inverse coupling procedure of Carter,[25] which is reliable in mildly separating flows.

The GRUMFOIL code of Melnik et al.[26,27] solves the full-potential equation in the interior of the unit circle that has been mapped from the infinite exterior of the airfoil. A modified Green's Lag-Entrainment boundary-layer method[24] is used as the viscous model, with the wall-transpiration model representing its influence on the potential flow. Higher-order corrections are also included to model the effects of wake curvature.[27] The semi-inverse coupling procedure of Carter[25] is employed to iterate between the inviscid and viscous solutions. The overall method appears comparable in accuracy to the VGK code in predicting general airfoil flows, as indicated by the comparisons of Lock and Williams.[19]

One of the most recent developments is the ISES code of Drela and Giles,[7,8] which solves the conservative steady Euler equations on an intrinsic streamline grid. A lagged-dissipation integral method represents the laminar and turbulent boundary layers and wake and uses an e^n-type formulation[28-30] to determine the transition points. The overall model permits accurate simulation of low-Reynolds-number airfoil flows dominated by transitional separation bubbles. Rather than iterating between the viscous and inviscid solvers, all of the discrete equations (Euler equations, integral boundary-layer equations, transition location equations, viscous/inviscid matching conditions) are solved simultaneously by a global Newton method. This solution technique is very fast and reliable, even in substantially separated flows. About 20–30 min of CPU time is required for a grid-independent viscous transonic solution on a MicroVAX II (0.1 Mflop) machine, and about 10 min is needed per point in a polar sweep. The polar sweep is faster due to the quadratic convergence of the Newton method from one point to the next. The ISES code has proven to be a reliable prediction method for transonic and low-Reynolds-number airfoil flows, accurately predicting transonic drag rise[31] and separation bubble behavior down to a Reynolds number of about 200,000.[32]

Inverse Methods

Current inverse methods are based on a wide range of basic formulations. The simplest methods are those based on a small-disturbance assumption. A prime example is of course Glauert's thin airfoil theory.[2] An inverse small-disturbance method valid for transonic flows is presented by Steger and Klineberg.[33] Although these methods can be useful for giving the designer a feel for the effects of geometry on surface pressures and vice versa, they cannot deal adequately with blunt leading edges—a region that has a very strong influence on airfoil behavior in real viscous flows. For this reason, the small-disturbance methods by themselves are poorly suited for airfoil design work, and general-body algorithms are needed.

All of the general two-dimensional inverse formulations developed to date can be loosely grouped into either the full-inverse or the mixed-inverse categories (Fig. 12). In a full-inverse method, the entire airfoil geometry is calculated from a speed $q(s/s_{\max})$ (or, equivalently, pressure) distribution specified over the entire airfoil surface. The entire speed distribution influences the entire geometry and vice versa. In practice, this approach is most useful for preliminary design of a new airfoil. A mixed-inverse method allows the geometry to be specified over a part of the airfoil and the surface speed to be specified over the rest. This is most useful for local modifications to an existing airfoil with desirable geometric features elsewhere which are to be preserved. For example, tuning an airfoil to maximize performance is best done by dealing with one surface at a time while "freezing" the other surface, especially in the case of a transonic airfoil. A mixed-inverse method is ideal for such an application. For greatest flexibility, however, the airfoil designer has access to both full- and mixed-inverse methods.

An alternative to either the full- or mixed-inverse method is the optimization technique,[11] where the airfoil geometry is repeatedly perturbed by preselected mode functions until its pressure distribution matches the specified distribution sufficiently closely (or some other objective is achieved). A standard direct calculation is performed to determine the sensitivity of the surface pressure distribution to each perturbation. Although this has the advantage in that it dispenses with an inverse formulation altogether, it does so at a large computational cost. A good inverse method will be one to two orders of magnitude faster than an equivalent direct/optimization method. Primarily for this reason, direct/optimization methods have not been used in the industry in lieu of true inverse methods.

Full-Inverse Methods

An effective full-inverse approach for incompressible flows is the complex mapping method originally formulated by Mangler[34] and Lighthill[35] and implemented in various forms by several workers, notably Arlinger[36] and Eppler and Somers.[12] A complex mapping function $z(\zeta)$ whose derivative has the form

$$\frac{\mathrm{d}z}{\mathrm{d}\zeta} = \left(1 - \frac{1}{\zeta}\right)^{(1-\epsilon)} \exp\left\{\sum_{n=0}^{\infty} C_n \zeta^{-n}\right\} \tag{1}$$

transform the unit circle in the ζ plane to an airfoil in the z plane. With $\pi\epsilon$ being the trailing-edge angle, the complex constants C_n uniquely define the airfoil shape and hence the surface speed. It must be noted that many specific forms of the mapping function are possible, the present form being chosen for illustration purposes. On the surface of the airfoil, where $\zeta = e^{i\omega}$, the conjugate velocity $u - iv$ is given by

$$u - iv = ie^{-i\frac{\omega}{2}(1+\epsilon)}(e^{i\frac{\omega}{2}} - e^{-i\frac{\omega}{2}})(e^{i(\frac{\omega}{2}-\alpha)} + e^{-i(\frac{\omega}{2}-\alpha)}) \exp\left\{\sum - C_n\zeta^{-n}\right\} \quad (2)$$

with α the angle of the freestream flow. Specifying the surface speed $q = |u - iv|$ on the airfoil then uniquely determines the coefficients C_n, which can be calculated by Fourier-analyzing the logarithm of Eq. (2) if the infinite series is truncated and ϵ is specified. The corresponding airfoil shape can then be calculated from Eq. (1) by straightforward numerical integration.

A very important result of Lighthill's theory is that the specified surface speed distribution $q(\omega)$ cannot be arbitrarily prescribed, but must satisfy

$$\int_0^{2\pi} \ell n \frac{q(\omega)}{q_\infty} \left\{ \begin{matrix} 1 \\ \cos\omega \\ \sin\omega \end{matrix} \right\} d\omega = 0 \quad (3)$$

where ω is the circle plane angle. These three integral constraints are equivalent to

$$\mathrm{Re}(C_0) = 0, \qquad \mathrm{Re}(C_1) = 1, \qquad \mathrm{Im}(C_1) = 0 \quad (4)$$

which require that the flowfield far from the airfoil be unstretched by the mapping (freestream speed be unchanged) and that the airfoil be closed. The imaginary part of C_0 simply rotates the entire airfoil and flowfield and hence is quite arbitrary. In practice, once all of the C_n coefficients are calculated from a specified input speed distribution $q(\omega)$, the C_0 and C_1 coefficients must be reset to their required values [Eq. (4)]. Naturally, this modification will produce a speed distribution different from that specified, but this is inevitable in any well-posed inverse method. If only small changes are made to an existing distribution that already satisfies the Lighthill constraints [Eq. (3)], then the difference between the specified and resulting speed distributions will be small in practice.

Tranen[37] has applied the gist of the circle-mapping formulation to transonic flows. The compressible nonconservative potential equation is mapped into the interior of the unit circle and solved there using a finite-difference method. Dirichlet airfoil surface boundary conditions obtained from an integration of a specified surface speed distribution on an approximate airfoil are imposed:

$$\Phi = \int q(s)\, \mathrm{d}s + C \quad (5)$$

The resulting mass flux through the approximate airfoil surface is used to calculate a new updated contour. The mapping is recalculated and the overall process repeated until the airfoil contour does not change significantly from one iteration to the next. Tranen[37] adjusts the otherwise arbitrary constant of integration in Eq. (5) to obtain trailing-edge closure. However, as pointed out by Volpe and Melnik,[38] this formulation leads to a formally ill-posed Dirichlet problem that cannot converge in the general case. Specifically, it does not address Lighthill's first constraint in Eq. (3), which does not allow the speed on the airfoil and the freestream speed to be both arbitrarily prescribed. The formulation of Volpe and Melnik, which is similar to that of Tranen, does eliminate this inconsistency by addressing all three constraints. Three shape functions weighted by three free parameters are added to the prescribed speed distribution, and the parameters are adjusted automatically during the calculation to enforce closure and consistency with the freestream speed. The method is shown to converge to an arbitrarily small tolerance, which indicates that it is well-posed.

Another notable formulation for the compressible full-inverse problem is that of Daripa and Sirovich.[39] The full-potential equation is transformed into the potential stream function plane, with the flow angle θ and Prandtl-Meyer function $v \equiv \int \sqrt{(1 - M^2)}/q \, dq$ chosen as the dependent variables. This is then further transformed into the interior of the unit circle, where it is solved using boundary conditions on v, which is uniquely related to the prescribed suface speed q. The method is limited to subsonic flows.

A distinct class of full-inverse methods is that based on the hodograph transformation[20,40] with a solution scheme for the potential and stream-function based on integration along complex characteristics. The airfoil geometry is extracted from this solution as a postprocessing step. Although efficient computationally, these methods can be difficult to use, as the required inputs are not directly related to physical speed and arc length on the airfoil. The method of Bauer et al.,[20] however, does accept speed distributions in terms of arc length and has been used as a design tool in industry. One of its reported disadvantages, however, is that the designer does not have firm control over the freestream Mach number, which is adjusted by the solver to satisfy certain consistency requirements in the formulation. For transonic airfoils, whose performance always critically depends on the freestream Mach number, this is a very undesirable feature.

Mixed-Inverse Methods

Most mixed-inverse methods that are documented in the literature formulate the problem in physical space rather than some mapped space such as the unit circle plane. An exception is the method of Woods,[41] who solves the mixed-inverse problem in a plane analytically mapped from the physical domain. As in Lighthill's theory, three integral constraints on the prescribed speed distribution arise. One constrains the potential on the airfoil to be consistent with the freestream potential, and the other two enforce geometric regularity at the points joining the specified and "free" segments of the airfoil. The method uses the Karman-Tsien tangent-gas approximation to treat subcritical compressible flows.

Carlson[42] solves the transonic mixed-inverse problem on a Cartesian mesh, with the geometry near the leading edge being prescribed. The leading-edge radius is adjusted to obtain trailing-edge closure. As pointed out by Volpe and Melnik,[38] this scheme has the same flaw as Tranen's method; namely, it does not address Lighthill's first constraint and hence cannot converge to the specified speed distribution within an arbitrarily small tolerance. Also, enforcing trailing-edge closure by varying the leading-edge radius is poorly suited to actual airfoil design practice, where firm control of the leading-edge radius is often necessary.

The ISES code,[8,43] described earlier as a viscous analysis method, also solves the mixed-inverse problem. The steady Euler equations used to represent the inviscid flow are discretized on a streamline-based finite-volume grid that evolves during the solution. A standard analysis problem results when each streamline on each airfoil side is fixed (or offset by the displacement thickness in viscous cases), and the surface pressures are calculated as a result. A mixed-inverse problem results if the surface pressures are specified over the inverse part of the airfoil as shown in Fig. 12, and the local surface streamline position (and hence the airfoil shape) is calculated as a result. The flowfield interior and far-field boundary conditons remain identical to those in the analysis case. To allow for integral constraints on the surface pressure distribution, it is expressed in the form

$$p(\sigma) = p_{\text{spec}}(\sigma) + A_1 f_1(\sigma) + A_2 f_2(\sigma) \tag{6}$$

where σ is the fractional surface arc length over the inverse segment, f_1 and f_2 are specified shape functions, and A_1 and A_2 are free parameters determined as part of the solution. The conditions of geometric continuity (or regularity) at the segment endpoints serve to implicitly determine A_1 and A_2. As a consequence, the final surface pressure $p(\sigma)$ will not in general match the specified $p_{\text{spec}}(\sigma)$ as expected in any inverse method. Curvature continuity at the segment endpoints can also be ensured by adding two additional degrees of freedom in the specified pressure expression [Eq. (6)]. These are then determined by enforcing a smoothness constraint on the geometry or the surface pressure. Of course, the more constraints that are imposed on the inverse problem, the greater the discrepancy will be between the specified and final surface pressures. This is a feature of all inverse methods and must be accepted.

The ISES code has met with wide acceptance in the industry, partly due to its speed and perfect compatibility with the code's viscous analysis mode. The transonic transport airfoil design example presented earlier demonstrated the code's mixed-inverse mode, with viscous effects fully included. A design calculation is guaranteed to be an exact solution to an analysis problem, which avoids extra analysis runs and dispenses with the need to redesign specifically to correct for viscous displacement effects.

Some mention must be made of the fictitious gas method of Sobieczky and Seebass,[44] which is aimed specifically at redesigning existing airfoils to achieve a transonic shock-free condition. Both two- and three-dimensional versions have been developed. Nevertheless, this method has not experienced wide use, possibly because airfoils that are shock-free at their design point are somewhat conservative and may have poor off-design performance, as shown in the transonic airfoil example. Hence, it is desirable to specify at least a weak shock at the design point of a transonic airfoil. Methods such as that of Volpe and Melnik,[38] and the ISES code, permit shock waves to be specified in the surface speed distribution.

Integrated Analysis/Design System

For efficient use of the computational methodologies reviewed in this chapter, they must be accessed through a reasonably user-friendly interface system. All tedious input data set generation should be automated and simplified to the point where the designer is not distracted from the design task. Since many design decisions are made on the basis of perceived trends (rate of drag rise with Mach number, sharpness of stall, etc.), output in graphic form should be used whenever possible. Sensing trends by viewing columns of numbers is simply not effective.

Naturally, the "ideal" design system is largely a matter of personal choice. This author's opinion of what constitutes an effective system is embodied in the XFOIL code,[6] whose logical organization is shown schematically in Fig. 13. This code was used in the laminar sailplane airfoil design example presented earlier, with the figures being basically screen dumps of the plots seen by the designer. It incorporates all its design and analysis elements in one unified menu-driven program. A common internal data representation is used by all facilities, eliminating the necessity of input data set conversion for communication between the various modules. Output is performed almost exclusively via graphics, and surface speed distributions for the mixed- and full-inverse solvers are specified in terms of surface arc length via a screen graphics cursor. This combination of features gives a code a good "What if ...?" capability, which allows the designer to quickly and easily make modifications with a wealth of input choices and to evaluate the results.

The XFOIL code's analysis solver is based on a panel method incorporating a Kármán-Tsien compressibility correction. The viscous layer formulation is identical to that of the ISES code, and a global Newton method is used to solve the overall viscous/inviscid equation set. This formulation is quite fast, permitting a high-resolution viscous point solution to be interactively calculated in at most a few minutes on a MicroVAX II machine. The ISES code is too slow for this type of execution on the MicroVAX and hence is configured for easy batch execution. It is supported by separate, interactive inverse input and plotting output programs, all of which reference a common disk solution save file. With a 100-Mflop supercomputer, or even a 10-Mflop minisupercomputer, a unified organization of ISES similar to that of XFOIL would be perfectly feasible.

Conclusions

This chapter has discussed the computational elements required for effective airfoil design and analysis systems. Design examples of two very different airfoils were presented to convey the complicated nature of the decisions that must be made in the course of the airfoil design process. The examples make it clear that airfoil design requires a fast viscous analysis method.

A brief review has also been presented of the currently most popular zonal viscous analysis methods. It is concluded that Navier-Stokes methods are at present unsuitable for routine airfoil design work due to their extremely demanding computational requirements. Current inverse methods and their applicability to airfoil design have also been discussed.

Finally, the author's personal view of what constitutes an effective overall airfoil design system has been presented. The importance of giving the designer flexibility in airfoil modification and employing graphics-based input and output has been stressed.

Acknowledgment

Support for the preparation of this manuscript has been given by the MIT Dean of Engineering Office through the Carl Richard Soderberg Faculty Development Chair.

References

[1]Rubbert, P., and Goldhammer, M., "CFD in Design: An Airframe Perspective," AIAA Paper 89-0092, Jan. 1989.

[2]Glauert, H., *Elements of Airfoil and Airscrew Theory*, Cambridge Univ. Press, Cambridge, UK, 1937.

[3]Liebeck, R. H., "A Class of Airfoils Designed for High Lift in Incompressible Flow," *Journal of Aircraft*, Vol. 10, Oct. 1973, pp. 610–617.

[4]Wortmann, F. X., "Aerofoil Design for Man Powered Aircraft," *Second International Symposium on the Technology and Science of Low Speed and Motorless Flight*, Massachussetts Inst. of Technology, Cambridge, MA, 1974.

[5]Drela, M., "Low-Reynolds Number Airfoil Design for the MIT Daedalus Prototype: A Case Study," *Journal of Aircraft*, Vol. 25, Aug. 1988, pp. 724–732.

[6]Drela, M., "XFOIL: An Analysis and Design System for Low Reynolds Number Airfoils," *Conference on Low Reynolds Number Aerodynamics*, Univ. of Notre Dame, Notre Dame, IN, June 1989.

[7]Drela, M., and Giles, M. B., "Viscous-Inviscid Analysis of Transonic and Low Reynolds Number Airfoils," *AIAA Journal*, Vol. 25, Oct. 1987, pp. 1347–1355.

[8]Giles, M. B., and Drela, M., "Two-Dimensional Transonic Aerodynamic Design Method," *AIAA Journal*, Vol. 25, Sept. 1987, pp. 1199–1206.

[9]Pfenninger, W., "Untersuchen über Reibungsverminderungen an Tragflügeln insbesondere mit Hilfe von Grenzschichtabsaugung," Institut für Aerodynamik ETH, Zürich Mitteilung, Rept. 13, 1946.

[10]Horstmann, K. H., and Quast, A., "Reduction of drag by Means of Pneumatic Turbulators," German Aerospace Research Establishment, Göttingen, FRG, Rept. DFVLR-FB-81-33, 1982.

[11]Vanderplaats, G. N., "An Efficient Algorithm for Numerical Airfoil Optimization," AIAA Paper 79-0079, Jan. 1979.

[12]Eppler, R., and Somers, D. M., "A Computer Program for the Design and Analysis of Low-Speed Airfoils," NASA TM-80210, Aug. 1980.

[13]Lighthill, M. J., "On Displacement Thickness," *Journal of Fluid Mechanics*, Vol. 4, 1958, pp. 383–392.

[14]Maksymiuk, C. M., and Pulliam, T. H., "Viscous Transonic Airfoil Workshop Results Using ARC-2D," AIAA Paper 87-0415, Jan. 1987.

[15]Coakley, T. J., "Numerical Simulation of Viscous Transonic Airfoil Flows," AIAA Paper 87-0416, Jan. 1987.

[16]Matsushima, K., Obayashi, S., and Fujii, K., "Navier-Stokes Computations of Transonic Flow Using the LU-ADI Method," AIAA Paper 87-0421, Jan. 1987.

[17]Holst, T. L., "Viscous Transonic Airfoil Workshop Compendium of Results," *Journal of Aircraft*, Vol. 25, Dec. 1988, pp. 1073–1087.

[18]Johnson, D. A., and King L. S., "A Mathematically Simple Turbulence Closure Model for Attached and Separated Turbulent Boundary Layers," *AIAA Journal*, Vol. 23, Nov. 1985, pp. 1684–1693.

[19]Lock, R. C., and Williams, B. R., "Viscous-Inviscid Interactions in External Aerodynamics," *Progress in Aerospace Sciences*, Vol. 24, 1987, pp. 51–171.

[20]Bauer, F., Garabedian, P., Korn, D., and Jameson, A., "Supercritical Wing Sections I, II, III," *Lecture Notes in Economics and Mathematical Systems*, Springer-Verlag, New York, 1972, 1975, 1977.

[21]Nash, J. F., and MacDonald A. G. J., "The Calculation of Momentum Thickness in a Turbulent Boundary Layer at Mach Numbers Up to Unity," Aeronautical Research Council, London, R & M Rept. 963, 1987.

[22]Wigton, L. B., and Holt, M., "Viscous-Inviscid Interaction in Transonic Flow," AIAA Paper 81-1003, 1981.

[23]Ashill, P. R., Wood, R. F., and Weeks, D. J., "A Semi-Inverse Version of the Viscous Garabedian and Korn Method (VGK)," Royal Aircraft Establishment, Farnborough, Hampshire, UK, TR-87002, 1987.

[24]Green, J. E., Weeks, D. J., and Brooman, J. W. F., "Prediction of Turbulent Boundary Layers and Wakes in Compressible Flow by a Lag-Entrainment Method," Aeronautical Research Council, London, R & M Rept. 3791, 1977.

[25]Carter, J. E., "A New Boundary Layer Inviscid Iteration Technique for Separated Flow," AIAA Paper 79-1450, July 1979.

[26]Melnik, R. E., Chow, R. R., and Mead, H. R., "Theory of Viscous Transonic Flow Over Airfoils at High Reynolds Number," AIAA Paper 77-680, June 1977.

[27]Melnik, R. E., "Turbulent Interactions on Airfoils at Transonic Speeds—Recent Developments," "Conference on Computation of Viscous-Inviscid Interactions," AGARD CP-291, 1980.

[28]Smith, A. M. O., and Gamberoni, N., "Transition, Pressure Gradient, and Stability Theory," Douglas Aircraft Co., Long Beach, CA, Rept. ES 26388, 1956.

[29]Van Ingen, J. L., "A Suggested Semi-Empirical Method for the Calculation of the Boundary Layer Transition Region," Dept. of Aerospace Engineering, Delft Univ. of Technology, The Netherlands, Rept. VTH-74, 1956.

[30]Gleyzes, C., Cousteix, J., and Bonnet, J. L., "Theoretical and Experimental Study of Low Reynolds Number Transitional Separation Bubbles," *Conference on Low Reynolds Number Airfoil Aerodynamics*, Univ. of Notre Dame, Notre Dame, IN, 1985.

[31]Liebeck, R. H., "Low Reynolds Number Airfoil Design for Subsonic Compressible Flow," *Conference on Low Reynolds Number Airfoil Aerodynamics*, Univ. of Notre Dame, Notre Dame, IN, June 1989.

[32]Evangelista, R., McGhee R. J., and Walker, B. S., "Correlation of Theory to Wind-Tunnel Data at Reynolds Numbers Below 500,000," *Conference on Low Reynolds Number Airfoil Aerodynamics*, Univ. of Notre Dame, Notre Dame, IN, June 1989.

[33]Steger, J. L., and Klineberg, J. M., "A Finite-Difference Method for Transonic Airfoil Design," AIAA Paper 72-679, Sept. 1972.

[34]Mangler, K. W., *Design of Airfoil Sections*, Jahrbuch Der Deutscher Luftfahrforschung, Göttingen, FRG, 1938.

[35]Lighthill, M. J., "A New Method of Two-Dimensional Aerodynamic Design," Aeronautical Research Council, London, R & M Rept. 2112, June 1945.

[36]Arlinger, G., "An Exact Method of Two-Dimensional Airfoil Design," SAAB Corp., SAAB Lintöping, Sweden, TN-67, Oct. 1970.

[37]Tranen, T. L., "A Rapid Computer Aided Transonic Airfoil Design Method," AIAA Paper 74-0501, June 1974.

[38]Volpe G., and Melnik, R. E., "The Design of Transonic Airfoils by a Well-Posed Inverse Method," *Proceedings of the International Conference on Inverse Design Concepts in Engineering Sciences*, Univ. of Texas, Austin, TX, 1984.

[39]Daripa, P., and Sirovich, L., "An Inverse Method for Subcritical Flows," *Journal of Computational Physics*, Vol. 63, April 1986, pp. 311–328.

[40]Boerstoel, J. W., and Huizing, G. H., "Transonic Shock-Free Airfoil Design by an Analytic Hodograph Method," AIAA Paper 74-539, June 1974.

[41]Woods, L. C., "The Design of Two-Dimensional Aerofoils with Mixed Boundary Conditions," *Quarterly Journal of Applied Mathematics*, Vol. 13, No. 2, 1955, pp. 139–146.

[42]Carlson, L. A., "Transonic Airfoil Analysis and Design Using Cartesian Coordinates, *Journal of Aircraft*, Vol. 13, May 1976, pp. 349–357.

[43]Drela M., and Giles, M. B., "ISES: A Two-Dimensional Viscous Aerodynamic Design and Analysis Code," AIAA Paper 87-0424, Jan. 1987.

[44]Sobieczky, H., and Seebass, A. R., "Supercritical Airfoil and Wing Design," *Annual Reviews of Fluid Mechanics*, Vol. 16, 1984, pp. 337–363.

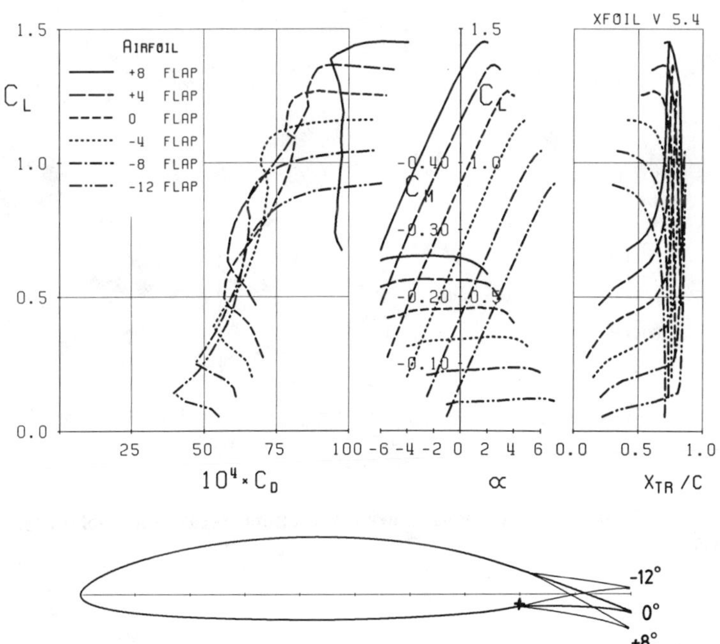

Fig. 1 Polars for laminar sailplane airfoil over range of flap settings [$Re = 10^6/\sqrt{(C_L)}$].

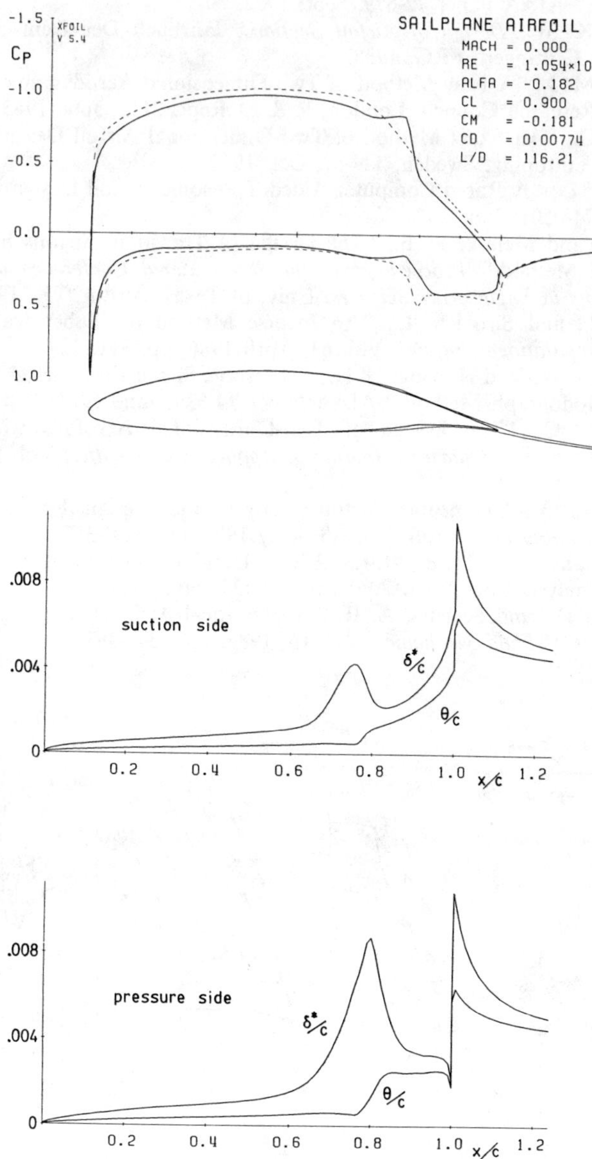

Fig. 2 C_p, δ^*, and θ distributions showing substantial separation bubbles for 0-deg flap case.

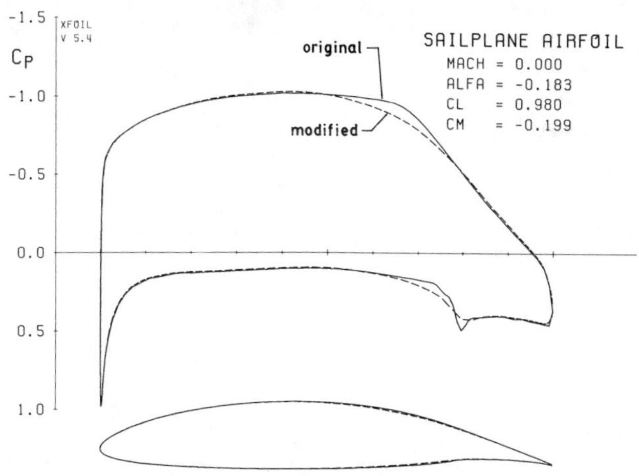

Fig. 3   Original and modified inviscid C_p distributions for reduction of separation bubble drag.

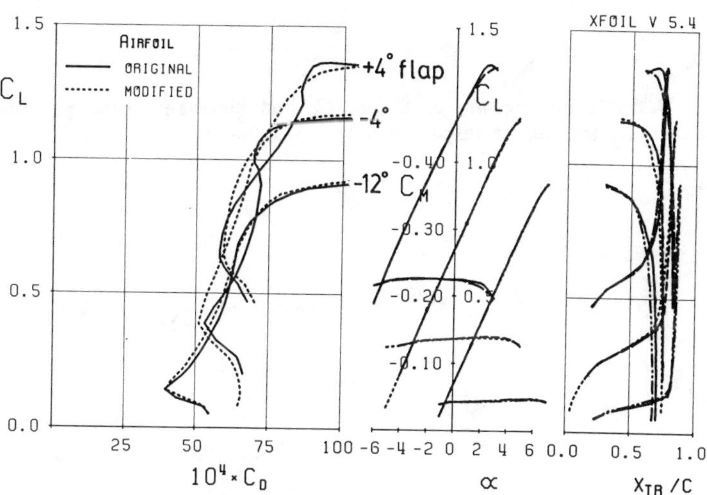

Fig. 4   Polars for original (solid line) and modified (dashed line) laminar sailplane airfoil $[Re = 10^6/\sqrt{(C_L)}]$.

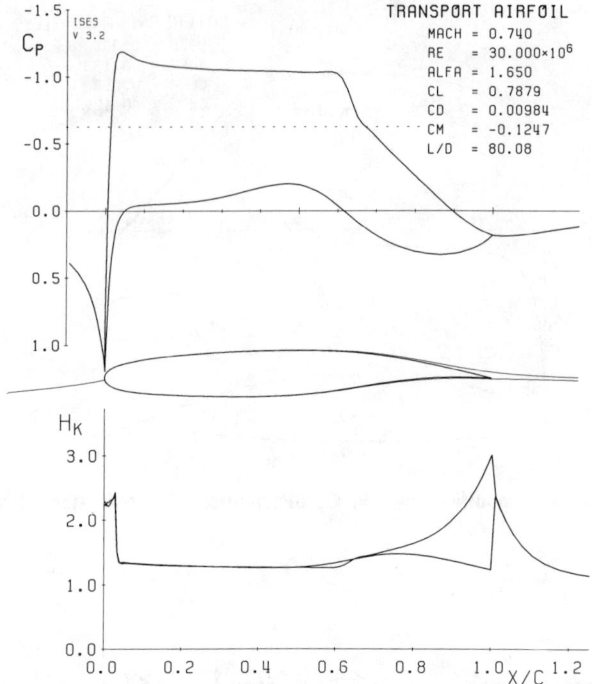

Fig. 5 Design point pressure coefficient C_p and kinematic shape parameter H_k distributions for transonic transport airfoil.

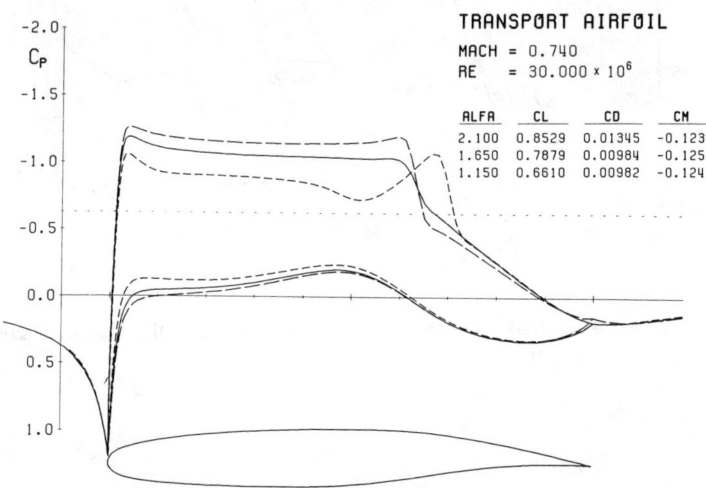

Fig. 6 Transonic transport airfoil C_p distribution for range of lift coefficients at design Mach number of 0.74.

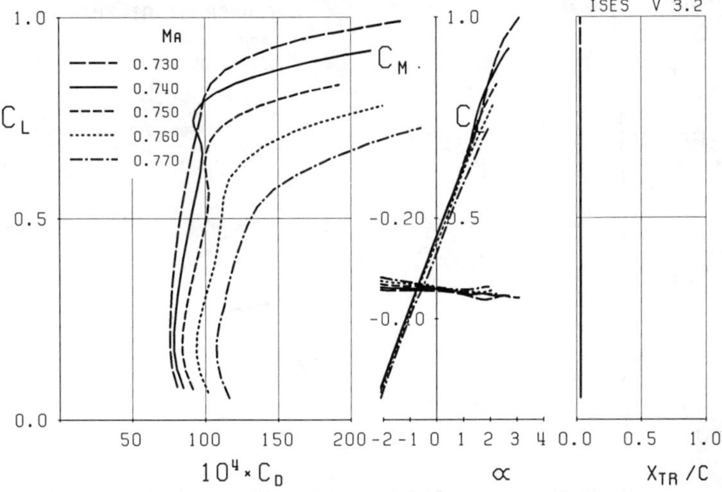

Fig. 7 **Calculated polars for transonic transport airfoil near drag divergence** $Re = 30 \times 10^6$.

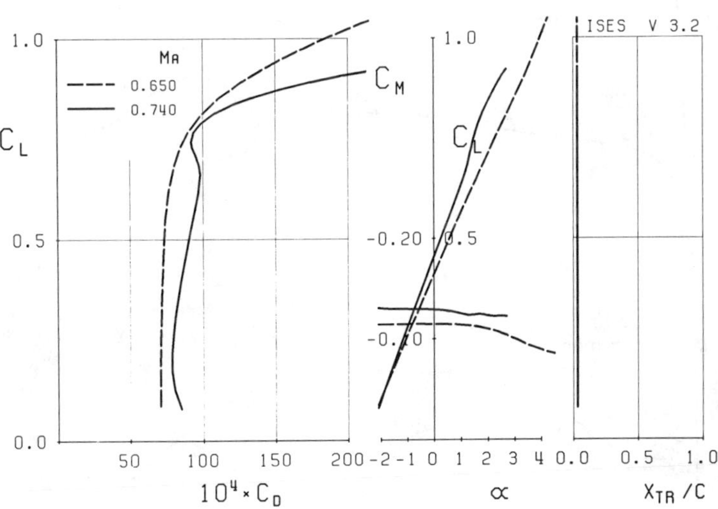

Fig. 8 **Calculated polars for transonic transport airfoil at the design Mach number** $M = 0.74$ **and a lower Mach number** $M = 0.65$ ($Re = 30 \times 10^6$).

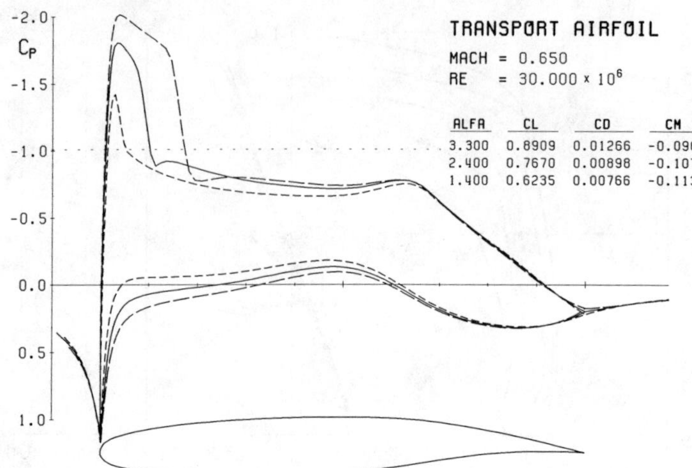

Fig. 9 C_p distributions for range of lift coefficients at lower Mach number $M = 0.65$.

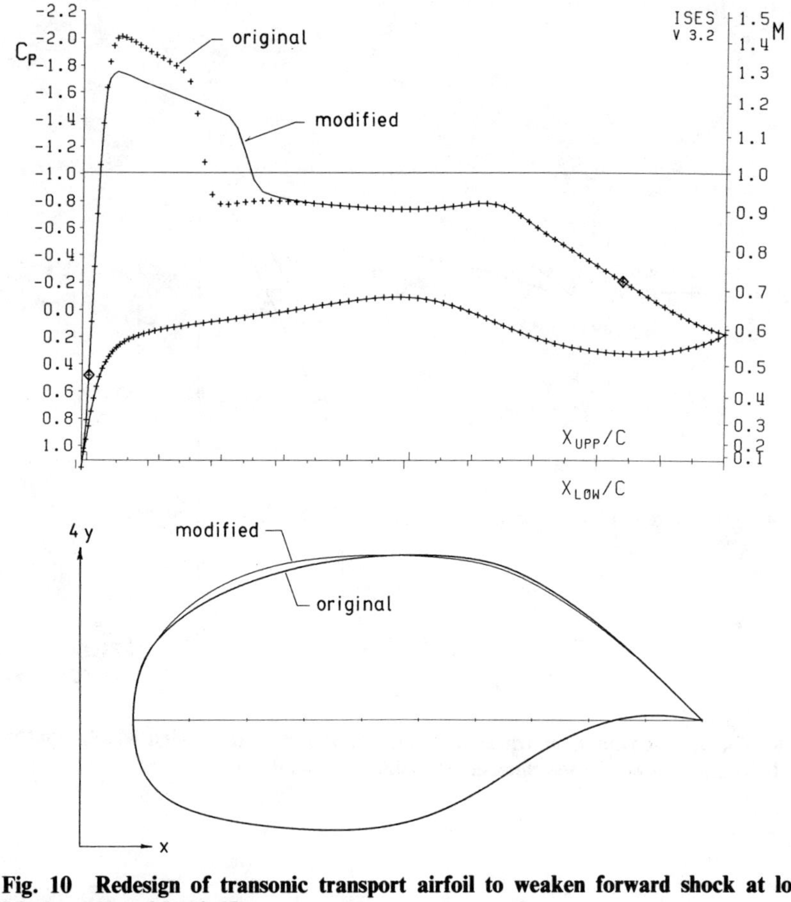

Fig. 10   Redesign of transonic transport airfoil to weaken forward shock at lower Mach number $M = 0.65$.

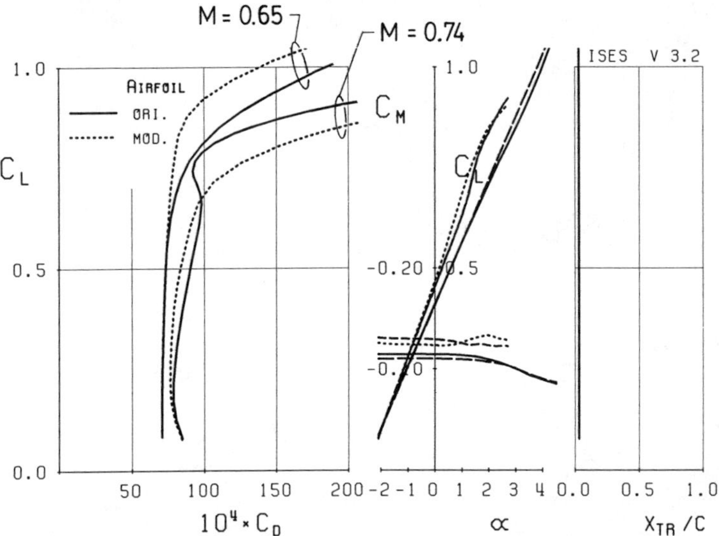

Fig. 11   Drag polars of original (solid line) and modified (dashed line) transonic transport airfoil.

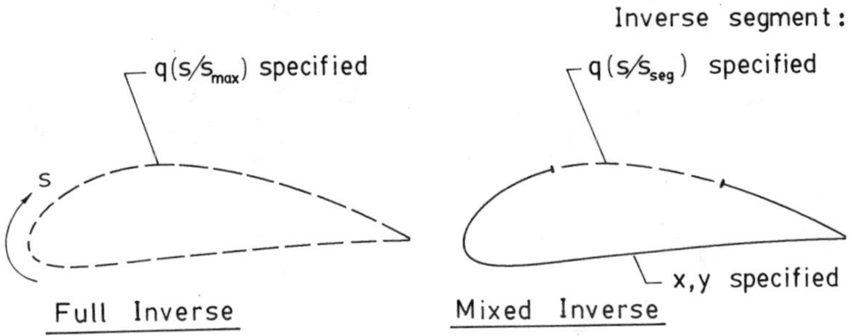

Fig. 12   Full-inverse and mixed-inverse problems.

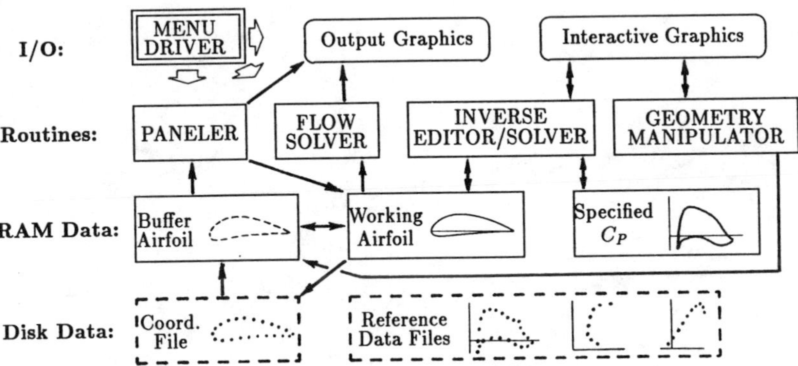

Fig. 13   XFOIL airfoil design/analysis system.

Chapter 7

Inverse Airfoil Design: A Classical Approach Updated for Transonic Applications

G. Volpe*

Grumman Corporate Research Center, Bethpage, New York

Introduction

THE problem of designing airfoil profiles has aroused considerable theoretical interest for well over half a century because of the tremendous practical implications. The aerodynamic performance of an aircraft can be greatly enhanced by tailoring the airfoil to its specific commercial or military requirements. An aircraft design usually calls for the wing profile to exhibit specified lift and/or drag characteristics, a particular lift distribution, or a specific velocity distribution that would provide a measure of control on the behavior of the boundary layer. In these cases, the airfoil design problem is reduced to the specification of a desired speed or pressure distribution.

Mangler[1] and Lighthill[2] discussed this "inverse" problem of airfoil theory for the case of incompressible flow and proposed various analytical solutions. Their methodology was refined and adapted for application on large and small computers by successive researchers.[3-7] However, the difficulty in extending the methodology developed for incompressible flow to the transonic regime eventually gave rise to a number of alternate methods. In a pure inverse-type method of airfoil design—such as those of Mangler and Lighthill—the speed (or pressure) distributions desired on the surface of the profile are specified along with the magnitude and direction of the freestream. In contrast to the direct problem in which the shape of the airfoil profile is specified and the surface speed is computed through a solution of a Neumann-type problem, the inverse problem does not necessarily have a solution. A solution to the inverse problem exists only if a certain constraint between the freestream speed and the surface speed is satisfied. In incompressible flow, which can be described by Laplace's equation, this can easily be shown and the constraint can be expressed in closed form. If, in addition, it is required that the airfoil profile be closed (or have a particular trailing-edge thickness), two additional constraints appear. They can also be expressed in closed form for incompressible flow. The existence of these constraints has been known since the work of Mangler and Lighthill, and Woods has extended the analysis to subcritical

Copyright © 1989 by G. Volpe. Published by the American Institute of Aeronautics and Astronautics, Inc., with permission.
*Senior Staff Scientist.

compressible flows of a Kármán-Tsien-type gas. The work of these authors indicated that a specified surface speed distribution had to be altered in such a manner as to satisfy the three constraints in order to guarantee a solution. In their methods, as in the refinements that followed,[4-7] the approach was to prescribe the surface speed distributions with three free parameters whose values were to be adjusted to satisfy the three constraints.

An obvious advantage of inverse methods is the control the designer has over the force characteristics of the airfoil profile and over the boundary-layer development on its surface, a control gained through the pressure (speed) distribution that is specified. This control can still be retained when making the changes that might be necessary to satisfy the constraints. The introduction of the three free parameters can be arranged such that an "ideal" speed distribution is modified only over selected segments of the airfoil surface. Desired characteristics of a speed distribution (e.g., "rooftops," Stratford-type pressure recovery, rear or front loading) can be retained with little or no modifications.

The formulation of an inverse method at supercritical speeds has been problematic because of the lack of closed-form expressions for the three constraints. The existence of constraints for the transonic design problem was intuitively true because the incompressible design problem was a subset of the more general compressible problem. The lack of a clear understanding of the nature of the first constraint was the main source of the difficulties. A number of alternate approaches to airfoil design were thus developed, with each approach having its particular advantages and disadvantages. A number of methods, such as those of Hicks et al.,[8] Davis,[9] McFadden,[10] Tranen,[11] and Carlson[12] can be classified as "direct" methods. In these methods the solutions (presure and/or force characteristics) for the flow over some arbitrary initial airfoil contour are compared with a desired set of values for the pressure distribution or forces. The differences between the "target" and "current" characteristics are used in some rational way to modify the airfoil profile in the hope of reducing these differences. The process obviously has to be iterated. One advantage of such methods is that a realistic airfoil profile is always obtained at every step of the iteration. The biggest disadvantage, however, is the lack of a guarantee that the iteration will converge with the differences between computed and target values reduced to arbitrarily small levels. The question of the existence of an airfoil solution for a particular target pressure distribution is skirted in these methods, and, in fact, they will not converge for arbitrarily prescribed pressure distributions that do not satisfy the three constraints. This approach to airfoil design is best suited to applications where the target pressure distribution is a "small" modification of the one computed over the initial profile.

Another approach pioneered by Sobieczky et al.[13] is built around the concept of the fictitious gas in regions of supercritical flow. This approach is ideally suited to the redesign of an existing contour in a way that the new contour will have shock-free flow. This technique is much easier to implement than hodograph methods for the design of shock-free airfoils such as those of Garabedian et al.[14] and Boerstoel.[15] However, in neither of these classes of methods does the user have control of the pressure distribution. Such control can be exercised only with an inverse method.

A formulation of the inverse problem for airfoil design at transonic speeds was finally given by Volpe and Melnik.[16] They devised a method in which the first constraint could be satisfied by treating the value of the freestream speed as a free parameter that is determined as part of the solution of the Dirichlet problem. Alternatively, the target surface speed distribution could be scaled keeping the freestream speed fixed. The two options are equivalent in practice since the quantity of interest is the ratio of surface to freestream speed. The discovery of a means by which the first constraint might be satisfied essentially completed the extension of the methodology of Mangler and Lighthill to transonic flow. The method was refined in a later work by Volpe.[17] The rest of this article provides a discussion of this method and will illustrate a number of applications. In the method, three free parameters are adjusted numerically to drive the values of the freestream speed and trailing-edge gap dimensions to pre-scribed values. The trailing-edge parameters are introduced in such a way that each mainly affects only one of the constraints. This permits the formulation of a diagonal-type iterative scheme in which the three parame-ters can be determined from three uncoupled one-dimensional relaxation methods. The result is a robust method for the design of airfoils that generates speed distributions in close approximation to an arbitrarily prescribed ideal.

Formulation of the Inverse Design Problem

The problem being addressed is the construction of the airfoil profile, which has a surface speed distribution q_0 equal to some desired function F everywhere along its arc length s. This is to be measured clockwise around the airfoil contour starting at the lower surface trailing-edge point. The airfoil's coordinates, x,y, can be parameterized as functions of s. A feature of practical airfoil contours is that the trailing edge be either closed or have a very small gap. Thus, a requirement on the to-be-determined airfoil is that the upper- and lower-surface trailing-edge points be separated by prescribed distances Δx and Δy. The horizontal gap Δx is usually set to zero, and the vertical gap Δy is set to zero (a closed airfoil) or to a small positive number. The freestream is also defined by prescribing values for the freestream velocity q_∞, temperature, and pressure (or density). These, in turn, determine the freestream Mach number M_∞. In incompressible flow, of course, it is only necessary to specify the velocity in order to identify the freestream uniquely. Our formulation applies in its entirety if we specify a surface-pressure distribution instead of a surface speed, since the two are uniquely related. Formally, then, the problem is to determine the airfoil profile of a specified trailing-edge thickness corresponding to the speed distribution

$$q_0/q_\infty = F(s/s_{\text{max}}) \tag{1}$$

Without loss of generality, s_{max} can be set equal to one; q_0 is taken as positive in the clockwise direction around the airfoil. The strategy followed is to iteratively modify some initial contour until the desired speed distribution q_0 is achieved. This initial contour can be mapped into the unit circle by the unique conformal transformation

$$\frac{dz}{d\zeta} = \left(1 - \frac{1}{\zeta}\right)^{1-\epsilon} e^{P+iQ} \tag{2}$$

where $z = x + iy$ and $\zeta = re^{i\omega}$ are the coordinates in the physical and mapped planes, respectively, and $\epsilon\pi$ is the included trailing-edge angle. This equation can be separated into its real and imaginary parts. Thus, on $r = 1$,

$$\frac{ds}{d\omega} = \left(2 \sin\frac{\omega}{2}\right)^{1-\epsilon} e^{P} \tag{3}$$

$$\theta = \frac{1}{2}(1+\epsilon)(\pi - \omega) - \frac{\pi}{2} + Q \tag{4}$$

where θ is the local slope of the airfoil. Q is the Fourier series

$$Q = \sum_{n=0}^{N} (A_n \sin n\omega - B_n \cos n\omega) \tag{5}$$

and P is its conjugate series. Because θ is known as a function of s, the coefficients of the series can be found by standard Fourier analysis as described in Ref. 4. With this mapping procedure, the leading terms of the series are related to the trailing-edge gap $(\Delta x, \Delta y)$ by

$$A_1 = -\left(\frac{\Delta x}{2\pi}\right)\sin B_0 - \left(\frac{\Delta y}{2\pi}\right)\cos B_0 + (1-\epsilon) \tag{6a}$$

$$B_1 = -\left(\frac{\Delta x}{2\pi}\right)\cos B_0 + \left(\frac{\Delta y}{2\pi}\right)\sin B_0 \tag{6b}$$

In the case of incompressible flow, if q_0 were defined in terms of the circle plane coordinate ω, then the airfoil contour corresponding to q_0 could be constructed immediately. The complex potential w for the incompressible flow past the circle (to which the unknown airfoil is mapped) is

$$w = q_\infty\left(\zeta e^{-i\alpha} + \frac{e^{i\alpha}}{\zeta}\right) + i\Gamma \log\zeta \tag{7}$$

where α is the angle of attack of the freestream, and Γ is the circulation around circle. The value of Γ is obtained by requiring that the velocity $dw/d\zeta$ be zero at the point on the circle corresponding to the airfoil's trailing edge. Now

$$\frac{ds}{d\omega} = \left|\frac{dz}{d\zeta}\right| = \left|\frac{dw}{d\zeta}\right|\left|\frac{dw}{dz}\right|^{-1} \tag{8}$$

and, in the physical plane, $|dw/dz| = q_0$. The right-hand side of Eq. (8) is then completely known as a function of ω. Using Eqs. (3) and (5) the coefficients of the series of the transformation can be evaluated. Once $dz/d\zeta$ has been expressed as a trigonometric series in ω, the airfoil coordinates, x and y, are found by integration.

If q_0 is prescribed as a function of arc length s, some iteration will be required in this procedure, since $|dw/dz|$, which is needed in Eq. (8), will not be known as a function of ω until after $s(\omega)$ is found. However, the design of the airfoil is still a straightforward process.

For compressible flow of a perfect gas, the flow past the circle cannot be expressed in closed form. However, it can be computed numerically, and a similar iterative procedure can be formulated for compressible flow.

Specifically, the procedure will be as follows: An initial airfoil contour is mapped into the unit circle, and the flow around the circle is solved subject to the conditions on the circle boundary that the tangential speed be the required total speed. The passage from $q_0(s)$ to $q_0(\omega)$ is done using the current $s(\omega)$. In this flow the circle boundary is not necessarily a streamline, and the departure of the boundary from a streamline can be used to find a correction to the aircraft contour. Using the new metric the process can then be repeated.

The following sections will describe a numerical scheme for computing the compressible flow in the circle plane and a method for updating the airfoil contour.

Transonic Flowfield Solution

The infinite flowfield around a unit circle can be transformed into the finite region inside the circle. The modulus of the transformation of the physical plane, z, to the inside of the circle is then written as

$$h = \frac{1}{r^2}\left|\frac{dz}{d\zeta}\right| \tag{9}$$

The term $dz/d\zeta$ is, of course, the quantity we seek since it describes the transformation of the unknown airfoil profile into the circle. It should be recalled that the transformation is conformal everywhere except at the airfoil's trailing edge where the metric $h = 0$.

The flow in the circle plane is assumed to be governed by the continuity equation

$$\frac{\partial}{\partial\omega}(\rho U) + r\frac{\partial}{\partial r}(\rho V) = 0 \tag{10}$$

where U and V are the transformed circle plane velocity components in the r and ω directions, respectively. For irrotational flow they can be expressed as gradients of a potential function ϕ; thus,

$$U = \phi_\omega, \qquad V = r\phi_r \tag{11}$$

As long as shock waves in the flow remain of moderate strength, the preceding assumptions will not introduce significant errors in the flow solution. The density ρ is found from the speed of sound a through the relation

$$\rho = (M_\infty^2 a^2)^{1/(\gamma - 1)}$$

where γ is the ratio of specific heats. In turn, a can be evaluated through the energy relation

$$a^2 = a_0^2 - \left(\frac{\gamma - 1}{2}\right)(u^2 + v^2)$$

where a_0 is the stagnation speed of sound, and u and v are the velocity components in the physical plane. These are related to the components in the circle plane by

$$u = \frac{rU}{h}, \qquad v = \frac{rV}{h} \tag{12}$$

It can be seen that the flow within the circle cannot be computed if h is not known. The assumed initial shape for the airfoil provides the initial estimate for h. In the limit of M_∞ going to zero, Eq. (10) reduces to Laplace's equation. Then, the solution for the flow within the circle is independent of the mapping metric h and is given by Eq. (7).

The mapping introduces singularities at infinity, but they can be removed by subtracting from the potential its behavior in the far field. As discussed by Ludford,[18] the solution in the far field is made up of a uniform stream plus a circulatory component. The potential functions describing these terms are known. Thus, one can define a reduced potential function:

$$G = \phi - q_\infty\left(r + \frac{1}{r}\right)\cos(\omega + \alpha) - E\tan^{-1}[\sqrt{(1 - M_\infty^2)}\tan(\omega + \alpha)] \tag{13}$$

where E is a circulation constant. This reduced potential is continuous and single-valued everywhere.

At infinity $(r = 0)$ $G = G_\infty$, a constant that can be set at zero in direct (Neumann) problems, but which must be determined as part of the solution in inverse (Dirichlet) problems by extrapolating from the interior of the flowfield.

For the direct (analysis) problem, q_∞ is usually set to unity, and the boundary conditions demand that $v = 0$ at the surface. The solution for the flowfield is computed numerically by discretizing the flowfield in conservation form along a polar coordinate mesh. The set of difference equations that approximates Eq. (10) is solved for the discrete values of the reduced potential G at the nodes of the computational mesh by an approximate factorization multigrid scheme similar to the one described by Jameson.[19] The value of the circulation constant E is determined from the Kutta condition, which requires that u be finite at the trailing edge. Since $h = 0$ at the trailing edge, U must be made to vanish at this point. In this direct problem the surface speed $q_0(s) \equiv u(s)$ is computed from the potential function G.

For the inverse design problem the boundary conditions at $r = 1$ are imposed on u rather than v. Using the known functional relation between s and ω for the current contour, the target speed distribution q_0 can then be expressed as a function of ω. Then, at the boundary in the circle plane we set $u_0 = u(\omega)$ equal to $q_0[s(\omega)]$. Of course, this would be true if h were the true mapping metric; in general, it is not. Hence, the boundary is not necessarily a streamline of the flow. In other words, v is not necessarily zero at the boundary. The flowfield is computed subject to the boundary condition $u_0 = q_0[s(\omega)]$ at $r = 1$ by a numerical scheme identical to the one used for the direct problem. The Dirichlet boundary conditions are implemented by integrating q_0 around the airfoil to find the reduced potential G at the boundary points using Eqs. (11–13). A constant of integration G_0 can be prescribed arbitrarily. The numerical problem that has boundaries at both $r = 0$ and $r = 1$ is well-posed, since the value at the inner boundary, G_∞, is determined as part of the solution. The circulation constant is determined by integrating u_0 around the full boundary. In general, in the Dirichlet problem there is a net mass flow emitted from the boundary. To allow for this, a source term $\sigma \log r$ is substracted from the potential leading to a new reduced potential $\bar{G}$ defined by

$$G = \bar{G} + \sigma \log r$$

The far-field boundary condition will then have the additional term

$$\frac{\sigma}{2} \log[1 - M_\infty^2 \sin^2(\omega + \alpha)]$$

The source term has a role similar to that of the circulation term for the Neumann problem. The value of σ is determined by setting V equal to 0 at the trailing edge. The introduction of the source term guarantees compatibility of surface and far-field boundary conditions during the iteration process. This source term vanishes as the design process converges to its final contour.

A nonzero normal velocity v will, in general, be computed on the circle boundary. This value of v can be used to modify the original airfoil contour.

Contour Modification

A nonzero normal velocity v at the circle boundary implies that the actual streamline is (to first order) rotated from the boundary by an angle of magnitude

$$\delta\theta = \tan^{-1}\left(\frac{v}{u}\right) \tag{14}$$

The mapping of a new streamline contour, $\tilde{z}$, to the unit circle is accomplished by a mapping similar to Eq. (2):

$$\frac{d\tilde{z}}{d\zeta} = \left(1 - \frac{1}{\zeta}\right)^{1-\epsilon} e^{\tilde{P} + i\tilde{Q}}$$

which implies relations similar to Eqs. (4) and (5). Letting

$$\tilde{Q} = \sum_{n=0}^{N} (\tilde{A}_n \sin n\omega - \tilde{B}_n \cos n\omega)$$

and $A'_n = \tilde{A}_n - A_n$ and $B'_n = \tilde{B}_n - B_n$, it follows that

$$\delta\theta - \frac{1}{2}\delta\epsilon(\pi - \omega) = \sum_{n=0}^{N} (A'_n \sin n\omega - B'_n \cos n\omega) \tag{15}$$

where

$$\delta\epsilon = \frac{1}{\pi}(\delta\theta_1 - \delta\theta_{2N})$$

and $\delta\theta_1$ and $\delta\theta_{2N}$ are the values of $\delta\theta$ at the two sides of the trailing-edge point. Since $\delta\theta$ is known as a function of ω, A'_n and B'_n can be evaluated by a standard Fourier analysis of Eq. (15). The new components of the series of the transformation, $\tilde{A}_n$ and $\tilde{B}_n$, can be formed. The new derivative of the arc length, $d\tilde{s}/d\omega$, is calculated in turn, as well as

$$\frac{d\tilde{x}}{d\omega} = -\frac{d\tilde{s}}{d\omega}\cos\tilde{\theta}, \qquad \frac{d\tilde{y}}{d\omega} = -\frac{d\tilde{s}}{d\omega}\sin\tilde{\theta}$$

The actual ordinates of the new airfoil contour are then obtained by integration. This new airfoil provides a new approximation for the metric h and a new relation $s(\omega)$, which are needed to set up a new Dirichlet problem in the circle plane. This process can be repeated until a desired tolerance in the maximum value of v/u is reached. At this point the tangential velocity is equal to the total target speed prescribed on the surface, and the target speed will have been modified appropriately to satisfy the constraints.

Constraints in Incompressible Flow

The question that must be asked at this point is whether an airfoil solution exists for an arbitrarily prescribed speed distribution. For incompressible flow, Mangler and Lighthill showed that, in fact, a solution exists only if certain integral constraints are satisfied by q_0, and this can be demonstrated as follows.

Since a lifting flow over a circle can be reduced to the nonlifting, symmetric flow as shown in Refs. 2 and 20, it is sufficient to consider the nonlifting case in order to simplify the discussion. The mapping between the z and ζ planes must have the form

$$z = \zeta + \sum_{n=0}^{\infty} a_n \zeta^{-n}$$

if the flow in the far field is to remain unscaled. Here, the a_n are complex constants. Therefore,

$$\frac{dz}{d\zeta} = 1 - \sum_{n=1}^{\infty} n a_n \zeta^{-(n+1)}$$

From Eq. (7),

$$\frac{dw}{d\zeta} = q_\infty \left(1 - \frac{1}{\zeta^2} \right)$$

Hence, combining the last two equations, one obtains

$$\frac{dw}{dz} = q_\infty \left(1 + \sum_{n=2}^{\infty} b_n \zeta^{-n} \right) \tag{16}$$

Since $dw/dz = q_0 e^{-i\theta}$, it follows that

$$\log\left| \frac{q_0}{q_\infty} \right| = \sum_{n=2}^{\infty} b_n' \zeta^{-n} \tag{17}$$

As pointed out by Lighthill and Thwaites, $\log|q_0/q_\infty|$ is an analytic function in the domain outside the circle (it fails to be analytic at stagnation points on the circle where $q_0 = 0$). Therefore, it can be expanded in a Fourier series on the circle itself. However, from Eq. (17) we see that the series cannot have terms of zero or first order. In fact, q_0 must be such that

$$\int_0^{2\pi} \log\left| \frac{q_0}{q_\infty} \right| \left\{ \begin{array}{c} 1 \\ \cos\omega \\ \sin\omega \end{array} \right\} d\omega = 0 \tag{18}$$

These are the three integral constraints that the prescribed speed distribution must satisfy for an airfoil solution to exist. These three constraints have arisen from the requirements that the airfoil be closed and from the imposition of a value on the freestream. It can be safely assumed that similar constraints exist also at supercritical speeds. The preceding discussion indicates that the prescribed speed distribution should contain, in general, three adjustable parameters to guarantee that the constraints may be satisfied. Thus, the surface speed distribution should be prescribed in the form

$$q_0/q_\infty = F(s/s_{\max}; p_1, p_2, p_3) \tag{19}$$

where p_1, p_2, and p_3 are the three parameters that are found as part of the solution. For compressible flow ($M_\infty \neq 0$), Eq. (18) is no longer an adequate expression for the constraints. One must then formulate alternative means of evaluating the parameters while still ensuring that the constraints are satisfied. The particular functional forms chosen to introduce the parameters will, of course, affect the class of airfoil solutions that can be obtained. These will be discussed later.

An Alternate Look at the Constraints

In order to formulate a well-posed inverse design procedure that would be valid at compressible speeds, the nature of the constraints must be re-examined since the earlier derived closed-form expressions are valid only for incompressible flow. It is logical to do this in the context of the computational method that has been outlined. The two constraints that arose because of the required trailing-edge gap are of a geometric nature. Hence, one can set up a procedure in which, by monitoring the trailing-edge gap size, the target speed can be modified in order to drive the gap's dimensions to its specified values. The first constraint creates a problem because there is no single physical quantity that reflects the constraint. This first condition is a statement of "compatibility" between the prescribed surface speed and the freestream speed. If the latter is also being prescribed, as is usually the case, the surface speed prescription has to be modified for ιne constraint to be satisfied. If the freestream speed is not specified, in the case of incompressible flow its value can be found from Eq. (18). In the absence of a closed-form expression, which would be valid at compressible speeds, the problem is to define a procedure whereby either the surface speed or the freestream might be changed to bring about compatibility.

Consider the incompressible flow over a circle again. As mentioned earlier, it will be sufficient to consider the nonlifting symmetric flow. The general solution for the flow on the outside of a circle of unit radius can be represented in the form

$$G = a_0 + a_1 r \cos\omega + \sum_{n=1}^{N} \frac{b_n}{r^n} \cos n\omega$$

with N being a sufficiently large number.

This is the most general solution to Laplace's equation that yields a uniform freestream flow in the far field ($r \to \infty$). Hence,

$$\frac{\partial G}{\partial r} = \left(a_1 - \frac{b_1}{r^2}\right)\cos\omega - \sum_{n=2}^{N} n\,\frac{b_n}{r^{n+1}}\cos n\omega \tag{20}$$

and

$$\frac{1}{r}\frac{\partial G}{\partial \omega} = -\left(a_1 + \frac{b_1}{r^2}\right)\sin\omega - \sum_{n=2}^{N} \frac{nb_n}{r^{n+1}}\sin n\omega \tag{21}$$

It follows that the total velocity

$$q = \left(G_r - \frac{i}{r}G_\omega\right)e^{-i\omega}$$

must be of the form

$$q = a_1 - \sum_{n=2}^{N} b_n'\zeta^{-n} \tag{22}$$

In the far field, as $\zeta \to \infty$, $q = q_\infty$. Hence, $a_1 = q_\infty$, and

$$q = q_\infty - \sum_{n=2}^{N} b_n'\zeta^{-n} \tag{23}$$

which reflects the result expressed in Eq. (17). Thus, if the flow over the circle is determined with the condition that $q = q_0(\omega)$ on the boundary $r = 1$, the expansion of $q_0(\omega)$ in a series

$$q_0(\omega) = c_0 + \sum_{n=1}^{N} c_n e^{-in\omega} \tag{24}$$

reveals restrictions on q_0 that echo the constraints described by Lighthill and Mangler. In particular, $c_0 = q_\infty$, the first constraint. Also, one can see that $c_1 = 0$, implying two additional restrictions on the speed, since c_1 is a complex constant. The rest of this section will concentrate on the first constraint.

It is interesting to note that Eq. (23) can be factored into the form

$$q = q_\infty\left(1 - \frac{1}{\zeta^2}\right)\left[1 - \sum_{n=2}^{N} b_n''\zeta^{-n}\right] \tag{25}$$

Since the first part on the right-hand side of this expression represents the solution for the flow over the unit circle, the expression in brackets formally gives the mapping $|d\zeta/dz|$, which generates the airfoil corresponding to $q = q_0(\omega)$ (assuming that q_0 satisfies the constraints).

At this point it should be mentioned that a proper speed distribution for an airfoil should contain at least one zero corresponding to the leading-edge stagnation point and two zeroes if the trailing edge is not cusped. Hence, the zeroes of q should match the zeroes of the flow over the circle if the metric is to be free of singularities with the possible exception of the trailing edge.

In the procedure described in the previous sections, the airfoil's contour is to be found by successively modifying some starting profile. The modifications are to be guided by the solution of a Dirichlet problem. Consider now such a Dirichlet problem in the case of incompressible flow. Boundary conditions are now imposed on the tangential velocity and can be expressed as $\partial G/\partial \omega = g(\omega)$. Expanding $g(\omega)$ in a trigonometric series, one obtains

$$\left(\frac{\partial G}{\partial \omega}\right)_{r=1} = c_0' + \sum_{n=1}^{N} c_n' \sin\omega \tag{26}$$

A comparison of Eqs. (26) and (21) leads to $N+1$ conditions for the constant a_1 and the N Fourier coefficients b_n. However, c_0' is identically 0 since

$$\int_0^{2\pi} \frac{\partial G}{\partial \omega} \, \mathrm{d}\omega = 0$$

Thus, there are really only N conditions, and a_1, the term representing the freestream, can be specified!

After the values of the b's have been determined the speed normal to the boundary can be found from Eq. (20). The total speed on the boundary, as anywhere else in the flowfield, must have the form given in Eq. (22). Thus, the solution to the Dirichlet problem yields a flow in which the freestream and the speed on any closed path that can be drawn within the flowfield are automatically compatible; there is no "first" constraint. Not all paths are of interest, however. Only the contours that pass through branch (stagnation) points of the flow can yield airfoil-like profiles. The location of the stagnation points of the flow computed from the solution of the Dirichlet problem depends on the value assigned to the freestream term, a_1. Regardless of the value assigned to a_1, an airfoil contour can be traced from the stagnation points, but the speed distribution on the contour is not necessarily equal to the one prescribed, or even close to it. The first constraint has been removed at the expense of retaining control over the speed on the airfoil, which is an undesirable result.

If the circle boundary were to be truly a streamline, G_ω would be identically equal to the sought-after total speed. If the branch points of the flow were to be on the circle boundary, it would be reasonable to expect that the streamline passing through them would be "close" to the circle and the total speed on the streamline would be close to G_ω. The contour-perturbation process described earlier may have a chance to work in such a situation. This closeness can be brought about by choosing a_1 in such a way that the branch points do fall exactly on the circle boundary. Since points where G_ω is already zero are already specified, it is natural to enforce these points to be the stagnation points of the flow. This is guaranteed by forcing G_r to be zero at these two points by appropriately choosing values for the freestream and for the mass-flow term σ. Since the formulation already called for σ to be chosen in such a way as to make $G_r = 0$ at the trailing edge, the procedure just outlined has reinstated the constraint between the freestream and the prescribed surface speed. The advantage consists in the fact that the constraint is now satisfied by making $G_r = 0$ at the leading-edge stagnation point rather than through an integral relation. This new approach to the constraints generates a profile with a speed distribution that automatically satisfies the integral expression without invoking it explicitly. In this new procedure, as in the classical one, the freestream speed can be kept at a specified value by introducing free parameters in the specified surface speed, as in Eq. (16); p_1 in Eq. (16) can be adjusted to return a_1 to its desired value.

In summary, an airfoil design procedure that satisfies the first constraint and is equivalent to the Mangler/Lighthill method can be formulated as follows. The design strategy calls for the computation of a flowfield about a circle on which the boundary conditions are such that the tangential velocity u (equivalent to G_ω in incompressible flow) is set equal to the target speed distribution $q_0(\omega;p_1)$ (trailing-edge closure is still being neglected). At convergence, u must be equal to q_0. If the value of p_1 is adjusted in such a way that $v = 0$ at the point on the circle where $u = 0$, this point will be a stagnation point for the flow. Hence, the streamline representing an airfoil-like contour must pass through that point. The speed distribution along that streamline is not equal to the target speed, but it automatically satisfies the freestream speed constraint. The mapping metric h can then be updated as previously described. It is worth mentioning at this point that setting $v = 0$ where $u = 0$ ensures that the ratio expressed in Eq. (14) remains finite at all times. With the new metric one can set up another Dirichlet problem that has a streamline passing closer to the circle, and the speed along it will be closer to the target speed. If one repeats this process until the circle itself becomes a streamline, the speed distribution on the streamline will then be equal to the specified target and will satisfy the constraint. Hence, adjusting p_1 in such a way that $v = 0$ at $u = 0$ at all times leads to the design of the airfoil that corresponds to $q_0(\omega;p_1)$, and the value of p_1 is the value that ensures satisfaction of the freestream speed constraint.

This technique for statisfying the constraint can be applied at supercritical speeds as well as for incompressible flow. The constraints imposed by the trailing-edge closure requirements can be accounted for by monitoring the trailing-edge gap during the iteration process and adjusting the two additional parameters, p_2 and p_3, in Eq. (16).

Enforcement of Constraints

The discovery of a method that ensures that the first constraint can be satisfied at compressible speed opens the way to the formulation of schemes whereby the necessary freedom can be introduced in the speed distribution.

For the remainder of this paper, it will be assumed that the target speed is of the form

$$\frac{q_0}{q_\infty} = f_1(s;p_1)[f_0(s) + f_2(s;p_2) + f_3(s;p_3)] \tag{27}$$

where $f_0(s)$ represents the ideal target speed distribution that, in practice, is usually a tabulated function. The functions f_1, f_2, and f_3 are introduced to modify the ideal target in order to satisfy the three constraints. In general, it is desirable to localize the effect of f_1, f_2, and f_3 so that the resulting surface speed will be close to the ideal speed distribution, $f_0(s)$, over most of the airfoil surface. Since in transonic flow it is not possible to relate p_1, p_2, and p_3 to the three constraints in closed form, a numerical search for the parameters must be made. The search is greatly facilitated by choosing f_1, f_2, and f_3 in such a way that each significantly affects only one of the constraints. One would then have three one-dimensional searches for p_1, p_2, and p_3. In Ref. 21 the sensitivity of a designed airfoil contour to various changes in the target speed distribution is reported. These results have guided the definition of f_1, f_2, and f_3 in Eq. (27). Three separate schemes will be described, but they hardly exhaust the number of possibilities and many more can be constructed.

Satisfaction of the first constraint is guaranteed by adjustment of p_1. By definition, f_1 causes a scaling of surface speed (q_0/q_∞). In scheme 1, f_1 is set equal to the constant p_1. This results in a scaling that is uniform along the airfoil. In this case p_1 can be looked upon as a scaling on either q_0 or q_∞. In the latter case one would essentially have q_∞ floating, and it would be determined as part of the solution. As discussed earlier, the value of p_1 is chosen to guarantee that the specified leading-edge stagnation point will truly be a branch point of the flow.

Control over Δy, the vertical separation between the upper- and lower-surface trailing-edge points, can be exercised by defining

$$f_2 = p_2 \sin(\tfrac{8}{3}\omega), \qquad \omega \leqslant \tfrac{3}{4}\pi \tag{28}$$

Outside of this range f_2 is set to zero. Here, the ordinate ω in the computational plane has been substituted for the arc length s. It is more convenient to use ω rather than s, and the formulation of the problem is not affected by this substitution. The function f_3 is the hardest to define. The horizontal separation between the two trailing-edge points, Δx, is affected primarily by the location of the leading-edge stagnation point. As shown in Ref. 20, a small shift in this stagnation point along the surface of the airfoil, on the order of 2% of the chord length, can alter the horizontal gap by 5–6%. It should be pointed out that a 2% shift in the stagnation point along the surface is hardly noticeable when viewed as a shift along the chord. In order to maintain a loose coupling among p_1, p_2, and p_3, the shift must be accomplished without altering the local velocity gradients. This can be accomplished by shifting the functional dependence of q_0 on s

locally, near the leading edge. This shift can be expressed by

$$f_3(s) = f_0(s') - f_0(s)$$

with

$$s' = s - p_3 h(s)$$

where

$$h(s) = \frac{1}{2}\left\{1 - \cos\left[\frac{\pi}{\Delta s}(s - s_T + 2\Delta s)\right]\right\}, \qquad s_T - 2\Delta s \leqslant s \leqslant s_T - \Delta s$$

$$= 1, \qquad s_T - \Delta s \leqslant s \leqslant s_T + \Delta s$$

$$= \frac{1}{2}\left\{1 + \cos\left[\frac{\pi}{\Delta s}(s - s_T + \Delta s)\right]\right\}, \qquad s_T + \Delta s \leqslant s \leqslant s_T + 2\Delta s$$

Elsewhere, $h(s)$ is zero. The point s_T denotes the location where $f_0(s)$ is zero in the leading-edge region, and Δs is some appropriate distance, typically 2.5% of the total arc length. This form for f_3 shifts the leading-edge stagnation point smoothly without introducing any "wiggles" in the target speed distribution and, in addition, has hardly any effect on the values of p_1 and p_2. This form for f_3 is common to all the three schemes described here.

A second scheme for modifying the target distribution uses a different definition for f_2 in Eq. (27). The expression given in Eq. (28) alters the target speed distribution only on the lower surface of the airfoil. It would therefore be unsatisfactory if one were trying to design a symmetric airfoil. An alternative form for f_2 is

$$f_2 = p_2\left(1 - \frac{\omega}{\omega_1}\right), \qquad \omega \leqslant \omega_1$$

$$= p_2\left(\frac{2\pi - \omega}{\omega_1} - 1\right), \qquad \omega \geqslant 2\pi - \omega_1 \qquad (29)$$

This function symmetrically alters the magnitude of the speed in the neighborhood of the trailing edge. In the present computational scheme, the speed takes on opposite signs on the upper and lower surfaces, accounting for the sign difference between the two parts of Eq. (29); ω_1 is typically taken as $\pi/3$. A third scheme is formulated by substituting for $f_1 = p_1$ in scheme 1 the function

$$f_1 = \sqrt{1 + p_1 \sin^2\left(\frac{\omega}{2}\right)}$$

which concentrates the scaling in the front half of the airfoil.

Regardless of which scheme is used, the three parameters are adjusted periodically during the solution of the Dirichlet problem that precedes each contour modification. At the end of every sweep of the flowfield, q_∞ and σ are determined by forcing v to be zero both at the leading-edge point where u is zero and at the trailing edge. The factor p_1 is then adjusted to scale q_∞ back to its specified value, and the flowfield is swept again. The value of the normal component of velocity at the leading-edge stagnation point, v, goes to zero quite fast (due to the continuous resetting of p_1). When v is below a given tolerance (typically 10^{-5} to 10^{-6}), estimates are made of the values that A_1 and B_1, the first-order terms of the series in Eq. (5), would have if the airfoil were modified at that stage. These values are compared with the values they should have for the airfoil to have the desired trailing-edge gap dimensions, as given by Eq. (6). The differences between the current and desired values, δA_1 and δB_1, are then used to change p_2 and p_3, respectively. The change in p_2 is made proportional to δA_1, and the change in p_3 is proportional to $(-\delta B_1)$. Since p_1 is introduced as a multiplier, a change in the surface boundary conditions due to a new p_1 can be transmitted through the entire flowfield by scaling the entire potential field. Using this procedure we can update p_1 after each multigrid sweep of the flowfield without seriously affecting the convergence rate of the numerical scheme.

This procedure is not possible with p_2 and p_3; therefore, they are updated infrequently. However, the method of false position can be used to accelerate convergence of p_2 and p_3. The flowfield is assumed to be converged when all of the residuals at all of the flowfield node points are below a specified tolerance, and v at the leading-edge stagnation point together with δA_1 and δB_1 are below their respective tolerances. At this point the airfoil contour is modified, and another Dirichlet problem is set up. There is no need to analyze the new airfoil contour with this procedure. A direct analysis is made at the very end of the calculation just to check our results.

To ensure convergence of the design process, it is necessary to under-relax the changes to the contour shape. Thus, only a fraction of the changes suggested by Eq. (14) are actually taken in the early design cycles. After several contour modifications the factor can be increased. The tangential velocity $u(\omega)$ at the boundary, which is interpolated from the desired $q_0 = F(s)$, is also under-relaxed when a new design cycle is started.

Results

A considerable number of airfoil contours have been designed by the method over a wide range of speed (or pressure) distributions, including cases in which shock waves were present in the flowfield. All of the examples that will be presented in this section have been computed on a mesh containing 192 points in the circumferential direction and 32 points in the radial direction. Five mesh levels were used in the multigrid sequence. In each case the angle of incidence of the designed contour was set at zero. In the present formulation the angle of attack can be specified; different choices for the angle result in different orientations of an otherwise identical contour within the given coordinate system.

A strong test of the system is illustrated in the redesign of the Korn airfoil using as a target the pressure distribution computed on the profile at $M_\infty = 0.750$, $\alpha = 0.5$ deg. At these flow conditions a shock is present in the flow on the upper surface, as can be seen in Fig. 1. Using the distribution given by the circles in Fig. 1 and using the NACA 0012 airfoil as a starting contour (see Fig. 2), the Korn airfoil is recovered exactly in about a dozen iterations of the airfoil shape. A measure of the convergence rate of the procedure is given by Fig. 3, which gives the maximum value of $|v/u|$ at each cycle. The program was run through 30 cycles, but usually no changes in the shape can be noticed once the maximum value of $|v/u|$ has been reduced below 0.01. The pressure distribution computed on the redesigned profile is given by the solid line in Fig. 1, and it is practically identical to the specified target (the symbols). Obviously, since the target was a direct solution for the flow over a known profile, it satisfied the three constraints of the inverse problem, and it should have generated an airfoil solution without the need of modifications. The actual values computed for the three parameters were (scheme 1 was used in this example) $p_1 = 1.000023$, $p_2 = 0.000040$, and $p_3 = -0.000668$, all well within the specified numerical tolerances. The slight differences near the shock are due to the fact that values for the target distribution were computed by central difference formulas at midpoints of the mesh, and the values associated with the recomputed profile were computed at node points. The Korn profile is recovered exactly, without any wiggles. In Fig. 4 the slope distribution of the computed profile in the leading-edge region is compared with the values of the original contour. Even in the vicinity of the shock, the redesigned profile is as smooth as the original Korn airfoil, as can be seen in Fig. 5. An interesting exercise is to use the pressure distribution in Fig. 1 as the target for designing airfoils at a freestream Mach number other than 0.75. This was tried with freestream Mach numbers of 0.730 and 0.770. As the results of Figs. 5 and 6 show, the recompression through the shock implied by the target distribution can no longer be achieved over a smooth profile. Thus, even though the "modified" target distributions for these two cases differ only slightly (see Fig. 6) from the original target, the designed airfoil profiles have dramatically different upper surfaces. As Fig. 5 shows, the profile designed for $M_\infty = 0.730$ has a convex (to the flow) corner underneath the shock, whereas the profile obtained for $M_\infty = 0.770$ has a concave corner at that point. This is consistent with expectations. For the $M_\infty = 0.730$ case, the specified shock is too strong. The opposite is true for the case at $M_\infty = 0.770$. The vertical extent of the supersonic regions in these several cases is of some interest, as can be seen in Fig. 7. The corners for the designs in Fig. 5 are real features of the airfoil solutions and are not due to numerical inaccuracies.

Evidence for the equivalence between the present design procedure and the one described for incompressible flow can be generated by trying to design a profile using as a target speed the distribution obtained by the analysis at a low Mach number of a known profile, which we will call q_1, multiplied by some arbitrary factor, p. Since q_1 automatically satisfies Eq. (19), (pq_1) cannot satisfy the constraints. Straightforward application of

Eq. (19) suggests that, for a contour to exist, either q_∞ must be scaled by the same factor, or p must be scaled back to one. The numerical procedure described in this paper accomplishes this same result as shown in Fig. 8. The Korn airfoil was analyzed at $M_\infty = 0.100$, $\alpha = 1.7$ deg, and the resulting speed distribution was scaled to provide a target. The scaling factor was assigned a value in the range of $0.2 \leqslant q_\infty \leqslant 2$. As was expected, using the previously mentioned scheme 1, the multiplier was scaled back to one in each case. As can be seen in Fig. 8, the scaling is accomplished almost entirely within the first design cycle—before any airfoil updates. Also, as expected, the resulting profile was the Korn airfoil.

These examples demonstrate the robustness and self-consistency of the numerical scheme. The remaining examples illustrate the use of the various schemes for satisfying the constraints in the design of airfoils. Rarely, if ever, are pressure distributions with shocks in the flowfield prescribed. At supercritical speeds, "shockless" airfoils are usually the goal. A reasonable target might be the distribution depicted by the symbols in Fig. 9. The freestream Mach number in this case is 0.800, and, again, the exercise is to design a closed airfoil using scheme 1 to make any necessary changes in the target distribution. The modified target speed distribution is given by the solid line in Fig. 9.

The shift in the location of the stagnation point should be noticed in this figure. The shift is achieved smoothly and makes it possible to close the x gap in the airfoil. The designed airfoil is depicted in Fig. 10, along with the computed pressure distribution. This pressure distribution is the result of a direct solution of the flowfield over the designed airfoil contour, and it agrees to three decimal places with the pressure distribution that corresponds to the target speed distribution (the solid line in Fig. 9). This airfoil solution is obtained regardless of the airfoil contour initially prescribed to start the iteration procedure. In Fig. 11 the designed airfoil contour is compared with four different starting shapes: the Korn airfoil, the NACA 0012, the NACA 0002, and, finally, a "needle"—two straight lines joined at the trailing edge and at the leading edge tangent to a semicircle of radius equal to 0.25% of the chord.

It is satisfying to note that the values of p_1, p_2, and p_3 are identical regardless of the starting shape (i.e., the modified target speed distribution is the same in all cases). Apparently, by decoupling the three parameters, we have ensured that only a single set of values exists that satisfies the three constraints. It is possible that, if the three parameters had been coupled, more than one set of values might exist that would satisfy the constraints. Even though we have no formal proof of this, decoupling appears to guarantee a unique solution as well as making the search simpler and faster. The convergence rate of the method for the various "starter" profiles is given in Fig. 12, which depicts the maximum value of $|v/u|$ as a function of design cycle. Again, after 10–12 cycles it is difficult to distinguish any changes in the airfoil shape. Typically, we run the code to a level where the maximum $|v/u|$ is 0.001 or smaller. A converged solution generally requires 4–5 min on a Cray-1M computer and about 20 min on the IBM/3081 machine.

The pressure distribution depicted in Fig. 10 appears to have very desirable features; in particular, the "plateau" region on the upper surface suggests the absence of a shock. However, a very large drag ($C_D = 0.0232$) is present even at the design point. If we look at the Mach number contours in Fig. 13, we see that, although there is no shock at the airfoil surface itself, a very strong shock is present off the surface. The contours represent increments of 0.01 in Mach number, and only contours for values greater than the freestream are shown. At off-design conditions, the shock reaches the surface. Several authors have observed this feature. A smooth recompression along the surface does not necessarily mean that the flow-field is shockless. Thus, the airfoil shown in Fig. 10 is impractical because of its high drag.

A truly shockless closed airfoil is depicted in Fig. 14, along with the computed pressure distribution (i.e., modified target) and the original, unmodified target. Note the low computed drag ($C_D = 0.0005$) of this airfoil. The computed isomach pattern in Fig. 15 shows that the flow over this airfoil is truly shock-free, and at off-design points only a weak shock develops. This case was computed using scheme 2 described earlier. It should also be noted that in this case the modifications made to the ideal target pressure distribution are considerably larger than those that resulted in the previous case. The changes on the lower surface reflect mostly the effect of f_2, as given by Eq. (29). An example of an airfoil designed using scheme 3 is shown in Fig. 16. Note, in this case again, the very low value for the drag and the considerable lift coefficient. The modifications to the ideal target that should be noticed apart from the scaling are concentrated near the trailing edge.

A very interesting profile designed to an unusual pressure distribution is depicted in Fig. 17. The airfoil was designed for laminar flow (remember that the present method is purely inviscid) to a distribution devised by Pfenninger[22] for $M_\infty = 0.766$. It is only one of a series of airfoils designed for such purposes. The scheme used was scheme 1. Since the ideal pressure distribution was based on the considerable personal experience of its designer, minor modifications were needed to generate the airfoil solution. The computed Mach number contours are depicted in Fig. 18. Notice the shallowness of the supersonic region compared to its length. This airfoil exhibits very low drag for a considerable range of flow conditions around its design point.

As mentioned earlier, the method will generate airfoil contours of arbitrary trailing-edge thickness. The contour shown in Fig. 19 has a trailing-edge thickness equal to 2% of its chord. Like the previous example, this represents an interesting design that, in addition to front loading, has a long and shallow supersonic flow region. A final set of examples depicts airfoils designed at a moderate freestream Mach number ($M_\infty = 0.675$). All were designed to have a trailing-edge thickness equal to 1% of the chord. The two examples shown in Figs. 20 and 21 were designed to original target pressure distributions that differed only in the leading-edge region. The resulting airfoils both have a substantial thickness: maximum values are 12.5 and 13.7% of the chord, respectively. The contour in Fig. 22 is not as thick (11% of chord), but it generates considerably more lift. The very shallow supersonic region present on this profile at its design point is of interest in Fig. 23.

Summary

The preceding sections have described the classical approach to airfoil design introduced by Mangler and Lighthill and have presented an extension of the methodology to transonic flow. The problem imposes constraints on the speed distribution to which the airfoil is to be designed. An effort has been made to illustrate the most elusive of the constraints—the one relating the target surface speed to the freestream speed—and to interpret it within the context of the numerical scheme presented. All of the constraints on the speed are accounted for in the formulation. The method is therefore well posed both theoretically and numerically. It is also quite general in the sense that the ideal specified speed distribution, represented by $f_0(s)$, is general, and an airfoil solution will always be found by modifying the target speed in order to satisfy the constraints. Also, the initial airfoil contour needed to start the procedure need not be close to the final contour to achieve convergence.

The particular forms presented for f_1, f_2, and f_3 are by no means exhaustive or even necessarily best. They do, however, provide the freedom needed to satisfy the constraints automatically, without user intervention, and to introduce only a loose coupling among their respective multipliers, making their evaluation simpler and computationally cost effective. Other forms for f_2 and f_3 are, of course, possible, although the search for p_1, p_2, and p_3 might be more difficult. Most alternative formulations for introducing free parameters will probably require a multidimensional search for the parameters. However, techniques exist for optimizing this search. Also, following the approaches of Arlinger, Strand, and Polito, it would be possible to develop formulations that would keep changes to a minimum. Additional free parameters could conceivably be introduced to prevent crossovers of the upper and lower surfaces of the airfoil, a possibility not ruled out by the present formulation. In its present form, however, the approach presented is a reliable and efficient method for the design of airfoil profiles of given trailing-edge thicknesses at transonic speeds.

References

[1]Mangler, W., "Die Berechnung eines Tragflügelprofiles mit Vorgeschriebener Druckverteilung," Jahrbuch der Deutschen Luftfahrtanschung, 1938.

[2]Lighthill, M. J., "A New Method of Two-Dimensional Aerodynamic Design," Aeronautical Research Council, London, R&M 2112, 1945.

[3]Woods, L. C., "Aerofoil Design in Two-Dimensional Subsonic Compressible Flow," Aeronautical Research Council, London, R&M 2845, March 1952.

[4]Van Ingen, J. L., "A Program for Airfoil Section Design Utilizing Computer Graphics," AGARD Short Course Notes, 1969.

[5]Arlinger, B., "An Exact Method of Two-Dimensional Airfoil Design," SAAB, Linköping, Sweden, TN-67, 1970.

[6]Strand, T., "Exact Method of Designing Airfoils with Given Velocity Distribution in Incompressible Flow," *Journal of Aircraft*, Vol. 10, Nov. 1973, pp. 651–659.

[7]Polito, L., "Un Metodo Esatto per il Progetto di Profili Alari in Corrente Incompressibile Aventi un Prestabilito Andamento della Velocità sul Contorno," Universita degli Studi di Pisa, Pisa, Italy, Rept. 42, 1974.

[8]Hicks, R. M., Vanderplaats, G. N., Murman, E. M., and King, R. R., "Airfoil Section Drag Reduction at Transonic Speeds by Numerical Optimization," NASA TMX-73097, Feb. 1976.

[9]Davis, W. M., "Technique for Developing Design Tools from the Analysis Methods of Computational Aerodynamics," AIAA Paper 79-1529, 1979.

[10]McFadden, G. N., "An Artificial Viscosity Method for the Design of Supercritical Airfoils, *Courant Inst. of Mathematical Science*, New York Univ., Research and Development Rept. C00-3077-158, July 1979.

[11]Tranen, T. L., "A Rapid Computer Aided Transonic Airfoil Design Method," AIAA Paper 74-501, 1974.

[12]Carlson, L. A., "Transonic Airfoil Analysis and Design Using Cartesian Coordinates," *Journal of Aircraft*, Vol. 13, May 1976, pp. 369–356.

[13]Sobieczky, M., Fung, K. Y., and Seebass, A. R., "A New Method for Designing Shock-free Transonic Configurations," AIAA Paper 78-1114, 1978.

[14]Bauer, F., Garabedian, P., and Korn, D., "Supercritical Wing Sections," Springer-Verlag, New York, 1972.

[15]Boerstoel, J. W., and Muizing, G. H., "Transonic Shock-Free Airfoil Design by an Analytic Hodograph Method," AIAA Paper 74-439, 1974.

[16]Volpe, G., and Melnik, R. E., "The Role of Constraints in the Inverse Design Problem for Transonic Airfoils," AIAA Paper 81-1233, June 1981.

[17]Volpe, G., "On the Design of Airfoil Profiles for Supercritical Pressure Distributions," *Proceedings of Second International Conference on Inverse Design Concepts and Optimization in Engineering Sciences*, edited by G. S. Dulikravich, Univ. of Texas at Austin, 1987, pp. 487–505.

[18]Ludford, G. S., "The Behavior at Infinity of the Potential Function of a Two-Dimensional Subsonic Compressible Flow," *Journal of Mathematics and Physics*, Vol. 30, 1951, pp. 117–130.

[19]Jameson, A., "Acceleration of Transonic Potential Flow Calculations on Arbitrary Meshes by the Multiple Grid Method," AIAA Paper 79-1458, 1979.

[20]Thwaites, B., *Incompressible Aerodynamics*, Oxford Univ. Press, Oxford, UK, 1960.

[21]Volpe, G., "The Inverse Design of Closed Airfoils in Transonic Flow," AIAA Paper 83-504, Jan. 1983.

[22]Pfenninger, W., Viken, J. K., Vemuru, C. S., and Volpe, G., "All Laminar SC LEC Airfoils with Natural Laminar Flow in the Region of the Main Wing Structure," AIAA Paper 86-2625, Oct. 1986.

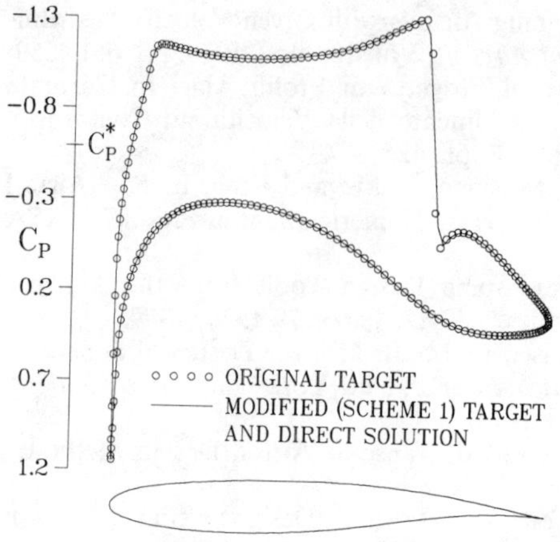

Fig. 1 Target pressure distribution (computed on Korn airfoil at 0.5 deg); redesigned Korn airfoil and direct solution ($M_\infty = 0.750$, $\alpha = 0$ deg, $C_L = 0.8334$, $C_D = 0.0063$).

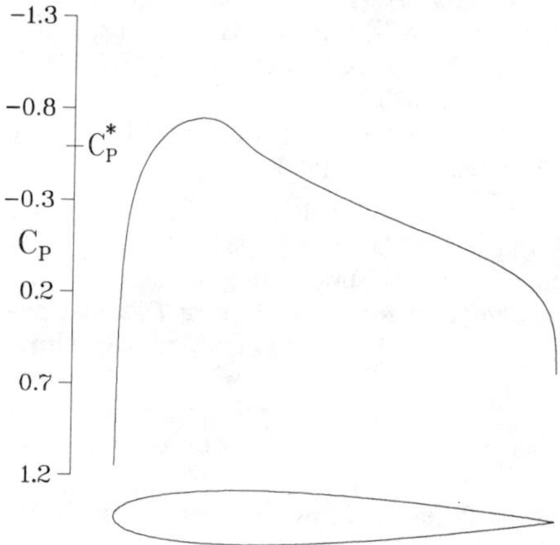

Fig. 2 Starter airfoil contour (NACA 0012) and computed pressure distribution ($M_\infty = 0.750$, $\alpha = 0$ deg, $C_L = 0.0$, $C_D = 0.0001$).

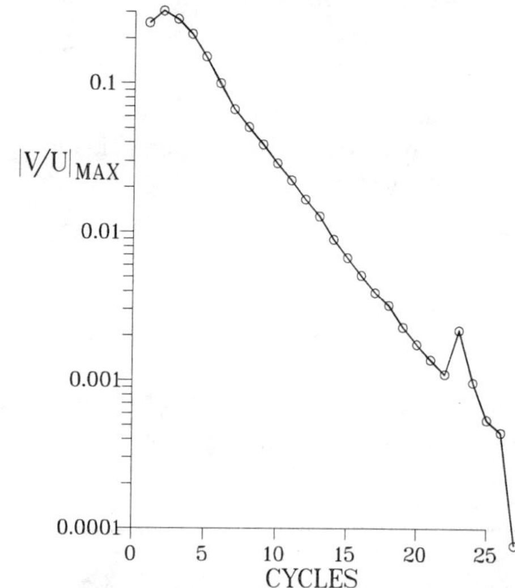

Fig. 3 Convergence history of maximum velocity ratio; Korn airfoil redesign ($M_\infty = 0.750$, $\alpha = 0$ deg).

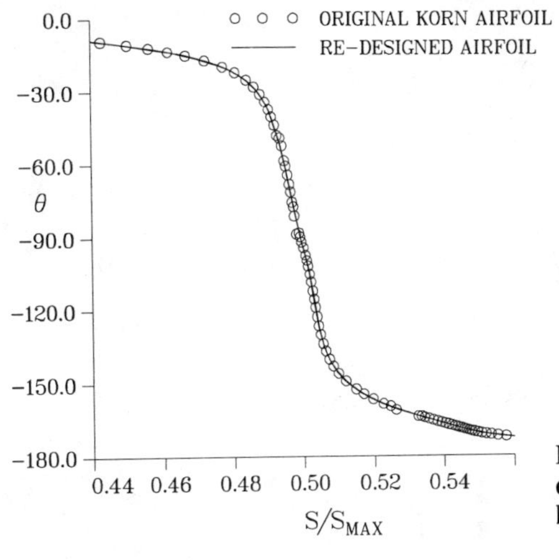

Fig. 4 Computed and actual slope distribution for Korn airfoil design in leading-edge region.

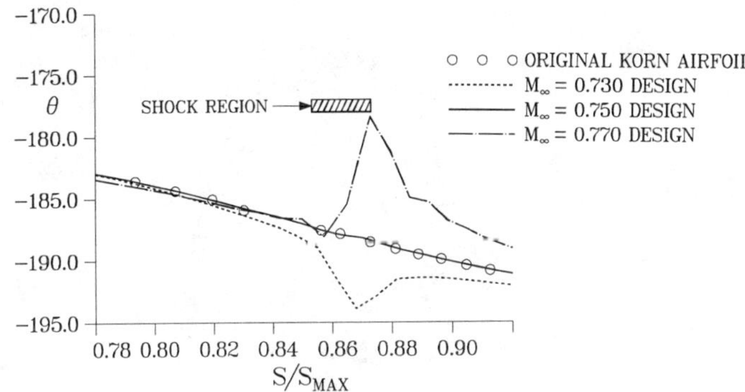

Fig. 5 Slope distribution in shock-wave region for Korn airfoil and for airfoils designed to Korn pressure distribution in Fig. 1 at various freestream Mach numbers.

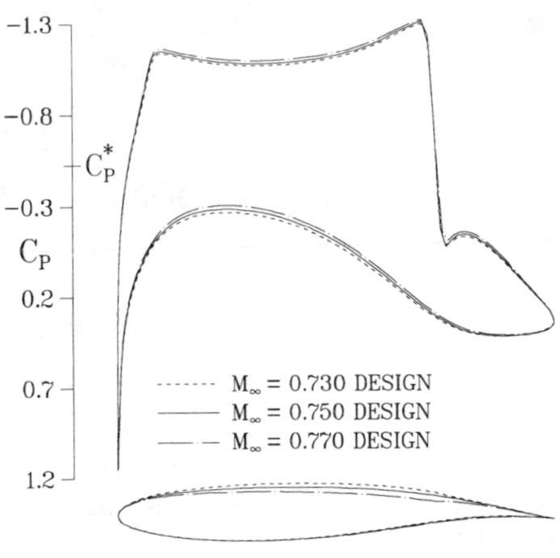

Fig. 6 Direct solutions (modified target distributions) for airfoils designed to Korn pressure distributions from Fig. 1 ($\alpha = 0$ deg). $M_\infty = 0.770$: $C_L = 0.8359$, $C_D = 0.0026$. $M_\infty = 0.750$: $C_L = 0.8334$, $C_D = 0.0063$. $M_\infty = 0.770$: $C_L = 0.8316$, $C_D = 0.0167$.

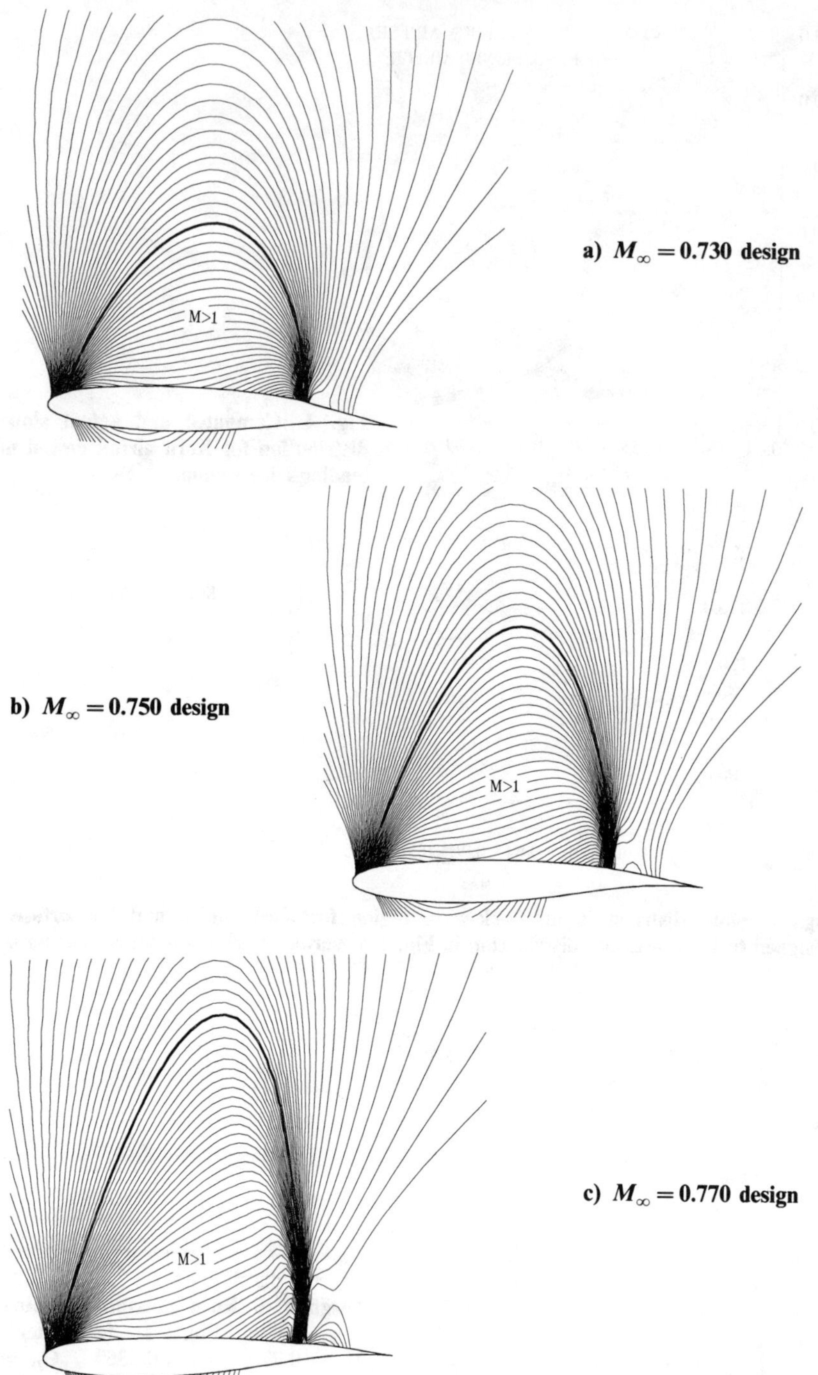

Fig. 7   Design point isomachs for airfoils designed to Korn pressure distributions from Fig. 6 ($\alpha = 0$ deg); contours shown at 0.01 intervals.

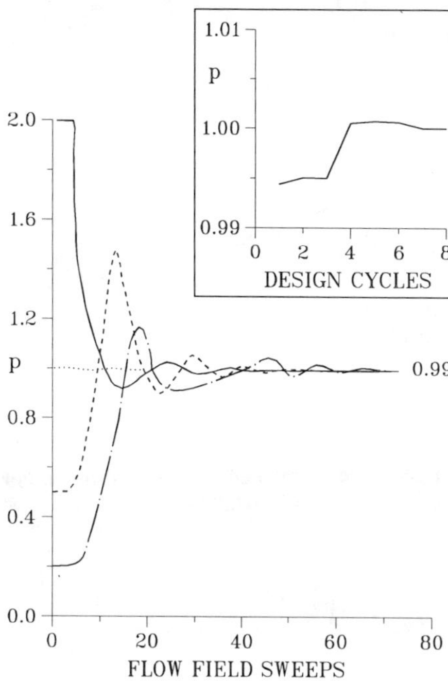

Fig. 8 Convergence history of scale factor for Korn airfoil design ($M_\infty = 0.100$).

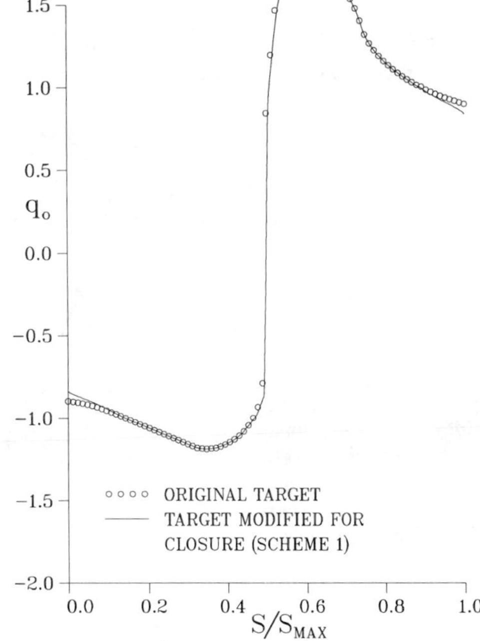

Fig. 9 Original and modified target speed distributions; shockless case ($M_\infty = 0.800$, $\alpha = 0$ deg).

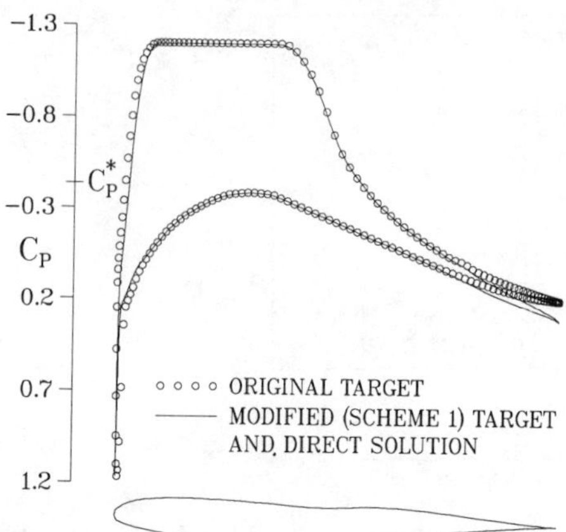

Fig. 10 Designed contour, original target, and computed pressure distribution; shockless case ($M_\infty = 0.800$, $\alpha = 0$ deg, $C_L = 0.4801$, $C_D = 0.0232$).

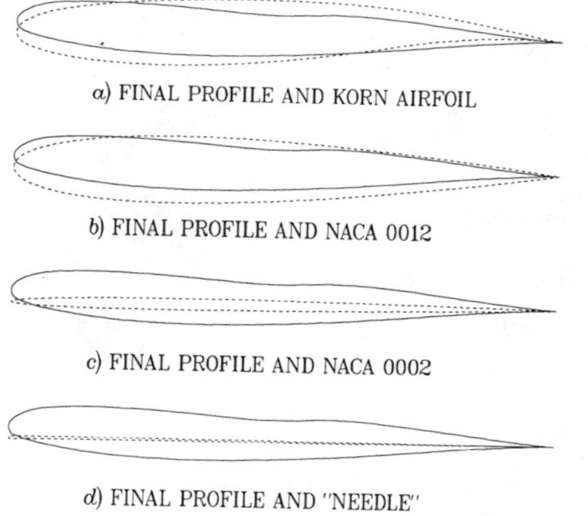

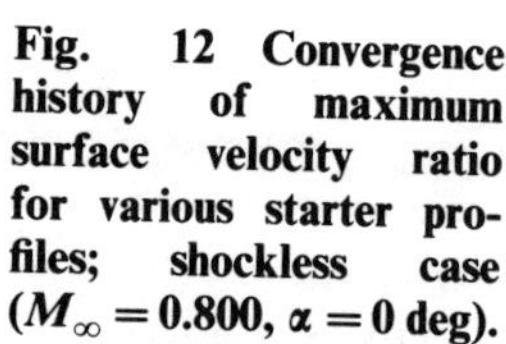

Fig. 11 Final airfoil profile (solid line) compared with starting profiles (dashed line).

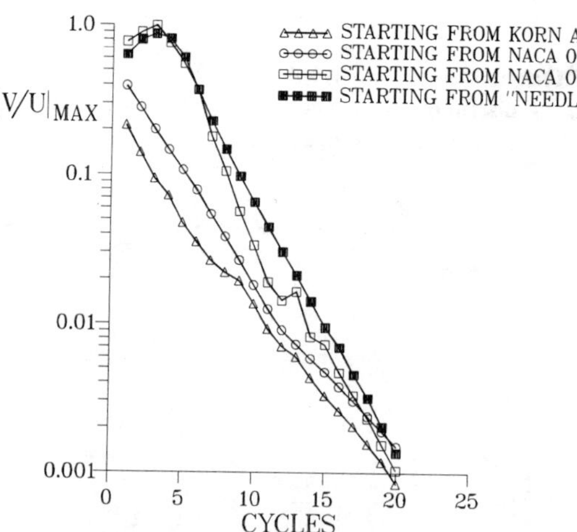

Fig. 12 Convergence history of maximum surface velocity ratio for various starter profiles; shockless case ($M_\infty = 0.800$, $\alpha = 0$ deg).

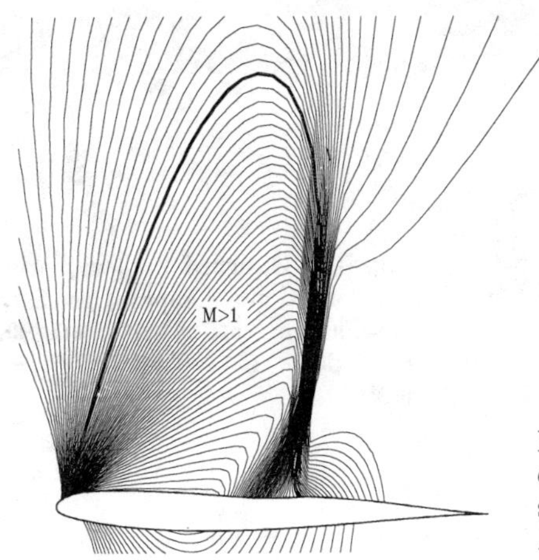

Fig. 13 Design point isomachs; shockless case ($M_\infty = 0.800$, $\alpha = 0$ deg); contours shown at 0.01 intervals beginning with $M = 0.810$.

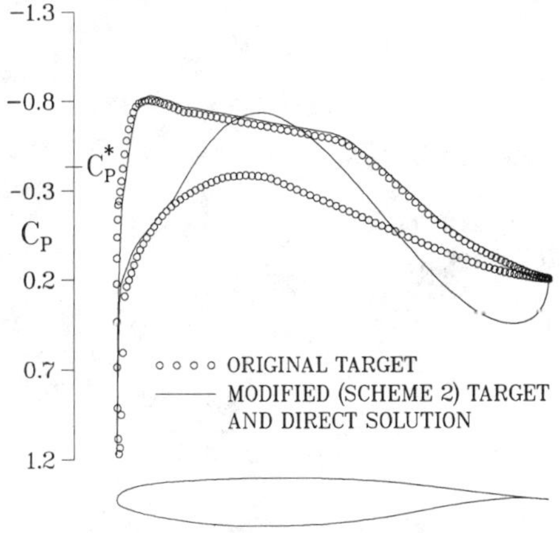

Fig. 14 Designed contour, original target, and computed pressure distribution; case 2 ($M_\infty = 0.800$, $\alpha = 0$ deg, $C_L = 0.2673$, $C_D = 0.0005$).

Fig. 15 Design point isomachs; case 2 ($M_\infty = 0.800$, $\alpha = 0$ deg); contours shown at 0.01 intervals beginning with $M = 0.810$.

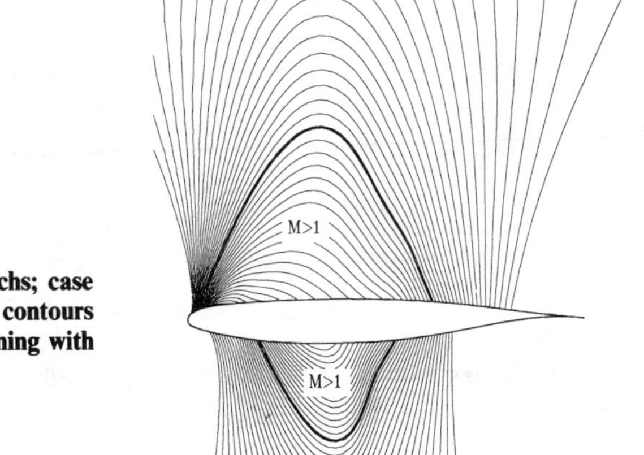

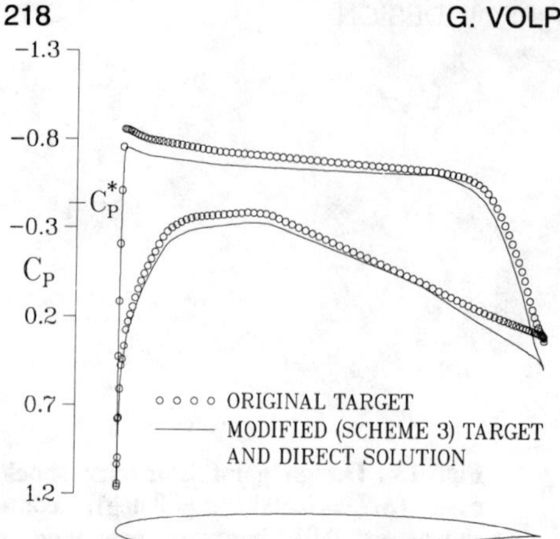

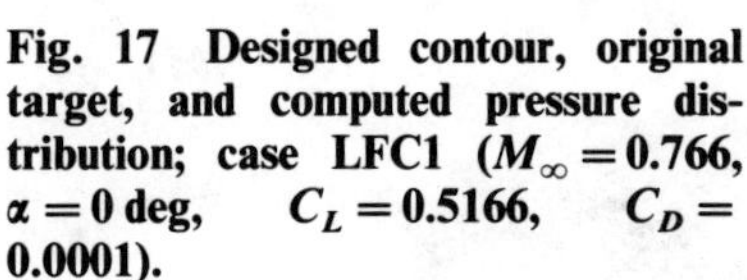

Fig. 16 Designed contour, original target, and computed pressure distribution; case 3 ($M_\infty = 0.800$, $\alpha = 0$ deg, $C_L = 0.4972$, $C_D = 0.0002$).

Fig. 17 Designed contour, original target, and computed pressure distribution; case LFC1 ($M_\infty = 0.766$, $\alpha = 0$ deg, $C_L = 0.5166$, $C_D = 0.0001$).

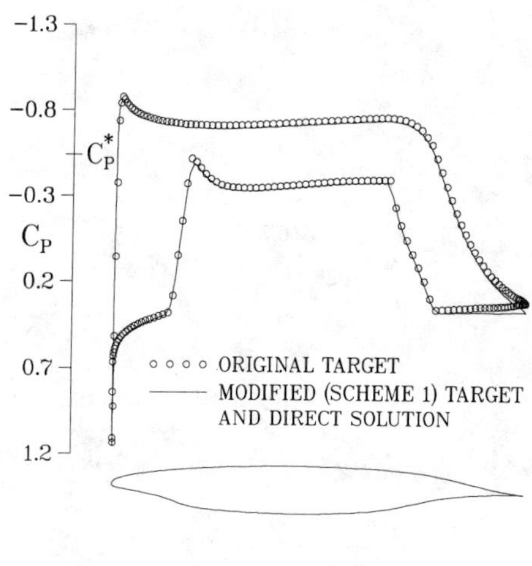

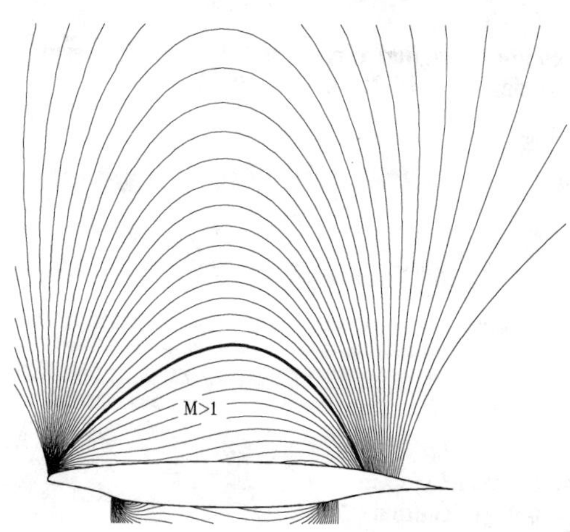

Fig. 18 Design point isomachs; case LFC1 ($M_\infty = 0.766$, $\alpha = 0$ deg); contours shown at 0.01 intervals beginning with $M = 0.770$.

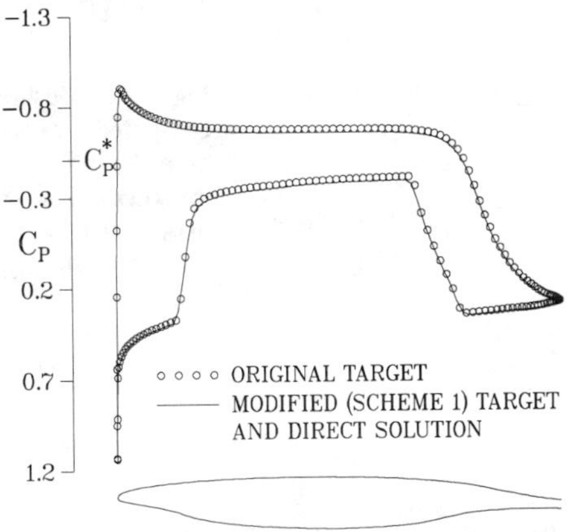

Fig. 19   Designed contour, original target, and computed pressure distribution; case LFC2, 2% trailing-edge thickness ($M_\infty = 0.775$, $\alpha = 0$ deg, $C_L = 0.4805$, $C_D = 0.0001$).

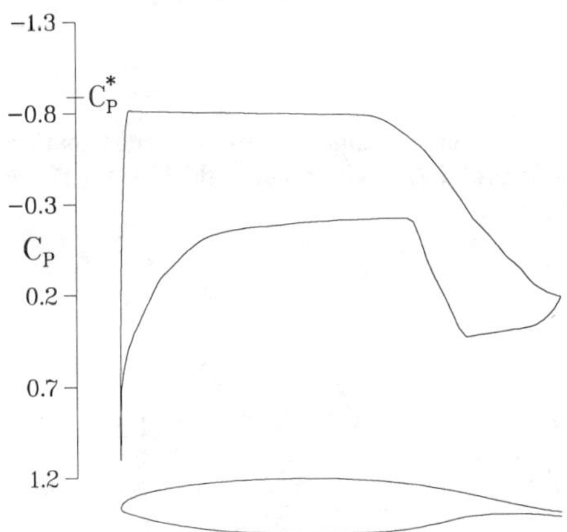

Fig. 20   Designed contour and computed pressure distribution; case AMSS-8, 12.5% maximum thickness, 1% trailing-edge thickness ($M_\infty = 0.675$, $\alpha = 0$ deg, $C_L = 0.6121$, $C_D = 0.0001$).

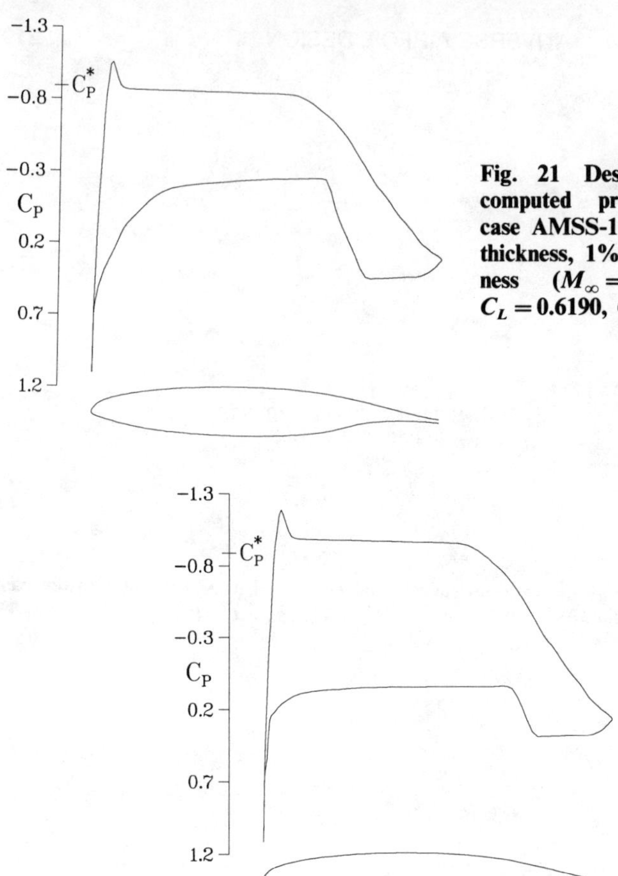

Fig. 21 Designed contour and computed pressure distribution; case AMSS-10, 13.7% maximum thickness, 1% trailing-edge thickness ($M_\infty = 0.675$, $\alpha = 0$ deg, $C_L = 0.6190$, $C_D = 0.0001$).

Fig. 22 Designed contour and computed pressure distribution; case AMSS-11, 11.0% maximum thickness, 1% trailing-edge thickness ($M_\infty = 0.675$, $\alpha = 0$ deg, $C_L = 0.8540$, $C_D = 0.0001$).

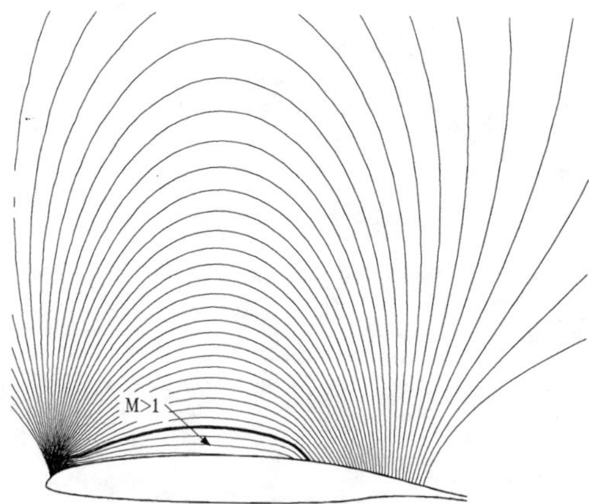

Fig. 23 Design point isomachs; case AMSS-11 ($M_\infty = 0.675$, $\alpha = 0$ deg); contours shown at 0.01 intervals beginning with $M = 0.680$.

Chapter 8

Innovation with Computational Aerodynamics: The Divergent Trailing-Edge Airfoil

P. A. Henne*
*Douglas Aircraft Company, McDonnell Douglas Corporation,
Long Beach, California*

Introduction

THE aerodynamic design of airfoil sections continues today as an elegant yet practical engineering design problem. It is elegant in that the solution of any airfoil design problem, however complex, is nothing more than a two-dimensional closed contour. It is practical in that many, if not most, problems in aerodynamics involve the generation of lift, drag, and moment by airfoil sections.

The airfoil design problem has received much attention in experimental, theoretical, and computational studies. Numerous volumes and documents have been compiled to summarize such studies. Reference 1 stands as a classic example of such documentation and provides an excellent summary of early airfoil research efforts. Early theoretical studies of airfoil design led to the decomposition of airfoil geometric characteristics into thickness, camber, and angle of attack, as shown in Fig. 1. Much of the early design studies dealt with the proper combination thickness shapes and camber shapes necessary to achieve some aerodynamic design goal. It is interesting to point out that few studies were focused on details associated with the airfoil trailing edge. Trailing-edge effects normally were of concern only when mechanical variations for high-lift systems or control systems were introduced.[1,2]

Traditionally, airfoil closure has been accomplished by using trailing-edge included angles typically in the 5- to 15-deg range in conjunction with essentially zero trailing-edge thickness. Early trailing-edge design principles were that the trailing edge must be sharp to avoid drag penalties, and the trailing-edge included angle, as defined by ϵ in Fig. 1, must be positive to allow for practical manufacture. These principles caused most early airfoil trailing-edge geometries to be wedge shaped.

More recently, work by Whitcomb[3] and others to develop the supercritical airfoil has shown the possibility of using a thin-trailing-edge geometry

---

Copyright © 1989 by the American Institute of Aeronautics and Astronautics, Inc. All rights reserved.

*Manager, Advanced Programs, Aerodynamics and Acoustics Subdivision; currently, Chief Design Engineer, MD-80 Program.

with near-parallel trailing-edge surfaces to produce a superior transonic airfoil section. The thin-trailing-edge and high-aft-camber geometry provides for significant aft lift loading on the supercritical airfoil, as indicated in Fig. 2. The aft loading represents a large increment in lift and contributes a major part of the transonic performance improvement associated with the supercritical airfoil.

Early analysis and critique of the supercritical airfoil revealed several adverse characteristics. First, the thin aft section introduced concern about structural penalties of the thin, highly loaded trailing-edge configuration. This concern resulted in a challenge to one of the early airfoil closure principles, namely, that of trailing-edge sharpness. Subsequent research, summarized in Ref. 4, provides the conclusion that a small amount of trailing-edge thickness or bluntness can be incorporated with little or no drag penalty. Such a trailing-edge thickness, less than 1% chord, is small but nevertheless alleviates concern for practical manufacturability. Applications of the supercritical airfoil have generally made use of this trailing-edge bluntness approach to achieve essentially parallel upper and lower surfaces in a practical design. Second, adverse viscous boundary-layer effects have been found to be more significant for highly aft-loaded airfoils.[5] The aft loading leads to more adverse pressure gradients and correspondingly more rapid boundary-layer growth near the trailing edge on both airfoil surfaces. A significant amount of the aft camber is effectively lost due to boundary-layer decambering near the upper-surface trailing edge and in the lower-surface concave region. As a result of these adverse characteristics, the full inviscid benefit of the supercritical airfoil is not obtained in practice.

Fundamental studies accomplished at McDonnell Douglas have identified a new type trailing edge that provides superior aerodynamic performance compared to the previous types of closures. This new trailing-edge concept utilizes three fundamental characteristics. First, a finite trailing-edge thickness is required. Second, strongly divergent trailing-edge upper and lower surfaces are introduced. Third, the surface curvature on the lower surface increases rapidly as the trailing edge is approached. The combination of these three characteristics produces an unconventional airfoil trailing edge, as shown in Fig. 3. This type of airfoil is referred to as a divergent trailing edge (DTE) airfoil (U.S. patent no. 4,858,852).[6-8] This chapter presents the original development of the DTE airfoil concept as well as computational analyses and wind-tunnel test results for airfoils and wings utilizing such trailing edges.

Origin of the Concept

The origin of this new concept in airfoil design is attributed to a two-part design study using computational aerodynamics that was performed in 1981. The first part of the 1981 study involved an analysis of the blunt trailing-edge effects reported in Ref. 4. The transonic airfoil analysis method of Bauer, Garabedian, and Korn, referred to as Pgm H,[9] was used to calculate the aerodynamic characteristics of the airfoil sections reported in Ref. 4. The Pgm H method utilizes an integral boundary-layer solution in conjunction with the profile drag formula of Squire and Young to evaluate viscous drag characteristics. This combination is quite reasonable for the drag created by airfoil upper and lower surfaces. However, a major shortcoming of the original Pgm H was the inability of the numerical scheme to evaluate base drag. In order to address the base drag characteristics, an semiempirical base drag formula was developed. The formula for incremental base drag coefficient is

$$C_{d_{\text{base}}} = -C_{p_{\text{base}}} \frac{q_{te}}{q_0} \frac{\Delta Z_{te}}{C}$$ (1)

where $C_{p\,\text{base}}$ is the base pressure coefficient, referenced to local potential-flow conditions at the trailing edge. The q_{te}/q_0 is the local to freestream dynamic pressure ratio. The $\Delta Z_{te}/C$ is the nondimensional trailing-edge thickness parameter, base thickness divided by airfoil chord. The key to this formula is determining the base pressure coefficient. The correlations of Nash[10] are used for this purpose. The interesting facet of this approach is that, for trailing-edge thickness values less than about 4%, the Nash data indicate a linear variation of base pressure coefficient with trailing-edge thickness. This linearity substituted into Eq. (1) results in

$$C_{d_{\text{base}}} \sim \left(\frac{\Delta Z_{te}}{C} \right)^2$$ (2)

The quadratic variation of the base drag increment with trailing-edge thickness, as indicated by Eq. (2), leads to the expectation that the base drag penalty for small trailing-edge thicknesses might indeed be negligible.

This formula, when combined with Pgm H calculations, provided the drag comparison shown in Fig. 4. In this figure the measured drag variation with trailing-edge thickness from Ref. 4 is compared to the calculated variation. The calculated drag has been shifted lower by several counts to match the data point for the thinnest trailing edge. The dashed line indicates the basic profile drag variation stemming from the Squire and Young calculation. It reveals an interesting but modest reduction in drag with increasing trailing-edge thickness. This effect is related to an increase in local velocities near the trailing edge as a result of the additional local thickness. The quadratic base drag increment added to this basic level is also shown. The calculated variation in drag with trailing-edge thickness matches the measured results remarkably well for drag data. The comparison indicates that the true minimum drag occurs at a small but nonzero trailing-edge thickness. It also indicates that trailing-edge thicknesses on the order of 0.5% can be utilized with drag penalties as low as one drag count. This 1981 base drag study reinforced the notion of Ref. 4 concerning the advantage of using a blunt trailing edge.

The second part of the 1981 study involved the evaluation of potential trailing-edge modifications of an existing transport wing geometry. The original wing section in this case predated the supercritical airfoil and has a conventional positive included-angle closure. The design objective for the study was to introduce some level of aft camber near the trailing edge of this section and make a corresponding improvement in the transonic lift and drag characteristics. In other words, the author sought to find a modest change to the trailing-edge region that would achieve some measure of the supercritical airfoil performance without having to redesign the entire wing.

The trailing edge of the original airfoil section is shown in Fig. 5. Three possible trailing-edge modifications are also shown in Fig. 5. Although all three shapes would add aft camber effect to the airfoil section, the small reversed wedge with no chord extension was the most intriguing configuration. The wedge added to the lower surface was intriguing in several respects. First, the base drag study had indicated that the base drag for such a wedge would be acceptable if its size were kept small. Second, of the three approaches the added lower-surface wedge was the most easily accomplished for wind-tunnel model testing. Third, the added wedge seemed like a natural, high-speed version of the Zaparka flap[11] or the Gurney flap,[12] which had been demonstrated to produce surprisingly large increments in lift[12] at low-speed conditions.

A preliminary computational assessment of this configuration was made to evaluate the benefit of such a modification. The computations, shown in Figs. 6–8, revealed a strong camber effectiveness at transonic normal airfoil design conditions. "Normal" in this context implies perpendicular conditions associated with simple sweep theory. The calculated pressure distributions for the original and modified airfoils at constant angle of attack are shown in Fig. 6. The normal Mach number is 0.74. The effect of the trailing-edge modification is a substantial increase in circulation, as can be noted by the change in velocity everywhere around the airfoil. The calculated increment in lift at constant angle of attack is 0.14, a large change in lift for transonic airfoil design. In Figs. 7 and 8 the calculations were performed at the same lift coefficient, which means that the angle of attack of the modified airfoil is less than that of the original airfoil. The results shown in the figures indicate that the circulation effect and angle-of-attack effect combine to give a redistribution of the airfoil loading and a reduction of shock strength at constant lift coefficient. The reduced shock strength leads to reduced drag, as indicated in Figs. 7 and 8.

The calculated performance improvement of the airfoil in the transonic range was impressive enough that a transonic wind-tunnel test was conducted to verify the results. The testing was accomplished in the National Aeronautical Establishment (NAE) High Reynolds Number Two-Dimensional Test Facility[13] in Ottawa, Canada. The testing did verify the calculated results, as shown in Figs. 9–11. The measured pressure distributions shown in Figs. 9 and 10, at constant angle of attack and constant lift coefficient, respectively, indicate the same character as the original computations. The calculated lift and drag increments due to the trailing-edge modification are remarkably well substantiated by the test data, as indicated in Fig. 11. This simple trailing-edge modification improved the airfoil lift capability and delayed the drag rise onset to higher lift coefficients.

The airfoil performance improvement identified for this small modification was gratifying, but, more importantly, the effectiveness of this modification caused a re-examination of trailing-edge design approaches.

Development of the DTE Airfoil Concept

Computational Development

The 1981 test substantiated the computationally predicted effectiveness of small variations in airfoil contour near the trailing edge. In order to exploit this effect a large computational study of various approaches to trailing-edge contouring was accomplished. The key to the study was to recognize that a thick trailing edge was acceptable, a thick trailing edge admits closures with negative included angles, and sharp curvature near the trailing edge is acceptable. This recognition is nothing less than discarding traditional closure principles. The results of this large computational effort led to the development of the DTE concept.

The context of the trailing-edge contouring study was advanced airfoils for transport aircraft application. Consequently, the starting point was a well-developed supercritical airfoil section. As indicated in Fig. 12, a baseline or original airfoil was used to initiate a series of parametric variations. The variations shown in the figure represent lower-surface contour changes very near the trailing edge. Succeedingly higher curvature is introduced near the trailing edge. Two constraints must be observed in this process. First, it is important to maintain sufficient structural thickness in this region to make a practical design. Second, the lower-surface contouring can create adverse pressure gradients sufficient to cause lower-surface boundary-layer separation. This condition needs to be avoided in the operational lift range of the airfoil section.

Results shown in Fig. 13 are presented to generalize some of the computational study efforts. In this figure total section drag is plotted vs trailing-edge included angle for a large number of airfoils derived from a baseline airfoil DLBA 186. The normal Mach number of 0.74 and the normal lift coefficient of 0.8 are typical design conditions for advanced airfoil designs. Each data point represents a discrete airfoil configuration that was computationally designed and analyzed in detail. The objective of the study was to determine the best trailing-edge closure for minimum drag without the traditional trailing-edge design constraints. As indicated, the trailing-edge included angles of interest are negative and represent upper and lower surfaces that strongly diverge from each other (hence the name divergent trailing edge). At any given trailing-edge angle some variation is noted with different lower-surface contours. Several constraint limits are also indicated in the figure. Two lower-surface boundary-layer separation boundaries are shown. The Nash-Macdonald method[14] of calculation is known to be conservative in the determination of separation. Therefore, this boundary offers some design margin, for example, to lower lift coefficients. The second separation limit is a Cebeci method of calculation.[15] This calculation of separation is typically more accurate. The third constraint is the structural thickness limit. Airfoils beyond this line have a minimum thickness point that is not sufficient for practical structure for the typical transport application. Determination of this constraint is dependent on the application and the aggressiveness of the structural designer. The results shown in the figure indicate that drag reductions on the order of

11% were obtained for airfoil sections in the -20 to -25-deg range. Such a performance improvement at an airfoil design point is quite substantial and was unexpected.

The airfoil geometry that is indicated as DLBA 243 is compared to the baseline DLBA 186 in Fig. 14. A closeup view of the the trailing edge is also shown. In this case the upper surfaces of the two airfoils are identical. The lower surfaces differ predominantly near the trailing edge where the DTE has been cut into the contour of the original airfoil. It is apparent in the blowup view that the upper- and lower-surface divergence is substantial, and there is a rapid increase in lower-surface curvature as the trailing edge is approached. However, in observing the view of the complete airfoil it is hard to recognize that a significant change has been made to the baseline airfoil.

Calculated pressure distributions for these two airfoils are compared in Figs. 15 and 16. Figure 15 illustrates the pressure distributions at 0.5 Mach number, whereas Fig. 16 is the same comparison at the design point Mach number of 0.74. The lift coefficient for both is 0.8. It is clear in these comparisons that the airfoil loading near the trailing edge has been substantially altered. A very large incremental C_p is maintained to essentially the trailing edge. In other words, the lower-surface high pressure is held to be very near the trailing edge, and the upper-surface low pressure is also maintained to the trailing edge. The upper-surface adverse gradient is reduced, whereas the lower-surface gradient is increased. The character of the DTE provides an uncoupling of the upper surface from the lower surface. The flow along each surface can be designed to produce a more balanced upper-surface and lower-surface level of work at the same operating condition. The upper-surface pressure distribution can be designed to terminate at one C_p level, whereas the lower-surface pressure distribution can be designed to terminate at a completely different C_p level. This uncoupling of the two surfaces is facilitated by the high curvature and divergence in the geometry of the trailing edge. The lower-surface flow is required to go through very rapid turning and acceleration to affect flow closure beyond the airfoil physical trailing edge. The flowfield pattern shown in Fig. 17 is hypothesized for the flow closure downstream of the DTE.

The trailing-edge C_p that is achieved by the DTE airfoil is between 0.1 and 0.15 more negative than the original airfoil. This effect is due to the combination of the trailing-edge thickness and divergence. If this variation in trailing-edge pressure level is considered in canonical pressure distribution form,[16] the ability to recompress and terminate the airfoil suction surfaces at a much lower pressure offers numerous airfoil design opportunities.

This implementation of the DTE introduces a significant increase in circulation at a given angle of attack. As a result, the comparison at constant lift coefficient shown in Fig. 16 indicates a significant reduction in shock strength associated with reduced angle of attack. It should also be noted that the airfoil pitching-moment coefficient for this implementation is approximately 0.04 more negative.

The calculated drag rise characteristics of the two airfoils are compared in Fig. 18. The advantage of the DTE at the design point is roughly a 0.0014 reduction in drag coefficient, an 11% reduction in drag. Such a reduction is considered quite remarkable. At lower lift coefficients the difference is small. This drag comparison indicates that the DLBA 243 type of implementation provides an airfoil design with higher lift or camber effectiveness; that is, low drag levels are maintained to higher lift levels. The special contouring and final divergence of the DTE produces the increased camber effectiveness in the airfoil design.

Airfoil designers recognize that increased camber effectiveness can be utilized to reduce compressibility drag at a given lift, to increase lift at a given angle of attack, to increase section thickness at given lift and drag, to increase drag divergence Mach number at given lift and drag, or to produce some combination of these preceding improvements. As a result, the DTE effectiveness can be used in several ways to develop an airfoil design.

In the previous example the DTE was used to minimize drag at a given lift. A second example is included to illustrate a different implementation of the DTE. In this example the camber effectiveness of the DTE is traded for substantial improvements in the airfoil thickness in the flap region, that is, improved airfoil geometric characteristics to aid structural design considerations. The DLBA 238 DTE airfoil was designed to be aerodynamically equivalent to an original airfoil designated DLBA 032. In this design the trailing-edge thickness, maximum thickness, and upper-surface contour were constrained to be the same as the original DLBA 032 section. The effective camber increase of the DTE was offset by increasing airfoil depth from approximately 50 to 95% chord along the lower surface. The DLBA 238 and DLBA 032 section geometries are compared in Fig. 19. The substantial increase in airfoil depth in the flap spar region, 70–80% chord, is quite evident. The DLBA 238 is roughly 30% thicker in this region.

The calculated pressure distributions for DLBA 238 and DLBA 032 are shown in Fig. 20 for the design point conditions. The upper-surface pressure distribution is nearly the same for both airfoils. The only notable exception is the slightly lower pressure along the upper surface near the trailing edge. However, considerable difference is noted in the lower-surface pressure distributions. The increased velocities from approximately 56 to 94% relate roughly to the region of increased airfoil depth. The lower-surface recompression gradient has been made nearly linear.

Figure 21 illustrates the calculated drag rise characteristics of the two sections at 0.6 and 0.8 lift coefficient. The characteristics are matched at the design point and within computational (and experimental) accuracy elsewhere. Nearly identical lift and moment characteristics were also calculated. If the geometric changes illustrated in Fig. 19 are considered in light of the performance equivalence shown in Fig. 21, the camber loss from 50 to 95% chord is offset by the camber increase from 95 to 100% chord. The ratio of these chordwise extents is a measure of the surprising effectiveness of the DTE concept.

Experimental Evaluation

The DLBA 243 and DLBA 238 DTE airfoils, as well as their baseline sections, were tested in a wind tunnel to experimentally verify the concept and to establish an experimental data base. The wind-tunnel tests were accomplished at the NAE facility. Special fabrication and inspection precautions were taken to ensure that the details of the DTE were physically reproduced at model scale.

The measured pressure distributions for DLBA 186 and DLBA 243 are shown in Figs. 22 and 23. The comparison in Fig. 22 is at 0.5 Mach number. If Fig. 22 is compared to Fig. 15, the measured pressure distributions reproduce the calculated pressure distributions quite well. The same is true of the design point data shown in Figs. 23 and 16. More importantly, if details of Figs. 22 and 23 are compared to Figs. 15 and 16, it can be noted that the incremental effects of the DTE implementation are well substantiated in the measured results. The second set of comparisons, illustrated in Figs. 24 and 25, is at a higher lift coefficient at 0.74 Mach number. The calculated and measured pressure distributions compare very well for both sections. At these high-lift conditions, the DTE advantage is maintained; the shock strength is weaker at the same lift. This advantage translates into significantly higher buffet lift coefficients for this particular DTE implementation.

The measured drag rise characteristics for DLBA 186 and DLBA 243 are illustrated in Fig. 26. At the design point, these data indicate a drag reduction of nearly 0.0020. If these measured results are compared to the calculations shown in Fig. 18, the measured design point drag improvement is slightly better than expected. At lower lift coefficients the drag characteristics are essentially the same, as was expected. Perhaps as important is the continued increase in drag improvement at lift coefficients higher than 0.8. At a lift coefficient of 0.85, the DLBA 243 drag characteristics are still remarkably good.

These measured aerodynamic characteristics match the predicted characteristics quite well. The incremental performance improvement of the DTE is validated by these measurements. The magnitude of the improvement in aerodynamic performance is considerable and represents a significant step in technology beyond the supercritical airfoil.

The second airfoil series, DLBA 032 and DLBA 238, was also successfully tested. The design point pressure distributions are illustrated in Fig. 27. If Fig. 27 is compared to Fig. 20, the alteration of the lower-surface recompression by the DTE implementation is clearly indicated. The similarity in the upper-surface pressure distributions is also recognized.

The DLBA 238 airfoil was designed to have matched lift and drag characteristics at the design point by trading the DTE camber effectiveness for substantially increased airfoil aft thickness. The measured lift-curve and drag polar characteristics shown in Fig. 28 confirm that the design trade was successful. The drag polar comparison reveals that the DLBA 238 DTE airfoil actually has slightly less drag at the 0.8 lift coefficient design point. Additionally, the lift curve indicates that at a constant angle of attack the DTE airfoil produced slightly higher lift than the original airfoil. This incremental lift indicates that a small portion of the increased camber effectiveness of the DTE was retained and not traded away. The lift-curve comparison illustrates this increment is essentially preserved through buffet onset (lift-curve-break) and to high-speed stall.

The measured drag rise characteristics for DLBA 238 and DLBA 032 are presented for 0.8 lift coefficient in Fig. 29. This comparison, as in the previous figure, indicates that the design trade was accomplished conservatively. A residual aerodynamic improvement remains for the DTE configuration at the Mach numbers near the design point.

The second set of airfoils utilizing a different implementation of the DTE concept represents an independent confirmation of the fundamental advantage of the DTE. The interesting aspect of this second set is the introduction of the design trade in conjunction with the DTE. The airfoil designer can contemplate the DTE as an additional degree of design freedom or flexibility not previously recognized.

The basic aerodynamic traits associated with the DTE include increased camber effectiveness, reduced upper-surface trailing-edge recompression pressure level, increased lower-surface recompression pressure very near the trailing edge, and aft loading carried essentially to the trailing edge. With regard to the latter characteristics, it is interesting to consider the DTE and the Kutta condition. Clearly, the upper- and lower-surface pressure distributions are matched to some condition, perhaps equivalence, near the trailing edge. However, the gradients in pressure near the DTE lower surface are necessarily very large in both the chordwise and normal directions. Such gradients render definition of the Kutta condition a bit more speculative for the DTE than for previous configurations.

New Wing Application of DTE Airfoils

In addition to airfoil design and testing, several wing development efforts have also been conducted in order to address any issues that might arise from a three-dimensional, swept-wing design application of the DTE concept. In one such development a high-performance supercritical wing configuration, previously developed for an advanced commercial transport application, was used as a baseline configuration. This configuration, designated W_7, was designed computationally using the methods outlined in Ref. 5. These methods included the Douglas version of Jameson's FLO22 and the Inverse Transonic Wing Design Method.[17] The design Mach number of the wing was 0.77. The baseline W_7 was well developed computationally to include tailoring for high-speed, cruise efficiency; low-speed, high-lift conditions; low-speed, stall characteristics; high-speed, buffet lift conditions; high-speed, pitch characteristics; surface curvature producibility considerations; and nacelle/pylon integration. This configuration had also been experimentally validated. In short, the aerodynamic definition was complete and ready for full-scale development.

This wing was computationally redesigned with an implementation of the DTE concept. The implementation used for this wing, designated W_{15}, was the same as that for the DLBA 243 airfoil. The objective of the design was to minimize drag at a given design lift. Constraints were imposed to maintain constant planform, constant spanload distribution, and constant spanwise distribution of airfoil maximum thickness, and to retain all of the tailored features of the baseline wing configuration. The planform for the baseline W_7 wing and the W_{15} DTE wing is shown in Fig. 30. With the imposed constraints the inviscid induced drag is held constant. Hence, the objective was to implement DTE airfoil sections into the design and achieve a reduction primarily in shock drag for the wing.

The tail-off design lift coefficient range for this wing is 0.55–0.6. The calculated improvement in compressibility drag, defined as the variation in drag above the level at $M = 0.5$, is illustrated in Fig. 31. The calculated drag coefficient improvement at 0.77 Mach number and 0.55 lift coefficient is 0.0005. At 0.6 lift coefficient, the DTE improvement is 0.0010. This latter increment is roughly 3% of aircraft drag for the original transport aircraft application. This predicted improvement is significant and prompted a wind-tunnel test to validate the results.

A wind-tunnel test of both wings was conducted in the Rockwell International 7-ft trisonic wind tunnel in El Segundo, California. Both wings were mounted to a representative fuselage. The test Reynolds number was approximately 3.6 million based on wing mean aerodynamic chord (MAC). At this subscale Reynolds number, care was taken to set transition on the wing upper and lower surfaces so that high-Reynolds-number drag characteristics could be simulated.[5] This transition position also ensured that the lower-surface boundary layer was not overly thickened in the region of the DTE. The model installation is shown in Fig. 32.

The measured DTE improvement in compressibility drag is presented in Fig. 33. At the $M = 0.77$ design conditions the drag increments are quite close to the computations illustrated in Fig. 31. Furthermore, the trend of continued improvement at higher lift coefficients, which was identified in the DLBA 243 airfoil characteristics, is also evident in this wing comparison. In fact, the W_{15} DTE wing drag rise character at the very high lift coefficient of 0.7 is still remarkably well behaved.

The 0.5 Mach number measured drag polars are compared in Fig. 34. At low lifts a small drag penalty is indicated. At lift coefficients in the design range, a small improvement is noted. This improvement is probably due to improved viscous drag due to lift. Such a trend can be noted in the airfoil characteristics shown in Fig. 26.

Measured and calculated incremental pitching-moment effects for the DTE wing are illustrated in Fig. 35. The measured increment is larger than the calculated increment by about 0.01. The measured increment in moment between the two wings would cause the DTE wing performance to deteriorate by as much as 1% in terms of increased trim drag if no c.g. management system is utilized. If a contemporary c.g. management system is utilized, this penalty is reduced.

Figure 36 illustrates the measured and calculated improvement in buffet onset lift coefficient for the DTE wing. An improvement of approximately 0.05 in lift coefficient is noted for both the computational point and the measurements at cruise Mach number.

Finally, wing/body lift-to-drag ratio (L/D) effects are compared in Fig. 37. The L/D characteristics at $M = 0.77$ indicate that at peak L/D an improvement of over 4% has been achieved with implementation of the

DTE concept on this wing configuration. Such an improvement in aerodynamic efficiency is quite substantial. Early studies of supercritical airfoil technology[18] identified aerodynamic efficiency improvements of 5–10% for application of high-aspect-ratio, supercritical wings compared to lower-aspect-ratio, conventional wings. If the conventional wing aspect ratio was equivalent to that of the supercritical wing, then about half of that advantage is lost, as indicated in Fig. 38. In comparison, the improvement measured for the DTE wing is roughly the same magnitude as the residual improvement associated with the supercritical wing. Hence, it has been demonstrated that the DTE airfoil represents a significant step in airfoil technology beyond that achieved with the supercritical airfoil.

Derivative Wing Application of DTE Airfoils

The DTE airfoil concept has also been applied to the modification of an existing wing with conventional airfoil sections.[7] From an engineering development and manufacturing standpoint, one advantage offered by a such a derivative wing over an "all-new" configuration is the ability to preserve existing structure, systems, and tooling. The motivation to preserve commonality with a baseline configuration creates numerous design constraints on the definition of a derivative wing.

The aircraft configuration for this study includes increased capacity and gross weight from a baseline configuration. Additional wing area is required due to the increase in gross weight. The existing wing has a planform area of 1209 ft^2, whereas the derivative wing requires 1294 ft^2. A planform drawing, including a tabulation of wing reference quantities, is given in Fig. 39. Trailing-edge chord extension is used to provide the required area increase for the derivative wing. The chord extension ranges from 4.2% at the side of fuselage to approximately 14.7% near the wing tip. The discussion in this section will refer to the baseline wing as W15F and the modified (DTE derivative) wing as W16A. The W15F designation is not related to the W_{15} new wing configuration in the previous section.

A reduction in wing compressibility drag and an increase in the wing buffet lift coefficient were both primary performance goals in the derivative wing design. An improvement in the wing buffet capability will generally result in an increase to the initial cruise altitude capability of the aircraft. In this design activity the objective was to satisfy the transonic performance goals while respecting the numerous structural constraints. In other words, the objective was to define a modest change to the trailing-edge region of a conventional wing that would achieve some of the performance improvements typical of those demonstrated by high-performance wing designs. The design study provided a natural opportunity to assess the performance benefits resulting from the use of DTE technology with a conventional wing as the starting point. The question was whether the significant performance improvements associated with the DTE airfoil would be realized with the imposition of many design restrictions.

In the design of the modified wing, both the wing structural box section and the airfoil contour forward of the flap rear spar were constrained to the existing definitions. Using the flap rear spar as the forward extent of the modification, the airfoil contour was "frozen" forward of a chordwise fraction of 0.82, as indicated in Fig. 40. The use of the existing wing structural box section effectively fixed the spanwise wing twist distribution.

The baseline aircraft uses a manual aileron control system. In order to preserve this system in the derivative, an aileron hinge moment constraint was enforced. As would be expected by reference to the calculated two-dimensional pressure distributions (Fig. 16), the aft loading of the DTE airfoil creates an increase in control surface hinge moments. As a result, the DTE airfoil cannot easily be used across the span of a manual aileron. Therefore, the DTE concept was limited to a partial span application in the derivative wing. The design restriction constrained the application of the DTE airfoil to the inboard 63.9% of the wing semispan. This represented a significant design constraint. However, by utilizing a trailing-edge extension, some airfoil recontouring was accomplished on the aileron. A gradual introduction or transition to the DTE concept was used on the inboard wing and a return to a basic chord extension was enforced at the inboard edge of the aileron.

Since the DTE airfoil will generate a significant increase in circulation (lift increase at a constant angle of attack) over a conventional airfoil, noticeable changes in the spanwise loading will result from a partial span application of the DTE airfoil. Wing-induced drag, low-speed stall characteristics, and wing buffet onset characteristics all have some degree of dependence on the spanwise wing loading. Therefore, the design of the derivative wing included consideration of the potential implications created by changes to be spanwise wing loading.

The spanwise distribution of the trailing-edge thickness for wing W16A and the spanwise variation in the trailing-edge included angle for both wings are given in Figs. 41 and 42, respectively. Both the trailing-edge thickness and the trailing-edge included angle show a gradual transition to the DTE airfoil on the inboard wing. At a semispan fraction of 0.303, the wing design has reached full application to the DTE concept. The comparisons also clearly show the return to a more conventional airfoil across the span of the aileron. The spanwise limitations to the DTE concept have effectively constrained "full" application of the DTE airfoil to a region between 30 and 53% of the wing span.

An airfoil comparison for wings W15F and W16A for a semispan fraction of 0.303 is shown in Fig. 43. The characteristic use of finite-trailing edge thickness, the strongly divergent upper and lower surfaces near the trailing-edge, and the high rate of change of lower surface curvature near the trailing edge are all very evident in this comparison.

Figure 44 compares the calculated three-dimensional pressure distributions for wings W15F and W16A. The solutions are for the design point Mach number of 0.76 and wing lift coefficient of 0.55. The comparison clearly shows the dramatic change to the wing pressure distributions that results from the trailing-edge modification. The calculated pressure distributions also point to an increase in the wing nose-down pitching moments for wing W16A, compared to wing W15F.

As expected by the qualitative interpretation of the calculated pressure distributions, a reduction in the compressibility drag of the modified wing

(W16A) relative to the baseline (W15F) was calculated. The calculated drag rise characteristics are given in Fig. 45. The calculated compressibility drag improvement increases with a corresponding increase in the wing lift coefficient. Compared to wing W15F, the calculated compressibility drag reduction for wing W16A is negligible at a wing lift coefficient of 0.45. The calculated drag improvement increases to a drag reduction of 13 counts at a lift coefficient of 0.60 and for the design Mach number of 0.76. At the design point of 0.76 Mach number and 0.55 lift coefficient, the trailing-edge modification is responsible for a 0.0007 reduction in the calculated compressibility drag. This represents an approximately 2% reduction to the aircraft cruise drag.

The combination of the calculated chordwise pressure distributions and a buffet onset prediction method identified a 0.12 increase in the wing buffet lift coefficient for the modified wing compared to the original wing. This is considered to be a relatively large increase in the wing buffet capability. The baseline wing for this design study is essentially of conventional technology (compared to a supercritical wing or a DTE wing). The relatively large improvement in the calculated buffet onset capability is due to a combination of both the DTE concept and the basic trailing-edge chord extension. When a high-performance supercritical wing is used as the starting point for the DTE concept, as described in the previous section, the calculated increase in the buffet lift coefficient is approximately 0.05.

The calculated improvements in the wing buffet capability and the reduction in compressibility drag for the modified wing were of sufficient magnitude to justify a wind-tunnel evaluation of the characteristics. A wind-tunnel test was conducted in the Rockwell trisonic wind tunnel for the purpose of validating the calculated performance improvements provided by wing W16A. Model wing geometries were built for both wing W15F and wing W16A. The test Reynolds number based on the wing mean aerodynamic chord was approximately 4.5 million, compared to the full-scale value of 24 million. To better simulate the high-Reynolds-number drag characteristics, transition was fixed on both the upper and lower surface of the wing.

The measured drag rise characteristics for each wing are compared in Fig. 46. The measured drag improvement was five counts at a Mach number of 0.76 and a wing lift coefficient of 0.55. The measured drag improvement agrees reasonably well with the calculated value of seven drag counts. As the wing lift coefficient is increased, the wind-tunnel data show an increase in the compressibility drag improvement for wing W16A compared to wing W15F. A similar trend was predicted by the calculated characteristics. At a wing lift coefficient of 0.60, a 12-count compressibility drag improvement was measured for W16A. This compares very well with the calculated value of 13 drag counts. The measured compressibility drag-improvement for the wing W16A increases from 5 counts to approximately 12 counts as the wing lift coefficient changes from 0.55 to 0.60. Both the measured and the calculated compressibility drag reduction for the modified wing essentially double as the wing lift coefficient increases from 0.55 to 0.60. Such a trend is consistent with the results for the all-new DTE wing configuration in the previous section.

The measured wing buffet lift coefficients for wings W15F and W16A are compared in Fig. 47. The experimental data confirm the predicted 0.12 improvement in the buffet lift coefficient for the modified wing at the design Mach number of 0.76. An increase in the buffet lift coefficient of this magnitude is considered to be a rather significant improvement beyond the baseline wing.

As previously mentioned, the aft lift loading created by the DTE airfoil will create an increase in the wing nose-down pitching moments. At the design point, the increment in the measured wing/body pitching moments is approximately -0.089 increased nose-down moment for wing W16A. The measured pitching-moment term includes both the increase in the wing nose-down pitching moment and differences in the fuselage pitching-moment contribution due to variation in the fuselage angle of attack at the design point. This latter effect is the result of the original wing-to-body rigging constraint. If the derivative wing could be rotated on the fuselage to maintain the same variation in wing lift coefficient with angle of attack as the baseline, approximately half of the moment effect would be eliminated. In any event, this increment is considered to be a rather large increase in the wing/body nose-down pitching moment and contributes to an increase in aircraft trim drag.

Figure 48 contains a comparison of the measured L/D for wings W15F and W16A. On a peak L/D basis, wing W16A shows a 5% improvement relative to the wing W15F. As the lift coefficient is increased, the L/D comparison clearly shows the increase in the relative performance benefits of W16A.

The L/D values given in Fig. 48 do not include a trim drag penalty associated with the increased nose-down pitching moments for wing W16A. If a c.g. management system were not used on the aircraft, the actual improvement in peak L/D may reduce to approximately 2%. This conclusion is dependent on the aircraft configuration and does not represent a general conclusion. For the particular derivative under consideration in this study, the improvement in peak L/D can be returned to approximately 4% with the use of a c.g. management system. Any conclusions regarding an assumed trim drag penalty are strongly dependent on the specific aircraft configuration. However, it is safe to say that the transonic improvements provided by the trailing-edge modification are not nullified by an excessive trim drag penalty.

Summary

A new concept in airfoil design has been developed. The concept, referred to as the DTE airfoil, utilizes unconventional trailing-edge closure geometric characteristics. These characteristics include a finite trailing-edge thickness, strongly divergent trailing-edge upper and lower surfaces, and high surface curvature on the lower surface at or near the trailing edge.

Two different implementations of the DTE in airfoil designs have been demonstrated. In the first case, DLBA 243, the increased camber effectiveness of the DTE was utilized to develop an airfoil design with much superior transonic lift and drag characteristics than was previously attainable. In the second case, DLBA 238, the DTE camber effectiveness was traded for improved geometric characteristics while aerodynamic performance was held essentially constant. Both airfoils were computationally designed and tested in a wind tunnel to validate the DTE concept. Both examples were quite successful in achieving the design objectives. The two different approaches for the use of the DTE also encourage the airfoil designer to consider yet other applications. The ability of the DTE to decouple the upper- and lower-surface pressure distributions near the trailing edge offers new possibilities for the designer.

The confidence established by the airfoil development naturally led to wing design applications to ensure that any three-dimensional design issues were addressed for a DTE application. An all-new wing design was computationally developed utilizing a DTE implementation very similar to that used in DLBA 243. The calculated aerodynamic performance improvement was large and was confirmed in subsequent wind-tunnel testing. The DTE performance improvement compared to a contemporary supercritical wing is similar in magnitude to the improvement originally associated with the supercritical wing. The DTE airfoil concept has also been applied to a derivative wing development in which numerous design constraints were imposed to preserve existing wing structure and systems. Such constraints limit the ability to take full advantage of the DTE airfoil. Nevertheless, calculated and subsequently measured improvements in compressibility drag and buffet lift coefficient were obtained in the design effort.

The DTE airfoil represents a significant step in airfoil aerodynamic technology. It is clear that this concept provides the aerodynamicist with an additional degree of design freedom and flexibility previously unrecognized. The inception and development of the DTE was enabled by the existence and application of well-developed computational aerodynamics methods. The original idea of small changes near the trailing edge was emphatically demonstrated by computational results. Development of the DTE-concept was accomplished as a result of airfoil trailing-edge design investigations utilizing several computational aerodynamics methods as the basic tools. It is very unlikely that the DTE airfoil would have been discovered without the availability of such methods. The computational aerodynamic development of the DTE led to the recognition that traditional airfoil closure principles could be abandoned. In doing so, airfoil performance limitations inherent in the traditional principles are also abandoned. Hence, innovation with computational aerodynamics methods has been demonstrated and will no doubt continue in the future.

References

[1]Abbott, I. H., and Von Doenhoff, A. E., *Theory of Wing Sections*, Dover, New York, 1959.

[2]Hoerner, S. F., *Fluid-Dynamic Lift*, Hoerner Fluid Dynamics, Brick Town, NJ, 1975.

[3]Whitcomb, R. T., "Review of NASA Supercritical Airfoils," *Proceedings of the Ninth International Congress of the International Council of the Aeronautical Sciences (ICAS)*, 1974.

[4]Harris, C. D., "Wind Tunnel Investigation of Effects of Trailing Edge Geometry on a NASA Supercritical Airfoil Section," NASA TM-X-2336, 1971.

[5]Henne, P. A., Dahlin, J. A., and Peavey, C. C., "Applied Computational Transonics—Capabilities and Limitations," *Progress in Astronautics and Aeronautics: Transonic Aerodynamics*, Vol. 81, AIAA, New York, 1982.

[6]Henne, P. A., and Gregg, R. D. III, "A New Airfoil Design Concept," AIAA Paper 89-2201, July 1989.

[7]Gregg, R. D. III, Hoch, R. W., and Henne, P. A., "Application of Divergent Trailing Edge Airfoil Technology to the Design of a Derivative Wing," Society of Automotive Engineers Paper 892288, 1989.

[8]Henne, P. A., and Gregg, R. D. III, United States Patent No. 4,858,852, 1989.

[9]Bauer, F., Garabedian, P., Korn, D., and Jameson, A., *Supercritical Wing Sections II*, Lecture Notes in Economics and Mathematical Systems, Vol. 108, Springer-Verlag, New York, 1975.

[10]Nash, J. F., "A Review of Research on Two-Dimensional Base Flow," British Aeronautical Research Council, London, RM-3323, 1963.

[11]Zaparka, E. F., United States Patent No. 19,412, 1935.

[12]Liebeck, R. L., "Design of Subsonic Airfoils for High Lift," *Journal of Aircraft*, Vol. 15, Sept. 1978, pp. 547–561.

[13]Ohman, L. H., "The National Aeronautics Establishment High Reynolds Number 15 × 60 Inch Two Dimensional Test Facility," National Research Council, Canada, LTR-HA-4, April 1970.

[14]Nash, J. F., and Macdonald, A. G. J., "The Calculation of Momentum Thickness in a Turbulent Boundary Layer at Mach Numbers up to Unity," British Aeronautical Research Council, London, CP-963, 1967.

[15]Cebeci, T., and Bradshaw, P., *Momentum Transfer in Boundary Layers*, McGraw-Hill, New York, 1977.

[16]Smith, A. M. O., "High-Lift Aerodynamics," *Journal of Aircraft*, Vol. 12, June 1975, pp. 501–530.

[17]Henne, P. A., Inverse Transonic Wing Design Method," *Journal of Aircraft*, Vol. 18, Feb. 1981, pp. 121–127.

[18]Steckel, D. K., Dahlin, J. A., and Henne, P. A., "Results of Design Studies and Wind Tunnel Tests of High Aspect Ratio Supercritical Wings for an Energy Efficient Transport," NASA CR-159332, Oct. 1980.

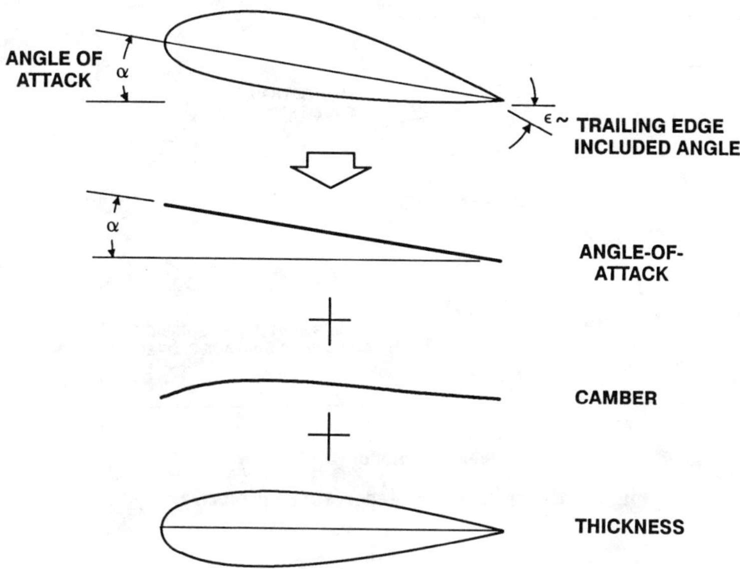

Fig. 1   Classical airfoil decomposition into angle of attack, camber, and thickness.

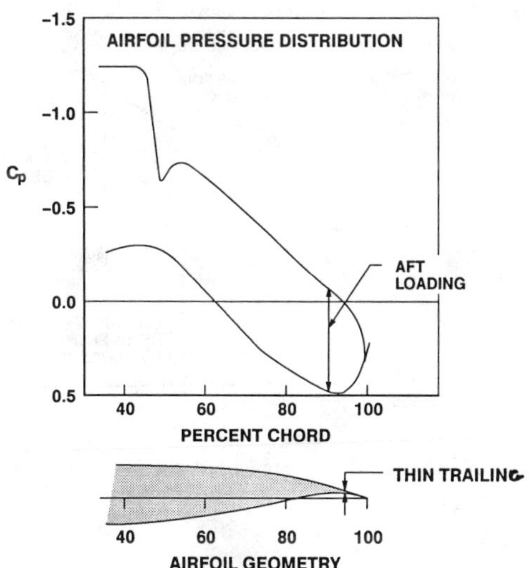

Fig. 2   Supercritical airfoil aft-loading and trailing-edge geometry.

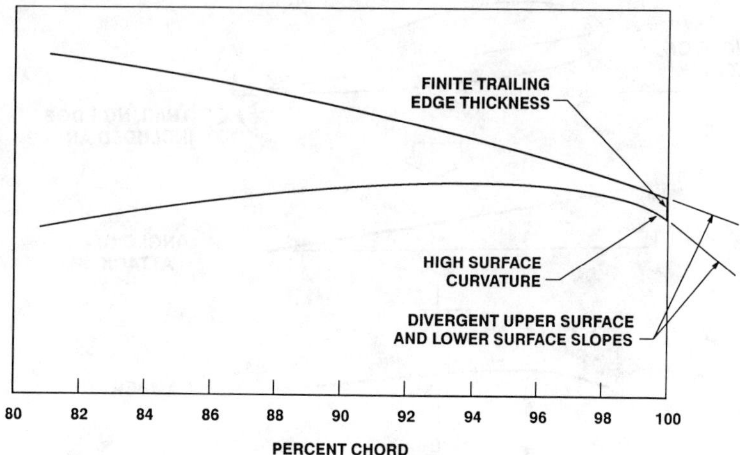

Fig. 3   Divergent trailing-edge airfoil geometry.

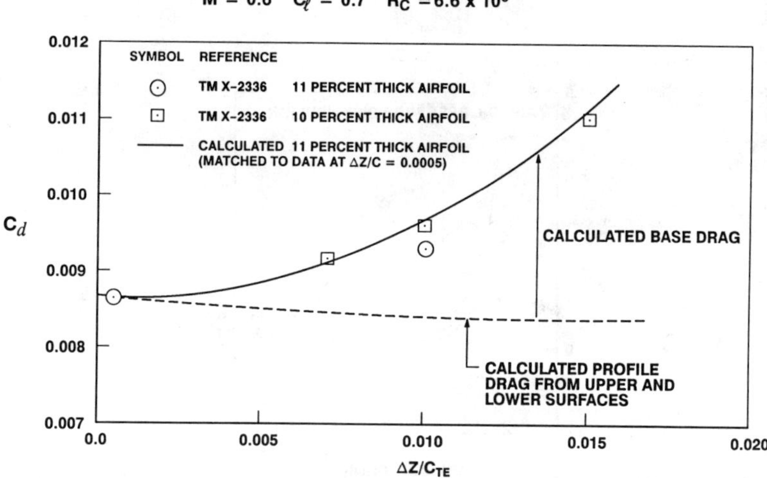

Fig. 4   Comparison of calculated and measured section drag variation with trailing-edge thickness.

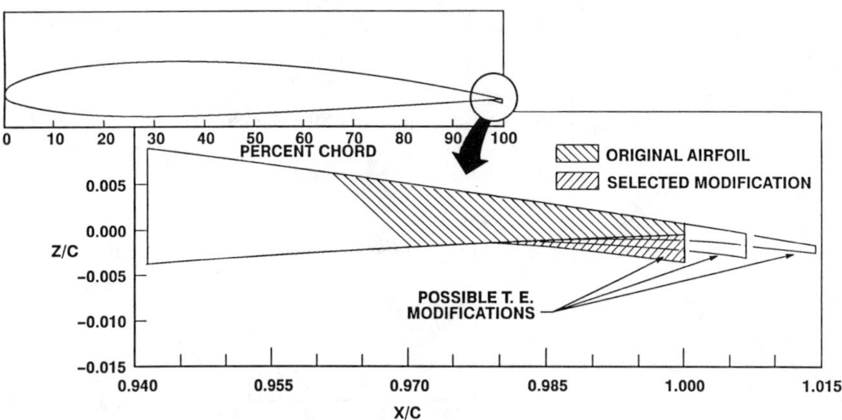

Fig. 5 Possible trailing-edge modifications for DSMA 661 airfoil, 1981 study case.

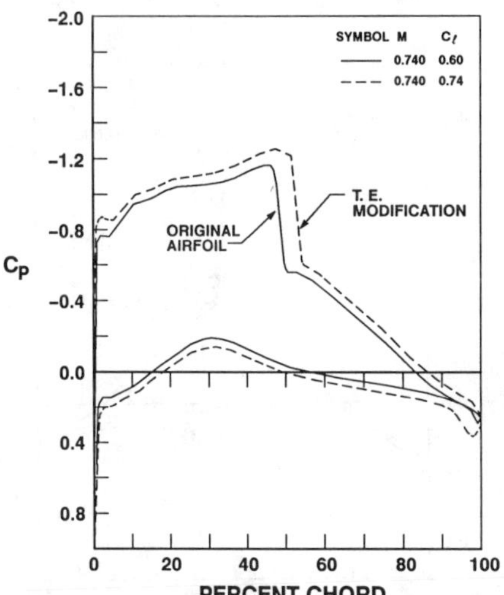

Fig. 6 DSMA 661 trailing-edge modification calculated chordwise pressure distribution at 2.2-degrees angle of attack.

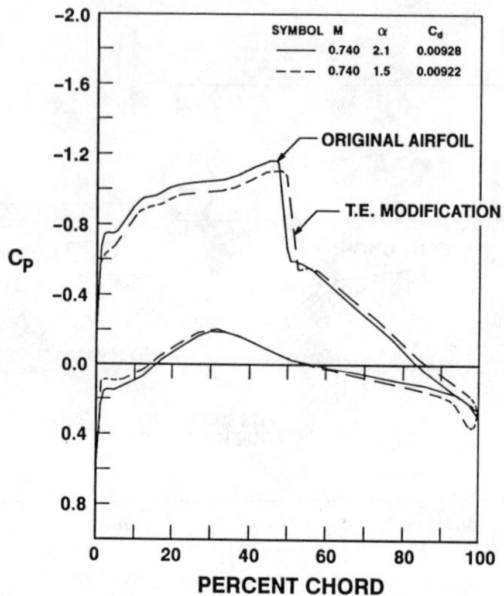

Fig. 7   DSMA 661 trailing-edge modification calculated chordwise pressure distribution at 0.6 lift coefficient.

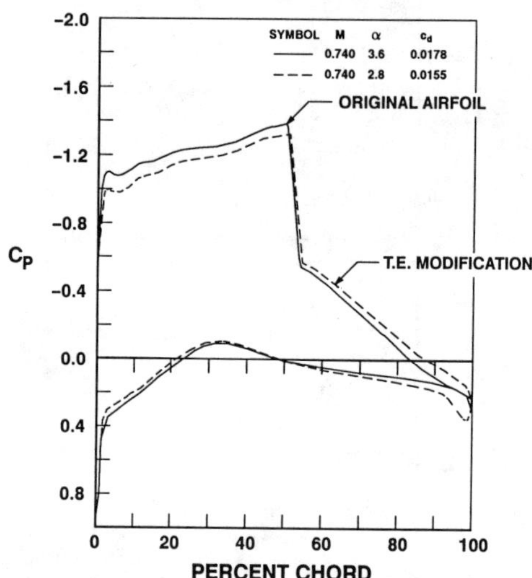

Fig. 8   DSMA 661 trailing-edge modification calculated chordwise pressure distribution at 0.8 lift coefficient.

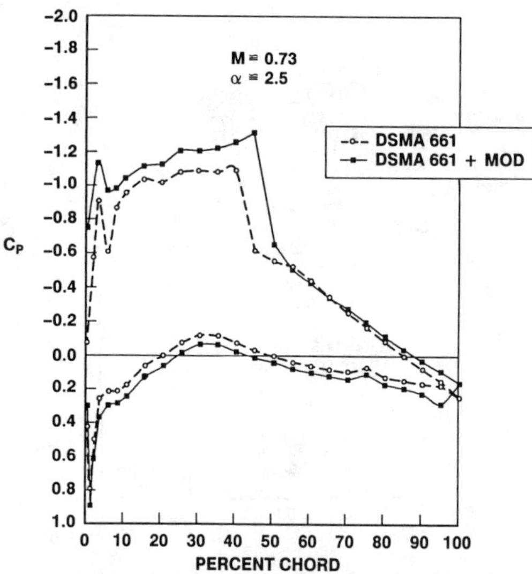

Fig. 9   Measured effect of trailing-edge modification on DMSA 661 pressure distributions at constant angle of attack.

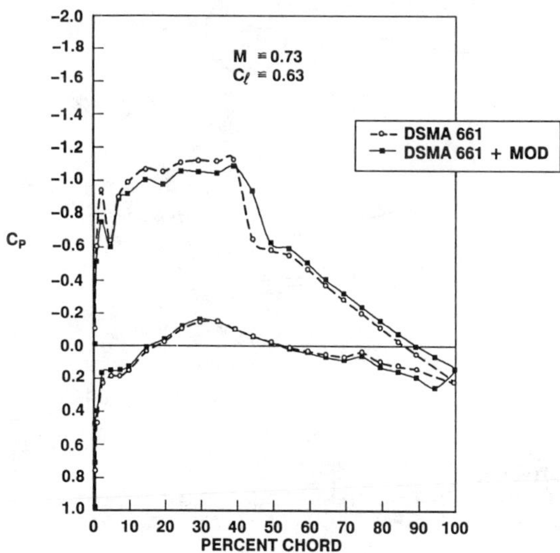

Fig. 10   Measured effect of trailing-edge modification on DMSA 661 pressure distributions at constant lift.

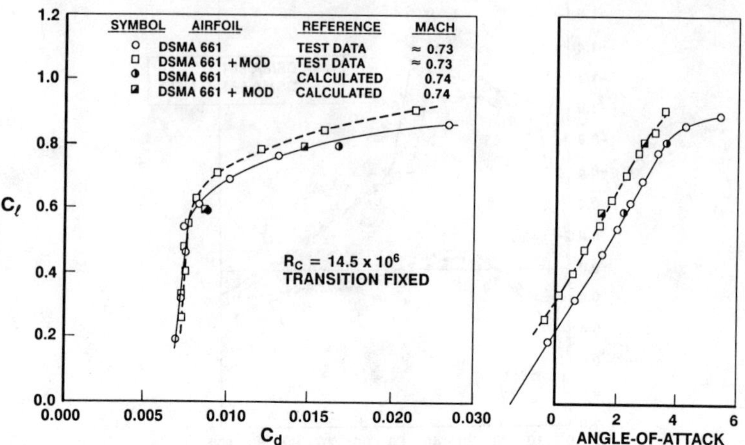

Fig. 11   Comparison of calculated and measured lift and drag characteristics for DSMA 661 modification.

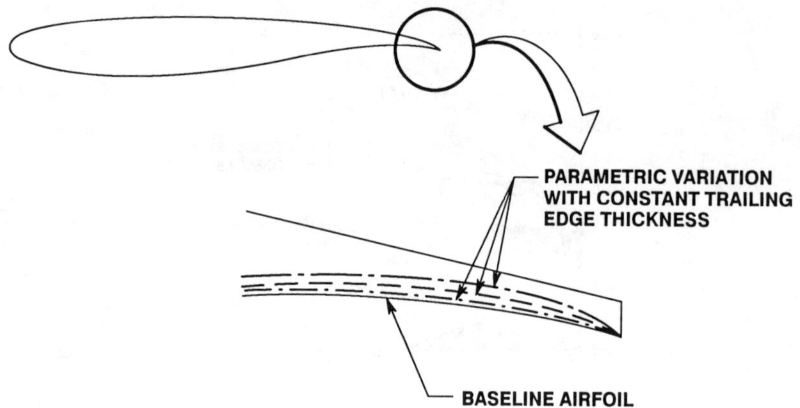

Fig. 12   Divergent trailing-edge airfoil concept development.

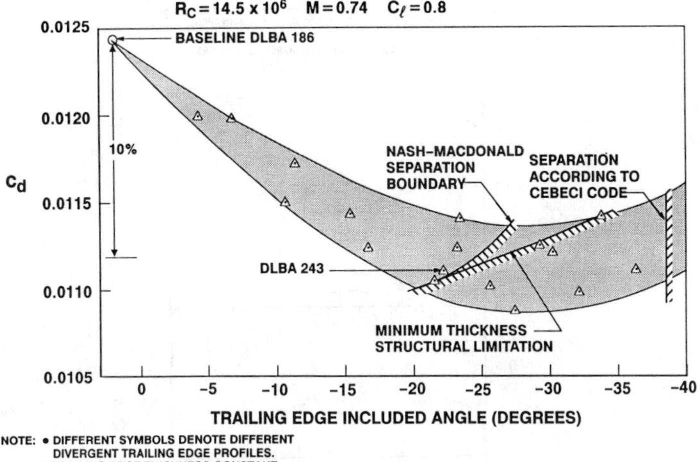

Fig. 13 **Effect of divergent trailing-edge profiles on calculated airfoil total drag coefficient.**

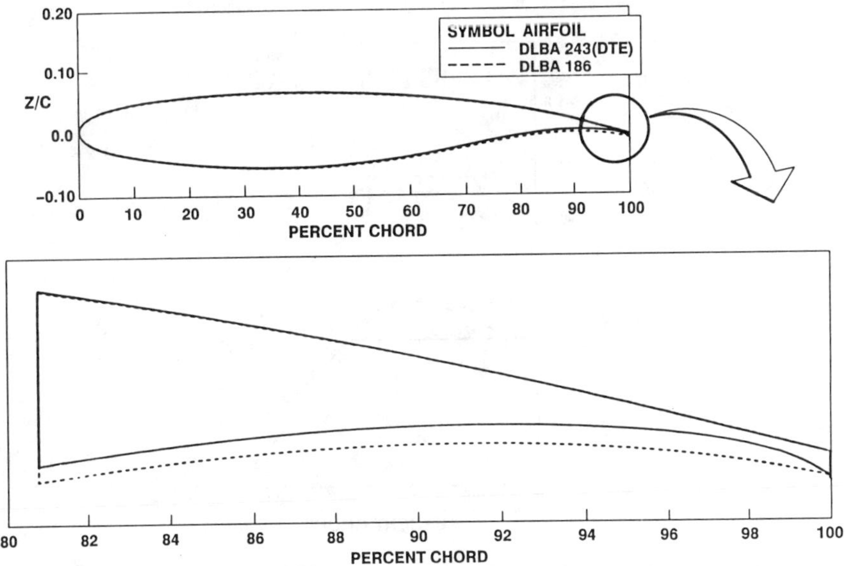

Fig. 14 **Comparison of DLBA 243 and DLBA 186 airfoil sections.**

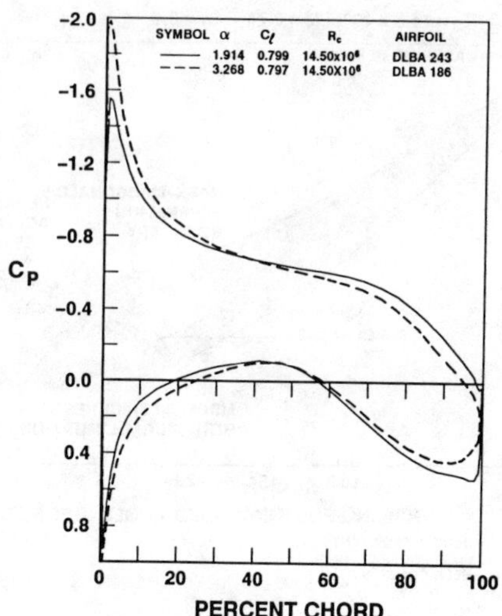

Fig. 15   Comparison of DLBA 243 (DTE) and DLBA 186 calculated chordwise pressure distributions at 0.5 Mach number.

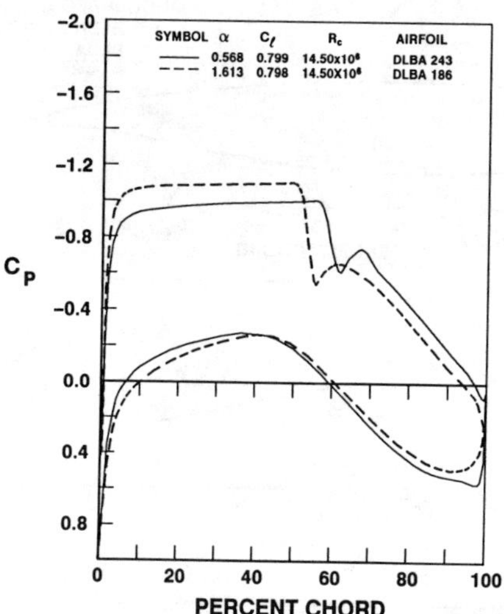

Fig. 16   Comparison of DLBA 243 (DTE) and DLBA 186 calculated chordwise pressure distributions at 0.74 Mach number.

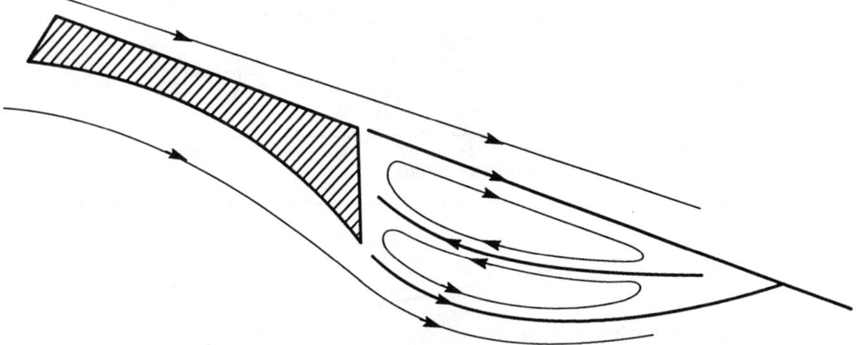

Fig. 17   Hypothesized flow closure for divergent trailing-edge airfoil.

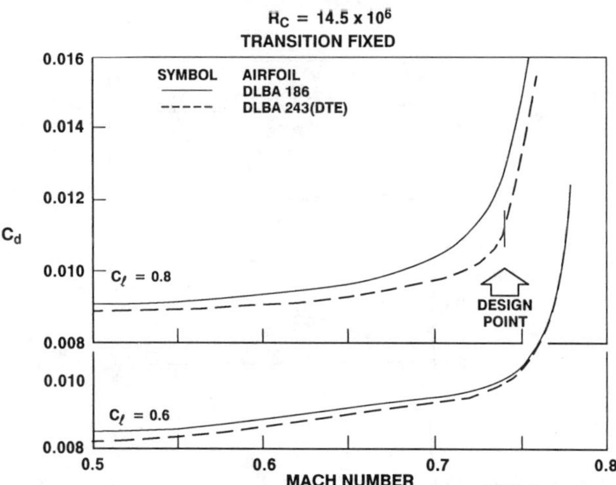

Fig. 18   Comparison of DLBA 243 and DLBA 186 calculated drag rise characteristics.

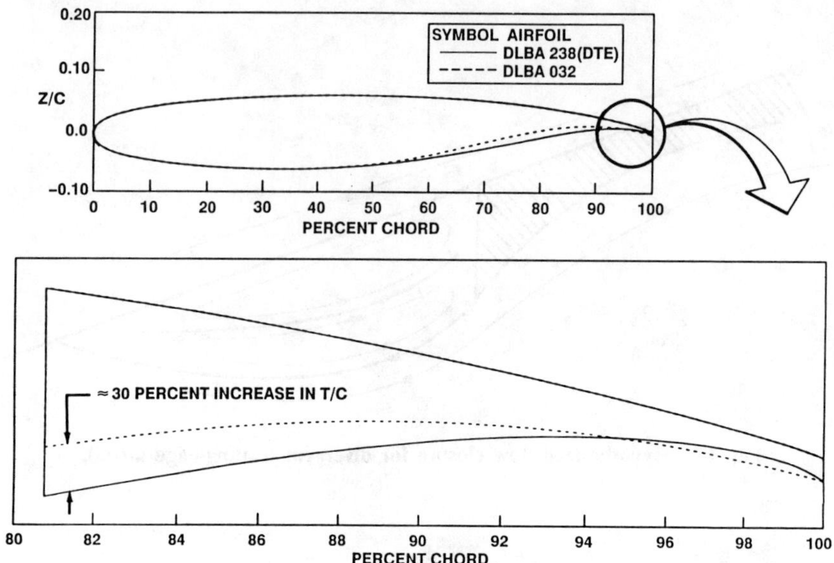

Fig. 19   Comparison of DLBA 238 and DLBA 032 airfoil sections.

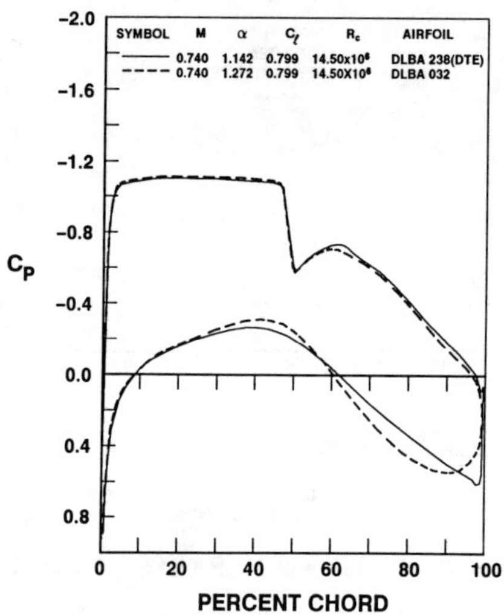

Fig. 20   Comparison of DLBA 238 and DLBA 032 calculated chordwise pressure distributions.

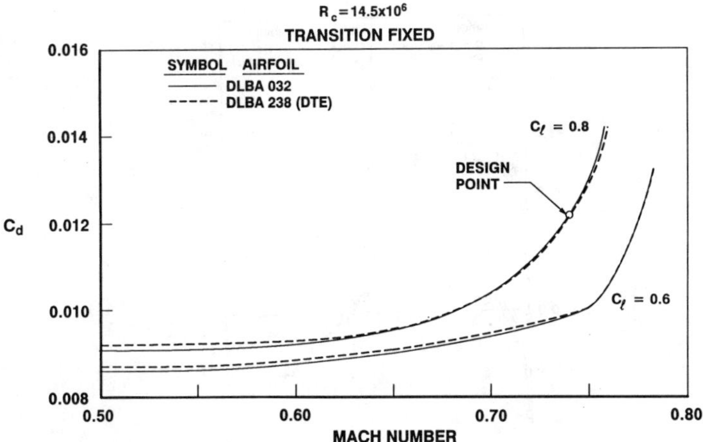

Fig. 21   Comparison of DLBA 238 and DLBA 032 calculated drag rise characteristics.

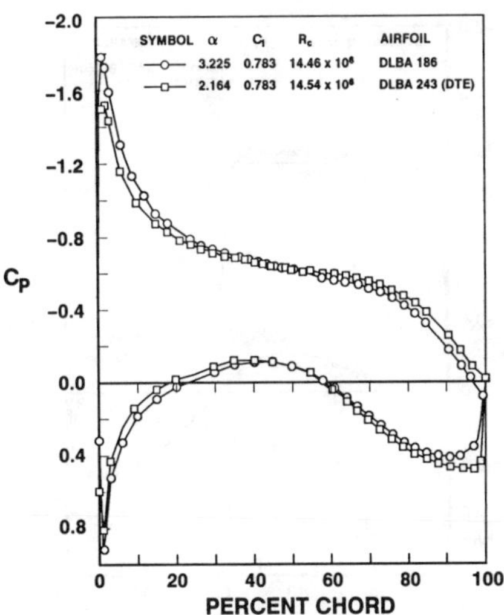

Fig. 22   Comparison of DLBA 243 and DLBA 186 measured chordwise pressure distributions at 0.5 Mach number.

P. A. HENNE

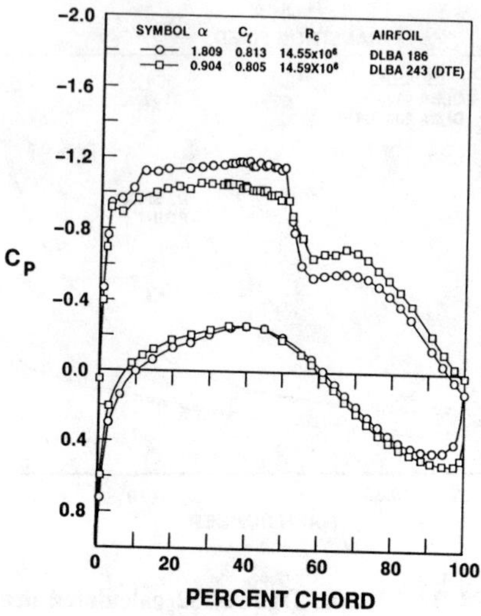

Fig. 23 Comparison of DLBA 243 and DLBA 186 measured chordwise pressure distributions at 0.74 Mach number.

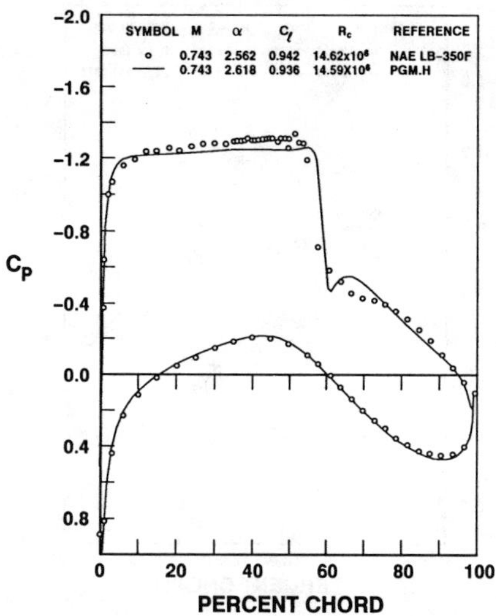

Fig. 24 Comparison of experimental and calculated DLBA 186 chordwise pressure distributions at 0.74 Mach number and 0.94 lift coefficient.

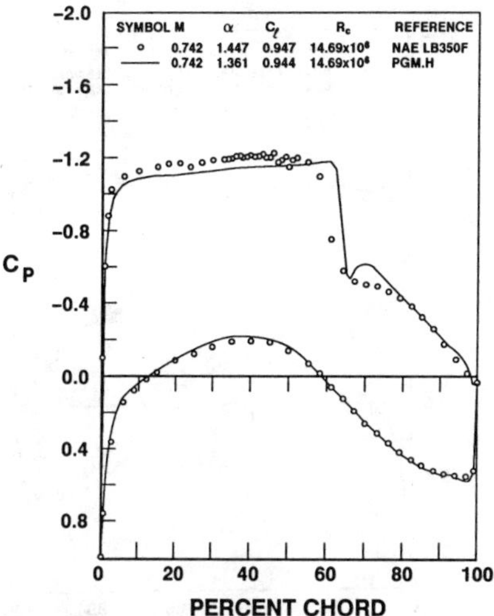

Fig. 25 Comparison of experimental and calculated DLBA 243 (DTE) chordwise pressure distributions at 0.74 Mach number and 0.95 lift coefficient.

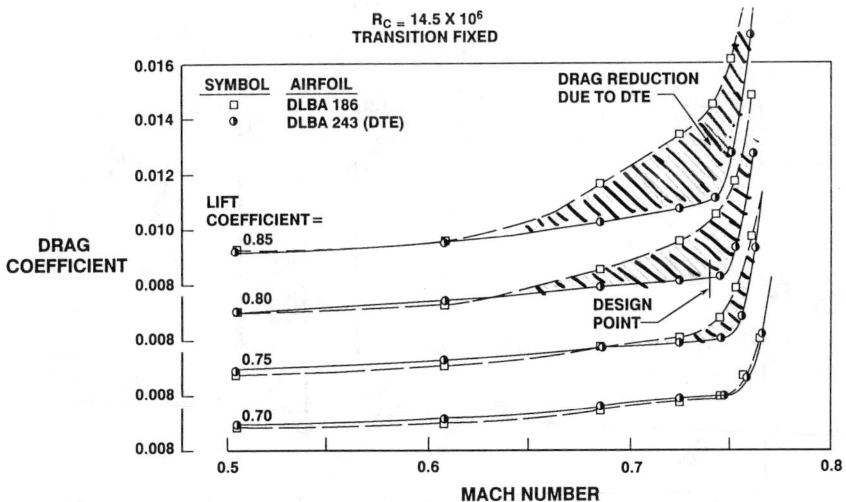

Fig. 26 Measured drag characteristics for DLBA 243 and DLBA 186.

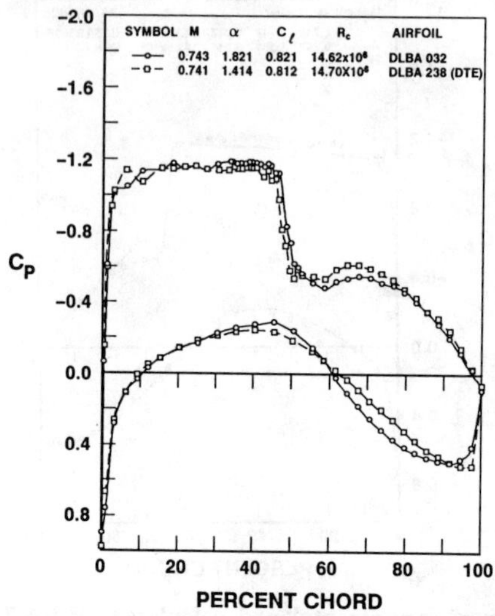

Fig. 27   Comparison of DLBA 238 and DLBA 032 measured chordwise pressure distributions.

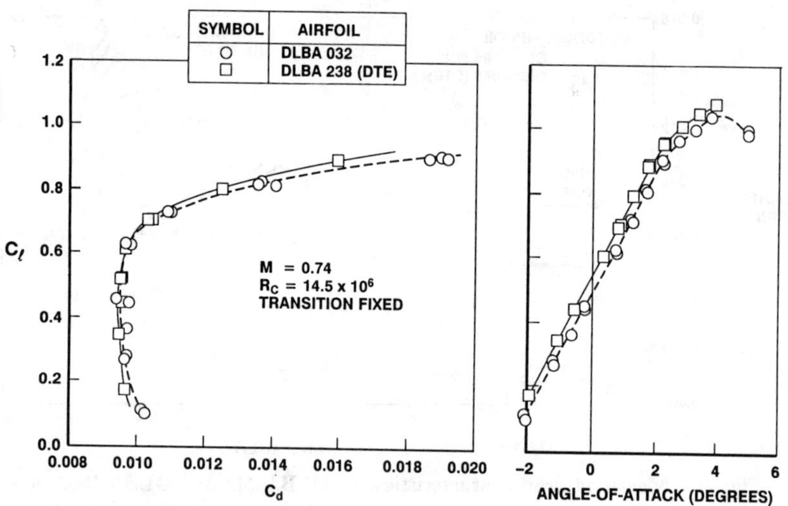

Fig. 28   Comparison of DLBA 238 and DLBA 032 measured lift-curve and drag polar characteristics.

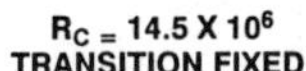
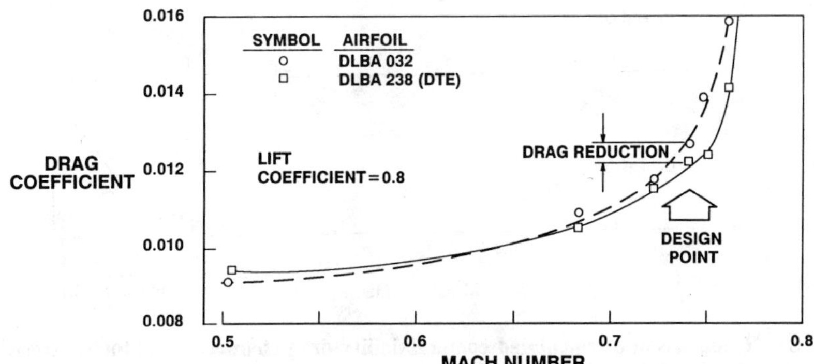

Fig. 29   Measured drag characteristics for DLBA 238 and DLBA 032.

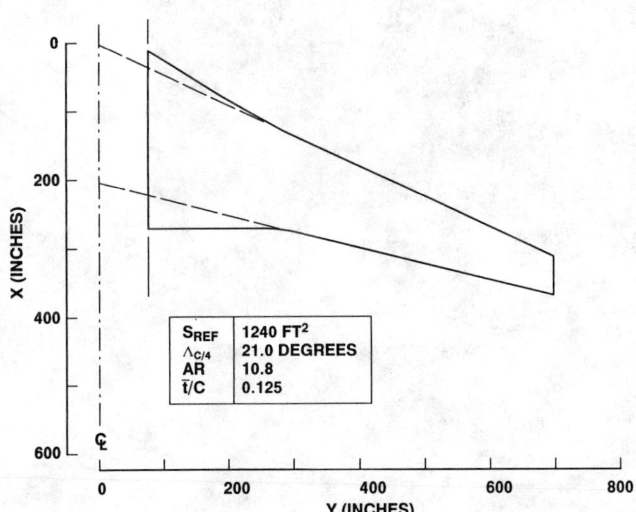

Fig. 30   Planform definition for wings W_7 and W_{15}.

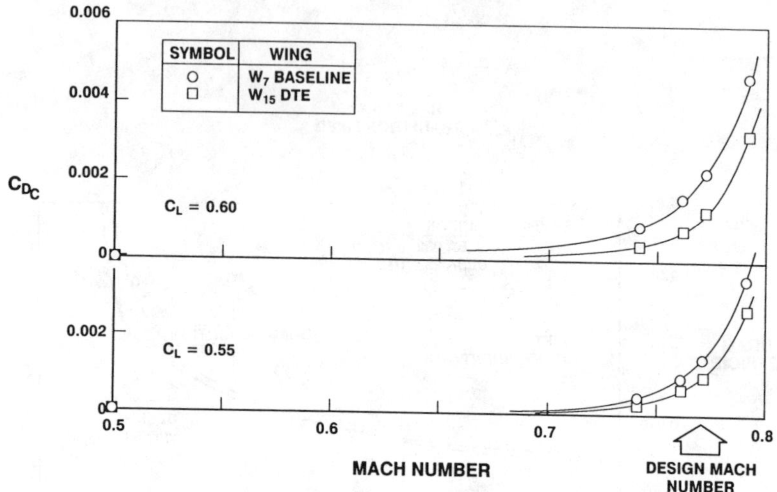

Fig. 31   Comparison of calculated compressibility drag characteristics for supercritical wing W_7 and DTE wing W_{15}.

Fig. 32 W_7 and W_{15} model installation in the Rockwell International 7-ft transonic wind tunnel.

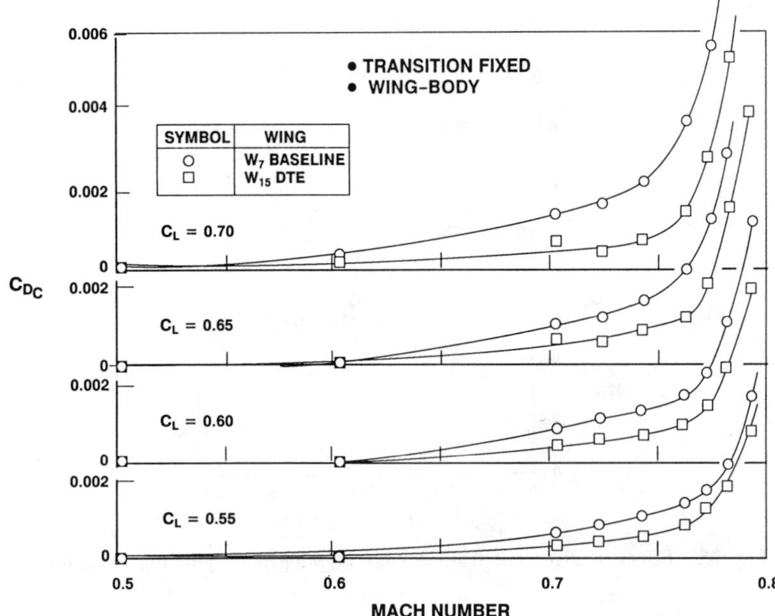

Fig. 33   Comparison of measured compressibility drag characteristics for supercritical wing W_7 and DTE wing W_{15}.

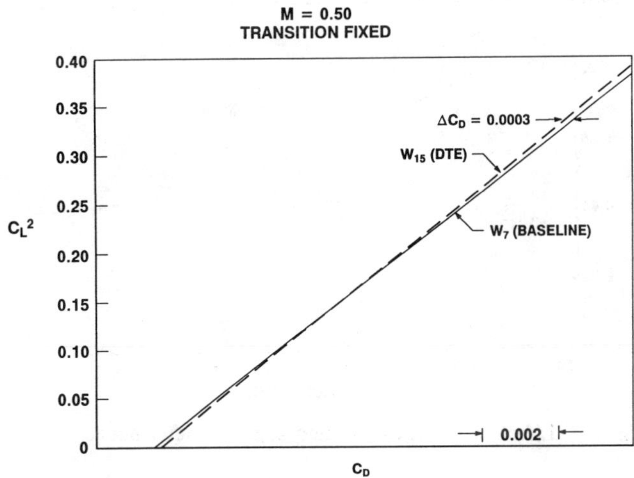

Fig. 34   Measured low-Mach-number drag polars.

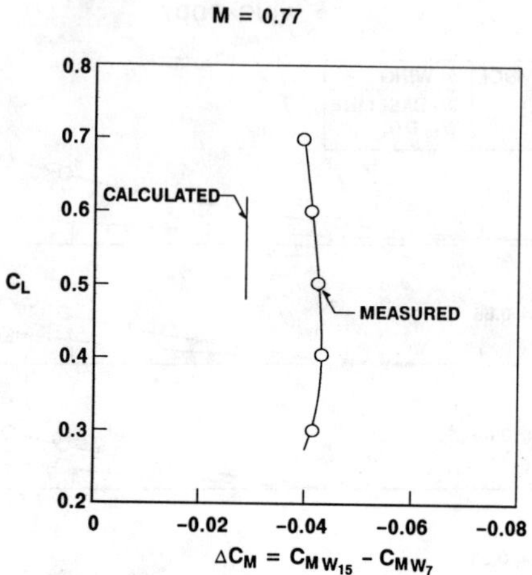

Fig. 35   Effect of divergent trailing edge on wing/body pitching moments.

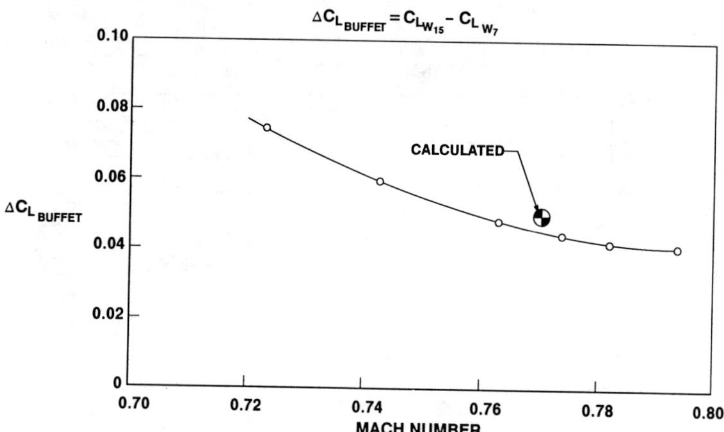

Fig. 36   Effect of divergent trailing edge on buffet boundary.

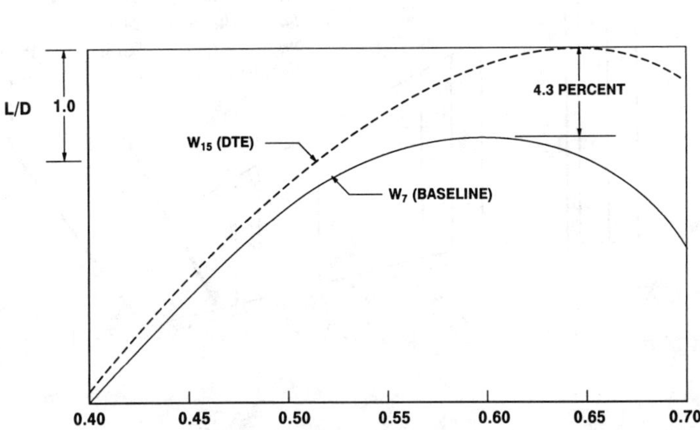

Fig. 37   Comparison of measured wing body L/D characteristics.

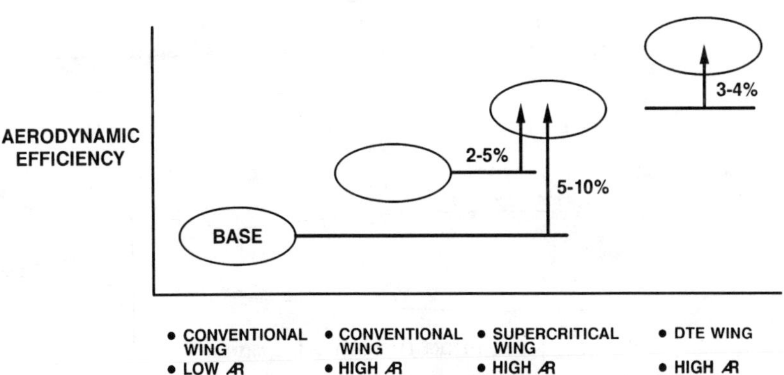

Fig. 38   Airfoil technology progression.

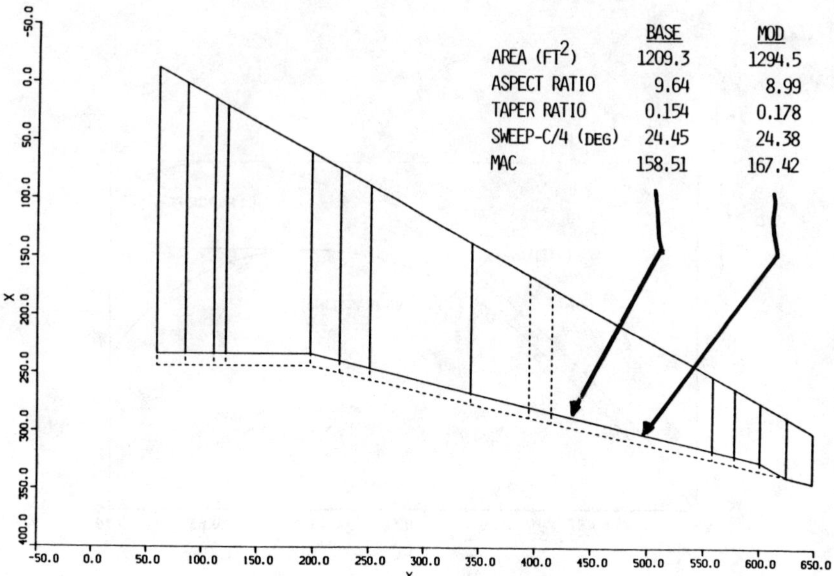

Fig. 39   Planform comparison of the baseline wing and the modified wing.

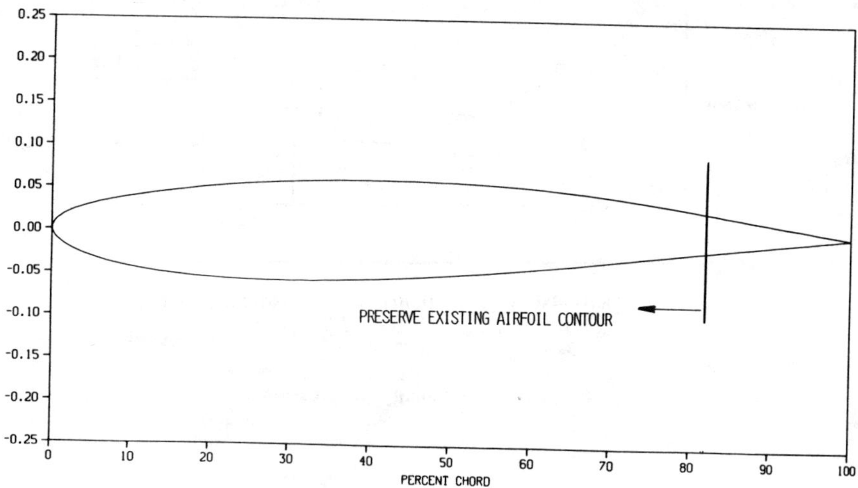

Fig. 40   Extent of airfoil modification.

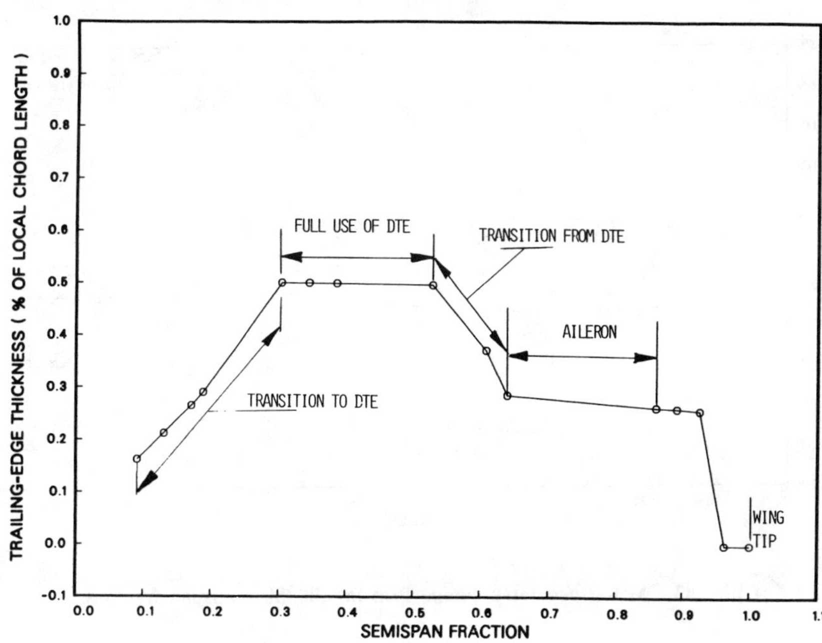

Fig. 41   Wing W16A (DTE) trailing-edge thickness.

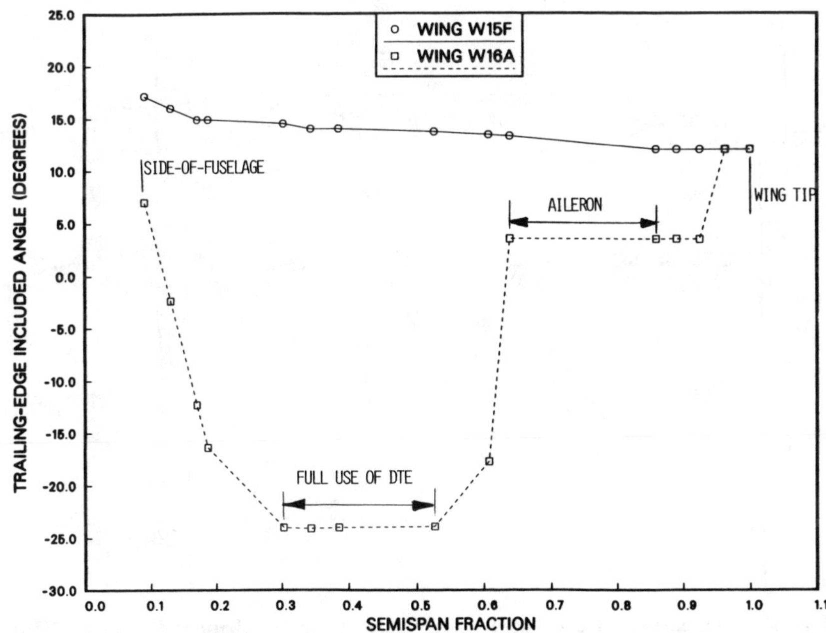

Fig. 42   Comparison of wings W15F and W16A trailing-edge included angle.

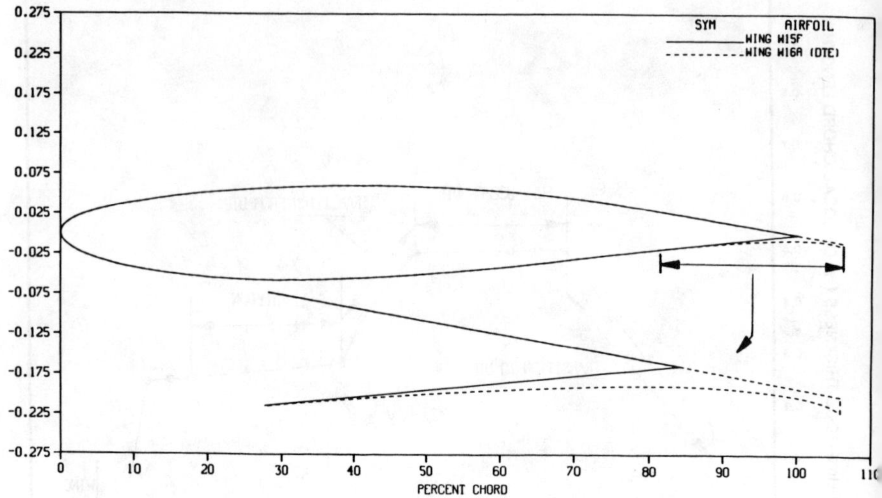

Fig. 43　Airfoil geometry comparison for 30.3% semispan fraction.

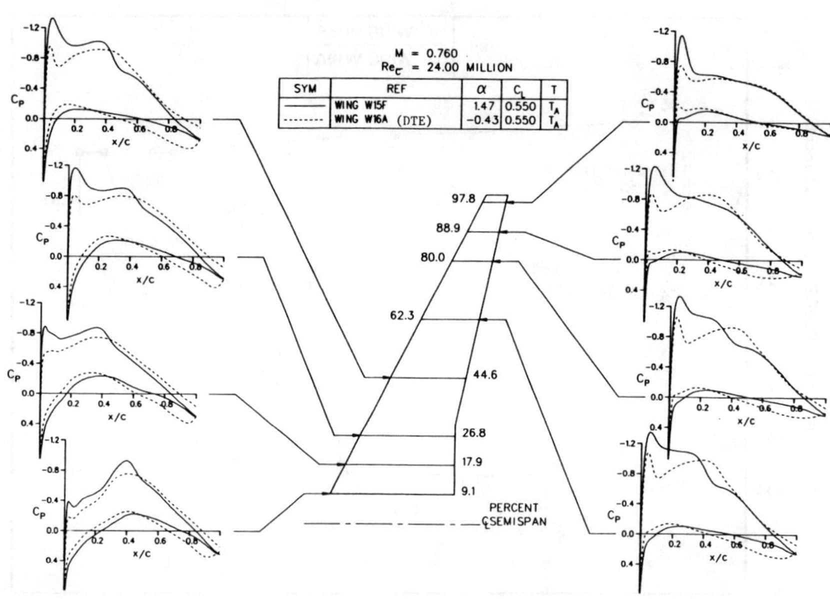

Fig. 44　Comparison of calculated chordwise pressure distributions for wings W15F and W16A.

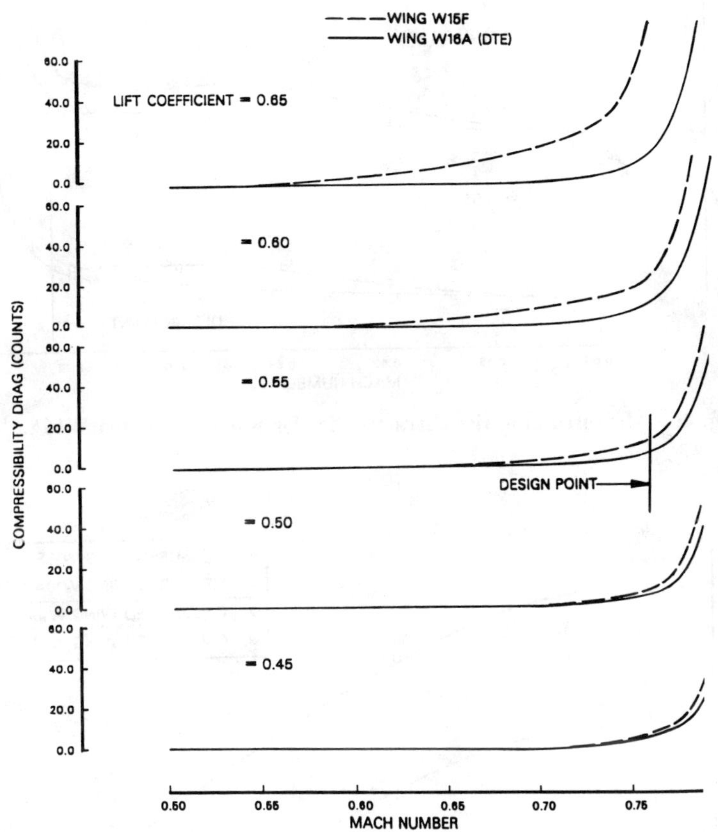

Fig. 45 Calculated drag rise characteristics for wings W15F and W16A.

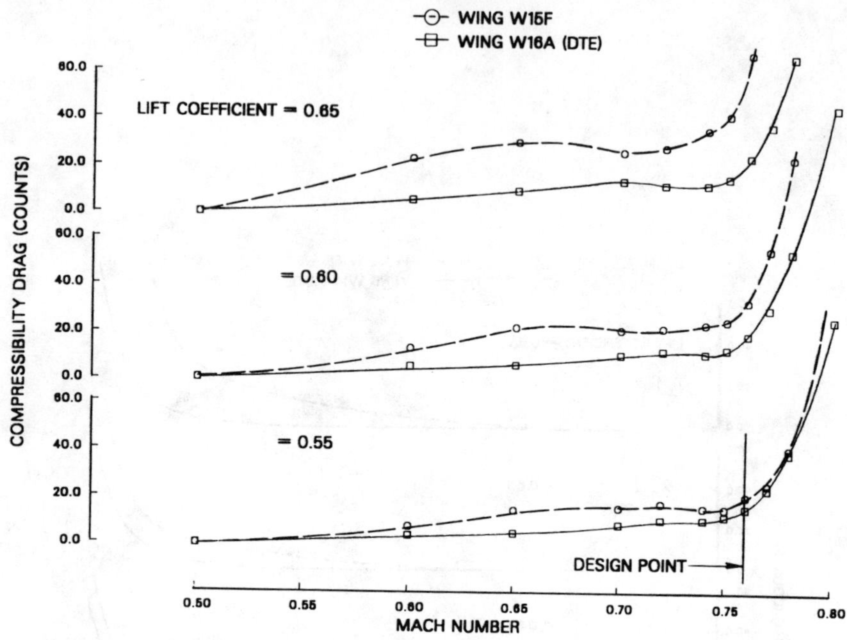

Fig. 46 Measured drag rise characteristics for wings W15F and W16A.

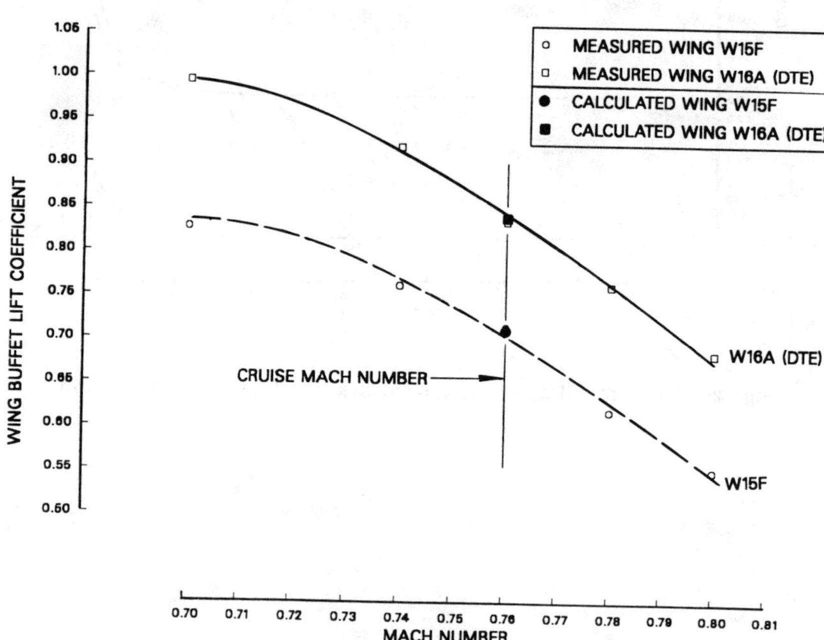

Fig. 47 Comparison of calculated and measured buffet-onset boundaries of wings W15F and W16A.

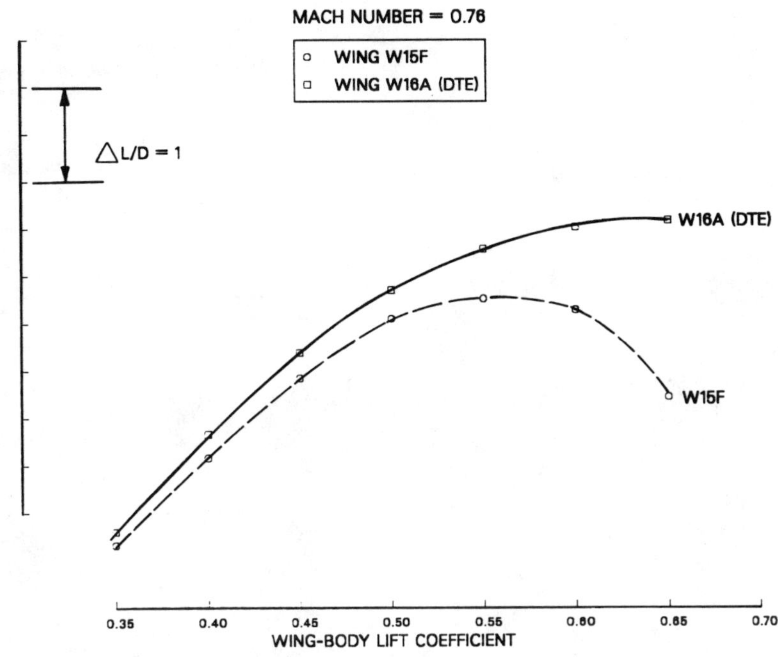

Fig. 48   Measured lift/drag characteristics for wings 15F and W16A.

Chapter 9

Euler and Potential Computational Results for Selected Aerodynamic Configurations

R. M. Hicks,* S. E. Cliff,* J. E. Melton,† R. G. Langhi,‡ A. M. Goodsell,*
D. D. Robertson,§ and S. A. Moyer§
NASA Ames Research Center, Moffet Field, California

Nomenclature

C = chord
C_D = drag coefficient
C_{D_0} = drag coefficient at zero lift
C_L = lift coefficient
C_M = pitching-moment coefficient
C_P = pressure coefficient
M = Mach number
Re = Reynolds number
X = chordwise distance
α = angle of attack
η = span station in fractions of span

Subscripts
∞ = freestream conditions
th = theoretical

Introduction

THE development of computational fluid dynamics (CFD) has passed through two phases over the past 30 years. The initial phase, lasting from the late 1950s to the early 1970s was a period of exploration and slow growth. Expectations for the future of CFD were modest, and wind-tunnel testing dominated the design process. The second phase, which lasted until the early 1980s was a period of rapid growth in algorithm and grid

---

Copyright © 1989 by the American Institute of Aeronautics and Astronautics, Inc. No copyright is asserted in the United States under Title 17, U.S. Code. The U.S. Government has a royalty-free license to exercise all rights under the copyright claimed herein for Governmental purposes. All other rights are held by the copyright owner.

*Aeronautical Engineer, Advanced Aerodynamic Concepts Branch.
†Aeronautical Engineer, Applied Aerodynamics Branch.
‡Programmer-Analyst, Sterling Federal Systems, Inc.
§Aeronautical Engineer, U.S. Airforce.

topological complexity. This was a period of great expectation for the future of CFD and a time when the imminent demise of wind tunnels was predicted. A third phase has been entered in which CFD limitations are more apparent and code advancement has been substantially replaced by code refinement. Many wind tunnels are being refurbished, and a few new tunnels are being planned or built.

Most of the computational results presented at technical symposia and in the technical literature during the early phase of development described the flow about two-dimensional or axisymmetric configurations with attached flow. This was a period when panel codes and finite-difference approximations of the potential equation were state of the art. Modest claims were made for the value of CFD as a design tool. The second phase was a time when exaggerated claims for the capabilities of CFD were made, and technical presentations often contained flowfield calculations for separated flows about simple geometries or for attached flow for complete aircraft. Finite-volume approximations to the full-potential, Euler, and Navier-Stokes equations began to appear, and finite-difference approximations became less popular. Most of the experiment/CFD correlations presented were good, with the calculations sometimes appearing to be a fairing of the experimental data. Flowfield computations for complete aircraft were presented by code developers after many months of code manipulation. Such calculations show the ultimate value of CFD, but tend to give false hope to the aircraft designer, who is led to believe that such codes can be used to *design* arbitrary aircraft configurations. When an engineer attempts to use these codes to design a configuration different from the one used to demonstrate the code, the idiosyncrasies and limitations of the code become apparent. The Euler and Navier-Stokes codes are not as robust as potential-flow codes and require a level of user skill beyond that of most design engineers. The use of such codes often requires frequent consultation with the author of the code because the numerical methods used to solve the fluid dynamic equations contain many parameters that must be adjusted to fit each aerodynamic configuration and/or flow condition. Many aircraft designers are specialists in aerodynamics but have little training in numerical analysis and fluid dynamics, making the application of Euler and Navier-Stokes codes to practical design problems difficult at best. Knowledge of numerical analysis is not sufficient to guarantee successful code application if the user is not also the code developer, since most CFD codes use little structured programming and include little documentation, making reprogramming during the design process nearly impossible. The Euler and Navier-Stokes codes require more computer memory and more CFD time than potential-flow codes. These are the primary reasons that the use of the Euler and Navier-Stokes codes in aircraft design has been limited to a class of problems that is dominated by vortical flow.

Before one attempts a design problem, the flowfield characteristics and flight conditons should be examined to determine the least complicated fluid dynamic equations that will describe the flow about the configuration being considered. If the potential equation is deemed adequate, there is little justification for using the Euler equations. A commercial transport wing flowfield is adequately described by the full-potential equation for the transonic cruise condition. The aerodynamic characteristics of highly swept wings are more accurately described by the Euler or Navier-Stokes equations because of the presence of rotational flow. Separated flows not initiated by abrupt geometric change (e.g., a sharp leading edge) are not predicted accurately by the most advanced CFD codes, even in two dimensions. The Navier-Stokes equations are valid for separated flow, but the numerical models used to solve the equations do not, in general, calculate separation correctly. The maximum lift coefficient for an arbitrary airfoil section is no closer to predictability today than it was 10 years ago. The absolute level of drag is another quantity that cannot be determined accurately for an arbitrary aircraft configuration.

In spite of the limitations and inaccuracies of current CFD codes, the aircraft designer can eliminate many hours of wind-tunnel testing by judiciously using CFD during the design process. The designer should calibrate the code against data for configurations that are similar to the one being designed, to ascertain whether the code is capable of calculating the absolute level of the aerodynamic forces and moments or can merely produce the correct trends with variation in flow condition and/or geometry. CFD can be used to enhance the utility of an experiment in many ways (e.g., the placement of pressure taps on a wind-tunnel model can be made less arbitrary by calculating the regions with large pressure gradients prior to model design).

The examples presented in this chapter will address a selection of successes and failures of CFD. Experiment/CFD correlations involving full-potential and Euler computations of the aerodynamic characteristics of four commercial transport wings and two low-aspect-ratio, delta-wing configurations will be shown. The examples will consist of experiment/CFD comparisons for aerodynamic forces, moments, and pressures. Navier-Stokes calculations will not be considered in this chapter. The computational results discussed in the following sections are representative of the level of accuracy that can be obtained without reprogramming the codes used in this study. Coding changes would require the active participation of the code authors and therefore would not be indicative of the level of accuracy that can be obtained from advanced CFD codes by a typical aircraft designer. An effort was made to determine optimum values for grid density, grid distribution, artificial dissipation, Courant, Friedrichs, Lewy (CFL) number, enthalpy damping, and multigrid scheme for each flow condition and configuration analyzed during this study.

Commercial Transport Wing/Body Configurations

Model and Test Description

A planform view of the large semispan model with the dimensions indicated in inches is shown in Fig. 1. This model was tested in the NASA Ames Research Center 11-ft transonic wind tunnel. The model was mounted approximately 6 in. above the tunnel floor, which brought the wing tip to a position only 2 ft from the tunnel ceiling. The model was manufactured from stainless steel and had eight chordwise rows of pressure taps. Each row had 23 upper-surface and 12 lower-surface orifices, giving a total of 280 surface pressure measurements on the wing. An installation photograph showing a rear quartering view of the model is shown in Fig. 2. Pressure rails (clearly visible in Fig. 2) were attached to both walls in an effort to assess the wall interference of the large semispan model, which had 2.8% blockage. The test was conducted over a Mach number range of 0.4–0.86. The Reynolds numbers, based on mean aerodynamic chord, varied from 1.4 to 5.1 million at Mach 0.40 and 2.3 to 8.5 million at Mach 0.86. Transition was fixed at 5% chord on the upper surface and 10% chord on the lower surface for most test conditions. Limited transition-free runs were also included. Four wings were tested on a common body with identical planforms but different airfoil sections and twist distributions.

Experiment/CFD Correlations

The primary purpose of this study was to evaluate an inviscid Euler code for the design and analysis of commercial transport wings. In actual practice there is little justification for using Euler solvers in transport wing design since the flight regime of primary interest is transonic cruise and experience has shown that the full-potential equation with a simple quasi-three-dimensional integral boundary-layer correction is adequate for this purpose.[1] However, Euler flow solvers have proliferated in recent years, driven by academic competition; hence, it is of interest to the aircraft designer to determine the range of applicability of such codes. This section will evaluate an Euler code with C-H- and H-O-grid topologies for transport wing design. (Two later sections of this chapter will evaluate the same code with the H-O-grid topology for design of low-aspect-ratio delta wings.) This section will also include limited computational results from a fully conservative potential-flow code and a nonconservative potential-flow solver. The conservative code is coupled to a three-dimensional, finite-difference, boundary-layer code, and the nonconservative code includes a quasi-three-dimensional, integral boundary-layer correction. The Euler code used for this study was originally developed by Jameson et al.[2] and was subsequently modified by personnel of the Lockheed Georgia Company. The code, designated FLO57, will be evaluated by addressing two issues. The first is the code's capability of determining quantitative aerodynamic forces, moments, and pressures; the second is the code's capability of ordering the four transport wings correctly with respect to drag level. The latter capability is the best that can be expected from most CFD codes.

The common planform showing the distribution of defining sections is shown in Fig. 3. Each of the four wings had 23 defining stations from root to tip. The most inboard section shown in the figure is inside the body and was obtained by extrapolating the sections toward the centerline. This was done to evaluate two different techniques of describing the wing to the C-H grid-generation code. Experience has shown that some grid generators give smoother cell distribution near the wing/fuselage intersection if a defining section inside the body is included in the wing definition. The results of this study showed that little difference could be detected in the computational results from including the additional section inside the body. Hence, all results presented were obtained from grids generated from the 23 defining stations begining at the wing root rather than from the 24 defining stations shown in the figure. The wing sections for the four wings analyzed during this study are shown in Fig. 4. Note the difference in lower-surface contours between wing A and wings B, C, and D from $\eta = 0.147$ to $\eta = 0.501$. Beyond $\eta = 0.501$ the wing sections become similar with minor differences in thickness.

Two different grid topologies were evaluated with FLO57. Most of the computational results shown in this section were obtained from an elliptic grid generator that produced C contours in the wraparound streamwise direction and H contours in the spanwise direction. This grid generation code was developed by personnel at the Lockheed Georgia Company. The distribution of cells on the wing surface is uniform in the spanwise direction and is clustered near the leading and trailing edges (Fig. 5). The grid had 151 points in the wraparound streamwise direction, with 121 points on the surfaces of the wing. The cell distribution in the plane of the wing is shown in Fig. 6. Note that the cells are skewed along the nose of the body—a well-known deficiency with C-H grids for long forebodies. The spacing between grid lines beyond the tip increases more rapidly than desired for optimum computational results. Also note the large grid cells adjacent to the centerplane upstream of the nose. These large cells, along with the skewed cells along the nose, could give questionable aerodynamic quanti-ties on the nose of the body. The C-H grid has 31 points in the spanwise and vertical directions, with 25 points on the wing and body. The cell distribution on the centerplane, including the half-body, is shown in Fig. 7. The large cells adjacent to the aft section of the fuselage and the high-aspect-ratio cells along the aft body are another reason for inaccurate body forces. The outer boundary is fairly close to the body, resulting in further uncertainties in the computational results. A limited amount of time was available to conduct this study; therefore, different outer-boundary loca-tions were not examined.

An algebraic grid generator was used to produce the H-O grid topology evaluated during this study.[3] The H- and O contours are in the streamwise and spanwise directions, respectively. The H-O wing-surface grid has clustering near the root and tip, with fewer points on the surface than the C-H grid has (compare Figs. 5 and 8). Note that the C-H grid has greater leading- and trailing-edge clustering than the H-O grid does. The grid cell distribution and shape in the plane of the wing adjacent to the body is more regular for the H-O grid than for the C-H grid (compare Figs. 6 and 9). The outer boundary of the H-O grid is considerably farther from the body than that of the C-H grid. The H-O grid also has more clustering near the nose of the body than the C-H grid does, resulting in better definition of the nose flowfield. The H-O grid distribution on the centerplane has more clustering and more uniform cell shape than the C-H grid (compare Figs. 7 and 10). Both grid topologies have regions of cell distortion that are largely unavoidable. An H-H grid topology avoids most of the problems associated with H-O and C-H grids, but requires substantial clustering near blunt leading edges for accurate prediction of pressure peaks. The computational results will show that in spite of the large differences between the H-O and C-H grid topologies the predicted surface pressures are similar.

The experimental and theoretical surface pressure distributions for wing A are shown in Fig. 11. The theoretical pressures were obtained from FLO57 with the C-H grid topology (designated FLO57C in the figures). Results will be shown for Mach numbers of 0.40, 0.70, and 0.80 and theoretical angles of attack of 0 and 2 deg, except as noted. (The phrase "theoretical angle of attack or $\alpha_{th} = X$ deg" refers to the angle of attack associated with the computational results.) Computational results were also obtained for angles of attack of -2 and 4 deg but will be shown on the force and moment plots only. The computational results are the best solutions obtainable from FLO57 after several grid densities and distributions were examined. The experimental and theoretical pressure distributions are compared for closely matched lift coefficients and Mach numbers in this section. Pressure comparisons will be shown at eight spanwise stations from 15% to 95% semispan. Note the self-scaling vertical axis when comparing pressure distribution plots. The experiment/CFD comparisons for $M_\infty = 0.40$ and $\alpha_{th} = 0$ deg are shown in Fig. 11a. The experimental upper and lower surface pressures are depicted by circular and square symbols, respectively. The computational and experimental pressures correlate fairly well at most span stations, except that the lower-surface pressures near the leading edge show small differences. The overall correlation is somewhat better near the root ($\eta = 0.15$) than near the tip ($\eta = 0.95$). The aft loading is predicted accurately by FLO57 at all span stations because of moderate adverse pressure gradients, thin boundary layers, and attached flows. The experiment/CFD correlation for $\alpha_{th} = 2$ deg is slightly worse than at the lower angle of attack, particularly near the leading edge on the upper surface of the outboard section of the wing (Fig. 11b).

The experiment/CFD pressure comparisons for wing A at Mach 0.70 are shown in Figs. 11c and 11d. Again, the correlation is good at both angles of attack with the exception of the tip region at $\alpha_{th} = 2$ deg (Fig. 11d). The shock position and strength are calculated accurately at most span stations. The shocks are weak, and good trailing-edge pressure recovery is observed at this Mach number; thus, the inviscid approximation is valid. The experiment/CFD pressure comparisons for Mach 0.80 are shown in Figs. 11e and 11f. The fact that an upper-surface shock wave near midspan is predicted to be too far aft with too much strength at $\alpha_{th} = 0$ deg indicates that a boundary-layer displacement surface should be included in the calculation (Fig. 11e). The pressure recovery at the trailing edge indicates attached flow at all span stations at $\alpha_{th} = 2$ deg, but the shock wave has sufficient strength to cause a thick turbulent boundary layer over the aft region of the wing. The decambering effect of a displacement surface would result in a weaker shock with a more forward position and better experiment/CFD correlation.

Wing A pressure distributions calculated by FLO57 with C-H- and H-O-grid topologies are compared for matched angle of attack at Mach 0.80 in Fig. 12 (FLO57 with the H-O grid topology is designated FLO57O in the figures). The C-H and H-O grids have 145,111 and 131,805 points, respectively. The shape of the pressure distributions calculated by the two grid topologies exhibit minor differences. For example, shock strength and position are different at 2-deg angle of attack over the inboard region of the wing (Fig. 12b). The computational results from the two grids show greater differences over the inboard region of the wing where the difference in grid cell distribution between the two topologies is largest. The span station nearest the tip might be expected to show large grid effects since this region of the wing is difficult computationally. The effect of grid topology on wing pressures is shown for only one Mach number since the differences were small.

A single calculation using an H-H-grid topology was generated for wing A by using the fast multigrid, isolated-wing, Euler code FLO60. H-H grids use upper and lower blocks and therefore require more computer memory than C-H and H-O grids. A grid with dimensions of $96 \times 16 \times 16$ was chosen for efficient use of the Cray XMP-48. Additional computations were not performed because body and boundary-layer effects were not included in the solutions, and the experiment/CFD correlation obtained is worse than for FLO57 (compare Figs. 11 and 13).

An attempt was made to determine the relative importance of viscous and body effects by conducting experiment/CFD correlations with an isolated-wing code with a boundary-layer correction. The code chosen for this part of the study was FLO22NM: a full-potential code with an integral boundary-layer subroutine and a C-H-grid topology with 192 streamwise, 24 vertical, and 32 spanwise points. FLO22NM was originally developed by Jameson and Caughey[4] and was subsequently modified by McDonnell Douglas Aircraft Company to include a boundary-layer correction. The boundary-layer displacement surface is computed iteratively along with the potential solution and is added stripwise to the wing surfaces. The boundary-layer displacement surface was updated five times during convergence. Computational pressures obtained from FLO22NM are compared with test pressure distributions in Fig. 14 for Mach numbers of 0.40 and 0.80. Comparisons are shown for theoretical angles of attack of 1 and 4 deg at both Mach numbers. These angles of attack were chosen because the lift coefficients match those for FLO57 (Fig. 11). The experimental and theoretical pressures are compared for closely matched lift coefficient, Mach numbers, and Reynolds number. The experiment/CFD comparisons for Mach 0.40 are good at both test conditions. The shapes of the pressure distributions including the aft loading are predicted accurately by FLO22NM at Mach 0.40. The experiment/CFD correlation for FLO22NM is slightly better than that for FLO57 at both lift coefficients. The angle of attack required for a given lift coefficient is greater for FLO22NM than that for FLO57 because FLO22NM neglects the body and includes viscosity, whereas FLO57 includes the body and neglects viscosity. FLO22NM predicts pressures more accurately near the tip than FLO57 does at Mach 0.40 (compare Figs. 11 and 14). The FLO22NM experiment/CFD correlation is also better than that for FLO57 at Mach 0.80. The shock strength and location are predicted more accurately by FLO22NM. It appears that an isolated-wing potential-flow code with a quasi-three-dimensional boundary layer gives more accurate predictions of transport wing pressures with less computer resources than does a wing-body Euler code.

The wing A experimental aerodynamic force and moment coefficients are compared with computational coefficients obtained from FLO57 with the C-H grid in Fig. 15. The FLO57 lift-curve slope is greater than the experimental slope at all Mach numbers because of neglected viscous effects. The inviscid drag prediction is less than the experimental drag, but the difference is less than expected because of inaccurate integration of surface pressures. The only favorable comment that can be made regarding the drag predictions is that the shape of the polar at Mach 0.70 appears to be correct—a result that may be fortuitous. The pitching moment curves predicted by FLO57 exhibit more static stability than do the test results at all Mach numbers, which is consistent with the pressure distributions shown earlier. The irregularities in the theoretical pitching moment curve for Mach 0.40 will be explained after the computational results for wings B and C have been discussed. The explanation will be more convincing when additional data have been presented. The smoother cell distribution of the H-O grid compared with the C-H grid results in a more regular pitching moment curve at Mach 0.40 (Fig. 16).

The FLO22NM force and pitching-moment computations are compared with test data and FLO57 computations for wing A at Mach 0.40 and 0.80 in Fig. 17. The FLO22NM computational lift-curve slopes are less than the experimental and FLO57 computational slopes at both Mach numbers, which is consistent with the missing body lift and the decambering of the wing caused by the boundary-layer displacement surface. The drag predicted by FLO22NM is lower than that of FLO57 because of inaccurate integration of surface pressures for the latter, but the polar shapes using FLO22NM are in better agreement with the test data, which indicates that its shock and induced-drag calculations are more accurate. The FLO22NM pitching moments are more negative than those of FLO57 and experiment because the destabilizing body moments are neglected.

Pressure distributions computed by FLO57 are compared with experimental pressures for wing B in Fig. 18. The lower-surface pressure contours for wing B differ from those for wing A because of greater lower-surface thickness over most of the span (compare Fig. 11a with Fig. 18a). The increased flow acceleration along the lower surface is evident at the four inboard span stations and is predicted accurately by FLO57 (Fig. 18). The experiment/CFD correlation is good at both angles of attack at Mach 0.40. The computational pressures are more positive than the experimental pressures near the leading edge of the wing near the tip at Mach 0.70 and $\alpha_{th} = 2$ deg. The experiment/CFD correlation is acceptable at all span stations for both angles of attack at Mach 0.70. The experiment/CFD correlation is unacceptable when shock waves are present on the wing at Mach 0.80. The theoretical shock strength and position is usually greater than the experimental values because viscous effects are neglected.

The effect of grid topology on the pressure distributions for wing **B** was examined at 2-deg angle of attack for Mach numbers of 0.40 and 0.70. The results are shown in Fig. 19. The computed pressures from the two topologies are similar except for minor differences near the leading edge at most span stations.

The wing **B** experimental and computational force and moment co-efficients are compared in Fig. 20. The degree of correlation of experiment

Table 1   Initial and final residuals in (   ) iterations (FLO57C)

		Mach number		
		0.4	0.7	0.8
		Wing A		
	-2	$0.1801 + 2$	$0.3189 + 2$	$0.3785 + 2$
		$0.9109 - 1$ (3000)	$0.9414 - 2$ (3000)	$0.1645 - 1$ (2800)
	0	$0.1804 + 2$	$0.3182 + 2$	$0.3778 + 2$
		$0.2704 - 2$ (1250)	$0.1379 - 1$ (3000)	$0.5188 - 2$ (1740)
	2	$0.1796 + 2$	$0.3180 + 2$	$0.3778 + 2$
		$0.4085 - 1$ (3000)	$0.8194 - 3$ (3000)	$0.4325 - 2$ (3000)
	4	$0.1799 + 2$	$0.3184 + 2$	$0.3785 + 2$
A		$0.4492 - 2$ (1250)	$0.2235 - 1$ (3000)	$0.5241 - 2$ (1740)
n		**Wing B**		
g	-2	$0.1740 + 2$	$0.3072 + 2$	$0.3524 + 2$
l		$0.1915 - 2$ (2000)	$0.1680 - 1$ (3000)	$0.1610 - 1$ (3000)
e	0	$0.1732 + 2$	$0.3057 + 2$	$0.3507 + 2$
o		$0.1389 - 1$ (2000)	$0.9693 - 2$ (3000)	$0.1079 - 1$ (3000)
f	2	$0.1726 + 2$	$0.3047 + 2$	$0.3495 + 2$
a		$0.7426 - 2$ (2000)	$0.9363 - 3$ (3000)	$0.5453 - 2$ (3000)
t	4	$0.1716 + 2$	$0.3043 + 2$	$0.3490 + 2$
t		$0.4513 - 1$ (2000)	$0.7306 - 2$ (3000)	$0.6130 - 2$ (3000)
a		**Wing C**		
c	-2	$0.1732 + 2$	$0.3054 + 2$	$0.3507 + 2$
k		$0.8751 - 2$ (2800)	$0.1141 - 1$ (2800)	$0.2068 - 1$ (2800)
	0	$0.1723 + 2$	$0.3045 + 2$	$0.3489 + 2$
		$0.1558 - 1$ (2800)	$0.3092 - 2$ (2800)	$0.8481 - 2$ (2800)
	2	$0.1717 + 2$	$0.3034 + 2$	$0.3477 + 2$
		$0.9070 - 2$ (2800)	$0.2282 - 2$ (2800)	$0.8982 - 2$ (2800)
	4	$0.1714 + 2$	$0.3029 + 2$	$0.3471 + 2$
		$0.3568 - 2$ (2800)	$0.3023 - 1$ (2800)	$0.5526 - 2$ (2800)
		Wing D		
	-2	$0.1732 + 2$	$0.3073 + 2$	$0.3525 + 2$
		$0.1647 - 1$ (2800)	$0.1922 - 1$ (2800)	$0.1205 - 1$ (2800)
	0	$0.1724 + 2$	$0.3057 + 2$	$0.3506 + 2$
		$0.2324 - 1$ (2800)	$0.1560 - 1$ (2800)	$0.1853 - 1$ (3000)
	2	$0.1718 + 2$	$0.3046 + 2$	$0.3494 + 2$
		$0.2396 - 1$ (2800)	$0.1082 - 2$ (2800)	$0.1867 - 2$ (3000)
	4	$0.1715 + 2$		$0.3488 + 2$
		$0.1793 - 1$ (2800)		$0.2026 - 2$ (2800)

with theory is similar to that shown for wing A (compare Figs. 15 and 20). The irregular shapes of the force and moment curves noted for wing A was also observed for wing B. In particular, the pitching-moment curve for wing B at Mach 0.40 exhibits irregularities similar to those shown for wing A. The wing forces and moments in the presence of the body and the body forces and moments in the presence of the wing were analyzed independently in an effort to explain the poor pitching moment results. The wing forces and moments in the presence of the body were smooth with no irregularities. However, the fact that the body forces and moments exhibited a break in the curves showed that the body is responsible for the irregular force and moment curves (Fig. 21). The irregular moment curves for wing A and B will be shown to correlate with variations in code convergence level when wing C is discussed.

The effect of grid topology on the forces and moments for wing B at Mach numbers of 0.70 and 0.80 is shown in Fig. 22. The computational lift and pitching moment curves from the two grids agree well in slope and magnitude. The drag polars are similar in shape, but the C-H grid gives a larger value of drag at all test conditions because of inaccurate surface pressure integrations. The fact that neither grid topology gives good correlation with the test data shows that the inviscid Euler equations with the grids used here are of questionable value for design and analysis of commercial transport wings.

The experimental and theoretical pressure distributions for wing C are shown in Fig. 23. The wing section contours for wing C are similar to those for wing B; hence, the experiment/CFD correlations for the two wings should be comparable. The predicted shock strength and position appear to be slightly better at some test conditions for wing C than for wing B (compare Fig.18f with 23f).

The computational and experimental aerodynamic forces and moments for wing C are compared in Fig. 24. The experiment/CFD correlation is comparable to that for wings A and B except that the wing C theoretical results are more regular at Mach 0.40. In particular, the pitching-moment curve for wing C does not have the sinusoidal shape of the curves for wings A and B. The body and wing planform is identical for all four wings; thus, it is unlikely that the unusual pitching moment characteristics are due to geometric differences between the wings. Examination of the convergence trends for wings A, B, and C provides a plausible explanation for the unusual moment curves. The initial and final average residuals for the computational results obtained from FLO57 with the C-H grid are shown in Table 1. The numbers shown in parentheses are the total number of iterations for each solution. The order-of-magnitude reduction in average residual for each solution is shown in Table 2. Note that the convergence level as a function of angle attack exhibits a wave pattern similar to the pitching-moment wave pattern for wing A at Mach 0.40 (Fig. 15a). Also note that the more fully converged solutions give pitching-moment coefficients that are nearest to the test data (Figs. 15a and 20a). The pitching-moment and convergence wave patterns for wing B are reversed from those of wing A, and the trends are consistent for each wing. The convergence levels for wing C at Mach 0.40 show less waviness, which is consistent with the shape of the pitching-moment curve (Fig. 24a). The order-of-magnitude reduction in values shown in Table 2 for Mach 0.70 and 0.80 show little variation with angle of attack, which is consistent with the pitching-moment curves for these test conditions. It appears that the irregular pitching-moment curves are related to varying levels of convergence for the computational results. All attempts to improve convergence for the cases in question by varying input parameters and running additional iterations were unsuccessful.

The effect of grid topology on the aerodynamic force and moment coefficients for wing C is shown in Fig. 25. The H-O-grid lift curves correlate somewhat better with the test data than those of the C-H grid at Mach 0.40 and 0.70. The opposite is true at Mach 0.80. The C-H-grid pitching moments correlate somewhat better with the experimental data than the H-O-grid moments, but neither prediction is acceptable. The H-O-grid drag coefficients are less than those of the C-H grid, primarily because of inaccurate surface pressure integrations for the latter. In summary, the drag and moment predictions for wing C are poor, and the lift-curve calculation is acceptable. However, accurate lift-curve prediction can be achieved by far simpler methods than a three-dimensional Euler code.

Experimental and theoretical pressure distributions for wing D are compared in Fig. 26. The wing D experimental pressures near the leading edge over the outboard section of the wing are slightly more negative than those of wing C (compare Figs. 23c and 26c). These pressures were not predicted accurately by FLO57; thus, the experiment/CFD correlation was somewhat poorer for wing D than for wing C. However, the differences are small, and the trends and conclusions remain the same as for the three wings discussed earlier. The experiment/CFD correlation for wing D at Mach 0.80 and $\alpha_{th} = 2$ deg is worse than for wing C at the same flow conditions (compare Fig. 23f with 26f). Note that the predicted shock position is less accurate for wing D over the outboard region of the wing at this Mach number. The experimental data show that wing D has less drag than wing C for most lift coefficients at Mach 0.70 and 0.80, and FLO57 predicts lower drag for wing C. Such results indicate that FLO57 has limited value for transport wing design when used with the grid generators discussed in this chapter.

The experimental pressure distributions for wing D are compared with computational pressures from FLO22NM in Fig. 27. The experiment/CFD correlation for wing D is similar to that shown for wing A at Mach 0.40 (compare Fig. 14 with Fig. 27). The pressure distribution predictions for the two wings, including the effects of small geometric differences in the leading-edge region, are acceptable.

The aerodynamic force and moment coefficients predicted by FLO57 are compared with experimental quantities for wing D in Fig. 28. The theoretical lift-curve slope is greater than the experimental slope at all Mach numbers because of neglected boundary-layer effects. The computational drag is clearly too large near zero lift, where the flow is shock-free at Mach

Table 2   Total change in the average residual in (   ) iterations in orders of magnitude (FLO57C)

		Mach number		
		Wing A		
		0.4	0.7	0.8
	-2	2.30 (3000)	3.53 (3000)	3.36 (2800)
	0	3.82 (1250)	3.36 (3000)	3.66 (1740)
A	2	2.64 (3000)	4.59 (3000)	3.94 (3000)
n	4	3.60 (1250)	3.15 (3000)	3.86 (1740)
g		**Wing B**		
l	-2	3.96 (2000)	3.26 (3000)	3.34 (3000)
e	0	3.10 (2000)	3.50 (3000)	3.51 (3000)
o	2	3.37 (2000)	4.51 (3000)	3.81 (3000)
f	4	2.55 (2000)	3.62 (3000)	3.75 (3000)
a		**Wing C**		
t	-2	3.29 (2800)	3.43 (2800)	3.23 (2800)
t	0	3.04 (2800)	3.99 (2800)	3.61 (2800)
a	2	3.28 (2800)	4.12 (2800)	3.59 (2800)
c	4	3.68 (2800)	3.00 (2800)	3.80 (2800)
k		**Wing D**		
	-2	3.02 (2800)	3.20 (2800)	3.47 (2800)
	0	2.87 (2800)	3.29 (2800)	3.28 (3000)
	2	2.86 (2800)	4.45 (2800)	4.27 (3000)
	4	2.98 (2800)		4.24 (2800)

0.40 and 0.70 primarily because of inaccurate surface pressure integrations. The computational pitching-moment coefficients show excessive static stability consistent with an inviscid theory. The body forces, in the presence of the wing, for Mach 0.40 show that the irregular shape of the wing/body pitching-moment curve is caused by inaccurate body forces (Fig. 29). However, the pitching-moment curve for wing D is more regular than those for wings A and B (Figs. 15 and 20). Note that the shape of the moment curve for wing D corresponds to the trend in convergence level with angle of attack shown in Table 2. This provides more evidence to support the hypothesis that the irregular moment curves are related to convergence-level variations.

The influence of grid topology on the experiment/CFD correlation for wing D at Mach numbers of 0.70 and 0.80 is shown in Fig. 30. The lift-curve slopes computed by use of the H-O and C-H grids are greater than the experimental slopes at both Mach numbers. The pitching-moments calculated by use of the C-H grid correlate more closely with experiment than those of the H-O grid for wing D, whereas the reverse is true for wing A (compare Figs. 16 and 30). The pitching moment predictions from both grid topologies are poor for the two wings, giving further evidence that the Euler computational methods described in this chapter have limited use in transport-wing design and analysis.

The experimental forces and moments for wing D are compared with computations from FLO22NM and FLO57 with the C-H grid for Mach 0.40 and 0.80 in Fig. 31. The lift-curve slope predicted by FLO22NM is less than experiment for Mach 0.40 because FLO22NM neglects body lift and includes viscous decambering of the wing, whereas the FLO57 lift-curve slope is too high because FLO57 neglects viscosity. The agreement between experimental and theoretical lift curves for FLO22NM at Mach 0.80 is fortuitous. The shape of the FLO22NM drag polars correlate somewhat better with experiment than those of FLO57, which indicates better shock and induced-drag prediction for the latter. The FLO22NM pitching moment curve exhibits excess static stability consistent with an isolated-wing calculation. When the body forces and moments calculated by FLO57 are added to the isolated-wing forces and moments from FLO22NM, the experiment/CFD correlation improves as shown in Fig. 32. This correlations is the best obtained for the four transport wings evaluated during this study. It is apparent that both viscosity and body effects should be included in the computations to obtain acceptable force and moment predictions. Even a quasi-three-dimensional boundary-layer correction appears adequate for attached-flow conditions.

Experimental pressure distributions are compared with FLO28BL pressure distributions for wing D at Mach 0.80 and $C_L = 0.55$ in Fig. 33. FLO28BL is a fully conservative wing-body potential-flow code[5] coupled with a three-dimensional finite-difference boundary-layer program. The experiment/CFD correlation is acceptable at all span stations. Small discrepancies are observed at the outboard stations, particularly near the tip. Note that the FLO28BL and FLO22NM experiment/CFD correlations are similar, indicating that acceptable pressure predictions can be achieved for flight conditions with attached shockless or weak-shock flow without including body effects (compare Fig. 27d with Fig. 32). The C-H grid used with FLO28BL has dimensions of $217 \times 25 \times 33$, giving a total of 179,025 points. The cell distribution on the center plane is shown in Fig. 34. Note that the cells are more uniformly distributed and more regular in shape than those of the C-H grid used with FLO57 (compare Fig. 7 with Fig. 34). The outer boundary of the FLO28BL grid is considerably farther from the body than that of FLO57 with the C-H grid.

An important part of this code evaluation was to determine whether FLO57 can predict 1) which of the four wings has the lowest drag and 2) the order of the wings with respect to drag. A CFD code that cannot predict drag increments resulting from small geometric changes has limited use in aerodynamic design. Experimental drag polars for the four wings at Mach 0.80 are shown in Fig. 35. The range of lift coefficients shown is appropriate for commercial transports operating at or near the transonic cruise condition. The experimentally determined order of the wings with respect to drag is fairly constant with lift coefficient, with wing A having the lowest drag and wing C the highest. The theoretical drag polars show that the wing with lowest drag is dependent on lift coefficient, and wing A does not have the lowest drag at any lift coefficient, in contrast to the experimental results (Fig. 36). These drag comparisons show that FLO57 with the C-H grid described here may not be capable of predicting drag as a function of small geometric change and hence may be unacceptable for transport-wing design. An assessment of the ability of the full-potential code FLO22NM to predict the correct order of wings A and D with respect to drag is shown in Fig. 37. Both experiment and theory show that wing A has lower drag than wing D has. However, the drag differences are small, and only two wings were considered; hence, it is difficult to draw meaningful conclusions from this part of the study.

Concluding Remarks

1) Experiment/CFD pressure correlations for a wing/body Euler code were acceptable for attached subsonic flow and transonic flow with weak shocks on a commercial transport wing.

2) Experiment/CFD pressure correlations for a wing/body Euler code were unacceptable for transonic flow with shock waves of moderate strength on a commercial transport wing.

3) Experiment/CFD force and moment correlations for a wing/body Euler code were unacceptable for subsonic or transonic flight conditions with attached flow for commercial transport wings.

4) The wing/body Euler code was not capable of predicting the order of the four commercial transport wings with respect to drag level.

5) Computed Euler results were little effected by changing from a C-H- to an H-O-grid topology.

6) A nonconservative, isolated-wing, full-potential code with a quasi-three-dimensional integral boundary-layer correction gave better correlation with experimental pressures than a wing/body Euler code without viscous corrections.

7) A full-potential wing-body code coupled to a three-dimensional finite-difference, boundary-layer program gave satisfactory agreement with experimental pressures for a typical transonic cruise condition of a commercial transport.

Generic Fighter

Experimental Studies

In an effort to increase the understanding of vortical flows at transonic speeds, Erickson and Rogers[6] conducted a series of experimental investigations into the behavior of a generic fighter model at transonic conditions. The model has a cropped delta wing with a leading-edge sweep of 55 deg, an aspect ratio of 1.8, and a taper ratio of 0.2. The wing sections were modified NACA 65A005 airfoils with sharp leading edges. A chine with wedge cross section was added to the forebody 0.5 in. above the wing. A planform view, a side view, and two cross-sectional views of the model are shown in Fig. 38. The model has a total of 80 upper-surface static pressure taps located at 30, 40, 50, 62.5, and 75% of the distance along the wing centerline chord, as depicted in Fig. 38. The wing was mounted on a generic fuselage that accommodated a four-module Scanivalve and a six-component balance.

Table 3   Computational run summary for generic fighter

Mach	α	Configuration	Iterations	CPU time (h)
0.60	4.0	W/B	1200	3.4
	8.0		1650	4.72
	12.0		2100	6.01
	16.0		2300	6.58
	20.0		2800	8.01
0.60	4.0	W/B/C	2000	5.72
	8.0		2000	5.72
	12.0		1800	5.15
	16.0		1800	5.15
	20.0		2400	6.87
0.80	5.0	W/B	1600	4.58
	8.0		2100	6.01
	12.0		3050	8.73
	16.0		3200	9.16
	20.0		3500	10.01
0.80	8.0	W/B/C	2300	6.58
	12.0		2500	7.15
	16.0		2400	6.87
	20.0		2900	8.30

The model was tested in the 7×10-ft transonic wind tunnel of the David Taylor Naval Ship Research and Development Center, at Mach numbers between 0.40 and 0.95 and at angles of attack between 0 and 22 deg. The effects of Mach number, angle of attack, and leading-edge flap deflection on the wing–upper-surface pressure distributions were studied. The model was also tested in the 6×6-ft wind tunnel of NASA Ames Research Center at Mach numbers between 0.40 and 1.8 and at angles of attack of 0–24 deg. The purpose of the Ames test was to determine the flowfield changes over the wing resulting from placing a chine at various locations on the forebody.[7]

Experiment/CFD Correlations

The grid generated about the generic fighter has an H-O topology that allows good leading-edge resolution. The grid was generated using an elliptic solver written by Melton and Langhi.[3] This code solves the Laplace equation in two dimensions in order to smoothly wrap a grid around the body at specific longitudinal locations. This grid is then algebraically redistributed in the direction normal to the wing surface to provide clustering specified by the user near the wing surface and also to provide a smooth transition between grid planes in the streamwise direction.

The grid used in the computations has a total of 426,790 points. There are 134 points in the streamwise direction, 94 of which are on the body. There are 49 points from the surface to the outer boundary and 65 points circumferentially around the wing/body configuration. The grid extends 6.5 centerline chordlengths upstream and 6.5 centerline chordlengths downstream of the body, and 7 semispanlengths radially to the outer boundary. Examples of the grid for the wing/body model are given in Figs. 39a and 39b. The chine was modeled as a flat plate. A grid plane at the chine trailing edge is shown in Fig. 39c.

FLO57 solutions were obtained at Mach numbers of 0.60 and 0.80 for both the wing/body and wing/body/chine configurations. Table 3 contains the computational run schedule. A minimum of 1200 iterations were run at the low angles of attack and 3000–3500 iterations at higher angles to achieve a reduction of at least three orders of magnitude in the average residual. The lift, drag, and moment histories generally show convergence after about 1000 iterations while the residuals are still decreasing. All solutions were obtained on the Cray-2 at NASA Ames Research Center. The memory requirement was approximately 16 megawords, and the CPU time required for 1000 iterations was approximately 10,300 s, resulting in 4–10 h for each solution. Computed pressure, lift, drag, and pitching-moment coefficients for each Mach/alpha combination were compared to experimental results at 30, 40, 50, 62.5, and 75% chord stations. A discussion of the results for each combination of Mach number and configuration will be presented later.

Mach = 0.60, Wing/Body

The flow is attached with no leading-edge vortex at $\alpha = 4$ deg. FLO57 pressure distributions compare well with experimental results (Figs. 40a–40d).

The experimental results show that a vortex has begun to form near the leading edge at $\alpha = 8$ deg. The computational results correlate poorly with experimental results as shown in Figs. 41a–41d. The Euler results show the formation of a weaker vortex inboard of the experimental vortex. Since the vortex is in the formative stage, the Euler solution may be sensitive to grid density and leading-edge resolution.

The vortex strength increases and the location moves inboard at $\alpha = 12$ deg (Figs. 42a–42d). The Euler results predict the strength and position of the primary vortex accurately at 40% chord. The secondary separation and other viscous effects can be seen in the experimental pressure distributions further aft on the wing. The secondary separation moves the primary vortex inboard and creates a low-pressure region extending outboard from the core of the primary vortex. These flow characteristics cannot be modeled with the Euler equations. FLO57 predicts a stronger primary vortex that lies outboard of the experimental vortex at 50 and 62.5% chord. At 75% chord the Euler results show a flattening of the vortex core, which suggests that the vortex may be bursting. Further evidence of vortex breakdown was given by numerical simulations of particle paths showing recirculation within the core.

Vortex bursting is apparent in both computational and experimental pressure distributions at $\alpha = 16$ and $\alpha = 20$ deg. The vortex structure is no longer maintained, and the pressure distributions become unsteady aft of the breakdown position (Figs. 43a–43d and 44a–44d). Numerical behavior believed associated with vortex bursting is depicted by the unstable and oscillatory moment coefficient histories shown in Fig. 45. FLO57 predicts the burst location ahead of the experimental position at both angles of attack. The comparison between Euler and experimental results improves at stations where experiment and computations indicate that breakdown has occurred. The existence of a cross-flow shock is shown in the experimental pressure distributions, but not in the FLO57 results (Fig. 43).

Comparisons between the experimental and computational force coefficients show that FLO57 predictions are accurate until approximately $\alpha = 16$ deg, where the vortex appears to burst over a large portion of the wing (Fig. 46). Moment predictions do not compare well with experiment above $\alpha = 8$ deg. However, the angle at which the moment-curve slope changes is accurately predicted by FLO57.

Mach = 0.60, Wing/Body/Chine

The comparisons between computational and experimental results improved for the wing/body/chine configuration. The pressure distributions for all angles of attack are shown in Figs. 47–51. The flow over the forebody without the chine is strongly affected by viscous separations that cannot be modeled in an Euler simulation. The dominant vortex from the chine is captured by the Euler code, allowing a more representative computational model of the forebody flowfield approaching the apex of the main wing.

Both Euler and experimental results show that the chine delays the onset of instabilities which lead to vortex breakdown. This is clearly evident by comparing the moment histories in Fig. 52 with those from the wing/body configuration shown in Fig. 45. The location of vortex breakdown given by the computational pressure distributions is ahead of that shown by the experimental data. FLO57 does not predict the strengthening of the crossflow shock by the chine. Increased grid resolution may be required to capture the shock.

The lift, drag, and moment curves are given in Fig. 53. The Euler code overpredicts the increment in lift due to the chine. The drag differences at low-lift conditions and the change in slope of the moment curve were accurately predicted.

Mach = 0.80, Wing/Body

The experiment/CFD correlations at $M_\infty = 0.80$ are similar to those at $M_\infty = 0.60$. The pressure distributions for $M_\infty = 0.80$ are shown in Figs. 54–58. The experimental and computational pressure distributions agree well at $\alpha = 5$ deg over the entire wing since the flow is fully attached. The computational pressure distributions are in poor agreement with the wind-tunnel data when the vortex begins to form near the leading edge at $\alpha = 8$ deg. An increase in Mach number causes the vortex burst location to move forward on the wing above $\alpha = 12$ deg. The Euler simulation predicts bursting prematurely, as observed at $M_\infty = 0.60$. The Euler solutions show poor agreement with experimental results at both $\alpha = 16$ and $\alpha = 20$ deg.

The FLO57 results do not show a distinct vortex at either angle of attack and have flat spanwise pressure distributions. The moment histories show random behavior for both angles of attack with a large increase in nose-down pitching moment (Fig. 59). The computational lift, drag, and moment curves show good agreement with test data for angles of attack below approximately 12 deg (Fig. 60). The Euler predictions become less accurate when vortex bursting becomes more extensive.

Mach = 0.80, Wing/Body/Chine

Placing the chine on the forebody above the wing surface improves the correlation of experiment with CFD and delays the onset of vortex burst. The pressure distributions are shown in Figs. 61–64. The moment histories given in Fig. 65 are smoother than those from the wing/body. The increase in lift and the decrease in drag are overpredicted by FLO57 (Fig. 66). The shape of the moment curve was accurately predicted by the Euler code.

Concluding Remarks

The Euler code FLO57 has been used to calculate transonic flow over two configurations of a generic fighter model. Results were computed at Mach numbers of 0.60 and 0.80 for selected angles of attack between 4 and 20 deg.

The Euler code predicts the aerodynamic quantities accurately for attached-flow conditions but less accurately when the leading-edge vortex is forming. This may be due to the sensitivity of FLO57 to grid resolution and geometry modeling near the leading edge. However, after the vortex is established, the Euler results show good agreement with experimental data until vortex breakdown occurs in the computations. The Euler code appears to predict bursting prematurely. The reason for this is unknown. When the experimental results indicate vortex bursting, the experiment/CFD correlation improves.

The effect of adding the chine to the forebody is to delay the onset of vortex breakdown both numerically and experimentally and to improve Euler predictions. The decoupled chine vortex interacts favorably with the main wing flowfield, increasing the effective leading-edge sweep.[8] The experiment/CFD correlations improve with the addition of the chine, which may result from fixed forebody separation at the chine, providing a dominant flow feature for the Euler code to capture. Without the chine, the Euler code has difficulty modeling the unsteady flow separations on the forebody, especially at high angles of attack.

Delta Wing

Study Objectives and Model Geometry

Wind tunnel-testing requirements for instrumentation, propulsion-simulation hardware, and support systems are often in conflict for physical space inside a wind-tunnel model. As a result, forces and moments are sometimes derived from integrated experimental pressure data. This method is usually suspect because of the relatively small number of pressure measurements available over complex three-dimensional surfaces. The first part of this section presents a comparison between CFD predictions and wind-tunnel data for surface pressure distributions and longitudinal forces and moments. The effects of numerical discretization were explored by using three computational grids of varying density. The large number of wind-tunnel pressure taps available on this model allowed for colored surface pressure contour comparisons between the experimental and computational results. This technique was found to be useful for highlighting small discrepancies in wing pressure distributions that were sometimes overlooked when viewing traditional pressure distribution plots. The second part of this section investigates the feasibility of integrating experimental pressures to determine the forces and moments acting on a wind-tunnel model. An integrated approach for determining forces and moments using wind-tunnel data and CFD predictions is described.

The wind-tunnel model was a full-span delta wing with 62.5-deg leading-edge sweep. A plan view showing pressure tap locations is given in Fig. 67. The simple geometry and large number of pressure taps (287) made it ideal for both CFD code validation and surface pressure integration. To accommodate other test requirements, the model was supported near the wing tips. As a result, there was a fairly large gap between the tenth and eleventh chordwise rows of pressure taps. In addition, a few geometric modifications were made to the wind tunnel model prior to the present test to assist in the pressure-integration study and to ensure tunnel compatibility. The shroud required for sting mounting also covered a small portion of the upper surface and some of the associated pressure taps. Outside of these regions, the model had an excellent pressure tap distribution on both the upper and lower surfaces. The geometry used for the CFD analysis and post-test pressure integrations was obtained in a panel-code-type input format. It consisted of three networks containing the forward-wing region, the aft-wing region, and the sting shroud. There were a total of 12 span stations, with 27 chord points used to define each airfoil.

The model was tested in the Ames 11×11-ft transonic wind tunnel over a wide range of Mach numbers and model attitudes. This section will present only the longitudinal data obtained at $M_\infty = 0.8$, $Re/\text{ft} = 2.5$ million; $M_\infty = 0.7$, $Re/\text{ft} = 2.5$ million; and $M_\infty = 0.4$, $Re/\text{ft} = 2.0$ million.

Experiment/CFD Correlations

Grid Generation

An H-O-grid topology was chosen for this study because it can handle wings with low taper ratios or pointed tips and provides for an efficient clustering of points along the leading edge. The grid-generation process can be broken down into three steps. In the first step, a commercial CAD/CAM system was used to generate cross sections of the model at specified longitudinal locations.[9] These cross sections were then redistributed in order to provide grid clustering in regions of high curvature. Finally, the surface grid at each cross section was extended into the three-dimensional space surrounding the body. In order to produce a smooth grid, a two-dimensional elliptic grid-generation routine was used to create each grid plane. A cross-sectional view of one of the fine grid planes is shown in Fig. 68. Since the grids are created at successive longitudinal stations, the sting can be easily incorporated. Three grids of varying density were used to investigate the sensitivity of the computations to different discretizations. The three model upper-surface grids are shown in Fig. 69. The grid crossover that occurs near the trailing edge in these planform views is not a true crossover, but a wrapping of grid lines around the double-valued sting surface. Information about the three grids is given in Table 4.

Flow Solver

The FLO57 CFD code integrates the Euler equations using Jameson's four-stage explicit Runge-Kutta algorithm.[2] The original program used in this study had been modified by numerous authors; further modifications were necessary to incorporate the H-O-grid topology. Details of the Euler solution strategy and arguments for the applicability of Euler codes to delta-wing configurations are given in Refs. 10 and 11.

The extensive pressure instrumentation allowed detailed surface pressure comparisons to be made over the entire span of the model. The lift, drag, and moment coefficients were also compared to the experimental data. Three grids of varying density were used to provide an estimate of the effect of grid density on solution characteristics. All solutions were run to a minimum of three orders of magnitude convergence in the density residual. Lift, drag, moment, and the number of supersonic points vs iteration were also monitored to ensure converged results. Table 4 provides information about the computational resources required for the FLO57 analysis. The computational run schedule is given in Table 5. The predicted and experi-

Table 4   Grid information and memory requirements for FLO57—delta wing

Grid	i dim	j dim	k dim	Surface points	Total points	Memory MW	CPU s/case
Coarse	67	21	43	45 × 43	60,501	2	3100
Medium	89	29	57	57 × 57	147,117	4	7800
Fine	113	37	73	73 × 73	305,213	8	18,000

Table 5   CFD solutions—delta wing

Mach	α	Coarse	Medium	Fine	PANAIR
0.4	−4,−2,0,2,4,6,8,10	X	X		X
0.7	−4,−2,0,2,4,6,8,10	X	X		X
0.8	−4,−2,0,2,4,6,8,10	X	X	X	X

mental surface pressure comparisons are shown in Figs. 70–72, and the resulting force and moment data are presented in Figs. 73–75. Included with the experimental data and FLO57 predictions are the results of an analysis using a panel code.[12] The agreement between the FLO57 computations and wind-tunnel pressure distributions is generally good except in regions where significant viscous effects were present. The pressure distributions of Figs. 70–72 show the ability of the Euler code to predict the details of the inboard pressure distributions, including the magnitude and extent of the leading-edge expansion spikes. At the trailing edge, FLO57 overestimates the amount of pressure recovery. The wind-tunnel pressure distributions do not recover to the same level as FLO57 because of boundary-layer displacement effects and trailing-edge separation.

The detailed mechanics of the leading-edge vortex formation and the secondary separations of the boundary layer caused by resulting adverse spanwise pressure gradients are also incorrectly modeled by the inviscid Euler code. The effect of these modeling differences on the pressure distributions becomes significant as the vortex strength increases along the leading edge. These vortex/boundary-layer interactions result in an increasingly poorer C_p comparison toward the tip. The result of neglecting the physics involved in these important flowfield interactions is clearly seen in the outboard pressure distributions of Figs. 70b, 71b, and 72b. Increasing the angle of attack strengthens the vortex and increases its interactions with the boundary layer. Although the overall pressure agreements between the computations and wind-tunnel tests were very good, the slope and sign of the moment curves predicted by FLO57 did not correlate well for any of the three Mach numbers investigated, as seen in Figs. 73–75. However, these correlations are better than those shown in the section on commercial transport wing/body configurations. The origin of the difference between the wind-tunnel moment curve and the FLO57 and PANAIR predictions was difficult to identify. The fact that this difference persisted even at low lift conditions was especially troubling, because the effects of incorrect vortex modeling, viscosity, and shock waves were not expected to be large for small angles of attack. Studying individual chordwise pressure distributions did not yield any obvious explanations. In order to study the decambering effect of the boundary layer, a turbulent displacement thickness was calculated and added to the upper and lower surfaces for the medium grid, but it did not appreciably change the FLO57 moment slope. A thorough check of the wind-tunnel data-reduction system was also made.

The origin of the difference between the FLO57 and wind-tunnel moment curves was eventually discovered by comparing the overall computed and experimental pressure distributions. This was accomplished by preparing colored planform views of the model that combined experimental and computational predictions. One-half of each surface of the wing was colored with the experimental pressures, and the other half by the predicted computational pressures.[13] Each wind-tunnel pressure tap was assigned a surrounding area on the wing, and these areas were colored by the C_p measured at the tap. Each surface cell in the FLO57 grid was similarly colored by the pressure predicted at its center, making a direct comparison between the experiment and computations possible. Although these colored planform comparisons are more qualitative than the traditional chordwise plots, they tend to accentuate overall trend differences. This can be seen by studying Plates 4 and 5 (see the color section), which correspond to α sweeps at $M_\infty = 0.40$ and 0.80. Upon close review, it can be seen that the upper-surface pressure distributions predicted by FLO57 tend to have a slightly lower pressure extending further aft on the wing. This additional aft loading, when integrated over the entire span, produces a more negative pitching moment and exists even at low lift conditions. Further review of the plots of individual chordwise pressure distributions shows this difference to be small, but to extend over the entire span. It would appear that geometric inaccuracies in the wing geometry network files, not viscous effects, are the major source of the computational moment discrepancy. The effect of the grid density on surface C_p distributions did not appear to be very large, but did increase somewhat near the wingtips. The lift, drag, and moment were also barely affected by grid size. Additional expensive, large-grid solutions were not obtained at $M_\infty = 0.4$ and 0.7 since the three grids gave similar results at $M_\infty = 0.8$.

Experimental Pressure Integration

The experimental pressure-integration scheme used the simple midpoint rule, where an area and unit normal vector were associated with each tap location. In order to compare the resulting integrated data with the balance data, a skin-friction estimate was added to the drag computations. Tap areas were determined by placing a panel around each pressure tap. Sides of these panels were positioned at the midpoint between adjacent taps. The unit normal vectors assigned to each area were the unit normals to the surface calculated at the centroid of each panel. These centroids were generally close to the tap locations. Since the sting shroud was not pressure instrumented, it was not modeled in the pressure-integration scheme. Instead, the panel areas of the closest spanwise taps were extended to the model centerline (Fig. 76). The integration scheme might have been improved by determining an average unit normal vector over the surface of a panel instead of using that found at the centroid. In addition, a higher-order numerical integration method could have been used.[14] The midpoint rule was used for this study because it was straightforward and simplified location and correction of errors in the integration scheme.

Skin-Friction Estimation

Skin-friction increments were applied only to the drag since their effect on pitching moment was assumed to be negligible. The incompressible, flat-plate, two-dimensional skin-friction estimates used were[15]

$$C_{F_{\text{Turb}}} = 0.074/Re^{0.2}$$

$$C_{F_{\text{Lam}}} = 1.328/\sqrt{Re}$$

Boundary-layer transition was forced near the leading edges of the wings by a transition strip set 1.0 in. streamwise from the leading edge. The drag increment due to this strip was shown to be insignificant. In addition, drag sensitivity to the turbulent transition location (in case the transition strip failed to trip the boundary layer) was analyzed and was found to be negligible. The above coefficients were based on the transition strip location and mean chord.

Integrated Force and Moment Comparisons

Force and moment comparisons of balance, integrated pressures, and integrated pressures plus skin friction are shown in Figs. 77–79. The lift curve derived from the integrated pressures agrees well with the balance data for all three Mach numbers. This is expected, since the distribution of integration panels projected onto a plane perpendicular to the lift axis is good. The only gaps occur at the sting shroud and just inboard of the two wingtip rows. The pitching-moment data also agree quite well. However, there is a general trend for the integrated pressures to consistently give a more positive pitching moment than the balance data. This trend becomes significant at higher angles of attack. A possible reason for this discrepancy is removal of the sting shroud from the integration scheme. The shroud extended 2.9 in. past the trailing edge. At zero angle of attack the surface of the shroud is nearly parallel to the freestream and does not contribute to the frontal area of the model. At higher angles of attack the aft end of the shroud becomes visible to the oncoming flow and exerts a nose-down moment not captured by the experimental pressure integration. The balance drag polar and the integrated pressure drag polar do not agree as well as the lift and moment curves at the three Mach numbers. The integrated pressures overestimate the drag by a constant increment over the range of lift coefficients, but give the correct shape of the polars. Analysis of this drag shift identified a number of possible causes:

1) inaccurate wind-tunnel data;
2) errors in the integration scheme;
3) inaccurate skin-friction estimate;
4) errors in geometry used in integration; and
5) inadequate number of pressures on the model surface.

These causes were analyzed and are discussed in the following sections.

Inaccurate Wind-Tunnel Data

The wind-tunnel data were thoroughly examined, and all data and corrections appear to be accurate within the limits of the instrumentation ($\Delta C_D \leq \pm 0.0003$).

Errors in the Integration Scheme

Possible errors in the integration scheme were evaluated by examining the unit normal vectors and panel areas assigned to each pressure tap. These values were compared with normal vectors and panel areas previously determined by the test engineer. The earlier values were not used for this study because the manner in which the areas around the sting shroud were computed was unclear. All of the values compared very well with only a few minor differences. These differences were not large enough to explain the drag shift found in the integration.

Inaccurate Skin-Friction Estimate

The analysis of errors in the skin-friction estimate was limited to the turbulent region since only a very small portion of flow over the model's surface was laminar. The turbulent estimate is a crude incompressible value from Prandtl's 1/7th-power velocity profile.[15] This relation is the classical turbulent skin-friction estimate generally used in conceptual design. A method that incorporates compressibility was proposed by Sivells and Payne.[16] Of course, the largest difference between the two methods is found at $M_\infty = 0.80$ where $C_{F_{turb}} = 0.00316$ for the classical Prandtl method and 0.002926 for Sivells and Payne. Here the skin-friction coefficient is reduced by 7.5%. However, a reduction in skin friction on the order of 50% would be necessary to match the balance. Also, as opposed to using a Reynolds number based on mean chord, the planform was divided into eight chordwise strips of the same width. The skin friction for each strip based on local chord was determined. The total was then derived by summing the skin friction over the strips. The error associated with this was on the order of 1%. Therefore, any error in skin friction would be due to three-dimensional and pressure-gradient effects that were not considered during this study.

Errors in Integration Geometry

The most obvious geometric difference between the balance measurement and the integration scheme was the sting shroud. As mentioned earlier, there were no pressures available for integration over the surface of the shroud. Instead, the sting shroud was excluded from the surface integration, and each pressure tap adjacent to the shroud was assigned an area extending to the model centerline. The balance drag includes skin friction and pressure drag; hence, a method of determining the effect of removing the sting shroud from the integration can be assessed. First, the increment in skin friction due to differences in wetted area with and without the sting shroud is calculated. The difference in total drag is $C_D = 0.0002$. This is on the order of the resolution of the balance. Second, the effect of removing the shroud from the integration of pressure forces is estimated. The surface of the sting shroud has no aft-facing area, and the final balance forces included base and cavity corrections. Therefore, the pressure on the sting shroud did not contribute to the final balance drag. In comparison, the geometry used for the pressure integration has an aft-facing surface area in the region occupied by the shroud. This additional aft-facing area in the integration geometry contributes to the axial force. Therefore, if the axial force due to this area is removed from the integration, a reasonable estimate of the effect of the shroud can be determined.

Another method for determining the effect of removing the shroud on the integration of pressures can be determined from computations. The FLO57 solution included the shroud; hence, the surface grid can be modified so that the cells adjacent to the shroud can be extended to the model centerline as the experimental panels were. Then an estimate of the drag differences with and without the shroud may be obtained. Results from both methods are presented in Fig. 80 for $M_\infty = 0.80$. Values found using the experimental pressures are combined with the data in Fig. 81. The figure demonstrates that the shroud correction improves the correlation between the balance and integration curves, but a sizable shift in drag still remains unexplained. No other major geometric differences are present to explain this drag shift.

Inadequate Number of Pressures on the Model Surface

Accurate drag calculation by surface pressure integration requires a large number of panels with significant leading- and trailing-edge clustering to resolve large pressure gradients. This wind-tunnel model was chosen for this study because of its large number of pressure taps and its simple shape. However, the drag still could not be estimated accurately. Economic and instrumentation limitations restrict the number of pressure taps that may be incorporated into a model. Increasing the number of taps on a wind-tunnel model to obtain drag by surface pressure integration is not practical. A higher-order numerical integration or curve-fitting technique could be used. However, no curve-fitting method will capture regions without taps, such as leading-edge suction, or, in this case, the sting shroud. This section demonstrates a method to correct the integrated pressures without increasing the number of taps. The pressure integration used a discrete pressure and a unit normal vector for each panel. The integration method more accurately approximates a continuous pressure field as the number of integration panels increases. The difference between the force computed from the discretized integration method and that from a continuous pressure field acting on infinitesimal areas is defined to be the discretization error. The most dense computational grid and the experimental integration panels at one location on the model are compared in Fig. 82. The FLO57 surface grid was assumed to approximate a continuous pressure field, and the discretization error was quantified in the following ways:

1) The pressure at each tap location predicted by FLO57 was determined. Tap pressures were obtained by interpolating from the surrounding cell centers.

2) These interpolated pressures were combined with the experimental tap areas and unit normals and were then integrated.

3) Drag polars from step 2 and from the original FLO57 solution quantify the discretization error as shown in Fig. 83.

Therefore, the discretization error associated with the experimental pressure integration scheme can be determined by relying on an estimate from the computational solution. The discretization error found in Fig. 83 can be expected to apply to the experimental pressure integration if FLO57 gives an accurate estimate of the flowfield. The integration results improve when this error is removed (Fig. 83). When the skin-friction estimate is added, the agreement with the balance data is good (Fig. 84). It is important to note here that the discretization error described earlier includes the effect of the sting shroud because the FLO57 results were obtained from geometry that included the shroud. Therefore, this error-estimation method is also capable of capturing geometric differences between the integration geometry and the actual model configuration. Unfortunately, the method does not retain the same accuracy for all test conditions. The discretization error found from FLO57 at high lift coefficients does not match the increment needed to correct the experimental integration at $M_\infty = 0.40$ (Fig. 85). However, at lower angles of attack the correction works well. The deviation at high angles of attack is probably due to inaccurate computational results at those test conditions. The computational drag polars show greater drag due to lift than the experimental data do at $M_\infty = 0.40$. When the discretization error determined from the computational solution is removed from the experimental pressure integration at $M_\infty = 0.40$ as shown in Fig. 85, the final result shows an increase in drag due to lift that is too large. However, the zero-lift drag coefficient matches the balance data closely. Therefore, discretization-error estimation using a computational solution is limited to conditions where the flowfield is accurately modeled by the CFD code. The discretization error found from computations is essentially the integration error due to the inability to interpolate pressures accurately between taps. The success of the method results from using the computational solution as a "higher-order" integrator. Also, the integrated experimental pressure data were easily corrected for the sting shroud since the computational geometry included the shroud.

Concluding Remarks

The Euler code FLO57 has been compared with experimental data at three Mach numbers and at several angles of attack. The code gave accurate results over the forward regions of the model, but, as expected, poor experiment/CFD correlations were observed near the trailing edge because of neglected viscosity. In addition, the CFD solution was used to improve the feasibility of using experimentally measured pressures to obtain quantitative forces and moments acting on a wind-tunnel model. The computational solution was used to correct the discretization error resulting from a finite number of pressure taps, thus giving improved values for drag. Experiment and CFD can be used to their mutual enhancement. Proper experimental validation of CFD codes is necessary to determine the conditions under which computations may be expected to give satisfactory results. In addition, CFD solutions may be used to assist the experimentalist before a test by improving the conceptual design, by indicating locations of high pressure gradients for improved pressure tap placement and by projecting the test condition limitations resulting from balance design limits. Furthermore, as demonstrated in this study, CFD may be used after a test to improve the quality and resolution of the experimental data.

Acknowledgments

The authors thank Francis Enomoto, Ron Fegenbush, and Paul Keller for their help with the computational geometry and PANAIR analyses;

Rodney Bailey for his encouragement and many insightful suggestions; Jeff Trosin for his help with the data acquisition and plotting; and John Schreiner and Gary Erickson for their assistance in interpreting the wind-tunnel data.

References

[1]Henne, P. A., and Hicks, R. M., "Wing Analysis Using a Transonic Potential Flow Computational Method," NASA TM-78464, July 1978.

[2]Jameson, A., Schmidt, W., and Turkel, E., "Numerical Solutions of the Euler Equations by Finite Volume Methods Using Runge-Kutta Time Stepping Schemes," AIAA Paper 81-1259, 1981.

[3]Melton, J. E., and Langhi, R. G., *Surface Grid Generation for Advanced Transport Configurations, Numerical Grid Generation in Computational Fluid Mechanics '88* Pineridge, Swansea, Wales, UK, 1988.

[4]Jameson, A., and Caughey, D. A., "Numerical Calculation of the Transonic Flow Past a Swept Wing," COO-3077-140, ERDA Math and Computations Lab., New York Univ., NY, June 1977.

[5]Caughey, D. A., and Jameson, A., "Numerical Calculation of Transonic Potential Flow About Wing-Fuselage Combinations," AIAA Paper 77-677, June 1977.

[6]Erickson, G. E., and Rogers, L. W., "Experimental Study of the Vortex Flow Behavior on a Generic Fighter Wing at Subsonic and Transonic Speeds," AIAA Paper 87-1262, Jan. 1987.

[7]Erickson, G. E., Rogers, L. W., Schreiner, J. A., and Lee, D. G., "Subsonic and Transonic Vortex Aerodynamics of a Generic Forebody Strake-Cropped Delta Wing Fighter," AIAA Paper 88-2596, June 1988.

[8]Private communication with Gary Erickson.

[9]Enomoto, F., and Keller, P., "Using a Commercial CAD System for Simultaneous Input to Theoretical Aerodynamic Programs and Wind-Tunnel Model Construction," NASA CP-2272, April 1983.

[10]Goodsell, A., Madson, M., and Melton, J., "TranAir and Euler Computations of a Generic Fighter Including Comparisons with Experimental Data," AIAA Paper 89-0263, Jan. 1989.

[11]Newsome, R. W., and Kandil, O. A., "Vortical Flow Aerodynamics—Physical Aspects and Numerical Simulation," AIAA Paper 87-0205, Jan. 1987.

[12]Carmichael, R. L., and Erickson, L. L., "PANAIR—A Higher Order Panel Method for Predicting Subsonic or Supersonic Linear Potential Flows About Arbitrary Configurations," AIAA Paper 81-1255, 1981.

[13]Hermstad, D. L., "RAID User's Guide," Sterling Federal Systems, Palo Alto, CA, TN: 33, Rev. 2, 1989.

[14]Rice, J. R., *Numerical Methods, Software, and Analysis: IMSL Reference Edition*, McGraw-Hill, New York, 1983, pp. 186–215.

[15]Schlichting, H., *Boundary-Layer Theory*, McGraw-Hill, New York, 1979, pp. 140, 640.

[16]Sivells, J. C., and Payne, R. G., "A Method of Calculating Turbulent-Boundary-Layer Growth at Hypersonic Mach Numbers," Arnold Engineering Development Center, Tullahoma, TN, AEDC-TR-59-3, ASTIA Doc. AD-208774, 1959.

 R. M. HICKS ET AL.

Fig. 1 Commercial transport wing/body.

88.07

12.75

200.22

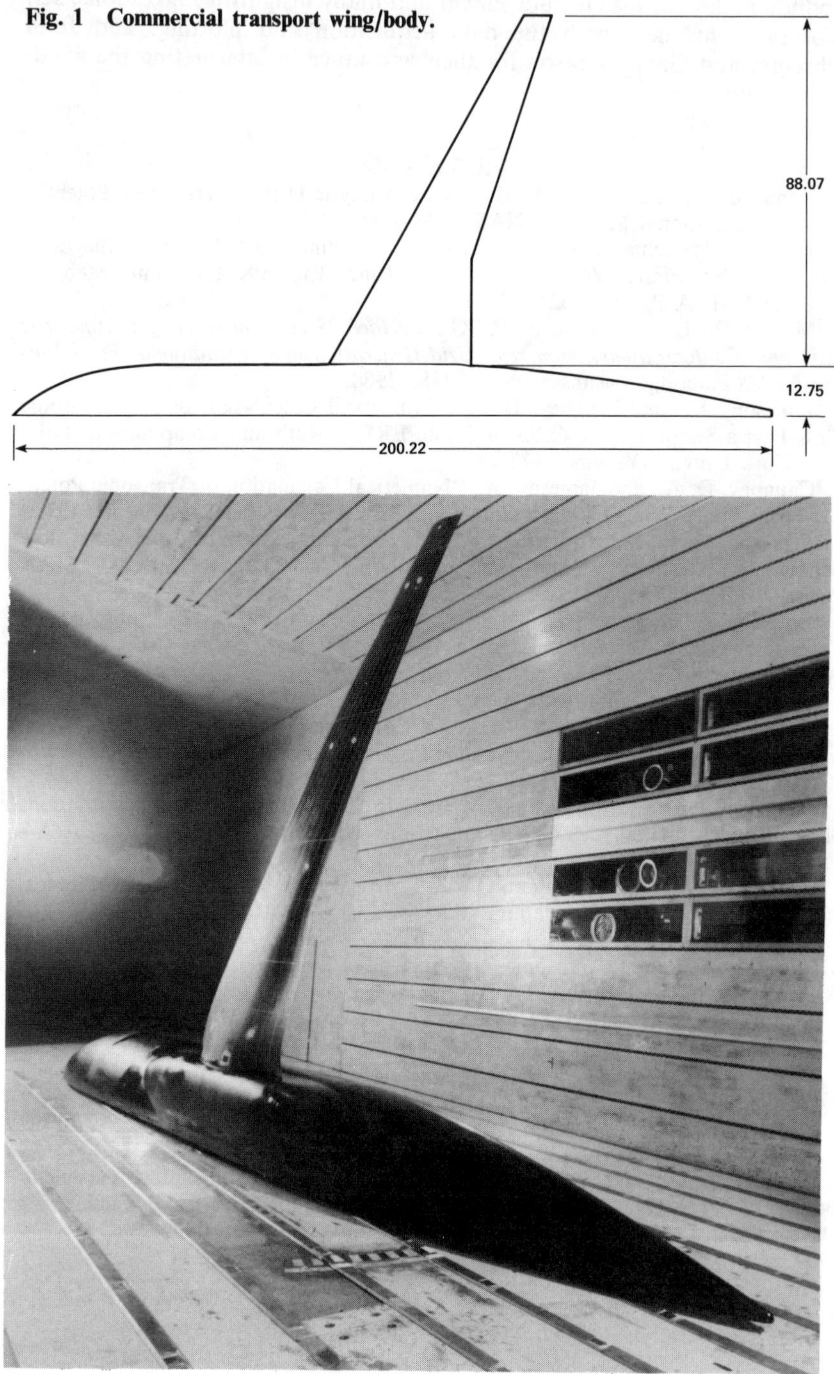

Fig. 2 Installation photograph, NASA Ames 11-ft transonic wind tunnel.

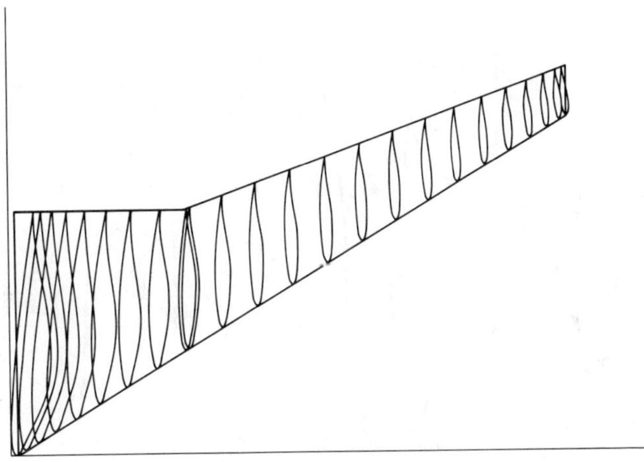

Fig. 3   Wing planform with defining sections.

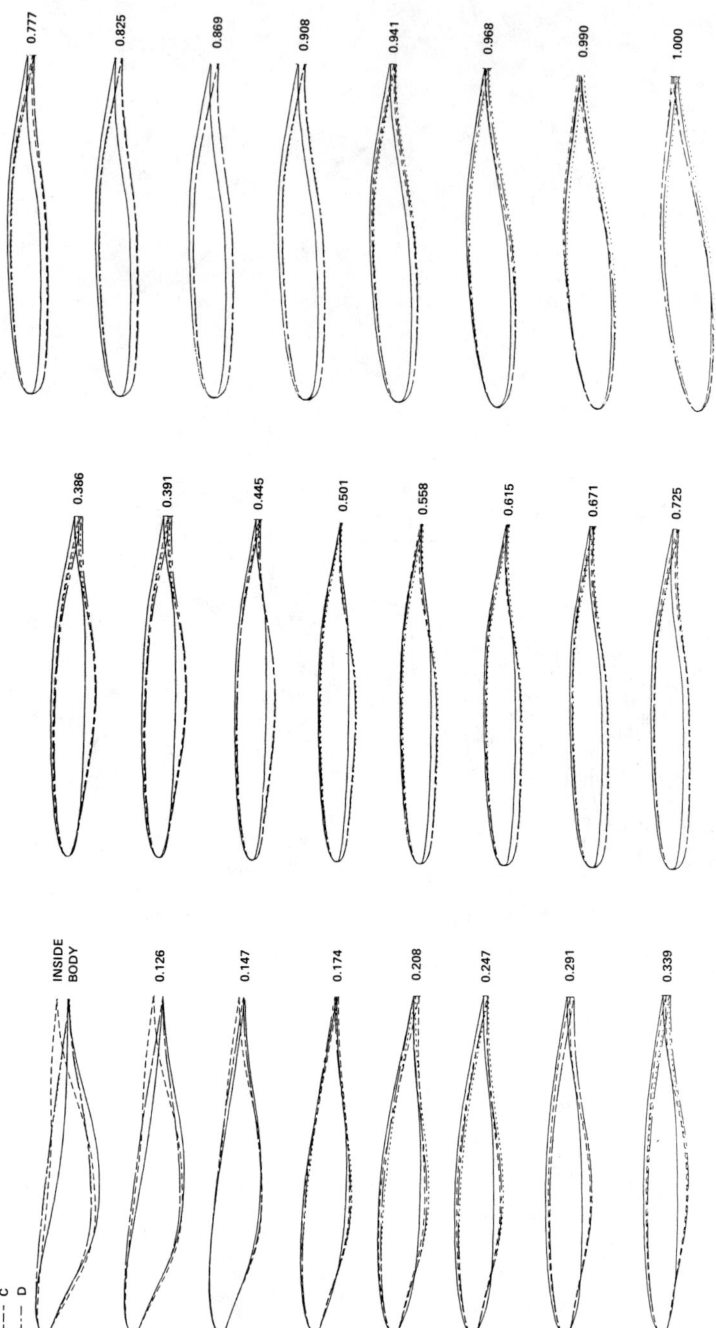

Fig. 4   Wing sections.

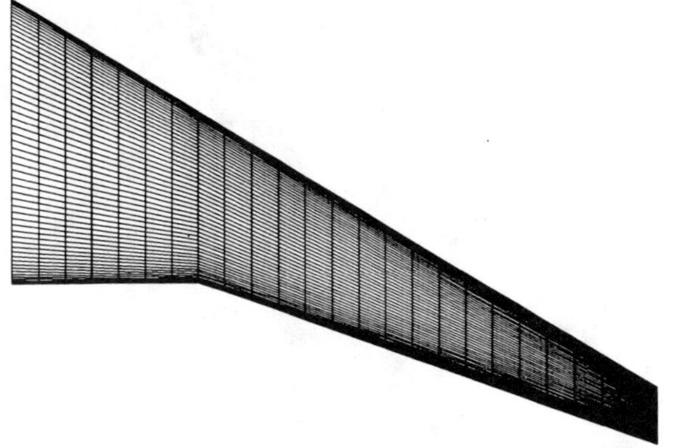

Fig. 5   C-H-grid distribution on wing surface, 151 × 31 × 31 points.

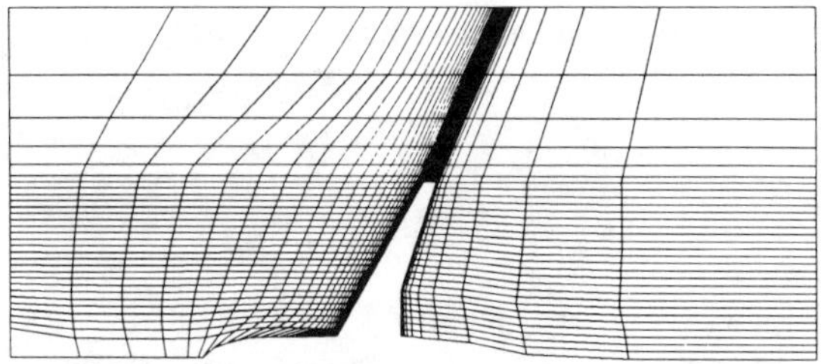

Fig. 6   C-H-grid distribution in plane of wing, 151 × 31 × 31 points.

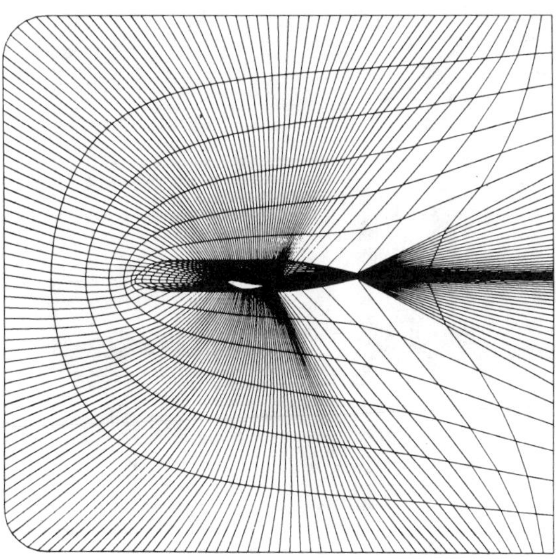

Fig. 7   C-H-grid distribution on centerplane, 151 × 31 × 31 points.

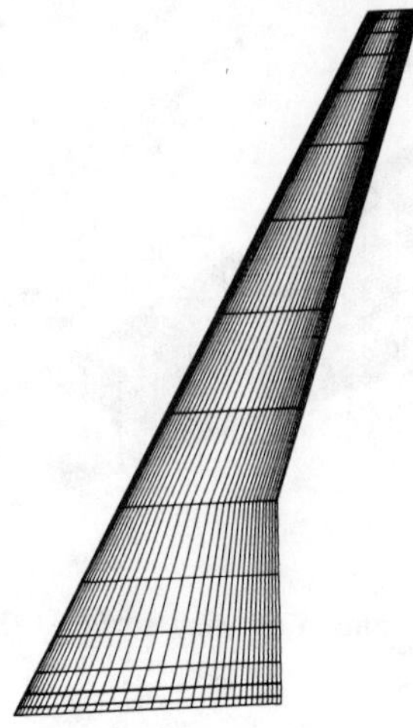

Fig. 8   H-O-grid distribution on wing surface, $101 \times 29 \times 45$ points.

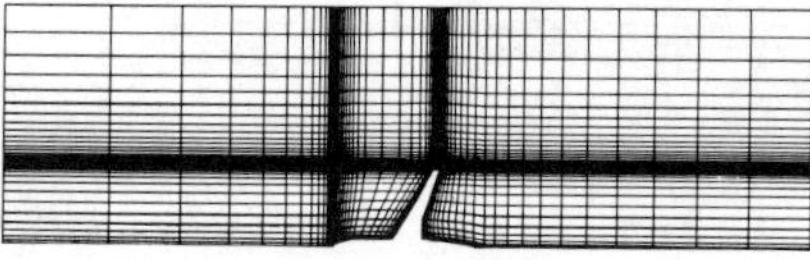

Fig. 9   H-O-grid distribution in plane of wing, $101 \times 29 \times 45$ points.

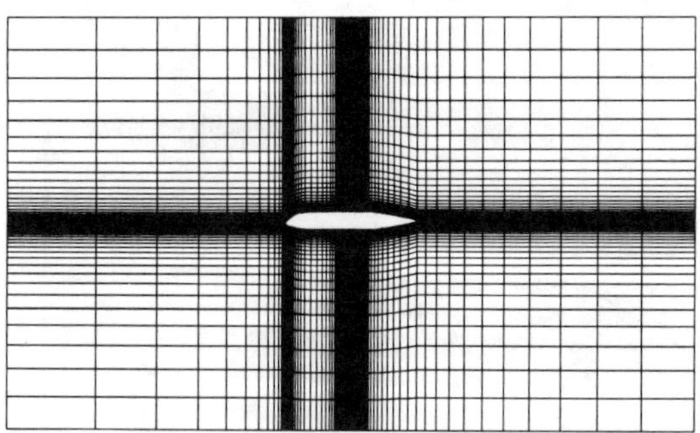

Fig. 10   H-O-grid distribution on centerplane, $101 \times 29 \times 45$ points.

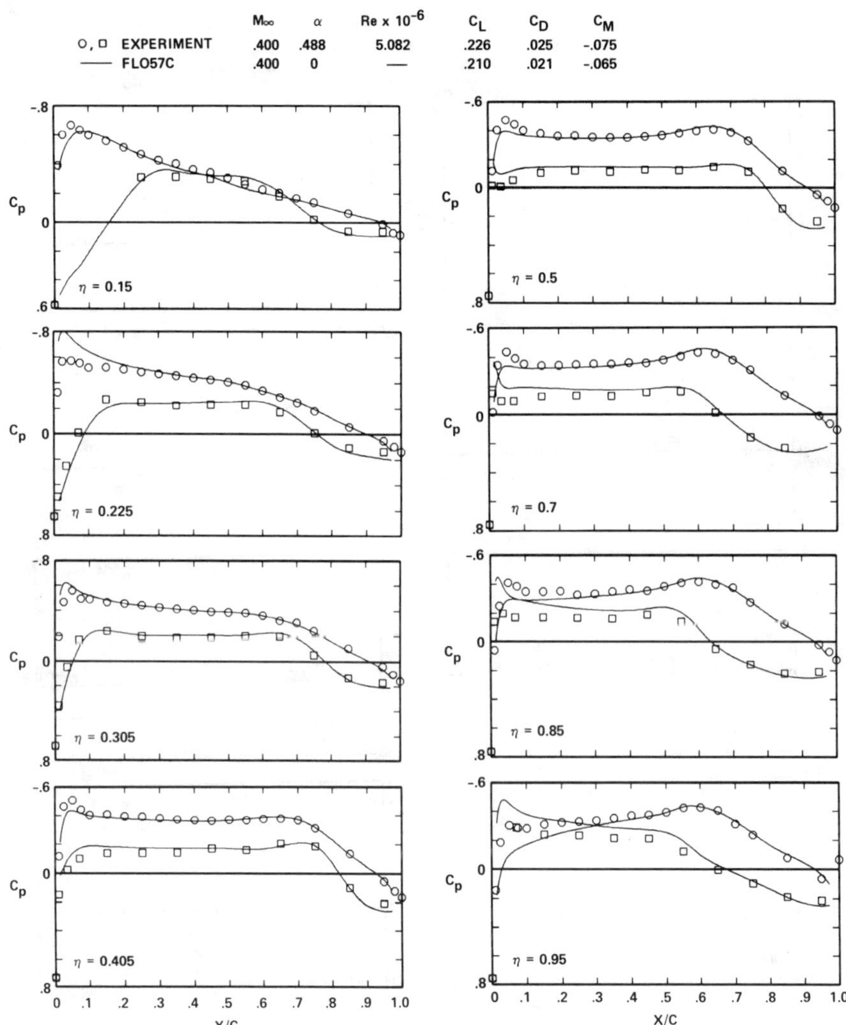

Fig. 11a   Experiment-CFD pressure distribution comparison for wing A, FLO57C.

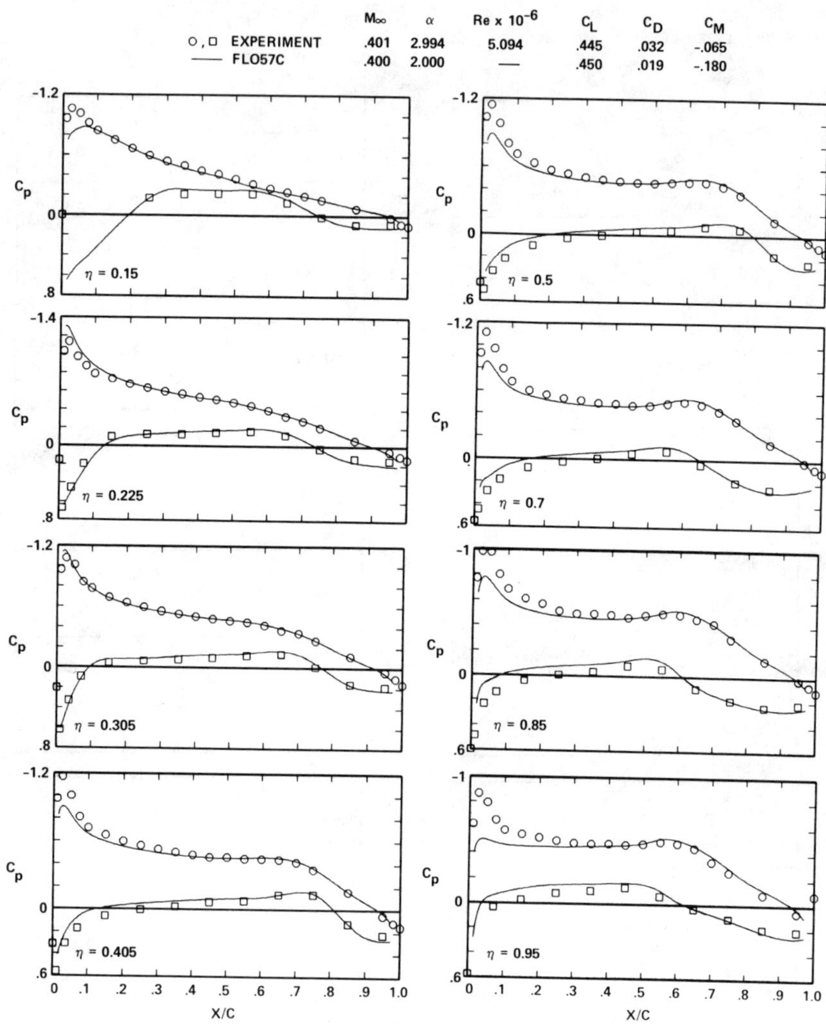

Fig. 11b (cont'd) Experiment-CFD pressure distribution comparison for wing A, FLO57C.

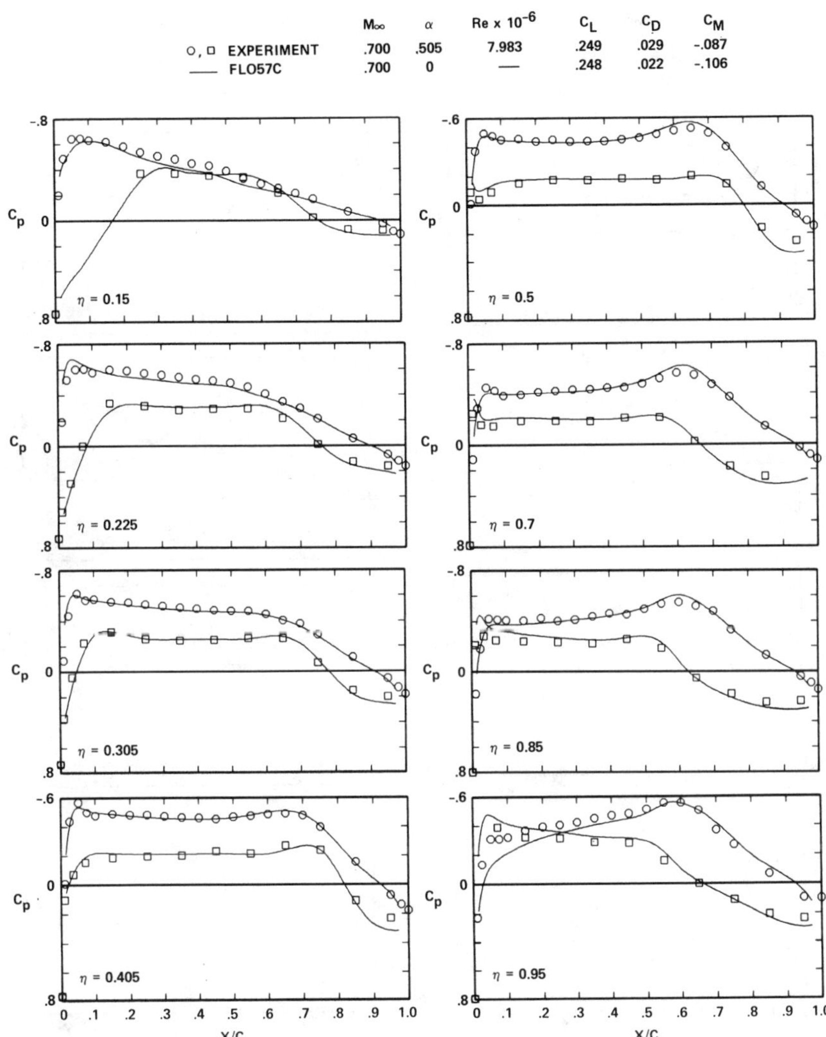

Fig. 11c (cont'd) Experiment-CFD pressure distribution comparison for wing A, FLO57C.

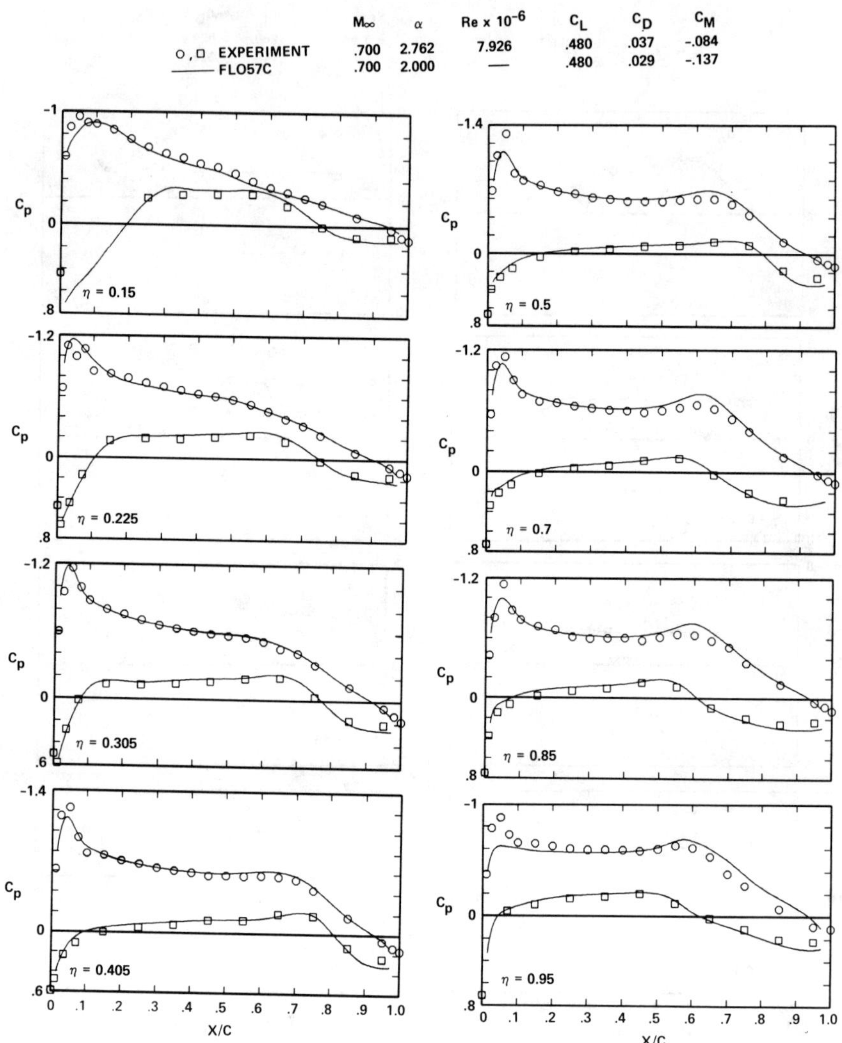

Fig. 11d   (cont'd) Experiment-CFD pressure distribution comparison for wing A, FLO57C.

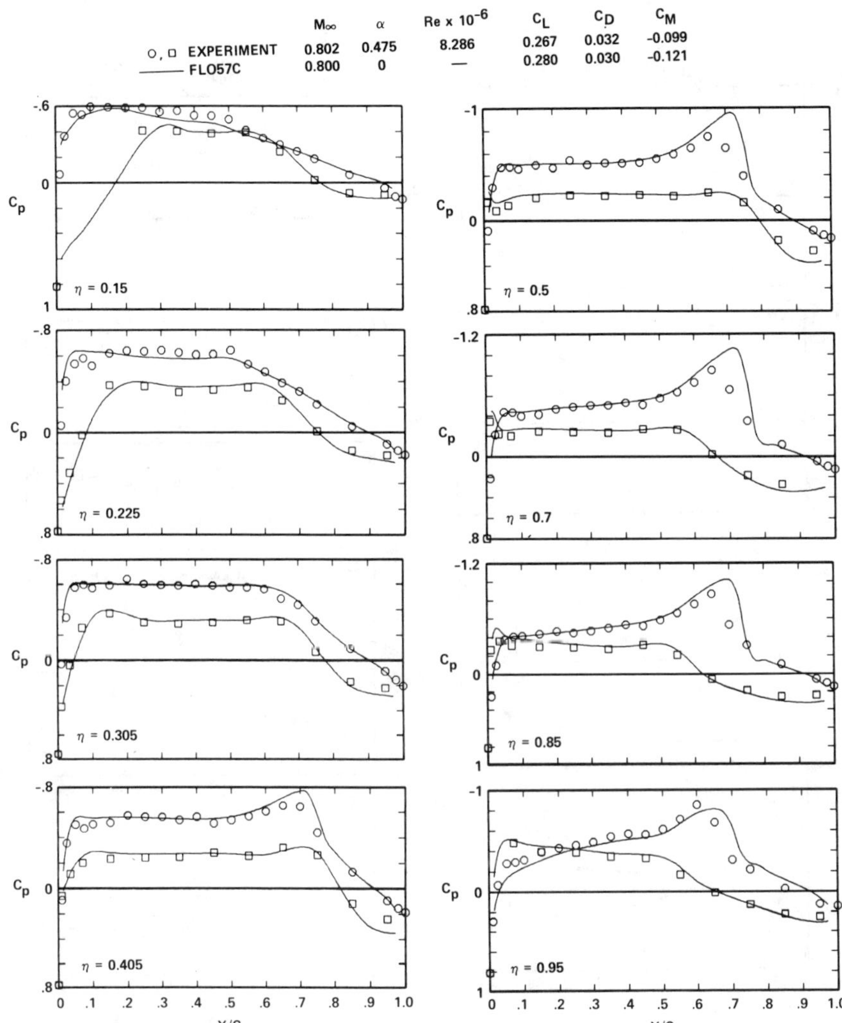

Fig. 11e (cont'd) Experiment-CFD pressure distribution comparison for wing A, FLO57C.

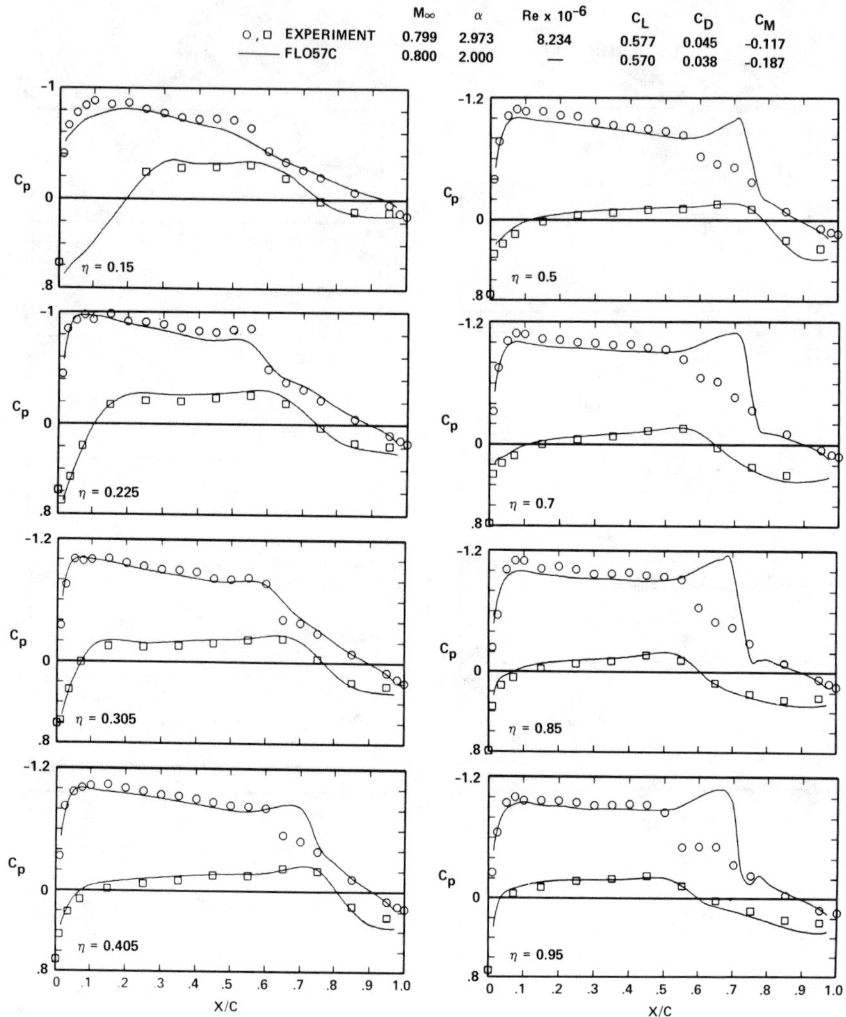

Fig. 11f   (cont'd) Experiment-CFD pressure distribution comparison for wing A, FLO57C.

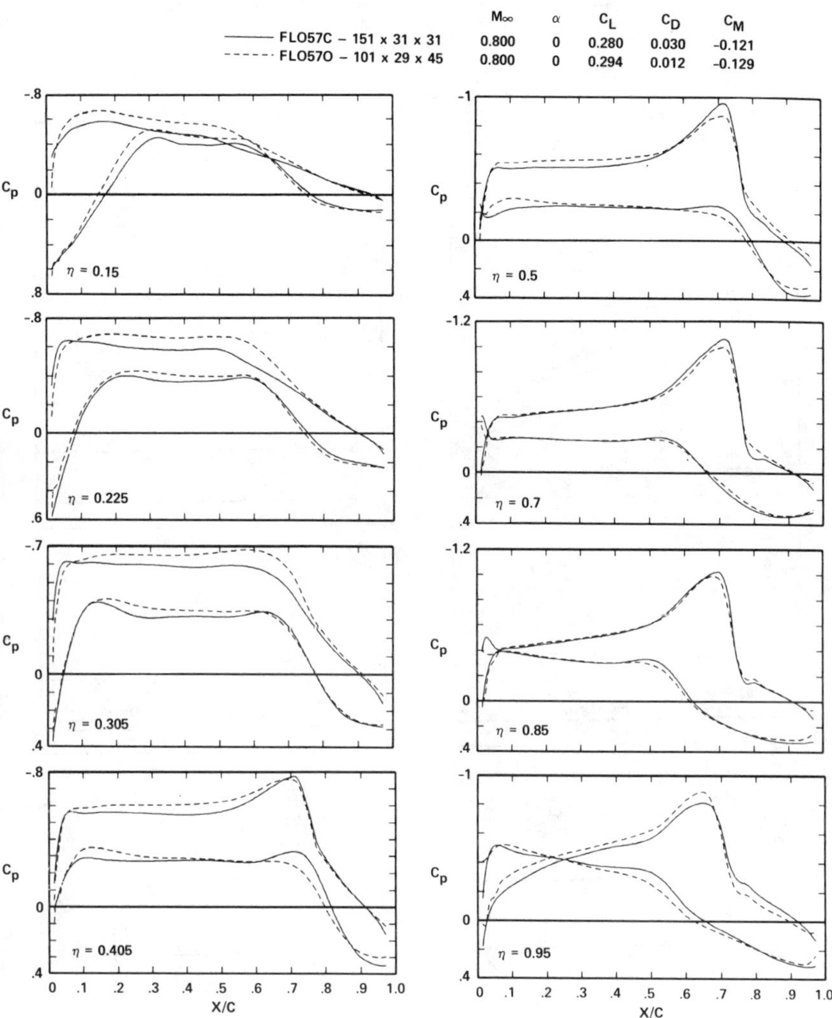

Fig. 12a   Pressure distribution comparison for wing A, FLO57C and FLO57O.

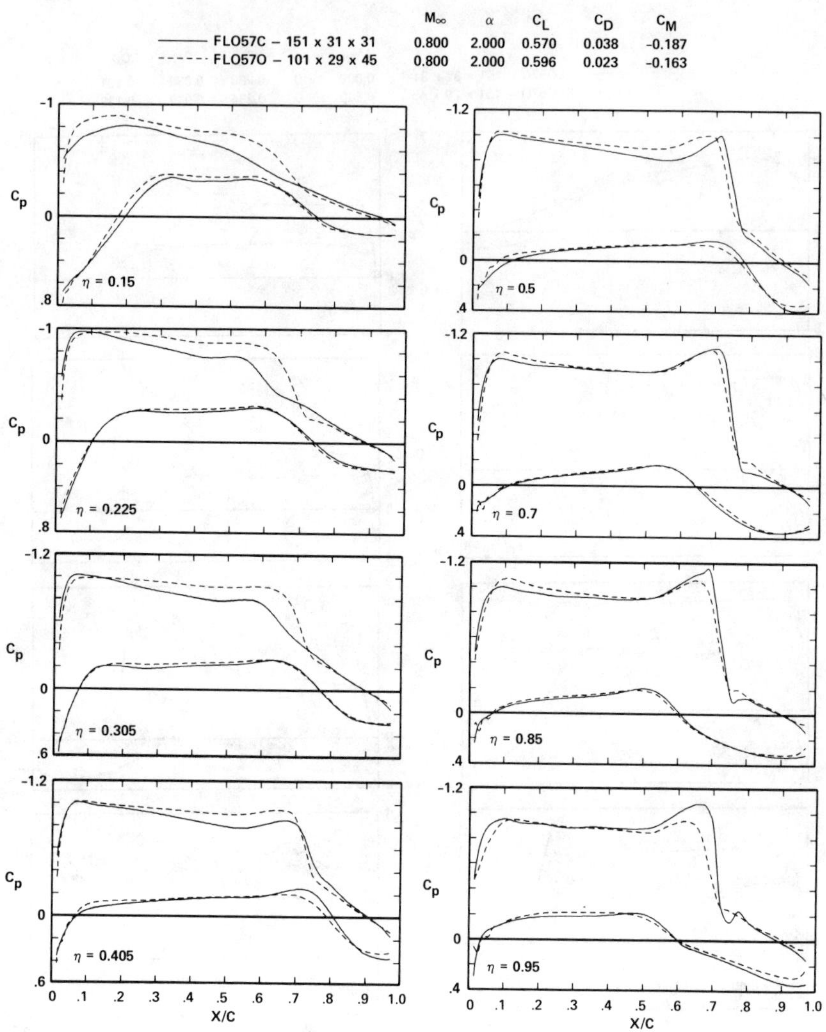

Fig. 12b (cont'd) Pressure distribution comparison for wing A, FLO57C and FLO57O.

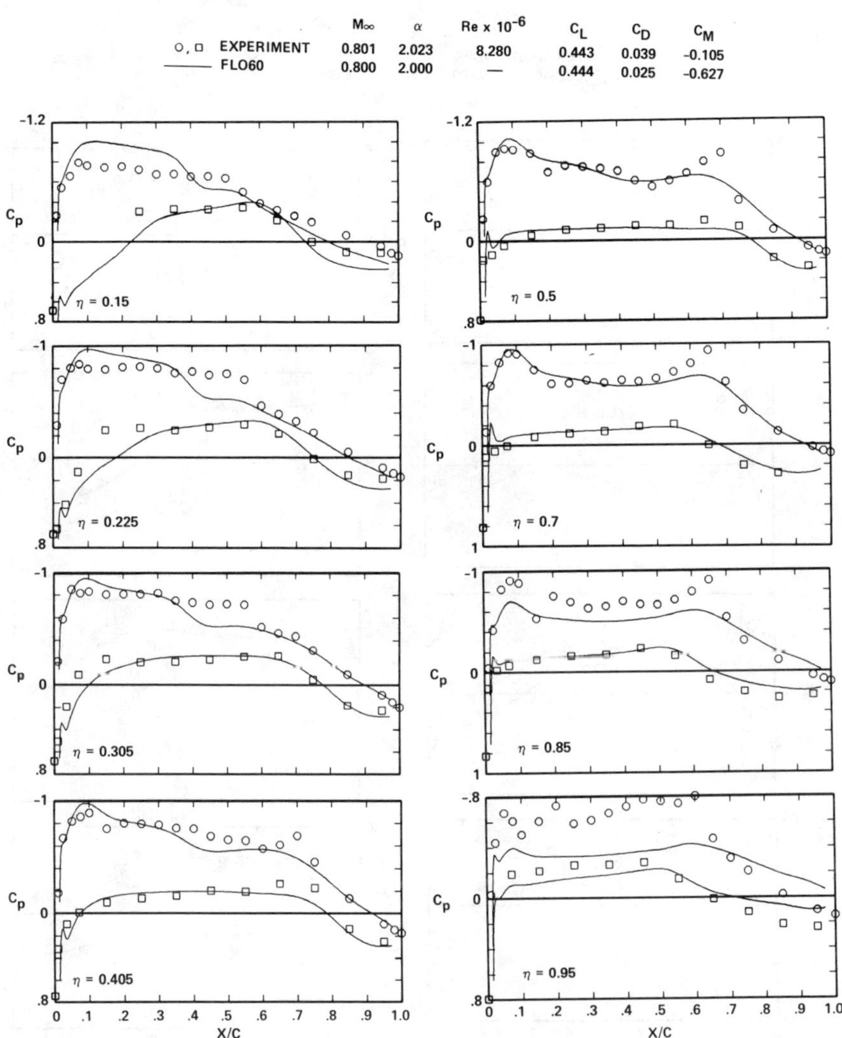

Fig. 13 Experiment-CFD pressure distribution comparison for wing A, FLO60.

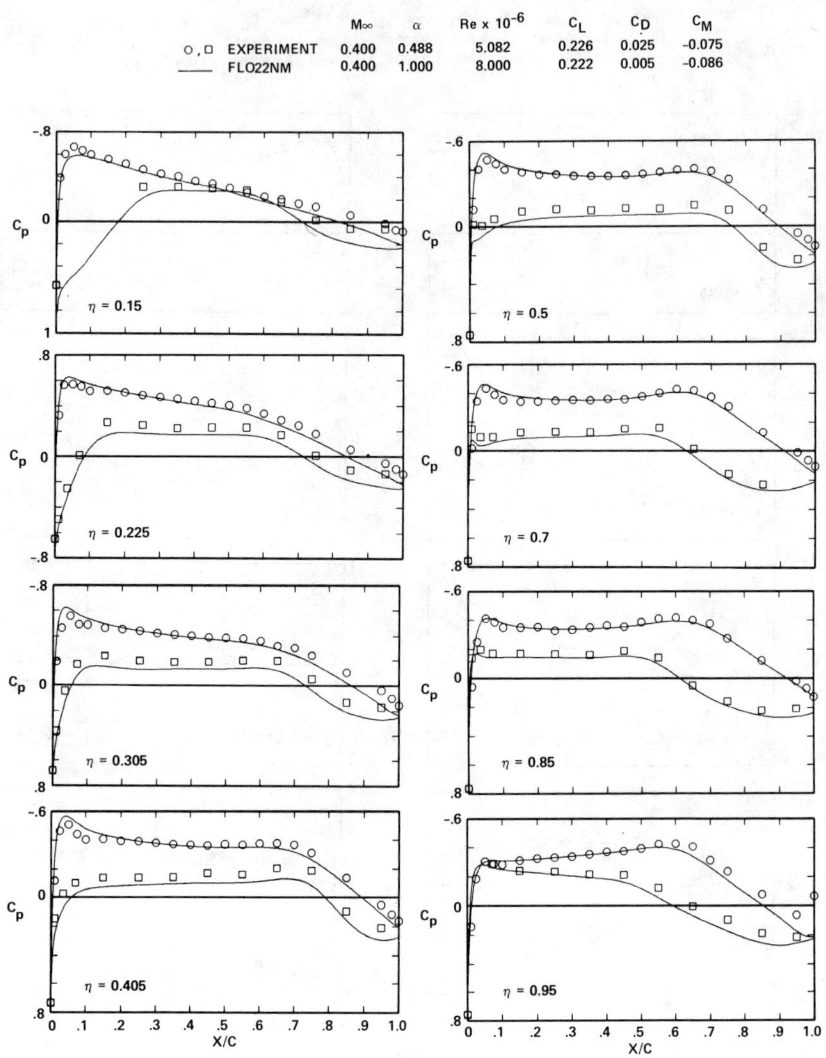

Fig. 14a Experiment-CFD pressure distribution comparison for wing A, FLO22NM.

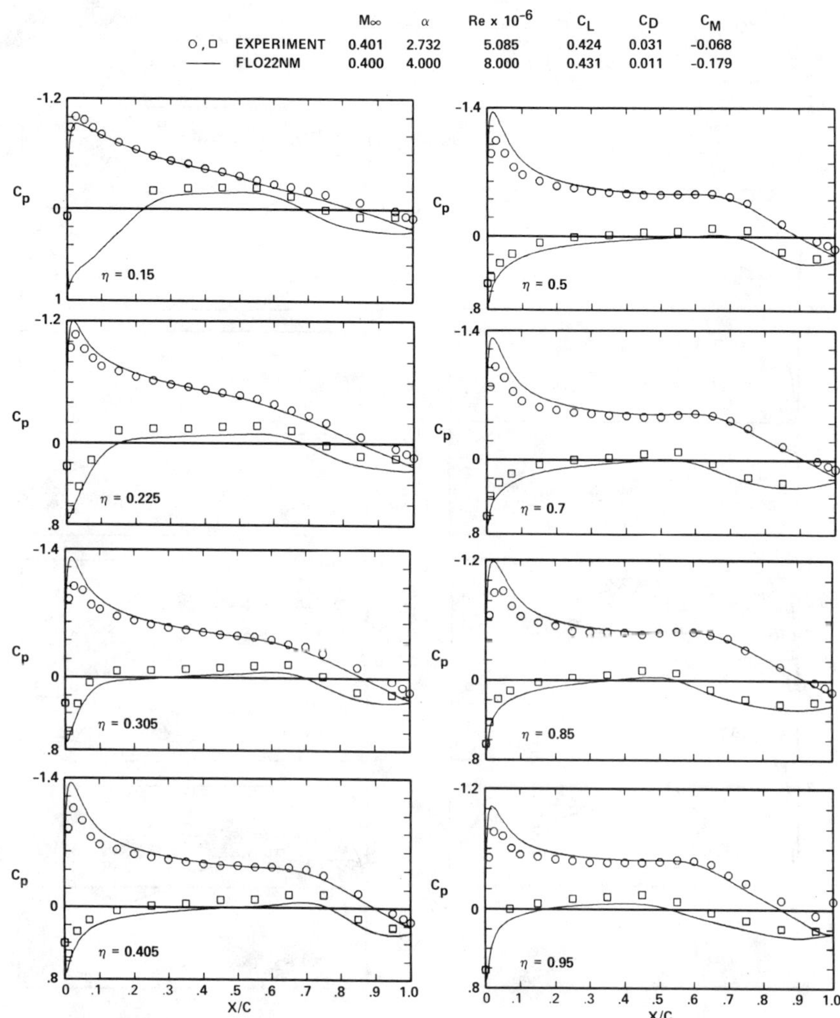

Fig. 14b   (cont'd) Experiment-CFD pressure distribution comparison for wing A, FLO22NM.

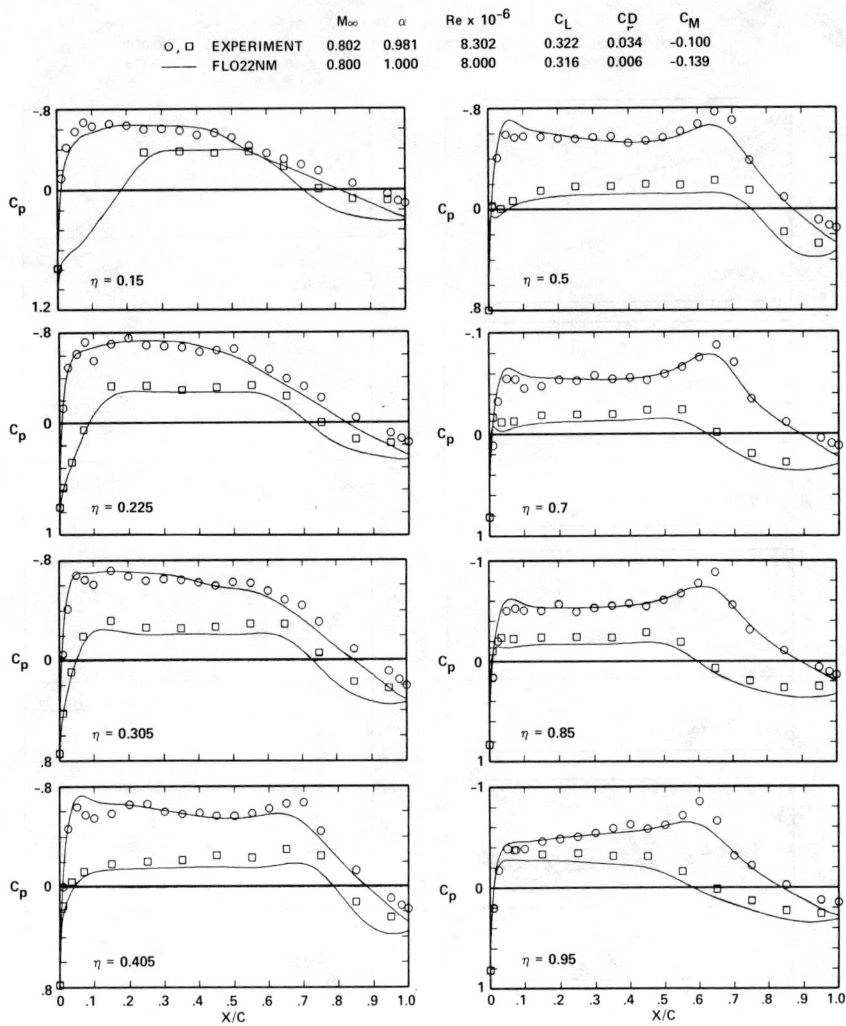

Fig. 14c   (cont'd) Experiment-CFD pressure distribution comparison for wing A, FLO22NM.

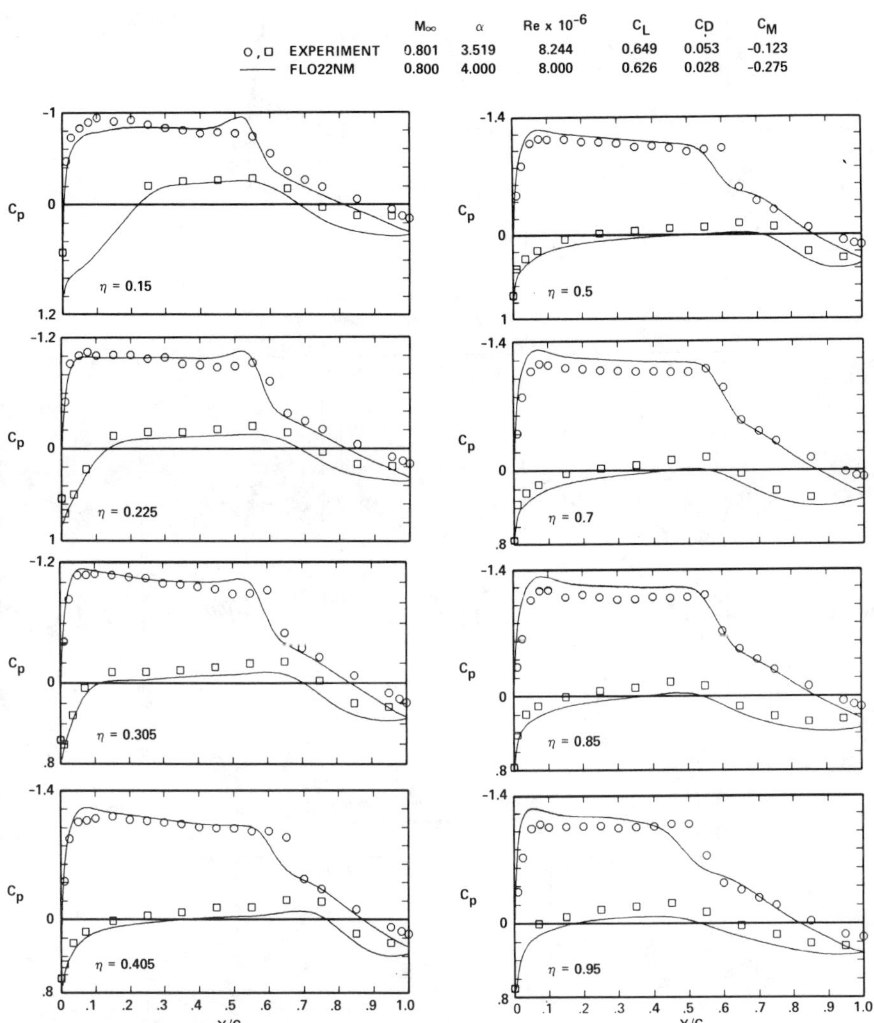

Fig. 14d (cont'd) Experiment-CFD pressure distribution comparison for wing A, FLO22NM.

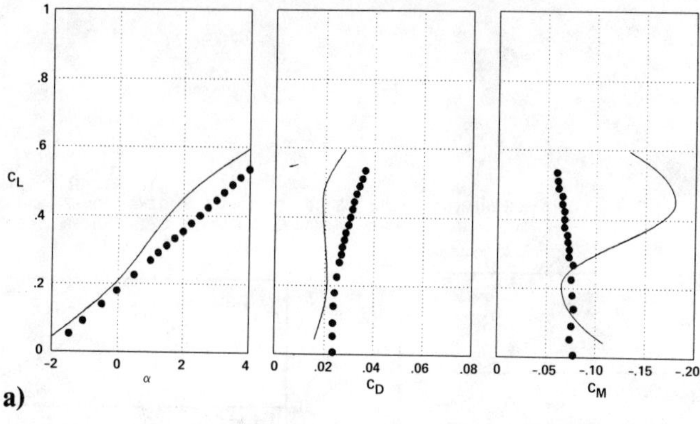

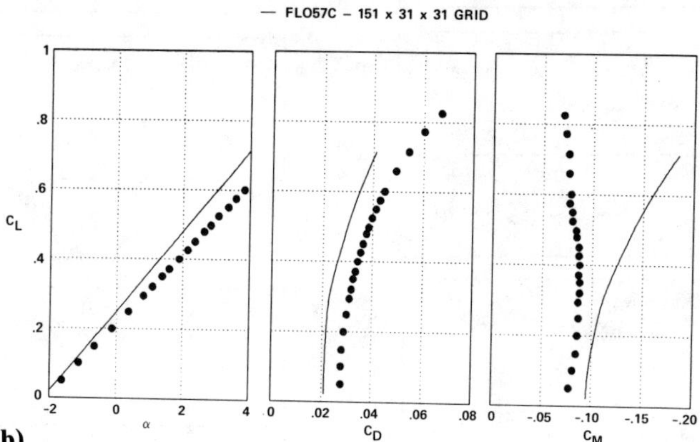

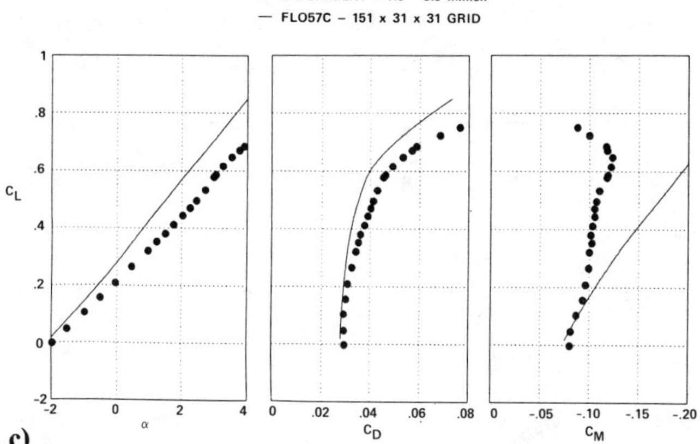

Fig. 15 Experiment-CFD force and moment comparison for wing A, FLO57C: a) $M = 0.40$; **b)** $M = 0.70$; **c)** $M = 0.80$.

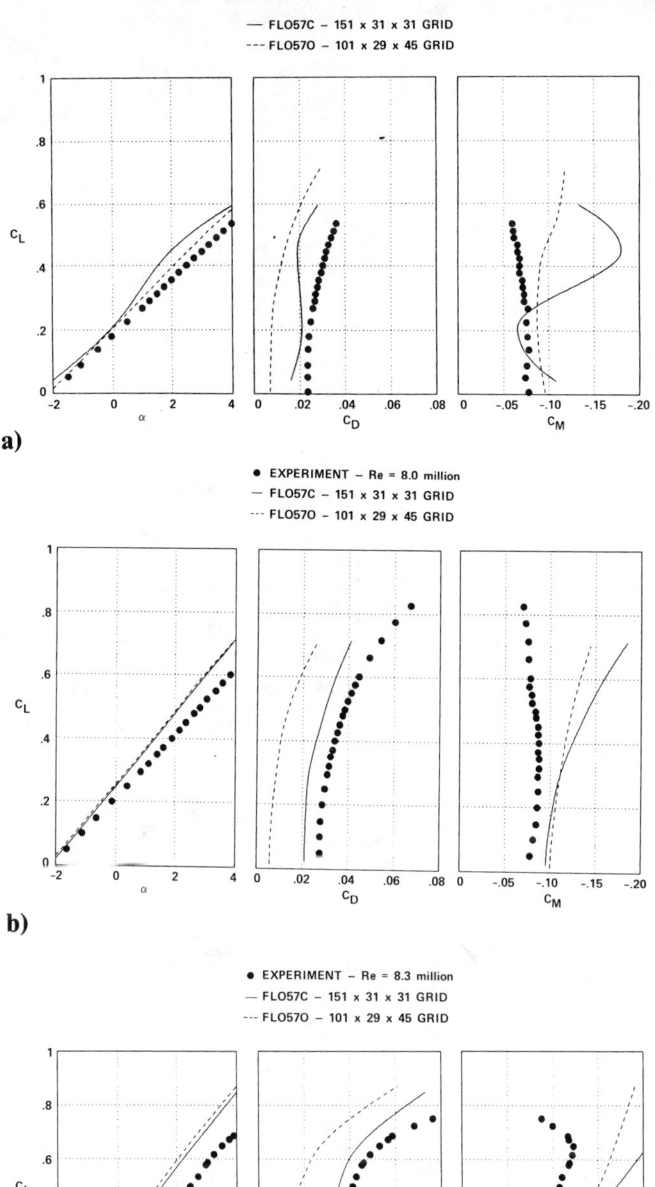

Fig. 16   Experiment-CFD force and moment comparison for wing A, FLO57C and FLO57O: a) $M = 0.40$; b) $M = 0.70$; c) $M = 0.80$.

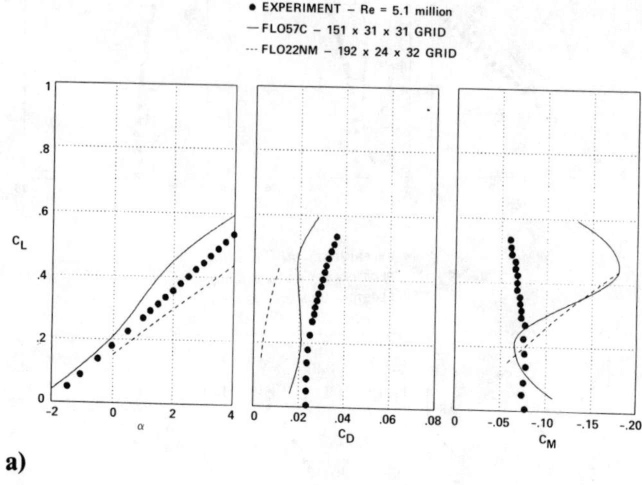

a)

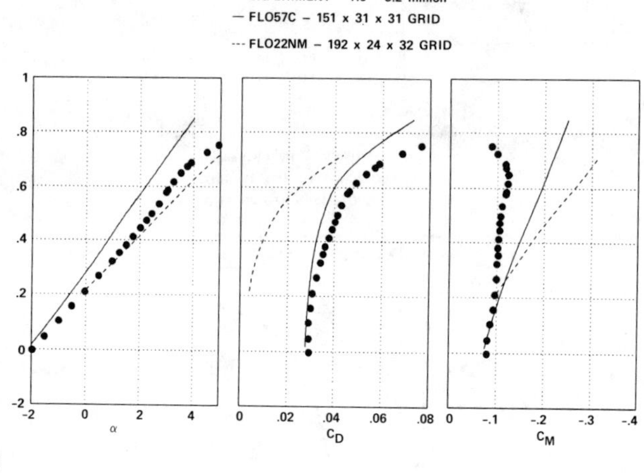

b)

Fig. 17　Experiment-CFD force and moment comparison for wing A, FLO57C and FLO22NM: a) $M = 0.40$; b) $M = 0.80$.

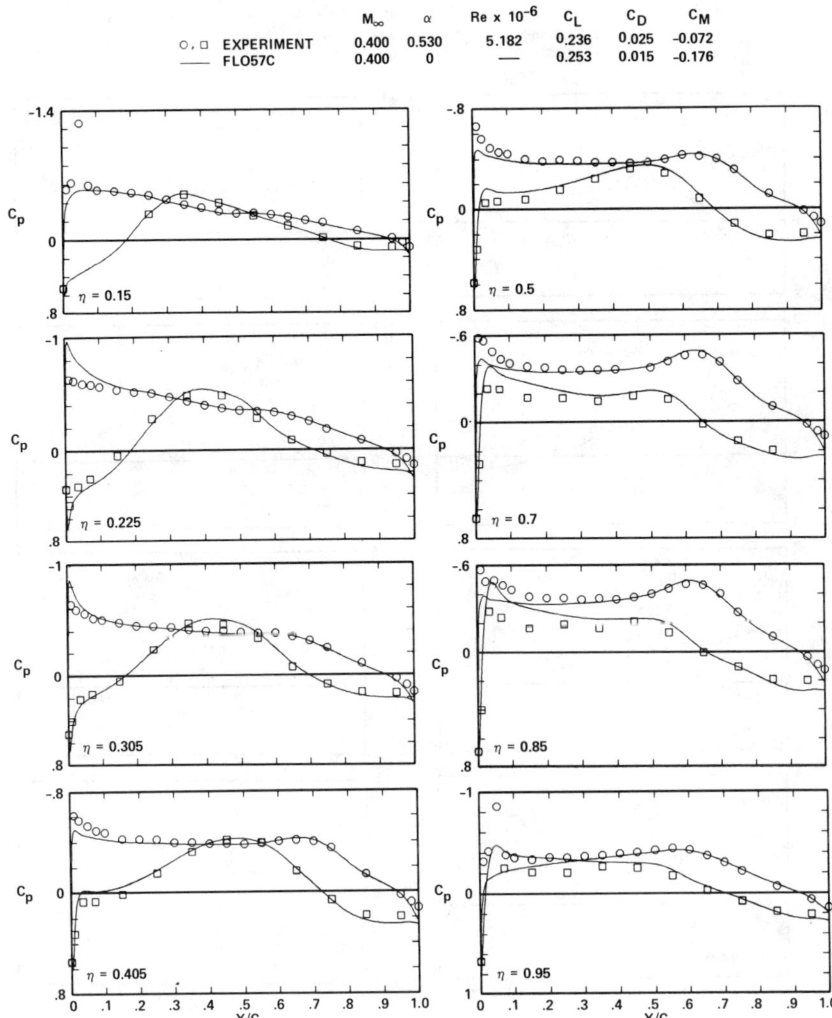

Fig. 18a Experiment-CFD pressure distribution comparison for wing B, FLO57C.

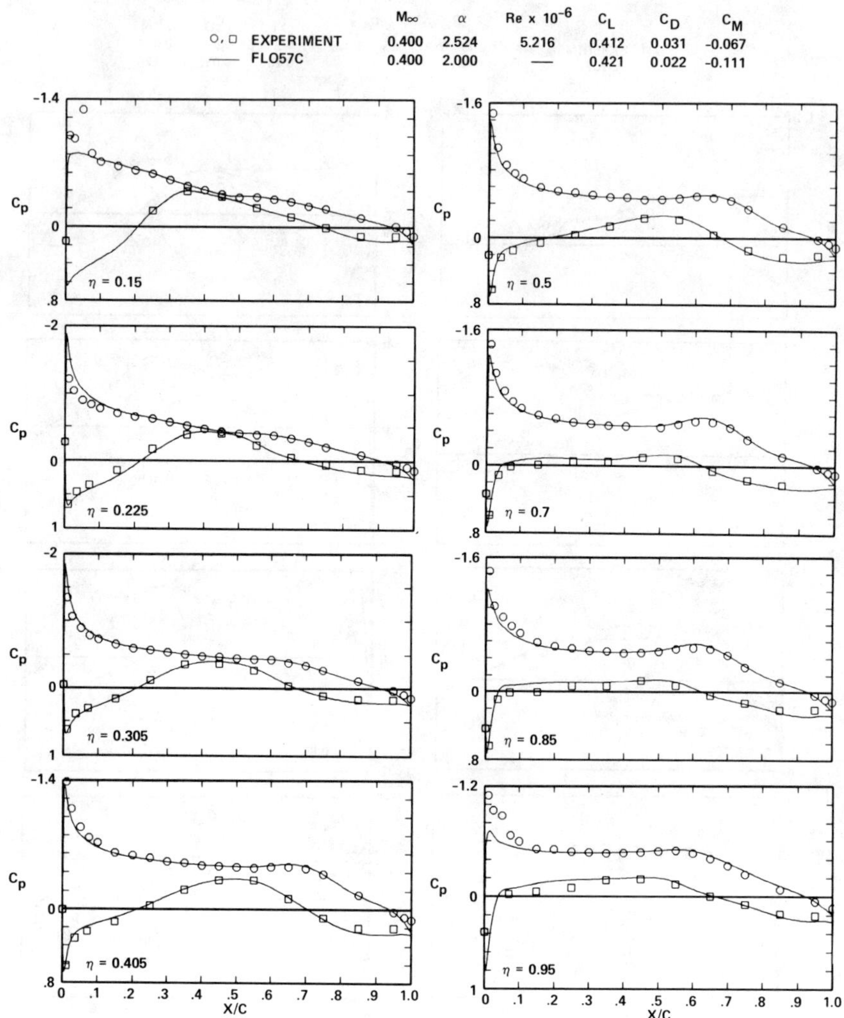

Fig. 18b   (cont'd) Experiment-CFD pressure distribution comparison for wing B, FLO57C.

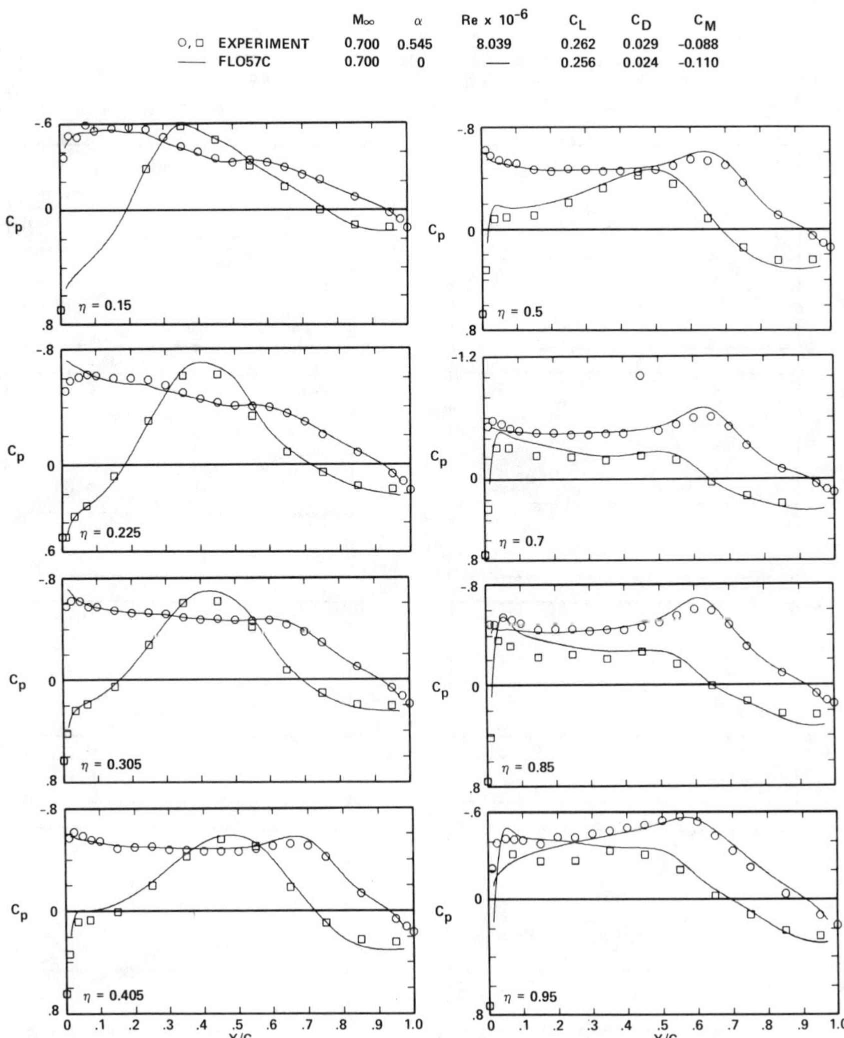

Fig. 18c   (cont'd) Experiment-CFD pressure distribution comparison for wing B, FLO57C.

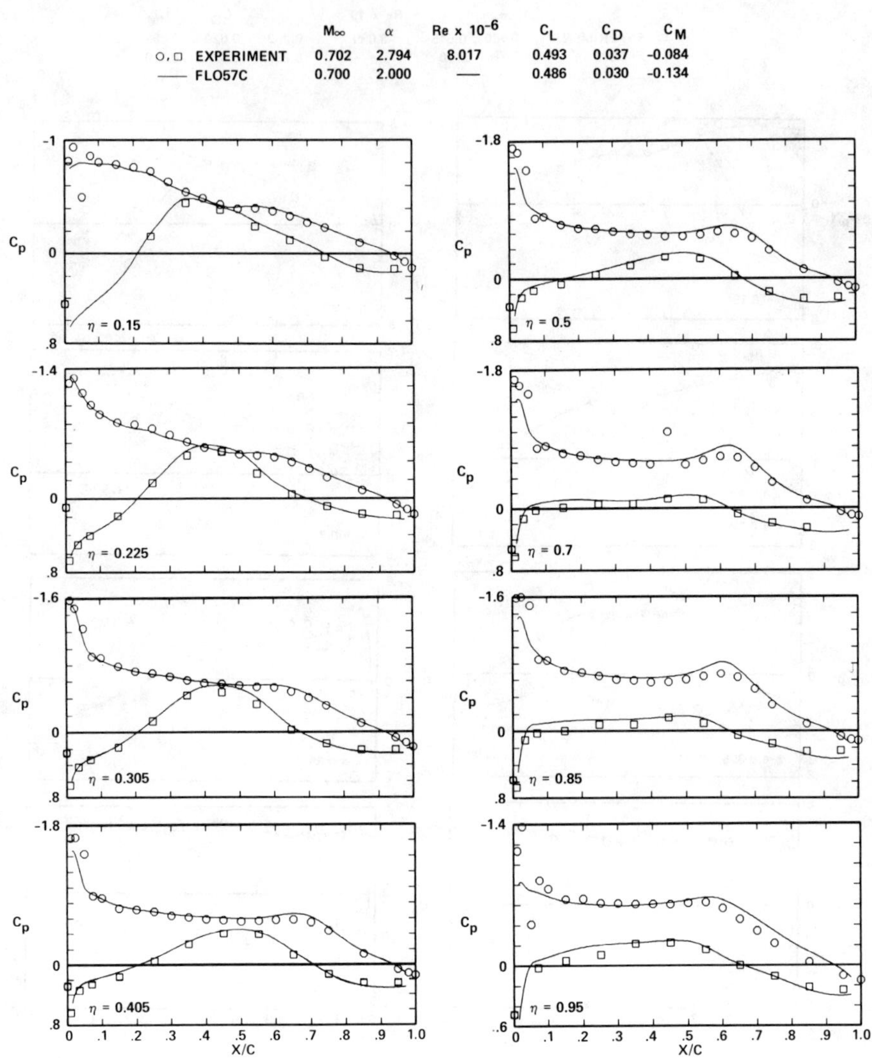

Fig. 18d　(cont'd) Experiment-CFD pressure distribution comparison for wing B, FLO57C.

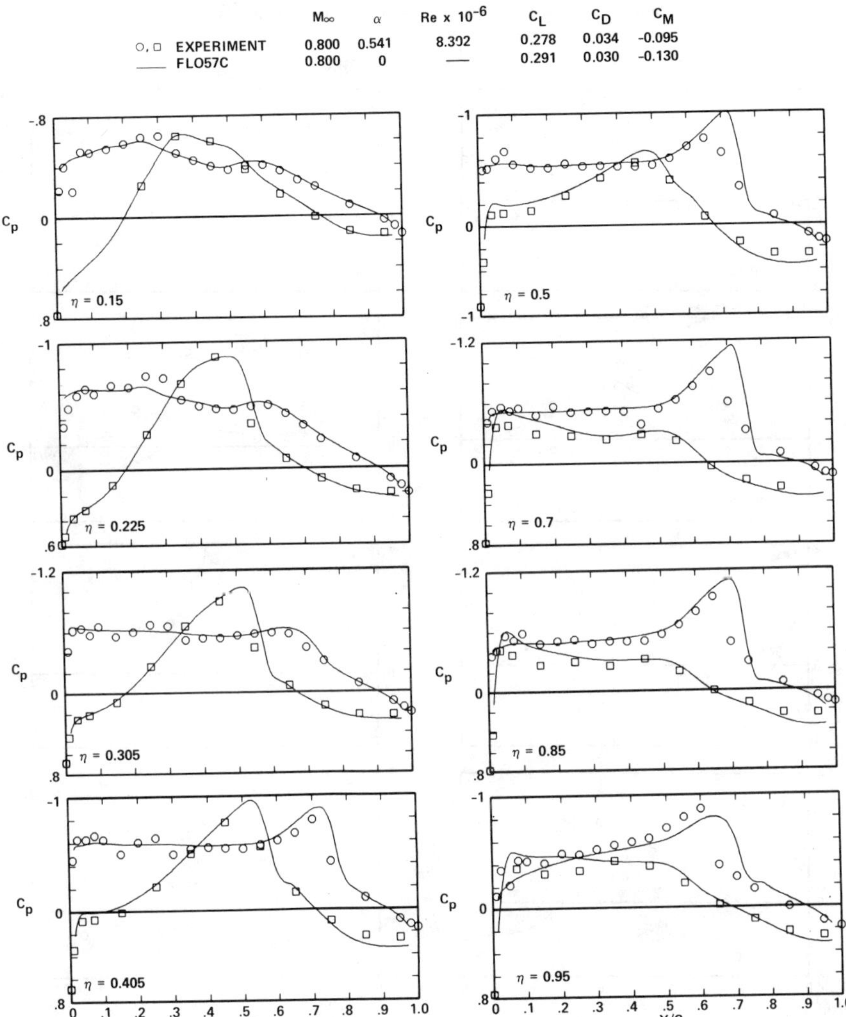

Fig. 18e   (cont'd) Experiment-CFD pressure distribution comparison for wing B, FLO57C.

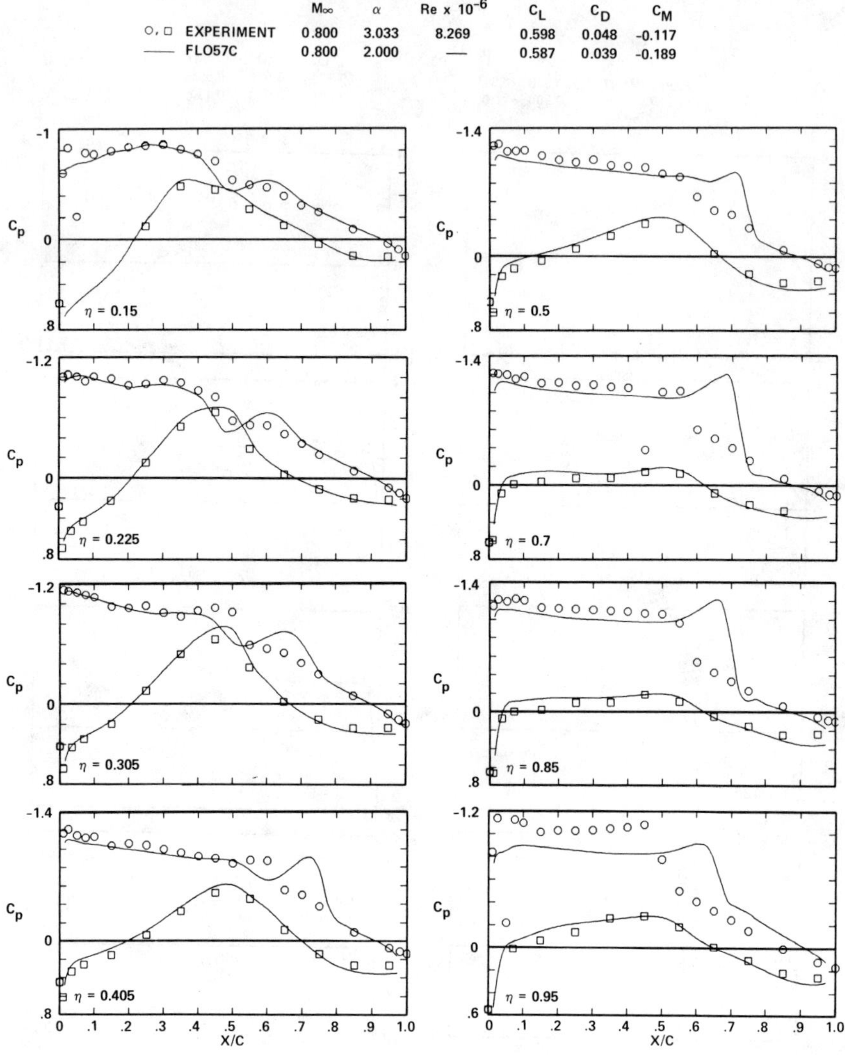

Fig. 18f   (cont'd) Experiment-CFD pressure distribution comparison for wing B, FLO57C.

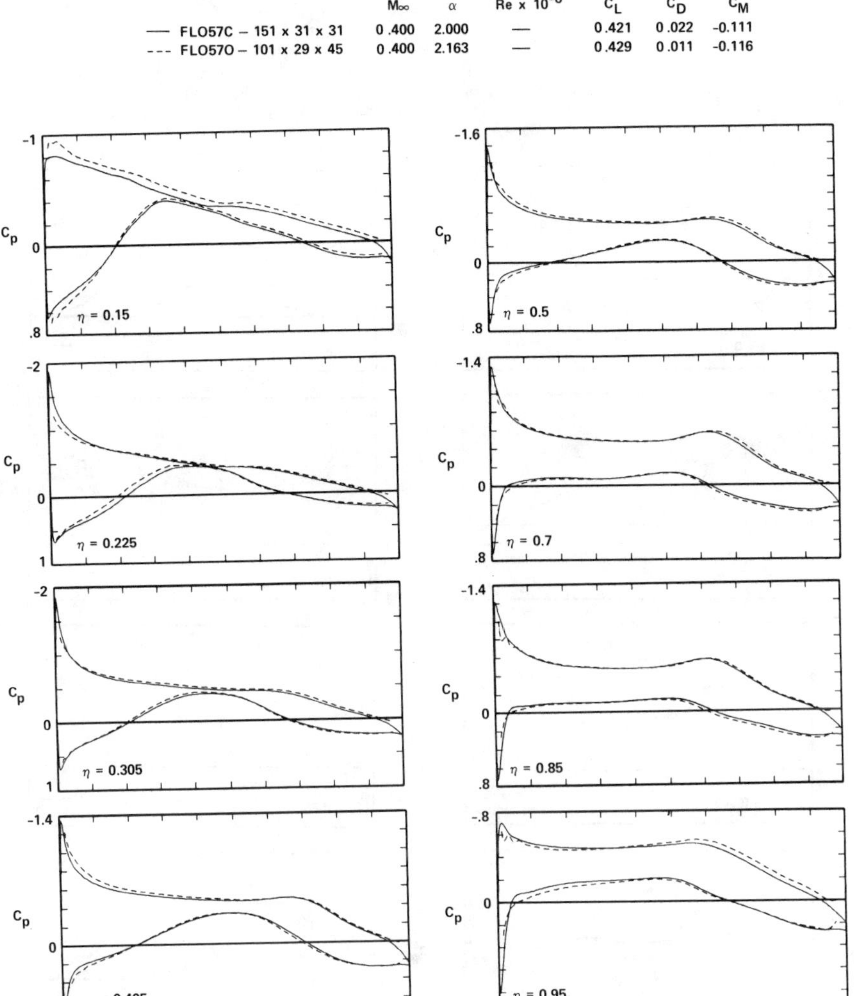

Fig. 19a   Pressure distribution comparison for wing B, FLO57C and FLO57O.

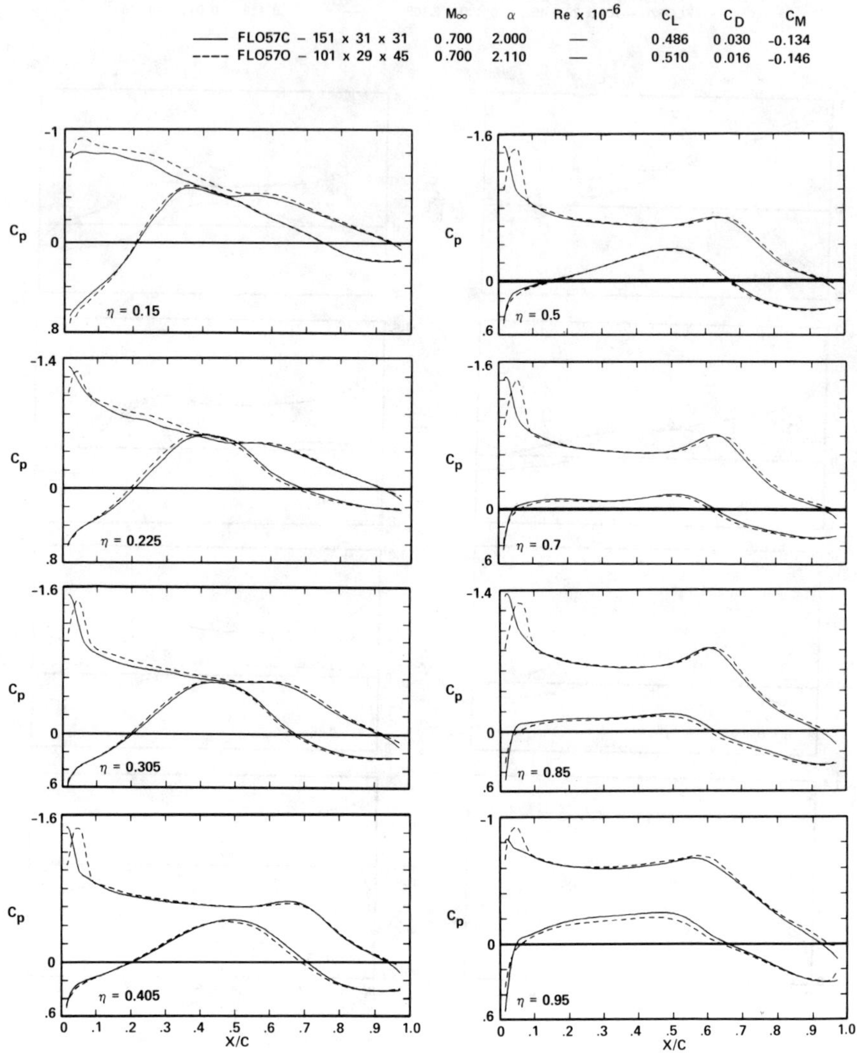

Fig. 19b (cont'd) Pressure distribution comparison for wing B, FLO57C and FLO57O.

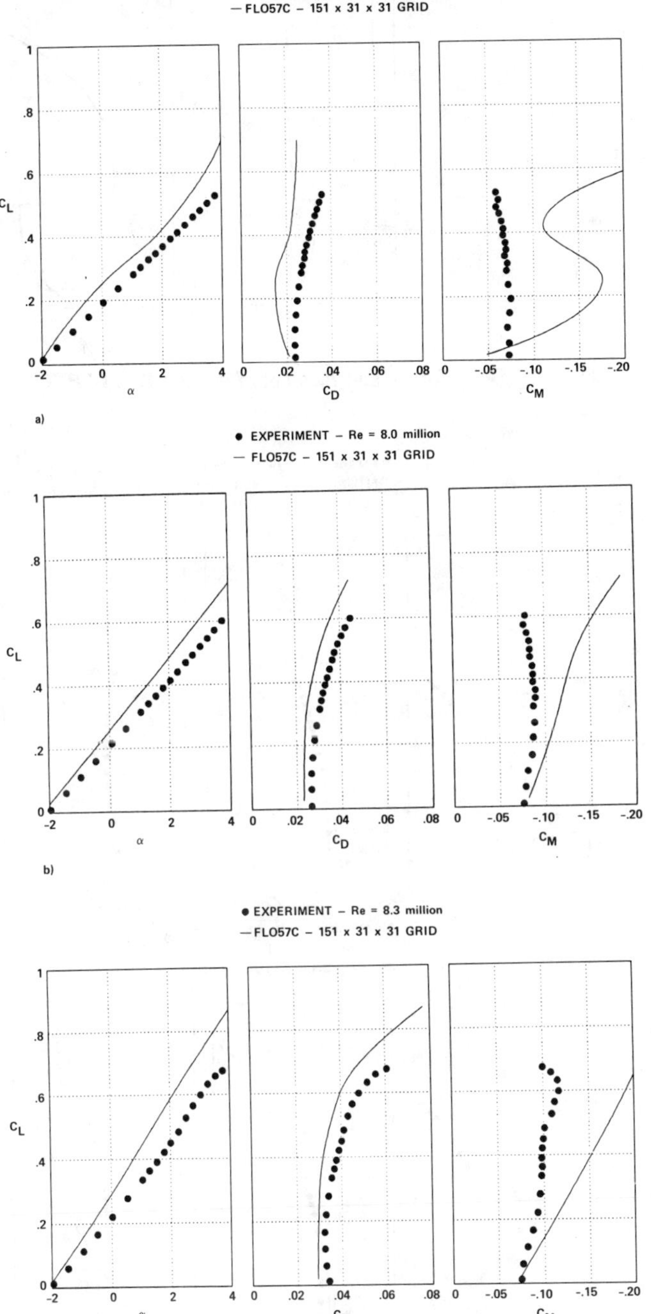

Fig. 20 Experiment-CFD force and moment comparison for wing B, FLO57C: a) $M = 0.40$; b) $M = 0.70$; c) $M = 0.80$.

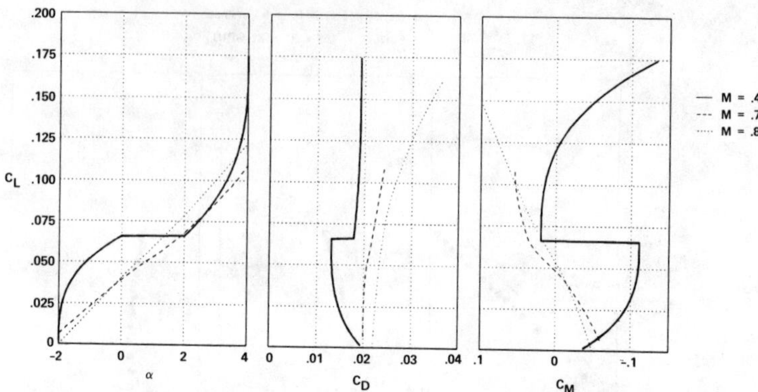

Fig. 21   Body forces and moments for wing B, FLO57C.

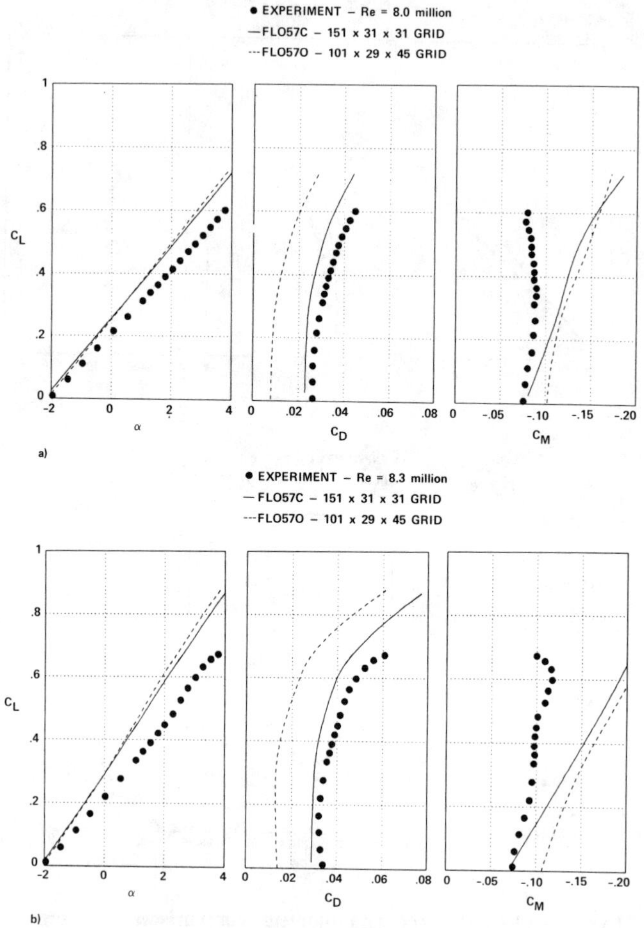

Fig. 22   Experiment-CFD force and moment comparison for wing B. FLO57C and FLO57O: a) $M = 0.70$; b) $M = 0.80$.

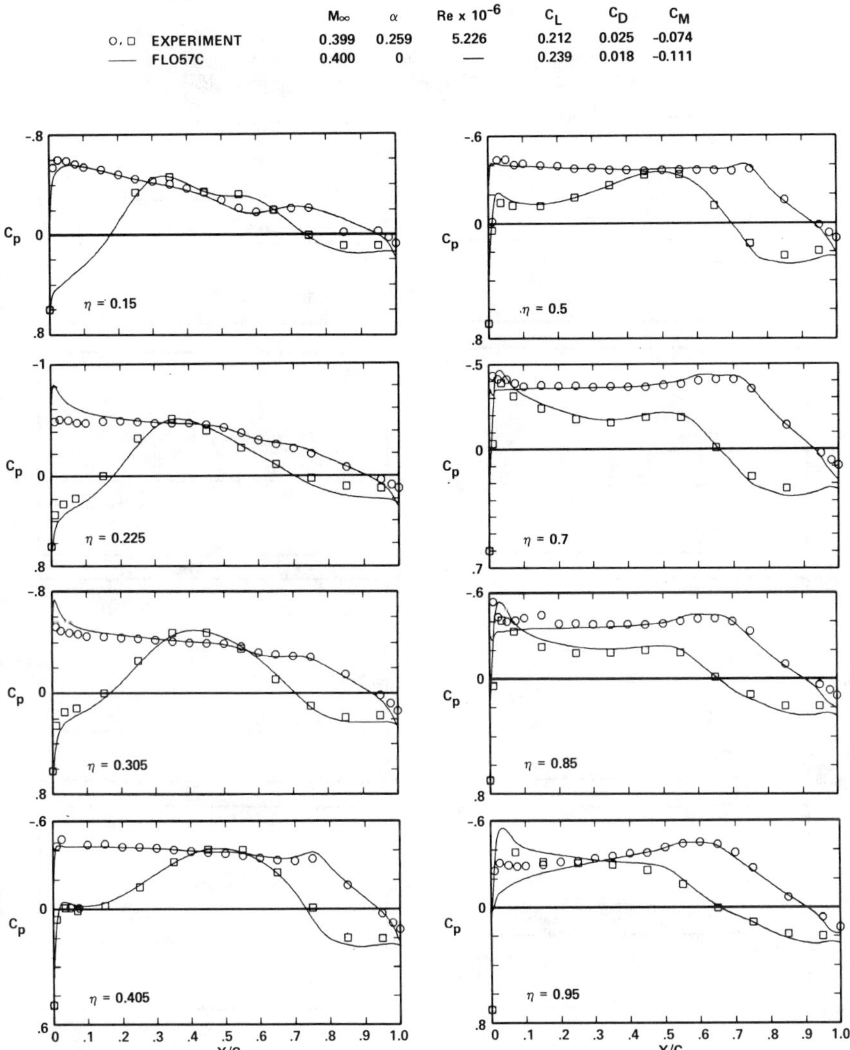

Fig. 23a   Experiment-CFD pressure distribution comparison for wing C, FLO57C.

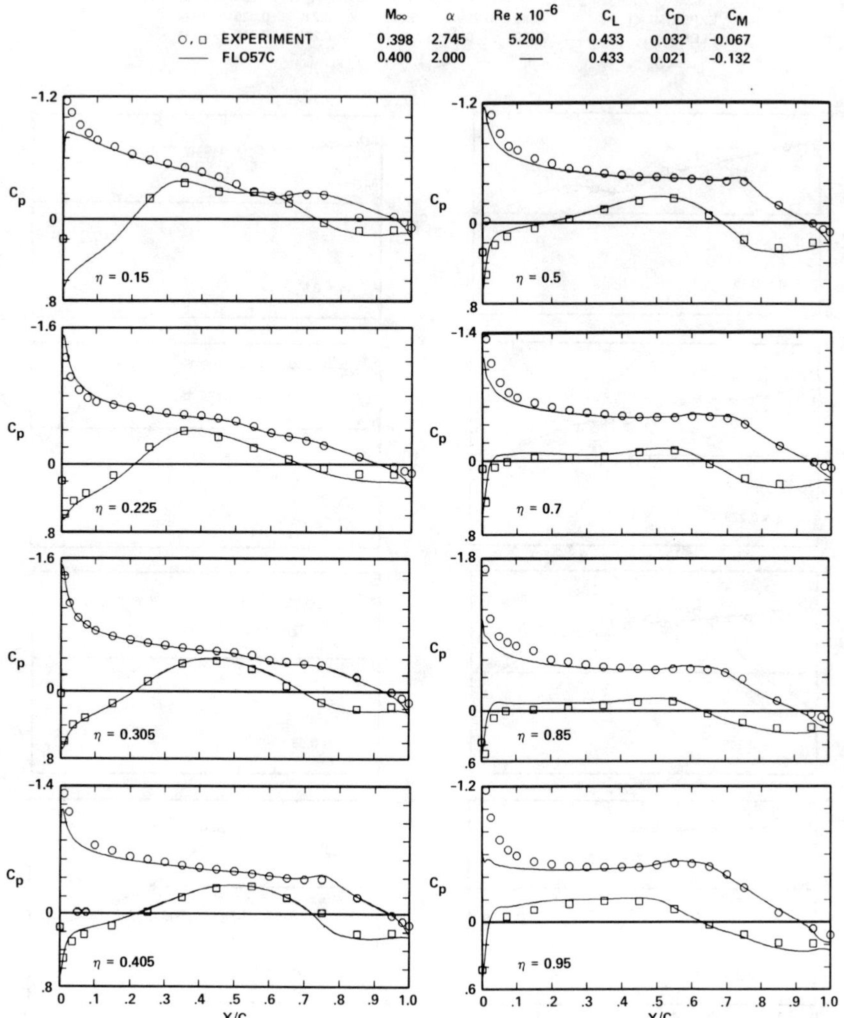

Fig. 23b   (cont'd) Experiment-CFD pressure distribution comparison for wing C, FLO57C.

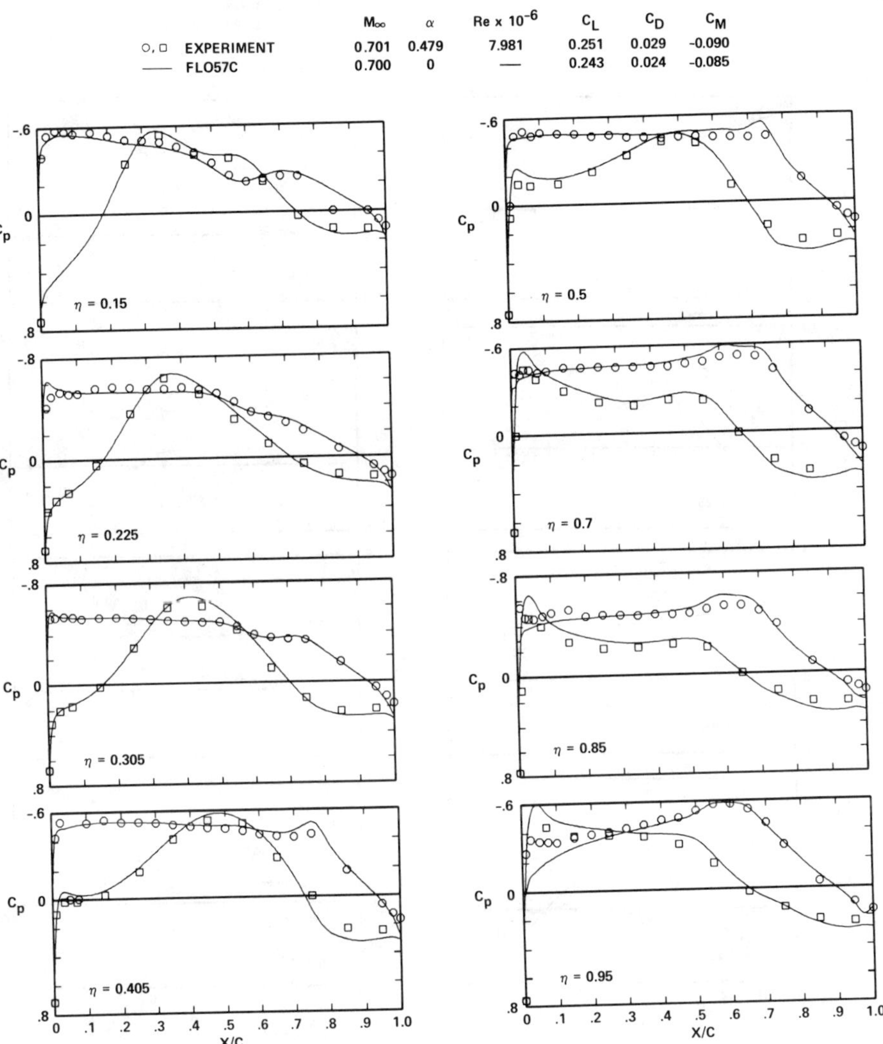

Fig. 23c   (cont'd) Experiment-CFD pressure distribution comparison for wing C, FLO57C.

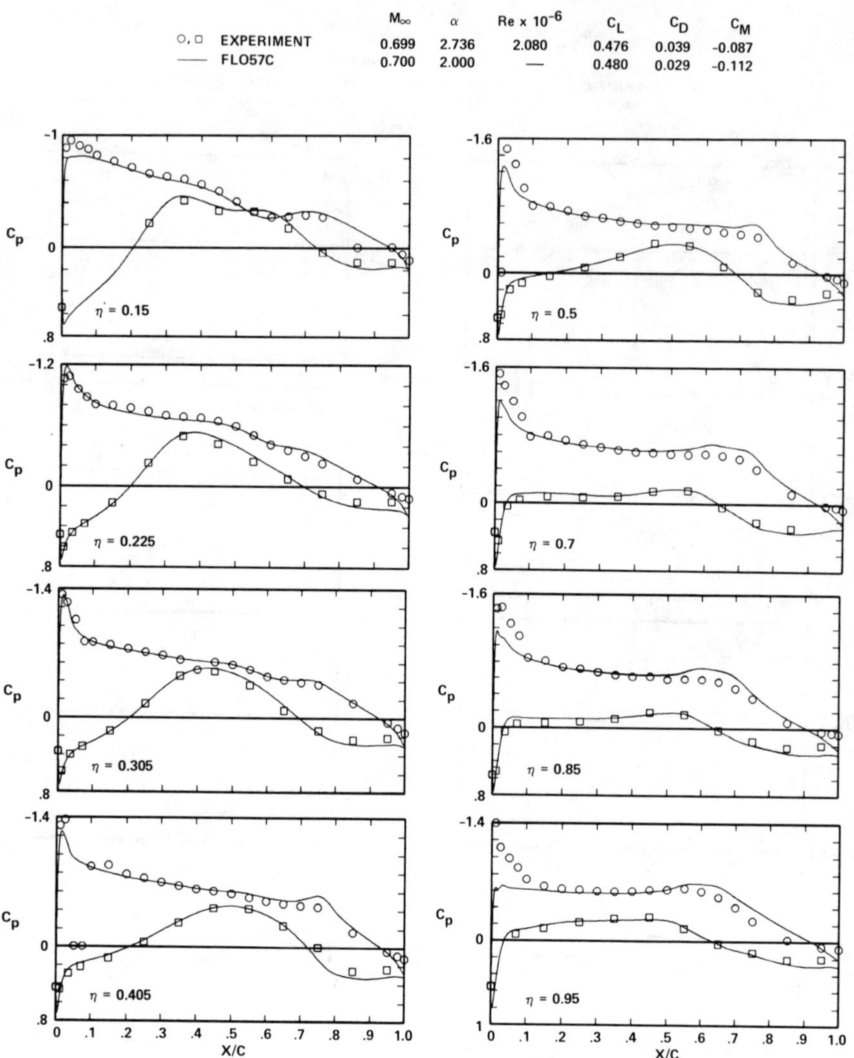

Fig. 23d   (cont'd) Experiment-CFD pressure distribution comparison for wing C, FLO57C.

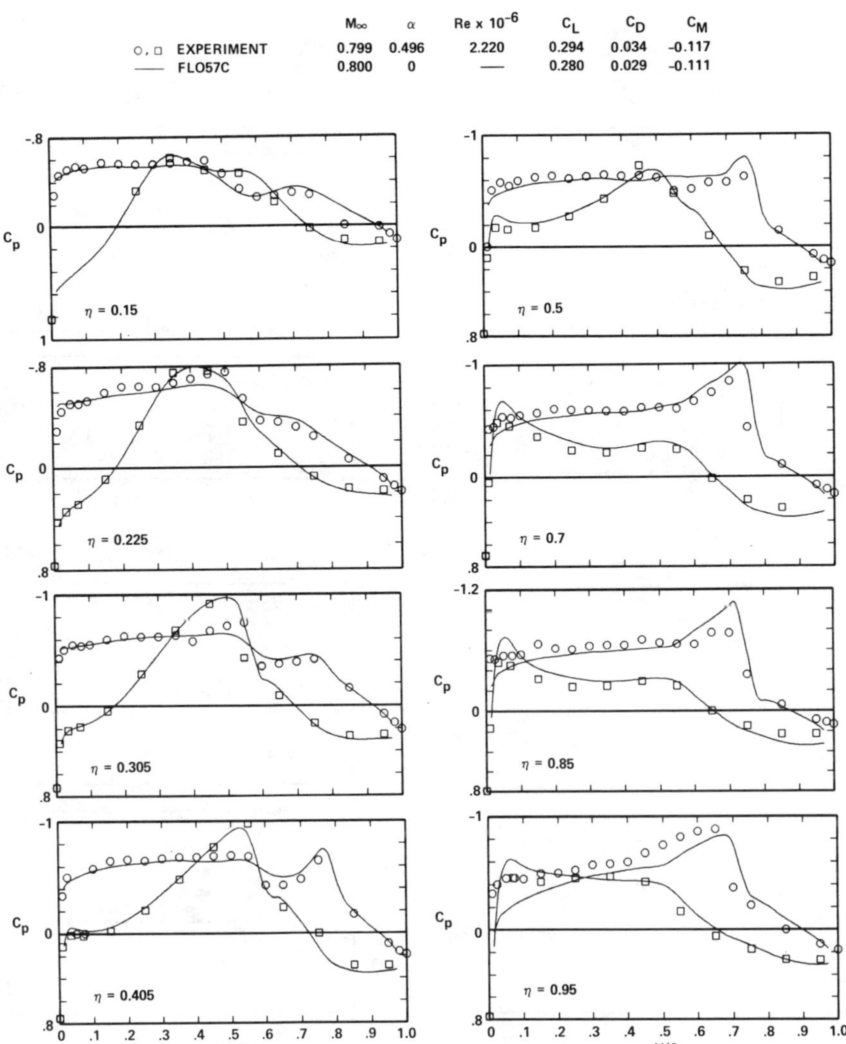

Fig. 23e   (cont'd) Experiment-CFD pressure distribution comparison for wing C, FLO57C.

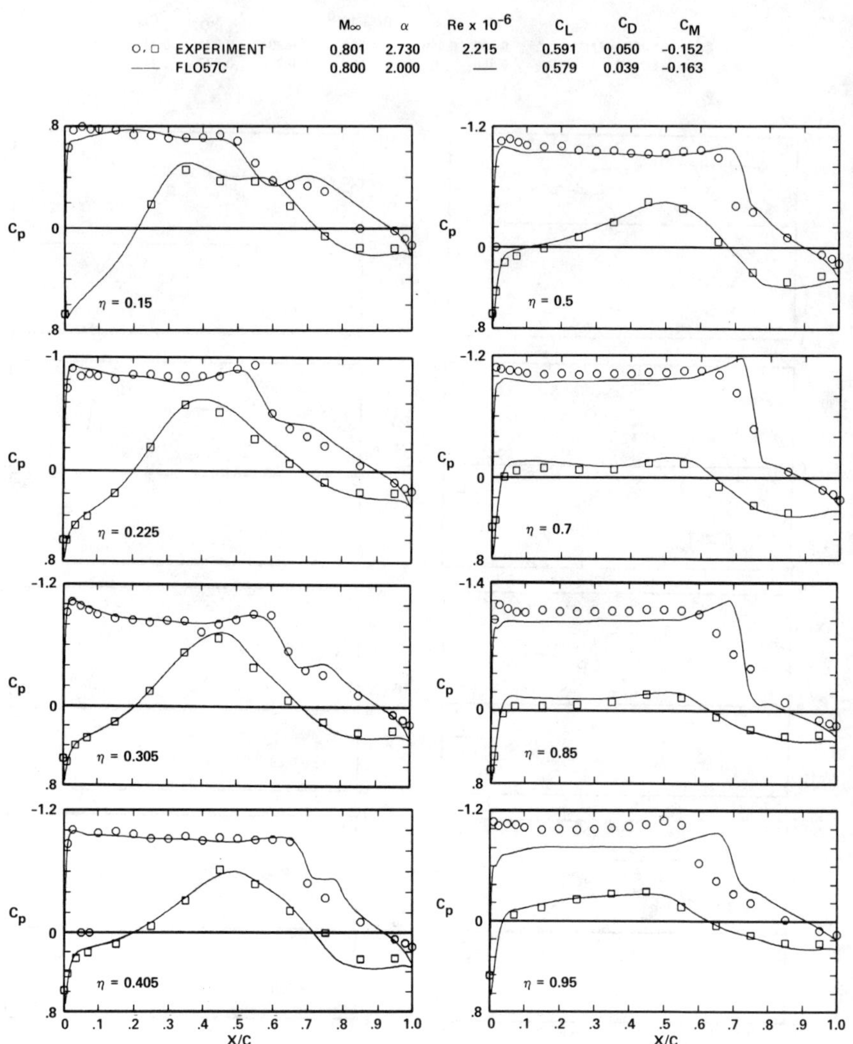

Fig. 23f (cont'd) Experiment-CFD pressure distribution comparison for wing C, FLO57C.

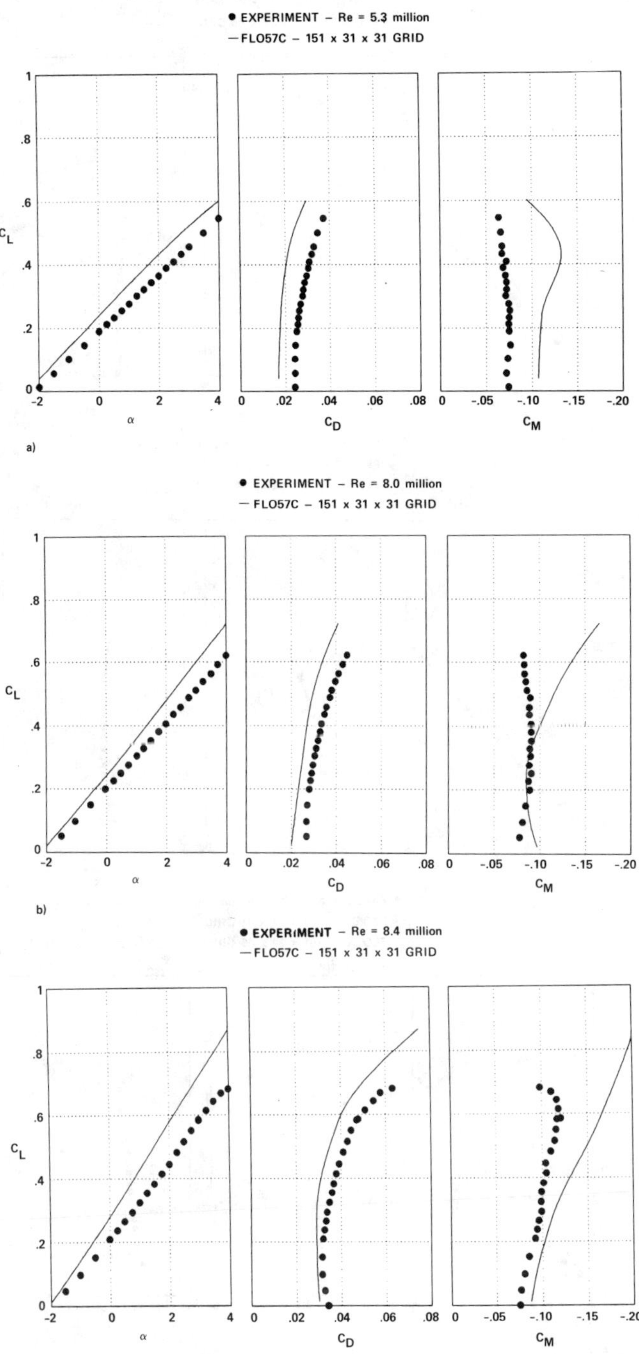

Fig. 24 Experiment-CFD force and moment comparison for wing C, FLO57C: a) $M = 0.40$; b) $M = 0.70$; c) $M = 0.80$.

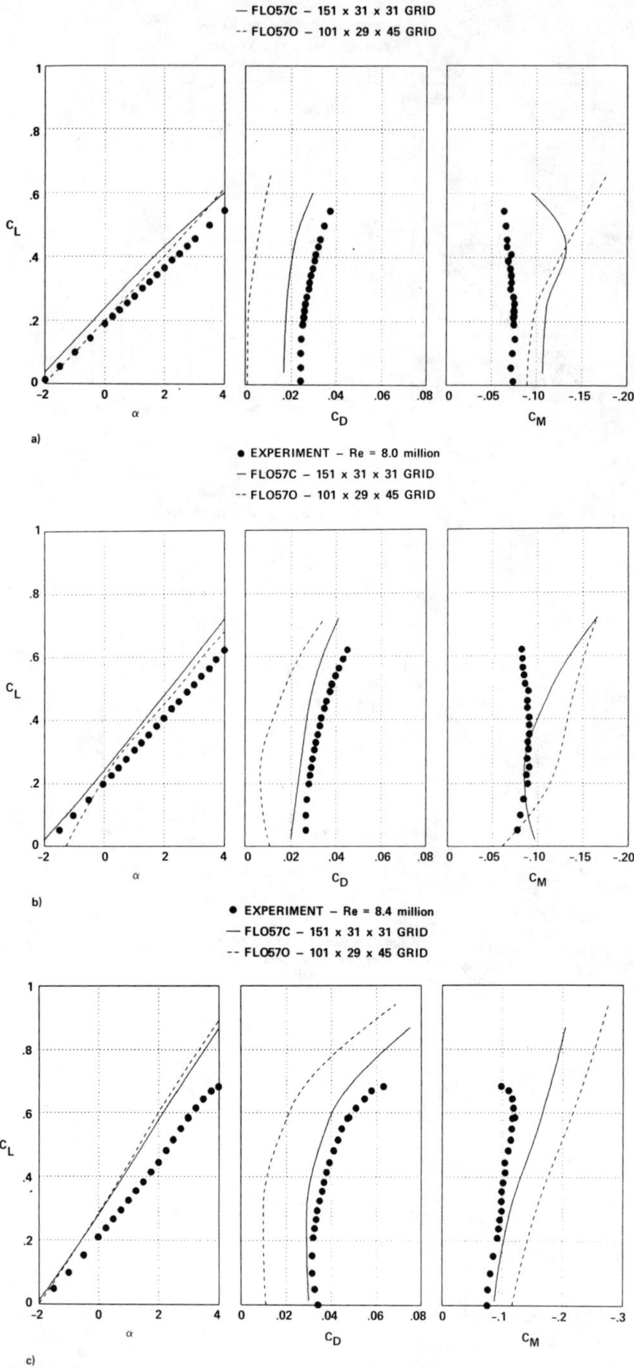

Fig. 25   Experiment-CFD force and moment comparison for wing C, FLO57C and FLO57O: a) $M = 0.40$; b) $M = 0.70$; c) $M = 0.80$.

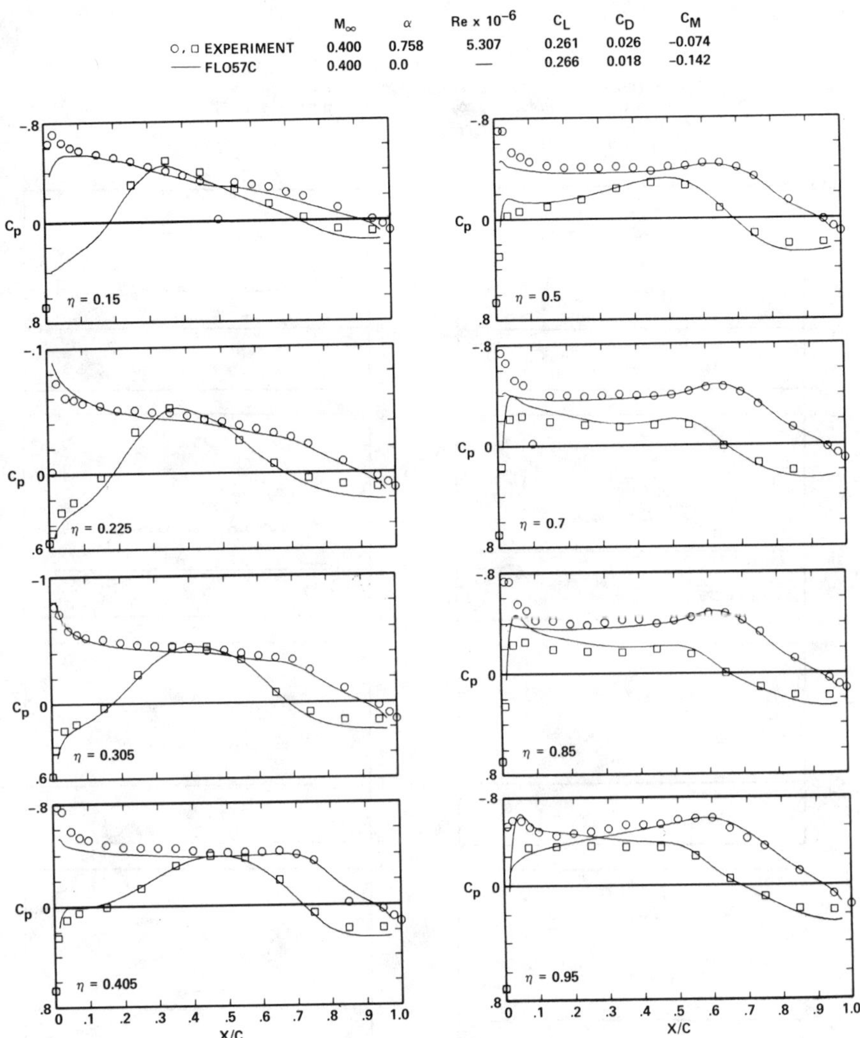

Fig. 26a   Experiment-CFD pressure distribution comparison for wing D, FLO57C.

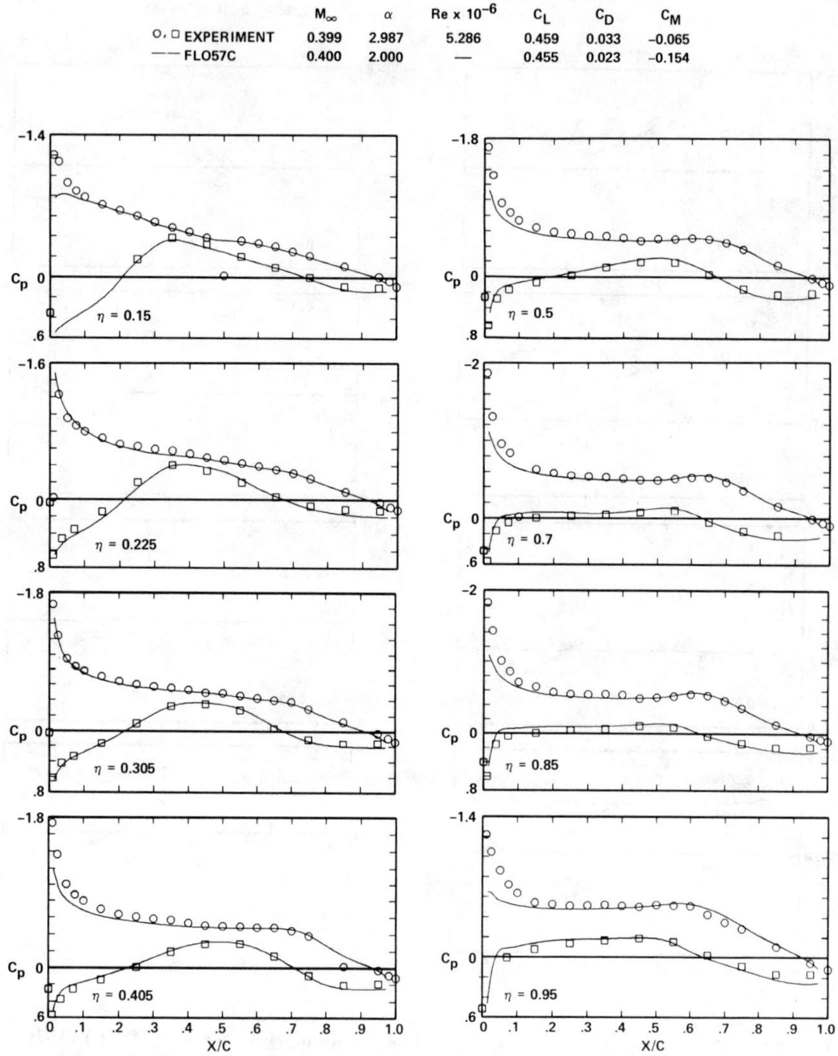

Fig. 26b   (cont'd) Experiment-CFD pressure distribution comparison for wing D, FLO57C.

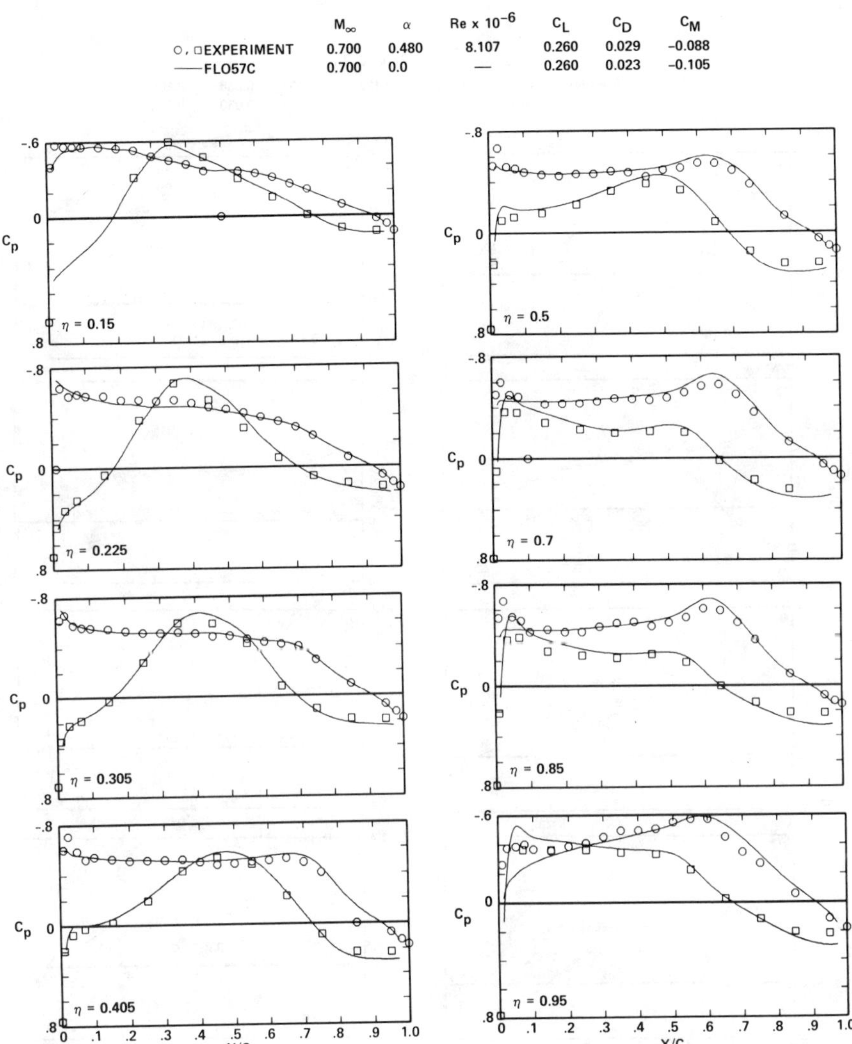

Fig. 26c (cont'd) Experiment-CFD pressure distribution comparison for wing D, FLO57C.

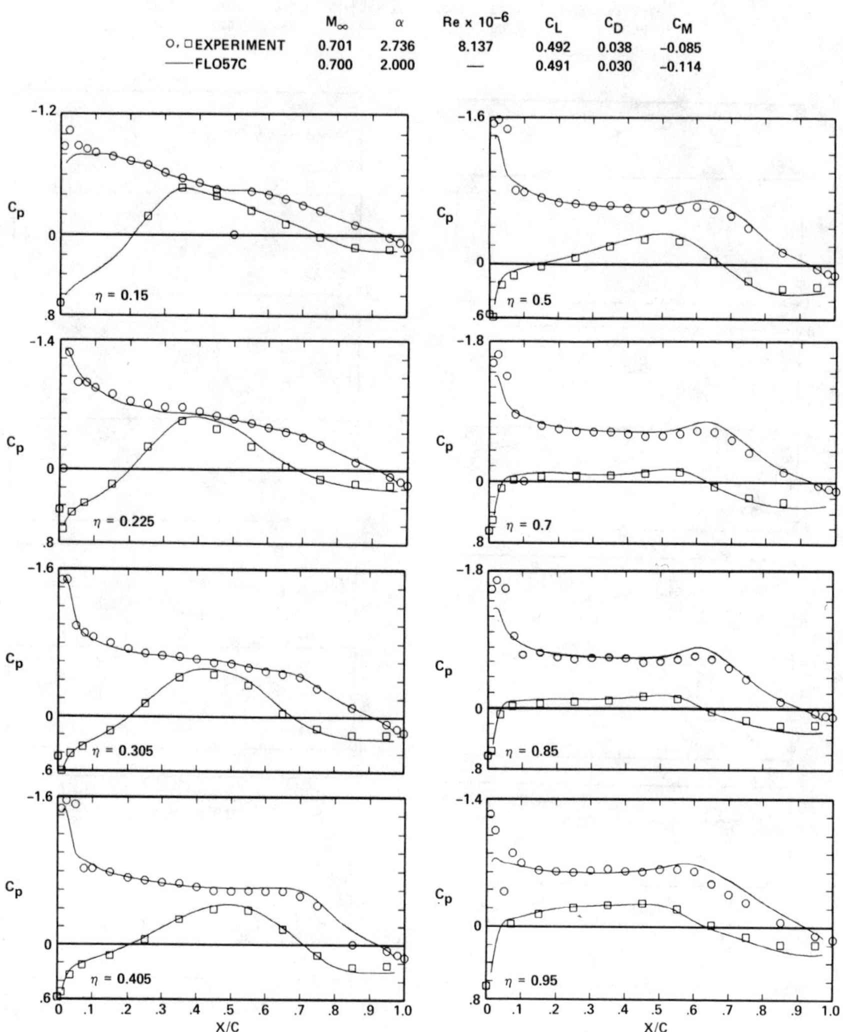

Fig. 26d   (cont'd) Experiment-CFD pressure distribution comparison for wing D, FLO57C.

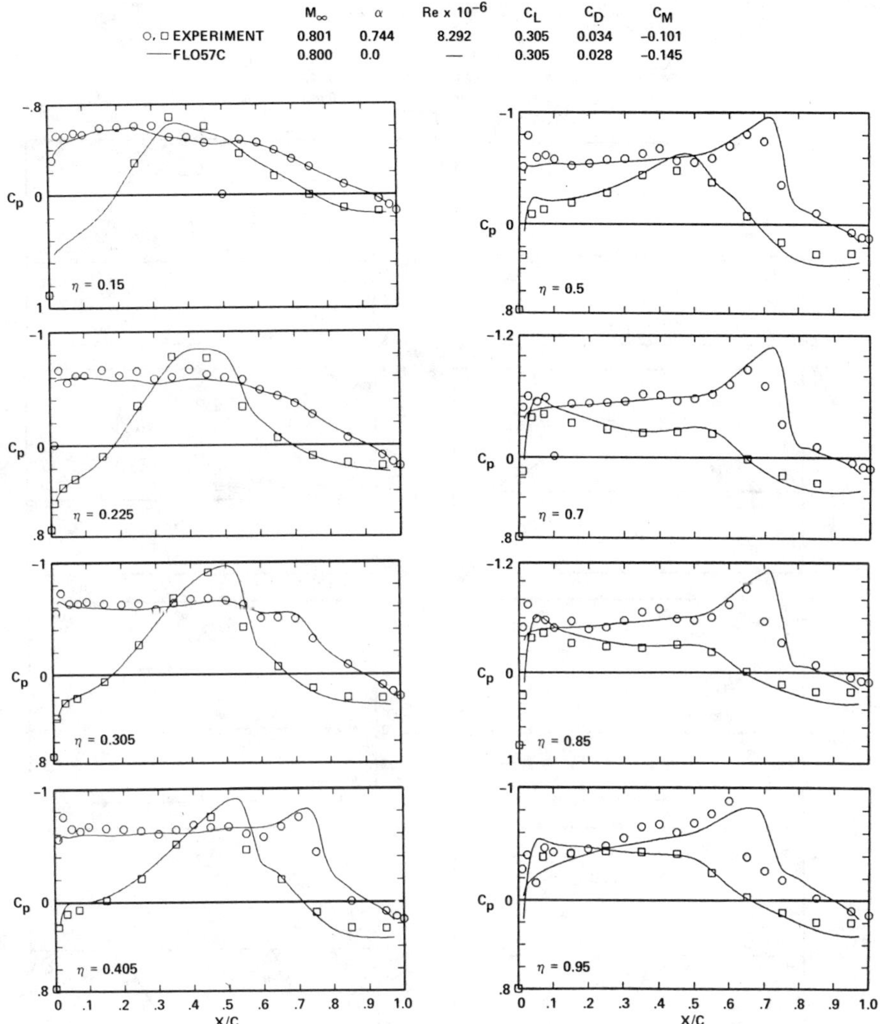

Fig. 26e   (cont'd) Experiment-CFD pressure distribution comparison for wing D, FLO57C.

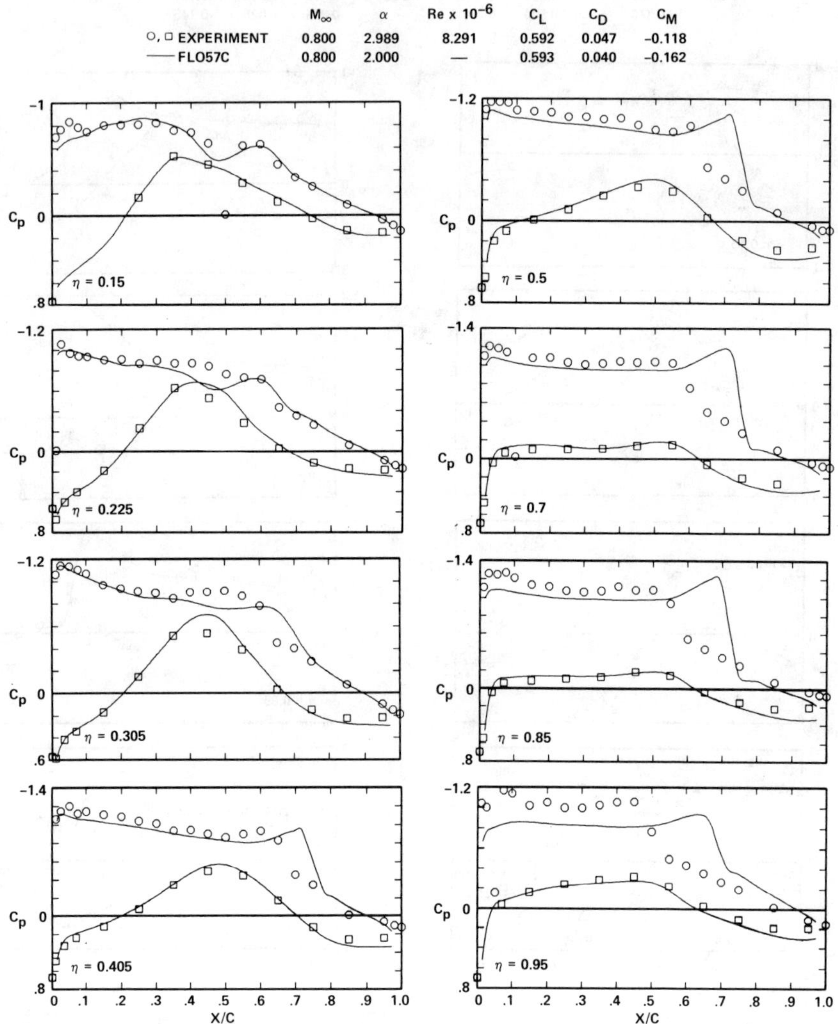

Fig. 26f (cont'd) Experiment-CFD pressure distribution comparison for wing D, FLO57C.

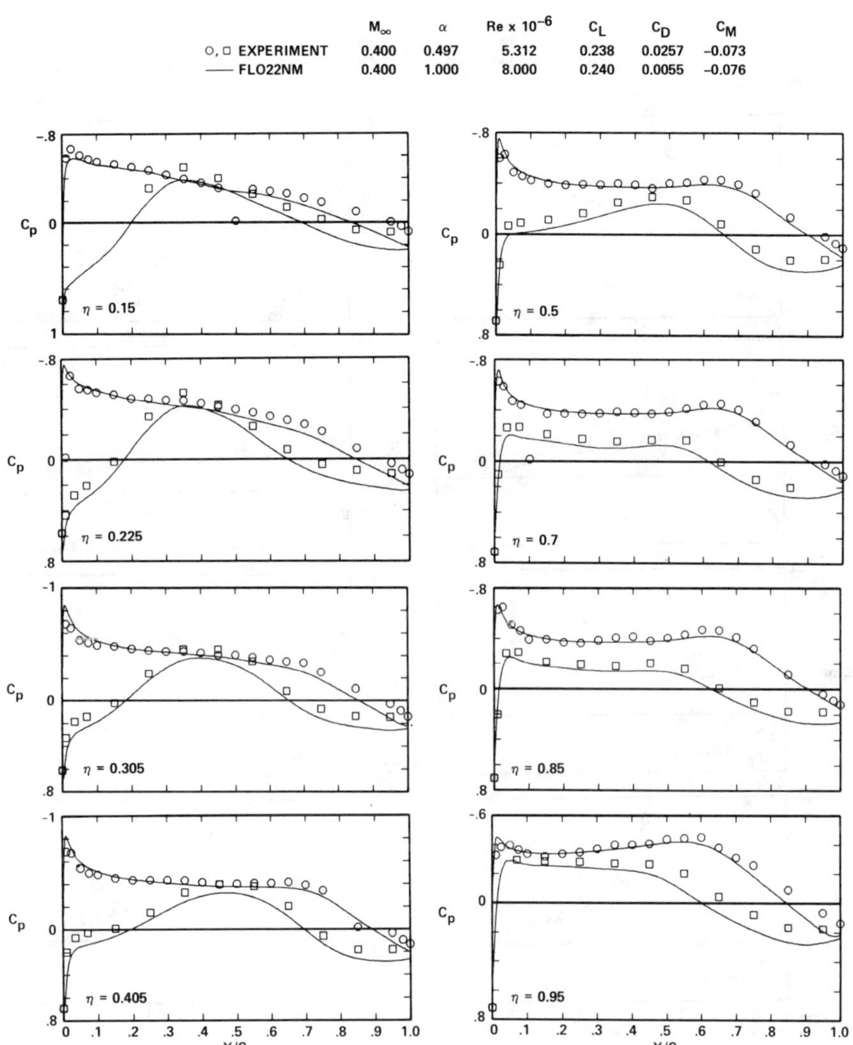

Fig. 27a   Experiment-CFD pressure distribution comparison for wing D, FLO22NM.

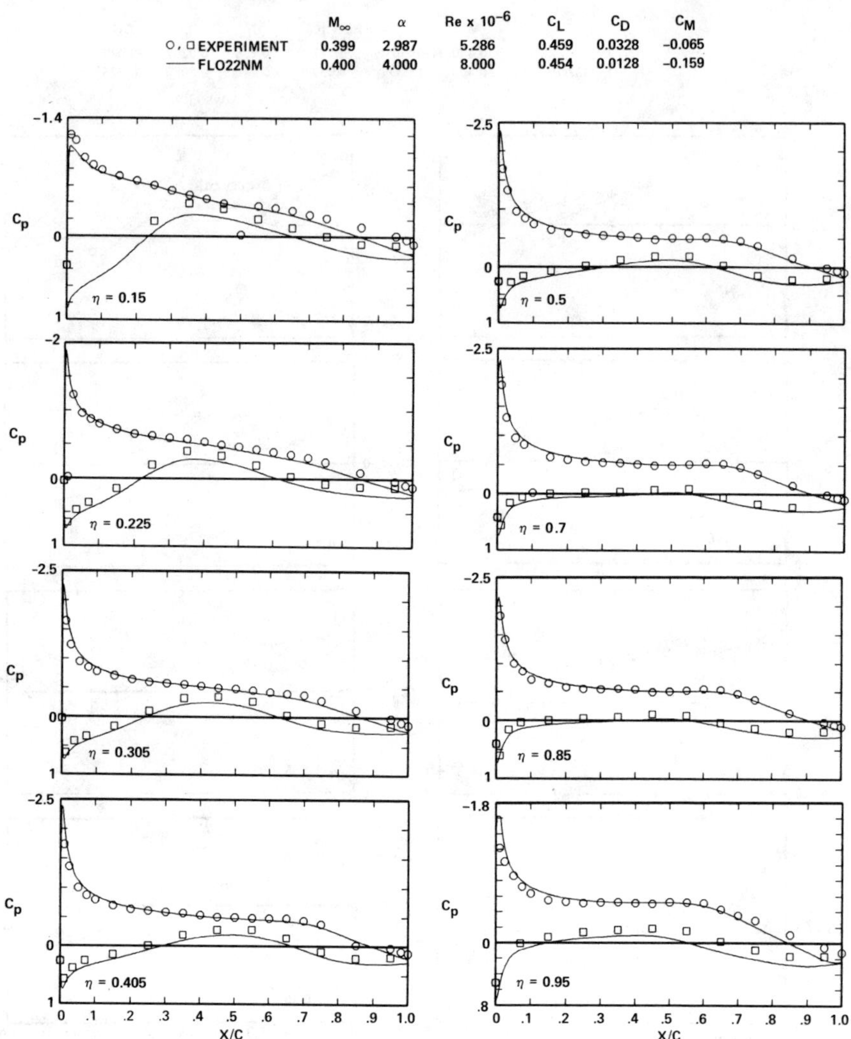

Fig. 27b   (cont'd) Experiment-CFD pressure distribution comparison for wing D, FLO22NM.

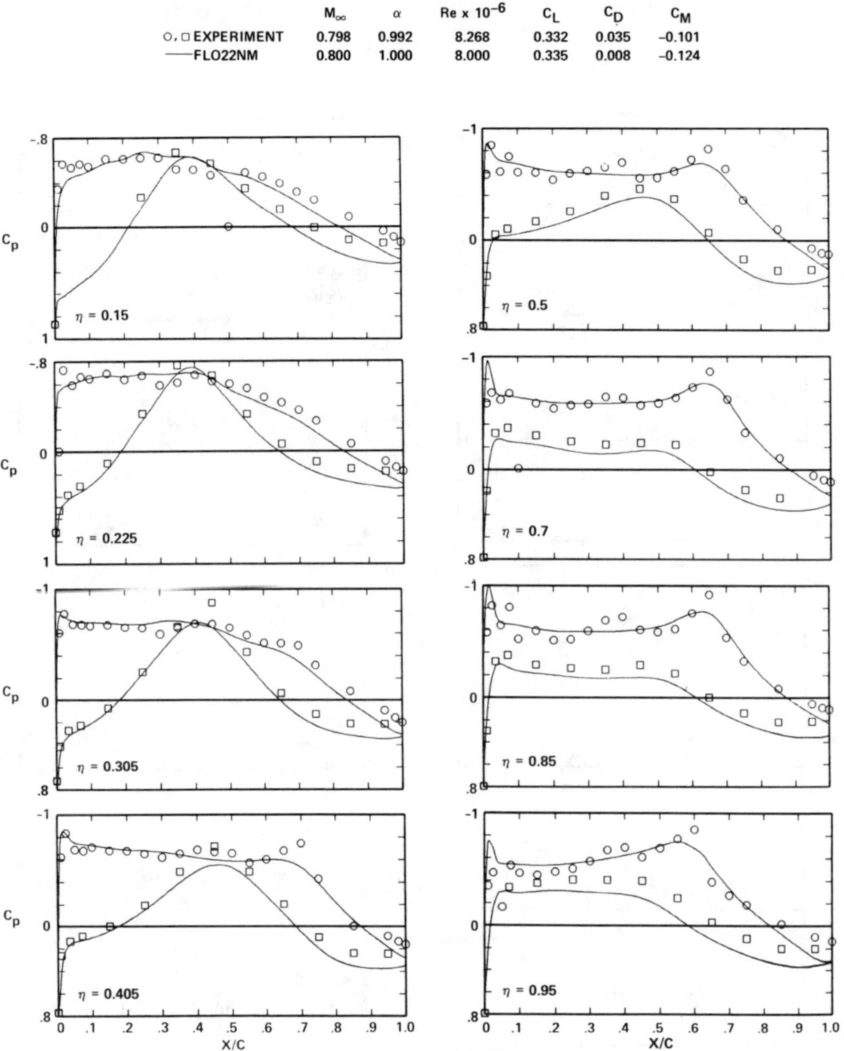

Fig. 27c   (cont'd) Experiment-CFD pressure distribution comparison for wing D, FLO22NM.

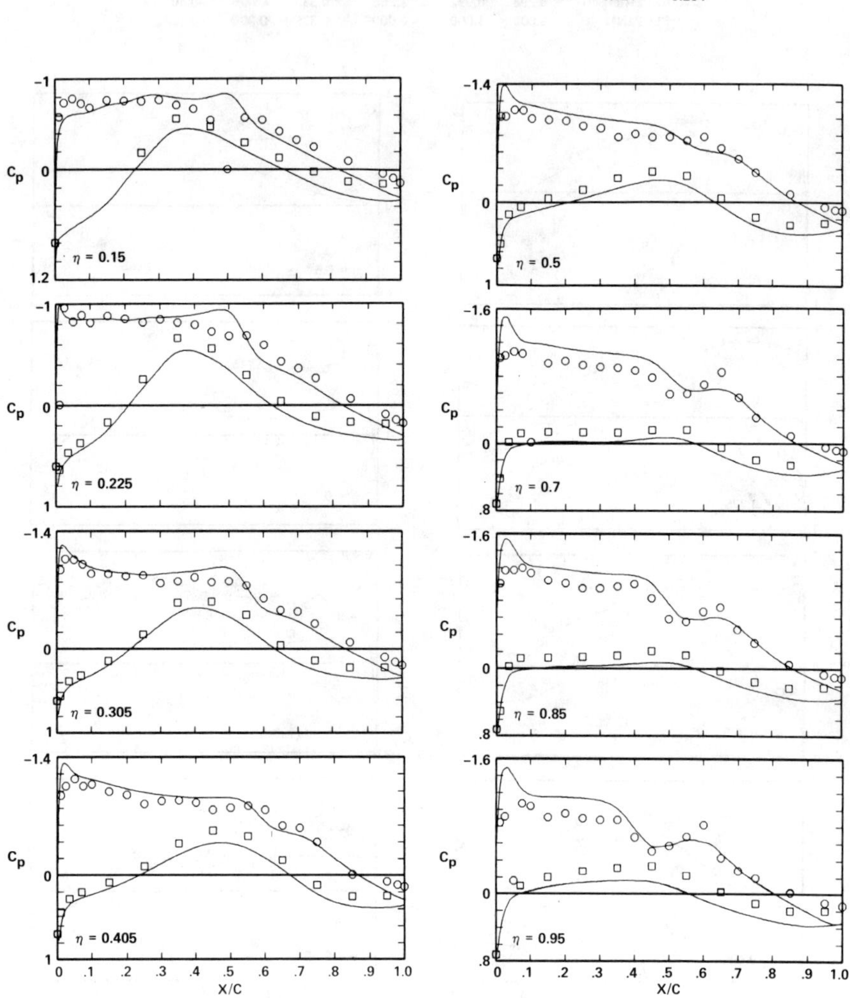

Fig. 27d (cont'd) Experiment-CFD pressure distribution comparison for wing D, FLO22NM.

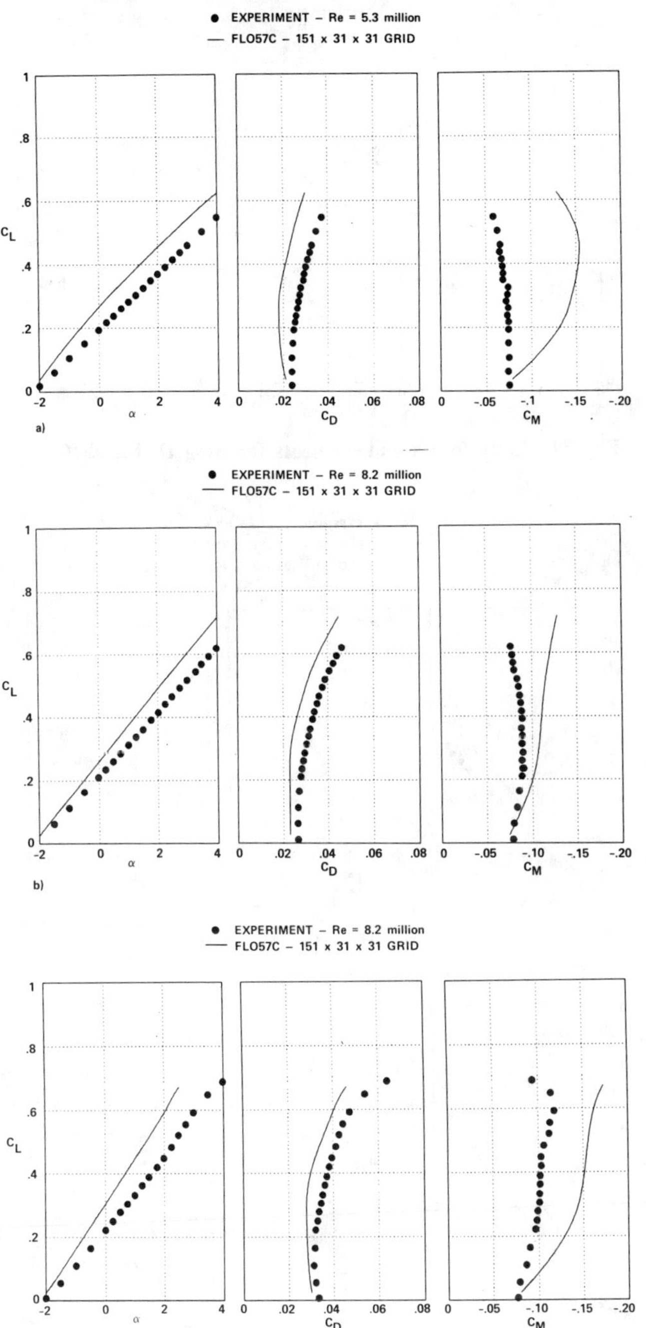

Fig. 28 Experiment-CFD force and moment comparison for wing D, FLO57C. a) $M = 0.40$; b) $M = 0.70$; c) $M = 0.80$.

R. M. HICKS ET AL.

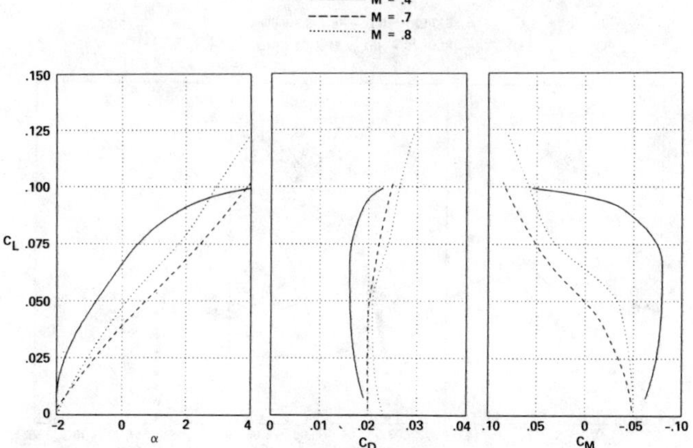

Fig. 29   Body forces and moments for wing D, FLO57C.

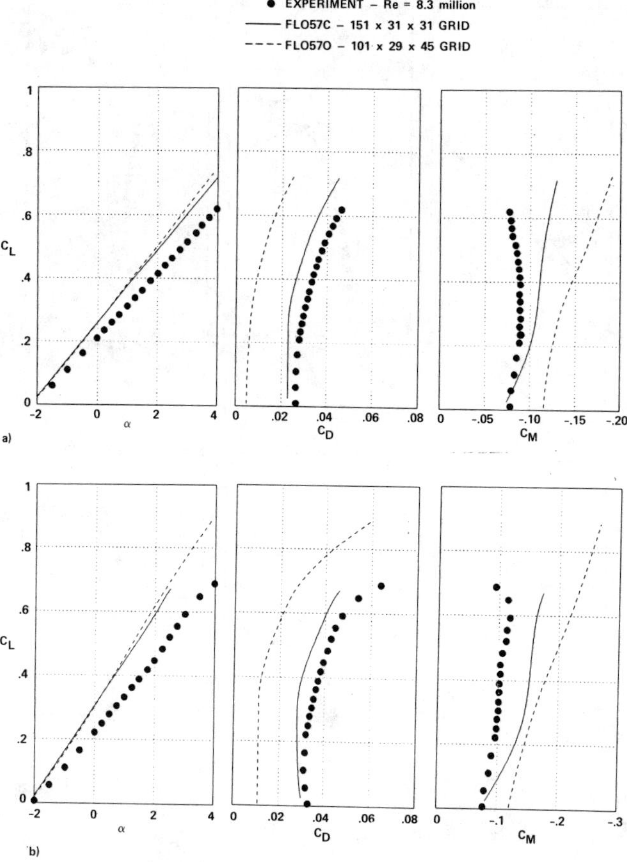

Fig. 30   Experiment-CFD force and moment comparison for wing D, FLO57C and FLO57O: a) $M = 0.70$; b) $M = 0.80$.

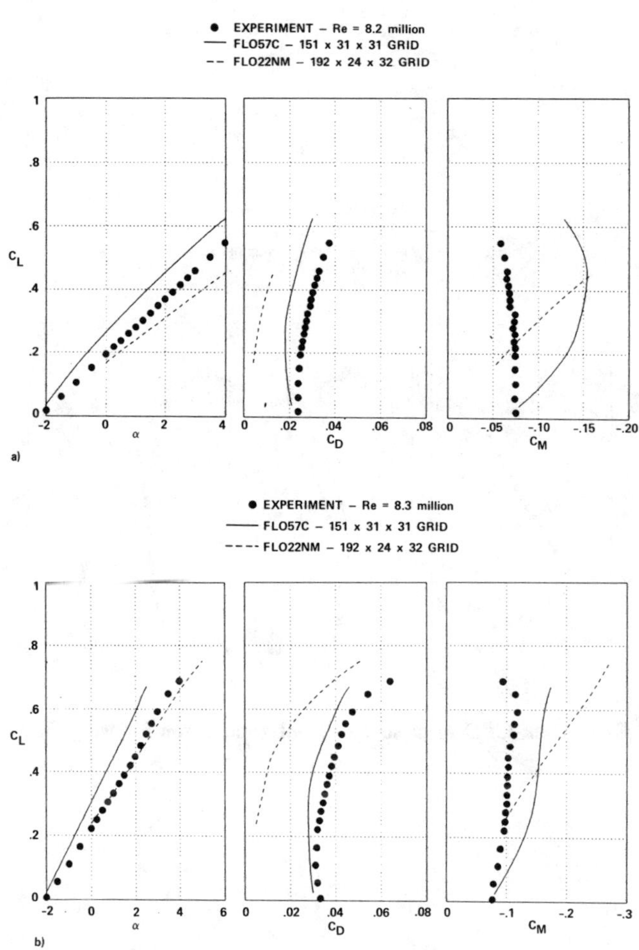

Fig. 31   Experiment-CFD force and moment comparison for wing D, FLO22NM: a) $M = 0.40$; b) $M = 0.80$.

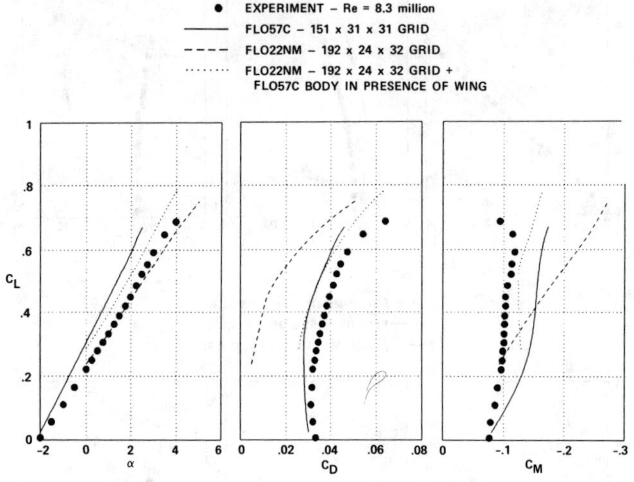

Fig. 32   Experiment-CFD force and moment comparison for wing D, $M = 0.80$.

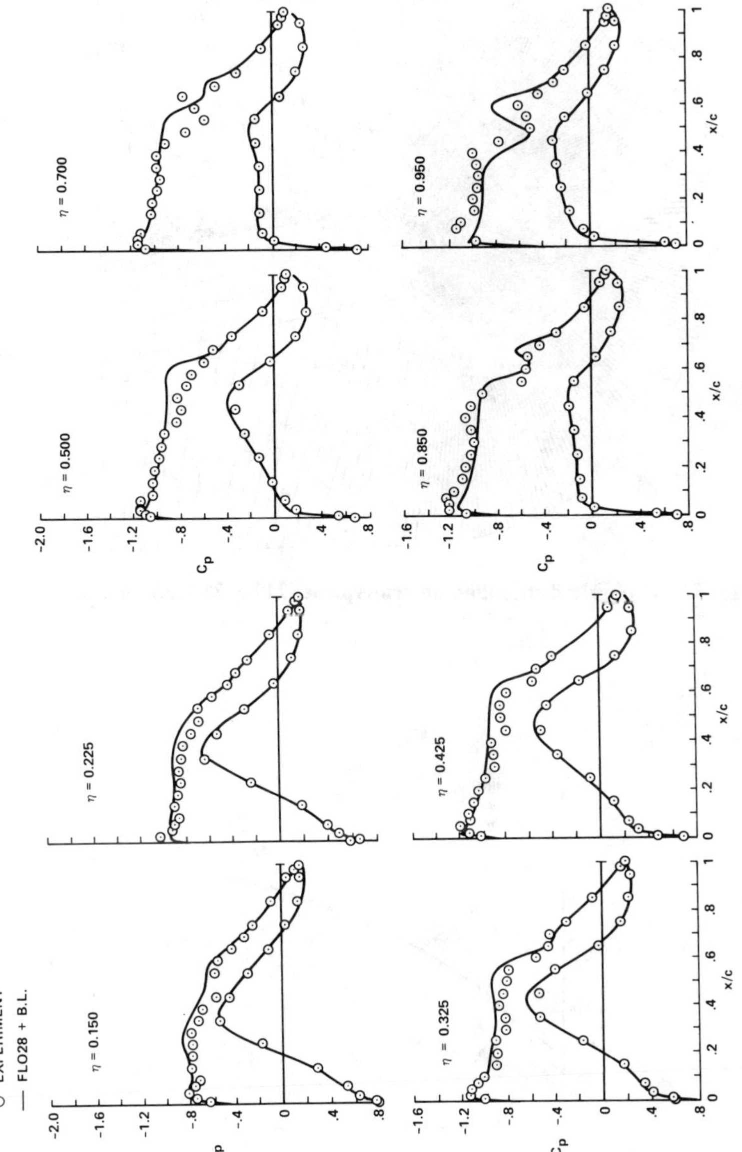

Fig. 33 Experiment-CFD pressure distribution comparison for wing D, $M = 0.80$, $C_L = 0.55$, FLO28 with boundary-layer correction.

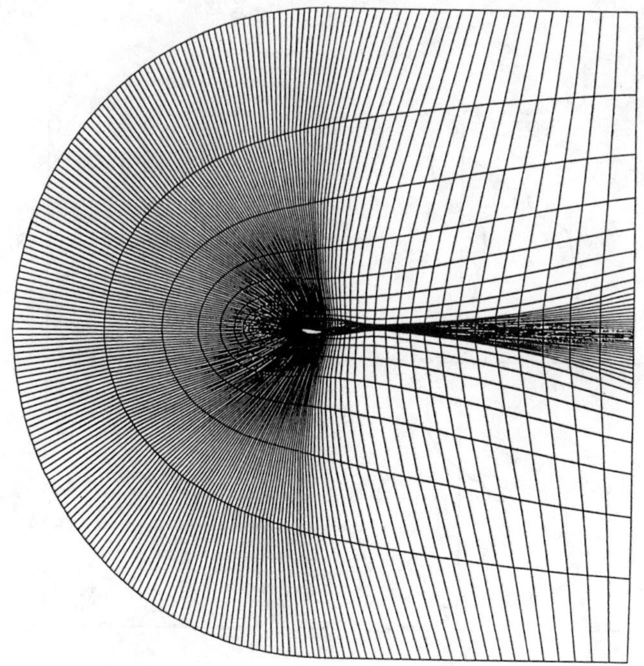

Fig. 34 C-H-grid distribution on centerplane, 217 × 25 × 33 points.

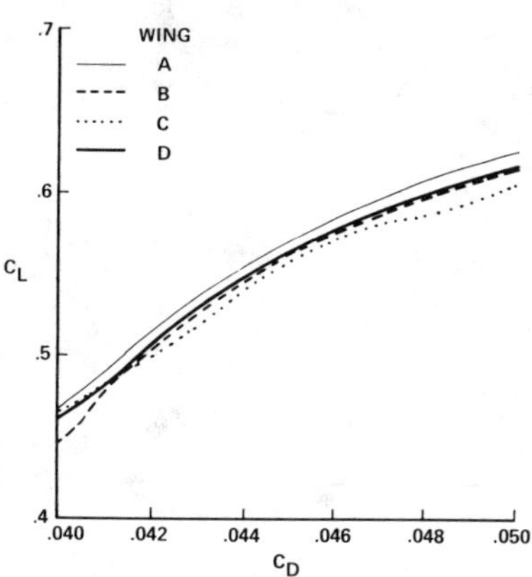

Fig. 35 Experimental drag comparison for transport wings, $M = 0.80$, $Re = 8.3 \times 10^6$.

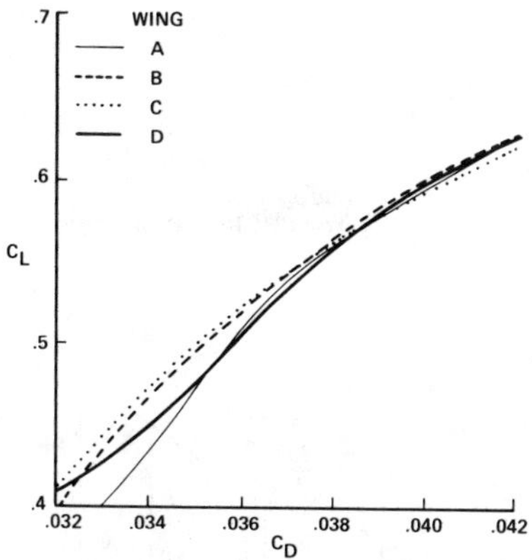

Fig. 36   Computational drag comparison for transport wings, $M = 0.80$, FLO57C.

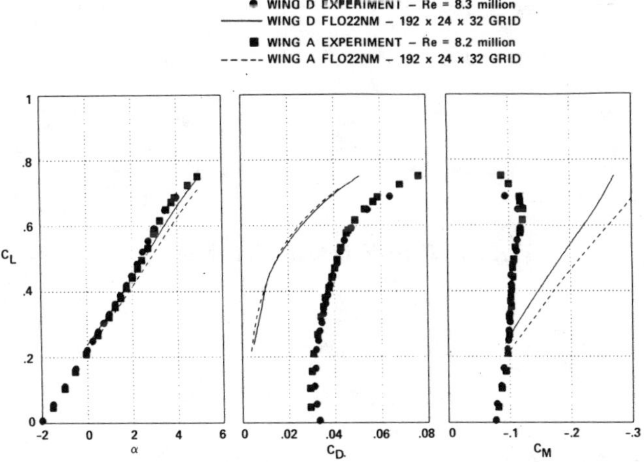

Fig. 37   Experiment-CFD force and moment comparison for wing A and wing D, $M = 0.80$, FLO22NM.

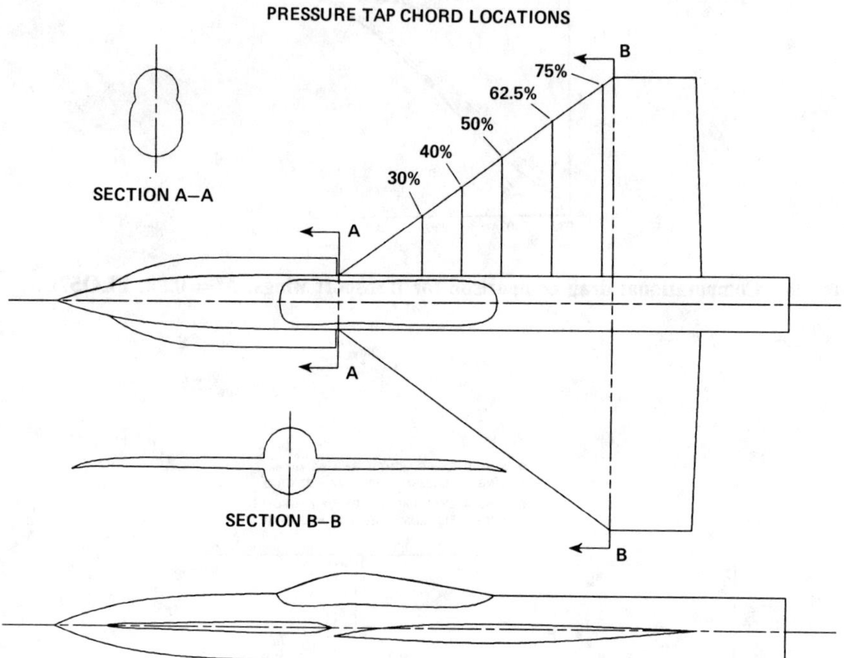

Fig. 38 Generic fighter with chine.

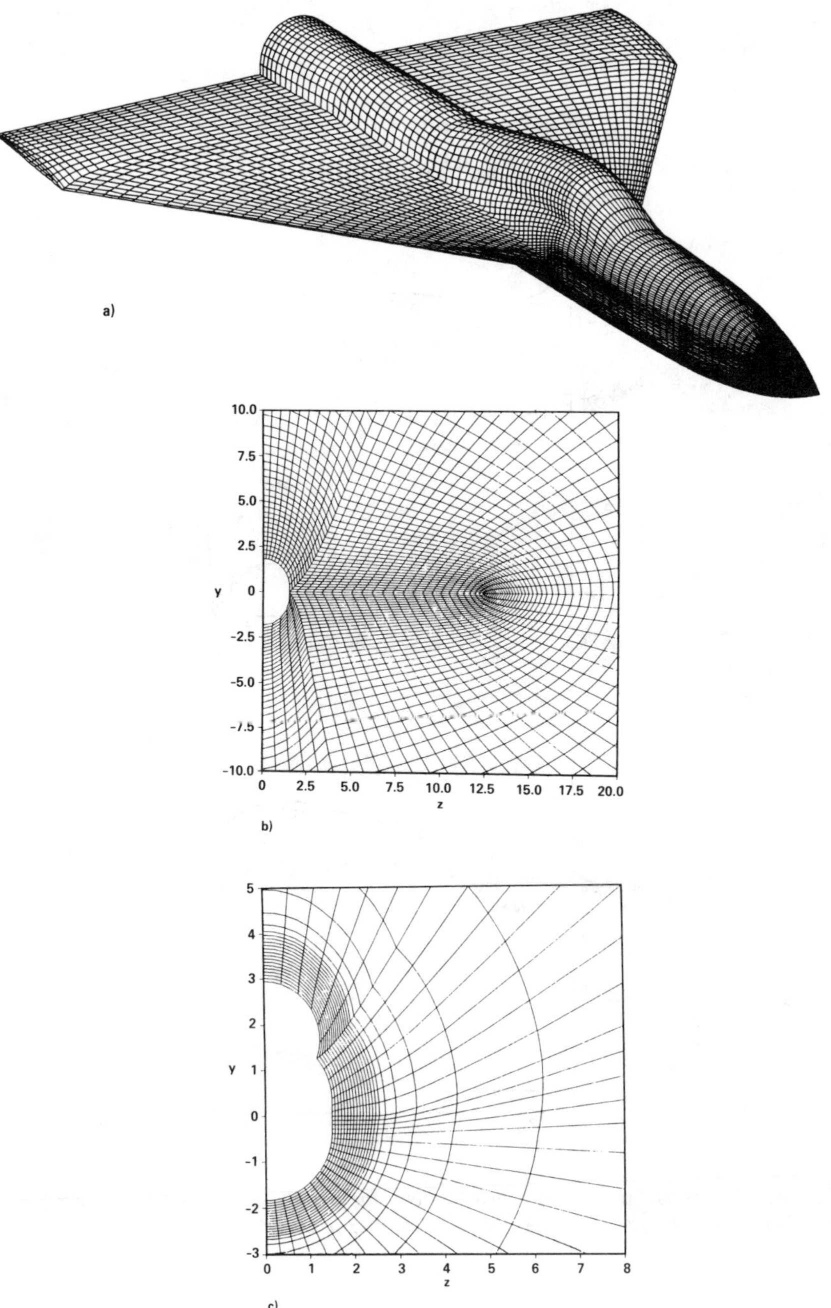

Fig. 39 Generic fighter wing/body grid, $134 \times 49 \times 65$ points: a) surface grid; b) wing trailing-edge plane; c) chine trailing-edge plane.

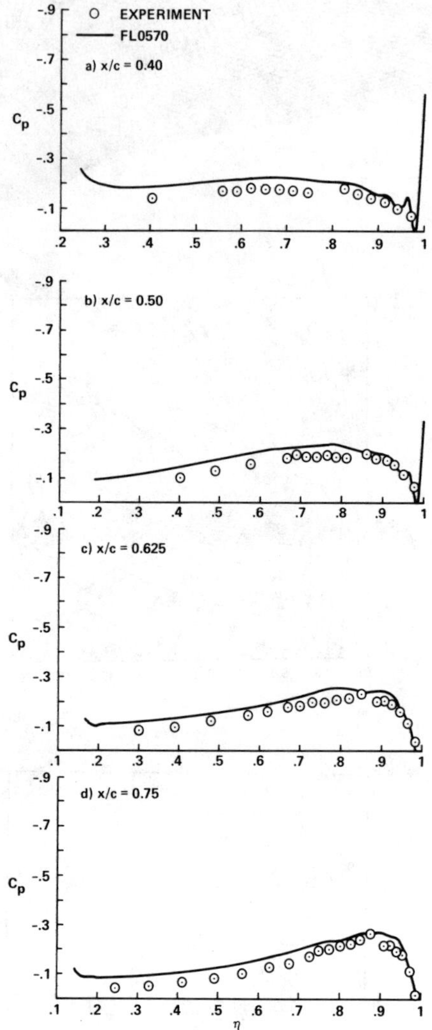

Fig. 40   Upper-surface pressure distributions for wing/body, $M = 0.6$, $\alpha = 4.0$ deg: a) $x/c = 0.40$; b) $x/c = 0.50$; c) $x/c = 0.625$; d) $x/c = 0.75$.

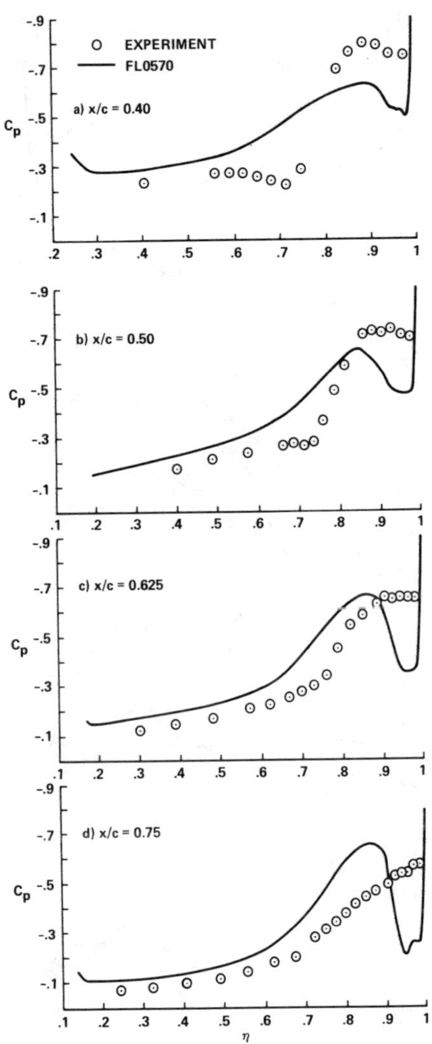

Fig. 41 Upper-surface pressure distributions for wing/body, $M = 0.6$, $\alpha = 8.0$ deg: a) $x/c = 0.40$; b) $x/c = 0.50$; c) $x/c = 0.625$; d) $x/c = 0.75$.

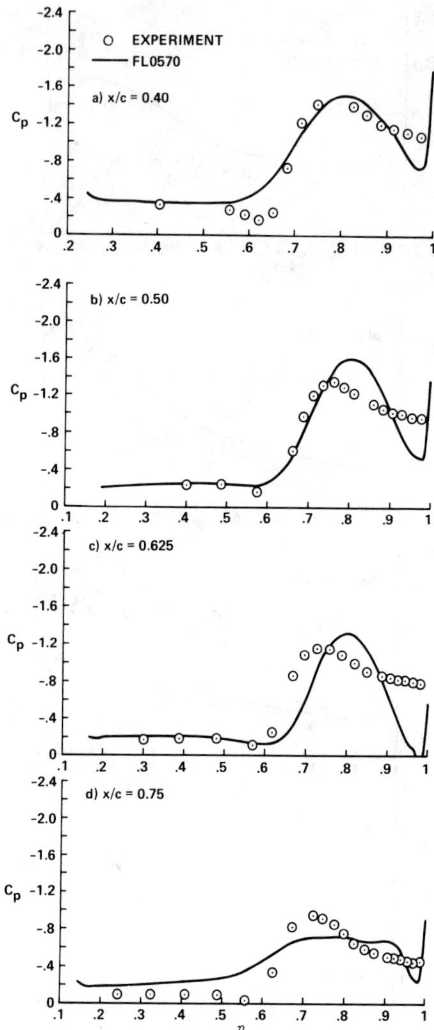

Fig. 42 Upper-surface pressure distributions for wing/body, $M = 0.6$, $\alpha = 12.0$ deg: a) $x/c = 0.40$; b) $x/c = 0.50$; c) $x/c = 0.625$; d) $x/c = 0.75$.

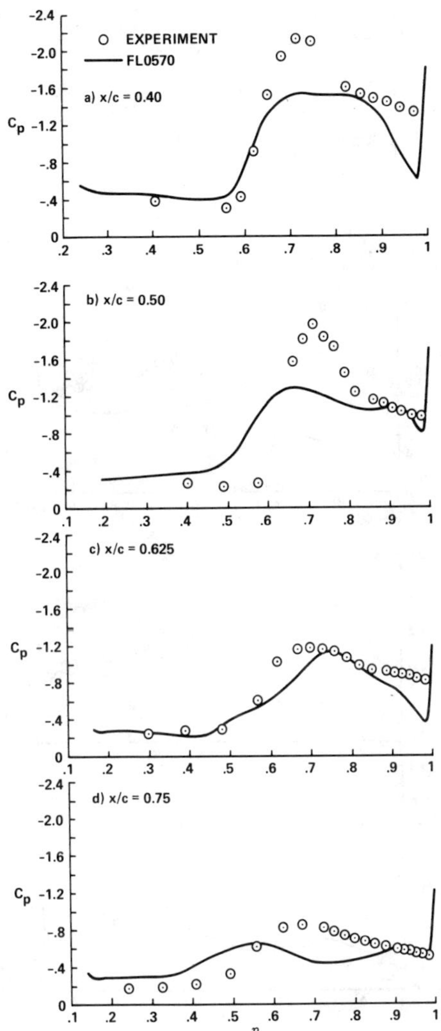

Fig. 43 Upper-surface pressure distributions for wing/body, $M = 0.6$, $\alpha = 16.0$ deg: a) $x/c = 0.40$; b) $x/c = 0.50$; c) $x/c = 0.625$; d) $x/c = 0.75$.

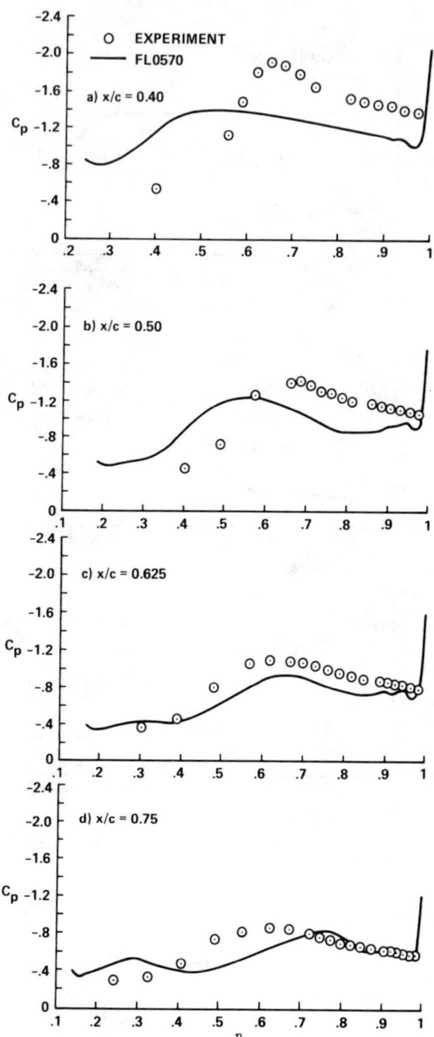

Fig. 44 Upper-surface pressure distributions for wing/body, $M = 0.6$, $\alpha = 20.0$ deg: a) $x/c = 0.40$; **b)** $x/c = 0.50$; **c)** $x/c = 0.625$; **d)** $x/c = 0.75$.

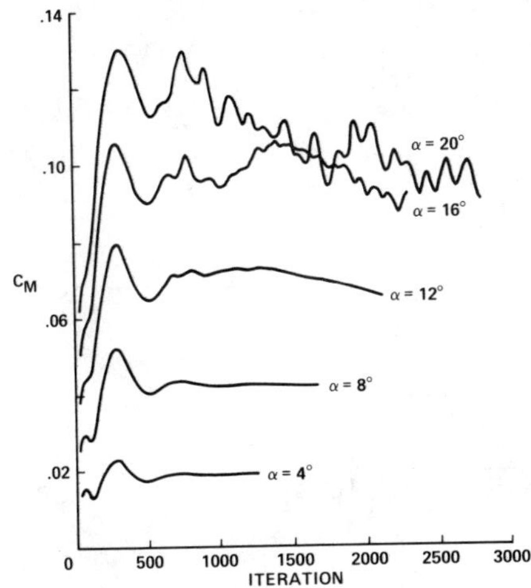

Fig. 45　Pitching-moment history for wing/body, $M = 0.60$.

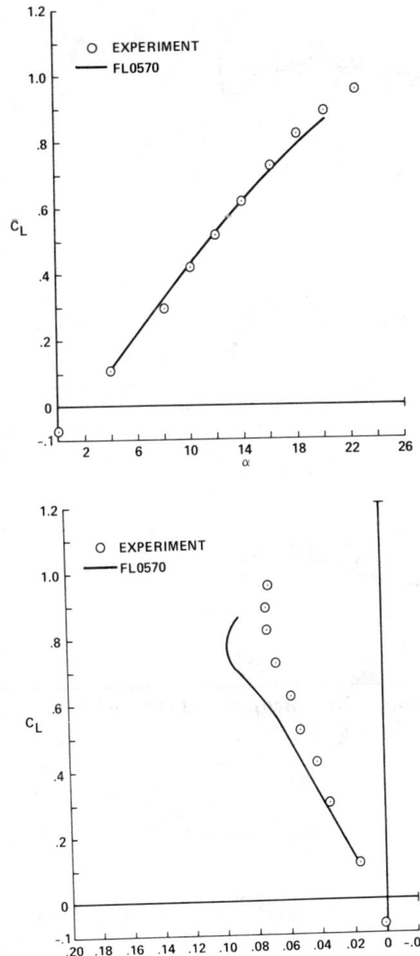

Fig. 46　Experiment-CFD forces and moments for wing/body, FLO57O, $M = 0.60$.

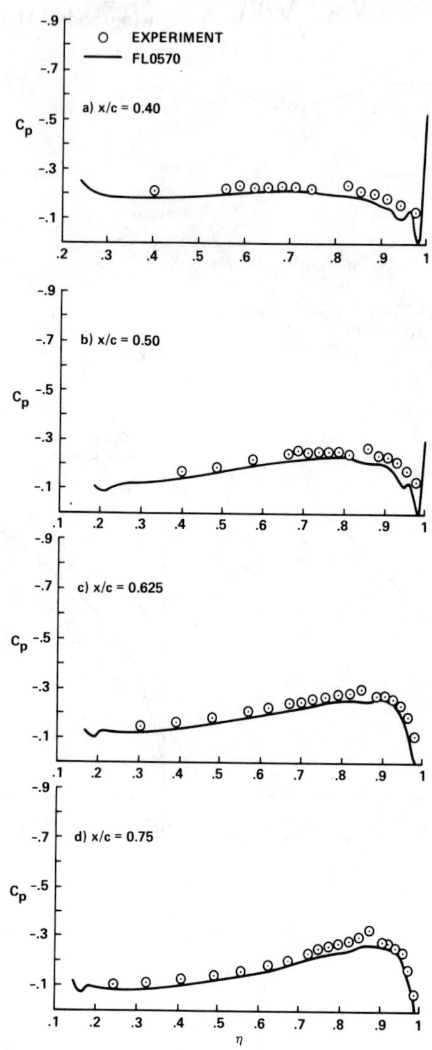

Fig. 47 Upper-surface pressure distributions for wing/body/chine, $M = 0.6$, $\alpha = 4.0$ deg: a) $x/c = 0.40$; b) $x/c = 0.50$; c) $x/c = 0.625$; d) $x/c = 0.75$.

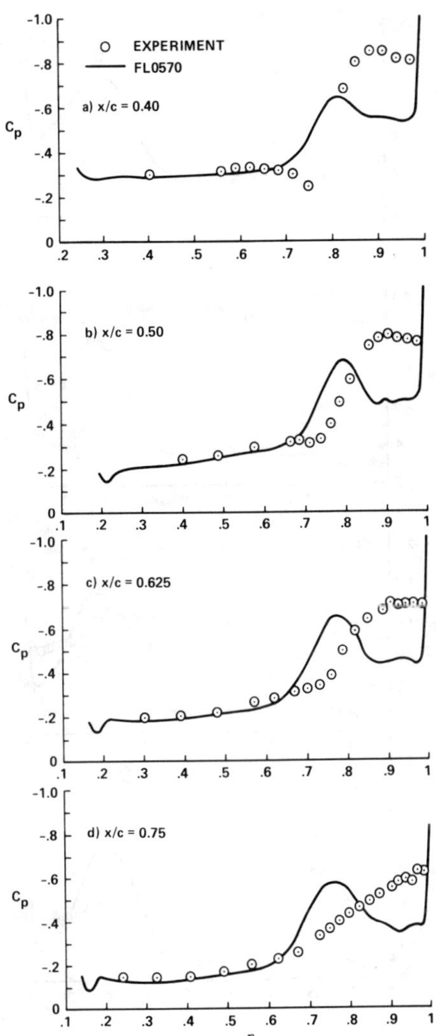

Fig. 48 Upper-surface pressure distributions for wing/body/chine, $M = 0.6$, $\alpha = 8.0$ deg: a) $x/c = 0.40$; b) $x/c = 0.50$; c) $x/c = 0.625$; d) $x/c = 0.75$.

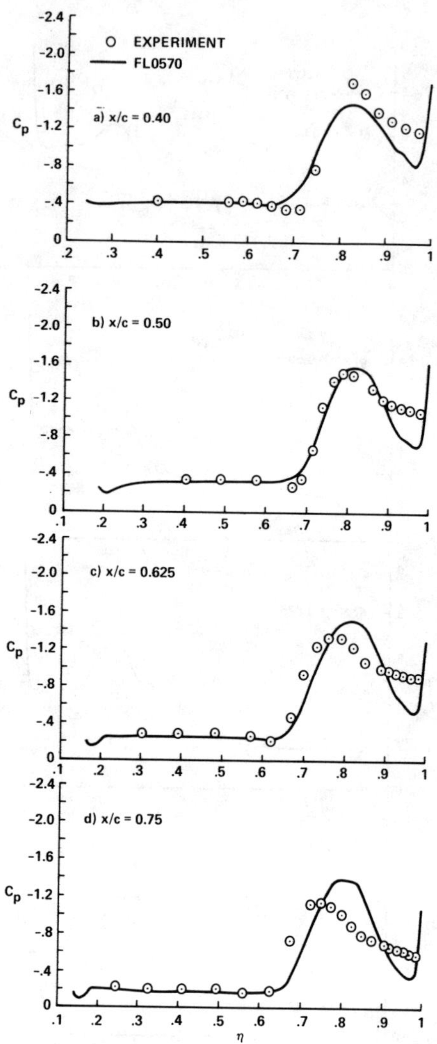

Fig. 49 Upper-surface pressure distributions for wing/body/chine, $M = 0.6$, $\alpha = 12.0$ deg: a) $x/c = 0.40$; b) $x/c = 0.50$; c) $x/c = 0.625$; d) $x/c = 0.75$.

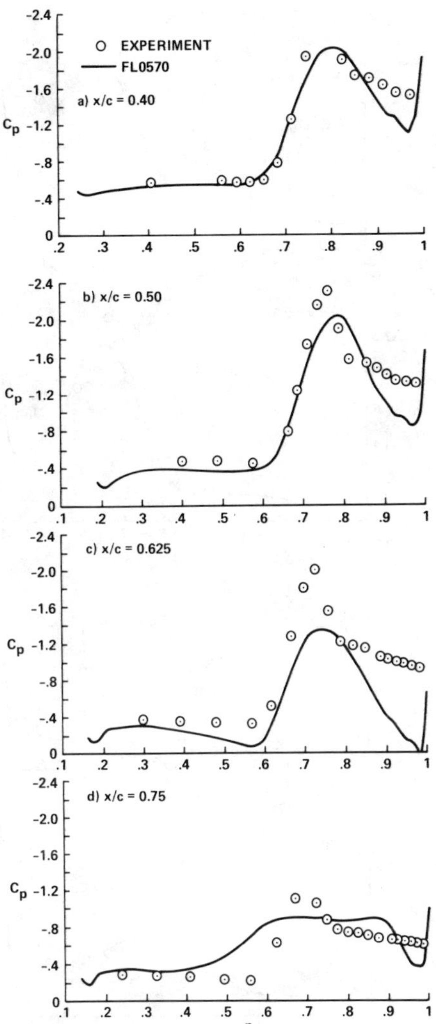

Fig. 50 Upper-surface pressure distributions for wing/body/chine, $M = 0.6$, $\alpha = 16.0$ deg: a) $x/c = 0.40$; b) $x/c = 0.50$; c) $x/c = 0.625$; d) $x/c = 0.75$.

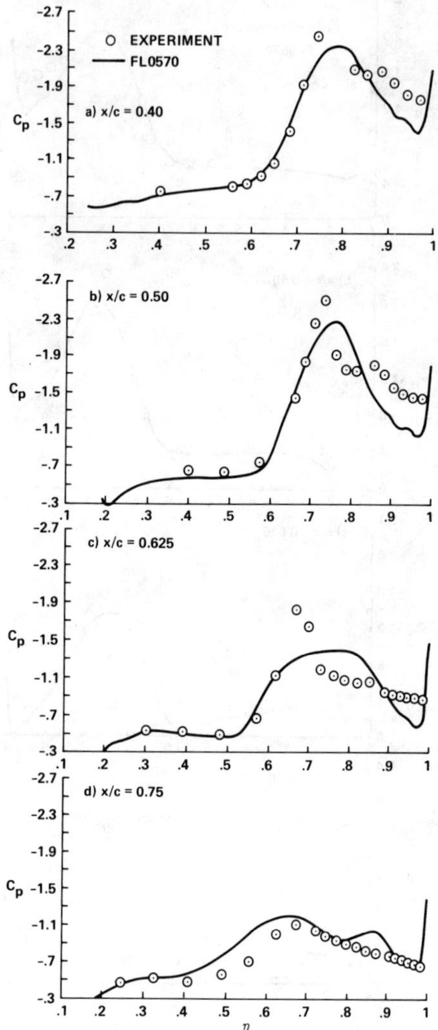

Fig. 51 Upper-surface pressure distributions for wing/body/chine, $M = 0.6$, $\alpha = 20.0$ deg: a) $x/c = 0.40$; b) $x/c = 0.50$; c) $x/c = 0.625$; d) $x/c = 0.75$.

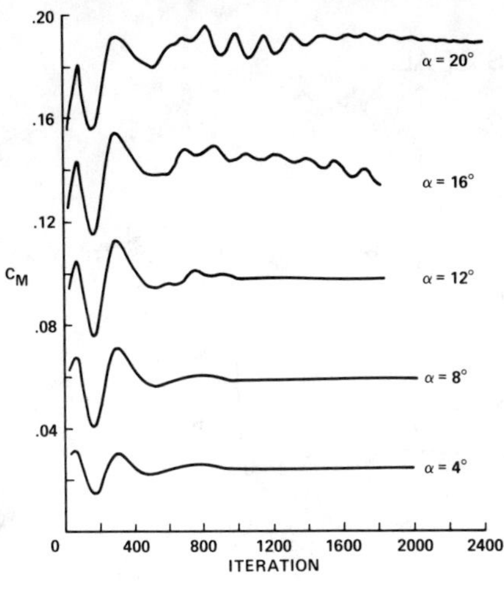

Fig. 52 Pitching-moment history for wing/body/chine, $M = 0.60$.

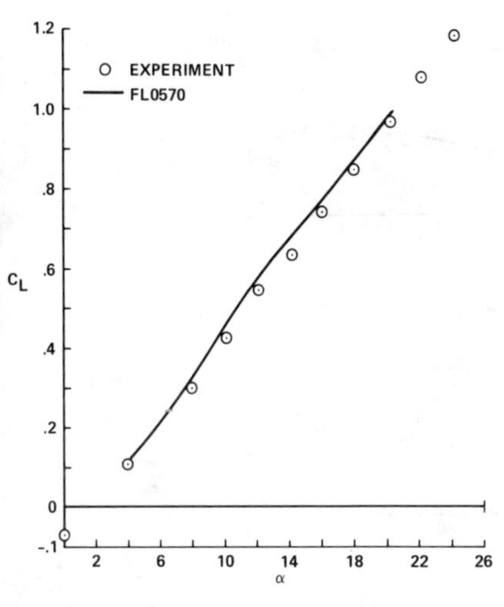

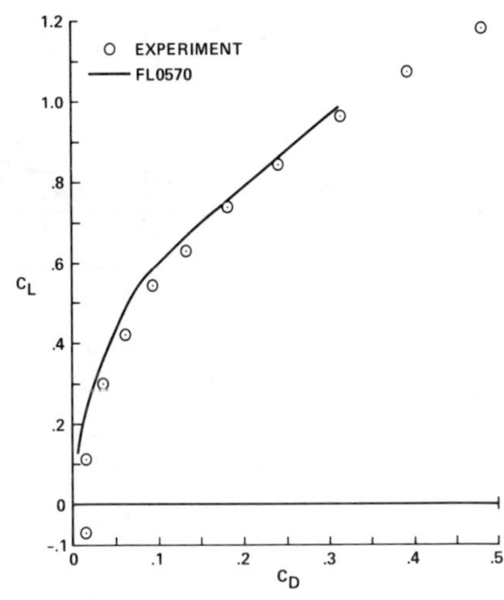

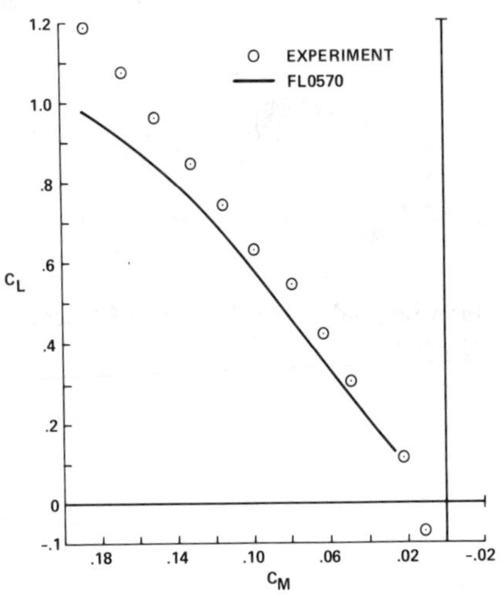

Fig. 53 Experiment-CFD forces and moments for wing/body/chine, $M = 0.60$.

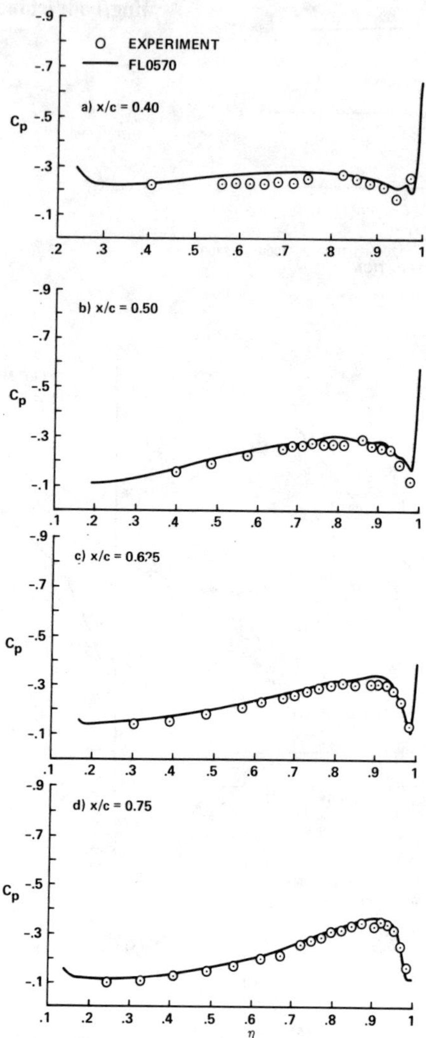

Fig. 54   Upper-surface pressure distributions for wing/body, $M = 0.80$, $\alpha = 5.0$ deg: a) $x/c = 0.40$; b) $x/c = 0.50$; c) $x/c = 0.625$; d) $x/c = 0.75$.

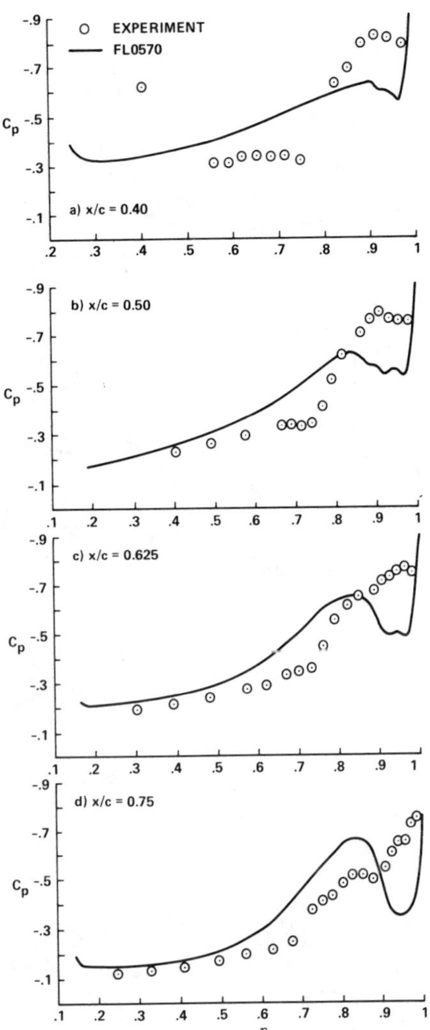

Fig. 55   Upper-surface pressure distributions for wing/body, $M = 0.80$, $\alpha = 8.0$ deg: a) $x/c = 0.40$; b) $x/c = 0.50$; c) $x/c = 0.625$; d) $x/c = 0.75$.

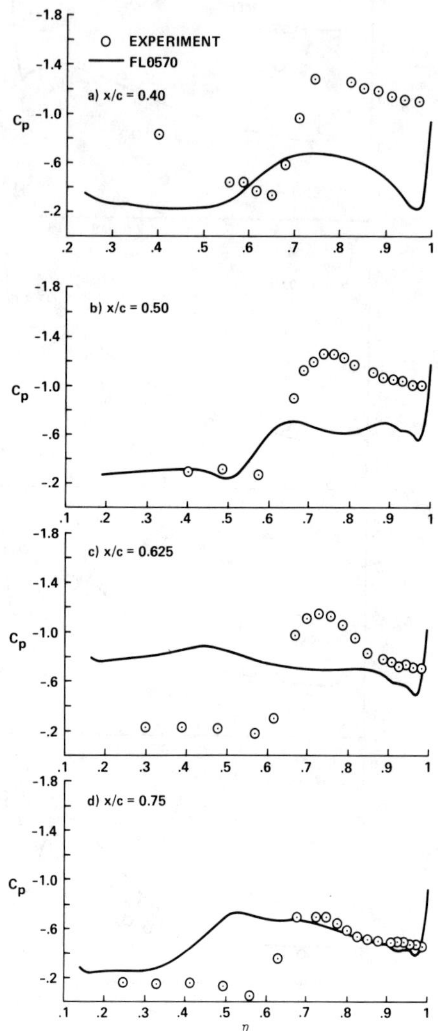

Fig. 56   Upper-surface pressure distributions for wing/body, $M = 0.80$, $\alpha = 12.0$ deg: a) $x/c = 0.40$; b) $x/c = 0.50$; c) $x/c = 0.625$; d) $x/c = 0.75$.

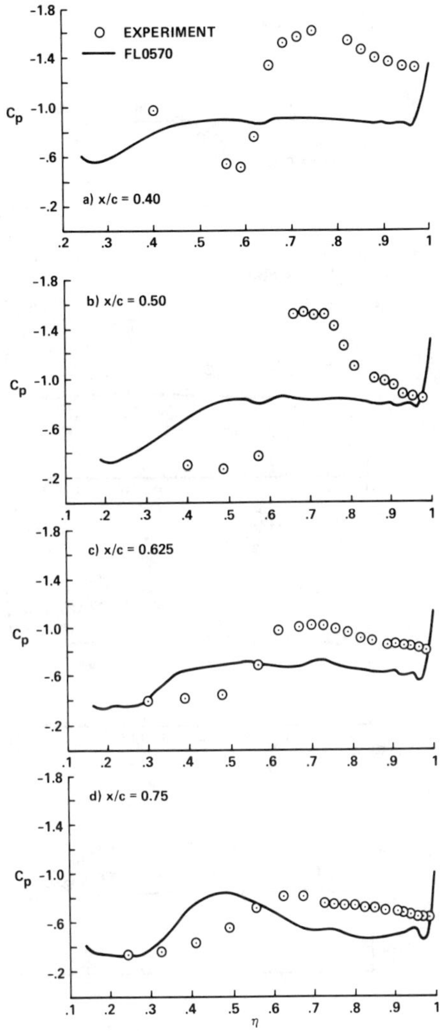

Fig. 57   Upper-surface pressure distributions for wing/body, $M = 0.80$, $\alpha = 16.0$ deg: a) $x/c = 0.40$; b) $x/c = 0.50$; c) $x/c = 0.625$; d) $x/c = 0.75$.

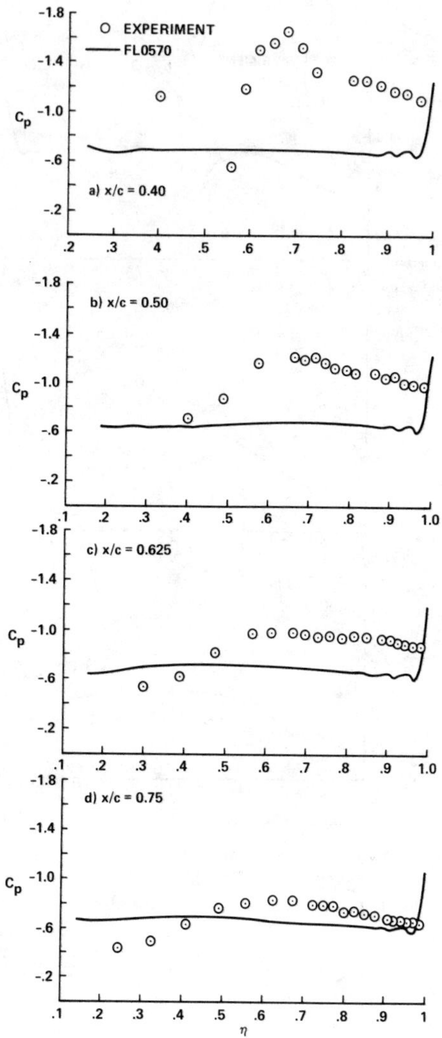

Fig. 58   Upper-surface pressure distributions for wing/body, $M = 0.80$, $\alpha = 20.0$ deg: a) $x/c = 0.40$; b) $x/c = 0.50$; c) $x/c = 0.625$; d) $x/c = 0.75$.

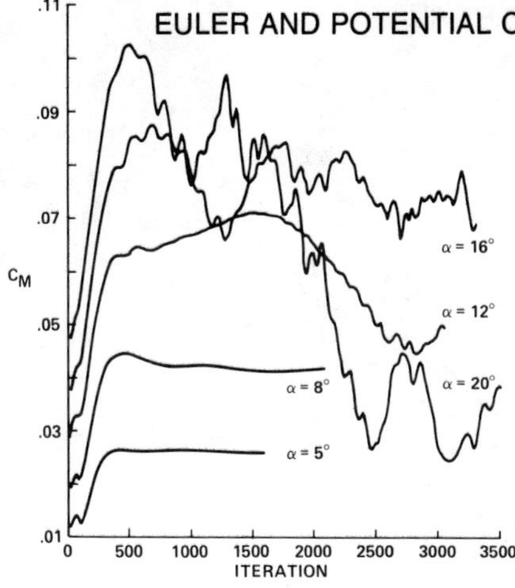

Fig. 59 Pitching-moment history for wing/body, $M = 0.80$.

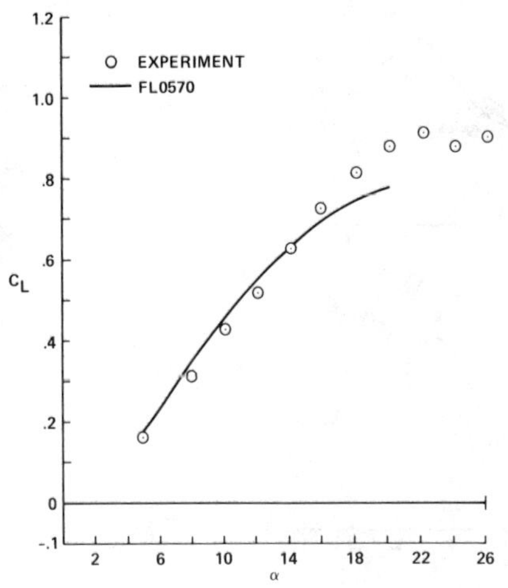

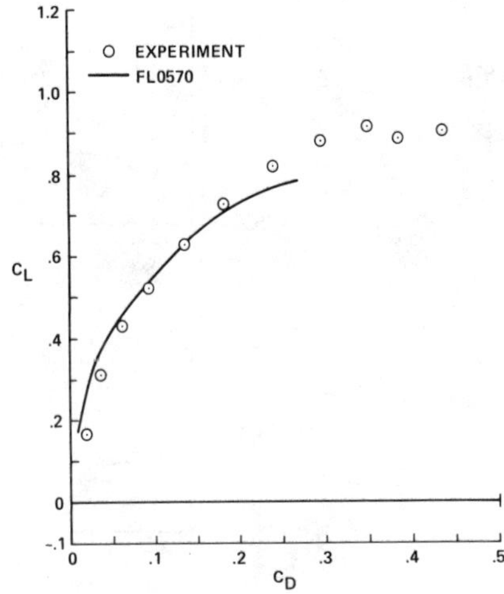

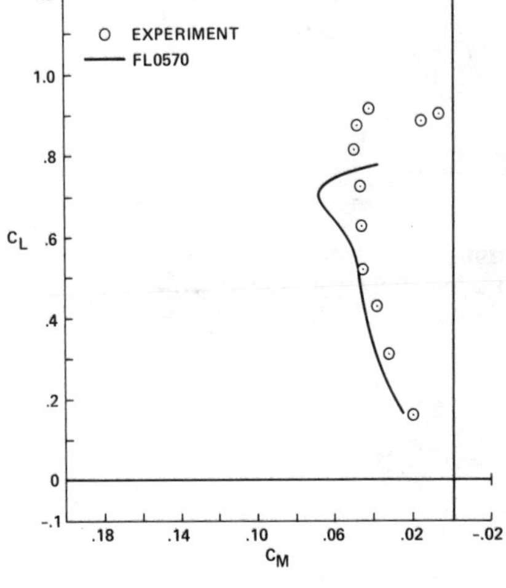

Fig. 60 Experiment-CFD forces and moments for wing/body, $M = 0.80$.

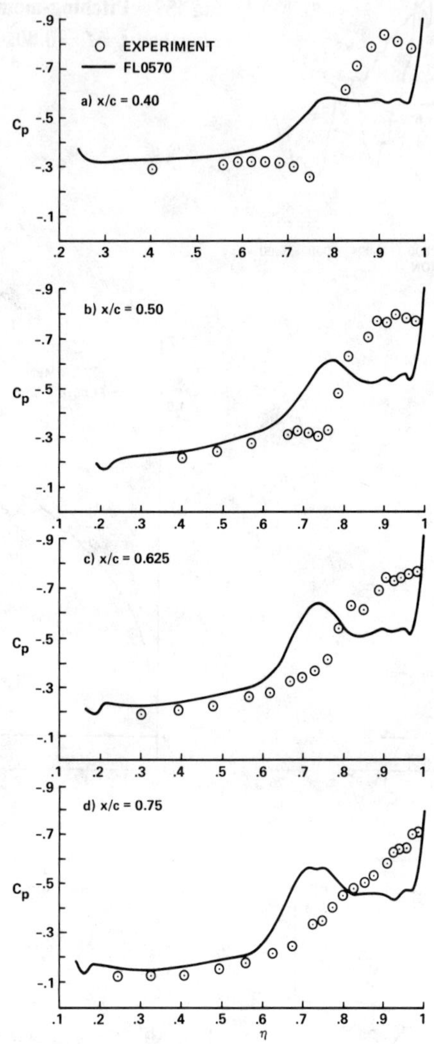

Fig. 61 Upper-surface pressure distributions for wing/body/chine, $M = 0.80$, $\alpha = 8.0$ deg: a) $x/c = 0.40$; b) $x/c = 0.50$; c) $x/c = 0.625$; d) $x/c = 0.75$.

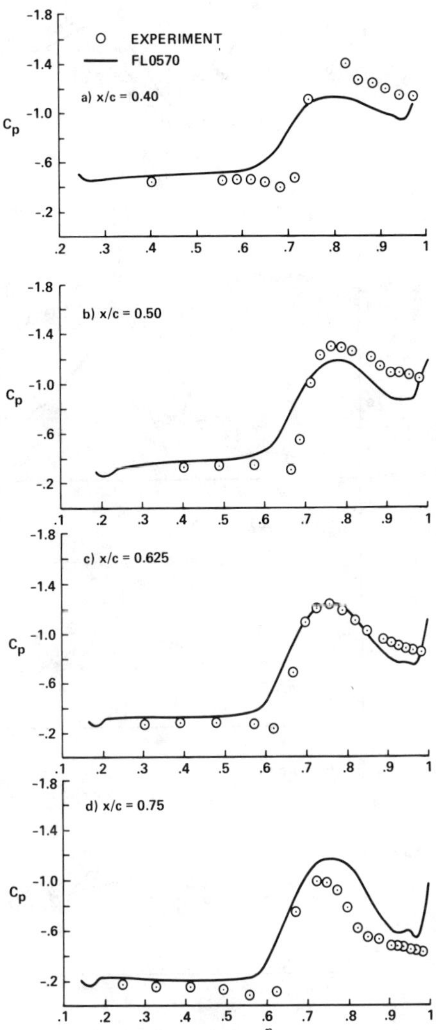

Fig. 62 Upper-surface pressure distributions for wing/body/chine, $M = 0.80$, $\alpha = 12.0$ deg: a) $x/c = 0.40$; b) $x/c = 0.50$; c) $x/c = 0.625$; d) $x/c = 0.75$.

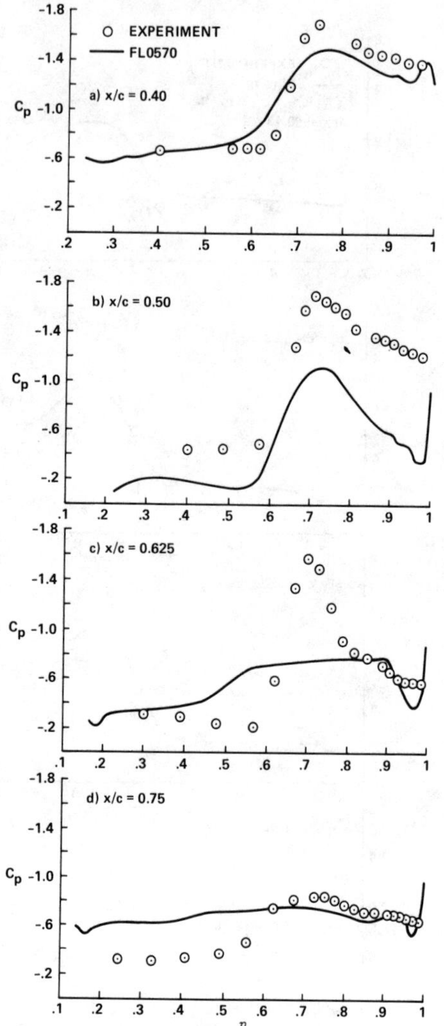

Fig. 63 Upper-surface pressure distributions for wing/body/chine, $M = 0.80$, $\alpha = 16.0$ deg: a) $x/c = 0.40$; b) $x/c = 0.50$; c) $x/c = 0.625$; d) $x/c = 0.75$.

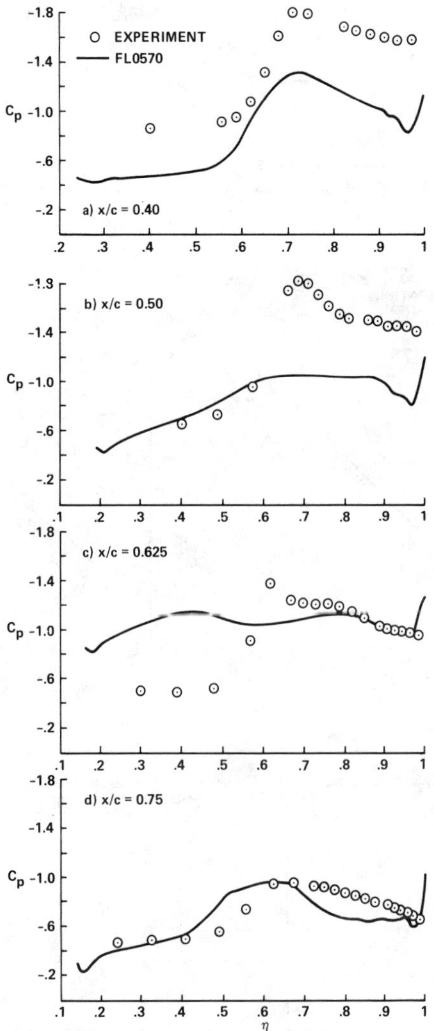

Fig. 64 Upper-surface pressure distributions for wing/body/chine, $M = 0.80$, $\alpha = 20.0$ deg: a) $x/c = 0.40$; b) $x/c = 0.50$; c) $x/c = 0.625$; d) $x/c = 0.75$.

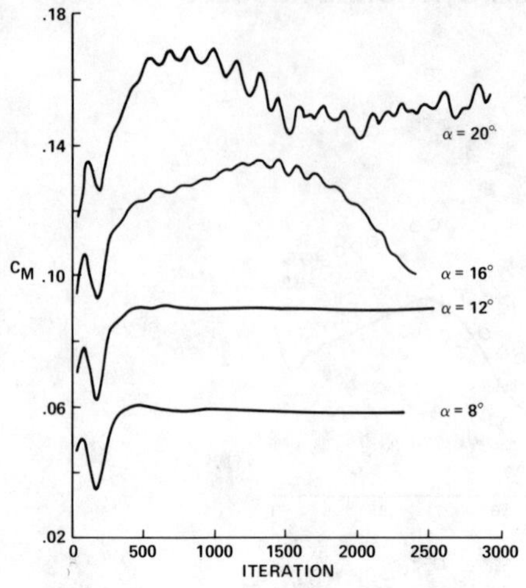

Fig. 65 **Pitching-moment history for wing/body/chine, $M = 0.80$.**

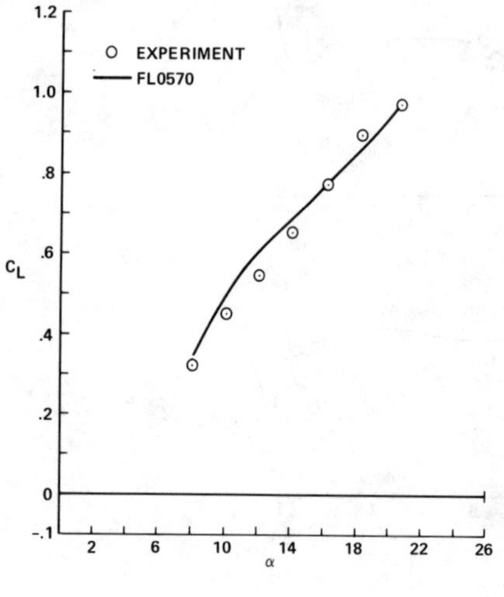

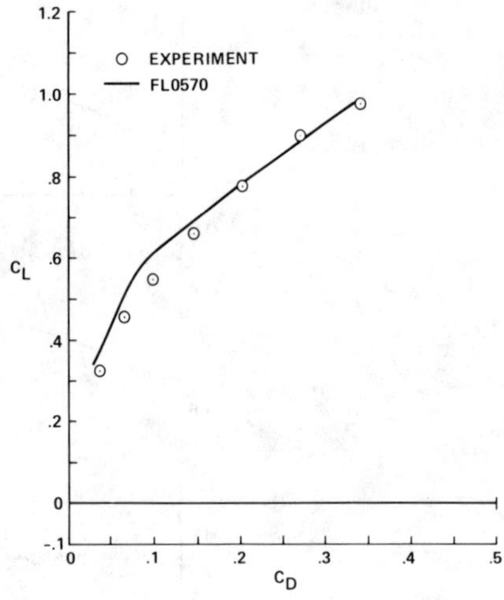

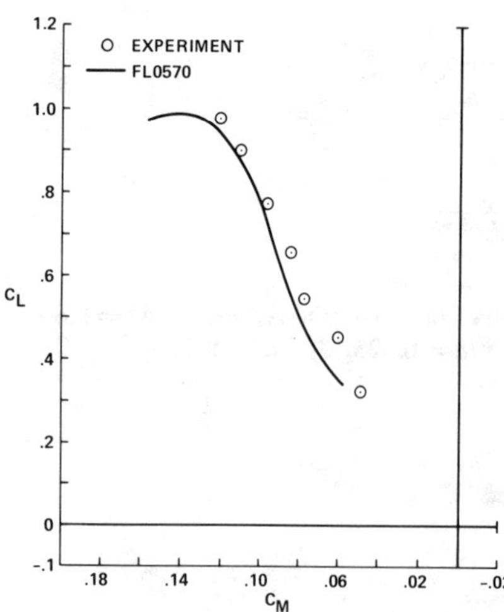

Fig. 66 **Experiment-CFD forces and moments for wing/body/chine, $M = 0.80$.**

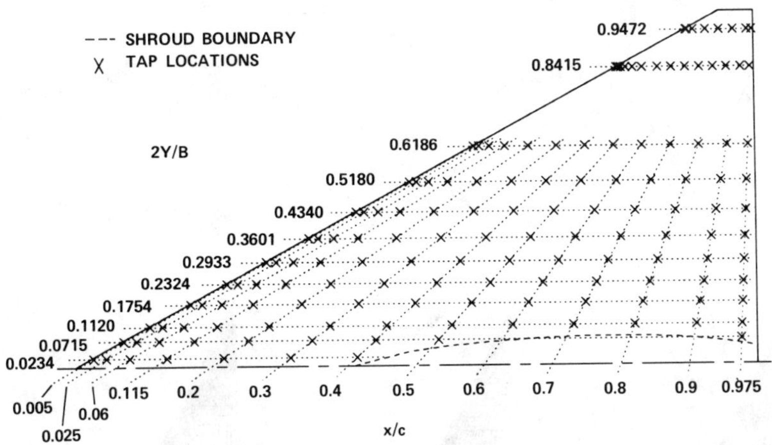

Fig. 67   Upper-surface plan view showing pressure tap locations.

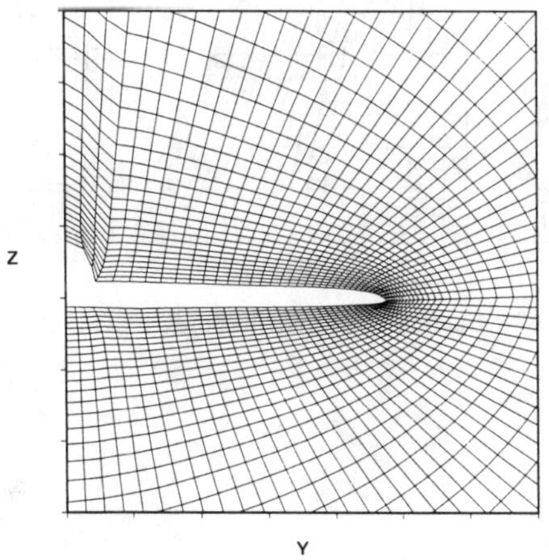

Fig. 68   Typical grid plane near trailing edge.

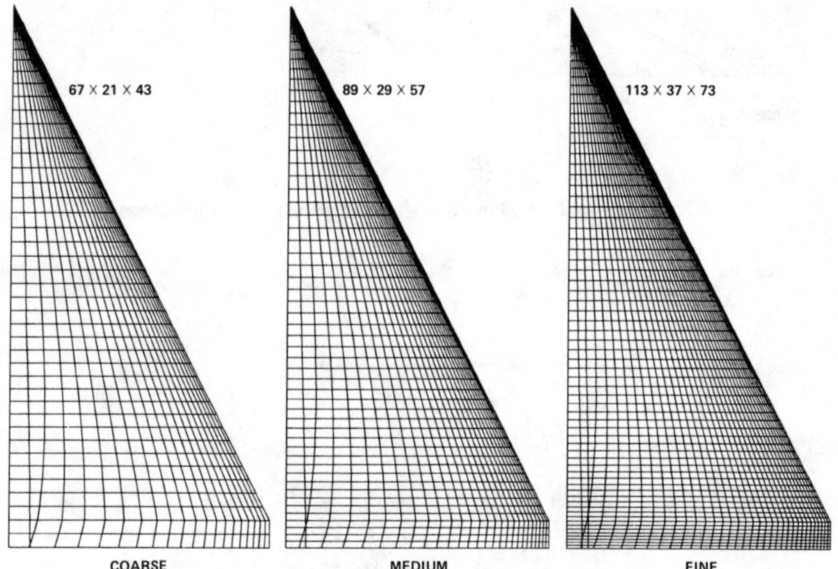

Fig. 69   Surface grid distributions, FLO57O.

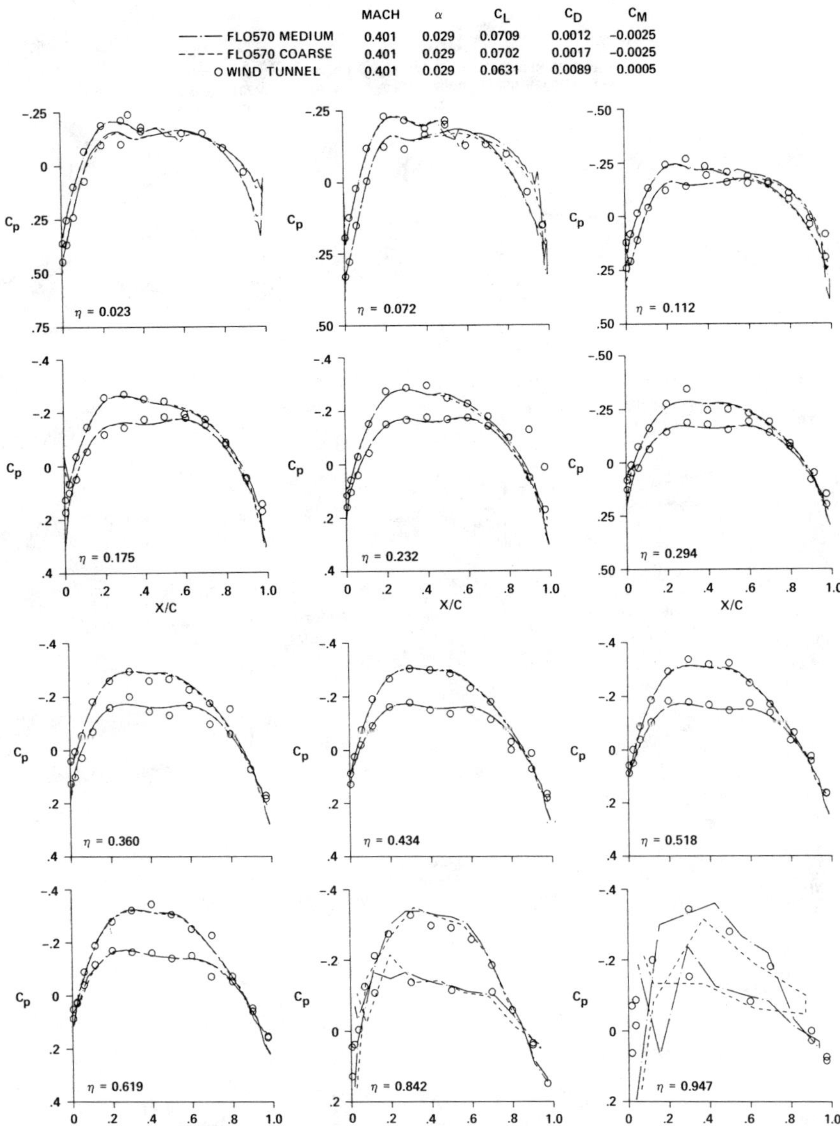

Fig. 70a Experiment-CFD pressure distribution comparison for delta wing, $M = 0.40$.

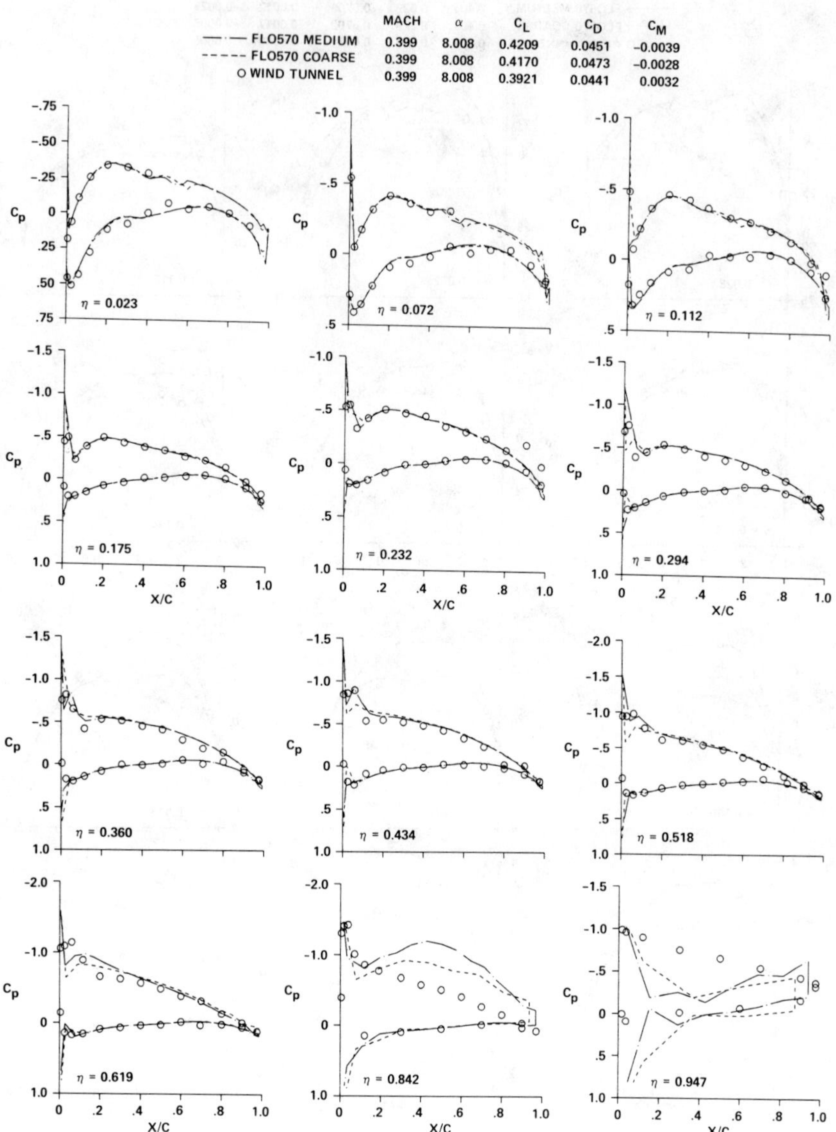

Fig. 70b   (cont'd) Experiment-CFD pressure distribution comparison for delta wing, $M = 0.40$.

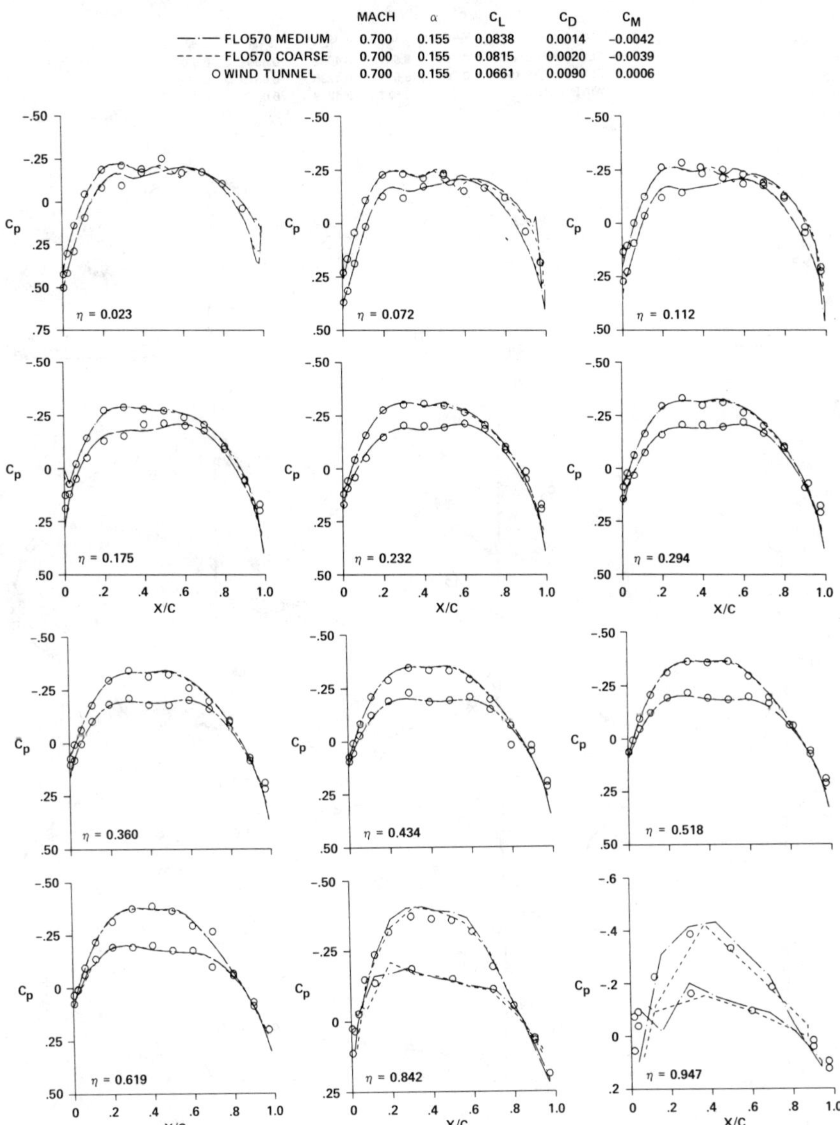

Fig. 71a Experiment-CFD pressure distribution comparison for delta wing, $M = 0.70$.

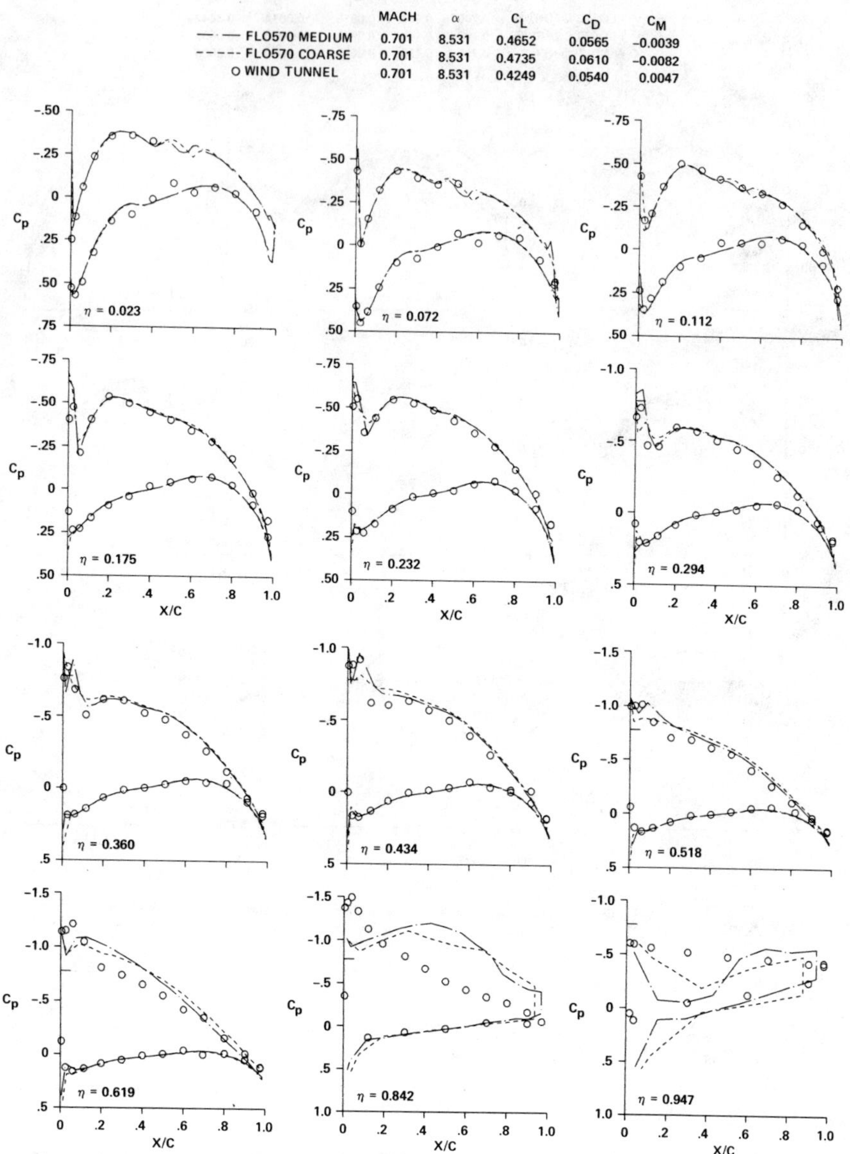

Fig. 71b   (cont'd) Experiment-CFD pressure distribution comparison for delta wing, $M = 0.70$.

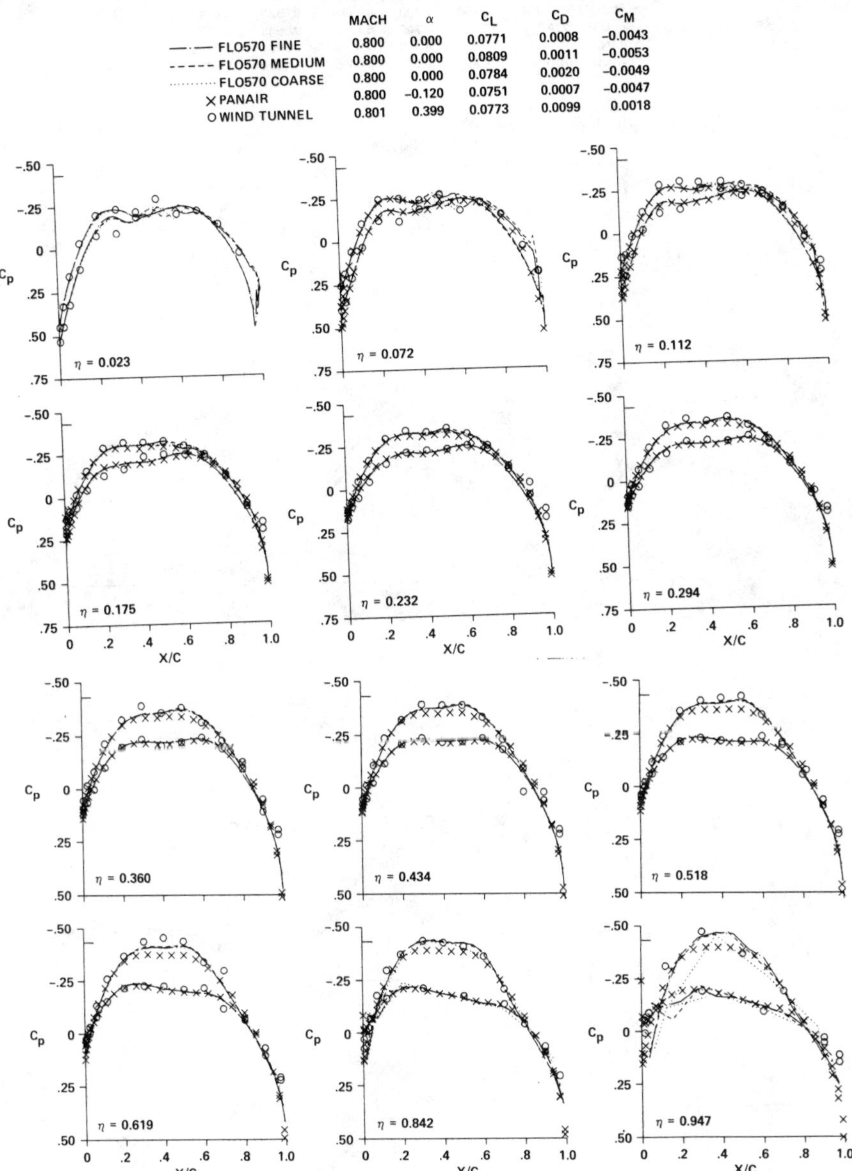

Fig. 72a Experiment-CFD pressure distribution comparison for delta wing, $M = 0.80$.

R. M. HICKS ET AL.

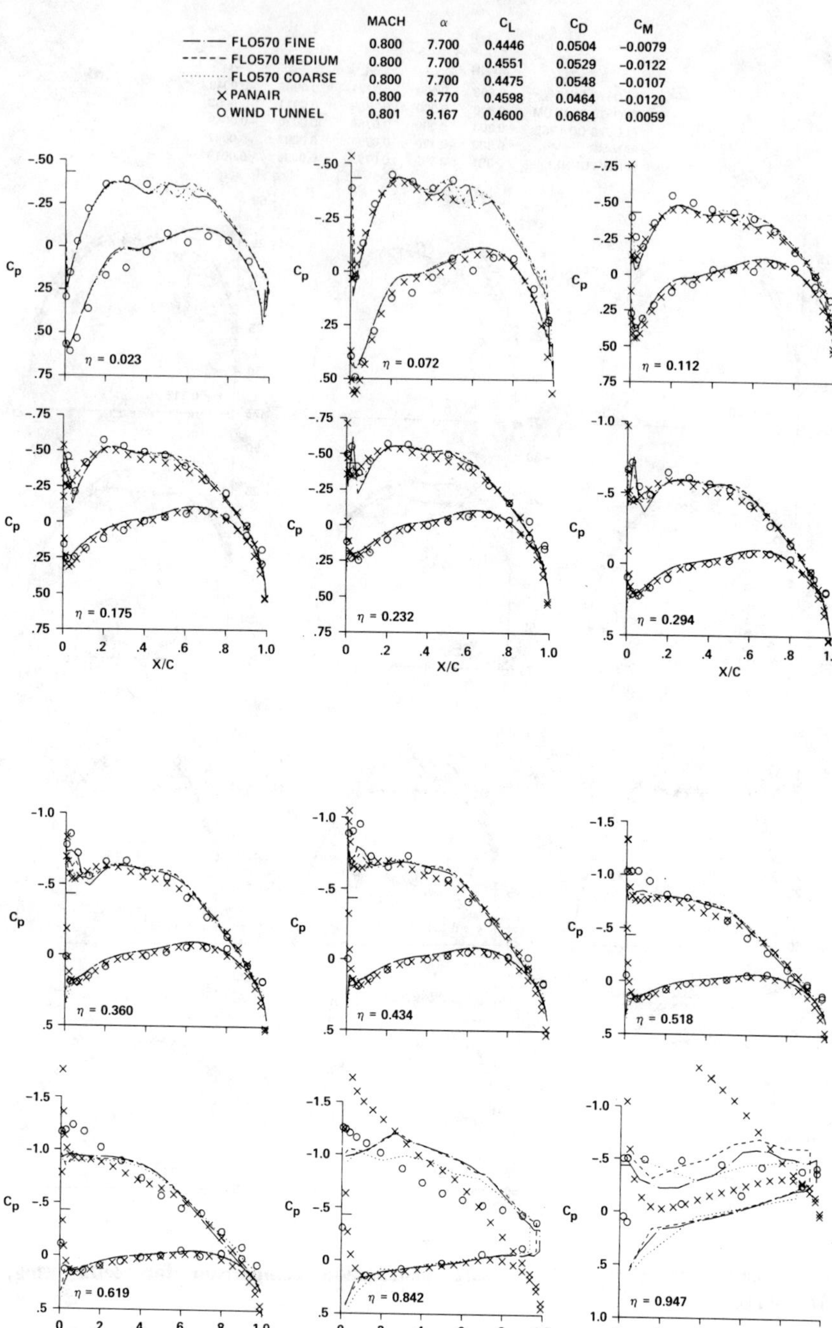

Fig. 72b (cont'd) Experiment-CFD pressure distribution comparison for delta wing, $M = 0.80$.

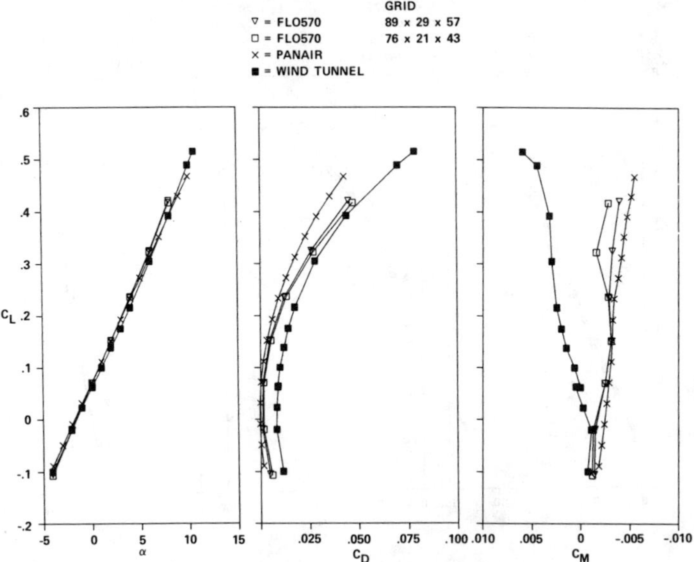

Fig. 73   Experiment-CFD force and moment comparisons for delta wing, $M = 0.40$.

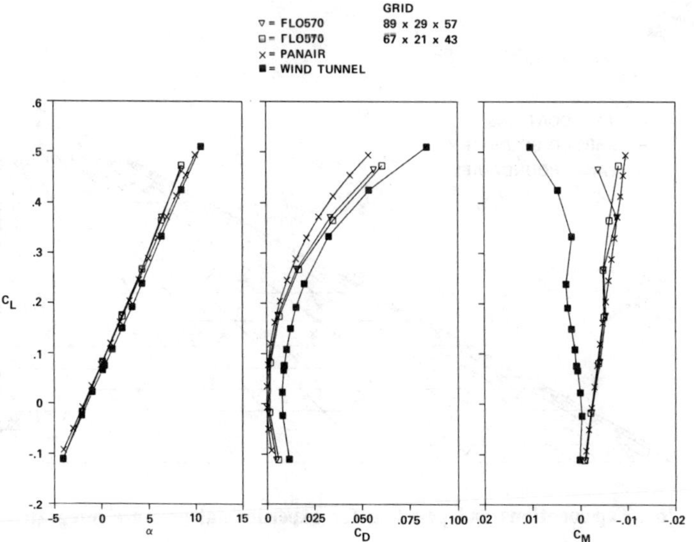

Fig. 74   Experiment-CFD force and moment comparisons for delta wing, $M = 0.70$.

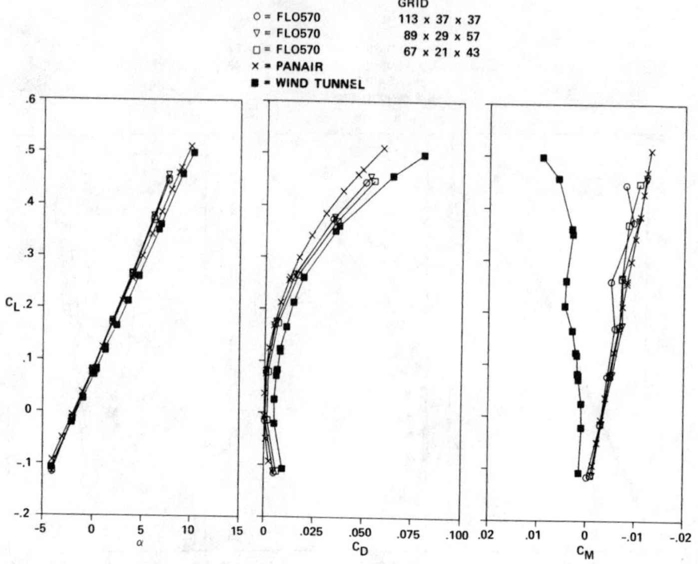

Fig. 75   Experiment-CFD force and moment comparisons for delta wing, $M = 0.80$.

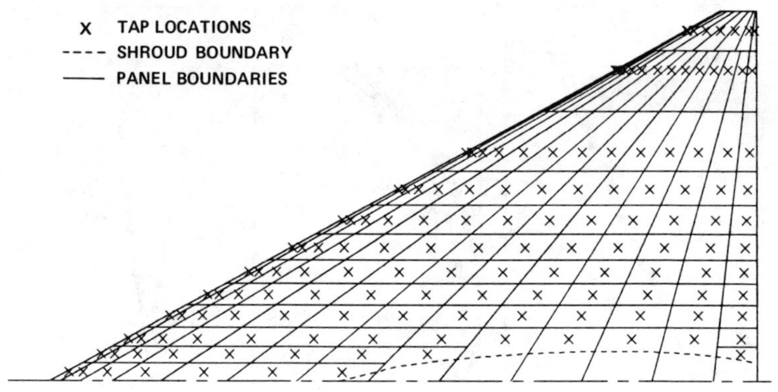

Fig. 76   Tap locations and paneling for experimental pressure integration.

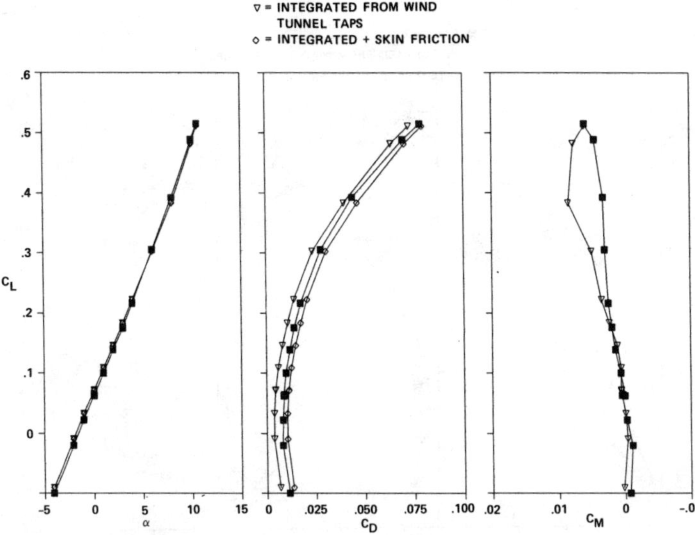

Fig. 77 Experimental forces and moments for delta wing, $M = 0.40$.

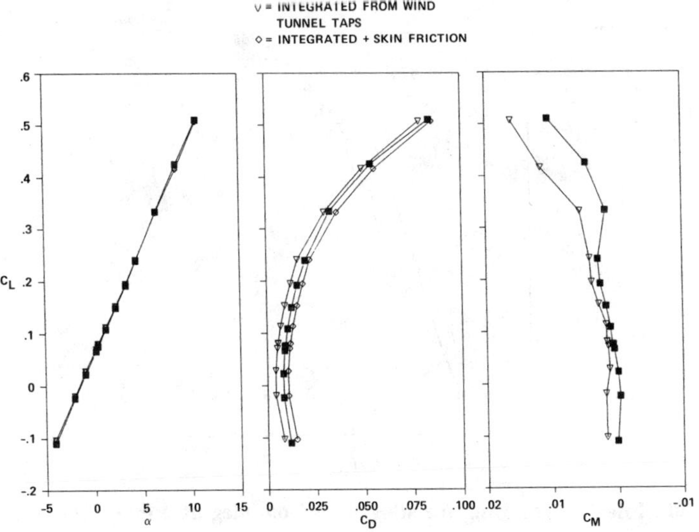

Fig. 78 Experimental forces and moments for delta wing, $M = 0.70$.

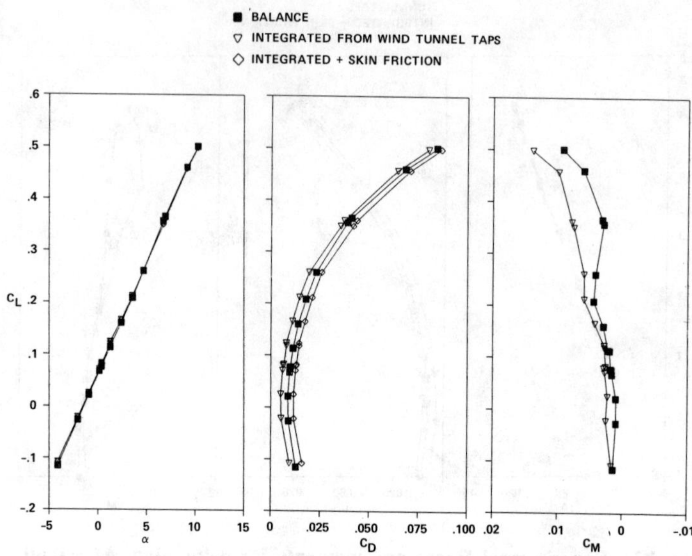

Fig. 79   Experimental forces and moments for delta wing, $M = 0.80$.

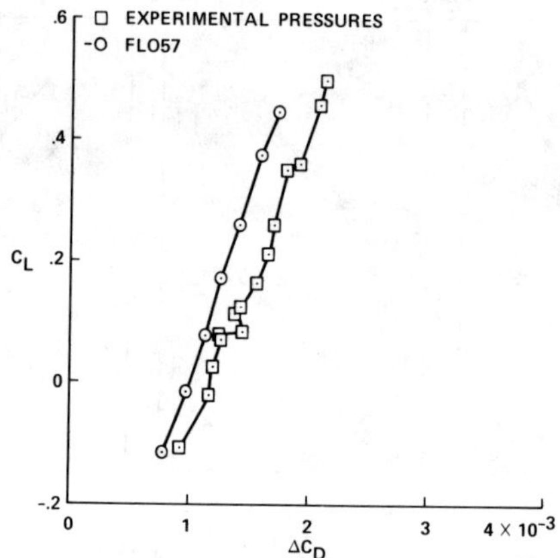

Fig. 80   Effect of removing the sting shroud on integrated drag, $M = 0.80$.

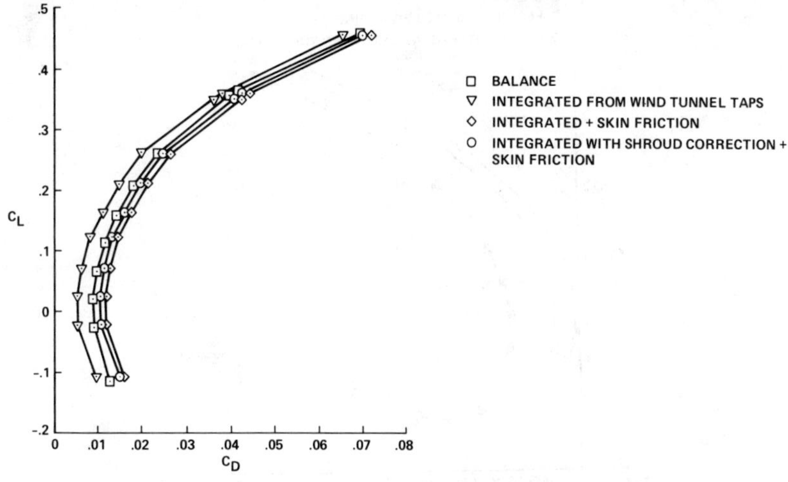

Fig. 81   Shroud increment removed from the $M = 0.80$ drag polar.

Fig. 82   Computational and experimental grid comparison, FLO57O.

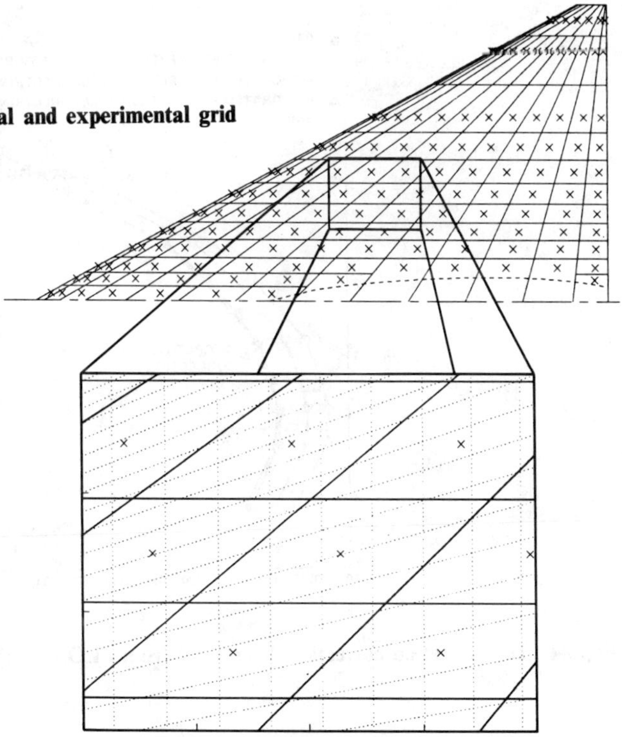

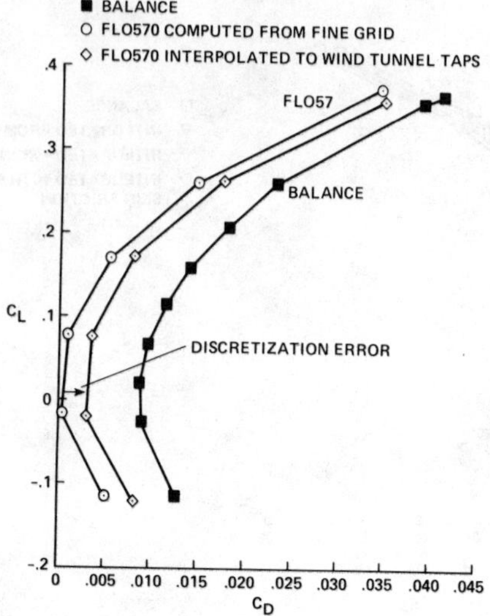

Fig. 83　Discretization effect of drag calculated from FLO57O, $M = 0.80$.

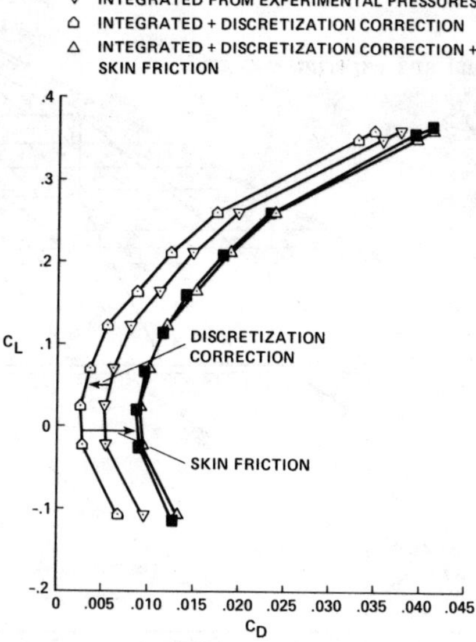

Fig. 84　Discretization correction determined from FLO57O applied to experimental
data, $M = 0.80$.

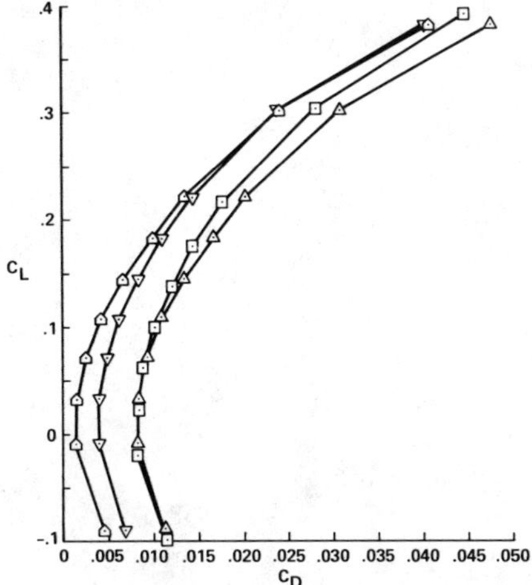

Fig. 85 Discretization correction determined from FLO57O applied to experimental data, $M = 0.40$.

Part 4.   High-Lift Systems

Chapter 10

Computational Aerodynamics Applied to High-Lift Systems

G. W. Brune* and J. H. McMasters*
The Boeing Company, Seattle, Washington

Nomenclature

b = wing span
c = basic cruise wing or airfoil chord
c_d, c_l, c_m = two-dimensional (section) drag, lift, and pitching moment coefficients, respectively, force/qc and moment/qc^2
C_D, C_L, C_M = three-dimensional configuration drag, lift, and pitching-moment coefficient, force/qS and moment/qSc
C_p = pressure coefficient, $\Delta p/q_\infty = (p_{local} - p_\infty)/q_\infty$
H = boundary-layer form parameter, δ^*/θ
L/D = ratio of lift to drag
M = Mach number
p = static pressure
q = dynamic pressure
Re = Reynolds number
S = wing area
s = arc length
t = airfoil thickness
V = total velocity
x = longitudinal coordinate
y = lateral coordinate
α = angle of attack
δ_f = flap-deflection angle
δ^* = boundary-layer displacement thickness
θ = boundary-layer momentum thickness
η = nondimensional spanwise wing station, $2y/b$
Λ = sweep angle measured at wing quarterchord

Subscripts
c = based on chord length
eff = effective viscous condition
exp = experimental value

Copyright © 1989 by the American Institute of Aeronautics and Astronautics, Inc. All rights reserved.
*Principal Engineer, Aerodynamics Engineering.

f = flap
geo = geometric value
max = maximum
min = minimum
∞ = freestream conditions
r = point at which pressure recovery to freestream conditions begins on an airfoil
visc = viscous

Superscripts
$(\char`\^)$ = adjusted or scaled quantity
$(*)$ = critical section

Introduction

HIGH-LIFT system performance has a major influence on the sizing, economics, and safety of any airplane configuration. Because of the combination of complicated geometry and fluid physics associated with the high-lift systems of even relatively simple aircraft types, the development of such systems has historically been an experiment intensive process. The primary objective, into the 1970s, was to maximize the maximum lift coefficient of a practical, economically viable, and manufacturable system. Since that time, the design problem has become more complex, with attention focused on high-lift systems that not only provide good maximum lift performance for landing approach but also good take off lift-to-drag ratio characteristics as well. In addition, the system must be as simple and economical as possible while retaining necessary performance levels.

The advent of computational tools capable of *partially* dealing with the problems of separated flows and viscous vortex interactions that are encountered in the analysis and design of high-lift systems is a relatively recent event. Only a limited number of applications of such methods to practical airplane development efforts have been reported in the literature.

Given the wide range of existing aircraft types, the overall performance of which depend to some degree on the performance of their high-lift systems, the focus of the present discussion must be limited. Thus, only unpowered lift concepts will be considered, and these will be further restricted to systems appropriate to conventional take off and landing subsonic aircraft (e.g., civil and military transports).

Because of the rather specialized nature of high-lift aerodynamics in general, and with respect to computational methods in particular, a supplement to the basic computational fluid dynamics (CFD) methods description in the early sections of this volume is considered necessary. This introductory material also includes a very brief review of basic aerophysics issues of importance, some of which are almost unique to the high-lift analysis and design. The CFD application examples described later have been drawn from the Boeing experience of the two authors, although corresponding examples should be citable by our colleagues in industry and various government facilities.

Two-Dimensional Airfoil Methods

Multielement Airfoil Problem

An accurate calculation of the flow over multielement airfoils designed for use on transport airplanes is presently an unsolved problem, even though much progress has been made by code developers in industry and government research centers. This may come as a surprise to some readers, since the flow is two dimensional and freestream Mach numbers are low, typically ranging from 0.1 to 0.4. Reynolds numbers of interest, based on velocity of an undisturbed uniform freestream and airfoil reference chord, are usually between 1–2 and 40 million.

Consideration of a few examples of typical multielement high-lift airfoils (Fig. 1) demonstrates vividly that geometric complexity is one of the main reasons why even two-dimensional computations are such a formidable problem. These complex geometries produce complex viscous flows at most flight conditions. The problem is illustrated in Fig. 2, which shows a modern four-element airfoil with a slotted leading-edge flap and a vane main type trailing-edge flap in a landing configuration. The low-speed designer has the task of analyzing this airfoil over a large angle-of-attack range, say, from -2 to 25 deg.

Many of the flow phenomena found on multielement airfoils are presently not well understood, in particular, those associated with the physics of merging shear layers, separated flows, and boundary-layer transition and hence are not accurately modeled in most high-lift airfoil codes. The lack of an adequate experimental data base with sufficiently detailed and accurate measurements over the full range of Reynolds number of interest is partly responsible for this situation.

Flow Physics

Studying the physics of multielement airfoils in more detail, one can compile the following list of flow features not found on cruise airfoils (see Butter and Williams[1] for a similar discussion):

1) The flow region surrounding multielement slotted airfoils is multiply connected, which complicates the topological laws governing viscous separated flows and even makes the calculation of inviscid flow a difficult task.

2) Limited regions of transonic flow may appear on the upper surface of the leading edge of highly loaded flapped airfoils, even though the freestream Mach number is low.

3) Wakes of upstream airfoil elements often merge with boundary layers on the surfaces of downstream elements. The resulting turbulent shear layer is referred to as a confluent boundary layer.

4) The region of viscous flow above the surface(s) of trailing-edge flaps is relatively thick, particularly on landing configurations, often resulting in flow separation even near normal operating conditions (i.e., well before maximum lift is attained).

5) Streamline curvature and its effect on turbulent flow development is significant.

6) Viscous wakes frequently develop in the presence of the strong pressure gradients of adjacent airfoil elements. Off-the-surface wake flow reversal can occur if pressure gradients are adverse.

7) Coves at the trailing edges of the main airfoil and flaps are formed when trailing-edge flaps are deployed. Cove flow is always separated, as is the flow on the lower surface of a leading-edge device (either slat or Krueger flap) at low angles of attack.

8) Laminar boundary layers on downstream elements will become turbulent when merging with the wake of an upstream element at the start of a confluent boundary layer. This type of boundary-layer transition may occur in addition to natural transition of attached boundary layers and transition in laminar short bubbles.

Confluent boundary layers and flow separation are at the center of the difficulties that computer code developers and designers of high-lift airfoils must deal with. These flow phenomena are closely related to each other and will be discussed in more detail from an engineering point of view, emphasizing their effect on airfoil characteristics.

Confluent Boundary Layers

These viscous layers form when wakes from upstream elements merge with boundary layers on the surfaces of downstream elements. Since the viscous wakes of airfoils are always turbulent in the range of Reynolds numbers of interest here, confluent boundary layers are also turbulent. The formation of these boundary layers depends on the size of the gap between neighboring airfoil elements and on flight condition, i.e., on angle of attack, Reynolds number, and Mach number. We can identify confluent boundary layers by studying mean flow velocity or, even better, total pressure profiles on the upper surface of a typical slotted multielement airfoil (Fig. 3).

Achievable airfoil lift depends strongly on the size of the gap between airfoil elements, as demonstrated by the data of Foster et al.,[2] measured on a two-element airfoil with a slotted trailing-edge flap at a constant angle of attack (Fig. 4). The actual measured lift data show an optimum gap size, whereas the inviscid lift of this airfoil decreases almost linearly with increasing gap. For gap sizes larger than optimum, lift losses are due to ordinary thin boundary layers, laminar and/or turbulent. For smaller gaps, part of the observed viscous lift loss is due to the confluent boundary layer forming on the upper flap surface.

On multielement airfoils optimized for maximum lift, the gap between the main wing section and the trailing-edge flap is large enough to delay merging of wing wake and flap boundary layer until both shear layers have reached the flap trailing edge. However, on optimum airfoils with a leading-edge flap the gap between the latter and the main wing section is usually such that a confluent boundary layer forms on the upper wing surface well upstream of the wing trailing edge.

Massive Flow Separation

Three physically distinct types of flow separation can be identified on multielement airfoils: 1) laminar separation bubbles with turbulent reattachment, also found on single airfoils; 2) separation in coves and on the lower surface of leading-edge devices; and 3) massive flow separation on one or more airfoil surfaces. The latter type of flow separation also occurs on single airfoils where it produces different lift curves and stall patterns depending on the origin of separation (see Gault[3]). Single-element airfoil flow separation can start at the leading and/or trailing edge and is affected by the presence of laminar bubbles, but only involves ordinary boundary layers. If trailing-edge stall prevails, maximum lift of single airfoils increases with Reynolds number and decreases with Mach number due to a process of ordinary turbulent boundary-layer separation that is relatively well understood and hence can be accurately predicted by existing computer codes. As we will see later, the maximum lift of multielement airfoils does not follow such a predictable trend and, in particular, does not necessarily increase with increasing Reynolds number.

Contrary to the case of single-element airfoils, the mechanism causing massive flow separation on multielement airfoils is not well understood at this time. However, we know that the interaction between the wake of an upstream element and boundary layers on downstream elements plays a central role. This interaction can be inviscid, in which case the wake flow affects flap pressures and consequently flap boundary layers, but it can also be viscous turbulent when confluent boundary layers are present.

A striking example of the effect of flow separation on high-lift airfoil performance is provided by Woodward et al.,[4] whose study of scale effects in low-speed, high-lift aerodynamics has greatly enhanced our understanding of the flow physics of multielement airfoils. Figure 5 shows results from a wind-tunnel study in which gap and overlap between the main airfoil and a single-slotted trailing-edge flap were optimized at a fixed flap angle of 40 deg to achieve the highest possible maximum lift. Results of this study are best understood by observing the effect of gap size on lift at a constant value of overlap, say, -0.5% chord. Varying gap size produces very different lift curves at otherwise constant test conditions such as Reynolds number, Mach number, and flap angle, as shown in the upper plot. The lower plot shows gap/overlap effects on maximum lift coefficient.

Small gaps produce a conventional lift-curve A characterized by strong interactions between the wake of the main airfoil and flap, which leads to a reduction in the flap suction peak and hence to attached flow at low angles of attack. Increasing angle of attack ultimately results in stall.

Intermediate gaps reduce the interactions between wake and flap, thereby increasing the flap suction peak, and consequently lead to flap boundary-layer separation at low angles of attack. The corresponding lift-curve B is therefore lower than A at low incidences. However, increasing angle of attack to intermediate values will result in a thickening of the wake, which in turn lowers the flap suction peak, and keeps the boundary layer attached until it eventually separates at rather large incidence angles. The resulting maximum lift exceeds that of lift-curve A.

At the largest gaps, wake/flap interactions are never strong enough to keep the flap boundary layer attached. Hence, lift-curve C is significantly lower than the others throughout the angle-of-attack range.

For some configurations and flap settings Reynolds number can have an equally strong effect on wake/flap interactions[5] (Fig. 6). Increasing the Reynolds number from 2.8 to 12 million for the airfoil shown with a vane main type trailing-edge flap causes a lift loss at lower angles of attack due to a decrease in wake/flap interactions. This low-angle-of-attack phenomenon, called inverse Reynolds number effect, does not increase lift with increasing Reynolds number as expected for a single airfoil.

Survey of Existing Methods

The development of two-dimensional high-lift airfoil methods is still an ongoing process; thus, any attempt to compile a comprehensive list of existing methods is a risky task. In any case, the present survey is not complete, but includes a sufficient number of methods of various types to illustrate how far the development of multielement airfoil methods has progressed. The list only includes viscous methods for the analysis of multielement slotted airfoils. Frequently, such methods evolved from single airfoil codes, but methods limited to single airfoils, no matter how advanced, are not contained in the list. Most referenced design methods are inviscid.

Excluded from the list are several inviscid, fully compressible methods that apparently had been developed with future addition of viscous flow effects in mind. These are the methods of Klevenhusen[6] and Oskam[7] based on the transonic full-potential equation and of Wigton,[8] who uses the Euler equations.

Table 1 groups existing airfoil methods in four categories that are defined later. Within each category the table indicates the main features of the flow model. The term "confluent boundary layers" refers to attached confluent boundary-layer flow, "massive flow separation" means the use of a model for upper surface flow separation allowing the computation of maximum lift, and the term "transonic flow" indicates the use of nonlinear compressible flow equations instead of simplified compressibility corrections.

Category I: Coupled Attached-Flow Methods

The codes listed use panel methods exclusively for the calculation of the outer inviscid flow. All boundary layers are assumed to be attached, but provisions are usually made for short laminar bubble separation with turbulent reattachment. Boundary-layer integral methods are preferred by most developers. A coupled solution of inviscid and viscous flows is achieved using the classical iteration technique of successive substitutions, which is adequate for these weak interactions. The development of these methods was essentially complete about 10 years ago, with many of them are still in use today.

Category II: Coupled Separated-Flow Methods

Here, only those methods are included that employ coupled inviscid/viscous solution procedures. All but one of the methods employ a panel technique to calculate inviscid flow with some correction for Mach number effects as in methods of the previous category. All methods feature models

Table 1 Existing computational methods for multielement airfoils

Category	Developers	Confluent boundary layers	Massive flow separation	Transonic flow	Ref.
I Coupled attached-flow methods	Stevens, Goradia, Braden	✓	—	—	9
	Seebohm, Newman	—	—	—	10
	Morgan	✓	—	—	11
	Dvorak, Woodward	✓	—	—	12
	Brune, Manke	✓	—	—	13
	Oskam	—	—	—	14
	Green, Newling	✓	—	—	15
II Coupled separated flow methods	Jacob, Steinbach	—	✓	—	16
	Henderson	—	✓	—	17
	Mani	✓	✓	—	18
	LeBalleur, Neron	—	✓	—	19
	Olson	✓	✓	—	20
	Butter, Williams	✓	✓	—	1
	King, Williams	—	✓	✓	21
III Navier-Stokes methods	Shima	✓	✓	✓	22
	Mavriplis	✓	✓	✓	23, 24
	Fritz	✓	✓	✓	25
	Rice, Schnipke, Cornelius, Normansell	✓	✓	✓	26
IV Design and optimization methods	Olson	✓	✓	—	20
	Henderson			—	17
	Bristow, Hawk		Inviscid methods	—	27
	Narramore, Beatty			—	28

✓ Included; —— not included.

for massive flow separation intended to predict maximum lift. Jacob and Steinbach,[16] Henderson,[17] Mani,[18] and Olson[20] use potential-flow singularities to model the large displacement effect of flow separation in which details of the viscous recirculating flow within the separated flow region are not accounted for. A very different separated flow model, the so-called viscous defect formulation due to LeBalleur[29] and East,[30] is used by Butter and Williams,[1] LeBalleur and Neron,[19] and King and Williams.[21] The latter is unique among the methods in this category since it provides a truly compressible solution for the inviscid flow based on the full-potential equation coupled to integral boundary-layer methods.

Category III: Navier-Stokes Methods

Development of these methods, generally based on Reynolds averages of the Navier-Stokes equations with closure by a suitable turbulence model, is still in its infancy. Presently, methods of this kind are research codes under development and, as far as we know, have not yet found their way into production use. Formidable problems associated with grid generation, turbulence modeling of separated flows, transition from laminar to turbulent flow, and numerics must be overcome before they routinely can be applied in an airplane design environment. Nevertheless, these methods have already produced some very impressive results and, in the long run, appear to have the greatest potential for accurately predicting high-lift airfoil flows of general interest.

Category IV: Design and Optimization Methods

In principle, shape design of an airfoil element for a given distribution of surface pressure can be accomplished by repeated application of analysis codes. However, some methods have been developed that allow the user to design all or part of a new airfoil shape in a single application. Multielement airfoil design is treated explicitly by several code developers, e.g., Henderson,[17] Bristow and Hawk,[27] and Narramore and Beatty,[28] as an inviscid problem solved by a panel method. Olson[20] addressed the related problem of optimizing slat and flap deflections required to achieve maximum lift while keeping the shape of all airfoil elements fixed.

Airfoils Below Stall

Presently, we can predict airfoil performance quite accurately for attached-flow conditions. Consider the airfoil shown in Fig. 7, which was tested at the National Aerospace Laboratory (NLR) in Holland by van den Berg[31] in an atmospheric pressure tunnel at a low chord Reynolds number of 2.5 million. Both airfoil components were designed and positioned to avoid flow separation at low to moderate angles of attack. Very detailed force, pressure, and boundary-layer data were acquired for the validation of high-lift codes employing attached-flow methods, and these have been predicted theoretically by many investigators with good success. Here, the theoretical surface pressures calculated by van den Berg and Oskam[14,32] are shown to give an excellent match with experiment. Boundary-layer data computed for the upper wing surface of the same airfoil (Fig. 8) also agree favorably with measured data.

Oskam[14] reported that the referenced older version of his code failed to converge at angles of attack higher than 6 deg due to a reversal of the viscous wing wake and turbulent boundary-layer separation on the upper flap surface. He consequently improved his code, as described in a publication coauthored with Laan and Volkers,[33] by adding a viscous wake model to account for off-the-surface pressure recovery and wake flow reversal. This is probably the first successful attempt at modeling the inviscid interactions between a viscous wing wake and flap pressures correctly. Figure 9 shows upper flap surface pressure peaks of the NLR airfoil decreasing in a physically realistic manner with increasing incidence.

In general, the application of airfoil codes employing attached-flow methods should be limited to high-lift airfoils with flap angles not exceeding about 15 deg (typical of maximum takeoff conditions) and moderately high angles of attack. The lift data of Fig. 10 emphasize this point. Here, it is shown that the theoretical predictions of the GA(W)-1 data measured by Wentz and Seetharam[34] using an attached-flow method fail at the geometry and flight conditions that cause massive flow separations.

It is usually more difficult to calculate the performance of airfoils that are not optimized and consequently feature large viscous flow regions with confluent boundary layers and strong viscous wake/flap interactions. An example of such a nonoptimum configuration is shown in the experimental data of Figs. 11 and 12 consisting of surface pressures and global airfoil characteristics of a four-element airfoil. The main objective of this investigation, which was conducted at Boeing by Brune and Sikavi,[35] was to obtain a detailed map of the viscous flow on the upper surface of a transport high-lift airfoil with takeoff flaps including profiles of boundary-layer mean flow and Reynolds stresses. In addition, surface pressures, lift, drag, and pitching moments were measured. A large leading-edge flap of the Krueger type was chosen for this study and positioned to obtain a strong confluent boundary layer. The use of two different high-lift airfoil theories[13,17] produced rather inaccurate surface pressures and unacceptable predictions of global airfoil data, even though trailing-edge flaps were in a typical takeoff position and the boundary layers stayed attached for much of the angle-of-attack range.

Most multielement airfoil methods of categories I and II are basically incompressible methods in which compressibility effects are accounted for by relatively simple corrections to an incompressible solution. However, high-lift designers have long recognized the need for modeling compressibility more correctly.[36] Even though transonic flow is a very localized phenomenon limited to the leading-edge region, it can have a strong influence on performance. Compressible high-lift methods capable of dealing with transonic flow effects and the geometric complexities of high-lift airfoils are just beginning to appear in the literature. An example of a true compressible flow result is shown in Fig.13, computed by Mavriplis[24] using a multigrid method for the Euler equations on an unstructured and adaptive grid. The leading-edge region of the main airfoil element features a strong shock with a local Mach number of 1.5 just ahead of it. Clearly, such a strong shock will have a dominant effect on the boundary-layer development of this airfoil and must be accounted for.

Maximum Lift

The maximum amount of lift that an airfoil can generate is limited by flow separation on the upper surface. Hence, category I methods that assume attached flow will not be able to compute maximum lift; thus, the methods of categories II and III are required.

The methods of Jacob and Steinbach[16] and of Henderson[17] use potential-flow artifices to model massive flow separation. The latter is an extension of the well-known method of Jacob for single airfoils, reported by Schlichting.[37] Viscous recirculating flow is not computed in detail, and the main role of attached boundary-layer models is to predict the onset of separation. Airfoil stall predictions including maximum lift by both methods are very accurate for the cases shown in Figs. 14 and 15. For comparison, inviscid flow results are also shown to underline the magnitude of the lift losses due to flow separation.

Notice that the results of Fig. 14 do not contain poststall predictions. It is difficult to go beyond maximum lift since the separated-flow regions become very large and in some cases cause numerical convergence problems. In addition, the validity of such models, based on the assumption of steady flow, becomes questionable, since postairfoil stall can be highly unsteady.

A Navier-Stokes solution for the lift of a multielement airfoil through stall (Fig. 16) has been obtained by Shima.[22] Viscous terms are calculated based on the thin-layer approximation, and turbulence modeling uses the Baldwin and Lomax[38] approach. With these approximations, Shima was able to obtain an accurate value for maximum lift, but he does not match lift level or slope very well at lower angles of attack. As imperfect as these predictions might appear at this time, Shima has made a significant step in the right direction toward producing a Navier-Stokes code of engineering utility.

Several accurate calculations of high-lift airfoil stall have been presented by the code developers, but we nevertheless consider the accurate prediction of maximum lift under a broad range of conditions an unsolved problem. As indicated, some very ingenious methods have been devised, yet the modeling presently used does not contain sufficient physical realism to adequately account for all aspects of the complex physics affecting stall. Viscous/inviscid interaction methods have been continously improved, as documented by Lock and Williams,[39] LeBalleur,[29] and others. However, Navier-Stokes methods for high lift, even though they are still in their early stages of development, as seen in the preliminary data of Shima, are in our view conceptually better suited to deal with the complexities of separating confluent boundary layers and other flow phenomena found on these airfoils.

Drag

In principle, there are three different ways to calculate airfoil drag. First, one can do the most obvious and integrate surface pressures and skin friction, a procedure that works very well for calculating lift. However, drag is very small in comparison to lift, so that even small uncertainties in the calculated surface pressures may cause relatively large drag errors. Second, encouraged by the reasonable success in predicting drag of single airfoils using the Squire and Young formula,[40] one is tempted to ignore the assumptions made in deriving this formula and apply it to each element of the multielement airfoil separately to obtain the appropriately summed total drag. Several researchers have tried this approach without much success.[13,17]

The third and theoretically most appealing approach is to calculate the total momentum deficit in the wake as a measure of drag. Olson et al.[41] have studied this method theoretically and experimentally on a two-element airfoil (Fig. 17) and have reported a significant mismatch between a computed and measured drag polar which they attribute entirely to drag errors (rather than a combination of lift and drag mispredictions). Their results are representative of the drag prediction accuracy that one can expect to get from high-lift airfoil codes at the present time. Calculating wake momentum deficit more accurately will require improvements in viscous flow methodology and primarily better computations of viscous wakes and confluent boundary layers. As discussed by Lock and Williams,[39] conventional parabolic boundary-layer methods have difficulties in calculating wake development due to significant cross-stream pressure gradients, streamline curvature affecting turbulence, and streamwise pressure gradients changing rapidly from favorable to adverse.

However, the problem is even more complex than described since the mismatch between theoretical and experimental drag is also due to large measurement uncertainties. Drag of two-dimensional airfoil models in wind tunnels is usually measured with a wake rake whose readings frequently depend strongly on spanwise position. Spanwise variations in drag of up to 10% arise from three-dimensional disturbances in boundary layers and wakes. Such variations cause the uncertainty in the wake rake drag measurement shown in Fig. 12.

Design and Optimization

The importance of inverse methods for designing and optimizing airfoil surface shapes and relative position of all airfoil elements need not be emphasized, yet very few publications deal with these tasks. The problem of high-lift airfoil design is closely related to the design of single airfoils so that only the differences are pointed out here. Each element of a high-lift airfoil is strongly affected by its neighbors and is not simply imbedded in an inviscid uniform flowfield as is a single airfoil. All airfoil design codes listed under category IV of Table 1 assume inviscid flow to simplify the modeling of the interactions between neighboring elements. An example is given in Fig. 18 illustrating the design of a trailing-edge flap in the presence of three other airfoil elements, published by Narramore and Beatty.[28]

Similiar results and design procedures have been presented by Henderson[17] and Bristow and Hawk.[27]

It should be emphasized that the design of a high-lift airfoil is usually constrained to match the shape of the cruise airfoil when all flaps are retracted. This only allows a partial shape design for most elements, or the other alternative of simultaneously designing cruise and high-lift airfoil.

The low-speed aerodynamicist, however, has other (limited) options for improving high-lift airfoil performance since he can modify flap size and setting parameters. Chord length, angle, gap, and overlap of each leading- and trailing-edge flap can be varied to some extent to achieve the desired performance. This process, referred to as optimization of flap size and setting, is more difficult than airfoil shape design because viscous flow methods rather than inviscid approximations are required to optimize most parameters. An example of this requirement is the gap optimization for a two-element airfoil intended to achieve the highest value of lift at a fixed angle of attack. As shown in Fig. 4, an inviscid method would completely close the gap between the two elements, whereas the measured optimum has a nonzero value of about 2% of the airfoil reference chord, which was correctly predicted by the Navier-Stokes method of Shima[22] (Fig. 19).

Three-Dimensional Methods

Flow Physics of Three-Dimensional High-Lift Systems

Our present understanding of the physics of three-dimensional high-lift flows is rather incomplete, partly because of the lack of detailed flow data on realistic low-speed airplane configurations over a sufficiently wide range of Reynolds numbers. During the development phase of a new airplane, numerous wind-tunnel tests are routinely conducted with transport models in high-lift configuration aimed at optimizing and recording low-speed performance. However, such investigations contribute relatively little to our basic understanding of the flow physics, since data are usually limited to the measurement of forces and surface pressures and do not include detailed boundary-layer and wake information. Such data are also highly configuration specific, and the configurations tested are generally complicated.

The nature of the basic flow around three-dimensional high-lift configurations is complicated further by the geometric complexities of practical airplane features (Fig. 1), which include such items as part-span slotted flaps and their supporting flap tracks, wing and flap sweep, variations in geometry across the span, and the fuselage and engine nacelles. As in the case of two-dimensional high-lift airfoils, we include here a list of several flow problems that are unique to three-dimensional high-lift systems and that are not found on the geometrically simpler cruise configurations (see also Butter[42] for a similar discussion):

1) High-lift configurations feature regions of viscous separated flows that are very complex and that may be topologically quite different from two-dimensional separated flows found on high-lift airfoils. An example is

given in Fig. 20, which shows the three-dimensional separation pattern of a wing with landing flaps at a very high angle of attack. The pattern shown consists of skin-friction lines or limiting surface streamlines. Such data must be supplemented by off-the-surface flowfield measurements (which are seldom available) and force data to arrive at a proper aerodynamic interpretation.

2) Separation in wing and flap coves as well as on the lower surface of a leading-edge flap also may be different from that on two-dimensional airfoils due to spanwise flow when significant wing sweep is involved.

3) High-lift systems of transport airplanes feature three-dimensional confluent boundary layers in which the merging of wakes and boundary layers may be affected by crossflow in ways not yet fully understood. Recently, Moghadam and Squire[43] have published the very first investigation of three-dimensional confluent boundary layers generated by swept wing wakes merging with the boundary layer on a flat plate.

4) Very strong vortices are shed from the edges of part-span flaps (Fig. 21), which are frequently more powerful than the tip vortices associated with unflapped wings.

5) There are significant three-dimensional viscous interference effects between neighboring airplane components. An example is the flow in the wing/fuselage junction, which, in the presence of a highly deflected trailing-edge flap, can be very different from that of a cruise configuration.

6) Because of the spanwise variation in loading of a practical high-lift system, significant spanwise variations in compressibility effects are also to be expected.

7) Each element of a swept high-lift system has its own attachment line (corresponding to the stagnation point in two dimensions). Transition from a laminar to a turbulent attachment line flow depends on sweep angle, Reynolds number, and velocity gradient in the leading edge region and may heavily influence downstream boundary-layer development. Hence, maximum lift is also affected and may vary with Reynolds number in the manner postulated by Woodward, et al.[4] (Fig. 22).

Three-Dimensional High-Lift Methods

Presently there is no truly three-dimensional CFD method available to compute the viscous separated flow of a transport airplane in a high-lift configuration. High-lift system designers must therefore be content at present with lifting surface and panel methods that are generally used in conjunction with two-dimensional airfoil codes in various ways that are described later. Such quasi-three-dimensional viscous methods have been developed on a more or less formal basis to provide an interim capability until a full three-dimensional capability emerges. The following discussion provides a brief outline of three-dimensional high-lift codes in use at Boeing at the present time. Codes with similar capabilities are in production use throughout the industry.

Lifting Surface and Panel Methods

Three-dimensional inviscid methods for arbitrary nonplanar wing/body geometries are generally based on a solution of the Prandtl-Glauert equation. The lifting surface methods in use at Boeing were developed by Goldhammer[44] and Feifel[45] for the analysis and design of high-lift systems. The term "lifting surface" refers to the main assumption of methods of this type, which is that the effects of wing thickness are small and can be neglected.

There are two reasons why lifting surface methods have found widespread use in high-lift aerodynamics. First, it is relatively easy to prepare geometry input data for even the most complex flapped wing geometry, since the details of wing/flap thickness need not be specified. The second and more important reason is that many of the predictions made with these methods are surprisingly accurate, benefitting from a partial cancellation of wing thickness and attached viscous flow effects that are both neglected by lifting surface methods. However, lifting surface methods do not calculate wing surface pressures and therefore cannot be coupled with boundary-layer codes that need such pressures as input.

Panel methods can accurately calculate inviscid flow and can be coupled with attached boundary methods if desired. The principal method of this kind at Boeing is A502/PAN AIR,[46] which is a general three-dimensional boundary-value problem panel method solver for the Prandtl-Glauert equation. It is applicable to the analysis of inviscid linear flows over vehicles in subsonic or supersonic flight. Several features distinguish A502/ PAN AIR from other panel methods, including higher-order numerics, continuity of doublet strength and geometry, a very general user-specified boundary condition equation, and the implementation of the Kutta condition. As discussed by Tinoco et al.,[47] all these features are absolutely essential for the analysis of complex high-lift systems.

VSAERO, which was developed by Maskew and Dvorak[48] for the analysis of very general three-dimensional configurations (Fig. 23), is another panel method in widespread use throughout the industry. It employs lower order singularity distributions and accounts for the effect of wake rollup.

The three-dimensional inviscid methods described are indispensable tools for the analysis of takeoff configurations at moderate angles of attack where most boundary layers stay attached. They do not predict maximum lift or profile drag but have been found to give accurate estimates for span load and induced drag at takeoff conditions. The larger trailing-edge flap deflections required for landing cause massive flow separation that cannot be predicted by lifting surface or panel methods without a model for viscous separated flows.

Emerging New Methods

Lifting surface and panel methods do not correctly account for compressibility effects and hence are not valid for high-lift systems with local transonic flow regions. The new TRANAIR code[49] that is presently being developed at Boeing avoids this limitation by solving the transonic full-potential equation. Complex configuration geometries are defined by surface panels similar to those used by A502/PAN AIR, even though the method is a true field method rather than a panel method. At this time,

CFD code developers are evaluating the **TRANAIR** prediction capability for high-lift applications.

We are not aware of any development effort at this time based on the Navier-Stokes equations applicable to high-lift wing/body configurations with slotted part span flaps.

Quasi-Three-Dimensional Viscous Methods

In the absence of a full three-dimensional viscous flow analysis capability for high-lift configurations, a few methods have been developed to deal with the problem in a quasi-three-dimensional fashion. There are basically two different ways of analyzing configurations in a quasi-three-dimensional manner. One approach is to couple a panel method or a field method with a simplified boundary-layer analysis performed along external streamlines or some other suitable two-dimensional strip. The other approach is to combine a two-dimensional airfoil code with a simple three-dimensional potential flow method assuming that neighboring airfoil sections affect each other only through the inviscid pressure field and that all three-dimensional viscous flow effects can be neglected. This assumption specifically excludes spanwise boundary-layer flow on wings and flaps.

Several researchers have developed computerized methods for formally correlating two-dimensional airfoil data with three-dimensional wing data in order to compute three-dimensional wing lift over the entire angle-of-attack range including stall. These methods are nonlinear, since, in general, airfoil lift changes nonlinearly with angle of attack. The problem has a relatively simple solution for straight wings requiring only a minor modification of classical lifting line theory[50,51] but is more difficult for swept wing/body configurations. Levinsky[52] appears to have published the first method for swept configurations that also accounts for unsteady aerodynamic effects. He utilized a panel method with a single chordwise panel combined with airfoil data for streamwise wing sections. Another method has been presented by Jacob,[53] who coupled Truckenbrodt's lifting surface procedure with his own airfoil code. This method accounts for changes of effective angle of attack as well as effective camber changes but apparently does not handle slotted flaps.

A quasi-three-dimensional viscous high-lift method recently developed at Boeing shares many features with that of Levinsky[52] but is intentionally limited to steady flow problems. It has been applied to realistic transport airplanes in high-lift configuration and has been used to successfully compute aerodynamic parameters without recourse to an empirical high-lift data base. Figure 24 shows the span load calculation of the Boeing 757-200 with takeoff flaps as an example. The high-lift airfoil data were obtained from a viscous flow code valid for attached flow which models both ordinary and confluent boundary layers. Hence, the three-dimensional predictions shown also include a rough estimate of confluent boundary-layer charactertistics. Nacelle effects are not modeled, however, causing the observed discrepancy between computed and measured span load data in the nacelle region.

The underlying assumptions of these quasi-three-dimensional methods restrict their application to those portions of a configuration where the flow is approximately two dimensional. They were developed to provide an interim capability but may have value in aiding preliminary design efforts even after a truly three-dimensional viscous-separated flow method becomes available.

Some Applications of Computational High-Lift Design Methodology

As the preceeding discussions have demonstrated, much work remains to be done before the computational methods needed to perform a fully analytic design of a modern high-lift system can be considered as relatively satisfactory as the methodology available at present to the designer of transport-type aircraft at subsonic and transonic *cruise* conditions.

However, this does not mean that the designer of high-lift systems has been unable to share in the riches that a computational capability can bring to the aerodynamic design process. What has been necessary so far is 1) the devotion of considerable resources to the development of adequate, if imperfect, two-dimensional viscous and three-dimensional inviscid computational methods; 2) exploration of semiempirical rules that allow practical implementation of the computational methods; and 3) determination of the proper complementary roles of computations and experiment in present and future airplane development efforts. For the past 15 years, Boeing Commercial Airplanes has supported such a combined effort, and the following discussion of applications of computational methods will be presented from this perhaps parochial point of view. However, we believe that this perspective is broadly representative of the state of affairs both here and abroad, and, where appropriate literature exists, these paralleling efforts are cited.

The initial effort to build a modern high-lift computational capability at Boeing recognized the facts that, due to the state of the art in both computer hardware and software and the combined physical and geometric complexity of high-lift aerodynamics, a fully developed three-dimensional viscous flow computational capability would be beyond reach for an extended period of time. Thus, the initial method development effort focused on three basic topics: 1) two-dimensional multielement airfoils in a viscous flow, 2) three-dimensional wings/bodies in inviscid flow, and 3) two-dimensional to three-dimensional correlation rules. It should also be noted here that, from the outset, a heavy emphasis was placed on providing suitable inverse (design) capabilities in the two-dimensional codes and inverse/optimization capabilities in the three-dimensional codes. In addition, throughout the development of these methods, there has been a heavy emphasis on applications to practical, aircraft-project-oriented design problems.

The initial development effort resulted in the availability of two computer programs for multielement airfoils developed by Brune and Manke[13] and Henderson[17] [subsonic analysis section system (SASS)]. As a further adjunct to the latter method, an auxiliary inverse boundary-layer (IBL) method was developed[54,55] for pressure distribution design. This IBL method has proved to be an important aid to the high-lift design process described presently. The primary code used to deal with three-dimensional high-lift problems is the lifting surface theory code of Goldhammer,[44] which employs the distributed vorticity method (DVM).

In addition to these three basic methods, which have been available for production use since 1980, a version of the A502/PAN AIR (higher-order panel method) code has become available for high-lift analysis purposes in these cases requiring more detailed information (e.g., chordwise pressure distributions) than can be provided by lifting surface theory. Such analyses also have been further extended by combining the output of A502/PAN AIR with the three-dimensional boundary-layer code of McLean[56] to provide viscous flow information for attached flow conditions. However, at this time, the high-lift A502/PAN AIR and three-dimensional boundary-layer codes are not automatically coupled.

With this suite of computer codes in hand, it has been possible to devise the computational/experimental design process outlined in Fig. 25. Broadly similar schemes have also been reported by Wedderspoon[57] at British Aerospace, Szodruch and Schnieder[58] at Messerschmitt-Bölkow-Blohm, and others. In order to make such design schemes work, it is first necessary to develop some semiempirical recipes for connecting the results of two-dimensional viscous and three-dimensional inviscid analyses.

High-Lift Flow Correlation and Prediction Techniques

The problem of establishing rational methods for connecting the results of three-dimensional potential flow with two-dimensional viscous flow analyses has been an important part of the high-lift research effort at Boeing. As a major part of this effort, it has been necessary to establish the correlation between two-dimensional multielement section characteristics with the corresponding sections on three-dimensional wings as influenced by sweep, induced angle-of-attack and camber effects, and spanwise components of boundary-layer flow.

As an example of early correlation methodology work, it was found that "simple sweep theory" type corrections to two-dimensional results, which are rigorously valid only for thin wings of constant chord and infinite aspect ratio, should be replaced by the more theoretically correct method due to Lock,[59] which explicitly accounts for taper and finite aspect ratio effects.

With the advent of the DVM lifting surface theory program, Goldhammer[44] was able to achieve a substantial advance in correlation/prediction methodology. For the first time, it became possible to reliably obtain potential-flow results for high-lift configurations representative of actual transport aircraft. An example of what Goldhammer achieved with combined use of the DVM and SASS programs to predict high-lift wing-body characteristics beyond the linear portion of the lift curve is demonstrated in Fig. 26. The assumptions made in this example are that airfoil section characteristics dominate the lift behavior of the wing and that, even in cases where the flow may be locally separated, spanwise boundary-layer flow effects can be neglected.

Use of the DVM program for cases of highly deflected part- or full-span flaps, with separation at normal operating conditions, leads to a substantial overprediction of lift. However, by analyzing "critical two-dimensional sections" of the wing (located at "peaks" in the span loading distribution) using the SASS program with its separated-wake modeling capability, an "effective" viscous flap-deflection angle can be determined, as shown in Fig. 26.

When this new effective viscous flap deflection is input to the potential flow analysis (DVM), the result is a dramatic improvement in the test-theory comparison of span loading shown in Fig. 26.

Perhaps a more remarkable outcome of this sort of analysis was the fact that applications of the DVM program to a variety of other transport-type configurations showed a repeatable (to first order) correlation between effective and geometric flap deflections for a given number of elements in the high-lift system. This led to the tentative construction of the graph relating effective and geometric flap deflections shown in Fig. 26. These relations hold only for standard wind-tunnel level Reynolds number, although a comparable set could be constructed for flight levels.

The general success of the two-dimensional/three-dimensional correlation methodology has led to the more recent quasi-three-dimensional viscous method described in the earlier three-dimensional theory section. It has also provided the foundation for the preliminary design level prediction method reported by Murillo and McMasters.[60] The following application examples further demonstrate the overall methodology.

Redesign of the Boeing 747 High-Lift System

The first example was selected because it demonstrates the way in which the general design methodology (Fig. 25), with its strong reliance on the use of inverse methods, was used to evaluate a complex design problem. The problem posed consisted of the following: Given the wing of the existing Boeing 747, is it possible to simplify the triple-slotted flap/variable camber Krueger high-lift system without degrading the approach speed? Further constraints were the following:

1) The cruise aerodynamic configuration must remain unaltered.
2) Major structural modification outside the flaps would not be allowed.
3) Handling characteristics should not be degraded.

As shown in Fig. 27, the baseline geometry was first analyzed in potential flow using the DVM lifting surface theory. This yielded the span loads at various values of lift coefficients. From the span loads, the critical two-dimensional sections were identified and evaluated using the two-dimensional multielement airfoil code SASS. These results, corrected for sweep are shown in Fig. 28. Additional information obtained from these analyses are viscous flow pressure distributions and details of the boundary-layer characteristics. At this point one has the basic data necessary to begin the redesign effort.

Using the two-dimensional IBL method, improved viscous flow pressure distributions are derived for the various elements of the multielement airfoil ensembles. These design point pressure distributions are then used in the inverse mode of the SASS program to generate new airfoil geometries.

These new geometries are a combination of revised surface contours and/or modified flap gap, overlap, and deflection relationships.

With the new geometry established, these sections were further analyzed in the SASS program to obtain full section lift curves, including the effects of flow separation, and thus maximum lift. Typical final results are shown in Fig. 29. As a final step, since the resulting new geometry differed substantially from the baseline case, a re-evaluation of the total high-lift systems in the DVM program could be conducted to ensure that a "converged" design solution had been achieved.

An interesting result of the present example is shown in Fig. 29. The new double-slotted flap does indeed yield the same section maximum lift coefficient as the baseline triple-slotted system. However, more important at a given angle of attack, the lift coefficient is lower for the redesigned system than for the baseline. This means that the net result of integrating this revised section into the system would be that the aircraft would have to approach at higher angles of attack to maintain the same approach speeds. The solution to this problem (e.g., to lengthen the landing gear to avoid tail strikes) would violate the initially imposed constraints. For this reason, the results of this particular exercise remain of largely academic interest, and the new configuration was not tested. However, the example does demonstrate clearly the power of the new analytic approach to high-lift design.

Transport Aircraft Takeoff Lift-to-Drag Ratio Improvement

The second application to be discussed, and one for which wind-tunnel data exist, concerns an investigation of possible improvements in takeoff L/D of a twin turbofan-powered, swept-wing transport aircraft configuration operating out of high-altitude airfields. Takeoff from such airports, particularly under high ambient temperature conditions, is typically limited by second segment climb L/D characteristics. The basic problem posed consisted of the following: Given a baseline cruise airplane configuration (i.e., cruise wing geometry completely specified) and high-lift system, establish the upper bound of takeoff L/D attainable by 1) allowing "unlimited" chord extension of a leading-edge slat, and/or 2) possible recontouring of the slat.

The approach to the design problem was to first analyze the baseline configuration using the three-dimensional DVM lifting surface theory to obtain predicted sectional and total aerodynamic characteristics of the wing. Configuration geometry and span loadings thus obtained are shown in Fig. 30.

Next, two-dimensional section coordinates were extracted from the baseline wing at "critical" spanwise stations as shown in Fig. 30. These sections were then used as starting airfoils for the redesign investigation. In the subsequent analysis, the two-dimensional inverse boundary-layer method was used to design improved pressure distributions on the forward portion of the sections as the "slat" chord was increased, while the baseline pressure distribution (and hence baseline airfoil contour) was maintained on the aft portion of the wing. Based on the optimized pressure distributions thus obtained, the airfoil sections were reshaped in the design region. Some typical pressure distributions and the overall result of the design effort are shown in Fig. 31. The study showed that recontouring the leading-edge slat proved to be the most important contribution toward improving performance. Simple chord increase (Fowler motion of the slat) also provided some L/D gains, but only at unrealistically high extensions. In addition to establishing the performance gain to be had by slat recontouring, a practical scheme for mechanizing the required geometry change was devised.

A medium extension version of this final configuration was subsequently selected for direct comparison wind-tunnel testing with the baseline. The comparison between wind-tunnel test results and computed L/D at takeoff conditions is shown in Fig. 32. Note that the analysis did not include nacelles, whereas the wind-tunnel model had nacelles installed. In the region of design point lift coefficients, the test verified the analysis of the baseline.

However, at higher lift coefficients the analysis predicted some separation that was alleviated in the test due to favorable nacelle interference. Predicted performance of the new slat was fully verified by the test. Specifically, the new slat resulted in a takeoff L/D increase from 5% (sea level, standard conditions) to 7% (high altitudes, 33°C). Additionally, maximum lift coefficient at the various takeoff flap settings increased by increments of 0.12–0.16.

Transport Aircraft Maximum Lift Performance Improvement

The next application example to be discussed is of interest for several reasons:

1) Both wind-tunnel and flight-test validation results exist.

2) The full computational methodology previously described was applied to a difficult flow problem involving a complex airplane geometry.

3) Although the computational methods alone were inadequate to cope with the full problem, when used to augment and guide carefully conducted wind-tunnel testing, they provided the crucial element in achieving a difficult aerodynamic goal.

4) A potentially powerful approach to partially circumventing some major limits of conventional low-Reynolds-number testing in high-lift system development was demonstrated. This approach can only be pursued efficiently by application of computational techniques.

The problem confronted in this example was as follows: Given a well-proven and thoroughly tested (both in the wind tunnel and in flight) baseline transport airplane powered by four low-bypass-ratio turbofan engines in "slender" nacelles, the objective was to retrofit this basic airframe with four large-diameter high-bypass-ratio turbofan engines, with minimum modification to the rest of the airframe and without an off-design (i.e., low-speed) performance penalty. The new nacelles were apparently compatible with the baseline airframe, provided the nacelle struts of the new installation were shorter than those of the baseline, resulting in the nacelles being placed in closer proximity to the wing. Wind-tunnel tests comparing the baseline and retrofit airplanes showed no low-speed performance penalty. Corresponding flight tests showed a 10% loss in airplane maximum lift capability. The comparison results are shown in Fig. 33 and are dreadful. Furthermore, based on low-Reynolds-number wind-tunnel force data alone, there appears to be no obvious experimentally derivable aerodynamic fixup, and a major puzzle is presented.

The puzzle regarding the cause of the lift loss was rather easily solved by additional wind-tunnel testing with particular emphasis placed on carefully documented flow visualization. Nacelle-on and nacelle-off tests clearly showed that flow separation occurred on the sides of the large-diameter nacelles at high angles of attack and high flap deflection conditions, leading to the formation of large vortices that flowed streamwise over the wing. Although the section characteristics of the wing were very strongly Reynolds number scale dependent, the paths and strength of the nacelle shed vortices were almost scale independent. Furthermore, under certain conditions, the vortices interacted in a complex and unfavorable way with the boundary layer on inboard sections of the wing downstream. The upshot of all this was that, at wind-tunnel Reynolds numbers, the maximum lift characteristics of the wing were dominated by the outboard section characteristics. At flight-level Reynolds numbers, the outboard wing sections benefited from the increased Reynolds number so that maximum lift performance was limited by the unfavorable inboard wing boundary-layer/nacelle vortex interaction. Thus, the two configurations, both with identical wings and high-lift systems, exhibited almost equal maximum lift performance in the wind tunnel, but not at flight conditions. Subsequent analysis of the wing using the procedure illustrated in Fig. 25 further validated (and clarified) this diagnosis, as shown in Fig. 34.

Thus, the puzzle was solved, but the problem was not. Having now clearly observed that the wind-tunnel model, which carefully duplicated the full-scale geometry of the proposed configuration, could not duplicate the necessary flow phenomena, the traditional approach would have been to embark on an expensive and time-consuming flight-test program, with the fear that a substantial revision of the baseline high-lift system might prove to be the only satisfactory solution. However, with the availability of computational tools, a quite different approach became feasible.

This approach was merely to implement the old idea of attempting to simulate the full-scale aerodynamics, rather than the full-scale geometry in defining the parts of the wind-tunnel model. Although conceptually appealing, this course is almost impossible to follow economically, unless one has sufficiently powerful computational tools with design (inverse) capability.

In the case under discussion, the aerodynamic simulation was rather crude but extremely effective. Having determined both by flow visualization and analysis that the stall characteristics at low Reynolds number were driven by outboard wing section characteristics, it was a straightforward procedure to design an alternative, nonstandard, leading-edge device that could be fitted to the outboard wing of the wind-tunnel model. (Note: At no time was it intended to fit such an alternative leading-edge device to the full-scale wing.) In this way, the modified outboard wing of the model behaved at wind-tunnel Reynolds number very much like the full-scale wing did in flight, and the nacelle vortex/wing boundary-layer interactions determined the stall in the wind tunnel.

Having now radically adjusted the wing stall patterns in the wind tunnel, we now turn our attention to necessary modifications of the nacelles, to improve the maximum lift performance. Relying on the flow-visualization work done previously, a simple fix was found which, at full-scale conditions required no change to the baseline high-lift system. Subsequently, the airplane with nacelles modified with the addition of "vortex control devices" (vortex generating vanes attached to the upper inboard sides of the nacelle cowls) was flown with the results shown in Fig. 35.

References

[1]Butter, D. J., and Williams, B. R., "The Development and Application of a Method for Calculating the Viscous Flow about High-Lift Airfoils," AGARD CP-291, Aug. 1980.

[2]Foster, D. N., Irwin, H. P. A. H., and Williams, B. R., "The Two-Dimensional Flow around a Slotted Flap," Royal Aircraft Establishment, Farnborough, England, UK, R&M 3681, Sept. 1970.

[3]Gault, D. E., "A Correlation of Low Speed Airfoil Section Stalling Characteristics With Reynolds Number and Airfoil Geometry," NACA TN-3963, 1957.

[4]Woodward, D. S., Hardy, B. C., and Ashill, P. R., "Some Types of Scale Effect in Low-Speed High-Lift Flows," International Council of the Aeronautical Sciences Paper 88-4.9.3, 1988.

[5]Morgan, H. L., Jr., Ferris, J. C., and McGhee, R. J., "A Study of High-Lift Airfoils at High Reynolds Numbers in the Langley Low Turbulence Pressure Tunnel," NASA TM-89125, July 1987.

[6]Klevenhusen, K. D., "A Calculation Method for Multielement Airfoils in Subsonic and Transonic Potential Flow," AIAA Paper 80-0340, Jan. 1980.

[7]Oskam, B., "Transonic Panel Method for the Full Potential Equation Applied to Multi-Component Airfoils," National Aerospace Lab., NLR, The Netherlands, Rept. MP-83030 U, May 1983.

[8]Wigton, L. B., "Application of MACSYMA and Sparce Matrix Technology to Multielement Airfoil Calculations," AIAA Paper 87-1142, June 1987.

[9]Stevens, W. A., Goradia, S. H., and Braden, J. A., "Mathematical Model for Two-Dimensional Multi-Component Airfoils in Viscous Flow," NASA CR-1843, July 1971.

[10]Seebohm, T., and Newman, B. G., "A Numerical Method for Calculating Viscous Flow Round Multiple-Section Aerofoils," *Aeronautical Quarterly*, Vol. 26, Pt. 3, Aug. 1975, pp. 176–188.

[11]Morgan, H. L., Jr., "A Computer Program for the Analysis of Multielement Airfoils in Two-Dimensional Subsonic, Viscous Flow," NASA SP-347, March 1975.

[12]Dvorak, F. A., and Woodward, F. A., "A Viscous/Potential Flow Interaction Analysis Method for Multi-Element Swept Wings," NASA CR-2476, Vol. 1, Nov. 1974.

[13]Brune, G. W., and Manke, J. W., "An Improved Version of the NASA/Lockheed Multi-Element Airfoil Analysis Computer Program," NASA CR-145323, March 1978.

[14]Oskam, B., "A Calculation Method of the Viscous Flow Around Multi-Component Airfoils," National Aerospace Lab., NLR, The Netherlands, TR-79097 U, Sept. 1980.

[15]Green, M. J., and Newling, J. C., "A Theoretical Method for 2D High-Lift Flow Prediction," Royal Aeronautical Society Symposium on High-Lift Aerodynamics, Cambridge, UK, 1986.

[16]Jacob, K., and Steinbach, D., "A Method for Prediction of Lift for Multi-Element Airfoil Systems with Separation," AGARD CP-143, April 1974.

[17]Henderson, M. L., "A Solution to the 2-D Separated Wake Modeling Problem and Its Use to Predict $C_{L_{max}}$ of Arbitrary Airfoil Sections," AIAA Paper 78-156, Jan. 1978.

[18]Mani, K. K., "A Multiple Separation Model for Multi-Element Airfoils," AIAA Paper 83-1844, July 1983.

[19]LeBalleur, J. C., and Neron, M., "Calcul d'écoulements visqueux décollés sur profils d'ailes par une approche de couplage," AGARD CP-291, Paper 11, Aug. 1980.

[20]Olson, L. E., "Optimization of Multi-Element Airfoils for Maximum Lift," NASA CP-2045, Vol. 1, March 1978.

[21]King, D. A., and Williams, B. R., "Developments in Computational Methods for High-Lift Aerodynamics," *Aeronautical Journal*, Vol. 92, No. 917, Aug./Sept. 1988, pp. 265–288.

[22]Shima, E., "Numerical Analysis of Multiple Element High-Lift Devices by Navier Stokes Equation Using Implicit TVD Finite-Volume Method," AIAA Paper 88-2574-CP, June 1988.

[23]Mavriplis, D. J., "Accurate Multigrid Solution of the Euler Equations on Unstructured and Adaptive Meshes," Institute for Computer Applications in Science & Engineering Rept. 88-40, June 1988.

[24]Mavriplis, D. J., "Adaptive Mesh Generation for Viscous Flows Using Delaunay Triangulation," 2nd International Conference on Numerical Grid Generation in CFD, Miami Beach, FL, Dec. 1988.

[25]Fritz, W., "Numerical Simulation of 2-D Turbulent Flow Fields with Strong Separation," International Council of the Aeronautical Sciences Paper 88-4, 6.4, 1988.

[26]Rice, J. G., Schnipke, R. J., Cornelius, D. K., and Normansell, M. D., "Navier-Stokes Computation of a Typical High-Lift Airfoil System," 4th International Symposium on Science and Engineering on Cray Supercomputers, Minneapolis, MN, Oct. 1988.

[27]Bristow, D. R., and Hawk, J. D., "A Mixed Analysis and Design Surface Panel Method for Two-Dimensional Flows," AIAA Paper 82–0022, Jan. 1982.

[28]Narramore, J. C., and Beatty, T. D., "An Inverse Method for the Design of Multielement High-Lift Systems," AIAA Paper 75-879, June 1975.

[29]LeBalleur, J. C., "Numerical Viscid-Inviscid Interaction in Steady and Unsteady Flows," Second Symposium on Numerical and Physical Aspects of Aeronautical Flows, Long Beach, CA, Jan. 1983.

[30]East, L. F., "A Representation of Second-Order Boundary Layer Effects in the Momentum Integral Equation and in Viscous-Inviscid Interaction," Royal Aircraft Establishment, Farnborough, UK, TR-81002, 1981.

[31]Van den Berg, B., "Boundary Layer Measurements on a Two-Dimensional Wing with Flap," National Aerospace Lab., NLR, The Netherlands, TR 79009 U, Jan. 1979.

[32]Van den Berg, B., and Oskam, B., "Boundary Layer Measurements on a Two-Dimensional Wing with Flap and a Comparison with Calculations," National Aerospace Lab., NLR, The Netherlands, Rept. MP-79034 U, Sept. 1979.

[33]Oskam, B., Laan, D. J., and Volkers, D. F., "Recent Advances in Computational Methods to Solve the High-Lift Multi-Component Airfoil Problem," National Aerospace Lab., NLR, The Netherlands, Rept. MP-84042 U, April 1984.

[34]Wentz, W. H., Jr., and Seetharam, H. C., "Development of a Fowler Flap System for a High Performance General Aviation Airfoil," NASA CR-2443, 1974.

[35]Brune, G. W., and Sikavi, D. A., "Experimental Investigation of the Confluent Boundary Layer of a Multielement Low Speed Airfoil," AIAA Paper 83-0566, Jan. 1983.

[36]Smith, A. M. O., "High-Lift Aerodynamics," *Journal of Aircraft*, Vol. 12, June 1975, pp. 501–530.

[37]Schlichting, H., *Boundary Layer Theory*, 7th ed., McGraw-Hill, New York, 1979.

[38]Baldwin, B. S., and Lomax, H., "Thin Layer Approximation and Algebraic Model for Separated Turbulent Flows," AIAA Paper 78-257, Jan. 1978.

[39]Lock, R. C., and Williams, B. R., "Viscous-Inviscid Interactions in External Aerodynamics," *Progress in Aerospace Sciences*, Vol. 24, Aug. 1986, pp. 51–171.

[40]Squire, H. B. and Young, A. D., "The Calculation of the Profile Drag of Airfoils," British Aeronautical Research Council, RM-1838, 1938.

[41]Olson, L. E., James, W. D., and McGowan, P. R., "Theoretical and Experimental Study of the Drag of Multielement Airfoils," AIAA Paper 78-1223, July 1978.

[42]Butter, D. J., "Research into Multi-Element Airfoil Aerodynamics," Third Symposium on Numerical and Physical Aspects of Aerodynamic Flows, Long Beach, CA, Jan. 1985.

[43]Moghadam, A., and Squire, L. C., "The Mixing of Three-Dimensional Turbulent Wakes and Boundary Layers," *Aeronautical Journal*, May 1989, pp. 153–161.

[44]Goldhammer, M. I., "A Lifting Surface Theory for the Analysis of Nonplanar Lifting Systems," AIAA Paper 76-16, Jan. 1976.

[45]Feifel, W. M., "Optimization and Design of Three-Dimensional Aerodynamic Configurations of Arbitrary Shape by a Vortex Lattice Method," *Vortex Lattice Utilization*, NASA, SP-405, 1976.

[46]Magnus, A. E., "PAN AIR Computer Program for Predicting Subsonic or Supersonic Linear Potential Flows About Arbitrary Configurations Using a Higher Order Panel Method," NASA CR-3251, Vol. 1, Theory Doc. (Version 1.0), 1980.

[47]Tinoco, E. N., Ball, D. N., and Rice, F. A. II, "PAN AIR Analysis of a Transport High-Lift Configuration," *Journal of Aircraft*, Vol. 24, March 1987, pp. 181–187.

[48]Maskew, B., and Dvorak, F. A., "Predicting Dynamic Separation Characteristics of General Configurations," AIAA Paper 86-1813, June 1986.

[49]Johnson, F. T., James, R. M., Bussoletti, J. E., Woo, A. C., and Young, D. P., "Transonic Rectangular Grid Embedded Panel Method," AIAA Paper 82-0953, June 1982.

[50]McVeigh, M. A., and Kisielowski, E., "A Design Summary of Stall Characteristics of Straight Wing Aircraft," NASA CR-1646, June 1971.

[51]Anderson, J. D., Jr., Corda, S., and Van Wie, D. M., "Numerical Lifting Line Theory Applied to Drooped Leading-Edge Wings Below and Above Stall," *Journal of Aircraft*, Vol. 17, Dec. 1980, pp. 898–904.

[52]Levinsky, E. S., "Theory of Wing Span Loading Instabilities Near Stall," AGARD CP-204, Sept. 1976.

[53]Jacob, K., "Advanced Method for Computing the Flow Around Wings With Rear Separation and Ground Effect," German Aerospace Research Establishment DFVLR, FB 86-17, Jan. 1986.

[54]Henderson, M. L., "Inverse Boundary Layer Technique for Airfoil Design," *Advanced Technology Airfoil Research*, Vol. 1, NASA CP-2045, Pt. 1, March 1978.

[55]McMasters, J. H., and Henderson, M. L., "Low-Speed Single-Element Airfoil Synthesis," *Science and Technology of Low-Speed and Motorless Flight*, NASA CP-2085, Pt. 1, March 1979.

[56]McLean, J. D., "Three-Dimensional Turbulent Boundary Layer Calculations for Swept Wings," AIAA Paper 77-3, Jan. 1977.

[57]Wedderspoon, J. R., "The High-Lift Development of the A320 Aircraft," International Council of the Aeronautical Sciences Paper 86-2.3, 1986.

[58]Szodruch, J., and Schnieder, H., "High-Lift Aerodynamics for Transport Aircraft by Interactive Experimental and Theoretical Tool Development," AIAA Paper 89-0267, Jan. 1989.

[59]Lock, R. C., "Equivalence Law Relating Three- and Two-Dimensional Pressure Distributions," British Aeronautical Research Council, London, R&M 5346, May 1962.

[60]Murillo, L. E., and McMasters, J. H., "A Method for Predicting Low-Speed Aerodynamic Characteristics of Transport Aircraft," *Journal of Aircraft*, Vol. 21, March 1984, pp. 168–174.

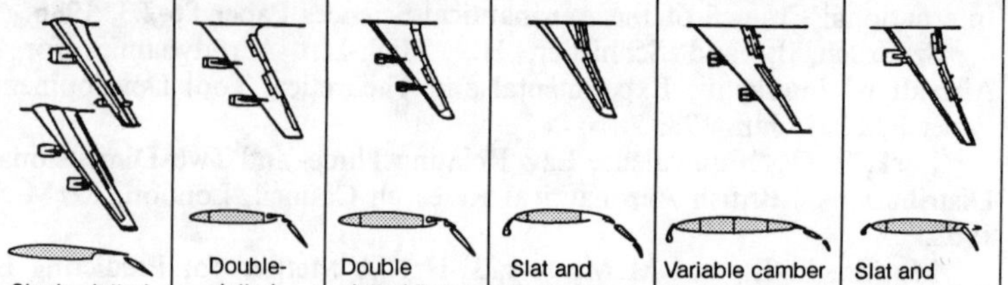

Type	B-47/B-52	367-80/ KC-135	707-320/ E-3A	727	747/E-4A	767
First flight	1947/1952	1954	1962	1963	1969	1981
Planform						
Typical airfoil	Single-slotted fowler flap	Double-slotted flap	Double slotted flap and Krueger leading edge	Slat and triple-slotted flap	Variable camber Krueger and triple-slotted flap	Slat and single-slotted flap
$C_{L_{max}}$	1.8	1.78	2.2	2.79	2.45	2.45

Fig. 1   Trends in Boeing transport high-lift system development.

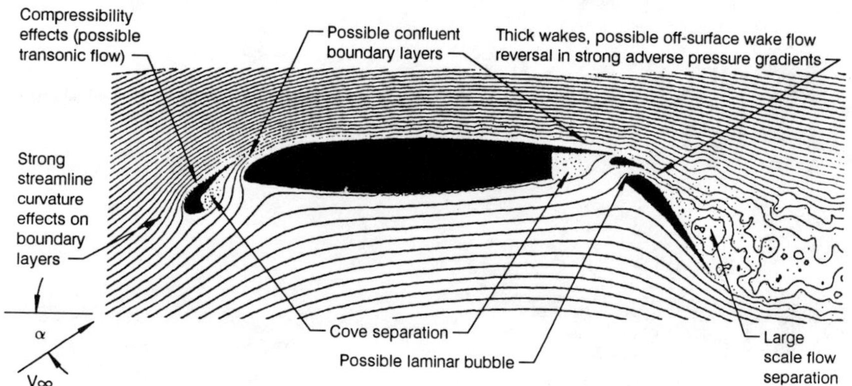

Fig. 2   Illustration of multielement airfoil problem (adapted from computation by Spalart, NASA Ames).

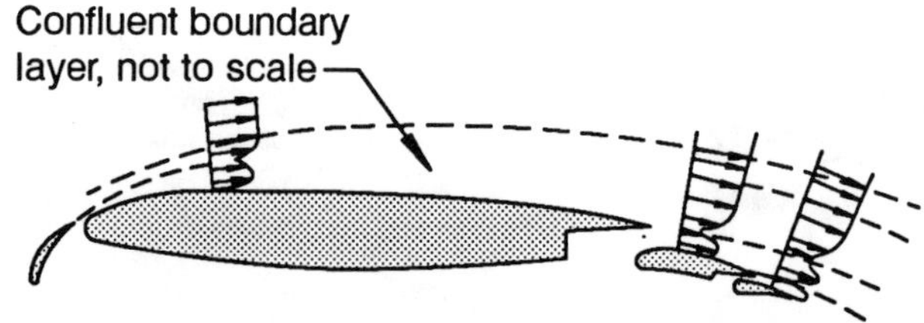

Fig. 3   Confluent boundary layer on upper surface of slotted multielement airfoil.

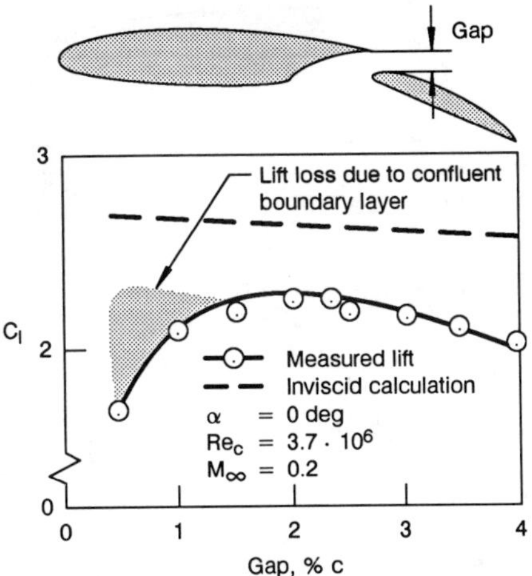

Fig. 4   Effect of confluent boundary layer on airfoil lift (data from Foster et al.[2]).

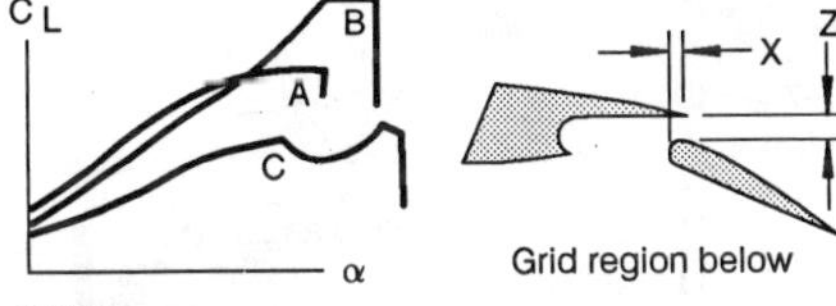

Notes:
A   Conventional lift curve.
B   Flap separations improving at high incidence.
C   Excessive separations, lift curve collapse.

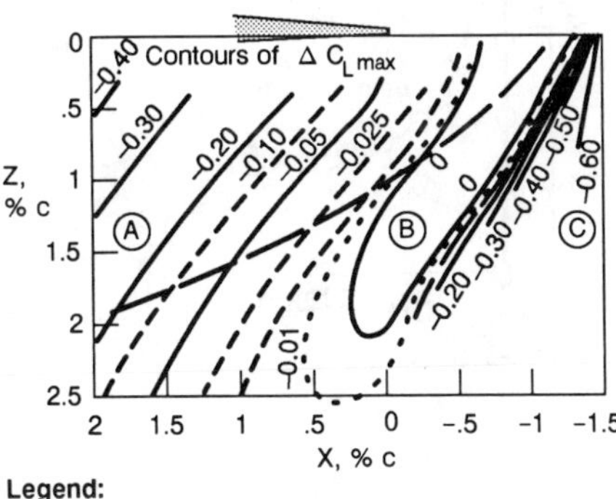

Legend:
‐ ‐ ‐ Flap flow development indicated by boundaries

Fig. 5   Effect of flap position on airfoil lift (from Woodward et al.[4]).

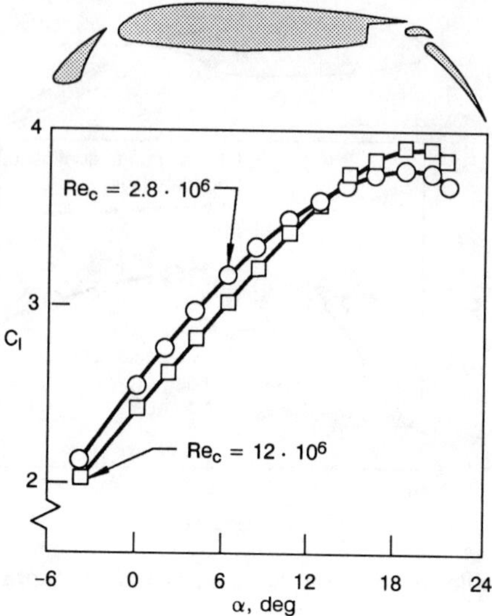

Fig. 6   Inverse Reynolds number effect.[5]

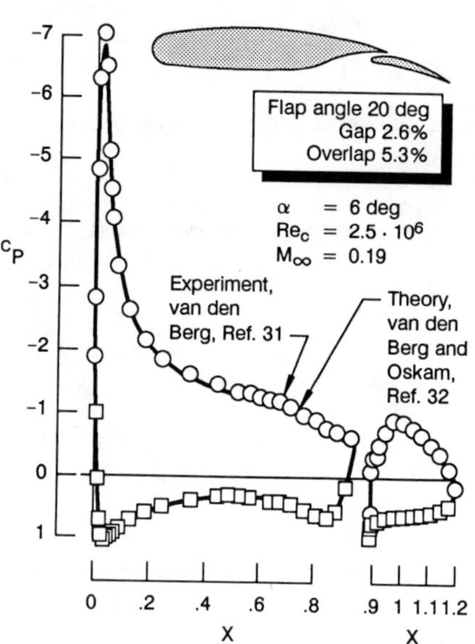

Fig. 7   Prediction of surface pressures of NLR airfoil.

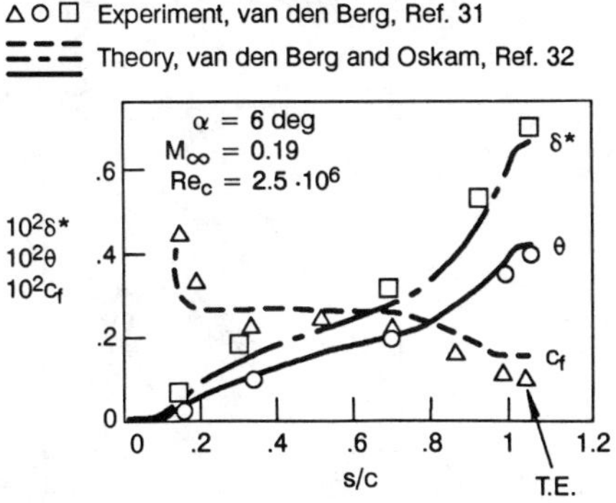

Fig. 8 Boundary-layer parameters on upper wing surface of NLR airfoil.

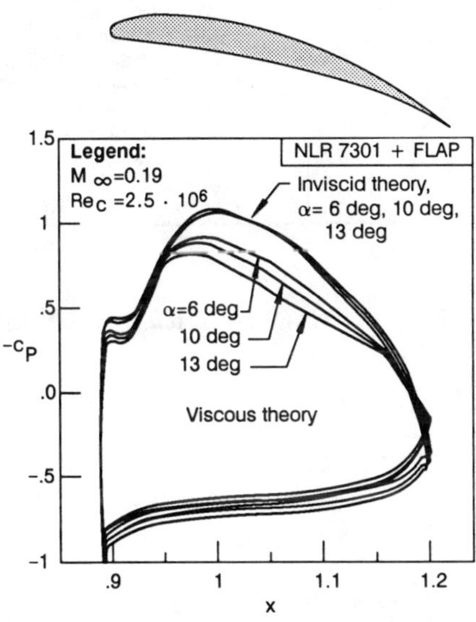

Fig. 9 Theoretical predictions of viscous and inviscid flap pressures by Oskam et al.[33]

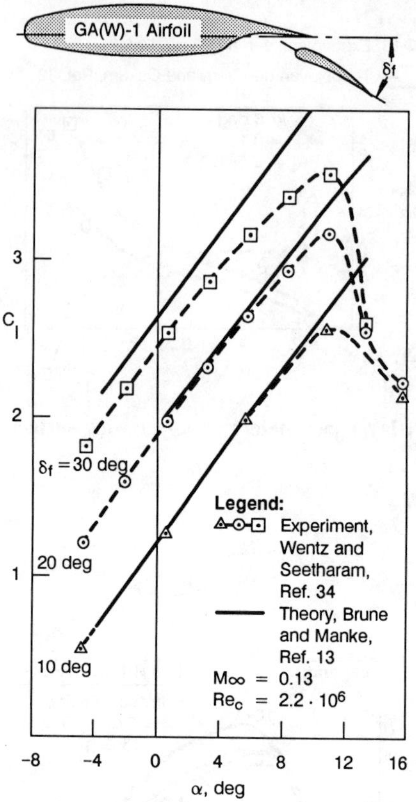

Fig. 10   Lift changes of GA(W)-1 airfoil with flap deflection.

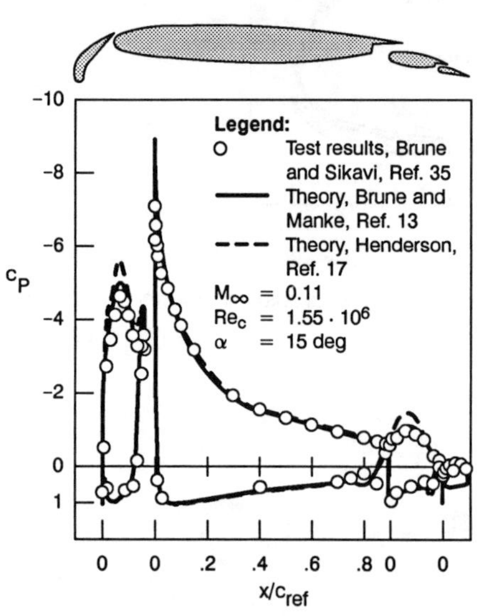

Fig. 11   Surface pressure of airfoil with takeoff flap setting.

Note: Krueger leading edge flap positioned to produce strong confluent boundary layer on main element.

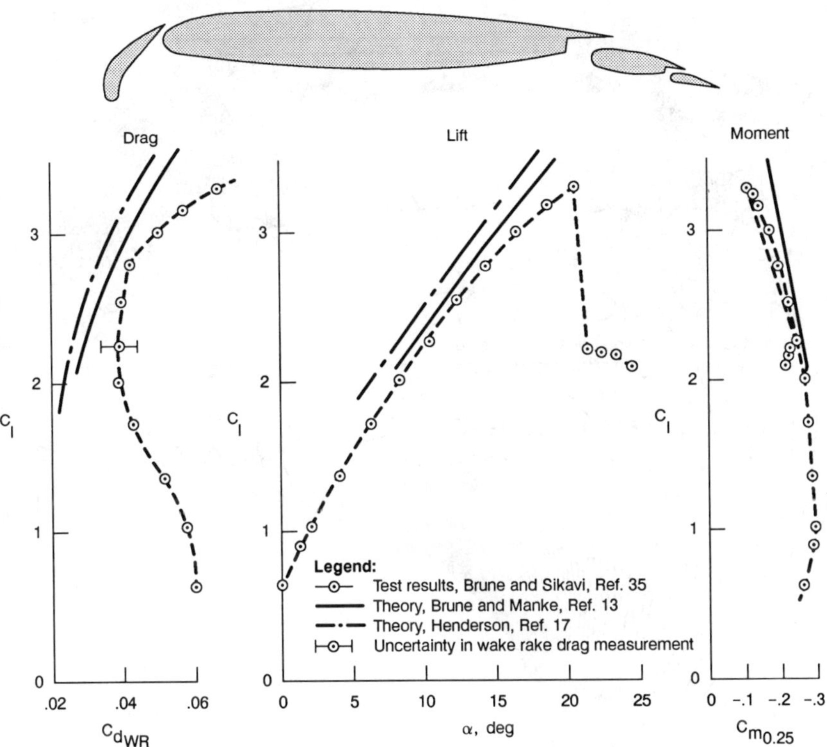

Fig. 12 Characteristics of airfoil with strong confluent boundary-layer effect.

a) Unstructured and adaptive mesh

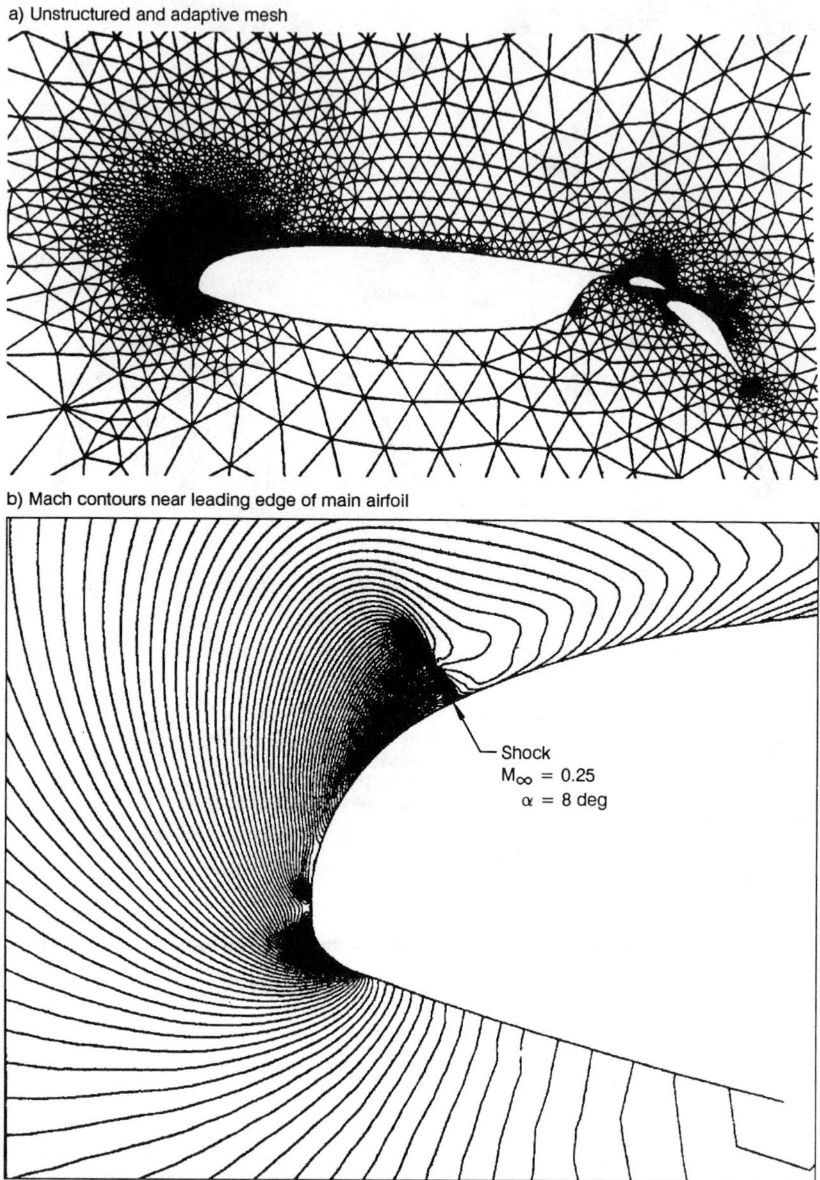

Fig. 13   Euler solution of Mavriplis[24] with local transonic flow: a) unstructured and adaptive mesh; b) Mach contours near leading edge of main airfoil.

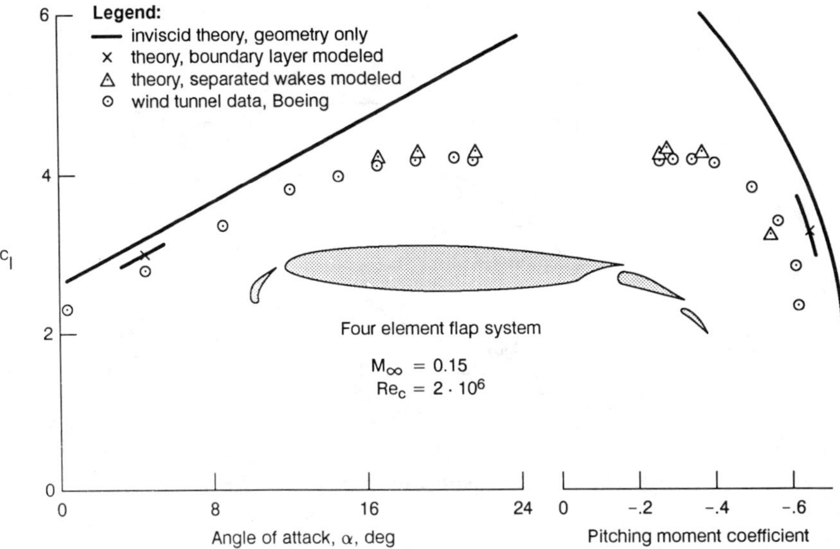

Fig. 14 Airfoil stall predictions of Henderson.[17]

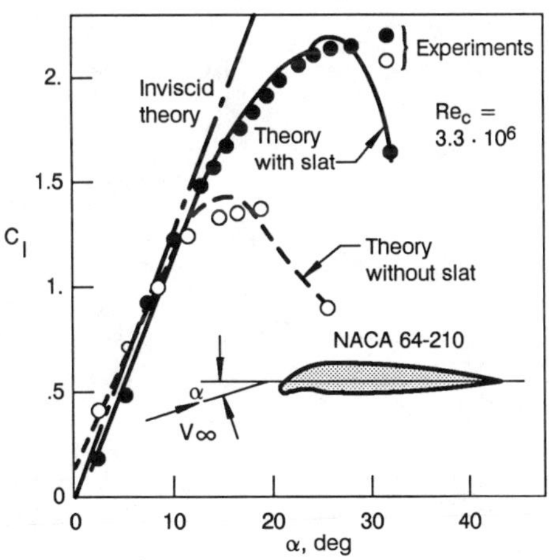

Fig. 15 Airfoil stall predictions of Jacob and Steinbach.[16]

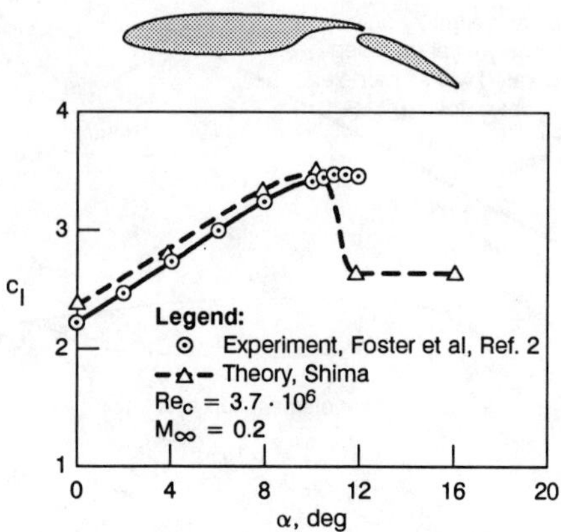

Fig. 16　Navier-Stokes stall calculations of Shima.[22]

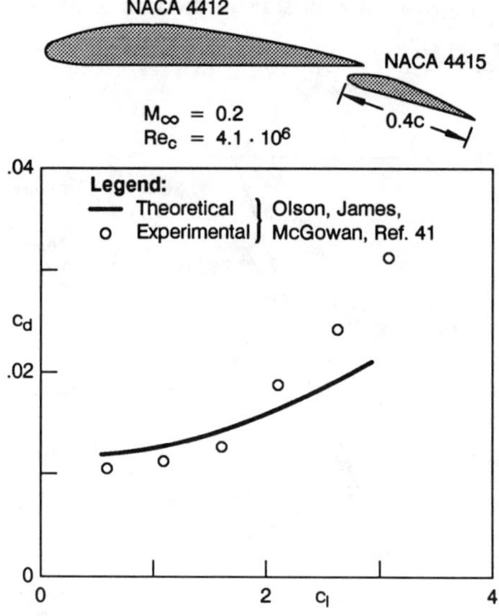

Fig. 17　Drag prediction based on the wake momentum defect method.

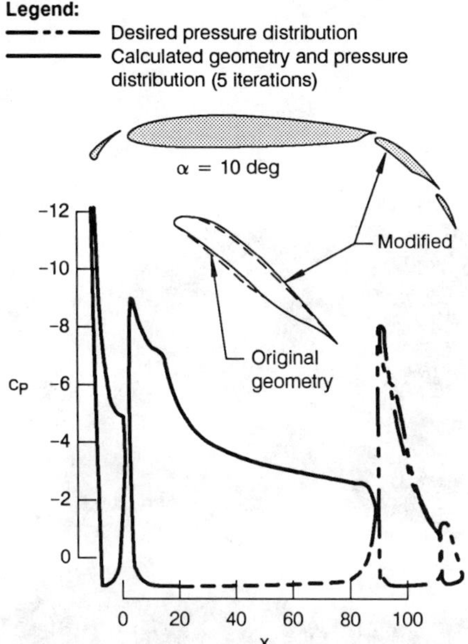

Fig. 18   Inviscid flap shape design by Narramore and Beatty.[28]

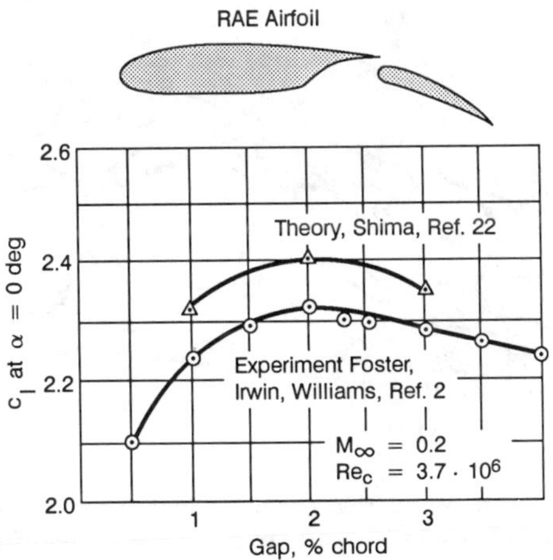

Fig. 19   Gap optimization at fixed angle of attack.

a) China clay photo of separation pattern

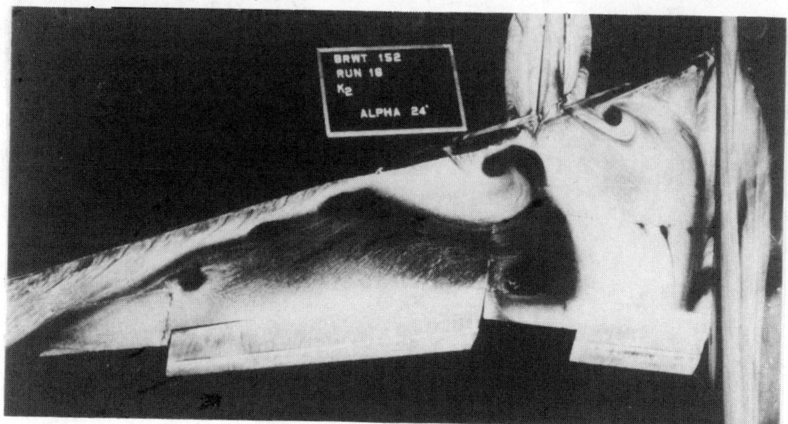

b) Interpretation of wing surface streamlines using critical point theory

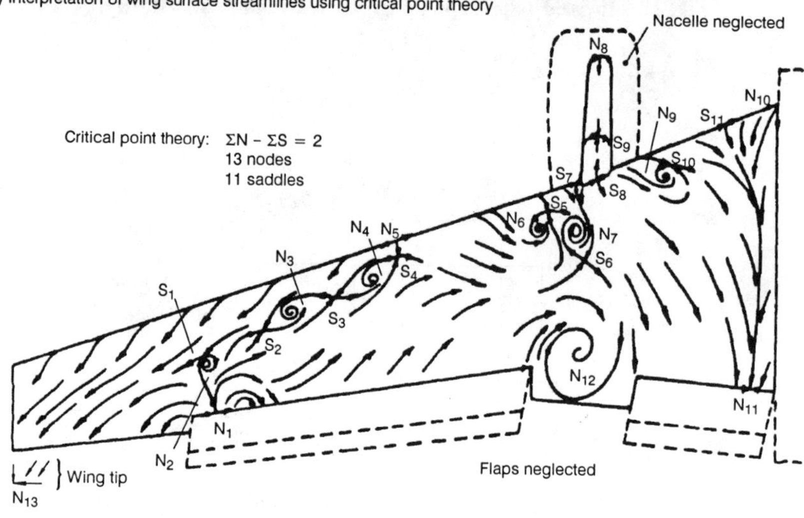

Fig. 20　Separated flow on transport wing in high-lift configuration at $\alpha = 24$ deg: a) China clay photo of separation pattern; b) interpretation of wing surface streamlines using critical point theory.

Fig. 21 Isobar pattern in wake of transport airplane model in high-lift configuration (courtesy of J. P. Crowder, Boeing Commercial Airplanes).

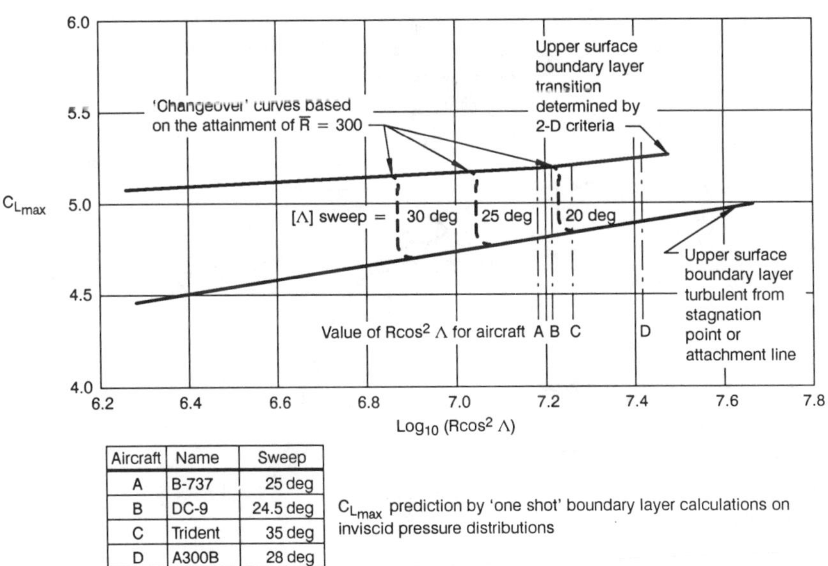

Aircraft	Name	Sweep
A	B-737	25 deg
B	DC-9	24.5 deg
C	Trident	35 deg
D	A300B	28 deg

Fig. 22 $C_{L_{max}}$ variation with Reynolds number predicted by Woodward et al.[4]

Fig. 23   Paneling of a complex high-lift configuration in VSAERO.

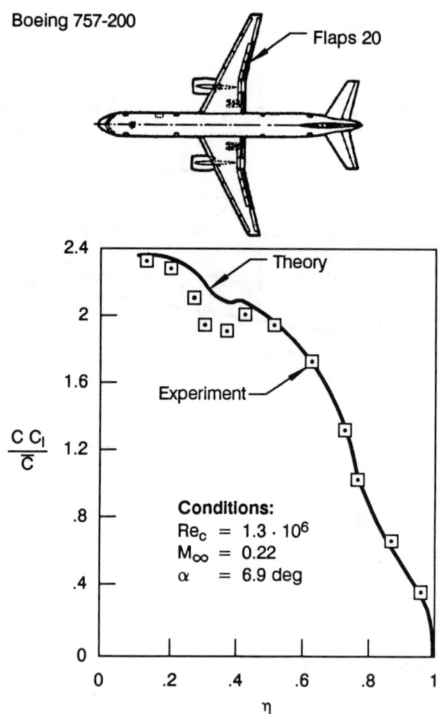

Fig. 24   Quasi-three-dimensional viscous calculation of spanload of transport airplane in takeoff configuration.

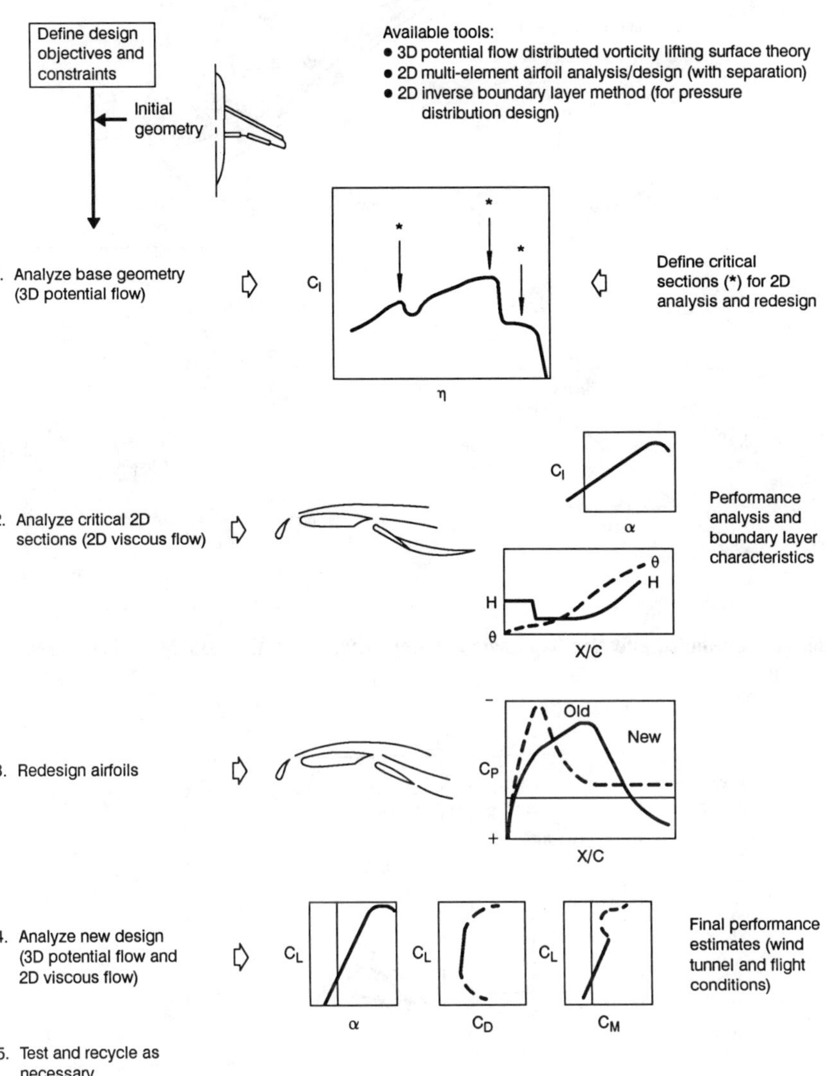

Fig. 25   High-lift analytic design procedure.

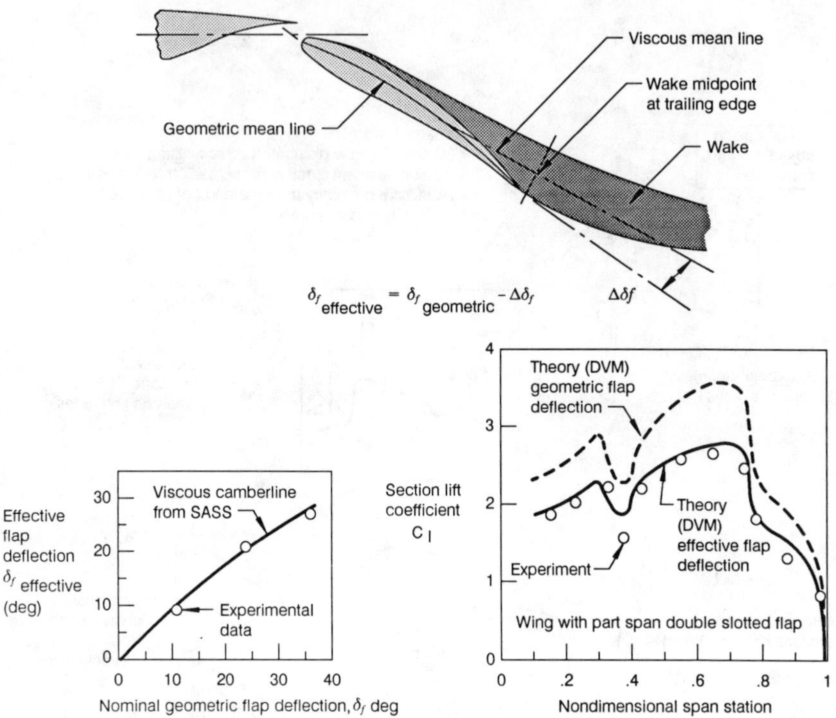

Fig. 26   Predicted effective flap angle and influence of effective flap deflection on span loading.

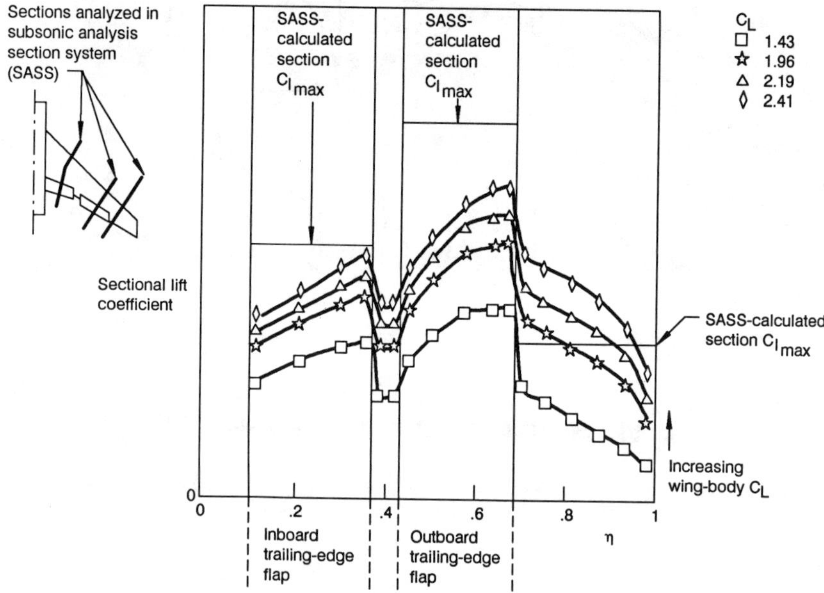

Fig. 27   Span loading on baseline 747-200, flaps 30, calculated with SASS.

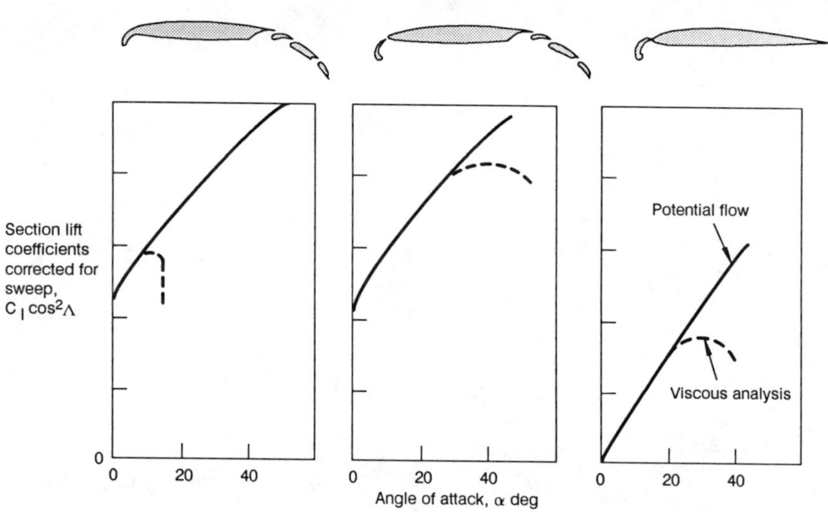

Fig. 28 SASS-calculated sectional lift curves, $\delta_f = 30$ deg.

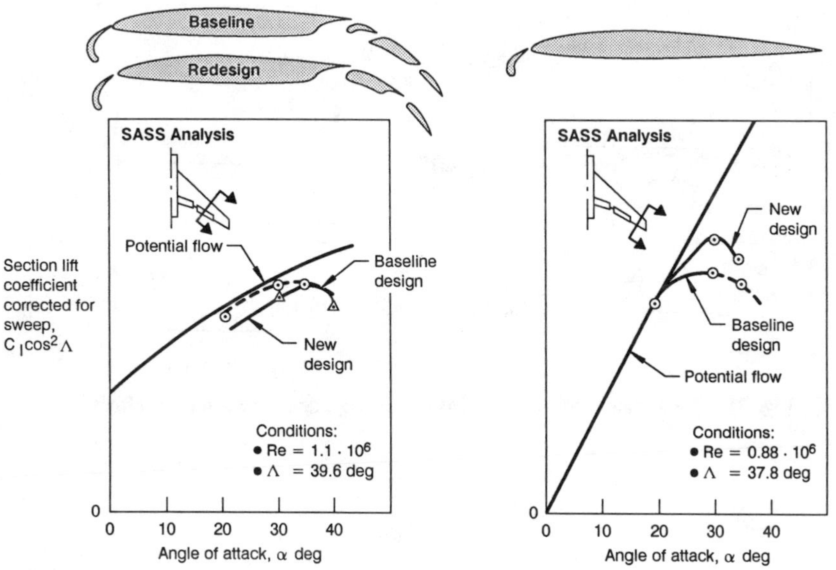

Fig. 29 SASS analysis of baseline and redesigned flaps.

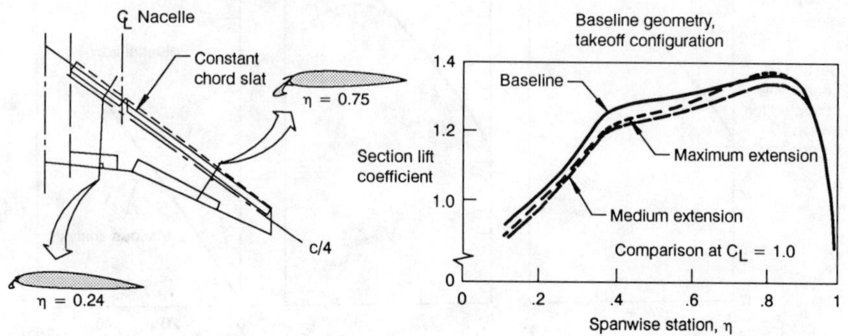

Fig. 30   Baseline geometry and computed span loading for slat redesign problem.

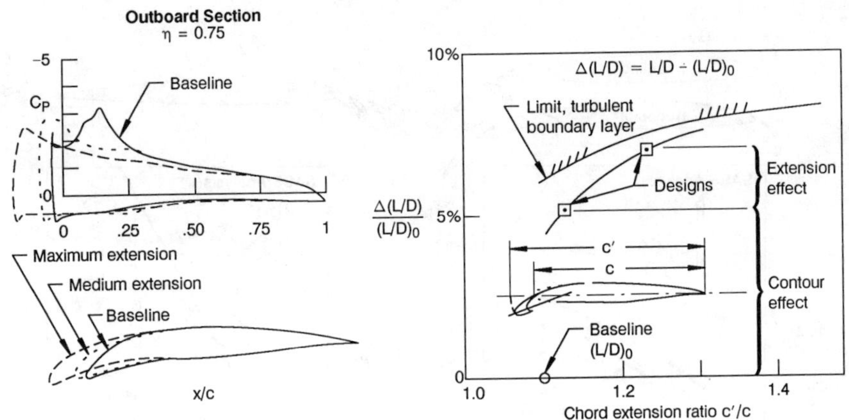

Fig. 31   Predicted results of improved leading-edge device design effort.

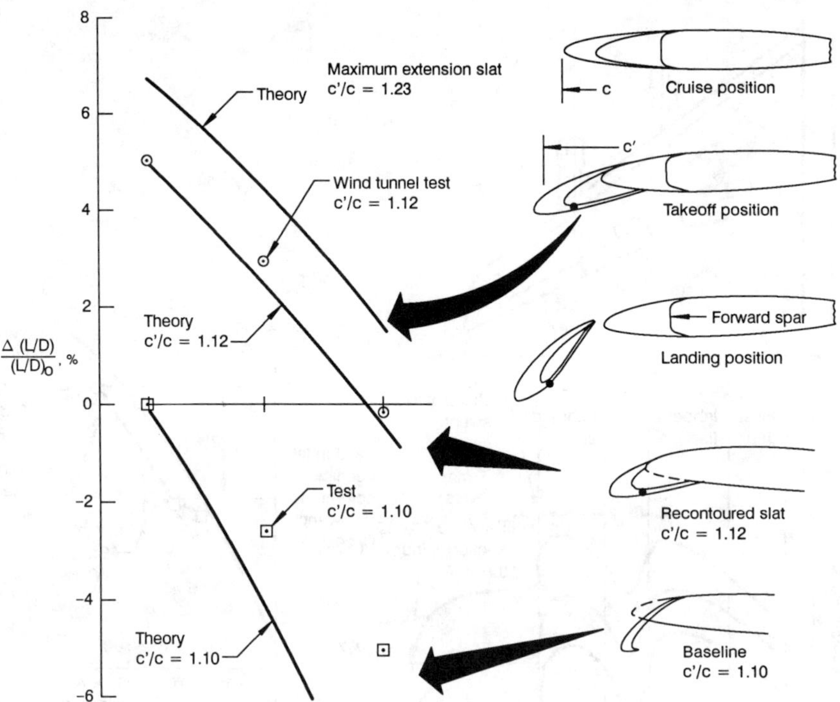

Fig. 32   Test-theory comparison of redesigned leading-edge device.

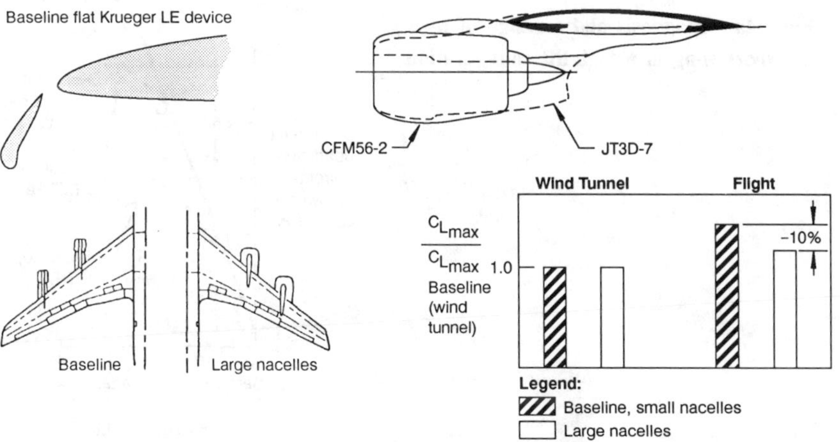

Fig. 33   Nacelle influence on $C_{L_{max}}$.

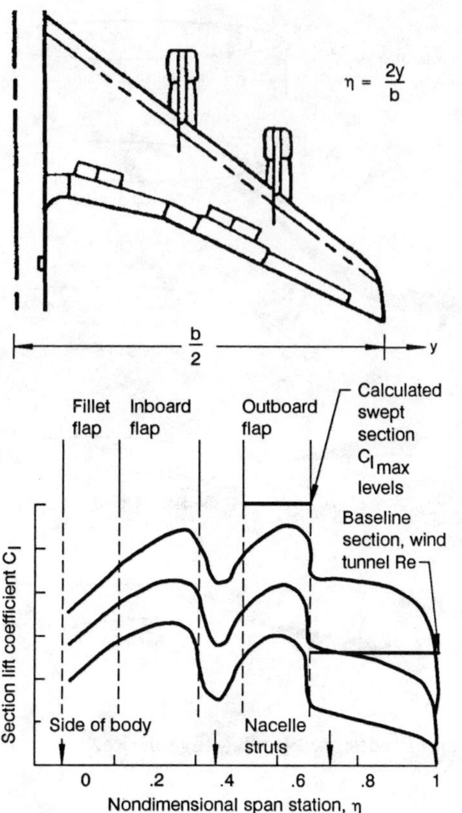

Wing span load distributions (δ_f, approach setting) for various values of wing lift coefficient lifting surface theory predictions

Fig. 34a   Diagnosis and redesign of a transport wing in a high-lift configuration.

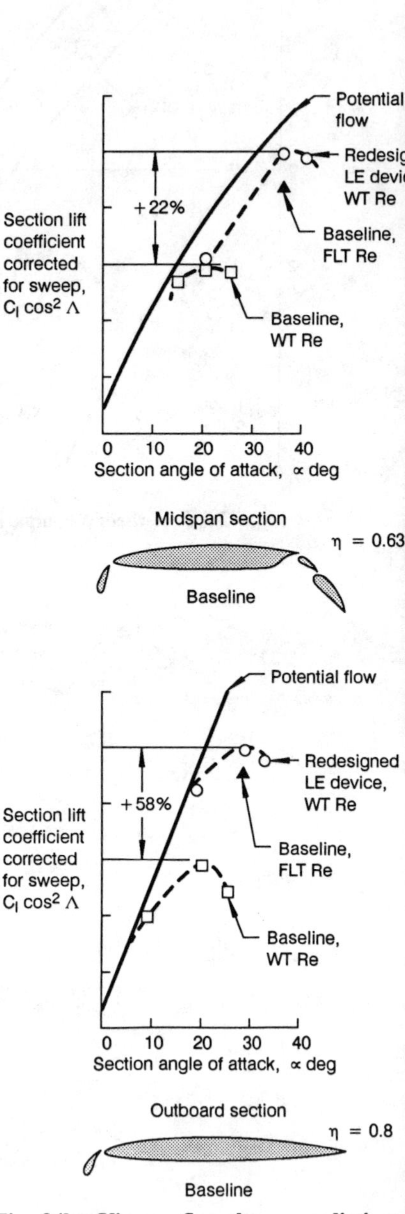

Fig. 34b   Viscous flow theory predictions.

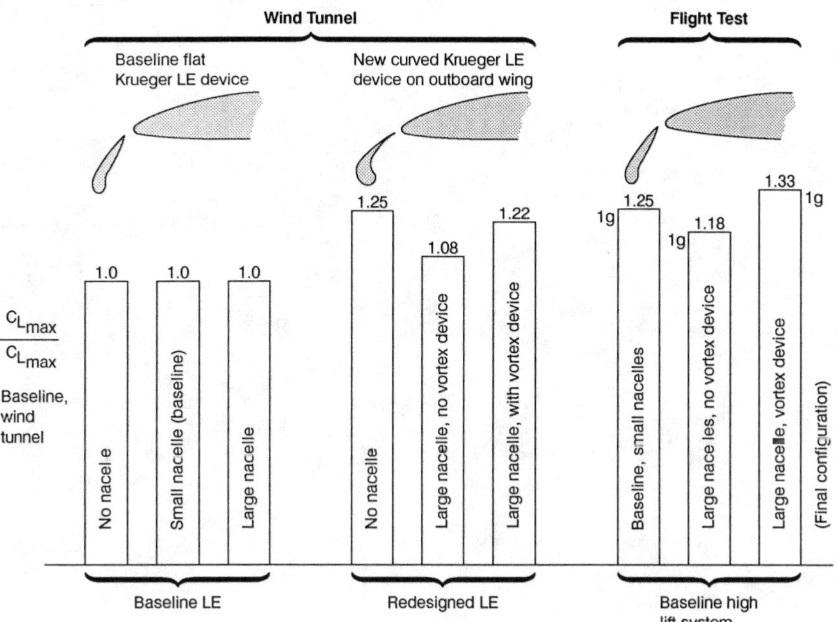

Fig. 35   Solution to $C_{L_{max}}$ problem.

Chapter 11

Development of High-Lift Wing Modifications for an Advanced Capability EA-6B Aircraft

Edgar G. Waggoner*
NASA Langley Research Center, Hampton, Virginia

Introduction

TREMENDOUS progress has been made during the past decade in computational fluid dynamics. Advances in computer hardware have spawned code and algorithm developments that are changing the complexion of applied computational aerodynamics. The availability of supercomputers has diminished the impact that operational speed and memory once had on code development. However, one must recognize that, as one passes from the arena of code or algorithm development to code application for practical configurations, the issues change. In the real world of finite resources and time constraints, there is a constant and continual tradeoff between geometric realism, modeling of the flow physics, and computational resources. One must be cognizant of issues related to code calibration and validation, reliability and robustness, and throughput requirements. Even though computational capabilities have expanded greatly, these issues still impose constraints on the use of computational fluid dynamics (CFD) methods to solve practical problems.

There are a significant number of aerodynamic problems an aerodynamicist faces for which transonic potential flow solutions are not only adequate but are often the only solutions available. Extended small-disturbance and full-potential equation solvers have been relied upon extensively at NASA Langley Research Center to provide reliable guidance in support of several developmental efforts. During these efforts opportunities have arisen to develop extensions to the basic analysis methods for specific applications. Wind-tunnel testing of scaled models and subsequent analyses and comparison with flight data have identified a need to predict aeroelastic effects[1] and wind-tunnel wall interference effects.[2] Laminar flow research on flight vehicles has emphasized the need for automated wing design.[3,4] These extensions have been reported in the literaturae and at technical meetings. These include an automated design procedure for airfoil sections and wings that modifies a contour to develop a specified pressure distribu-

Copyright © 1989 by the American Institute of Aeronautics and Astronautics, Inc. No copyright is asserted in the United States under Title 17, U.S. Code. The U.S. Government has a royalty-free license to exercise all rights under the copyright claimed herein for Governmental purposes. All other rights are reserved by the copyright owner.
*Head, Subsonic Aerodynamics Branch, Applied Aerodynamics Division.

tion. Also, an interactive structural and aerodynamic analysis method has been developed to predict the effect of static aeroelastic deflections on configuration aerodynamics. In addition, a method has been developed to predict the interference effect of various types of wind-tunnel walls on the aerodynamics of a model in a wind tunnel. In these cases necessity has indeed been the mother of invention. Although this chapter will not address these research efforts, the interested reader will find the cited references informative.

Over the past three to four years, several applications of computational methods to practical transonic aerodynamic problems have been addressed at NASA Langley. These have involved design, modifications, and/or experimental support for various developmental or research programs. Specific applications have included wing modifications to improve the high-lift capability of the EA-6B,[5] a laminar flow glove design to support the F-14 variable-sweep transition flight experiment,[3] and the design of a laminar flow wing for the Gulfstream Peregrine.[6] Although the experiences encountered during application of potential flow analysis and design codes during these efforts is of interest, it was felt that, within the spirit of this book, a more meaningful and lasting chapter would discuss one of these applications in detail rather than present an overview of the work. The following discussion will be devoted to the efforts involved in wing modifications for the Navy's EA-6B Prowler. The discussion will present an overview of the computational and experimental effort, brief descriptions of the codes used, details of the computational design procedures and experimental verification efforts, and summary observations.

EA-6B High-Lift Wing Modification

NASA Langley Research Center has been involved in a cooperative program with the Navy and Grumman aircraft systems to improve the maneuvering capability of the EA-6B aircraft. This cooperative program is one facet of the Navy's Advanced Capability Program (ADVCAP) for the EA-6B. The EA-6B aircraft is an electronics countermeasures aircraft that evolved from the A-6 attack aircraft. The aircraft is a four-place, twin engine configuration with a midfuselage mounted wing. The wing is swept back 25 deg at the quarterchord and has an aspect ratio of 5.3 (Fig. 1). Wing contours are based on the NACA 64 series airfoils. The EA-6B and the A-6 configurations are somewhat different. The EA-6B has an extended fuselage (54.5 in.), a pod-shaped fairing on top of the vertical tail, and an engine with an approximately 20% increase in sea-level thrust. However, the planforms, section contours, and aerodynamics of the wings for the two aircraft are very similar. One of the most obvious differences between the two aircraft is their weights. The A-6 has a maximum landing weight of 36,000 lb, and the EA-6B has a maximum landing weight of 45,500 lb. This coupled with virtually the same wing characteristics results in significantly reduced stall maneuver margins. For example, at 2 g and 250 knots in a 60-deg banked turn the velocity above stall speed is reduced almost 60% for the EA-6B relative to the A-6.

One objective of this phase of the maneuver improvement program was to investigate the design of relatively simple wing modifications that could be incorporated into the aircraft design to improve the low-speed, high-lift capability. This was to be accomplished with minimum degradation in high-speed cruise performance. A variety of two- and three-dimensional low-speed and transonic computational techniques were employed during the design effort. Modifications to the leading-edge slat and trailing-edge flap contours were evaluated computationally and the results verified experimentally during extensive two- and three-dimensional testing in several experimental facilities at NASA Langley. Modifications were defined that yielded significantly increased maximum lift at low speed with minimal impact on high-speed cruise performance.

Computational Techniques

Four computer codes were used in the design and evaluation of the wing modifications during this study. Three-dimensional transonic analyses were performed using a full-potential analysis code (TAWFIVE)[7] and an extended small-disturbance analysis code (WBPPW),[8] each of which has been verified extensively at the NASA Langley Research Center. These codes are widely used throughout the industry to provide reliable solutions of transonic, three-dimensional flowfields. The NYU airfoil analysis code[9] was used to provide two-dimensional transonic analysis. The two-dimensional low-speed MCARF code[10] was used to study changes in maximum lift coefficient due to various modifications. Salient features of the computational techniques are presented in the following discussion.

TAWFIVE Analysis Code

A computer code for the Transonic Analysis of a Wing and Fuselage with Interacted Viscous Effects (TAWFIVE) was used in the study. The code utilizes the interaction of an inviscid and a viscous flow solver to obtain transonic flowfield solutions about wing-fuselage combinations. The outer inviscid flowfield is solved using a conservative, finite-volume, full-potential method based on FLO-30 by Caughey and Jameson.[11] No modifications were made to the internal grid-generation algorithm in FLO-30, which is a body-fitted, sheared, parabolic coordinate system.

Viscous effects are computed using a compressible integral method that calculates three-dimensional boundary layers for wings. The code has the capability of computing laminar or turbulent boundary layers with the method of Stock[12] and Smith,[13] respectively. An important addition to the code is Streett's treatment of the wake.[14] The wake model used in FLO-30 was replaced with a model that satisfies flow tangency on the wake displacement body and the pressure jump condition resulting from wake curvature. These changes in the code can make significant differences in results obtained on various configurations.[14]

WBPPW Analysis Code

The Wing-Body-Pod-Pylon-Winglet (WBPPW) code, developed by Charles Boppe[8] of Grumman Aerospace Corporation, is characterized by a unique grid-embedding technique that provides excellent flowfield resolution about various configuration components. The code solves for the flowfield about a wing-fuselage configuration that can include engine pods or stores, wing pylons, and wing-tip-mounted winglets at transonic speeds. Using finite-difference approximations, a modified small-disturbance potential-flow equation is iteratively solved in a system of multiple embedded grids. The modifications to the classical small-disturbance equation are in the form of extra terms, which, when added to the equation, provide more accurate resolution of shock waves with large sweep angles and a better approximation of the critical velocity where the full-potential equation changes from elliptic to hyperbolic type.

The computational space used in the method is filled with a relatively crude global grid system. This grid is stretched to planes corresponding to infinity in all directions. The global grid basically serves two purposes. It provides the proper representation of the effects of the configuration on the far field and, conversely, the effects of the far-field conditions on the flowfield near the configuration. In addition, the crude grid provides the channels of communication between the various embedded fine grids.

Fine grid regions around components of interest are embedded into the global continuous grid. The fine grids are distributed along the wing span and, if desired, may also encompass the fuselage, engine pods or stores, pylons, and/or a winglet. Within the fine grids, the resolution is much enhanced relative to the global grids. This allows for greater resolution in areas where flowfield gradients are large.

Viscous effects are approximated in the code by coupling a modified Bradshaw boundary-layer computation to the finite-difference potential-flow solution. The modified method provides a technique to extend a two-dimensional boundary-layer calculation to account for the first-order sweep effects.[15] The viscous effects are incorporated in the solution by adding the boundary-layer displacement slopes to the wing surface slopes. This modifies the wing surface to an equivalent "fluid" wing shape, which is then analyzed by the potential flow code.

NYU Airfoil Analysis Code

The New York University airfoil analysis code written by Bauer et al.[9] is used extensively by many researchers to provide transonic two-dimensional viscous analysis of airfoils. The inviscid solution solves for the steady, isentropic, irrotational flow about an airfoil contour. Viscous corrections are provided by adding the turbulent displacement thickness to the airfoil surface. There is no laminar boundary layer calculated by the code. The momentum thickness is initialized at the transition point, which can be set arbitrarily. By use of the turbulent boundary-layer method of Nash and Macdonald,[16] the boundary-layer characteristics are computed from the results of the potential-flow analysis and the airfoil geometric characteristics.

MCARF Two-Dimensional High-Lift Code

The two-dimensional subsonic panel code, MCARF,[10] is used to evaluate low-speed, high-lift characteristics. The inviscid solution is found using a distributed vortex concept to solve Laplace's equation. Compressibility effects are represented through the use of the Karman-Tsien correction. Viscous effects are included by iteratively updating the airfoil shape to include the displacement thicknesses from both laminar and turbulent boundary-layer calculations. A laminar boundary layer is calculated using the basic approach of Cohen and Reshotko.[17] The instability criteria of Schlichting-Ulricht-Granville[18,19] is used in the determination of the transition location, and Morgan's method[20] is used to predict laminar stall. Two turbulent boundary-layer models are used. The first, an approximate method developed by Goradia, is used in the early inviscid-viscous iteration cycles; the second model, based on the method of Nash,[21] is more accurate and is used for the final cycles.

Computational Design Problem

The objective of the reported study was to increase the low-speed maximum lift capability of the EA-6B aircraft with minimal degradation in high-speed performance. The approach was to make minor modifications to the wing contour that could easily be incorporated into the wing. The modifications were constrained to the leading-edge slat and trailing-edge flap elements of the wing. The extent of the wing planform over which the modifications were allowed is shown in Fig. 1. Note that the span of the slat and flap elements encompass approximately half of the wing span. By constraining the modifications to only these regions on the wing, hardware requirements for experimental wind-tunnel verification of the modifications involved fabricating only new slat and flap elements, a relatively rapid and inexpensive process. Similarly, full-scale verification of the modifications would be enhanced by the simple, straightforward wing modifications. Two design conditions were defined for the computational study. The low-speed condition was defined as $M = 0.2$. The objective at this condition was to maximize the lift coefficient available from the wing in the cruise configuration, that is, without deflecting the slat and flap elements. The transonic, high-speed design condition was level flight at $M = 0.8$ and 30,000 ft for a 50,000-lb aircraft.

The computational design was to be performed with available analysis methods described previously. The primary objective was to increase the low-speed, high-lift capability for the wing. However, the only method available to predict this characteristic was the MCARF code, a two-dimensional analysis method. It was felt that an integrated two- and three-dimensional design and analysis approach could be employed for this configuration. The initial feasibility of this approach was established by comparing two- and three-dimensional analyses of the configuration at equivalent conditions. The wing/body configuration was analyzed in the TAWFIVE and WBPPW codes at the transonic cruise design condition. In addition, streamwise airfoils at the inboard and outboard edge of the slat and flap system were analyzed in the NYU airfoil analysis code at equivalent conditions. The equivalent conditions were derived from the flight conditions and span loadings available from the three-dimensional analysis and application of simple sweep theory corrections based on the quarterchord wing sweep. The airfoil analyses were conducted at $M = 0.722$ and a corrected lift coefficient based on the corresponding sectional lift coefficient from the three-dimensional analysis. Comparisons of pressure distributions for the inboard and outboard span locations are presented in Fig. 2.

For the inboard span location the comparisons are strikingly similar. There are minor differences in the pressure coefficients in the midchord region. At the outboard station the shock is located slightly forward of the three-dimensional analysis. However, the objective is merely to determine the relative magnitude of three-dimensional effects on the characteristics of the pressure distributions. The similarity of the characteristics indicate that taper and finite span effects are not significant. Therefore, two-dimensional computations could be used to predict three-dimensional trends.

The design process would be characterized as a computational "cut-and-try" procedure. That is, modifications were made by hand to the wing section, smoothed with an automated smoothing routine, and analyzed with the appropriate computational techniques. Although it was not attempted because of the time limitations on the project, the process lends itself to automation. This would allow the design space to be investigated more quickly and efficiently than with the manual procedure employed. An optimization procedure such as that employed in the TRO-3D code of Davis et al.[22] lends itself to this application. The procedure could be coupled with each of the analysis codes for an integrated evaluation of candidate modifications.

The inboard airfoil was initially addressed in the design process. The trailing-edge modifications were designed, analyzed, and evaluated as the first step in the design process. The trailing-edge modifications were designed to increase the loading over the rear of the airfoil and to reduce the upper surface pressure gradients to delay trailing-edge separation. After an acceptable design was established for the trailing edge of the inboard airfoil, the leading-edge modifications were addressed. These modifications were analyzed and evaluated in conjunction with the trailing-edge modification. The leading-edge modifications were designed to reduce the maximum flow velocities over the leading edge at low-speed, high-angle-of-attack flight conditions. After both the leading and trailing edges were designed for the inboard section, the outboard section was addressed. This approach allowed the incremental increase in the stall angle of attack observed in the inboard design to be integrated into the design process. Modifications to the outboard section evolved from the successful modifications to the inboard section.

The criteria used to evaluate the modifications were based on the low-speed maximum lift and the transonic drag available from the two-dimensional analyses. Typically, a modification would be analyzed in the MCARF code to determine the incremental maximum lift coefficient relative to the baseline airfoil or a modified airfoil. Concurrently, comparisons were made between the wave drag levels predicted in the NYU airfoil analysis code at both transonic cruise and off-design flight conditions in an incremental manner. This allowed simple trade studies between low-speed and transonic performance for the candidate designs to be performed during the design process.

Initial Two-Dimensional Design and Experimental Verification

In order to increase the loading over the rear part of the airfoil and to reduce the local velocity gradients on the upper surface near the trailing edge, both the upper and lower surface trailing-edge modifications were designed with concave regions (Fig. 3a). This resulted in a relatively thick (1% of the local chord) trailing edge. The leading-edge radius was increased, resulting in a thicker, more blunt leading edge than the baseline geometry (Fig. 3b). A summary of results from the design of the inboard span location is presented in Table 1. The data are summarized for the baseline airfoil, the baseline plus the trailing-edge modification (Mod 1 TE), and the baseline plus the trailing- and leading-edge (Mod 1 LE) modifications. The data presented are the predicted stall angle and maximum lift coefficient from the MCARF code at $M = 0.2$ and $Re = 7$ million. In addition, drag coefficients corresponding to three-dimensional Mach numbers of 0.8 and 0.85 for level flight at 30,000 ft are presented. The drag predictions are from the NYU airfoil code. Note that the differences in predicted drag were small for the three configurations. It is also interesting to observe that the predicted increase in maximum lift coefficient for the leading-edge modification was significantly greater than the increase resulting from the trailing-edge modification. Overall, the predicted increment in maximum lift coefficient resulting from the modifications was 26% of the baseline, and the predicted stall angle increased 2 deg.

The modifications to the outboard span location are presented in Fig. 4. The modifications are similar to the inboard modifications with one exception. The leading edge was drooped slightly in order to increase the outboard stall angle the same increment (2 deg) as the inboard to maintain the three-dimensional stall progression pattern. A summary of predicted results for the outboard design modifications is presented in Table 2. The trends follow those observed for the inboard design. Maximum lift increased 28%, and the stall angle increased 2 deg at $M = 0.2$ and $Re = 4$ million. The drag predictions show a 16-count reduction in drag at the overspeed Mach number. An upper limit of the maximum lift increase was predicted by integrating the predicted sectional increases over the portion of the wing span encompassed by the slat/flap system. This yielded an untrimmed, maximum lift increment of 16% for the wing-body configuration.

Experimental verification of the designs was accomplished through a series of wind-tunnel tests conducted in NASA Langley's National Transonic Facility (NTF), 7×10-ft high-speed tunnel, 16-ft transonic tunnel, low-turbulence pressure tunnel (LTPT), and 6×28-in. transonic tunnel. Because of the time-critical nature of the program, two- and three-dimensional testing was conducted concurrently. Initial data on the modifications were obtained on a 1/16-scale model of the EA-6B in the 7×10 high-speed

Table 1   Predicted two-dimensional inboard design results

| | a_{stall}, deg | $C_{l_{\max}}$ | C_D | |
| | M = 0.2 | | M = 0.722 | M = 0.767 |
Configuration	Re = 7 million		C_l = 0.40	C_l = 0.35
Baseline	12.0	1.40	0.0063	0.0077
Mod 1 TE Baseline LE	11.5	1.53	0.0064	0.0073
Mod 1 TE Mod 1 LE	14.0	1.77	0.0064	0.0072

Table 2   Predicted two-dimensional outboard design results

| | a_{stall}, deg | $C_{l_{\max}}$ | C_D | |
| | M = 0.2 | | M = 0.722 | M = 0.767 |
Configuration	Re = 4 million		C_l = 0.60	C_l = 0.53
Baseline	11.5	1.33	0.0080	0.0119
Mod 1 TE Baseline LE	10.5	1.39	0.0075	0.0103
Mod 1 TE Mod 1 LE	13.5	1.70	0.0075	0.0103

tunnel. The data presented in Figs. 5 and 6 are for the baseline and modified configurations with seven stores installed (six wing-mounted and one fuselage-mounted). The experimental data in Fig. 5 show that the maximum lift at $M = 0.4$ increased 15% for the modified configuration. This compares favorably with the computational prediction of 16%. In addition, the drag data in Fig. 6 show that drag levels were reduced at lift coefficients above approximately 0.3 at the high-speed cruise design Mach number ($M = 0.8$). However, evident in the data were two somewhat disturbing facts. First, the data showed that the modifications to the leading edge had virtually no influence on the maximum lift. This is contrasted with the two-dimensional predictions presented in Tables 1 and 2, which showed a larger increment resulting from the leading-edge modifications than the trailing-edge modifications. Second, the drag penalty observed at minimum drag was approximately 28 counts. This penalty is attributed to the increase in base drag resulting from the thickened trailing edge. Although this was not a design condition, it corresponds to a transonic, low-altitude penetration mission for the aircraft. Hence, the design objectives were met, but the degradation in transonic, low-altitude level flight was undesirable and bordered on unacceptable. Before additional data were obtained, a catastrophic fan blade failure in the 7×10 high-speed tunnel precluded further investigations during this initial tunnel entry.

Trailing-Edge Redesign

As previously stated, the drag penalty at the transonic low-lift condition was attributed to the increase in base area associated with the trailing-edge modifications. A simple modification that reduced the base area 40% was proposed. The modification, which is shown in Fig. 7, extended the local surface slope on the upper surface at 88% local chord to the trailing edge. The inboard section with the proposed modified trailing edge (Mod 2 TE) was analyzed in the two-dimensional analysis codes to determine what, if any, effects would be predicted. The analysis from the MCARF code predicted only a very slight effect on maximum lift coefficient.

Experimental Verification

Because of the time constraints associated with the modification project, two- and three-dimensional experimental investigations were performed concurrently. This discussion will consider the two-dimensional investigations first and then address the three-dimensional investigations in the next section.

Two-Dimensional Investigations

Airfoil tests were conducted in NASA Langley's LTPT and 6×28 high-speed tunnel. The force data available from the tests in the LTPT[23] were primarily used to assess the effects of the various modifications on low-speed maximum lift coefficient. Data presented in Fig. 8a demonstrate the variation in maximum lift coefficient resulting from the leading- and trailing-edge modifications to the inboard design section as a function of Mach number. The data correspond to $Re = 10$ million based on chord with transition fixed at 15 and 5% on the upper and lower surfaces, respectively. In Fig. 8b the experimental data in incremental form are compared to predictions from the MCARF code at $M = 0.2$ and 0.3. Note the rather rapid decrease in the effectiveness of the leading-edge modifications as the Mach number increases. The incremental effects are predicted quite adequately by the code. However, the absolute magnitude of maximum lift was inconsistently predicted by the code. The maximum lift coefficient for the baseline section was overpredicted by 0.21 at $M = 0.2$ and underpredicted by -0.13 at $M = 0.3$. In each case the criteria used to establish the maximum lift was the local flow in the leading-edge region reaching sonic velocity. It is interesting to note that these data clearly show an increase in incremental maximum lift resulting from the leading-edge modifications at a Mach number comparable to the test conditions in the 7×10 tunnel test on the full configuration. The effect of the redesigned trailing edge on the maximum lift coefficients appears to be minimal, as predicted by the MCARF code.

Drag measurements from the 6×28 wind-tunnel test are presented in Fig. 9. These data were computed from wake-rake total and static pressures using the method of Pankhurst and Holder.[24] These data, presented as a function of Mach number for a series of normal force coefficients, show a significant drag reduction associated with the redesigned trailing edge (Mod 2 TE). In addition, the redesigned trailing edge does not appear to degrade the drag divergence Mach number improvement available from the original modified inboard section. One may infer from these data that the section with the redesigned trailing edge has significantly improved performance from a drag standpoint over the original trailing-edge modification.

Comparisons of experimental surface pressure distributions from the 6×28 wind-tunnel test with predictions from the NYU analysis code are presented in Figs. 10 and 11. Comparisons are presented for both the baseline and modified inboard section at two-dimensional conditions near the high-speed cruise and overspeed conditions. Mach number, Reynolds number, transition location, and lift coefficient from the experiment were matched in the computational predictions. The results demonstrate excellent agreement between the experimental and computations. Note particularly the predictions in the leading-edge region for the upper and lower surfaces for both Mach numbers. Also notice that the shock locations are predicted slightly aft of the experiment data for the lower Mach number (Fig. 10) and show excellent agreement at the higher Mach number (Fig. 11).

Three-Dimensional Investigations

Three-dimensional testing in the NTF at NASA Langley provided further verification of the computational design and insight into understanding the data obtained during testing in the 7×10 high-speed tunnel. Recall that the low-speed data from the 7×10 entry showed no effect on maximum lift resulting from the modifications to the leading edge of the wing (Fig. 5). The only data obtained that included the leading-edge modifications was on a configuration that also included the installation of seven external stores. There were various rationale proposed to explain the anomaly observed. These included flowfield effects resulting from the installed stores, model fidelity, and three-dimensional effects.

Data from the NTF are presented in Fig. 12 for the EA-6B wing body at $M = 0.3$ and two different Reynolds numbers. The Reynolds number variation resulted from obtaining data on the configuration at 1 and 4 atm total pressure. The data at the lower (1.4 million) Reynolds number showed virtually the same characteristics as the data from the 7×10 test. That is, all of the increase in maximum lift coefficient is associated with the modification to the trailing edge. However, the data obtained at the higher (5.4 million) Reynolds number demonstrate a measurable increase in maximum lift resulting from the leading-edge modification. These data demonstrate a maximum lift increase of approximately 11% compared with the upper-limit prediction of 16% from the two-dimensional analysis. The MCARF analysis yielded approximately the same percentage increase in maximum lift over the Mach range where analyses were conducted ($M = 0.2$–0.3).

Materials that had been used in the model fabrication were not compatible with the cryogenic environment available in the NTF. Hence, the method used to obtain data at elevated Reynolds numbers was to operate the tunnel at increased total pressure. Because of load limitations on the model, 5.4 million was the maximum test Reynolds number. However, based on these limited results, it was felt that at full-scale Reynolds numbers the leading-edge modification would be even more effective.

Data at transonic conditions verified the approach taken to redesign the original trailing-edge modification. Drag polars are presented in Fig. 13 for the baseline wing-body configuration and the original and redesigned trailing-edge modifications. The data show a 16-count drag penalty associated with the original trailing-edge modification (Mod 1 TE) at $M = 0.8$ and 0.10 lift coefficient. The polar for the original modification is rotated such that a reduction in drag relative to the baseline is evident at lift coefficients greater than 0.46. The redesigned trailing edge offered a reduction in the low-lift drag penalty to approximately 10 counts. In addition, at lift coefficients greater than 0.4, drag levels were reduced relative to the baseline configuration. Furthermore, trimmed drag data for the complete configuration with seven external stores show a 5-count drag penalty at sea-level maximum velocity and drag levels reduced over a substantial portion of the aircraft's flight envelope compared with the baseline configuration.

The drag reduction does not come without a small reduction in low-speed maximum lift. Comparisons of untrimmed lift as a function of angle of attack are presented in Fig. 14 for the baseline and the two trailing-edge modifications. These data show a decrease of 0.04 in the maximum lift coefficient associated with the reduced trailing-edge thickness at $M = 0.3$. The trailing-edge modification still offers a significant increase in maximum lift compared to the baseline configuration.

Force and moment data comparing the baseline wing-body configuration and the proposed leading- and trailing-edge modifications are presented in Figs. 15–17. Lift data in Fig. 15 at $M = 0.3$ show the increase in maximum lift coefficient resulting from the modifications discussed previously. The wing-body drag polars presented in Fig. 16 show a 17-count drag penalty at minimum drag. The modified configuration demonstrates reductions in drag levels for lift coefficients greater than 0.4. The pitching moment data presented in Fig. 17 show that the longitudinal stability characteristics are not significantly affected by the modifications. However, the zero-lift pitching moment is shifted -0.03 at $M = 0.8$. This would result in a small penalty in trim drag.

The final set of comparisons presented are computational predictions of the wing surface static pressure distributions (Fig. 18) and span loadings (Fig. 19) for the baseline wing-body configuration and the wing-body plus leading- and trailing-edge modifications. The comparisons are from analyses with the TAWFIVE code at $M = 0.8$ and lift coefficients of approximately 0.32. To obtain the same lift coefficients from the analysis code required a difference in angle of attack of 0.7 deg between the configurations. This compares with an observed difference in the experimental data of 0.65 deg at these conditions. The most striking difference in the pressure distributions results from the trailing-edge modification. The modified configuration carries a significant portion of the load over the aft 20% of the chord. This increases the torsional loads carried through the wing box.

The leading-edge modification allows a more rapid flow expansion over the first 5% of the chord. Although it only has limited practical applicability in an operational environment, the modified pressure distributions would support natural laminar flow at these cruise conditions from the leading edge to the midchord region. It might be possible to exploit this if a new leading-edge slat were incorporated into the actual aircraft. This could allow laminar flow to be achieved from the leading edge to the gap between the slat and the main wing, offering further drag reductions.

Computational predictions of the spanwise loading distributions at comparable lift coefficients are presented in Fig. 19 for the subject configurations. As expected, the loadings increased over that portion of the span where the modifications were made. This results in an increase in the root bending moment of 1.1% at the transonic design condition, $M = 0.8$, and a lift coefficient of 0.32.

Summary

Improvements to increase the maneuver performance of the EA-6B aircraft have been addressed in a complementary computational design and experimental validation study. State-of-the-art low-speed and transonic computational techniques were used to design modifications to the wing leading-edge slat and trailing-edge flap. Computational predictions were used to evaluate candidate modifications. The objective of the modifications was to maximize low-speed high-lift performance while minimizing transonic drag increases. Experimental testing of the modifications was necessary to verify the computational predictions and design philosophy. Experimental data were also necessary to assess interactions and aerodynamic effects for which accurate and reliable predictions are not available (such as the effect of installed stores on high-lift performance and base drag resulting from the thick trailing edge). Data showed a significant increase in usable maximum lift at low speeds on the wing-body configuration. At transonic conditions, the modifications yielded drag reductions over a large portion of the aircraft's operational envelope.

Several specific observations were made during the program and are delineated as follows:

1) The integrated two- and three-dimensional design philosophy was successful. Modifications defined based on airfoil analyses met the objective (increased maximum lift) within the design constraint (transonic drag) and were successfully integrated into the aircraft configuration.

2) The MCARF code provided useful guidance in evaluation of candidate leading- and trailing-edge modifications. The code was most successfully used when the results were evaluated in an incremental manner and judgments on the candidate modifications were made on these increments.

3) The drag constraint was successfully integrated into the design. Evaluations from the NYU code predicted trends observed in the experimental data on the modified sections and the total configuration.

4) Conclusions based on observations made on the low-Reynolds-number data from the three-dimensional test would not have indicated a benefit in maximum lift resulting from the leading-edge modification. However, at higher Reynolds numbers, the data indicated measurable benefits associated with the leading-edge modification.

The modifications identified during this study can significantly improve the cruise maneuver lift capability of the EA-6B aircraft. These modifications, along with others that have been identified during the joint program, will enhance the airplane mission effectiveness by improving the EA-6B maneuver and stall margins in the cruise configuration.

Concluding Remarks

The role of CFD in applied aerodynamics is becoming increasingly important. The preceding discussion identified several extensions to potential flow analysis codes that have been developed at NASA Langley. The need for these extensions was identified from developmental and research programs involving configurations that were to go through full-scale development and flight testing. During these programs CFD was relied on extensively in what will become a classical context. Preliminary assessment of the feasibility of various ideas was explored. Experimental test programs were augmented through the use of CFD for both flight and wind-tunnel testing. Enhanced understanding of flow phenomena was made possible through interrogation of the details available from the computational solutions.

Specific extensions to the analysis codes referenced in the previous discussion included automated design to a specific surface pressure distribution, inclusion of aeroelastic effects, and modeling of various types of wind-tunnel wall boundary conditions. The automated design and aeroelastic codes were developed in a modular fashion. This allowed different aerodynamic or structural analysis codes to be included in the systems as new or updated codes became available. The design and aeroelastic systems have been exercised with a variety of aerodynamic analysis codes. The wall interference approach is useful for a priori prediction of model in wind-tunnel effects or prediction of wall interference effects in the absence of wall pressure data.

Specific applications referenced included the design of a wing for the Gulfstream Peregrine that generated extensive runs of natural laminar flow at its transonic design condition. This design was accomplished within a restrictive set of aerodynamic, structural, and temporal constraints. An outer panel glove was designed for the F-14 to support the Variable Sweep Transition Flight Experiment. The glove was designed computationally, verified during wind-tunnel testing, modified computationally, and successfully installed and flown on the F-14. The comparisons of wind-tunnel and flight experimental data with the computations were excellent. Finally, reported herein are details of the modifications to the wing slat and flap contours for the EA-6B which were designed computationally to increase the low-speed, high-lift capability with minimal degradation to transonic performance. These designs were verified through a series of wind-tunnel tests at low speed and transonic conditions. Results were significant enough for the Navy to commit to a flight-test program involving the modifications while a parallel effort involves detail design to incorporate the modifications into the fleet.

This summary of code work and modifications indicates that there is indeed a place in applied CFD for potential flow solvers. For those who labor in applied computational aerodynamics, the ability to understand the applicability and limitations of all classes of computational methods is important. Understanding when and how to apply methods ranging from vortex-lattice to full Navier-Stokes is critical to efficiently solve the variety of problems that the researcher continuously faces.

Acknowledgment

I wish to acknowledge the contributions of Dennis Allison and Bill Sewall of NASA Langley for the experimental data on the EA-6B. In addition, Bill Gato and Matt Masiello of Grumman Aircraft lent valuable insight into the operational impact associated with the computationally defined wing modifications. Finally, I would like to thank Bud Bobbitt of NASA Langley for his helpful suggestions and support throughout the entire design process.

References

[1]Campbell, R. L., "Calculated Effects of Varying Reynolds Number and Dynamic Pressure on Flexible Wings at Transonic Speeds," NASA CR-2327, April 1984, pp. 309–327.

[2]Phillips, P. S., and Waggoner, E. G., "A Transonic Wind Tunnel Wall Interference Prediction Code," AIAA Paper 88-2538, June 1988.

[3]Waggoner, E. G., Campbell, R. L., and Phillips, P. S., "Computational Wing Design in Support of an NLF Variable Sweep Transition Flight Experiment," AIAA Paper 85-4074, Oct. 1985.

[4]Campbell, R. L., and Smith, L. A., "A Hybrid Algorithm for Transonic Airfoil and Wing Design," AIAA Paper 87-2552-CP, Aug. 1987.

[5]Waggoner, E. G., and Allison, D. O., "EA-6B High Lift Wing Modifications," AIAA Paper 87-2360-CP, Aug. 1987.

[6]Campbell, R. L., Waggoner, E. G., and Phillips, P. S., "Design of a Natural Laminar Flow Wing for a Transonic Corporate Transport," AIAA Paper 86-0314, Jan. 1986.

[7]Melson, N. D., and Streett, C. L., "TAWFIVE: A User's Guide," NASA TM-84619, Sept. 1983.

[8]Boppe, C. W., "Aerodynamic Analysis for Aircraft With Nacelles, Pylons, and Winglets at Transonic Speeds," NASA CR-4066, April 1987.

[9]Bauer, F., Garabedian, P., Korn, D., and Jameson, A., "Supercritical Wing Sections II," Vol. 108, *Lecture Notes in Economics and Mathematical Systems*, Springer-Verlag, New York, 1975.

[10]Stevens, W. A., Goradia, S. H., and Braden, J. A., "Mathematical Model for Two-Dimensional Multi-Component Airfoils in Viscous Flow," NASA CR-1843, July 1971.

[11]Caughey, D. A., and Jameson, A., "Recent Progress in Finite Volume Calculations for Wing-Fuselage Combinations," AIAA Paper 79-1513, July 1979.

[12]Stock, H. W., "Integral Method for Calculation of Three-Dimensional Laminar and Turublent Boundary Layers," NASA TM-75320, 1978.

[13]Smith, P. D., "An Integral Prediction Method for Three-Dimensional Compressible Turbulent Boundary Layers," Royal Aircraft Establishment, Farnborough, Hampshire, UK, Rept. 3739, 1974.

[14]Streett, C. L., "Viscous-Inviscid Interaction for Transonic Wing-Body Configurations Including Wake Effects," AIAA Paper 81-1266, July 1981.

[15]Mason, W. H., MacKenzie, D., Stern, M., Ballhaus, W. F., and Frick, J., "An Automated Procedure for Computing the Three-Dimensional Transonic Flow Over Wing-Body Combinations, Including Viscous Effects," Vol. 1, Air Force Flight Dynamics Lab. Wright-Patterson AFB, OH, Rept. AFFDL-TR-122, Oct. 1977.

[16]Nash, J. F., and Macdonald, A. G. J., "The Calculation of Momentum Thickness in a Turbulent Boundary Layer at Mach Numbers up to Unity," Aeronautical Research Council, London, CP-963, 1967.

[17]Cohen, C. B., and Reshotko, E., "The Compressible Laminar Boundary Layer with Heat Transfer and Arbitrary Pressure Gradient," NACA TR-1294, 1956.

[18]Schlichting, H., and Ulrich, A., "Zur Berechnung Laminar-Turbulent," *Jahrbuch der Deutscher Luftfahrtforschung*, No. 1, 1942, pp. 8–35.

[19]Granville, P. S., "The Calculation of the Viscous Drag of Bodies of Revolution," David Taylor Model Basin, Rept. 849, July 1953.

[20]Morgan, H. L., "High-Lift Flaps for Natural Laminar Flow Airfoils," NASA/SAE/AIAA/FAA Laminar Flow Aircraft Certification Workshop, Wichita, KS, April 15–16, 1985.

[21]Nash, J. F., "A Practical Calculation Method for Compressible Turbulent Boundary-Layers in Two-Dimensional and Axisymmetric Flows," Lockheed-Georgia, Marietta, GA, RM, ER-9428, Aug. 1967.

[22]Davis, W. H., Aidala, P. V., and Mason, W. H., "A Study to Develop Improved Methods for the Design of Transonic Fighter Wings by the Use of Numerical Optimization," NASA CR-3995, 1986.

[23]Sewall, W. G., McGee, R. J., Ferris, J. C., "Wind Tunnel Test Results of Airfoil Modifications for the EA-6B," AIAA Paper 87-2359, Aug, 1987.

[24]Pankhurst, R. C., and Holder, D. W., *Wind-Tunnel Technique*, Pitman, London, 1965.

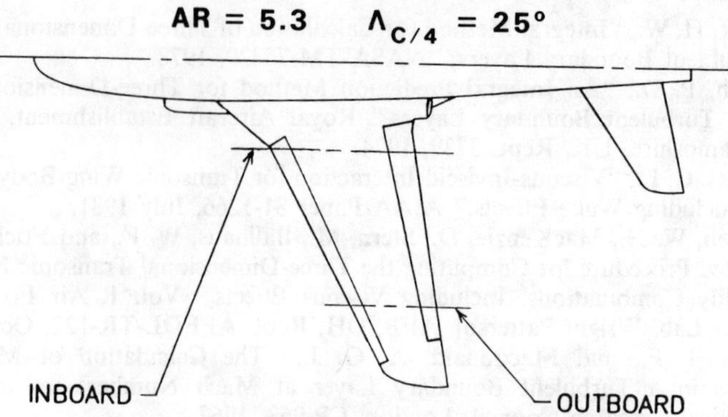

Fig. 1   Planform sketch of EA-6B configuration.

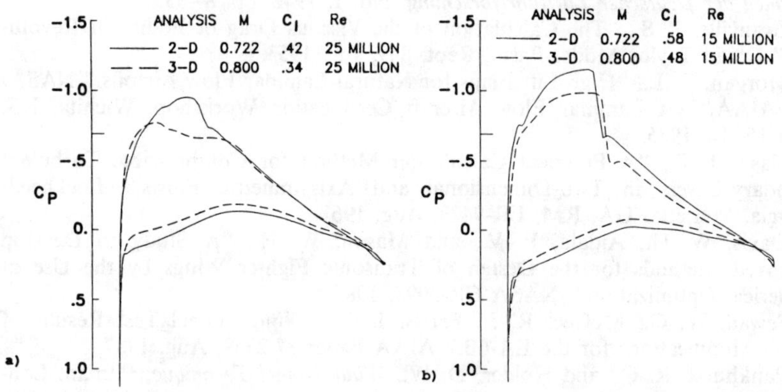

Fig. 2   Pressure distributions from two- and three-dimensional analyses at equivalent conditions for wing-body configuration: a) inboard span station; b) outboard span station.

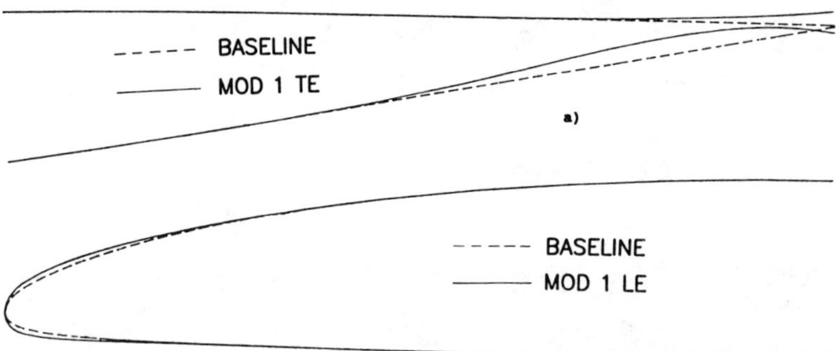

Fig. 3   Airfoil geometry at the inboard span station: a) trailing-edge geometry; b) leading-edge geometry.

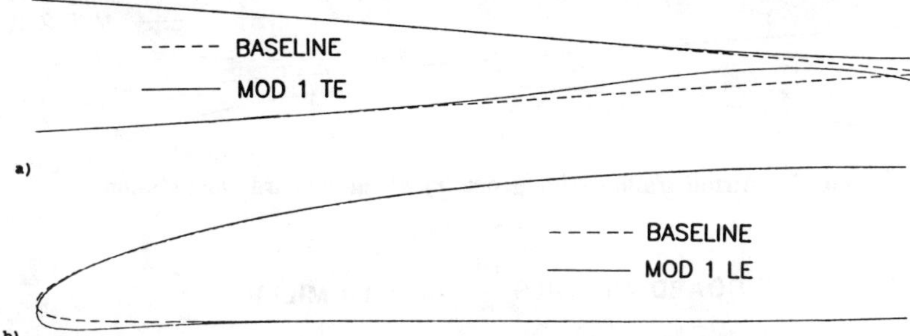

Fig. 4 Airfoil geometry at the outboard span station: a) trailing-edge geometry; b) leading-edge geometry.

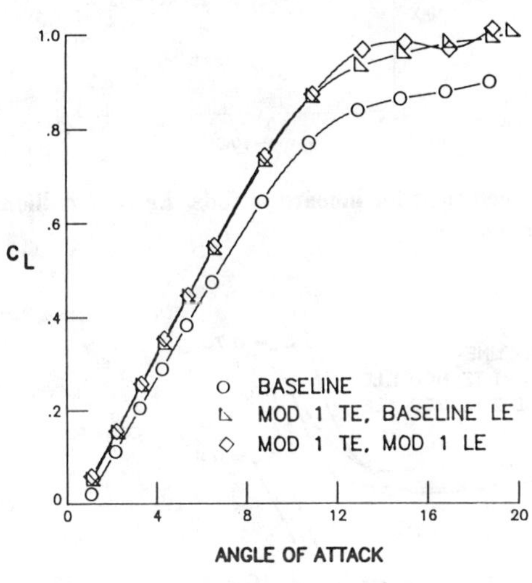

Fig. 5 Variation of configuration lift coefficient with angle of attack.

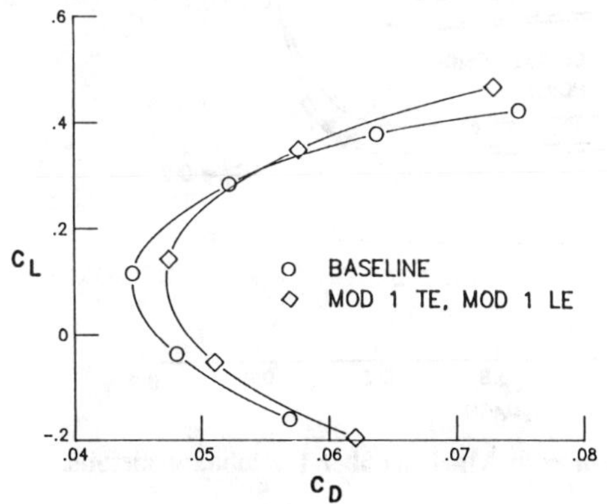

Fig. 6 Variation of configuration lift coefficient with drag coefficient.

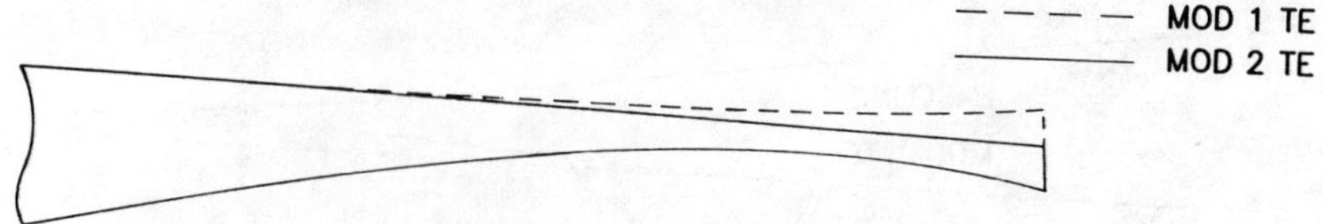

Fig. 7   Airfoil trailing-edge geometry at the inboard span station.

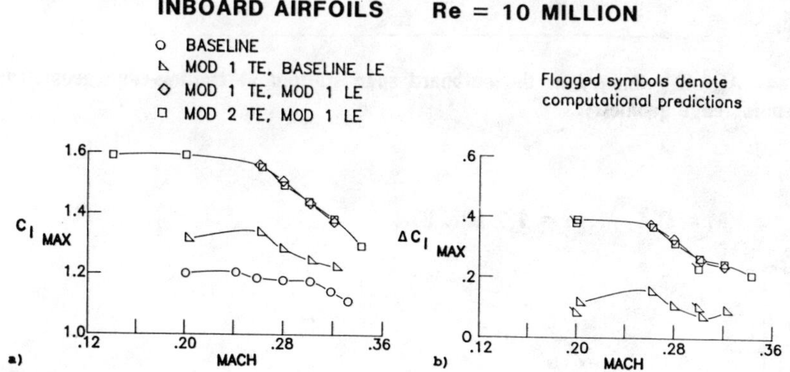

Fig. 8   Comparison of maximum lift coefficient for inboard airfoils, $Re = 10$ million:
a) experiment; b) incremental comparisons.

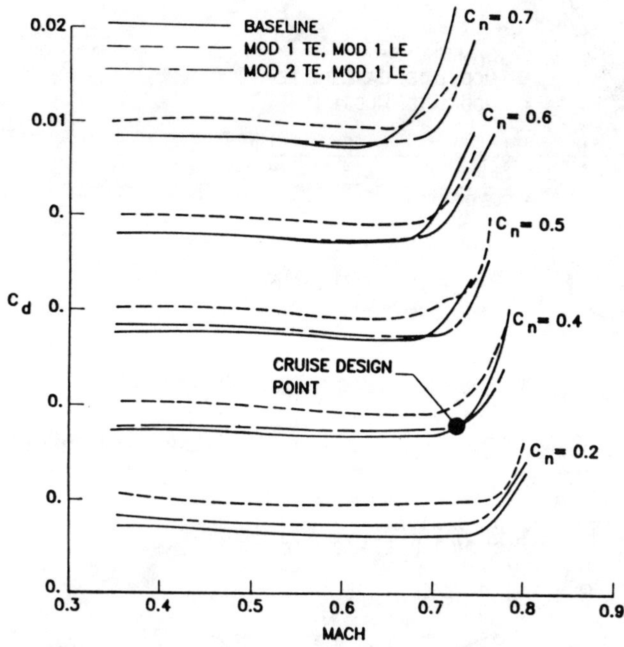

Fig. 9   Variation of drag coefficient with Mach number for inboard airfoils.

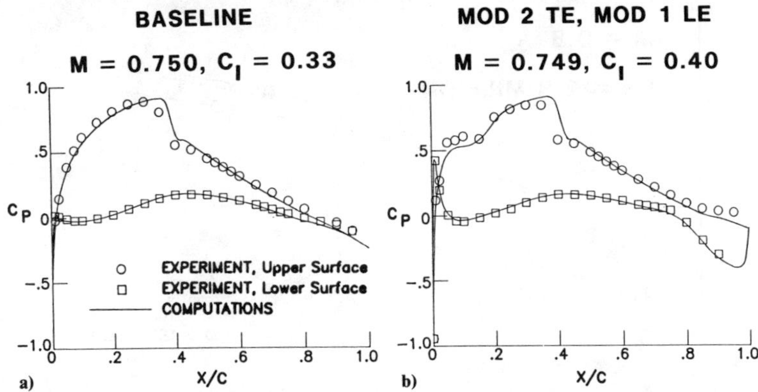

Fig. 10   Comparison of computational and experimental airfoil pressure distributions, $Re = 10$ million: a) baseline; b) Mod 2 TE, Mod 1 LE.

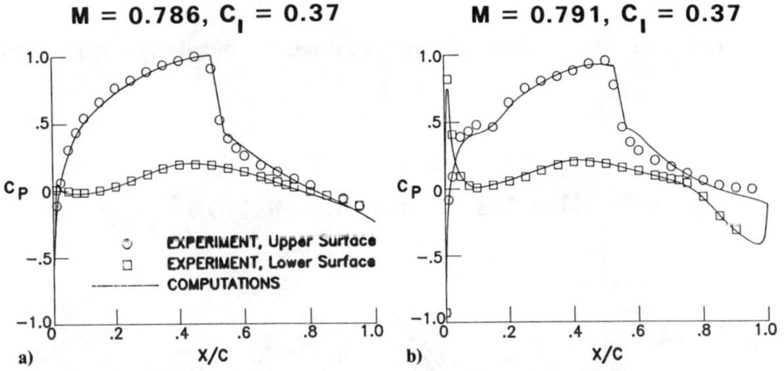

Fig. 11   Comparison of computational and experimental airfoil pressure distributions, $Re = 10$ million: a) baseline; b) Mod 2 TE, Mod 1 LE.

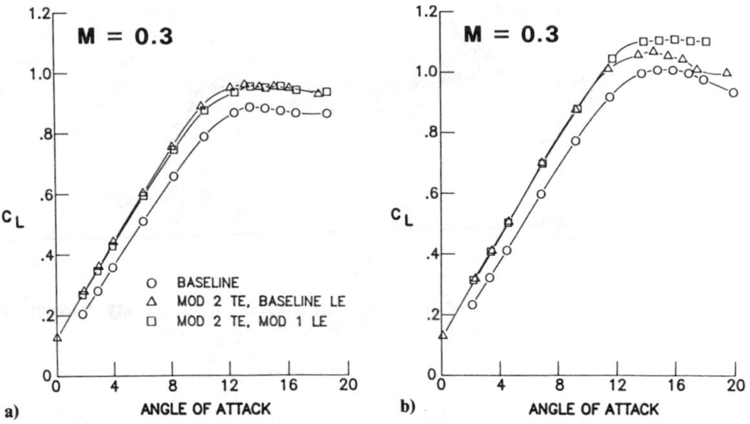

Fig. 12   Variation of lift coefficient with angle of attack for wing-body configuration: a) $Re = 1.4$ million; b) $Re = 5.4$ million.

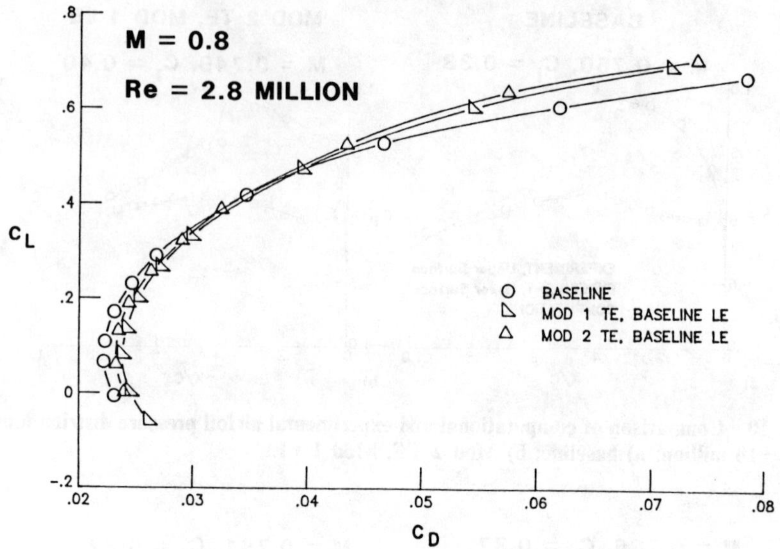

Fig. 13   Variation of lift coefficient with drag coefficient for wing-body configuration.

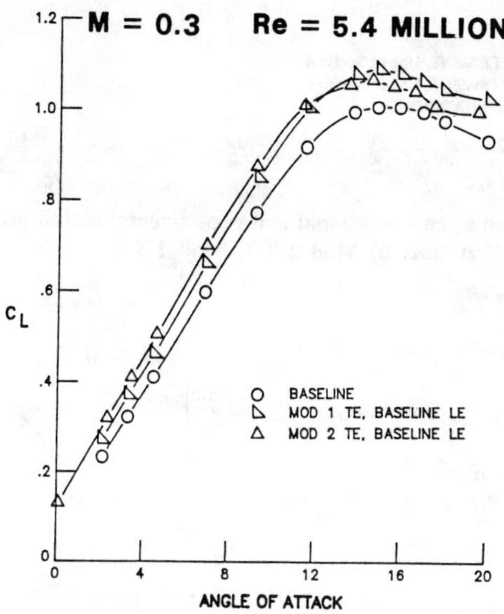

Fig. 14   Variation of lift coefficient with angle of attack for wing-body configuration.

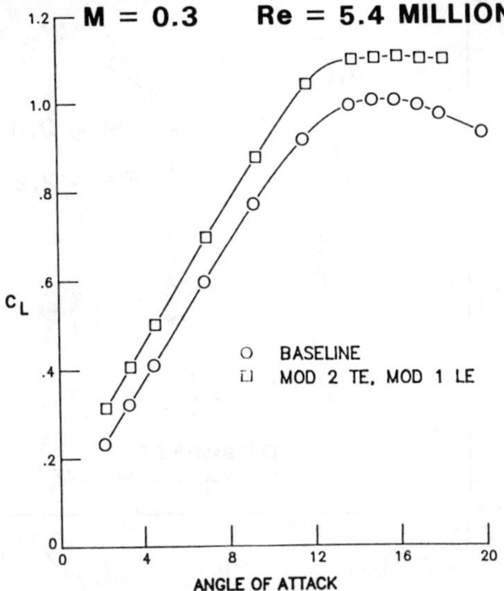

Fig. 15   Variation of lift coefficient with angle of attack for wing-body configuration.

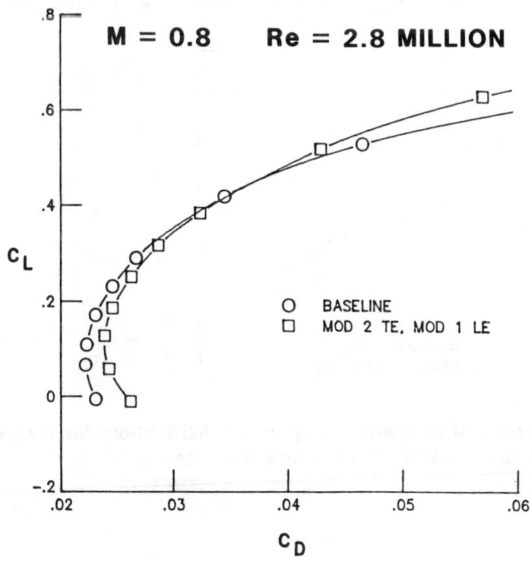

Fig. 16   Variation of lift coefficient with drag coefficient for wing-body configuration.

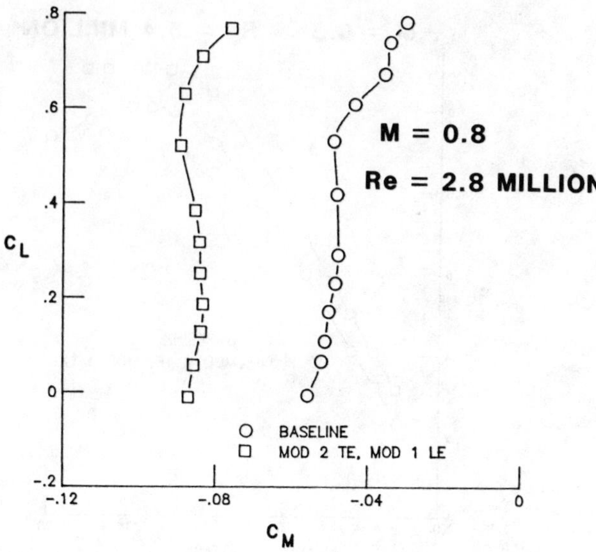

Fig. 17 Variation of lift coefficient with pitching moment coefficient for wing-body configuration.

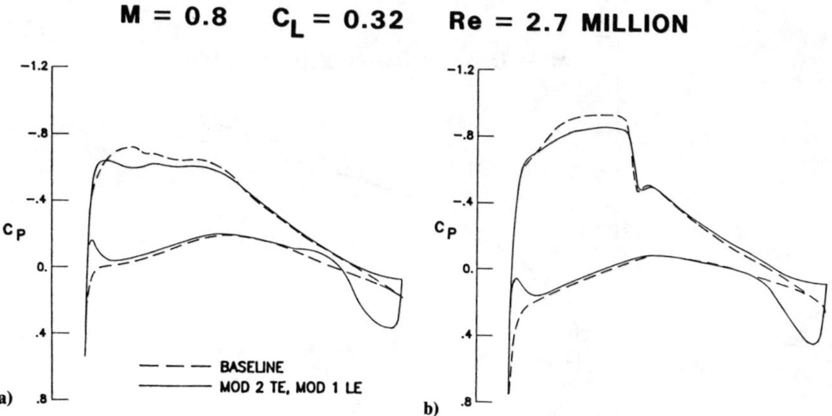

Fig. 18 Comparison of computational pressure distributions for wing-body configuration: a) Inboard span station; b) Outboard span station.

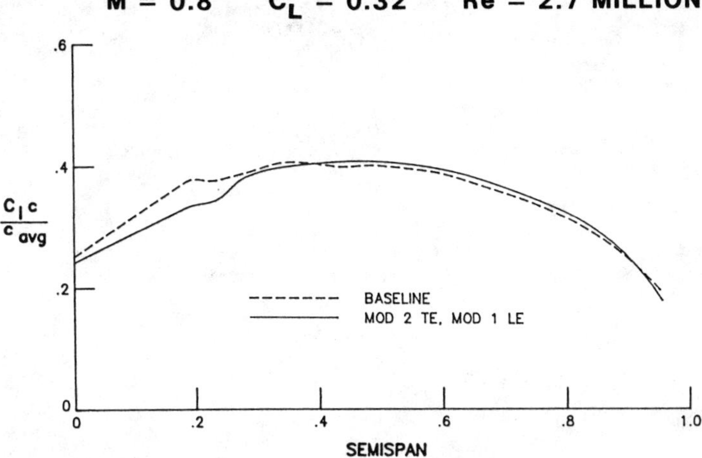

Fig. 19 Comparison of predicted spanwise loading distributions for wing-body configuration.

Part 5.   Propulsion Systems

Chapter 12

Navier-Stokes Methods for Internal and Integrated Flow Prediction

Raymond R. Cosner*
McDonnell Douglas Corporation, St. Louis, Missouri

The Problem

A KEY requirement in aircraft design is the prediction of internal flows. Generally, these flows are associated with the propulsion system (inlets, diffusers, and nozzles). Occasionally, internal flow analysis also is needed to design the environmental control system, cockpit or avionics cooling, or ground test facilities (i.e., wind-tunnel components). This discussion will be confined to issues associated with fighter aircraft propulsion systems: the inlet, diffuser and nozzle, and their integration with the external aerodynamic vehicle.

The major engineering problems in internal flow are associated with energy loss mechanisms (shock waves and viscosity), viscous layer behavior, unsteadiness, and nonuniformity. Some common issues are encountered in both inlet-diffuser analysis and exhaust system analysis. Some significant differences between the two areas can also be identified.

Inlet-Diffuser Design Considerations

For inlet-diffuser design, the parameters to be predicted are, in order of importance, the following:

1) Area-averaged total pressure recovery at the engine face.

2) A measure of nonuniformity in the total pressure distribution at the engine face ("distortion"), both steady state and transient.

3) Information on the distribution of other flow properties, such as static pressure or Mach number, at the engine face.

The flowfield interactions that must be modeled include the following:

1) nonuniform flow approaching the inlet;

2) viscous losses through boundary layers (possible separated flow or strong secondary flow);

3) shock losses and shock–boundary-layer interactions;

4) sustained adverse pressure gradient; and

Copyright © 1989 by the American Institute of Aeronautics and Astronautics, Inc. All rights reserved.
*Principal Technical Specialist, Computational Fluid Dynamics.

5) substantial turning of the duct, accompanied by major changes in cross-sectional shape.

A major focus of inlet-diffuser design is to control these losses through methods such as boundary-layer removal (distributed suction, "bleed", or open slots), boundary-layer energization (tangential blowing), vortex generators, and innovative contouring of surfaces. An effective engineering computational fluid dynamics (CFD) tool must be applicable to each of these control methods.

Exhaust System Design Considerations

Analysis of exhaust systems poses some of the same problems as inlet-diffuser systems. Unique issues are also present:

1) The flow at the analysis start plane (often the turbine exit) is highly nonuniform, i.e., turbulent and swirling. Often two streams of air are present at the start plane (i.e., flow from the engine core and the bypass).

2) Frequently, wall heating rate is a key parameter of interest.

3) Various wall cooling schemes are used, such as addition of cool air at the wall.

4) For ejector nozzles modeling the entrainment in a non-self-similar free shear layer is critical.

5) Generally, the internal pressure gradients are favorable.

6) In some applications the ducts may split prior to reaching the nozzle exit to the ambient air.

CFD Requirements

In several respects the requirements for CFD analysis are far more severe for internal flow than for external flow. The most difficult problem of all, of course, is integrated flow (external flow coupled with internal flow). Most of the important engineering information depends on viscous fluid interactions. Inviscid analysis methods therefore are of little value unless they are coupled with a viscous analysis.

Boundary-layer separation and strong secondary flows are very common problems in inlet-diffuser systems and to a lesser degree in exhaust systems. Patched viscous-inviscid analysis methods have had little success in modeling these complex three-dimensional flows, except where they have been "tuned" to a specific type of configuration.

The preferred analysis tool is a three-dimensional simulation based on the Reynolds-averaged Navier-Stokes equations. In several organizations the need to analyze internal flow has triggered investment in Navier-Stokes technology. A variety of tools have been used and will be reviewed later.

Internal flow analysis is characterized also by a wide range of geometric complexity, particularly in the integrated problem (external plus internal flow). The response from the CFD community has been the development of multizone solution methods.

The cost of internal flow analyses often is very high compared with external flows. For a confined steady internal flow, the mass flow must be uniform at every station in the duct (except, of course, if mass addition or removal is employed). Computationally, the mass flow can be specified and controlled only through boundary conditions at the inflow and outflow boundaries. (Some other techniques will be mentioned later.)

The transient results in an evolving internal flow solution must interact globally since the net flow of mass, momentum, and energy is constrained throughout the duct. Information-bearing waves must pass through the complete computational domain many times to attain a global balance of these properties. Furthermore, the sustained adverse pressure gradient in diffusers means that complex and sensitive viscous interactions are usually present. These phenomena lead to slow convergence rates and high run times. For similar problems (same number of grid points, same Mach numbers, etc.), a solution for internal flow can easily require an order of magnitude more computing time than an external flow solution.

The sustained adverse pressure gradients and strong secondary flows also expose weaknesses in turbulence models. In some cases the flow may be fundamentally unsteady as a result of these interactions. (Normally, the engineering team will design to avoid this.)

Available Technology

The key issues for internal flow are addressed most effectively using CFD methods based on the Reynolds-averaged Navier-Stokes equations (hereafter, they will be referred to as Navier-Stokes methods). Current problems can be addressed effectively using single-zone and multizone methods with boundary conditions appropriate for internal flow. The PARC3D code,[1] among others, has been used effectively for these applications.

Many practical problems in internal flow involve split flow paths, bifurcated ducts, inlet bleed, secondary air removal or addition, and other complex flow paths. To treat these problems, most engineering organizations have modified existing codes developed for simpler problems, or else they have developed new methods. Therefore, in general, standard accepted tools cannot be said to exist for these applications.

The situation is even less clear when the problem of external-internal flow coupling is considered. Unfortunately, this area is often far more important from an engineering perspective than are the areas of purely external or purely internal flow. Generally, different grid topologies are optimal for different regions of the problem. Multizone methods seem essential for these coupled requirements, unless the problem can be decomposed in some manner (e.g., by restricting the method to supersonic flow so that upstream flow of information can be ignored).

Here, too, some codes are available from government organizations that have served as the starting point for developing capability. The compressible Navier-Stokes (CNS) code from NASA Ames Research Center has been applied to coupled flow problems on the F-16 aircraft.[2] In many cases engineering organizations have developed new methods to perform these analyses.[3-5]

The well-known problems in acquiring geometry and constructing grids are compounded for inlet and nozzle integration studies. No standard methods have been developed to relieve this bottleneck, although promising efforts are being pursued both under government sponsorship and as proprietary industrial research.

The picture is more encouraging in the area of postprocessing tools. Many of the standard graphics tools such as PLOT3D[6] are available with multizone capability and can be used without modification to display the required data. However, additional tools are needed to compute performance parameters such as engine face recovery and distortion and corrected air flow.

In general much basic knowledge and several operational tools are available to form a basis for predicting complex aeropropulsion flowfields. However, the methods in the public domain by themselves do not provide the complete required analysis methodology. Therefore, all aircraft companies have, to some degree, developed proprietary methods for these analyses.

These methods tend to be similar in their overall features, since they have been developed to solve similar problems. The approach used at McDonnell Aircraft Company (MCAIR) for aeropropulsion integration studies will be described as an example of current practice.

Diffusers

An obvious application for internal flow prediction tools is to model subsonic flow through diffusers. A sample diffuser installation is illustrated in Fig. 1. The requirement is to predict viscous losses in the duct resulting from boundary-layer growth.

This application often places high demands on CFD codes. For example, one purpose of the diffuser is to decelerate the flow ahead of the engine. Thus, the boundary layers on the diffuser walls are subjected to a sustained adverse pressure gradient, which frequently leads to flow separation.

Another purpose of the duct, of course, is to connect the inlet with the engine. In many cases the inlet and engine axes are offset, and the diffuser must turn the flow. The cross-sectional shape of the duct can change from an arbitrary shape to circular at the engine face. Often the location of the diffuser must be compromised for vehicle integration concerns. For example, the duct may be displaced from the aerodynamic preferred location to accomodate aircraft structure, landing gear, or other components. All of these effects lead to complex boundary-layer behavior, including strong secondary flows, three-dimensional accelerations and compressions, and corner flows.

For some diffuser designs the secondary flows are weak and the boundary layer remains attached. Simple analysis methods based on inviscid core flows with a boundary-layer correction can be valuable.

In many cases, though, the assumptions of boundary-layer theory are violated. The boundary layer can change rapidly in the streamwise direction, normal pressure gradients can be large, the walls may be curved, and separation or strong secondary flow can be dominant features. For these reasons CFD methods based on the Reynolds-averaged Navier-Stokes equations have gained acceptance for these applications.

At MCAIR, the preferred CFD codes for this application are called X3D and NASTD. The features of these codes were mentioned previously. Grids are generated with ZGRID,[3] a zonal method based on three elliptic partial differential equations. These tools have capabilities that are common throughout organizations concerned with aircraft design.

A representative set of experimental data for current diffuser concepts was developed by the Air Force in the early 1980s.[7] The B19 diffuser from this study has been used to validate CFD predictions. The B19 diffuser geometry is illustrated in Fig. 2. This design has an area ratio of 1.7 with a square entrance and a circular exit. The duct is short, with a length of 3.5 diam (engine face diameter) and an offset of 1.5 diam. Because of the short length and high offset, substantial secondary flow develops in the duct.

The results for this diffuser are shown in Fig. 3. The most interesting element is seen in the contours of total pressure, which show that the boundary layer is separated for more than half the duct length along one wall. For an engineering study these predictions would be used to guide a redesign of the duct to improve the flow quality.

Exhaust Nozzles

Exhaust duct and nozzle analysis is generally similar to diffuser analysis. However, one difference is that the pressure gradient in an exhaust duct is generally negligible or favorable; hence, strong viscous interactions are uncommon.

One common requirement is to predict heating rates, to guide the design so that the walls can be cooled adequately. Heating rate predictions have their own subtleties and unique requirements, generally related to additional grid density near the surface. The fluid is usually depleted air and hydrocarbon combustion products, although the altered chemical composition is generally ignored.

Another complication is the inflow boundary condition. For diffusers the inflow often is taken to be uniform. However, the flow behind a turbofan engine cannot be simplified in that way. At the exhaust duct inflow station (generally, just aft of the engine), the flow consists of a hot core surrounded by cool flow from the fan. These different flows must be represented. The cooler fan flow will pose no heating problem to the designer, but the downstream impingement of hot core flow on walls must be predicted. The inflow is swirling and highly nonuniform as a result of its passage through several stages of turbomachinery. Generally, these factors are not modeled.

Wind-Tunnel Installation

Another challenging problem is to model the flow over a model installed in a wind tunnel. The model must be installed on a support system, which may modify the flow over the model in unexpected ways. CFD tools are also needed to determine the difference between flow through unconfined air and flow through a wind tunnel confined by walls. The walls often are perforated with a many small holes, or they may be slotted. At best, the holes or slots are modeled only in their overall impact on the flowfield.

The challenges to CFD analysis arise mainly in two areas: 1) providing the geometric capabilities to analyze the model, the support system, and the wind-tunnel surfaces; and 2) developing adequate models of the gross interactions near nonsolid wind-tunnel walls. These concerns are most acute for transonic testing. Aspects of these issues are reviewed in Refs. 8–10.

Coupled External/Internal Flow

The coupling between external and internal flow, aeropropulsion integration, is a key concern in fighter aircraft design. High interest is seen in analyzing both the inlet and exhaust nozzle regions. This interest is often strongest in the transonic range, where the most complex and sensitive interactions occur.

Recent trends have been toward close integration of the inlets and nozzles with other airframe components (wings, fuselage, etc.). This leads to high demands for grid-generation tools and complex fluid dynamic models.

For both inlet and nozzle flowfields, natural unsteadiness often occurs in the flows. To an increasing degree, aircraft designers are finding that this unsteadiness must be predicted. Inlet transient phenomena such as buzz and hammershock often define aircraft performance limits. Large-amplitude acoustic fields about cavities and nozzles can cause structural fatigue of aircraft components; consequently, these interactions must be predicted.

Inlet Flowfields

Inlet design and analysis is a critical area in fighter configuration development. The complexity of inlet flowfields, combined with their importance, has been the key consideration in setting the direction for CFD research at several organizations. For example, at MCAIR this was the original stimulus for extending Navier-Stokes methods to multizone schemes (in 1981).

Many different design concepts have been evaluated for fighter inlets. Inlets can be integrated with the aircraft in many different locations. Therefore, one emphasis in CFD analysis must be to develop tools that can be used for radically different classes of geometries. Multizone methods have been developed as a part of this approach. Another part of the solution is to use interactive graphics and powerful workstations to assist the engineer in developing grids for arbitrary geometries.

Multizone Navier-Stokes methods can be regarded as an enabling technology for practical inlet analyses. The general approach is to break a complex computational domain into several simpler subdomains (usually called "blocks" or "zones"). Grids are generated separately in each zone with the use of relatively simple methods. Together, the zones span the required domain. Existing flow solvers must be modified to use this approach. Flowfield boundary conditions are communicated between zones during the solution process to provide solution continuity and accuracy. Another benefit of the zonal approach is that a large problem can be broken into several smaller problems. This can facilitate the use of computers with smaller real memories, or coarse-grained parallel processing.

Multiple computational grids can be created and used in many different ways. The approach at MCAIR, which is described in Ref. 3, is based on multiple nonoverlapping grids, with no required relationship between grid points across zone boundaries. By avoiding a requirement for continuity of grid lines across zone boundaries, maximum versatility is achieved. Another attractive method is the chimera approach,[9] in which grids are constructed with a high level of overlap.

For most inlet integration problems the analysis model at a minimum must consist of the fuselage forward of the inlet (the forebody), the inlet itself (both internal and external surfaces), and the external aircraft near and immediately aft of the inlet. In some cases additional components such as canards must be modeled. Generally, the internal flow domain is extended downstream beyond the inlet near field, since the need is for data on flow properties at the engine face. The zonal computation for a subsonic inlet integration study is presented in Fig. 4.

Transonic inlet flows are highly three dimensional, but two-dimensional methods often are valuable for predicting supersonic flows. Sample two-dimensional methods will be reviewed first.

Multizone methods have been used extensively for two-dimensional inlet-analysis. At MCAIR, the preferred code is called Factored Algorithm, Navier-Stokes Implicit (FANSI). This code is described in Refs. 5 and 11. This code implements a finite-volume formulation of the Beam-Warming algorithm, with several options for upwind and total variational diminishing schemes. The Baldwin-Lomax turbulence model is employed. For supersonic inlet studies four zones are commonly used. These zones are depicted in Fig. 5.

Several aspects of inlet performance are of interest and are illustrated in Fig. 6. Foremost is the area-averaged total pressure ("recovery") at the engine face. This property is strongly affected by boundary-layer buildup and separation.

At supersonic speeds, recovery depends strongly on the strength of the inlet shock system. Simple tools based on familiar relationships for two-dimensional shocks are used extensively in the early stages. However, the location of the terminal normal shock depends on a complex interaction and is not estimated accurately with simple methods.

The inlet lip poses a dilemma for the designer. High performance at subsonic speeds over a range of conditions is achieved with a blunt, rounded lip to minimize local flow separation. However, the goal of minimum supersonic drag leads to sharp inlet lips. CFD methods are used extensively to balance these competing requirements.

Often, active control of the boundary layer is used to enhance inlet performance. The boundary layer may be removed, completely or partially, using open slots or bleed through porous walls. Inlet CFD tools must be capable of modeling these mechanisms.

These requirements are all present in most supersonic inlet studies. A representative solution is presented in Fig. 7. For this Mach 2.0 case, two oblique shocks are present, followed by a terminal normal shock ahead of the inlet lip. A key interest is the local flow over the inlet cowl lip, which in this case is seen to be well behaved. Excellent agreement between predicted and measured surface pressure is seen on the compression ramps; the accuracy is somewhat degraded near the normal shock.

CFD is commonly used to optimize the inlet design over a range of flight conditions. One interest is to design the bleed system using tools such as FANSI. To maximize system performance, the designer wants to remove the smallest amount of air required to prevent adverse boundary-layer interactions. With CFD tools, the location and rate of mass removal can be optimized.

Two-dimensional Navier-Stokes tools such as FANSI are very popular among engineering design groups. However, for some supersonic concepts and for most subsonic and transonic designs, two-dimensional analysis is not acceptable. In these cases, inlet flowfields are highly three dimensional. Two three-dimensional multizone Navier-Stokes codes have been developed at MCAIR to fill this need: X3D (velocity-splitting method[3]) and NASTD (three-dimensional extension of FANSI[4]). Other codes with similar capabilities have been developed by government organizations and other industry groups. The most commonly used multizone Navier-Stokes code probably is PARC3D.[1]

Three-dimensional codes are being used extensively for forebody and inlet integration studies. One example is presented in Fig. 8 for a study at Mach 0.67, $\alpha = 10$ deg. Good agreement is demonstrated between CFD predictions (using X3D) and test data for the pressure distribution inside the lower lip. This is a key application, since at higher angles of attack the flow will separate first at this location.

Nozzle Flowfields

The same tools used for inlet analyses are generally applicable to nozzle design studies. Several viscous-dominated interactions must be predicted; these are illustrated in Fig. 9. For most nozzle concepts boundary-layer prediction on internal surfaces poses no severe challenges. However, for ejector nozzles or flows with multiple streams the viscous mixing and entrainment must be modeled carefully.

The behavior of the external nozzle boundary layer can be very complex and difficult to predict. The highest interest is at transonic speeds, where the interactions are at peak complexity and sensitivity. The boundary layer tends to separate easily due to recompression approaching the nozzle exit. The external flow interacts strongly with the jet as it emerges from the internal nozzle. This jet interaction depends on both the shape of the jet (an inviscid interaction) and the viscous near-field entrainment of the free shear layer. These interactions can have a major impact on overall vehicle drag. In some cases the external flow is participating in a complex unsteady interaction coupled with the jets and the internal flow.

Two-dimensional predictions are widely used for engineering studies; three-dimensional tools are being developed and used, but their acceptance is impeded by high cost. Transonic prediction of complex viscous transonic nozzle-jet interactions is the most costly CFD analysis performed currently for engineering objectives at MCAIR, with the sole exception of hyersonic studies involving finite-rate chemistry models. For this three-dimensional problem computation times often exceed 10 h on a Cray XMP. Furthermore, with the grid densities used at this cost, grid independence has not been established for flows with boundary-layer separation.

For some applications two-dimensional methods can be used effectively, as long as one always bears in mind their limitations. The most obvious application is to optimize the external contour of two-dimensional nozzles at supersonic speeds. The predictions can be qualitatively valuable, and, depending on the problem, they even can be quantitatively correct. However, at subsonic and transonic speeds, a two-dimensional prediction of installed nozzle external flows can be dangerously misleading.

The validity of a two-dimensional prediction for internal flows is difficult to define. However, the low cost of this approach in support of engineering trade studies has made two-dimensional viscous analysis very popular. Careful cross checking with available test data or three-dimensional predictions is essential to avoid being misled.

Two-dimensional methods have been used to assist development of ejector nozzle systems. In an ejector nozzle a hot primary flow with high total pressure is mixed with a cool low-energy secondary flow inside the nozzle. The secondary flow is usually used to shield nozzle surfaces from the hot primary flow. Some thrust augmentation benefits also may be achieved.

A relatively small amount of engineering data is available to guide the integration of ejector nozzles with an aircraft propulsion system. The performance of this type of nozzle depends strongly on the rate of entrainment between the primary and secondary flows; that is, high entrainment is desired.

From a computational perspective the key issue is turbulence modeling in the initial region of a free shear layer. Often the primary flow is transonic, whereas the secondary flow is at a very low speed. Multizone methods have been used to model the primary, secondary, and combined flow passages. One recent solution is illustrated in Figs. 10 and 11. This CFD modeling approach contains several significant approximations in both the geometric and fluid dynamic models. However, this level of analysis (CFD), has proven far more effective than previous empirical methods for designing ejector nozzles.[12]

One of the most challenging problems in aircraft design is the prediction of afterbody flowfields for the purpose of drag reduction. The behavior of afterbody flows is closely related to the propulsion system, through both the exhaust jet properties and through the variation of external and internal nozzle contours as the propulsion power level is varied. At transonic speeds, the boundary layer often is separated, usually with dramatic circumferential variations. Unsteady flow is usually present, involving complex interactions between the boundary layer, the inviscid flow, and the jets. The flowfield is often dependent on interactions with surrounding aircraft structure: tail surfaces, nearby structures such as booms and fairings, and open gaps (required for movable nozzle flaps).

CFD models for these flows are subject to several approximations: neglecting unsteady flows, omitting geometric features, utilizing turbulence models with known deficiencies, etc. In many cases the factors that degrade predictive accuracy are not understood. The accuracy achieved in afterbody flow predictions varies widely.

One example of this was found in the study described in Ref. 13. Several afterbody-nozzle configurations were analyzed for the advanced nozzle concepts (ANC) vehicle, which was designed and tested by MCAIR under USAF sponsorship. This vehicle features a twin-engine nozzle system, with the engines submerged in the fuselage and the nozzles integrated with the afterbody. Both axisymmetric (circular) and two-dimensional (rectangular) nozzles were integrated with this model and tested.[14]

With relatively simple CFD models good predictions of afterbody drag were obtained for several axisymmetric nozzle shapes over a range of Mach numbers.[13] For the two-dimensional nozzles transonic predictions were less consistent and generally less accurate. As would be expected, predictions for both nozzle types were much better at supersonic speeds.

Integrated Vehicles

For many years analysis of a complete, integrated aircraft configuration was the ultimate objective of CFD groups in government and industry. Since about 1985, several groups have attained the capability to perform Euler or Navier–Stokes analyses for near-complete configurations (fuselage, wing, tails, inlets, nozzles, and internal flows excluding the engine). Some compromises in geometric modeling remain necessary, but the omissions and adjustments often are not obvious.

The most prominent of these efforts has been the work at NASA Ames Research Center to model the F-16 aircraft. A U.S. Air Force group obtained solutions for a high-speed lifting-body configuration. Jameson and Baker using unstructured grid methods, have obtained Euler solutions for complete transport and fighter configurations. Several other groups also have achieved similar capabilities for important classes of vehicles.

The reaction to these pioneering demonstrations has been surprisingly subdued. This is due in part to the high cost (both computational and labor) of these solutions to date. Because of the large cost and time required, these solutions have not yet had an impact on the engineering design process. Engineering CFD analysis remains focused on optimizing single aircraft components and supporting local integrations of a few aircraft elements.

Ultimately, near-complete vehicle analyses will become common in the engineering design practice. For this to happen major advances must occur in several aspects of CFD technology. Major improvements are yet needed in reducing the cost of analysis (both computing and labor costs) and improving the versatility and reliability of analysis methods.

Assessment

The nature of most internal flow problems is such that inviscid methods provide little useful information to the engineer. Current engineering usage of CFD is based on both two-dimensional and three-dimensional Navier-Stokes methods. For most inlet studies engineering tools use some type of multiple grid approach to provide the needed geometric capability.

As with all CFD applications, the use of these tools is impeded by the high cost of analysis. A single CFD solution often has little value; engineering design groups need several solutions to assess the impact of proposed geometry changes over a range of flow conditions. Thus, from the perspective of the design engineer or manager, a solution requiring 1 h of Cray XMP CPU time can be *very* expensive.

For Navier-Stokes methods many two-dimensional solutions and all three-dimensional solutions exceed this threshold of pain. To gain acceptance at this cost, the solutions must apply to the exact geometry and must provide high-confidence answers to the engineering questions. Current work in developing improved tools must lead to reduced cost. A tremendous range of technical methods is being examined to meet this requirement. These include both new numerical methods and improved strategies for using current methods. The impressive advances in computing power do not generally provide near-term improvement, since most facilities upgrade their supercomputers at long intervals.

The problems in internal flow and aeropropulsion integration have two unifying characteristics: they feature a wide range of complex geometries and are dominated by turbulent viscous interactions. A third characteristic also should be considered: the frequent presence of unsteady flow interactions. These unifying problems are the driving force behind current CFD development requirements:

1) Fast, versatile grid-generation tools are required.

2) Improved turbulence models are needed for secondary flow, separated flow, and the near-field (non-self-similar) region of free shear layers.

3) Current methods must be validated to show that engineering accuracy is achieved for unsteady predictions.

References

[1]Cooper, G. K., Jordan, J. L., and Phares, W. J., "Analysis Tool for Application to Ground Testing of Highly Underexpanded Nozzles," AIAA Paper 87-2015, 1987.

[2]Flores, J., Reznick, S. G., Holst, T. L., and Gundy, K. L., "Transonic Navier-Stokes Solutions for a Fighter-Like Configuration," AIAA Paper 87-0032, Jan. 1987.

[3]Cosner, R. R., "Integrated Flowfield Analysis Methodology for Fighter Inlets," AIAA Paper 85-3071, Oct. 1985.

[4]Bush, R. H., "A Three Dimensional Zonal Navier-Stokes Code for Subsonic Through Hypersonic Propulsion Flowfields," AIAA Paper 88-2830, 1988.

[5]Bush, R. H., "External Compression Inlet Predictions Using an Implicit, Upwind, Multiple Zone Approach," AIAA Paper 85-1521, July 1985.

[6]Bunig, P. G., and Steger, J. L., "Graphics and Flow Visualization in Computational Fluid Dynamics," AIAA Paper 85-1507, July 1985.

[7]Lee, C. C., and Price, W. A., "Subsonic Diffusers for Highly Survivable Aircraft," AFWAL TR-86-3025 (AFWAL/FIMM, Wright-Patterson Air Force Base, OH), May 1986.

[8]Kraft, E. M., and Ritter, A., "Advances at AEDC in Treating Transonic Wind Tunnel Wall Interference," AIAA Paper ICAS-86-1.6.1, London, Sept. 1986.

[9]Donegan, T. L., Benek, J. A., and Erickson, J. C., "Calculation of Transonic Wall Interference," AIAA Paper 87-1432, June 1987.

[10]Phillips, P. S., and Waggoner, E. G., "A Transonic Wind Tunnel Wall Interference Prediction Code," AIAA Paper 88-2538-CP, 1988.

[11]Bush, R. H., Vogel, P. G., Norby, W. P., and Haeffele, B. A., "Two Dimensional Numerical Analysis for Inlets at Subsonic Through Hypersonic Speeds," AIAA Paper 87-1751, June/July 1987.

[12]Rhodes, J. A., and Croxford, J. E., "Zonal Modelling of Flows Through Multiple Inlets and Nozzles," AIAA Paper 89-0005, Jan. 1989.

[13]Mace, J. L., and Cosner, R. R., "Analysis of Viscous Transonic Flow Over Aircraft Forebodies and Afterbodies," AIAA Paper 83-1366, June 1983.

[14]Hiley, P. E., and Bowers, D. L., "Advanced Nozzle Integration for Supersonic Strike Fighter Application," AIAA Paper 81-1441, July 1981.

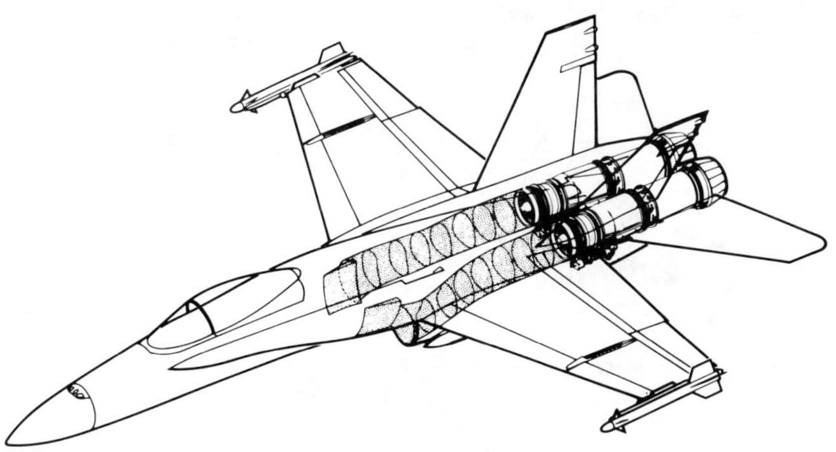

Fig. 1 Sample diffuser installation.

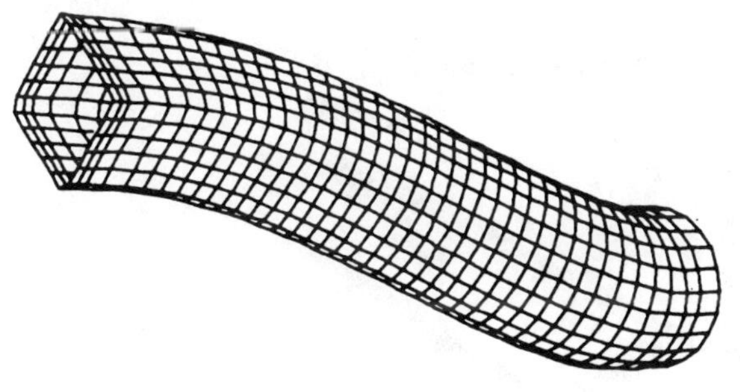

Fig. 2 B19 diffuser geometry.

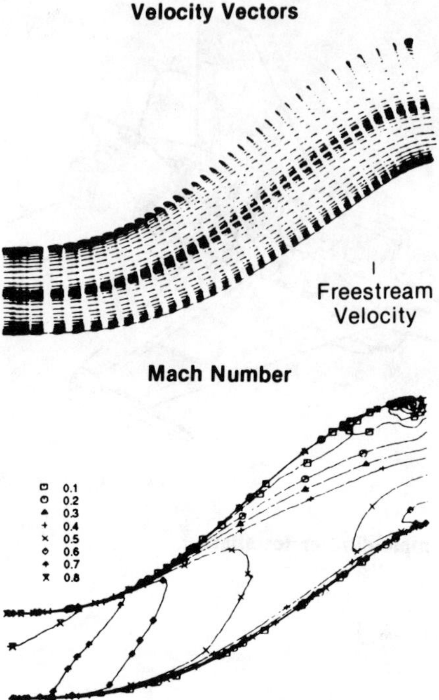

Fig. 3   B19 diffuser flowfield.

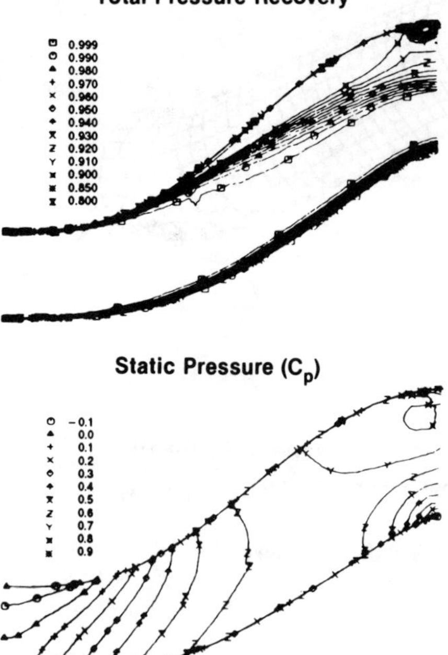

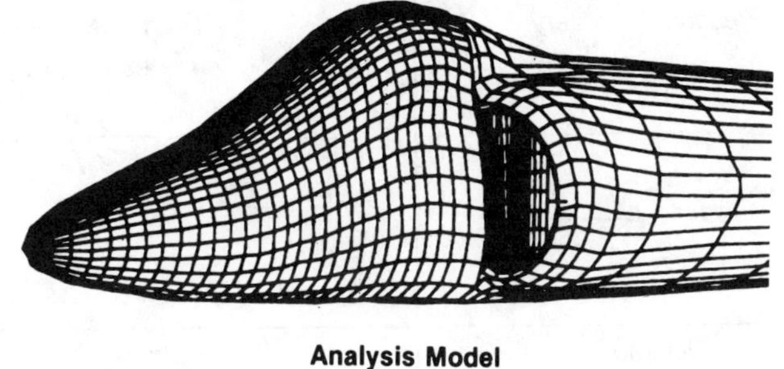

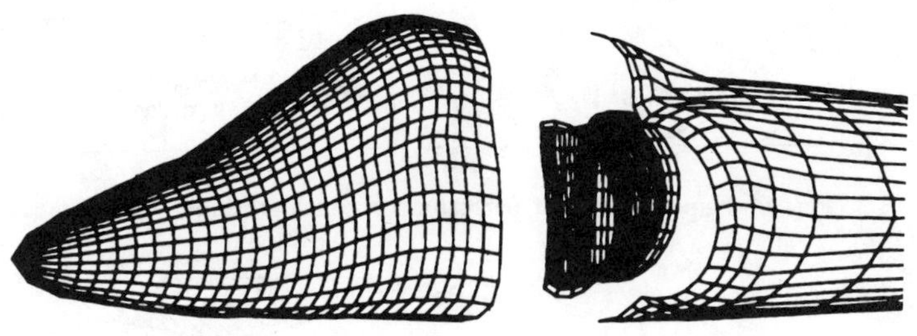

Fig. 4   AV-8B inlet computational domain (3 zones).

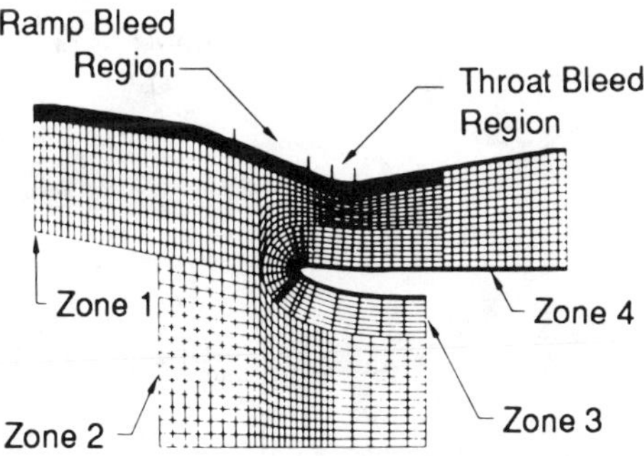

Fig. 5   FANSI four-zone inlet layout.

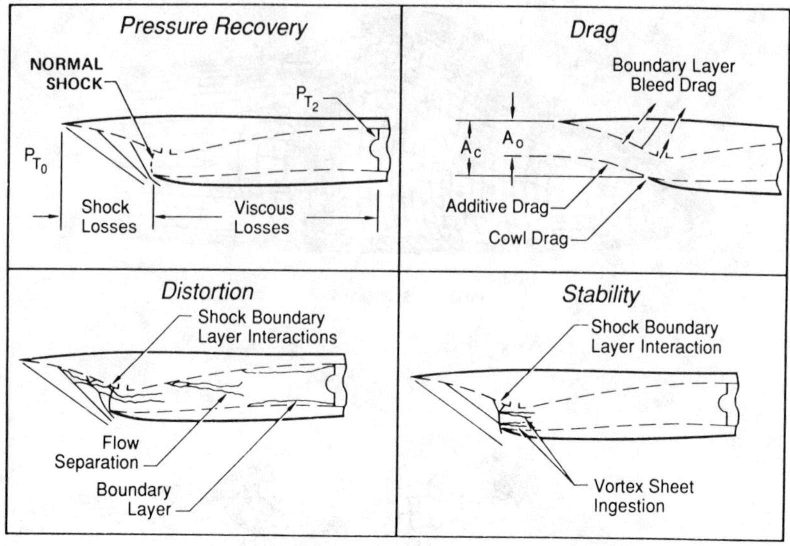

Fig. 6   Supersonic inlet performance considerations.

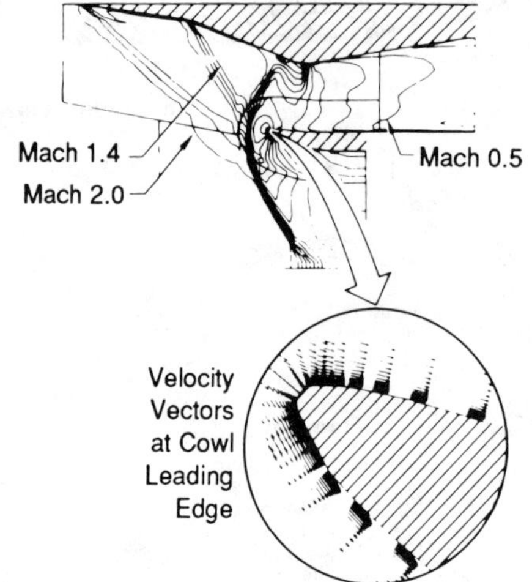

Fig. 7   FANSI solution of two-dimensional inlet.

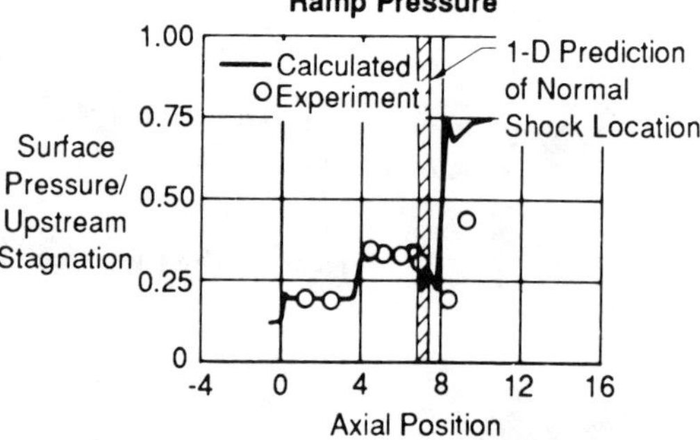

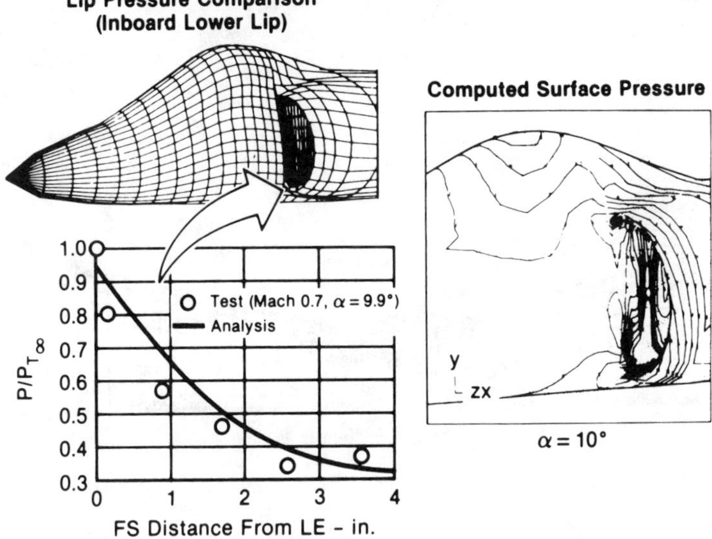

Fig. 8   AV-8B X3D forebody-inlet solution.

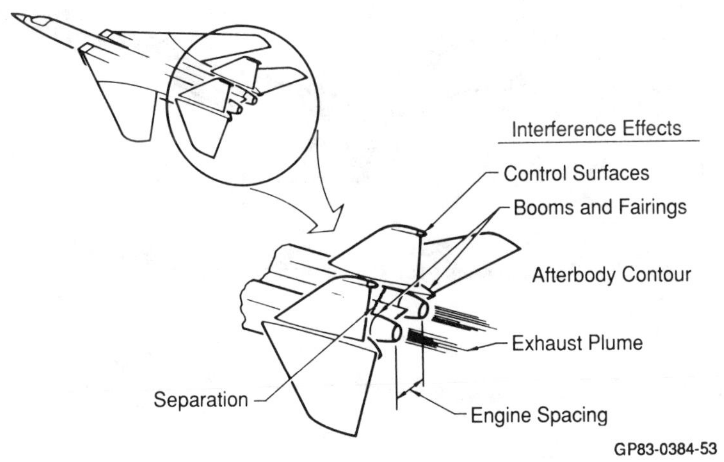

Fig. 9   Nozzle design and analysis issues.

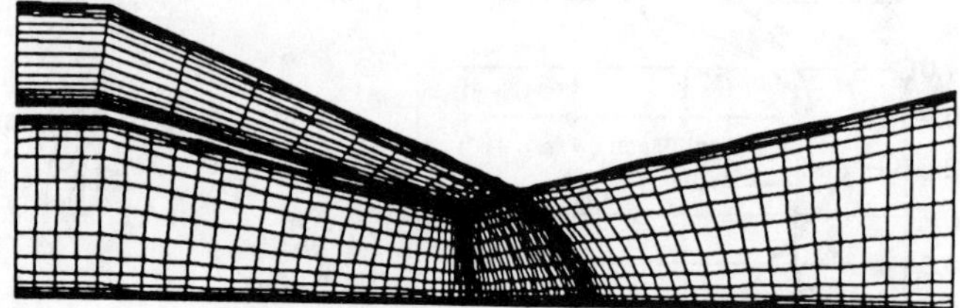

Fig. 10 Grid for ejector nozzle.

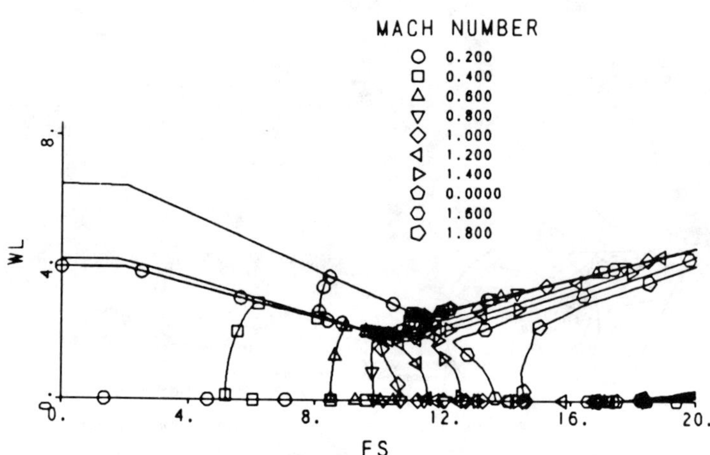

Fig. 11 Mach contours for ejector nozzle solution.

Chapter 13

Computational Analysis of Rotor-Stator Interaction in Turbomachinery Using Zonal Techniques

Nateri K. Madavan*
Sterling Federal Systems, Palo Alto, California
and
Man Mohan Rai†
NASA Ames Research Center, Moffett Field, California

Introduction

MOST techniques for the numerical prediction of flows in turbomachinery have been based on the approximation that the flow in each stator or rotor row is steady in the respective frames of reference; i.e., the spacing between adjacent stator and rotor rows is assumed to be far enough that there is no interaction between them. An extensive body of numerical results exists in the literature dealing with a wide variety of two- and three-dimensional cascade geometries, both planar and annular, stationary and rotating (see, for example, Refs. 1–5). Such analyses have gained widespread acceptance in the industrial community and are now being routinely used in the design process.

In reality, however, flows in turbomachinery are unsteady. The unsteadiness arises from the interaction of the downstream airfoils with the wakes and passage vortices generated upstream, from the motion of the rotors relative to the stators (potential effect), and from vortex shedding at blunt airfoil trailing edges. The unsteady interaction effects impact the aerodynamic, thermal, and structural performance of the airfoils and become increasingly significant as the distance between successive stator and rotor rows is decreased. The problem of rotor-stator interaction is already having an effect on existing designs, where axial gaps between adjacent airfoil rows of one-fourth to one-half of the airfoil chord are common; these effects are bound to be more severe in future designs with higher loading, lower aspect ratios, and reduced overall length.[6] Thus, a need clearly exists for analytical tools that treat the rotor and stator airfoils as a system and provide

---

Copyright © 1989 by the American Institute of Aeronautics and Astronautics, Inc. No copyright is asserted in the United States under Title 17, U.S. Code. The U.S. Government has a royalty-free licence to exercise all rights under the copyright claimed herein for governmental purposes. All other rights are reserved by the copyright owner.
*Principal Analyst.
†Research Scientist, Applied Computational Fluids Branch.

information regarding the magnitude and the impact of the unsteady effects. Such information can be very helpful in optimizing performance and in evaluating new design efforts.

In recent years, considerable progress has been made toward fulfilling this need, spurred by the development of improved computational methodologies and increased computer hardware capabilities (both speed and memory). In particular, the development of zonal techniques (that allow treatment of flowfields where some regions move relative to others) and robust high-accuracy algorithms for the numerical solution of the Navier-Stokes equations has contributed significantly to this progress. The application of such techniques to the problem of rotor-stator interaction forms the subject matter of this chapter.

Zonal Techniques

From a computational point of view, one major difficulty in simulating rotor-stator flows arises because of the relative motion of the rotor and stator airfoils. A single stationary grid that wraps around both the rotor and stator would have to distort considerably to accommodate the motion of the rotor and could result in inaccurate calculations. For small values of the axial gap between the rotor and stator airfoils, such an approach may even be altogether impractical. A second factor contributing to the complexity of grid generation is the necessity to cluster grid points in regions where the dependent variables and their gradients change rapidly (selective grid refinement).

The previously mentioned problems can be overcome by resorting to a zonal approach where the region of interest is divided into several geometrically simpler subregions (or zones). Grids can then be independently generated for each zone using existing grid generation schemes. Figure 1 which shows the multiply-connected flow region associated with a combination of three airfoils, serves to illustrate the approach. Such a region would be difficult to discretize with a single grid; the task becomes Herculean if the two airfoil rows are in relative motion. The division of the flowfield into 13 zones or subdomains, as shown in Fig. 1, results in simple four-sided regions that can be easily discretized. Relative motion between the two rows can also be accommodated by containing the stationary parts in one (set of) subregion(s) and the moving parts in another (set of) subregion(s). Clearly, the zonal approach has several advantages:

1) The grid for any arbitrary region can be generated in a simple, straightforward manner (since the zones are geometrically simple).

2) Flow regions requiring grid refinement can be isolated in separate zones, and the required number of grid points can be introduced in these zones (if necessary this can be done adaptively as the solution progresses in time).

3) The zonal approach facilitates the use of different equation sets in different zones; hence, subsets of the complete equations can be used in regions of the flow where the solution of the complete equations is not warranted.

4) The zonal approach also facilitates the use of block-processing schemes (wherein only the data corresponding to certain regions of the flowfield are required to reside in the core memory of the computer; the remaining data can be stored on disk or tape).

Two different zonal techniques are currently in use. The first is the patched-grid approach where the different zones come together along common boundaries.[7] Figure 2a shows the use of patched grids in discretizing the area between an inner circle and an outer square. In this approach information transfer between zones occurs at the interfaces that separate the zones. The second technique is the overlaid-grid technique.[8] Figure 2b shows the use of overlaid grids in discretizing the region of interest. The two zones do not have a common boundary, but instead overlap one another. In this case information transfer between zones takes place in the overlap region.

Both the patched- and overlaid-grid approaches have their relative advantages and disadvantages. Disadvantages of overlaid grids include 1) a problem in n spatial dimensions requires interpolation in n dimensions (in order to transfer information from one grid onto another), whereas patched grids require only an $(n-1)$ dimensional interpolation; and 2) maintaining global conservation seems somewhat more difficult with overlaid grids (this would be important in cases where the region of interest contained flow discontinuities). However, the fact that zones can (and should) overlap results in a certain amount of flexibility in generating grids. A second advantage of overlaid grids may lie in the ease with which grids can be moved relative to each other (for calculations involving bodies in relative motion). This would be especially important in cases where the motion of the bodies is not predetermined but is calculated as part of the solution.

The boundary conditions used to transfer information from one grid to another must satisfy several requirements before they can be used effectively. For example, the boundary conditions must be 1) numerically stable, 2) spatially and temporally accurate, 3) easily applicable in generalized coordinates, and 4) conservative so that flow discontinuities can move from one grid to another without any distortion. The conservative property, although desirable, is not required in the case of flows without discontinuities.

In recent years, zonal boundary conditions that meet the requirements specified earlier have been developed and rigorously tested for a wide variety of flow problems. One such technique that has been developed by Rai[7,9] will be discussed in this chapter.

Integration Schemes

Another factor contributing to the progress being made in the prediction of three-dimensional unsteady flows in turbomachinery is the development of robust algorithms for solving the Euler and Navier-Stokes equations. Notable among these are the conservative upwind schemes based on the work of Steger and Warming,[10] Roe,[11] Chakravarthy and Osher,[12] and others. Complex flowfields containing flow discontinuities (shocks, slip surfaces, etc.) can now be simulated almost routinely. The conservative nature of these schemes allow for accurate resolution of flow discontinuities. It must be borne in mind, however, that these improvements in the quality of the numerical solution are obtained at the expense of programming simplicity. The accuracy or resolution of such schemes has been improved through highly accurate total variation diminishing (TVD) approximations of the convective terms coupled with proper differencing of the dissipative terms.[13,14] In this chapter we discuss zonal techniques for turbomachinery flow computations in the context of such schemes.

Scope of Present Study

The focus of this chapter is the application of zonal techniques to the prediction of the rotor-stator interaction in turbomachinery. The basic concepts involved in transferring information between grids for the patched- and overlaid-grid approaches are discussed in the context of turbomachinery flowfields. The implementation of these techniques in the framework of a time-accurate two- or three-dimensional Navier-Stokes algorithm to predict rotor-stator interaction is outlined. Results obtained from application of these techniques to the flow in idealized as well as realistic turbine configurations are presented and compared to experimental data where available.

Some disclaimers are in order here. The purpose of this chapter is to familiarize the reader with the salient features of the zonal approach (as it pertains to zones in relative motion) and demonstrate its applicability in computing two- and three-dimensional unsteady flows in turbomachinery. To this end, we focus primarily on the work of Refs. 15–18 since it embodies the current state of the art. This chapter is not a review of the development of zonal techniques; neither is it a review of the current status of unsteady flow prediction in turbomachinery. Thus, the large body of literature that exists dealing with the development and application of zonal techniques is not discussed. (References 19–25 constitute a brief list of relevant work, but this list is by no means comprehensive.) By the same token, other research efforts in predicting rotor-stator interaction are not discussed, except for the brief mention later. The interested reader should consult these references for additional details.

Pioneering work in predicting rotor-stator interaction was conducted by Erdos et al.[26] as far back as 1977. However, the subject has only very recently become the focus of increasing attention. Fourmaux[27] and Giles[28] have calculated two-dimensional rotor-stator interactions using the Euler equations; Giles uses a novel concept of "time-inclined" computational planes to treat difficulties encountered when the stator-rotor pitch ratio is not a ratio of two small integers. Lewis et al.[29] solve the quasi-three-dimensional, inviscid equations, and Jorgenson and Chima[30,31] have used an explicit Runge-Kutta algorithm to solve the quasi-three-dimensional Euler and thin-layer Navier-Stokes equations. Three-dimensional calculations have been presented in Refs. 32 and 33. Reference 34 presents results for flow in a compressor stage obtained using a shearing-grid technique, where a single grid wraps around the stator and rotor and shears in order to allow relative motion between them. The data from the sheared grid are interpolated onto an undistorted initial grid at points in time when the distortion exceeds a certain tolerance level.

Integration Schemes

The zonal boundary schemes discussed in this section were developed for use in conjunction with both implicit and explicit integration schemes. Our focus here is on implicit schemes; the fine grids required to resolve boundary-layer properties result in rather restrictive time-step limitations for viscous calculations using explicit schemes. The integration schemes discussed here are set in an iterative framework, where the nonlinear equations are solved at each time step.[35] (The schemes revert to the conventional, noniterative, implicit type when the number of iterations at each time step is restricted to one.) There are several advantages to using an iterative approach.

First, some desirable properties, such as monotonicity of the solution to the nonlinear equations, that may be lost upon linearization can be restored through iteration at each time step. Second, various approximations can be made in the linearization of the equations in order to reduce the computational effort; these linearization errors can be eliminated during the iteration process. Third, in multidimensional problems, where approximate factorization is usually resorted to in order to avoid large-bandwidth matrix inversions, factorization errors can also be driven to zero. Fourth, the iterative approach has been found to enhance stability, allowing the use of larger time steps than with noniterative schemes. When used in conjunction with zonal techniques, the iterative approach has an added advantage in that it considerably simplifies the application of implicit zonal boundary conditions. The primary disadvantage of the iterative approach is the added computations that the iterations require. This is offset to some extent by the larger step sizes that are possible with the iterative scheme.

The integration scheme used is discussed in this section, both for the two- and three-dimensional case. Only a brief outline is given; further details are available in Refs. 15 and 17. Note that, although the fluxes are evaluated here using Roe's[11] scheme, other methods (Osher, Steger-Warming, etc.) can be (and have been) easily incorporated (see, for example, Ref. 15). Furthermore, in the schemes described, approximate factorization is used to split the multi-dimensional operators into single-dimensional operators (which are then solved efficiently using block-tridiagonal elimination methods). An alternative approach would be the use of a relaxation strategy.[36] The zonal-boundary techniques discussed here can also be applied in a relaxation-type setting. The reader is referred to Ref. 37 for details.

Integration Scheme in Two Spatial Dimensions

To describe the integration scheme we consider the unsteady, Navier-Stokes equations in two dimensions:

$$Q_t + E_x + F_y = R_x + S_y \qquad (1)$$

where

$$Q = \begin{pmatrix} \rho \\ \rho u \\ \rho v \\ e \end{pmatrix}, \quad E = \begin{pmatrix} \rho u \\ p + \rho u^2 \\ \rho u v \\ (e + p)u \end{pmatrix}, \quad F = \begin{pmatrix} \rho v \\ \rho u v \\ p + \rho v^2 \\ (e + p)v \end{pmatrix} \tag{2a}$$

$$R = \begin{pmatrix} 0 \\ \tau_{xx} \\ \tau_{xy} \\ \beta_x \end{pmatrix}, \quad S = \begin{pmatrix} 0 \\ \tau_{xy} \\ \tau_{yy} \\ \beta_y \end{pmatrix} \tag{2b}$$

and

$$\tau_{xx} = 2\mu u_x + \lambda(u_x + v_y) \tag{3a}$$

$$\tau_{xy} = \mu(u_y + v_x) \tag{3b}$$

$$\tau_{yy} = 2\mu v_y + \lambda(u_x + v_y) \tag{3c}$$

$$\beta_x = u\tau_{xx} + v\tau_{xy} + \gamma\mu Pr^{-1}\bar{e}_x \tag{3d}$$

$$\beta_y = u\tau_{xy} + v\tau_{yy} + \gamma\mu Pr^{-1}\bar{e}_y \tag{3e}$$

$$\lambda = -2\mu/3 \tag{3f}$$

$$\bar{e} = p/[\rho(\gamma - 1)] \tag{3g}$$

Making the independent variable transformation

$$\tau = t \tag{4a}$$

$$\xi = \xi(x,y,t) \tag{4b}$$

$$\eta = \eta(x,y,t) \tag{4c}$$

and the thin-layer approximation,[38] we obtain

$$\tilde{Q}_\tau + \tilde{E}_\xi + \tilde{F}_\eta = Re^{-1}\tilde{S}_\eta \tag{5}$$

where

$$\tilde{Q} = Q/Ja \tag{6a}$$

$$\tilde{E}(Q,\xi) = (\xi_t Q + \xi_x E + \xi_y F)/Ja \tag{6b}$$

$$\tilde{F}(Q,\eta) = (\eta_t Q + \eta_x E + \eta_y F)/Ja \tag{6c}$$

$$Ja = \xi_x \eta_y - \eta_x \xi_y \tag{6d}$$

The symbols (ξ_t, ξ_x, ξ_y) and (η_t, η_x, η_y) denote the metrics of the transformation, Ja denotes the Jacobian of the transformation, and Re is the Reynolds number. The vector $\tilde{S}$ is given by

$$\tilde{S} = \begin{pmatrix} 0 \\ K_1 u_\eta + K_2 \eta_x \\ K_1 v_\eta + K_2 \eta_y \\ K_1 [Pr^{-1}(\gamma-1)^{-1}(a^2)_\eta + (q^2/2)_\eta] + K_2 K_3 \end{pmatrix} \tag{7}$$

where

$$K_1 = \mu(\eta_x^2 + \eta_y^2)$$

$$K_2 = \mu(\eta_x u_\eta + \eta_y v_\eta)/3$$

$$K_3 = u\eta_x + v\eta_y$$

$$q^2 = u^2 + v^2$$

and Pr denotes the Prandtl number. Equations (5) and (7) assume that the body surface is a constant η line (in incorporating the thin-layer approximation).

The factored, iterative, implicit algorithm is given by

$$\left[I + \frac{\Delta\tau}{\Delta\xi}(\nabla_\xi \tilde{A}_{j,k}^+ + \Delta_\xi \tilde{A}_{j,k}^-) \right]^p$$

$$\times \left[I + \frac{\Delta\tau}{\Delta\eta}(\nabla_\eta \tilde{B}_{j,k}^+ + \Delta_\eta \tilde{B}_{j,k}^- - Re^{-1}\delta_\eta \tilde{M}) \right]^p (\tilde{Q}_{j,k}^{p+1} - \tilde{Q}_{j,k}^p)$$

$$= -\Delta\tau \left(\frac{\tilde{Q}_{j,k}^p - \tilde{Q}_{j,k}^n}{\Delta\tau} + \frac{\hat{E}_{j+\frac{1}{2},k}^p - \hat{E}_{j-\frac{1}{2},k}^p}{\Delta\xi} \right.$$

$$\left. + \frac{\hat{F}_{j,k+\frac{1}{2}}^p - \hat{F}_{j,k-\frac{1}{2}}^p}{\Delta\eta} - \frac{\hat{S}_{j,k+\frac{1}{2}}^p - \hat{S}_{j,k-\frac{1}{2}}^p}{Re\,\Delta\eta} \right) \tag{8}$$

where

$$\tilde{A}^\pm = \left(\frac{\partial \tilde{E}}{\partial \tilde{Q}} \right)^\pm \tag{9a}$$

$$\tilde{B}^\pm = \left(\frac{\partial \tilde{F}}{\partial \tilde{Q}} \right)^\pm \tag{9b}$$

$$\tilde{M} = \frac{\partial \tilde{S}}{\partial \tilde{Q}} \tag{9c}$$

and Δ, ∇, and δ are forward-, backward-, and central-difference operators, respectively. The quantities $\hat{E}_{j+\frac{1}{2},k}$, $\hat{F}_{j,k+\frac{1}{2}}$, and $\hat{S}_{j,k+\frac{1}{2}}$ are numerical fluxes consistent with the physical fluxes $\tilde{E}$, $\tilde{F}$, and $\tilde{S}$. In Eq. (8), $\tilde{Q}^p$ is an approximation to $\tilde{Q}^{n+1}$. When $p = 0$, $\tilde{Q}^p = \tilde{Q}^n$ and when Eq. (8) is iterated to convergence at a given time step, $\tilde{Q}^p = \tilde{Q}^{n+1}$. Note that, because the left-hand side of this equation can be driven to zero at each time step (by iterating to convergence), linearization and factorization errors can be driven to zero during the iteration process. As mentioned earlier, the scheme reverts to a conventional, noniterative scheme when the number of iterations is restricted to one. When second-order accuracy in time is required, the term $(\tilde{Q}^p_{j,k} - \tilde{Q}^n_{j,k})$ on the right-hand side of Eq. (8) must be replaced by $(1.5\tilde{Q}^p_{j,k} - 2.0\tilde{Q}^n_{j,k} + 0.5\tilde{Q}^{n-1}_{j,k})$ in addition to iterating to convergence. Typically, three or four iterations per time step are sufficient to reduce the residual by an order of magnitude or more.

The numerical fluxes $\hat{E}$ and $\hat{F}$ can be evaluated in many different ways, the different choices leading to different schemes. For the sake of illustration, we consider the scheme due to Roe.[11] The numerical fluxes for the spatially third-order accurate scheme are evaluated as

$$\hat{E}_{i+\frac{1}{2},j} = \tfrac{1}{2}[\tilde{E}(Q_{i,j},\xi_{i+\frac{1}{2},j}) + \tilde{E}(Q_{i+1,j},\xi_{i+\frac{1}{2},j})]$$

$$+ \tfrac{1}{6}[\Delta E^{+}(Q_{i-1,j},Q_{i,j},\xi_{i+\frac{1}{2},j}) - \Delta E^{+}(Q_{i,j},Q_{i+1,j},\xi_{i+\frac{1}{2},j})]$$

$$+ \tfrac{1}{6}[\Delta E^{-}(Q_{i,j},Q_{i+1,j},\xi_{i+\frac{1}{2},j}) - \Delta E^{-}(Q_{i+1,j},Q_{i+2,j},\xi_{i+\frac{1}{2},j})] \qquad (10)$$

Note that, in Eq. (8), an approximate linearization is used that results in the left-hand side corresponding to the Steger-Warming split-flux scheme. In order to preserve the block-tridiagonal matrix structure of the left-hand side, the terms that contribute to the higher-order spatial accuracy ($\Delta E^{\pm}$) are evaluated at the pth level and hence appear only on the right-hand side. The flux differences $\Delta E^{\pm}$ using the method of Roe are given by

$$\Delta E^{\pm}(Q_{i,j},Q_{i+1,j},\xi_{i+\frac{1}{2},j}) = \tilde{A}^{\pm}(Q_{i+\frac{1}{2},j},\xi_{i+\frac{1}{2},j}) \times (Q_{i+1,j} - Q_{i,j}) \qquad (11)$$

The dependent variables necessary to evaluate $\tilde{A}^{\pm}$ at the intermediate point $(i+\frac{1}{2},j)$ are evaluated from the following expressions:

$$u_{i+\frac{1}{2}} = \frac{u_i\sqrt{\rho_i} + u_{i+1}\sqrt{\rho_{i+1}}}{\sqrt{\rho_i} + \sqrt{\rho_{i+1}}} \qquad (12a)$$

$$v_{i+\frac{1}{2}} = \frac{v_i\sqrt{\rho_i} + v_{i+1}\sqrt{\rho_{i+1}}}{\sqrt{\rho_i} + \sqrt{\rho_{i+1}}} \qquad (12b)$$

$$h_{i+\frac{1}{2}} = \frac{h_i\sqrt{\rho_i} + h_{i+1}\sqrt{\rho_{i+1}}}{\sqrt{\rho_i} + \sqrt{\rho_{i+1}}} \qquad (12c)$$

$$h = \frac{e + p}{\rho} \qquad (12d)$$

The subscript j has been left out of the expressions in Eqs. (12) for convenience. The eigenvalues of the Jacobian matrix $\tilde{A}$ need to be calculated in the process of calculating the matrices $\tilde{A}^{\pm}$. These eigenvalues are modified as outlined in Ref. 39 to prevent expansion shocks. The numerical flux $\hat{F}_{i,j+\frac{1}{2}}$ is evaluated in a similar manner.

The viscous flux vector $\hat{S}_{i,j+\frac{1}{2}}$ is evaluated using central differences, that is,

$$\hat{S}_{i,j+\frac{1}{2}} = \hat{S}[Q_{i,j+\frac{1}{2}}, (Q_\eta)_{i,j+\frac{1}{2}}, \eta_{i,j+\frac{1}{2}}] \tag{13a}$$

$$Q_{i,j+\frac{1}{2}} = \tfrac{1}{2}(Q_{i,j} + Q_{i+1,j}) \tag{13b}$$

$$(Q_\eta)_{i,j+\frac{1}{2}} = Q_{i,j+1} - Q_{i,j} \tag{13c}$$

Integration Scheme in Three Spatial Dimensions

To describe the integration scheme in three dimensions, we consider the unsteady, three-dimensional Navier-Stokes equation

$$Q_t + E_x + F_y + G_z = R_x + S_y + T_z \tag{14}$$

where

$$Q = \begin{bmatrix} \rho \\ \rho u \\ \rho v \\ \rho w \\ e \end{bmatrix}, \qquad E = \begin{bmatrix} \rho u \\ p + \rho u^2 \\ \rho uv \\ \rho uw \\ (e+p)u \end{bmatrix}$$

$$F = \begin{bmatrix} \rho v \\ \rho uv \\ p + \rho v^2 \\ \rho vw \\ (e+p)v \end{bmatrix}, \qquad G = \begin{bmatrix} \rho w \\ \rho uw \\ \rho vw \\ p + \rho w^2 \\ (e+p)w \end{bmatrix} \tag{15a}$$

$$R = \begin{bmatrix} 0 \\ \tau_{xx} \\ \tau_{xy} \\ \tau_{xz} \\ \beta_x \end{bmatrix}, \qquad S = \begin{bmatrix} 0 \\ \tau_{yx} \\ \tau_{yy} \\ \tau_{yz} \\ \beta_y \end{bmatrix}, \qquad T = \begin{bmatrix} 0 \\ \tau_{zx} \\ \tau_{zy} \\ \tau_{zz} \\ \beta_z \end{bmatrix} \tag{15b}$$

and where

$$\tau_{xx} = 2\mu u_x + \lambda(u_x + v_y + w_z) \tag{16a}$$

$$\tau_{xy} = \mu(u_y + v_x) \tag{16b}$$

$$\tau_{xz} = \mu(u_z + w_x) \tag{16c}$$

$$\tau_{yx} = \tau_{xy} \tag{16d}$$

$$\tau_{yy} = 2\mu v_y + \lambda(u_x + v_y + w_z) \tag{16e}$$

$$\tau_{yz} = \mu(v_z + w_y) \tag{16f}$$

$$\tau_{zx} = \tau_{xz} \tag{16g}$$

$$\tau_{zy} = \tau_{yz} \tag{16h}$$

$$\tau_{zz} = 2\mu w_z + \lambda(u_x + v_y + w_z) \tag{16i}$$

$$\beta_x = u\tau_{xx} + v\tau_{xy} + w\tau_{xz} + \gamma\mu Pr^{-1}\bar{e}_x \tag{16j}$$

$$\beta_y = u\tau_{yx} + v\tau_{yy} + w\tau_{yz} + \gamma\mu Pr^{-1}\bar{e}_y \tag{16k}$$

$$\beta_z = u\tau_{zx} + v\tau_{zy} + w\tau_{zz} + \gamma\mu Pr^{-1}\bar{e}_z \tag{16l}$$

$$\lambda = -2\mu/3 \tag{16m}$$

$$\bar{e} = p/[\rho(\gamma - 1)] \tag{16n}$$

The independent variable transformation

$$\tau = t \tag{17a}$$

$$\xi = \xi(x,y,z,t) \tag{17b}$$

$$\eta = \eta(x,y,z,t) \tag{17c}$$

$$\zeta = \zeta(x,y,z,t) \tag{17d}$$

is then applied to Eq. (14). For high-Reynolds-number flows with one of the coordinates in the transformed coordinate system (ξ,η,ζ) corresponding to the body surface, one usually makes the thin-layer assumption; that is, the viscous terms evaluated as derivatives in the directions tangent to the body surface are assumed to be negligible in magnitude.[38] Extension of this concept to thin layers in all three directions is presented in Ref. 40. In the present study the viscous terms are retained in two coordinate directions: in the direction normal to the hub surface (the ζ direction) and the direction normal to the airfoil surfaces (the η direction). The transformed equations now take the form

$$\tilde{Q}_\tau + \tilde{E}_\xi + \tilde{F}_\eta + \tilde{G}_\zeta = Re^{-1}(\tilde{S}_\eta + \tilde{T}_\zeta) \tag{18}$$

where
$$\tilde{Q} = Q/Ja$$

$$\tilde{E}(Q,\xi) = (\xi_t Q + \xi_x E + \xi_y F + \xi_z G)/Ja \tag{19a}$$

$$\tilde{F}(Q,\eta) = (\eta_t Q + \eta_x E + \eta_y F + \eta_z G)/Ja \tag{19b}$$

$$\tilde{G}(Q,\zeta) = (\zeta_t Q + \zeta_x E + \zeta_y F + \zeta_z G)/Ja \tag{19c}$$

The formulas for the metrics of the transformation $(\xi_t,\xi_x,\xi_y,\xi_z)$, $(\eta_t,\eta_x,\eta_y,\eta_z)$, and $(\zeta_t,\zeta_x,\zeta_y,\zeta_z)$ and the Jacobian of the transformation are given in Ref. 38. The vector $\tilde{S}$ is given by

$$\tilde{S} = \begin{bmatrix} 0 \\ K_1 u_\eta + K_2 \eta_x \\ K_1 v_\eta + K_2 \eta_y \\ K_1 w_\eta + K_2 \eta_z \\ K_1[Pr^{-1}(\gamma-1)^{-1}(a^2)_\eta + (q^2/2)_\eta] + K_2 K_3 \end{bmatrix} \tag{20}$$

where

$$K_1 = \mu(\eta_x^2 + \eta_y^2 + \eta_z^2) \tag{21a}$$

$$K_2 = \mu(\eta_x u_\eta + \eta_y v_\eta + \eta_z w_\eta)/3 \tag{21b}$$

$$K_3 = u\eta_x + v\eta_y + w\eta_z \tag{21c}$$

$$q^2 = u^2 + v^2 + w^2 \tag{21d}$$

The vector $\tilde{T}$ can be obtained by replacing η with ζ in Eqs. (20) and (21). The factored, iterative, implicit, third-order accurate scheme is given by

$$\left[I + \frac{\Delta\tau}{\Delta\xi}(\nabla_\xi \tilde{A}^+_{i,j,k} + \Delta_\xi \tilde{A}^-_{i,j,k}) \right]^p$$

$$\times \left[I + \frac{\Delta\tau}{\Delta\eta}(\nabla_\eta \tilde{B}^+_{i,j,k} + \Delta_\eta \tilde{B}^-_{i,j,k} - Re^{-1}\delta_\eta \tilde{M}) \right]^p$$

$$\times \left[I + \frac{\Delta\tau}{\Delta\zeta}(\nabla_\zeta \tilde{C}^+_{i,j,k} + \Delta_\zeta \tilde{C}^-_{i,j,k} - Re^{-1}\delta_\zeta \tilde{N}) \right]^p (\tilde{Q}^{p+1}_{i,j,k} - \tilde{Q}^p_{i,j,k})$$

$$= -\Delta\tau \left(\frac{\tilde{Q}^p_{i,j,k} - \tilde{Q}^n_{i,j,k}}{\Delta\tau} + \frac{\hat{E}^p_{i+\frac{1}{2},j,k} - \hat{E}^p_{i-\frac{1}{2},j,k}}{\Delta\xi} \right.$$

$$+ \frac{\hat{F}^p_{i,j+\frac{1}{2},k} - \hat{F}^p_{i,j-\frac{1}{2},k}}{\Delta\eta} + \frac{\hat{G}^p_{i,j,k+\frac{1}{2}} - \hat{G}^p_{i,j,k-\frac{1}{2}}}{\Delta\zeta}$$

$$\left. - \frac{\hat{S}^p_{i,j+\frac{1}{2},k} - \hat{S}^p_{i,j-\frac{1}{2},k}}{Re\,\Delta\eta} - \frac{\hat{T}^p_{i,j,k+\frac{1}{2}} - \hat{T}^p_{i,j,k-\frac{1}{2}}}{Re\,\Delta\zeta} \right) \tag{22}$$

where

$$\tilde{A}^{\pm} = \left(\frac{\partial \tilde{E}}{\partial \tilde{Q}}\right)^{\pm} \tag{23a}$$

$$\tilde{B}^{\pm} = \left(\frac{\partial \tilde{F}}{\partial \tilde{Q}}\right)^{\pm} \tag{23b}$$

$$\tilde{C}^{\pm} = \left(\frac{\partial \tilde{G}}{\partial \tilde{Q}}\right)^{\pm} \tag{23c}$$

$$\tilde{M} = \frac{\partial \tilde{S}}{\partial \tilde{Q}} \tag{23d}$$

$$\tilde{N} = \frac{\partial \tilde{T}}{\partial \tilde{Q}} \tag{23e}$$

The numerical fluxes for the third-order Roe scheme are evaluated as

$$\hat{E}_{i+\frac{1}{2},j,k} = \tfrac{1}{2}[\tilde{E}(Q_{i,j,k},\xi_{i+\frac{1}{2},j,k}) + \tilde{E}(Q_{i+1,j,k},\xi_{i+\frac{1}{2},j,k})]$$

$$+ \tfrac{1}{6}[\Delta E^{+}(Q_{i-1,j,k},Q_{i,j,k},\xi_{i+\frac{1}{2},j,k}) - \Delta E^{+}(Q_{i,j,k},Q_{i+1,j,k},\xi_{i+\frac{1}{2},j,k})]$$

$$+ \tfrac{1}{6}[\Delta E^{-}(Q_{i,j,k},Q_{i+1,j,k},\xi_{i+\frac{1}{2},j,k}) - \Delta E^{-}(Q_{i+1,j,k},Q_{i+2,j,k},\xi_{i+\frac{1}{2},j,k})] \tag{24}$$

The flux differences $\Delta E^{\pm}$ using the method of Roe are given by

$$\Delta E^{\pm}(Q_{i,j,k},\, Q_{i+1,j,k},\xi_{i+\frac{1}{2},j,k}) = \tilde{A}^{\pm}(Q_{i+\frac{1}{2},j,k},\xi_{i+\frac{1}{2},j,k}) \times (Q_{i+1,j,k} - Q_{i,j,k}) \tag{25}$$

The dependent variables necessary to evaluate $\tilde{A}^{\pm}$ at the intermediate point $(i+\frac{1}{2},j,k)$ are evaluated from the following expressions:

$$u_{i+\frac{1}{2}} = \frac{u_i\sqrt{\rho_i} + u_{i+1}\sqrt{\rho_{i+1}}}{\sqrt{\rho_i} + \sqrt{\rho_{i+1}}} \tag{26a}$$

$$v_{i+\frac{1}{2}} = \frac{v_i\sqrt{\rho_i} + v_{i+1}\sqrt{\rho_{i+1}}}{\sqrt{\rho_i} + \sqrt{\rho_{i+1}}} \tag{26b}$$

$$w_{i+\frac{1}{2}} = \frac{w_i\sqrt{\rho_i} + w_{i+1}\sqrt{\rho_{i+1}}}{\sqrt{\rho_i} + \sqrt{\rho_{i+1}}} \tag{26c}$$

$$h_{i+\frac{1}{2}} = \frac{h_i\sqrt{\rho_i} + h_{i+1}\sqrt{\rho_{i+1}}}{\sqrt{\rho_i} + \sqrt{\rho_{i+1}}} \tag{26d}$$

$$h = \frac{e+p}{\rho} \tag{26e}$$

The subscripts j and k have been left out of the expressions in Eq. (26) for convenience. The numerical fluxes $\hat{F}_{i,j+\frac{1}{2},k}$ and $\tilde{G}_{i,j,k+\frac{1}{2}}$ are also evaluated in a similar manner. As in the two-dimensional case, the viscous flux vectors $\hat{S}_{i,j+\frac{1}{2},k}$ and $\hat{T}_{i,j,k+\frac{1}{2}}$ are evaluated using central differences.

Zonal Boundary Conditions

The use of zonal techniques result in several zonal boundaries where boundary conditions must be specified. The requirements that these boundary conditions must meet before they can be effectively used have been stated earlier in the Introduction. In this section, both conservative and nonconservative zonal boundary schemes are briefly outlined. Details can be found in Refs. 7 and 9. The conservative scheme described is restricted to calculations using patched grids. A conservative scheme applicable to overlaid grids can be developed, but is not considered here. The nonconservative scheme described here, however, is applicable to both patched and overlaid grids; the only constraint on its use is that the flow be free of discontinuities. Although easily extendable to three dimensions, the conservative zonal boundary scheme is described here for the two-dimensional case. This two-dimensional zonal-boundary scheme can also be used in three-dimensional calculations where the grids are generated by stacking two-dimensional grids in the spanwise direction (thus resulting in two-dimensional patch boundaries).

Conservative Patched Zonal-Boundary Scheme

For the sake of simplicity, we develop the conservative patch-boundary condition for the inviscid case. Consider the two curvilinear grids used to discretize the flow region shown in Fig. 3. The line AB represents the patch boundary that separates the two grids used to discretize the given region. Let l and m be the indices used in the ξ and η directions, respectively, in zone 1, and let j and k be the corresponding indices for zone 2. Let n represent the time step for both zones. A superscript within parentheses will denote the zone to which a given quantity belongs; for example, $\Delta \tau^{(1)}$ denotes the marching step size in zone 1.

Establishing two independent variable transformations (one for each grid),

$$\tau^{(i)} = t \tag{27a}$$

$$\xi^{(i)} = \xi^{(i)}(x,y,t) \tag{27b}$$

$$\eta^{(i)} = \eta^{(i)}(x,y,t) \tag{27c}$$

and applying these transformations to the Euler equations in two spatial dimensions, we obtain

$$\tilde{Q}^{(i)}_{\tau^{(i)}} + \tilde{E}^{(i)}_{\xi^{(i)}} + \tilde{F}^{(i)}_{\eta^{(i)}} = 0 \tag{28}$$

where

$$\tilde{Q}^{(i)} = Q/Ja^{(i)} \tag{29a}$$

$$\tilde{E}^{(i)}(Q,\xi^{(i)}) = (\xi_t^{(i)}Q + \xi_x^{(i)}E + \xi_y^{(i)}F)/Ja^{(i)} \tag{29b}$$

$$\tilde{F}^{(i)}(Q,\eta^{(i)}) = (\eta_t^{(i)}Q + \eta_x^{(i)}E + \eta_y^{(i)}F)/Ja^{(i)} \tag{29c}$$

$$Ja^{(i)} = \xi_x^{(i)}\eta_y^{(i)} - \eta_x^{(i)}\xi_y^{(i)} \tag{29d}$$

Let the conservative difference schemes used to integrate Eq. (28) be given by

$$\frac{\Delta\tilde{Q}_{l,m}^{(1)}}{\Delta\tau^{(1)}} + \frac{\hat{E}_{l+\frac{1}{2},m}^{(1)} - \hat{E}_{l-\frac{1}{2},m}^{(1)}}{\Delta\xi^{(1)}} + \frac{\hat{F}_{l,m+\frac{1}{2}}^{(1)} - \hat{F}_{l,m-\frac{1}{2}}^{(1)}}{\Delta\eta^{(1)}} = 0 \tag{29e}$$

in zone 1, and,

$$\frac{\Delta\tilde{Q}_{j,k}^{(2)}}{\Delta\tau^{(2)}} + \frac{\hat{E}_{j+\frac{1}{2},k}^{(2)} - \hat{E}_{j-\frac{1}{2},k}^{(2)}}{\Delta\xi^{(2)}} + \frac{\hat{F}_{j,k+\frac{1}{2}}^{(2)} - \hat{F}_{j,k-\frac{1}{2}}^{(2)}}{\Delta\eta^{(2)}} = 0 \tag{30}$$

in zone 2. The terms $\hat{E}^{(i)}$ and $\tilde{F}^{(i)}$ are, once again, numerical fluxes consistent with the transformed fluxes $\tilde{E}^{(i)}$ and $\tilde{F}^{(i)}$, respectively. For an explicit integration scheme these numerical fluxes are evaluated at the nth time level, whereas for an implicit scheme these fluxes are evaluated at the $(n+1)$th time level.

Consider first the explicit case. The patch-boundary treatment in this case consists of the following steps:

1) Integrate the dependent variables at grid points (of both the grids) that do not belong to the patch boundary using Eqs. (29) and (30).

2) Integrate the dependent variables at the patch-boundary points of one of the zones (say, zone 2 of Fig. 3) using a scheme that conserves fluxes across the patch boundary.

3) Obtain the dependent variables at the patch-boundary points of the other zone (say, zone 1 of Fig. 3) such that the dependent variables are continuous along the patch boundary.

The implementation of the first step of the zonal scheme is straightforward. The implementation of the second step is described later. Assume that the patch-boundary points of zone 2 are to be updated using the finite-difference scheme of Eq. (30). This calculation requires the use of the fluxes $\hat{F}_{j,\frac{1}{2}}^{(2)}$. These fluxes have to be calculated such that global conservation is maintained. A typical cell ($RSTU$) of a patch-boundary point ($j,1$) is shown in Fig. 3. The points R and S are midpoints of the cells they lie in, and the points T and U are obtained as follows: The constant j lines of zone 2 are extrapolated into zone 1 to intersect the line CD (CD corresponds to $m = m_{\max} - \frac{1}{2}$ in zone 1 and to $k = \frac{1}{2}$ in zone 2). The intersection points have the indices ($j,\frac{1}{2}$). Point T is midway between the points ($j+1,\frac{1}{2}$) and ($j,\frac{1}{2}$), point U is midway between points ($j,\frac{1}{2}$) and ($j-1,\frac{1}{2}$). The global conservation property can be shown to be satisfied if the following relationship is satisfied:

$$\frac{1}{2}(\hat{F}^{(2)}_{1,\frac{1}{2}} + \hat{F}^{(2)}_{j_{\max},\frac{1}{2}}) + \sum_{j=2}^{j_{\max}-1} \hat{F}^{(2)}_{j,\frac{1}{2}}$$

$$= \frac{1}{2}\left(\hat{F}^{(1)}_{l,m_{\max}-\frac{1}{2}} + \hat{F}^{(1)}_{l_{\max},m_{\max}-\frac{1}{2}}\right) + \sum_{l=2}^{l_{\max}-1} \hat{F}^{(1)}_{l,m_{\max}-\frac{1}{2}} \tag{31}$$

A close examination of Eq. (31) shows each side of this equation is nothing but a discrete form of the line integral of the numerical flux $\hat{F}$ along the line CD in Fig. 3 while the equation itself represents flux conservation across the patch boundary. Equation (31) is only a necessary condition and is not sufficient to define the fluxes $\tilde{F}^{(2)}_{j,\frac{1}{2}}$ in a physically meaningful way. Assume that the $\hat{F}^{(2)}_{j,\frac{1}{2}}$ are obtained by interpolating the $\hat{F}^{(1)}_{l,m_{\max}-\frac{1}{2}}$, that is,

$$\tilde{F}^{(2)}_{j,\frac{1}{2}} = \sum_{l=p}^{q} N_{j,l} \hat{F}^{(1)}_{l,m_{\max}-\frac{1}{2}} \tag{32}$$

where the $N_{j,l}$ are interpolation coefficients, and p and q define the set of fluxes of zone 1 that will be used in the interpolation. We now describe a simple way of obtaining the interpolation coefficients $N_{j,l}$ such that Eq. (31) is automatically satisfied. Let the line CD in Fig. 4 correspond to the line CD in Fig. 3. The dots represent the grid points of zone 1 and the crosses represent those of zone 2. A running parameter s is established along the line CD. The quantity s represents the distance of a point from the point C along the curve CD. Representative numerical values of $\hat{F}^{(1)}_{l,m_{\max}-\frac{1}{2}}$ are plotted on the positive y axis. Assume a piecewise constant variation of the numerical fluxes $\hat{F}^{(1)}_{l,m_{\max}-\frac{1}{2}}$ along CD; that is, $\hat{F}^{(1)}_{l,m_{\max}-\frac{1}{2}}$ is constant between $s^{(1)}_{l-\frac{1}{2}}$ and $s^{(1)}_{l+\frac{1}{2}}$. Consider a point of zone 2, $(j, 1)$. The $\hat{F}^{(2)}_{j,\frac{1}{2}}$ are now calculated from

$$\tilde{F}^{(2)}_{j,\frac{1}{2}}(Q,\eta^{(2)}_{j,\frac{1}{2}}) = \int_U^T \frac{\hat{F}^{(1)}_{l,m_{\max}-\frac{1}{2}}(Q,\eta^{(1)}_{l,m_{\max}-\frac{1}{2}})\,\mathrm{d}s}{s^{(1)}_{l+\frac{1}{2}} - s^{(1)}_{l-\frac{1}{2}}}$$

$$= \sum_{l=p}^{q} N_{j,l} \hat{F}^{(1)}_{l,m_{\max}-\frac{1}{2}}(Q,\eta^{(1)}_{l,m_{\max}-\frac{1}{2}}) \tag{33}$$

where the values of $N_{j,l}$ are given by

$$N_{j,l} = \begin{cases} 0 & \text{if} \quad s^{(1)}_{l+\frac{1}{2}}, s^{(1)}_{l-\frac{1}{2}} \le s^{(2)}_{j-\frac{1}{2}} \\ 0 & \text{if} \quad s^{(1)}_{l+\frac{1}{2}}, s^{(1)}_{l-\frac{1}{2}} \ge s^{(2)}_{j+\frac{1}{2}} \\ (s_r - s_l)/(s^{(1)}_{l+\frac{1}{2}} - s^{(1)}_{l-\frac{1}{2}}) & \text{otherwise} \end{cases} \tag{34}$$

and

$$s_r = \min(s^{(2)}_{j+\frac{1}{2}}, s^{(1)}_{l+\frac{1}{2}})$$

$$s_l = \max(s^{(2)}_{j-\frac{1}{2}}, s^{(1)}_{l-\frac{1}{2}})$$

The use of Eqs. (33) and (34) in calculating the numerical fluxes $\hat{F}_{j,\frac{3}{2}}^{(2)}$ identically satisfies the discrete conservation condition given by Eq. (31). The simple expressions of Eq. (34) are valid only for a piecewise constant variation of the $\hat{F}_{l,m_{max}-\frac{1}{2}}^{(1)}$. A piecewise linear or any other variation would result in different formulas for the interpolation coefficients $N_{j,l}$. Equation (34), in an indirect manner, also yields the endpoints in the interpolation, p and q (p and q include only that set of fluxes of zone 1 that are multipled by a nonzero interpolation coefficient for the given flux of zone 2). The treatment of the points $(1,1)$ and $(j_{max},1)$ and the calculation of the metrics used in the integration of the zonal points can be found in Ref. 7.

The final step of the patch-boundary scheme consists of updating the patch-boundary points of zone 1 such that the dependent variables are continuous across the patch boundary. This is very simply done by an interpolation process. The dependent variables at the patch-boundary points of zone 1 are obtained by interpolating the updated dependent variables at the patch-boundary points of zone 2. The results presented in this study were obtained using a linear interpolation procedure.

The conservative patched-boundary scheme previously described can be extended to the implicit, iterative integration schemes used in this study. In such a setting, the points on the patch boundary are integrated using a modified form of the iterative algorithm [Eq. (8)]. For example, the patch-boundary point $(j,1)$ of zone 2 is integrated using the following equation:

$$\left[I + \frac{\Delta\tau}{\Delta\xi}(\nabla_\xi(\tilde{A}_{j,1}^+)^p + \Delta_\xi(\tilde{A}_{j,1}^-)^p) \right]$$

$$\times \left[I + \frac{\Delta\tau}{\Delta\eta}(\Delta_\eta(\tilde{B}_{j,1}^-)^p + \sum_{l=p}^{q} N_{j,l}^{n+1}(\tilde{B}_{l,m_{max}}^+)^p(Ja_{j,1}^{(2)}/Ja_{l,m_{max}}^{(1)})^p) \right]$$

$$\times [(\tilde{Q}_{j,1}^{(2)})^{p+1} - (\tilde{Q}_{j,1}^{(2)})^p] = \text{RHS of Eq. (8)} \tag{35}$$

Evaluating the right-hand side of this equation requires the numerical flux $\hat{F}_{j,\frac{3}{2}}^{(2)}$. This flux is obtained using the conservative interpolation procedure described for the explicit patch-boundary scheme. Equation (35) also requires the interpolation coefficients to be evaluated at the $(n+1)$th time step. This is a simple matter if the movement of the patches relative to one another is known a priori (which is the case when the velocity of the rotor blades is specified). The coefficients $N_{j,l}^{n+1}$ can be replaced by $N_{j,l}^p$ if the motion is not predetermined but is coupled to the fluid characteristics.

The implicit patched-grid scheme thus comprises the following steps:

1) Integrate the dependent variables at all of the grid points of zone 2, using Eq. (8) in conjunction with the implicit patch-boundary condition Eq. (35) (only one iteration).

2) Interpolate the newly obtained values of $[(Q_{j,l}^{(2)})^{p+1} - (Q_{j,1}^{(2)})^p]$ to yield a new set of $[(Q_{l,m_{max}}^{(1)})^{p+1} - (Q_{l,m_{max}}^{(1)})^p]$.

3) Integrate the dependent variables at the grid points of zone 1, using Eq. (8) (only one iteration) and the most recent values of $[(\tilde{Q}^{(1)}_{l,m_{max}})^{p+1} - (\tilde{Q}^{(1)}_{l,m_{max}})^{p}]$ (a Dirichlet type of boundary condition).

4) Interpolate the values of $(Q^{(2)}_{j,1})^{p+1}$ to obtain the values of $(Q^{(1)}_{l,m_{max}})^{p+1}$ along the patch boundary (discard the ones obtained as a result of the integration procedure in step 3).

5) If the maximum values of the magnitudes of all $[(Q^{(i)})^{p+1} - (Q^{(i)})^{p}]$ is less than a prescribed tolerance limit, go to the next intergration step; if not, go back to step 1 and iterate.

It must be mentioned that approximate linearization of the fluxes is used to obtain Eq. (35). Proper linearization of the fluxes in the vicinity of the patch boundary results in the incremental change in the dependent variables on the two sides of the patch boundary becoming coupled and prevents factorization. However, the approximations made are corrected through the iteration process.

Several additional details of the patched-grid scheme have not been included here; these details can be found in Refs. 7 and 9.

Nonconservative Zonal Boundary Scheme

The nonconservative treatment of the zonal interfaces is conceptually and computationally far simpler than the conservative treatment of such boundaries. The procedure described later is applicable to both patched and overlaid grids. The only requirement is that there be at least a half-grid-cell overlap between neighboring zones at the boundary where these conditions are applied. All of the points on the zonal boundaries are integrated by using the equation

$$(\tilde{Q}^{p+1} - \tilde{Q}^{p})_{zb} = 0 \tag{36a}$$

where the subscript zb refers to the points on a zonal boundary. This is followed by an explicit, corrective interpolation procedure at the end of each iteration wherein the values of Q^{p+1} along the zonal boundary are obtained from interpolating the dependent variables of the neighboring grid in which the zonal boundary lies. For the results presented here, a linear interpolation over triangles was used.

It is important to note that Eq. (36a) is not the same as that given by the following equation:

$$(\tilde{Q}^{n+1} - \tilde{Q}^{n})_{zb} = 0 \tag{36b}$$

Equation (36a) [in addition to Eq. (8) in the interiors of zones] together with the postupdate correction mentioned earlier allows $(\tilde{Q}^{n+1} - \tilde{Q}^{n})_{zb}$ to assume its right value when the iteration process is carried to convergence. Both time accuracy and a spatial accuracy consistent with the order of the interpolation scheme are maintained at the zonal boundaries by using Eq. (36a).

Natural Boundary Conditions

The treatment of the various internal boundaries that result from the zonal approach has so far been discussed. We now turn to the conditions that must be specified at the natural boundaries of the flow domain. Figure 5 shows the computational domain and the associated boundaries for a typical single-stage turbine configuration. This domain is representative of the ones considered in this study. The natural boundaries include the stator inlet and the rotor exit, the stator and rotor airfoil surfaces, and, for the three-dimensional case, the hub and outer casing surfaces. For the sake of generality, we consider the boundary conditions for the three-dimensional case. The two-dimensional case follows easily.

Airfoil Surface Boundary Condition

A no-slip boundary condition, an adiabatic wall condition, and a zero normal derivative of the pressure condition is imposed at all points on the airfoil surfaces. Note that in the case of the rotor, the no-slip condition implies zero relative, not absolute velocity. The pressure derivative condition, the adiabatic wall condition, and the equation of the state together yield

$$\frac{\partial \rho}{\partial n} = 0$$

$$\frac{\partial e}{\partial n} = u\frac{\partial \rho u}{\partial n} + v\frac{\partial \rho v}{\partial n} + w\frac{\partial \rho w}{\partial n}$$

where n is the direction normal to the airfoil surface. These boundary conditions are implemented in an implicit manner by using the following equation to update the grid points on the airfoil surfaces:

$$C(\tilde{Q}_{i,1,k}^{p+1} - \tilde{Q}_{i,1,k}^{p}) + D(\tilde{Q}_{i,2,k}^{p+1} - \tilde{Q}_{i,2,k}^{p}) = 0 \tag{37}$$

where

$$C = \begin{bmatrix} 1 & 0 & 0 & 0 & 0 \\ 0 & 1 & 0 & 0 & 0 \\ 0 & 0 & 1 & 0 & 0 \\ 0 & 0 & 0 & 1 & 0 \\ 0 & \alpha & \beta & \gamma & 1 \end{bmatrix}, \quad D = \begin{bmatrix} \theta & 0 & 0 & 0 & 0 \\ 0 & 0 & 0 & 0 & 0 \\ 0 & 0 & 0 & 0 & 0 \\ 0 & 0 & 0 & 0 & 0 \\ 0 & \alpha\theta & \beta\theta & \gamma\theta & \theta \end{bmatrix}$$

and

$$\theta = -\frac{Ja_{i,2,k}}{Ja_{i,1,k}}$$

$$\alpha = -u_{\text{wall}}$$

$$\beta = -v_{\text{wall}}$$

$$\gamma = -w_{\text{wall}}$$

Equation (37) is an implicit, spatially first-order-accurate implementation of the no-slip (adiabatic wall) condition (first-order accurate because the zero normal derivative condition is implemented using a two-point forward difference). A second-order-accurate, three-point, forward-difference corrector step is also implemented after each time step. It should be noted that Eq. (37) requires the grid to be orthogonal at the airfoil surfaces and the Jacobians of the transformation $Ja_{i,1,k}$ and $Ja_{i,2,k}$ to be independent of τ.

Endwall Boundary Condition

The no-slip boundary condition, the adiabatic wall condition, and the pressure derivative condition are also imposed on the hub and outer casing surfaces in a manner similar to that employed at the airfoil surfaces.

Stator Inlet Boundary Condition

For subsonic inlet conditions, four quantities must be specified at this boundary. The four chosen for this study are a Riemann invariant, the entropy, and the radial and tangential flow velocities, that is,

$$R_1 = u + \frac{2c}{\gamma - 1} \tag{38a}$$

$$s = \frac{p}{\rho^{\gamma}} \tag{38b}$$

$$v_{\text{inlet}} = 0 \tag{38c}$$

$$w_{\text{inlet}} = 0 \tag{38d}$$

The fifth quantity (which is necessary to update the points on this boundary) is also a Riemann invariant,

$$R_2 = u - \frac{2c}{\gamma - 1} \tag{39}$$

and is extrapolated from the interior. These boundary conditions are appropriate for inviscid flow. In reality, the incoming flow contains the hub boundary layer and the outer casing boundary layer. This problem was overcome by imposing the no-slip boundary condition only at those grid points of the hub and outer casing that were a small distance downstream of the inlet boundary (typically two grid points downstream of the inlet boundary), thus making the two boundary layers originate downstream of the inlet boundary. The preceding approximation is justified if the incoming boundary layers are extremely thin. However, these boundary conditions can easily be replaced by experimentally observed inlet distributions or by an additional zone that comprises the entrance to the stage.

Rotor Exit Boundary Condition

For subsonic exit conditions with no reverse flow (i.e., no flow into the stage) only one quantity has to be specified at the exit boundary; the other four quantities can be extrapolated from the interior. The quantity to be specified can be picked either to reflect all pressure waves as in the case of an open-end duct or to transmit all pressure waves as in the case of an infinite duct. In the subsonic calculations presented here, the pressure-reflection boundary condition was chosen to simulate experimental conditions.

In two dimensions, the pressure-reflection condition implies specifying a constant pressure at the rotor exit. In the three-dimensional case of this study a constant pressure is imposed at midspan. The pressure at all other radial locations at the rotor exit are obtained from the radial equilibrium condition,

$$\frac{\partial p}{\partial r} = \frac{\rho v_t^2}{r} \tag{40}$$

where v_t is the tangential velocity, and r is the radius measured from the center of the hub. All five dependent variables are extrapolated implicitly from the interior, using instead of Eq. (22), the equation

$$(\tilde{Q}^{p+1}_{i_{\max},j,k} - \tilde{Q}^{p}_{i_{\max},j,k}) - \frac{Ja_{i_{\max}-1,j,k}}{Ja_{i_{\max},j,k}} (\tilde{Q}^{p+1}_{i_{\max}-1,j,k} - \tilde{Q}^{p}_{i_{\max}-1,j,k}) = 0 \tag{41}$$

A postupdate correction is then made to the pressure, using Eq. (40).

Periodicity Boundary Condition

Turbomachines are designed with unequal airfoil counts in the stator and rotor rows in order to minimize vibration and noise. A complete simulation including all of the airfoils in the stator and rotor rows would be prohibitively expensive. The approach used here is to assume that the ratio of the number of stator to rotor airfoils is unity or the ratio of two small integers. In order to invoke this approximation, a rescaling of either the stator or the rotor geometry is required (to keep blockage effects the same). Since the stator and rotor pitches (or composite pitches when multiple airfoils are considered) are now equal, the periodicity condition can be imposed in a straightforward manner.

Other approaches to handling this boundary condition without resorting to rescaling the geometry have been discussed in the literature. Erdos et al.[26] suggested the use of a time-lagged boundary condition. However, this approach imposes an additional constraint of temporal periodicity on the solution that may not be appropriate for viscous computations. Giles[28] inclines the computational grid in time to automatically satisfy the lagged boundary condition.

Results

The preceding sections have dealt with the underlying concepts in incorporating the zonal approach within the framework of a Navier-Stokes algorithm. In this section, we demonstrate the capabilities of this approach through applications to both idealized and realistic rotor-stator configurations. Both two- and three-dimensional results are presented and compared to experimental data where available.

Demonstration Euler Calculation

To demonstrate the capabilities of the zonal boundary scheme outlined in the previous sections, we first consider an idealized rotor-stator configuration where both the rotor and stator are simply circular arc airfoils; the axial gap between the rotor and stator is 35% of the chord length. The freestream Mach number is 1.5. For this calculation, viscous effects were neglected and the unsteady Euler equations were solved. The numerical fluxes were evaluated using the scheme of Chakravarthy and Osher,[12] and flux limiters were used to render the scheme TVD in each spatial dimension as in Ref. 37. Additional details of the calculation can also be found in Ref. 37.

Figure 6 shows the two-zone patched grid used to discretize the region of interest. The grid in zone 2 is stationary (as is the aft airfoil), and the grid in zone 1 is fixed to the first airfoil that is moving vertically downward. Although the grid lines are continuous at the patch boundary in Fig. 6, a discontinuity in these grid lines will develop as the first airfoil and zone 1 move downward. The patch-boundary points of zone 1 will slip past the patch-boundary points of zone 2.

Periodic boundary conditions are imposed on the upper and lower boundaries of both patches. Supersonic inlet conditions are imposed on the left boundary of zone 1, and supersonic exit boundary conditions are imposed on the right boundary of zone 2. The conservative patch-boundary condition is used at the boundary separating the two patches. Details of the inviscid surface boundary condition used can be found in Ref. 37.

The calculation was initially performed with both airfoils stationary. After 50 integration steps a downward velocity (the magnitude of the velocity corresponding to a Mach number of 0.1 with respect to freestream conditions) was imparted to the forward airfoil, and the calculation was continued until the solution became periodic in time. The calculation was performed at a CFL number of approximately 2.0. At this CFL number, 250 integration steps were required for each cycle (one cycle corresponds to the motion of the upper boundary of zone 1 from its current position to the position occupied currently by the lower boundary of zone 1). Approximately three cycles were required to eliminate the initial transients and to establish a solution that was periodic in time.

Figure 7 shows pressure contours at various positions of the forward airfoil (with respect to the aft airfoil) as it moves downward. These contours were obtained after the initial transients had subsided. Although the calculation was performed with only two airfoils, for the sake of clarity

Figs. 7a–7f depict four airfoils; information regarding the additional airfoils being obtained from the periodicity condition. Figure 7a shows contours at $t = 0.0$ (one cycle corresponds to $t = 1.0$). The downward motion of the forward airfoil results in an effective angle of attack that in turn results in an attached oblique shock on the lower side and a weak expansion fan on the upper side at the leading edge of the first airfoil. The interaction of this shock with the adjacent forward airfoil is clearly seen. A second weak attached shock is also evident at the trailing edge (lower side) of the forward airfoil. At the position $t = 0.0$ the leading-edge shock associated with the second airfoil is detached. It is seen impinging on the surface of the adjacent airfoil.

The interaction of the trailing-edge shock of the first airfoil and the leading-edge shock of the second airfoil is also clear from Fig. 7a. This interaction area moves downward as the first airfoil moves downward. This in turn results in the leading-edge shock of the second airfoil attaching and detaching periodically from the leading edge. Figures 7b–7f depict this attachment/detachment process. In Fig. 7b the shock is beginning to reattach. In Fig. 7c it is a weak attached shock, and in Fig. 7d it is a strong attached shock (beginning to detach again). It is completely detached in Fig. 7e, and, finally, in Fig. 7f the contours are identical to those in Fig. 7a, thus demonstrating the accuracy of the present technique in calculating periodic flows.

An important feature in Figs. 7a–7f is that the contours are continuous, even slope continuous, across the patch boundary along which grid points from the two grids are slipping past each other. This high degree of continuity results from the conservative nature of the patched-grid scheme and the manner in which continuity of dependent variables is enforced across the patch boundary. Another interesting feature is that the captured shocks are almost oscillation-free. The absence of large oscillations is because of the TVD nature of the integration scheme.

Figure 8 shows the pressure variation with time at midchord on the lower surface of the aft airfoil. This pressure variation corresponds to the fourth and fifth cycles. Clearly, the pressure is periodic in time.

One aspect of rotor-stator configurations that is not represented in the present results is the effect of the aft airfoil on the forward airfoil (the supersonic nature of the flow does not permit such an interaction). However, the patch-boundary conditions presented earlier will accurately simulate such an interaction, were it present. This fact is borne out by the results presented in the next subsections for a low-speed axial turbine stage.

Two-Dimensional Rotor-Stator Interaction Calculations

The geometry considered here is a low-speed, single-stage axial turbine that has been extensively tested by Dring and co-workers.[6,41–44] The geometry consists of 22 stator airfoils and 28 rotor airfoils. Figure 9 is a perspective view of the configuration showing the pressure side of the stator and the suction side of the rotor. In Fig. 9 the outer casing has been removed for clarity. The stator airfoil extends from the hub to the outer casing and is attached at both ends. The rotor airfoil is attached only at the hub; there is a clearance region between the rotor tip and the outer casing. The hub is in two sections; the section to which the stator airfoil is attached is stationary, whereas the section to which the rotor airfoil is attached rotates.

The nonconservative zonal boundary conditions discussed earlier were used in these calculations, since the flow is entirely subsonic and free of flow discontinuities (the maximum Mach number in the system is less than 0.30).

For the two-dimensional calculations, the rotor-stator geometry corresponding to the midspan location was considered. The multizone grid used to discretize the region consists of four zones, as shown in Fig. 10. The first zone contains the stator (stator inner zone) and is discretized with an O grid. The second zone contains the rotor (rotor inner zone) and is also discretized with an O grid. These grids were generated using an elliptic grid generator of the type developed in Ref. 45. Note that, for the purpose of clarity, Fig. 10 does not show all the grid points used in the calculation.

The stator and rotor inner zones are patched into the outer zones, zones 3 and 4, respectively, as shown in Fig. 10. The grids for these outer zones were generated using an algebraic grid generator. Note that the inner boundary of zone 3 corresponds to the outer boundary of zone 1, and, similarly, the inner boundary of zone 4 corresponds to the outer boundary of zone 2. This interface boundary commonality of the inner and outer stator (and rotor) zones facilitates information transfer between these zones. The outer zones abut each other along the patch boundary $ABCD$ and slip past each other as the rotor airfoil moves downward.

An interesting feature of zones 3 and 4 as seen in Fig. 10 is that they do not align with each other. The segment AB of zone 4 does not seem to align with any part of the patch boundary of zone 3, and, similarly, the segment CD of zone 3 does not seem to align with any part of the patch boundary of zone 4. However, the periodicity boundary condition can be used to solve this problem, the result being that the segment AB is matched with the segment CD.

As mentioned earlier, the airfoil geometry consists of 22 stator airfoils and 28 rotor airfoils. An accurate simulation of this configuration would require at least 11 stator airfoils and 14 rotor airfoils, thus making the computation extremely expensive. The rescaling strategy described earlier was used to reduce the number of airfoils. Two separate calculations were performed to evaluate the implications of this strategy. The first corresponded to a one-rotor/one-stator case where the rotor was enlarged by a factor (28/22), and it was assumed that there were 22 stator and rotor airfoils. The second corresponded to a four-rotor/three-stator case where the stator was enlarged by the factor (22/21), and it was assumed that there were 21 stators and 28 rotors. Note that the pitch-to-chord ratio of the airfoils was not changed during the rescaling process.

The calculations were performed at a constant time-step value of 0.04 (the corresponding maximum CFL number in the grid system was approximately 250). Between five and seven cycles (a cycle corresponds to the motion of the upper boundary of zone 4 from its current position to the position occupied currently by the lower boundary of zone 4) were required to get rid of the initial transients and establish a solution that is periodic in time. For more details, the reader is referred to Refs. 15 and 16.

The dependent variables were nondimensionalized with respect to the inlet pressure (p_∞) and density (ρ_∞). This yields

$$u_\infty = M_\infty\sqrt{\gamma}$$

$$v_\infty = 0 \quad \text{(inlet flow is axial)}$$

where M_∞ is the inlet Mach number. The Riemann invariants that are prescribed at the inlet are determined using the dependent variables defined earlier. The rotor velocity is determined from the desired flow coefficient (0.78 in this case) and the inlet axial velocity (u_∞). An iterative procedure is required to obtain the correct rotor velocity (corresponding to a flow coefficient of 0.78) when the solution becomes periodic in time. This is because the boundary conditions specified at the inlet are not the dependent variables themselves, but the Riemann invariants.

The Reynolds number for the calculation was chosen to be 100,000/in. The Baldwin-Lomax turbulence model[46] was used to determine the eddy viscosity; the kinematic viscosity was calculated using Sutherland's law.

In the figures that follow there are several comparisons made with experimental data. Several important differences between the rotor-stator configuration simulated here and the experiment should be remembered when these comparisons are evaluated. First, the airfoil geometry used in the numerical calculations only approximates that used in the experiment, because of the rescaling strategy employed here. Second, the geometry rescaling requires the modification of the Reynolds number to simulate conditions equivalent to the experiment; it is not clear how this modification should be effected. Third, the axial gap between the airfoils in the experiment was 15% of the chord length. The calculations reported here were performed using an axial gap that was 15% of the average chord length of the modified configuration. Finally the current analysis is only two dimensional in nature, whereas the actual flow is three dimensional. Results from three-dimensional calculations will be presented later in this section.

One-Rotor/One-Stator Calculation

Figure 11 shows the time-averaged pressure coefficient (C_p) as a function of the axial distance along the stator. The pressure coefficient is defined as

$$C_p = \frac{p_{\text{avg}} - (p_t)_{\text{inlet}}}{\frac{1}{2}\rho_{\text{inlet}}\omega^2}$$

where p_{avg} is the static pressure averaged over one cycle, $(p_t)_{\text{inlet}}$ is the average total pressure at the inlet, ρ_{inlet} is the average density at the inlet, and ω is the rotor velocity. Clearly, there is good agreement between theory and experiment. A small separation bubble was found near the trailing edge on the pressure side of the stator in the numerical results. This is seen as a sharp dip and rise of C_p toward the trailing edge on the pressure side. The experimental data also indicates such a separation. However, the magnitude of the pressure fluctuation obtained numerically may be suspect because the turbulence model was not tailored to yield accurate estimates of the eddy viscosity in such regions.

Figure 12 shows the time-averaged C_p distribution for the rotor. The agreement is good except on the suction side of the rotor toward the trailing edge $(4.0 \le x \le 7.0)$. As in the case of the stator, a small separation bubble (seen as a spatial fluctuation in pressure) was noticed on the trailing-edge circle.

The amplitude of the temporal pressure fluctuation is a measure of the unsteadiness of the flow. Figure 13 shows pressure amplitudes $\tilde{C}_p$ on the surface of the stator plotted as a function of the axial distance. The quantity $\tilde{C}_p$ is defined as

$$\tilde{C}_p = \frac{p_{\max} - p_{\min}}{\frac{1}{2}\rho_{\text{inlet}}\omega^2}$$

where $p_{\max}$ and $p_{\min}$ are, respectively, the maximum and minimum pressures that occur over a cycle at a given point. The numerical amplitude distribution shows most of the qualitative features that are found in the experimental results. However, the numerical data seem to form a wider large-amplitude region than that found experimentally. In addition, the predicted peak is to the left of the experimental peak, and the pressure amplitude minimum on the suction side seen in the experimental results $(x = -2.4)$ is absent in the calculated results. These distortions are because of a combination of reasons that are discussed later.

The current calculation assumes an equal number of stator and rotor airfoils and were obtained using a fixed exit static-pressure (totally reflective) boundary condition. An acoustic analysis[47] shows that in such a situation every harmonic in time (if one were to perform a Fourier decomposition of the unsteady pressures in the region between the stator and rotor) results in a propagating wave in the axial direction. In the experimental configuration there are 22 stator airfoils and 28 rotor airfoils. This results in only the higher harmonics in time giving rise to propagating waves; the lower harmonics give rise to decaying signals. Since the higher harmonics are much smaller in magnitude, the unsteady pressures that reach the exit boundary are much smaller in the case of the experiment. The reflective exit boundary condition used in the calculation reflects the relatively large calculated pressure waves that reach the exit boundary back into the system, thus distorting the unsteady pressures everywhere. This problem is further discussed later.

Figure 14 shows the $\tilde{C}_p$ distribution on the rotor. The agreement between theory and experiment is not as good as in the case of the stator. The suction side amplitude peak is shifted to the left of the experimental one. A sizable portion of the pressure side peak toward the trailing edge is due to strong pressure waves being reflected back from the exit boundary. The stator pressure amplitude distributions tend to be predicted better because the rotor airfoils partially shield the stator airfoils from the reflected pressure waves (reflected off the exit boundary). However, the numerical data shown in Fig. 14 do predict all of the qualitative features shown by the experiment.

Nonreflecting boundary conditions can, and have been, developed[16] that prevent reflections of the pressure waves from the inlet and exit boundaries. However, these result in only limited improvements to the predicted unsteady pressure field, since the pressure signals being generated by the one-rotor/one-stator system are inherently different from those that would be generated by a multirotor/multistator system. The right approach is to perform a multirotor/multistator calculation (the reflective properties of the exit boundary condition will be relatively less important for more realistic rotor and stator airfoil counts). The results from such a calculation are discussed later.

Three-Stator/Four-Rotor Calculation

A second calculation was performed with three stator airfoils and four rotor airfoils.[16] Figure 15 shows the computational grid that was used. The ratio of the number of rotor and stator airfoils (4/3) is much closer to the experimental ratio (28/22) than is the ratio obtained for the one-rotor/one-stator case. Consequently, the amount of rescaling is also much smaller. In addition, it can be shown that only the third harmonic results in a propagating pressure signal. The first two harmonics result in decaying pressure signals. Since the third harmonic is much smaller in magnitude than the first two, the pressure signals reaching the exit boundary can also be expected to be much smaller in magnitude. Hence, the reflective properties of the exit boundary condition play a smaller role in determining unsteady pressures on the airfoils.

Although not shown here, the time-averaged pressures on the stator and rotor obtained from the three-stator/four-rotor calculation were almost identical to those obtained from the single-stator/single-rotor calculation and presented in Figs. 11 and 12. Thus, for the rotor-stator configuration considered here, the single-blade approximation does not seem to affect the time-averaged pressure distribution.

Figure 16 shows the pressure amplitude distribution obtained on the stator. Clearly, there is a marked improvement over the numerical results depicted in Fig. 13. The positions of the experimental and numerical peaks are in much better agreement. The numerical data show an amplitude minimum on the suction side just as the experimental data do. However, some differences exist between the numerical results and the experimental data on the suction side of the stator; this may be because of the small difference between experimental and numerical rotor/stator pitch ratios or because of three-dimensional effects.

Figure 17 shows pressure amplitudes on the rotor. As in the case of the stator, the numerical results are considerably closer to the experimental results with three stators and four rotors than those obtained with one rotor and one stator (Fig. 14). The pressure side peak of Fig. 14 has decreased in magnitude, resulting in much better comparisons between theory and experiment. Similar improvements can also be seen on the suction side of the rotor. However, just as in the case of the stator, there are still some quantitative differences between theory and experiment.

Three-Dimensional Rotor-Stator Interaction Calculations

Three-dimensional calculations have also been carried out for the rotor-stator configuration of Fig. 9. The results described here were obtained using the single-stator/single-rotor assumption. These calculations account for endwall effects due to the hub and outer casing, the leakage effects in the tip clearance region, and the overall three dimensionality of the flow. Additional details can be found in Refs. 17 and 18.

The computational grid used in the three-dimensional calculations consists of a sequence of two-dimensional grids (shown in Fig. 18) that are stacked together in the radial direction (from hub to tip). The grids shown in Fig. 18 are similar to the grids described earlier in Fig. 10 for the two-dimensional calculations, except that the inner stator and rotor zones do not conform to the inner boundaries of the outer zones, but instead overlap into the outer zone interiors. Furthermore, these grids lie on a cylinder of constant radius (measured from the center of the hub) and not on a plane (as in the case of the two-dimensional calculations). Again, for the sake of clarity, Fig. 18 does not show all the grid points used in the calculation.

Figure 18 shows also an additional grid in the region interior to the rotor airfoil. This grid is used only in the rotor tip clearance region (corresponding to the last 0.4% of the span). It is essentially an O grid with the innermost O grid line collapsing into a double-valued curve. As before, several of these interior grids are stacked in the radial direction to form the grid in the tip clearance region. The set of innermost O grid lines forms a surface along which the grid transformation is undefined and therefore requires special treatment during the solution process. Note that the stator, on the other hand, does not need any additional grids since it extends all the way from the hub to the casing.

The radial locations of the stacked two-dimensional grids are the same for both the rotor and stator zones. This leads to two-dimensional interface boundaries and thus reduces the interface logic by almost an order of magnitude.

As indicated earlier, grid points are densely packed close to the airfoil surfaces in zones 1 and 2 to resolve the viscous effects. Viscous effects are also predominant in the boundary layers associated with the hub and outer casing. The stacking of the two-dimensional grids in the radial direction reflects the presence of these two boundary layers; the two-dimensional grids are packed densely near the hub and outer casing.

All calculations were performed at a constant time step value of about 0.04 (this translates into 2000 time steps per cycle). Approximately five cycles were required to eliminate the initial transients and establish a solution that was periodic in time (a cycle corresponds to the motion of the rotor through an angle equal to $2\pi/N$, where N is the number of stator or rotor airfoils). A modified version of the Baldwin-Lomax turbulence model[40] was used to determine the eddy viscosity.

Several differences between the experiment and the two-dimensional calculations discussed earlier have been pointed out. These differences should again be kept in mind when evaluating the three-dimensional results. In addition, the following points should be noted. First, the tip clearance used in the present calculation is 0.4% of the span compared to the 1.4% value in the experiment. A preliminary calculation with a clearance of 1.4% resulted in separated flow on the suction side of the rotor in the last 40% of the chord in the vicinity of the clearance region. Past experience indicates that the Baldwin-Lomax turbulence model does not predict eddy viscosities accurately in separated regions. For this reason the clearance was reduced until the flow reattached. Second, the geometry rescaling was effected only in two directions (not in the spanwise direction), thus altering the aspect ratio of the scaled airfoil. Last, it must be noted that the experimental data presented in these figures were obtained with an axial gap of 50%,[41] whereas the numerical data were obtained with an axial gap of 15%. However, the results of Ref. 6 indicate that the axial gap has negligible effect on time-averaged stator surface pressures and at most a weak effect on time-averaged rotor surface pressures. Hence, the following comparisons between computations and experiment are, to a large extent, valid (subject to the geometric and other approximations discussed).

The following figures show results obtained from both a coarse-grid as well as a fine-grid calculation with twice as many gridpoints as the former. The fine-grid calculation provided enhanced resolution in the spanwise direction that captured better the endwall and tip clearance effects. In the figures, the fine-grid calculation is denoted by the solid line, the coarse-grid calculations are denoted by the dashed line, and the symbols represent the experimental data.

Figures 19a–19c show experimental and numerical time-averaged pressure coefficients, C_p, on the stator at 2.0, 50.0 and 98.0% of the span, respectively. The comparison between the computations and experiment is good all the way from the hub to the tip. As in the case of the two-dimensional results, a small separation bubble can be seen on the trailing-edge circle of the stator in the numerical results. The computed three-dimensional results at midspan are almost identical to the two-dimensional results presented earlier. The coarse- and fine-grid results are almost identical to each other at the midspan location (Fig. 19b) but differ slightly at the hub (Fig. 19a) and at the outer casing (Fig. 19c). Although both results are of comparable quality at the hub (Fig. 19a), the fine-grid results at the casing (Fig. 19c) are in slightly better agreement with the experiment than the course-grid results. This improvement is probably due to the improved resolution of the casing passage vortex obtained with the fine grid.

Figures 20a–20c show the time-averaged C_p distributions on the rotor airfoil at 2.0, 50.0 and 87.5% of the span, respectively. The increasing effect of the tip-leakage vortex on the rotor surface pressures in the region close to the rotor tip is evident from the lower pressure values at about 70% chord (Fig. 20c) on the suction side. Figures 20a–20c show reasonably good agreement between theory and experiment. The coarse- and fine-grid results lie close to each other and within the range of the experimental data except at the hub (Fig. 20a). The fine-grid results at the hub are in better agreement with the experiments because of the improved resolution of the passage vortex. Note that comparisons between the numerical results and the experiment at values of span larger than 87.5% would be inappropriate and have not been made because the tip-leakage vortex in the experiment is much stronger than the one calculated (because of the larger clearance between the rotor and the casing in the experiment).

Figure 21 presents the pressure amplitude distribution ($\tilde{C}_p$) obtained from the three-dimensional calculations on the stator at the midspan location. As in the case of the two-dimensional single-stator/single-rotor calculations, both the coarse- and fine-grid calculations predict a wider large-amplitude region than that found experimentally; the predicted peaks are also shifted to the left of the experiment. As discussed earlier, these distortions are a direct consequence of the single-stator/single-rotor approximation.

The predicted $\tilde{C}_p$ distribution on the rotor at the midspan location is shown in Fig. 22. The agreement between the numerical results and the experiment is not as good as in the case of the stator. A multistator, multirotor calculation should result in much better agreement with experiment.

Figures 23a–23c present Mach number contours at various instants in time (corresponding to various positions of the rotor with respect to the stator). The data in these figures correspond to the midspan location. Although the calculation was performed with only one stator and one rotor airfoil, for the sake of clarity these contour plots depict several airfoils. The information regarding the additional airfoils is obtained from the periodicity condition. The thickening of the boundary layer from the leading to the trailing edges of the rotor and stator airfoils is evident. The wakes associated with both airfoils and the interaction of the rotor airfoil with the stator wakes are also clearly observed in the figures. Figures 23a–23c show contour lines that are smooth and continuous across the various zonal boundaries used in the calculation except for small discontinuities in the region of the boundary between the stator and rotor. Although not shown here, these discontinuities are more easily discernible in plots of instantaneous pressure contours. Efforts are currently underway to remove these discontinuities through the use of higher-order interpolation to transfer information between the stator and rotor grids.

The nature of the flow close to the surfaces of the rotor and stator airfoils and the hub surface can be better understood from the pattern of limiting streamlines on these surfaces. These streamline patterns not only yield information regarding the flow close to the surface but also reflect the presence of flow features such as shocks and vortices in the interior of the region under consideration. The results presented in the following paragraphs were obtained by releasing particles on the grid surface just above the hub or airfoil surface and then allowing these particles to move according to the time-averaged velocity field. The motion of these particles is restricted to the grid surface on which they were originally released (i.e., the out-of-surface component of velocity is not used to move the particle).

All of the results presented in the following figures were obtained from the fine-grid calculations.

Figure 24 presents limiting streamlines on the stator-hub surface. The flow is uniformly axial in the inlet region and then aligns itself to the stator passage as it moves through it. The flow separates as it approaches the leading edge of the stator, with the flow separation resulting in the formation of a horsehoe vortex. One leg of the vortex wraps itself around the leading edge of the stator and then impinges on the suction side of the stator at about 35% of the chord. The other leg of the vortex enters the passage and impinges on the suction side of the next stator. Figure 24 also shows a saddle point of separation with two attachment lines and two separation lines. The first attachment line starts at the saddle point and meets the airfoil surface on the leading-edge circle on the pressure side of the stator. The second attachment line starts at the saddle point and extends upstream toward the inlet boundary. The separation lines extend from the saddle point into adjacent passages. The horseshoe vortex is formed between the separation lines and the airfoil surface. These results agree qualitatively with the experimental results of Ref. 48 (obtained for a cascade geometry).

Figure 25 shows limiting streamlines on the stator pressure side. The flow is seen to be almost two dimensional everywhere except in a small region near the endwalls in the vicinity of the leading edge (where the incoming boundary layer separates and forms a horseshoe vortex). These flow features are again in close agreement with those reported in Ref. 48.

Figure 26 shows limiting streamlines on the stator suction surface. The upward motion of the fluid particles near the hub (0–15% span) is caused by the passage vortex generated by the neighboring stator airfoil. The upper half of the figure is qualitatively a mirror image of the lower half. The passage vortex associated with the casing rotates in a direction counter to the hub passage vortex. Hence, it induces a radially inward flow on the stator suction surface that extends almost all the way to midspan. The casing passage vortex, just like the hub passage vortex, is pulled toward the hub because it contains low total pressure fluid. However, unlike the hub passage vortex, which gets confined to the hub region because of its downward motion, the casing passage vortex gets elongated and thus affects a greater portion of the stator surface. The midspan flow on the suction side is largely two dimensional.

Figure 27 depicts limiting streamlines on the rotor-hub surface. Note that rotor-relative velocities are used to generate these and the following figures. As in the case of the stator, there is evidence of leading-edge separation and the formation of a horseshoe vortex. The secondary flow in the rotor passage is much stronger than that in the stator passage. The rotor passage flow is not aligned with the passage but has a large tangential component. Hence, it can be expected that the passage vortex impinges on the rotor suction side at a large angle, thus creating a local stagnation region. This would explain the higher pressures seen at midchord on the suction side of the rotor (Fig. 20a). The saddle point of separation, the attachment, and detachment lines are clearly discerned in this figure. Another feature that can be seen is a line of separation on the suction side that extends from about 30% chord all the way to the trailing edge of the rotor.

Figure 28 shows limiting streamlines on the rotor airfoil pressure surface. A radial outflow can be discerned in the first 30% of the chord from the leading edge. Beyond this the flow is almost two dimensional. This had been demonstrated in Ref. 42 to be because of the "relative-eddy" effect. This eddy is the axial component of the relative vorticity in the flow when viewed from the rotating frame of reference. Additional details regarding the relative-eddy effect can be found in Ref. 42. The numerical results are in fair agreement with experimental flow visualization results shown in Fig. 29.[42]

The rotor suction surface limiting streamlines are shown in Fig. 30. As in the case of the stator, the endwall passage vortices induce strong radial components in the surface velocity field. The figure clearly shows the two lines of separation caused by the endwall passage vortices; the two separation lines move toward each other and the midspan region. The spanwise location along which the radial component of the velocity is zero is at about 55% span. These observations are in good agreement with the experimental flow visualization studies shown in Fig. 31.[42] The sudden change in curvature in the upper separation line in Fig. 31 can be discerned in the experimental results also, albeit somewhat less readily.

The limiting streamlines presented earlier demonstrate the effect of the stator- and rotor-generated passage vortices on the flow. A better understanding of the vortex structure can be obtained from contours of total pressure at the exit to the stator and the exit to the rotor. The passage vortices contain fluid from the endwall boundary layers and the airfoil surface boundary layers and therefore represent low total pressure regions (compared to freestream total pressure values). The difference in vortex total pressures and freestream total pressure enables one to clearly observe passage vortex structure by studying total pressure contours at appropriate locations.

Figure 32 present the time-averaged total pressure contours, in the absolute frame of reference, at the exit to the stator (8.8% of the stator chord behind the stator trailing edge). Three distinct entities are clearly seen in the figures: the hub passage vortex, the casing passage vortex, and the stator wake. The hub passage vortex is seen to be confined to a region close to the hub, whereas the casing passage vortex is elongated and occupies a relatively larger portion of the span. The reason for the difference in structure between the two vortices has already been discussed. Figure 33 shows the experimentally observed[44] total pressure contours obtained at the same location. The vortex structure predicted by the three-dimensional calculation is seen to be in close agreement with experiment. Note that Fig. 32 also shows the endwall boundary layers in the form of highly clustered total pressure contours at the lower and upper boundaries.

The results previously given demonstrate the validity and the capabilities of the zonal approach in the prediction of turbomachinery rotor-stator interaction. These results, however, are only a representative sample of those presented in Refs. 15–18. The reader is referred to these studies for more details.

Areas of Current and Future Research

In order to keep the subject matter tractable, we have refrained from discussing one aspect of the computational approach, namely, turbulence modeling, despite its extreme importance. Turbulence modeling for unsteady turbomachinery flowfield predictions is a major problem area and needs to be addressed in future work. All of the results presented here are obtained using an algebraic turbulence model. Experience indicates that such algebraic turbulence models do not accurately predict eddy viscosities in regions of separation and in wakes without requiring extensive "fine-tuning" of the model. Hence, it is important to use more generally applicable turbulence models such as the two-equation models. The use of such models may enable more accurate prediction of heat transfer and losses. Turbulence modeling in the tip clearance region poses additional problems. Currently, the values of eddy viscosity in the tip clearance region are set to zero. Again, more generally applicable turbulence models, such as the one of Ref. 49, may yield realistic eddy viscosity values for the clearance region.

The results of the two-dimensional calculations presented here clearly demonstrate the necessity of using the correct (or at least a close approximation to the correct) stator and rotor airfoil counts in order to accurately predict the unsteady rotor-stator flowfield. Efforts are currently underway to extend the capabilities of the three-dimensional code to handle an arbitrary number of stator and rotor airfoils. Efforts are also being directed toward extensions to handle multiple stator-rotor stage configurations.[50]

One major current drawback to simulating unsteady three-dimensional flows in complex geometries such as turbomachinery is the large amount of computer time required to obtain accurate solutions. Until these computing times are reduced by an order of magnitude or more, three-dimensional, multirotor/multistator codes will only serve as research tools used to understand the basic flow mechanisms and not enter the design stage. Various algorithmic improvements that will substantially reduce run times by permitting larger time steps are currently being investigated.

Inlet and exit boundary conditions affect the transient solution in the interior of the domain. Although nonreflective boundary have been developed in Refs. 7 and 22, it is not clear whether these boundary conditions will result in more accurate simulations in all cases. This is because in real situations the first stage may be followed by another stage, or the flow may exit through a nozzle to the ambient or it may encounter bends in the piping, to mention just a few of the possibilities. Computing entire systems such as multistage turbomachinery with the appropriate upstream and downstream conditions will require far more computing power than currently available. Hence, the area of inlet and exit boundary conditions still requires development. Proper treatment of these boundaries is mandatory for accurate prediction of the far-field acoustics of the system (for noise prediction and minimization studies).

Summary

Zonal techniques have been successfully used in the computational analysis of rotor-stator interaction in turbomachinery. Unsteady, thin-layer Navier-Stokes codes, in both two and three dimensions, have been developed to predict such interaction, based on a zonal approach that incorporates a combination of patched and overlaid grids and allows for relative grid motion. The capabilities of the zonal approach have been demonstrated by simulating both idealized and realistic rotor-stator configurations. Results obtained from both two- and three-dimensional calculations for a single-stage, low-speed axial turbine configuration show good agreement with experimental data.

Acknowledgments

Some of the work described here was funded by the Office of Naval Research. The authors would also like to thank Paul Kelaita of Sterling Federal Systems for providing the graphic support for the project.

References

[1]Delaney, R. A., "Time-Marching Analysis of Steady Transonic Flow in Turbomachinery Cascades Using the Hopscotch Method," *Journal of Engineering for Power*, Vol. 105, April 1983, pp. 272–279.

[2]Davis, R. L., Ni, R.-H., and Carter J. E., "Cascade Viscous Flow Analysis Using the Navier-Stokes Equations," AIAA Paper 86-0033, Jan. 1986.

[3]Subramanian, S. V., Bozzola, R., and Povinelli, L. A., "Computation of Three-Dimensional, Rotational Flow Through Turbomachinery Blade Rows for Improved Aerodynamic Design Studies," American Society of Mechanical Engineers, Paper 86-GT-26, 1986.

[4]Chima, R. V., "Development of an Explicit Multigrid Algorithm for Quasi-Three-Dimensional Viscous Flows in Turbomachinery," NASA TM-87128, 1986.

[5]Weinberg, B. C., Yang, R. J., McDonald, H., and Shamroth, S. J., "Calculations of Two- and Three-Dimensional Transonic Cascade Flow Fields Using the Navier-Stokes Equations," *Journal of Engineering for Gas Turbines and Power*, Vol. 108, Jan. 1986, pp. 93–102.

[6]Dring, R. P., Joslyn, H. D., Hardin, L. W., and Wagner, J. H., "Turbine Rotor-Stator Interaction," *Journal of Engineering for Power*, Vol. 104, Oct. 1982, pp. 729–742.

[7]Rai, M. M., "A Conservative Treatment of Zonal Boundaries for Euler Equation Calculations," *Journal of Computational Physics*, Vol. 62, Feb. 1986, pp. 472–503.

[8]Benek, J. A., Steger, J. L., and Dougherty, F. C., "A Flexible Grid Embedding Technique with Application to the Euler Equations," AIAA Paper 83-1944, July 1983.

[9]Rai, M. M., "An Implicit Conservative Zonal Boundary Scheme for Euler Equation Calculations," *Computers and Fluids*, Vol. 14, No. 3, 1986, pp. 295–319.

[10]Steger, J. L., and Warming, R. F., "Flux Vector Splitting of the Inviscid Gas Dynamic Equations with Application to Finite Difference Methods," *Journal of Computational Physics*, Vol. 40, April 1981, pp. 263–293.

[11]Roe, P. L., "Approximate Riemann Solvers, Parameter Vectors, and Difference Schemes," *Journal of Computational Physics*, Vol. 43, Oct. 1981, pp. 357–372.

[12]Chakravarthy, S. R., and Osher, S., "Numerical Experiments with the Osher Upwind Scheme for the Euler Equations," *AIAA Journal*, Vol. 21, Sept. 1983, pp. 1241–1248.

[13]Chakravarthy, S. R., and Osher, S., "High Resolution Applications of the Osher Upwind Scheme for the Euler Equations," AIAA Paper 83-1943, July 1983.

[14]Chakravarthy, S. R., Szema, K.-Y., Goldberg, U. C., and Gorski, J. J., "Application of a New Class of High Accuracy TVD Schemes to the Navier-Stokes Equations," AIAA Paper 85-0165, Jan. 1985.

[15]Rai, M. M., "Navier-Stokes Simulations of Rotor-Stator Interaction Using Patched and Overlaid Grids," *Journal of Propulsion and Power*, Vol. 3, Sept. 1987 pp. 387–396.

[16]Rai, M. M., and Madavan, N. K., "Multi-Airfoil Navier-Stokes Simulations of Turbine Rotor-Stator Interaction," AIAA Paper 88-0361, Jan. 1988.

[17]Rai, M. M., "Unsteady Three-Dimensional Navier-Stokes Simulations of Turbine Rotor-Stator Interaction," AIAA Paper 87-2058, June 1987.

[18]Madavan, N. K., Rai, M. M., and Gavali, S., "Grid Refinement Studies of Turbine Rotor-Stator Interaction," AIAA Paper 89-0325, Jan. 1989.

[19]Hessenius, K. A., and Pulliam T. H., "A Zonal Approach to Solution of the Euler Equations," AIAA Paper 82-0969, June 1982.

[20]Usab, W. J., and Murman, E. M., "Embedded Mesh Solutions of the Euler Equation Using a Multiple-Grid Method," AIAA Paper 83-1946-CP, July 1983.

[21]Lombard, C. K., and Venkatapathy, E., "Implicit Boundary Treatment for Joined and Disjoint Patched Mesh Systems," AIAA Paper 85-1503-CP, July 1985.

[22]Berger, M. J., "On Conservation at Grid Interfaces," Institute for Computer Applications in Science & Engineering Rept. 84-43, Sept. 1984.

[23]Atta, E., "Component Adaptive Grid Interfacing," AIAA Paper 81-0382, Jan. 1981.

[24]Allmaras, J. D., and Baron, J. R., "Embedded Mesh Solutions of the 2-D Euler Equations: Evaluation of Interface Formulations," AIAA Paper 86-0509, Jan. 1986.

[25]Walters, R. W., Thomas, J. L., and Switzer, G. F., "Aspects and Application of Patched Grid Calculations," AIAA Paper 86-1063, May 1986.

[26]Erdos, J. I., Alzner, E., and McNally, W., "Numerical Solution of Periodic Transonic Flow Through a Fan Stage," *AIAA Journal*, Vol. 15, Nov. 1977, pp. 1559–1568.

[27]Fourmaux, A., "Unsteady Flow Calculation in Cascades," American Society of Mechanical Engineers Paper 86-GT-178, 1986.

[28]Giles, M. B., "Stator/Rotor Interaction in a Transonic Turbine," AIAA Paper 88-3093, July 1988.

[29]Lewis, J. P., Delaney, R. A., and Hall, E. J., "Numerical Prediction of Turbine Vane-Blade Interaction," AIAA Paper 87-2149, 1987.

[30]Jorgenson, P. C. E., and Chima, R. V., "An Explicit Runge-Kutta Method for Unsteady Rotor/Stator Interaction," AIAA Paper 88-0049, Jan. 1988.

[31]Jorgenson, P. C. E., and Chima, R. V., "An Unconditionally Stable Runge-Kutta Method for Unsteady Flows," AIAA Paper 89-0205, Jan. 1989.

[32]Koya, M., and Kotake, S., "Numerical Analysis of Fully Three-Dimensional Periodic Flows Through a Turbine Stage," *Journal of Engineering for Gas Turbines and Power*, Vol. 107, Oct. 1985, pp. 945-952.

[33]Chen, Y. S., "3-D Stator-Rotor Interaction of the SSME," AIAA Paper 88-3095, 1988.

[34]Gibeling, H. J., Weinberg, B. C., Shamroth, S. J., and McDonald, H., "Flow through a Compressor Stage," Scientific Research Associates, Glastonbury, CT, Rept. R86-910004-F, May 1986.

[35]Rai, M. M., and Chakravarthy S. R., "An Implicit Form for the Osher Upwind Scheme," *AIAA Journal*, Vol. 24, May 1986, pp. 735–743.

[36]Chakravarthy, S. R., "Relaxation Methods for Unfactored Implicit Upwind Schemes," AIAA Paper 84-0165, Jan. 1984.

[37]Rai, M. M., "A Relaxation Approach to Patched-Grid Calculations with the Euler Equations," *Journal of Computational Physics*, Vol. 66, No. 1, Sept. 1986, pp. 99–131.

[38]Pulliam, T. H., and Steger, J. L., "On Implicit Finite-Difference Simulations of Three-Dimensional Compressible Flow," *AIAA Journal*, Vol. 18, Feb. 1980, pp. 159–167.

[39]Yee, H. C., Warming, R. F., and Harten, A., "Implicit Total Variation Diminishing (TVD) Schemes for Steady-State Calculations," NASA TM-84342, March 1983.

[40]Hung, C. M., and Buning, P. G., "Simulation of Blunt-Fin-Induced Shock-Wave and Turbulent-Layer Interaction," *Journal of Fluid Mechanics*, Vol. 154, 1985 pp. 163–185.

[41]Joslyn, H. D., and Dring, R. P., "Experimental and Analytical Study of Turbine Temperature Profile Attenuation," Air Force Office of Scientific Research Contract F49620-86-C-0020, Aug. 1988.

[42]Dring, R. P., and Joslyn, H. D., "The Relative Eddy in Axial Turbine Rotor Passages," American Society of Mechanical Engineers Paper 83-GT-22, 1983.

[43]Dring, R. P., and Joslyn, H. D., "Measurements of Turbine Rotor Blade Flows," *Journal of Engineering for Power*, Vol. 103, No. 2, April 1981, pp. 400–405.

[44]Sharma, O. P., Butler, T. H., Joslyn, H. D., and Dring, R. P., "An Experimental Investigation of the Three-Dimensional Unsteady Flow in an Axial Flow Turbine," AIAA Paper 83-1170, June 1983.

[45]Steger, J. L., and Sorenson, R. L., "Automatic Mesh-Point Clustering Near a Boundary in Grid Generation with Elliptic Partial Differential Equations," *Journal of Computational Physics*, Vol. 33, No. 3, Dec. 1979, pp. 405–410.

[46]Baldwin, B. S., and Lomax, H., "Thin Layer Approximation and Algebraic Model for Separated Turbulent Flow," AIAA Paper 78-257, Jan 1978.

[47]Tyler, J. M., and Sofrin, T. G., "Axial Flow Compressor Noise Studies," *SAE Transactions*, Vol. 70, 1970, pp. 309–332.

[48]Langston, L. S., Nice, M. L., and Hooper, R. M., "Three-Dimensional Flow Within a Turbine Cascade Passage," *Journal of Engineering for Power*, Vol. 99, Jan. 1977, pp. 21–28.

[49]Chien, K.-Y., "Predictions of Channel and Boundary-Layer Flows with a Low-Reynolds-Number Turbulence Model," *AIAA Journal*, Vol. 20, Jan. 1982, pp. 33–38.

[50]Gundy-Burlet, K. L., and Rai, M. M., "Two-Dimensional Computations of Multi-Stage Compressor Flows Using a Zonal Approach," AIAA Paper 89-2452, 1989.

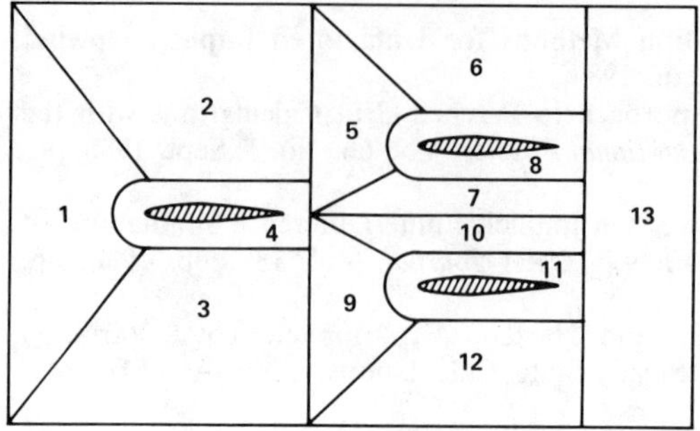

Fig. 1   Zoning of the multiply-connected region associated with three airfoils.

a) Example of patched grids

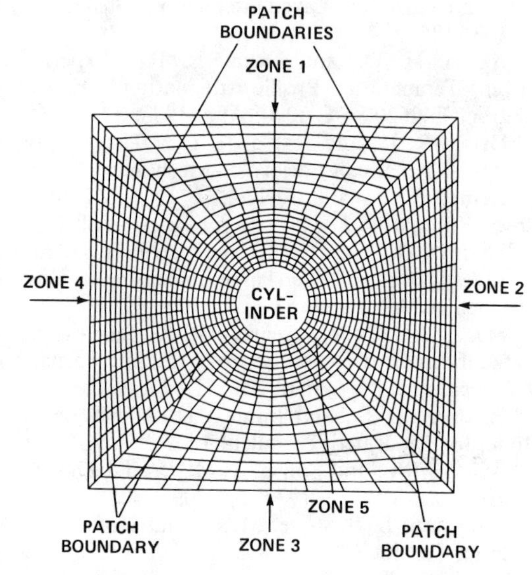

b) Example of overlaid grids

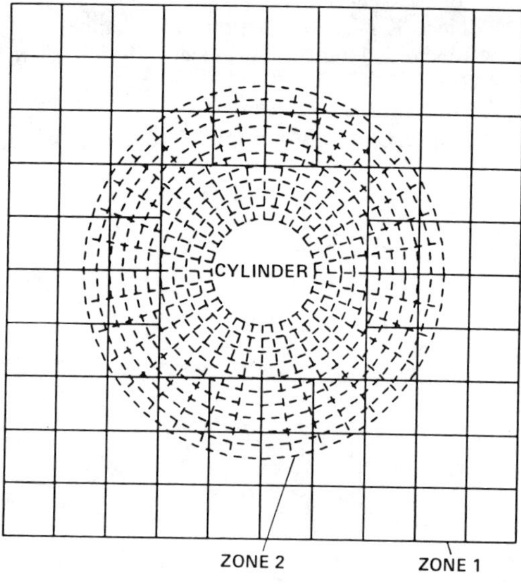

Fig. 2   Types of zonal grids used in finite-difference calculations.

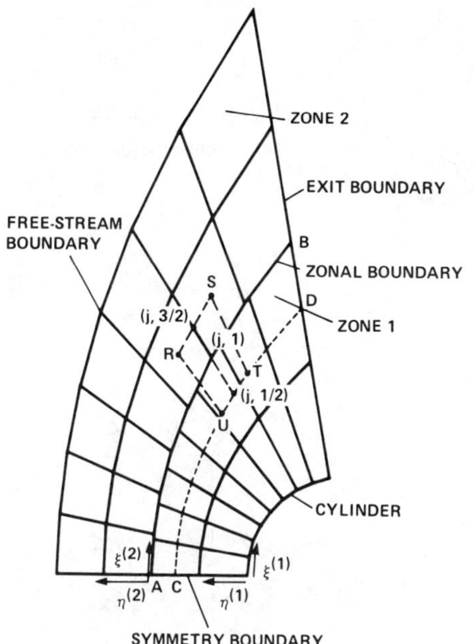

Fig. 3   Two-zone grid to illustrate the conservative, patch-boundary scheme.

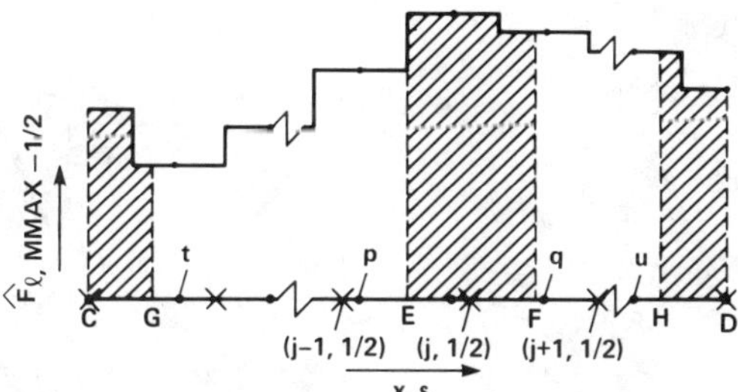

Fig. 4   Piecewise constant variation of the numerical flux $\hat{F}$ as a function of s.

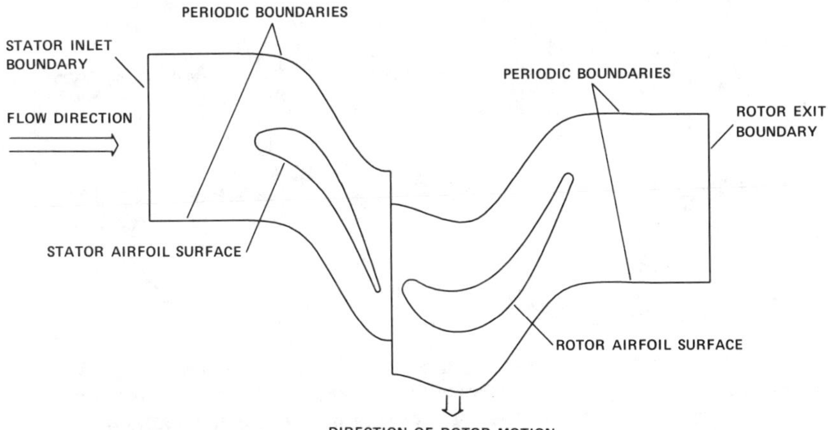

Fig. 5   Natural boundaries for a typical single-stage turbine configuration.

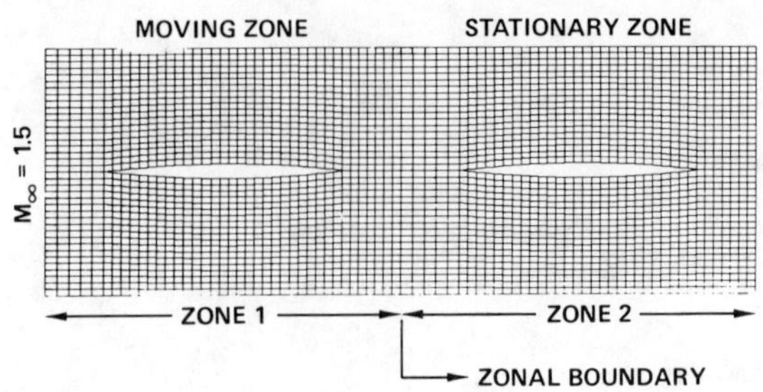

Fig. 6 Two-patch grid for the double-airfoil calculation.

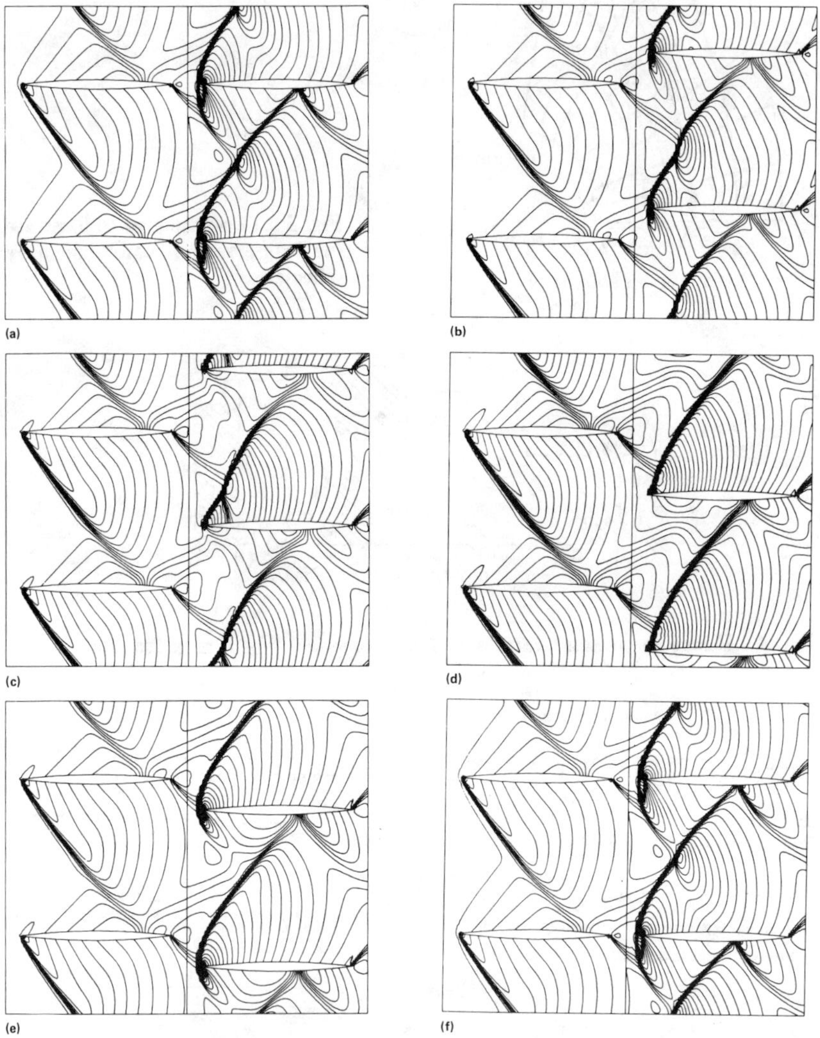

Fig. 7 Pressure contours at various time intervals after convergence to a time-periodic solution: a) pressure contours after 4.0 cycles, b) pressure contours after 4.2 cycles; c) pressure contours after 4.4 cycles; d) pressure contours after 4.6 cycles; e) pressure contours after 4.8 cycles; f) pressure contours after 5.0 cycles.

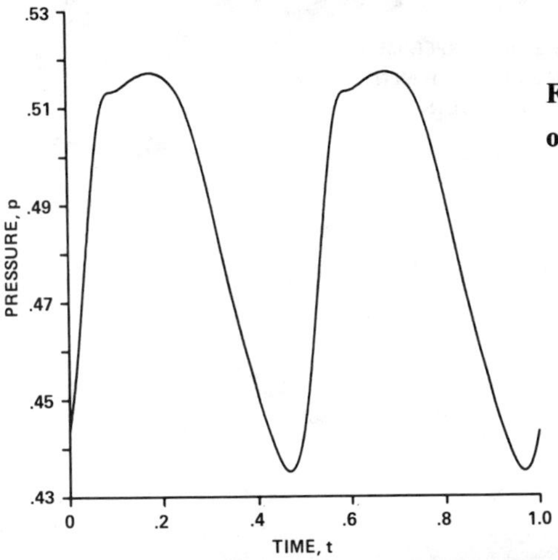

Fig. 8   Pressure history at midchord on the lower surface of the aft airfoil.

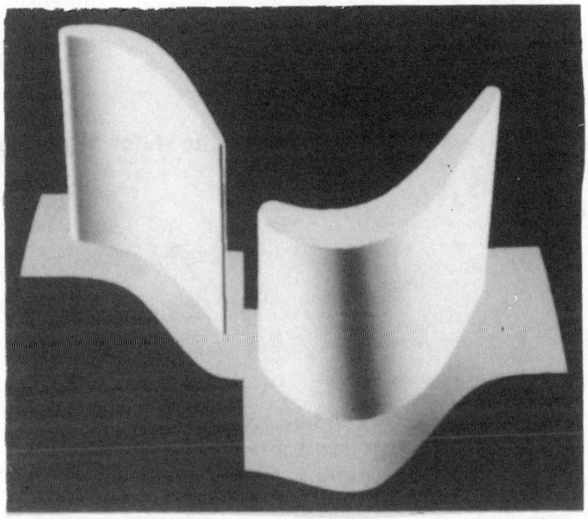

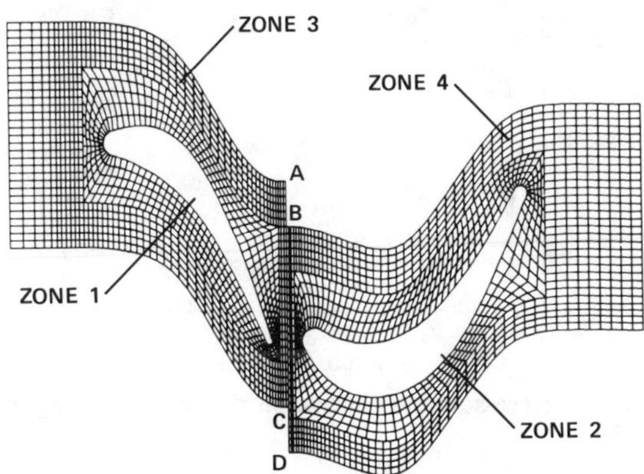

Fig. 9   Perspective view of the rotor-stator geometry (showing pressure side of stator, suction side of rotor).

Fig. 10   Multizone grid used in the two-dimensional calculations.

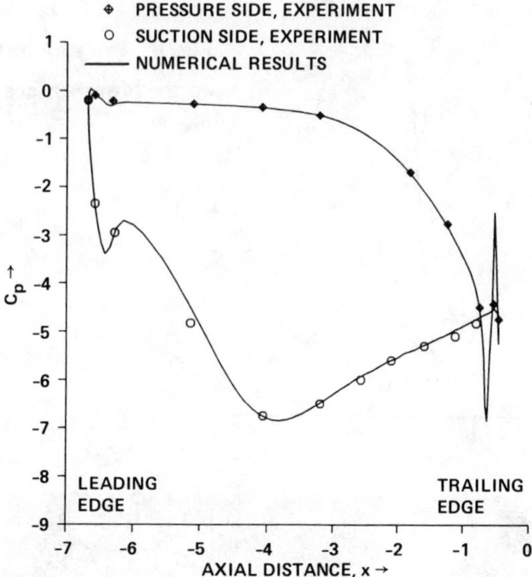

Fig. 11   Time-averaged pressure distribution on the stator at midspan, single-stator/single-rotor calculation.

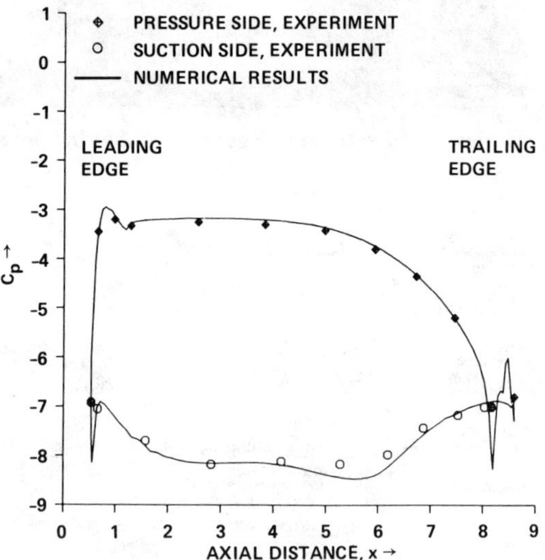

Fig. 12   Time-averaged pressure distribution on the rotor at midspan, single-stator/single-rotor calculation.

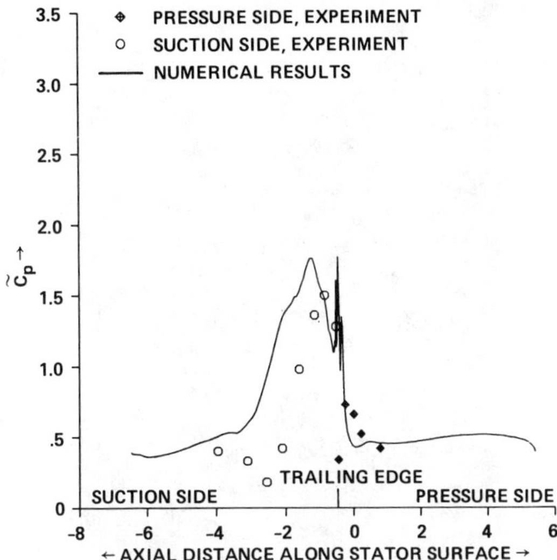

Fig. 13 **Pressure amplitude distribution on the stator at midspan, single-stator/ single-rotor calculation.**

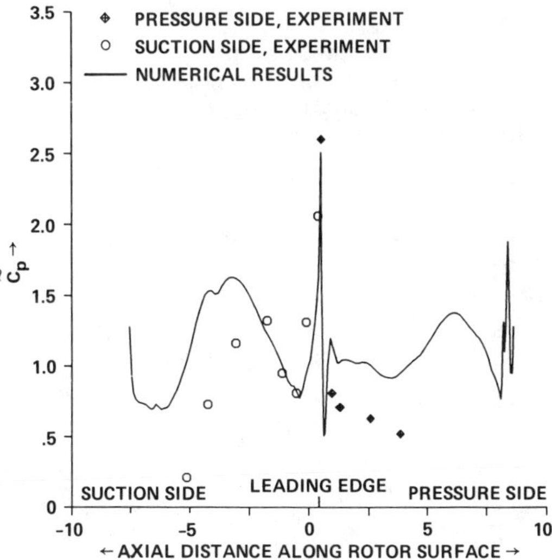

Fig. 14 **Pressure amplitude distribution on the rotor at midspan, single-stator/single-rotor calculation.**

Fig. 15   Multizone grid used in the two-dimensional multiblade calculations.

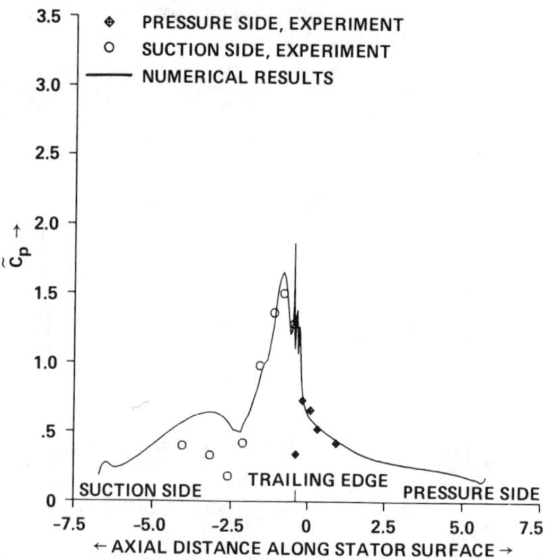

Fig. 16   Pressure amplitude distribution on the stator at midspan, three-stator/four-rotor calculation.

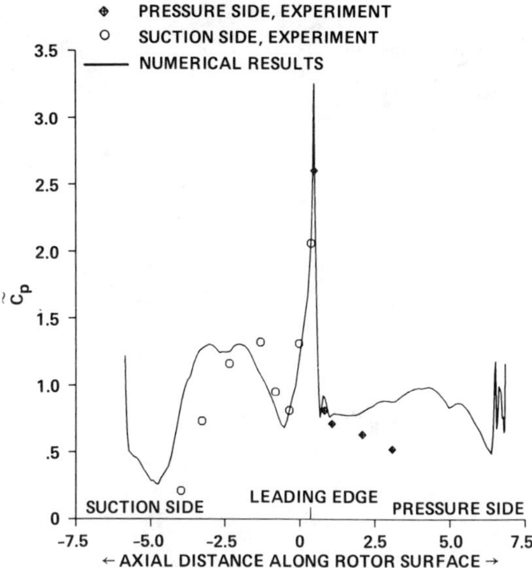

Fig. 17 **Pressure amplitude distribution on the rotor at midspan, three-stator/four-rotor calculation.**

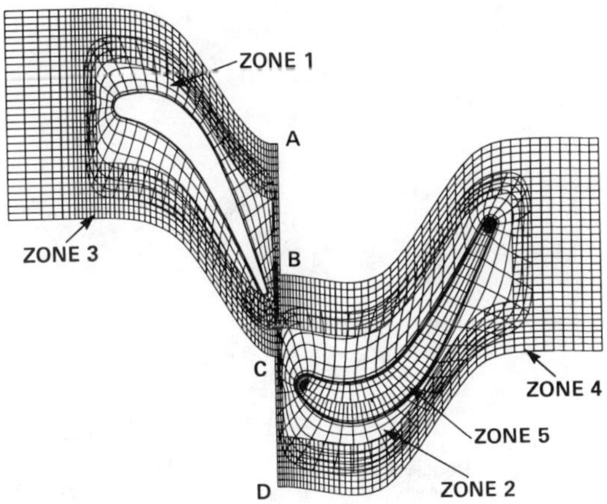

Fig. 18 **Multizone grid used in the three-dimensional calculations.**

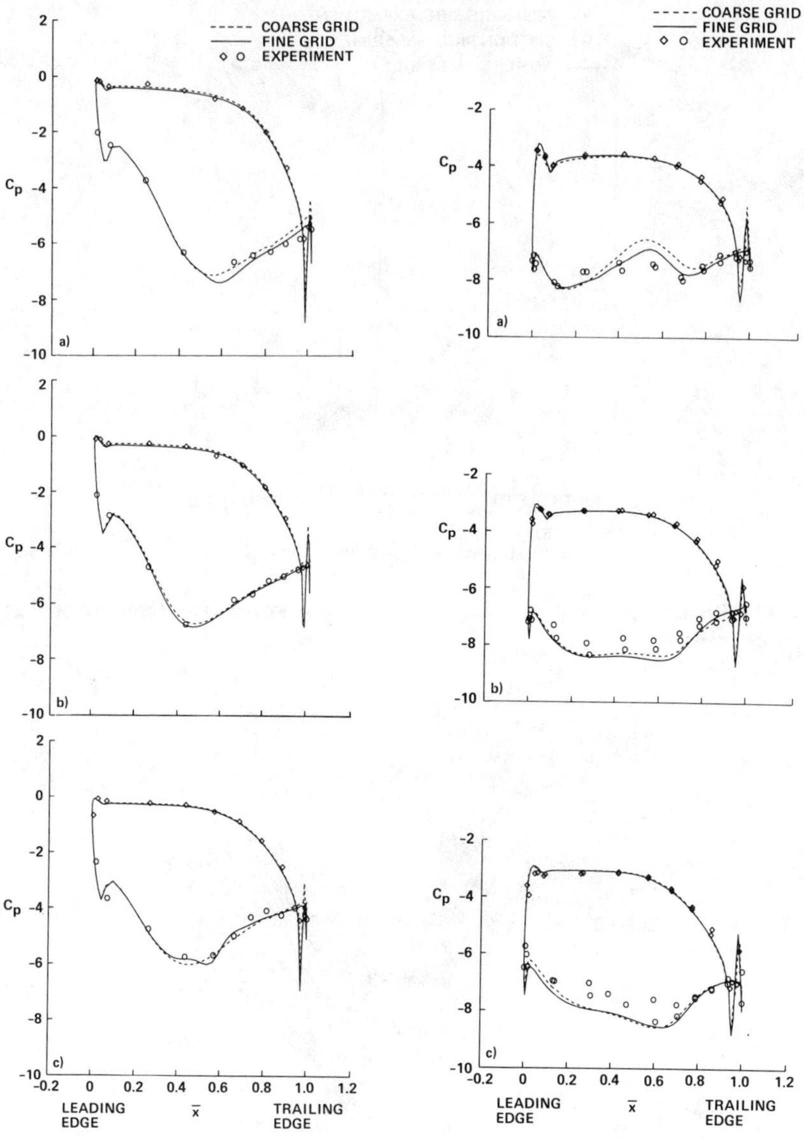

Fig. 19 Spanwise variation of time-averaged pressure distributions on the stator, three-dimensional single-stator/single-rotor calculation: a) 2.0% span; b) 50.0% span; c) 98.0% span.

Fig. 20 Spanwise variation of time-averaged pressure distributions on the rotor, three-dimensional single-stator/single-rotor calculation: a) 2.0% span; b) 50.0% span; c) 87.5% span.

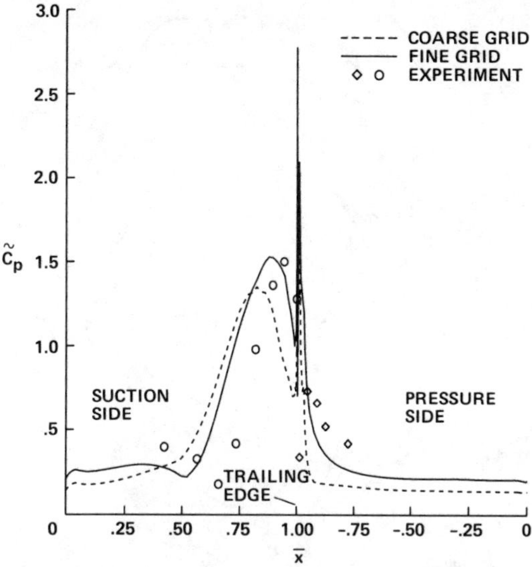

Fig. 21   Pressure amplitude distribution on the stator at midspan, three-dimensional single-stator/single-rotor calculation.

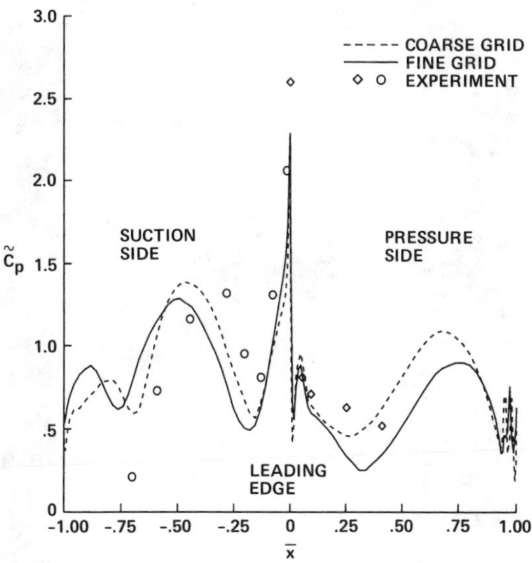

Fig. 22   Pressure amplitude distribution on the rotor at midspan, three-dimensional single-stator/single-rotor calculation.

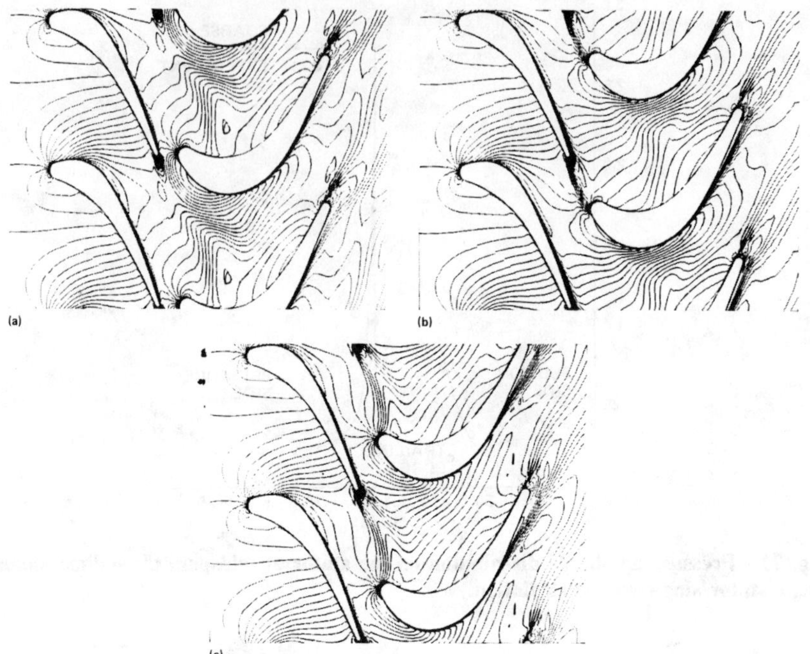

Fig. 23   Instantaneous Mach number contours; a) time instant $t = 0.0$; b) time instant $t = 0.33$; c) time instant $t = 0.66$.

Fig. 24   Time-averaged limiting streamlines on the stator hub.

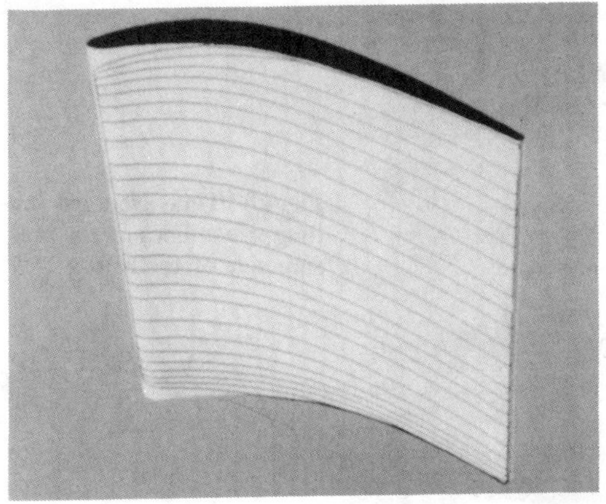

Fig. 25 Time-averaged limiting streamlines on the pressure side of the stator.

Fig. 26 Time-averaged limiting streamlines on the suction side of the stator.

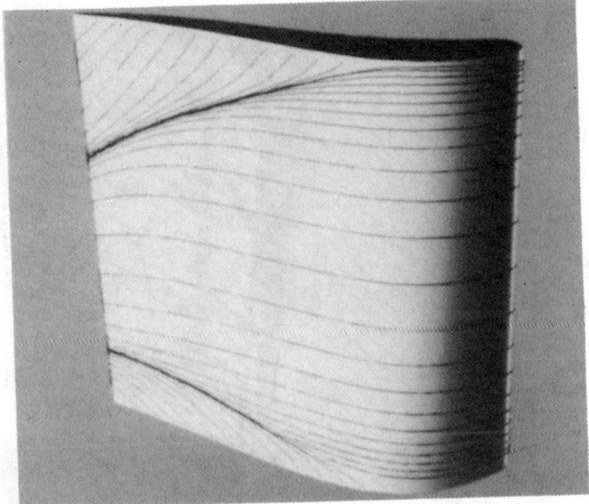

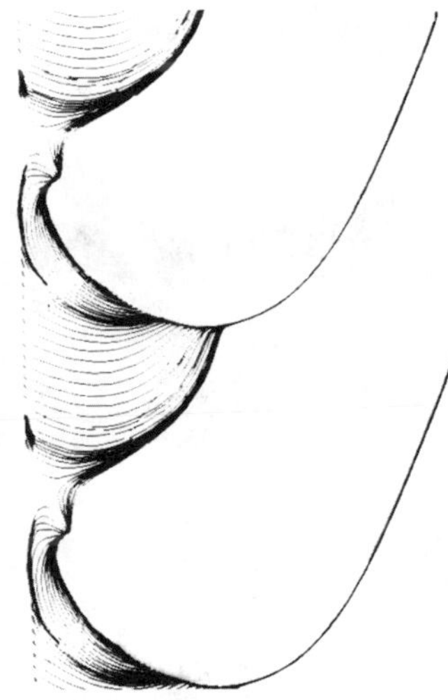

Fig. 27 Time-averaged limiting streamlines on the rotor hub.

Fig. 28 Time-averaged limiting streamlines on the pressure side of the rotor.

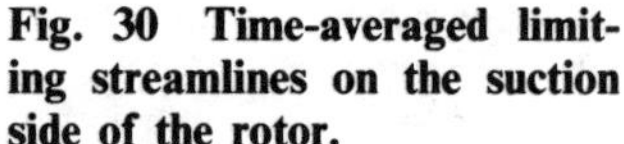

Fig. 29   Experimental visualization of rotor pressure side flow.

Fig. 30   Time-averaged limiting streamlines on the suction side of the rotor.

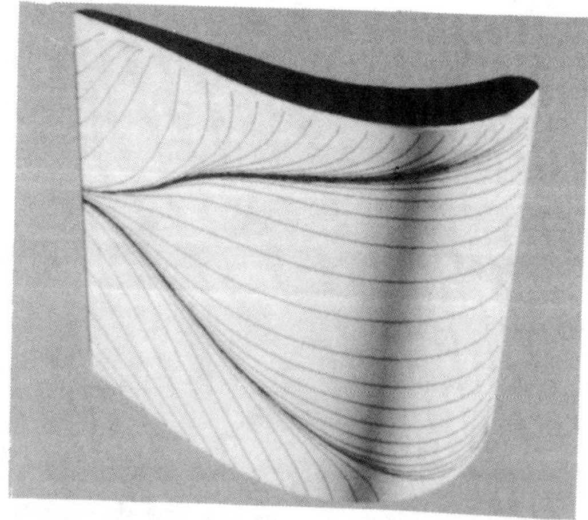

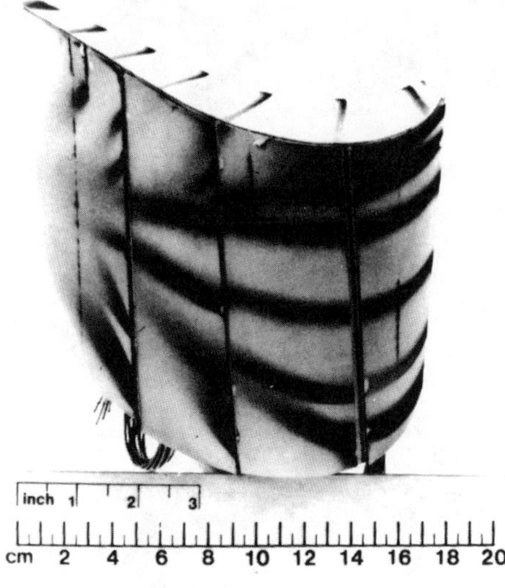

Fig. 31   Experimental visualization of rotor suction side flow.

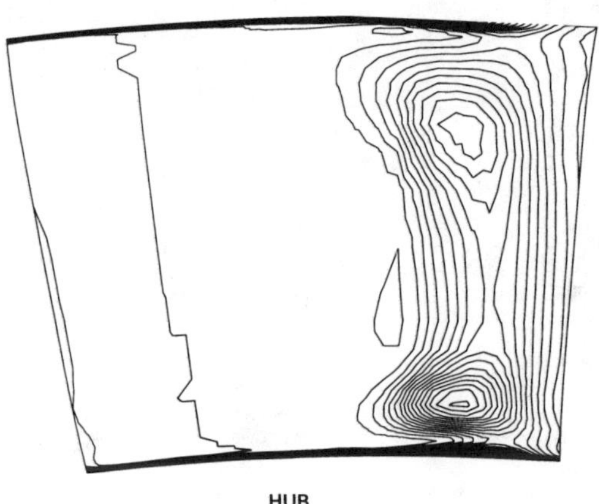

Fig. 32   Total-pressure contours at the exit to the stator airfoil (in a plane perpendicular to the hub axis).

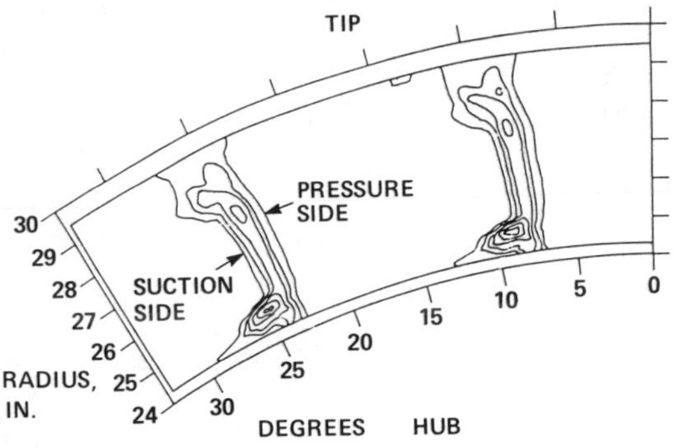

Fig. 33   Experimental total-pressure contours at the exit to the stator airfoil (in a plane perpendicular to the hub axis).

Part 6. Rotors

Euler/Navier-Stokes Calculations of the Flowfield of a Helicopter Rotor in Hover and Forward Flight

Ramesh K. Agarwal* and Jerry E. Deese†
McDonnell Douglas Corporation, St. Louis, Missouri

Introduction

AS computing power has progressively increased over the past decade, the development of computational codes for the flowfield of a helicopter rotor in hover and forward flight has progressed from the solution of the transonic small-disturbance equation,[1] to the full-potential equation,[2,3] and more recently to that of the Euler and Navier-Stokes equations.[4,5]

Modern helicopter rotor flowfields are characterized by transonic shocks, complex vortical wakes, and blade-vortex interactions (Fig. 1). The Euler equations accurately describe wave drag, shock position, and shock pressure rise, and also admit vortical solutions.

An Euler/Navier Stokes code, designated MDROTH, was recently developed at McDonnell Douglas Research Laboratories for calculating the flowfield of a multibladed helicopter rotor in hover and forward flight. The code computes the flowfield by solving the three-dimensional Euler/Navier-Stokes equations in a rotating coordinate system on body-conforming curvilinear grids around the blades. Equations are recast in absolute-flow variables so that the absolute flow in the far field is uniform but the relative flow is nonuniform. The equations are solved for the absolute-flow variables by employing Jameson's finite-volume explicit Runge-Kutta time-stepping scheme.[6] This formulation is essential for calculating rotary-wing flowfields. References 7–9 describe the details of the methodology. Inclusion of viscous effects makes it possible to simulate complex flow phenomena, such as flow separation on retreating blades, tip-vortex rollup and its interaction with rotating blades, and transonic shock−boundary-layer interaction. Viscous effects are modeled by employing the thin-layer approximation to the Reynolds-averaged Navier-Stokes equations in conjunction with the Baldwin-Lomax algebraic turbulence model.

Copyright © 1989 by the American Institute of Aeronautics and Astronautics, Inc. All rights reserved.

*Principal Scientist, Research Laboratories.
†Scientist, Research Laboratories.

The Euler/Navier-Stokes equations admit vortical solutions that allow the transport of vorticity, rollup of the wake following the blade, and convection of the tip vortex past the blade; however, at the present state of development, limitations are still associated with finite-difference or finite-volume computations of the Euler or Navier-Stokes equations for predicting the development of vortical wakes. The key problem is false diffusion of vorticity due to truncation error and artificial viscosity in the numerical algorithm employed in the flow solver. Substantial grid refinement and high-order-accurate differencing may alleviate this problem, but computational power beyond the Cray 2 class of supercomputers will be required. Presently, rotor-wake effects are modeled in the form of a correction applied to the geometric angle of attack along the blades. This correction is obtained by computing the local induced downwash with the prescribed wake analysis programs HOVER and B-TRIM of McDonnell Douglas Helicopter Company or CAMRAD of NASA Ames Research Center. HOVER, B-TRIM, and CAMRAD are based on the lifting-line theory and provide estimates of the induced downwash and the induced angle of attack.

The code has been fully vectorized (98%) for optimum performance on a Cray XMP or a Cray 2. The code has also been microtasked to optimally utilize four processors of a Cray XMP/48.

Calculations are presented for the flowfield of a model helicopter rotor in hover at various collective pitch angles using both the Euler and Navier-Stokes codes. For a rotor in forward flight, calculations are performed for the flowfield of a model rotor, the OLS rotor, the 500-E rotor, and the Apache AH-64 rotor using the Euler code. Comparisons with the experimental data are presented where possible. All of the calculations reported in this paper have been performed on a Cray XMP or a Cray 2 (at NASA Ames Research Center).

Governing Equations

Compressible Reynolds-averaged Navier-Stokes equations (Euler equations are a subset of these equations) are formulated in a Cartesian reference frame rotating with angular velocity Ω of the rotor. The equations are recast in the absolute-flow variables so that the absolute flow in the far field is uniform but the relative flow is nonuniform; we solve for the absolute-flow variables. This formulation allows more accurate calculation of the fluxes in the finite-volume method and is essential for obtaining accurate solutions on nonuniform grids. Next, the equations are transformed in the body-conforming curvilinear coordinate system (ξ,η,ζ) and are simplified by employing the thin-layer approximation.

Let p_∞, ρ_∞, H_∞, μ_∞, and κ_∞ denote the pressure, density, total enthalpy, molecular viscosity, and molecular thermal conductivity, respectively, in the freestream; and let $\mathscr{L}$ be a typical body dimension (blade chord, for example) in the flow. Then the nondimensionalized pressure p, the density ρ, the absolute velocity components (u,v,w), the absolute total energy and total enthalpy (e,h), the laminar and turbulent viscosity coefficients (μ_ℓ,μ_t), the laminar and turbulent conductivity coefficients (κ_ℓ,κ_t), the independent spatial variables (ξ,η,ζ), and the time variables t are nondimensionalized by p_∞, ρ_∞, $\sqrt{(p_\infty/\rho_\infty)}$, p_∞/ρ_∞, μ_∞, κ_∞, $\mathcal{L}$, and $L\sqrt{(\rho_\infty/p_\infty)}$, respectively.

Let (u,v,w), (u_r,v_r,w_r), and $(u_\Omega,v_\Omega,w_\Omega)$ denote the absolute, relative, and rotational velocity components, respectively, in the rotating Cartesian coordinate system (x,y,z) shown in Fig. 1. The following relationships hold among the various velocity components:

$$u = u_r - u_\Omega; \qquad v = v_r - v_\Omega; \qquad w = w_r - w_\Omega \qquad (1)$$

For a rotor that rotates with a uniform angular velocity Ω about the y axis (as shown in Fig. 1),

$$u_\Omega = -\Omega z, \qquad v_\Omega = 0, \qquad \text{and} \qquad w_\Omega = \Omega x \qquad (2)$$

The governing equations in a body-conforming rotating coordinate system then can be written as

$$\frac{\partial W}{\partial t} + \frac{\partial L}{\partial \xi} + \frac{\partial M}{\partial \eta} + \frac{\partial N}{\partial \zeta} + U_\Omega \frac{\partial W}{\partial \xi} + V_\Omega \frac{\partial W}{\partial \eta} + W_\Omega \frac{\partial W}{\partial \zeta} = T + \frac{\sqrt{\gamma} M_\infty}{Re_\infty} S \qquad (3)$$

where

$$W = J \begin{bmatrix} \rho \\ \rho u \\ \rho v \\ \rho w \\ \rho e \end{bmatrix}, \qquad L = J \begin{bmatrix} \rho U \\ \rho u U + \xi_x p \\ \rho v U + \xi_y p \\ \rho w U + \xi_z p \\ \rho U h \end{bmatrix}$$

$$M = J \begin{bmatrix} \rho V \\ \rho u V + \eta_x p \\ \rho v V + \eta_y p \\ \rho w V + \eta_z p \\ \rho V h \end{bmatrix}, \qquad N = J \begin{bmatrix} \rho W \\ \rho u W + \zeta_x p \\ \rho v W + \zeta_y p \\ \rho w W + \zeta_z p \\ \rho W h \end{bmatrix} \qquad (4)$$

with

$$U = \xi_x u + \xi_y v + \xi_z w, \qquad V = \eta_x u + \eta_y v + \eta_z w \qquad (5a)$$

$$W = \zeta_x u + \zeta_y v + \zeta_z w \qquad (5b)$$

$$U_\Omega = \xi_x u_\Omega + \xi_y v_\Omega + \xi_z w_\Omega, \qquad V_\Omega = \eta_x u_\Omega + \eta_y v_\Omega + \eta_z w_\Omega \qquad (5c)$$

$$W_\Omega = \zeta_x u_\Omega + \zeta_y v_\Omega + \zeta_z w_\Omega \qquad (5d)$$

$$J = x_\xi y_\eta z_\zeta + x_\zeta y_\xi z_\eta + x_\eta y_\zeta z_\xi - x_\xi y_\zeta z_\eta - x_\eta y_\xi z_\zeta - x_\zeta y_\eta z_\xi \qquad (5e)$$

$$T = J \begin{bmatrix} 0 \\ -\rho\Omega w \\ 0 \\ \rho\Omega u \\ 0 \end{bmatrix} \qquad (6a)$$

$$S = J \begin{bmatrix} 0 \\ m_1 u_\eta + m_2 \eta_x \\ m_1 v_\eta + m_2 \eta_y \\ m_1 w_\eta + m_2 \eta_z \\ m_1 \dfrac{\gamma\kappa_{\text{eff}}}{\mu_{\text{eff}}(\gamma-1)Pr}(p/\rho)_\eta + (q^2/2)_\eta + m_2 V \end{bmatrix} \qquad (6b)$$

with

$$m_1 = \mu_{\text{eff}}(\eta_x^2 + \eta_y^2 + \eta_z^2) \qquad (7a)$$

$$m_2 = \mu_{\text{eff}}(\eta_x u_\eta + \eta_y v_\eta + \eta_z w_\eta)/3 \qquad (7b)$$

$$q^2 = (u^2 + v^2 + w^2) \qquad (7c)$$

$$\mu_{\text{eff}} = \mu_\ell + \mu_t, \qquad \kappa_{\text{eff}} = \kappa_\ell + \kappa_t, \qquad \kappa_t = \mu_t/Pr_t \qquad (7d)$$

In Eqs. (7), μ_ℓ and k_ℓ are assumed to vary with temperature according to the empirical relations of Worsøe-Schmidt and Leppert.[10]

$$\mu_\ell = (p/\rho)^{0.67} \qquad \text{and} \qquad \kappa_\ell = (p/\rho)^{0.71}$$

The eddy viscosity μ_t in Eqs. (7) is calculated by employing the simple algebriac turbulence model of Baldwin and Lomax.[11]

Also, the velocity components (u,v,w), the pressure p, the total energy e, and the total enthalpy h are related through the equation of state for a perfect gas as follows:

$$e = \frac{p}{(\gamma-1)\rho} + \tfrac{1}{2}(u^2 + v^2 + w^2) \qquad (8)$$

$$h = e + \frac{p}{\rho} \qquad (9)$$

Let

$$F = L\hat{e}_\xi + M\hat{e}_\eta + N\hat{e}_\zeta$$

and

$$U_\Omega = U_\Omega\hat{e}_\xi + V_\Omega\hat{e}_\eta + W_\Omega\hat{e}_\zeta$$

where $(\hat{e}_\xi, \hat{e}_\eta, \hat{e}_\zeta)$ are the unit vectors in the (ξ, η, ζ) coordinate system. Equation (3) can be written as

$$\frac{\partial W}{\partial t} + \nabla \cdot F + U_\Omega \cdot \nabla W = T + \frac{\sqrt{\gamma} M_\infty}{Re_\infty} S \tag{10}$$

We solve Eq. (10) with the finite-volume method described later. In Eqs. (3) and (10), taking $S \equiv 0$ gives the Euler equations.

Finite-Volume Method

To apply the finite-volume method, Eq. (10) is written in the integral form

$$\frac{\partial}{\partial t} \iiint_V W \, dV + \iint_{\partial S} \hat{n} \cdot F \, dS + \iint_{\partial S} U_\Omega \cdot \hat{n} W \, dS = \iiint_V (T + rS) \, dV \tag{11}$$

for a domain V with a bounding surface ∂S, where $r = \sqrt{(\gamma)} M_\infty / Re_\infty$, and $\hat{n}$ denotes the unit outward normal vector to the surface element dS.

The physical domain is divided into hexahedral cells that map into unit cubes in the computational domain denoted by the subscripts (i,j,k). Assuming that the dependent variables are known at the center of each cell, we obtain a system of ordinary differential equations by applying Eq. (11) separately to each cell. These equations have the form

$$\frac{d}{dt} (J_{ijk} W_{ijk}) + P_{ijk} + Q_{ijk} = J_{ijk}(T_{ijk} + rS_{ijk}) \tag{12}$$

where J_{ijk} is the cell volume, P_{ijk} represents the net absolute flux out of the cell, and Q_{ijk} is the rotational flux out of the cell. The calculation of J_{ijk}, P_{ijk}, and Q_{ijk} for a hexahedral cell is described in Ref. 7. We rewrite Eq. (12) as

$$\frac{d}{dt} (J_{ijk} W_{ijk}) + \mathscr{F}_{ijk} = 0 \tag{13}$$

The finite-volume scheme [Eq. (13)] constructed in this manner reduces to a central-difference scheme on a Cartesian mesh and is second-order-accurate, provided that the mesh is sufficiently smooth and without any abrupt changes in cell shape and volume. Scheme (13) is not dissipative and therefore allows undamped oscillations at odd and even mesh points.

To suppress the tendency for odd- and even-point decoupling and to prevent the appearance of oscillations in regions containing severe pressure gradients near shock waves and stagnation points, the finite-volume scheme is augmented by the addition of artificial dissipative terms. Equation (13) is replaced by

$$\frac{d}{dt} (J_{ijk} W_{ijk}) + \mathscr{F}_{ijk} - D_{ijk} = 0 \tag{14}$$

where D_{ijk} denotes the dissipative terms that are generated by dissipative fluxes. Jameson et al.[6] have established that an effective form of dissipative terms for flows with discontinuities is a blend of second and fourth differences with coefficients that depend on the local pressure gradient. Dissipative terms are constructed as follows:

$$D_{ijk} = (D_x + D_y + D_z)W_{ijk} \tag{15}$$

where

$$D_x W_{ijk} = d_{i+\frac{1}{2},j,k} - d_{i-\frac{1}{2},j,k} \tag{16}$$

The dissipative flux $d_{i+\frac{1}{2},j,k}$ is defined as

$$d_{i+\frac{1}{2},j,k} = \frac{J_{i+\frac{1}{2},j,k}}{(\Delta t_\ell)_{i+\frac{1}{2},j,k}} [\epsilon^{(2)}_{i+\frac{1}{2},j,k} \Delta_x W_{ijk} - \epsilon^{(4)}_{i+\frac{1}{2},j,k} \Delta_x^3 W_{i-1,j,k}] \tag{17}$$

where Δ_x denotes the forward-difference operator $\Delta_x = (W_{i+1,j,k} - W_{i,j,k})$, and $\epsilon^{(2)}$ and $\epsilon^{(4)}$ are adaptive coefficients defined later. We define

$$v_{ijk} = \left| \frac{p_{i+1,j,k} - 2p_{ijk} + p_{i-1,j,k}}{p_{i+1,j,k} + 2p_{ijk} + p_{i-1,j,k}} \right| \tag{18}$$

$$\epsilon^{(2)}_{i+\frac{1}{2},j,k} = \kappa^{(2)} \max(v_{i+1,j,k}, v_{ijk}) \tag{19}$$

and

$$\epsilon^{(4)}_{i+\frac{1}{2},j,k} = \max\{0, [\kappa^{(4)} - \epsilon^{(2)}_{i+\frac{1}{2},j,k}]\} \tag{20}$$

where typical values of the constants $\kappa^{(2)}$ and $\kappa^{(4)}$ are $\kappa^{(2)} = 1/4$ and $\kappa^{(4)} = 1/256$. The expression $(\Delta t_\ell)_{ijk}$ is also defined in Ref. 7. The terms $D_y W_{ijk}$ and $D_z W_{ijk}$ in Eq. (15) are calculated in an analogous manner.

The scaling $J/\Delta t$ in Eq. (17) conforms to the inclusion of the cell volume in the dependent variables of Eq. (14). Since Eq. (17) contains undivided differences, it follows that, if $\epsilon^{(2)} = 0(\Delta x^2)$ and $\epsilon^{(4)} = 0(1)$, then the added terms are of $0(\Delta x^3)$, as will be the case in the regions where the flow is smooth. Near a shock wave $\epsilon^{(2)} = 0(1)$, and the scheme behaves locally as a first-order-accurate scheme.

Time-Stepping Scheme

The classical fourth-order Runge–Kutta scheme is used to integrate Eq. (14). Suppressing the subscripts (i,j,k), we can write Eq. (14) as

$$\frac{dW}{dt} + \frac{1}{J}[\mathscr{F}(W) - D(W)] = 0 \tag{21}$$

**Table 1   Performance evaluation of helicopter rotor code MDROTH
on Cray XMP/48**

Test case: NACA 0012 untwisted, untapered, two-bladed rotor; $M_t = 0.52$,
$\alpha = 0$ deg, aspect ratio $= 6$

Mesh used: 97 (chordwise) $\times$ 33 (wing normal) $\times$ 21 (spanwise)

Main memory required: 2 million words

Processing rate on one processor: 2.1×10^{-5} CPU s/mesh point/iteration

98% of total CPU time spend in two key subroutines, EULER and FILTER

Theoretical speedup on 4 processors (Amdahl's law)

$$SP = \frac{\text{CPU time}}{\text{wall clock time}} = \frac{1}{\dfrac{(0.98)}{4} + (1 - 0.98)} = 3.77$$

Actual speedup achieved after microtasking $= 3.73$

Processing rate on 4 processors: 6.3×10^{-6} wall clock s/mesh point/iteration

At time level n, we set

$$W^{(0)} = W^n$$

$$W^{(1)} = W^{(0)} - \alpha_1 \frac{\Delta t}{J} [\mathscr{F} W^{(0)} - D W^{(0)}]$$

$$W^{(2)} = W^{(0)} - \alpha_2 \frac{\Delta t}{J} [\mathscr{F} W^{(1)} - D W^{(0)}]$$

$$W^{(3)} = W^{(0)} - \alpha_3 \frac{\Delta t}{J} [\mathscr{F} W^{(2)} - D W^{(0)}]$$

$$W^{(4)} = W^{(0)} - \alpha_4 \frac{\Delta t}{J} [\mathscr{F} W^{(3)} - D W^{(0)}]$$

and

$$W^{n+1} = W^{(4)} \tag{22}$$

In scheme (22), the dissipative terms are frozen at their values in the first
stage, thus avoiding expensive computations for each stage. The choice of
coefficients determines the characteristics of the scheme. A good choice of
the coefficients is $\alpha_1 = 0.6$, $\alpha_2 = 0.6$, $\alpha_3 = 1$, and $\alpha_4 = 1$. The scheme is
fourth-order-accurate in time and is stable for Courant numbers $\leqslant 2\sqrt{(2)}$.
The scheme has the property that, if $[\mathscr{F} W^n - D W^n] = 0$, then $W^{(1)} = W^{(0)}$,
$W^{n+1} = W^n$, and the steady-state solution is $W^n - D W^n = 0$, independent

of time step Δt. Thus, a variable time step (Δt_ℓ) determined by the bound on the local Courant number can be used to accelerate convergence to steady state without altering the steady state. To obtain time-accurate calculations of unsteady flow phenomena, we determine the allowable time step by the minimum value of the Courant number in the field.

Residual Smoothing

The stability range of the explicit scheme given by Eq. (22) is extended by employing implicit residual smoothing. The residual smoothing is applied to the product

$$(1 - \epsilon_i \delta_x^2)(1 - \epsilon_j \delta_y^2)(1 - \epsilon_k \delta_z^2)\bar{R}_{ijk} = R_{ijk} \tag{23}$$

where $R_{ijk} = [\mathcal{F}(W_{ijk}) - D(W_{ijk})]$ is the residual before smoothing, and $\bar{R}_{ijk}$ is the new residual. It has been shown by linear stability analysis that stability can be obtained for any Courant number λ, provided the smoothing parameter satisfies

$$\epsilon \geqslant \frac{1}{4}\left[\left(\frac{\lambda}{\lambda^*}\right)^2 - 1\right]$$

where λ^* is the Courant number limit for the unsmoothed scheme given by Eq. (22). However, in practice, for the three-dimensional thin-layer Reynolds-averaged Navier-Stokes equations, it has not been possible to extend the stability limit substantially. In implementing Eq. (23), it is only necessary to solve a sequence of tridiagonal equations for scalar variables; thus, the residual smoothing requires a relatively small amount of additional computational effort per time step. Comparatively, other implicit schemes usually need to solve a coupling system with a much more costly block-tridiagonal solver.

Optimization and Microtasking

The code MDROTH has been fully vectorized for optimum performance on a single-processor Cray XMP and Cray 2. It has also been microtasked on a four-processor Cray XMP/48 for reducing the wall clock time by judicious use of various Cray software techniques.[8] The actual speedup in wall clock time achieved by microtasking ($=3.73$) is very close to the theoretical speedup possible ($=3.77$). Table 1 shows the performance of the microtasked code for a typical test run.

Boundary Conditions

Special considerations are required to determine the fluxes in Eq. (13) at the boundaries. Far-field boundary conditions are applied at a finite distance from the blade surface, whereas flow tangency (with Euler solver) or no-slip boundary condition (with Navier-Stokes solver) is applied at the blade surface.

The normal pressure gradient at the blade surface is estimated from the condition that $\partial(\rho V_n)/\partial t = 0$, where V_n is the normal-velocity component. The pressure at the wall is then calculated from the pressure at the adjacent cell centers with the known value of $\partial p/\partial n$.

The treatment of the far-field boundary condition is based on the introduction of Riemann invariants for a one-dimensional flow normal to the boundary. Let subscripts ∞ and e denote far-field values and values extrapolated from the interior cells adjacent to the boundary, respectively, and let V_n and a be the velocity component normal to the boundary and the speed of sound, respectively.

If the flow in the far field is subsonic, we introduce fixed and extrapolated Riemann invariants,

$$R_\infty = V_{n\infty} - \frac{2a_\infty}{\gamma - 1}$$

and

$$R_e = V_{ne} + \frac{2a_e}{\gamma - 1}$$

corresponding to incoming and outgoing waves. These invariants may be added and subtracted to obtain

$$V_n = \tfrac{1}{2}(R_e + R_\infty)$$

and

$$a = \tfrac{1}{4}(\gamma - 1)(R_e - R_\infty)$$

where V_n and a are the actual normal velocity component and the speed of sound to be specified in the far field, respectively. At an outflow boundary, the tangential velocity component and entropy are extrapolated from the interior, whereas at an inflow boundary they are specified as having far-field values. These four (two extrapolated, two specified) quantities provide a complete definition of the flow in the far field.

If the flow is supersonic in the far field, all of the quantities are specified at the inflow boundary, and they are all extrapolated from the interior at an outflow boundary.

For a computational region large enough to enclose the entire flowfield about the rotor and its wake, the preceding description of the boundary conditions is complete. However, for a computational region covering only a portion of the rotor disk, the effects caused by the other rotor blades and the flow outside the finite computational region need to be incorporated into the boundary conditions in one of several ways. In the present work these effects are incorporated simply by employing the transpiration velocity technique. In this method the velocity at the blade surface is set equal to the downwash velocity instead of zero. The downwash velocity is given by

$$v_{\text{wake}} = u_{\Omega} \tan\alpha_{\text{eff}} \tag{24}$$

where α_{eff} is the effective angle-of-attack distribution along the blade obtained from a helicopter wake code such as CAMRAD, HOVER, or B-TRIM.

Computed Examples

Computations are performed for the flowfield of a model helicopter rotor, the OLS rotor, the 500-E rotor, and the Apache AH-64 rotor in hover and forward flight at various collective pitch angles.

Hover Flowfield Calculations (Euler and Navier-Stokes Solutions)

These calculations have been performed for the flowfield of a model helicopter rotor. The rotor has two blades, of aspect ratio equal to 6, which are untwisted, untapered, and have the NACA 0012 airfoil section. The experiments on this rotor have been performed by Caradonna and Tung[12] at NASA Ames Research Center for a range of blade tip Mach numbers M_t and collective pitch angles θ_c. Euler calculations are performed on a $97 \times 33 \times 21$ mesh on a Cray XMP/48, and the Navier-Stokes computations are performed on a $97 \times 33 \times 33$ mesh on a Cray 2. A typical Euler computation requires 2 million words of main memory and 2.1×10^{-5} s of CPU time per mesh point for each iteration, and a typical Navier-Stokes calculation requires 4 million words of main memory and 5.6×10^{-5} s of CPU time per mesh point for each iteration. The fully vectorized version of the code decreases CPU time by a factor of 3, and the microtasked version of the code on four processors of the Cray XMP/48 provides an additional time-gain factor of approximately 3.73 in wall clock time.

For the Euler calculations, a six-order-of-magnitude reduction in the residual for density was obtained in approximately 800 time steps at a Courant number of 2. A converged solution required approximately 19 min of CPU time on the Cray XMP/48.

For the Navier-Stokes calculations, on the other hand, a four-order-of-magnitude reduction in the residual for density was obtained in approximately 2500 time steps. The residual for density is defined as the root-mean-square value of $\partial\rho/\partial t$ (calculated as $\Delta\rho/\Delta t$ for the complete time step).

Nonlifting Case:

$$M_t = 0.52, \qquad Re_c = 2 \times 10^6, \qquad \theta_c = 0 \text{ deg}$$

Figure 2 shows the surface pressure distribution on the nonlifting blade at different spanwise locations. The agreemnt between the Euler, the thin-layer Reynolds-averaged Navier-Stokes calculations, and the experimental data is excellent. As expected, both the Euler and the thin-layer Navier-Stokes calculations gave the same result for the pressure distribution.

Lifting Cases:

$$\text{1) } M_t = 0.44, \qquad Re_c = 2 \times 10^6, \qquad \theta_c = 8 \text{ deg}$$

$$\text{2) } M_t = 0.877, \qquad Re_c = 4 \times 10^6, \qquad \theta_c = 8 \text{ deg}$$

$$\text{3) } M_t = 0.612, \qquad Re_c = 2.67 \times 10^6, \qquad \theta_c = 8 \text{ deg}$$

As mentioned in the Introduction, for these calculations, the rotor-wake effects are modeled in the form of a correction applied to the geometric angle of attack of the blades. The free-wake analysis program HOVER of McDonnell Douglas Helicopter Company is used to calculate the induced downwash v_i. The results of calculations using HOVER are shown in Fig. 3. The induced angle of attack is calculated from the relation $\alpha_i = \tan^{-1}(v_i/\Omega r)$, where Ω is the rotational velocity of the rotor, and r is the local radial distance of the blade section from the rotor axis. As can be seen from Fig. 3, the slope of the curve v_i as a function of r remains constant except near the blade tip. For 95% of the blade span α_i can be taken as approximately 3.8 deg for all cases 1–3. The effective angle of attack thus becomes $\alpha_e = \theta_c - \alpha_i = 4.2$ deg. Figures 4–6 show the surface pressure distribution on the blade at various spanwise stations for cases 1–3, respectively. The agreement between the thin-layer Reynolds-averaged Navier-Stokes calculations and the experimental data is satisfactory. However, we do not show the results of Euler calculations; they are almost identical to Navier-Stokes calculations for the pressure distributions.

Forward-Flight Flowfield Calculations (Euler Solutions)

These calculations have been performed for the model NACA 0012 rotor, the OLS rotor, the 500-E rotor, and the Apache AH-64 rotor. All of the computations have been performed on a $97 \times 33 \times 33$ mesh on a Cray XMP/14 and a Cray 2. A typical calculation for the entire rotor half-plane from $\psi = 0$ to $\psi = 180$ deg requires approximately 6 h of CPU time; approximately 80 iterations are required for 1-deg movement in the azimuth direction to obtain a converged solution.

Figures 7 and 8 show a comparison of the computed pressure distributions on the model helicopter rotor in forward flight with experimental data[13] for a location near the blade tip; and

1) tip Mach number $M_t = 0.7$, advance ratio $\mu = 0.3$, and collective rotor pitch angle $\theta_c = 0$ deg; and

2) tip Mach number $M_t = 0.8$, advance ratio $\mu = 0.2$, and collective rotor pitch angle $\theta_c = 0$ deg, respectively.

In both the cases, the computed results show good agreement with the experimental data.

Figure 9 shows the effective angle of attack along the OLS rotor blade at various azimuth angles computed using CAMRAD for a tip Mach number $M_t = 0.633$, advance ratio $\mu = 0.298$, and collective pitch angle $\theta_c = 0$ deg. Figure 10 shows the computed pressure distributions on the OLS rotor blade for a location near the blade tip. The computations were performed with Euler and full-potential codes, and the results are compared with the in-flight data. Again, the Euler calculations show good agreement with the experimental data.

Figure 11 shows the effective angle of attack along the 500-E rotor blade at various azimuth angles computed using CAMRAD for a tip Mach number $M_t = 0.58$, advance ratio $\mu = 0.312$, and collective pitch angle $\theta_c = 0$ deg. Figure 12 shows the pressure distributions, using the Euler code, computed on the 500-E rotor blade for a location near the blade tip. Comparison with the experiment could not be made because experimental data were unavailable. However, Fig. 13 compares the lift coefficient C_L along the blade at $\psi = 90$ deg obtained with the Euler code and the full-potential code. Reasonable agreement is obtained. Plate 6 (see the color section) shows the Mach number distribution on the blade upper surface at various rotor angles.

Figure 14 shows the pressure distributions, using the Euler code, computed on the upper and lower surface of the AH-64 rotor blade for a location near the blade tip for a tip Mach number $M_t = 0.65$ and advance ratio $\mu = 0.326$. Again, comparison with the experiment could not be made because experimental data have not been available. For this rotor Mach contours on the blade upper surface are displayed at various azimuth angles in Plate 7. An enlarged view of the Mach contours near the tip at $\psi = 90$ deg is shown in Plate 8. A significant region of supersonic flow is apparent on the upper surface.

Conclusions

A three-dimensional Euler/thin-layer Reynolds-averaged Navier-Stokes solver has been developed for calculating transonic flow about rotor blades of a helicopter in hover and forward flight. The computational code is found to be robust, efficient, and accurate for both the steady and unsteady flowfield calculations. With the computing power currently available on the Cray XMP/48, Cyber 205, or Cray 2 class of machines, wake modeling is still needed for accurate calculation of aerodynamic loads. For this reason, a free-wake analysis program has been integrated with the Euler/thin-layer Navier-Stokes solver.

The code is fully vectorizable on a Cray supercomputer. The code can be easily multitasked/microtasked on a four-processor Cray XMP/48 for reducing the wall clock time by judicious use of various Cray software techniques. The actual speedup in wall clock time achieved by microtasking $(=3.73)$ is very close to the maximum theoretical speedup possible $(=3.77)$.

Acknowledgments

This research was conducted under the McDonnell Douglas Independent Research and Development Program.

The authors gratefully acknowledge the assistance of M. Shu of McDonnell Douglas Research Laboratories with programming on the Cray 2. The authors would also like to thank B. Charles and D. S. JanakiRam of

McDonnell Douglas Helicopter Company and N. L. Sankar of Georgia Institute of Technology for many useful discussions and assistance in various aspects of this work.

All of the calculations reported in this paper were performed on the Cray 2 at NASA Ames Research Center. The access to NAS (Numerical Aerodynamic Simulation) Processing System Network and the award of computational time are gratefully acknowledged.

References

[1]Caradonna, F. X., and Isom, M. P., "Subsonic and Transonic Potential Flow over Helicopter Blades," *AIAA Journal*, Vol. 10, 1972, pp. 1606–1612.

[2]Arieli, R., and Tauber, M. E., "Computation of Subsonic and Transonic Flow about Lifting Rotor Blades," AIAA Paper 79-1667, 1979.

[3]Strawn, R. C., and Caradonna, F. X., "Numerical Modeling of Rotor Flows with a Conservative Form of the Full-Potential Equation," AIAA Paper 86-0079, 1986.

[4]Roberts, T. W., and Murman, E. M., "Solution Method for a Hovering Helicopter Using the Euler Equations," AIAA Paper 85-0436, 1985.

[5]Sankar, L. N., Wake, B. E., and Lekoudis, S. G., "Solution of the Unsteady Euler Equations for Fixed and Rotary Wing Configurations," AIAA Paper 85-0120, 1985.

[6]Jameson, A., Schmidt, W., and Turkel, E., "Numerical Solution of the Euler Equations by Finite Volume Methods Using Runge-Kutta Time-Stepping Schemes," AIAA Paper 81-1259, 1981.

[7]Agarwal, R. K., and Deese, J. E., "Euler Calculations for Flowfield of a Helicopter Rotor in Hover," *Journal of Aircraft*, Vol. 24, 1987, pp. 231–238.

[8]Agarwal, R. K., and Deese, J. E., "An Euler Solver for Calculating the Flowfield of a Helicopter Rotor in Hover and Forward Flight," AIAA Paper 87-1427, 1987.

[9]Agarwal, R. K., and Deese, J. E., "Navier-Stokes Calculations of the Flowfield of a Helicopter Rotor in Hover," AIAA Paper 88-0106, 1988.

[10]Worsøe-Schmidt, P. M., and Leppert, G., "Heat Transfer and Friction for Laminar Flow of a Case in a Circular Tube at High Heating Rates," *International Journal of Heat and Mass Transfer*, Vol. 8, 1965, p. 1281.

[11]Baldwin, B. S., and Lomax, H., "Thin-Layer Approximation and Algebraic Model for Separated Turbulent Flows," AIAA Paper 78-257, 1978.

[12]Caradonna, F. X., and Tung, C., "Experimental and Analytical Studies of a Model Helicopter Rotor in Hover," *Vertica*, Vol. 5, 1981, pp. 149–161.

[13]Caradonna, F. X., Desopper, A., and Tung, C., "Finite-Difference Modeling of Rotor Flow Including Wake Effects," 8th European Rotorcraft Forum, Aix-en-Provence, France, Aug. 1972.

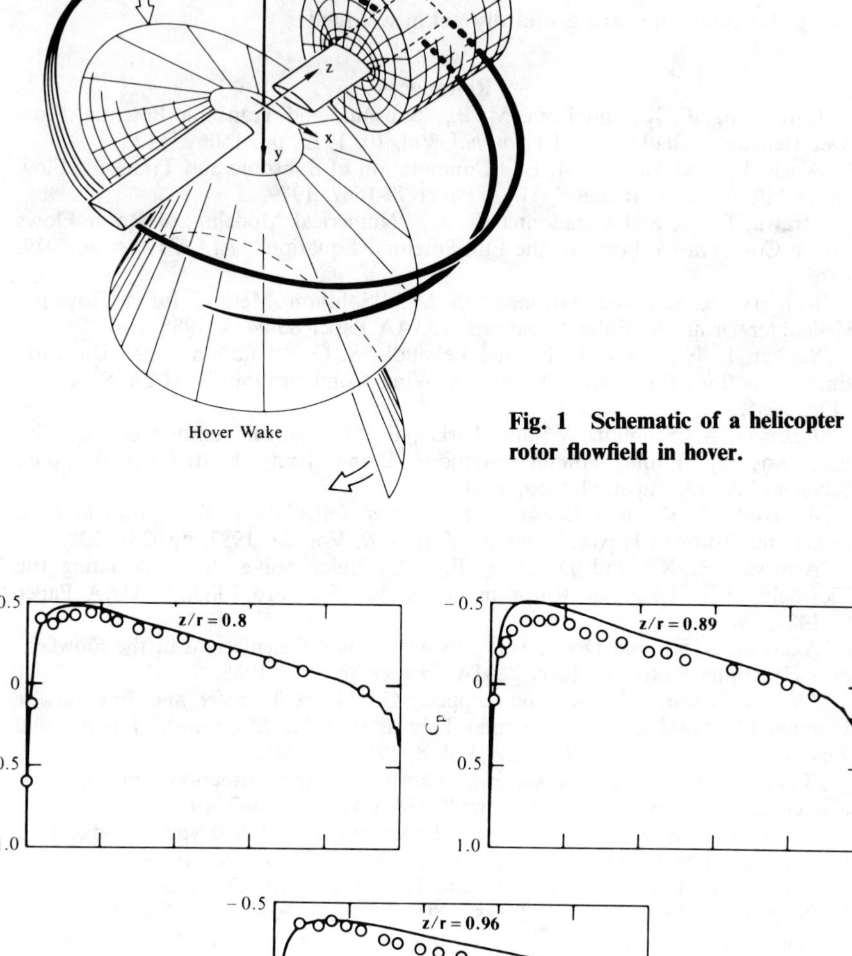

Fig. 1 Schematic of a helicopter rotor flowfield in hover.

Fig. 2 Surface pressure distributions on a nonlifting rotor in hover: untwisted, untapered, NACA 0012 blade, $M_t = 0.52$, $\theta_c = 0$ deg, AR = 6.0, $97 \times 33 \times 21$ mesh.

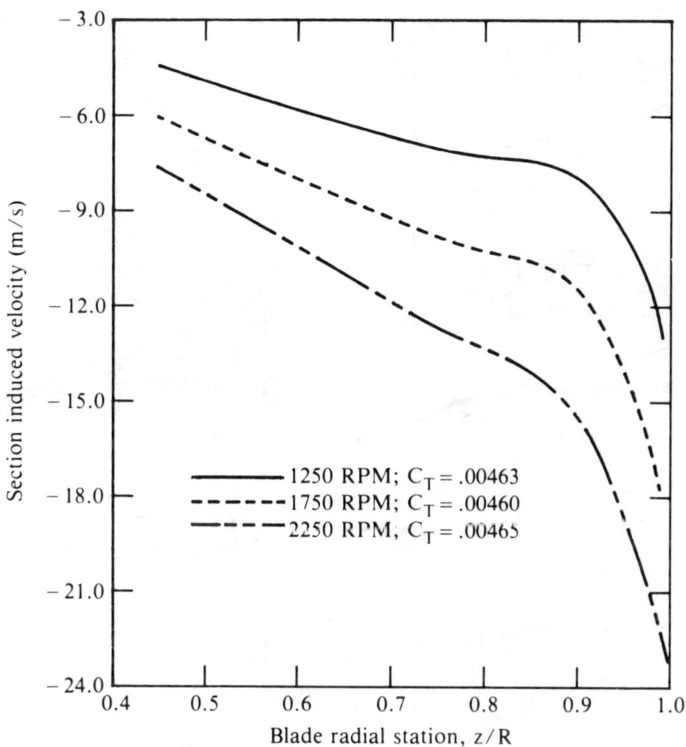

Fig. 3 Radial-induced velocity distributions on a two-bladed hovering rotor obtained with the free-wake analysis program HOVER of McDonnell Douglas Helicopter Company; untwisted, untapered, NACA 0012 blade, $AR = 6.0$, $\theta_c = 8$ deg, $R = 1.143$ m, $C = 0.1905$ m.

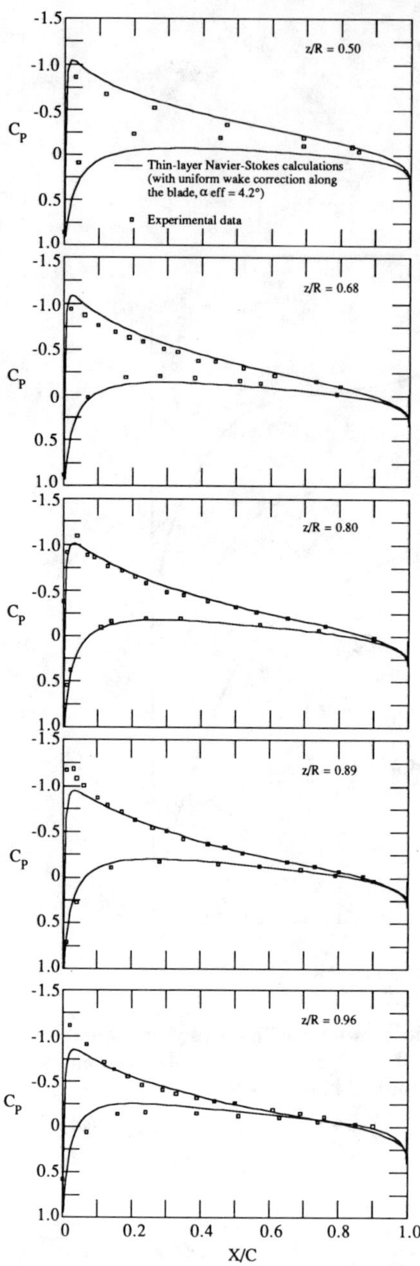

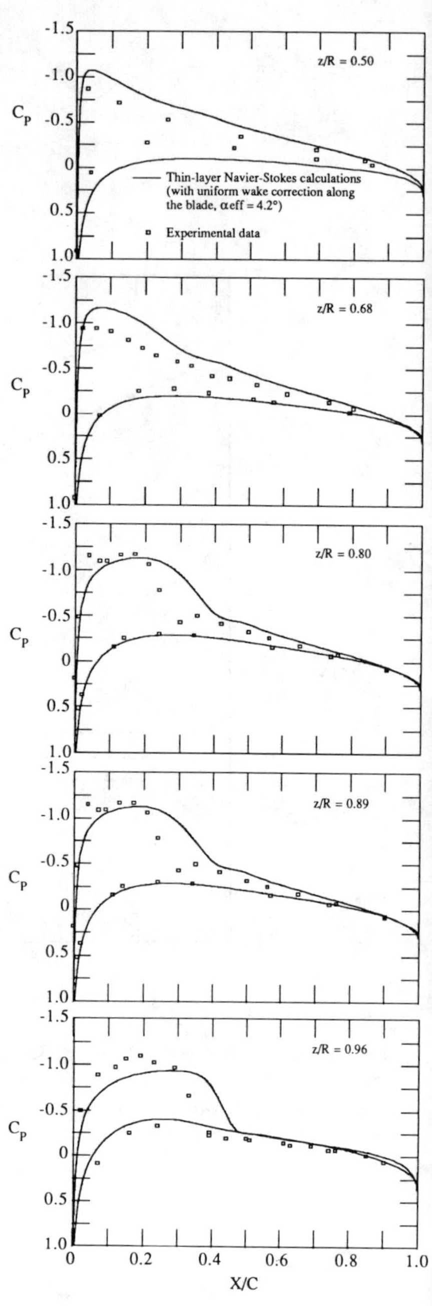

Fig. 4 Surface pressure distribution on a lifting rotor in hover; untwisted, untapered, NACA 0012 blade, $M_t = 0.44$, $Re_c = 2 \times 10^6$, $\theta_c = 8$ deg, AR = 6.0, $97 \times 33 \times 33$ mesh.

Fig. 5 Surface pressure distribution on a lifting rotor in hover; untwisted, untapered, NACA 0012 blade, $M_t = 0.877$, $Re_c = 4 \times 10^6$, $\theta_c = 8$ deg, AR = 6.0, $97 \times 33 \times 33$ mesh.

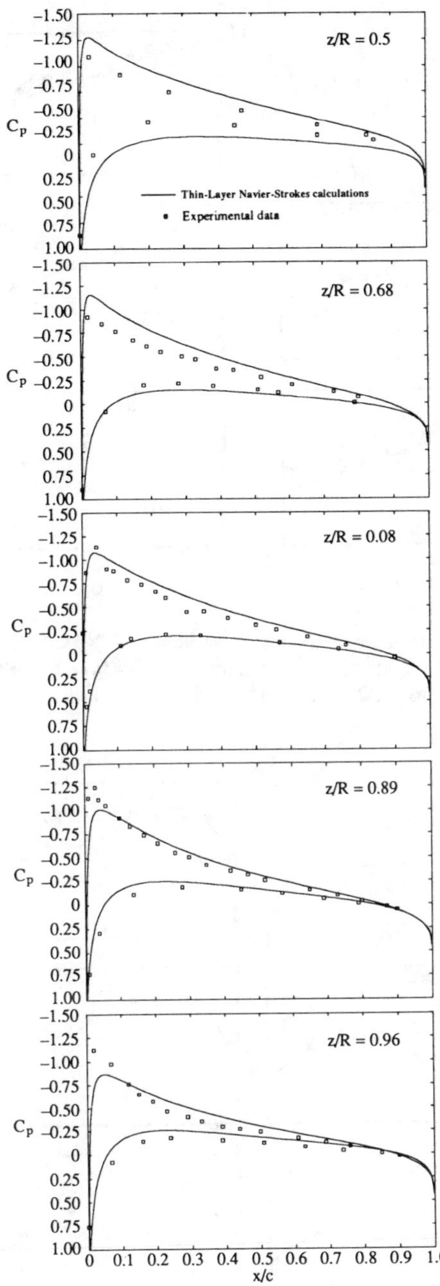

Fig. 6 Surface pressure distributions on a lifting rotor in hover; untwisted, untapered, NACA 0012 blade, $M_t = 0.612$, $Re_c = 2.67 \times 10^6$, $\theta_c = 8$ deg, AR $= 6.0$, $97 \times 33 \times 33$ mesh.

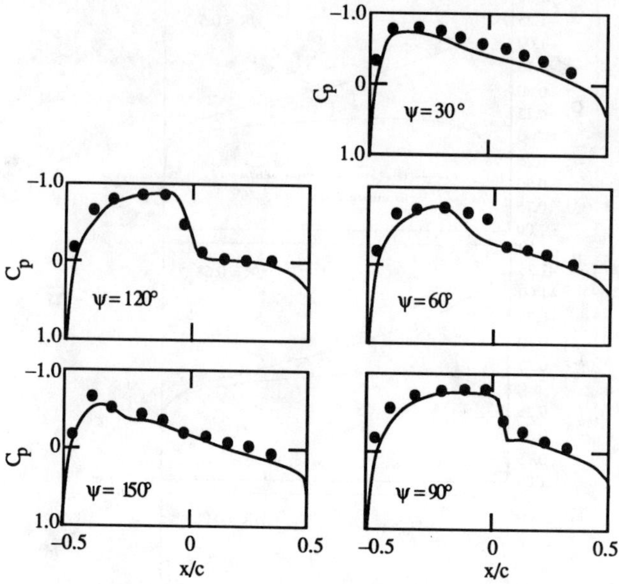

Fig. 7   Pressure distribution on a model rotor blade of NACA 0012 airfoil section and AR = 6 in forward flight; tip Mach number $M_t = 0.7$, advance ratio $\mu = 0.3$, collective pitch $\theta_c = 0$, 96 × 32 × 32 mesh. Solid lines, Euler calculations; circles, experimental.

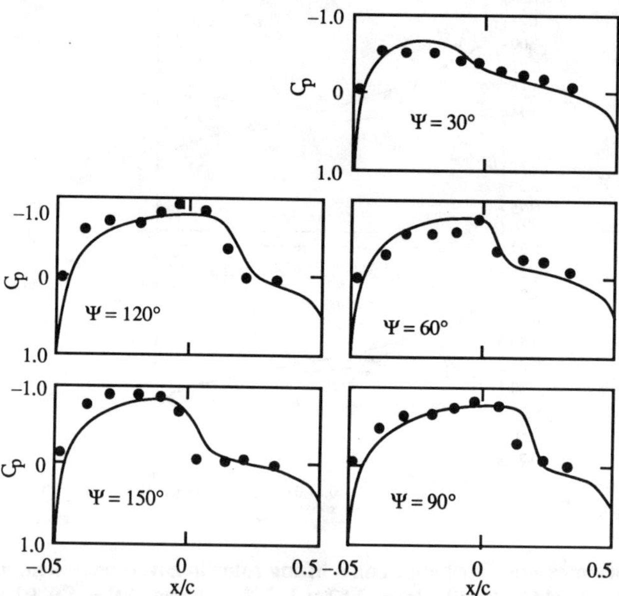

Fig. 8   Pressure distribution on a model rotor blade of NACA 0012 airfoil section and AR = 6 in forward flight; tip Mach number $M_t = 0.8$, advance ratio $\mu = 0.2$, 96 × 32 × 32 mesh. Symbols as in Fig. 7.

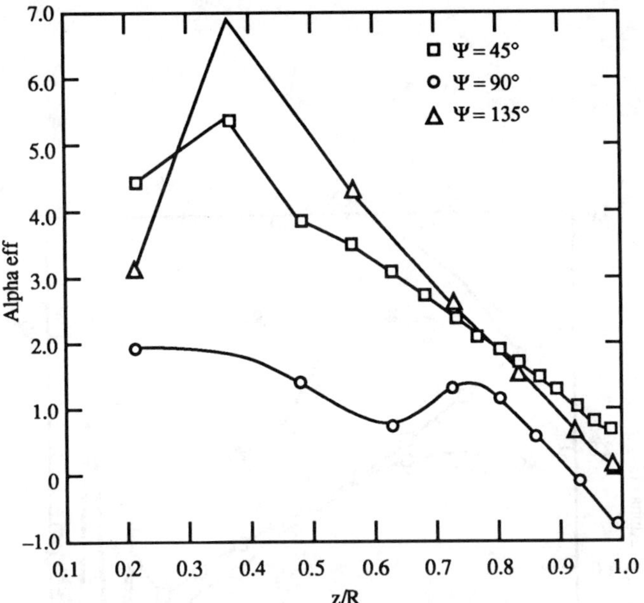

Fig. 9   Effective angle of attack along the OLS rotor blade at various azimuth angles computed using CAMRAD; tip Mach number $M_t = 0.633$, advance ratio $\mu = 0.298$, collective pitch $\theta_c = 0$.

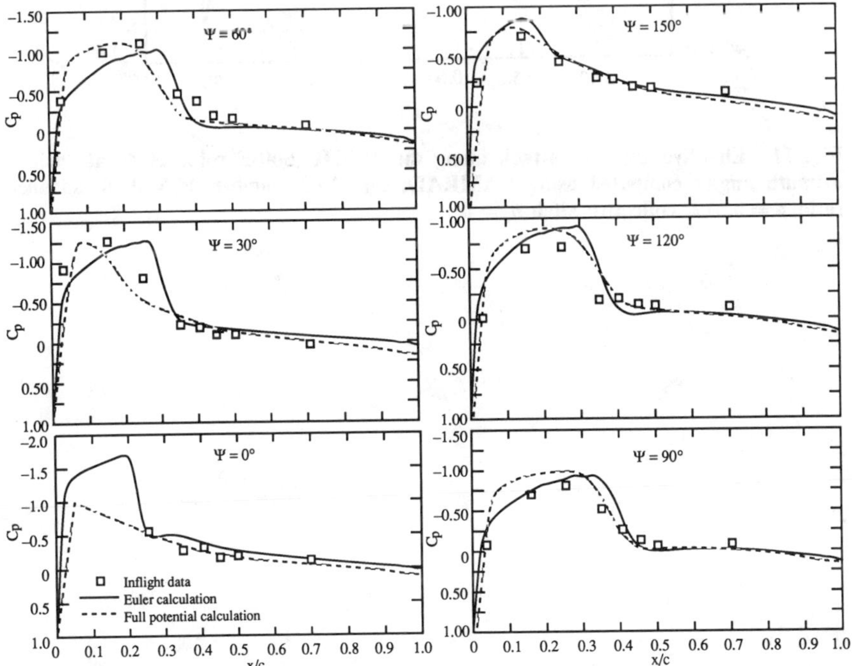

Fig. 10   Pressure distribution on the upper surface of the OLS rotor blade in forward flight at 95% span location; tip Mach number $M_t = 0.663$, advance ratio $\mu = 0.298$, collective pitch $\theta_c = 0$, $96 \times 32 \times 32$ mesh.

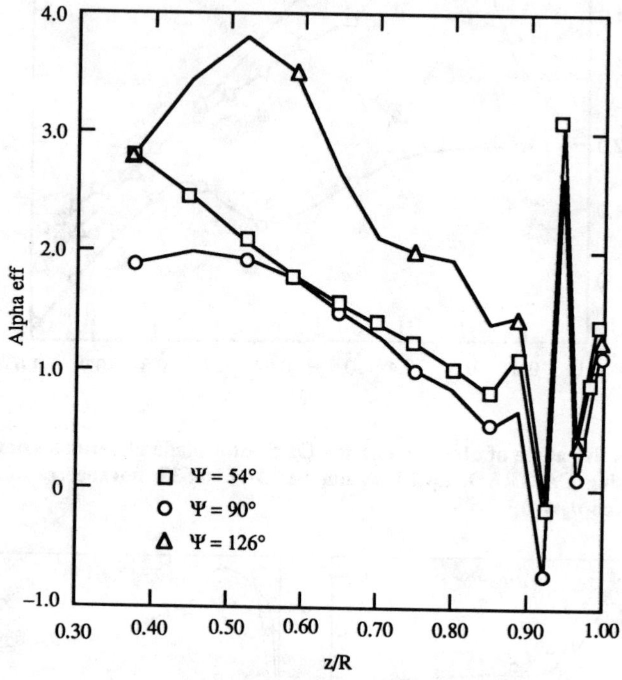

Fig. 11 Effective angle of attack along the MDHC 500-E rotor blade at various azimuth angles computed using CAMRAD; tip Mach number $M_t = 0.58$, advance ratio $\mu = 0.312$, collective pitch $\theta_c = 0$.

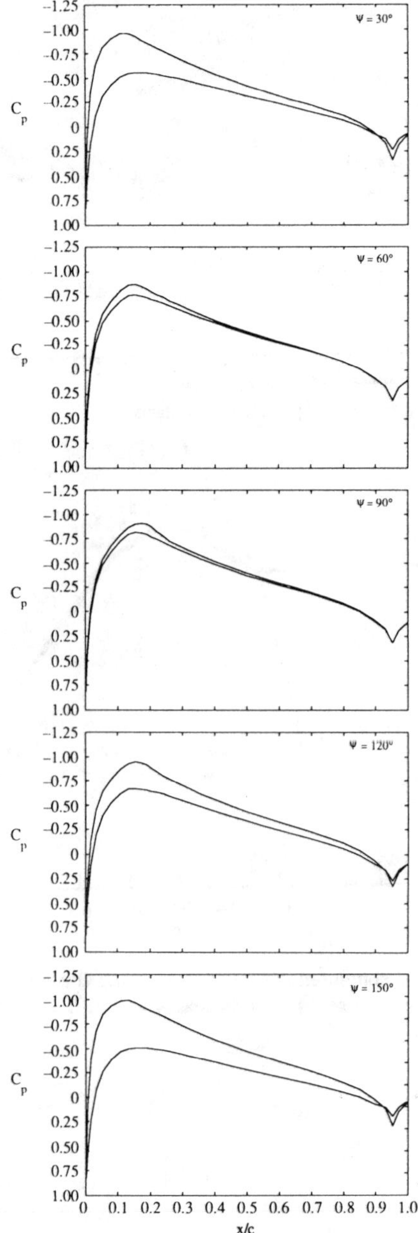

Fig. 12 Computed pressure distributions on upper and lower surfaces of the MDHC 500-E rotor blade in forward flight at 95% span location; $M_t = 0.58$, $\mu = 0.312$, $\theta_c = 1.2$ deg, $96 \times 32 \times 32$ mesh.

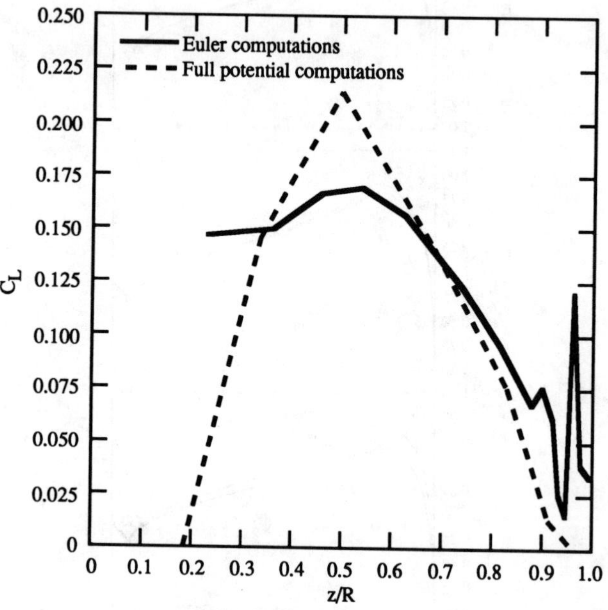

Fig. 13   Comparison of computed lift coefficient along the blade of a 500-E rotor at $\psi = 90$ deg azimuth location at $M_t = 0.58$ and $\mu = 0.312$.

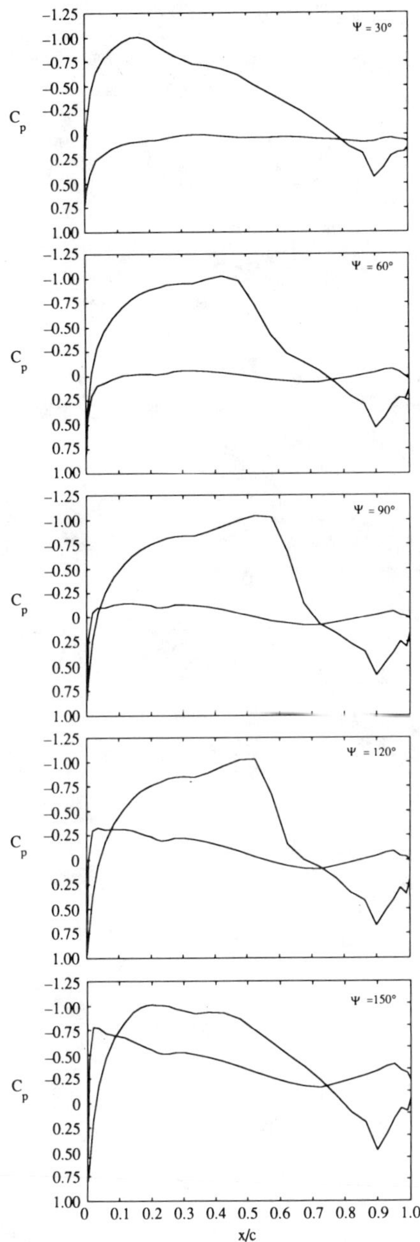

Fig. 14 Computed pressure distributions on upper and lower surface of the AH-64 rotor blade in forward flight at 95% span location; $M_t = 0.65$, $\mu = 0.326$, $96 \times 32 \times 32$ mesh.

Part 7.   Complex Configurations

Chapter 15

CFD Applications to Complex Configurations: A Survey

Edward N. Tinoco*
The Boeing Company, Seattle, Washington

Introduction

COMPUTATIONAL fluid dynamics (CFD) is playing an ever-increasing role in the design, analysis, and support of air vehicles.[1-4] The value of CFD in industry is in its application. CFD has joined the wind tunnel and flight test as a principal technology for aerodynamic design. CFD is used to lend understanding to the flow phenomena about a given geometry and to aid in the design of a piece of hardware, whether it be an all new wing or a minor modification to an existing configuration. Airplanes are flying today that may not have been deemed practical if not for the insight provided by CFD studies. On the basis of CFD analysis, modifications such as the addition of new engines, antennas, or sensor pods have been incorporated into existing airframes with little or no wind-tunnel testing before flight. CFD has also been used in lieu of flight testing to assist in the aircraft certification process. For many of these applications, it is no longer possible to focus the CFD analysis to a simple subset of the aircraft configuration, such as just the wing or the wing body.

Most aerodynamic configurations are geometrically complex. Commercial aircraft, for example, feature nacelles, struts, empennage, flap track fairings, flaps, slats, etc., in addition to wings and fuselages. The geometry of a military aircraft can be even more complex, featuring the preceding items plus a variety of external stores and sensors. The flow about these configurations may include shock waves, engine exhaust interactions, strong leading-edge expansions, leading-edge vortices, and significant areas of boundary-layer separation. Most of the aircraft geometry and an increasing level of the flow physics must be considered to produce useful engineering solutions. To aid in the design of such complex configurations, the engineering community needs computational tools that faithfully represent the dominant flow physics and the required geometric complexity and are capable of providing reliable solutions in a timely, affordable manner. This is a difficult and as yet unfilled order by any single method. The approach in industry has been to assemble a collection of CFD tools that meet the preceding requirements to varying degrees.

Copyright © 1989 by the American Institute of Aeronautics and Astronautics, Inc. All rights reserved.
*Principal Engineer, Computational Fluid Dynamics Laboratory.

In this collection are the "production" tools that are in wide use throughout the industry by a variety of CFD users. These are well-documented codes that have been specialized to a certain extent and can be run by the nonexpert CFD users. These codes are generally available through COSMIC, DOD, NASA, academia, or the individual developers and will run on computers ranging from high-end personal computers to the latest supercomputers. Just about all linear panel methods and most full-potential codes could be considered production tools. Some production codes include coupled boundary-layer solutions to account for some of the viscous effects.

There are "expert user" tools that may have more general and advanced capabilities than the production codes, but are not so well developed, require special skill to run successfully, and nearly always require a supercomputer. A common feature of expert user tools is the need for complex field grid generation. Most codes based on the Euler or Navier-Stokes formulation are classed as expert user codes.

CFD today covers a wide range of capabilities in terms of computational flow physics and geometric complexity. Figure 1 illustrates the boundary in terms of the complexity of flow physics and configuration geometry that encompass what we believe is practical in industry on a routine basis, and what has been demonstrated by the research community. We realize that these boundaries are continuously being challenged here and abroad. Solutions for complete aircraft configurations based on the thin-layer, Reynolds-averaged Navier-Stokes equations have been achieved by several researchers. However, just because this level of complexity in the formulation of the flow physics can now be considered, it does not necessarily follow that it should. Several factors must be considered when deciding what level of flow physics formulation should be applied to a specific problem. These not only include the level of physics necessary to adequately simulate the relevant flow phenomena, but also the level of accuracy and reliability of the numerical method, the resources available in terms of time and labor to set up the analysis, the computer resources required, and the level of confidence in the solutions.

The choice of what level of flow physics formulation to use on a particular problem requires the user to have some idea of what physics is relevant. For attached flows, inviscid solvers with coupled boundary-layer codes may be adequate for most problems. These methods will generally give a good indication if there are regions of separated flow. Inviscid solutions may be adequate for many applications if it is known that viscous effects are not significant. For some problems only the Navier-Stokes formulation can adequately represent the dominant physics. The level of flow physics is also influenced by the expected accuracy and reliability of the numerical method. Experience has shown that methods based on the linear potential and full-potential formulations can be routinely converged to a higher level of accuracy than has been practical with methods based on the Euler or Navier-Stokes formulation.

This chapter will focus on the application of CFD to complex configurations and will try to show not only what can be done with CFD analysis but also how CFD can influence the aircraft development and support process. This chapter will restrict itself to CFD applications that are in practical use today (1989). As such, the focus will mainly be on inviscid methods, although some examples will include an accounting of significant viscous effects.

General Aircraft Characteristics

Computational methods have been used to predict overall stability characteristics of complex aircraft configurations. Much of the applications to date have been based on linear panel methods such as A502/PAN AIR, VSAERO, QUADPAN, HISSS, and MACAERO.[5-13] The availability of panel methods, their ability to model complex configurations without getting involved in flowfield grid generation, and their relative economy favors the use of these methods for many applications. Most panel methods can solve for several flight conditions (i.e., angle of attack and sideslip) in a single analysis. Some modern linear panel methods, such as A502/PAN AIR, HISSS, and QUADPAN, can use the same paneling model for both subsonic and supersonic analysis. Agreement with experimental data is generally good except for conditions with significant transonic and/or viscous effects. Panel methods may frequently be used in the preliminary design phase, and, as the design of a configuration is refined, it may then become necessary to resort to the more complete computational methods for more detailed analysis.

Aerodynamic Characteristics

Use of linear panel methods dates back to the late 1960s. Panel methods with linearized boundary conditions[14,15] were used extensively for configuration development and aeroelastic analysis in the U.S. supersonic transport program. Figure 2 shows a comparison of computational results with experimental data across the Mach number range of 0.3–2.7. Excellent agreement with experimental data is seen for the prediction of lift-curve slope. The prediction of aerodynamic center shows the correct trend with Mach number but is too far aft. These methods worked relatively well for very slender configurations with thin wings. Overall lift distributions were well predicted, but detail pressure distributions were not done well.

Panel methods with exact "on the surface" boundary conditions improved the prediction of surface pressure distributions, but only slightly improved the prediction of overall forces.[16-18] An advanced fighter configuration for which comparisons with experimental data are available,[19] is shown in Fig. 3. Longitudinal characteristics comparisons for a Mach number range of 0.4–1.4 are also presented in Fig. 3. The computational results are from the HISSS code,[13] a higher-order panel method that is very similar to A502/PAN AIR.[5-7] The comparisons, which were made at a low angle of attack before the onset of leading-edge vortical flow, show good agreement between the computational results and experimental data. Semiempirical methods based on "suction analogy" can improve agreement with experiment for cases with leading-edge vortex flows.[20,21]

Lateral-directional stability characteristics have always proved to be difficult to predict computationally. A NASA study[22] found that A502/

PAN AIR generally provided accurate predictions at moderate angles of attack for a complete delta wing/body configuration with either single or twin vertical tails. Comparisons with experimental data for two of the vertical tail configurations tested are shown in Fig. 4. Comparisons are for yawing moment per unit sideslip, rolling moment per unit sideslip, and side force per unit sideslip as a function of angle of attack are shown for Mach numbers of 1.60 and 2.86. The agreement of the computed stability derivatives as a function of angle of attack with the experimental data is very good.

These types of calculations are normally used in the preliminary design stage to support the basic layout of the aircraft or to guide the incorporation of significant aerodynamic modifications. Lockheed used QUAD-PAN[11] to guide the placement of a large radome on a P-3 aircraft (Fig. 5). The objective was to position the radome so that the aircraft aerodynamic center was not changed. VSAERO[10] has been run on a high-end personal computer to support the design and analysis of innovative light aircraft,[23] (Fig. 6). MACAERO[8,9] was used to compute the effectiveness of leading- and trailing-edge flaps on an F-18-type fighter configuration. The resulting computed results are in good agreement with experimental data, as shown in Fig. 7. Linear panel methods are generally adequate for these types of analyses.

More complete analysis with higher-order methods (full-potential, Euler, Navier-Stokes) are certainly possible, but the greater level of resources required (man hours for grid generation, computer CPU, and memory) generally limits their use in the initial preliminary design phase. A notable exception may be codes that employ marching schemes for analysis of inviscid supersonic flows.[24-26] These codes solve the full-potential or the Euler equations on a series of cross-plane grids (usually constant x planes) marching downstream from the nose of the configuration. The grid-generation process is restricted to a series of two-dimensional cross planes, which is certainly more "doable" than a full three-dimensional surface-fitted grid generation. An example of the cross-sectional grids for the analysis of an advanced short takeoff and vertical landing (ASTOVL) configuration is illustrated in Fig. 8 (Ref. 26). A comparison of computational results from two linear methods and a full-potential marching code[24] with wind-tunnel data for a supersonic cruise interceptor is shown in Fig. 9. The results from the full-potential code are in substantially better agreement with the test data than are the results from the linear methods. Solution times for these nonlinear marching methods are on the order of a few minutes on a Cray XMP.

Control Effectiveness

As a configuration is further refined, computational methods may be used to supplement wind-tunnel tests in the investigation of more detailed aerodynamic characteristics. In a study to develop improved methods for predicting full-scale flight characteristics,[27] CFD predictions from several codes were compared with wind-tunnel- and flight-test-derived stabilizer and elevator effectiveness of the Boeing 767 and 757 transports. The two CFD codes evaluated were the A502/PAN AIR panel method and A488,[28-31] a full-potential transonic code coupled with a three-dimensional boundary-layer code on the lifting surface. Test-theory comparisons between the inviscid panel method and low-speed wind-tunnel results for the Boeing 767-200 are shown in Figs. 10 and 11. Although the results are quite reasonable, the effect of neglecting viscosity is very evident. CFD predictions of the stabilizer and elevator effectiveness are compared to wind-tunnel and flight-test values in Fig. 12. Again, the trends predicted by the CFD methods are reasonable, although the absolute level differs substantially from the flight-test and wind-tunnel results. When the viscous effects on the stabilizer are accounted for via a coupled three-dimensional boundary layer (viscous effects on the flow over the body are neglected), the CFD results agree better with the flight-test data.

The discrepancies between the wind-tunnel and flight-test data also can be seen in Fig. 12. Scale corrections based on historical data can be applied to the wind-tunnel results. However, those corrections are not nearly as large as the discrepancies found. As a result of this study, a prediction method for full-scale flight characteristics that utilizes both CFD and wind-tunnel techniques while avoiding their respective shortcomings was developed and is presented in Ref. 27. The method essentially reduces the classical problem of extrapolating wind-tunnel data to flight conditions to one of interpolating between wind-tunnel data and an upper bound established by inviscid CFD results. CFD predictions of the stabilizer and elevator effectiveness for a Boeing 757-200 transport are compared to wind-tunnel and flight-test values in Fig. 13. In addition, results from the interpolation between the CFD results and wind-tunnel data are also presented. The interpolated estimates are in very good agreement with the full-scale flight-test results. Similar agreement was also obtained for the interpolated results for the 767-200 transport. This example shows how CFD can be used to complement a wind-tunnel test to obtain a better estimate of full-scale flight characteristics. This same procedure can also be applied to improve estimates of aileron or rudder effectiveness.

Viscous Effects on General Configurations

In the previous section we have seen that we can use inviscid CFD methods together with subscale wind-tunnel results to improve preflight estimates of important aircraft stability and control characteristics. Practical engineering applications frequently demand that viscous effects be considered without the benefit of prior experiment, or to greater detail than can be accounted for by semiempirical interpolation schemes with test data. Solutions for complete aircraft configurations based on the thin-layer Reynolds-averaged Navier-Stokes equations have been achieved by several researchers.[32-36] Again, however, just because this level of complexity in the formulation of the flow physics can now be considered, it does not necessarily follow that it should. Applying the Navier-Stokes formulation to all problems is not only not practical today, but may not satisfy other necessary criteria such as timeliness, accuracy, reliability, availability of computing resources, and management confidence.

Inviscid/Coupled Boundary-Layer Methods

For attached flows, inviscid solvers with coupled boundary-layer codes may be adequate for most problems. The coupled boundary-layer/inviscid solver approach[37–40] offers a very practical way of dealing with viscous effects on complex configurations. There are significant limitations, though 1) boundary-layer calculations are generally limited to a portion of the configuration; 2) the calculations are not valid for corner flows; and 3) separated flows cannot be adequately modeled. However, these methods will generally give a good indication if there are regions of separated flow. For attached flows on high-aspect-ratio wings, viscous boundary-layer effects will be considerably more significant for transonic flows than for subcritical flow.

A comparison of computational results with experimental data on a wing-body configuration is shown on Figs. 14 and 15. Two sets of computational results are shown. One set of results is from a purely inviscid full-potential solution using FLO28.[29] The second set is from the same inviscid solver iteratively coupled to a three-dimensional finite-difference boundary-layer code, A488. The comparison on Fig. 14 is for Mach 0.70 where the flow is essentially subcritical everywhere. The agreement with experimental data is quite good, and except for improved agreement with experiment near the wing trailing edge, there is little difference between the computational results. Figure 15 presents a similar comparison on the same configuration at Mach 0.86 where there is considerable transonic flow. Here only the coupled computational results are in good agreement with the experimental data. For transonic flows the inviscid/viscous coupling is very necessary. The inviscid-alone results are in considerable disagreement with the experimental data and would not be adequate for wing-design work.

Coupled solutions have been applied to a variety of configurations. They are used extensively along with an inverse design method in wing design. (At Boeing, these methods have been used in excess of 2000 times a year for the past several years.) Figure 16 shows a comparison of the A488 coupled results with wing pressure distributions measured in flight on a Boeing 737-300. The computational model consisted of the wing, body, strut, and nacelle. The wing definition included the estimated aeroelastic twist for the condition flown. Although the character of the pressure distribution on the wing changes dramatically across the span, the computational results agree reasonably well with the measured data. This is just one of the several solutions calculated by the airplane support staff (not the research staff) in support of a flight loads program. The flight conditions analyzed ranged from subcritical to well beyond the normal cruise Mach number.

Another application of this method for the 737 was to get a quick assessment of the drag impact of an out-of-contour fairing that would be necessary if aluminum were used instead of titanium for the landing gear beam in the heavier weight 737-400 derivative. A quick decision on using aluminum instead of titanium saved considerable development flow time and money. A488, along with its inverse design counterpart, A555,[40] was used to design a modified leading-edge section for a hybrid laminar flow control (HLFC) flight experiment to be conducted as a joint NASA/Boeing experiment. A 17-ft spanwise leading-edge section ahead of the front wing spar was designed to provide a specific pressure distribution in flight for the experiment (Fig. 17). Adding to the complication of the design was the placement of the section starting just outboard of the engine strut nacelle location. This requires that the CFD method accurately account for the presence of the strut nacelle on the adjacent leading-edge wing pressure distribution. The design has been committed to flight without a prior high-speed wind-tunnel test. The experiment is scheduled to fly in 1990.

Figure 18 shows the results of another application of A488. Computational wing surface pressure distributions are compared to flight measured data for a Grumman F-14 with the wing set at the 20-deg sweep position. These analyses were done in support of a variable-sweep boundary-layer transition flight experiment.[41] With the exception of the innermost station, the computational results agree well with the measured data. The discrepancy at the most inboard station is believed to be due to the inability of the method, with its single block C-grid topology, to adequately represent the fuselage and engine inlet. The surface grid used for the analysis is shown in Fig. 19. The general cross section of the fuselage is represented, but the engine inlet has been gridded over. This example illustrates the need for methods capable of handling more general and complex geometries.

Coupled solutions (with an Euler solver)[42,43] have also been applied to wing-body-winglet development. The importance of viscous effects on winglet aerodynamic characteristics is shown in Fig. 20. A comparison of inviscid and coupled solutions with test data is for the outboard part of a wing and for the winglet. The boundary-layer solutions were carried out on both the wing and winglet during the solution inviscid/viscous iterations. The coupled results are in reasonable agreement with the test data, whereas the inviscid results greatly overpredict the loading on the winglet at this transonic condition. The correct loading on the outboard part of the wing and the winglet are essential to achieve the drag-reduction benefits of the winglet. If the winglet design and orientation had been based solely on the inviscid results, the proper loadings and anticipated drag reduction would not have been achieved. This same approach was used to evaluate three winglet designs in a two-week period prior to wind-tunnel model manufacture. Two of the three designs had acceptable transonic characteristics, but the third had to be modified. In the subsequent wind-tunnel test, all three winglets met their design objectives. This is an example of how the timely use of CFD can result in use of the wind tunnel for validation and performance testing rather than for cut-and-try development.

Although we have no direct comparisons of a coupled full-potential or Euler/boundary-layer solutions with thin-layer Reynolds-averaged Navier-Stokes solutions, we shall make the following observations. Although the full-potential A488 coupled code lacks some of the physics of a Navier-Stokes code, it can do a better job with what it has. The inviscid solver can be converged to greater accuracy than is practical with the Navier-Stokes method. The finite-difference three-dimensional boundary-layer code uses a much denser grid within the boundary layer (30–40 grids normal to the surface) than is practical with current Navier-Stokes solvers. There is a significant interaction between the shock and the boundary layer that neither the boundary-layer code alone nor the Navier-Stokes code can adequately handle. The A488 coupled method includes a semiempirical shock/boundary-layer interaction model[31] that significantly improves the solution and agreement with test data. For attached flows on wings of transport type and similar configurations for which the boundary-layer theory assumptions are valid, coupled solutions will continue to offer the preferred solution approach for most problems.

Parabolized Navier-Stokes

As useful as the coupled solutions are for some circumstances, there are a variety of configurations and flow conditions for which they are not. The flows over many fighter-type configurations do not lend themselves to the coupled inviscid/boundary-layer methods. Navier-Stokes solutions over complete fighter-type configurations have been calculated with varying degrees of success and at great cost in terms of grid-generation flow time and computer resources.[33,34,36] One approach that can be applied in supersonic flows which can handle complex geometries and account for some of the dominant viscous effects and is practical in terms of grid-generation flow time and computer resources is the parabolized Navier-Stokes (PNS) method.[44] This method is based on the steady thin-layer Reynolds-averaged Navier-Stokes equations with the approximation that the pressure across the subsonic portion of the boundary layer is constant. The resulting equations can be marched in the downsteam direction when the inviscid portion of the flow is supersonic. The grid-generation problem is reduced to a series of two-dimensional cross planes, as in the case of the inviscid codes previously mentioned. This leads to a significant reduction in the computing resources and, even more importantly, reduces the grid generation from a complex three-dimensional problem to a series of simpler two-dimensional problems.

The PNS method was applied to the model 350 supersonic fighter configuration shown in Fig. 21.[45,46] A comparison of computed results with experimental data is shown in Fig. 22 for the basic wing-body configuration for a supersonic high-lift condition. The agreement with the test data is quite good for this extreme condition. A comparison that shows the influence of the canard is shown in Fig. 23. A solution was also carried out including the underwing engine nacelles. These comparisons with limited experimental data are shown in Fig. 24. Calculated pressure distributions are shown for the configuration with and without the underwing nacelles. Again the agreement with the test data is good. The PNS method is a powerful tool for the supersonic design of advanced supersonic configurations. The grid generation is simplified to a series of two-dimensional problems, and the computing and memory requirements, though moderate compared to solving the Navier-Stokes equations without the parabolic assumption, are still formidable, requiring 4–5 h on a Cray X-MP.

Wakes and Flowfields

Detailed knowledge of the flowfields surrounding a configuration is of interest in the design of an aircraft. Of particular interest are flowfield effects on engine inlets; the effect of the wing trailing wake on the stability and control characteristics of aircraft and on any aft-mounted propulsion system; and flowfield effects on nearby objects, i.e., weapon separation, or the expulsion of any other object from an aircraft.

For most aircraft configurations with high-aspect-ratio wings at moderate angles of attack, boundary-layer flow separation and the resultant vortex or wake shedding are confined to the wing trailing edge. The shed wake is convected downstream and tends to roll up into two or more vortex cores. The details of the wake generally have only a minor influence on the generating surface from which they were shed, but can have significant effects on trailing components. In potential-flow methods, both in linear panel methods and in nonlinear full-potential methods, the wake position is specified as part of the problem definition. The wake imparts a potential jump across its surface and, in essence, satisfies the requirement that the pressure jump across its surface be zero, usually only to first order. A wake is not impermeable. Ideally, it should be aligned with the local flow, but, because it is specified as part of the problem definition, it rarely is. Some computational methods[10,47,48] attempt to iterate or relax the wake to satisfy both the requirements that the pressure jump across the wake be zero and that the wake be aligned with the local flow. In practice, these methods have been successful in treating only a limited range of configuration geometries. Wake relaxation is an inherently unstable process. In our opinion, a truly robust potential-flow analysis code for complex geometries capable of automatically relaxing the wake does not exist yet.

In the Euler or Navier-Stokes formulations the shed wake becomes a natural part of the solution. In these methods the wakes are naturally convected downstream. However, this does not automatically guarantee a better solution than the simplified approach taken in potential methods. Dissipation and lack of adequate grid resolution in the flowfield may result in a less accurate simulation of wake and downwash effects.

747-300 Flight Comparison

Several examples of the ability of potential methods to adequately simulate the trailing wake behind a complex aircraft configuration follow.[49] The first example does not present direct comparisons between calculated and experimental results, but does illustrate the ability of panel methods to produce solutions that look right. Photographic data were found for a 747-200 in a landing flaps configuration. These data were acquired as part of NASA's investigation into trailing-wake turbulence.[50] Streamlines were calculated from the wing and flap trailing edges of an existing panel model for a 747-300 at the same flap setting. (The only significant external aerodynamic difference between a 747-200 and a 747-300 is the length of the upper deck cab.) Figure 25 shows a comparison of streamline shapes from A502/PAN AIR for the 747-300 flaps 30 model, with flight-test vortex imaging results in which smoke bombs were mounted on the outboard flap edges and on the wing tip. The computed results and the flight-test photographic data agree remarkably well. The lack of the inboard wake constriction is thought to be related to the lack of the horizontal tail and engines in the panel model.

737-300 Wind-Tunnel Comparison

The second example features A502/PAN AIR flowfield results from a 737-300 wing-body-strut-nacelle model with takeoff flaps (15 deg) compared with five-hole probe wind-tunnel data. The A502/PAN AIR model, available from a previous investigation[51] (see Chapter 10, Brune and McMasters). The model featured a slotted leading-edge slat, simplified single-slotted trailing-edge flaps, and a flow-through nacelle. Figure 26 (see also Plate 10 in the color section) shows the panel geometry, the computed streamlines, and the location of the off-body survey plane at which the wind-tunnel measurements were taken. The circles shown on the survey network indicate the intersection of the computed streamlines with the survey plane. Figure 27 shows the total pressure contours at the survey plane measured with a five-hole pressure probe. The dark circles shown on the plot are the computed streamline intersections with this survey plane. The agreement between the computed results and the experimental measurements is good considering the complexity of the configuration. The biggest discrepancy appears to be in the region aft of the engine. The modeling of the actual geometry may have been oversimplified in this region. The computational model has a gap between the inboard and outboard trailing-edge flaps. In the airplane and in the wind-tunnel model, the aft portion of the combination nacelle strut and flap track fairing eliminates the gap and seals against the flap ends. The loss of lift in the computational model in that region results in the trailing streamlines not being deflected down as much as measured in the wind tunnel.

727-100 GE/Boeing UDF Demonstrator

A practical application of CFD calculations of wake and flowfield effects was realized shortly after the Boeing/General Electric 727-100 UDF demonstrator airplane, shown in Fig. 28, first flew. Initial flight tests revealed a strong vibration as the UDF power levels were increased. An initial suspicion was that the wing trailing wake was being deflected into the propeller disk by engine power effects. To evaluate the feasibility of this hypothesis, a very simple propeller model was implemented with the 727-100 configuration in an A502/PAN AIR panel method analysis. The propeller model was essentially a disk with a specified uniform inflow related to an intentionally high thrust coefficient based on momentum theory. The calculated streamlines showed that the wake deflections were small and remained away from the propeller disk. Further confidence in the calculations is provided in Fig. 29, in which streamline locations predicted by the panel method for the power-off case are compared to the wing wake image from a total pressure survey obtained in the wind tunnel with the same configuration.

These studies showed that, although increasing the power of the UDF did deflect the wing wake upward toward the propeller disk, the changes were small, and the resultant wake position was still below the propellers. This study provided a better understanding of the flowfield and allowed engineers to eliminate the idea of the wing wake hitting the propfan blades as a source of the problem. Once engineers were free to focus on the actual cause of the vibration, it was found to be related to the resonance frequency of the forward and aft propellers of the counter-rotating propfan. These CFD results were of value because they were available within a very few days to the engineers confronted with the vibration problem.

Aft Propfan Aircraft Body Design

Another application that made use of the knowledge of the flowfield around the aft portion of an aircraft fuselage came during the development of a commercial transport configuration with aft-mounted propfans.[2] The task was to install a propfan engine on the aft body of an all new medium-sized transport. Prior to any powered wind-tunnel testing, CFD was used to developed the normal details of a body-mounted engine. Although there were locally supersonic flows to deal with at the prescribed freestream Mach of 0.80, the installation of the unpowered nacelle was accomplished without much difficulty using the linear potential A502/PAN AIR method.

The powered installation proved to be a much more formidable task. In the initial transonic wind-tunnel testing, significant installation losses occurred. In addition, significant dynamics occurred, and the propfan blades of the powered model simulator began to break from fatigue. An A502/PAN AIR analysis of the flowfield indicated that significant excursions in flow angularity were being presented to the propfan blades. It was theorized that this nonuniform onset flow was causing the dynamics, blade fatigue, and possibly even the performance losses. The aft body was redesigned to reduce the nonuniformity of the flowfield at the propfan station. Figure 30 compares the initial and final geometries and the flowfields obtained by this CFD analysis/design exercise. Subsequent wind-tunnel testing of the two geometries demonstrated that the design change solved all the problems experienced by the initial configuration. Measurement of the flowfield with a rake of five-hole flow angularity probes confirmed that the CFD predicted changes due to body geometry did, in fact, occur.

CFD provided the level of detail necessary for the design engineers to explore the flowfield, find a potential problem, and then fix it. Whether the solution would have been found solely on the basis of exhaustive wind-tunnel testing is not clear, but the fact that CFD was able to identify the problem and lead to a solution in only one design iteration is impressive.

Drain Mast Certification

CFD calculations of the flowfield around the drain masts on the 767-300 and 737-400 were used in place of flight test during the FAA certification of these aircraft. The drain masts in question are the efflux of "gray" water from the aircraft lavatories. Gray water is the used water from the hand wash basins. This water can be dumped overboard during flight, but must not impinge on the aircraft surface where it could freeze into a chunk of ice. Certification procedures usually require that a colored dye be expelled from the drain masts while a chase airplane observes whether the fluid impinges on the aircraft surface. Both the 767-300, shown in Fig. 31, and the 737-400 are longer fuselage derivatives of existing aircraft that had been certified using the flight-test procedure. CFD results from existing computational models of the aircraft in question were used to show that, although the fuselage lengths had changed, the flowfields in the vicinity of the drain masts were similar enough to the certified aircraft to justify certification without an additional flight test.

Airbus A320 Ground Effect Testing

Flowfield calculations about an Airbus A320 in landing configuration, shown in Fig. 32, were used to adjust and check experimental flow conditions of landing tests conducted in the ONERA S1MA wind tunnel.[52] A 4728-panel model (per plane of symmetry) was used to calculate the inviscid streamlines on the ground plane. Blowing through a slit in the wind-tunnel floor is used to simulate the ground effect in landing. The blowing was adjusted so that the streamline directions visualized by small flags just outside the wind-tunnel floor boundary layer coincided with those of the calculation. This application is just another example of the complementary nature of CFD and the wind tunnel.

Weapons Store Behavior in Complex Aircraft Flowfields

It has been demonstrated that the linear panel method such as A502/PAN AIR can predict the behavior of a weapon store in the presence of a complex aircraft configuration at both subsonic and supersonic speeds.[17,53,54] At subsonic speeds the predictions of store forces and moments are generally excellent, whereas at transonic and supersonic speeds the forces and moments will be less accurate in the presence of strong shocks. These types of calculations have been carried out on a variety of weapons, including those shown in Fig. 33. An example of the accuracy that can be expected from the prediction of normal force and pitching moment on a weapon traversing beneath the carriage aircraft is taken from Ref. 17. A transverse of the planar wing weapon (PWW) along a waterline just below the fuselage of a Grumman supersonic tactical aircraft configuration (STAC) provided a relatively severe test of a linear method to predict the forces and moments of the weapon in close proximity to the delivery aircraft. The data shown are for a freestream Mach number of 1.95. Normal force and pitching moment on the PWW along the traverse are presented in Fig. 34. The computed results are in reasonable agreement with the wind-tunnel measurements. One obvious discrepancy is the apparent phase shift in the force and moment predictions in the region where the inlet shock intersects the traverse. This shift results because linear theory assumes that flow disturbances propagate along straight Mach waves, whereas the physical process involves finite shock waves that are less inclined to the freestream. Nonlinear full-potential, Euler, or Navier-Stokes methods do not share this limitation. However, these methods generally present formidable challenges in the field grid generation that is required for solution. An exception would be the full-potential finite-element TRANAIR method[55-57] (also see Chapter 19, Madson and Erickson). The panel method type of user interface allows the analysis about the most complex configurations (Fig. 35, for example) without the attendant grid-generation problems. Other approaches include use of the CHIMERA scheme of embedded, overlapping surface-fitted grids along with an Euler solver[58] and use of embedded Cartesian and cylindrical grids with a refined transonic small-disturbance solver.[59]

Engine Airframe Integration

One area of aircraft development that has benefitted greatly from CFD application has been that of engine/airframe integration on transport configurations. The quest for improved propulsive efficiencies for transport aircraft has led first to the development of large-diameter, high-bypass-ratio turbofan engines and more recently to unducted-fan, or propfan, engines. The installation of these engines can lead to large aerodynamic interference penalties if sufficient care is not taken in the design and development of these installations. This chapter will focus on the general installation effects, including jet exhaust and propfan power effects. Engine inlet design and the calculation will not be covered. References 60–63 discuss some of the aspects of CFD application to inlet design and flowfields.

Turbofan Integration

Early implementations of high-bypass-ratio engines, C-5A, B-747, DC-10, L-1011, A300, and even more recent designs have placed the wing-mounted engines on pylons well away from the wing to avoid excessive installation drag penalties. Empirical attempts to develop new low-drag nacelle installations closer to the wing have not been too successful, largely because of the lack of detailed understanding of the nature of the interference-drag-producing phenomena. A coordinated computational and wind-tunnel study in the mid-1970s[37,64] led to the understanding of the impact of nacelle integration on induced drag and the need to avoid or minimize supercritical flow on the wing lower surface in the region of the nacelle/strut. With the aid of this understanding, guidance from CFD, and refinement in the wind tunnel, the new nacelle installations illustrated in Fig. 36 were developed.

Exhaust Jet Modeling

The computational representation of an engine installation must include some representation of the jet exhaust or plume. Jet exhausts usually feature higher total pressures and temperatures than freestream, mixed subsonic, and supersonic regions, possibly having strong interactions with the boundary layers, mixing layers, and regions of separated flow. Such flows are properly the subject of analysis using the Navier-Stokes formulation. Unfortunately, it is not yet practical or desirable to use these more complete methods in the analysis of complex wing/body/nacelle configurations on a routine basis.

The Euler and Navier-Stokes equations have been shown to be sufficient for modeling the physics of most exhaust flows.[65,66] For many installations viscous effects can be ignored, and an exhaust simulation based on the potential or Euler formulation can be used. The potential-flow formulation is inherently limited to regions of a constant total temperature and pressure and at first glance does not seem well suited for modeling exhaust flows. The simplest approach to exhaust flow modeling in the past has been to treat the exhaust plume as a solid body as illustrated in Fig. 37.[64] The shape of the exhaust plume can be chosen to maintain a constant exit area, or derived from a more complete isolated axisymmetric calculation such as Navier-Stokes, Euler, or a multistream tube curvature method. This model has been previously used successfully in panel methods[67,68] and in full-potential transonic analysis.[68,69] A major disadvantage of this model is that the solution is only as good as the choice of the solid plume shape, and the exhaust plume geometry must change to reflect changes in the exhaust fan nozzle pressure ratio or core cowl geometry. Transpiration velocities can be specified on the plume to try to simulate pressure ratio changes. A recent approach updates the solid plume shape by iterating between the three-dimensional installed analysis and an axisymmetric solution for the exhaust plume that is a function of the static conditions around the circumference of the installed nacelle exhaust nozzle exit.[70]

Another approach to potential exhaust plume modeling is also illustrated in Fig. 37. Here the exhaust plume is modeled as a wake rather than a hard surface. This also allows the geometry internal to the exhaust to be modeled. In a typical high-bypass-ratio turbofan engine, this would include the core cowl and a portion of the fan duct. Since the exhaust plume boundary is modeled as a wake, with a boundary condition of zero pressure jump, its position is not as critical to the solution as that of a solid surface. It is usually adequate to have the wake maintain a fan flow having a constant cross-sectional area. Thrust effects are modeled by treating the flow inside the exhaust plume as if it were operating at a different total pressure than freestream. Low-order panel methods generally specify a linear variation in potential as a function of the exhaust pressure ratio in the downstream direction of the exhaust plume surface, which is equivalent to satisfying the zero static pressure jump condition across the plume surface to first order.

A higher-order panel method, A502/PAN AIR, and the full-potential TRANAIR method have a somewhat more sophisticated implementation, satisfying the zero static pressure jump condition across the plume wake surface to the second-order pressure relationship in A502/PAN AIR or the isentropic pressure relationship in TRANAIR and allowing both the total pressure and total temperature within the exhaust plume to be specified. Figure 38 shows the pressure distribution on the fan cowl and core cowl surface of a turbofan type engine as calculated by the linear potential A502/PAN AIR and by an Euler code. The results shown are for two specified fan-nozzle pressure ratios, one for which the exhaust total pressure is equal to freestream total, the other for a higher exhaust total pressure that results in a flow that is still subcritical. The linear potential results are essentially the same as those from the Euler code. For a supercritical exhaust the linear potential methods would not be able to model the transonic character of the flow. However, a full-potential method can capture the transonic character of a supercritical flow. Results from TRANAIR are compared to those from a Navier-Stokes method, PARC2D,[71] and to experimental data in Fig. 39. The data are from a static exhaust nozzle test. The Navier-Stokes and TRANAIR results match the experimental test data reasonably well. The TRANAIR solution took advantage of an adaptive grid-refinement technique to adequately capture the multiple shock waves in the fan nozzle flow over the core cowl. The level of the pressure peaks matches that from the experiment for the multiple shocks over the core cowl. The spacing of the multiple shocks is not predicted as well, possibly due to the lack of precise knowledge of the boundary conditions from the experiment.

Installed Turbofan Analysis with Potential Methods

All of the early studies and most of the computational work, which was responsible for the successful designs illustrated in Fig. 36, were based on subcritical panel methods. Subcritical methods were the first to possess the geometric fidelity to accurately represent the complete wing/body/strut/nacelle configuration. These methods could assess the impact of the installation on induced drag and could give an indication of the presence of supercritical flow, but could not provide quantitative details of the supercritical flow. References 37 and 64 illustrate the use of a first-order panel method in determining the interference drag mechanisms of a wing-strut-mounted nacelle and some of the concepts of designing installations with low interference drag. Further applications of linear panel methods to engine/airframe integration can be found in Refs. 67, 68 and 72.

Subsequent developments in CFD have led to methods that possess the geometric fidelity to accurately represent the complex geometry and to resolve the details of the supercritical flow that might be present in the region of the engine installation. Transonic small-disturbance (TSD) methods can indicate the correct trends, but inherently lack the required geometric precision for underwing-strut-mounted nacelles through their use of linearized wing boundary conditions.[38,73,74] However, through the use of multiple embedded grids (Fig. 40), TSD methods have been used successfully in the analysis of modern business jets with body-aft-mounted engines.[75,76]

Full-potential methods, such as the A488 finite-volume analysis method with coupled boundary layer on the wing, have been used for the analysis of a variety of transport configurations with underwing-strut-mounted nacelles. A key feature in the usability of A488 for this type of analysis is the "black box" elliptic grid-generation method used to generate a surface-fitted C-type grid.[77,78] For a general strut/nacelle installation, it is usually difficult to produce an exact surface-fitted grid with smooth, well-distributed mesh spacing in the field. The approach taken in A488 involves a simplification of the nacelle inlet geometry and a relaxation of the requirement that the grid lines lie along the corners formed by the nacelle/strut intersection and along the nacelle keel line, as illustrated in Fig. 41. The tight clustering of grid lines close to the wing allows adequate grid resolution in the region between the wing lower surface and the nacelle. References 68 and 69 detail some of the applications of the method and illustrate the use of the solid exhaust plume modeling technique in simulating engine installation details such as engine primary core cowl geometry effects and engine power effects. A comparison of results from A488 and flight surface pressure measurements was previously shown in Fig. 16.

The complexities of surface-fitted grid generation can be avoided through the use of Cartesian grid methods. Reyhner's approach[79] features a finite-difference full-potential method on a Cartesian grid. The program uses a sequence of grids of increasing density and a multigrid strategy to improve the efficiency of the solution. Figure 42 shows the grid cross section through a nacelle installation. Dense grid spacing is maintained in the region of primary interest and then stretched to the far-field boundaries. The exhaust plume is modeled as a solid boundary, whose shape can be iteratively adjusted in response to the external pressures imposed by the surrounding airframe.[70] Figures 43 and 44 show a comparison of computed results with wind-tunnel data on a NASA high wing transport configuration with underwing-strut-mounted nacelles. The wing surface pressure distribution data shown on Fig. 43 are for a wing buttline just inboard of the nacelle strut. The comparisons presented on Fig. 44 are for the inboard and outboard sides of the installed nacelle. The agreement with test data is very good considering the lack of viscous modeling the potential solution.

Although still in the developmental stage (early 1989), the finite-element full-potential TRANAIR code[55-57] is already seeing use in turbofan engine installation studies. The ability to handle exhaust streams of different total pressures and temperatures and the panel-method-like configuration definition with almost automatic field grid generation make it a usable method for analyzing transonic flows, including engine exhaust plumes, over complex configurations. TRANAIR uses the same configuration input format as that of the A502/PAN AIR panel method. It can and has used the same computational models for analysis. However, since the nature of transonic flow makes it much more sensitive to both surface curvature and slope than a subcritical flow, a typical TRANAIR model will use a significantly greater number of panels for the surface discretization than a typical panel method. Twin engine configurations have featured 10,000–15,000 panels, with up to 24,000 panels for a four-engine configuration. Figures 45–47 illustrate the level of geometric detail possible in the configuration descriptions. Figure 48 illustrates the refined field grid in the region of the strut nacelle of a twin engine aircraft. The program automatically refines the grid near the configuration surface and in regions of specified interest. The user need only define a few parameters that control the levels of refinement in each specified region. Automatic flowfield refinement has recently been implemented into the code, but was not sufficiently checked out enough to be used for any of the following cases presented here.

As part of the TRANAIR validation, comparisons with experimental data were made on several configurations with installed underwing turbofan nacelles.[80,81] A comparison of wing surface pressure distributions on the 747-200 configuration shown in Fig. 45 is presented in Fig. 49. The computational representation included a complete modeling of the flow-through nacelles used on the wind-tunnel model. A low supercritical Mach number, $M = 0.80$, was chosen for the comparisons to minimize the effects of the viscous interactions in the real flow. The comparisons are along wing pressure rows just inboard and outboard of the two nacelle locations. As reference, test data are also presented for the configuration without the struts and nacelles. The computational results are in good agreement with the experimental data. The lack of viscous modeling in the computational results is most evident in the trailing-edge recovery pressures. The agreement with experimental data at other wing stations was equally as good.[80]

An analysis was also made of the 737-400 twin engine transport, shown in Fig. 46. Figure 50 shows TRANAIR results for the 737-400 compared to experimental flight data on the wing surface from a 737-300. The only significant external difference between the two aircraft is the body length. The level of agreement is surprising considering that the TRANAIR solution is inviscid. The area of greatest discrepancy was near the leading edge just outboard of the nacelle. However, there is some doubt as to the quality of the experimental measurements in this area. The flight measurements could have been contaminated by possible problems with the leading-edge slat sealing to the wing, or by problems with the strip-a-tubing used for the measurements. Figure 51 shows TRANAIR results compared to results from the full-potential A488 code and with experimental flight data on the wing surface just inboard of the nacelle strut. TRANAIR solution is shown in Plate 9 (see the color section). The A488 results, which include viscous coupling and which are in good general agreement with flight-test measurements across the span of the wing, previously seen in Fig. 16, do not match the flight data as well as the TRANAIR results do next to the strut nacelle. The grid topology of the A488 code did not allow the detail definition of the nacelle strut that extends over the leading edge of the wing. Furthermore, the exhaust plume in the A488 analysis was modeled as a solid surface, rather than as an exhaust stream. The benefits of the improved geometric capability and exhaust modeling in TRANAIR are clearly illustrated in this comparison.

The exhaust plume modeling in TRANAIR also yields a solution inside the exhaust stream. The local Mach number distribution at two circumferential positions on a core cowl of an installed short-duct turbofan nacelle are shown in Fig. 52. The distributions are for both cruise thrust and idle conditions. The influence of the nearby airframe on this installation, although not severe, is clearly evident in these results. TRANAIR has been used for the analysis of a variety of engine/airframe installations, including both short- and long-duct turbofan nacelles. In one instance, the wind-tunnel test of a proposed installation was cancelled on the basis of the TRANAIR analysis so that a better installation could be developed.[81]

Installed Turbofan Analysis with a Euler Code

Installed engine/airframe analysis has also been carried out using codes based on the Euler formulation. A twin engine transport with underwing nacelle (Fig. 53) was analyzed with both TRANAIR and a multiblock, surface-fitted grid, Euler code.[42,82,83] The Euler code is a refinement of the finite-volume Jameson FLO57[84] and is similar to the TEAM code.[85,86] The surface discretization for the twin engine transport Euler solution was identical to that used for the TRANAIR analysis with the exception that the fuselage was not included. The field grid was made up of 27 blocks, as illustrated in Fig. 54. A portion of the grid in the region of the strut is shown in Fig. 55. A total of 600,000 field grid points were used for the solution. A comparison of the results from the Euler code with those from TRANAIR in the region of the strut nacelle show them to be essentially identical (Fig. 56). The run costs for the two codes are similar; however, the surface-fitted field grid generation for the Euler code took 2–3 months!

Transports with engines mounted on the upper surface by the wing, utilizing upper-surface blowing for high lift during landing and takeoff, have been analyzed by several researchers[87–89] who used finite-volume Euler codes that require a structured surface-fitted grid. Again, the grid-generation process for these complex geometries may take weeks or even months of preparation. Reference 89 presents results for the ASKA, a Japanese research aircraft developed by the National Aerospace Laboratory (NAL). The surface gridding for the ASKA solution is shown in Fig. 57. Results from this calculation are shown in Fig. 58 compared to wind-tunnel data on a model with turbopowered engine simulators.

Propfan Installations

The significant reduction in fuel consumption predicted for propfan propulsion systems has stimulated interest in high-subsonic propfan airplanes. To understand propfan engine and nacelle installation effects on airplane aerodynamic performance, computational methods that can simulate a wing/body/nacelle/propeller flowfield, including power effects, are needed.

Propfan Analysis with Potential Methods

The ability of panel methods to predict the flowfields about complex configurations and their relative ease of use has led to their use in the development of aft-mounted propfan aircraft. The use of a panel method to guide the recontouring of the aft fuselage of one aft propfan configuration has already been discussed in the section on wakes and flowfields. Improved propfan modeling[90] has been combined with a general panel method to study low-speed flight conditions at high power loadings. Figures 59 and 60, from Ref. 91, show the paneling of a McDonnell-Douglas configuration and the resulting solutions compared to experimental data on the aft fuselage. The comparisons illustrate the effects of engine power and the ability of the panel method to model them.

A full-potential method with coupled boundary layer and an Euler correction method using the Clebsch transformation to simulate propeller power effects[92] is being used to guide the design of the engine pylon for transonic conditions on the McDonnell-Douglas aircraft. Comparisons with experimental data on the engine pylon shown in Fig. 61 illustrate the importance of including the fuselage boundary layer in the computational modeling.

Potential methods can also be applied to configurations with tractor-mounted propfans. However, the need to define the slipstream wakes as part of the geometry input and the indirect manner by which swirl must be simulated do not make them methods of choice. References 93–96 present some successful applications of linear (panel method), TSD, and full-potential methods, in which the swirling propeller slipstream was simulated using a linearized transpiration boundary condition.

Propfan Analysis with Euler Codes

The Euler technology has also been applied to the analysis of an advanced propfan-powered aircraft,[82,83] shown in Fig. 62. For a low-speed, high angle-of-attack case, the resulting moments and side force are shown in Plate 11 (see color section). A primary concern for a configuration with aft-mounted propfan engines is the power-induced interference effects on the aircraft's aerodynamics. Although many of these characteristics can be investigated in the wind tunnel with powered propfan simulators, some flight conditions cannot be properly simulated. Small and powerful enough propfan simulators for testing at transonic conditions on a full model in yaw were not available to us at the time of this analysis. The asymmetric effects of a failed engine at cruise could only be investigated by computational methods. (This may still be the case today.) A full-configuration analysis at both high and low speeds and at yaw with various combinations of thrust on the right and left side engines was carried out to examine these issues. A grid of approximately 600, 000 cells was used. The CPU time on a Cray XMP for a solution was approximately 2 h. The power-induced effects at high speed on the pressure distributions on the pylon and nacelle upper surfaces are illustrated in Fig. 63. The low-speed case for which there are experimental data was to be used to validate the computational results. The Euler code would then be used to support the estimation of the engine-out handling characteristics of the aircraft. Development of this particular aircraft was canceled before this line of analysis could be completed.

The ability of the Euler formulation to capture trailing vortex wakes, propeller slipstreams, and rotational flow make it desirable for the analysis of configurations with tractor-mounted propfans. An Euler propfan code[97,98] was applied to a NASA wing-mounted tractor configuration (Fig. 64). To simulate the propeller power and swirl effects, the total pressure and the swirl angle distributions downsteam of the propeller obtained from experiments were used as input for the disk boundary conditions. The total temperature needed for the disk simulation was calculated from the total pressure distribution using the isentropic relation. The computational results presented are from a solution that utilized only 50, 000 field grid cells, which is minimal for an adequate solution. Figure 65 shows computed and measured wing-surface pressure distributions on the propeller-off configuration. The presence of the nacelle, which causes a high-pressure peak inboard of the nacelle, has been well simulated. Figure 66 shows the results of the propeller-on configuration with the propeller rotating upward inboard of the nacelle. The swirl effects causing an increase of the angle of attack inboard of the nacelle and resulting in an increase of the shock strength have been correctly predicted.

These types of Euler analyses were used in a preliminary design study of the feasibility of installing wing-mounted tractor propfan engines on existing turbofan-powered aircraft. By use of a code with coupled boundary-layer analysis,[42] it was possible to investigate the effects of nacelle design and propfan slipstream on the aerodynamic characteristics of the wing. CFD methods have become an integral part of the engine/airframe integration process whether used in the preliminary design or in the detail design phase.

Concluding Remarks

It is now possible to analyze the flow about complex aerodynamic configurations using CFD methods that range in formulation complexity from inviscid linear potential panel methods running on high-end personal computers to solutions of the Reynolds-averaged Navier-Stokes equations utilizing the most modern supercomputers. It is no longer a question of which code or formulation may have the capability of handling the complex geometry of a complete aircraft, but rather which is most appropriate for the particular task. As previously stated, several factors must be considered when deciding what level of flow physics formulation should be applied to a specific problem. These not only include the level of physics necessary to adequately simulate the relevant flow phenomena, but also the level of accuracy and reliability of the numerical method, the resources available in terms of time and labor to set up the analysis, the computer resources required, and the level of confidence in the solutions. In industry, timeliness, accuracy, cost, and confidence are key factors in the choice of method for analysis. Timeliness and management confidence are paramount before CFD will be used in industry. The balance between cost (including labor to set up the problem) and accuracy becomes a function of need and budget.

The engineering development process is usually schedule-driven. Rarely does a project have months available to set up the grid generation, or paneling, for a CFD solution. All too often, by the time the grid has been prepared, the configuration geometry has changed. It is difficult to imagine how it will be possible to develop a timely method for structured surface-fitted grid generation for arbitrary geometries in the near-term. Reference 99 describes many of the developments and trends in three-dimensional grid generation for both structured and unstructured grids. Reyhner's Cartesian grid method, TRANAIR, and some of the recent advances in unstructured grid methods[100,101] may provide an answer to the timeliness problem. However the field grid generation problem is handled, there still remains the problem of the surface definition and discretization (paneling) for complex configurations. Complex configurations are complex in shape; simple geometry programs are just not adequate. Engineering computer-aided design (CAD) programs and special geometry programs such as AGPS[102] are now being used to speed the surface discretization process.

Before CFD can be seriously used, there must be management confidence in the results. Confidence rarely comes without a demonstration of repeated successes. As impressive as some of today's Navier-Stokes solutions on complete aircraft configurations are, it is going to take further refinement and many more solutions to develop the accuracy and that confidence. It is difficult to develop that track record when each solution takes tens of hours on today's fastest supercomputers. Routine use of Navier-Stokes codes will have to await the next generation of supercomputers. Recent progress in development of an efficient multigrid acceleration technique for solving the three-dimensional Navier-Stokes equations[103] may speed the time when they can be routinely applied to flows about complex configurations.

CFD has become a part of the aerodynamic design process, alongside the wind tunnel and flight test. It can give insight that is impossible to get from experimental means alone. CFD can also be used to generate geometry that will yield a desired pressure distribution or other payoff function, a function that testing cannot produce. CFD will not replace the wind tunnel in the next 10 or 20 years, but together the two will lead to the superior aerodynamic designs in less time than is possible with either alone.

Acknowledgment

This paper is based in part on work conducted for the Boeing Commercial Airplanes and Boeing Advanced Systems Independent Research and Development Program. Many of the CFD programs discussed were developed with government funding. The sponsoring agencies are acknowledged in the cited references.

References

[1]Rubbert, P. E., and Tinoco, E. N., "Impact of Computational Methods on Aircraft Design," AIAA Paper 83-2060, Aug. 1983.

[2]Rubbert, P. E., and Goldhammer, M. I., "CFD in Design: An Airframe Perspective," AIAA Paper 89-0092, Jan. 1989.

[3]Smith, C. W., Braymen, W. W., Bhateley, I. C., and Londenberg, W. K., "The Application of CFD for Military Aircraft Design at Transonic Speeds," NASA CP 3020, Vol. 1, Pt. 1, April 1989.

[4]Miranda, L. R., "Transonic and Fighter Aircraft: Challenges and Opportunities in CFD," NASA CP-3020, Vol. 1, Pt. 1, April 1989.

[5]Moran, J., Tinoco, E. N., and Johnson, F. T., "User's Manual—Subsonic/Supersonic Advanced Panel Pilot Code," NASA CR-152047, Feb. 1978.

[6]Magnus, A. E., and Epton M. A., "PAN AIR—A Computer Program for Predicting Subsonic or Supersonic Linear Potential Flows About Arbitrary Configurations Using A Higher Order Panel Method," Vol. 1, Theory Document (Version 1.0), NASA CR-3251, April 1980.

[7]Saaris, G. R., "A502H User's Manual—PAN AIR Technology Program for Solving Problems of Potential Flow About Arbitrary Configurations," Boeing Document D6-54703, Feb. 1990 (also available through NASA Ames Research Center, Aerodynamics Division).

[8]Bristow, D. R., and Hawk, J. D., "Subsonic Panel Method for Designing Wing Surfaces from Pressure Distribution," NASA CR-3713, July 1983.

[9]Hawk, J. K., and Bristow, D. R., "Subsonic Surface Panel Method for Airframe Analysis and Wing Design," AIAA Paper 83-0341, Jan. 1983.

[10]Maskew, B., "Prediction of Subsonic Aerodynamic Characteristics—A Case for Low-Order Panel Methods," *Journal of Aircraft*, Vol. 19, Feb. 1982, pp. 157–163.

[11]Youngren, H. H., Bochard, E. E., Coopersmith, R. M., and Miranda, L. H., "Comparison of Panel Method Formulations and its Influence on the Development of QUADPAN, an Advanced Low Order Method," AIAA Paper 83-1827, July 1983.

[12]Hess, J. L., and Friedman, D. M., "An Improved Higher-Order Panel Method for Three-Dimensional Lifting Flow," U.S. Naval Air Development Center, Warminster, PA, Naval Air Development Center, NADC Rept. 79277-60, Dec. 1981.

[13]Fornasier, L., "HISSS—A Higher-Order Subsonic/Supersonic Singularity Method for Calculating Linearized Potential Flow," AIAA Paper 84-1646, June 1984.

[14]Middleton, W. D., and Carlson, H. W., "Numerical Method of Estimating and Optimizing Supersonic Aerodynamic Characteristics of Arbitrary Planform Wings," *Journal of Aircraft*, Vol. 2, No. 4., July–Aug. 1965, pp. 261–265.

[15]Woodward, F. A., Tinoco, E. N., and Larsen, J. W., "Analysis and Design of Supersonic Wing-Body Combinations, Including Flow Properties in the Near Field, Part 1—Theory and Application," NASA CR-73106, Aug. 1967.

[16]Tinoco, E. N., Johnson, F. T., and Freeman, L. M., "The Application of a Higher Order Panel Method to Realistic Supersonic Configurations," AIAA Paper 79-0247, Jan. 1979.

[17]Cenko, A., "PAN AIR Applications to Complex Configurations," AIAA Paper 83-0007, Jan. 1983.

[18]Siclari, M., Visich, M., Cenko, A., Rosen, B., and Mason, W., "An Evaluation of NCOREL, PAN AIR and W12SC3 for the Prediction of Pressures on a Supersonic Maneuver Wing," AIAA Paper 84-0218, Jan. 1984.

[19]Fornasier, L., and Heiss, S., "Application of HISSS Panel Code to a Fighter-Type Aircraft Configuration at Subsonic and Supersonic Speeds," AIAA Paper 87-2619, Aug. 1987.

[20]Polhamus, E. C., "A Concept of the Vortex Lift of Sharp-Edge Delta Wings Based on a Leading-Edge-Suction Analogy," NASA TN D-3767, Dec. 1966.

[21]Lamar, J. E., "Extension of Leading-Edge-Suction Analogy to Wings with Separated Flow Around the Side Edges at Subsonic Speeds," NASA TR R-428, Oct. 1974.

[22]Lamb, M., Sawyer, W. C., and Thomas, J. L., "Experimental and Theoretical Supersonic Lateral-Directional Stability Characteristics of a Simplified Wing-Body Configuration with a Series of Vertical-Tail Arrangements," NASA TP-1878, Aug. 1981.

[23]Lednicer, D., "A VSAERO Analysis of Several Canard Configured Aircraft," Society of Automotive Engineers Rept. 881485, SAE SP-757, Oct. 1988.

[24]Szema, K. Y., and Shankar, V., "Nonlinear Computation of Wing-Body-Vertical Tail-Wake Flows at Low Supersonic Speed," AIAA Paper 84-0427, Jan. 1984.

[25]Szema, K. Y., and Shankar, V., "Validation of a Full Potential Method for Combined Yaw and Angle of Attack," AIAA Paper 86-1834, June 1986.

[26]Szema, K. Y., Chakravarthy, S. R., Pan, D., Bihari, B. L., Riba, W. T., Akdag, V. M., and Dresser, H. S., "The Application of a Unified Marching Technique for Flow over Complex 3-Dimensional Configurations across the Mach Number Range," AIAA Paper 88-0276, Jan. 1989.

[27]Reichenbach, S. H., and McMasters, J. H., "A Semiemperical Interpolation Technique for Predicting Full-Scale Flight Characteristics," AIAA Paper 87-0427, Jan. 1987.

[28]Caughey, D. A., and Jameson, A., "Recent Progress in Finite Volume Calculations for Wing-Fuselage Combinations," AIAA Paper 79-1513, July 1979.

[29]Yu, N. J., and Rubbert, P. E., "Acceleration Schemes for Transonic Potential Flow Calculations," AIAA Paper 80-0338, Jan. 1980.

[30]McLean, J. D., and Randall, J. L., "Computer Program to Calculate Three-Dimensional Boundary Layer Flows over Wings with Wall Mass Transfer," NASA CR-3123, Feb. 1978.

[31]McLean, J. D., and Matoi, T. K., "Shock/Boundary-Layer Interaction Model for Three-Dimensional Transonic Flow Calculations," *Turbulent Shear Layer/ Shock Wave Interactions*, Springer-Verlag, Berlin, 1986.

[32]Takanashi, S., Obayashi, S., Matsushima, K., and Fujii, K., "Numerical Simulation of Compressible Viscous Flows Around Practical Aircraft Configurations," AIAA Paper 87-2410CP, Aug. 1987.

[33]Flores, J., Reznick, S. G., Holst, T. L., and Gundy, K., "Navier-Stokes Solutions for a Fighter-Like Configuration," AIAA Paper 87-0032, Jan. 1987.

[34]Huband, G. W., Rizzetta, D. P., and Shang, J. J., "The Numerical Simulation of the Navier-Stokes Equations for an F-16 Configuration," AIAA Paper 88-2507-CP, June 1988.

[35]Buning, P. G., Chiu, I. T., Obayashi, S., Rizk, Y. M., and Steger, J. L., "Numerical Simulation of the Integrated Space Shuttle Vehicle in Ascent," AIAA Paper 88-4359, Aug. 1988.

[36]Agarwal, R., Deese, J., Johnson, J., and Steinhoff, J., "Euler/Navier-Stokes Calculations for Flow Over a Complete Aircraft," AIAA Paper 89-2221, Aug. 1989.

[37]DaCosta, A. L., "Application of Computational Aerodynamics Methods to the Design and Analysis of Transport Aircraft," International Council of the Astronautical Sciences Paper 78.B2-01, Sept. 1978.

[38]Boppe, C. W., "Aerodynamic Analysis for Aircraft with Nacelles, Pylons, and Winglets at Transonic Speeds," NASA CR-4066, April 1987.

[39]Melson, N. D., and Streett, C. L., "TAWFIVE: A User's Guide," NASA TM-84619, Sept. 1983.

[40]Tinoco, E. N., "Transonic CFD Application at Boeing," NASA CP-3020, Vol. 1, Pt. 1, April 1989.

[41]Rozendaal, R. A., "Variable Sweep Transition Flight Experiment (VSTFE)-Parametric Pressure Distribution Boundary Layer Stability Study and Wing Glove Design Task," NASA CR-3992, June 1986.

[42]Chen, H. C., Yu, N. J., Marcum, D. L., Kao, T. J., and Kusunose, K., "Coupled Euler/Boundary Layer Analysis of a Complete Aircraft with Viscous Modeling on the Lifting Surface," AIAA Paper 87-2614-CP, Aug. 1987.

[43]Yu, N. J., Chen, H. C., Chen, A. W., and Wittenberg, K. R., "Grid Generation and Flow Analysis for Wing/Body/Winglet Configurations," AIAA Paper 88-2548CP, June 1988.

[44]Schiff, L., and Steger, J., "Numerical Simulation of Steady Supersonic Viscous Flow," NASA TP-1749, 1981.

[45]Chaussee, D., Blom, G., and Wai, J., "Numerical Simulation of Steady Supersonic Flow over a Generic Fighter Configuration," 6th GAMM Conference on Numerical Methods in Fluid Dynamics, Göttingen, FRG, Sept. 1985.

[46]Wai, J. C., Blom, G., and Yoshihara, H., "Calculations for a Generic Fighter at Supersonic High-Lift Conditions," AGARD CP-412, April 1986.

[47]Johnson, F. T., Tinoco, E. N., Lu, P., and Epton, M. A., "Three-Dimensional Flow over Wings with Leading-Edge Vortex Separation," *AIAA Journal*, Vol. 18, April 1980, pp. 367–380.

582　　　　　　　　　　　　　　E. N. TINOCO

[48]Hoeijmakers, H. W. M., and Vaatstra, W., "A Higher-Order Panel Method Applied to Vortex Sheet Roll-Up," *AIAA Journal*, Vol. 21, April 1983, pp. 516–523.

[49]Wittenberg, K. R., Reichenbach, S. H., Kao, T. J., and Tinoco, E. N., "CFD Wake and Flowfield Predictions," AIAA Paper 87-2618, Aug. 1987.

[50]Burnham, D., "B-747 Vortex Alleviation Flight Tests: Ground Based Sensor Measurements," DOT-FAA-RD-81-99, Feb. 1982.

[51]Tinoco, E. N., Ball, D. N., and Rice, F. A., II, "PAN AIR Analysis of a Transport High-Lift Configuration," *Journal of Aircraft*, Vol. 24, March 1987, pp. 181-187.

[52]Viviand, H., Lecomte, C., and Morice, P., "Computational Fluid Dynamics in France," AIAA Paper 87-1131CP, June 1987.

[53]Cenko, A., Tinoco, E. N., Dyer, R. K., and DeJongh, J., "PAN AIR Applications to Weapons Carriage and Separation," *Journal of Aircraft*, Vol. 18, Feb. 1981, pp. 128–134.

[54]Cenko A., Tinoco, E. N., and Tustaniwskyj, J., "PAN AIR Applications to Mutual Interference Effects Due to Close Prolixity," AIAA Paper 84-0217, Jan. 1984.

[55]Samant, S. S., Bussoletti, J. E., Johnson, F. T., Burkhart, R. H., Everson, B. L., Melvin, R. G., and Young D. P., "TRANAIR: A Computer Code for Transonic Analysis of Arbitrary Configurations," AIAA Paper 87-0034, Jan. 1987.

[56]Samant, S. S., Bussoletti, J. E., Johnson, F. T., Melvin, R. G., and Young, D. P., "Transonic Analysis of Arbitrary Configurations Using Locally Refined Grids," *Proceedings of the 11th International Conference on Numerical Method in Fluid Dynamics*, 1988.

[57]Johnson, F. T., Samant, S. S., Bieterman, M. B., Melvin, R. G., Young, D. P., Bussoletti, J. E., and Madson, M. D., "Application of the TRANAIR Rectangular Grid Approach to the Aerodynamic Analysis of Complex Configurations," AGARD Fluid Dynamics Panel Specialists' Meeting, May 24, 25, 1989, Loen, Norway, to be published.

[58]Fox, J., Donegan, T., Jacocks, J., and Nichols, R., "Computation of the Euler Flow Field Produced by a Transonic Aircraft with Stores," AIAA Paper 89-2219, Aug. 1989.

[59]Rosen, B. S., "External Store Carriage Loads Prediction of Transonic Speeds," AIAA Paper 88-0003, Jan. 1988.

[60]Cosner, R. R., "Transonic Propulsion System Integration Analysis at McDonnell Aircraft Company," NASA CP-3020, Vol. 1, Pt. 2, April 1989.

[61]Boppe, C. W., "Elements of Computational Engine/Airframe Integration," *Progress in Astronautics and Aeronautics: Numerical Methods for Engine-Airframe Integration*, Vol. 102, AIAA, New York, 1986, pp. 127–168.

[62]Sokhey, J. S., "Three-Dimensional Flow Analysis of Turboprop Inlet and Nacelle Configurations," *Progress in Astronautics and Aeronautics: Numerical Methods for Engine-Airframe Integration*, Vol. 102, AIAA, New York, 1986, pp. 279–304.

[63]Lahti, D. J., Dietrich, D. A., Stockman, N. O., and Faust, G. K., "Application of Computational Methods to the Design of Large Turbofan Engine Nacelles," *Progress in Astronautics and Aeronautics: Numerical Methods for Engine-Airframe Integration*, Vol. 102, AIAA, New. York, 1986, pp. 305–334.

[64]Gillette, W. G., "Nacelle Installation Analysis for Subsonic Transport Aircraft," AIAA Paper 77-102, Jan. 1977.

[65]Perry, K. M., and Forester, C. K., "Numerical Simulation of Multi-Stream Nozzle Flows," AIAA Paper 79-1549, July 1979.

[66]Chen, H. C., Kusunose, K., and Yu, N. J., "Flow Simulations for Detailed Nacelle-Exhaust Flow Using Euler Equations," AIAA Paper 85-5003, Oct. 1985.

[67]Chen, A. W., and Tinoco, E. N., "PAN AIR Applications to Aero-Propulsion Integration," *Journal of Aircraft*, Vol. 21, March 1984, pp. 161–167.

[68]Tinoco, E. N., and Chen, A. W., "CFD Applications to Engine/Airframe Integration," *Progress in Astronautics and Aeronautics: Numerical Methods for Engine-Airframe Intergration*, Vol. 102, AIAA, New York, 1986, pp. 219–255.

[69]Tinoco, E. N., and Chen, A. W., "Transonic CFD Applications in Engine/Airframe Integration," AIAA Paper 84-0381, Jan. 1984.

[70]Colehour, J. L., Farquhar, B. W., and Reyhner, T. A., "Analytical Methods for Subsonic Propulsion System Integration," *Proceedings of the 9th International Symposium on Air Breathing Engines*, Athens, Greece, AIAA, Washington, DC, 1989, pp. 1064–1069.

[71]Cooper, G. K., "The PARC Code: Theory and Usage," Arnold Engineering Development Center, Tullahoma, TN, TR-87-24, Oct. 1987.

[72]Clark, D. R., Maskew, B., and Dvorak, F. A., "The Application of a Second-Generation Low-Order Panel Method (the VSAERO Program) to Powerplant Installation Studies," *Progress in Astronautics and Aeronautics: Numerical Methods for Engine-Airframe Integration*, Vol. 102, AIAA, New York, 1986, pp. 256–278.

[73]Waggoner, E. G., "Computational Analysis for an Advanced Transport Configuration with Engine Nacelles," AIAA Paper 83-1851, July 1983.

[74]Wai, J. C., and Yoshihara, H., "Transonic Perturbation Analysis of Wing-Fuselage-Nacelle-Pylon Configuration with Powered Jet Exhausts," AIAA Paper 82-0255, Jan. 1982.

[75]Kafyeke, F., Piperni, P., and Robin, S., "Application of KTRAN Transonic Small Disturbance Code to the Challenger Business Jet Configuration with Winglets," Society of Automotive Engineers Rept. 881483, SAE SP-757, Oct. 1988.

[76]Chandrasekharan, R. M., Murphy, W. R., Taverna, F. P., and Boppe, C. W., "Computational Aerodynamic Design of the Gulfstream IV Wing," AIAA Paper 85-0427, Jan. 1985.

[77]Yu, N. J., "Grid Generation and Transonic Flow Calculations for Three-Dimensional Configurations," AIAA Paper 80-1391, July 1980.

[78]Yu, N. J., "Transonic Flow Simulations for Complex Configurations with Surface Fitted Grids," AIAA Paper 81-1258, June 1981.

[79]Reyhner, T. A., "Three-Dimensional Transonic Potential Flow About Complex Three-Dimensional Configurations," NASA CR-3814, July 1984.

[80]Saaris, G. R., Gilkey, R. D., Smit, K. L., and Tinoco, E. N., "Transonic Analysis of Complex Configurations Using TRANAIR Program," Society of Automative Engineers Rept. 892289, Sept. 1989.

[81]Chen, A. W., Curtin, M. M., Carlson, R., and Tinoco, E. N., "TRANAIR Applications to Engine/Airframe Integration," AIAA Paper 89-2165CP, Aug. 1989.

[82]Yu, N. J., Kusunose, K., Chen, H. C., and Summerfield, D. M., "Flow Simulations for a Complex Airplane Configuration Using Euler Equations," AIAA Paper 87-0454, Jan. 1987.

[83]Kusunose, K., Marcum, D. L., Chen, H. C., and Yu, N. J., "Transonic Analysis for Complex Airplane Configurations," AIAA Paper 87-1196, June 1987.

[84]Jameson, A., Schmidt, W., and Turkel, E., "Numerical Solutions of the Euler Equations by Finite Volume Methods Using Runge-Kutta Time-Stepping Schemes," AIAA Paper 81-1259, April 1981.

[85]Raj, P., and Long, L. N., "An Euler Aerodynamic Method for Leading-Edge Vortex Flow Simulation," NASA CP-2416, Oct. 1985.

[86]Raj, P., and Brennan, J. E., "Improvements to an Euler Aerodynamic Method for Transonic Flow Analysis," AIAA Paper 87-0040, Jan. 1987.

[87]Narain, J. P., "Inviscid Flow Predictions for Complex Aircraft Configurations Using Euler Equations," AIAA Paper 87-2616, Aug. 1987.

[88]Birchelbaw, L., Atta, E., Rubertus, C., Whipkey, R., and Cornelius, K., "Euler Analysis and Correlations for Realistic Upper Surface Blowing Configurations," AIAA Paper 87-2271, Aug. 1987.

[89]Swada, K., and Takanashi, S., "A Numerical Investigation on Wing/Nacelle Interferences of USB Configurations," AIAA Paper 87-0455, Jan. 1987.

[90]Valarezo, W. O., and Hess, J. L., "Time-Averaged Subsonic Propeller Flowfield Calculations," AIAA Paper 86-1897CP, June 1986.

[91]Vernon, D. F., and Hughes, J. P., "Aerodynamic Integration of Aft-Mounted UHB Propulsion System," AIAA Paper 87-2920, Sept. 1987.

[92]Chen, L. T., Yu, K. C., and Dang, T. Q., "A Transonic Computational Method for an Aft-Mounted Nacelle/Pylon Configuration with Propeller Power Effect," AIAA Paper 89-0560, Jan. 1989.

[93]Boctor, M. L., Clay, C. W., and Watson, C. F., "An Analysis of Prop-Fan/Airframe Aerodynamic Integration," NASA CR-152186, Oct. 1978.

[94]Boppe, C. W., and Rosen, B., "Computation of Propfan Engine Installation Aerodynamics," International Council of the Astronautical Sciences Paper 85-5.5.3, Sept. 1984.

[95]Samant, S. S., et al., "Prop-Fan Nacelle Installations at Transonic Speeds," NASA CR-16376, Oct. 1982.

[96]Samant, S. S., Yu, N. J., and Rubbert, P. E., "Transonic Flow Simulation of Prop-fan Configurations," AIAA Paper 83-0187, Jan. 1983.

[97]Yu, N. J., Samant, S. S., and Rubbert, P. E., "Flow Predictions for Propfan Configurations Using Euler Equations," AIAA Paper 84-1645, June 1984.

[98]Yu, N. J., and Chen, H. C., "Flow Simulations for Nacelle-Propeller Configurations Using Euler Equations," AIAA Paper 84-2143, Aug. 1984.

[99]Baker, T. J., "Developments and Trends in Three-Dimensional Mesh Generation," NASA CP-3020, Vol. 1, Pt. 1, April 1989.

[100]Jameson, A., and Baker, T. J., "Improvements to the Aircraft Euler Method," AIAA Paper 87-0452, Jan. 1987.

[101]Lohner, R., Morgan, K., and Peraire, J., "Improved Adaptive Refinement Strategies for the Finite Element Aerodynamic Configurations," AIAA Paper 86-0499, Jan. 1986.

[102]Snepp, D. K., and Pomeroy, R. C., "A Geometry System for Aerodynamic Design," AIAA Paper 87-2902, Sept. 1987.

[103]Vatsa, V. N., and Wedan, B. W., "Development of an Efficient Multigrid Code for 3-D Navier-Stokes Equations," AIAA Paper 89-1791, June 1989.

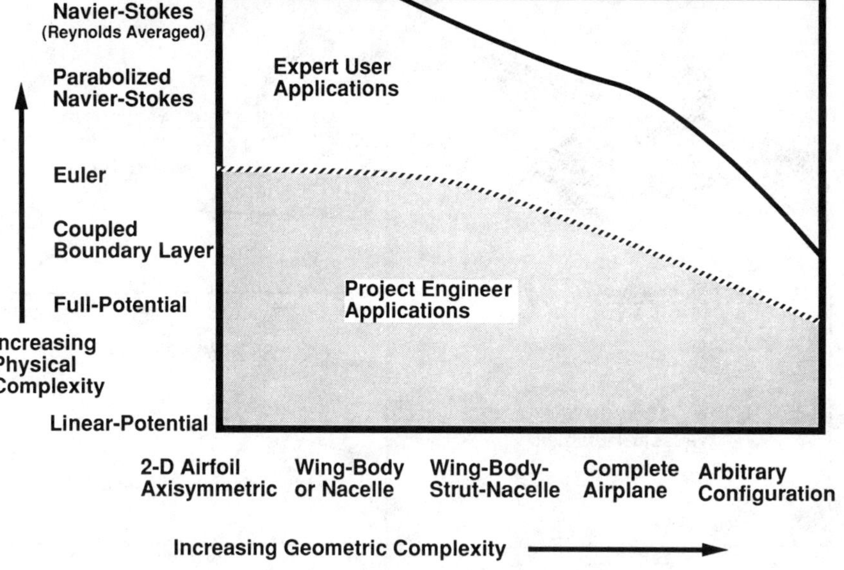

Fig. 1   State of the art in practical CFD applications, 1989.

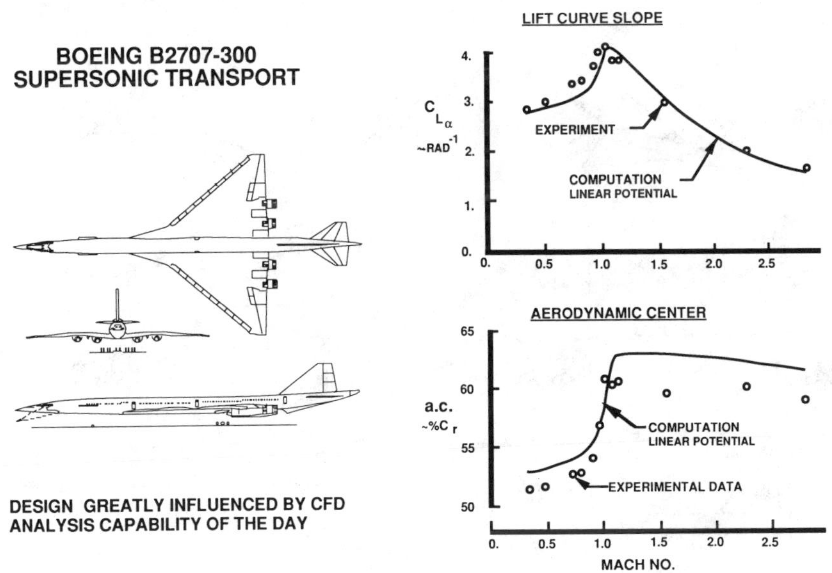

Fig. 2   CFD vs test, circa 1968–1970. C_{L_α}, lift coefficient; C_r,   a.c., aerodynamic center.

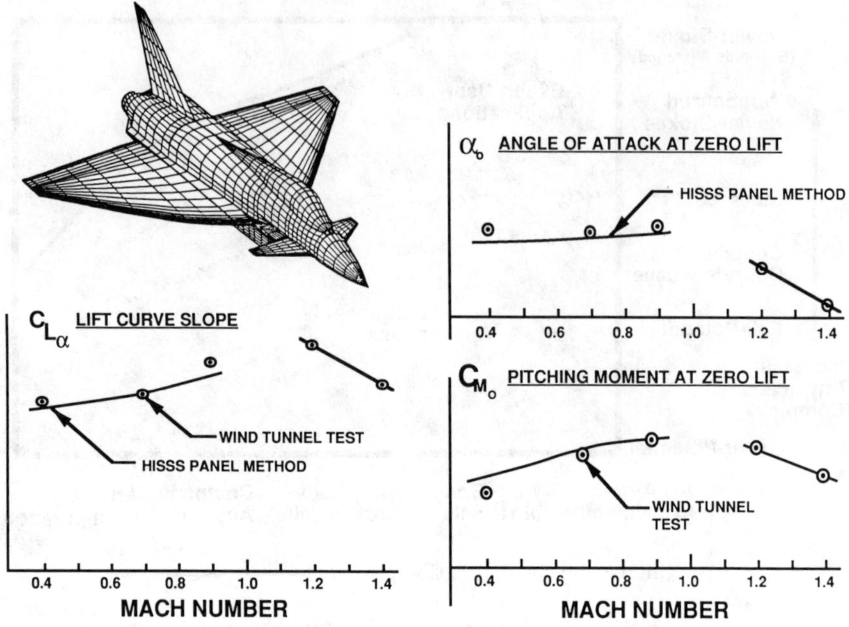

Fig. 3   Longitudinal characteristics of an advanced supersonic fighter: HISSS predictions.

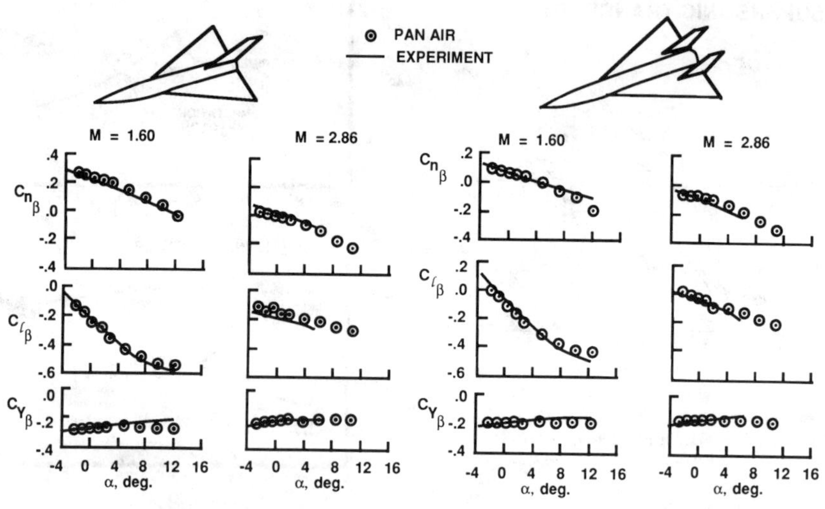

Fig. 4   Lateral-directional stability characteristics: A502/PANAIR. C_{n_β}, normal force coefficient; C_{ℓ_β}, C_{Y_β}, side force coefficient.

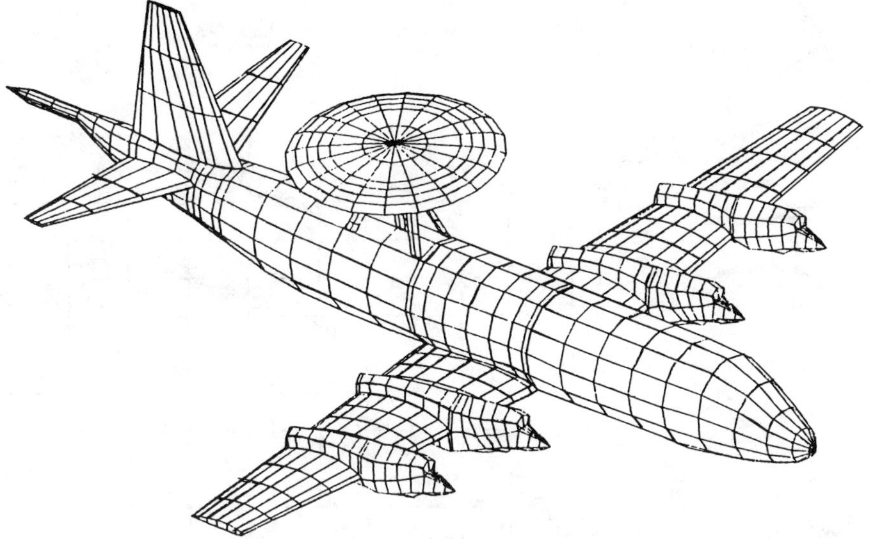

Fig. 5   P-3 AEW QUADPAN model.

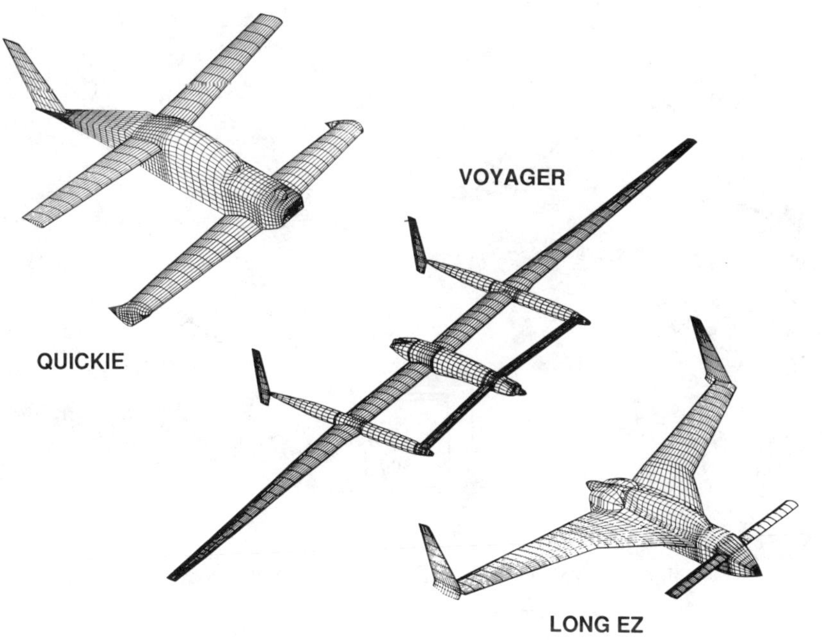

Fig. 6   Innovative aircraft: VSAERO panel models.

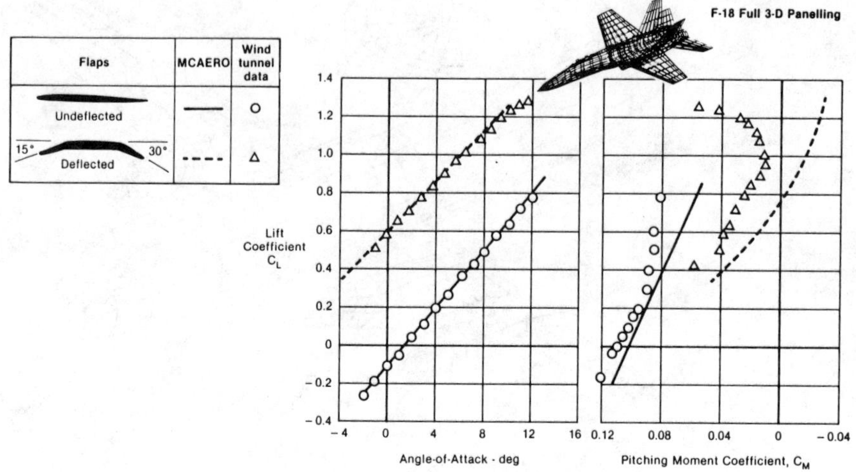

Fig. 7   Longitudinal characteristics: MCAERO predictions.

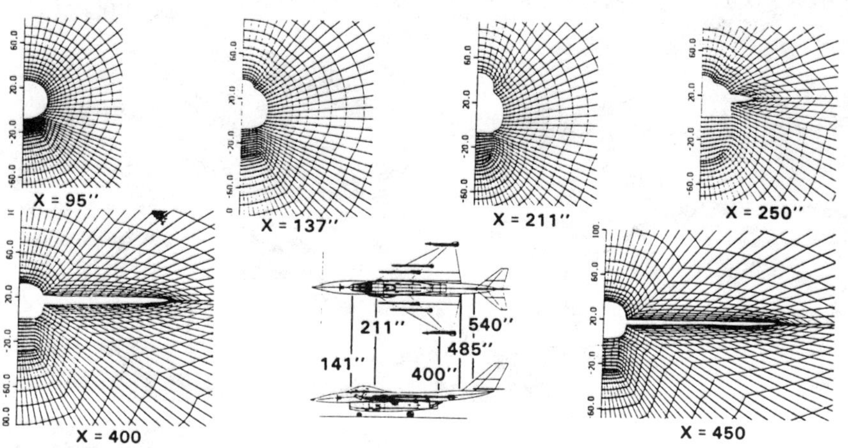

Fig. 8   Cross-sectional grids: EMTAC Euler code.

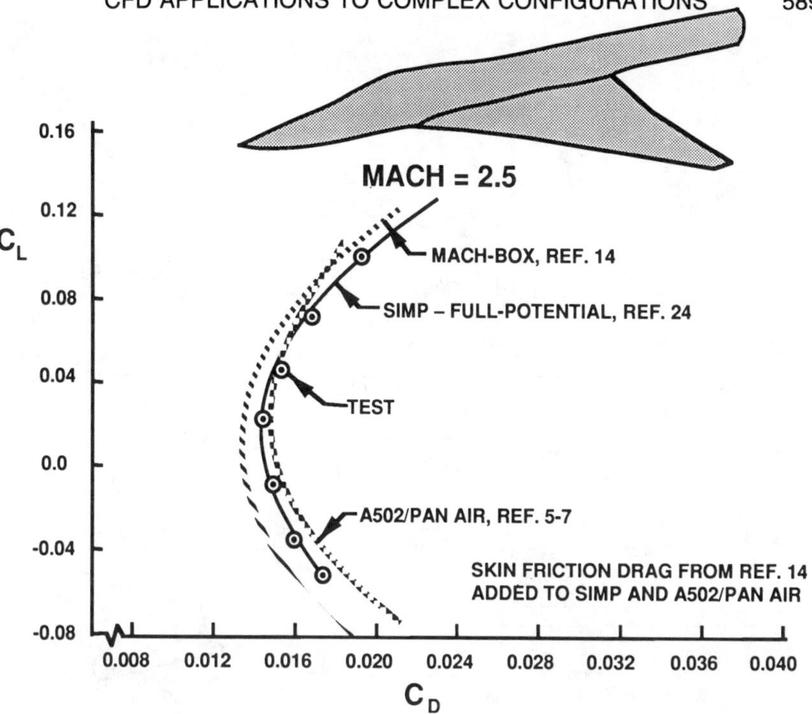

Fig. 9 Drag predictions of a supersonic fighter interceptor. C_D, drag coefficient.

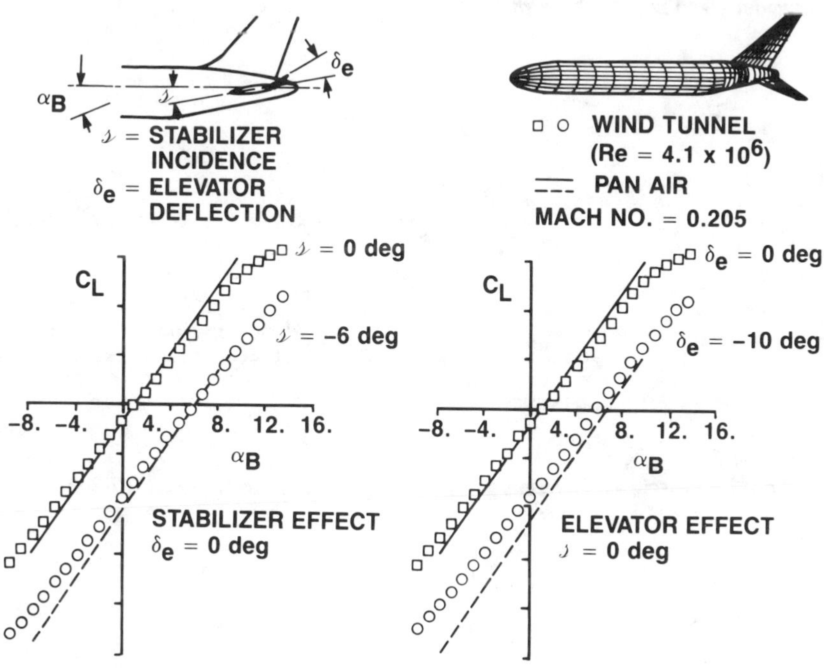

Fig. 10 A502/PAN AIR prediction of Boeing 767 stabilizer and elevator effectiveness.

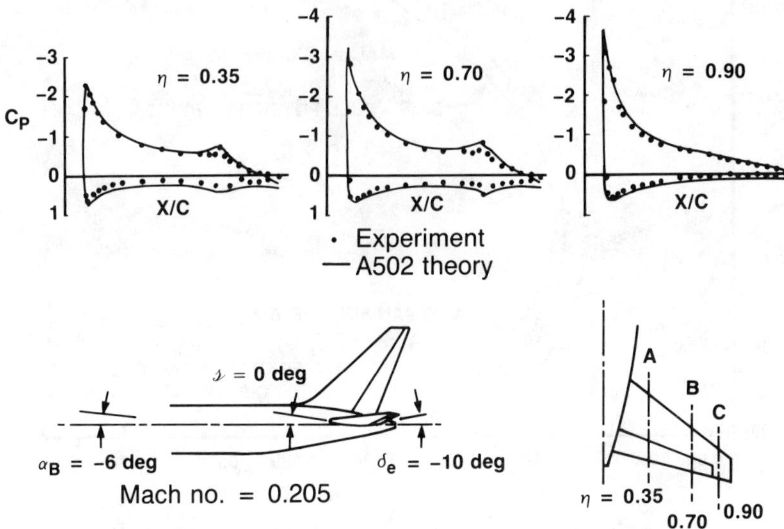

Fig. 11 A502/PAN AIR pressure distribution test-theory comparison for a 767 horizontal stabilizer. X/C, chord fraction.

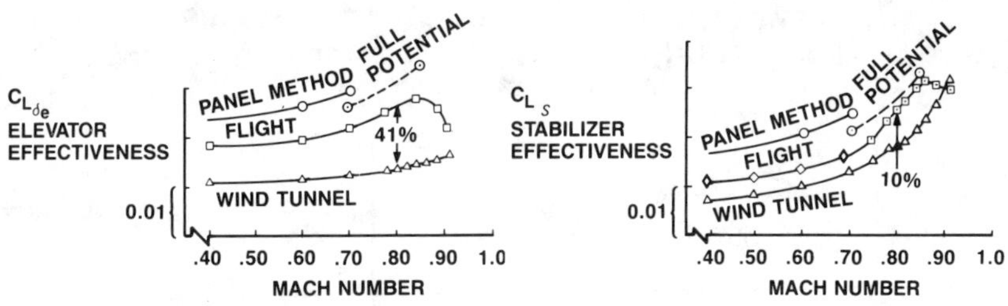

Fig. 12 Inviscid CFD results for stabilizer/elevator effectiveness: 767-200.

- Semiempirical interpolation using a
  viscous CFD based curve fit

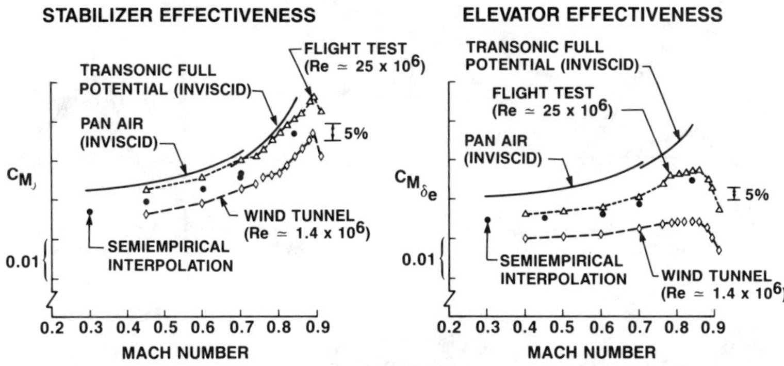

Fig. 13 Predictions using semiempirical interpolation: 757-200.

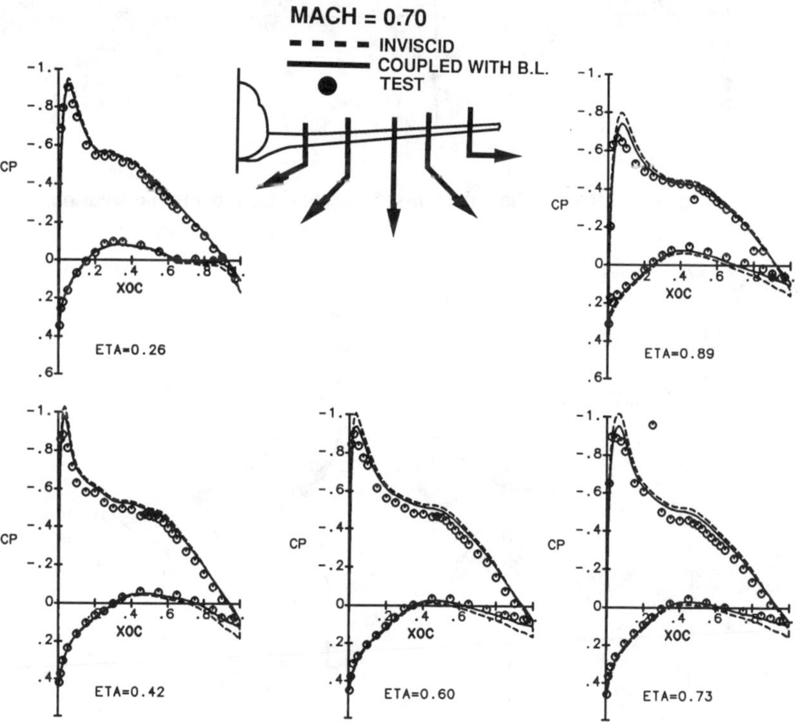

Fig. 14 Effect of boundary-layer coupling on a subcritical solution. C_P, pressure coefficient; xoc chord fraction; ETA, span fraction.

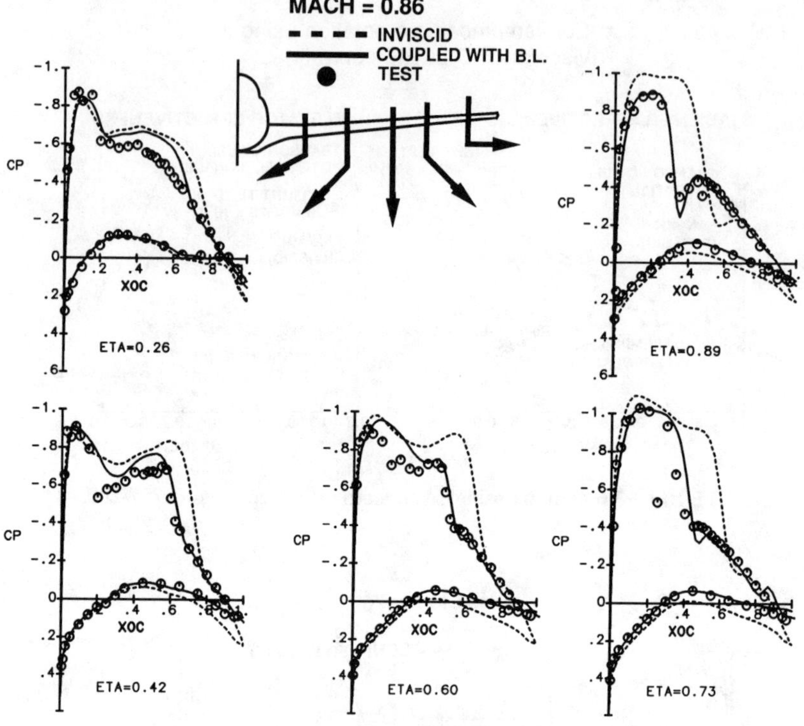

Fig. 15   Effect of boundary-layer coupling on a transonic solution.

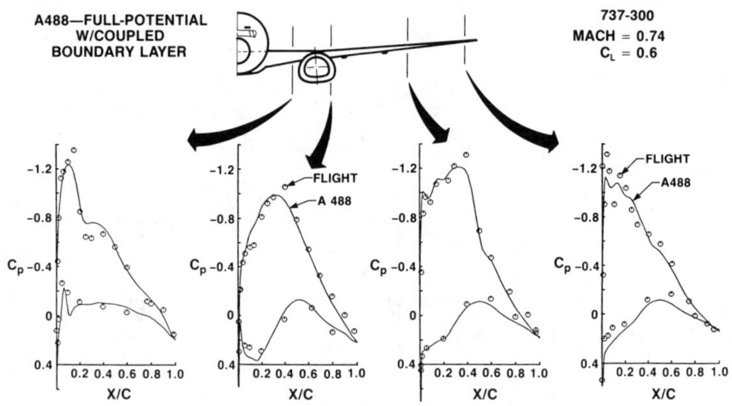

Fig. 16   Flight pressure distributions: 737-300.

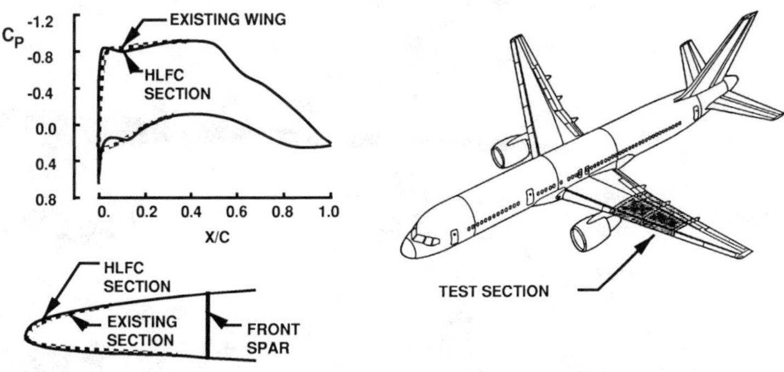

Fig. 17   757 wing leading-edge modification for the HLFC flight experiment: a CFD design.

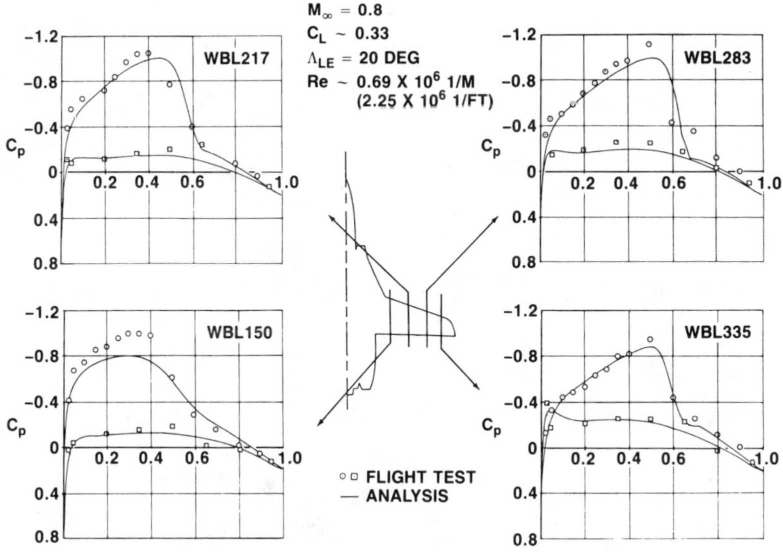

Fig. 18   Comparison with flight pressure distributions: Grumman F-14A. WBL, wing buttline.

Fig. 19   Surface grid for F-14A in A488 code.

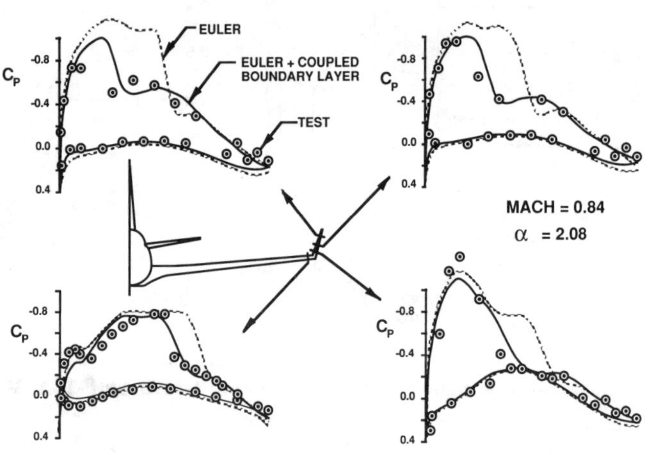

Fig. 20   Test-theory comparisons for a wing-body-winglet: Euler.

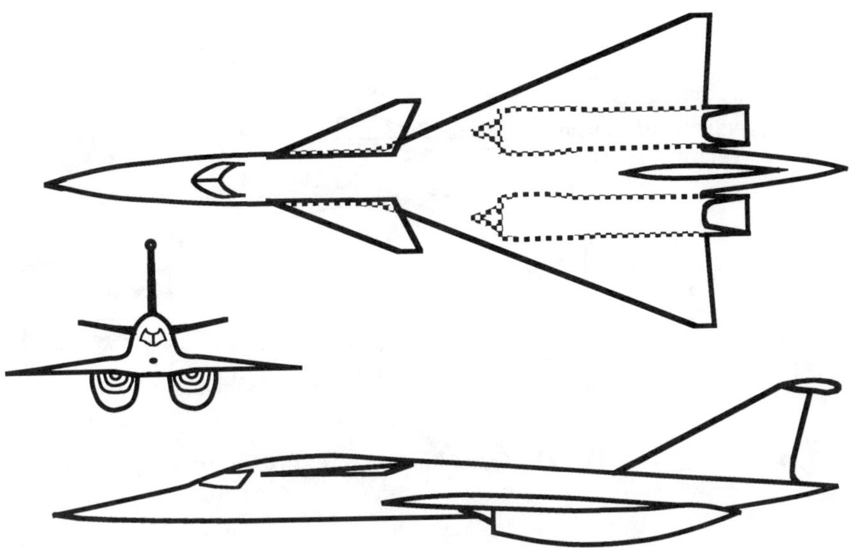

Fig. 21   Generic fighter model-350.

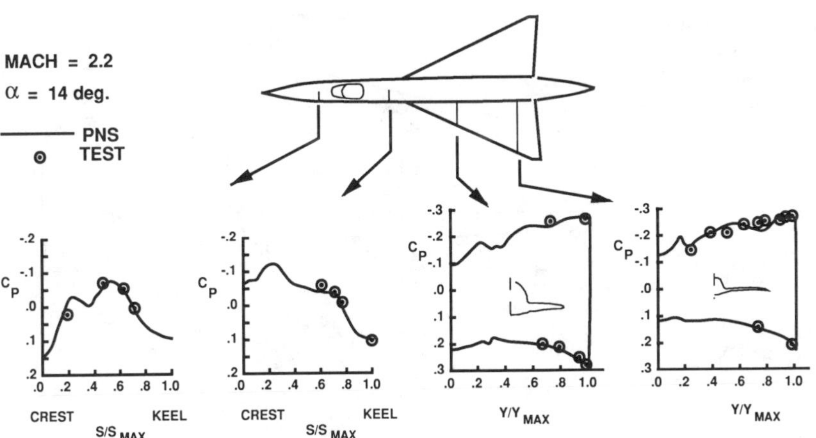

Fig. 22   Test-theory comparison of wing and fuselage pressures: PNS.

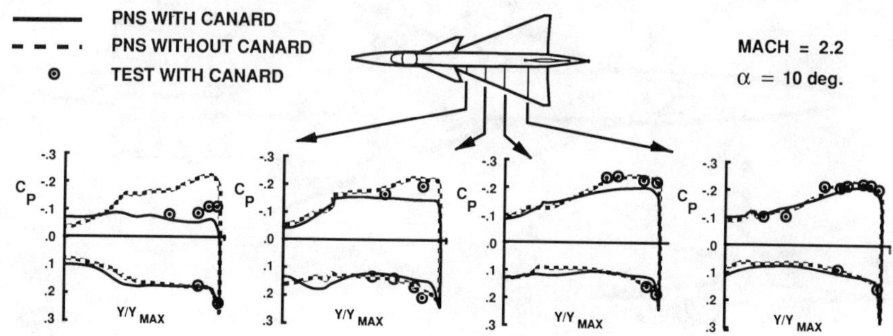

Fig. 23   Canard interference: PNS.

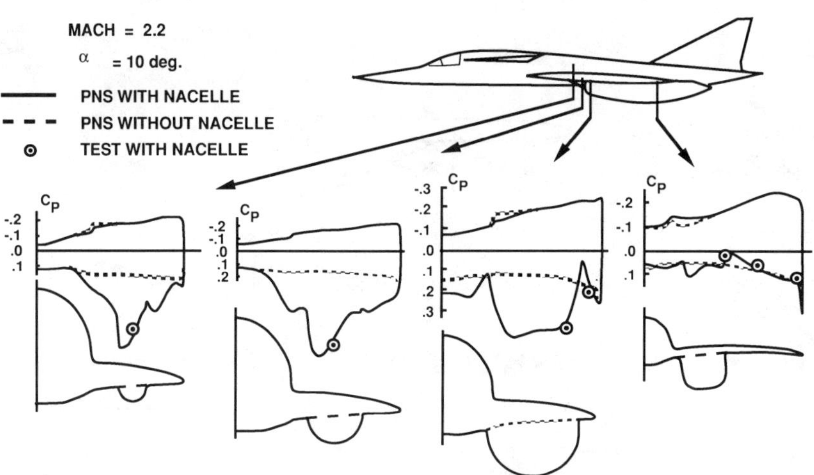

Fig. 24   Nacelle interference: PNS.

Theoretical to Flight Test Comparison
Wake Vortex Effects

B747 Wake Vortex Characteristics
NASA Flight Test—1979
Flaps 30, wt = 607,000 lb, 150 kn

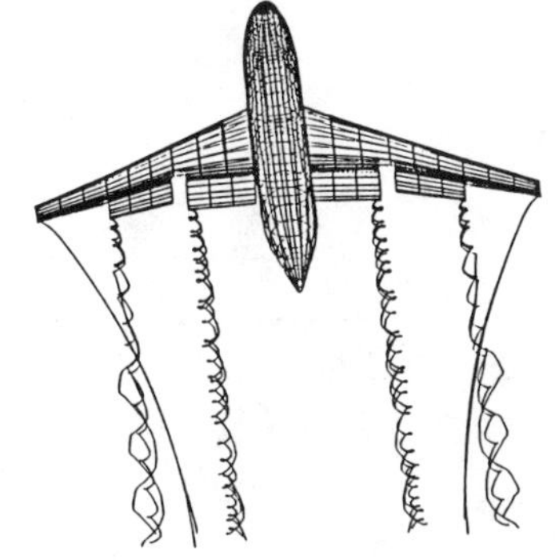

747-300 PANAIR Model, Flaps 30
Mach = 0.23, alp = 7.5 deg, tail-off

Fig. 25   CFD to flight comparison of wake vortex effects.

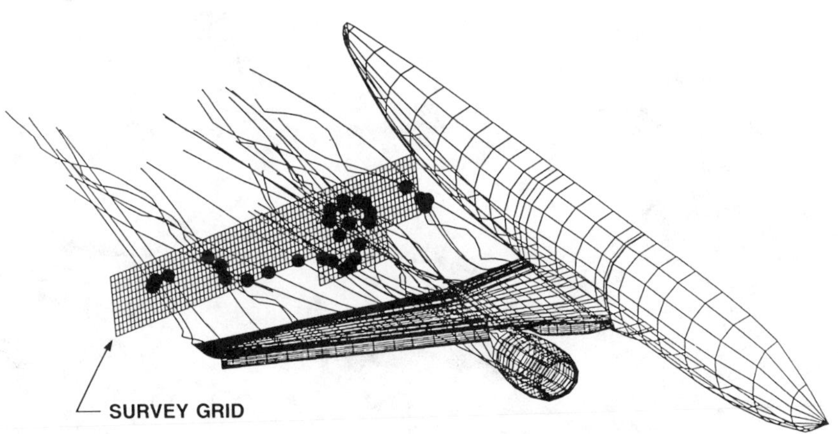

● PANAIR streamline
intersection with
survey plane

Fig. 26   A502/PAN AIR panel model and streamline calculations.

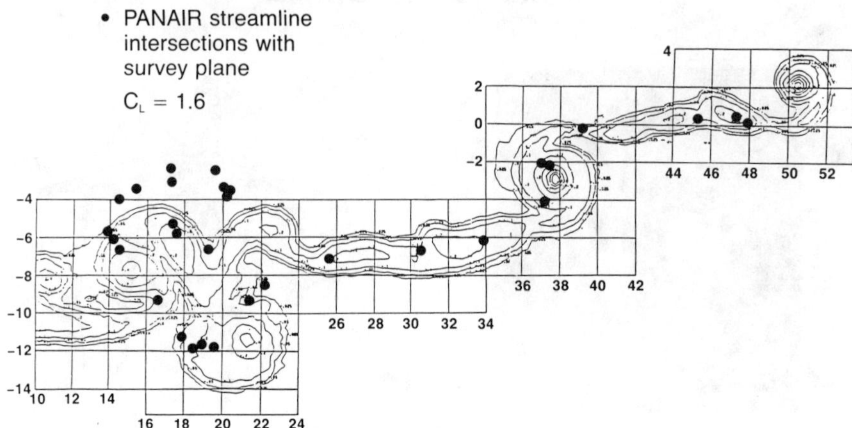

Total pressure wake contours from flow-through nacelle test

Fig. 27 A502/PAN AIR streamline intersections with survey plane.

Fig. 28 Boeing/GE UDF demonstrator.

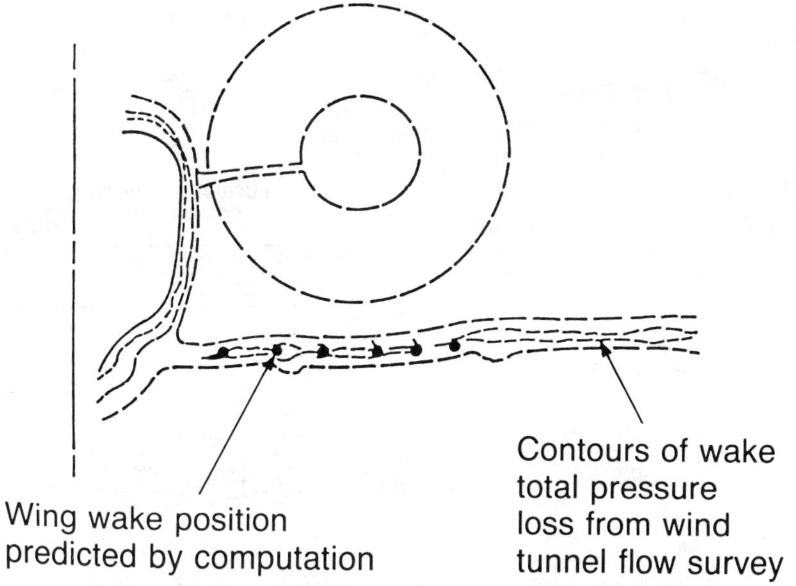

Fig. 29　Streamline prediction comparison with experiment.

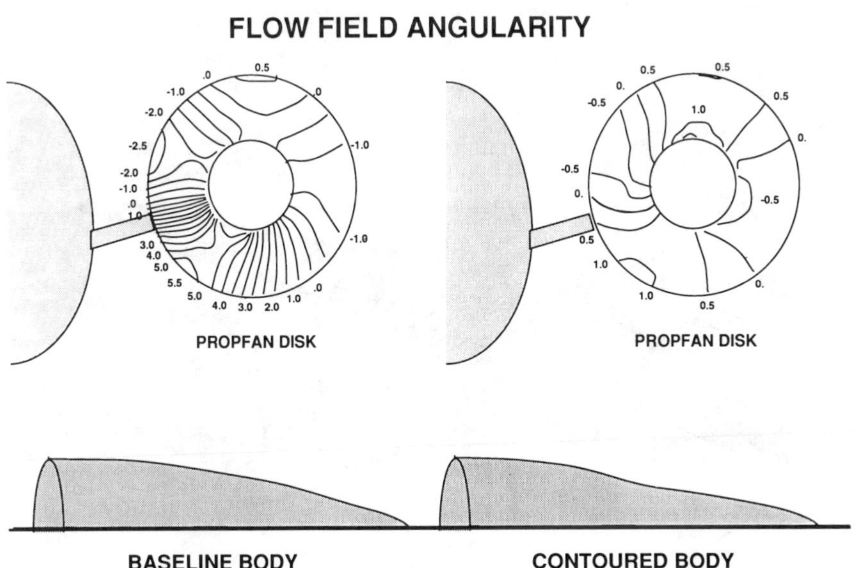

Fig. 30　Body contour effect on prop-plane flowfield angularity: a CFD design.

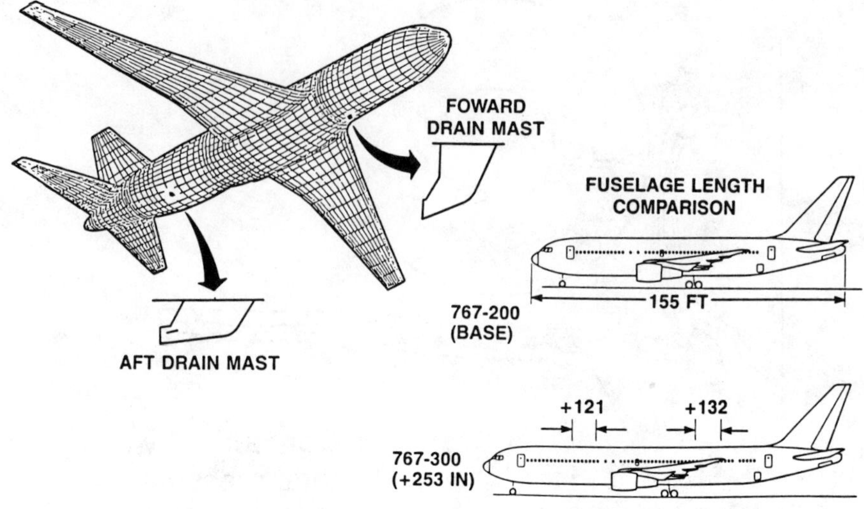

Fig. 31   CFD certification of drain mast in lieu of flight test.

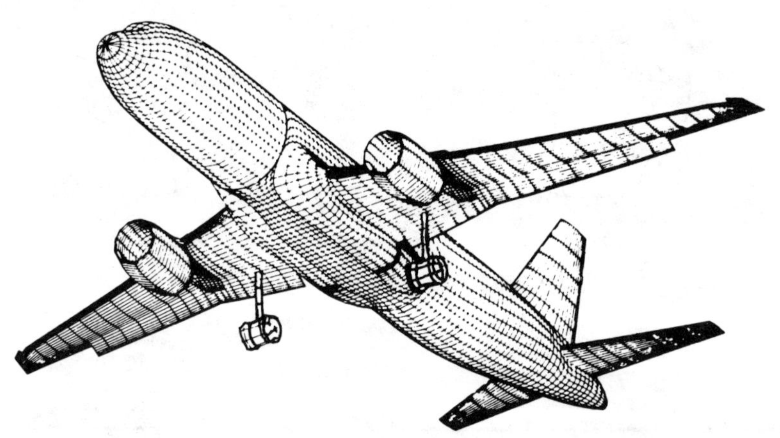

Fig. 32   CFD support of A320 wing tunnel ground effect testing.

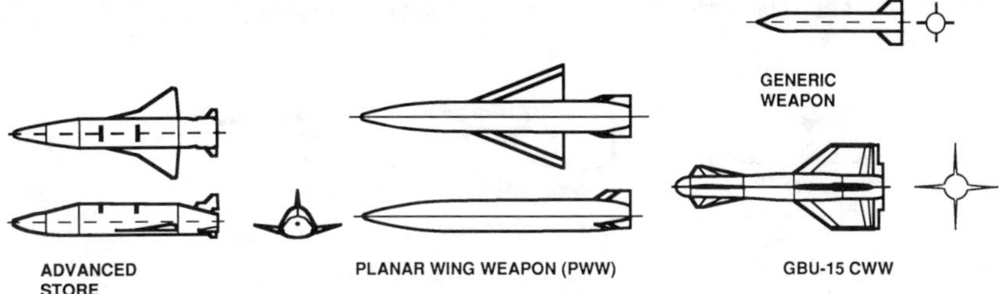

Fig. 33 Aircraft weapon stores.

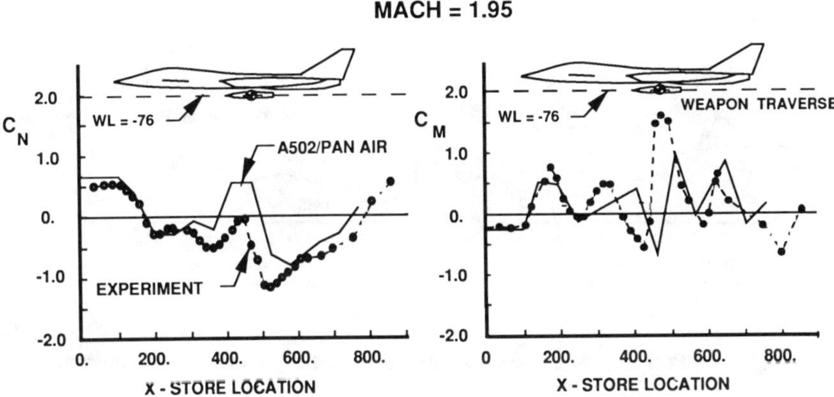

Fig. 34 Normal force and pitching-moment prediction: planar wing weapon in STAC flowfield.

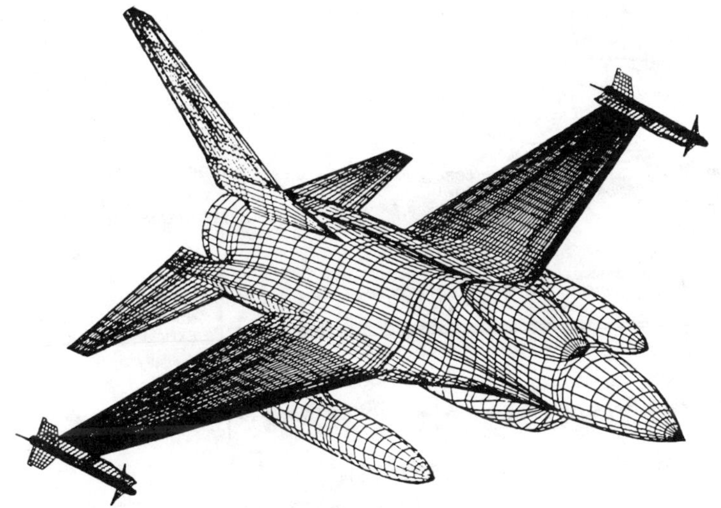

Fig. 35 TRANAIR model: F-16A aircraft with tanks and missiles.

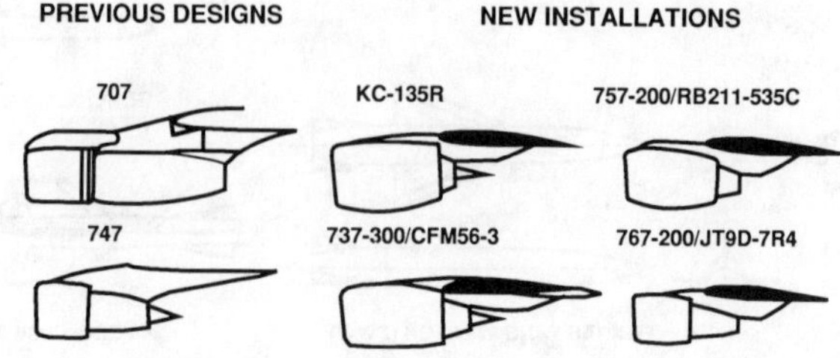

Fig. 36   Computationally derived close-coupled nacelle positions.

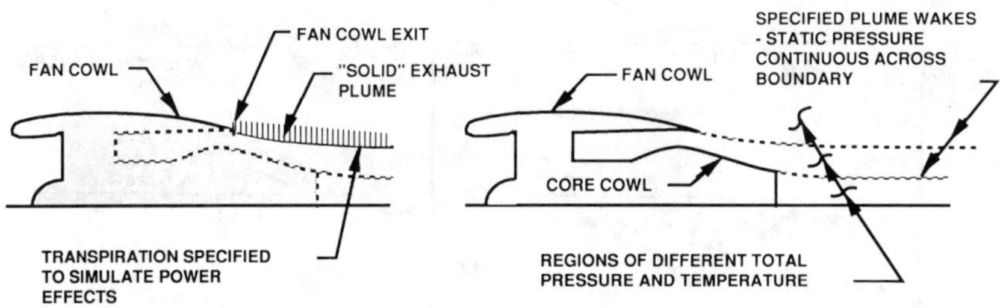

Fig. 37   Potential-flow computational models for jet exhaust simulation.

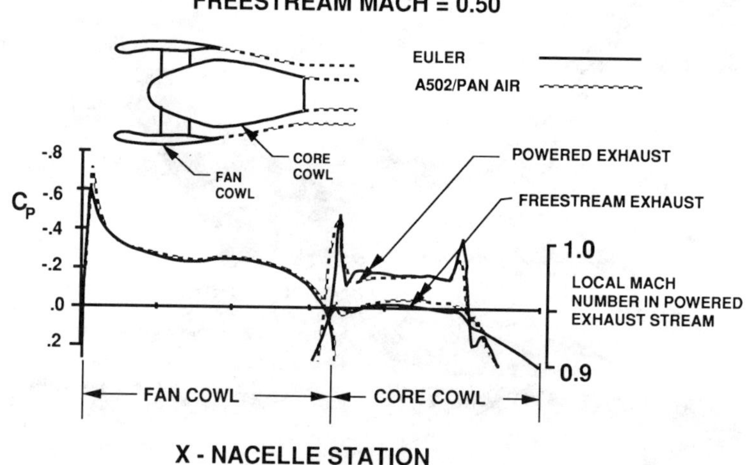

Fig. 38   Subcritical exhaust flows: A502/PAN AIR vs Euler.

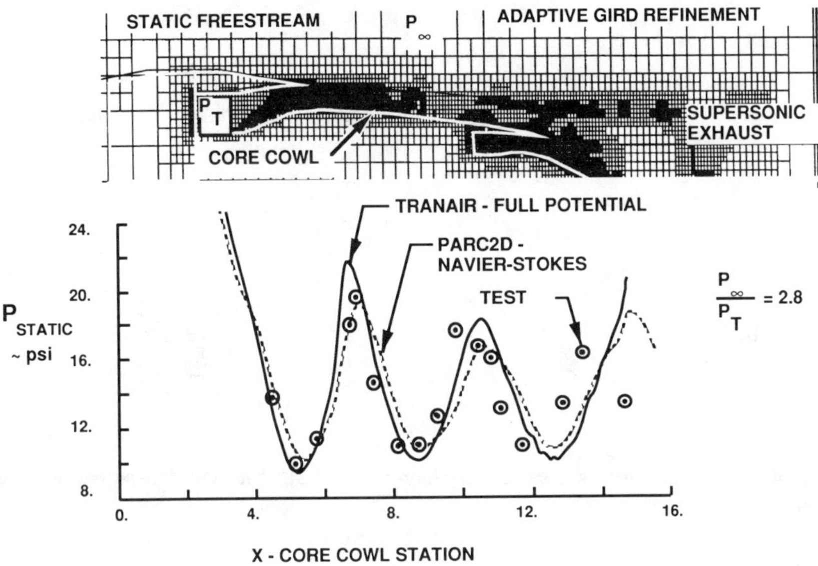

Fig. 39   Supercritical exhaust flows: comparison with experimental data.

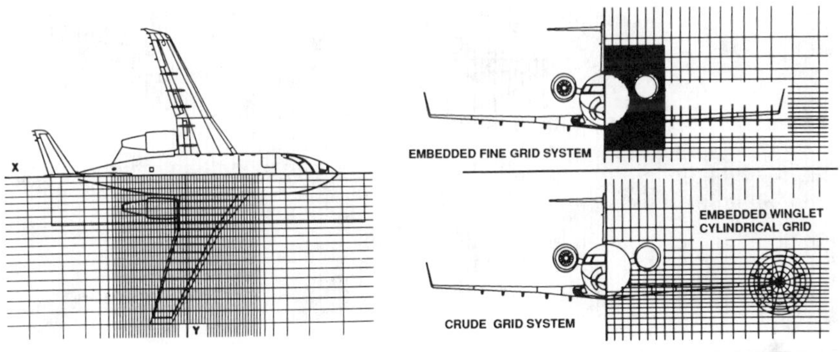

Fig. 40   Grid structure for extended transonic small-disturbance code.

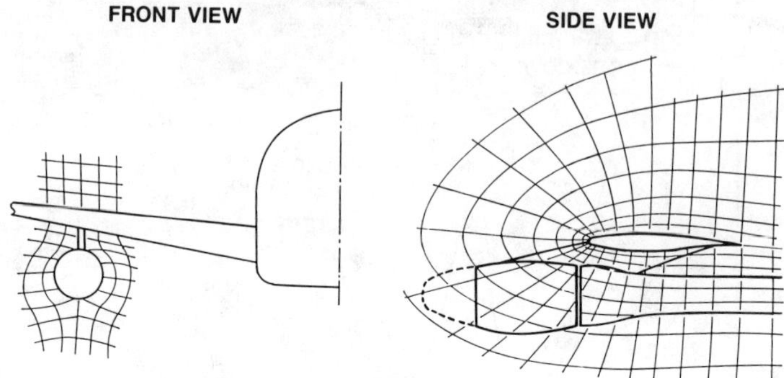

Fig. 41   Grid structure around under-wing-mounted strut nacelle full-potential code: A488.

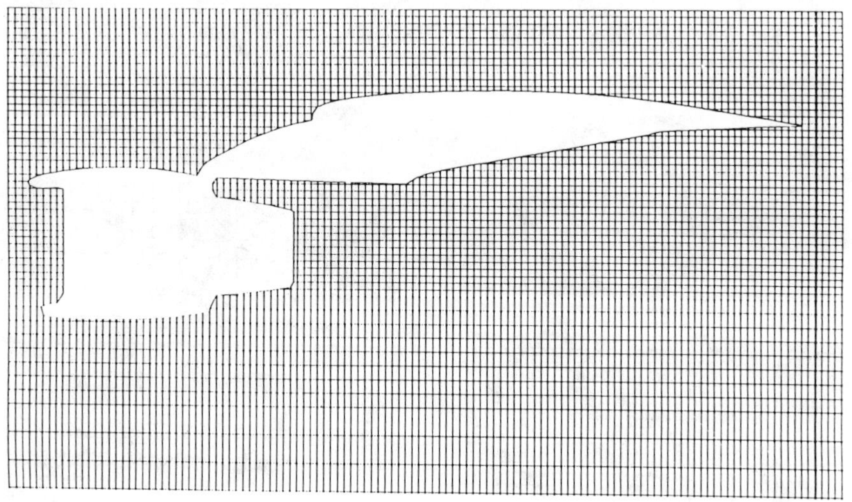

Fig. 42   Grid structure around under-wing-mounted strut nacelle for Cartesian grid full-potential code.

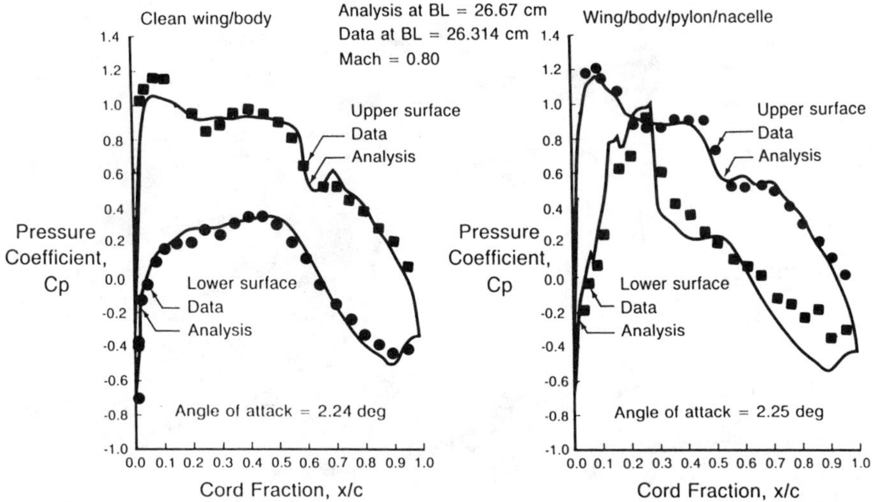

Fig. 43 Wing surface pressure distribution for NASA high wing transport with under-wing strut nacelle: Cartesian grid full-potential code.

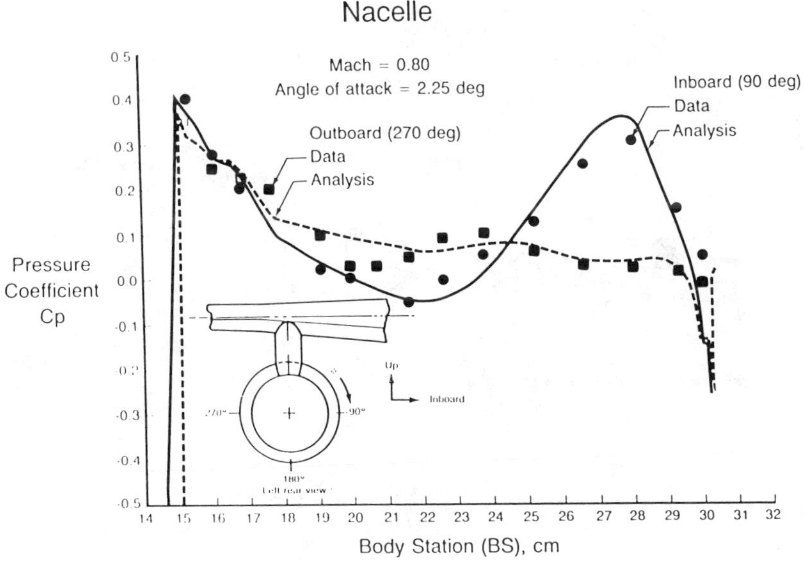

Fig. 44 Nacelle surface pressure distribution for NASA high wing transport with under-wing strut nacelle: Cartesian grid full-potential code.

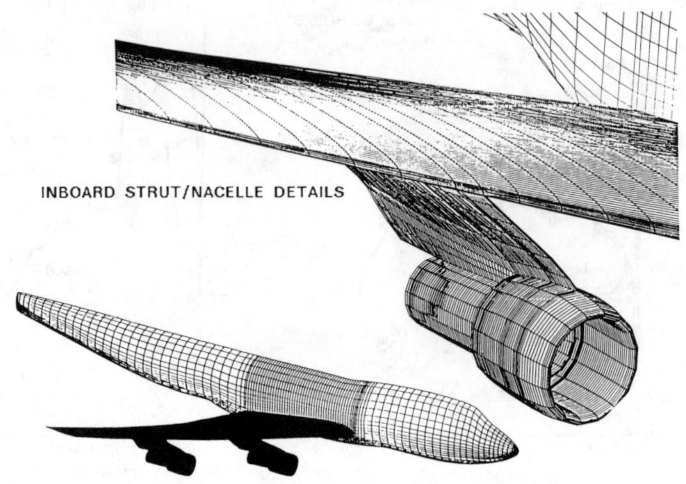

Fig. 45 TRANAIR model for 747-200 wing body strut nacelle model.

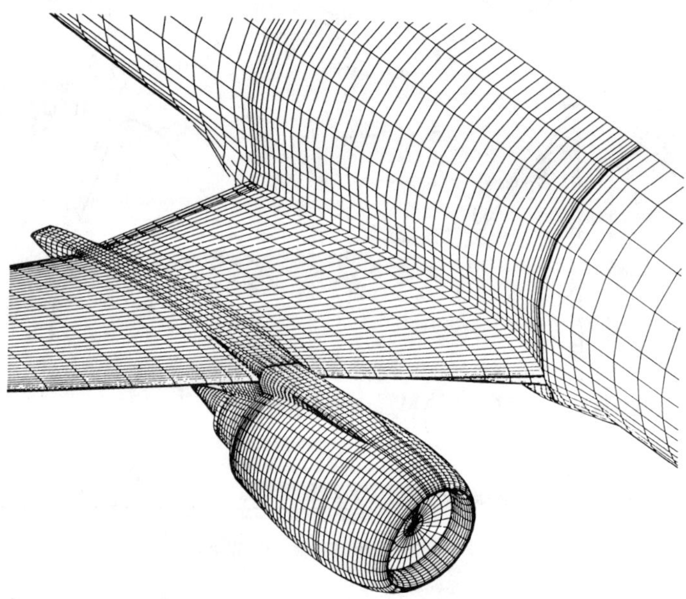

Fig. 46 TRANAIR model for 737-400: nacelle closeup.

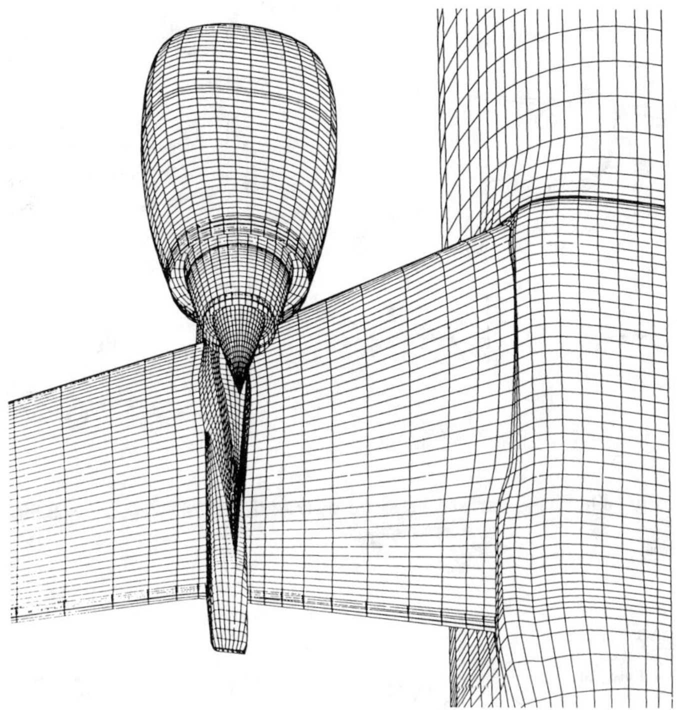

Fig. 47   TRANAIR model for 737-400: nacelle closeup, bottom view.

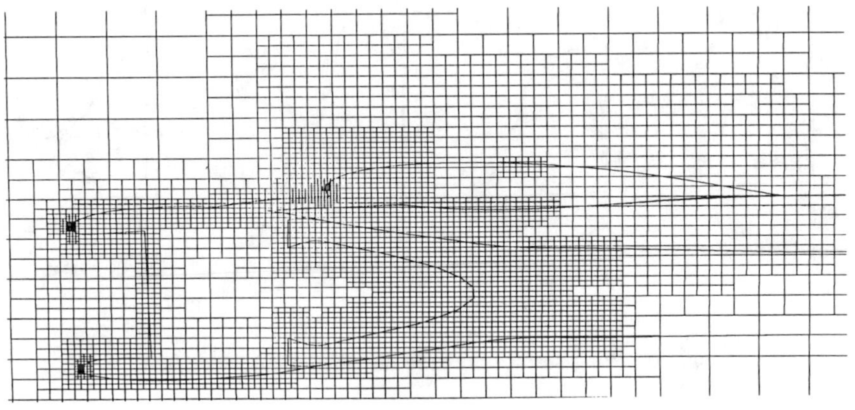

Fig. 48   TRANAIR field grid detail: z-x plane.

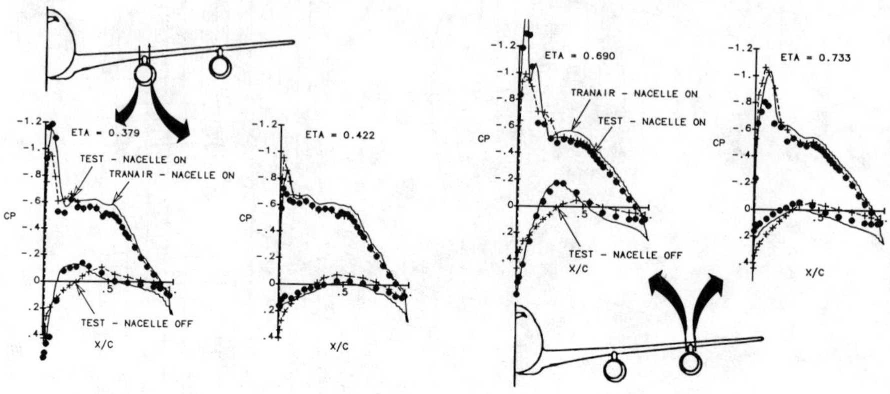

Fig. 49   Wing surface pressure distributions: 747-200, $M = 0.80$.

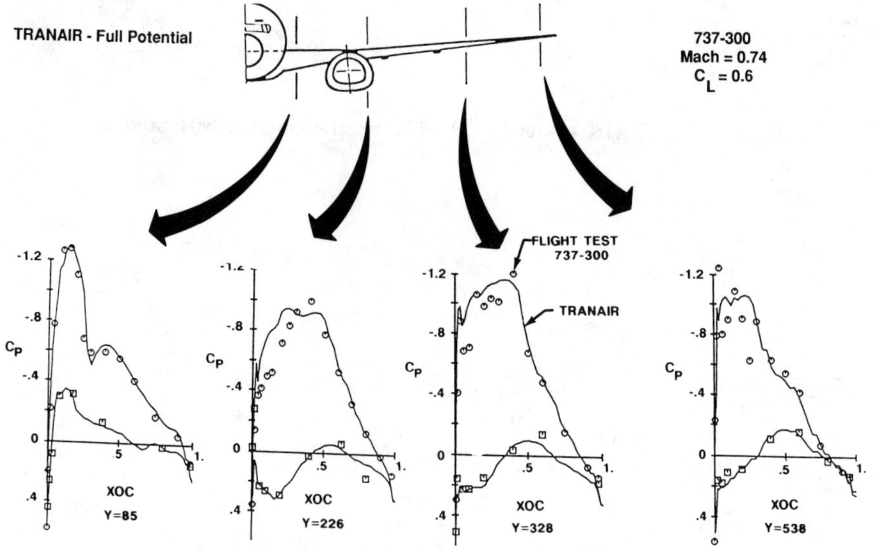

Fig. 50   Wing surface pressure distributions: 737-300/400.

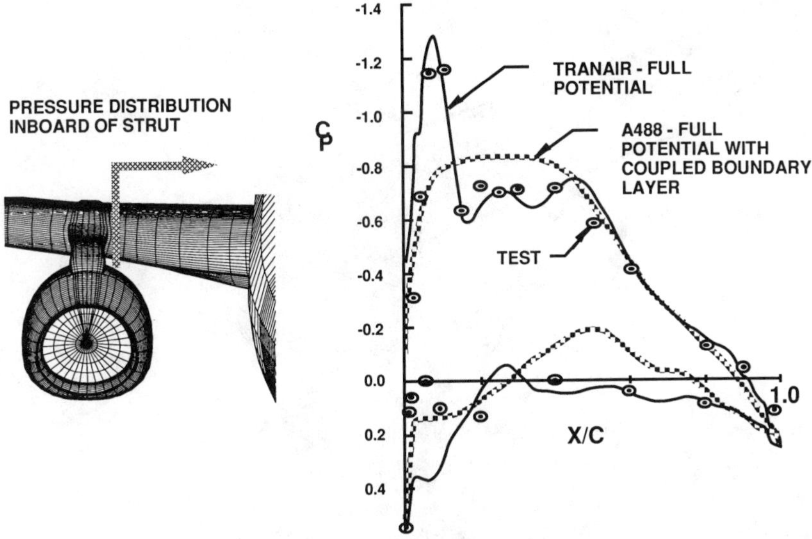

Fig. 51   Wing pressure distribution adjacent to engine installation.

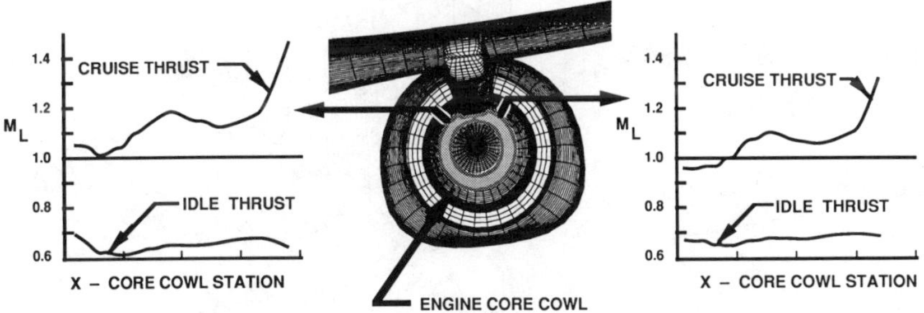

Fig. 52   Installed nacelle core cowl Mach number distributions: effect of engine thrust.

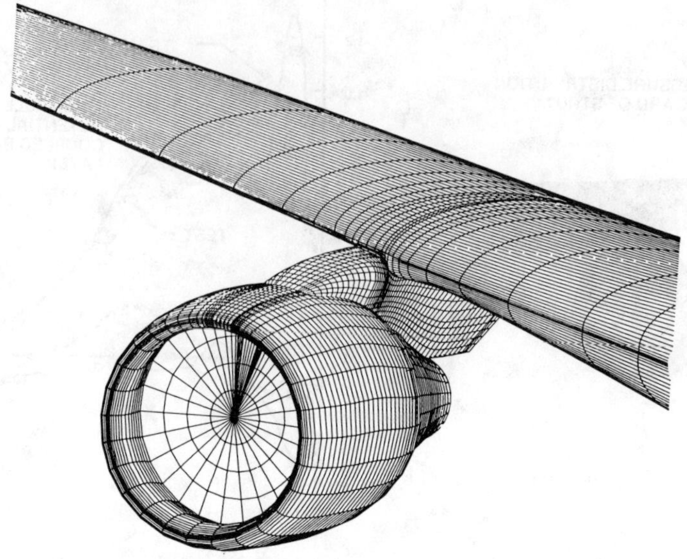

Fig. 53   Surface grid for TRANAIR and multiblock Euler codes.

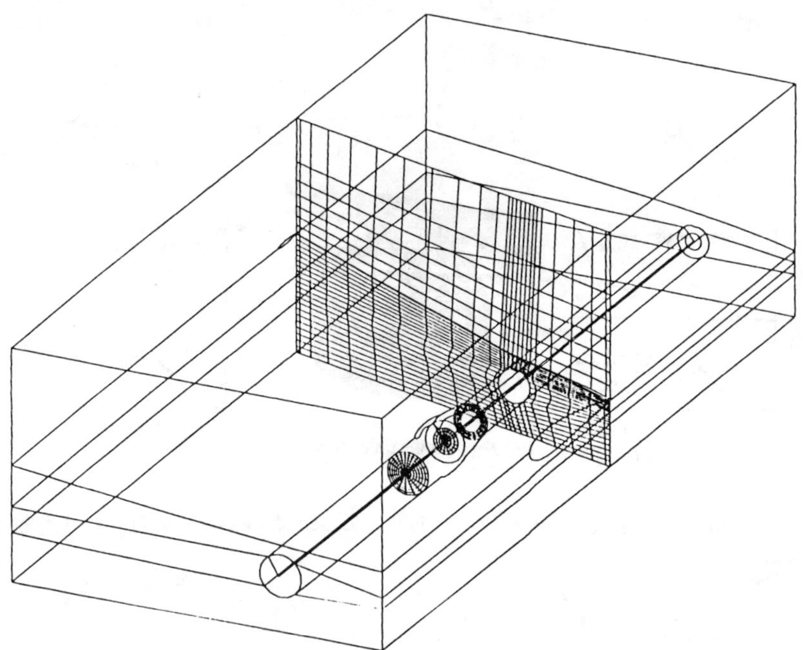

Fig. 54   Block structure for wing strut nacelle.

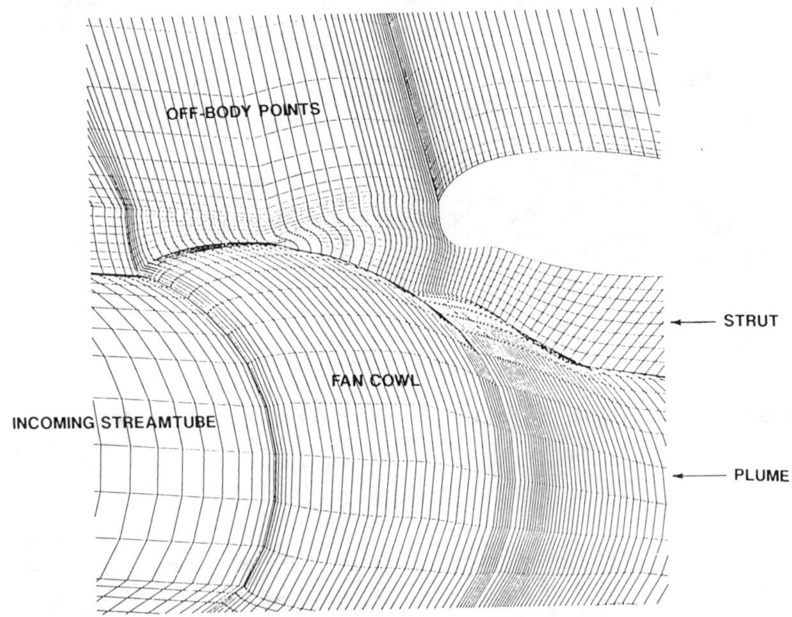

Fig. 55   Multiblock grid detail for wing strut nacelle.

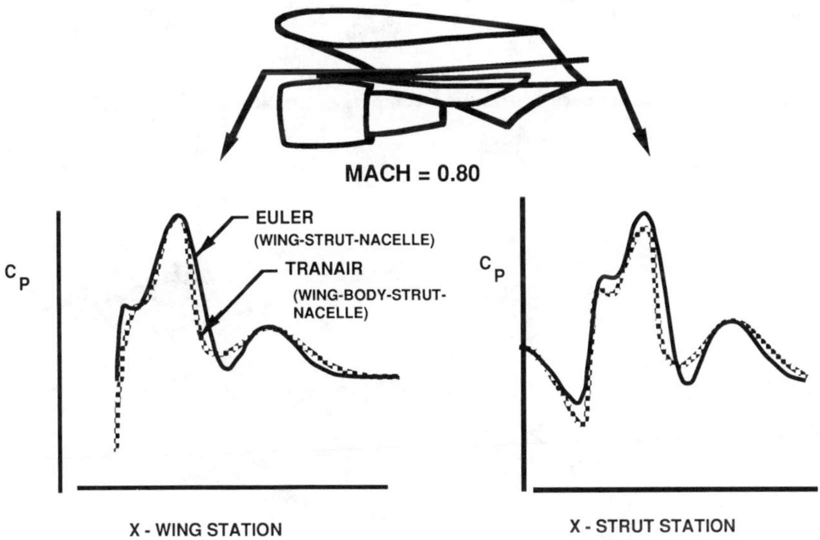

Fig. 56   Surface pressure distribution comparison: TRANAIR vs multiblock Euler.

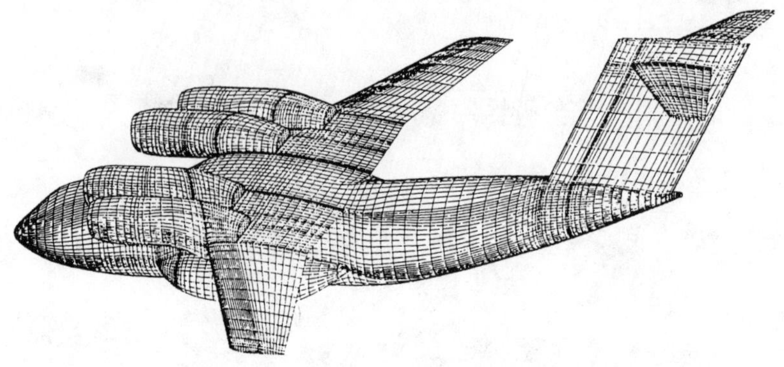

Fig. 57 Surface grid for ASKA complete aircraft configuration: multiblock Euler.

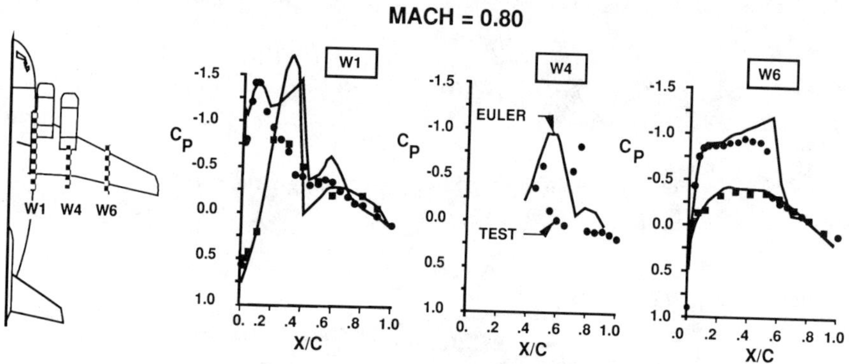

Fig. 58 Surface pressure distribution comparison: multiblock Euler vs test.

Fig. 59 Paneling for complete propfan configuration: PROP85.

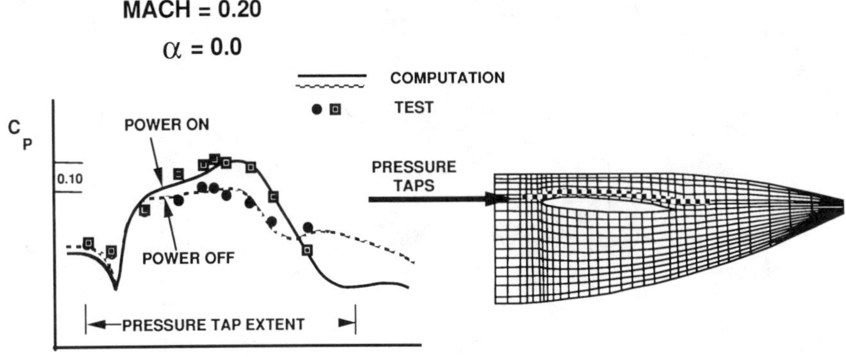

Fig. 60   Aft body surface pressure distribution comparison: effect of propfan power.

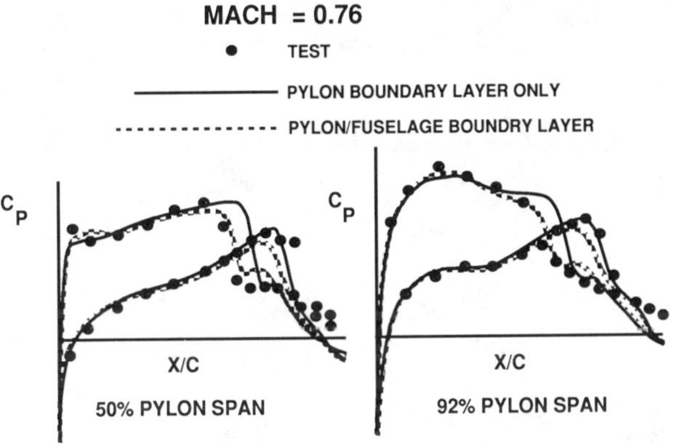

Fig. 61   Nacelle pylon surface pressure distribution comparison: effect of fuselage and pylon boundary layer.

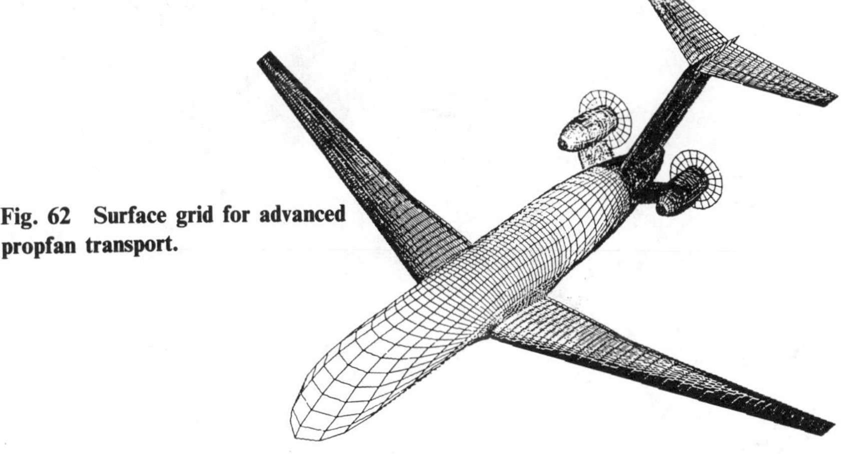

Fig. 62   Surface grid for advanced propfan transport.

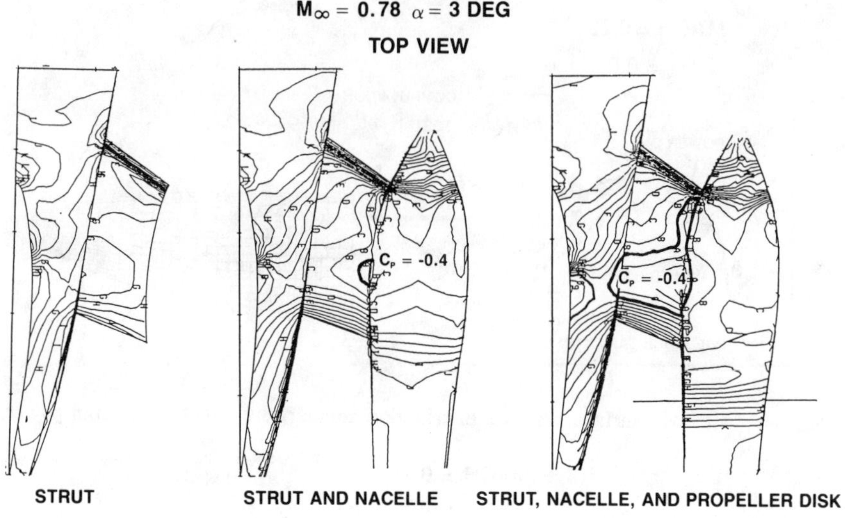

Fig. 63   **Effect of propfan on aft body strut nacelle: multiblock Euler.**

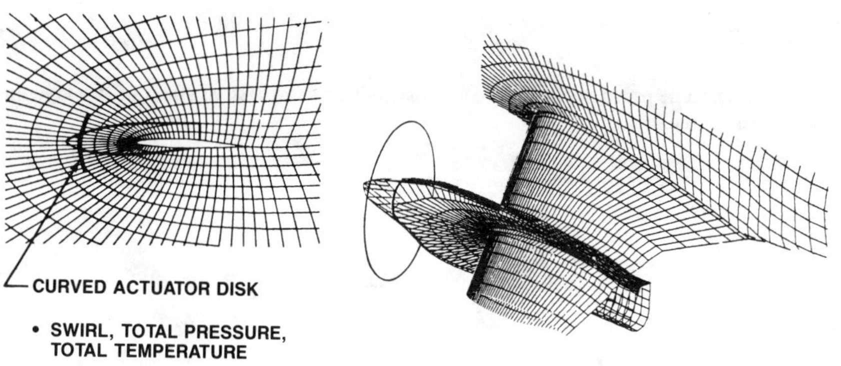

Fig. 64   **Grid structure for wing-mounted turboprop configuration.**

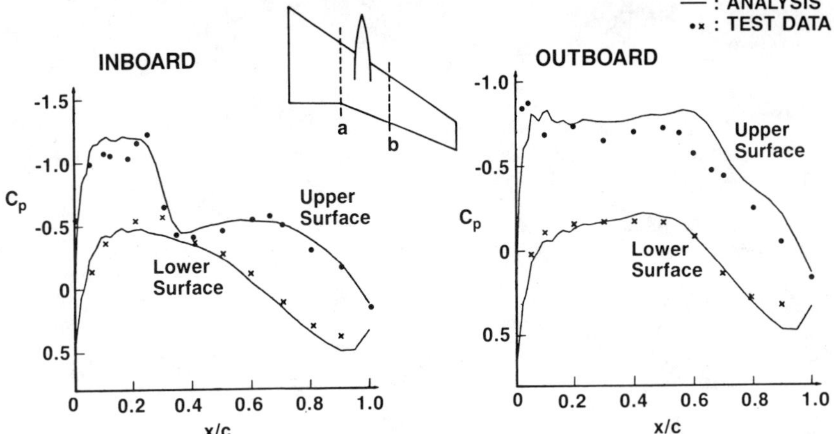

Fig. 65   Wing surface pressure distribution for a turboprop configuration: propeller off, multiblock Euler.

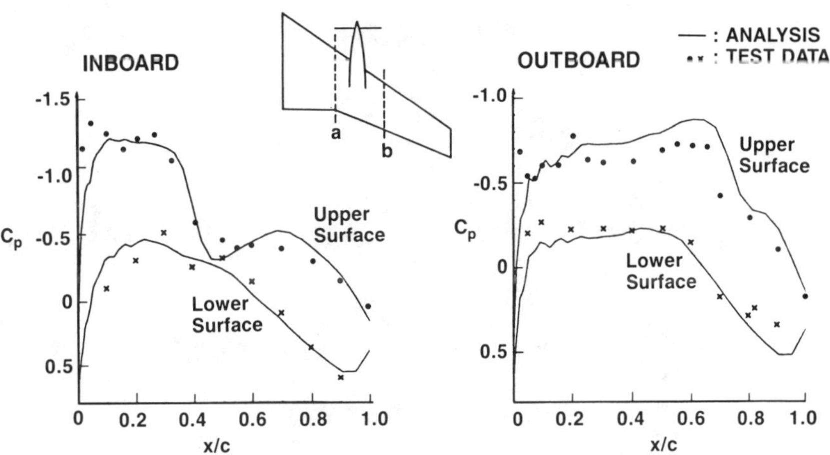

Fig. 66   Wing surface pressure distribution for a turboprop configuration: propeller on, multiblock Euler.

Computational Aerodynamics Applied to General Aviation/Business Aircraft

Neal J. Pfeiffer*
Beech Aircraft Corporation, Wichita, Kansas

Introduction

THE companies that manufacture general aviation or business aircraft are using computational aerodynamics tools more frequently than they have in the past to improve products and to design better new airplanes. Engineering budgets in these companies tend to be oriented toward specific aircraft projects, with little funding for research or methods development. In general, these companies can only support very small computational aerodynamics staffs that must work on specific applied aerodynamic problems. Given these constraints, it is obvious that these companies find it hard to commit to writing complex aerodynamics computer programs. Instead, these companies must depend to a large extent on government-sponsored research to produce the necessary aerodynamic computer codes. Fortunately, a large variety of programs are available to compute the flow past airfoils, wings, and full configurations. In addition to the government-sponsored programs, there are commercially available programs that often fill specific needs.

Geometry: The First Hurdle

The computational aerodynamics program user quickly learns that simply obtaining the appropriate code to do the desired analysis is only the beginning of the solution. Often the most time-consuming portion of the solution is the generation of a geometry model for the code to use. Typically, the geometry definition resides on a computer-aided design (CAD) system that stores the geometry as mathematical surfaces formed from parametric curves. Most aerodynamics programs, however, utilize data files composed of ordered points distributed across the surfaces. A method of translating the CAD geometry definition into a aerodynamic program input data file must be developed. To make things more difficult,

Copyright © 1989 by the American Institute of Aeronautics and Astronautics, Inc. All rights reserved.
*Staff Engineer, Aerodynamics.

there is a large variety of CAD systems and a large number of aerodynamic programs. In most cases, the transformation of the data from one system to the other is not a trivial exercise and may not have been done for this specific combination of CAD system and aerodynamic code formats before. Only recently have comprehensive geometry interface programs, such as I3G[1] and AGPS,[2] been developed by the larger airframe manufacturers.

Computational Aerodynamics Usage

Even before the geometry is fully defined, the required aerodynamics analysis or design program is chosen. The user must understand the limitations of the modeling used by the program and carefully review the results. Users should be especially careful when releasing results for a new type of geometry.

Computational aerodynamics is used in four key ways for general aviation aircraft. It is used to assist in the design, development, certification, and modification of an aircraft.

Design

The traditional design of high-aspect-ratio wings has been a two-dimensional airfoil design approach. This method is fairly simple and has a large variety of ready tools. A more recent approach method has been to treat the design completely in three dimensions. Programs for this method hold a great deal of promise and should become more widely used as these programs mature.

Combined Two- and Three-Dimensional Method

Basic configuration design work for business airplanes typically starts with the general airplane sizing and quickly moves to the wing characteristics. Often the three-dimensional wing problem is reduced to two-dimensional by defining appropriate airfoil sections and lift requirements. For wings with little or no sweep, it is easy to define the airfoil sections and lift. For swept wings, however, one should use sweep-taper theory[3] to generate curved airfoil sections that are normal to the local span lines and adjust the lift based on the effective sweep angle.

Viscous effects are usually modeled during airfoil design and analysis to better represent the airfoil characteristics, especially at low and high speeds. The low-speed characteristics of an airfoil depend on the growth of the boundary layer and when and how it separates from the surface. For high-speed airplanes, the viscous effects can cause large changes in shock position and strength.

The Eppler code[4,5] has been used widely for the design of subsonic airfoil sections. It uses a potential-flow method with boundary-layer corrections to generate an airfoil shape from an indirect definition of the shape of the pressure distribution. Initially, the method tends not to be very intuitive, but with some experience, it works quite well. It is possible to use the defining parameters to design for a range of conditions.

Some transonic design has been done with a full transonic potential flow code coupled to an optimization program.[6] This pair of programs uses the potential-flow code in the analysis mode to run multiple solutions to check for an optimum solution within set constraints. This technique allows for the design of an airfoil using lift, drag, pitching moment, and thickness as the primary design variables. It makes the procedure very intuitive to work with, but not as fast as other methods. For example, an airfoil can be designed by solving for the maximum lift-to-drag ratio with constraints on thickness and pitching moment. It would be possible to extend this technique to include off-design conditions directly in the optimization.

Some limited work has been done with TRANDES.[7] This transonic program uses a prescribed nose shape and pressure distribution over the rest of the chord length to design an airfoil. Since TRANDES does not design the leading-edge region, it must be used in conjunction with a program such as the Eppler code to define the desired leading-edge characteristics.

The ISES code[8] has probably the most potential of any of the current transonic airfoil design programs. It uses a unique method of coupling an Euler solver with the boundary-layer equations to obtain a direct solution of both. This code designs an airfoil to a prescribed pressure distribution. The generation of this target distribution requires expertise on the part of the aerodynamicist who runs the code.

There are two factors that often flavor the design of small airplane airfoil sections. First, most of these airplanes are built without powered control systems. This means that using aft cambered airfoils where there is a control surface could prove to be very challenging. Second, leading-edge high-lift devices are nearly nonexistent on this class of airplane. This means that, in order to get satisfactory high-lift characteristics, the airfoil leading-edge shape tends to be different from that for an airfoil with a leading-edge slat.

All of the previous programs have been used for the design of single-element airfoils. After a candidate airfoil has been designed, the airfoil should be checked for off-design conditions. These would include a range of angles of attack and control surface or flap deflections. Control surface or flap deflections also might well include gaps or slots.

The Eppler code has analysis capabilities and can be used to study unslotted control surface or flap deflections. It is very easy to use in the analysis mode and again yields reasonable results for subsonic cases. The mainstay of the transonic airfoil analysis has been the Bauer-Garabedian-Korn-Jameson program.[9] This program exists in many versions, with the latest ones having a good boundary-layer capability with arbitrary transition. The ISES code can also be used for airfoil analysis. This program has been shown to produce good results for a wide range of Reynolds numbers and has been used effectively for such diverse vehicles as man-powered airplanes and jet transports.

Cases with slotted flaps are normally analyzed with the NASA multiele-ment code.[10] One should spend a good deal of time working with this code in order to feel confident in using the results for high-angles-of-attack cases. Since this is where the analysis is most important for a flapped airfoil, the boundary-layer results from this program should be examined closely to check for possible problems.

After airfoils have been designed and analyzed at various off-condition points, they can be integrated into a trial wing shape. A panel method such as VSAERO[11] for subsonic cases or a full-potential method such as FLO-22[12] for transonic cases could be used to analyze the wing. Panel methods may be used for quick studies of some transonic cases. The analysis is run at high subsonic Mach numbers, and the general wing loading and possible interferences can be studied.

Business aircraft usually have wing-mounted nacelles or nacelles mounted very close to the wing on the aft fuselage. In either case, the effects on the wing can be considerable. Wing-alone analysis must therefore be followed by wing-body-nacelle analysis to get a better picture of the flow. VSAERO analysis is normally done for these cases because of the capability that it has to model complex configurations. For transonic cases, FLO-30[13] wing-body analysis or WIBCO[14] modeling follows the VSAERO computations to estimate the compressibility effects.

Canard configurations require that the entire configuration be modeled. A wing-alone analysis is of very limited use since the canard will greatly alter the shape of the span load.

Three-Dimensional Alone Method

A number of three-dimensional design methods have been under investigation. A method being developed at NASA Langley Research Center[15] appears to be interesting for business jet configurations with aft fuselage nacelles placed close to the wing. This method would allow the direct tailoring of the wing for the close proximity of the nacelle by defining target pressure distributions. Techniques of this general type could provide a faster and better way to design wings. As with some of the two-dimensional methods, there remain questions of what the best target pressure distributions should be.

Development

Once a basic configuration has been settled upon, there is an almost immediate need to establish the aerodynamic loads so that the detailed structural design can commence. Panel method analysis has been used to generate aerodynamic loads for complete configurations. These results have typically been validated by wind-tunnel or flight testing, or by comparison to similar validated analysis.

The project groups involved with system design and development can have questions about where to put inlets and exits to duct air for cooling avionics or for air conditioning. A three-dimensional potential flow model can often provide just the flow information that they require.

Once the airplane has reached the flight stage, it is possible that some flight characteristic may not be as anticipated. Computational methods can often be used to analyze the situation and give the needed additional information to identify the source of the discrepancy.

Assist Certification

Panel methods have been used to generate aerodynamic loads. The computed aerodynamic loads along with the required validation documentation become part of the certification of the airplane.

The calculation of ice shapes has been done by using computational methods. An icing program, LEWICE,[16] was developed under the funding of the NASA Lewis Research Center and the FAA technical center. This program is capable of calculating the droplet trajectories and impingements on two-dimensional airfoil sections. It then models the resulting ice accretion. For a swept wing, sweep-taper theory may be used to make a reasonable two-dimensional simplification.

A flight condition of interest is identified and a three-dimensional analysis is done for a baseline configuration, one with an angle-of-attack increment, and one with an elevator increment. A trim solution can then be found and analyzed. The trimmed analysis gives the three-dimensional pressure distribution and section lift at any point of interest on the airplane. For a straight wing the two-dimensional section and lift coefficient are directly available. For a swept wing a curved airfoil section is generated at the point of interest by sweeping out an arc that is tangent to the local span lines. This two-dimensional section is analyzed at a lift coefficient that is based on the effective sweep angle. The LEWICE program does the two-dimensional analysis of the airfoil and builds the ice shape. These shapes can then be used to build foam models of the expected ice for certification flight testing of the airplane.

Assist Modification

The current Beech product line has been very evolutionary. For example, the outboard wing panels on the Barons, the King Airs, and the 1900 commuter airliners share the same basic planform and airfoil sections as the original Bonanza wing that was designed during the 1940s. These wings use NACA five-digit sections based on the 230 mean line, which has good turbulent flow characteristics. These sections can even sustain modest amounts of laminar flow on well-maintained airplanes without ice-protection boots. The similarity between these airplanes goes even beyond the wing to include the fuselage. The major driver for all of these airplanes was to maintain similar basic tooling and minimize manufacturing costs.

Even with this stability, computational design work has been done on these airplanes to develop modifications. Typical areas of study are fairings, leading-edge cuffs, wing tips, winglets, and flap design. As with most endeavors, many more modifications are analyzed than are actually put into production.

Applications

The Beech Starship is an airplane that has used computational aerodynamics throughout its development. Figure 1 shows the twin-turboprop-powered pusher, canard configuration of this airplane.

The initial design work was begun in two dimensions. Airfoil sections were generated by using an airfoil design computer code of the same general type as the Eppler airfoil program. The airfoil camber lines were then used to define a thin-sheet geometry that was analyzed using a modified NASA Lamar-Margason vortex lattice program.[17] All of this work was done on a small personal computer by an outside consultant.

When development work within Beech started on the Starship, VSAERO was obtained to analyze the entire configuration and generate aerodynamic loads. VSAERO was chosen to do this analysis because of the ability to handle complex geometry, the ability to generate relaxed wake shapes, the perturbation potential formulation, and the commercial availability of the program.

In a conventional airplane, much of the analysis or design work can be done with the tail off. The tail is typically placed more than a wing chord length behind the wing, and the lift generated by the tail is small compared to that of the wing. This means that the tail does not greatly affect to basic wing aerodynamics. Furthermore, the wing produces a predictable downwash field so that the tail aerodynamics may be readily predicted.

For an aerodynamically stable canard airplane, however, the lift of the canard is a major portion of the total airplane lift. This lift produces a sizable downwash over only the inboard portion of the aft wing and significantly lowers the wing lift in this area. A separate wing and canard analysis scheme is still possible, but would require considerable care and iteration to handle the mutual interaction of the two surfaces. The current three-dimensional panel methods are capable of analyzing such a configuration where the lift of one surface depends so greatly on the lift of another surface.

Margason et al.[18] and Strang et al.[19] have each presented a basic comparison of panel method results with experiment for conventional configurations. Comparisons were made for several panel methods using lift and pitching moment plots.

A wind-tunnel investigation was carried out to provide force and surface-pressure data to compare with the VSAERO analysis results for the Starship configuration. A pressure-tapped 1/7-scale model was made and tested at the University of Washington 8 × 12-ft low-speed wind tunnel. Subsequently, an entry was made in the Boeing transonic wind tunnel to generate data for higher speeds up to dive. Computer runs were made for conditions that matched the wind-tunnel testing. The results were used for the internal evaluation of VSAERO and then later presented to the FAA as validation of the air loads used for the certification of the airplane.

The position of the canard wake with respect to the aft wing was of great concern during the initial development of the airplane. VSAERO analysis was done using the wake-relaxation option. It quickly became obvious that it was not difficult to make the canard wake careen through the aft-wing gridding. This caused large gradients in the basic solution of the aft-wing potential flow and rendered the pressure calculations useless there. Even with the unusable aft-wing solution, the basic track of the canard wake agreed reasonably well with wind-tunnel experiments for a range of conditions. For low angles of attack most of the canard wake passes slightly below the aft wing with the tip vortex at or slightly above the aft wing. For moderate angles of attack, the canard wake moves to the upper side of the aft wing.

A brief study was made of the Starship configuration without nacelles in order to investigate the affect of canard wake position. In one model the canard wake was placed so that it passed just slightly above the aft wing and was fixed to the intersection of the upper wing with the fuselage. In a second model the canard wake was placed slightly below the aft wing and was fixed to the intersection of the lower wing with the fuselage. Figure 2 shows that there is no significant difference in the span load with either placement, even for this moderate angle of attack.

The canard wake was placed below the aft wing for additional analysis. This was done to avoid the complexity of splitting the wake around either the aft wing or the nacelle.

Two geometrically accurate computer models of the airplane were constructed that included the wing fences just inboard of the elevons and all of the movable control surfaces. The first model was for the cruise configuration with the flaps nested and the forward wing swept aft. The second model was for the landing configuration with the flaps extended and the forward wing in the forward-sweep position to balance the pitching moment due to the flap. Again, the forward-wing wakes in both of these models extended back under the aft wing and met the fuselage at the lower aft wing to fuselage intersection. Figure 3 shows the basic cruise configuration model.

No attempt was made to fill in the small openings that were created when the control surfaces were deflected. The trailing wakes moved with the control surfaces and produced discontinuous wakes at control surface edges when deflected.

Conditions were tested and analyzed to define a baseline configuration and several perturbations. These perturbations included angles of attack and yaw, combined elevator and elevon deflection for pitch control, differential elevon deflection for roll control, and combined left and right rudder deflections.

The VSAERO program has the ability to calculate a strip boundary-layer analysis along a surface streamline. This provides a simplified means of calculating a boundary layer on a three-dimensional geometry. The version of the program that was available when this analysis was being done had problems identifying a starting point for the boundary-layer analysis on a swept wing. Since swept wings have an attachment line instead of a stagnation point, the code would occasionally try to turn along the attachment line near the leading edge and follow it upstream. Sometimes it would select a starting point somewhere along this line, but often it would not and would simply stop the calculations. The latest versions of the code have been modified to correct this problem, but as it was, the program was not reliable enough to use with viscous iterations. Therefore, all of the analysis for comparison with experiment was done without boundary-layer corrections.

Analysis results for the lift forces and pitching moments of the wind-tunnel model are presented in Figs. 4 and 5. These results are compared with the matching wind-tunnel data. The lift-curve plot shows the typical difference between inviscid analysis and measured flow. The nonlinearity in the lift-curve slope is caused by the boundary-layer thickening and eventual trailing-edge separation. The slope of the moment curve is nearly the same for the analysis and the experiment for the lower lift coefficients and departs as the lift is increased. This again is caused by the viscous effects.

Surface pressure distributions were also measured in both the wind tunnel and in flight. Figure 6 shows a comparison of analysis and wind-tunnel pressure distributions for a Mach number $M = 0.27$ case with no control deflections and 0-deg angles of attack and yaw. Figure 7 shows a similar case at $M = 0.6$. Both of these cases show good agreement between the analysis and the wind-tunnel results, but the FAA still had concerns pertaining to the scale effects for a novel configuration. This required that a flight test be carried out with external pressure belts. The company had an 85% scale proof of concept (POC) demonstrator vehicle flying at that time, and it was agreed that it was appropriate for this test.

A POC geometry model was analyzed using the angles of attack and yaw and the various control deflections that were measured by flight instrumentation. Two cases are presented here.

Figure 8 shows a comparison between analysis and POC results for a $M = 0.44$ case in straight and level flight. Figure 9 shows a similar comparison for a $M = 0.44$ case in a 2-g turn. The shapes of the pressure distributions compare favorably between the analysis and the measured results with the largest variation at the inboard canard station. During the POC development testing, a series of leading-edge cuffs were tested on the inboard portion of the canard. When the final shape was approved, measurements were made and used as a starting point for the final loft contours. The VSAERO model was built from the loft contours which may have differed slightly from the actual POC. This would cause a difference in the pressure distribution.

Computational aerodynamics was also used to study how the wing trailing-edge location between the fuselage and the nacelle would affect the total span load distribution. As the POC was being built, the trailing edge in this region was moved slightly upward from the original loft. VSAERO models were made for both configurations and were analyzed for trimmed conditions. These results showed that this slight upward movement caused a large decrease in the inboard wing lift with a mild increase over the outboard wing to balance the lift and moment.

In studying these results, it was obvious that a small droop of this short section of trailing edge would greatly lower wing loads and that this could be checked on the POC. The POC trailing edge in this region is about 3-ft wide. It was cut out and faired in 3.4 in. lower at the trailing edge. VSAERO analysis was done for both of these contours, and the resulting span loads for trimmed conditions at a constant airplane lift coefficient are presented in Fig. 10. The change in canard lift is fairly small, but the aft wing shows a very large shift in span load.

Photographic measurements of wing-tip deflection were made to compare the static position on the ground with various conditions in flight. The lowered trailing edge produced half the wing-tip deflection compared to the original trailing edge for straight and level trimmed flight at cruise speeds. For the critical loads case, this translates into approximately a 14% decrease in wing bending. Unfortunately, the POC lost approximately 7% in speed with this change. This large speed loss was totally unexpected. Blame was initially directed to diffuser losses due to flow separation in the channel formed by the fuselage, wing, and nacelle. However, flow visualization studies in this region showed no indications of separation.

When the panel method results for the two trimmed cases were more carefully examined, it was noticed that the integrated surface pressures for the entire airplane showed a drag difference that closely matched the measured speed difference. The details of the pressure integration indicated that the aft fuselage gave the largest contribution to the drag increment. The redistribution of the lift inboard induced a reduced pressure on the aft-facing taper of the fuselage, which caused an increase in the airplane drag.

The historical method for calculating induced drag has involved the integration of the far-field downwash. Surface pressure integration has been used recently with apparent success.[20]

In order to gain more confidence in the use of surface pressure integration for the calculation of induced drag, a quick study was undertaken. The Starship wind-tunnel model was analyzed for a range of angles of attack with zero control deflections. A drag polar of the following form was fit to the experimental data:

$$\text{Drag coefficient} = \text{parasite drag} + \text{induced drag}$$

where

$$\text{Induced drag} = \text{const} * (\text{lift coefficient}) ** 2$$

A similar fit was also made to the VSAERO analysis results. The "parasite drag" term from the panel method results represents the sum of method and discretization error. The parasite drag term was subtracted from both the wind-tunnel and analysis results to study just the lift-induced drag. Figure 11 shows the comparison of the analysis with the wind-tunnel results. The inviscid analysis method slightly underpredicts the measured induced drag, but this should be expected since viscous effects are lift dependent.

It appears that the surface pressure integration method of calculating induced drag can be useful, but must be done carefully. As the panel density is increased, the parasite drag should decrease, and induced drag should converge to a specific value. Surface pressure integration is probably best for evaluating the increments due to small localized geometry modifications while the majority of the model stays the same. It obviously does not give any useful information about parasite drag.

With the drag cause identified, a series of trailing edges were analyzed and then checked in the wind tunnel to evaluate the loads the drag impact.

The final configuration represents a tradeoff between the wing loads and the airplane drag.

Computational Aerodynamics Challenges

Aerodynamic analysis has not become standarized in the same way that structural analysis has become standardized. Aerodynamic computer analysis is fundamentally different from the current structural computer analysis in one important aspect. The aerodynamics engineer has to be concerned with nonlinear viscous and separation effects that are a regular occurrence. The structural engineer, on the other hand, often deals with elements that are constrained to have linear stress-strain relationships in order to have structural integrity. Because of these aerodynamic nonlinear viscous effects and the uncertainty in modeling turbulence, even the best current computational aerodynamics tools are inherently somewhat approximate.

To complicate matters, the computational methods tend to be most approximate for two key design problems. The first is maximum lift, and the second is cruise drag. The project engineer would like to have a good estimate of maximum lift in order to set a stall speed and all of the other parameters that stall speed controls. The maximum lift of an airfoil can be estimated with reasonable certainty, but the calculation of the maximum lift of a full flapped configuration is not possible with current computational aerodynamics. This means that stall speed cannot be directly calculated. For the cruise drag case, the viscous effects are better behaved and an integrated drag can be calculated. Although this drag may serve well for doing relative comparisons for small modifications, it is typically not accurate in an absolute sense.

Future Directions

Beech will continue to use two-dimensional methods because they are easy to use and can give reasonable estimates of the drag level and the maximum lift coefficient. There are standard ways to extend these results into three dimensions and provide a good first cut for general aviation airplanes. Three-dimensional panel methods will continue to be used heavily to analyze complex configurations. Higher-speed analysis will initially continue with either two-dimensional transonic methods being extended by sweep-taper theory or three-dimensional small-disturbance methods that can handle more complex configurations. Attention is being given to the modern gridding programs, especially in following how they can be used in conjunction with new Euler codes. These methods are allowing more and more complex geometries to be studied with fewer assumptions. The TRANAIR program[21] development is being followed to evaluate it for possible future work.

Most three-dimensional methods are currently inviscid, and many cases could be better solved with viscosity included. The question of how viscosity should be handled remains split between those favoring the coupling of a potential or Euler code with boundary-layer analysis and those favoring a form of Navier-Stokes analysis. In the near term, one can hope that more will be done along both paths.

At this point, general aviation will continue to use analysis codes developed by others. Panel methods, small-disturbance transonic, and full-potential methods have been the mainstay in the past. As the newer methods mature and allow the calculation of more complex configurations, these methods will be acquired and used.

References

[1]Snepp, D. K., and Pomeroy, R. C., "A Geometry System for Aerodynamic Design," AIAA Paper 87-2902, 1987.

[2]LaBozzetta, W. F., "Interactive Graphics for Geometry Generation—A Program with a Contemporary Design," AIAA Paper 84-2389, Oct. 1984.

[3]Boppe, C. W., "Computational Aerodynamic Design: X-29, the Gulfstream Series and a Tactical Fighter," Society of Automotive Engineers Paper 851789, 1985.

[4]Eppler, R., "A Computer Program for the Design and Analysis of Low-Speed Airfoils," NASA TM-80210, 1980.

[5]Eppler, R., "Supplement to: A Computer Program for the Design and Analysis of Low-Speed Airfoils," NASA TM-81862, 1981.

[6]Hicks, R. M., "Transonic Wing Design Using Potential-Flow Codes—Successes and Failures," Society of Automotive Engineers Paper 810565, 1981.

[7]Carlson, L. A., "TRANDES: A Fortran Program for Transonic Airfoil Analysis or Design," NASA CR-3376, Dec. 1980.

[8]Drela, M., and Giles, M. B., "ISES: A Two-Dimensional Viscous Aerodynamic Design and Analysis Code," AIAA Paper 86-0424, 1986.

[9]Bauer, F., Garabedian, P., and Korn, D., "A Theory of Supercritical Wing Sections, with Computer Programs and Examples," *Lecture Notes in Economics and Mathematical Systems*, Springer-Verlag, New York, 1972.

[10]Stevens, W. A., Goradia, S. H., and Braden, J. A., "Mathematical Model for Two-Dimensional Multi-Component Airfoils in Viscous Flow," NASA CR-1843, 1971.

[11]Maskew, B., "VSAERO—Theory Document," NASA CR-4023, Nov. 1986.

[12]Jameson, A., Caughey, D. A., Newman, P. A., and Davis, R. M., "A Brief Description of the Jameson-Caughey NYU Transonic Swept Wing Computer Program—FLO 22," NASA TM-X-73996, 1976.

[13]Caughey, D. A., and Jameson, A., "Progress in Finite-Volume Calculations for Wing-Fuselage Combinations," AIAA Paper 79-1513, 1979.

[14]Boppe, C. W., "Aerodynamic Analysis for Aircraft With Nacelles, Pylons, and Winglets at Transonic Speeds," NASA CR-4066, April 1987.

[15]Campbell, R. L., and Smith, L. A., "A Hybrid Algorithm for Transonic Airfoil and Wing Design," AIAA Paper 87-2552, 1987.

[16]Ruff, G. A., and Berkowitz, B. M., "Users Manual for the NASA Lewis Ice Accretion Prediction Code (LEWICE)," NASA CR-185129, 1989).

[17]Margason, R. J., and Lamar, J. E., "Vortex-Lattice FORTRAN Program for Estimating Subsonic Aerodynamic Characteristics of Complex Planforms," NASA TN D-6142, Feb. 1971.

[18]Margason, R. J., Kjelgaard, S. O., Sellers, W. L., Morris, E. K., Walkley, K. B., and Shields, E. W., "Subsonic Panel Methods—A Comparison of Several Production Codes," AIAA Paper 85-0280, 1985.

[19]Strang, W. Z., Berdahl, C. H., Nutley, E. L., and Murn, A. J., "Evaluation of Four Panel Aerodynamic Prediction Methods (MCAERO, Pan Air, Quadpan, and VSAERO)," AIAA Paper 85-4092, 1985.

[20]Vijgen, P. M. H. W., Van Dam, C. P., and Holmes, B. J., "Sheared Wing-Tip Aerodynamics: Wind-Tunnel and Computational Investigations of Induced-Drag Reduction," AIAA Paper 87-2481, 1987.

[21]Samant, S. S., Bussoletti, J. E., Johnson, F. T., Burkhart, R. H., Everson, B. L., Melvin, R. G., and Young, D. P., "TRANAIR: A Computer Code for Transonic Analyses of Arbitrary Configurations," AIAA Paper 87-0034, Jan. 1987.

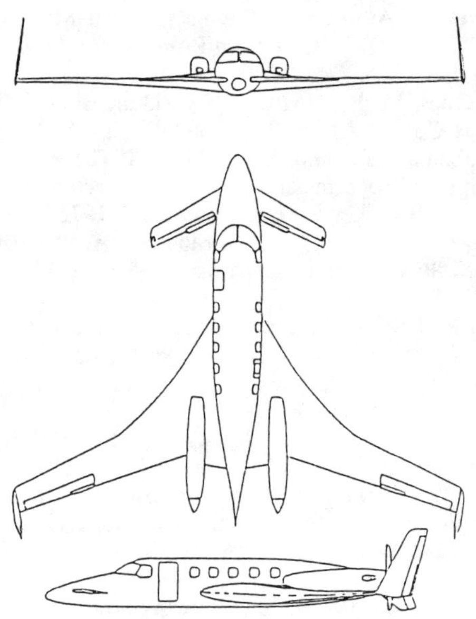

Fig. 1   Starship configuration.

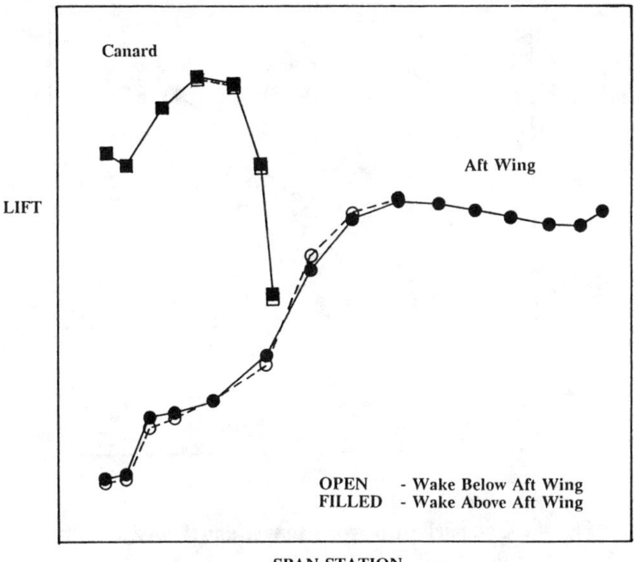

Fig. 2 Effect of wake placement on span load.

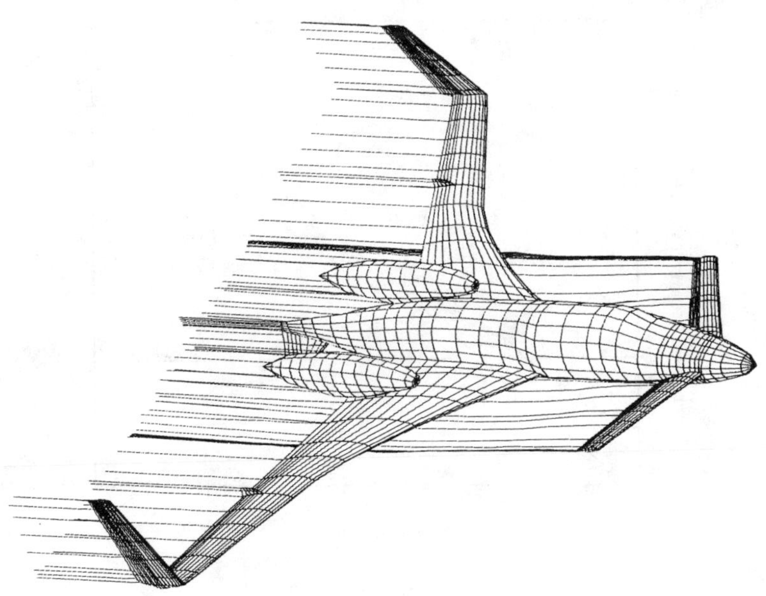

Fig. 3 Surface panels and wakes for cruise configuration.

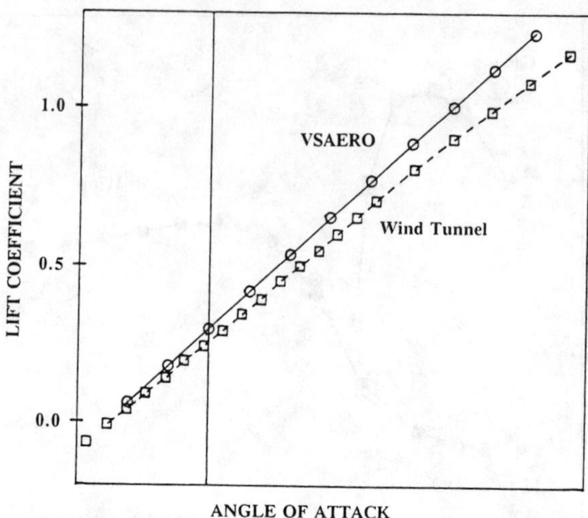

Fig. 4   Comparison of wind-tunnel and VSAERO lift.

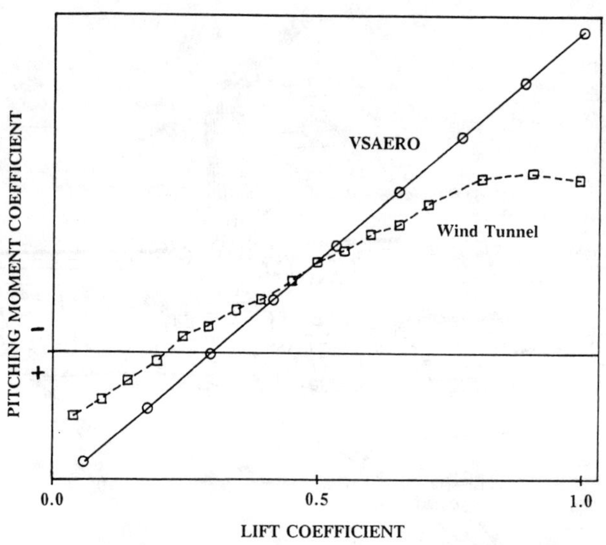

Fig. 5   Comparison of wind-tunnel and VSAERO pitching moments.

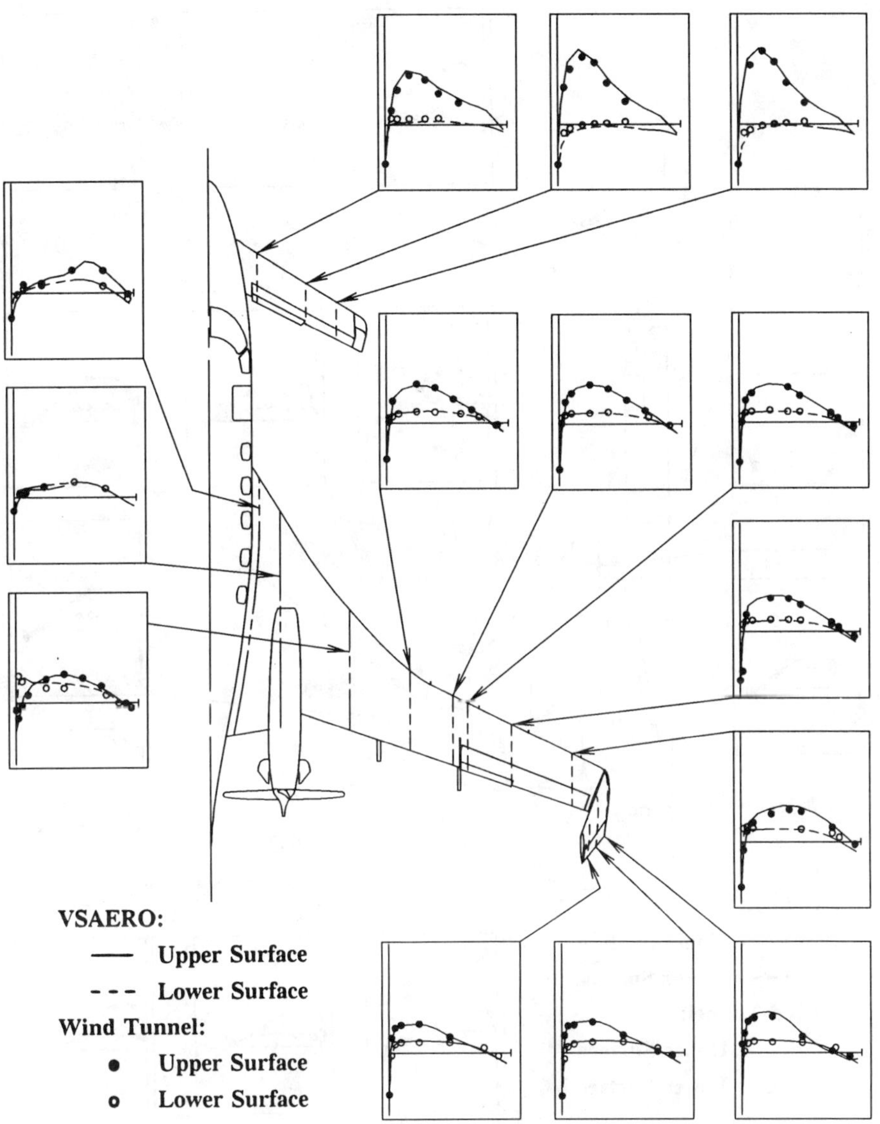

Fig. 6 Analysis to experimental pressure distribution comparison for a baseline, $M = 0.27$, wind-tunnel case.

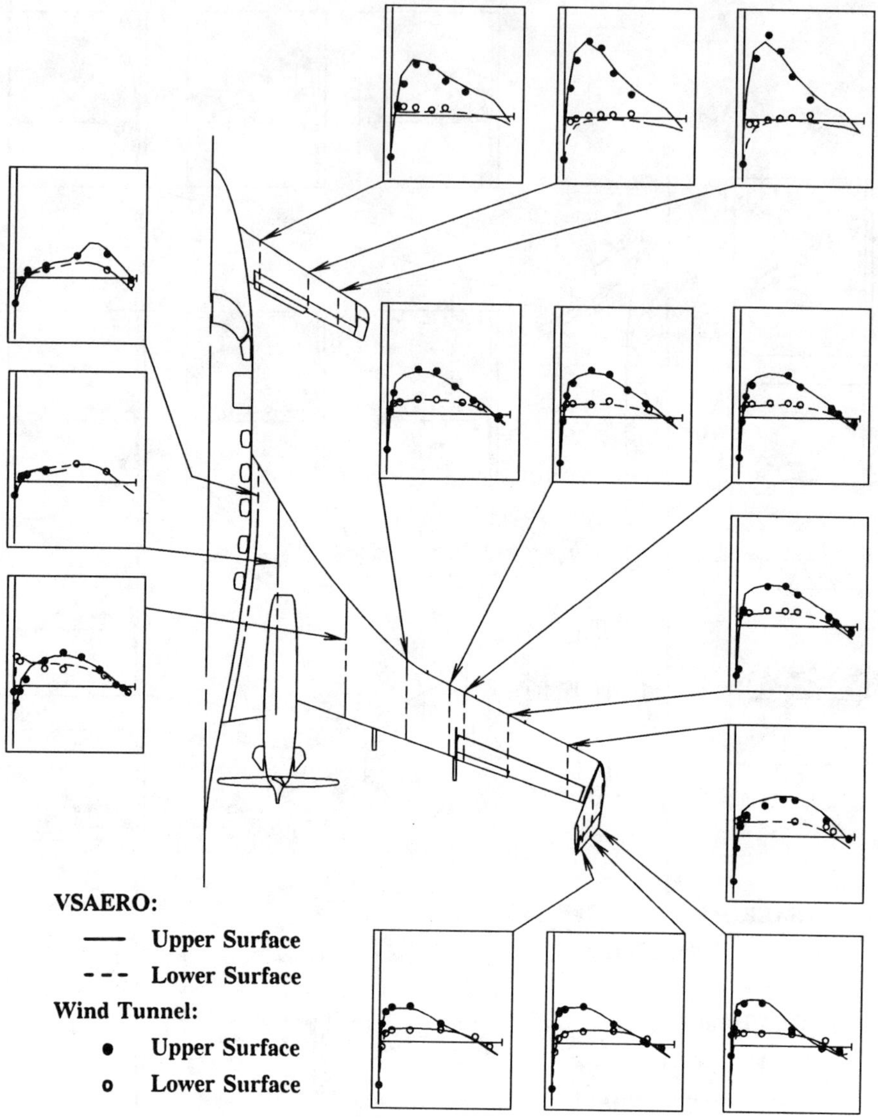

Fig. 7   Analysis to experimental pressure distribution comparison for a baseline, $M = 0.60$, wind-tunnel case.

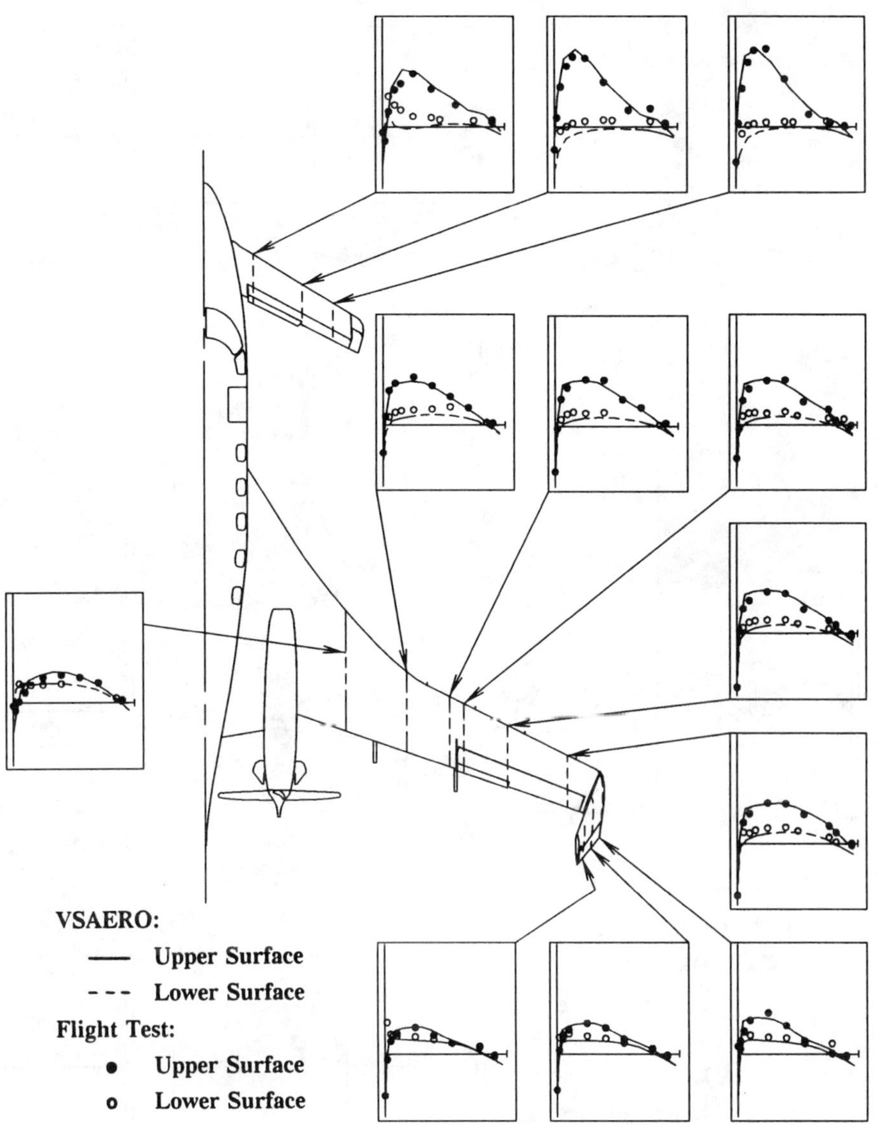

Fig. 8 Analysis to experimental pressure distribution comparison for a POC, $M = 0.44$, straight and level case.

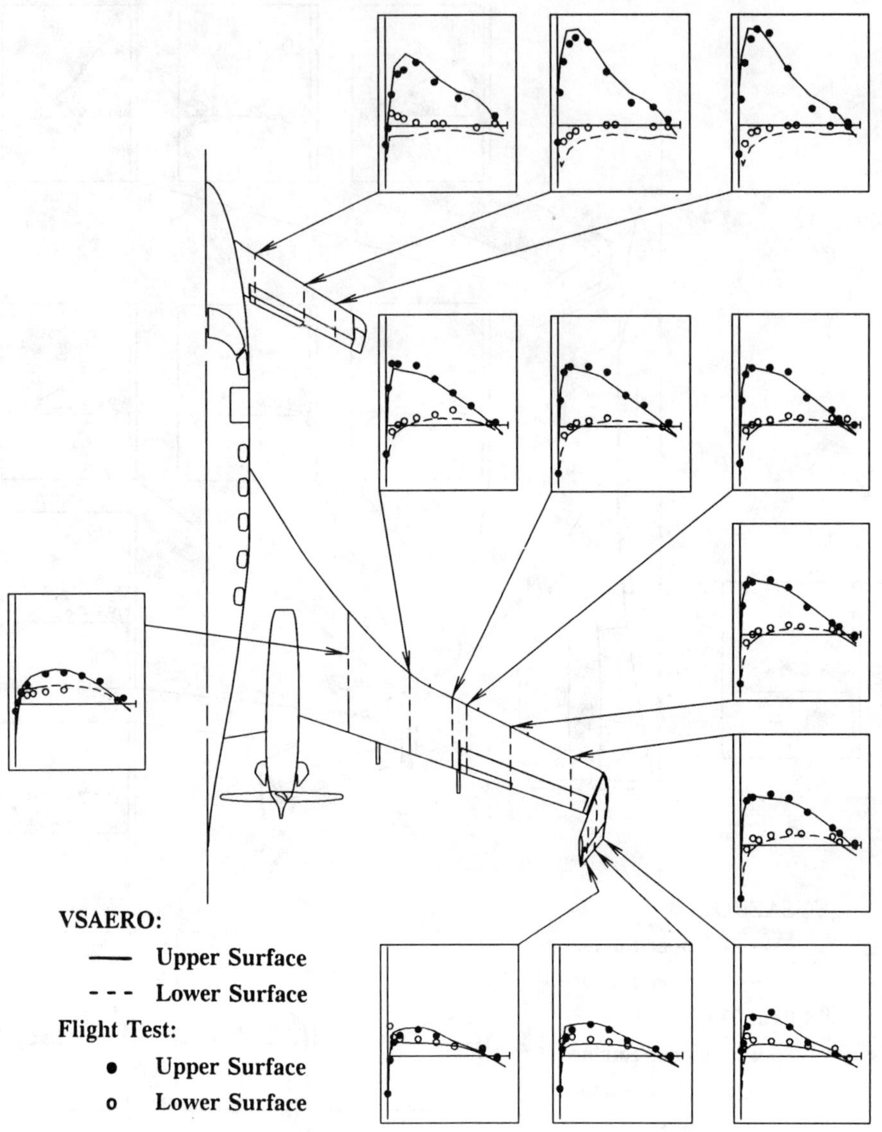

Fig. 9 Analysis to experimental pressure distribution comparison for a POC, $M = 0.44$, 2-*g* turn case.

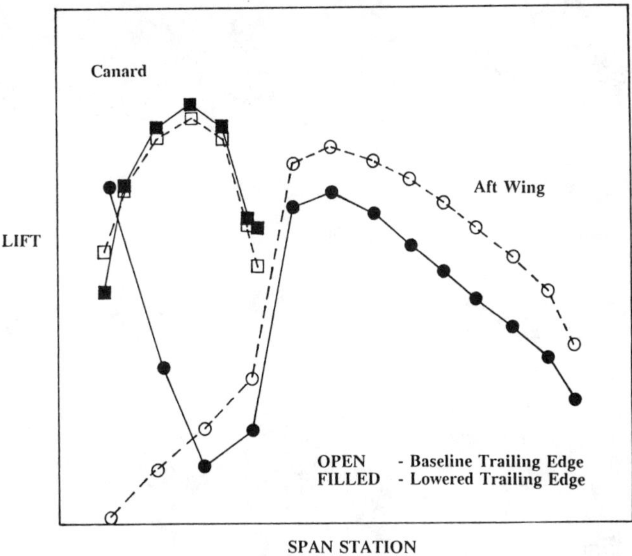

Fig. 10 Trimmed span load comparison for two inboard trailing-edge heights.

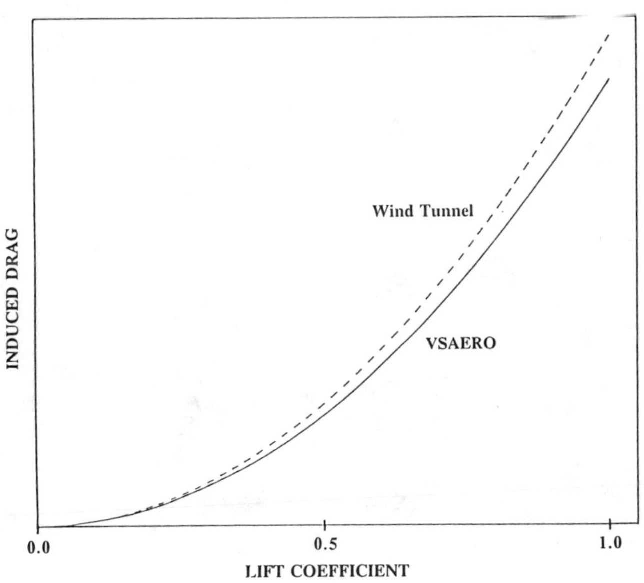

Fig. 11 Analysis vs wind-tunnel-induced drag.

Chapter 17

Application of Computational Fluid Dynamic Methods to the Component Aerodynamic Design of the V-22 Osprey Tilt Rotor Vehicle

J. C. Narramore [*]

Bell Helicopter Textron, Inc., Fort Worth, Texas

Nomenclature

C_d = airfoil section drag coefficient
C_l = airfoil section lift coefficient
C_m = airfoil section pitching moment coefficient about the quarterchord
C_p = pressure coefficient
C_t = rotor thrust coefficient
H_{32} = energy thickness form parameter (energy thickness/momentum thickness)
M = drag divergence Mach number
M_{DD} = freestream Mach number
MUE = boundary-layer transition "roughness parameter"
r/R = blade radial station/total blade radius
Re = Reynolds number based on chord length
Re_θ = Reynolds number based on momentum thickness
t/c = airfoil maximum thickness divided by chord length
x/c = airfoil chord location/chord length
$\Delta C_{l_{\max}}$ = increment in maximum lift coefficient above zero flap deflection maximum lift coefficient
σ_T = rotor thrust weighted solidity

Introduction

DURING the Joint Services Vertical Lift Aircraft (JVX) Program design studies leading to the development of the V-22 Osprey, emphasis was placed on producing excellent performance levels within the constraints of the mission requirements for this vehicle. In order to achieve this goal and to reduce design risks, computational fluid dynamic (CFD) methods were utilized to develop efficient aerodynamic contours. Numerous trade studies have been carried out by using state-of-the-art com-

---

Copyright © 1989 by Bell Helicopter Textron Inc. Published by the American Institute of Aeronautics and Astronautics, Inc., with permission.
*Group Engineer, Aerodynamic Technology Division.

putational aerodynamic methods to tailor the contour of aerodynamic components for optimum performance at critical design points using both iterative-direct and inverse techniques. Significant guidance in determining the lines for the V-22 were obtained through the use of these computational methods. Wind-tunnel test results of the aerodynamic designs and correlation to theoretical analyses have substantiated the benefits obtained from using computational methods for design of aerodynamic components.

Objective

This paper will give an overview of computational fluid dynamic methods that were used to develop the aerodynamic shapes for the V-22 tilt rotor vehicle. Because so many aspects of this design were impacted by CFD methods, these descriptions will necessarily be brief. However, it is hoped that the references given will lead the reader to greater detail in specific areas of interest.

Aerodynamic Design of the Blade Sections

Introduction to the Blade Design Problem

An effort to design improved airfoils for tilt rotor application has been carried out at Bell Helicopter. This was done to produce aerodynamic contours that would improve the maneuverability in helicopter mode, to improve the propeller efficiency in the airplane cruise mode condition, to allow for high speeds with no compressibility drag buildup on the rotor blades, and to improve the hover performance compared to the NACA 64 series airfoils on the XV-15. It was hoped that better performance for the V-22 could be achieved if there were significant airfoil performance gains

Table 1   Tilt rotor airfoil design goals and constraints

Radial station r/R	Design constraints		Aerodynamic design optimization goals			
	t/C	Incomp. C_m	Maneuver $(C_{l_{max}})$	Cruise (C_d)	Max. speed (M_{DD})	Hover (L/D_{max})
1.0	0.8	-0.02	1.35 at $M = 0.6$	0.006 at $C_l = 0.3$, $M = 0.75$	0.81 at $C_l = 0.3$	80.0 at $M = 0.65$
0.75	0.12	-0.03	1.40 at $M = 0.45$	0.006 at $C_l = 0.2$, $M = 0.65$	0.72 at $C_l = 0.2$	95.0 at $M = 0.5$
0.50	0.18	-0.05	1.50 at $M = 0.03$	0.007 at $C_l = 0.0$, $M = 0.57$	0.64 at $C_l = 0.0$	80.0 at $M = 0.3$
0.25	0.28	-0.12	1.35 at $M = 0.19$	0.018 at $C_l = 0.0$, $M = 0.51$	0.59 at $C_l = 0.0$	50.0 at $M = 0.2$

over the baseline XV-15 sections that provided good performance for this vehicle.

The objective during the design of the V-22 rotor airfoil sections was to take advantage of advances in airfoil design technology to develop optimized sections specifically for four different radial stations on a tilt rotor blade. In order to accomplish this goal, design criteria were established and prioritized based on the envelope of flight conditions of the blade. These airfoil performance goals were established to provide performance improvements at critical conditions in its flight envelope relative to XV-15 airfoils. Table 1 gives the design goals and constraints that were established for four radial stations on the blade. These criteria were then used to develop completely new advanced technology airfoils specifically for tilt rotors.[1]

Description of CFD Codes

The sections were designed using several computational fluid dynamic codes that are integrated into the aerodynamic design and analysis methodology (ADAM) system.[2] In order to develop the rotor airfoil contours, an inverse airfoil design method[3] was used to determine desired pressure distributions to be input into an inverse conformal mapping airfoil design routine. The resulting sections were evaluated using both panel methods including the Eppler program[4] and the North Carolina State University version of the Lockheed program[5] and potential methods including the Bell Helicopter laminar/turbulent version of the Bauer-Garabedian-Korn (BGK) program[6] and the GRUMFOIL program.[7] These codes were used to refine the candidate airfoils during early studies[8] that led to the development of the V-22 rotor airfoils.

The Bell Helicopter laminar/turbulent version of the Garabedian-Korn program was developed to allow for the computation of a displacement thickness due to laminar flow and to compute a transition location. The original transonic analysis code written by Bauer, Garabedian, Korn, and Jameson neglected the effect of a laminar boundary layer in the nose region of the airfoil. For sections operating at Reynolds numbers above 20 million, this is a valid assumption. However, for tilt rotor airfoil sections that typically operate in a range of Reynolds numbers from 1 to 5 million, this is not the case. Therefore, in an effort to improve the accuracy of this code for application at low Reynolds numbers, it was decided to add a laminar boundary-layer calculation.

The laminar boundary-layer calculation scheme used is the Walz method II.[9] This method uses an integal momentum and energy equation formulation. It can be applied to the computation of laminar and turbulent boundary layers in arbitrarily accelerated or retarded incompressible and compressible flows with or without heat transfer. For this enhancement to the transonic program only laminar calculations without heat transfer were used.

The transition from laminar flow to turbulent flow is determined by a method of Eppler.[10] In this method an empirical criterion for boundary-layer transition based on the momentum thickness Reynolds number is used. Transition is assumed to occur if

$$\ell n(Re_\theta) \geqslant 18.4H_{32} - 21.74 - 0.36 * (MUE - 3)$$

In this equation MUE is a "roughness factor" that may be used to model differing levels of roughness on the transition location. A roughness factor of $MUE = 3$ corresponds to natural transition on a smooth surface with no freestream turbulence. Increasing the value of MUE causes the transition to occur earlier on the surface of the airfoil. This roughness factor can be used to model different turbulence levels found in different wind tunnels.

In incorporating this method into the original code, the effective airfoil is modified in the laminar flow region by adding the displacement thickness to the airfoil and producing a new airfoil shape. This approach is consistent with the original code in the turbulent flow region. Figure 1 shows a comparison of the results for boundary-layer displacement thickness in the nose region of the upper surface on an airfoil from the original code to the BHTI version that includes laminar flow computations. The BHTI version of the code was run first to calculate a transition location and the laminar boundary-layer displacement. The original version of the code was then executed using the transition location computed previously as the input for this code. Figure 1 shows that the new version of the code models a laminar flow displacement effect and transition. The pressure distributions from the two codes for this case are given in Fig. 2. The differences in the pressure distribution are very small, and there are subtle differences in the drag and moment coefficients even with the same transition location.

Prior to the development of the BHTI version of this code, the start of the turbulent flow was usually arbitrarily set at 5% chord on the upper and lower surface. Figures 3 and 4 show the differences in drag and pitching moment, respectively, that occur when the original code is run with a turbulent boundary layer starting at 5% chord on the upper and lower surface and when the BHTI version is run with natural transition. The drag is slightly changed, and the pitching moment has a different characteristic at the high Mach numbers. This result emphasizes the importance of the laminar boundary-layer development and transition location on the airfoil performance characteristics.

Examples of Evaluations Performed

This and other airfoil analysis codes were used to evaluate the performance characteristics of candidate V-22 tilt rotor blade airfoil sections. Figures 5 and 6 show comparisons of the lift coefficient as a function of angle of attack and drag coefficient as a function of lift coefficient for two 12% candidate airfoils at the 0.75% radius station as predicted by the Eppler program.[4] In Figs. 7 and 8 the same comparison is made using the North Carolina State University program.[5] Although these codes predict different levels of lift and drag, the trends are similar from the two codes. In both cases the lift is higher for the V-22 airfoil compared to the XV-15 baseline airfoil. Figures 9 and 10 compare the drag coefficient and pitching-moment coefficient as a function of Mach number at a lift coefficient of 0.2 for these two candidate airfoils from the BHTI version of the Garabedian-Korn program. The same comparison is made in Figs. 11 and 12 using the GRUMFOIL code.[7] These figures indicate that the trends predicted by these two different codes are the same. The drag divergence Mach number of the V-22 airfoil is slightly lower than that for XV-15 airfoil, but the drag level below drag divergence is lower for the V-22 airfoil. Figures 10 and 12 show that both codes predict that the V-22 airfoil will have a smaller nose-down pitching moment than the baseline section. During the design studies that led to the selection of the V-22 airfoils, these types of comparisons were made in order to give confidence that the design goals could be met. Over 13 candidate airfoils for four radial stations on the V-22 rotor blade were extensively analyzed using two-dimensional CFD methods.

Wind-Tunnel Tests of Rotor Airfoils

Wind-tunnel models of 13 candidate sections were built and tested in the two-dimensional channel insert of the United Technologies Research Center's (UTRC) large subsonic wind tunnel in 1980.[11] Lift, drag, and pitching-moment aerodynamic performance data were acquired for a series of Mach numbers up to 0.81 at low angles of attack and at angles of attack up to 34 deg at low Mach numbers. The airfoils for the XV-15 tilt rotor were previously tested in this wind tunnel. The design goals were prioritized prior to the testing, and the wind-tunnel tests were used to select the V-22 airfoils based on this wind-tunnel test data. Analyses were carried out to determine the impact on total performance of the vehicle of each of the candidate sections. Based on evaluations carried out at both Bell Helicopter and Boeing Vertol, the sections designed using the subsonic inverse design technique in the ADAM system were selected for application on the blade.

In Fig. 13, the 12% thick airfoil developed in the inverse design method is compared to the baseline XV-15 airfoil of the same thickness. Several features that were considered during this design study are depicted around the new airfoil. The pressure distribution on the upper surface of the nose was developed so that no spikes in the pressure would be obtained at high angles of attack. On the upper surface at the design condition for the drag divergence, the pressure distribution was set so that no strong shock would be produced. Pressure in the recovery region was set to produce a favorable development of the turbulent boundary-layer momentum thickness. On the lower surface the pressure was set so that no pressure spikes would occur at low lift coefficients and the pressure would be favorable on the lower surface at the cruise design condition. Pressure recovery on the lower surface was set not only to produce favorable boundary layer but also to set the pitching moment to the same value as the baseline 64 series airfoil.

Figure 14 shows a comparison of the maximum lift coefficient as a function of Mach number for the selected 12% thick V-22 rotor airfoil compared to the results for the 12% thick XV-15 airfoil from the same wind tunnel. Maximum lift at a retreating blade condition of 40 knots was the first priority goal for the 12% thick rotor airfoil. Figure 14 shows that a significant increase in this performance parameter at its design point was obtained from the design developed using the inverse method. All of the design goals for the new airfoils were met or exceeded by the airfoils that were designed using the inverse design methodology.

Additional testing of these sections was carried out in the Boeing Supersonic Wind Tunnel (BSWT) facility in Seattle.[12] The contours tested in BSWT were very close to those tested at UTRC except for small modifications to meet the required V-22 trailing-edge thickness requirement. This trailing-edge thickness was achieved by subtracting a wedge from the surface of the airfoil. Specific comparisons of these results are given for the XN12 airfoil. Figure 15 shows a comparison of lift coefficient vs angle of attack at a Mach number of 0.45 for the XN12 airfoil as tested both at UTRC and in BSWT 592. Test BSWT 592 data yield a zero-lift angle of attack that is 0.5 deg more negative and a maximum lift that is 0.1 higher than the UTRC data. These variations may be attributed to a small camber increase introduced by the trailing-edge modification and a difference in the operating Reynolds number. An increase in camber would cause the negative shift in the zero-lift angle and contribute to an increase in maximum lift. The higher Reynolds number for the BSWT data (7.6 million vs 4.0 million) would also cause an increase in maximum lift.

Figure 16 compares the drag polars from the two tests at a Mach number of 0.45. The polars indicate that the UTRC drag levels for the smooth models were lower in the low lift region and higher at the high lifts. The lower minimum drag can be attributed to the lower turbulence level of the UTRC facility, which will allow more extensive laminar flow regions and hence more pronounced drag "buckets." The higher drag of the UTRC data at lift coefficients above 0.8 can be, at least in part, attributed to the lower Reynolds number and lower camber of the UTRC model. Some differences are also due to differences between test facilities.

The effect of Mach number on the drag at a lift coefficient of 0.0 is shown in Fig. 17. Because of the lower turbulence level in the test section, the drag level from the UTRC test is always lower than the BSWT 592 level. The drag divergence Mach number from the UTRC test is approximately 0.04 higher than that achieved at BSWT. These differences may be due to the wind-tunnel characteristics at high subsonic Mach numbers.

It should be noted that the BSWT wind-tunnel test results, which produced higher drag and lower drag divergence Mach numbers, were used to establish the section performance characteristic guarantees for the V-22 rotor blade. These data were condensed into lift, drag, and moment characteristics as a function of angle of attack and Mach number which were used in the performance simulation routines for the V-22.

The results given here point out some of the uncertainties surrounding the interpretation of wind-tunnel test results. Different wind tunnels almost always produce different performance characteristics for the same airfoil.[13] Quite frequently CFD methods can be used to help interpret the wind-tunnel data and make logical conclusions from the results.[14]

Correlation of Test and Analysis

After each of these wind-tunnel tests were completed, correlation of the predicted analysis to the wind-tunnel data was carried out. Examples of some of these correlations are presented here.

Figures 18–21 show the correlation of ADAM theoretical results for lift, drag, drag divergence, and pitching moment, respectively, to the BSWT tunnel test results on the 12% thick V-22 rotor airfoil (XN-12). In Fig. 18 the lift coefficient as a function of angle-of-attack comparison between theory and test at a Mach number of 0.40 is given. The theoretical results are from the Eppler program.[4] As can be seen the theoretical results correlate well to the BSWT test data at this condition. The drag coefficient as a function of lift coefficient comparison for this condition is given in Fig. 19. It shows that the theoretical drag is slightly higher than the value measured in the tunnel but gives a very good trend.

A comparison of drag as a function of Mach number at a lift coefficient of 0.0 is shown in Fig. 20 for the test and theory. The Bell Helicopter version of the Garabedian-Korn program was used to calculate the theoretical results. As shown in Fig. 20, the theoretical drag level corresponds closely to the BSWT results, but the theoretical drag divergence Mach number is higher than the BSWT test results indicate. It should be noted that the theory correlates better with the UTRC test results for drag divergence Mach number.

Figure 21 shows a comparison of theoretical and test results for the pitching moment as a function of Mach number at a lift coefficient of 0.0. It shows that the theoretical results are in general agreement with the BSWT test results, but the Mach number at which the pitching-moment divergence starts is overpredicted.

During the design of the V-22 airfoils, two-dimensional potential-flow codes represented the state of the art for production airfoil design and analysis methods. However, since that time, significant advances in airfoil design and analysis methods have been realized. Currently, Navier-Stokes codes are used to resolve issues about the validity of wind-tunnel test data in the highly nonlinear regions, to enhance test data, and to design new airfoil sections. At BHTI, two different Navier-Stokes airfoil analysis codes are currently being used. These include the ARC2D code[15] and the Georgia Institute of Technology two-dimensional Navier-Stokes code (GIT2DNS) that was originally developed by Sankar and Tang.[16] A new inverse design code has recently been developed that uses a residual-correction algorithm in conjunction with a two-dimensional Navier-Stokes code.[17] These codes are being correlated with the wind-tunnel data for the V-22 airfoils and other airfoils to determine the factors needed for adequate prediction.

Results of Blade Tests

The new airfoils were used, along with the planform and twist rate developed by momentum element method studies, to define the contour of the V-22 blade. Figure 22 illustrates the blade contour that resulted.

Many model tests have been conducted on the V-22 configuration.[18] These included, but were not limited to, unpowered drag and stability tests, powered aerodynamic interaction tests, aeroelastic stability tests, spin tests, and a large-scale scale rotor performance and wing download test. In all of these tests the rotor performance met or exceeded the design goals.

A large-scale experimental investigation was conducted at the NASA Ames Research Center Outdoor Aerodynamic Research Facility (OARF) to accurately measure the hover performance of the V-22 Osprey rotor.[19] Blades with a radius of 25 ft, representing a 0.658-scale model of V-22 blades, were tested at the OARF. The balance system used at this facility was designed to be very sensitive to rotor thrust and torque, with minimal interactions caused by other forces, moments, or thermal effects. This blade, which has the new airfoils distributed along the radius, showed significant figure of merit increases over the baseline XV-15 blade as shown in Fig. 23. This is a very good indication that the goal of increasing the performance of the V-22 airfoils relative to the XV-15 airfoils was accomplished.

Aerodynamic Design of the Wing/Flap Section

Wing Section Design Approach

The wing section was developed using the design and analysis capabilities afforded by the use of Bell Helicopter Textron's ADAM system. In order to minimize the number of iterations required to produce an airfoil with desired characteristics, several design capabilities have been developed in the ADAM system.

One approach taken in the ADAM airfoil design methodology is to assume generalized equations for the velocity distribution along the perimeter of the airfoil. Advantages of using equations for the velocity distribution stem from the fact that it minimizes the input (since only a few coefficients are used); allows fast evaluation of lift, moment, etc., by integrated equations; and thus minimizes the number of design iterations. The design method evaluates families of velocity distributions to determine which distribution on the airfoil will best satisfy the performance requirements. Once a pressure distribution is determined that satisfies the design requirements, the geometry of the airfoil contour is calculated using an inverse airfoil design method.

During the design of the airfoil for the V-22 wing, the ADAM subsonic design component was used[20] to develop the section coordinates. This design method had been substantiated by wind-tunnel testing[21] prior to the start of the V-22 design study.

Wing Airfoil Design Goals and Constraints

Airfoil design objectives and priorities were established for a new wing section based on anticipated tilt rotor wing flight conditions. Maneuverability in the transition corridor was the first priority goal. Therefore, since this parameter is limited by stall at low Mach numbers, the first priority goal for the new wing airfoil was to increase the maximum lift over the baseline NACA 64A223 airfoil used on the XV-15. The second priority for the new section was to produce low compressibility drag at a high-speed and high-altitude level flight condition. Successfully obtaining this goal would allow for efficient cruise at 30,000-ft altitude for speeds up to 300 KTAS. Third in priority was the drag value for a low-altitude cruise condition. It was established to allow for efficient cruise at sea level and 250 KTAS. To obtain the desired lift values on such a thick wing section, it was felt that rear loading was required. In order to ensure that the lower surface would not stall at low lift coefficients from this aft loading, a design goal was established for the theoretical separation location on the lower surface. A final priority was established for the pitching moment on the wing section at moderate Mach numbers. This was done to limit the trim drag at forward center-of-gravity conditions. Table 2 summarizes the aerodynamic design goals established for the V-22 wing section. Geometric goals and constraints were also established in order to provide structural and dynamic response fidelity. The thickness requirements were established to provide an efficient spar torque box based on a structural and aerodynamic optimization trade study.

In order to provide a very wide transition corridor, a maximum lift increment due to flap deflection goal was established for the wing section. The goal was to provide a maximum lift coefficient of 2.2, which would allow for a 60-knot-wide transition corridor at a 20-deg nacelle angle on the V-22.

Wing and Flap Design: Codes Used

The resulting 23% thick wing section was tailored specifically for the lift coefficients, Mach number-Reynolds number conditions, and geometric constraints appropriate for the V-22 wing using the incompressible inverse design approach. Desired pressure distributions were evaluated using the subsonic design methodology[3] and the best distribution chosen to serve an input to an incompressible inverse design code. Figure 24 gives an input pressure distribution and the resulting airfoil shape that was produced.

Table 2   Tilt rotor wing airfoil aerodynamic design goals

Priority	Design goal	Flight condition	C_l	$Re \times 10^{-6}$	M
1	$C_{l_{max}} = 1.67$	Transition maneuverability at sea level	——	11.8	0.197
2	$M_{DD} = 0.509$	300 KTAS at 30,000 ft	0.847	12.1	——
3	Low drag	250 KTAS at sea level	0.457	22.3	0.378
4	Min. separation on lower surface	305 KTAS at sea level	0.307	27.2	0.461
5	Pitching-moment coef. $< \lvert -0.06 \rvert$	250 KTAS at sea level	——	22.3	——

Having defined a final section, the characteristics of the airfoil were then assessed systematically at both Bell and Boeing Vertol by means of surface singularity and viscous transonic flowfield analysis techniques as was done with the blade airfoil designs.

The flap contour was designed using an iterative-direct approach in which sections were developed using computer-aided drafting techniques and evaluated using multielement aerodynamic analysis surface singularity methods (i.e., panel methods). During this process, the flap surface that is covered by the cove region when the flap is undeflected was modified until desired pressure distributions were produced on the flap.

The multielement analysis method used is based on the Hess-Neumann potential-flow method[22] with an intergal boundary-layer method added.[9] Figure 25 illustrates the type of design iterations that were carried out. The objective was to reduce the peak pressure on the wing flap and eliminate the predicted separation while producing a smooth pressure distribution on the flap. Because of the desire to simplify the structure, minimize the weight, and enhance roll control for the vehicle, a simple hinge flaperon configuration was designed. This configuration allows for flap deflections in the range from 25-deg trailing edge up to 67-deg trailing edge down above and below the undeflected position, respectively. Studies were performed to determine the best position for the hinge using experimental data and multielement analysis runs. The final wing section geometry and a depiction of the flaperon motion from up to 25 deg to down 67 deg is illustrated in Fig. 26.

Wind-Tunnel Tests

Two-dimensional wind-tunnel tests of the wing section and flap system were conducted. The airfoil section and two flap systems were tested in the Wichita State University 7×10 ft Walter H. Beech Memorial Low Speed Wind Tunnel. This test included evaluation of a 33% chord plain flap and a 31% chord slotted flap configuration at low Mach numbers. Figure 27 shows the slotted and plain flap configurations that were tested and gives an indication of the distribution of pressure taps on these models. A major objective of this test was to determine the effectiveness of a plain flap and a slotted flap for application on the V-22 wing. Figure 28 compares the lift coefficient as a function of flap deflection for several angles of attack. It shows that the increment in lift due to flap deflection for the plain flap is much less than that of the slotted flap at positive flap deflections (trailing edge down). At a wing angle of attack of zero the plain flap remains effective up to 10-deg flap deflection but then degrades compared to the slotted flap at higher flap deflections. However, at higher wing incidences the plain flap effectiveness is degraded at all flap deflections above 0 deg. This is due to the fact that at the high flap deflections and angles of attack the flow is separated from the plain flap at the hinge line while the slotted flap continues to produce attached flow as is depicted in Fig. 29. Maximum lift increments for the plain and slotted flap are shown in Fig. 30. The slotted flap with a simple hinge motion shows a significant improvement over the plain flap configuration. These results indicate that the flap design studies, which had been carried out prior to this test, generated a configuration that produced significant performance improvements over the baseline configuration.

During this wind-tunnel test flap track optimization studies were also carried out. Figure 31 shows that a maximum lift coefficient of 3.31 was obtained at the optimum flap position for 35-deg deflection for this 23% thick airfoil. This result is comparable to the maximum lift values achieved by much thinner airfoil sections with single-slotted flaps and is an indication of the ability of the computational methods to produce contours with good performance while satisfying stringent geometric constraints. There is a range of speeds that the tilt rotor can fly as a function of the nacelle angle. The low end of this range is affected by the stall speed of the wing. As the maximum lift coefficient of the wing/flap system is improved the corridor through which the V-22 can operate becomes larger. Figure 32 depicts the expansion of this corridor due to the development of a simple flap system that exceeded the design goals. These types of developments were the result of the application of computational aerodynamic techniques during the design of these contours.

Mach number effects on the wing airfoil performance were ascertained in the Boeing Seattle 12×36-in. supersonic wind tunnel (BSWT 592). Most of the wing airfoil testing was conducted at a Reynolds number corresponding to 0.15 scale with some data also being measured at a Reynolds number corresponding to 0.30 scale during this test. Figure 33 gives the maximum lift as a function of Mach number for the wing airfoil for these two Reynolds numbers. Adding a theoretical Reynolds number lift increment to the test results yields a full-scale maximum lift coefficient of 1.73 at a Mach number of 0.3. The maximum lift does not degrade due to compressibility effects up to a Mach number near 0.5. This Mach number penetration will enable high-load factor maneuvers at high altitudes and gross weights to be performed by this vehicle. In addition, this figure shows a comparison of the maximum lift coefficient as a function of Mach number from the BSWT test and the WSU test. This indicates that the two tunnels produced similar results for the first priority goal of high maximum lift coefficient.

Angle-of-attack sweeps at Mach numbers up to 0.67 were performed for the V-22 wing section during the BSWT test. The drag divergence Mach number as a function of lift coefficient is shown in Fig. 34. It shows that the second priority goal for high-speed and high-altitude flight was achieved. In addition, the results from computational methods are included and show the excellent agreement between computational methods and the test results for this design parameter. Drag coefficient as a function of lift coefficient from the measured data is given in Fig. 35. In order to display this data more effectively the zero in the drag axis is changed for each different Mach number. This effectively spreads the results out so that each Mach number may be evaluated independently. This figure shows that the third and fourth priority goals for the wing airfoil were achieved. Figure 36 gives the pitching-moment coefficient as a function of lift coefficient from the test data. As with the drag data, the zero for each Mach number slides to a new location to expand the data. This figure shows that the fifth priority goal of providing a pitching-moment coefficient of less than -0.06 was satisfied as a lift coefficient of zero for the low Mach numbers. It is felt that the ability to satisfy all of these design goals with the constraints on thickness that were applied can be attributed to the use of the computational fluid dynamic methods used during the aerodynamic design of this airfoil shape.

Correlation of Test and Analysis

Figure 37 shows lift as a function of angle of attack for the BSWT test data and the theoretical results from the Eppler program[4] at a Mach number of 0.4. For this 23% thick section with large camber, the Eppler theory overpredicts the lifting capability of the section. Since this program does not do any iteration to modify the pressure distribution due to displacement of the boundary layer, it frequently predicts the zero-lift angle to be more negative than measured results on highly cambered airfoils. Figure 38 shows the drag coefficient as a function of lift coefficient for the same condition. It shows a general agreement in the form and level of the drag between theoretical and measured data. The comparison of the predicted and measured pressure distribution for an angle of attack of 0 deg and Mach number of 0.4 is shown in Fig. 39. It indicates that the potential methods give reasonable trends for the pressure distribution but do not accurately predict the measured values.

A comparison of the pressure distribution on the wing with a 20-deg flap deflection computed by the Hess-Neumann program and the measured pressure distribution is given in Fig. 40. It shows that, although the panel method with boundary layer may be used to give the trends of the pressure distributions, the pressure distribution, especially on the flap, is not adequate.

A comparison of the test results to theory for drag as a function of Mach number at an angle of attack of 0 deg is shown in Fig. 41. It shows that the drag is accurately predicted up to the drag divergence Mach number, but after that the drag computed is lower than that measured. However, this correlation may be fortuitous, since the correlation of the pressure distribution, as shown in Fig. 42, shows that the sonic zone is missed and the lower-surface pressure is shifted at a Mach number of 0.60.

Results from the wind-tunnel tests indicate that the airfoil and flap system, which were designed using computational aerodynamic methods, met or exceeded the design goals. However, detailed correlation at off-design points indicates that the technology at that time did not provide adequate results in the nonlinear regions.

Results of Three-Dimensional Tests

Over 8000 h of wind-tunnel testing of eight different airframe models of the V-22 have been conducted. All of these tests have confirmed that the performance requirements for the V-22 are satisfied.

Additional CFD Successes on the V-22

Fuselage Fairing Aerodynamic Design

Tests of the initial V-22 airframe model indicated that regions of separation existed in the area forward of the wing-fuselage fairing. The contour of this fairing was originally designed to minimize the blade fold (rotation) area. Analysis of this configuration using computational methods indicated that the adverse gradients produced by the fairing caused boundary-layer separation ahead of the wing fairing.

In order to develop a new fairing shape, an iterative approach was developed in which the flow about the two-dimensional profile of the fuselage and fairing was interrogated for separation and modified until separation was eliminated in the forward region of the wing fairing. Figure 43 compares the original and final profiles, and Fig. 44 presents the two-dimensional pressures and separation location for these contours. A three-dimensional fairing was developed using the resulting profile shape. This fairing was analyzed using a three-dimensional panel method code[22] to determine the resulting three-dimensional pressure distribution and ensure that pressure gradients were not adversely changed. Plate 12 (see the color section) gives the panel model with the resulting pressure distribution superimposed on the panels. Subsequent wind-tunnel testing of the V-22 airframe model, with the new wing-fuselage fairing contour, indicated that separation was eliminated in the forward region of the fairing. Separation aft of the fairing was suppressed to much higher angles of attack than that experienced on the initial fairing shape. These results indicate that three-dimensional panel methods can be used to tailor detailed aerodynamic shapes that will provide improved performance.

Nacelle Inlet Aerodynamic Design

The stringent performance requirements and geometric constraints for the engine inlet of the V-22 posed a very challenging design task. This is a result of the fact that the inlet must work efficiently in helicopter mode, where the inlet is perpendicular to the general flow direction; in the transition mode; and in the airplane mode, where the inlet is parallel to the general flow direction. In addition, crosswind components will be encountered during flight. Three critical flight conditions were selected at which the inlet flow was computed for each configuration considered. Since it was obvious from the outset of the design study that configurations with large regions of separated flow would be unacceptable because of drag and/or reduced inlet recovery, the goal of eliminating predicted separation was established. In addition, constraints on the sand separation efficiency and acceptable geometry were imposed. The VSAERO[23] and particle trajectory[24] programs were used to evaluate configurations and to guide the contour modification process to the selected configuration for the final lines of the V-22 inlet.[25] Figure 45 shows a typical flowfield from the VSAERO program that was used for trajectory evaluation. Results of the wind-tunnel test of the configuration developed by using these computational methods indicated that the performance goals were achieved for the V-22 inlet.

Spinner Aerodynamic Design

The shape of the V-22 spinner was developed using the three-dimensional VSAERO panel method code as a design tool. The length and diameter of the V-22 spinner could not be varied because of the shipboard folding and storage requirements for this vehicle. However, the spinner contour was tailored using VSAERO in an iterative-direct design mode to minimize the pressure spike and resulting gradient over the spinner nose and to minimize the pressure recovery over the region of the blade openings. A 0.3-scale V-22 spinner wind-tunnel test was conducted at the 7×10-ft Vought Corporation low-speed wind tunnel. The test was conducted to investigate methods to reduce the V-22 spinner drag.[26] Surface pressures were also measured along the spinner which allowed comparison to the previously computed VSAERO results. Figure 46 shows a comparison of the measured pressure distribution and computed pressure distribution on the spinner shape at a spinner angle of attack of 14 deg. As can be seen in this figure, the VSAERO computed results correlate well with the measured values. In addition, the design goal of minimizing the pressure spikes on the nose of the spinner was achieved by this design. Results indicate good correlation between the test data and the VSAERO predictions and validate the use of the VSAERO code to aid in the detailed design of complicated components.

Airload Distributions

CFD methods were used in conjunction with other simulation and structural analysis methods during the loads analysis for the V-22. Flight-path simulations using the V-22 Generic Tilt Rotor Simulation Program (GTR) determined time histories of maneuvers from which peak load cases for the V-22 could be extracted. Airload pressure distributions were computed for these flight conditions using the VSAERO code for the wing, fuselage, nacelle, and empennage components. The results from these analyses were used as input to the NASTRAN structural analysis code[27] and as input to determine static loading distributions for structural testing.

Plate 13 (see the color section) shows a model that was developed to evaluate the empennage loads distributions for the critical conditions. The model was developed so that the rudders and elevators could be deflected and the pressure distributions and integrated aerodynamic moment and shear distributions computed for the cases could be analyzed. The balance of forces and moments of the VSAERO results and the GTR results is critical for input into the NASTRAN code. Therefore, angle of attack, sideslip angle, and control surface deflections in the VSAERO runs were set to provide the exact total load and moment computed from the GTR code. For the case shown here, the flow about the model was computed using VSAERO for an angle of attack of 8 deg, an elevator deflection of 20 deg, and a rudder deflection of 20 deg. In Fig. 47 the pressure distributions on the left horizontal and the center of the carry-through, beaver-tail structure is compared. It shows that the load is reduced in the center section region as would be expected. The shear and bending moment diagrams on the horizontal tail component for this case are shown in Figs. 48 and 49, respectively. These values were computed from the pressure distributions determined by the VSAERO code in the ADAM system LOADS postprocessor component.

Plate 14 (see the color section) shows a model that represents more of the total vehicle but necessarily has less detail than the model specifically designed to produce empennage loads distributions. In addition, it was run in a symmetrical mode so that only one side of the vehicle was modeled. This model was used to determine loads on the flaperons when deflected to negative flap angles and allowed for the deflection of the flaperons and elevators so that critical symmetrical load conditions could be evaluated. In this plate the pressure distribution computed for an angle of attack of 5 deg is represented on the panel surfaces as different color shades.

Other VSAERO models were used in several different applications to loads computations for the V-22 including the effect of the wing on rotor dynamic loads.[28] For this study, the inclusion of airframe-induced velocity from VSAERO results allowed the prediction of higher-frequency blade loads by an aeroelastic rotor analysis code.

In addition to transferring these data to the NASTRAN models for airload distributions, information from these types of runs was transferred to the test areas so that this data could be reflected in the static test article. For airload determinations on the V-22, the computational aerodynamic methods were an integral part of the design process.

Concluding Remarks

Computational fluid dynamic methods provide a tool that can greatly enhance the design process. These codes are currently being used mainly to aerodynamically tailor the lines of flight vehicles to produce specified performance characteristics and to provide aerodynamic loads information to structures groups. In the future, as computational methods for application to rotor craft vehicles improve, the use of these methods in stability and control, handling qualities, performance computation, and other areas will also increase. Because of the benefits that can be achieved by their use, these new methods are being incorporated into the aerodynamic design process at BHTI.

Acknowledgment

The author would like to thank Don Axley, Mike Farrell, Ralph Fricker, and Rick Reinesch for their technical assistance during the preparation of this paper.

References

[1]Narramore, J. C., "Airfoil Design, Test, and Evaluation for the V-22 Tilt Rotor Vehicle," *Proceedings of the 43rd Annual National Forum of the American Helicopter Society*, Alexandria, VA, May 1987.

[2]Narramore, J. C., and Yeary, R. D., "Airfoil Design and Analysis Using an Information Systems Approach," AIAA Paper 80-1444, July 1980.

[3]Narramore, J. C., "An Approach to Subsonic, Turbulent Flow Airfoil Design Using Mini-Computers," Society of Automotive Engineers, Paper 770479, 1977.

[4]Eppler, R., and Somers, D. M., "A Computer Program for the Design and Analysis of Low-Speed Airfoils," NASA TM-80210, Aug. 1980.

[5]Smetana, F. O., Summery, D. C., Smith, N. S., and Carden, R. K., "Light Aircraft Lift, Drag, and Moment Prediction—A Review and Analysis," NASA CR-2523, May 1975.

[6]Bauer, F., Garabedian, P., and Korn, D., *Supercritical Wing Sections II, A Handbook*, Lecture Notes in Economics and Mathematical Systems, Vol. 108, Springer-Verlag, New York, 1975.

[7]Mead, H. R., and Melnik, R. E., "GRUMFOIL: A Computer Program for the Viscous Transonic Flow Over Airfoils," Final Rept. for Langley Research Center, NASA, March 1980.

[8]Narramore, J. C., "Advanced Technology Airfoil Development for the XV-15 Tilt-Rotor Vehicle," AIAA Paper 81-2623, Dec. 1981.

[9]Walz, A., *Boundary Layers of Flow and Temperature*, edited and translated by Hans Jorg Oser, MIT Press, Cambridge, MA, 1969.

[10]Eppler, R., "Turbulent Airfoils for General Aviation," *Journal of Aircraft*, Vol. 15, Feb. 1978, pp. 93–99.

[11]Goepner, B. W., "Bell Two-Dimensional Airfoil Test," United Technologies Research Center, Rept. R80-935004, July 27, 1980.

[12]Jenks, M. D., and Narramore, J. C., "Final Report for the 2-D Test of Model 901 Rotor and Wing Airfoils (BSWT 592)," Bell-Boeing Rept. D901-99065-1, May 1984.

[13]McCroskey, W. J., "A Critical Assessment of Wind Tunnel Results for the NACA 0012 Airfoil," Paper 1, AGARD Fluid Dynamics Panel Symposium on Aerodynamic Data Accuracy and Quality: Requirements and Capabilities in Wind Tunnel Testing, Naples, Italy, 1987.

[14]Shevell, R. S., "Aerodynamic Anomalies: Can CFD Prevent or Correct Them?" *Journal of Aircraft*, Vol. 23, Aug. 1986, p. 641.

[15]Pulliam, T. H., and Steger, J. L., "Recent Improvements in Efficiency, Accuracy, and Convergence for Implicit Approximate Factorization Algorithms," AIAA Paper 85-0360, Jan. 1985.

[16]Sankar, N. L., and Tang, W., "Numerical Solution of Unsteady Viscous Flow Past Rotor Sections," AIAA Paper 85-0129, Jan. 1985.

[17]Malone, J. B., Narramore, J. C., and Sankar, L. N., "An Efficient Airfoil Design Method Using the Navier-Stokes Equations," AGARD FDP Specialists' Meeting on Computational Methods for Aerodynamic Design (Inverse) and Optimization, Loen, Norway, May 22–23, 1989.

[18]Rosenstein, H., and Clark, R., "Aerodynamic Development of the V-22 Tilt Rotor," AIAA Paper 86-2678, Oct. 1986.

[19]Felker, F. F., Maisel, M. D., and Betzina, M. D., "Full Scale Tilt Rotor Hover Performance," *Proceedings of the 41st Annual National Forum of the American Helicopter Society*, Alexandria, VA, May 1985.

[20]Narramore, J. C., "The Use of Small Computers in a Subsonic Airfoil Design Application," *Proceedings of the Advanced Technology Airfoil Conference, NASA CP-2046*, Vol. 2, March 1979, pp. 81-97.

[21]Narramore, J. C., Wentz, W. H., and Ostowari, C., "Evaluation of a Procedure for the Design of Low-Speed Airfoils: Analytical and Experimental Results," Wichita State University, Rept. AR79-5, Oct. 1979.

[22]Hess, J. L., and Smith, A. M. O., "Calculation of Potential Flow about Arbitrary Bodies," *Progress in the Aeronautical Sciences*, Vol. 8, Pergamon, New York, 1966.

[23]Maskew, B., "Program VSAERO, a Computer Program for Calculating Nonlinear Aerodynamic Characteristics of Arbitrary Configurations," NASA CR-166476, Nov. 1982.

[24]Breer, M., and Seidel, W., "Particle Trajectory Computer Program User's Manual," Boeing Document D3-9855-1, Rev. B, Dec. 1974.

[25]Paynter, G. C., "Application of CFD Design Technology in Development of the JVX Engine Inlet," *Proceedings of the 41st Annual National Forum of the American Helicopter Society*, Alexandria, VA, May 1985.

[26]Farrell, M. K., "Aerodynamic Design of the V-22 Osprey Proprotor," *Proceedings of the 45th Annual National Forum of the American Helicopter Society*, Alexandria, VA, May 1989.

[27]Agnihotri, A., Schuessler, W., and Marr, R., "V-22 Aerodynamic Loads Analysis and Development of Loads Alleviation Flight Control System," *Proceedings of the 45th Annual National Forum of the American Helicopter Society*, Alexandria, VA, May 1989.

[28]Schillings, J., and Reinesch, R., "Rotor/Airframe Interference on the V-22 Tilt Rotor," *Proceedings of the American Helicopter Society National Specialists' Meeting on Aerodynamics and Aeroacoustics, American Helicopter Society*, Alexandria, VA, Feb. 1987.

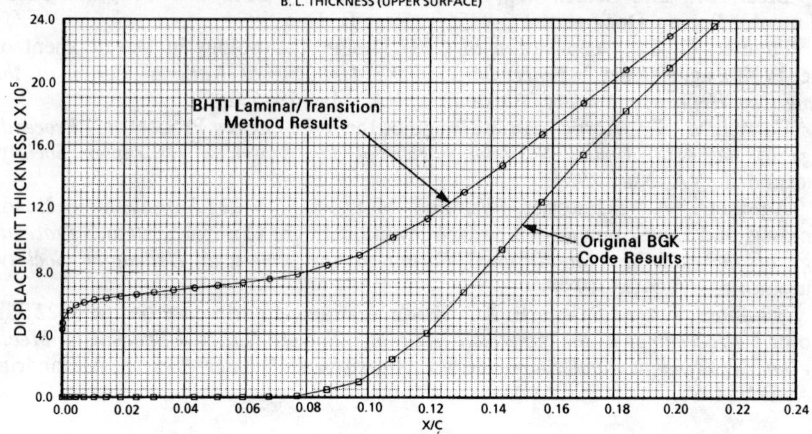

SYM	AIRFOIL	SOURCE	MACH	ALF	RE	CL	CD	CM	TRANU	TRANL
□	SFN12M2S	ARSN04	0.77	0.0	11.8	0.199	0.0111	-0.0282	0.11	0.03
O	SFN12M2S	ARSN05	0.77	0.0	11.8	0.199	0.0113	-0.0294	0.11	0.03

Fig. 1 BHTI laminar/transition method modifies displacement thickness in the nose region.

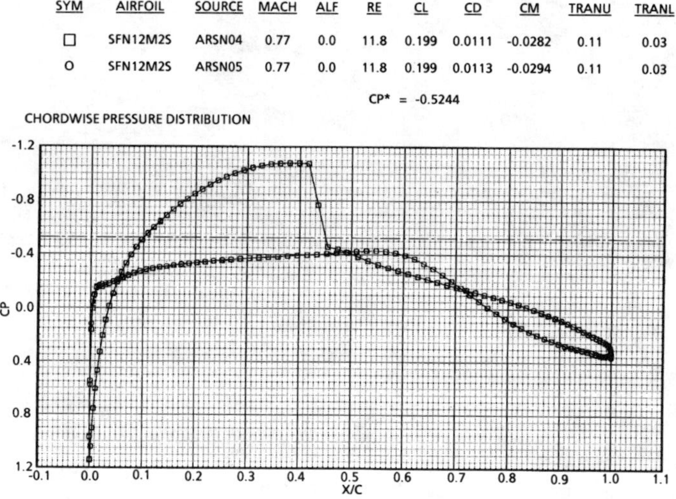

SYM	AIRFOIL	SOURCE	MACH	ALF	RE	CL	CD	CM	TRANU	TRANL
□	SFN12M2S	ARSN04	0.77	0.0	11.8	0.199	0.0111	-0.0282	0.11	0.03
O	SFN12M2S	ARSN05	0.77	0.0	11.8	0.199	0.0113	-0.0294	0.11	0.03

CP* = -0.5244

Fig. 2 Pressure distribution changes due to the laminar boundary-layer displacement are insignificant.

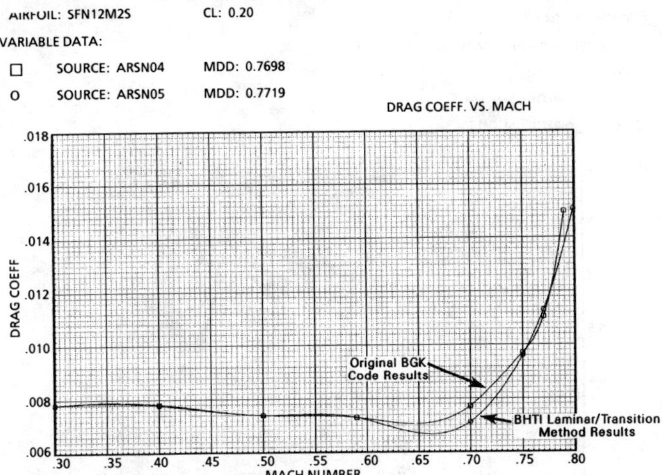

Fig. 3   Drag is slightly changed due to the natural transition and boundary-layer displacement.

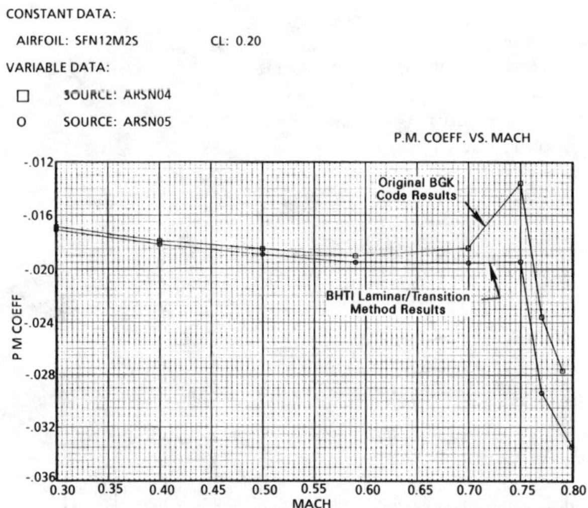

Fig. 4   Moment characteristics are changed due to natural transition and laminar boundary-layer displacement.

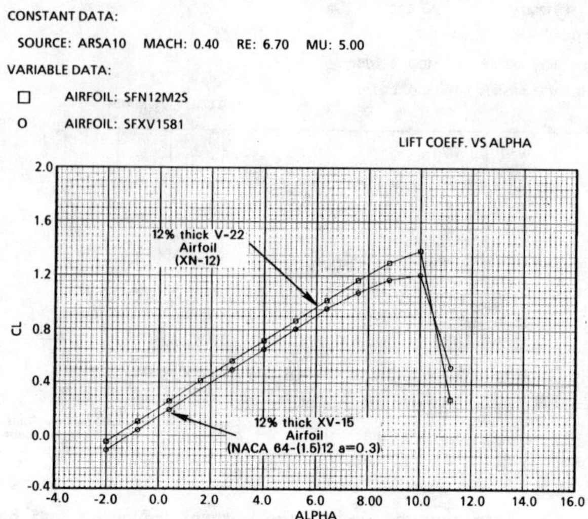

Fig. 5   Eppler code results indicate that the newly designed airfoil will have higher lift than the XV-15.

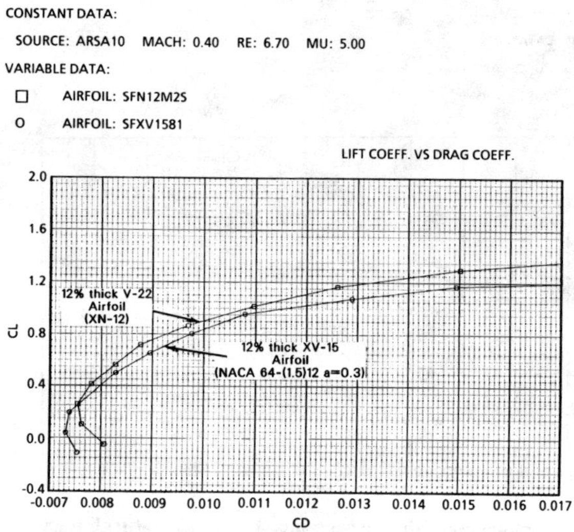

Fig. 6   Eppler code results indicate that the newly designed airfoil will have higher L/D than the XV-15 airfoil.

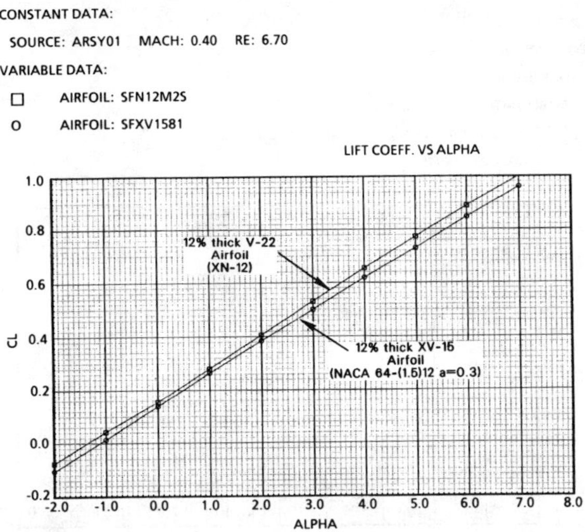

Fig. 7   North Carolina State University code results give a similar lift coefficient trend as that of the Eppler code.

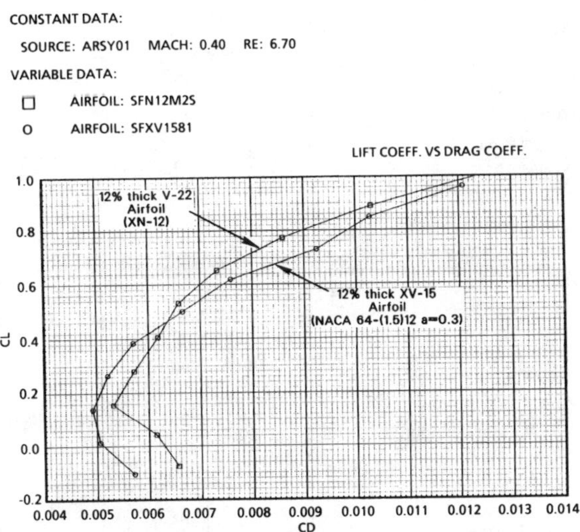

Fig. 8   North Carolina State University code results give a similar drag coefficient trend as that of the Eppler code.

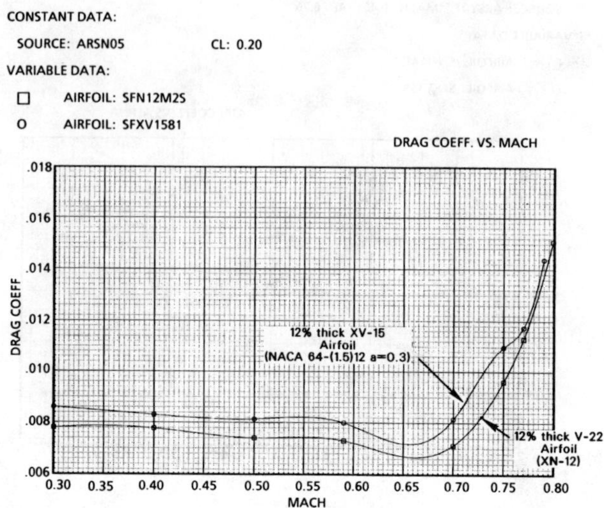

Fig. 9   BHTI-BGK code results indicate that the newly designed airfoil produces less drag than the baseline.

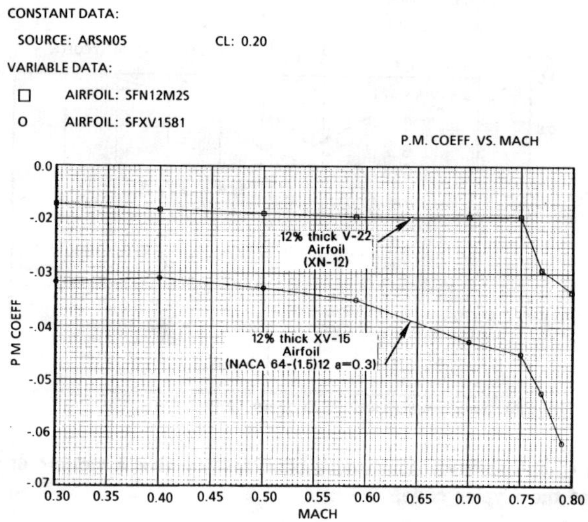

Fig. 10   BHTI-BGK code results indicate that the newly designed airfoil has less nose-down pitching moment.

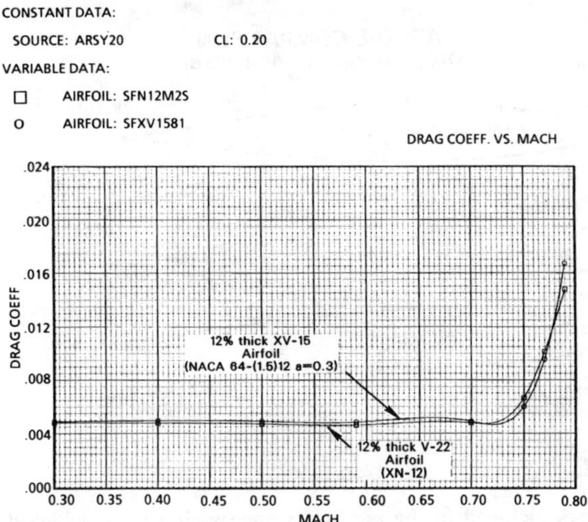

Fig. 11   GRUMFOIL code gives similar trends for drag as a function of Mach number.

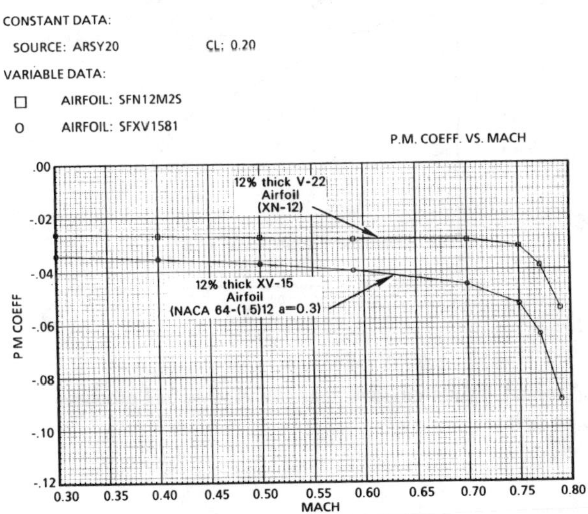

Fig. 12   GRUMFOIL code gives similar trends for pitching moment as a function of Mach number.

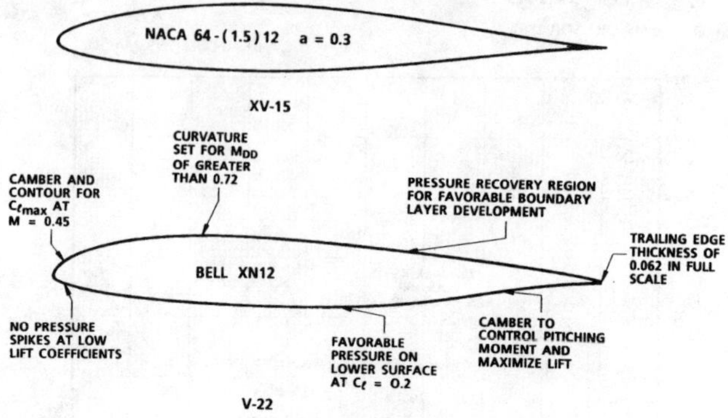

Fig. 13   Features included in the newly designed airfoil were achieved using inverse design methods.

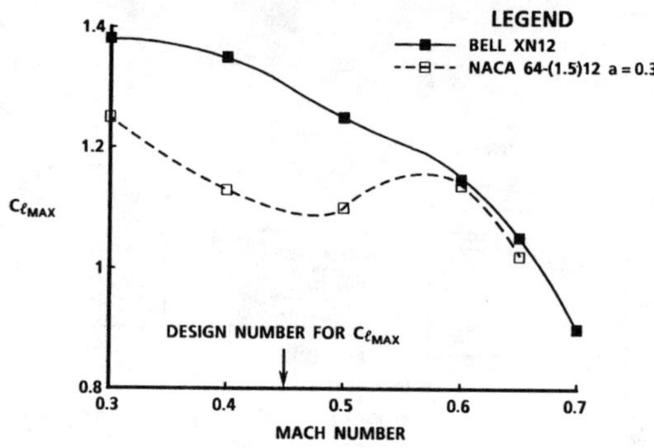

Fig. 14   Wind-tunnel results confirm that the new airfoil produces higher maximum lift at the design point.

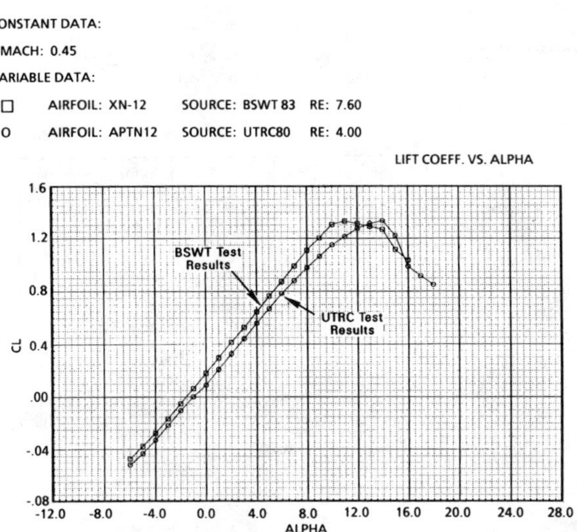

Fig. 15   Small differences between tunnel test results may be attributed to camber and Reynolds number differences.

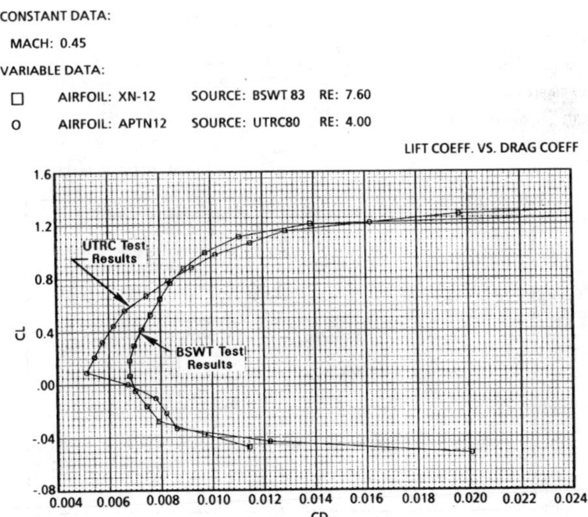

Fig. 16   UTRC wind tunnel has a lower turbulence level than the BSWT wind tunnel.

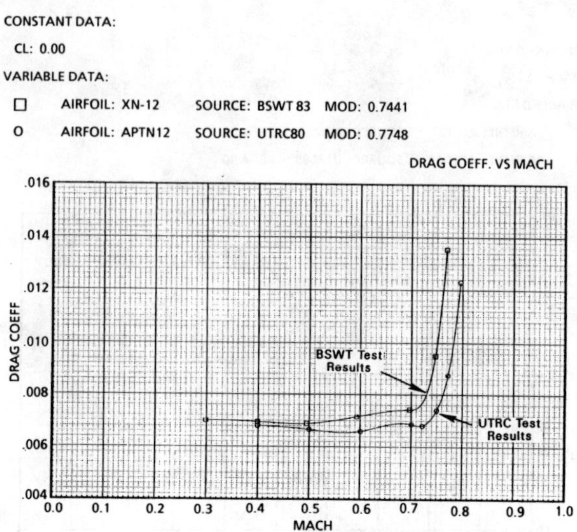

Fig. 17   Compressible results from the UTRC tunnel produce higher drag divergence than the BSWT tunnel.

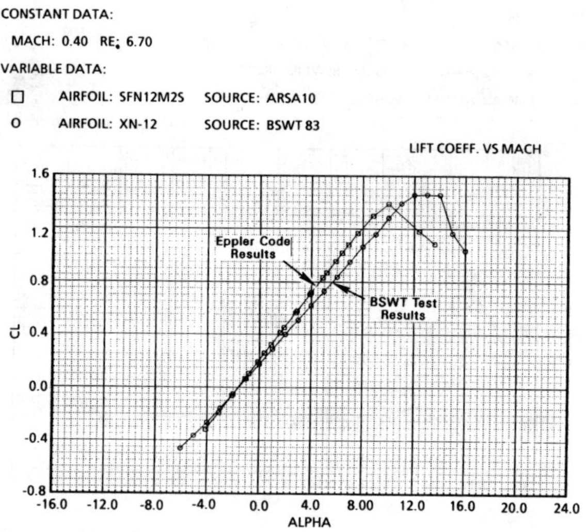

Fig. 18   Eppler code has an empirical method to compute the maximum lift coefficient.

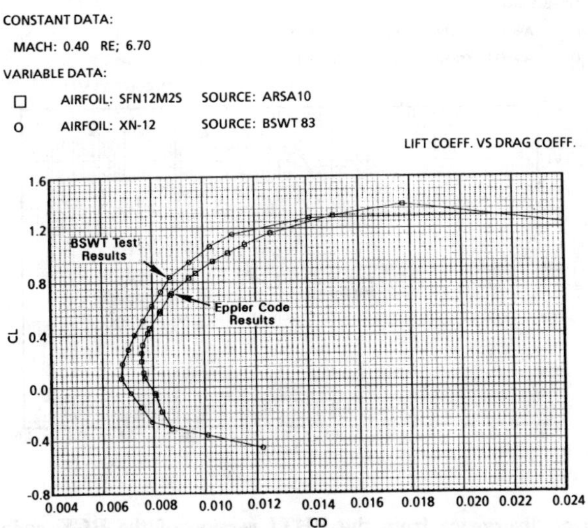

Fig. 19   Drag level from the Eppler code is slightly conservative.

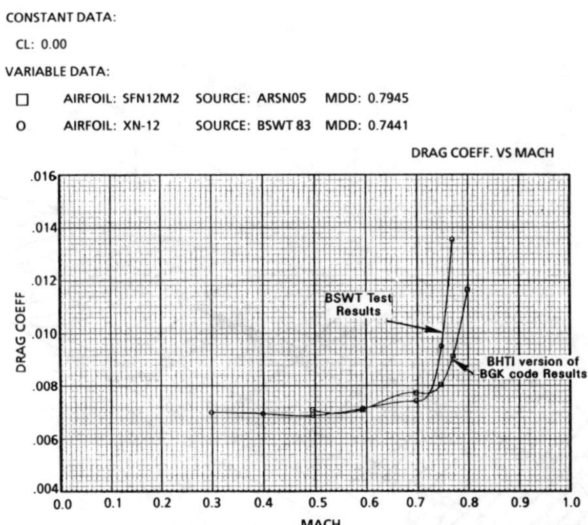

Fig. 20   Drag divergence from the BHTI version of the BGK code is higher than that measured at BSWT.

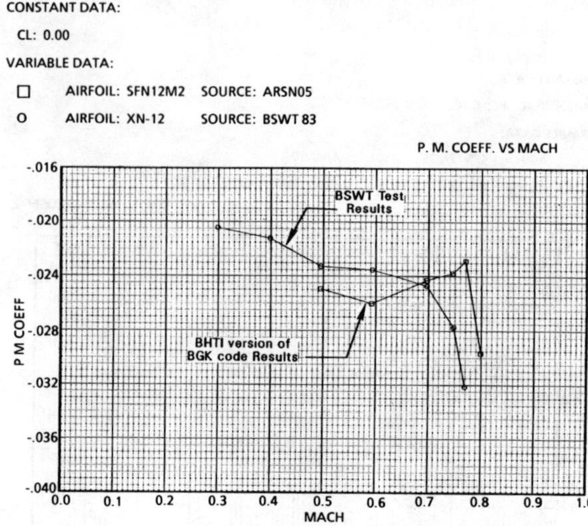

Fig. 21 Moment divergence from the BHTI version of the BGK code is different from that measured at BSWT.

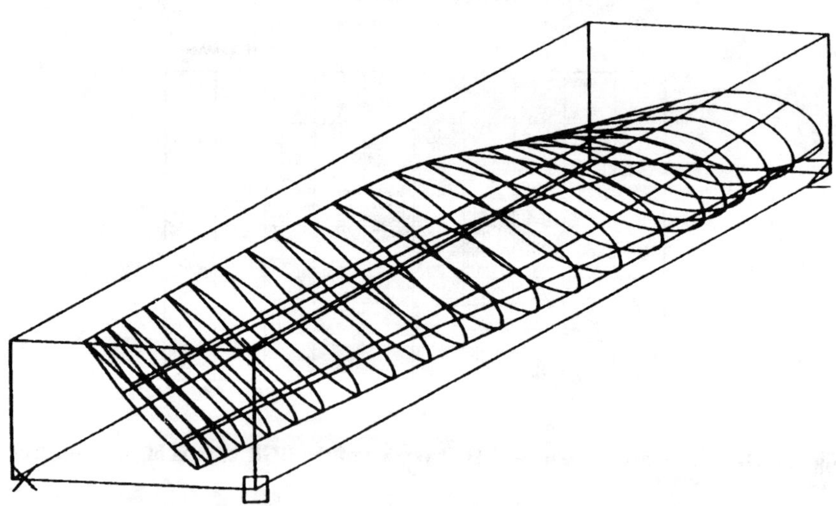

Fig. 22 Airfoils designed using the inverse method were used to define the V-22 rotor blade surface.

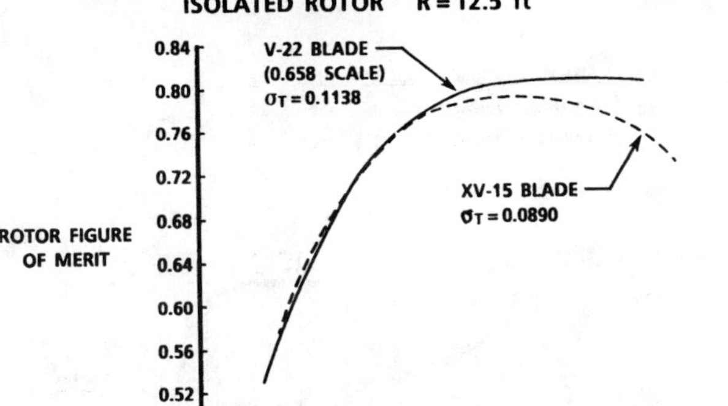

Fig. 23 V-22 blade performance exceeds the XV-15 blade performance at V-22 hover conditions.

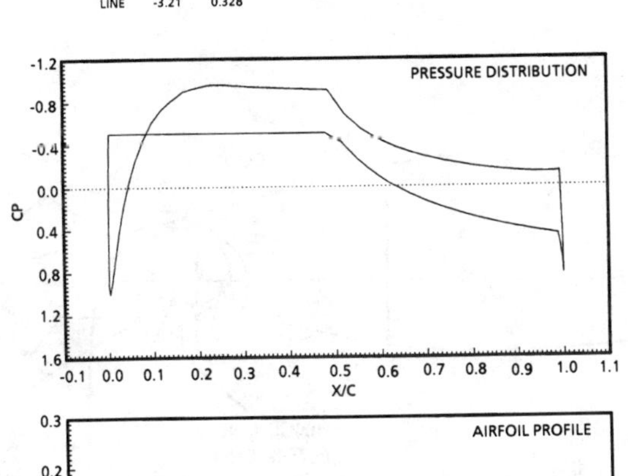

Fig. 24 Input pressure distribution produces a wing airfoil that is 23% thick.

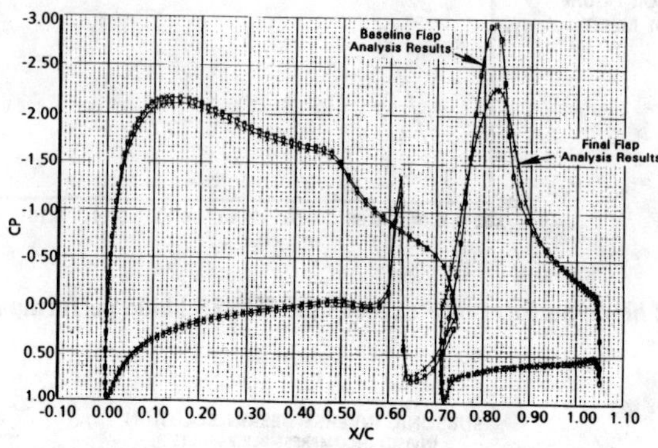

Fig. 25   Flap modifications were made to produce a more favorable pressure distribution at 15-deg deflection.

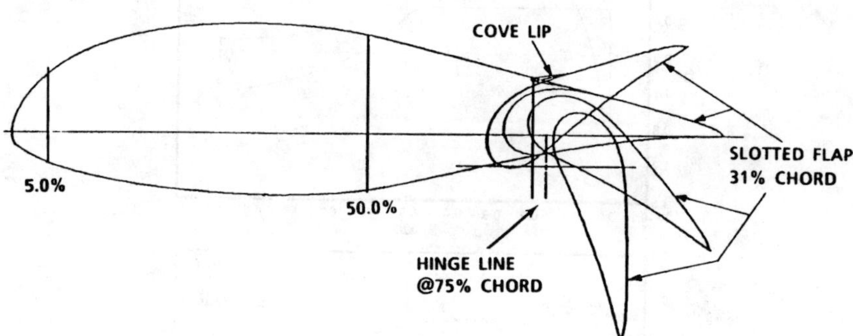

Fig. 26   A simple hinge flaperon system was developed for the V-22.

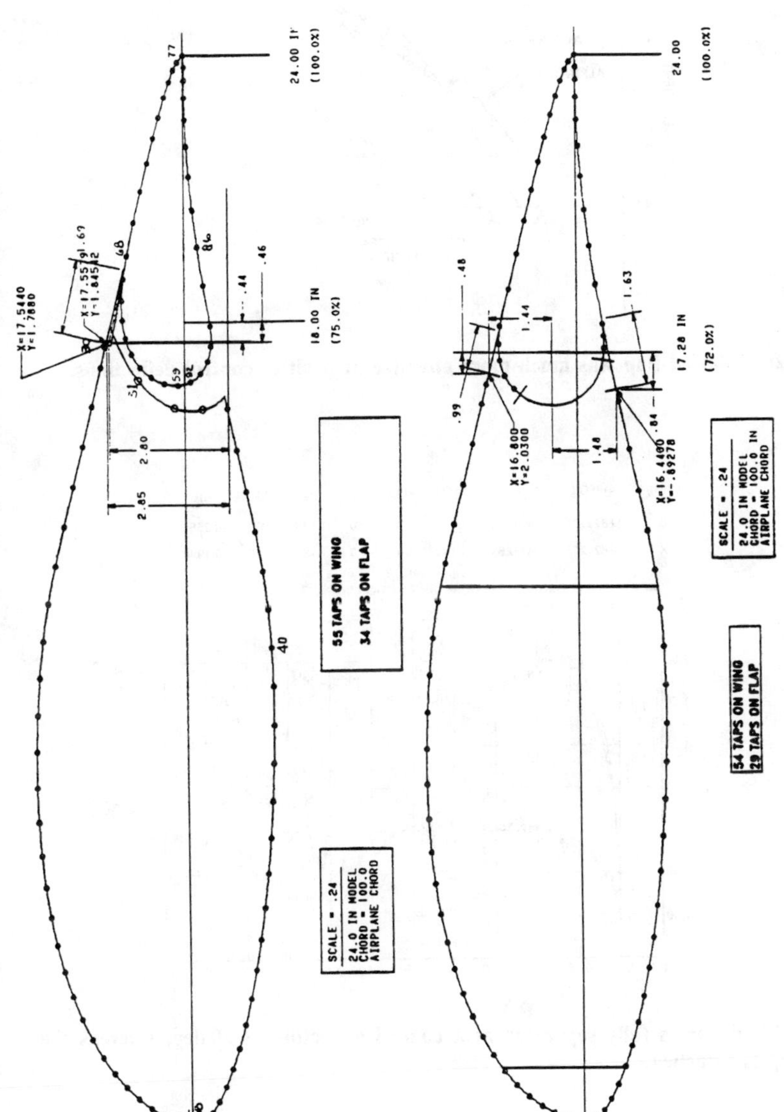

Fig. 27 Both slotted and plain flap configurations of the V-22 wing airfoil were tested.

Fig. 28　Slotted flap was much more effective at positive control deflections.

SYMB	AIRFOIL	SOURCE	MACH	ALF	RE	CL	CD	CM
0	SF821201	WSU 83	.013	-0.1	2.0	1.428	0.0319	-0.2561
X	SF821201	WSU 83	.013	-0.1	2.0	1.181	0.0503	-0.1990

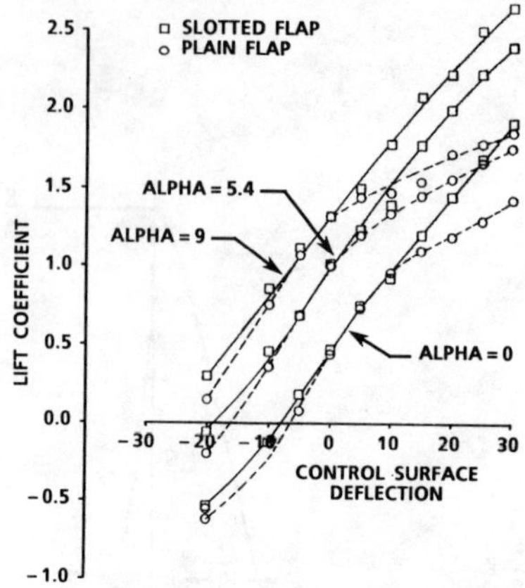

Fig. 29　Plain flap is fully separated at a control deflection of 20 deg, whereas the slotted flap is attached.

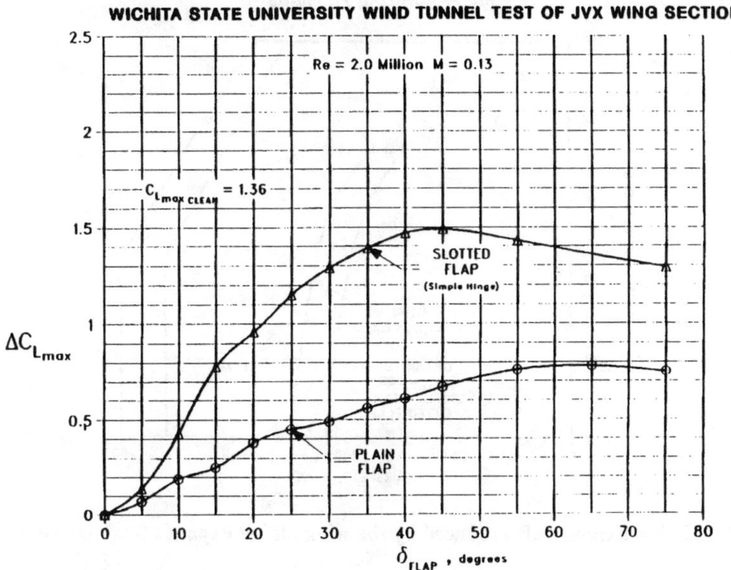

Fig. 30   Slotted flap produces much higher maximum lift increments than the plain flap.

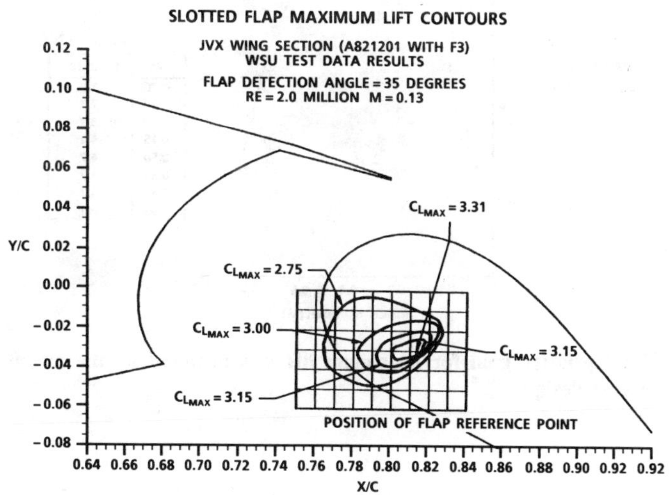

Fig. 31   Optimizing the flap location generated high maximum lift for the 23% V-22 wing with single-slotted flap.

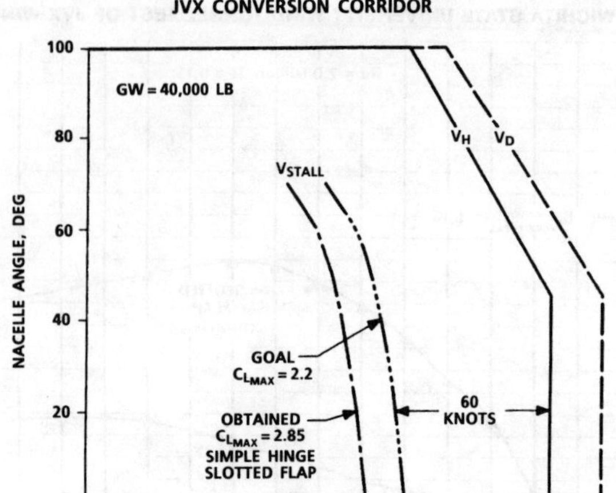

Fig. 32   High maximum lift produced by the wing airfoil expands the V-22 conversion corridor.

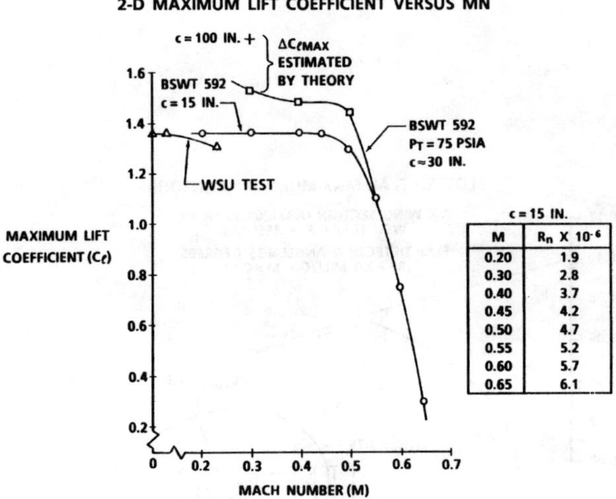

Fig. 33   First priority goal for maximum lift was achieved by the airfoil produced from an inverse design.

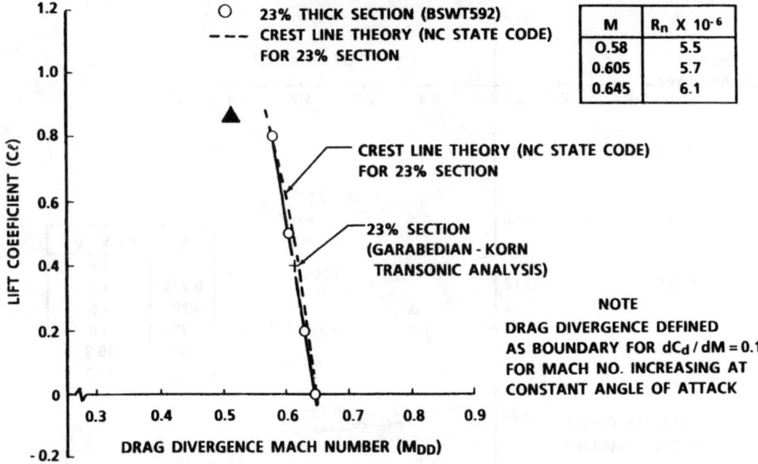

Fig. 34   Drag divergence goal for the V-22 wing airfoil was exceeded.

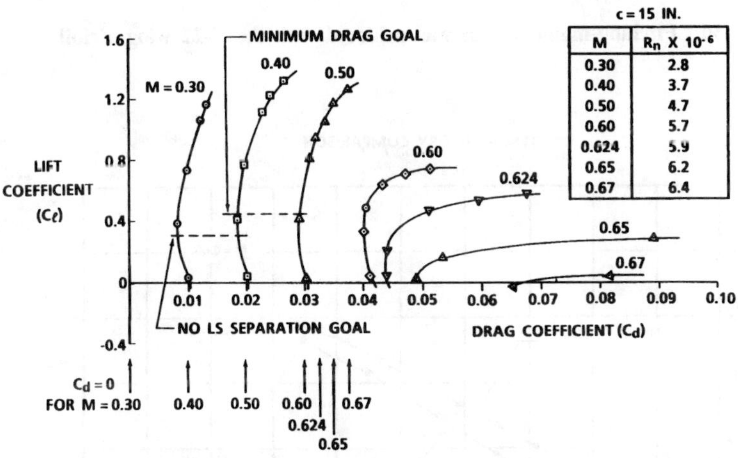

Fig. 35   Drag results indicated that there is no flow separation on the lower surface at low lift coefficients.

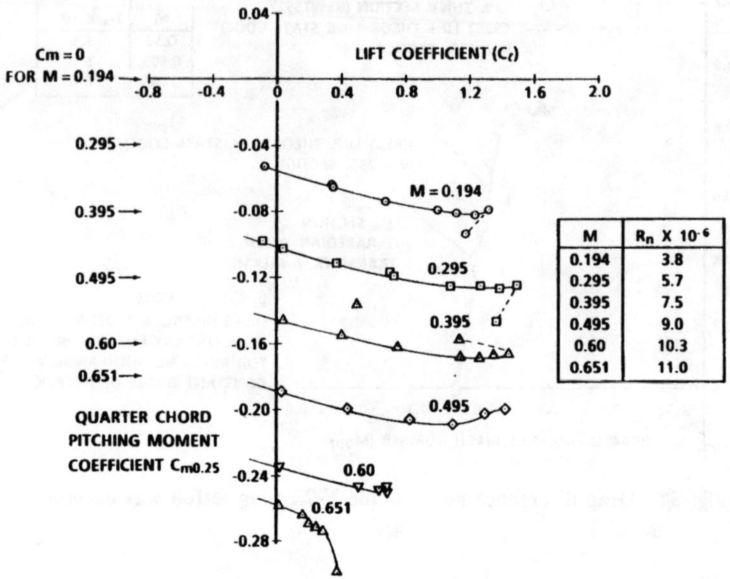

M	$R_n \times 10^{-6}$
0.194	3.8
0.295	5.7
0.395	7.5
0.495	9.0
0.60	10.3
0.651	11.0

Fig. 36　Pitching-moment goal was achieved with the V-22 wing airfoil.

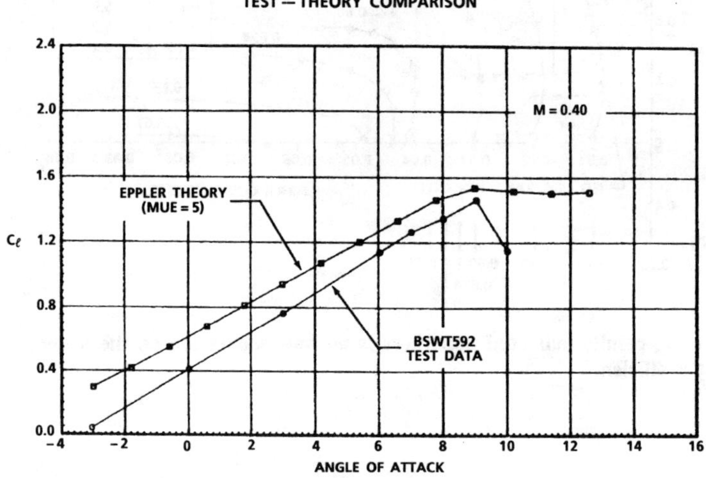

Fig. 37　Eppler code overpredicts the maximum lift capability of this 23% thick wing airfoil.

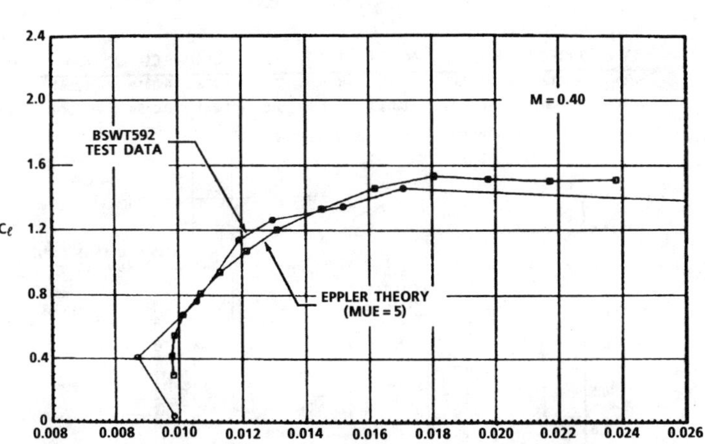

Fig. 38   Eppler code drag is in general agreement with the measured data for the 23% thick wing airfoil.

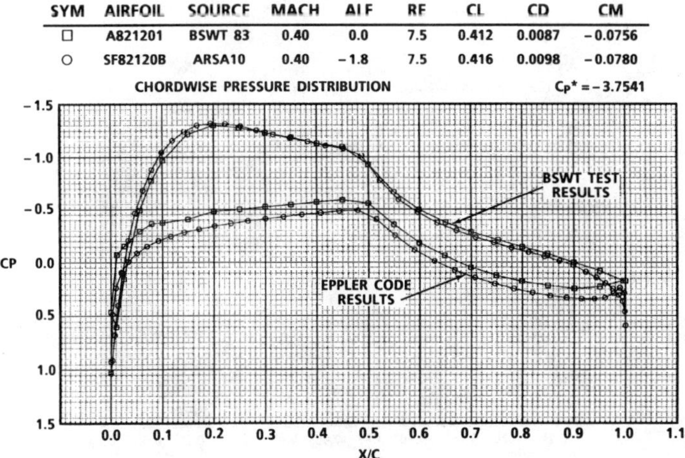

SYM	AIRFOIL	SOURCE	MACH	ALF	RE	CL	CD	CM
□	A821201	BSWT 83	0.40	0.0	7.5	0.412	0.0087	−0.0756
○	SF82120B	ARSA10	0.40	−1.8	7.5	0.416	0.0098	−0.0780

Fig. 39 Eppler code does not accurately predict the measured lower-surface pressure distribution.

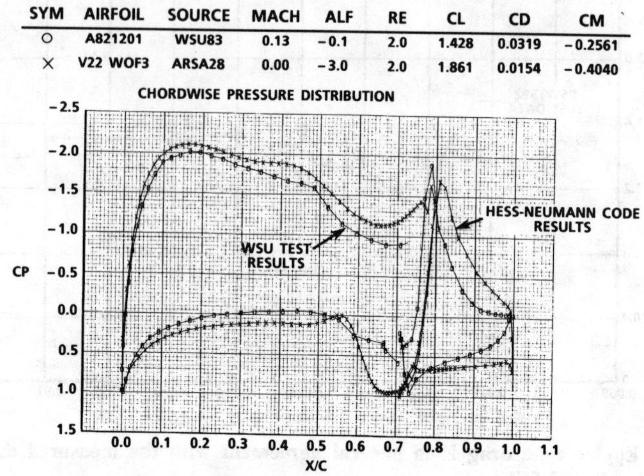

SYM	AIRFOIL	SOURCE	MACH	ALF	RE	CL	CD	CM
O	A821201	WSU83	0.13	−0.1	2.0	1.428	0.0319	−0.2561
×	V22 WOF3	ARSA28	0.00	−3.0	2.0	1.861	0.0154	−0.4040

Fig. 40   Simple panel method calculations are not adequate for multielement configurations.

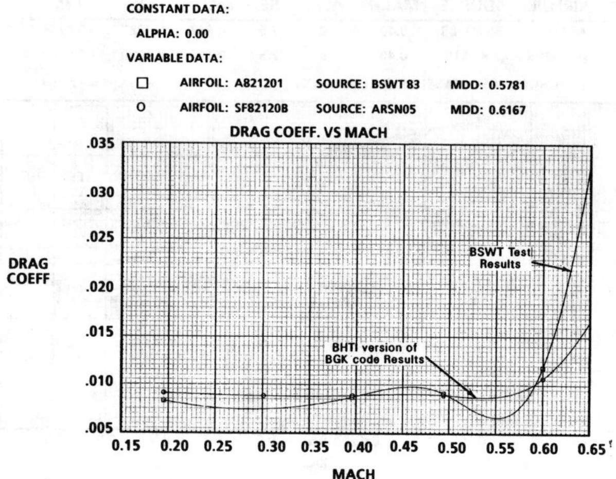

Fig. 41   Drag levels from the potential method do not correlate with test data past drag divergence.

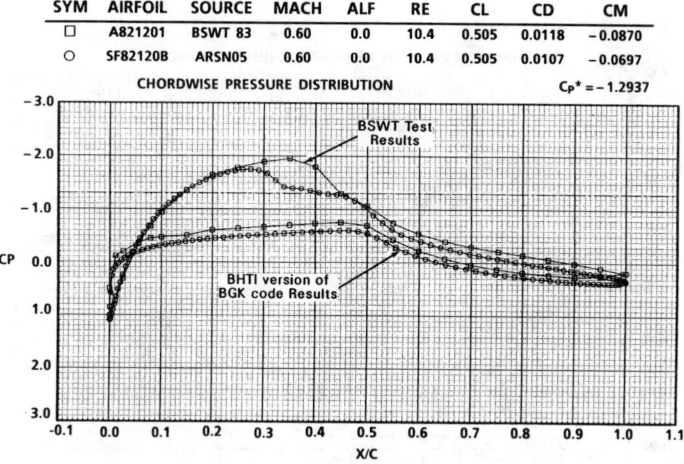

SYM	AIRFOIL	SOURCE	MACH	ALF	RE	CL	CD	CM
□	A821201	BSWT 83	0.60	0.0	10.4	0.505	0.0118	− 0.0870
○	SF82120B	ARSN05	0.60	0.0	10.4	0.505	0.0107	− 0.0697

Fig. 42 Potential method does not correlate well with test data for pressure distribution at this condition.

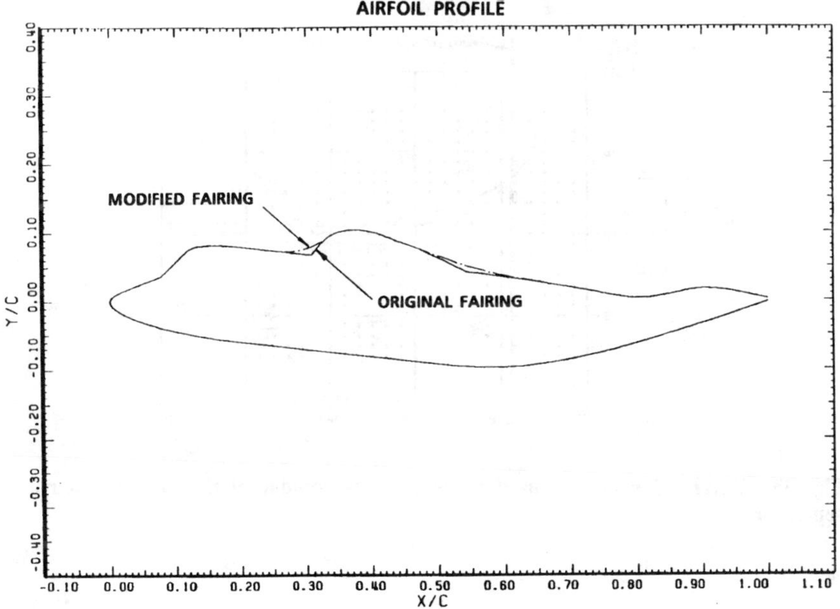

SYMBOL	CONFIGURATION	SOURCE
───	JVX OFM	AASY03
−.−.	JVX8 MF	AASY03

Fig. 43 Profile of the original V-22 fuselage had an abrupt slope change at the front of the wing fairing.

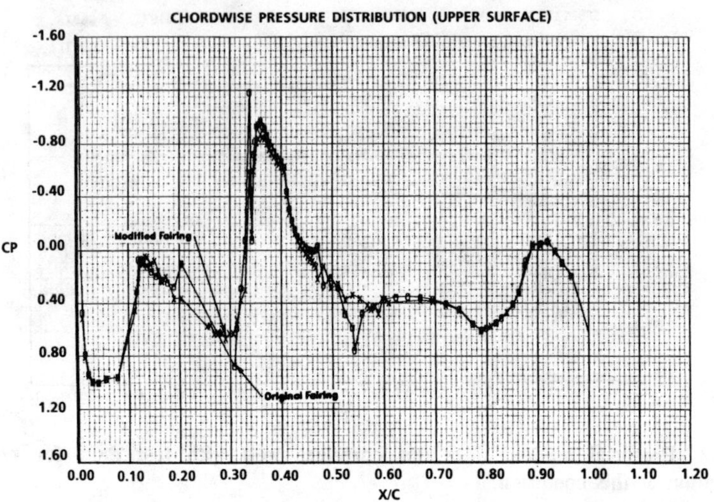

Fig. 44   Modified shape reduced the pressure spike and eliminated the calculated flow separation.

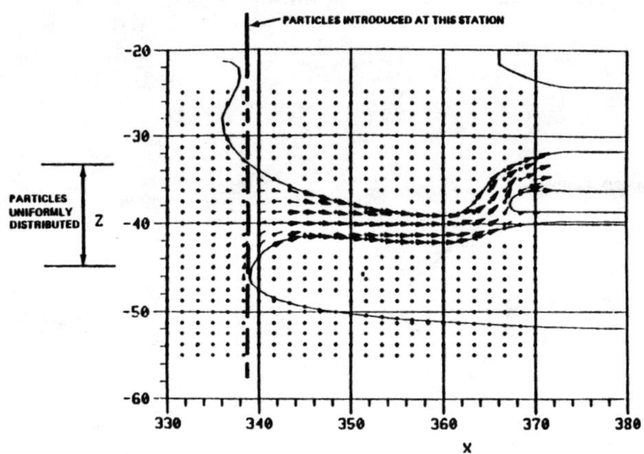

Fig. 45   VSAERO code was used to develop the contour of the inlet and particle separator.

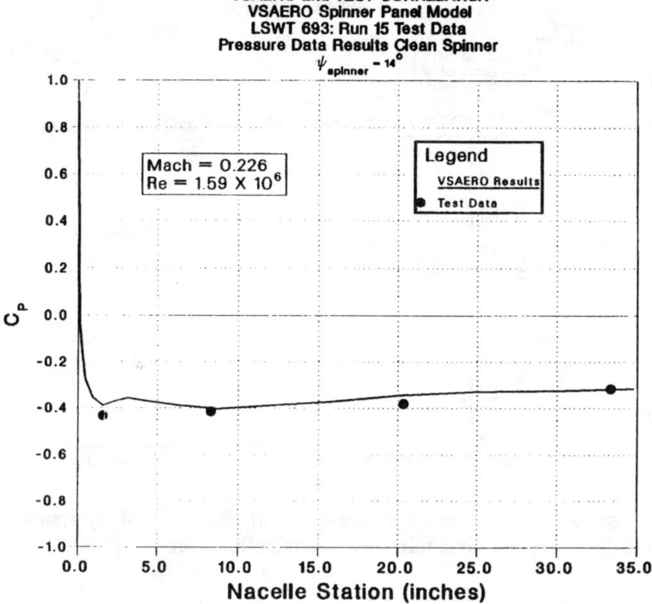

Fig. 46   VSAERO code was used to develop the contour of the spinner.

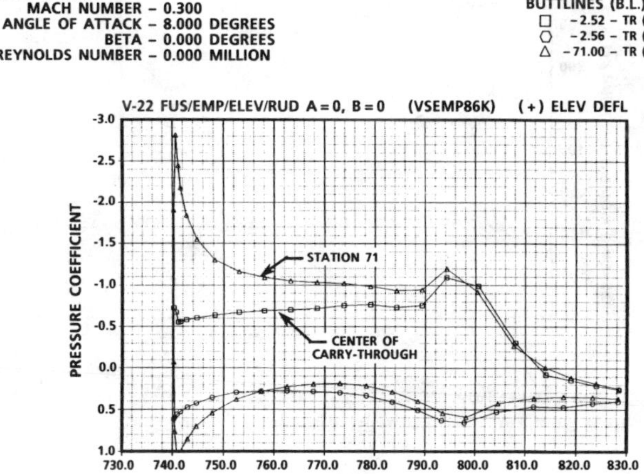

Fig. 47   Computed pressure distributions reflect the reduced load in the carry-through
region.

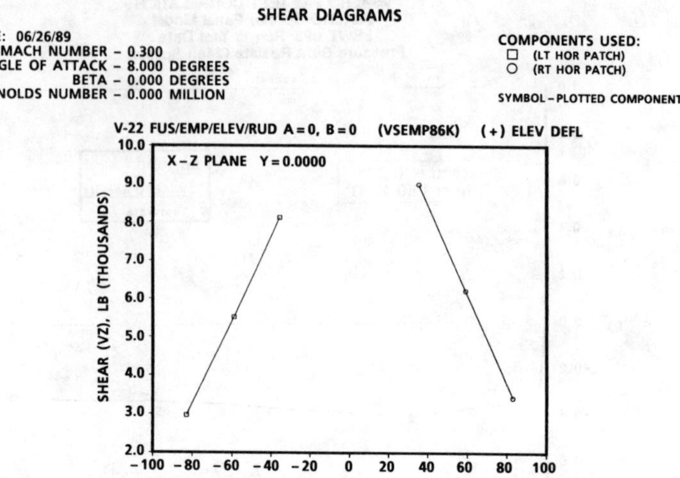

Fig. 48 Special postprocessing components of the ADAM system can integrate pressures to produce aerodynamic shear distribution plots.

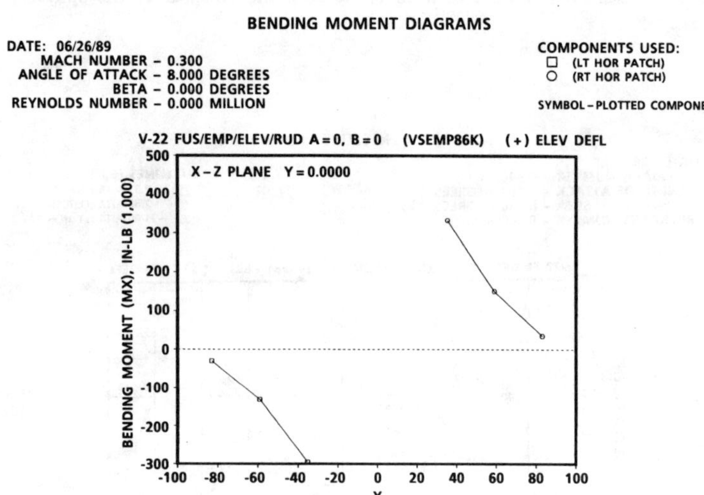

Fig. 49 Special postprocessing components of the ADAM system can integrate pressures to produce aerodynamic moment distribution plots.

Chapter 18

Aerodynamic Analysis Using Euler Equations: Capabilities and Limitations

Pradeep Raj*

Lockheed Aeronautical Systems Company, Burbank, California

Introduction

THE escalating cost of aircraft development continues to be a serious concern. One of the major factors contributing to cost escalation is the need to evaluate numerous geometric modifications to ensure that the final design will meet all required performance characteristics. Such an evaluation is very costly and time-consuming if conducted through wind-tunnel testing alone. Complementing the testing with computational simulations is widely recognized as one of the most promising approaches to reducing cost and time. Consequently, computational fluid dynamics (CFD), i.e., the simulation of a flow phenomenon by solving the fluid dynamic equations on a digital computer, is emerging as a crucial enabling technology for aircraft development.

A number of CFD codes are now available for aerodynamic analysis of aircraft configurations. These codes vary in the complexity of mathematical models employed and in their ability to model flow phenomena. The most sophisticated codes are based on the Reynolds-averaged Navier-Stokes (RANS) equations. Although the RANS codes are capable of simulating a wide variety of flow phenomena as demonstrated by some impressive computations on complex geometries,[1-3] their effectiveness for aircraft analysis is hampered primarily by two factors: 1) the codes typically require large .computational time and storage, and 2) results of complex-flow simulations are not reliable due to the inadequacies of turbulence modeling.[4,5] With ongoing advances in computers and numerical algorithms, coupled with an improved understanding of turbulence, it is only a matter of time before the RANS methods become more effective.

At present, the more widely used methods are based on the inviscid approximations to the Navier-Stokes equations. The boundary-integral methods, popularly known as panel methods,[6-9] solve the linearized potential-flow equations that represent the lowest-level inviscid approximation to

Copyright © 1989 by the American Institute of Aeronautics and Astronautics, Inc. All rights reserved.
*Computational Aerodynamics Technical Leader, Aerodynamics Department.

the Navier-Stokes equations. The panel methods are now relatively mature and quite effective in simulating purely subsonic or supersonic attached flows. For transonic flows, methods have been developed to solve either the transonic small-disturbance equations[10] or the full-potential equations.[11,12] However, the accuracy of their solutions deteriorates in the presence of strong shocks or regions of rotational flows.

The 1980s are notable for significant advances in numerical algorithms to solve the Euler equations[13-17] that represent the highest-level inviscid approximation to the Navier-Stokes equations. The Euler codes eliminate many limitations of the potential-flow codes since nonisentropic shocks and rotational flows can be a part of their solution. Of course, the precise structure of the computed shocks and the rotational-flow regions is inaccurate because the viscous terms are absent. However, the global solutions are found to be quite useful for engineering purposes as demonstrated by numerous investigations.[18-26] In spite of a substantial amount of validation effort to date, we do not yet have a full understanding of the capabilities and limitations of the Euler codes.

A complete and thorough assessment of a code's capabilities requires an evaluation of its performance using numerous criteria, such as the ability to handle complex geometries, the ability to simulate a wide range of flow conditions, the accuracy of solutions, the ease of use, the computational efficiency, and the robustness. Such an evaluation involves analysis of a large number of configurations and flow conditions. For Euler methods, the scope of such an evaluation is further expanded by the wide range of flow phenomena that can be modeled using them. Consequently, assessing the capabilities of an Euler code is a monumental yet crucially important task.

In this chapter, attention is focused on evaluating the capabilities of one representative Euler code, namely, the three-dimensional Euler/Navier-Stokes aerodynamic method, TEAM.[26] It has been developed by Lockheed under a U.S. Air Force contract. Although TEAM can solve the Navier-Stokes equations for laminar flows and the RANS equations for tubulent flows (using the well-known Baldwin-Lomax tubulence model), only its Euler solutions are considered here. Three criteria are used to assess TEAM's capabilities: configuration geometry, flow conditions, and solution accuracy.

Configuration Geometry

The ability of a code to handle complex geometries is clearly one of the most important criteria to be considered. TEAM's ability of analyzing a variety of configurations ranging from two-dimensional airfoils to three-dimensional aircraft configurations is demonstrated by the results included in this chapter.

Flow Conditions

The ability of a code to analyze a wide range of flow conditions, i.e., Mach numbers and attitude angles (pitch and yaw) is another important feature. TEAM's ability of simulating perfect-gas flow from subsonic through hypersonic Mach numbers is demonstrated by the results presented here. The code can also be used to simulate equilibrium real-gas effects at hypersonic speeds. Symmetrical as well as asymmetrical flight conditions can be analyzed. At present, the code cannot model unsteady aerodynamics; the freestream speed as well as the body orientation are assumed to be fixed at the start of the analysis.

Solution Accuracy

Evaluating the accuracy of the computed Euler solutions is one of the most important tasks. It is also a more difficult one due to the following three reasons: 1) The computed solutions are affected by numerous factors, such as, grid quality, grid density, discretization approximations, numerical implementation of boundary conditions, numerical dissipation, treatment of dissipation near the boundaries of the computational domain, truncation errors, and roundoff errors. Therefore, extensive studies are needed to evaluate the effect of the various factors on the accuracy of the computed solutions. 2) The lack of accurate analytical solutions of the nonlinear Euler equations, especially for transonic and rotational flows, renders the usual approach of comparing computed and analytical solutions ineffective. Such an approach has been quite effective in evaluating the accuracy of the linear potential-flow codes. 3) It is known a priori that the inviscid Euler equations cannot accurately model all aspects of the real flows. Therefore, evaluation of solution accuracy through a comparison of computed solutions with experimental data must be approached with caution.

In order to evaluate TEAM's capabilities subject to the criteria outlined earlier, computed solutions have been correlated with experimental data for a variety of geometries and flow conditions.[22-26] For many cases, sensitivity to grid density, numerical dissipation, and boundary condition treatment have also been evaluated. From an applications viewpoint—the one emphasized here—a large collection of such correlations is essential to assessing the capabilities and limitations of the code. In this chapter, solutions for six representative test cases are presented to illustrate the capabilities as well as limitations of the code. A brief discussion of TEAM's basic features precedes the presentation of the solutions. A few concluding remarks complete the chapter.

Basic Features of TEAM

The TEAM code is designed to compute steady-state solutions of the Euler equations representing conservation of mass, momentum, and energy, expressed in unsteady, integral form. The computational method, proposed by Jameson et al.,[13] employs finite-volume spatial discretization producing a set of ordinary differential equations that are integrated in time using a multistage time-stepping procedure. The finite-volume formulation requires that the region surrounding a given configuration be divided into an ordered set of hexahedral cells. Such a set, called a grid or a mesh, may be constructed in any convenient manner; only the Cartesian coordinates of the cell vertices are required to carry out the solution process. Since TEAM's algorithm is described in detail in Ref. 26, only the basic features are highlighted here.

Multiple Zones

The TEAM code can accommodate patched zonal grids of arbitrary topologies. Across zonal interfaces three classes of grid distributions are allowed: 1) one-to-one nodal point matching; 2) nodal points of one grid being an ordered subset of the other grid; and 3) mismatched nodal point distributions, as illustrated in Fig. 1. This feature affords considerable flexibility in analyzing complex configurations.

Spatial Discretization

A cell-centered finite-volume formulation is used to discretize the spatial terms. The flow variables are defined at cell centers and fluxes are computed at cell faces. A simple arithmetic average of the cell-center values is used to estimate the flow quantities at the faces. This formulation is equivalent to a central-difference approximation that is second-order accurate on smooth grids. The resulting semidiscrete equations have to be augmented by numerical dissipation terms (also known as numerical or artificial viscosity) in order to stabilize the solution process, to suppress odd-even decoupling, to capture shocks, and to minimize oscillations in regions of high gradients. In TEAM, two types of schemes, namely, adaptive and characteristic-based, are used to construct the dissipation terms. The adaptive-dissipation schemes employ differences of the flow variables directly, whereas the characteristic-based schemes require characteristic decomposition of the flow variables.

Adaptive Dissipation

Three adaptive dissipation schemes are included in TEAM: 1) standard adaptive dissipation (SAD); 2) modified adaptive dissipation (MAD); and 3) flux-limited adaptive dissipation (FAD). Both SAD and MAD schemes use blended first and third differences as proposed by Jameson et al.[13] The two schemes differ only in the scaling factors used for the difference terms resulting in overall lower levels of dissipation for the MAD scheme compared to the SAD scheme. The FAD scheme employs only third differences with appropriate flux limiters[27]; the first-difference terms are not used. For a scalar conservation law it becomes a total variation diminishing (TVD) scheme provided that the coefficients are appropriately chosen. The FAD scheme is capable of nonoscillatory shock capture.

All three schemes use two user-specified parameters, VIS-2 and VIS-4, to control the level of dissipation. For the SAD and MAD schemes the input VIS-4 parameter (divided internally by 64) scales the third-difference terms and controls the background dissipation needed to suppress high-frequency error components, whereas the VIS-2 parameter is used to minimize wiggles and overshoots near shocks and stagnation points. For the FAD scheme the value of VIS-4 determines a lower bound on dissipation and VIS-2 may be adjusted to ensure that there is enough dissipation to eliminate oscillations near shocks. Although the adaptive dissipation schemes are quite effective, difficulties may be encountered in simulating high-speed flows because these schemes do not accurately model the wave propagation in supersonic flows.

Characteristic-Based Dissipation

The characteristic-based dissipative terms provide an appropriate upwind bias for supersonic flows albeit at an expense of additional computations compared to the adaptive schemes. In TEAM the dissipative terms are constructed using the symmetric TVD formulation with Roe averaging of Gnoffo et al.[28] In its simplest form this formulation leads to a first-order-accurate upwind scheme. A second-order-accurate upwind scheme results from using a flux limiter. Both options are included in TEAM. The user can select either one by changing the value of an input parameter. The input value of the VIS-2 parameter serves as an eigenvalue limiter near the critical points, i.e., stagnation points, sonic points, and shocks. The input value of VIS-4 is used to scale the dissipative terms to provide enhanced stability.

Boundary Conditions

Appropriate nonreflecting boundary conditions based on Riemann invariants are used at the far-field boundaries. A no-normal-flow condition is imposed on the solid surface. The value of pressure on the surface is estimated using the normal momentum equation as described in Ref. 24. At an inlet face, Mach number, pressure ratio, speed ratio, or mass flow ratio (MFR) can be specified. At the exhaust face, nozzle pressure ratio (NPR), total temperature ratio, and exit flow direction need to be prescribed. At a zonal interface along which two zones are patched, the flux-conserving boundary conditions are used.

Time Marching

The set of ordinary differential equations resulting from the finite-volume spatial discretization is integrated in time using a multistage time-stepping scheme with options for three, four, or five stages. Convergence to a steady state is typically achieved in a few hundred time steps because the use of local time step, i.e., a different step size for each cell determined solely by the local stability restrictions, is augmented by enthalpy damping and implicit residual smoothing. This contrasts with thousands of steps typically required by a more conventional explicit scheme using a global minimum step size.

Analysis Process

The analysis starts with the flow variables, namely, density, three Cartesian components of momentum, and total energy, in all cells initialized to freestream conditions. This is equivalent to an impulsive start. Following each time step, the ratio of the current and the initial values of the average residual (root-mean-square value of the net mass flux) is checked to determine whether a steady state has been reached. The time marching stops if a prespecified convergence criterion is met or after a prescribed number of steps. The code has a provision to restart the solution process and perform additional steps if needed.

Analysis Results

In this section TEAM solutions are presented for six test cases: NLR 7301 airfoil, Lockheed-AFOSR Wing C, 74-deg delta wing, cone derived Mach 6 waverider, a canard-wing-body, and the advanced nozzle concept (ANC) fighter configuration. In all cases the flow was impulsively started, and no Kutta condition was explicitly applied. Shocks and rotational-flow regions were automatically captured. For all symmetrical-flight cases only half of the configuration was analyzed.

NLR 7301 Airfoil

Analysis of this airfoil was primarily motivated by the availability of an "exact" shock-free transonic solution of the Euler equations that could be used to evaluate the accuracy of the TEAM computations. The exact solution[29] obtained using the hodograph technique corresponds to a freestream Mach number (M_∞) of 0.721 and an angle of attack (α) of -0.194 deg. For this set of M_∞ and α the airfoil has been analyzed using several O-type grids with the coarsest having 2025 nodes and the finest having 51,681. For all analyses, the far-field boundary was located nearly 80 chords away based on the results of a previous study[30] where the effect of the far-field boundary location was investigated. No far-field vortex correction was used because its effect on the results was found to be negligible when the outer boundary was moved past 60 chords.

The aerodynamic lift, drag, and pitching-moment coefficients computed using the SAD, MAD and FAD schemes are compared with the hodograph predictions and the exact solutions (numerical integration of hodograph surface pressures) in Table 1. For these analyses, a grid with 11,809 nodes (49×241, i.e., 49 O curves each defined by 241 points) was used, and the values of VIS-2 and VIS-4 were fixed at 0.001 and 1.0,

Table 1   Sensitivity to numerical dissipation; NLR 7301
($M_\infty = 0.721$, $\alpha = -0.194$ deg, 49×241 O grid)

Numerical dissipation	C_L	C_D	C_{MP}
SAD	0.5973	0.000457	-0.1326
MAD	0.598	0.000265	-0.1325
FAD	0.593	0.00083	-0.132
Hodograph theory	0.5949	0.0	——
Exact[a]	0.5939	0.0005	-0.1298

[a]Numerical integration of hodograph data.

respectively. Note that the lowest value of the drag coefficient corresponds to the MAD scheme. This data, along with an examination of the computed pressure, Mach number, and total pressure loss distributions on the surface, confirmed the overall lower levels of numerical dissipation associated with the MAD scheme.[25] Consequently, the MAD scheme was selected to study the effect of grid density on the solution.

The solution for this airfoil was found to be quite sensitive to grid distribution.[25] For most grids but the coarsest, computed surface pressure and Mach number distributions agreed well with the exact solution, except in the region of transition from supersonic to subsonic flow, where the computed solutions exhibited a wiggle whose amplitude varied with the grid distribution. The amplitude of the wiggle increased as the number of nodes on the O curves increased from 161 to 241 to 321 while the number of O curves was held at 49. On the other hand, the amplitude decreased as the number of O curves increased from 33 to 49 to 65 to 81 while the number of nodes on each curve was held at 241.

The latest analysis used a 161×321 grid. A part of this grid is shown in Fig. 2, where every other grid line only is plotted for the sake of clarity. The corresponding flowfield represented as Mach contours in the vicinity of the airfoil is also shown in the same figure. The corresponding values of lift, drag, and pitching-moment coefficients are 0.5957, 0.0002, and -0.1325, respectively (see Table 1 for the exact solution). The computed surface pressure distribution is compared with the exact solution in the top part of Fig. 3, and the total pressure loss distribution on the surface is shown on the bottom part. The computed pressure distribution exhibits a small wiggle. On a large part of the airfoil, the values of the total pressure loss are less than 0.25%. Ideally, the total pressure loss should be zero. Further investigation of this case is continuing.

Lockheed-AFOSR Wing C

This wing, which is characteristic of a transonic maneuver fighter, with a leading-edge sweep of 45 deg, aspect ratio of 2.6, and a taper ratio of 0.3, is shown in Fig. 4. It was analyzed at its design conditions of $M_\infty = 0.85$ and $\alpha = 5$ deg using several C-H and C-O type grids. Results for one C-H grid having 106,425 nodes ($25 \times 129 \times 33$) are presented here. The grid had 33 H planes, each plane containing 25 C curves, each curve being defined by 129 nodes. The wing surface was covered with an 81×25 mesh. The results illustrate TEAM's ability to model transonic flows.

In Fig. 5 surface pressures computed using the SAD, MAD, and FAD schemes are compared with each other at five stations. For this study the value of VIS-4 was set at 2.0 and that of VIS-2 at 0.5, except for the MAD scheme, which required increasing VIS-2 to 1.0. In Fig. 6 the corresponding computed total pressure loss distributions are shown for four stations in a plane at the 50% semispan location. These results further confirm the lower levels of numerical dissipation associated with the MAD scheme compared to SAD and FAD schemes as reflected by the rearward movement of the shock and the smallest change in the total pressure. The relatively large values of the total pressure loss close to the surface may be attributed to errors associated with the boundary treatment of dissipation and implementation of solid surface boundary condition, which are being investigated further.

In Fig. 5, the computed solutions are also compared with two sets of measured data, one from a large-scale test[31] and the other from a small-scale test.[32] Note that the Reynolds number (10 million) and the Mach number (0.85) are the same for the two test data, but the angles of attack are different. The large-scale test data corresponds to $\alpha = 5$ deg and the small-scale test data to 5.9 deg. For these values the two sets of measured lift coefficients and the leading-edge pressure peaks match. All computed solutions agree well with the large-scale test data for the two inboard stations. For the outboard stations all solutions exhibit larger discrepancies that can be attributed to the absence of viscous effects in the TEAM computations. The experimental data[31] show localized flow separation on the outer parts of the wing near the tip. It is interesting to note that the differences between the two sets of measured data are relatively larger than the differences among the solutions computed using three different numerical dissipation schemes.

74-Deg Delta Wing

TEAM analysis of this wing, shown in Fig. 7, illustrates the ability of the code to model leading-edge vortices resulting from flow separation along sharp leading edges of a swept wing at moderate to high angles of attack. The wing was analyzed at $M_\infty = 0.3$ and $\alpha = 10, 20, 30, 35,$ and 40 deg, using an H-O type grid with 85,470 nodes ($35 \times 37 \times 66$). The grid had a total of 66 H planes, each with 35 O curves. Each O curve had 37 nodes. The computed lift, drag and pitching-moment coefficients are compared with the experimental data[33] in Fig. 8. The computed values for $\alpha = 10, 20,$ and 30 deg are in good agreement with the measured data, considering that the viscous effects are not included in the present computations.

For $\alpha = 35$ and 40 deg, converged solutions were not obtained. This does not imply that the solution process diverged, only that no steady-state solutions were obtained. The average residuals dropped approximately two to three orders of magnitude and then remained essentially unchanged. Similar lack of convergence was observed in other investigations using TEAM.[22,23] The lack of convergence could be traced to localized vortex instabilities. In the region of instability the flow variables changed continuously as the solution process advanced in time. As a consequence, the flow did not reach a steady state. The lift, drag, and pitching-moment coefficients continued to fluctuate with the pitching moment coefficient exhibiting the largest amplitude. For all aerodynamic coefficients the amplitudes of fluctuations were larger for the $\alpha = 40$-deg case than the $\alpha = 35$-deg case. Interestingly, the water vapor patterns[33] suggest vortex breakdown at 35- and 40-deg angles of attack. Since local time-stepping process was used in the present analysis, the unconverged solutions were not physically meaningful except perhaps in a qualitative sense. Time-accurate analysis is required for further study.

In Fig. 9 surface pressures computed using the SAD, MAD, and FAD schemes are compared with measured data for the $\alpha = 20$-deg case at four cross-plane stations: $X/C_r = 0.133, 0.4, 0.667$, and 0.93. Here X is measured from apex, and C_r is the root chord. The computed results do not exhibit significant differences among one another. However, the MAD scheme results show highest C_p peaks consistent with the other findings regarding its overall lower levels of dissipation. Note that the computed results fail to model the secondary vortices that are evident in the measured data.

The delta wing was also analyzed in sideslip at $\alpha = 20$ deg and $M_\infty = 0.3$. For this analysis a two-zone grid, one zone each for the left and right halves, was used. Each zone had 85,470 nodes. The cross-plane pressures for a sideslip angle (β) of 10 deg are correlated with experimental data in Fig. 10. Good correlation is obtained on the downwind side. On the upwind side the computed pressure distributions are more peaky than the measured ones, due most likely to the absence of secondary vortices (and other viscous effects) in the computations. For a higher value of β (25 deg), no converged solution was obtained. The analysis procedure exhibited a lack of convergence similar to the high-angle-of-attack symmetrical-flow cases. The experimental data[33] suggest vortex breakdown for sideslip angles greater than 20 deg.

Cone-Derived Mach 6 Waverider

Results for the waverider configuration demonstrate TEAM's ability to model hypersonic flows. The configuration, shown in Fig. 11, was analyzed at its design condition of $M_\infty = 6$ and $\alpha = 0$ deg using an H-O type grid. Solutions were also obtained at $+4$ and -4 deg. The grid contains a singular line emanating from the apex and terminating at the upstream boundary located slightly ahead of the apex. The downstream boundary was located at the base of the configuration. The grid contained 39 H planes, 37 of them on the configuration. Each plane contained 30 O curves, each curve being defined by 45 nodes, resulting in a total of 52,650 nodes.

In Fig. 12 the computed lift and drag coefficients are correlated with experimental data and theoretical results of Ref. 34. Overall trends are correct, but the computed values are underpredicted. Note that the values of the drag coefficient related to the TEAM analysis include an estimate of the skin friction drag given by Schindel[34] as 0.0076. The aerodynamic coefficients showed minimal sensitivity to the type of upwind scheme used, i.e., the first-order accurate or the second-order accurate.

The computed surface pressures for the design condition of $\alpha = 0$ deg are compared with measured values at $X/L = 0.8$ in Fig. 13. Here, X is the distance from the apex, and L is the length of the configuration. Good agreement between computed and measured data is seen on the upper surface and along the lower surface out to about 60%. The discrepancy past that is primarily due to the inviscid nature of the present computations, whereas the measured data strongly suggest shock-induced flow separation.

Canard-Wing-Body

Results for the canard-wing-body configuration not only demonstrate TEAM's ability to model transonic flow about a relatively complex configuration but also illustrate its ability to simulate canard-wing interaction. The configuration with canard in position 1 and wing in position 6 (see Fig. 14) was analyzed using a five-zone grid of H-H topology with 491,232 nodes. Analysis of the canard-off configuration used the same number of points. Each of the zones had 168 planes between the upstream and the downstream boundaries and 34 planes between the plane of symmetry and the outboard side boundary. The canard upper and lower surfaces were each defined by a 34×13 grid (34 nodes chordwise and 13 spanwise). The wing upper and lower surfaces were defined by a 38×22 grid.

In Plate 15 (see the color section), color-coded displays of surface pressure distributions on the canard-on and canard-off configurations are compared with each other at $M_\infty = 0.9$ and $\alpha = 4$ deg. The surface grids are also visible in this plate. The pressure distributions clearly illustrate the ability of the TEAM code to model the effect of canard wake in reducing the pressure peak on the inboard wing stations. Correlations of computed solutions and experimental data[35] at four stations on the wing are shown in Fig. 15. Good overall agreement is found with the exception of the shock location. The computed shock is located approximately 10% aft of the measured one. However, such a discrepancy is typical of inviscid simulations of flows exhibiting significant shock–boundary-layer interaction. For the present case effects of such an interaction is seen in the measured data but not simulated in the TEAM computations. The agreement between the computed results and measured data is much better at a lower value of the freestream Mach number (0.6) when the flow is largely subcritical as shown in Fig. 16.

ANC Fighter Configuration

The ANC fighter configuration, shown in Fig. 17, represents the most complicated geometry analyzed to date using the TEAM code. This configuration features close-coupled canards mounted on the outboard side of podded engine nacelles that are separated from the fuselage by straked-wing sections. The configuration was analyzed at $M_\infty = 1.19$ and $\alpha = 4.85$ deg using a 27-zone H-H type grid having 378,464 cells. The grid generation process is described in detail by Olling and Mani.[2] At the inlet face of the nacelle, the MFR value was set at 0.745. At the exhaust face of the nacelle, the NPR value was set at 7.622.

The computed surface pressure distributions are shown in color-coded display in Plate 16 (see the color section). The computational grid on the surface is also visible in this plate. The computed lift and moment coefficients agree well with the experimental data as reported in Ref. 2. The chordwise surface C_p distributions at two wing stations are shown in Fig. 18. For the inboard station, the computed values could be correlated with the available measured data.[36] The results demonstrate the ability of the TEAM code to compute the flowfield around a complete aircraft configuration.

Concluding Remarks

Over the past few years, considerable progress has been made in developing computational techniques to solve the Euler equations about complex configurations for a variety of flow conditions. However, the full range of capabilities of the Euler methods is not known. A thorough evaluation of the capabilities requires 1) conducting sensitivity studies to investigate the effect of various computational parameters, such as grid density, numerical dissipation, and treatment of boundary conditions, on the solutions; and 2) correlating computed solutions with experimental data to determine the types of flows that can be modeled and the associated levels of accuracy. Only then can an Euler code be considered validated for routine aerodynamic analysis. The enormity and the importance of the evaluation task should not be underestimated. With that in mind, this chapter should be considered a progress report rather than a final verdict on the capabilities of the TEAM code.

The TEAM code affords a cost effective means of obtaining steady-state solutions to the Euler equations. The code accommodates patched multizone structured grids of arbitrary topologies and thereby facilitates analysis of complex configurations. This was demonstrated by the solutions for six test cases ranging from an airfoil to a complete aircraft presented in this chapter. The ability to generate suitable grids about aircraft configurations in a timely manner holds the key to using TEAM effectively. Grid generation continues to be a challenging task and is considered a pacing item for effectively using the advanced CFD codes, including TEAM.

The results presented in this chapter demonstrated TEAM's ability to model subsonic through hypersonic flows. Correlations with experimental data illustrated that a good approximation to flows with shock waves and rotational flows was obtained in regions where the viscous effects were relatively weak. These results coupled with other published and unpublished data provide enough confidence in the abilities of the code for Lockheed to use it in developing new aircraft, such as the advanced tactical fighter. With ongoing investigations, a more comprehensive assessment of TEAM's capabilities for aerodynamic analysis will be forthcoming. A number of other developments, e.g., the use of unstructured tetrahedral grids, parallel computers, and solution-adaptive grid techniques, hold considerable promise in improving the effectiveness of the Euler codes for aerodynamic analysis of aircraft configurations.

Acknowledgments

Lockheed Aeronautical Systems Company, Burbank, CA, provided funding for preparing this chapter under an Independent Research and Development program. The author wishes to thank Josef Sikora, Charles Olling, Joseph Keen, Kurian Mani, and Steven Singer for their invaluable assistance in TEAM development and applications to geometries for which the results are presented in this chapter.

References

[1]Ghaffari, F., Luckring, J. M., Thomas, J. L., and Bates B. L., "Navier-Stokes Solutions about the F/A-18 Forebody-LEX Configuration," AIAA Paper 89-0338, Jan. 1989.

[2]Olling, C. R., and Mani, K. K., "Navier-Stokes and Euler Computations of the Flow Field around a Complete Aircraft," *Advanced Aerospace Aerodynamics*, Society of Automotive Engineers SP-757, Oct. 1988, pp. 231–242.

[3]Flores, J., and Chaderjian, N. M., "The Numerical Simulation of Transonic Separated Flow About the Complete F-16A," AIAA Paper 88-2506, June 1988.

[4]Holst, T. L., "Viscous Transonic Airfoil Workshop Compendium of Results," AIAA Paper 87-1460, June 1987.

[5]Lakshminarayana, B., "Turbulence Modeling for Complex Shear Flows," *AIAA Journal*, Vol. 24, Dec. 1986, pp. 1900–1917.

[6]Miranda, L. R., Elliott, R. D., and Baker, W. M., "A Generalized Vortex Lattice Method for Subsonic and Supersonic Flow Applications," NASA CR-2865, Dec. 1977.

[7]Magnus, A. E., Ehlers, F. E., and Epton, M. A., "PANAIR—A Computer Program for Predicting Subsonic or Supersonic Linear Potential Flow About Arbitrary Configurations Using a Higher-Order Panel Method," NASA CR-3251, April 1980.

[8]Youngren, H. H., Bouchard, E. E., Coopersmith, R. M., and Miranda, L. R., "Comparison of Panel Method Formulations and Its Influence on the Development of QUADPAN, an Advanced Low Order Panel Method," AIAA Paper 83-1827, July 1983.

[9]Johnston, C. E., Youngren, H. H., and Sikora, J. S., "Engineering Applications of an Advanced Low-Order Panel Method," Society of Automotive Engineers Paper 851793, Oct. 1985.

[10]Boppe, C. W., "Aerodynamic Analysis for Aircraft with Nacelles, Pylons, and Winglets," NASA CR-4066, April 1987.

[11]Caughey, D. A., and Jameson, A., "Progress in Finite-volume Calculations for Wing-Fuselage Combinations," *AIAA Journal*, Vol. 18, Nov. 1980 pp. 1281–1288.

[12]Raj, P., "A Multigrid Method for Transonic Wing Analysis and Design," *Journal of Aircraft*, Vol. 21, Feb. 1984, pp. 143–150.

[13]Jameson, A., Schmidt, W., and Turkel, E., "Numerical Solution of the Euler Equations by Finite-Volume Methods Using Runge-Kutta Time-Stepping Schemes," AIAA Paper 81-1259, 1981.

[14]Rizzi, A., "Damped Euler Equation Method to Compute Transonic Flow Around Wing-Body Combinations," *AIAA Journal*, Vol. 20, Oct. 1982, pp. 1321–1328.

[15]Jameson, A., and Baker, T. J., "Multigrid Solution of the Euler Equations for Aircraft Configurations," AIAA Paper 84-0093, Jan. 1984.

[16]Thomas, J. L., van Leer, B., and Walters, R. W., "Implicit Flux-Split Schemes for the Euler Equations," AIAA Paper 85-1680, July 1985.

[17]Chakravarthy, S. R., and Osher, O., "A New Class of High-Accuracy TVD Schemes for Hyperbolic Conservation Laws," AIAA Paper 85-0363, Jan 1985.

[18]Rizzi, A., and Eriksson, L. E., "Computation of Flow Around Wings Based on the Euler Equations," *Journal of Fluid Mechanics*, Vol. 148, Nov. 1984, pp. 45–71.

[19]Hitzel, S. M., and Schmidt, W., "Slender Wings with Leading-Edge Vortex Separation: A Challenge for Panel Methods and Euler Solvers," *Journal of Aircraft*, Vol. 21, Oct. 1984, pp. 751–759.

[20]Agarwal, R. K., and Deese, J. E., "An Euler Solver for Calculating the Flowfield of a Helicopter Rotor in Hover and Forward Flight," AIAA Paper 87-1427, June 1987.

[21]Yu, N. J., Chen, H. C., Chen, A. W., and Wittenberg, K. R., "Grid Generation and Flow Analyses for Wing/Body/Winglet Configurations," AIAA Paper 88-2548-CP, June 1988.

[22]Raj, P., Sikora, J. S., and Keen, J. M., "Free-Vortex Flow Simulation Using a Three-dimensional Euler Aerodynamic Method," *Journal of Aircraft*, Vol. 25, Feb. 1988, pp. 128–134.

[23]Raj, P., Keen, J. M., and Singer, S. W., "Applications of an Euler Aerodynamic Method to Free-Vortex Flow Simulation," AIAA Paper 88-2517-CP, June 1988.

[24]Raj, P., and Brennan, J. E., "Improvements to an Euler Aerodynamic Method for Transonic Flow Analysis," *Journal of Aircraft*, Vol. 26, Jan. 1989, pp. 13–20.

[25]Raj, P., "An Euler Code for Nonlinear Aerodynamic Analysis: Assessment of Capabilities," *Advanced Aerospace Aerodynamics*, Society of Automotive Engineers SP-757, Oct. 1988, pp. 215–230.

[26]Raj, P., Olling, C. R., Sikora, J. S., Keen, J. M., Singer, S. W., and Brennan, J. E., "Three-Dimensional Euler/Navier-Stokes Aerodynamic Method (TEAM), Vol. I: Computational Method and Verification," AFWAL-TR-87-3074, June 1989.

[27]Jameson, A., "A Nonoscillatory Shock Capturing Scheme Using Flux Limited Dissipation," Princeton Univ., Princeton, NJ, MAE Rept. 1653, 1983.

[28]Gnoffo, P. A., McCandless, R. S., and Yee, H. C., "Enhancements to Program LAURA for Computation of Three-Dimensional Hypersonic Flow," AIAA Paper 87-0280, Jan. 1987.

[29]"Test Cases for Inviscid Flow Field Methods," AGARD-AR-211, May 1985.

[30]Raj, P., "TEAM: A Three-Dimensional Euler Aerodynamic Method," AIAA Professional Study Series, Euler Solvers Workshop, Monterey, CA, Aug. 20–21, 1987.

[31]Keener, E. R., "Pressure-Distribution Measurements on a Transonic Low-Aspect Ratio Wing," NASA TM-86683, Sept. 1985.

[32]Hinson, B. L., and Burdges, K. P., "Acquisition and Application of Transonic Wing and Far-Field Test Data for Three-Dimensional Computational Method Evaluation," Air Force Office of Scientific Research TR-80-0421, March 1980.

[33]Wentz, W. H., Jr., "Effects of Leading-Edge Camber on Low-Speed Characteristics of Slender Delta Wings," NASA CR-2002, Oct. 1972.

[34]Schindel, L. H., "Design of High Performance Hypersonic Missiles," AIAA Paper 82-0391, Jan. 1982.

[35]Stewart, V. R., "Evaluation of a Propulsive Wing/Canard Concept at Subsonic and Supersonic Speeds," Vol. I, Rockwell International Rept. NR82H-85, Aug. 1983.

[36]Zilz, D. E., "Propulsion and Airframe Aerodynamic Interactions of Supersonic V/STOL Configurations," NASA CR-177343, Sept. 1985.

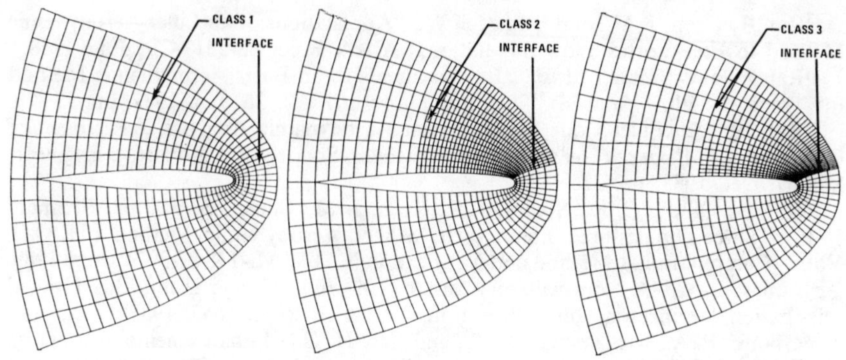

Fig. 1 Three classes of zonal interfaces of multizone grids for TEAM.

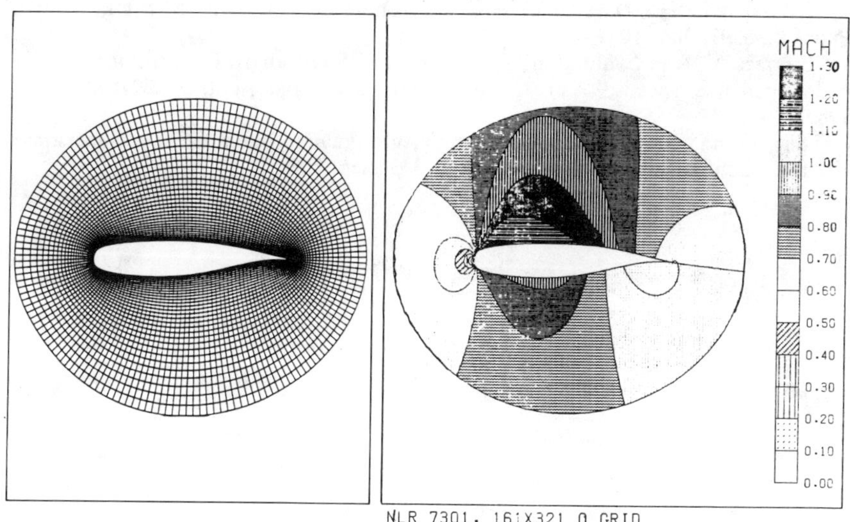

Fig. 2 Portion of grid and flowfield; NLR 7301 airfoil ($M_\infty = 0.721$, $\alpha = -0.194$ deg, 161×321 O grid).

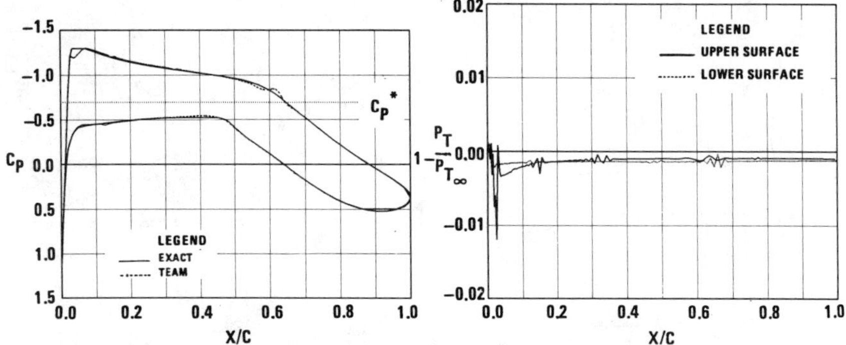

Fig. 3 Surface pressure and total pressure loss distributions; NLR 7301 airfoil ($M_\infty = 0.721$, $\alpha = -0.194$ deg, 161×321 O grid).

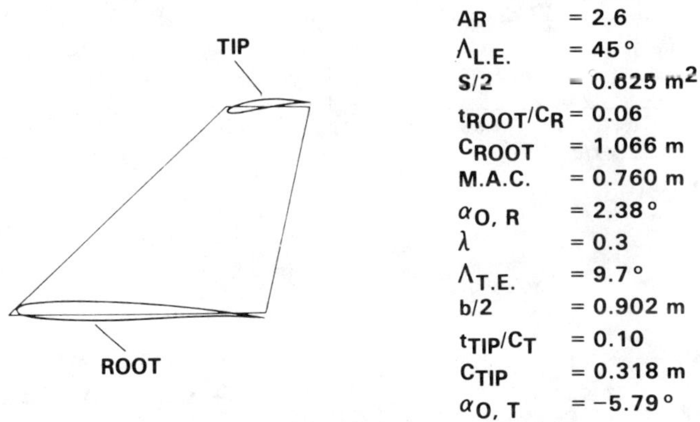

Fig. 4 Wing C geometric features.

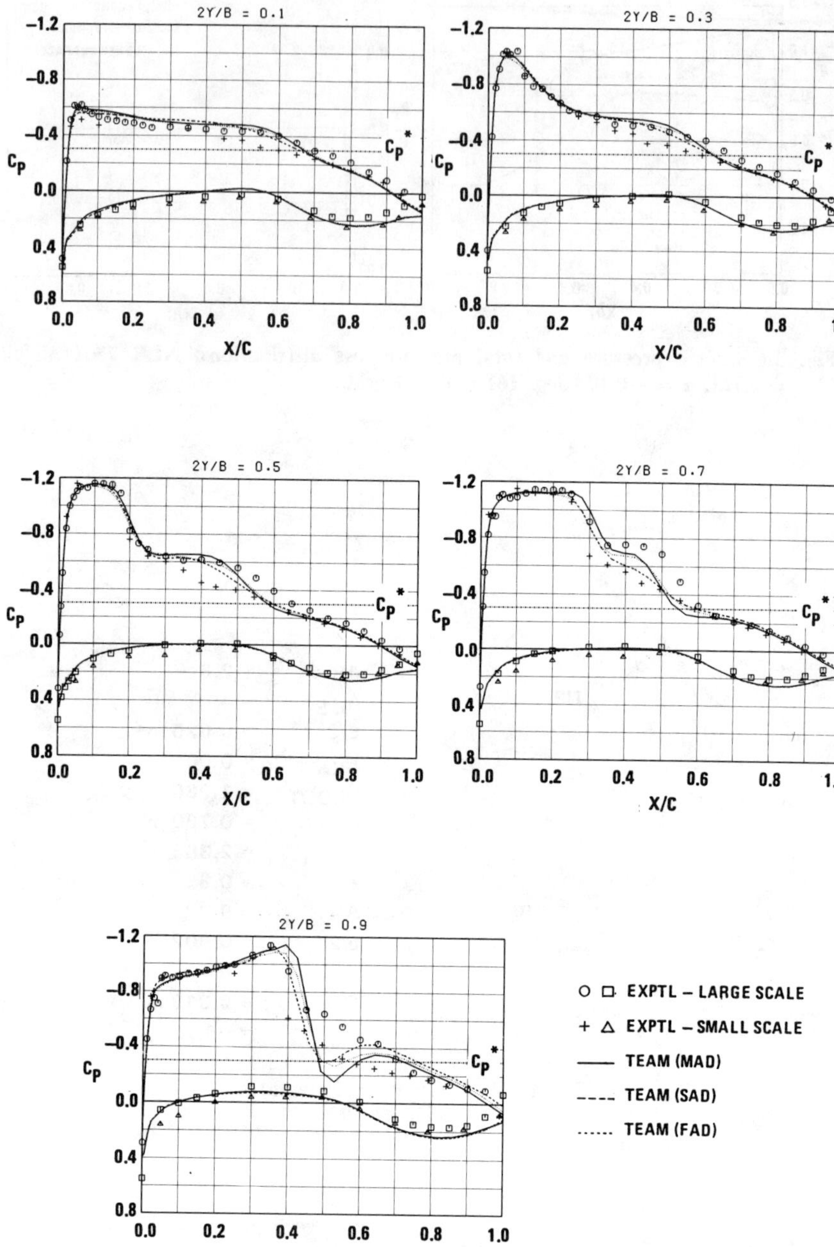

Fig. 5 Solution sensitivity to numerical dissipation and correlation with experimental data; wing C ($M_\infty = 0.85$, $\alpha = 5$ deg, $25 \times 129 \times 33$ C-H grid).

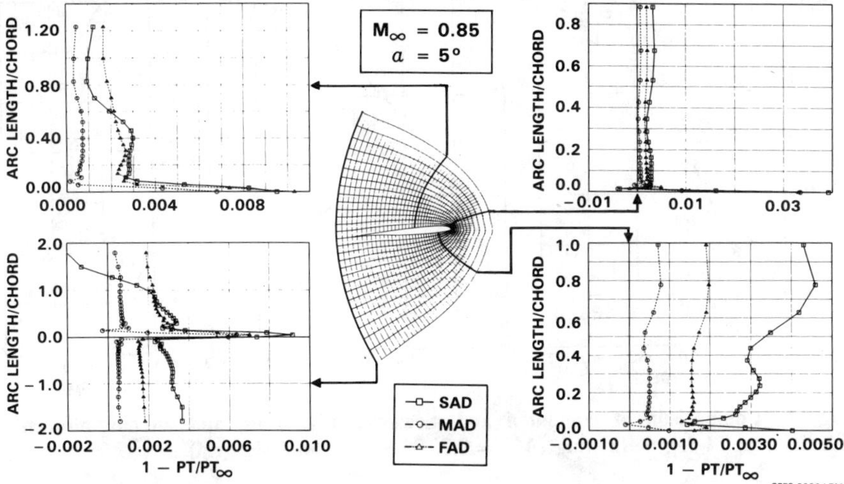

Fig. 6 Effect of numerical dissipation on total pressure loss variation in $2y/b = 0.5$ plane; wing C ($M_\infty = 0.85$, $\alpha = 5$ deg, $25 \times 129 \times 33$ C-H grid).

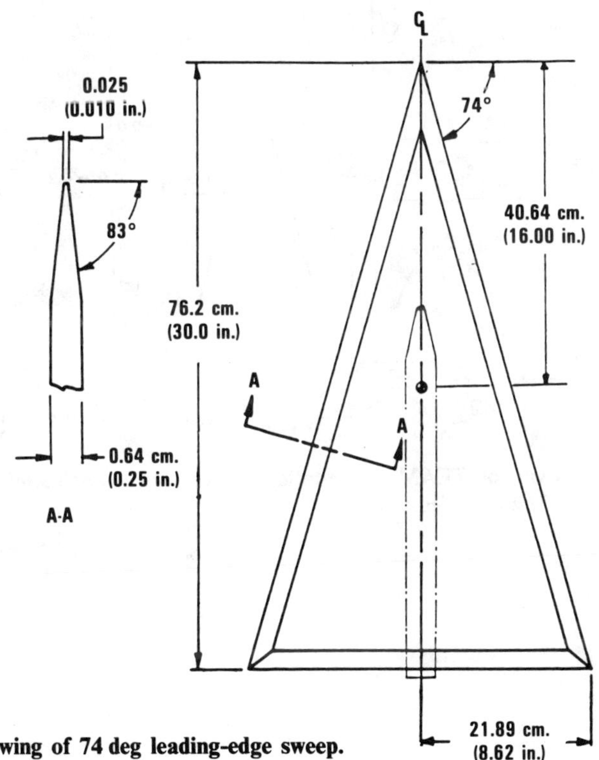

Fig. 7 Delta wing of 74 deg leading-edge sweep.

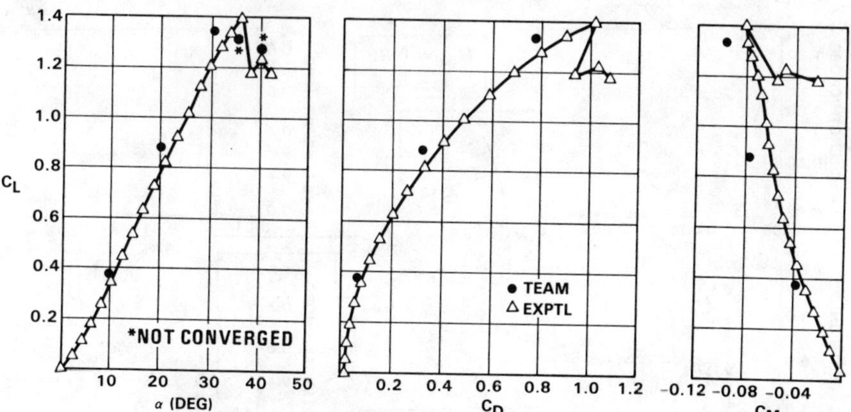

Fig. 8　Comparison of computed and measured lift, drag, and pitching-moment coefficients; 74-deg delta wing ($M_\infty = 0.3$, $35 \times 37 \times 66$ O-H grid).

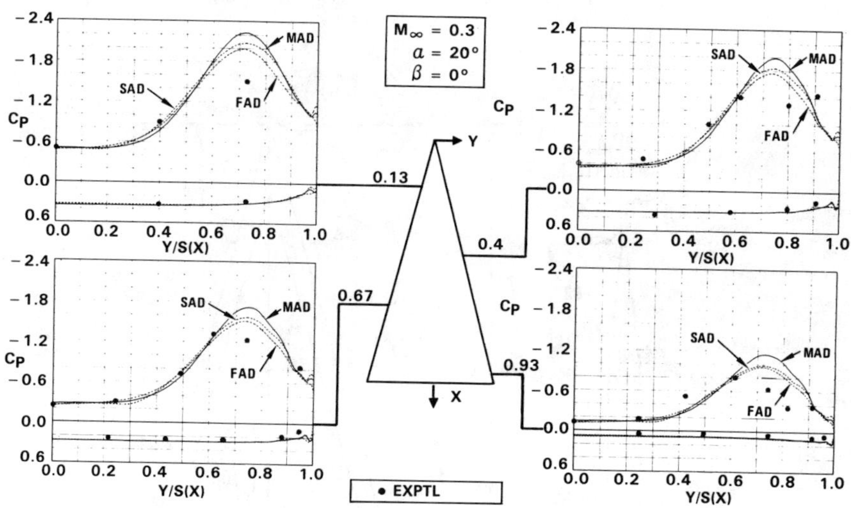

Fig. 9　Sensitivity of TEAM computations to numerical dissipation; 74-deg delta wing ($M_\infty = 0.3$, $\alpha = 20$ deg).

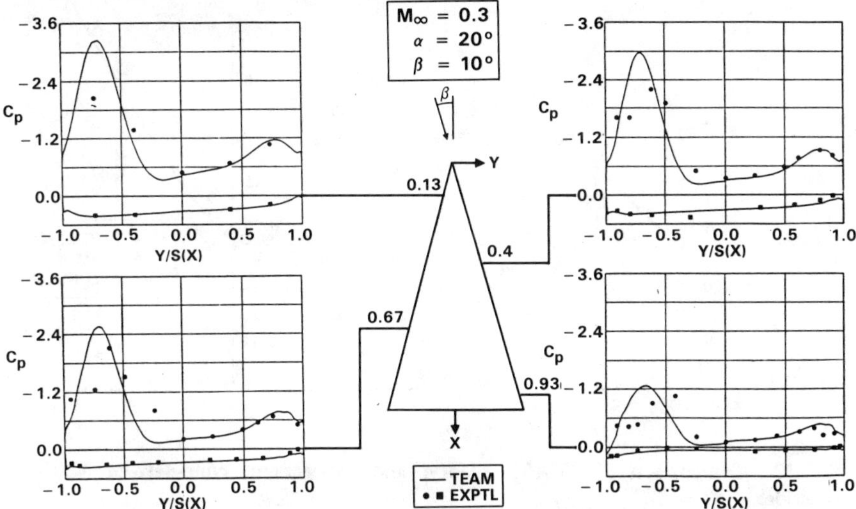

Fig. 10   Cross-plane surface pressure distribution; 74-deg delta wing ($M_\infty = 0.3$, $\alpha = 20$ deg, $\beta = 0$ deg, $35 \times 37 \times 66$ O-H grid).

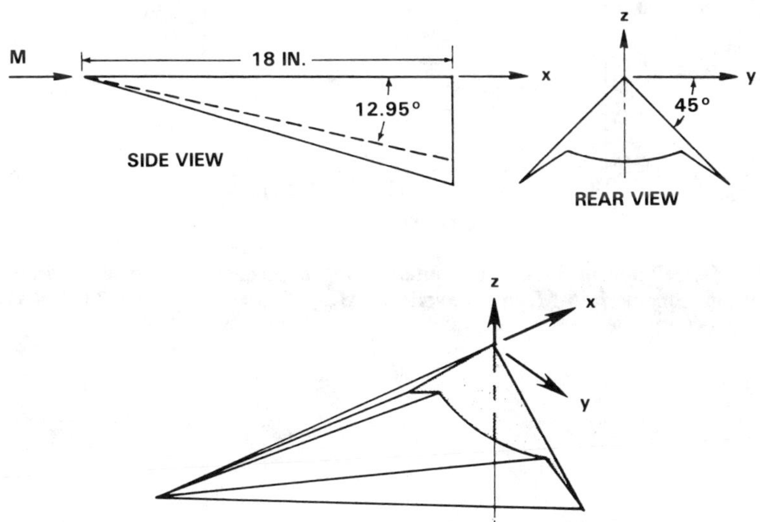

Fig. 11   Cone-derived Mach 6 waverider.

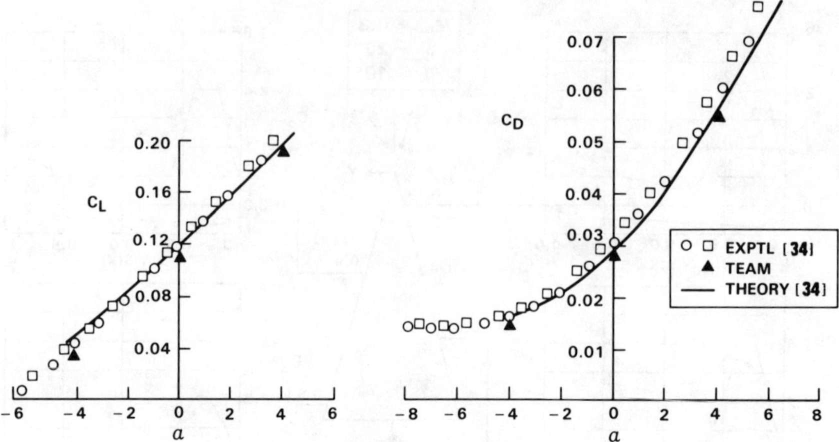

Fig. 12 Comparison of TEAM, theory, and experiment; cone-derived Mach 6 waverider ($M_\infty = 6$).

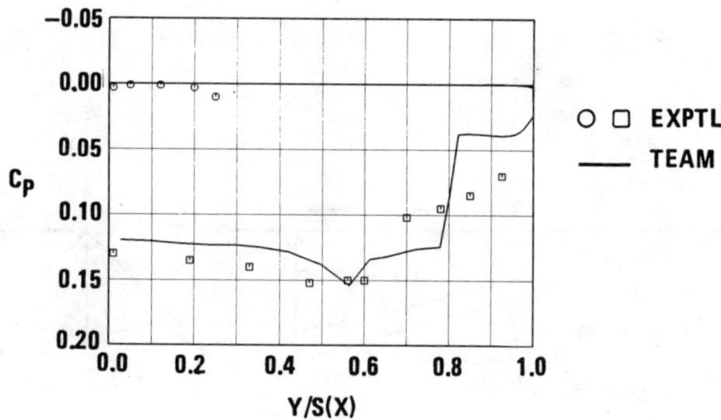

Fig. 13 Correlation of TEAM computation and experimental cross-plane pressure distribution; cone-derived Mach 6 waverider ($M_\infty = 6$, $\alpha = 0$ deg, $45 \times 30 \times 39$ O-H grid, $x/L = 0.8$).

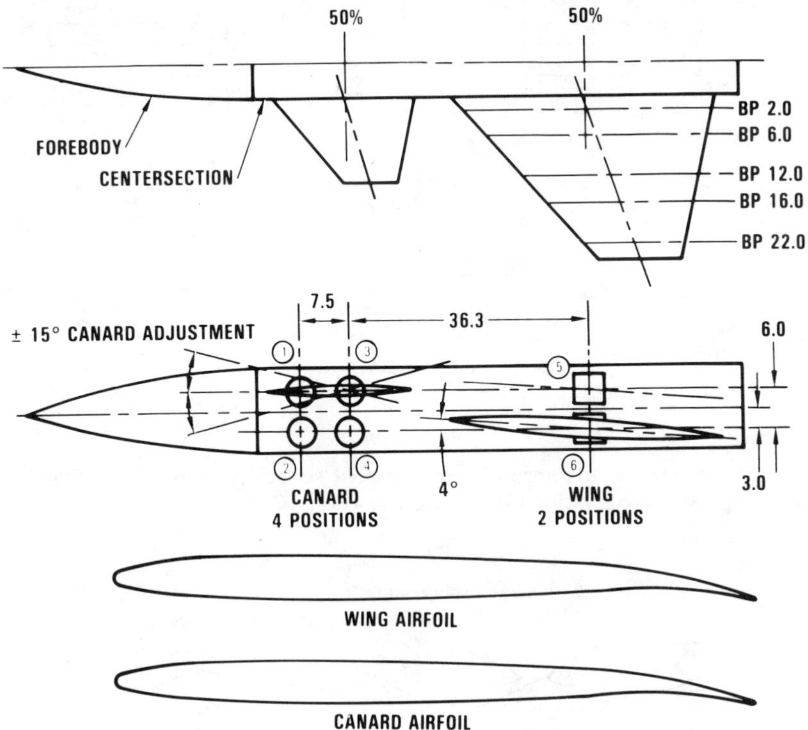

Fig. 14 Canard-wing-body configuration.

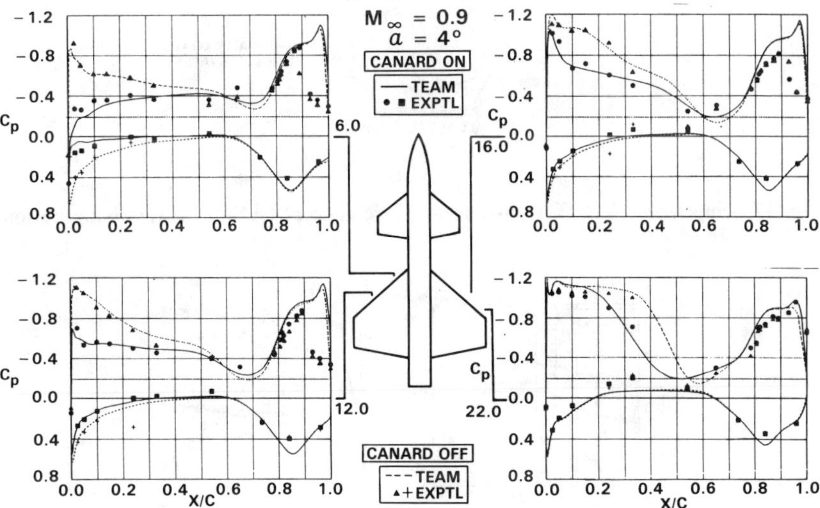

Fig. 15 Correlation of computed and measured surface pressure distributions; canard-wing-body configuration ($M_\infty = 0.9$, $\alpha = 4$ deg, $168 \times 84 \times 34$ H-H grid).

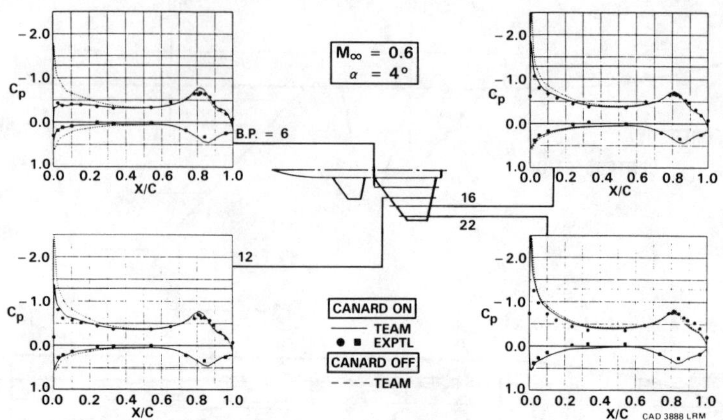

Fig. 16   Correlation of computed and measured surface pressure distributions; canard-wing-body configuration ($M_\infty = 0.6$, $\alpha = 4$ deg, $168 \times 84 \times 34$ H-H grid).

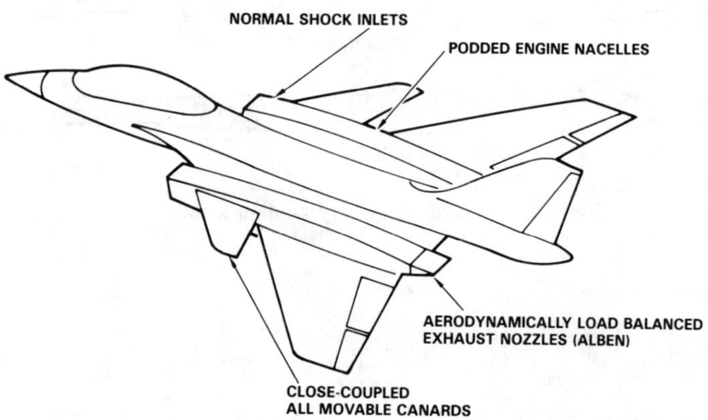

Fig. 17   Supersonic V/STOL advanced nozzle concept (ANC) fighter configuration.

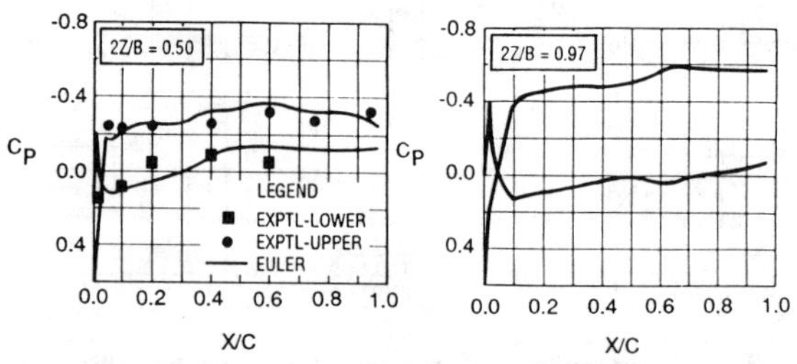

Fig. 18   Computed and measured chordwise pressure distributions on the wing; ANC fighter ($M_\infty = 1.2$, $\alpha = 4.85$ deg).

Chapter 19

Toward the Routine Aerodynamic Analysis of Complex Configurations

Michael D. Madson* and Larry L. Erickson*
NASA Ames Research Center, Moffett Field, California

Introduction

THE ability to quickly and routinely predict the aerodynamics of complete aircraft configurations is a common goal among designers and engineers. Unfortunately, a trade off has always existed between the generality of the geometry that could be modeled and the level of physics that could be applied to the problem. In general, the more complex the physics that is modeled, the simpler the geometry needs to be to generate a suitable surface-conforming flowfield grid. Although Navier-Stokes and Euler codes are able to handle increasingly complex geometries because of improvements in the grid-generation process, they are still not able to model truly complete configurations in a routine manner. Each new geometry demands a substantial amount of calendar time in the generation of the flowfield grid, and there is the need to "tweak" various parameters within the codes themselves in order to adjust them for each new geometry and flow condition. If an analysis of a series of candidate configurations is desired, each new configuration requires a newly defined flowfield grid.

The ability to quickly model and analyze a configuration is of great importance to the engineer, who is often met with very restrictive time constraints when asked to provide a prediction of the aerodynamics of a new configuration, or to calculate the effect of adding or removing a piece of geometry from an existing model. For many cases it is not necessary to model the complete physics of the problem in order to obtain a solution that provides important insight into the design of a particular configuration.

Linear-potential panel methods have been able to model arbitrary geometries for many years. One of these codes, PanAir,[1-5] is able to predict supersonic as well as subsonic results about these general geometries. Unfortunately, the simplifications of the actual physics necessary to run

---

Copyright © 1989 by the American Institute of Aeronautics and Astronautics, Inc. No copyright is asserted in the United States under Title 17, U.S. Code. The U.S. Government has a royalty-free license to exercise all rights under the copyright claimed herein for governmental purposes. All other rights are held by the copyright owner.
*Research Scientist, Aerodynamics Division.

these configurations are substantial. For subsonic cases, the flow is assumed to be inviscid and irrotational, and the perturbation velocities must be small compared to the freestream in order for the solution to be completely valid. The flow must be subsonic everywhere in the flowfield, thus, transonic flow problems cannot be solved. Supersonically, the geometry must be relatively sharp so that surface panels are not inclined to the freestream flow at angles greater than the Mach cone angle. Also, the flow must be supersonic everywhere in the flowfield, so that Mach numbers close to 1.0 are not generally valid. Still, linear-potential codes such as PanAir have been successfully applied to many complex configurations.[6-15] A tremendous degree of confidence in the linear-potential solution exists when the code is applied to a problem within the range of applicability of the method. Several PanAir models and their results are presented in the first part of this paper. These cases demonstrate the type of modeling generality afforded by panel methods, as well as some of the successes and shortcomings of these methods. Among the models presented are the NASA/ MCAIR 279-3, which is a concept for a 1990s vertical/short takeoff and landing (V/STOL) fighter/attack aircraft; a General Dynamics E-7 V/ STOL fighter/attack aircraft mounted in the NASA Ames 12-Ft Pressure Wind Tunnel, and a modified Boeing 747-SP model.

The main advantage of panel method codes is the complete modeling generality they allow. For example, to estimate the effect of a fuel tank on the aerodynamics of a configuration, two quick runs may be made: one for the configuration without the fuel tank and one with the fuel tank added. Although the magnitude of flow quantities may be slightly inaccurate due to viscous effects and other flow features not predicted by a potential method, the incremental aerodynamic effect of adding the fuel tank to the geometry should provide useful insight. The generation of the paneled models themselves is a relatively simple process, and the solution may be obtained directly once the geometry is modeled (i.e., no flowfield grid generation is required). However, the inability to solve for transonic flow about these geometries leaves an important hole in the analysis of many aircraft, since most of today's transport configurations and military aircraft fly in the transonic regime.

A computer code called TranAir has been developed that solves the full-potential equation for transonic flow about completely arbitrary geometries. Although the code itself is not a panel method code, it employs the surface-paneling technology of PanAir in the geometry definition. The paneled geometry is embedded in a flowfield grid consisting of a rectangular array of grid points. The uniform global grid may be locally refined in regions where flow properties are rapidly changing, such as regions where shocks exist, and around wing leading edges. This local refinement produces an unstructured global grid. Unlike panel method codes, TranAir solutions are not hampered by small-perturbation assumptions and may be applied to very blunt geometries in the transonic flow regime. Presented in this paper is a brief description of TranAir, including some of the input required to run the program, and the basics of the solution process. Also presented are results for several configurations, which serve to demonstrate TranAir's ability to compute both the subsonic and transonic aerodynamics about very general geometries. Geometries presented include the F-16A with wing-tip missiles and launchers and underwing fuel tanks; a generic fighter configuration with a cropped delta wing and forebody chine; and a model of the NASA Ames 12-Ft Pressure Wind Tunnel, including the 25:1 contraction ratio and the support strut and centerbody downstream of the test section. TranAir results have recently been presented for several other configurations.[16-21]

To summarize, this chapter discusses the ability to routinely compute the aerodynamics of complex aircraft configurations. Presented are subsonic and supersonic results for a linear-potential method and subsonic and transonic results for a full-potential method. The full-potential code, TranAir, is able to solve transonic flow problems about the same general geometries allowed by linear-potential methods. Solutions are compared with experimental data as well as with results from other computational fluid dynamics (CFD) codes when available. Some of the successes, as well as some of the shortcomings, of the two solution methods are discussed.

PanAir: Applications

MCAIR 279-3

A jointly sponsored NASA Ames and Navy program was initiated in 1980 to develop aerodynamic technology for post-1990 V/STOL fighter/attack aircraft. One of the configurations that arose from this program was the MCAIR 279-3[22] (referred to from here on as 279). The wind-tunnel model of the 279 installed in the NASA Ames 11×11-Ft Transonic Wind Tunnel is shown in Fig. 1. One of the goals of the project was to determine how well the aerodynamics of complex configurations could be computationally predicted. PanAir, one of the few panel codes that treats both subsonic and supersonic flow, was chosen as the analysis tool. The linear theory on which PanAir is based assumes that local flow velocities do not vary greatly from the freestream velocity (small-perturbation assumption). Consequently, the theory is particularly valid for very slender configurations, especially in the supersonic flow regime. But the 279 geometry is relatively nonslender. This investigation was undertaken[7,8] to determine the degree of success that could be achieved by applying PanAir to the relatively complex, nonslender geometry of the 279. PanAir results for $M_\infty = 0.6$ and 1.2 are presented.

PanAir Model

The 279 configuration employs a close-coupled canard/wing, has large inlets for a high-bypass-ratio engine, and has a maximum design Mach number of 2.0. The configuration also has four deflecting exhaust nozzles for thrust-vectoring capability similar to that of the Harrier aircraft. These nozzles were modeled for the case of zero nozzle deflection only. The PanAir surface-paneled definition of the 279 geometry is shown in Fig. 2. A total of 990 surface panels was used to describe the geometry. The

PanAir model did not include the sting shroud that was present on the wind-tunnel model (see Fig. 1b). Consequently, aerodynamic effects caused by the presence of the shroud on the wind-tunnel model was not represented in the computations. The absence of the shroud from the PanAir model was due to the unavailability of the shroud-geometry definition, and not from any PanAir modeling limitations. Several PanAir models were examined to determine the effect of canard-wake position and nozzle-exit flow modeling on the predicted aerodynamics.

Canard-Wake Position

Being a potential method, PanAir requires that wakes be defined for any surfaces from which lift may be generated (e.g., wing trailing edges). For many configurations, defining the wake as a flat sheet that travels downstream parallel to the x axis provides a sufficient wake model. For configurations in which a wake may pass over another lifting surface, the position and shape of the wake becomes more critical. Two canard-wake cases were defined in order to evaluate the effect of wake position (not shape) on the wing-lift distribution. The first case defined a canard wake that was parallel to the x axis in the body-axis system and remained unchanged with angle of attack. The second case defined a canard wake that was aligned with (i.e., parallel to) the freestream at each angle of attack. For simplicity, other wakes in the configuration remained unchanged for both canard-wake cases. Figure 3 shows the root sections of the canard and the wing, with the respective wakes defined parallel to the x axis in the body-axis system (canard-wake case A) and with the canard wake aligned with freestream angle of attack (canard-wake case B). The wing has a 9-deg anhedral (see Fig. 1a); thus, the vertical distance between the canard wake and the wing surface increases in the outboard direction. If the case A canard-wake definition is utilized through a wide range of angles of attack, the inboard wing circulation begins to be seriously affected, as shown in Fig. 4a. By aligning the canard wake with the angle of attack (case B), which is a rather simplistic modeling idea, the effect on the wing circulation can be readily observed (Fig. 4a). The slight change in circulation on the canard is caused more by the effect of increased lift on the close-coupled wing than from the rotation of the canard wake. An improved lift prediction at the higher angles of attack is also evident in Fig. 4b. Including the effect of wake rollup would also have an effect on the wing circulation.

Exhaust Modeling

The main difficulty in defining the computational model of the 279 was the modeling of the flow-through conditions at the nozzle exits for both the subsonic and supersonic cases that were analyzed. As seen in Fig. 1b, the nozzle geometry and its relationship to the rest of the aircraft geometry is quite complex. Four nozzle-exit flow models were studied. These models (Table 1) were combinations of using and not using base networks (panels that represent the nozzle exit planes with flow-through boundary conditions, i.e., freestream velocity at the exit planes) and wakes. All four models were studied for the supersonic case.

Table 1   Nozzle exit models for the 279-3 configuration

Model	Nozzle exit planes[a]	Nozzle wakes		
1	No	No		Supersonic
2	No	Yes		Supersonic
3	Yes	No	Subsonic	Supersonic
4	Yes	Yes	Subsonic	Supersonic

[a]Base networks.

For the subsonic case, the cases where no exit planes were used were not studied, since the geometry downstream of the exit planes affects the flow at (and even upstream of) the exit planes, thus disturbing the zero-perturbation potential-boundary condition imposed on the inside of the geometry. A view of the wake model used for cases requiring nozzle wakes is shown in Fig. 5. The boldface lines outline all wake networks, and the dashed lines outline the nozzle wakes (the presence of which was case dependent). All of the wakes shown extended much farther downstream, but are shown chopped off for clarity.

For the supersonic case, the best overall results were obtained by the two models with the nozzle wakes. The results for these two models were essentially identical. Model 4 was used to obtain the supersonic results, since the presence of base networks at the nozzle exits provided output data for important flow quantities at the exits.

For the subsonic case the opposite was true: the nozzle wakes adversely affected subsonic PanAir results for both lift and pitching moment. As a result of this information, model 3 was used to obtain the subsonic PanAir results for the 279 model.

Figure 6 shows the nozzle-exit geometry, along with the boundary conditions applied at the nozzle-exit planes for the subsonic and supersonic cases, and the nozzle-wake definitions for the supersonic case. It is not yet understood why similar wake models for the subsonic and supersonic cases yielded opposite results in terms of the degree of agreement with wind-tunnel data.

Subsonic Results

The complex geometric features of the 279 configuration presented a challenge to the linear-potential assumptions of PanAir. After resolving modeling questions involving canard-wake positioning and nozzle-exit-plane boundary condition modeling, PanAir results were generated for $M_\infty = 0.6$ and 1.2. The quality of the comparisons with experimental data for the two cases was mixed. PanAir results at $M_\infty = 0.6$ are shown in Fig. 7 and are plotted against wind-tunnel data obtained in the Ames 11 × 11-ft Transonic Wind Tunnel.[23,24] The lift prediction is in generally good agreement with wind-tunnel data, even up to 10 deg, where one might expect that the complexity of the actual flow would cause PanAir to provide a result that compares poorly with wind-tunnel data. Pitching-moment data for $M_\infty = 0.6$ were in reasonable agreement with the linear portion of the wind-tunnel data.

In PanAir, pressure coefficients can be calculated from the isentropic equation for flow of a perfect gas and from approximations based on small perturbation assumptions. For small-perturbation flow the second-order equation, in which only first- and second-order terms of the isentropic equation are retained, is nearly equivalent to the isentropic relation. The discrepancy in the isentropic and second-order results in Fig. 7 are an indication that there are regions of the configuration where the small-perturbation assumption inherent in linear-potential methods is being violated.

Supersonic Results

The results for $M_\infty = 1.2$ are shown in Fig. 8. The slope of the second-order lift curve is nearly parallel to the wind-tunnel data and is shifted slightly to the left. The slope of the isentropic lift curve is not in good agreement with wind-tunnel data. The pitching-moment data were scattered. The isentropic result was in fair agreement with wind-tunnel data, but second-order results were poor. The divergence between isentropic and second-order results indicates that the 279 geometry is not as slender as supersonic linear theory prefers and that the small-perturbation assumption was violated at this Mach number for the configuration.

E-7/12-Ft Tunnel Model

Models tested in the NASA Ames 12-Ft Pressure Wind Tunnel over an angle-of-attack range of 0–90 deg have been mounted on a floor strut that protrudes from a large support bump. In high-angle-of-attack tests (40–90 deg), for which the floor support was originally designed, the effects of the flow angularities produced by the bump are often negligible. This is not so for low-angle-of-attack tests (0–40 deg). Since there are no standard means for correcting test data for this bump effect, low-angle-of-attack testing with the bump was not recommended by the Ames wind-tunnel staff. However, because the floor support system provides a greater sideslip range than the conventional rear-sting support system, it was often used for low-angle-of-attack tests.

PanAir Model

An exploratory study of a technique for correcting balance forces and experimental pressures for combined wall and bump effects was performed in which the aircraft, wind-tunnel walls, and support-bump geometries were modeled for PanAir analysis. For this study,[6] the aircraft modeled was the General Dynamics E-7 short takeoff and vertical landing (STOVL) aircraft. A front view of the E-7 mounted in the NASA Ames 12-Ft Pressure Wind Tunnel is shown in Fig. 9. Notice that the model is not mounted along the vertical centerline of the tunnel. Also notice the "bump on the bump" on the floor support system. For the PanAir model of this installation, the model lies exactly on the vertical centerline so that the plane-of-symmetry option could be used to reduce the size of the problem, and the bump on the bump was not modeled. Also, the tunnel was modeled as a constant 12-ft section, and the sting support was not modeled. The PanAir definition of the E-7 model in the 12-ft tunnel and its symmetric image are shown in Fig. 10. A schematic side view of the PanAir model showing the tunnel and bump, the E-7 definition including wakes, and the position of two survey planes is shown in Fig. 11. Survey-plane networks (i.e., networks with zero-strength sources and doublets) are used to evaluate properties of the flowfield at the survey plane, without actually affecting the flowfield. For this case they were used to verify that the mass flow entering the tunnel was the same as the mass flow leaving the tunnel, i.e., that there were no leaks. The PanAir model consisted of 773 panels to define the E-7 model and 674 panels to model the tunnel walls. Also, 132 wake panels and 121 survey-plane panels were defined, for a total of 1700 panels. The study compared test data from the Ames 12-ft tunnel[25] and the General Dynamics low-speed tunnel,[26] with corresponding PanAir predictions for the E-7 aircraft.

Results

A series of PanAir runs was made at the wind-tunnel test Mach number of 0.2 and angles of attack ranging from 0 to 9 deg. Three sets of runs were obtained and will be referred to as the free-air case, the walls case, and the walls-plus-bump case. For the free-air case, the E-7 model was run without the presence of the wind-tunnel model or the floor bump. The walls case included the E-7 model and the walls of ther 12-ft tunnel without the floor bump. The walls-plus-bump case included the E-7 model and the walls and floor bump of the 12-ft tunnel. These descriptions are also applied to the corresponding wind-tunnel cases.

Free-air results for the lift coefficient (C_L) are shown in Fig. 12. The results from the Ames 12-ft tunnel include corrections to remove wall effects, but not to remove the unknown bump effects. The wind-tunnel data with wall corrections for the General Dynamics tunnel should be more representative of free-air data since there are no bump effects to influence the results. Between angles of attack of 1 and 7 deg, the PanAir lift-curve slope is within 1.5% of that obtained from test data in the General Dynamics tunnel. The magitude of the lift coefficient predicted by PanAir is on the order of 0.01 lower than that predicted by the General Dynamics test data. For the wind-tunnel model of the E-7, the back end of the fuselage was enlarged to accommodate the support sting. This modification was not reflected in the PanAir geometry. A preliminary analysis (non-PanAir) indicated that the increased wind-tunnel-model camber caused by the modification to the nozzle area produced an increase in C_L equivalent to that of a 0.3-deg increase in angle of attack. Assuming this estimate to be in the ballpark, the two wind-tunnel C_L curves are high by a ΔC_L value of 0.011. This adjustment would make the PanAir data coincide very well with the General Dynamics wall-corrected test data.

The lift curves predicted by PanAir for all three cases are shown in Fig. 13. The walls cause the lift-curve slope to increase slightly. The addition of the bump causes a noticeable shift in the zero-lift angle of attack resulting from the increase in the local angle of attack caused by the presence of the bump. The symbols in Fig. 13 denote the result of applying the standard wall-correction values from the Ames 12-ft tunnel[27] to the PanAir walls-case C_L values. This standard correction is used to convert tunnel-measured C_L vs α curves to free-air conditions. The standard correction appears to be essentially the same as that predicted by PanAir, since the symbols are very nearly centered on the PanAir curve for free-air results.

From Fig. 14 it is seen that the increment in C_L ($C_{L,\text{walls}+\text{bump}} - C_{L,\text{free air}}$) determined by the wind-tunnel measurements is about twice that predicted by PanAir. The differences between the PanAir predictions and wind-tunnel measurements for free air (Fig. 12) and for the effect of the walls alone (Fig. 13) are too small to account for this disagreement. Therefore, the discrepancy between the PanAir and wind-tunnel ΔC_L values seem to be related to the PanAir bump model.

Three factors were identified as being possible contributors to the discrepancies shown in Fig. 14. First is the possibility of flow separation from the bump. A quick PanAir study was performed to attempt to measure this effect. The case of "total" separation was modeled by extending the maximum thickness station of the bump all the way downstream in the tunnel. These results are shown in Fig. 14 for $\alpha = 0$ and 9 deg; they indicate that flow separation from the bump has a substantial effect at $\alpha = 0$ deg and a minimal effect at $\alpha = 9$ deg. This makes sense since, as the angle of attack increases, the model moves farther away from the bump, and the effect the bump has on the model tends to decrease.

Other factors considered to be possible contributors to the discrepancies shown in Fig. 14 were the neglect of the smaller bump on the large bump (see Fig. 9) and the absence of a boundary-layer effect in the solution. Both of these factors would tend to increase the effective size of the bump and cause an increase in the ΔC_L imparted on the model.

SOFIA Project

A feasibility study is currently being conducted at NASA Ames in the development of a 3-m-class airborne observatory for an improved astronomy-data-gathering capability.[28] This project, identified as the Stratospheric Observatory For Infrared Astronomy (SOFIA), has chosen a modified Boeing 747-SP as the platform for this 3-m telescope. An artist's concept of this configuration is shown in Fig. 15. The telescope is mounted in the fuselage directly behind the cockpit area. A large bay door slides open during flight to allow the telescope to "see" out of the aircraft. It is this large cavity created by the opening of the sliding door that is of concern to the project engineers.

It was desired that a wind-tunnel model be designed on as large a scale as possible in order to reduce the scale effects on the important experimental optical measurements. A proposal was made to build a wind-tunnel model in which the wings were removed outboard of the inboard nacelle, allowing the overall scale of the model to be increased. A question that arose was how removing the outboard wing sections at this station would affect the flow properties along the fuselage where the bay door would exist.

PanAir Model

This problem required the analysis of two different configurations. It was ideal for PanAir analysis because of PanAir's ability to quickly analyze different geometries. The modeling generality afforded by PanAir allowed for the generation of a 747-SP paneled definition from an existing PanAir definition of a 747-200 definition provided to NASA by Boeing. This particular model did not include definitions of the outboard nacelle or the empennage. The model was being used by Boeing to analyze the flowfield over the cab, and the absence of these particular pieces of geometry reduced the size of the problem without diminishing the accuracy of the solution in the region of interest. Since the area of concern for the SOFIA model was also the fuselage area forward of the wing leading edge, it was felt that this model of the 747-200 would be sufficient. The original nacelle definition provided by Boeing included all the details of a flow-through nacelle. Since the problem only required the volume effect of the presence of the nacelle, the internal definition of the nacelle was omitted from the problem. This also served to reduce the overall size of the problem and improved the turnaround time without affecting the accuracy of the solution along the fuselage. Instead, networks covering the inlet and exhaust openings of the nacelle were used, with appropriate flow-through boundary conditions at each opening.

Modifications were made to the fuselage in order to change it from a "200" configuration to the desired "SP" model. These changes were performed on a CAD/CAM machine and were based on drawings that specified the differences between the two models. This process took approximately 6 man-hours. The section of the wing to be removed for the reduced-span model was defined by separate networks of panels. Therefore, to move from the full-span to the reduced-span model required only the removal of the outboard wing networks from the input file and the addition of a new network to close off the tip of the reduced-span wing. The actual PanAir definitions of the 747-SP used for the analysis are shown in Fig. 16. The outboard wing networks are shown slightly removed from the rest of the wing so that the difference between the full-span and reduced-span PanAir models can be pointed out.

Results

Figure 17 shows a two-dimensional projection in the x-z plane of the PanAir three-dimensional streamline results produced for the two models at $M_\infty = 0.6$. A set of 13 streamline starting points was defined around the fuselage at FS 505. These streamlines were allowed to run to approximately FS 980. The actual location of the bay door is from FS 520 to FS 700. Originally, both cases were run at an angle of attack of 4 deg. These results showed that, by removing the outer portion of the wing, the local angle of attack on the fuselage was reduced. By raising the reduced-span angle of attack to 4.5 deg, which is the result shown in Fig. 17, the flowfield for the full-span case at 4 deg (shown in the figure) was essentially recreated in the region of interest along the fuselage. With this information, the project engineers were able to proceed with the manufacturing of the reduced-span wind-tunnel model of the 747-SP knowing that the flowfield patterns over the region of the fuselage where the bay door exists were not going to be greatly affected over the range of angles of attack that were of interest. The full-span flowfields along the fuselage forebody could be recreated in wind-tunnel tests by increasing slightly the angles of attack of the reduced-span model.

TranAir: Description

TranAir solves the full-potential equation for transonic flow about complex configurations. A detailed description of the inner workings of the TranAir program is beyond the scope of this paper. Several publications describe in depth the theory and algorithms employed by the code.[18,29–32] However, since the approach taken by TranAir to solve the transonic full-potential equation about very general geometries is unique, an overview of the solution method is presented here.

Input Requirements

The input required to run the code is designed to be simple and quick to generate. Most of the code comprising the input processor was taken directly from the pilot code version of the PanAir panel method code. The aircraft geometry is described in terms of logically organized networks of surface panels. Because the input format for TranAir is of the same general format as that for panel methods, the same surface-paneled definition of a geometry analyzed by a linear panel method can be used by TranAir to obtain transonic solutions. TranAir also allows the specification of very general boundary conditions.

Once the user has defined the geometry, most of the work is done. TranAir does not use surface-conforming grids in the solution process. Instead, it embeds the surface-paneled geometry in a rectangular array of flowfield grid points, some of which are internal to the volume defined by the surface panels. This flowfield grid is defined by the user, who specifies the dimensions of the grid and the number of grid points in each direction. An example of a paneled geometry embedded in a uniform rectangular grid is shown in Fig. 18. Asymmetric geometries are allowed, and the user may specify a sideslip angle. There are a few other optional input parameters that are discussed in the next paragraphs, since some discussion of the solution process is necessary in order for these parameters to make sense.

Solution Process

The solution begins with the construction of the flowfield grid for the problem based on the parameters specified by the user. This uniform grid may then be locally refined based on two factors: 1) local surface panel size and 2) user-specified "regions of interest." The user may specify a minimum and maximum number of refinements to be performed over the entire configuration. This helps to control "runaway" refinement, which sometimes occurs in areas where the surface panel size is very small relative to the geometry. The user may also specify three-dimensional regions of interest around portions of the geometry in which he wishes to more carefully control the refinement. An example might be to specify a region around the wing leading edge to make sure enough grid is present to accurately capture the leading-edge pressure peak.

A modification of an oct-tree data structure[33] has been developed that stores essentially all of the information for the refined grid. This structure typically requires storage equal to two times the number of unrefined grid boxes[32] and efficiently provides information such as node and element centroid location, box size and level, box adjacency, and the identity of boundary boxes.

Having produced a refined computational grid, the solver constructs finite-element operators on the grid. A trilinear basis function for the potential is associated with each element. The discretized operators are obtained using a variational principle[34] in a manner that is fully conservative and second-order accurate. This discretization yields a set of nonlinear algebraic equations that simulate the original full-potential partial differential equation. Operators in supersonic regions are altered using first-order artificial dissipation for shock capturing. The set of nonlinear algebraic equations is solved by an iterative process that is driven by an orthogonal direction algorithm called GMRES,[35,36] in conjunction with multiple preconditioners. The preconditioners consist of a fast Poisson solver, which is particularly effective for subcritical flow regions, and an incomplete factorization of the sparse matrix produced by linearizing the set of nonlinear algebraic equations. The sparse matrix is defined by selecting a "reduced set" of the entire system of equations and by closing the set through the imposition of a Dirichlet boundary condition. This reduced set of points consists of all element nodes that make up a refined box on the global grid, plus all element nodes in which the flow is supersonic. A nested-dissection ordering based on the physical location of the finite elements in the grid is used to order the sparse matrix system. The incomplete factorization is performed by using a drop tolerance when the sparse matrix is factored. Each element in the decomposition is compared with the main diagonal entry of the current row, and, if the ratio is smaller than the drop tolerance, the element is dropped. The preconditioners work more effectively in combination than would a single preconditioner by itself.

A significant advantage of the formulation is that the flowfield grid need only extend to a point where the flow is linear (Prandtl-Glauert flow), rather than to a point where the flow is essentially unperturbed. For cases in which there exist embedded regions of supersonic flow, the flowfield grid must be large enough to include the supersonic regions. The equation solved about the perimeter of the grid (the outer set of grid boxes) changes from the nonlinear full-potential equation to the linear Prandtl-Glauert equation. The ends of the computational grid can be relatively close to the configuration because the unknowns defined on the global grid consist of sources for the velocity potential. These sources exist on a grid that is theoretically infinite in extent, but the sources go to zero rapidly as a function of distance from the boundary surfaces. In fact, the sources are generally weak in the entire computational domain except near shocks, geometry boundaries, and wakes. The potential induced by specified sources is computed efficiently by a convolution integral of the sources with a discrete exterior Green's function.

Output Processing

Once the solution is completed, either by executing the maximum number of iterations (user-specified), or by reducing the residual below a certain value (also user-specified), the value of the potential at each grid point has been obtained. From these values, velocities in the flowfield and on the surface may be computed, from which the user may obtain forces and moments for the configuration and pressure coefficients, streamlines, and Mach contours both in the field and on the surface of the geometry.

TranAir: Applications

For the case of analyzing a complete aircraft configuration, generally one of two situations applies: 1) the only data that exist for the configuration are from a wind-tunnel test, in which case the results from the code are usually too late to have any impact on the design process; or 2) no data exist. When applying a code to a configuration for which no other data exist, the user needs a high degree of confidence in the solution, since design decisions will be made based on the results.

An example of a code in which users have a high degree of confidence is PanAir. The reasons for this confidence are 1) an understanding of the well-defined applicability of the code, i.e., linear, irrotational, inviscid flow; and 2) the huge history of successful applications of the code.[6-15] This type of confidence is desired for any new code, and TranAir is no exception. Two examples are presented here in which TranAir results were compared not with experimental results, but with codes for which a high degree of confidence existed in the solutions: PanAir and FLO28.[37]

Although comparing computational results with experimental or flight-test data yields some insight into how well a code is doing, it is not always the most desirable approach to take. Experimental results are generally influenced by such things as vortex lift, flow separation, and wind-tunnel wall effects. Few, if any, of these effects can be modeled or predicted by most codes. Thus, it is desirable to compare the results from a new code with results computed by existing codes for which a high degree of confidence exists. As long as there is an understanding of the limitations of the solution, a great deal of insight can be gained in such "code-on-code" comparisons.

ONERA M6 Wing

One of the cases modeled by TranAir for comparison with results from PanAir (subsonic) and full-potential FLO28 (transonic) was the ONERA M6 wing.[38] A top view of the paneled definition of the wing, along with the airfoil section is shown in Fig. 19. The PanAir model of this wing consisted of approximately 1800 surface panels.

Subsonic Results

The comparison with PanAir was made at $M_\infty = 0$ and $\alpha = 3.06$ deg. A two-dimensional cut of the TranAir refined grid at the plane of symmetry is shown in Fig. 20. As can be seen in Fig. 21, TranAir results were in excellent agreement with PanAir. In TranAir, a subcritical case can be solved using the Prandtl-Glauert equation, as it is in the linear-potential solution of PanAir, or it can be solved using the full-potential equation. The close agreement between the TranAir Prandtl-Glauert result and PanAir serves as a verification of TranAir's ability to accurately calculate subcritical flows. For panel method codes such as PanAir, the CPU time required to solve a problem tends to go up with approximately the cube of the number of panels. TranAir solution times tend to be a linear function of the number of grid boxes in the flowfield. Because of the efficiency with which TranAir solves the subcritical problem, run times tend to be faster than panel method solutions for models involving complex configurations.

Transonic Results

A transonic solution for the ONERA wing was obtained with TranAir and compared to the FLO28 conservative full-potential code. FLO28 utilizes surface-conforming grids in its solution, is applicable to wing/body configurations, and has an established reputation among the users of the code as being highly reliable. A two-dimensional cut of the TranAir refined grid generated for this case at the plane of symmetry is shown in Fig. 22. The comparison was made at $M_\infty = 0.84$ and $\alpha = 3.06$ deg, and the results are shown in Fig. 23. FLO28 results using both first- and second-order viscosity models are plotted against the results from TranAir, which uses a first-order viscosity model. It is not evident from the results in Fig. 23 that anything is gained by the second-order viscosity model. The first-order results from both codes are generally in excellent agreement.

This case serves as a verification of TranAir's ability to solve transonic flow problems accurately. Of course, TranAir is not being developed to solve problems like isolated wings (although it is certainly capable of solving such problems). The objective of the code is to solve transonic flow problems about very general geometries by maintaining the modeling generality afforded by the use of surface panels. The models that follow demonstrate the variety of configurations that may be analyzed by the code. Also included are discussions of the positives and negatives of full-potential solutions for each of the geometries.

Generic Fighter Model

One facet of CFD analysis that is not used as often as would seem to be desirable is the comparison of results from different CFD codes for geometries that have wind-tunnel and/or flight data available. This serves to cross validate both experimental and computational results for a particular configuration. An example of this idea was recently presented[21] in which TranAir full-potential results were compared with FLO57[39] Euler results for a generic fighter configuration.

Wind-Tunnel Model

A series of wind-tunnel investigations of the transonic behavior of the generic fighter were conducted[40,41] in a effort to increase the understanding of vortex flows at transonic speeds. A schematic of the wind-tunnel model definition of the generic fighter showing the location of the pressure taps on the upper-wing surface is presented in Fig. 24. The wing is a cropped delta wing with a 55-deg leading-edge sweep, an aspect ratio of 1.8, and a taper ratio of 0.2. The airfoils used in the definition of the wing were modified NACA 65A005 sections with sharp leading edges. The wind-tunnel model of the generic fighter also employed three different wedge-shaped chine models. For the computational study, the model in which the chine was placed 0.5 in. above the wing (the overall length of the wind tunnel model is 42 in.) was studied. This case will be referred to as the wing/body/chine case. Also studied was the wing/body case, in which no chine was present on the model.

Computational Models

The TranAir surface-paneled definition of the generic fighter wing/body/ chine model is shown in Fig. 25. A two-dimensional cut of the refined grid used in obtaining the TranAir solutions is shown in Fig. 26. This is a spanwise cut just upstream of the cropped tip of the delta wing. The TranAir definition of the wing/body geometry consisted of 1576 surface panels and 124 wake panels. The wing/body/chine case consisted of 1688 surface panels and 280 wake panels. The refined flowfield grid was defined by approximately 85,000 grid boxes.

The FLO57 surface grid for the wing/body geometry is shown in Fig. 27. The grids at the chine trailing edge and the wing trailing edge are presented in Figs. 28a and 28b, respectively. The flowfield grid consisted of 426,790 grid points for both cases. In the streamwise direction, 134 points were defined, 74 of which were on the body. A total of 65 points were defined circumferentially around the body, and 49 points were defined from the wing surface to the outer boundary. Further details on the TranAir and FLO57 models are presented in Ref. 21.

Results

The objective of the computational study was to evaluate the transonic solutions from both an Euler and a full-potential code over a range of angles of attack. Both codes are able to predict supersonic flows (shocks), and the Euler code is also able to predict vortex flows. The expectation was that both TranAir and FLO57 would yield accurate solutions at the lower angles of attack, where vortical flow was not a dominant feature in the flowfield. Also, it was expected that the CPU time required for the TranAir solution would be roughly an order of magnitude less than that required by FLO57 for subsonic cases, and a factor of three to five less for transonic cases. As the angle of attack increased, it was expected that TranAir predictions would be degraded because of the inability of the code to capture vortex flows. On the other hand, it was expected that FLO57 would do well at these higher angles of attack.

Results were obtained from both codes at $M_\infty = 0.6$ and 0.8 for both the wing/body and wing/body/chine case. The FLO57 results were obtained on the Cray-2 at the Ames Research Center, requiring 16×10^6 words of central memory and anywhere from 3.4 to 10.0 CPU h, depending on the freestream conditions. TranAir results were obtained on a Cray X-MP/48, also at Ames. Solutions required less than 2×10^6 words of central memory, and run times varied from 0.35 to 1.2 CPU h, again depending on the freestream conditions. Results presented here are for the low angles of attack. Results for higher angles (up to 12 deg for TranAir and up to 20 deg for FLO57) are presented in Ref. 21.

Subsonic Results

Figure 29 shows the pressure coefficient C_p results at $M_\infty = 0.6$ and $\alpha = 4$ deg for the upper wing surface of the wing/body case. The values of x/c in the figure are based on an imaginary wing chord at the centerline (see Fig. 24). The agreement between all three predictive methods (two computational methods and one experimental method) for this case is quite good. Results for the same freestream conditions for the wing/body/chine case are shown in Fig. 30. Again, the various comparisons are generally good. One difference is that, for the wing/body case, the computational methods tended to predict pressure coefficients that were slightly more negative than those measured in the wind tunnel. For the wind/body/chine case, the codes predicted pressure coefficients that were slightly less negative than those measured in the wind tunnel. Also, at the 40% streamwise station, the pressures computed by TranAir along the inboard portion of the wing were quite wavy. For this case a chine wake, defined parallel to the x axis in the body-axis system, passes over the wing very near to the surface. This wavy area predicted by TranAir is probably caused by an insufficient grid resolution in the region.

Transonic Results

Results for the wing/body case at $M_\infty = 0.8$ and $\alpha = 5$ deg are shown in Fig. 31. These comparisons again were generally very good. For the wing/body/chine case, wind-tunnel data at $M_\infty = 0.8$ started at an angle of attack of 8 deg. At this angle, a substantial amount of vortical flow is indicated in the wind-tunnel data. In previous high-angle-of-attack TranAir runs, poor predictions were generated; hence, no TranAir runs were made at this particular flow condition.

Basic F-16A Model

The full-configuration case that has been used as a testbed for TranAir is the F-16A geometry. This geometry was chosen as the benchmark for the TranAir code for three main reasons: 1) the basic geometry is complex, as are the various external stores that could be added to the original geometry; 2) the wind-tunnel model was extensively instrumented, and a large amount of tunnel data exist for the various configurations[42]; and 3) attempts were being made to solve the Euler and Navier-Stokes equations for the F-16A geometry. TranAir results for the F-16A configuration were first published in 1986.[16] Results from Euler[43] and Navier-Stokes[44] efforts have also been published.

Description of the Model

A surface-paneled definition of the basic F-16A geometry (no external stores) is shown in Fig. 32. There are several areas of this geometry that present a major challenge to CFD codes: the engine inlet and diverter area under the fuselage; the ventral fins; the canopy; the leading-edge strake; and the empennage geometry, which includes cutouts around the nozzle exit. For this case, the velocity of the engine inlet was specified to be approximately the freestream value. The F-16A paneled definition consisted of approximately 3500 surface panels, which represents a very large and time consuming case for most panel methods.

Results

It was desired to compare results for this configuration with PanAir results at a subsonic Mach number as a means of verifying the TranAir results. As can be seen in Fig. 33, TranAir and PanAir C_p results on the wing of the F-16A at $M_\infty = 0.6$ and $\alpha = 4$ deg compare very closely. Although not presented here, the results generated by both codes over the entire configuration were compared. These comparisons showed essentially the same results everywhere on the geometry. This case, like that of the ONERA M6 wing previously, serves as a verification of the accuracy with which TranAir solves a subcritical case. This case also demonstrates the type of geometric complexity that TranAir is capable of handling. The panel method solution took nearly 3600 CPU s on a Cray X-MP/48. The TranAir solution for this case took approximately 1800 CPU s on the same machine. Recent enhancements to TranAir have improved the efficiency of the code considerably, and the same job today would take closer to 1000 CPU s.

F-16A with Wing-Tip Missiles and Fuel Tanks

After TranAir results for the basic F-16A geometry were verified, additional components were added to the configuration. Another F-16A model was defined in which wing-tip missiles and launchers, and underwing fuel tanks were added to the original geometry. These new pieces of geometry were defined and paneled on a CAD/CAM machine based on original drawings. The process of generating these new pieces of geometry and including them in the TranAir definition of the F-16A took on the order of 24 man-hours. A picture of the paneled definition of this configuration is shown in Fig. 34. The new F-16A geometry now consists of over 7000 surface panels, which is far too large for most panel method codes to handle, and no other nonlinear code has been able to successfully model this type of geometric complexity. For this case, experimental data was the only available means of checking TranAir's results. Two two-dimensional cuts of the refined grid generated for this case are shown in Fig. 35. A spanwise cut along the wing and fuselage, which also shows the grid refinement around the fuel tank and the tip missile and launcher, is shown in Fig. 35a. A closeup streamwise cut through the fuel tank and wing assembly is shown in Fig. 35b.

Transonic Results

TranAir C_p results were obtained for the fully configured F-16A at $M_\infty = 0.9$ and $\alpha = 4$ deg and are shown plotted with experimental data in Fig. 36. TranAir is able to capture the shock on the upper surface of the wing, but is is approximately 10% downstream of the location reflected in the experimental data. This rearward shift in the shock location is characteristic of a conservative full-potential formulation.[45] This tendency should be considered by the engineer when he interprets the computational results from such a formulation in order to more accurately predict a shock location on a wind-tunnel model.

At the $\eta = 0.32$ station, which is inboard of the fuel tank location, the effect of the presence of the fuel tank can readily be observed. The pressures on the lower surface of the wing at this station are dramatically different than for the case without the fuel tank (Fig. 37). Unfortunately, only two lower-surface pressure taps exist on the wind-tunnel model between the 20 and 80% chord locations. Although the correlation between TranAir and experiment along the lower surface of the wing at this station appears to be quite poor, it is difficult to evaluate the experimental results based on so few data points. It is possible that viscous effects play a dominant role in the region bounded by the fuel tank, the wing, and the fuselage (see Fig. 35a), since a "channel" flow seems to exist. The large negative C_p value seen for the wind-tunnel data at $x/c = 0.4$ could reflect a bad data point for that particular run, or it could represent the beginning of a shock at that location. Notice that TranAir has predicted a shock along the lower surface at the 65% chord location.

Wing Tip Revisited

Wind-tunnel data for just a "plain" wing tip for the F-16A were not available. The launcher was present for all of the runs, and most of the runs also included the AIM-9 tip missile. When the paneled definition of the missile-and-launcher geometry was created, one of the previous F-16A runs was repeated so that the effect of the launcher and missile on the aerodynamics at the wing tip could be evaluated. Previous TranAir results obtained at the most outboard tap station ($\eta = 0.95$) had been ignored since the effect of the launcher and missile were not being captured, and comparisons with experiment were predictably poor. The influence of adding the launcher and missile to the F-16A model was evaluated by examining the pressures at the $\eta = 0.95$ station. Figure 38 shows the results of this comparison at $M_\infty = 0.6$ and $\alpha = 4$ deg. The TranAir results without the missile and launcher were obtained by the original version of TranAir, which utilized a uniform flowfield grid with no local refinement.[16,17] The uniform-grid version of the code was unable to resolve the geometry of the missile and launcher because of the relatively large size of the grid boxes in relation to the missile-and-launcher geometry. The grid-refinement capability alleviated this problem. Two effects are reflected in Fig. 38. The first is the addition of the launcher-and-missile geometry to the computational model, and the second is the addition of grid refinement to the solution process. The leading-edge suction peak obtained for the model with the launcher and missile is a reflection of the grid refinement rather than of the presence of the missile and launcher. Early results with the uniform-grid version of TranAir consistently missed leading-edge pressure peaks because of the relatively coarse grid resolution around the wing leading edge. However, the improved results over the rest of the chord, can be attributed to both the presence of the missile and launcher and to the improved resolution of the local flowfield resulting from grid refinement.

Wind-Tunnel Flow Simulation

One of the largely untapped areas ideal for CFD analysis is that of wind-tunnel modeling. Experimental results obtained in wind tunnels have wall and support effects that alter the flowfield over the wind-tunnel model. These effects are not represented in free-air computational results. The ability to include a definition of the wind-tunnel walls and support system in the computational model is desirable in that their effects on the wind-tunnel model can then be predicted. An example of this idea was described earlier in the paper with the General Dynamics E-7 Model mounted in the NASA Ames 12-Ft Pressure Wind Tunnel. The previous study was undertaken to evaluate the effects of the wind-tunnel walls and the high-angle-of-attack support bump on an installed wind-tunnel model. Since that time, a redesign of the test section in the 12-ft tunnel has been undertaken. Several modifications to the existing tunnel definition have been suggested, and geometric models have been generated and analyzed using PanAir. One of the studies involved altering the thickness of the existing low-angle-of-attack rear support strut in the wind tunnel to study the effects on the power required to drive the tunnel as well as the effect on the flowfield in the test section. The maximum strut thicknesses varied from 6 in. (the existing strut) to 10 in. and 14 in. Experimental data were obtained along the section of the tunnel wall directly across from the support strut, and the Mach number along the wall was calculated for each of the strut thicknesses.

Computational Model

A surface-paneled definition of the 12-ft tunnel was generated and is shown in Fig. 39. The paneled definition of the tunnel was based on blueprint drawings, as well as on direct measurements made in the actual tunnel. The tunnel was faithfully modeled from just downstream of the upstream screens, through a 25:1 contraction ratio to the test section, and back to the end of the diffuser section to the point just before the flow is turned in the actual wind tunnel. Also modeled were the support strut and the centerbody to which wind-tunnel models are attached. The flats that exist in the wind-tunnel test section were modeled, since they subtract substantially from the overall area in the test section. Figure 39a shows a side view of the entire paneled definition, including the support strut and centerbody in the test section. A detailed view of the support strut and centerbody is provided in Fig. 39b. A total of approximately 1200 surface panels was used to describe the complete model. The open ends of the tunnel were closed off by networks of panels upon which flow-through boundary conditions were applied. Two two-dimensional cuts of the TranAir refined grid along the tunnel centerline are shown in Fig. 40. The refined grid over the entire tunnel model is shown in Fig. 40a, and a closeup of the refined grid around the support strut in the test section is shown in Fig. 40b.

Results

PanAir was originally used as the computational tool to analyze the various candidate wind-tunnel test section models. TranAir at that point did not have the boundary condition generality necessary to model this problem. The code has since been supplied with a boundary condition generality almost as liberal as that available in PanAir. PanAir results compared reasonably well with experimental data at the lower test section Mach numbers. At $M_\infty = 0.55$, however, the disagreement between PanAir results and experimental results for Mach number along the tunnel wall was fairly dramatic for each of the three strut thicknesses, as can be seen in Fig. 41. Also shown in the figure are the shapes of the wind-tunnel wall and support strut. These geometries are to scale in the flow direction, but are slightly exaggerated in the vertical direction in the figure.

For the $M_\infty = 0.55$ case, TranAir was run for each of the three strut models, using the same paneled definitions and boundary conditions as were used by PanAir. The results for these cases are also presented in Fig. 41. TranAir did a better job of calculating the acceleration of the flow along the wall than did PanAir. TranAir agreement with experiment is quite good for each of the three strut thicknesses, except right at the point where the flow enters the diffuser area. At this location, TranAir overpredicts the acceleration for each of the three models. This is probably a result of boundary-layer and/or separation effects in the wind tunnel at the point where the flow enters the diffuser area.

Since TranAir solves the full-potential equation, the code's accuracy can be assessed by comparing its results with those of the exact inviscid one-dimensional gasdynamics equations for compressible flow.[46] The known wind-tunnel cross-sectional areas at the diffuser exit and test section stations and the desired test section Mach number of 0.55 require that the diffuser-exit Mach number be 0.124. The latter was input to TranAir as a boundary condition at the diffuser-exit station. The TranAir-computed wall Mach number at the test section was 0.548, in excellent agreement with the theoretical 0.55 value.

The exact inviscid test section to diffuser-exit station density and velocity ratios for this case are 0.870 and 4.314, respectively. This large velocity change between the two stations violates the small-perturbation assumption used in the linear Prandtl-Glauert equation solved by PanAir. The linearized expression for density inherent in the Prandtl-Glauert equation is given by

$$\frac{\rho}{\rho_{\text{ref}}} = 1 - M_{\text{ref}}^2 - M_{\text{ref}}^2 \frac{V}{V_{\text{ref}}}$$

This expression can be obtained by expanding the isentropic equation for density, Eq. (39) of Ref. 46, and retaining only the linear terms in velocity. It can also be obtained from PanAir's expression for mass flux, which in turn is based on Eq. (1.11.22) of Ref. 47. Using the exact values for Mach number and velocity at the two tunnel stations, this approximation to the density gives a density ratio of 0.812 if the test section station is used as the reference and 0.949 if the diffuser-exit station is the reference. Thus, the linearized density approximation gives a density ratio error of -6.7 or 9.1%, depending on which station is picked for the reference.

For incompressible flow, the small-perturbation restriction does not apply to the Prandtl-Glauert equation, and accurate results should be attained even for large velocity changes. This is demonstrated by arbitrarily selecting Mach numbers of 0.1 and 0.2. The velocity ratio for two stations having these Mach numbers is 1.99. The linearized density approximation gives a -0.5 and $+0.5$% error for the density ratio of these two stations, again depending on the reference station selected. Even though the velocity ratio is still not small, the density ratio error is small for this case because the flow is essentially incompressible at these low Mach numbers.

From these calculations it is evident that TranAir accurately solves the nonlinear full-potential equation, since it reproduces the one-dimensional gasdynamic equation results for compressible subcritical flow. The PanAir solution is in error for this problem because the Prandtl-Glauert equation is inappropriate for flows having such a large velocity change in a compressible-flow Mach number range.

The next step in the aerodynamic analysis process, the installation of a paneled definition of an aircraft model in the computational model of the 12-ft tunnel, is a straightforward process that could easily be performed in one day, assuming the paneled geometry already exists. Once a paneled definition of a wind tunnel is generated, it can be used repeatedly with different aircraft models in the performance of a sort of "computational" wind-tunnel test. Such an analysis would allow for the evaluation of wall and strut effects in extrapolating wind-tunnel results to free air.

Future Directions

Several enhancements to the existing version of TranAir are currently being implemented that will make the code more efficient, accurate, and user-friendly. A grid-sequencing capability, which solves for a series of grids from coarse to fine, has been implemented, and early testing has shown that the number of iterations required to converge a transonic solution can be reduced by approximately a factor of two. A solution-adaptive grid-refinement capability that is currently being implemented[48] will refine the initial grid according to conditions in the local flowfield. For instance, when evidence of a shock occurs, the code will automatically refine the flowfield grid in the region of the shock in order to capture it more accurately. This capability will release the user from having to know what the flowfield looks like ahead of time in defining grid-refinement regions. Although certain areas such as wing leading edges are known to have flow conditions that generally require a finer grid resolution, the location of shocks or other critical flow areas is not always immediately obvious to the user. Solution-adaptive grid refinement will provide a more efficient refined grid in terms of the distribution of the flowfield grid points. The current version of TranAir is limited to subsonic ($M_\infty < 1.0$) freestream Mach numbers, though local regions of supersonic flow are allowed. The ability to specify supersonic freestream Mach numbers will soon be available.

TranAir provides the capability for efficient solution of the full-potential equation for subsonic, transonic, and supersonic flow about arbitrary geometries. The enhancements to the code that are being implemented will improve this capability. Some observations are provided here on the kind of developments that would further enhance the usefulness of the code for analyses of the type that have been presented.

One possibility is the development of a "modified" full-potential code. This code would be able to capture vortex sheets from sharp leading edges and trailing edges of lifting surfaces, in addition to the already available capability of specifying regions of differing total pressure and total temperature. The ability to track the path of a vortex sheet would be enhanced by the solution-adaptive grid-refinement capability, which would allow the grid around the sheet to be sufficiently resolved in order to capture its path more effectively. This implementation would provide a "pseudo-Euler" solution and would be desirable in that the program would still be solving for only one unknown in the flowfield—the potential—rather than for the five-variable state vector solved for in the Euler equations. It would also prevent the formation of "false entropy" and spurious separation that are unwanted but ever-present features of numerical solutions of the Euler equations.

Another possible development would be to solve the Euler equations directly using the same geometry definition and flowfield grid structure as is currently utilized by TranAir. Again, the vortical flow predicted by the Euler codes would be effectively captured by the solution-adaptive grid-refinement capability. The major problem in solving the Euler equations is that the effectiveness of the method seems to be dependent on both the geometry and the flow conditions. For transport wings, full-potential codes generally do a better job than the Euler equations. For fighter-type configurations (delta wings), Euler equations tend to do a better job than full-potential codes because of their ability to predict rotational flow. Both methods suffer the shortcoming of solving only the inviscid problem, although it is possible to integrate a boundary-layer method into the solution. Convergence problems tend to occur with the Euler equations when the freestream Mach number is in the low subsonic range. However, the expertise gained in solving the Euler equations on an unstructured Cartesian grid could lead to the eventual solution of the Navier-Stokes equations. Theoretically, the unstructured nature of the flowfield grid could be maintained, although care would have to be taken in the modeling of the flowfield near the boundary of the geometry in order to resolve the boundary layer.

Finally, some form of boundary-layer coupling in TranAir would be a desirable development. Although this is not a trivial goal, the rewards to be gained would be substantial. Boundary layers affect the actual location of the shock on a body. If an inviscid code is applied to a transonic flow problem, some allowance needs to be made in calculating where the shock on an actual configuration might reside by correcting the shock location predicted by the inviscid code. Other examples where boundary-layer effects tend to be signficant are the problem of flow through a wind tunnel (an example of this problem was presented earlier) and the internal flow of a flow-through nacelle model.

Closing Remarks

Several examples have been presented in demonstrating the capability to analyze arbitrary configurations in the subsonic, transonic, and supersonic attached-flow regimes. Linear-potential panel methods are able to reliably predict the aerodynamics of geometries in the subsonic regime. A few panel methods are also able to analyze configurations in the supersonic flow regime. A code that can analyze both subsonic and supersonic flows about complex geometries is PanAir. Results from this code were presented for a variety of configurations.

For the transonic analysis of complex configurations, a full-potential code called TranAir was employed. Solutions for several different geometries were presented. The ability of TranAir to routinely solve for transonic flows about the variety of configurations presented here represents a significant aerodynamic-analysis capability. Surface-conforming flowfield grids are not used, significantly reducing the setup time to generate the input to a problem. Since TranAir is a potential method, run times and storage requirements are significantly lower than those typical of Euler and Navier-Stokes solvers. TranAir is expected to eventually replace PanAir as the main tool used for potential-flow analysis about complex configurations. It has shown the ability to match linear-potential results in the subsonic flow regime, generally at less cost because of the efficiency of the solution process. It will be able to solve supersonic freestream problems after the present code development is completed. Also, it solves problems in the transonic flow regime, which linear-potential codes such as PanAir are unable to do.

The future of analyzing complex configurations with nonlinear equations (full-potential, Euler, Navier-Stokes) appears to lie in the development of unstructured-grid codes. Most of the burden of generating a flowfield grid is then passed from the user to the computer, and the geometric modeling limitations typically encountered by surface-conforming grids are greatly reduced.

Acknowledgments

The authors wish to thank several individuals for their contributions to the work presented here. Paul Keller was responsible for the generation of most of the geometries. His contributions are greatly appreciated. Ron

Fegenbush provided the surface definitions of the generic fighter models. Ralph Carmichael provided the PanAir results for the Ames 12-Ft Pressure Wind Tunnel geometry. None of the TranAir results would be possible without the work of the members of the Boeing "TranAir Team," who have developed the code under NASA Contract NAS2-12513: Forrester Johnson, Satish Samant, Dave Young, John Bussoletti, Robin Melvin, and Mike Bieterman.

References

[1]Magnus, A. E., and Epton, M. A., "PAN AIR—A Computer Program for Predicting Subsonic or Supersonic Linear Potential Flows About Arbitrary Configurations Using a Higher Order Panel Method, Vol. I—Theory Document (Version 3.0)," NASA CR-3251, 1989.

[2]Sidwell, K. W., Baruah, P. K., and Bussoletti, J. E., "PAN AIR—A Computer Program for Predicting Subsonic or Supersonic Linear Potential Flows About Arbitrary Configurations Using a Higher Order Panel Method, Vol. II—User's Manual (Version 3.0)," NASA CR-3252, 1989.

[3]Sidwell, K. W. (ed.), " PAN AIR—A Computer Program for Predicting Subsonic or Supersonic Linear Potential Flows About Arbitrary Configurations Using a Higher Order Panel Methods, Vol. III—Case Manual (Version 1.1)," NASA CR-3253, 1981.

[4]Baruah, P. K., Bussoletti, J. E., Massena, W. A., Nelson, F. D., Purdon, D. J., and Tsurusaki, K., "PAN AIR—A Computer Program for Predicting Subsonic or Supersonic Linear Potential Flows About Arbitrary Configurations Using a Higher Order Panel Method, Vol. IV—Maintenace Document (Version 3.0)," NASA CR-3254, 1989.

[5]Carmichael, R. L, and Erickson, L. L., "PAN AIR—A Higher Order Panel Method for Predicting Subsonic or Supersonic Linear Potential Flows About Arbitrary Configurations," AIAA Paper 81-1255, June 1981.

[6]Snyder, L. D., and Erickson, L. L, "PAN AIR Prediction of NASA Ames 12-Foot Pressure Wind Tunnel Interference on a Fighter Configuration," AIAA Paper 84-0219, Jan. 1984.

[7]Madson, M. D., and Erickson, L. L., "Application of PAN AIR to an Advanced Supersonic Fighter/Attack Aircraft," AIAA Paper 85-4093, Oct. 1985.

[8]Madson, M. D., and Erickson, L. L., "PAN AIR Analysis of the NASA/MCAIR 279-3: An Advanced Supersonic V/STOL Fighter/Attack Aircraft," NASA TM-86838, 1986.

[9]Miller, S. G., and Youngblood, D. B., "Applications of USSAERO-B and the PANAIR Production Code to the CDAF Model, A Canard/Wing Configuration," AIAA Paper 83-1829, July 1983.

[10]Towne, M. C., Strande, S. M., Erickson, L. L., Kroo, I. M., Enomoto, F. Y., Carmichael, R. L., and McPherson, K. F., "PAN AIR Modeling Studies," AIAA Paper 83-1830, July 1983.

[11]Strande, S. M., Erickson, L. L., Snyder, L. D., and Carmichael, R. L, "PAN AIR Modeling Studies II: Side Slip Option, Network Gaps, Three-Dimensional Forebody Flow and Thick Trailing Edge Representation," AIAA Paper 84-0220, Jan. 1984.

[12]Ghaffari, F., "PAN AIR Application to the F-106B," NASA CR-178165, 1986.

[13]Tseng, W. W., Llorens, R., and Gardner, J., "Analysis of the Flow Field About a T-45 Using PANAIR," Naval Air Development Center, Rept. 84077-60, 1984.

[14]Goetz, A. R., Osborn, R. F., and Smith, M. L., "Wing-In-Ground-Effect Aerodynamic Predictions Using PANAIR," AIAA Paper 84-2429, Oct.–Nov. 1984.

[15]Chen, A. W., and Tinoco, E. N., "PAN AIR Applications to Aero-Propulsion Integration," *Journal of Aircraft*, Vol. 21, March 1984, pp. 161–167.

[16]Erickson, L. L., Madson, M. D., and Woo, A. C., "Application of the TranAir Full-Potential Code to Complete Configurations," International Council of the Astronautical Sciences Paper 86-1.3.5, 1986.

[17]Erickson, L. L., Madson, M. D., and Woo, A. C., "Application of the TranAir Full-Potential Code to the F-16A," *Journal of Aircraft*, Vol. 24, Aug. 1987, pp. 540–545.

[18]Samant, S. S., Bussoletti, J. E., Johnson, F. T., Burkhart, R. H., Everson, B. L., Melvin, R. G., Young, D. P., Erickson, L. L., Madson, M. D., and Woo, A. C., "TranAir: A Computer Code for Transonic Analyses of Arbitrary Configurations," AIAA Paper 87-0034, 1987.

[19]Madson, M. D., "Transonic Analysis of the F-16A with Under-Wing Fuel Tanks: An Application of the TranAir Full-Potential Code," AIAA Paper 87-1198, 1987.

[20]Madson, M. D., Carmichael, R. L., and Mendoza, J. P., "Aerodynamic Analysis of Three Advanced Configurations Using the TranAir Full-Potential Code," Pt. 2, Vol 1, NASA CP-3020, 1989, pp. 437–452.

[21]Goodsell, A. M., Madson, M. D., and Melton, J. E., "TranAir and Euler Computations of a Generic Fighter Including Comparisons with Experimental Data," AIAA Paper 89–0263, 1989.

[22]Hess, J. R., and Bear, R. L., "Study of Aerodynamic Technology for Single-Cruise-Engine V/STOL Fighter/Attack Aircraft, Phase I Final Report," NASA CR-166269, 1982.

[23]Gross, L. W., Hoef, J. A., Steese, C. W., and Voda, J. J., "Experimental Aerodynamic Characteristics of a Single-Engine V/STOL Fighter/Attack Aircraft (Flow-Through Model Test Results), Vol. I—Background and Analysis, Phase II Final Report," NASA CR-177434, 1987.

[24]Gross, L. W., Hoef, J. A., Steese, C. W., and Voda, J. J., "Experimental Aerodynamic Characteristics of a Single-Engine V/STOL Fighter/Attack Aircraft (Flow-Through Model Test Results), Vol. III—Appendix B: 11 Ft. × 11 Ft. Transonic Wind Tunnel Data, Phase II Final Report," NASA CR-177434, 1987.

[25]Foley, W. H., Albright, A. E., Powers, D. J., and Smith, C. W., "Study of Aerodynamic Technology for Single-Cruise-Engine V/STOL Fighter/Attack Aircraft, Vols. I and II—Phase II Final Report," NASA CR-177367, 1985.

[26]Elbers, W. K., "Model and Test Information Report, 1/9 Scale Single Engine V/STOL Model (E7)," General Dynamics Fort Worth Division Rept. FZT-453, 1982.

[27]Sivells, J. C., and Salmi, R. M., "Jet-Boundary Corrections for Complete and Semispan Swept Wings in Closed Circular Wind Tunnels," NACA TN-2454, 1951.

[28]Wiltsee, C., (ed.), "SOFIA (Stratospheric Observatory For Infrared Astronomy) Phase A System Concept Description," NASA PD-2001, 1987.

[29]Rubbert, P. E., et al., "A New Approach to the Solution of Boundary Value Problems Involving Complex Configurations," *Computational Mechanics—Advances and Trends*, American Society of Mechanical Engineers, New York, 1986, p. 49.

[30]Johnson, F. T., Samant, S. S., Bieterman, M. B., Melvin, R. G., Young, D. P., Bussoletti, J. E. and Madson, M. D., "Application of the TranAir Rectangular Grid Approach to the Aerodynamic Analysis of Complex Configurations," AGARD Specialists' Meeting on Application of Mesh Generation to Complex 3-D Configurations, Loen, Norway, 1989.

[31]Melvin, R. G., Bieterman, M. B., Young, D. P., Johnson, F. T., Samant, S. S., and Bussoletti, J. E., "Local Grid Refinement for Transonic Flow Problems," 6th

International Conference on Numerical Methods in Laminar and Turbulent Flow, Swansea, UK, 1989.

[32]Young, D. P., Melvin, R. G., Bieterman, M. B., Johnson, F. T., Samant, S. S., and Bussoletti, J. E., "A Locally Refined Rectangular Grid Finite Element Method" (to be published in the *Journal of Computational Physics*).

[33]Samet, H., "The Quadtree and Related Hierarchical Data-Structures," *Computing Surveys*, Vol. 16, No. 2, 1984.

[34]Bateman, H., "Irrotational Motion of a Compressible Inviscid Fluid," *Proceedings of the National Academy of Sciences*, Vol. 16, 1930, pp. 816–825.

[35]Saad, Y., and Schultz, M. H., "GMRES: A Generalized Minimal Residual Method for Solving Nonsymmetric Linear Systems," Yale Research Rept. YALEU/DCS/RR-254, 1983.

[36]Wigton, L. B., Yu, N. J., and Young, D. P., "GMRES Acceleration of Computational Fluid Dynamics Codes," AIAA Paper 85-1494, July 1985.

[37]Jameson, A., "Transonic Flow Calculations," Princeton Univ. Dept. of Mechanical and Aerospace Engineering Rept. 1651, 1983.

[38]Advisory Group for Aerospace Research and Development (AGARD), "Experimental Data Base for Computer Program Assessment: Report of the Fluid Dynamics Panel Working Group 04," AGARD-AR-138, 1979.

[39]Jameson, A., and Baker, T. J., "Solution of the Euler Equations for Complex Configurations," AIAA Paper 83-1929CP, July 1983.

[40]Erickson, G. E., and Rogers, L. W., "Experimental Study of the Vortex Flow Behavior on a Generic Fighter Wing at Subsonic and Transonic Speeds," AIAA Paper 87-1262, 1987.

[41]Erickson, G. E., Rogers, L. W., and Schreiner, J. A., "Subsonic and Transonic Vortex Aerodynamics of a Generic Forebody Strake-Cropped Delta Wing Fighter," AIAA Paper 88-2596, 1988.

[42]Hammond, D. G., "Wind Tunnel Data Report, 1/9-Scale F-16 Pressure Loads Test, AEDC 16'T Test TF-397 (Phase I and II), Vol. II of VI," General Dynamics Rept. 16PR135, 1976.

[43]Karman, S. L., Steinbrenner, J. P., and Kisielewski, K. M., "Analysis of the F-16 Flow Field by a Block Grid Euler Approach," 58th Meeting of the AGARD Fluid Dynamics Panel Symposium on Applications of CFD in Aeronautics, Aix-En-Provence, France, 1986.

[44]Flores, J., and Chaderjian, N. M., "The Numerical Simulation of Transonic Separated Flow About the Complete F-16A," AIAA Paper 88-2506, 1988.

[45]Holst, T. L., "Numerical Computation of Transonic Flow Governed by the Full-Potential Equation," NASA TM-84310, 1983.

[46]Ames Research Staff, "Equations, Tables, and Charts for Compressible Flow," NACA Rept. 1135, 1953.

[47]Ward, G. N., "Linearized Theory of Steady High-Speed Flow," Cambridge Univ. Press, Cambridge, UK, 1955.

[48]Bieterman, M. B., Bussoletti, J. E., Johnson, F. T., Melvin, R. G., Samant, S. S. and Young, D. P., "Solution Adaptive Local Rectangular Grid Refinement for Transonic Aerodynamic Flow Problems," 8th GAMM Conference on Numerical Methods in Fluid Mechanics, Delft, The Netherlands, 1989.

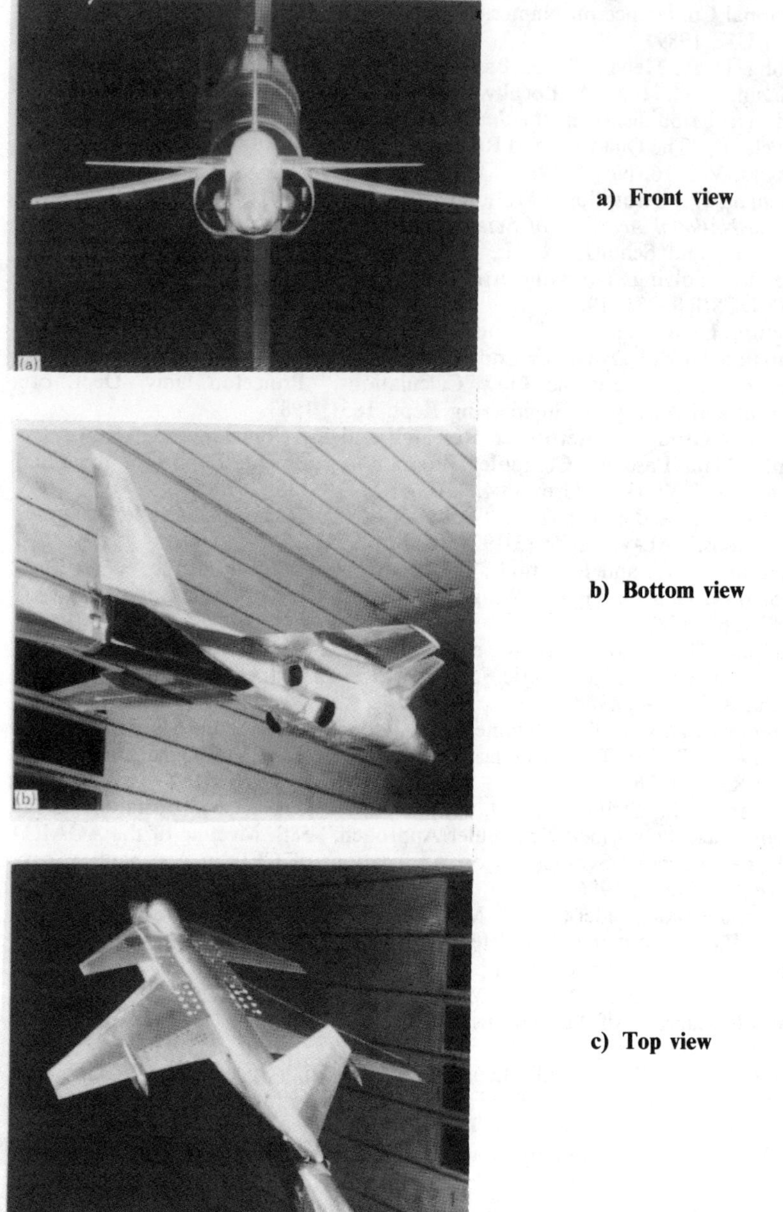

Fig. 1 Wind-tunnel model of NASA/MCAIR 279-3 installed in Ames 11 × 11-Ft Transonic Wind Tunnel.

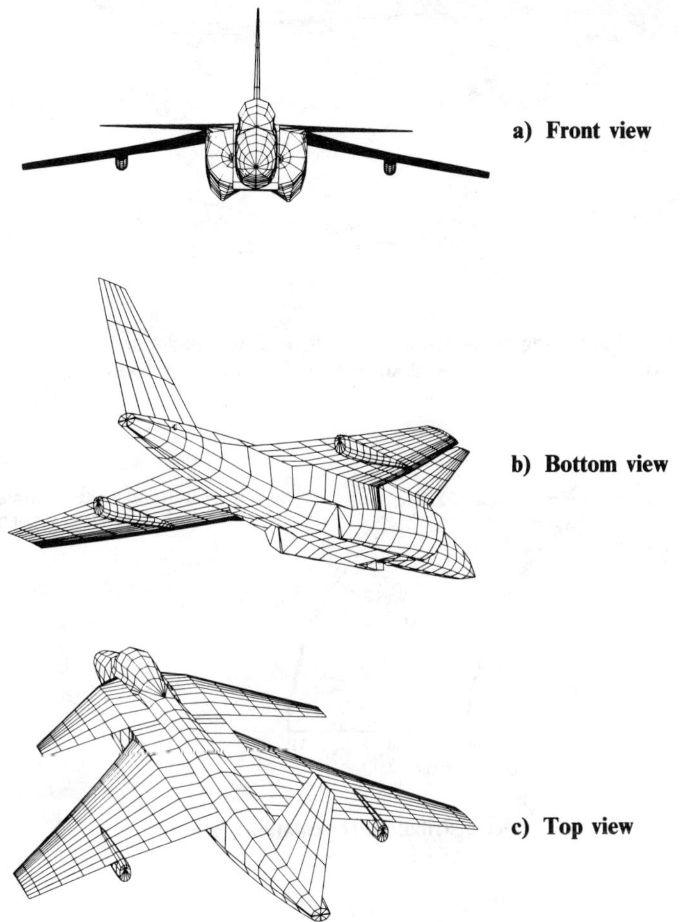

Fig. 2   PanAir paneling of the 279 configuration.

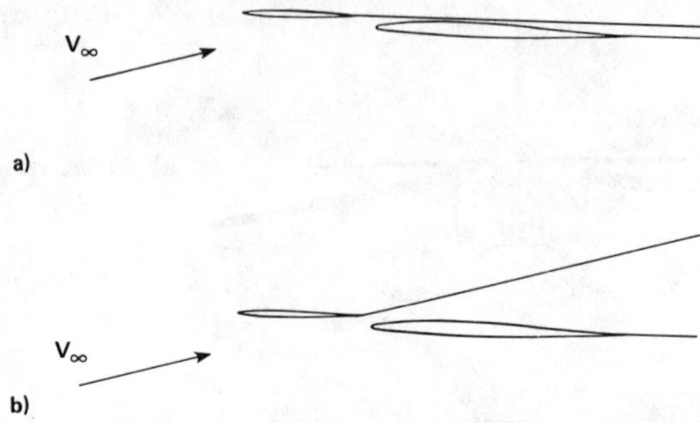

Fig. 3   Canard and wing root-sections for 279, with wakes defined off both surfaces. Canard wake aligned with a) *x* Axis of body-axis system; b) freestream.

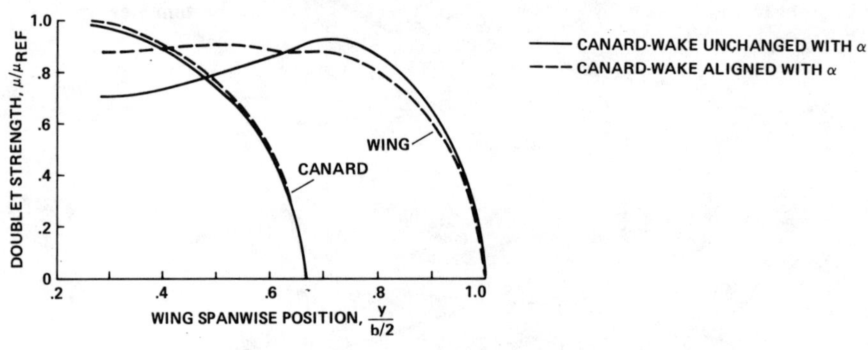

a) Doublet distribution (circulation), $\alpha = 10$ deg

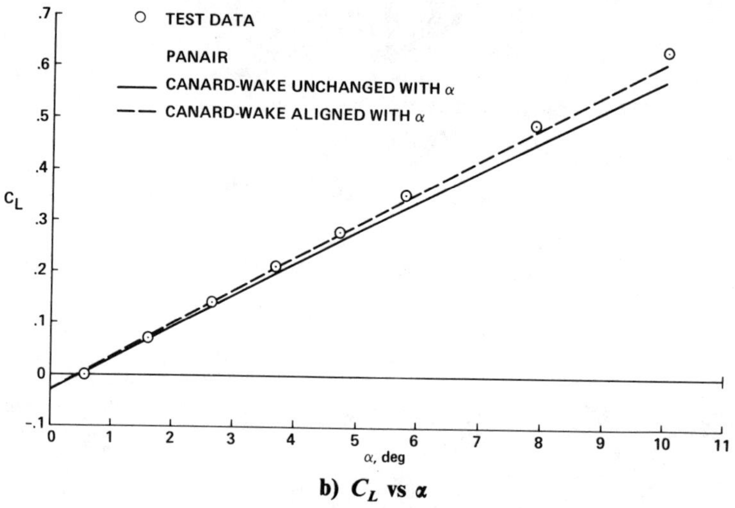

b) C_L vs α

Fig. 4   Effect of canard-wake location on PanAir results for the 279 ($M_\infty = 0.6$).

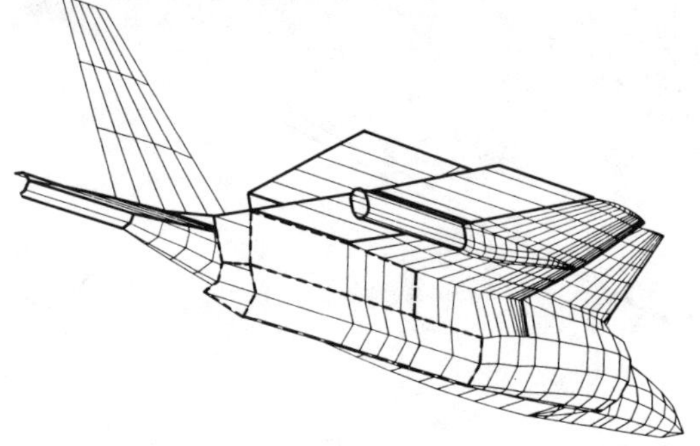

Fig. 5 Location of nozzle wakes for supersonic PanAir model of the 279.

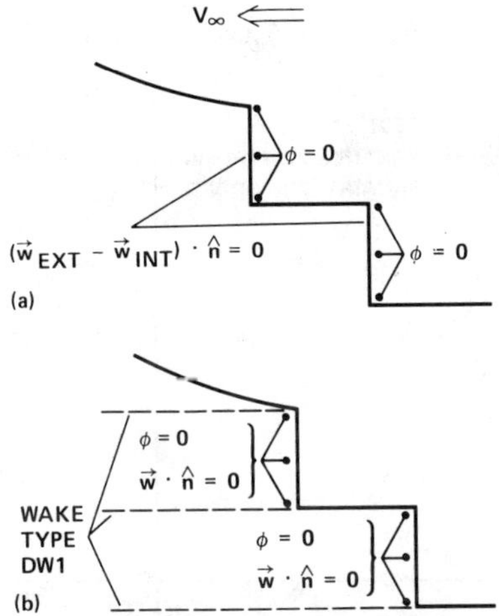

Fig. 6 Nozzle-exit plane models used to obtain PanAir results for the 279:
a) subsonic; b) supersonic.

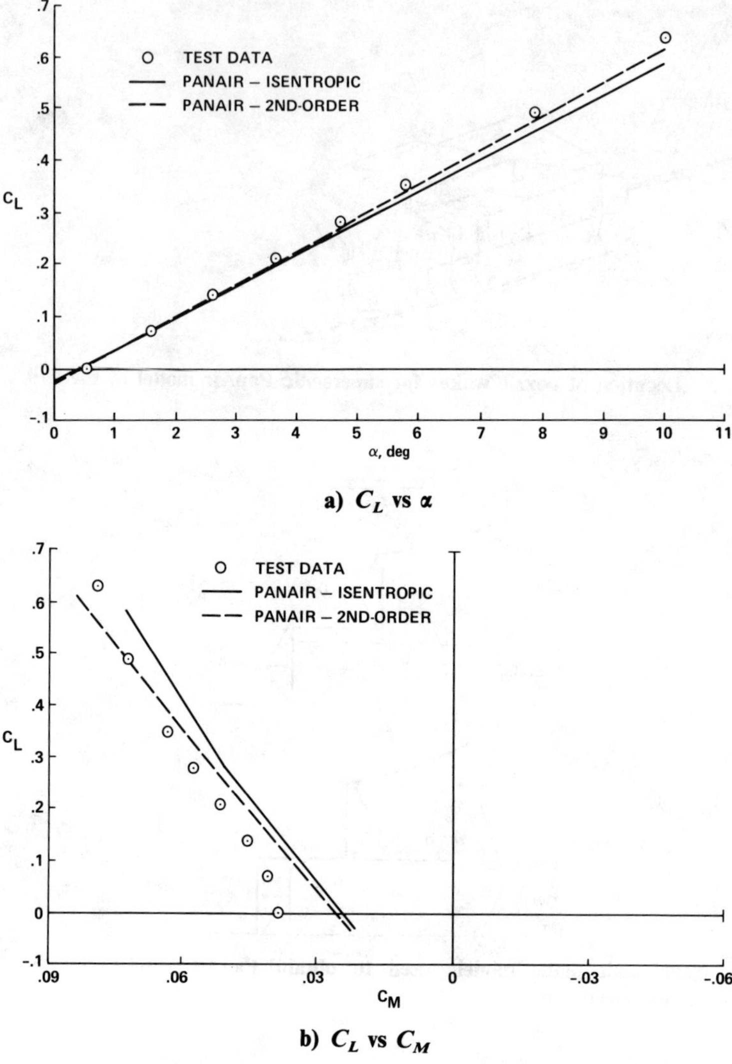

a) C_L vs α

b) C_L vs C_M

Fig. 7 Comparison of PanAir results and wind-tunnel test data for the 279 ($M_\infty = 0.6$).

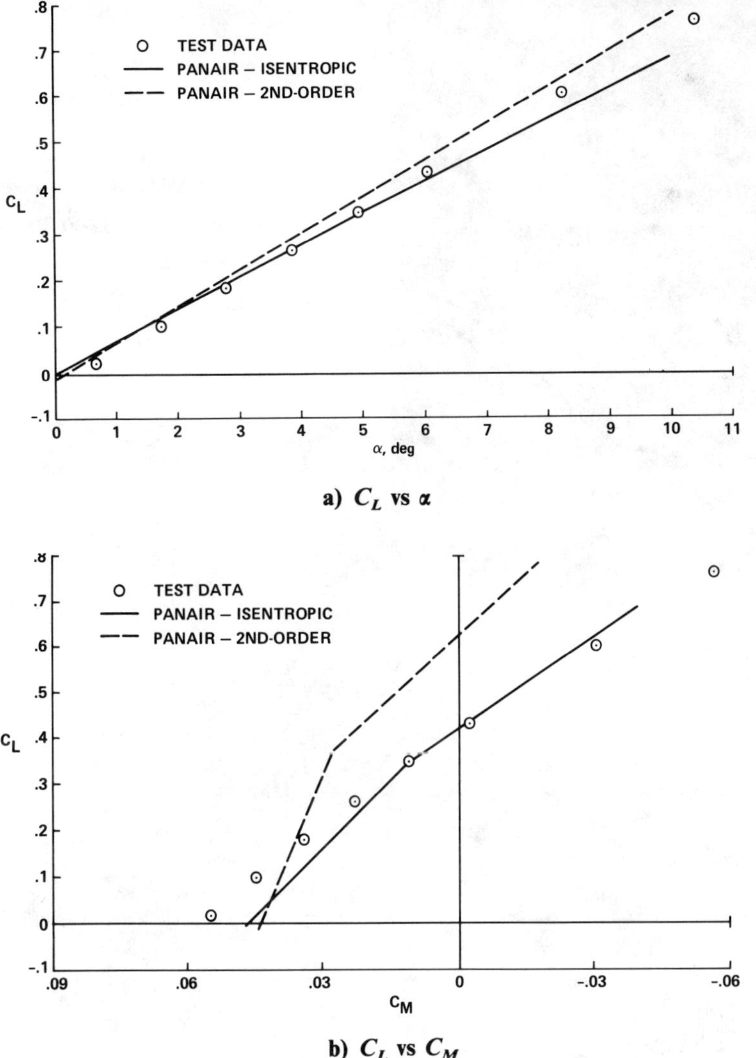

a) C_L vs α

b) C_L vs C_M

Fig. 8 Comparison of PanAir results and wind-tunnel test data for the 279 $(M_\infty = 1.2)$.

Fig. 9   NASA/GD E-7 model installed in Ames 12-Ft Pressure Wind Tunnel.

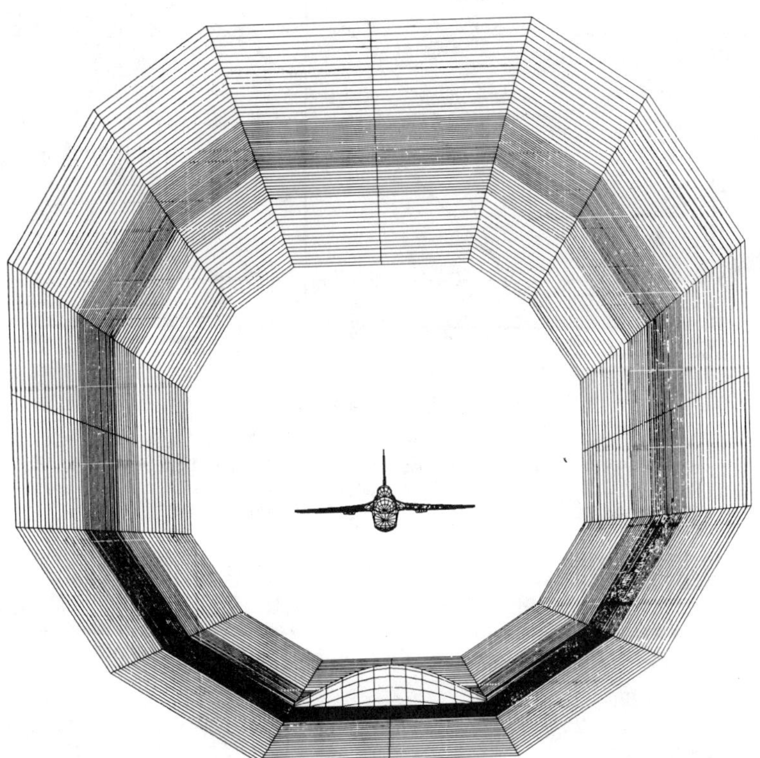

Fig. 10   PanAir model of E-7 geometry installed in Ames 12-Ft Pressure Wind Tunnel.

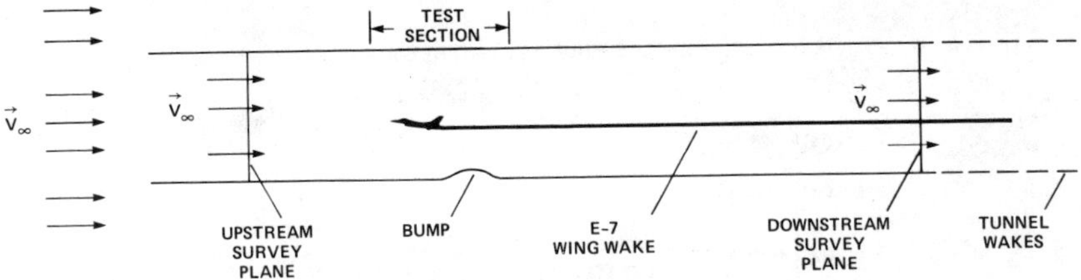

Fig. 11   Schematic of complete E-7/12-ft PanAir model at plane of symmetry.

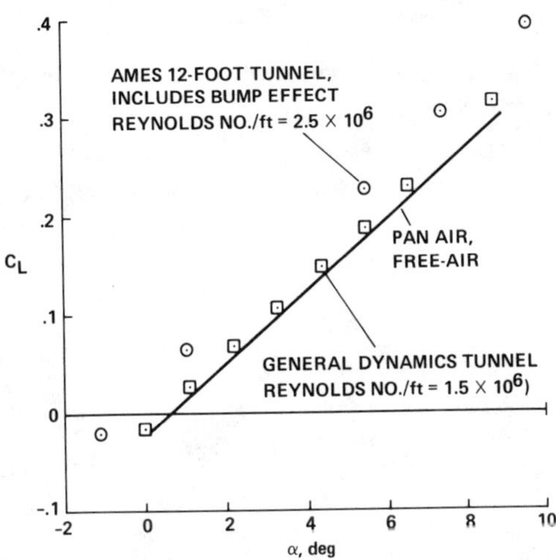

Fig. 12   Free-air E-7 lift curves ($M_\infty = 0.2$). Wind-tunnel curves include $\Delta\alpha$ corrections to remove wall effects.

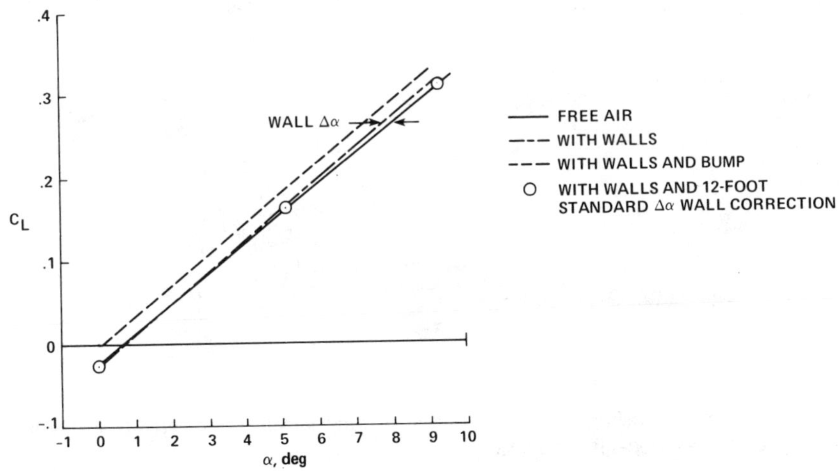

Fig. 13   PanAir-predicted lift curves for the E-7/12-ft mdoel ($M_\infty = 0.2$).

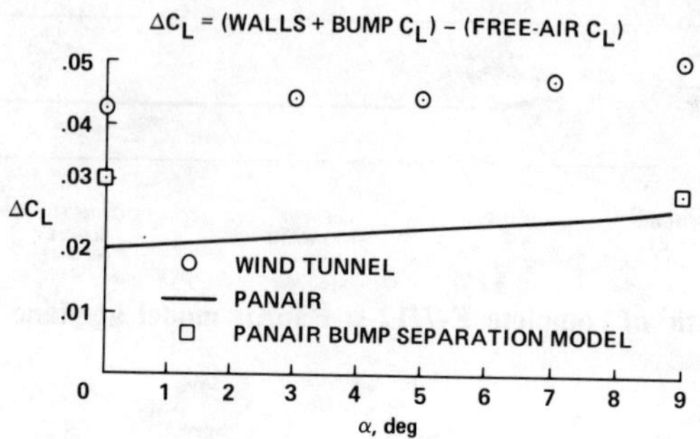

Fig. 14   PanAir and wind-tunnel increments in C_L owing to influence of wall and bump for the E-7/12-ft model.

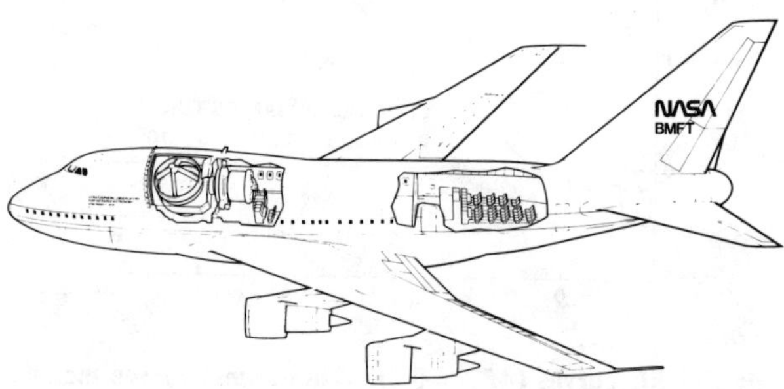

Fig. 15   Artist's concept of modified Boeing 747-SP for SOFIA project.

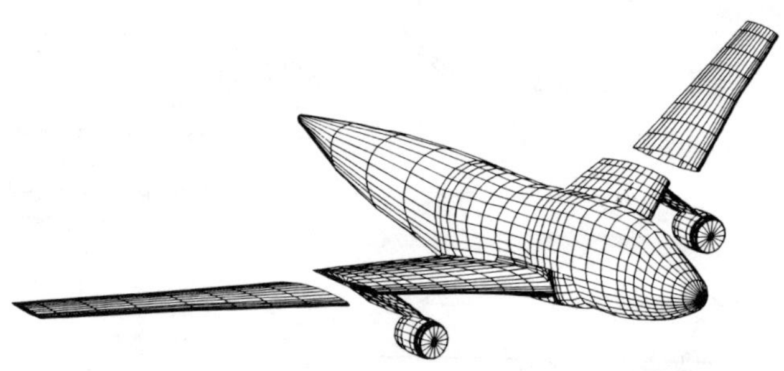

Fig. 16   PanAir paneling of 747-SP, outboard wing paneling separated to distinguish full-span and reduced-span models.

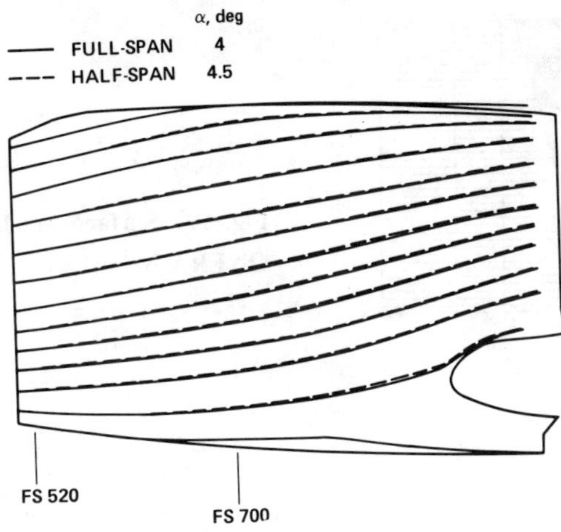

Fig. 17   PanAir streamlines along fuselage section of 747-SP.

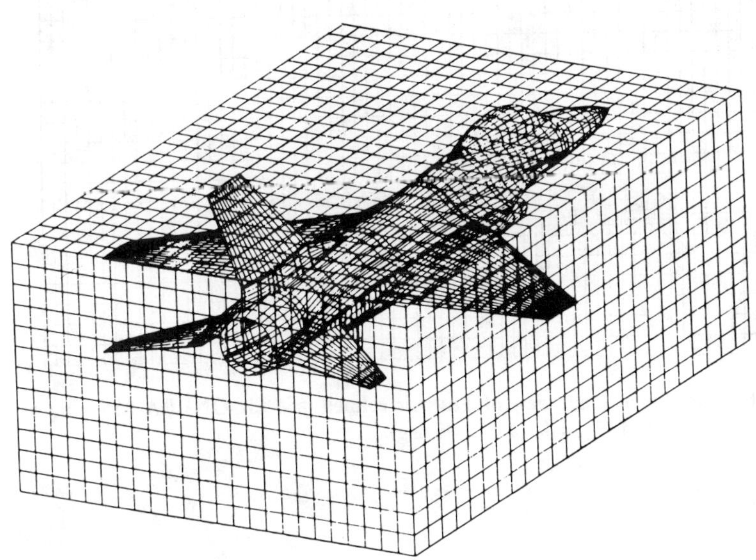

Fig. 18   Schematic of surface-paneled geometry embedded in uniform rectangular flowfield grid.

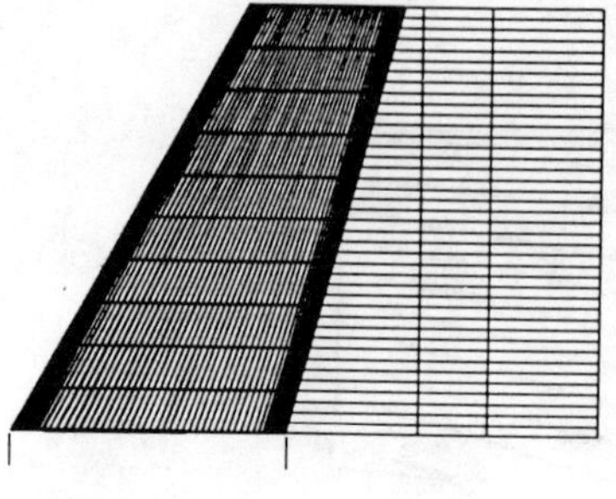

Fig. 19   Surface-paneled definition of ONERA M6 wing.

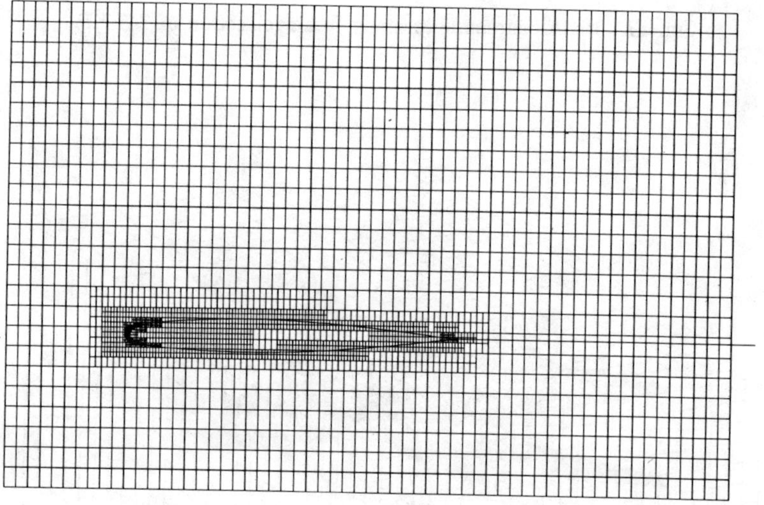

Fig. 20   Two-dimensional view of TranAir refined grid for ONERA M6 wing at the plane of symmetry ($M_\infty = 0$, $\alpha = 3.06$ deg).

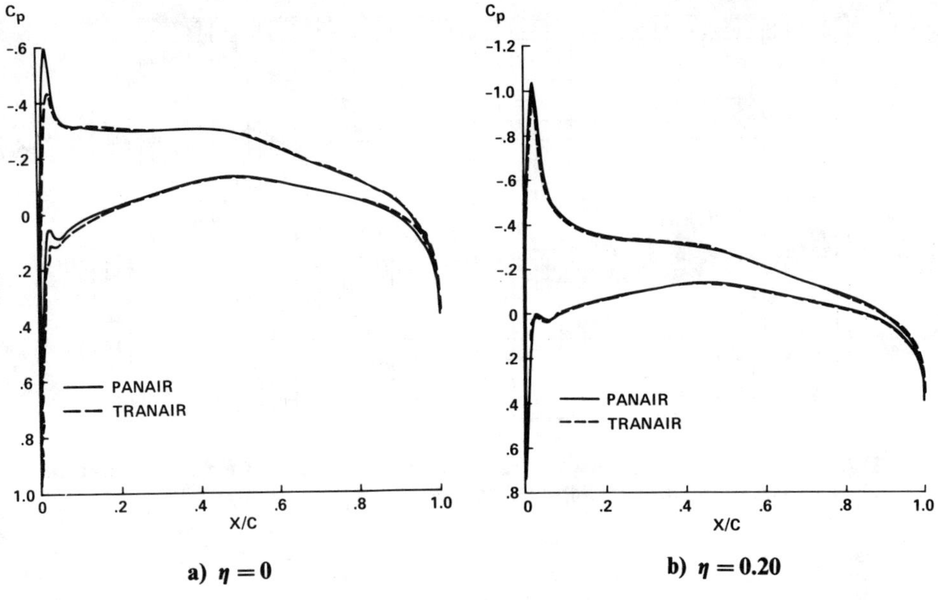

a) $\eta = 0$

b) $\eta = 0.20$

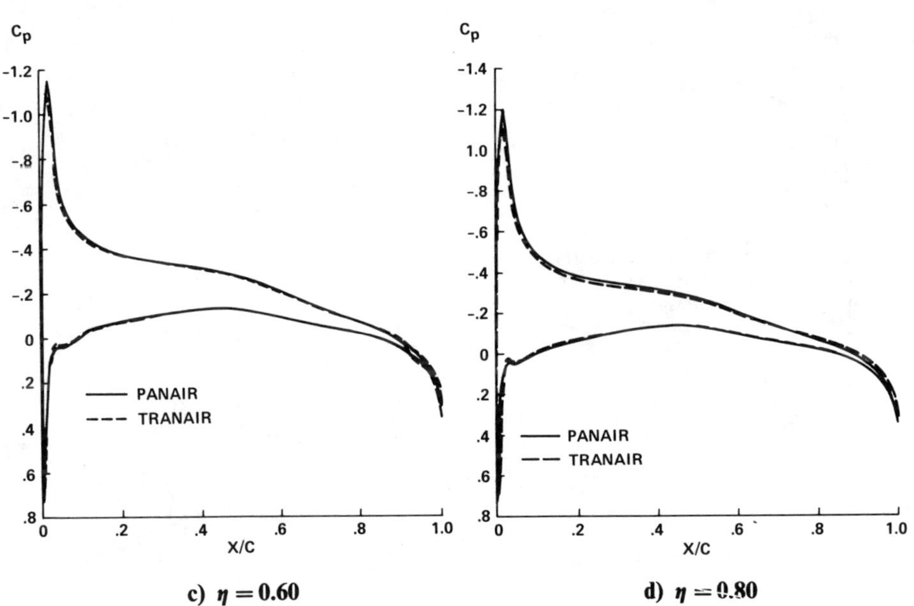

c) $\eta = 0.60$

d) $\eta = 0.80$

Fig. 21 PanAir and TranAir C_p results for ONERA M6 wing ($M_\infty = 0$, $\alpha = 3.06$ deg).

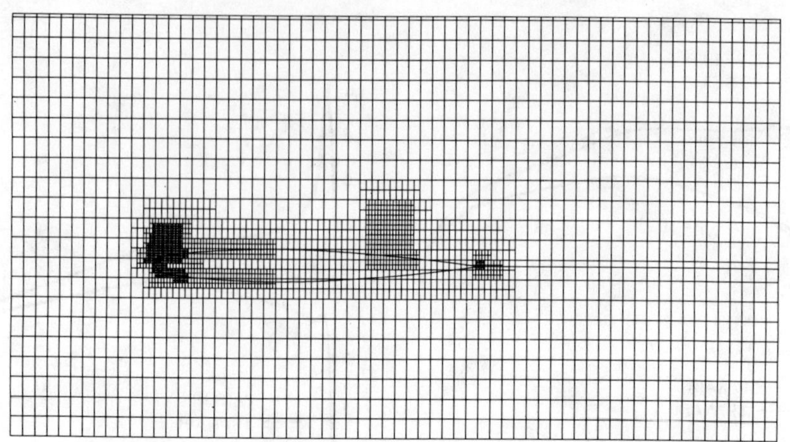

Fig. 22   Two-dimensional view of TranAir refined grid for ONERA M6 wing at the plane of symmetry ($M_\infty = 0.84$, $\alpha = 3.06$).

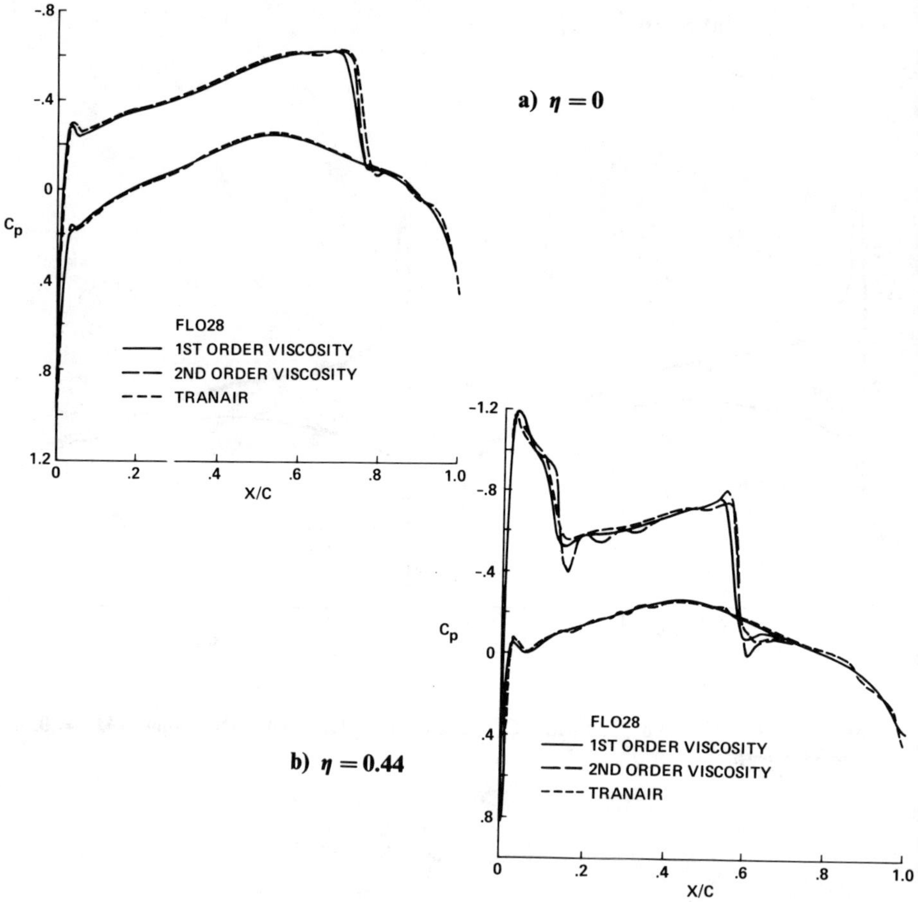

Fig. 23   FLO28 and Tranair C_p results for ONERA M6 wing ($M_\infty = 0.84$, $\alpha = 3.06$ deg).

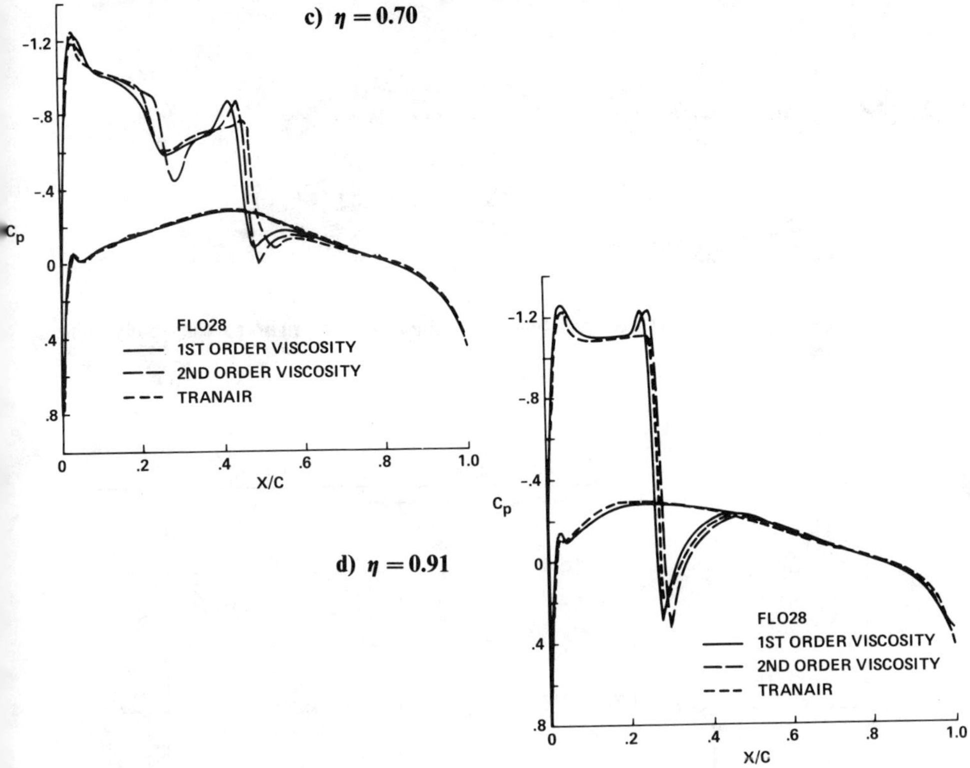

c) $\eta = 0.70$

d) $\eta = 0.91$

Fig. 23   (cont'd) FLO28 and Tranair C_p results for ONERA M6 wing ($M_\infty = 0.84$, $\alpha = 3.06$ deg).

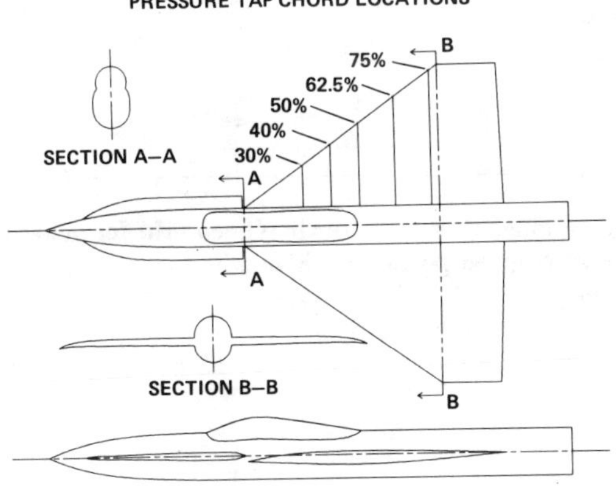

Fig. 24   Schematic of generic fighter with chine, including wind-tunnel tap-station locations.

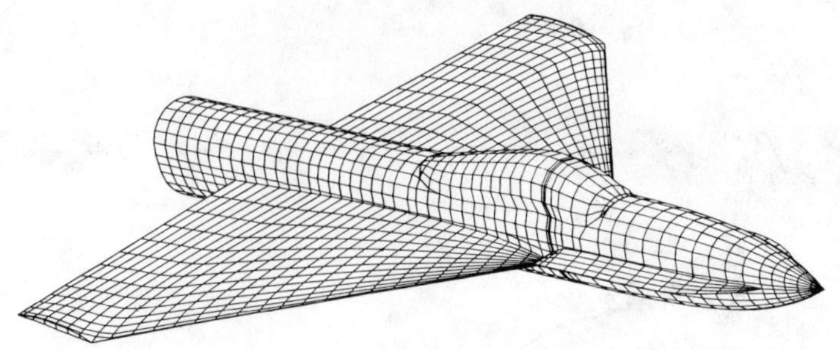

Fig. 25 TranAir surface paneling for the generic fighter wing/body/chine configuration.

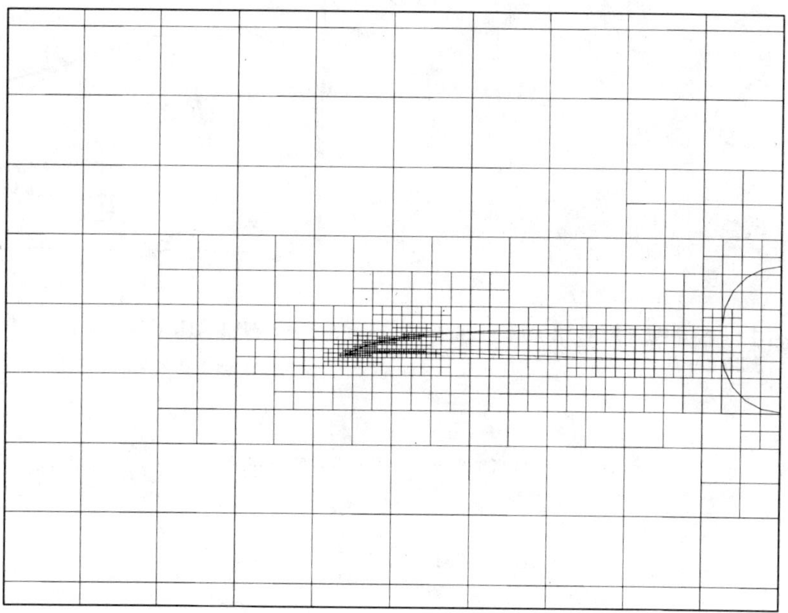

Fig. 26 Two-dimensional view of TranAir refined grid for generic fighter. Spanwise cut just upstream of cropped portion of delta wing.

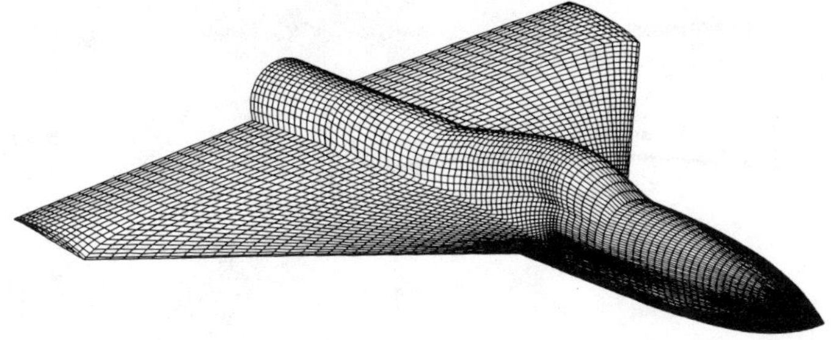

Fig. 27   FLO57 surface grid for the generic fighter wing/body configuration.

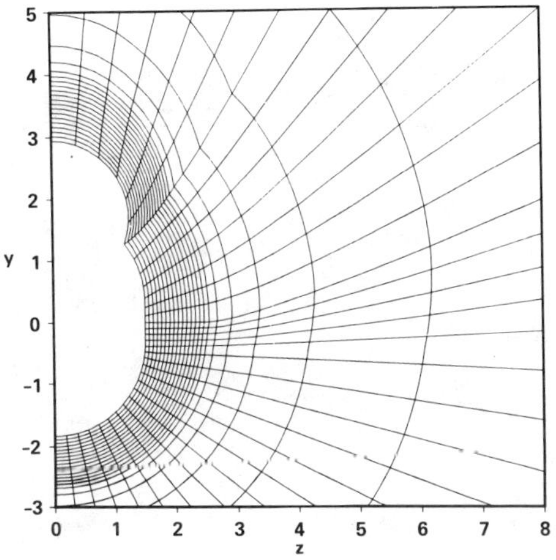

a) Grid at chine trailing edge

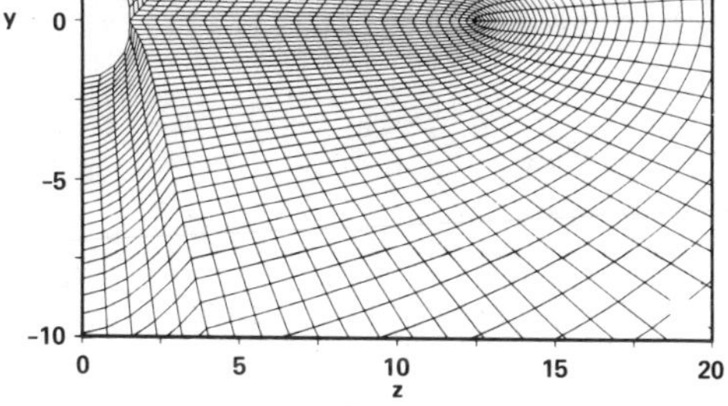

b) Grid at wing trailing edge

Fig. 28   FLO57 flowfield grid planes for the generic fighter.

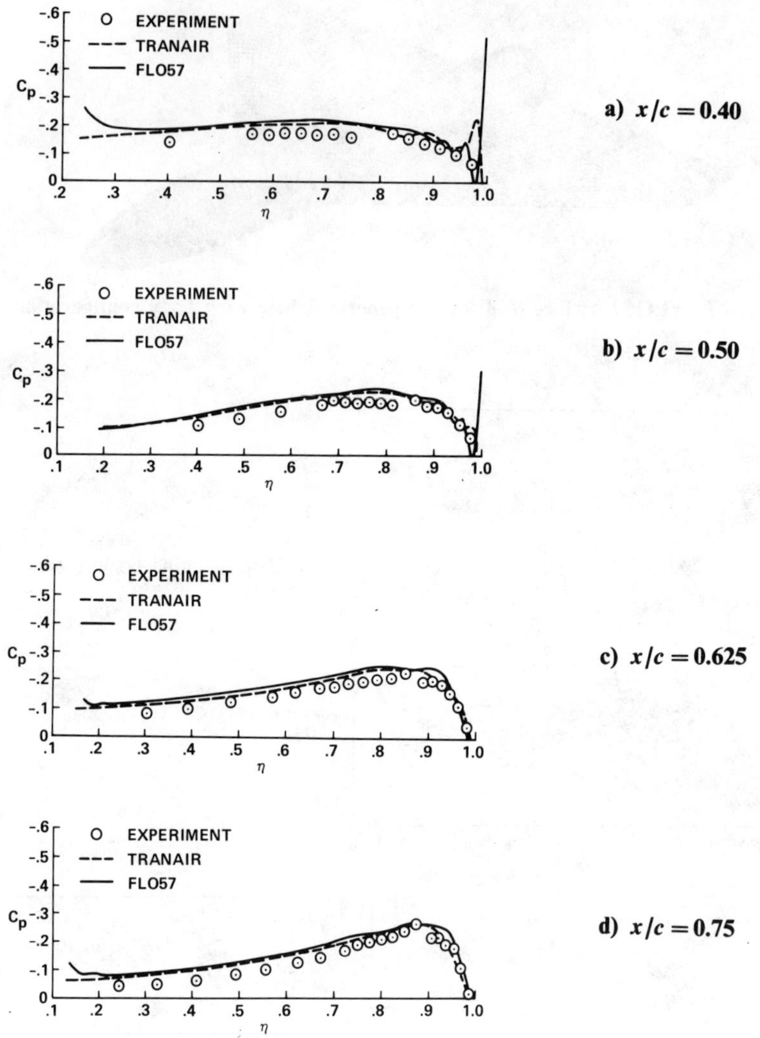

Fig. 29 C_p vs η results for the generic fighter wing/body configuration ($M_\infty = 0.6$, $\alpha = 4$ deg).

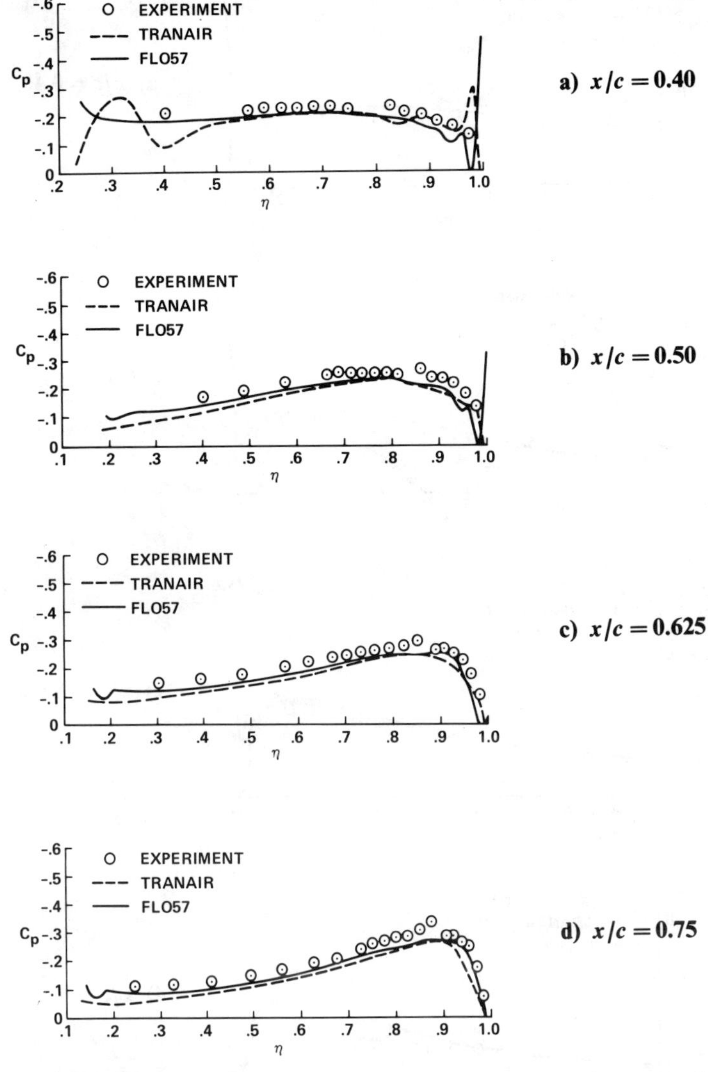

Fig. 30 C_p vs η results for the generic fighter wing/body/chine configuration ($M_\infty = 0.6$, $\alpha = 4$ deg).

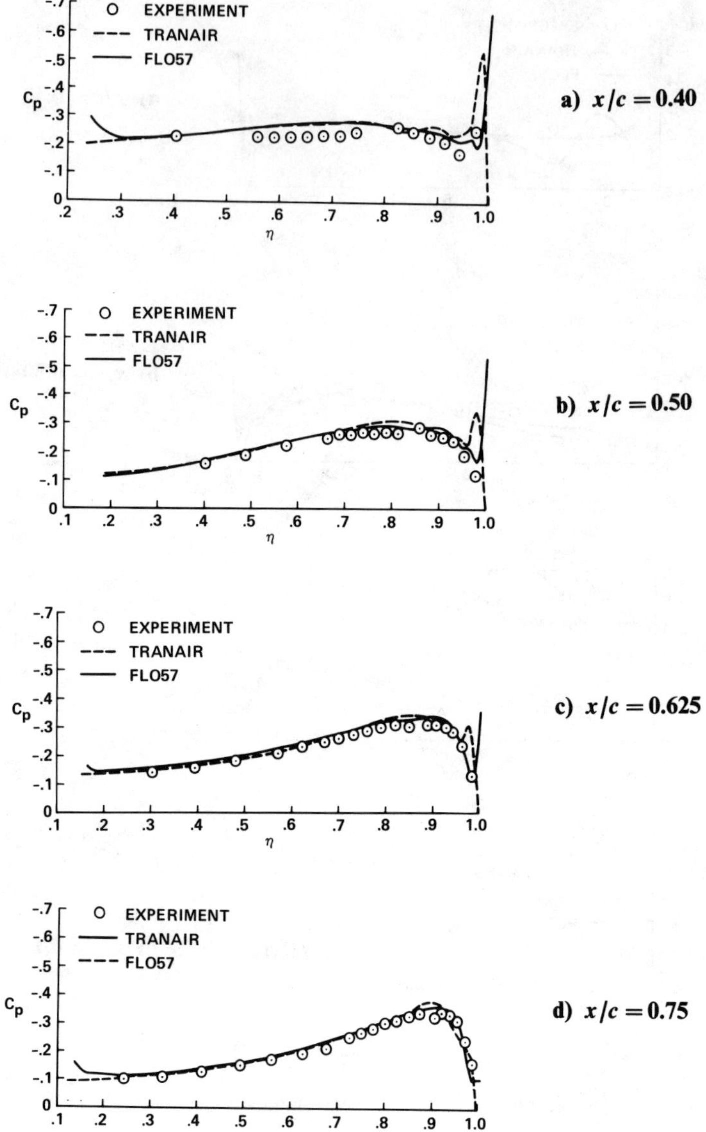

Fig. 31 C_p vs η results for the generic fighter wing/body configuration ($M_\infty = 0.8$, $\alpha = 5$ deg).

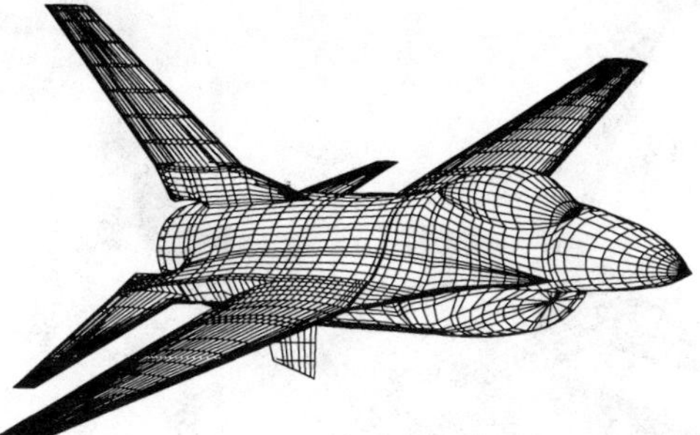

Fig. 32   Surface-paneled definition of F-16A geometry.

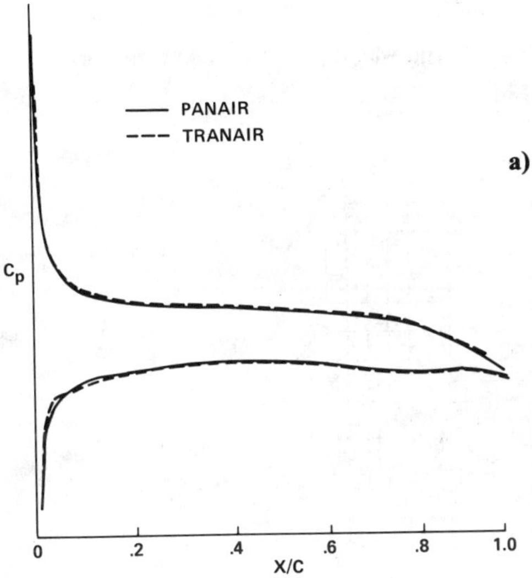

Fig. 33   PanAir and TranAir C_p vs x/c results for F-16A ($M_\infty = 0.6$, $\alpha = 4$ deg).

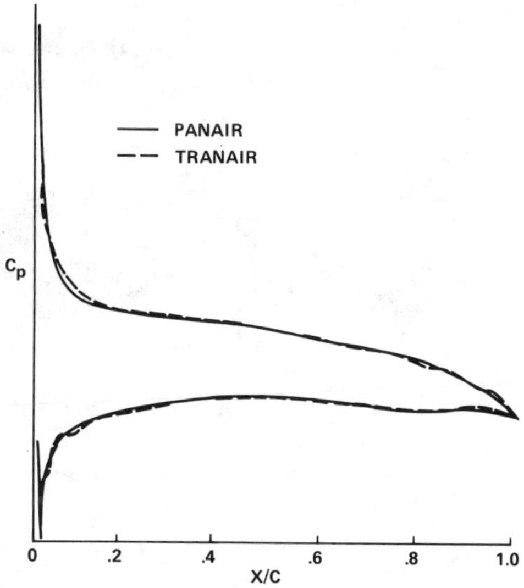

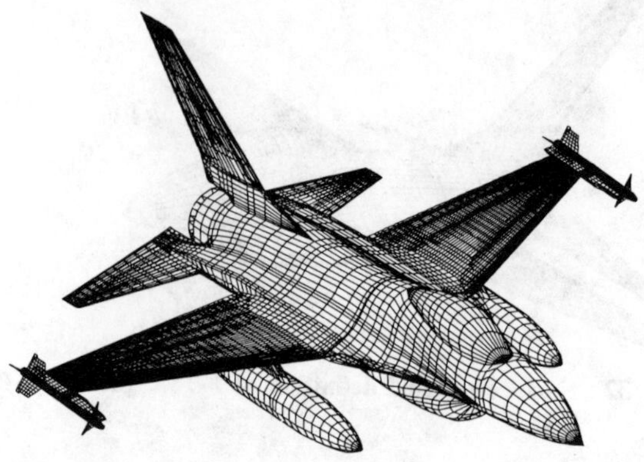

Fig. 34 Surface-paneled definition of F-16A with wing-tip missiles and launchers and underwing fuel tanks.

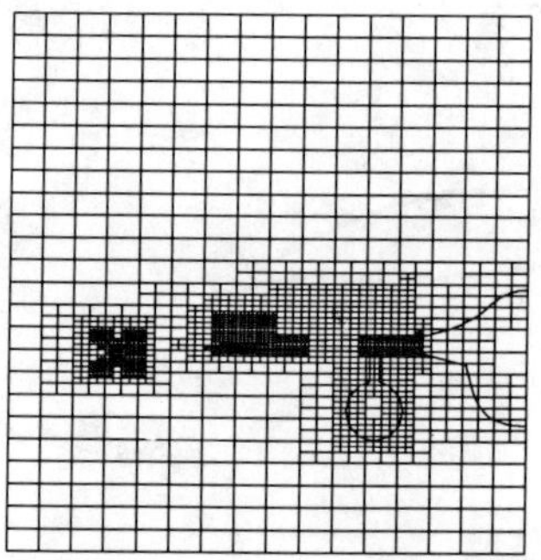

a) Spanwise cut through wing, fuel tank, and missile

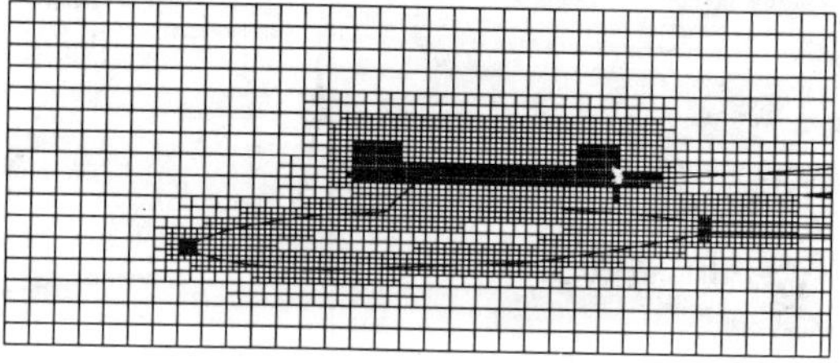

b) Streamwise cut through wing and fuel tank

Fig. 35 Two-dimensional view of refined grid for fully configured F-16A geometry.

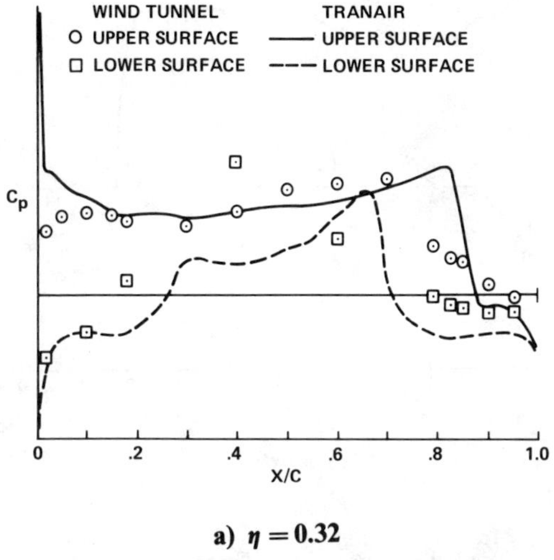

a) $\eta = 0.32$

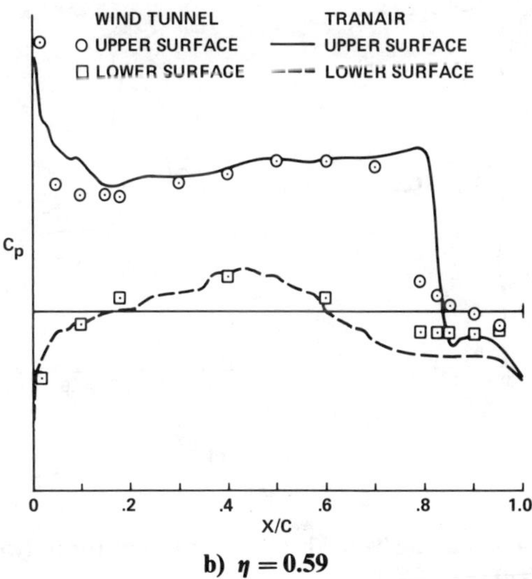

b) $\eta = 0.59$

Fig. 36 PanAir and TranAir C_p vs x/c results for fully configured F-16A ($M_\infty = 0.9, \alpha = 4$ deg).

(Fig. 36 continued on next page.)

M. D. MADSON AND L. L. ERICKSON

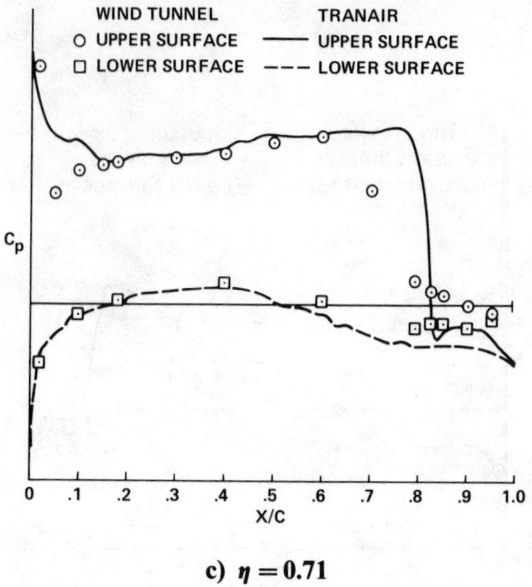

c) $\eta = 0.71$

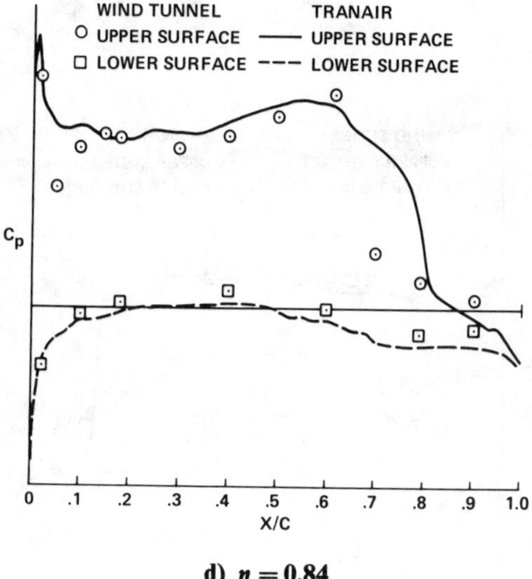

d) $\eta = 0.84$

Fig. 36 (cont'd.) **PanAir and TranAir C_p vs x/c results for fully configured F-16A ($M_\infty = 0.9$, $\alpha = 4$ deg).**

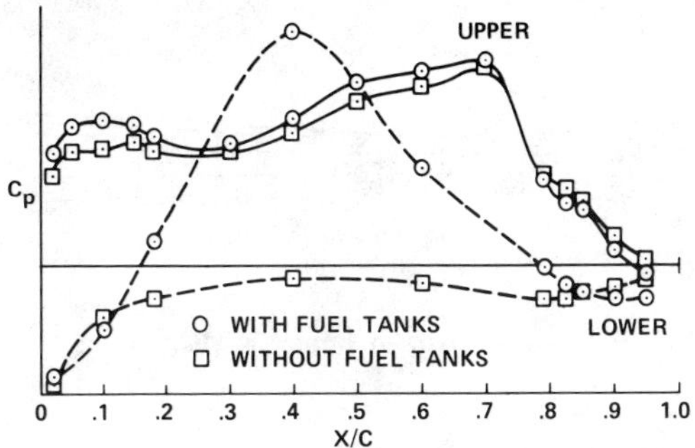

Fig. 37 Effect of fuel tank on experimental results of F-16A wing ($\eta = 0.32$, $M_\infty = 0.9$, $\alpha = 4$ deg).

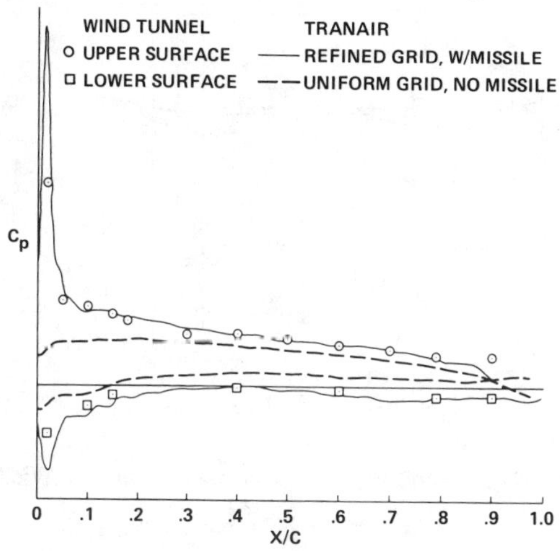

Fig. 38 Effect of missile geometry and grid refinement on TranAir C_p results for the F-16A wing ($\eta = 0.95$, $M_\infty = 0.6$, $\alpha = 4$ deg).

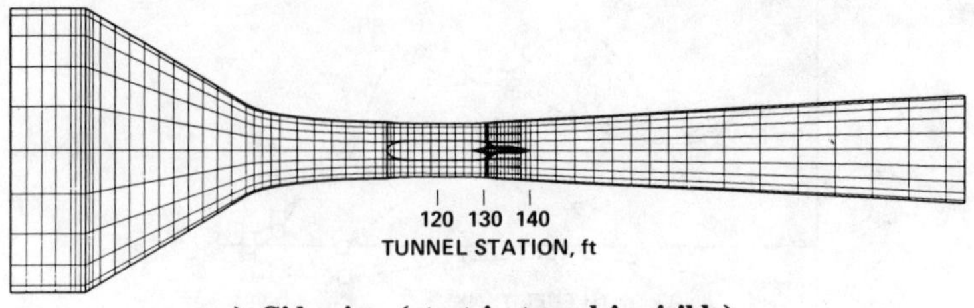

a) Side view (strut in tunnel is visible)

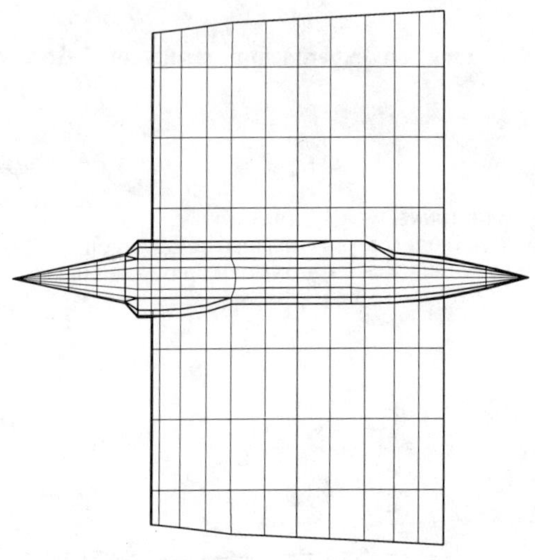

b) strut/centerbody assembly

Fig. 39   Surface paneling for the Ames 12-Ft Pressure Wind Tunnel.

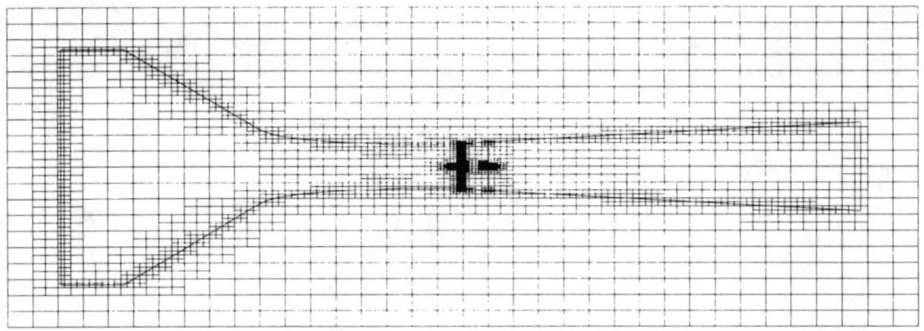

a) Side view along plane of symmetry

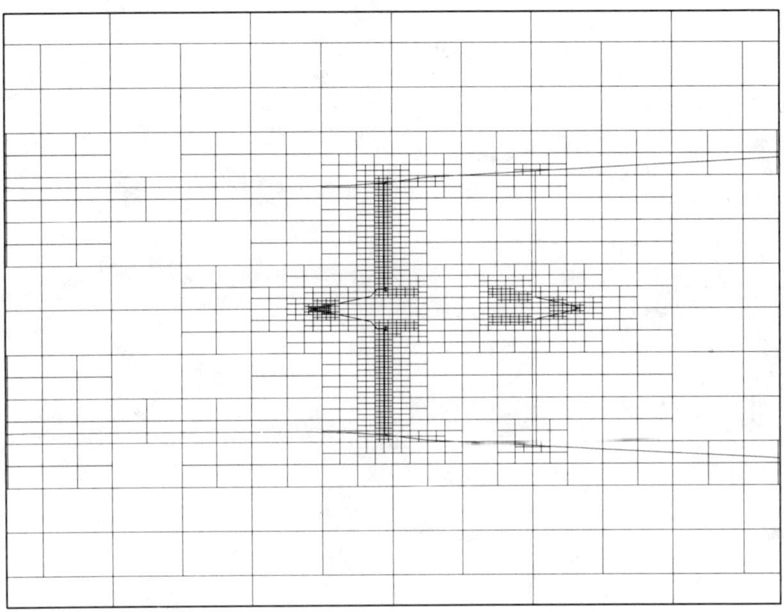

b) Closeup view of strut region

Fig. 40   Two-dimensional view of TranAir refined grid for the 12-ft tunnel model.

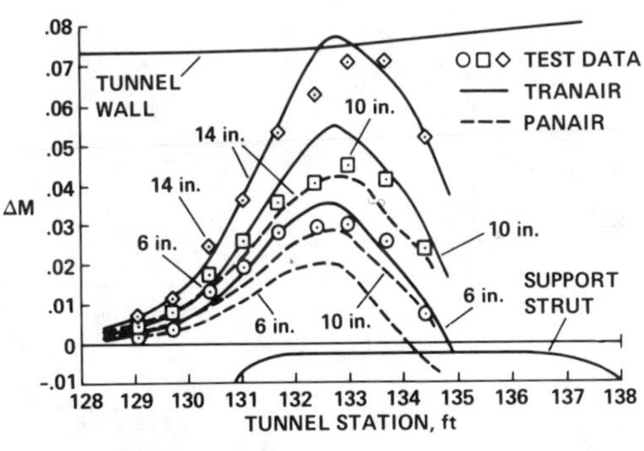

Fig. 41   PanAir and TranAir ΔM results for three strut thicknesses in the 12-ft tunnel model, $M_{\text{test-section}} = 0.55$.

Chapter 20

Computational Aerodynamic Simulation Experience

E. Bonner,* C. J. Woan,* and G. J. Sova*
*Rockwell International Corporation, North American Aircraft,
Los Angeles, California*

Nomenclature

AR   = aspect ratio
c = local chord
d = diameter
b = wing span
C_D = drag coefficient
$C_{L\alpha}$ = theoretical lift-curve slope per radian
C_p = static pressure coefficient, $(P - P\infty)/q\infty$
I = inviscid
M = Mach number
L/D = lift-to-drag ratio
NT = number of time steps
P = static pressure
P_t = total pressure
q = dynamic pressure
R_n = mean aerodynamic chord Reynolds number
S = suction parameter (see Fig. 5)
V = viscous
x,y,z = axial, lateral, vertical Cartesian coordinates
α = angle of attack
δ = flap or jet deflection angle
θ = twist or polar angle
Λ = sweep angle

Subscripts
F = friction
i = ideal
j = jet
t = tip
∞ = freestream

Copyright © 1989 by the American Institute of Aeronautics and Astronautics, Inc. All rights reserved.
*Member of Technical Staff, Computational Fluid Dynamics Group.

Introduction

SUBSTANTIAL advancements have occurred in numerical design and analysis since the first Transonic Perspective[1] at the Ames Research Center in the spring of 1981. The modeling of general wing-body arrangements is well developed at the full-potential/Euler level and has been coupled to boundary-layer equations to approximate the effects of viscosity. The interaction of multiple surfaces can be systematically treated. In short, most of the identified deficiencies at the time have been eliminated. Recently, modeling of extensively separated flows has been successfully demonstrated using Reynolds-averaged Navier-Stokes analysis.[2] This is encouraging and provides the impetus to numerically investigated increasingly complex design conditions.

One of the pacing technologies in numerical design and analysis is grid generation. The wide variety of geometry and flow gradients emphasizes multiblock/multigrid simulations in the interest of generality and computational efficiency through accelerated solution convergence. Grid size constraints place emphasis on adaptive strategies in order to resolve viscous regions, local interactions, etc.

The development of unified solution algorithms for nonlinear equations[3,4] allows consideration of subsonic (elliptic), transonic (mixed elliptic-hyperbolic), and supersonic (hyperbolic) inviscid/viscid steady and unsteady flows.

Discussion

Representative computational simulations will be presented to illustrate the current state of the art in numerical analysis. Results at subsonic, transonic, and supersonic speeds will be given since these conditions are commonly coupled vis-à-vis imposed geometric constraints.

Two general cases will be selected for detailed discussion. The first case addresses conceptual numerical design capability. The second case describes the simulation generality that is possible using unified solution algorithms. The discussion is concluded by citing a number of related developments to further define the scope and success of recent numerical efforts and the direction of future developments.

Advanced Concept

A multistage process can be used to achieve a conceptual aerodynamic design, defined here as an activity that is used to numerically screen and define the arrangement and flow characteristics prior to committing to a subscale test. The procedure is summarized in Fig. 1 in conjunction with the approximating equations used and proceeds from left to right. Linear theory is used to establish necessary far-field thickness and lifting constrained optimums and associated geometry. A transonic potential nonlinear analog is currently under development.[5] Linear results are necessary but not sufficient in nature since they do not explicitly deal with embedded shock waves and viscous effects. Full-potential and Euler-boundary-layer simulations are subsequently used to resolve the wave system and manage weak pressure/viscosity interactions. Finally, there is a design space in which viscous effects dominate to the extent that force and moment nonlinearities dictate structural and stability/control system requirements. These flows are typically extensively separated and are statically or dynamically coupled with the structure. Because of the complex nature of such conditions, the slowest progress from both a numerical and test standpoint has resulted for this class of problems. Aircraft design deficiencies can be commonly traced to a failure to deal with this consideration adequately.

In order to illustrate the process of Fig. 1, the following example[6] is presented for the tactical fighter concept of Fig. 2, which had transonic acceleration, supersonic cruise, and subsonic/transonic/supersonic maneuver design points. These diverse operating conditions were reconciled using quarterchord full-span deflectable leading- and trailing-edge wing flaps to provide variable camber and an aeroelastically tailored structure to increase nose-down twist with pitch angle. Subsonic and transonic maneuver design pretest expectations in terms of wing surface pressure coefficient distributions are presented on Figs. 3 and 4. Inviscid and viscous results are labeled I and V and are for 100 and 230 relaxation cycles, respectively. The secondary pressure peaks are due to the discrete deflections of the two-element leading-edge and single-element trailing-edge flaps. The leading-edge peak was not fully suppressed in order to reduce these local accelerations. Full-potential simulation with and without boundary-layer modeling indicated that trailing-edge-type separation would exist. At $M = 0.9$ a shock wave of increasing spanwise strength occurred downstream of the trailing-edge flap hingeline. The flow is separated at the foot of the shock in the outboard region and is partially a consequence of supersonic efficiency considerations that limit the fixed geometric camber between the flaps. The approach here is thus one of accepting trailing-edge separation and limiting its extent through adverse pressure gradient location. Post-test comparisons with measurements presented on Figs. 3 and 4 indicate that this objective was realized. Surface flow data (not shown) further corroborated the anticipated separation extent.

Measured performance results in terms of the aerodynamic lifting efficiency parameter, S, are presented on Fig. 5. Two cases are shown. The first design placed increased emphasis on acceleration/cruise, and the second placed increased emphasis on maneuver. Both were designed numerically. Theoretical upper-bound lifting efficiency corresponds to $S = 1$ and is associated with an elliptic span load. $S = 0$ corresponds to the zero suction drag of a flat plate of the same gross planform. The test-derived $M = 0.6$ variation with lift coefficient indicates the onset and growth of separation at subsonic conditions. Comparison with the transonic characteristics at $M = 0.9$ indicates the impact of the formation and strengthening of embedded shock waves and associated upper-surface separation. The test results compared to upper-bound level verify that the numerical design is operating efficiently over a broad band of operating lift coefficient at both subsonic and transonic conditions. A corollary result is that numerical pressure gradient/boundary-layer control through camber and twist management is an effective strategy. The impact of twist/camber variations at supersonic speeds is presented on Fig. 6. Also shown is the effect of transonic considerations associated with reduced leading-edge sweep and increased leading-edge radius, camber, and twist relative to an unconstrained supersonic design. The impact of the twist/camber changes between the first and second transonic design of Fig. 5 were secondary at this condition and consequently is not shown. Numerical pretest expectations are in good agreement with measurements in all cases. Comparison of full-potential and Euler predictions indicate minor rotational effects at the supersonic maneuver condition.

In summary, numerical design was very effective in developing high aerodynamic efficiency for an advanced concept over a broad Mach number/lift coefficient operating envelope and was realized with a minimal number (specifically two) of test entries. The impact of computational fluid dynamics on the design effort is summarized on Fig. 7, which compares the present advanced concept subscale test results with flight characteristics for representative inventory tactical aircraft of the same (twin engine) class. Transonic improvements are attributed to increased lifting efficiency and supersonic improvements to increased volumetric efficiency. These and similar results not presented here indicate that numerical aerodynamic performance design has progressed to a relatively mature status. Accurate prediction of transonic drag does, however, remain an area of research, particularly for conditions that have separation present.

Unified Simulations

The previous development made use of a number of analyses that utilized various specializations such as small perturbations, isentropic conditions, transonic or supersonic flows, inviscid or boundary layer, and nill or weak viscous interactions.

Recent development of unified solution algorithms[3,4] now make it possible to systematically remove these approximations using a single solver. Of even greater long-term consequence is the ability of this advancement to treat the class of flows associated with strong viscosity interactions, albeit approximately.

Several cases encompassing both external and internal flows will be presented to illustrate this capability. A common denominator for all these results is that a unified solver was used to generate them. Grid topology must, of course, be specialized to anticipate and capture high gradient regions of the problem(s) under investigation. Fully unified simulations consequently await integration of ongoing adaptive grid developments.

Research Wing-Body

A series of numerical computations will be compared to test results for the wing-body arrangement of Fig. 8 to illustrate the current capability to systematically simulate a wide variety of conditions encompassing subsonic, transonic, and supersonic conditions for both attached and separated flows using a unified solution algorithm and static grid topology. The analysis is capable of treating unsteady as well as steady flows. Only results for the latter will be presented. The geometry under consideration is undesigned;

consequently, the subject pertinent to aerodynamic development is how reliably the characteristics associated with this arrangement are numerically captured. This, of course, is a prerequisite to their modification through numerical design.

The multiblock six-zone H-grid topology of Fig. 9 was used for the numerical analysis. An isometric upper-half plane of the computational domain is presented in Fig. 9a. A typical cross-sectional cut is presented on Fig. 9b to further define the blocking and clustering. Vertically finer inner blocks immediately above and below the wing surface were used in recognition of the high-velocity gradients in such areas. Various zonal grid densities are tabulated in the upper-right-hand corner of the figure. Nominally 89,000 grid points were employed for an in core Cray XMP/14 simulation.

Euler results[7] for 6-deg angle of attack at $M = 0.9$ and 1.2 are presented in Fig. 10 for various wing span stations and fuselage polar angle locations of ± 15 deg above and below the plane of the wing where the wing-body interaction is strong. Comparison with measured surface pressure coefficient results is excellent, except for the wing tip at $M = 0.9$. Predictions for 20 and 60% span are equally successful[7] as those shown but have been omitted for brevity. At $M = 0.9$ a strong shock exists on the fuselage above the plane of the wing. Embedded leading- and trailing-edge shock exists on the suction side of the wing which increase in strength spanwise until they coalesce and further increase in strength until shock-induced tip separation occurs. Ample pretest warning is provided in this area that the shock is strong and that separation is a distinct possibility. From a design viewpoint, it can either be accepted or explicitly dealt with through redesign. The Euler numerical results typically required 20 min of Cray CPU time per case. It should be pointed out that a 33,660 grid point single zone C-H grid full-potential simulation at $M = 0.9$ produced similar prediction success in 80 CPU s. No important rotational flow effects consequently existed for this case. A similar conclusion at supersonic speeds was provided by the space-marching full-potential and Euler results of Fig. 6.

The multiblock grid of Fig. 9 was subsequently used for turbulent Navier-Stokes simulation[8] of the research wing-body arrangement. Typical impact of angle of attack on measured wing surface pressure coefficient characteristics is presented on Fig. 11. The inboard upper-surface loading increases with the leading-edge suction peak exhibiting pronounced broadening, which is indicative of the formation of a vortex in this region. The outboard wing exhibits a pronounced decrease in loading and a flat upper-surface pressure level typical of wing stall. Clearly, this is a relatively complex viscous-dominated flow that provides a difficult practical test for Reynolds-averaged Navier-Stokes simulation. Two cases were selected for evaluation ($M = 0.6$, $\alpha = 14$ deg, and $M = 0.9$, $\alpha = 10$ deg, corresponding to the highest angle of attack tested at each Mach number. Both exhibit the previously discussed inboard vortex/outboard separation behavior.

A zero-equation (Baldwin-Lomax) turbulence model was used in zones 1 and 4 of Fig. 9b for the viscous simulation. The numerical analysis is compared to measured surface pressure results on Figs. 12 and 13. The major features of the flow are well reproduced for both cases. In particular, the formation of the inboard wing vortex and the outboard wing stall are captured. Considering the relatively coarse grid being used to simulate the phenomena, the results are quite good and were not materially changed by subsequent doubling of the vertical grid in the zone immediately above the wing. It should be noted that it was necessary to carry molecular viscosity terms in zone 3 in order to achieve the $M = 0.6$ upper surface comparisons of Fig. 12. This consideration was not important for the $M = 0.9$ results of Fig. 13.

The potential for using Navier-Stokes analysis for structural design and high-angle-of-attack viscous-dominated nonlinear modeling appears promising. The Navier-Stokes solutions required between 1.5 and 2.5 Cray CPU h to nominally converge. Although this level of computer resources is considerable, it is not unreasonable from an aerodynamic design point of view. This is particularly true if such simulations are used discriminately and projected to advanced computers such as the Cray 3.

In summary, numerical simulation of a research wing-body arrangement was successfully demonstrated at subsonic, transonic, and supersonic speeds. Both attached and separated flow conditions were considered. These results were achieved with a unified solution algorithm and static multizone grid topology. The associated computer resources were consistent with advanced concept development activities. The viscous modeling success indicates that systematic consideration of strong viscous interaction design problems appears promising.

Research Inlet

The numerical analysis for the inlet arrangement and associated two-zone $83 \times 50 \times 11$ grid of Fig. 14 are presented to illustrate the simulation of a relatively broad range of air-induction system operating conditions. This case corresponds to one of those considered for the advanced concept of Fig. 2. Euler Mach number contours for $M = 2.15$ supercritical, near-critical, and subcritical operation (corresponding to mass flow ratios of 0.83, 0.823, and 0.40, respectively) are presented on Fig. 15 for polar angles of 90 (centerline), 63, and 0 deg. Transition from supersonic oblique shock to transonic mixed oblique/normal shock to subsonic normal shock internal flow is indicated as the mass flow is reduced. Progressive movement of the normal shock ahead of the cowl lip is further indicated for the subcritical condition with increasing polar angle. A rotational "Euler slip" layer is clearly visible downstream of the conical centerbody intersection with the normal shock at θ of 90 and 63 deg for a mass flow ratio of 0.4.

Cray XMP/14 computing time per solution time step was 4.45 s. The results were obtained with 400 time steps (29 min) for the supercritical case, 1090 time steps (1 h 21 min) for the near-critical case, and 760 time steps (56 min) for the subcritical case. For the supercritical analysis, the solution converged quickly to residuals of 10^{-4}. For the critical and subcritical cases, the solution convergence became relatively slow after the residuals reduced to 10^{-2}. This behavior may be due to the strong normal shocks at the cowl leading edge and the elongated cells used in the computations as shown in Fig. 14. The solution was terminated for these latter cases when the change in mass flow became small.

Euler predictions are compared to Rockwell test measurements for a near-critical operating condition at polar angles of 90 (centerline), 60, 30, and 8.4 deg on Fig. 16. Measured exit mass flow ratio is 0.818 compared to 0.823 for the present analysis. Agreement between predicted and measured centerbody surface static pressure ratio for θ of 90 and 60 deg is good. The peak pressure at 30 and 8.4 deg is underpredicted and may be due to the boundary-layer diverter that separates the test inlet and reflection plane plate which was neglected in the analysis. The degree of three dimensionality may be judged by the lateral flow gradients. Major features of the flow for this condition have been captured; consequently, the numerical simulation provides a point of departure for inlet design modifications, integration, etc.

Research Nozzle

Two-dimensional Reynolds-average Navier-Stokes nozzle analysis will be compared to static measurements for the thrust vectoring asymmetric arrangement of Fig. 17. Results for both under- and overexpanded operating conditions corresponding to internally attached and separated flow will be presented. The three-zone 100×50, 130×50, 100×50 grid of Fig. 17 was used for the turbulent zero-equation Navier-Stokes results of Figs. 18–20. Mach number contours indicate that the fluid dynamic throat ($M = 1.0$ contour line) is skewed relative to the physical throat. An asymmetric shock diamond pattern is visible in the jet plume.

Table 1   Related simulation results

Simulation	Formulation	Reference
Grid generation	Multiblock	AIAA Papers 88-0312, 88-0521
Multiple surface interactions	Full-potential	AIAA Paper 86-1795
Wing-body static elasticity	Full-potential	AIAA Paper 87-0707
Wing-body dynamic elasticity	Full-potential	AIAA Paper 87-1238 AIAA Paper 89-3468
Subcritical, critical, and supercritical inlet flow	Euler	Present
Attached and separated wing-body flow	Euler, RANS[a]	AIAA Paper 88-2547 NASA CP-3020, pp. 195–216
Boattail ramp and back step separation	RANS	NASA CP-2454, pp. 87–107
Attached and separated nozzle flow	RANS	AIAA Paper 88-2586
Cavity flow	RANS	AIAA Paper 87-0117

[a]RANS, Reynolds-averaged Navier-Stokes.

The computed results for a nozzle pressure ratio of 3 indicates that the turbulent boundary layer separates just downstream of the upper wall throat and slightly later than the test results. The overall comparison of wall static pressure is quite good considering the complexity of the flow. The comparison for the nozzle pressure ratio of 5 in the attached flow case is even better, as would be expected. The upper-surface exit oblique shock is clearly visible and sufficiently weak to preclude separation due to shock–boundary-layer interaction. A comparison of the gross thrust, pitching moment, and kinematic flow deflection angle is presented on Fig. 20. Nozzle thrust ratio (relative to ideal) is overpredicted for the attached flow case by 1% and underpredicted by 2% for the separated case. This is reasonably accurate considering the neglect of internal corners and side surfaces by the two-dimensional analysis. The moment and effective jet deflection angle are well predicted for both the attached and separated flow cases. A nominal Cray XPM/14 solution time of 1 h/case was required for a 4500 time step analysis.

Related Advancements

A series of developments will be cited to further define the scope and success of recent numerical efforts. Table 1 summarizes these activities. Detailed results are provided in the indicated references and will not be repeated here for brevity. Examination of the analyses and comparison with test measurements where pertinent establish that advancement is occurring in a variety of areas covering grid generation, multiple surface interactions, steady and unsteady aeroelasticity, and strongly shocked and separated flows. The research considered a hierarchy of fluid dynamic equations covering full-potential, Euler, and Reynolds-averaged Navier-Stokes approximations. Unification of the grid topology and flow solvers is being emphasized in order to reduce the effort associated with development of simulations covering a wide design space.

Future Directions

Computational aerodynamic simulation is becoming increasingly concerned with issues relating to refined resolution of complex highly three-dimensional flows. Increased effort is being directed at strategies that adjust the grid cell location, size, alignment, etc., as the solution evolves. Both structured and unstructured cells of fixed or growing number are under consideration. Determination of nonboundary properties such as high-accuracy shear-layer velocity distributions, cross-sectional flow profiles, and associated integrations etc. become nontrivial. This is a result of interpolations of cell point properties that satisfy conservation laws to derive characteristics that in general do not.

A second area of development is concerned with turbulence modeling for a Reynolds-averaged Navier-Stokes simulation. The use of algebraic length scales for zero- and one-equation models is restricted for problems that have multiple regions of high vorticity that may not be easily segregated. An example of this is two vertically separated wall shear layers with a mixing layer in the middle that is intersected by a strong curved shock. In this situation, a length scale model would have to simultaneously deal with four different vortex sheets. The use of two-equation K-ϵ models eliminate or at least reduce these difficulties.

From a numerical design point of view, progress is required to move beyond the commonly used parametric and/or "numerical filing" approach. This is difficult, considering the wide design space that is often encountered as evidenced by the advanced concept development discussed previously. Five different conditions spanning Mach numbers of 0.5, 0.9, 1.2 and 1.6 and lift coefficients of 1.13, 0.8, 0.1, and 0.1/0.32, respectively, were considered. A major constraint for the problem was that the geometry from 25 to 75% wing chord, although subject to design, is fixed thereafter for the various operating points. Application of numerical optimization, which has been relatively successful for single-point (cruise) supercritical flow wing design, is not clear and in fact may be of limited use.

A topic of special but increasing interest is associated with the growth of disturbances in a laminar boundary layer associated with the onset of transition, or, more accurately, the delay/suppression of the onset of transition. An extremely wide range of length scales is associated with the inviscid global flow and the very thin wall shear layers. The latter properties must be captured accurately to the second derivative level if small-amplitude stability analysis is used in the design process.

Clearly, advancements in numerical aerodynamic simulation are required in a number of areas to support future advanced design efforts.

Conclusions

Substantial progress is occurring in numerical simulation and design as a result of a computational fluid dynamic algorithm and high-speed digital computer advancements. Aerodynamic numerical performance design is in a relatively well-developed state, as evidenced by the ability to reliably develop efficient broad band characteristics at the conceptual design level. Recent aeroelastic and separated flow simulation results indicate that a capability to consider an important new range of design problems is emerging. Grid generation remains a pacing technology for numerical aerodynamic simulation efforts. It becomes increasingly demanding for complex geometry/viscous flow resolution. Overall grid size limits must be compatible with conceptual development activity in terms of time and resources available if advanced numerical analyses are to have the desired early design impact.

References

[1]Nixon, D. (ed.), *Transonic Aerodynamics, Progress in Astronautics and Aeronautics,* Vol. 8, AIAA, New York, 1981.

[2]Foughner, J. T. Jr. (ed.), "Transonic Symposium: Theory, Application, and Experiment," NASA CP-3020, April 1988.

[3]Shankar, V., and Chakravarthy, S., "Development and Application of Unified Algorithms for Problems in Computational Science," NASA CP-2454, March 1987, pp. 87–107.

[4]Chakravarthy, S., and Szema, K. Y., "Advances in Finite Difference Techniques for Computational Fluid Dynamics," *State of the Art Survey in Computational Mechanics*, edited by A. K. Noor and J. T. Oden (to be published by the American Society of Mechanical Engineers).

[5]Malmuth, N. D., and Cole, J. D., "Wave Drag Due to Lift for Transonic Airplanes," NASA CP-3020, April 1988, pp. 293–308.

[6]Bonner, E., "Nonlinear Aerodynamic Wing Design," NASA CR-3950, Dec. 1985.

[7]Woan, C. J., and Chakravarthy, S. R., "Transonic Euler Calculation of a Wing-Body Configuration Using a High Accuracy TVD Scheme," AIAA Paper 88-2547, June 1988.

[8]Bonner, E., "Transonic Aerodynamic Design Experience," NASA CP-3020, April 1988, pp. 195–216.

[9]Sova, G., "Applications of Navier-Stokes Analysis to Predict the Internal Performance of Thrust Vectoring Two-Dimensional Convergent-Divergent Nozzles," AIAA Paper 88-2586, June 1988.

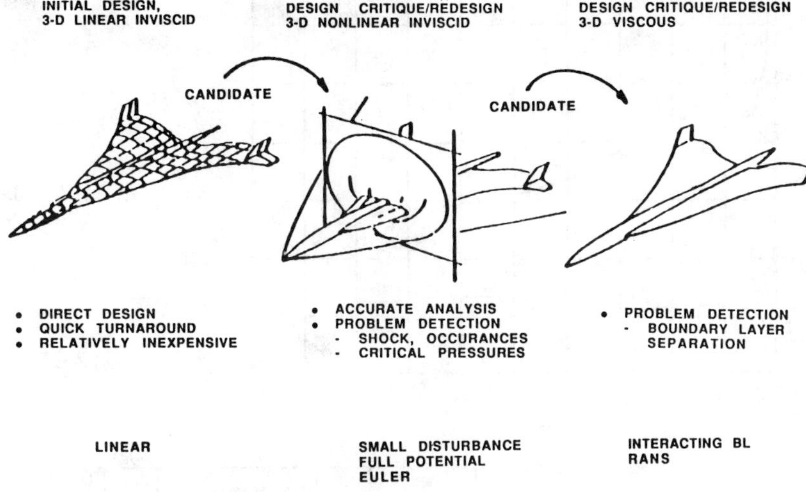

Fig. 1   Numerical design approach.

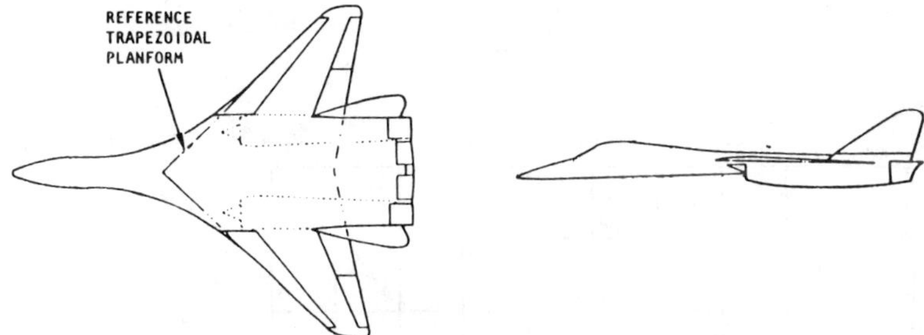

GEOMETRIC PARAMETER	WING (TRAP)	VERTICAL (TRAP)
ASPECT RATIO	3	1
TAPER RATIO	0.2	0.2
LE SWEEP	48°	55°
DIHEDRAL	0	70°
THICKNESS RATIO	0.04	0.04

Fig. 2   Advanced concept.

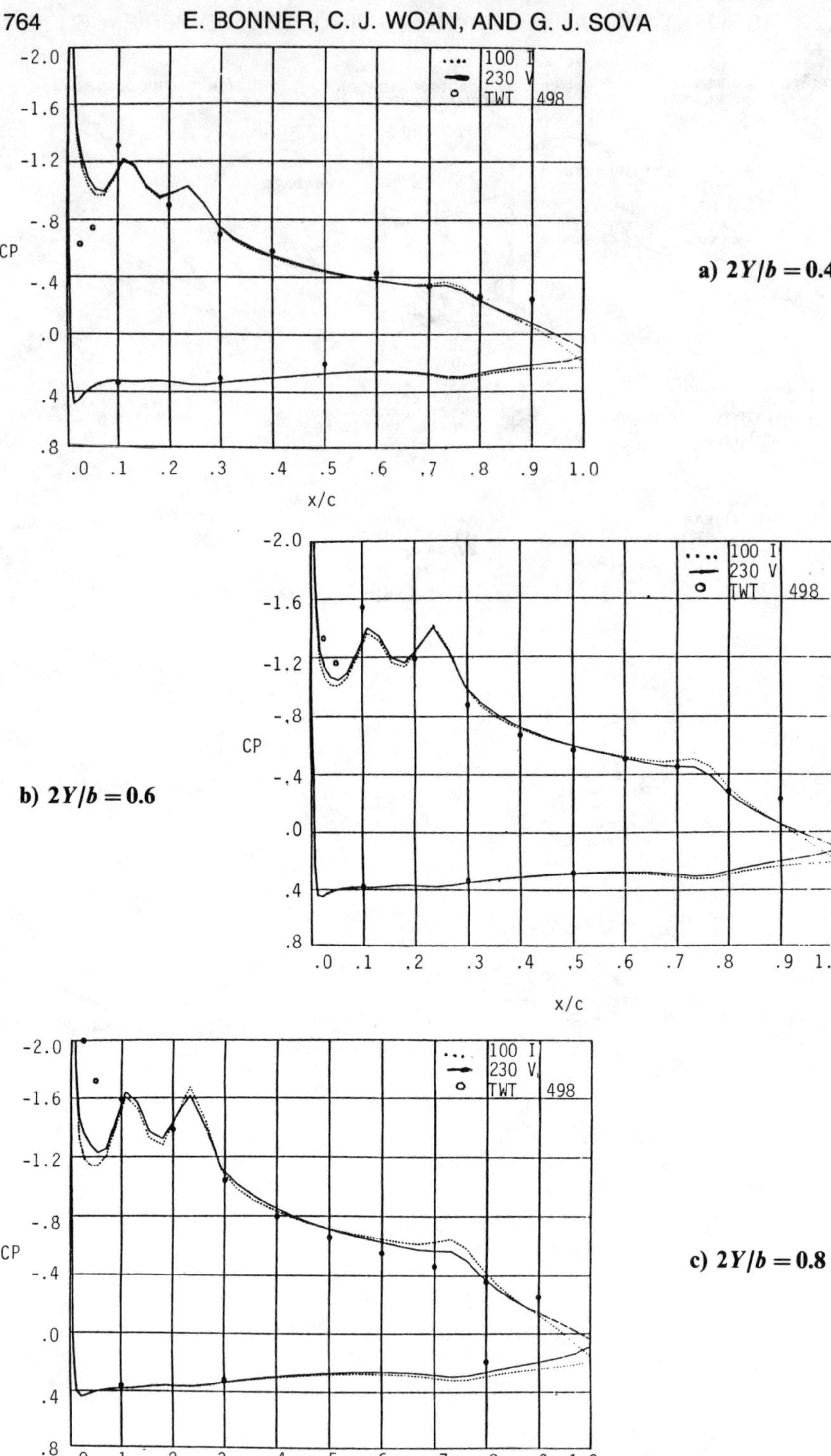

Fig. 3 Advanced concept subsonic wing pressure distributions at $M = 0.6$, $\alpha = 14$ deg, $Re = 5 \times 10^6$. TWT, Rockwell trisonic wind tunnel.

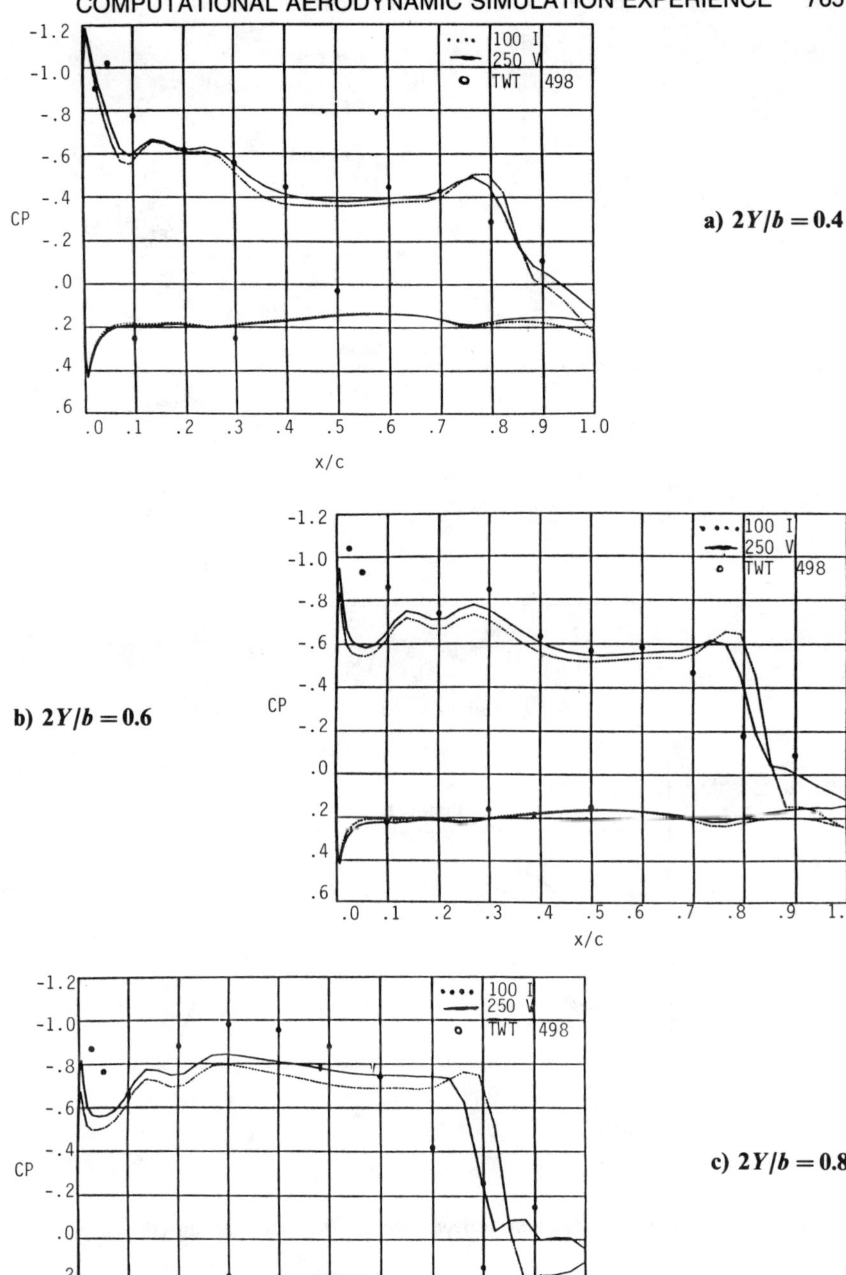

Fig. 4 Advanced concept transonic wing pressure distributions at $M = 0.9$, $\alpha = 8.25$ deg, $Re = 5 \times 10^6$.

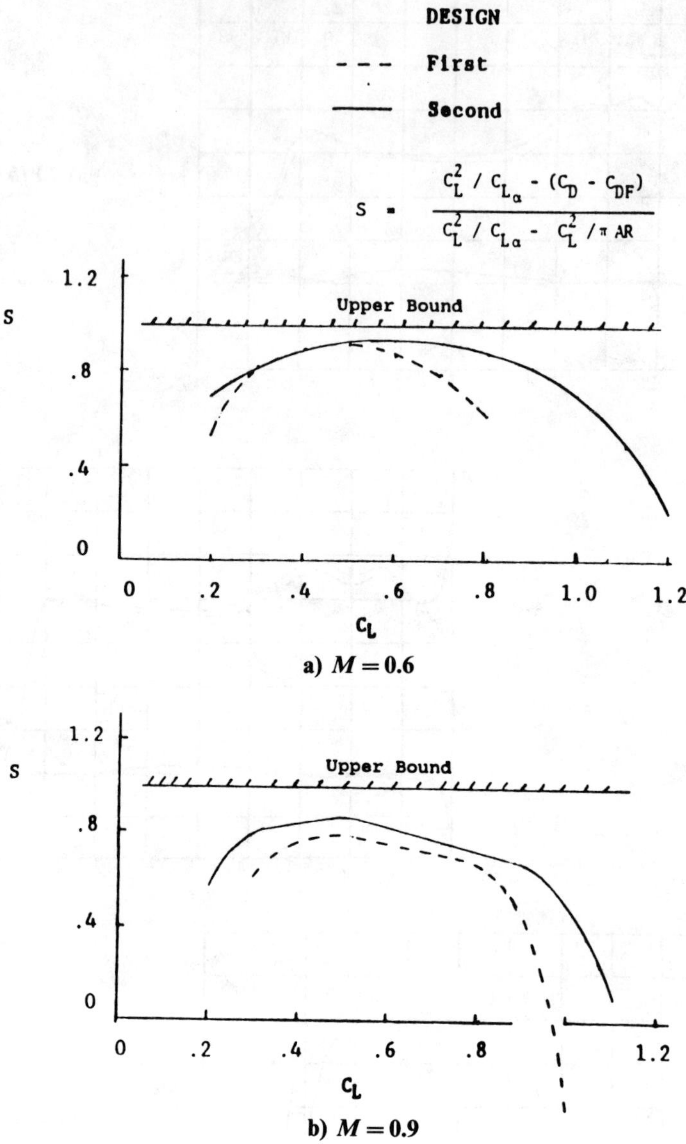

a) $M = 0.6$

b) $M = 0.9$

Fig. 5   Advanced concept test-derived subsonic/transonic lifting efficiency.

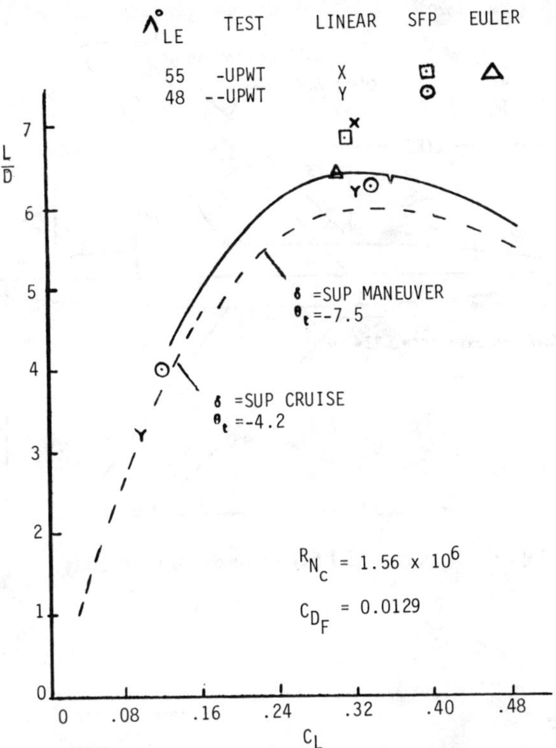

Fig. 6   Advanced concept measured and predicted aerodynamic efficiency at $M = 1.6$. UPWT, Langley unitary plan wind tunnel; SFP, supersonic full potential.

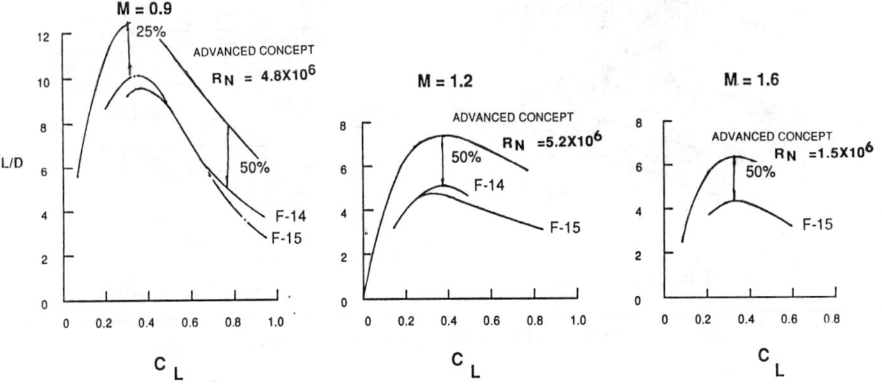

Fig. 7   Impact of numerical design on aerodynamic development.

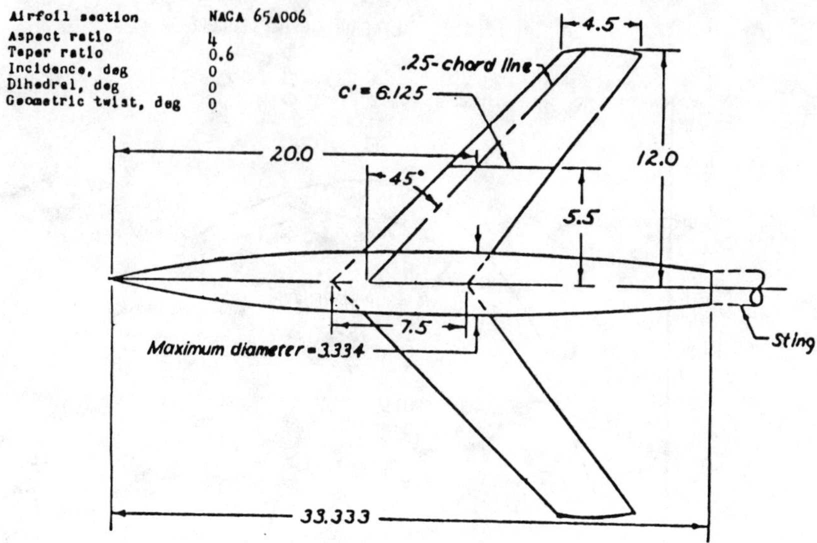

Fig. 8   NACA RM L51F07 research wing body.

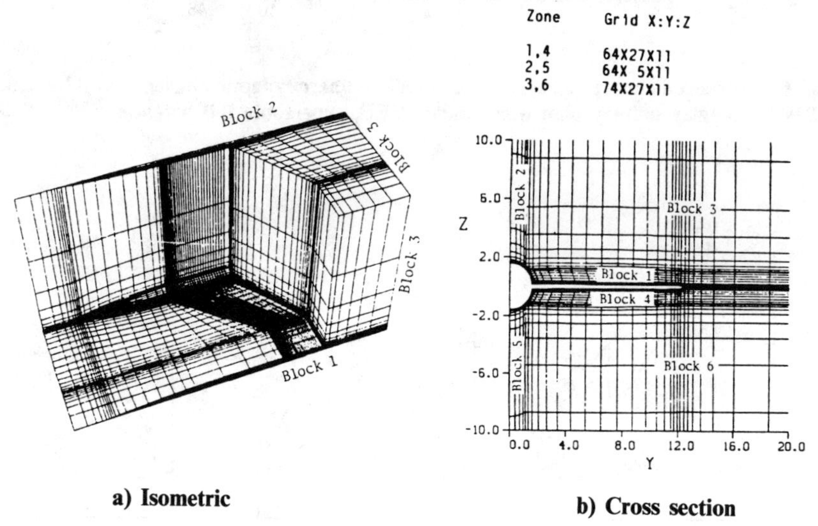

a) Isometric   b) Cross section

Fig. 9   Research wing-body multiblock computational grid.

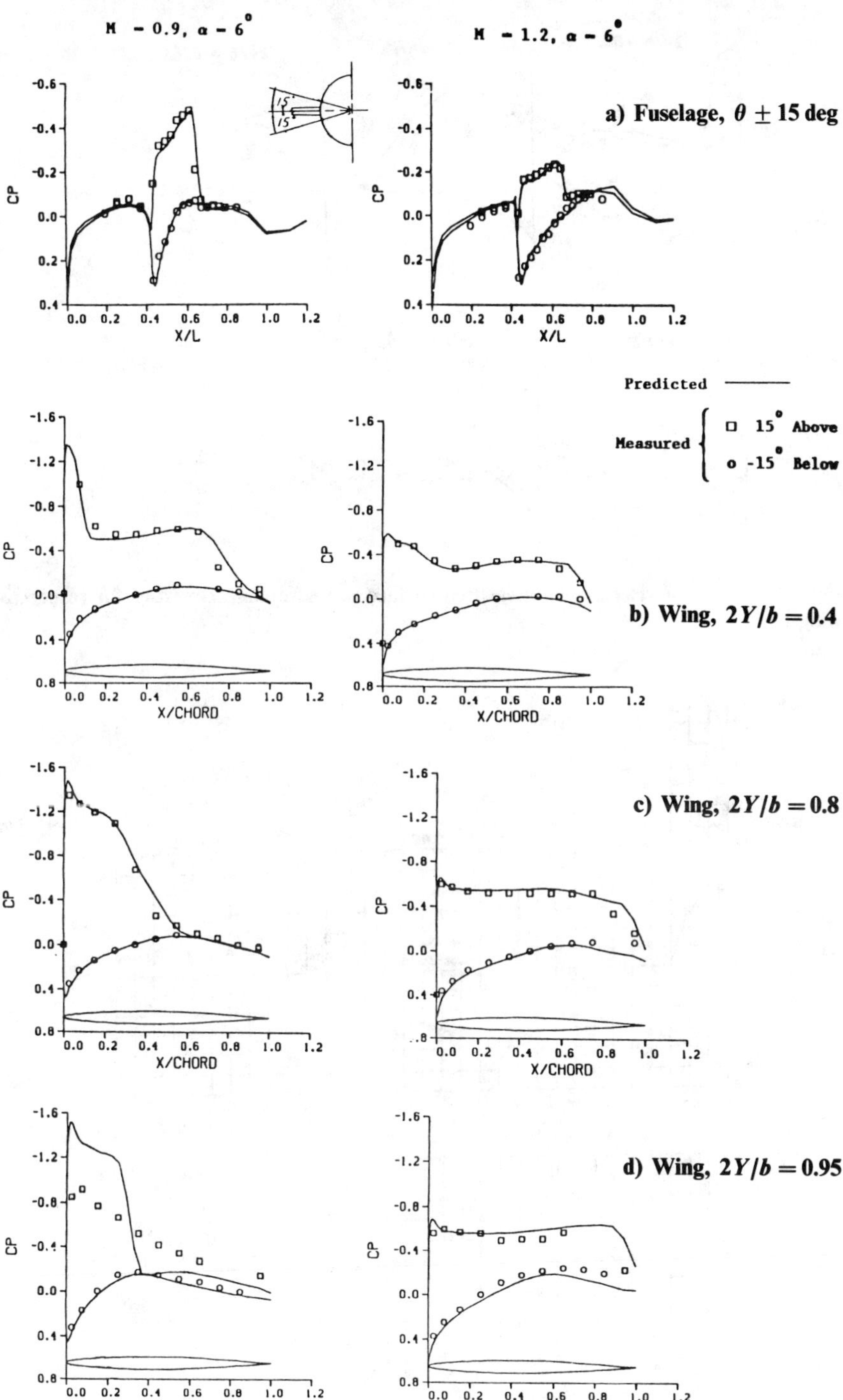

Fig. 10   Research wing-body unified TVD Euler solution at transonic and supersonic conditions.

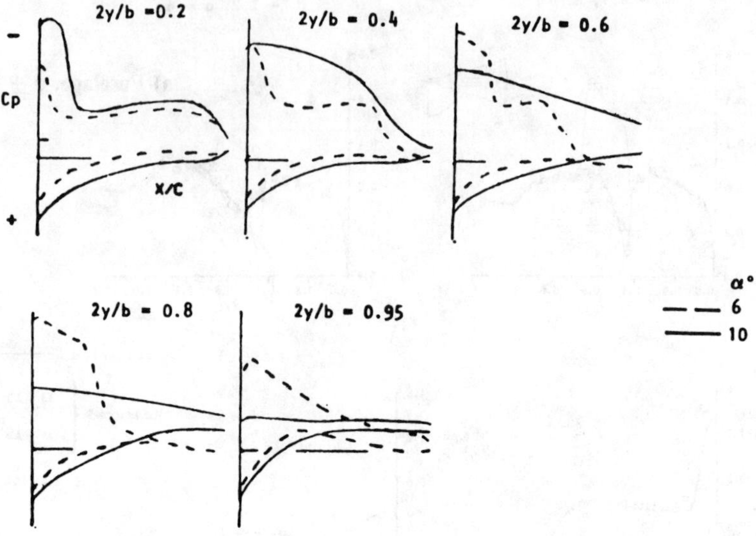

Fig. 11 Effect of viscosity on measured surface pressure characteristics for research wing-body.

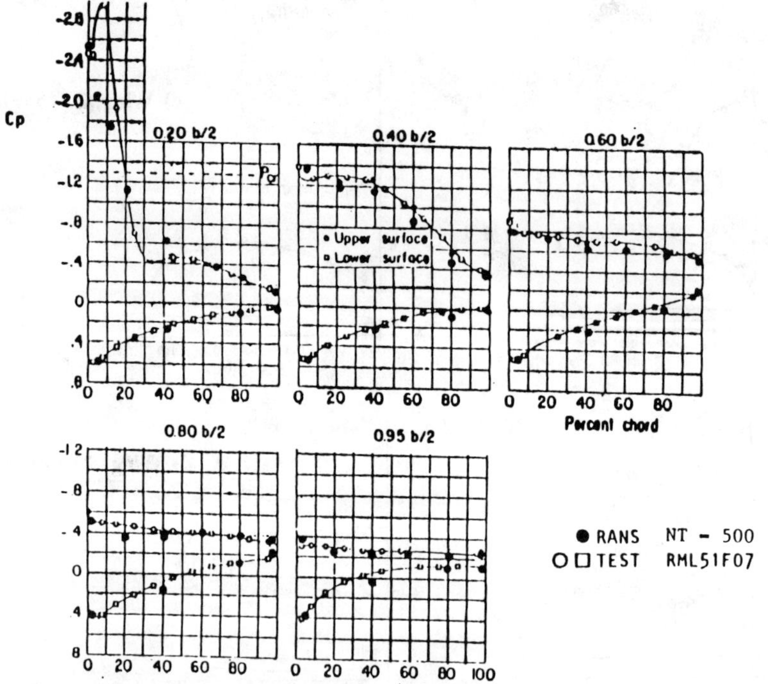

Fig. 12 Reynolds-averaged Navier-Stokes separated flow simulation for research wing-body at $M = 0.6$, $\alpha = 14$ deg, $Re = 1.74 \times 10^6$.

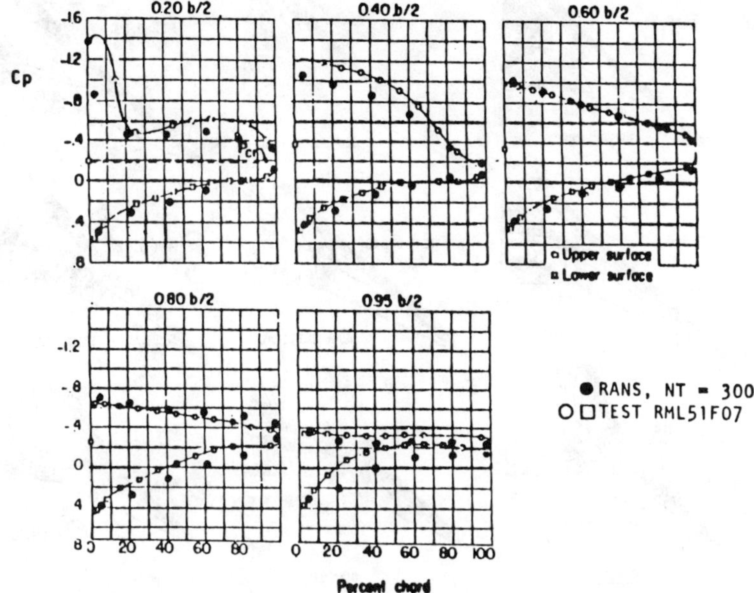

Fig. 13 Reynolds-averaged Navier-Stokes separated flow simulation for research wing-body at $M = 0.9$, $\alpha = 10$ deg, $Re = 2.1 \times 10^6$.

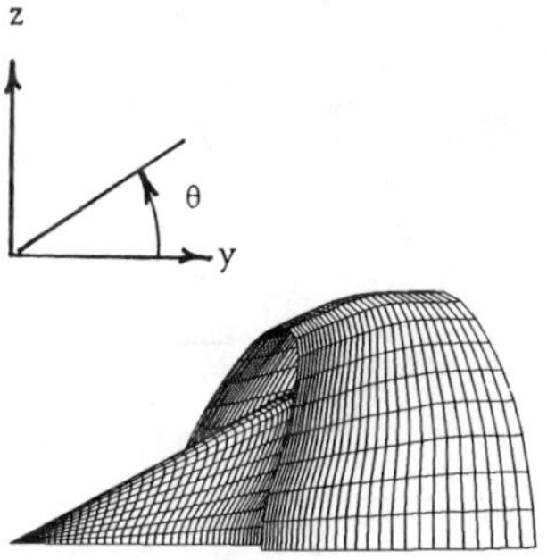

a) Surface grid

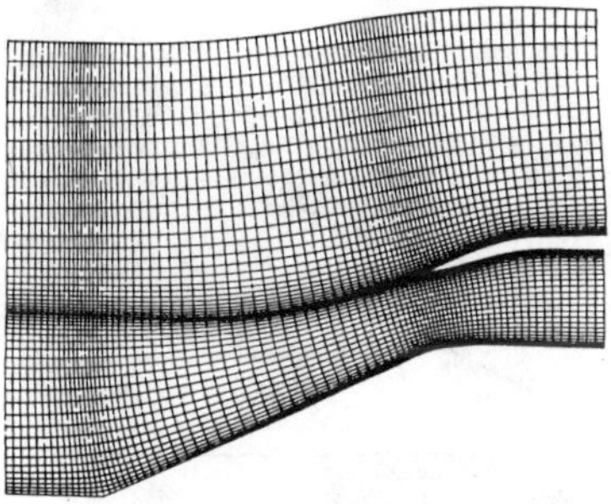

b) Sectional grid at $\theta = 90$ deg

c) Sectional grid at $\theta = 0$ deg

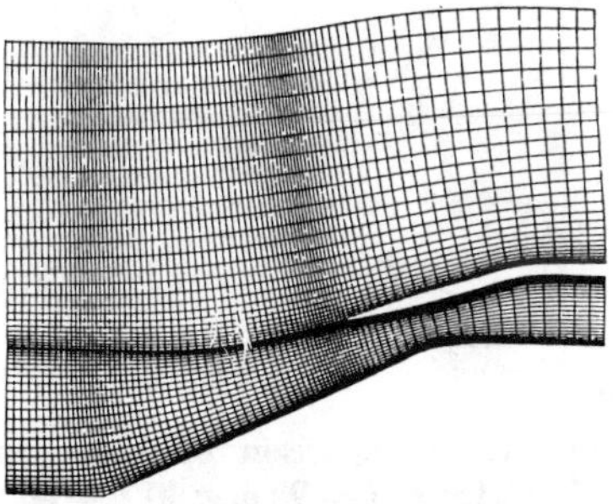

Fig. 14 Research supersonic inlet model.

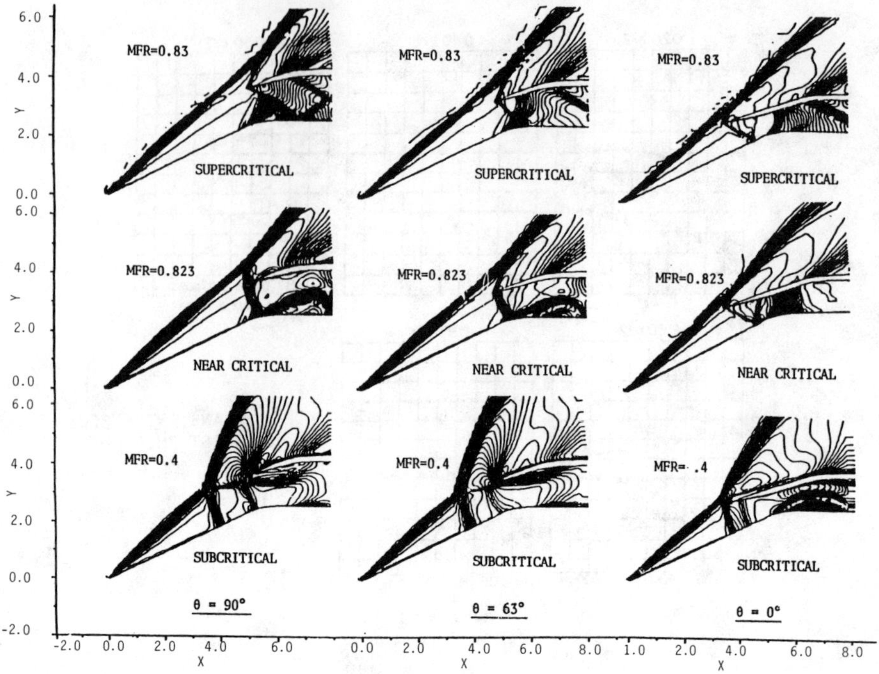

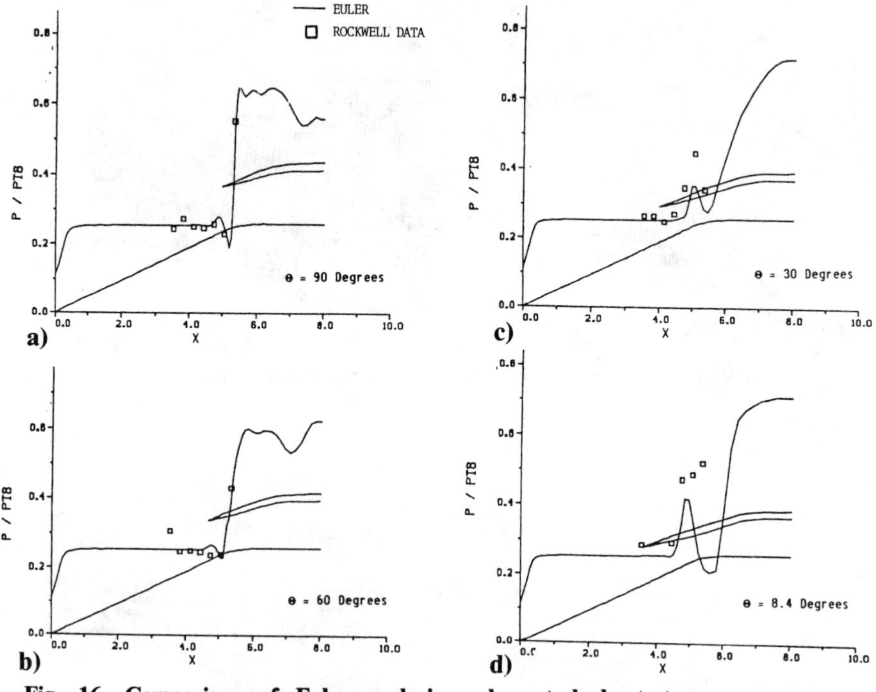

Fig. 15   Euler Mach number contours at $M = 2.15$.

Fig. 16   Comparison of Euler analysis and centerbody test measurements at $M = 2.15$: a) $\theta = 90$ deg; b) $\theta = 60$ deg; c) $\theta = 30$ deg; d) $\theta = 8.4$ deg.

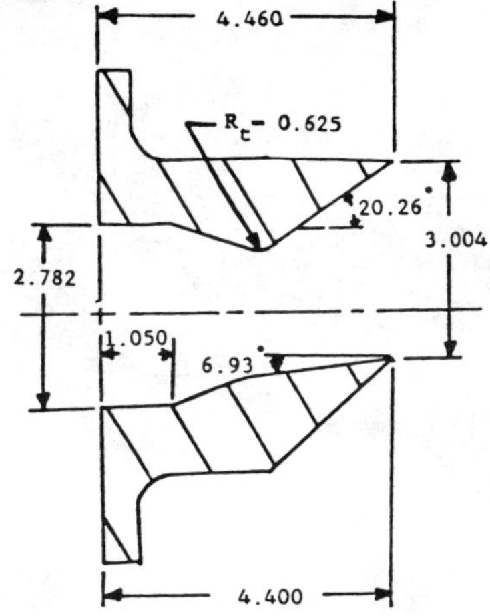

a) Geometry

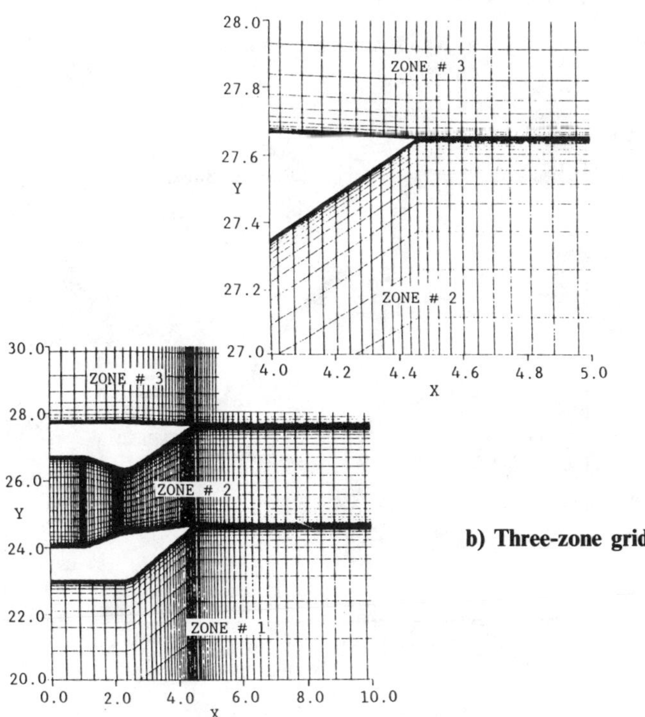

b) Three-zone grid

Fig. 17   Research two-dimensional thrust vectored nozzle model.

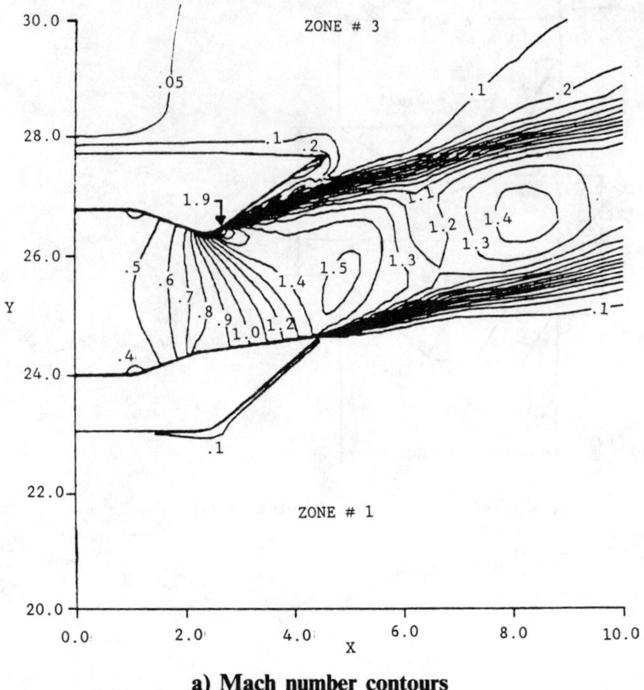

a) Mach number contours

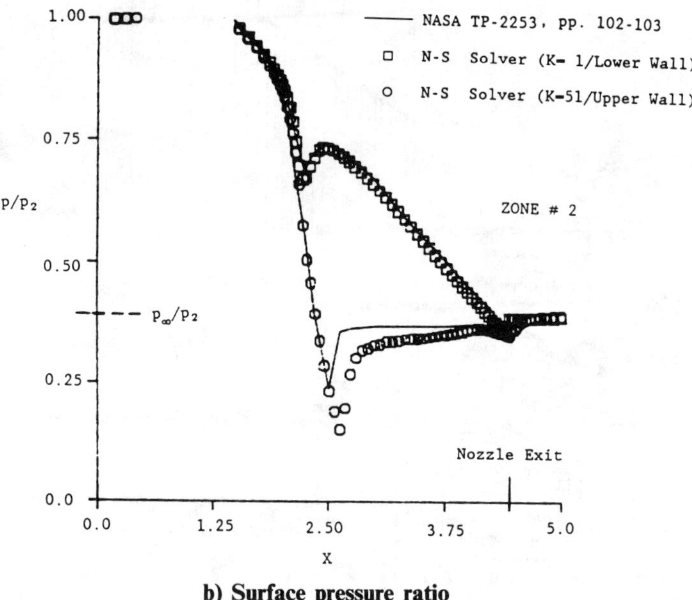

b) Surface pressure ratio

Fig. 18 Reynolds-averaged Navier-Stokes analysis for nozzle pressure ratio of 3.03.

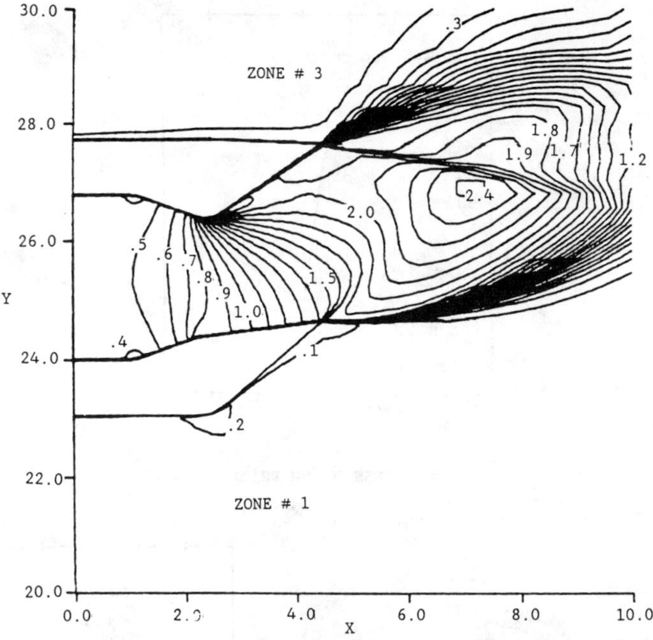

a) Mach number contours

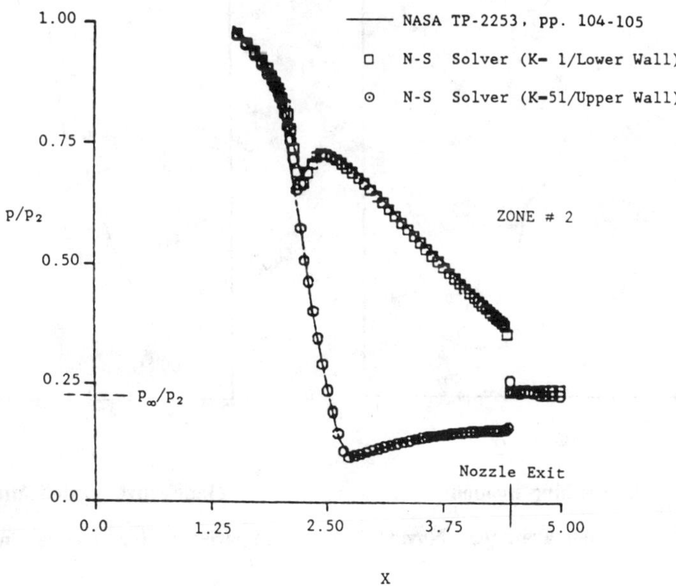

b) Surface pressure ratio

Fig. 19   Reynolds-averaged Navier-Stokes analysis for nozzle pressure ratio of 5.04.

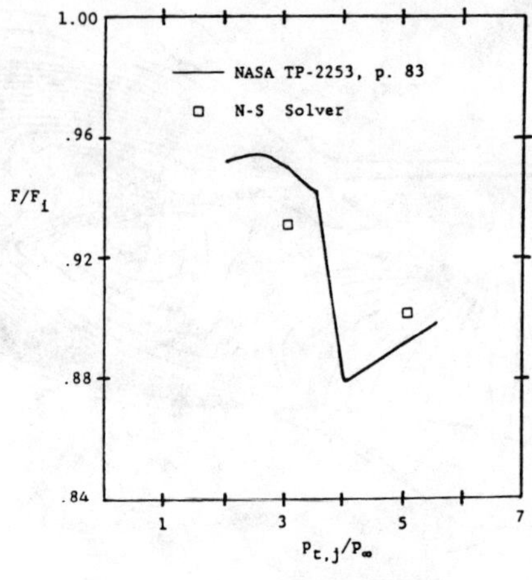

a) Gross thrust ratio

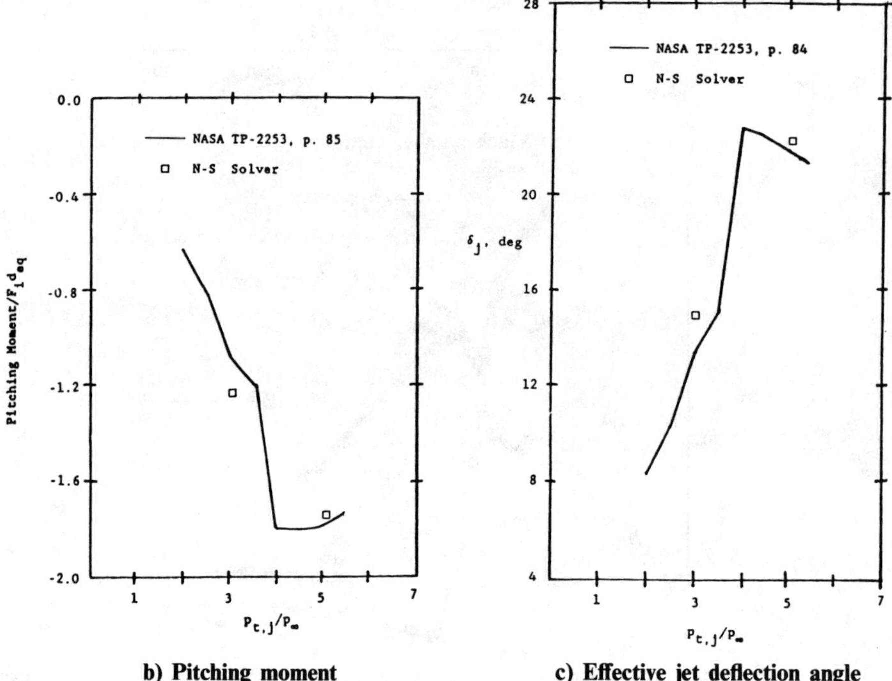

b) Pitching moment   c) Effective jet deflection angle

Fig. 20 Reynolds-averaged Navier-Stokes comparisons for nozzle integrated characteristics.

Chapter 21

Navier-Stokes Computations About Complex Configurations Including a Complete F-16 Aircraft

Terry L. Holst,* Jolen Flores†
NASA Ames Research Center, Moffett Field, California

Ünver Kaynak‡
Sterling Software, Palo Alto, California
and
Neal M. Chaderjian†
NASA Ames Research Center, Moffett Field, California

Introduction

COMPUTATIONAL fluid dynamics (CFD) algorithm and computer hardware developments have been rapidly advancing over the last decade. CFD computer codes have been used in many applications to improve the design of aircraft. For example, linear panel methods have been used to improve the design of complex aircraft configurations providing the flow does not contain shock waves or significant separation. Nonlinear potential methods have been used to remove the adverse characteristics caused by shock waves in the transonic regime with the limitation that vorticity and entropy production are not large. The problem of solving the Euler/Navier-Stokes equations about reasonably complete aircraft configurations is now the central issue facing the CFD research community, and, in fact, many significant results demonstrating the power of this approach have already been produced.

There are many reasons for pursuing this line of research. First, the large computer costs associated with the numerical solution of the Euler/Navier-Stokes equations in three dimensions is rapidly decreasing. This is a direct result of new improved algorithms and fast vector computers such as the Cray XMP and YMP, the Cray 2, and the CDC Cyber 205. Second, there

Copyright © 1989 by the American Institute of Aeronautics and Astronautics, Inc. No copyright is asserted in the United States under Title 17, U.S. Code. The U.S. Government has a royalty-free license to exercise all rights under the copyright claimed herein for Governmental purposes. All other rights are reserved by the copyright owner.
*Chief, Applied Computational Fluids Branch.
†Research Scientist, Applied Computational Fluids Branch.
‡Research Scientist; currently with Turkish Aerospace Industries (TAI), Ankara, Turkey.

is an increasing demand within the aircraft industry for applications codes which use the more complete Euler/Navier-Stokes formulation. This is a direct result of the success experienced by aircraft designers using the simpler potential methods mentioned earlier and the desire to use CFD methods for more complex flow simulations in which potential methods are not valid. Examples of flowfields that logically require the Euler/Navier-Stokes formulation are associated with propulsion/airframe integration, store separation, multiple lifting surface interactions, high-angle-of-attack flowfield prediction, performance boundary prediction, and flutter and buffet calculations where unsteady boundary-layer separation is present. In some of these cases CFD tools are vital because of difficulties associated with wind-tunnel and/or flight testing. A key example of this is the propulsion/airframe integration problem associated with the National Aerospace Plane (NASP) Project.

The literature indicates that research with the time-dependent Navier-Stokes formulation in three spatial dimensions has advanced rather rapidly in the last several years. A few examples of applications include the hypersonic aircraft computations of Shang and Scherr,[1] the unsteady turbomachinery applications in both two and three dimensions by Rai,[2-4] the rotorcraft applications by Srinivasan et al.[5] and Srinivasan and McCroskey,[6] the wing-fuselage and nearly complete transport aircraft computations of Fujii and Obayashi[7] and Obayashi et al.,[8] the jet with crossflow in ground effect work of Van Dalsem,[9] and the transonic wing computations of Vatsa.[10] A large number of Navier-Stokes computations have also been published in the high-angle-of-attack flow regime where the flowfields are often unsteady. Several examples of this type of computation include the hemisphere-cylinder work of Ying et al.,[11] the ogive-cylinder computations of Schiff et al.[12] and Degani and Schiff,[13] the delta-wing calculations of Thomas et al.,[14] the strake-delta-wing computations of Fujii and Schiff[15] and Fujii et al.,[16] and the vortical-flow computations over elliptical-body missiles of Newsome and Adams.[17]

Most of these later calculations utilized relatively fine grids consisting of 100,000–1,000,000 grid points. These grids are relatively fine based on the standard grid size from 10 years earlier, but still may not be fine enough to fully capture all of the pertinent mean-flow details associated with some of the more complex flowfields of interest. The primary pacing factor for these Navier-Stokes computations is computer speed and memory. Most of the more recent computations cited earlier required hours or even tens of hours of computer time and millions of words of memory on the most advanced computers available. The advent of these first pioneering calculations has set the stage for an aggressive assault on the problem of solving the Navier-Stokes equations about complex configurations including a complete aircraft. For a more in depth review of the field of three-dimensional Navier-Stokes computations for complex configurations, the interested reader is referred to the survey article in Ref. 18.

Another area of interest associated with complex configuration CFD simulations is grid generation. Generating a single, well-behaved grid about a reasonably complete aircraft for an inviscid simulation is a difficult task.

For viscous flow simulations coupled with the desire to handle truly general configurations, the task takes on enormous proportions. This fact has probably been the single most important reason for slow development of three-dimensional CFD applications about complex geometries (see Chapman[19]). One methodology that improves this situation is the use of zonal-grid techniques. With this approach grid zones can be added or deleted as the configuration changes, coarse or fine as the flowfield dictates, or viscous or inviscid as required. In a sense, the grid can be adapted to the expected solution.

Recently, the use of zonal- or blocked-grid approaches has become quite popular. Benek et al.,[20] Hessenius and Pulliam,[21] Rai,[22] Hessenius and Rai,[23] and Dougherty et al.[24] have used different types of zonal-grid approaches for solving the Euler equations. Benek et al.,[25] Rai,[2-4] and Buning et al.[26] have used different types of zonal schemes for solving the Navier-Stokes equations. With these approaches, topologically complicated geometries have been solved with several grid zones, which by themselves are not complicated. Therein lies the power of the zonal-grid approach. Additional information on zonal-grid schemes can be found in the review articles by Steger and Benek[27] and Thompson and Steger.[28]

In this chapter, Navier-Stokes solutions taken from several publications[29-40] are presented for a series of three-dimensional geometries. These geometries include two different wings and an essentially complete F-16A aircraft. The computer code used for all these applications is called the transonic Navier-Stokes (TNS) code and utilizes a zonal-grid approach with the numbers of zones varing from 4 to 54. The Euler equations are solved in zones away from all no-slip surfaces, and the thin-layer Navier-Stokes equations are solved in all zones immediately adjacent to all no-slip surfaces. For zones existing in "corners," that is, with no-slip boundary conditions on two different surfaces, a thin-layer formulation along both directions is employed. Using this philosophy a zonal construction with the appropriate set of boundary conditions can be created for almost any application.

Governing Equations and Numerical Approach

The basic governing equations, numerical algorithm, and turbulence model used in the TNS computer code have been taken from the Pulliam-Steger ARC3D flow solver.[41,42] This code solves either the Euler or the Reynolds-averaged Navier-Stokes equations in strong conservation law form. The Reynolds-averaged Navier-Stokes equations are simplified by using the standard thin-layer approximation. Two numerical algorithms are generally supported by the ARC3D flow solver, including an ADI algorithm that solves block-tridiagonal matrices along each coordinate direction due to Beam and Warming[43] and a diagonalized ADI algorithm that solves scalar pentadiagonal matrices along each coordinate direction due to Pulliam and Chaussee.[44] Both schemes use second-order-accurate central differencing of the spatial terms with added artificial dissipation to stabilize the time-integration process. Because the diagonal scheme is more computationally efficient than the block scheme, all results presented herein have been computed with the diagonal scheme. The turbulence model used for the majority of computations presented herein is the Baldwin-Lomax[45] algebraic eddy viscosity model. This model is used because it produces reasonable accuracy for attached flow cases and it is easy to implement. Additional information about the algorithm including convergence characteristics can be found in Refs. 30 and 35.

Zonal-Grid Approach

The grid zoning process for the wing geometry (TNS-WING code) will now be discussed in some detail. Additional aspects of the full F-16A topology will be discussed in subsequent sections but in less detail. The grid-generation process for the wing topological class begins with the generation of a single-zone base grid that includes the entire flowfield. This grid contains no viscous clustering and has a H-mesh topology in both the spanwise and chordwise directions. It can be generated from either of two approaches: the elliptic solver approach of Sorenson and Steger[46] or the parabolic solver approach of Edwards.[47] Both of these approaches solve partial differential equations to generate well-behaved finite-difference meshes. They have the capability of generating suitable grids about isolated-wing geometries with either free-air or wind-tunnel wall outer boundaries.

Once the base grid is generated, a zoning algorithm is used to divide the grid into separate zones. The first grid zone (grid 1) is the base grid itself, with a small block of grid points near the wing removed. The second grid zone (grid 2) occupies the space left open by the block of points removed from grid 1 with a small region of overlap (usually one or two grid cells on all boundaries). Grid 2 is constructed so as to contain twice as many grid points in each spatial direction as the original base grid. This refinement of grid 2 relative to the base grid is accomplished via cubic spline interpolation. Similar to grid 1, grid 2 has a small block of points removed near the wing.

The final two grid zones (grids 3 and 4) occupy the space left open by the block of points removed from grid 2, again with a small region of overlap included. Grids 3 and 4 are constructed so as to contain the same number of points in both the spanwise and chordwise directions as grid 2. However, the grid points in the normal direction are highly clustered in order to capture viscous effects on the wing surfaces. Grid 3 is designed to capture the upper wing surface viscous effects and grid 4 the lower wing surface viscous effects. The two outer inviscid grid zones are topologically represented in the computational domain as cubes with smaller cubes removed from the middle. The third and fourth viscous zones are topologically represented as simple cubes in the computational domain. This grid topology can be more adequately explained by an example.

Figure 1 shows a typical grid with outer-boundary positions specified to coincide with the position of the wind-tunnel walls from the NASA Ames High Reynolds Number Channel I.[48] The grid is plotted in perspective so that detail on the upper and lower wind-tunnel wall surfaces, the inflow and outflow planes, and the wing-symmetry plane are all visible. This grid, which is generated directly by the parabolic grid generation approach, becomes the outer, coarse grid zone (grid 1). The grid detail near the wing/symmetry plane juncture has been removed. It is in this region that grid zones 2–4 are located. A blowup of the grid in the vicinity of the wing showing detail of grid zones 2–4 is shown in Fig. 2. The wing geometry used in this case is composed of NACA 0012 cross sections, has a taper ratio of 1.0, 20 deg of leading-edge sweep, an aspect ratio of 3.0, and is rigged in the wind-tunnel wall grid at 2-deg angle of attack. This wing does not have any twist or dihedral. Note that the grids immediately adjacent to the wing surface (grid zones 3 and 4) are highly clustered in the normal direction and therefore are appropriate for a Navier-Stokes flow solver. Also note that the Navier-Stokes grids expand in thickness from the leading edge toward the trailing edge to better capture the growing. boundary layer.

The total number of grid points used in the grid of Figs. 1 and 2 is 149,071. The individual grid point breakdown for each zone is as follows: grid 1, $65 \times 20 \times 19 = 24,700$; grid 2, $69 \times 29 \times 21 = 42,021$; grid 3, $61 \times 27 \times 25 = 41,175$; and grid 4, $61 \times 27 \times 25 = 41,175$. This grid has been constructed to fit the geometry used in the experiment of Lockman and Seegmiller.[48] Another grid with exactly the same characteristics for zones 2–4 but with free-air boundaries specified for grid zone 1 has also been generated. The free-air grid is constructed from the wind-tunnel wall grid by simply adding several grid surfaces to the top and bottom boundaries of the wind-tunnel wall grid and to the side-wall boundary out board of the wing tip. Therefore, the wind-tunnel wall grid is an exact subset of the free-air grid. The grid immediately surrounding the wing geometry, including the inflow and outflow boundary positions, is exactly the same in both the free-air and wind-tunnel wall versions. The number of points in the free-air grid (166,621) is slightly larger than that for the wind-tunnel wall grid. This is solely due to the fact that grid zone 1 is slightly larger ($65 \times 26 \times 25 = 42,250$).

Since the free-air and wind-tunnel grids are identical in the vicinity of the wing, computed flow for the same conditions on both grids must be free of grid-induced perturbations near the wing. Any differences between the free-air and wind-tunnel wall cases must be attributed to changes in the outer-zone grid and the use of different outer-boundary conditions. This represents a better way of simulating wind-tunnel wall interference than approaches that require the generation of two grids which can be drastically different in the vicinity of the wing, especially when the wind-tunnel walls are close to the model being tested. In a later section, numerically computed flowfields taken from these two grids will be presented and compared with experiment.

Communication between the blocks is achieved via an interpolation procedure. The grid zones are carefully (but automatically) constructed to overlap by a specific number of cells, usually one or two cells. Then, information required at the boundary of one zone is interpolated from the interior of another zone. Because the grid zones are carefully constructed from a base grid so that surfaces requiring interpolation are coincident, the interpolation is greatly simplified. The most complicated zonal interface boundary condition for the wing topology involves only a series of one-dimensional linear interpolations. (For more general geometries, such as wing fuselage configurations, a somewhat more general and therefore more complicated interpolation scheme based on trilinear elements is used.) The situation involving a series of one-dimensional interpolations is illustrated in Fig. 3. In this hypothetical case, grid zone 1 is a coarse grid that interfaces with grid zone 2, a fine grid (see Fig. 3a). The idea is to obtain flowfield information from the interior of zone 1 and use it to satisfy the proper conditions on the boundary of grid zone 2 (surface ABCD). Since the grids of zones 1 and 2 have both been constructed to coincide with surface ABCD, this process is greatly simplified. A blowup of a typical interface surface (either ABCD or EFGH) is shown in Fig. 3b. Note that the points from both grid zones are shown and that the grid points from grid zone 1 represents a subset of the grid zone 2 points. Thus, the transfer of information from grid zone 1 to grid zone 2 is accomplished by a series of one-dimensional interpolations on the $\zeta = \text{const}$ surface.

The process of interpolating information back from zone 2 to zone 1, which is required for the EFGD interface surface, is even simpler than that in the first case. This is because the interpolation becomes simple "injection." That is, even though interpolation is carried out, the end result of this operation is the straight transfer of values from zone 2 to zone 1 without interpolation errors. This is because every grid point on the zone 1 EFGH boundary is concident with a grid point on the zone 2 EFGH boundary. This is readily seen by looking at Fig. 3b and noting that every zone 1 point (circles) has an exact matching zone 2 point (plus signs).

The example in Fig. 3 represents only one type of interface situation where the transverse grid spacing at the interface is 2:1. Any ratio is possible, including ratios as high as 30 or 40:1, which might correspond to viscous grid interfacing with an inviscid grid (see Ref. 29 for an example of this type). Of course, for large ratios such as this, the accuracy obtained by the fine grid will be greatly degraded locally at the interface by the interaction with the coarse grid. Nevertheless, in some cases this kind of interface is desired. It is also appropriate to interface the surface of one zone with only a portion of another zone. Thus, several zones may be required to fully supply the interface boundary conditions of a single surface from one zone. Many different interface situations are possible, since this methodology has a great deal of flexibility. The present zonal procedure is generally quite straightforward even as the number of interfaces becomes large. With the four-grid-zone topology for isolated wings, the total number of interface surfaces that must be handled each iteration is 28. For the more complex wing/fuselage and wing/fuselage/inlet topologies the number of interface surfaces approaches 100. Thus, it is quite important to automate the process with a convenient bookkeeping scheme, which is described in the next section.

Data Management

Once the grid is generated and divided into the proper zones, the flow solver is initiated. For the isolated-wing case, the iteration procedure starts in the outer Euler block (grid 1), proceeds to the inner Euler block (grid 2), and ends with the two Navier-Stokes blocks, first the upper block (grid 3) and then the lower block (grid 4). Only one iteration using a spatially varying time step is completed in each grid zone before passing to the next. Other iteration or time-stepping strategies could be used to improve convergence. For example, for the wing/fuselage and wing/fuselage/inlet topologies, the iteration procedure starts at the fore part of the fuselage in the boundary layer, proceeds downstream along the fuselage in the various viscous grid zones, and eventually ends in the outer Euler grids. This latter strategy makes somewhat more sense because it starts at the body where the boundary conditions that drive the solution are implemented. However, little testing to determine optimal strategies has been completed.

Only the flowfield solution (Q arrays), transformation Jacobian (Ja), and the turbulence model arrays (when appropriate) associated with a single-block reside in main memory at a time (for the present discussion implementation on a Cray XMP computer will be assumed). The information associated with the other blocks resides in extended storage. On the Cray XMP this device is called solid-state device (SSD). The SSD is utilized functionally in the same manner as standard rotating disk extended storage. However, the SSD extended storage is physically composed of semiconductor memory and therefore is much faster. Use of the SSD instead of disk greatly reduces input/output (I/O) wait time, and, for jobs that are normally I/O bound, this is a significant advantage.

NACA 0012 Wing Results

The next topic of discussion is centered around code calibration. For this activity it is appropriate to utilize a set of less complicated cases involving isolated-wing flowfields. The geometry used in this section involves an analytically defined wing for which there is a significant amount of experimental data. The test results include subcritical and transonic flow cases, cases with significant wind-tunnel wall interference, and cases with significant shock-wave-induced boundary-layer separation. Using these cases, the effects of artificial dissipation, boundary conditions, turbulence model, and tip geometry on solution accuracy will be investigated. Most of the results from this section are taken from Refs. 29 and 39. Additional isolated-wing results associated with a highly swept, tapered, and twisted wing will be presented in the next section.

Subcritical Nonlifting Airfoil Case

The first issue of interest is to assess the general correctness of the TNS-WING code including the accuracy associated with the turbulence and artificial dissipation models. To study this, a simple subcritical nonlifting case involving an attached airfoil flow is utilized. The test case consists of two-dimensional flow around an NACA 0012 airfoil at $M_\infty = 0.5$, $Re = 2.89 \times 10^6$, and $\alpha = 0$ deg. The TNS-WING program was run for a very-large-aspect-ratio wing making the flow at the symmetry plane essentially two dimensional. The computed and experimental pressure coefficient distributions are compared in Fig. 4. Additional results are included from the TRIVIA[49] computer code, which utilizes a full potential inviscid solver coupled to a finite-difference boundary-layer technique. The experimental data are taken from Thibert et al.[50] The agreement for this admittedly easy case is excellent, at least in terms of the surface pressure.

In Fig. 5, the turbulent boundary-layer profiles computed by the TNS-WING code are compared with appropriate boundary-layer profiles from the TRIVIA code for the case presented in Fig. 4. TRIVIA utilizes the same turbulence model as TNS-WING in solving the boundary-layer equations and has proven to be accurate for a broad range of airfoil flow conditions. The y^+ values of the first grid point off the surface are between 1 and 2 for the TRIVIA result and between 2 and 4 for the TNS-WING result. As can be seen from Fig. 5, the agreement is excellent at all profile locations. In order to achieve this level of agreement, the artificial dissipation used in the TNS-WING code had to be multiplied by the local Mach number squared (M^2). This reduced the standard level of artificial dissipation in the boundary layer, especially near the wall, and allowed the proper determination of the skin friction. Additional testing demonstrated the need to modify the artificial dissipation multiplier function to just M in regions of reversed flow. More information on the modified artificial dissipation scheme and the rationale for its use can be found in Ref. 39.

Effect of Boundary Conditions

The purpose of this section is to study the effect of the various physical boundary conditions on the computed results. One desirable feature of the TNS-WING program, as was already discussed, is that wind-tunnel wall-interference effects can be directly modeled, at least for solid-wall situations. The H-H mesh topology employed in the TNS-WING code allows the outer computational surfaces to exactly match the test section of the wind tunnel. An advantage of this gridding topology is shown in Fig. 6, where pressures computed utilizing both free-air and solid-wall boundary conditions are compared with experimental data from Lockman and Seegmiller.[48] This case involves nonlifting supercritical flow about an NACA 0012 wing with 20 deg of leading-edge sweep, a taper ratio of 1.0, and an aspect ratio of 3.0. The Reynolds number based on chord is 8 million, and the freestream Mach number is 0.795. The wind-tunnel walls are solid (not slotted or porous) and are slightly less than two chords above and below the wing. This is the same geometry presented in Figs. 1 and 2 (except that $\alpha = 0$ deg for the present case). As seen from Fig. 6, the case with wind-tunnel walls modeled is in much better agreement with the experimental data. For this calculation, a weak shock exists across the wing and is captured quite successfully.

The inflow and outflow boundary conditions are also important for wind-tunnel simulations, because wind-tunnel flows are basically channel flows and are known to be sensitive to these conditions.[51] This aspect of the problem was investigated by Coakley[52] for two-dimensional airfoil flows, and quite encouraging results were obtained by fixing the outflow pressure distribution. In order to assess the effect of inflow-outflow boundary conditions, another supercritical flow with a large separation from the Lockman and Seegmiller experiment[48] was computed. The flow conditions for this case include $M_\infty = 0.836$, $Re = 8 \times 10^6$, and $\alpha = 2$ deg. The original set of inflow and outflow boundary conditions, which will be called BC-1, were the following: freestream conditions at the inflow boundary and zeroth-order space extrapolation for the five dependent variables (density; x, y, and z components of the momentum; and the total energy per unit volume) at the outflow boundary. The computed wind-tunnel wall pressure coefficients using these boundary conditions are presented in Fig. 7 by the dashed lines. As is evident, the computed results do not compare well with those of the experiment.

Furthermore, a careful study of the flowfield revealed that fixing the inflow Mach number to freestream does not guarantee a smooth variation of the Mach number near the inflow boundary. Instead, it was found that there was a jump in the Mach number between the inflow plane and the first computational plane downstream of inflow. Hence, the freestream conditions are not properly satisfied at inflow. At the outflow plane, another inconsistency was detected. It was found that the coefficient of pressure in the computation goes to zero, whereas the experiment indicates a value of about $C_{p,\text{exit}} = -0.05$.

In order to improve this situation, characteristic-type boundary conditions similar to Coakley[52] and Rai[3] are implemented (designated as BC-2). At the inflow plane, a locally one-dimensional flow assumption is made. The theory of characteristics indicates that for a subsonic inflow there are two ingoing and one outgoing characteristic waves. Associated with the ingoing characteristics, two parameters are fixed, the freestream entropy and the R^+ Riemann invariant, such that

$$s_\infty = \frac{p_1}{\rho_1^\gamma}$$

$$R_\infty^+ = u_1 + \frac{2a_1}{(\gamma - 1)}$$

where the subscript 1 denotes the inflow plane. Associated with the outgoing characteristic, the R^- Riemann variable is extrapolated from inside as follows:

$$R_\text{ext}^- = u_1 - \frac{2a_1}{(\gamma - 1)}$$

Also, for a one-dimensional inflow,

$$v_1 = 0$$

$$w_1 = 0$$

completes the set of five conditions required for five unknowns. At the subsonic outflow plane, there is only one incoming characteristic; hence, the coefficient of pressure C_{pe} is fixed. The velocity components are extrapolated from inside, and the adiabatic energy equation is used to complete the set of five required conditions. The new boundary conditions (BC-2) are a dramatic improvement over the first set of boundary conditions (BC-1), as shown in Fig. 7 (the solid line result).

Effect of Turbulence Model

In order to study turbulence modeling effects, the transonic separated-flow case that was introduced in the last section is used. The flow conditions for this case are $M_\infty = 0.836$, $Re = 8 \times 10^6$, and $\alpha = 2$ deg. This case has the strongest shock wave and largest region of separated flow out of all the cases tested by Lockman and Seegmiller.[48] The effect of two different inflow and outflow boundary condition schemes (BC-1 and BC-2) on wind-tunnel wall pressures was already presented in Fig. 7, but the flow topology aspects were not presented so that they could be dealt with in this section. Since there is an intimate relation between the strength and location of the shock wave and the surface oil flow, it is natural to expect some changes in the simulated oil flow patterns (i.e., calculated skin-friction lines), as a result of the new inflow/outflow boundary condition scheme.

The investigation begins by studying the wing upper-surface skin-friction line patterns produced from the physical flowfield. The oil-flow picture from Lockman and Seegmiller[48] for the $M_\infty = 0.836$ case is presented in Fig. 8, followed by the postulated skin-friction lines in Fig. 9. The oil-flow pattern basically shows a mushroom-type separation with two counter-rotating vortices as a result of the impinging shock wave. The reverse-flow region extends across the entire span and from the shock wave to an area very close to the trailing edge. This case shows some oil accumulation near the wing root, which is indicative of a strong shock wave hitting the wall. This shock wave causes the oil streaks in the vicinity of the wing root to deflect about 15 deg after the shock foot.

The skin-friction topology of the mushroom-type separation is shown in Fig. 9. There is a line of separation that emanates from a saddle point of separation. On each side of the saddle point, there are counter-rotating vortices. The vortex on the left (wing-root side) spirals inward, as is evident from the thick oil deposit. However, the sense of rotation (inward or outward) of the vortex on the right (near the tip) is in question. It is depicted as spiraling inward in Fig. 9, but it may also be spiraling out because of the presence of a dark spot at the center, that is, the absence of oil. Also, a nodal point of separation and a saddle point near the tip-side vortex could be postulated. Another saddle point near the trailing edge from which the reverse flow emanates is needed to complete a possible skin-friction map. However, the reader is cautioned that this postulation is by no means final, and another author could choose a different combination of critical points as long as it did not violate topological rules.[53] Additional discussion on this postulated flowfield can be found in Refs. 35 and 39.

The effect of different inflow-outflow boundary conditions on the skin-friction field of the previously mentioned flow is presented in Figs. 10 and 11. Figure 10 shows the simulated oil flow with the BC-1 boundary conditions, and Fig. 11 shows the same flow with the BC-2 boundary conditions. In comparing these figures it is evident that there is a significant difference in the size of the separated flow regions. The separation region in Fig. 10 is quite small relative to that of the experimental result shown in Fig. 8; the size of the separation region shown in Fig. 11 is a better approximation. As far as the size of the separated zone is concerned, this is a significant improvement. Both of these solutions predict a separation at the root section of the wing that is absent in the experiment. As mentioned, the experiment shows a strong shock wave with a large oil streak deflection at the root. In the numerical scheme, the wing-root section is treated inviscidly, and this is probably not a proper modeling. The wind-tunnel wall boundary layers near the wing root section should be included in the simulation to obtain proper flow details in the vicinity of the root. See Vatsa[10] for more information on this point.

Although there is a significant improvement in the size of separation, the two computed results (Figs. 10 and 11) are qualitatively, or topologically, the same (at least in the outboard portion of the wing). There is a nodal point of separation from which a line of separation emanates and a line of reattachment where the flow reattaches through a saddle point. This is not in agreement with the experimental oil-flow picture. In order to claim qualitative agreement, a focus/saddle-point/focus sequence must exist on the line of separation. In an attempt to create the proper type of critical point sequence, we now turn the investigation to turbulence modeling.

As already mentioned, the Baldwin-Lomax[45] turbulence model is used in the TNS-WING code. It is known that this model, which is an equilibrium model, predicts too sharp an increase in the outer-layer eddy viscosity through the separation shock.[54] The increase of eddy viscosity from its local equilibrium value far from the wall leads to more turbulent mixing and to more shear stress to balance the adverse pressure gradient, suppressing the tendency toward separation. Rotta[55] concluded from experimental data that, when turbulent flow is perturbed from its local equilibrium state, a distance of about one order of magnitude greater than the boundary-layer thickness is required to attain a new equilibrium state. To account for the upstream turbulence history effects, Shang and Hankey[56] used a relaxation eddy viscosity model:

$$\mu_T = \mu_{Teq} - (\mu_{Teq} - \mu_{T0}) \exp\left(-\frac{x - x_0}{\lambda}\right)$$

for two-dimensional applications. The parameter μ_T is the turbulent dynamic eddy viscosity, μ_{Teq} is the local equilibrium eddy viscosity evaluated from the standard Baldwin-Lomax model, μ_{T0} is the eddy viscosity at a location x_0 just upstream of separation, and λ is the relaxation length. In the work of Kaynak and Flores[39] the Shang-Hankey relaxation formula is applied to three-dimensional problems using a "strip" approach; i.e., the two-dimensional formula is applied at each streamwise computational cross section of the wing. A good review of turbulence model relaxation techniques can be found in Hung.[54] Conceptually, the Shang-Hankey relaxation formula approximates the experimental observation that, in an abrupt disturbance of a turbulent flow, the Reynolds stress first remains nearly frozen at its initial value while it is being convected along streamlines and then exponentially approaches a new equilibrium state. In a numerical calculation the initial location of the disturbance from which the relaxation process is initiated, x_0, and a relaxation length scale, which describes the exponential decay of the eddy-viscosity distribution, λ, must be specified. There are two limiting cases that bound the relaxation length. For $\lambda = 0$ the turbulent eddy viscosity equals the local equilibrium value, and for $\lambda = \infty$ the initial value μ_{T0} is frozen and is used everywhere in the region $x > x_0$.

The relaxation turbulence model has been applied to the separated-flow case in the previous comparison. Different λ values were tried, and $\lambda = 40\,\delta_0$ was found to give the best agreement with the experimental oil-flow map. Here, δ_0 is the thickness of the boundary layer at x_0. In general, there is not an established rule for choosing the value of λ. It is a trial-and-error process that is undertaken with guidance from the experiment. The resulting computed skin-friction field using the relaxation model for the case discussed in Figs. 10 and 11 is presented in Fig. 12.

The remarkable feature of Fig. 12 is that it shows a topological change from a one node-separation configuration as shown in Figs. 10 and 11 to a focus/saddle-point/node sequence. The predicted inboard vortex (focus), a saddle point, and a node on the line of separation are topologically the same as the experimental pattern, because a node and a focus fall into the same category.[35] Besides the focus/saddle-point/node sequence, the saddle point of reattachment is also predicted. This simulation is a very important step toward predicting a mushroom-type separation shown in Fig. 8. The next item to be investigated is the effect of wing-tip geometry definition.

Effect of Geometry Representation

In the previous two sections, and especially in the last one, important steps were taken to reproduce the mushroom-type separation, and the relative importance of each issue has been pointed out. However, one of the counter-rotating vortices (the one near the tip) is still missing. The problem is not due to grid resolution, since a 1.1×10^6 mesh point solution presented in Ref. 39 showed no significant differences to the moderate grid calculations presented herein. Recent tip-vortex studies by Srinivasan et al.[5] have domonstrated that the shape of the wing-tip cap is extremely important in forming the tip vortex, as well as the tip-flow region. The shape of the tip in the Lockman and Seegmiller[48] experiment is a revolution of the NACA 0012 profile. Contrary to this, the tip geometry in the preceding computations is wedge-shaped.

In an attempt to improve the tip geometric modeling, a new grid was generated using an advanced version of the PAR3W parabolic grid generator,[47] which can generate wing grids with round tips. The resulting computation is presented in Fig. 13. This configuration finally fulfills expectations and captures the missing vortex near the tip, thereby predicting a mushroom-type separation zone. In Fig. 13 there is a line of separation emanating from a saddle point with two counter-rotating vortices on each side. Strictly speaking, the vortex on the right (near the tip) is a limit cycle. The particles inside the cycle spiral outward, whereas the ones from outside spiral inward, and all of these finally collapse on the limit cycle. This is structurally, as well as mathematically, a stable configuration.[57] As remembered, the sense of rotation of this vortex was questionable in the experiment because of the dark spot in the oil-flow picture (Fig. 8). Hence, this limit cycle may be an accurate representation of the experimental phenomenon.

Strong Shock Case

The last solution presented in this section consists of a free-air case with massive, shock-induced boundary-layer separation. This case was computed to ascertain the degree of robustness of the present algorithm and, in particular, the ability of the present zonal interface scheme to cope with large flow gradients. The geometry used is the same as that of the last two cases. The freestream Mach number and angle of attack have been arbitarily chosen to be 0.9 and 5 deg, respectively. Utilization of the wind-tunnel wall boundaries, as described in Figs. 1 and 2, produced a "choked" solution with a shock wave spanning the tunnel. After several hundred iterations and a moderately converged calculation, the solution (as expected) diverged. This, of course, was a consequence of the "fixed" upstream boundary conditions forcing more mass flow through the tunnel than the choked condition would allow.

The solution was repeated with free-air boundary conditions, and then convergence was easily achieved. Computed particle paths on the upper wing surface are displayed in Fig. 14. Note that approximately half of the upper wing surface is separated. The solution contains several interesting features including a separation saddle point, a focus, a reattachment saddle point, and a node. Note also that this computed solution has a stable sequence of critical points on the separation line, that is, a node followed by a saddle point and then a focus (see Ref. 53).

Two different perspective views of the three-dimensional particle paths are shown in Fig. 15. Figure 15a shows a view from outboard of the wing tip, and Fig. 15b shows a view from behind and above the wing. The dashed and solid lines are particle paths created by "releasing" particles just above the wing surface at $x/c = 0.05$ and 0.40, respectively. The height and three dimensionality of the separation zone are apparent in these figures. The dashed particle paths move along the wing surface until the separation line is encountered and then are deflected up over the separation bubble with a few of the dashed paths captured by the primary swirling flow at the center of the wing. The solid particle paths are more intimately involved with the two swirling pockets of flow and essentially serve to define these regions.

The position of the separation region relative to the zonal interface boundary can be viewed by plotting particle paths constrained to lie in spanwise cross-sectional planes. Two such plots are displayed in Fig. 16. Figure 16a shows cross-sectional particle paths for a semispan station of $2y/b = 0.66$, and Fig. 16b shows a blowup of the recirculating region at the same location. This station is close to the separation saddle point, as shown in Fig. 14. Inboard of about $2y/b = 0.20$ the separation region is small and does not approach the zonal boundary. However, at $2y/b = 0.66$ the separation region is large and easily extends above the zonal boundary from the Navier-Stokes region into the Euler region. Nevertheless, the solution looks qualitatively reasonable. From this figure it can be seen that the particle paths pass smoothly across the interface boundary with no function or slope discontinuities. Thus, the primary objective of doing this calculation was achieved. Despite the existence of a strong shear gradient across the explicitly updated interface boundary, the present approach is capable of predicting a stable solution that is reasonably free from interface boundary influence.

Wing C Results

In this section additional code calibration results are presented for a more sophisticated wing geometry involving camber, twist, spanwise section variation, taper, and a large amount of leading-edge sweep—the so-called wing C configuration. This will allow a more accurate assessment of the TNS-WING code for realistic geometric variations. As in the previous series of cases an extensive amount of experimental data are available for comparisons. These data have been taken from two different experiments run in two different wind tunnels and thus will allow not only computational/experimental comparisons, but also experimental/experimental comparisons. To begin, a discussion of the wing geometry and the two wing C experiments is presented.

Wing C Design and Testing

The planform view of wing C including the TNS-WING surface grid is displayed in Fig. 17. This wing has an aspect ratio of 2.6, a twist of 8.17 deg, a taper ratio of 0.3, and a leading-edge sweep of 45 deg. It is composed of supercritical airfoils with relatively thick sections and exhibits (at the design condition) moderate aft loading, mild shock waves, and a mild pressure recovery. The design condition selected was a Mach number of 0.85 with a lift coefficient of approximately 0.5 and an angle of attack of 5 deg. The mild shock wave chacteristic was obtained by limiting the leading-edge normal Mach numbers in the vicinity of the leading edge to a maximum of 1.2. This value is commonly accepted as the upper limit in preventing shock-induced flow separation on wings. More information on this geometry can be found in Refs. 58 and 59.

A small-scale, 0.261-m semispan model of wing C was tested in the Lockheed-Georgia wind tunnel (CFWT).[58] The test Reynolds number based on the mean aerodynamic chord was 10 million. Surface pressures were measured both on the wing and on the wind-tunnel walls. These data are presented in Ref. 58, and a comparison of these measurements with several three-dimensional transonic inviscid potential codes is presented in Refs. 58 and 60. The top and side walls of the three-dimensional test section have a variable porosity capacity of up to 10%. Most of the testing was conducted at a fixed-wall porosity of 4% to minimize wall interference effects. The influence of the wind-tunnel walls on the test data was explored computationally in Ref. 58. It was found that a Mach number correction of $\Delta M = -0.005$ and an angle-of-attack correction of $\Delta \alpha = 0.9$ deg were necessary to provide proper experimental/computational correlations at wing design conditions. Transition strips were located at a fixed distance from the leading edge equal to 5% of the mean aerodynamic chord on both the upper and lower surfaces of the wing.

A series of tests with a large-scale wing C geometry (0.90-m semispan) was performed in the NASA Ames Research Center 6-ft supersonic wind tunnel.[59] The model blockage ratio in the test section for this model was 1.3% at 0 angle of attack. Surface pressure measurements and oil-flow studies were made at the design angle of attack of 5 deg over a Mach number range of 0.25–0.96 and a Reynolds number range of $3.4–10 \times 10^6$. No Reynolds number effect on the results was reported. The lift interference from the wind-tunnel walls was reported to be small. This is because the leading-edge pressures of the experiment and computations in the correlations happened to agree with each other at the design angle of attack of 5 deg. Transition strips were installed at 4.5% of chord for this series of tests. Much of the discussion that follows in the next several sections is taken from Ref. 32.

Attached-Flow Cases

The first wing C flow computations using the TNS-WING code are attached-flow cases at or below the design Mach number and lift coefficient. Attached-flow calculations were completed for two supercritical cases: $M_\infty = 0.70$ and 0.82 at $\alpha = 5$ deg and $Re = 6.8 \times 10^6$. In all of these and subsequent wing C computations, the Reynolds number is based on the mean aerodynamic chord, and the turbulent calculations are started at the wing leading edge without any transition model. General features of these two cases are similar as far as the surface pressure and skin-friction fields are concerned. Therefore, only the latter case with $M_\infty = 0.82$, $\alpha = 5$ deg, and $Re = 6.8 \times 10^6$ will be presented.

The calculated pressure coefficients are compared with data from the small-scale and large-scale model tests in Fig. 18. Results at five different span stations ranging from 10% of semispan to 90% are presented. The comparison is very good in general, with the TNS-WING solution in slightly better agreement with the small-scale model data.[58] The discrepancy between the two sets of data is probably caused by differences in wall-interference effects, Reynolds number effects, or differences associated with the measurement of freestream Mach number and angle of attack. Overall the differences between computation and experiment are smaller than the differences between the two experimental results.

Separated-Flow Cases

The flow conditions for the first separated-flow case correspond to the design conditions: $M_\infty = 0.85$, $\alpha = 5$ deg, and $Re = 6.8 \times 10^6$. These conditions were intended to result in attached flow with a mild shock wave and a mild pressure recovery. But, for reasons discussed in Ref. 59, these conditions produced a small or "local" (as it is called by the author) flow separation.

The calculated pressure coefficient distribution for the wing C design case is compared with experimental data in Fig. 19. The angle of attack is 5 deg for the Keener data[59] and 5.9 deg for the Hinson and Burdges data.[58] As already discussed, this is in accordance with Hinson and Burdges' accounting for wind-tunnel wall effects. To check the influence of angle of attack on the solution, computations were performed at two angles of attack (5 and 5.9 deg). The computation at $\alpha = 5$ deg is generally in better agreement with the two experimental results, especially on the upper surface at the wing leading edge. Note that the differences between the $\alpha = 5$-deg computation and experiment and between the two experiments are generally on the same order. The experimental differences are probably associated with the uncertainties in the measurements of Mach number and angle of attack and with the assessment of the wall-interference effects associated with the experiments. The numerical simulation has difficulties associated with proper grid resolution at the wing tip and an inadequate turbulence model.

Figure 20 shows the wing-planform Mach number contours plotted at a location above the upper wing surface boundary layer for the case just discussed. The well-known lambda shock-wave pattern for swept wings is clearly displayed. The peak Mach number is just above 1.65 and exists near the wing leading edge at about mid-semispan. Overall, this pattern agrees well with the experimental pattern for this case given in Ref. 59.

The second separated case associated with wing C consists of a flow at off-design conditions: $M_\infty = 0.90$, $\alpha = 5$ deg and $Re = 6.8 \times 10^6$. The computed and experimental pressure coefficient distributions are compared in Fig. 21. The leading-edge and aft shock waves are stronger than those of the previous case, and the aft shock wave is further downstream. This is a consequence of the increased freestream Mach number. Despite the discrepancy in the shock location at the tip, the overall agreement between the experiment and computation for this difficult case is good. The disagreement at the tip is probably caused by coarse grid effects, i.e., lack of proper tip resolution, and separated flow turbulence model inadaquacies.

The strength and position of the transonic lambda shock wave is accurately predicted on the wing upper surface everywhere except at the tip where the shock position is in error by about 15%. The planform view of wing C showing this lambda-shaped shock wave is presented in Fig. 22. Note that the maximum Mach number on the upper surface is approximately 1.6 and is located near the leading edge at about mid-semispan. The Mach number contours in the spanwise cross-sectional plane at $2y/b = 0.77$ are shown in Fig. 23. The zonal boundaries are also shown in this plot. Note the smoothness with which the shock wave crosses the zonal interface boundary. This indicates that the communication between the zones is implemented in an acceptable manner. In addition, most of the other contours cross the zonal interface boundaries in a smooth, continuous fashion. Downstream in the wake where the fine grid that is used to capture viscous effects interfaces with a relatively coarse inviscid grid, the wake abruptly stops. This aspect of the solution exists because the coarse grid cannot retain the sharp wake velocity gradients, and, under the present grid topology constraints, this should be expected.

F-16A Results

In previous sections a number of relatively simple code calibration cases were presented and demonstrated a generally accurate aerodynamic prediction capability for wings. In this section, computations about a reasonably complete F-16A aircraft are presented, thus demonstrating the ability of the present approach to provide realistic simulations about full-scale aircraft. The grid-generation, grid zoning, flow-solver, and data management algorithms are all essentially the same as previously described, and, except for a brief description of the F-16A grid zone topology, they will not be discussed further. Additional details can be found in Flores and Chaderjian,[40] from which most of the material from this section is taken.

Geometry and Grids

Top and bottom perspective views of the F-16A geometry, as modeled in the present study, are shown in Figs. 24 and 25, respectively. In addition, geometric detail around the inlet is shown in Fig. 26. The complexity of this geometry can be clearly seen from these three figures. In particular, the forebody, canopy, leading-edge strake, wing, aft shelf region, horizontal and vertical tail surfaces, fuselage-mounted inlet, and inlet diverter regions are all true representations of the original F-16A geometry and are included in the present numerical simulation. Achieving appropriate clustering at all solid boundaries for this complicated geometry is extremely difficult with a single grid; however, it is a manageable task when zonal-grid methods are used. The first step in obtaining a zoned grid for the TNS code for the F-16A configuration is the generation of a base inviscid grid about the entire geometry. The base grid is generated by means of an elliptic-solver grid-generation method.[46] Plate 17 (see the color section) shows the general grid topology about the wing and fuselage. A polar-grid topology with the singular axis along the stream direction is used to efficiently distribute points on the fuselage. This necessitates an H-grid topology about the wing in both the spanwise and chordwise directions, exactly as utilized with the TNS-WING code. The polar grid collapses into a singular line forward of the fuselage nose, but causes no problem in the flow solver because of a special averaging boundary condition about this singularity.

Next, the base grid is fed into the zonal routine, which subsequently creates 27 zones by subdividing and enriching the original base grid. (For sideslip cases the resulting zonal grid contains a total of 54 zones.) This process is controlled by various parameters that exist in the zonal routine. The zones created are essentially of three types: 1) inviscid zones, 2) viscous zones with clustering on one face corresponding to wing or fuselage no-slip surfaces, and 3) viscous zones with clustering on two adjacent faces corresponding to the wing-fuselage juncture. Figures 27–29 and Plate 18 (see the color section) show various views of the zonal boundaries used in the F-16A zonal grid. Figure 27 shows a cross-sectional cut through the wing/fuselage combination. In this view the upper and lower viscous wing zones, the upper and lower fuselage viscous zones, the wing/fuselage juncture viscous zones, the viscous wing-tip zone, and the outer inviscid zones are all displayed. Figure 28 shows a cut through the wing at an approximately constant spanwise station. Note the similarity between this zonal scheme and the previous zonal topology for the TNS-WING code. In this F-16A application a viscous wake zone has been added that was absent in the wing cases. Plate 18 (see the color section) and Fig. 29 show detail about the grid zoning topology used to include the lower fuselage-mounted inlet. This topology consists of a total of three zones that are all viscous in nature. The first zone is triangular or wedge-shaped and covers the region immediately upstream of the inlet between the inlet face and the lower part of the fuselage. The second and third zones occupy the region between the lower fuselage and the upper inlet beside the inlet diverter and are easily recognized in Fig. 29.

The total number of grid points used in the F-16A simulation is approximately 522,000 for cases without sideslip, i.e., with a symmetry plane assumption, and approximately 1,044,000 points for cases with sideslip. The inviscid zones away from the geometry have as few as 5000 grid points, and the viscous zones near the body contain as many as 36,000 grid points. For typical transonic flow conditions involving a Reynolds number of 4.5 million based on the wing root chord, the viscous clustering used in this grid yields an average y^+ value for the first grid point off the wing surface of 3 and off the the rest of the aircraft of 6. Based on previous experience this level of refinement should provide good resolution of the boundary layer and for the wall skin friction.

Computed Results

The first computational result presented for the F-16A geometry is for $M_\infty = 0.9$, $\alpha = 6.0$ deg, and a Reynolds number based on root chord of 4.5 million. Figure 30 shows a comparison of computed surface pressure distributions with experiment[61] for this case at several different locations on the aircraft. In particular, comparisons are included for the main wing, the horizontal and vertical tails, the fuselage-inlet juncture, and the upper-surface symmetry plane. The quantity η denotes a local semispan position parameter where $\eta = 0.0$ corresponds to a root position and $\eta = 1.0$ corresponds to a tip position. The quantity ξ denotes a normalized length parameter aligned with the streamwise direction and is equal to 0.0 at the inlet face and 1.0 at the end of the inlet (when the inlet diverter has fully merged into the fuselage). For this calculation a Mach number correction of 0.02 was used to improve agreement in the main wing aft shock location. Use of this correction moved the aft shock location aft on the wing upper surface by about 10% of chord without significantly affecting the other surface pressures. Justification for this correction is presumably based on correction of wind-tunnel wall interference; however, there is insufficient information on the level of wall interference in Ref. 61 to fully assess this situation. The main wing computed pressures, which are shown in Figs 30a and 30b, are in good agreement with the experimental values. The upper-surface double shock pattern is clearly evident at both semispan stations. Overall the solution could be improved by using a finer grid, primarily at the leading edge, the wing tip, and the shock wave.

Figures 30c and 30d display C_p comparisons between the computed solution and experiment for two fuselage cross-sectional planes near the inlet region. Figure 30c shows a comparison at a streamwise location of $\xi = 0.67$, and Fig. 30d shows a comparison at $\xi = 0.37$. The dashed line is the local fuselage definition and shows the inlet, diverter, canopy, and leading-edge strake. Generally the computed pressure results are in good agreement with experiment. The sharp jump in pressure, which is exhibited in both the experimental and computational results, is caused by the expansion of flow around the strake leading edge.

The computed surface pressures on the horizonal and vertical tail surfaces are compared with experiment in Figs. 30e–30g. Generally, the comparisons for the vertical tail are good, but do not show a significant amount of loading as this set of results is for zero sideslip. The computed surface pressures for the horizontal tail show the proper trends but are in only fair agreement with experiment as too much suction is predicted over most of this lifting surface. Finally, the computed pressure distribution along the fuselage upper-surface centerline shown in Fig. 30h is in generally good agreement with the experimental pressure distribution. The expansion and recompression regions caused by the canopy and tail are somewhat underpredicted by the computation, which is probably a coarse grid effect. The pressure distribution downstream of the canopy on the fuselage immediately adjacent to the wing is in good agreement with experiment, including the compression caused by the wing aft shock at about 70% of the fuselage length.

A global picture of the F-16A upper-surface pressure field for the case just presented is displayed in Plate 19 (see the color section). This pressure map has been developed from the nine viscous grid zones that make up or define the aircraft upper surface. Note the compression-expansion-compression flow over the canopy and the lambda shock pattern on the main wing. The upper surface of the horizontal tail generally displays a higher pressure than the main wing upper surface, which is due to the downwash from the main wing effectively reducing the angle of attack at the tail leading edge. Despite the large number of grid zones that have been used to produce this figure, and therefore the large number of zonal-grid interfaces, the contour lines smoothly map out the F-16A upper-surface pressure field without function or slope discontinuities.

Plate 20 (see the color section) displays a simulated oil-flow pattern on the F-16A with emphasis on the upper inlet surface. Particle traces are released at every other grid point location in the streamwise and spanwise directions beginning at the inlet lip and diverter locations. These traces are restricted to the computational plane one grid point off the geometry surface. It can be seen from these traces that there exists a strong spanwise component of flow in the boundary layer. Particles released near the lip and diverted plate proceed in a spanwise direction until the particle paths are influenced by the outer flow and then abruptly are twisted and entrained by the streamwise flow. It should be pointed out that the discontinuity in the particle traces at the diverter/outer grid juncture is due to the plotting routine. The lower surface of the lower diverter grid zone corresponds to an i-j plane in computational space. The inlet surface of the outer grid zone corresponds to a j-k plane in computational space. The plotting routine cannot maintain continuity of particle traces across two different types of computational planes. A separation line appears near the lip of the inlet. Boundary-layer profiles of the u component of velocity in this region do in fact indicate reverse flow.

Plate 21 (see the color section) displays three-dimensional particle traces (i.e., they are not confined to any computational surface) that are released around the upper-inlet separation region. These particles are released at several streamwise stations beginning with the inlet lip. The resulting particle traces originating near the diverter region proceed downstream with a small spanwise component of velocity. The particle traces originating at the inlet lip go up and over the separation bubble. Additional particle traces originating in the separation region itself serve to better define the extent of the separation. It appears that the particles closer to the diverter and at higher elevations proceed in a spanwise direction and under those particles released farther outboard. Eventually, all particles are entrained in the outer fluid at about the same elevation and are swept downstream.

The computation just presented in Fig. 30 and Plates 19–21 required about 5000 iterations and about 25 h of CPU time on a single processor of the Cray XMP computer. During convergence the L_2 norm of the residual was reduced by approximately three orders of magnitude in all grid zones. The viscous grid zones tended to converge somewhat faster than the inviscid zones. The wing-fuselage junction region, which has viscous clustering in two directions, had to be run at a lower time step for stability reasons. When standard time steps were used in the wing-fuselage junction zone, stability problems were encountered at the interface between this zone and the zone directly above it.

Additional variations on the basic F-16A geometry are generally easy to analyze with the present zonal-grid approach. For example, to model a power-on condition, two additional grid zones were added downstream of the nozzle exit, one to model the upper plume and the second to model the lower plume. Previously, the plume had been modeled, as in the wind-tunnel simulation, with a solid sting that extended from the nozzle exit to an effective downstream infinity. Power-on computations were carried out by specifying boundary conditions at the nozzle exit plane. A nozzle exit to freestream static pressure ratio of 2, a freestream to exit temperature ratio of 5, and an exit Mach number of 1 were arbitrarily chosen for this purpose. Plate 22 (see the color section) shows static temperature contours from the nozzle symmetry plane just downstream of the nozzle exit. The temperature gradients at the hot-plume/freestream interface are clearly evident in this relatively easy plume computation. Additonal geometric complications using this zonal approach have also been studied, including the simulation of flow inside the F-16A inlet up to the compressor face. Details can be found in Flores and Chaderjian.[40]

In addition, an F-16A computation with 5 deg of sideslip has also been computed with the present approach. As previously mentioned, this involved a reflection of the zonal topology so as to fully define both sides of the aircraft and involved a total of 54 zones and over 1 million grid points. Plate 23 (see the color section) shows unrestricted particle traces that have been released at a fuselage cross section just downstream of the nose. The particles are forced around the canopy area and then travel aft on the aircraft to interact with the vertical tail. Most of the windward particle traces cross the fuselage centerline and actually pass the vertical tail on the leeward side.

Conclusions

Transonic flowfields about two wings and a reasonably complete aircraft have been computed using an Euler/Navier-Stokes formulation. In the present approach the flow domain is divided into several zones or blocks, which offers several distinct advantages over nonzonal techniques. The flowfields can be broken into grid zones that tend to adapt to the flowfields in a beneficial way. That is, regions of the flowfields with significant gradients can be resolved with dense grid zones, whereas regions in which the flow has small gradients can be resolved with coarse grid zones. In addition, the use of a zonal technique makes the problem of computing flowfields about complex geometries more tenable. If a new geometric feature is added or removed, grid zones can be added or removed to accommodate the change without completely redoing the basic grid topology. Finally, the use of a grid zonal technique, as utilized in the present study, represents an effective data management scheme. This is especially effective for large application problems implemented on computers with relatively small main memories. One grid zone is processed in main memory while the other grid zones remain on extended storage.

The present zonal-grid approach has been used to compute transonic flow solutions about various geometries with significant shock/boundary-layer interaction involving moderate to massive flow separation. Many numerical aspects associated with these flows have been investigated, including the effect of artifical dissipation, boundary conditions, grid refinement, turbulence modeling, and geometric modeling. The artificial dissipation had little influence on the pressure distribution but a large effect on the structure of the lower boundary-layer profile. The inflow and outflow boundary conditions, turbulence modeling, and geometric modeling had large effects on the topological aspects of the separated flowfield.

References

[1]Shang, J. S., and Scherr, S. J., "Navier-Stokes Solution of the Flow Field Around a Complete Aircraft," AIAA Paper 85-1509, July 1985.

[2]Rai, M. M., "Navier-Stokes Simulations of Rotor/Stator Interaction Using Patched and Overlaid Grids," *Journal of Propulsion and Power*, Vol. 3, No. 5, Sept.–Oct. 1987, pp. 387–396.

[3]Rai, M. M., "Unsteady Three-Dimensional Navier-Stokes Simulations of Turbine Rotor-Stator Interaction," AIAA Paper 87-2058, June 1987.

[4]Rai, M. M., and Madavan, N. K., "Multi-Airfoil Navier-Stokes Simulations of Turbine Rotor-Stator Interaction," AIAA Paper 88-0361, Jan. 1988.

[5]Srinivasan, G. R., McCroskey, W. J., Baeder, J. D., and Edwards, T. A., "Numerical Simulations of Tip Vortices of Wings in Subsonic and Transonic Flows," *AIAA Journal*, Vol. 26, Oct. 1988, pp. 1153–1162.

[6]Srinivasan, G. R., and McCroskey, W. J., "Navier-Stokes Calculations of Hovering Rotor Flowfields," *Journal of Aircraft*, Vol. 25, Oct. 1988, pp. 865–874.

[7]Fujii, K., and Obayashi, S., "Navier-Stokes Simulation of Transonic Flow over Wing-Fuselage Combinations," AIAA Paper 86-1831, June 1986.

[8]Obayashi, S., Fujii, K., and Takanashi, S., "Toward the Navier-Stokes Analysis of Transport Aircraft Configurations," AIAA Paper 87-0428, Jan. 1987.

[9]Van Dalsem, W. R., "Study of Jet in Ground Effect with Crossflow Using the Fortified Navier-Stokes Scheme," AIAA Paper 87-2279-CP, Aug. 1987.

[10]Vatsa, V. N., "Accurate Solutions for Transonic Viscous Flows over Finite Wings," AIAA Paper 86-1052, May 1986.

[11]Ying, S. X., Steger, J. L., Schiff, L., and Baganoff, D., "Numerical Simulations of Unsteady, Viscous, High-Angle-of-Attack Flows Using a Partially Flux-Split Algorithm," AIAA Paper 86-2179, Aug. 1986.

[12]Schiff L. B., Degani, D., and Gavali, S., "Numerical Simulation of Vortex Unsteadiness on Slender Bodies of Revolution at Large Incidence," AIAA Paper 89-0195, Jan. 1989.

[13]Degani, D., and Schiff L. B., "Numerical Simulation of the Effect of Spatial Distrubances on Vortex Asymmetry," AIAA Paper 89-0340, Jan. 1989.

[14]Thomas, J. L., Taylor, S. L., and Anderson, W. K., "Navier-Stokes Computations of Vortical Flows over Low Aspect Ratio Wings," AIAA Paper 87-0207, Jan 1987.

[15]Fujii, K., and Schiff, L. B., "Numerical Simulation of Vortical Flows Over a Strake-Delta Wing," AIAA Paper 87-1229, June 1987.

[16]Fujii, K., Gavali, S., and Holst, T., "Evaluation of Navier-Stokes and Euler Solutions for Leading-Edge Separation Vorticies," NASA TM-89458, June 1987.

[17]Newsome, R. W. and Adams, M. S., "Numerical Simulations of Vortical-Flow Over an Elliptical-Body Missile at High Angles of Attack," AIAA Paper 86-0559, Jan 1986.

[18]Holst, T. L., "Numercial Solution of the Navier-Stokes Equations about Three-Dimensional Configurations—A Survey," "Supercomputing in Aerospace," NASA CP-2454, March 1987, pp. 281–298.

[19]Chapman, D. R., "Computational Aerodynamics: Review and Outlook," *AIAA Journal*, Vol. 17, Dec. 1979, pp. 1293–1313.

[20]Benek, J. A., Steger, J. L., and Dougherty, F. C., "A Flexible Grid Embedding Technique with Application to the Euler Equations," AIAA Paper 83-1944, July 1983.

[21]Hessenius, K. A., and Pulliam, T. H., "A Zonal Approach to Solution of the Euler Equations," AIAA Paper 82-0969, June 1982.

[22]Rai, M. M., "A Conservative Treatment of Zonal Boundaries for Euler Equation Calculations," AIAA Paper 84-0164, Jan. 1984.

[23]Hessenius K. A., and Rai, M. M., "Applications of a Conservative Zonal Scheme to Transient and Geometrically Complex Problems," AIAA Paper 84-1532, June 1984.

[24]Dougherty, F. C., Benek, J. A., and Steger, J. L., "On Application of Chimera Grid Schemes to Store Separation," NASA TM-88193, Oct. 1985.

[25]Benek, J. A., Donegan, T. L., and Suhs, N. E., "Extended Chimera Grid Embedding Scheme with Application to Viscous Flows," AIAA Paper 87-1126-CP, June 1987.

[26]Buning, P. G., Chiu, I. T., Obayashi, S., Rizk, Y. M., and Steger, J. L., "Numerical Simulation of the Integrated Space Shuttle Vehicle in Ascent," AIAA Paper 88-4359-CP, Aug. 1988.

[27]Steger, J. L., and Benek, J. A., "On the Use of Composite Grid Schemes in Computational Aerodynamics," *Computer Methods in Applied Mechanics and Engineering*, Vol. 64, Nos. 1–3, Oct. 1987, pp. 301–320.

[28]Thompson, J. F., and Steger, J. L., "Three Dimensional Grid Generation for Complex Configurations—Recent Progress," AGARDograph 309, 1988.

[29]Holst, T. L., Kaynak, Ü., Gundy, K. L., Thomas, S. D., Flores, J., and Chaderjian, N., "Numerical Solution of Transonic Wing Flows Using an Euler/ Navier-Stokes Zonal Approach," AIAA Paper 85-1640, July 1985; also *Journal of Aircraft*, Vol. 24, Jan. 1987, p. 17.

[30]Flores, J., "Convergence Acceleration for a Three-Dimensional Euler/Navier-Stokes Zonal Approach," AIAA Paper 85-1495, July 1985.

[31]Holst, T. L., Thomas, S. D., Kaynak, Ü., Gundy, K. L., Flores, J., and Chaderjian, N. M., "Computational Aspects of Zonal Algorithms for Solving the Compressible Navier-Stokes Equations in Three Dimensions," International Symposium on Computational Fluid Dynamics, Tokyo, Sept. 1985.

[32]Kaynak, Ü., Holst, T. L., Cantwell, B. J., and Sorenson, R. L., "Numerical Simulation of Transonic Separated Flows Over Low Aspect Ratio Wings," AIAA Paper 86-0508, Jan. 1986; also *Journal of Aircraft*, Vol 24, Aug. 1987, pp. 531–539.

[33]Flores, J., Holst, T. L., Kaynak, Ü., Gundy, K., and Thomas, S. D., "Transonic Navier-Stokes Wing Solution Using a Zonal Approach: Part 1. Solution Methodology and Code Validation," AGARD Paper 30A, April 1986.

[34]Chaderjian, N. M., "Transonic Navier-Stokes Wing Solutions Using a Zonal Approach: Part 2. High Angle-of-Attack Simulation," AGARD Paper 30B, April 1986.

[35]Kaynak, Ü., Holst, T. L., and Cantwell, B. J., "Computation of Transonic Separated Wing Flows Using an Euler/Navier-Stokes Zonal Approach," NASA TM-88311, July 1986.

[36]Flores, J., Reznick, S. G., Holst, T. L., and Gundy, K. L., "Transonic Navier-Stokes Solutions for a Fighter-Like Configuration," AIAA Paper 87-0032, Jan. 1987.

[37]Reznick, S. G., and Flores, J., "Strake-Generated Vortex Interactions for a Fighter-Like Configuration," AIAA Paper 87-0589, Jan. 1987.

[38]Flores, J., Chaderjian, N. M., and Sorenson, R. L., "Simulation of Transonic Viscous Flow over a Fighter-Like Configuration Including Inlet," AIAA Paper 87-1199, June 1987.

[39]Kaynak, Ü., and Flores, J., "Advances in the Computation of Transonic Separated Flows over Finite Wings," AIAA Paper 87-1195, June 1987.

[40]Flores, J. and Chaderjian, N. M., "The Numerical Simulation of Transonic Separated Flow About the Complete F-16A," AIAA Paper 88-2506, June 1988.

[41]Pulliam, T. H., "Euler and Thin-Layer Navier-Stokes Codes: ARC2D, and ARC3D," Notes for the Computational Fluid Dynamics Users' Workshop, University of Tennessee Space Institute, Tullahoma, TN, March 1984.

[42]Pulliam, T. H. and Steger, J. L., "Implicit Finite-Difference Simulations of Three-Dimensional Compressible Flow," AIAA Paper 78-10, Jan. 1978; also *AIAA Journal*, Vol. 18, Feb. 1980, pp. 159–167.

[43]Beam, R., and Warming, R. F., "An Implicit Finite-Difference Algorithm for Hyperbolic Systems in Conservation-Law Form," *Journal of Computational Physics*, Vol. 22, Sept. 1976, pp. 87–110.

[44]Pulliam, T. H., and Chaussee, D. S., "A Diagonal Form of an Implicit Approximate Factorization Algorithm," *Journal of Computational Physics*, Vol. 39, No. 2, Feb. 1981, pp. 347–363.

[45]Baldwin, B. S., and Lomax, H., "Thin-Layer Approximation and Algebraic Model for Separated Turbulent Flows," AIAA Paper 78-257, Jan. 1978.

[46]Sorenson, R. L., and Steger, J. L., "Grid Generation in Three Dimensions by Poisson Equations with Control of Cell Size and Skewness at Boundary Surfaces," *Advances in Grid Generation-FED*, Vol. 5, edited by K. Ghia, American Society of Mechanical Engineers, New York, 1983.

[47]Edwards, T. A., "Noniterative Three-Dimensional Grid Generation Using Parabolic Partial Differential Equations," AIAA Paper 85-0485, Jan. 1985.

[48]Lockman, W. K., and Seegmiller, H. L., "An Experimental Investigation of the Subcritical and Supercritical Flow about a Swept Semispan Wing," NASA TM-84367, June 1983.

[49]Van Dalsem, W. R., and Steger, J. L., "Finite-Difference Simulation of Transonic Separated Flow Using a Full-Potential Boundary-Layer Interaction Approach," AIAA Paper 83-1689, July 1983.

[50]Thibert, J. J., Grandjacques, M., and Ohman, L. H., "NACA 0012 Airfoil: An Experimental Data Base for Computer Program Assessment," AGARD-AR-138, May 1979, pp. A1-1–A1-19.

[51]Mehta, U., and Lomax, H., "Reynold-Averaged Navier-Stokes Computations of Transonic Flows, The-State-of-the-Art," *Progress in Astronautics and Aeronautics: Transonic Aerodynamics*, Vol. 81, edited by D. Nixon, AIAA, New York, 1982, pp. 297–395.

[52]Coakley, T. J., "Numerical Methods for Gas Dynamics Combining Characteristic and Conservation Concepts," AIAA Paper 81-1257, June 1981.

[53]Tobak, M., and Peake, D. J., "Topological Structures of Three-Dimensional Separated Flows," AIAA Paper 81-1260, June 1981.

[54]Hung, C. M., "Development of Relaxation Turbulence Models," NASA CR-2783, Dec. 1976.

[55]Rotta, J. C., "Turbulent Boundary Layers in Incompressible Flow," *Progress in Aerospace Sciences*, Vol. 2, Pergamon, New York, 1962, p. 1.

[56]Shang, J. S., and Hankey, W. L., "Numerical Solution of the Navier-Stokes Equations for Compression Ramp," AIAA Paper 75-4, Jan. 1975.

[57]Andronov, A. A., Leontovich, E. A., Gordon, I. I., and Maier, A. G., "Theory of Bifurcations of Dynamic Systems on a Plane," NASA TT F-556, 1971.

[58]Hinson, B. L., and Burdges, K. P., "Acquisition and Application of Transonic Wing and Far-Field Test Data for Three-Dimensional Computational Method Evaluation, Vol. II. Appendix B, Experimental Data," Lockheed Georgia Co. AFOSR TR-80-0422, 1980.

[59]Keener, E. R., "Computational-Experimental Pressure Distributions on a Transonic, Low-Aspect-Ratio Wing," AIAA Paper 84-2092, Aug. 1984.

[60]Hinson, B. L., and Burdges, K. P., "An Evaluation of Three-Dimensional Transonic Codes Using New Correlation Test Data," AIAA Paper 80-0003, 1980.

[61]Reue, G. L., Doberenz, M. E., and Wilkins, D. D., "Component Aerodynamic Load from 1/9-Scale F-16A Loads Model," General Dynamics, Fort Worth, TX, Rept. 16PR316, May 1976.

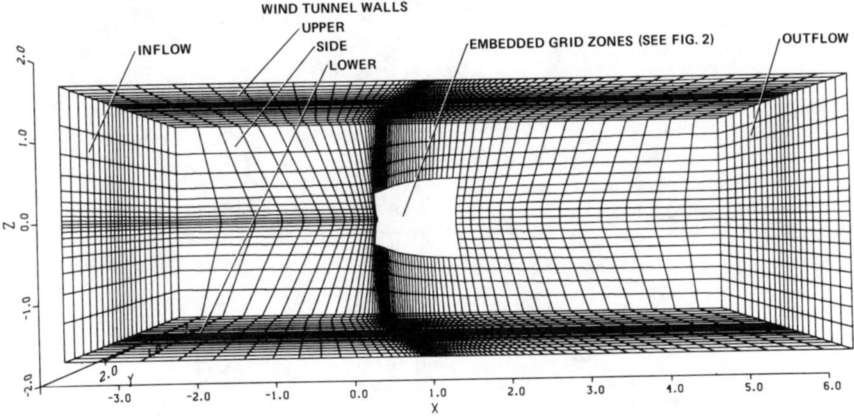

Fig. 1 Global finite-difference grid for the TNS-WING code showing detail on the wind-tunnel walls (taken from Holst et al.[29]).

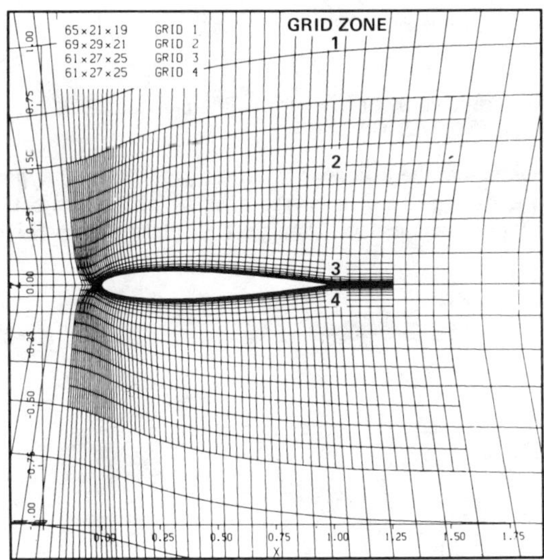

Fig. 2 Blowup of the embedded grid used in the TNS-WING code near the wing geometry (taken from Holst et al.[29]).

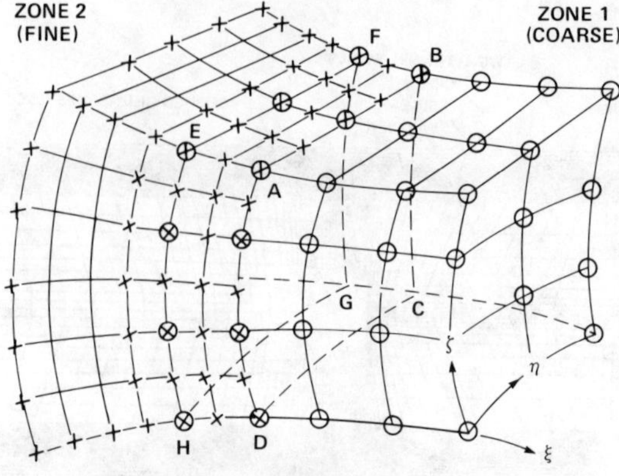

(a)   TWO-ZONE GRID SHOWING OVERLAP AT ABCD AND
      EFGH PLANES IN PHYSICAL SPACE

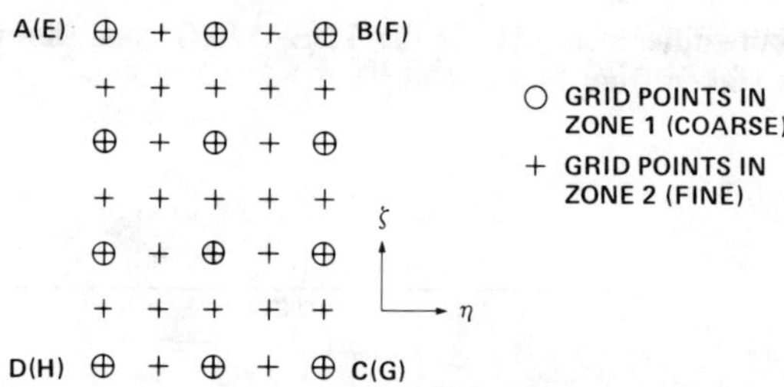

(b)   GRID POINT DETAIL IN THE OVERLAP REGION IN
      TRANSFORMED SPACE.

Fig. 3   Sketch illustrating the grid zone interface procedure: a) hypothetical grid zone interface between a fine grid and a coarse grid; b) planar projection of the ABCD (EFGH) grid zone interface surface from part 3a (taken from Kaynak et al.[35]).

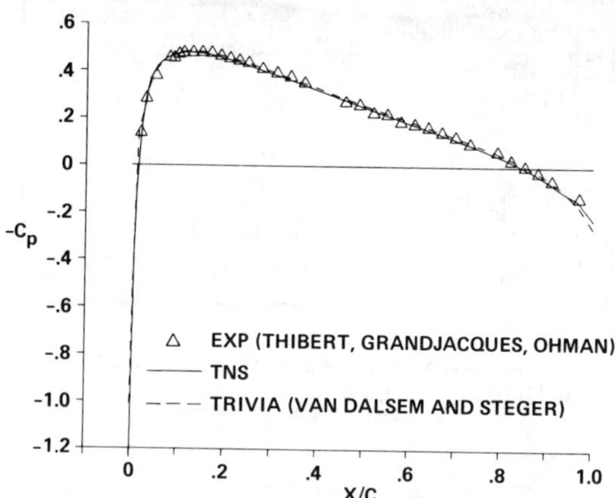

Fig. 4   Pressure coefficient comparison; NACA 0012 airfoil sections, $\Lambda = 0$ deg, large aspect ratio, $TR = 1.0$, $M_\infty = 0.5$, $\alpha = 0$ deg, $Re = 2.89 \times 10^6$ (taken from Holst et al.[29]).

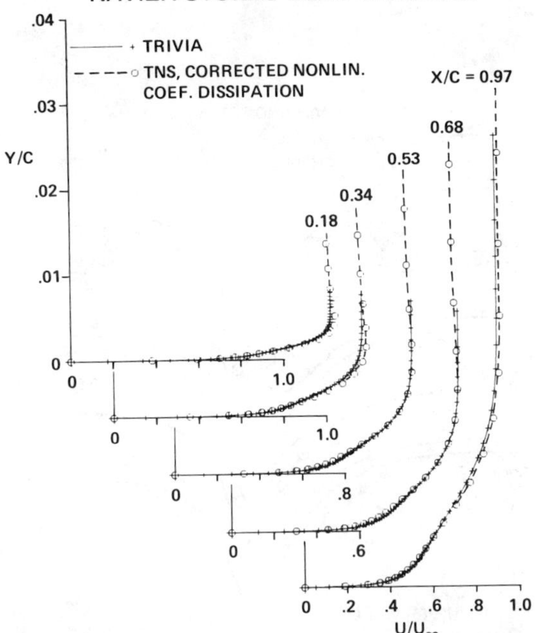

Fig. 5 Velocity profile comparisons from a boundary-layer code (TRIVIA) and the TNS-WING code using an improved artificial dissipation scheme; NACA 0012 airfoil sections, $\Lambda = 0$ deg, large aspect ratio, $M_\infty = 0.5$, $\alpha = 0$ deg, $Re = 2.89 \times 10^6$ (taken from Kaynak and Flores[39] and Holst et al.[29]).

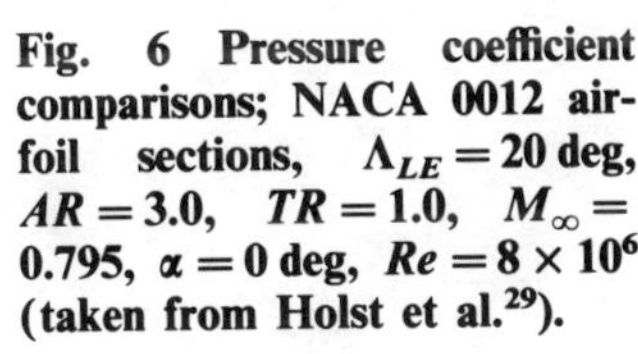

Fig. 6 Pressure coefficient comparisons; NACA 0012 airfoil sections, $\Lambda_{LE} = 20$ deg, $AR = 3.0$, $TR = 1.0$, $M_\infty = 0.795$, $\alpha = 0$ deg, $Re = 8 \times 10^6$ (taken from Holst et al.[29]).

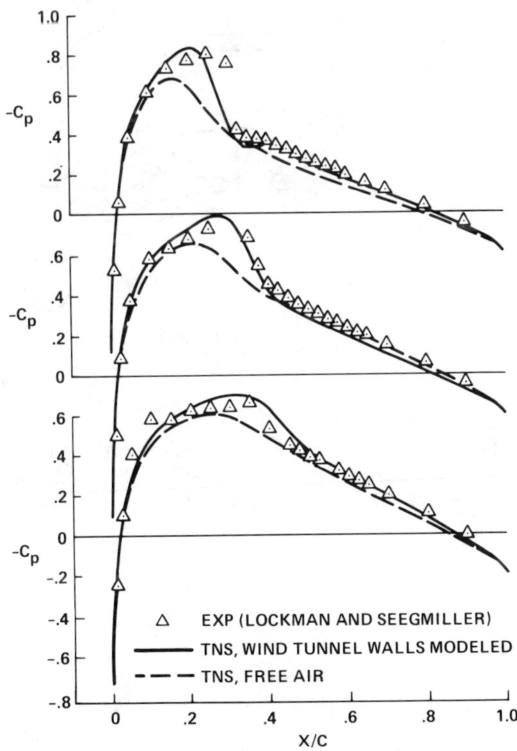

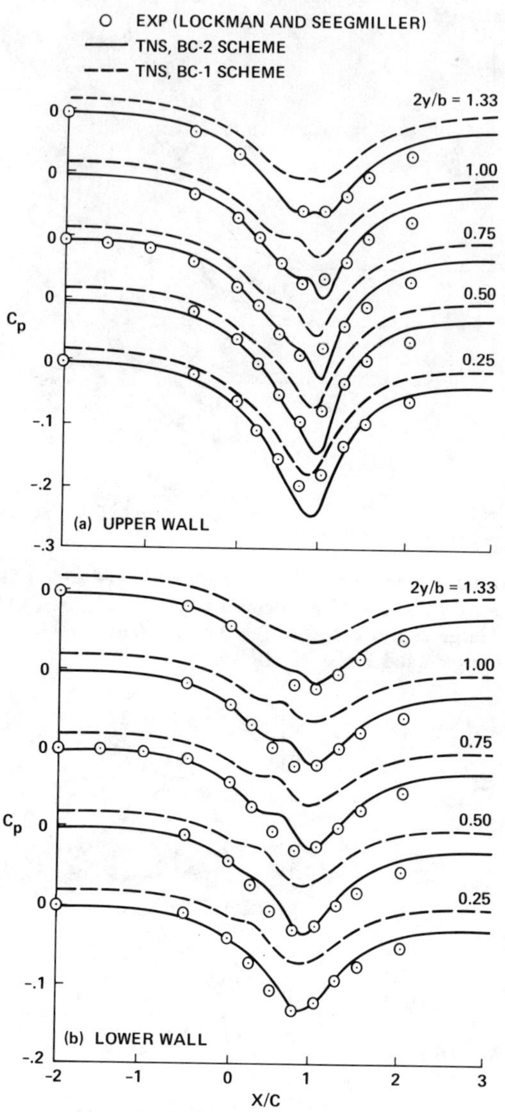

Fig. 7　Effect of wind-tunnel inflow-outflow boundary conditions on the wind-tunnel wall pressure coefficient; NACA 0012 airfoil sections, $\Lambda = 20$ deg, $AR = 3.0$, $TR = 1.0$, $M_\infty = 0.836$, $\alpha = 2$ deg, $Re = 8 \times 10^6$ (taken from Kaynak and Flores[39]).

Fig. 8 Experimental oil-flow pattern on the upper wing surface; NACA 0012 airfoil sections, $\Lambda = 20$ deg, $AR = 3.0$, $TR = 1.0$, $M_\infty = 0.836$, $\alpha = 2$ deg, $Re = 8 \times 10^6$ (taken from Lockman and Seegmiller[48]).

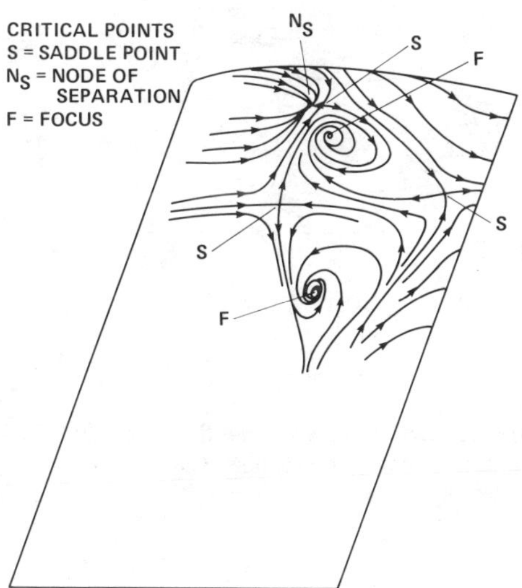

Fig. 9 Postulated skin-friction lines for the flow over the NACA 0012 wing of Fig. 8; $M_\infty = 0.836$ (taken from Kaynak and Flores[39]).

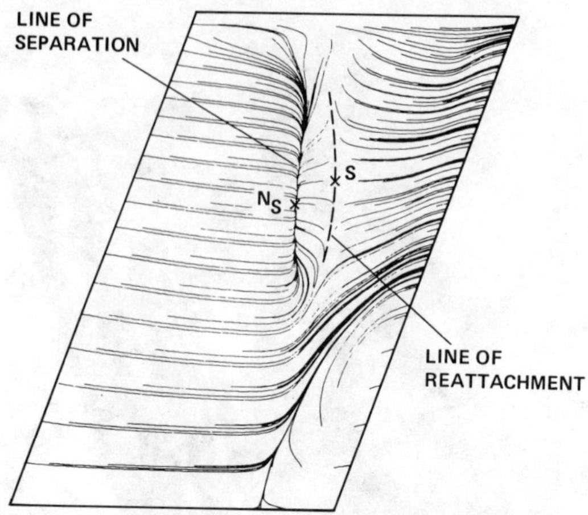

Fig. 10   Computed skin-friction lines for the flow over the NACA 0012 wing of Fig. 8 using the standard outer-boundary conditions (BC-1), the standard artificial dissipation scheme, and the baseline turbulence model; $M_\infty = 0.836$ (taken from Kaynak and Flores[39]).

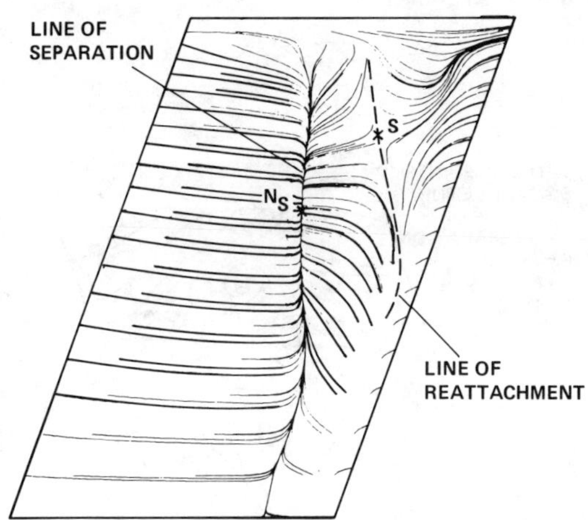

Fig. 11   Computed skin-friction lines for the flow over the NACA 0012 wing of Fig. 8 using the new outer-boundary conditions (BC-2), the new artificial dissipation scheme, and the baseline turbulence model; $M_\infty = 0.836$ (taken from Kaynak and Flores[39]).

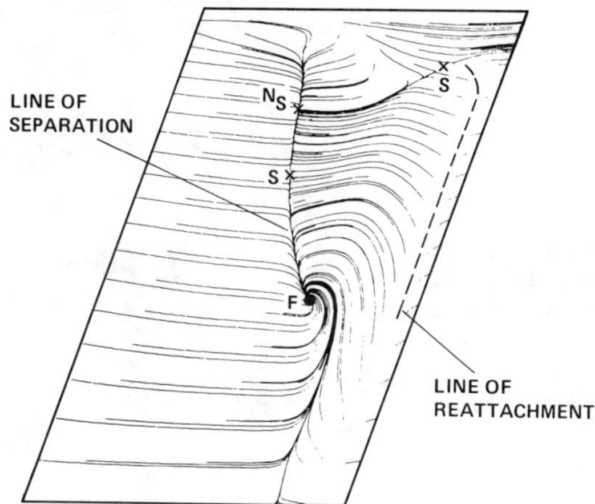

Fig. 12 Computed skin-friction lines for the flow over the NACA 0012 wing of Fig. 8 using the new outer-boundary conditions (BC-2), the new artificial dissipation scheme, and the modified turbulence model; $M_\infty = 0.836$ (taken from Kaynak and Flores[39]).

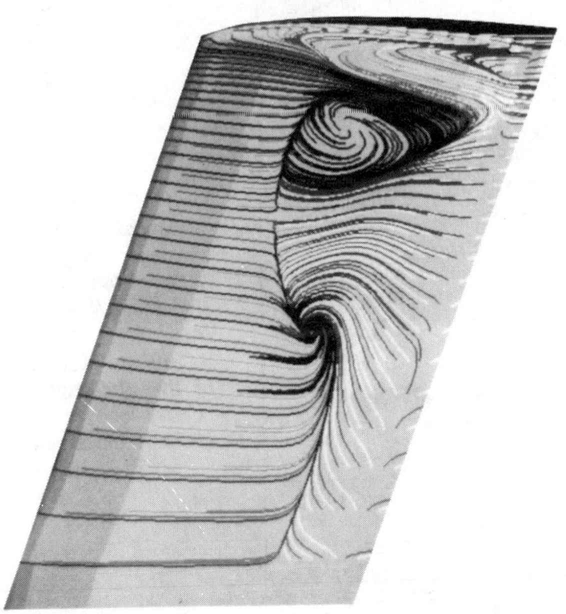

Fig. 13 Simulated oil flow over the NACA 0012 wing of Fig. 8, including the effects of the new artificial dissipation scheme, the new boundary condition scheme, the new turbulence model, and the new geometric modeling at the wing tip; $M_\infty = 0.836$ (taken from Kaynak and Flores[39]).

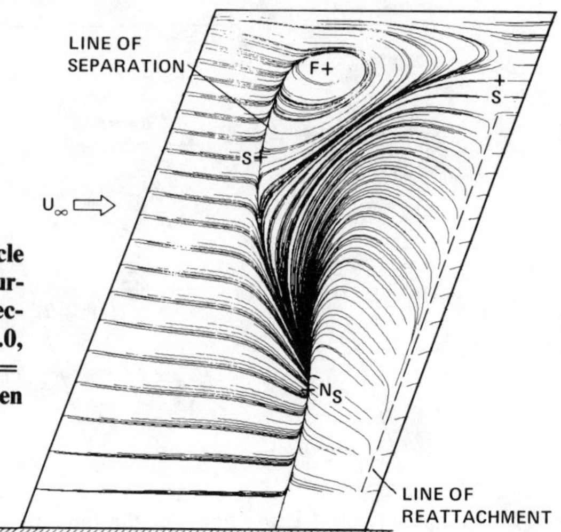

Fig. 14 Computed particle paths on the upper wing surface; NACA 0012 airfoil sections, $\Lambda_{LE} = 20$ deg, $AR = 3.0$, $TR = 1.0$, $M_\infty = 0.9$, $\alpha = 5$ deg, $Re = 8 \times 10^6$ (taken from Holst et al.[29]).

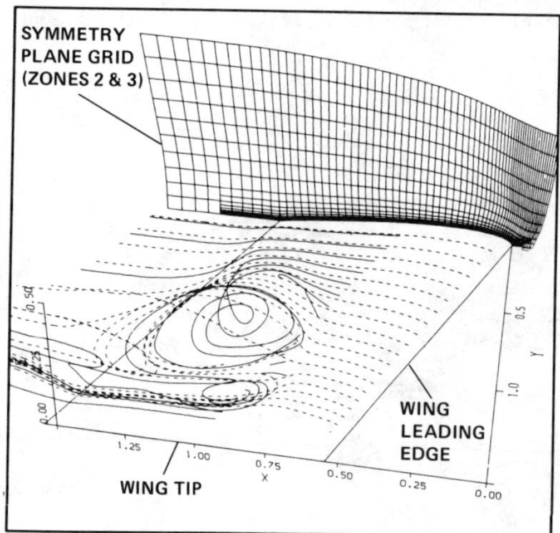

Fig. 15 Computed three-dimensional particle paths above the upper wing surface; NACA 0012 airfoil sections, $\Lambda_{LE} = 20$ deg, $AR = 3.0$, $TR = 1.0$, $M_\infty = 0.9$, $\alpha = 5$ deg, $Re = 8 \times 10^6$: a) view from outboard of the wing tip; b) view from behind and above the wing (taken from Holst et al.[29]).

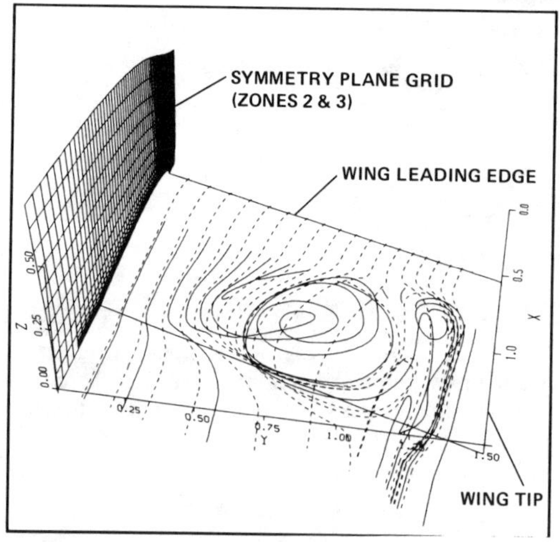

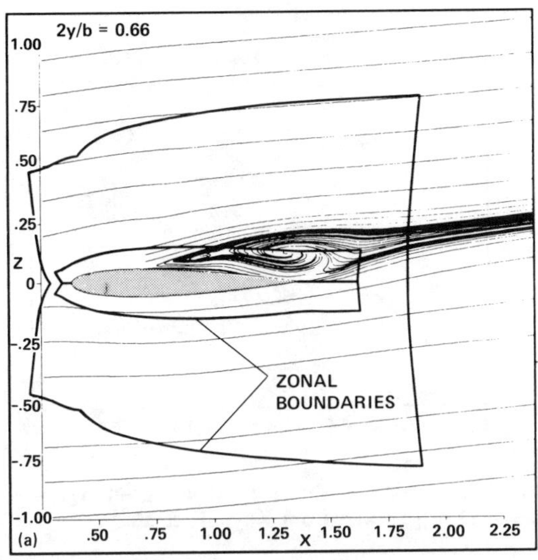

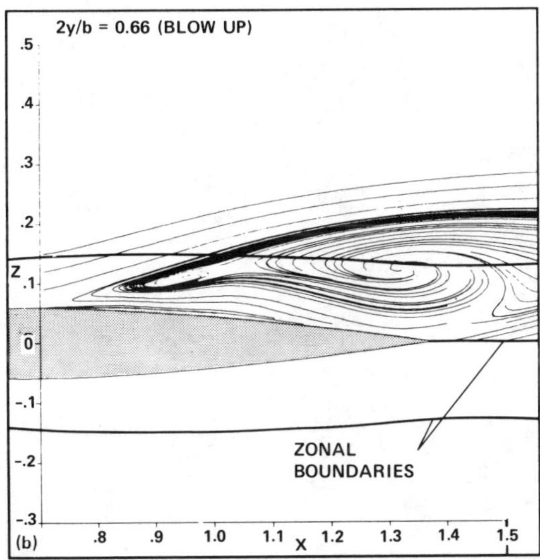

Fig. 16 Computed cross-sectional particle paths: NACA 0012 airfoil sections, $\Lambda_{LE} = 20$ deg, $AR = 3.0$, $TR = 1.0$, $M_\infty = 0.9$, $\alpha = 5$ deg, $Re = 8 \times 10^6$, $2y/b = 0.66$: a) normal view showing entire airfoil section; b) blowup of separated-flow region (taken from Holst et al.[29]).

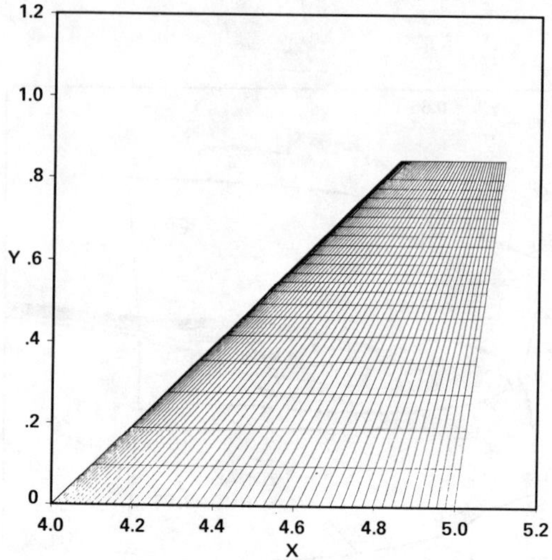

Fig. 17 Planform view of wing C showing surface grid; $\Lambda_{LE} = 45$ deg, $AR = 2.6$, $TR = 0.3$, $\alpha_{\text{TWIST}} = 8.17$ deg (taken from Kaynak et al.[32]).

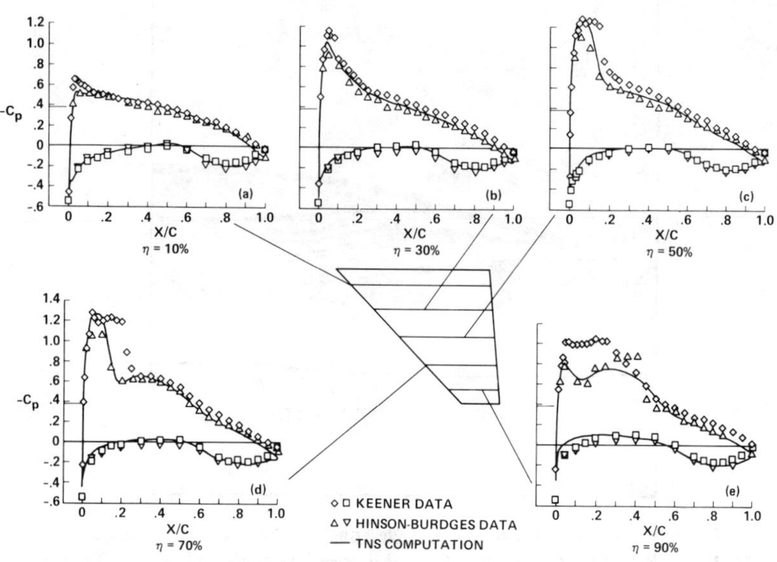

Fig. 18 Comparison of experimental and computed pressure coefficients for wing C; $M_\infty = 0.82$, $\alpha = 5$ deg, $Re_{\text{mac}} = 6.8 \times 10^6$ (taken from Kaynak et al.[32]).

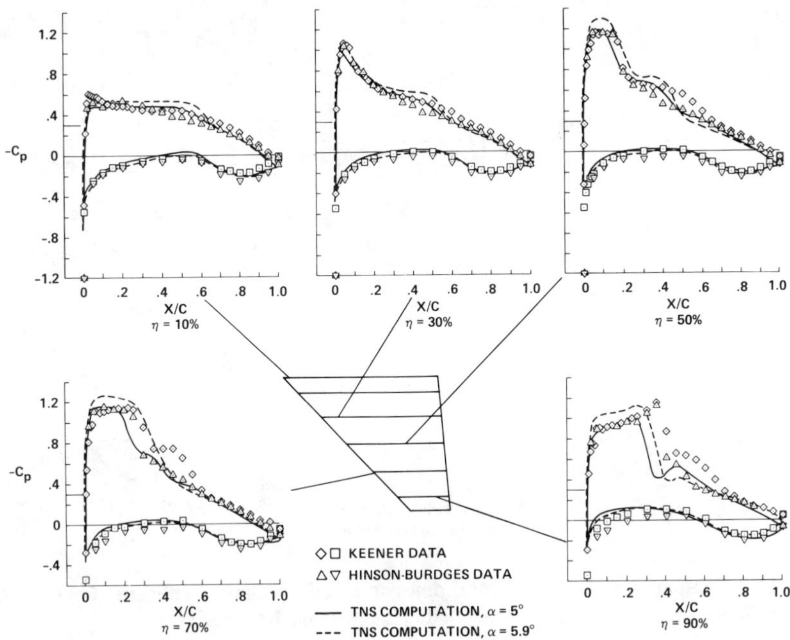

Fig. 19 Comparison of experimental and computed pressure coefficient for wing C; $M_\infty = 0.85$, $Re_{mac} = 6.8 \times 10^6$ (taken from Kaynak et al.[32]).

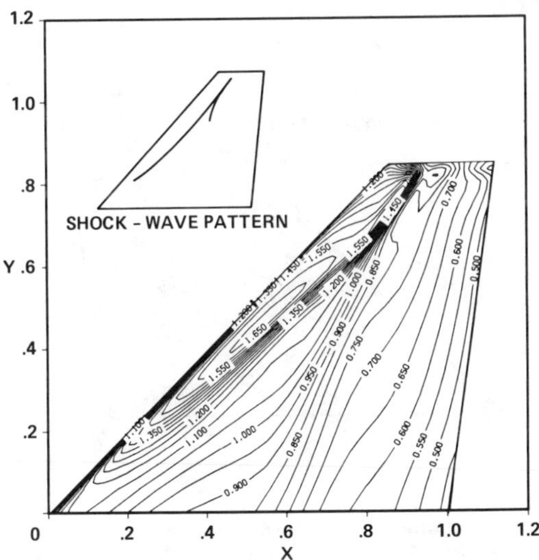

Fig. 20 Mach number contours just above the boundary layer on the upper surface of wing C; $M_\infty = 0.85$, $\alpha = 5$ deg, $Re_{mac} = 6.8 \times 10^6$ (taken from Kaynak et al[32]).

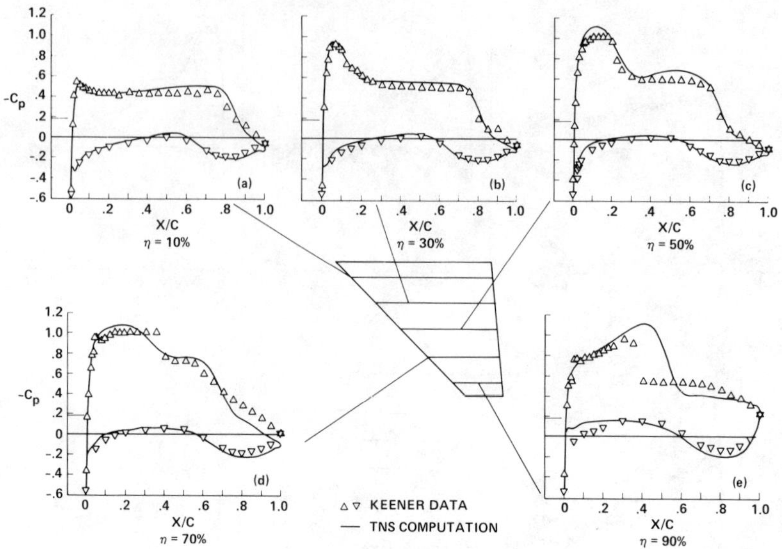

Fig. 21   Comparison of experimental and computed pressure coefficients for wing C; $M_\infty = 0.90$, $\alpha = 5$ deg, $Re_{\mathrm{mac}} = 6.8 \times 10^6$ (taken from Kaynak et al.[32]).

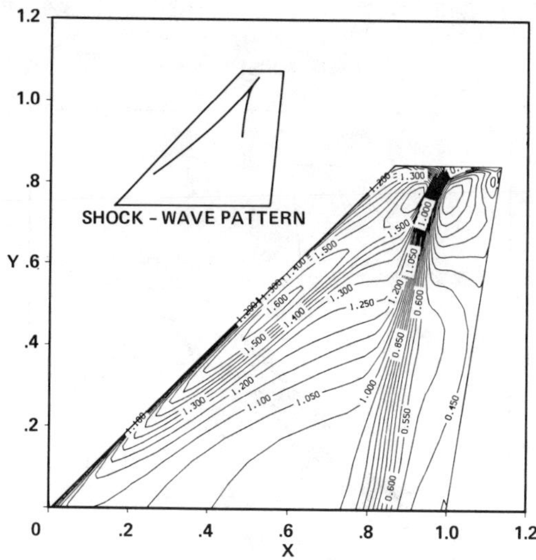

Fig. 22   Mach number contours just above the boundary layer on the upper surface of wing C; $M_\infty = 0.90$, $\alpha = 5$ deg, $Re_{\mathrm{mac}} = 6.8 \times 10^6$ (taken from Kaynak et al.[32]).

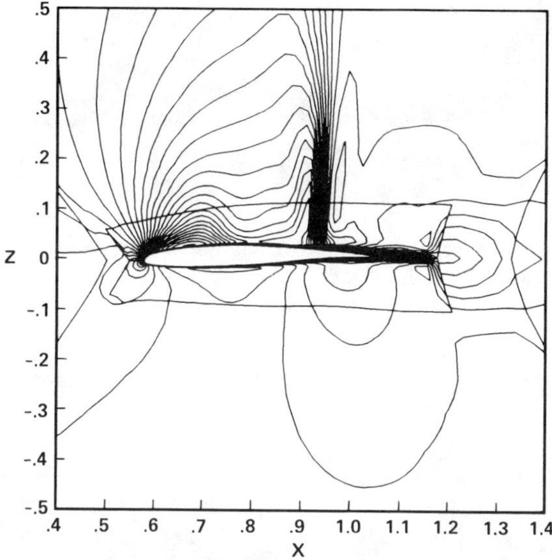

Fig. 23 Cross-sectional Mach number contours for wing C at $2y/b = 0.77$; $M_\infty = 0.90$, $\alpha = 5$ deg, $Re_{\mathrm{mac}} = 6.8 \times 10^6$ (taken from Kaynak et al.[32]).

Fig. 24 Top perspective view of the computationally modeled F-16A geometry (taken from Flores and Chaderjian[40]).

Fig. 25 Bottom perspective view of the computationally modeled F-16A geometry (taken from Flores and Chaderjian[40]).

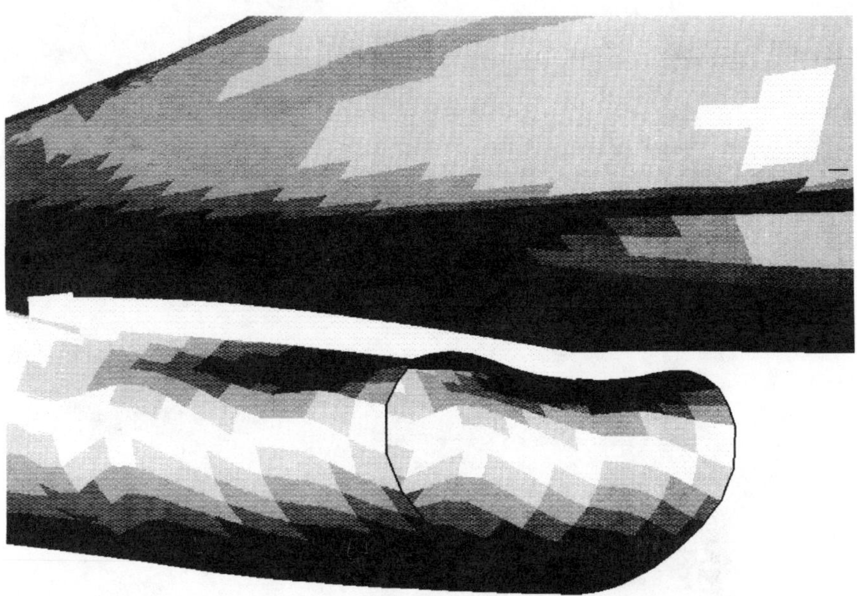

Fig. 26 F-16A geometry showing an expanded view of the inlet and diverter sections (taken from Flores and Chaderjian[40]).

Fig. 27 Grid zone topology at the wing/fuselage junction, F-16A geometry (taken from Flores and Chaderjian[40]).

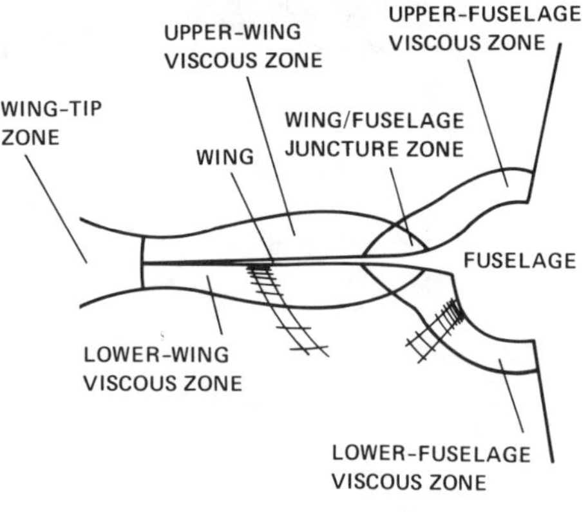

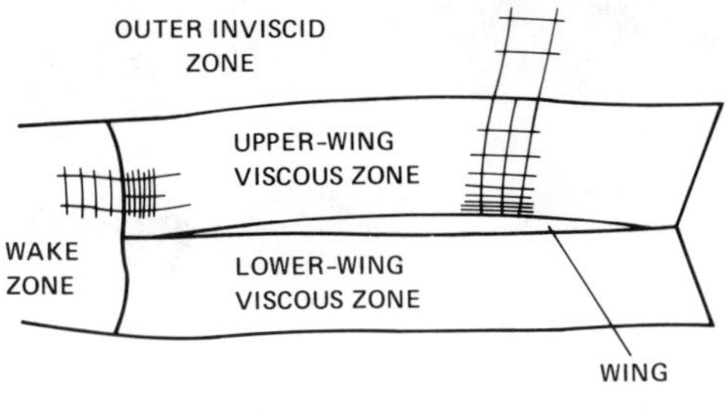

Fig. 28   Grid zone topology at a fixed spanwise cut through the wing, F-16A geometry (taken from Flores and Chaderjian[40]).

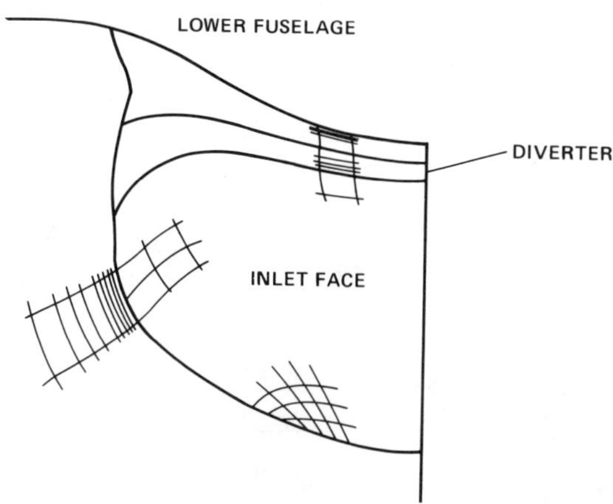

Fig. 29   Fuselage cross-sectional cut showing the grid zone topology at the inlet face with the inlet, upper diverter, and lower diverter grid zones highlighted (taken from Flores and Chaderjian[40]).

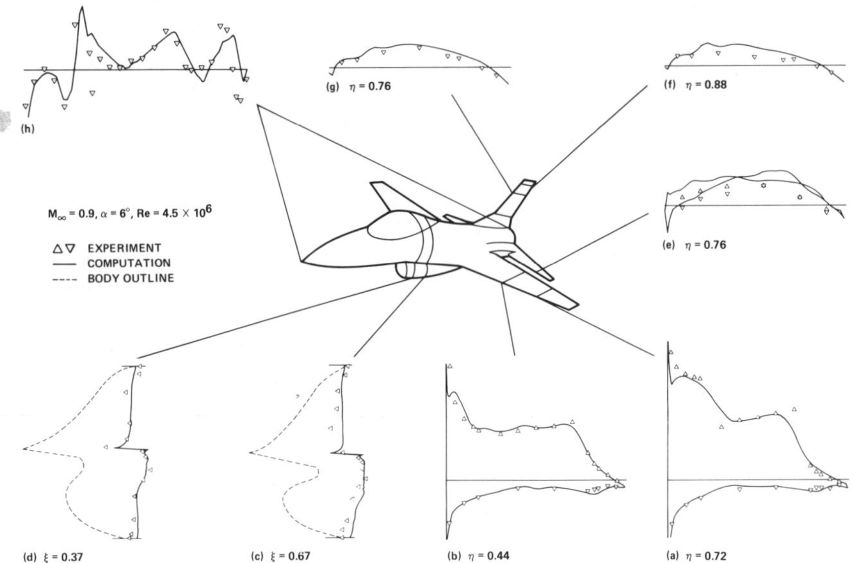

Fig. 30   Pressure coefficient comparisons at various locations on the F-16A surface, $M_\infty = 0.90$, $\alpha = 6.0$ deg, $Re_c = 4.5 \times 10^6$ (taken from Flores and Chaderjian[40]).

Chapter 22

Computational Fluid Dynamics Design Applications at Supersonic/Hypersonic Speeds

H. Robert Welge*
*Douglas Aircraft Company, McDonnell Douglas Corporation,
Long Beach, California*

Introduction

ADVANCED computational fluid dynamics (CFD) methods have had a significant impact on the design of aircraft flying at supersonic/hypersonic speeds. This chapter will review four cases to illustrate the impact. These cases are presented in the order of increasing complexity.

First the development of the external shape of an advanced vehicle configuration will be discussed. The introduction of computer-generated graphic displays to visualize and understand the complex flowfields, coupled with traditional and advanced CFD methods, were used to develop a new vehicle design concept that had significantly better cruise lift-to-drag ratios, under the same mission design constraints, than was possible using previously available analytical techniques.

The second example is concerned with sonic boom ground-level pressure signature prediction capability. Prior to the advanced nonlinear CFD solutions, sonic booms for vehicles flying at hypersonic speed could not be calculated. Existing theories were limited by linear assumptions that excluded hypersonic speeds and nonslender vehicles. Advanced CFD has removed these limitations. Analysis of more complex configurations and the potential for boom reduction through innovative vehicle shapes are now possible.

The third example describes the design of a wing at Mach 2.2. Two features were incorporated in the design that were not possible before CFD. These are large extents of laminar boundary-layer flow and increased leading-edge suction with subsequent improvements in drag due to lift. Only by virtue of advanced CFD and the details of the aerodynamic flow over the wing that these solutions delivered could these improvements have occurred.

Copyright © 1989 by the American Institute of Aeronautics and Astronautics, Inc. All rights reserved.
*Principal Specialist, High Speed Commercial Transport Business Unit.

The fourth and last example is the most complex. It deals with the design of a Mach 5.0 inlet. Inlet flows are very complex because, by design, they deal with strong shocks and viscous interactions, and in all cases the flows are very three dimensional. The inlet illustrated here was initially designed to be of the two-dimensional type using method of characteristics and boundary-layer codes in an iterative procedure. But, as will be shown, the flow is not two dimensional at all, and only by using three-dimensional Navier-Stokes solutions were the necessary flow details modeled and displayed so that improvements in the design could be made.

Case 1: Hypersonic Vehicle Design and Integration

For a vehicle to fly at hypersonic flight speeds the energy content of the fuel per pound of weight must be large enough to overcome the energy required for the vehicle drag and velocity. In this flight regime hydrogen fuel is a primary candidate to satisfy this requirement for sustained, long-range flight. The onboard storage of hydrogen with its extremely low density is a dominant factor in defining the shape of the aircraft. Hydrogen-fueled configurations require a high ratio of volume to surface area to provide adequate fuel storage, resulting in low vehicle "slenderness" or high drag. In addition, these speeds require large contraction and expansion area ratios in the inlet and nozzle of the propulsion system, necessitating the aircraft lower surface to act as the engine inlet and nozzle for successful integration of the propulsion system. In the past, to accommodate these constraints, configurations have typically used conventional wing-body arrangements with large, low slenderness fuselages where the hydrogen fuel is stored in the fuselage. A family of such configurations is shown in Fig. 1. These configurations have low lift-to-drag ratios due mainly to the low slenderness fuselage.

Prior to the implementation of CFD methods, the performance for hypersonic configurations was computed for simple idealized geometries or by correlation of existing experimental and empirical data. With this approach estimates of the total vehicle forces could be obtained, but no details of the configuration flowfields were known. It was impossible to determine the cause of the low performance of the configuration working with only the integrated lift and drag forces, and there were no data to guide further configuration development. Likewise, without a graphic display it was extremely tedious to match computed tabulated output values with their location on the configuration.

The solution to improved performance was twofold. First, the ability of the CFD methods to accurately compute detailed nonlinear surface pressures was combined with advanced color graphics postprocessing. Second, a procedure was employed that integrated the surface pressures to obtain the vehicle lift and drag forces. This procedure has proven accurate for design purposes in the supersonic/hypersonic speed region. To emphasize the distinction between regions of high pressure that contribute to lift and those that only contribute to drag, the graphic postprocessing was modified to display the local "drag" instead of pressure. This color-graphic postprocessing allows the user to rapidly identify areas of high drag by using the minds ability for pattern recognition to the fullest extent. The graphic display of the computed flow properties facilitates an understanding of the global flow characteristics so that development of modified geometries could be focused to the areas with the greatest potential for improvement.

By using the understanding obtained from flowfields of the combined CFD and graphic postprocessing, it was possible to identify for the family of wing-body configurations shown previously in Fig. 1 that the sides of the fuselage ahead of the wing/fuselage intersection were a region for modification to improve the total configuration performance. To illustrate the process, the results of a parametric study to improve the vehicle performance by optimally integrating the wing and fuselage into a single blended body will be shown. The blended body will provide the same fuel volume but can be tailored to provide lower drag at the same lift.

The study was made using two methods: 1) the Mark IV Supersonic/Hypersonic Arbitrary Body Program (HABP)[1] and 2) the Streamline Coordinate Riemann Axial Marching (SCRAM) Program.[2] The HABP program combines a number of supersonic and hypersonic flow computation methods with a powerful geometric modeling capability. The HABP program is a quasi-three-dimensional fully nonlinear analysis with a full three-dimensional geometric representation, yet many of the aerodynamic flow solution methods are based on two-dimensional approximations of the configuration or on the local geometry. The HABP program was used to screen the initial configurations, with its ease of use and low computational time used to an advantage. The SCRAM program is an advanced supersonic Euler analysis program. The SCRAM program does not have the geometric modeling capability of the HABP program, but provides a true three-dimensional aerodynamic flow solution and also computes the off-body flowfield properties. For this study SCRAM was used after development of selected configurations to verify the HABP computations and to determine the off-body flow properties near the engine inlet.

A typical wing-body configuration was used as a reference to determine the performance improvement of the blended-body configurations. The cause of the poor performance of the wing-body configurations can be determined from the computed aerodynamic pressure properties displayed in Plate 24 (see the color section). In this plate and subsequent plates high-pressure or drag regions are shown in red. To provide the required fuel volume in a reasonable length, the fuselage of a wing-body configuration must have a low fineness ratio (length/diameter) that creates a high vehicle drag. Furthermore, since the fuselage sides are nearly vertical, the positive pressures do not provide any beneficial lift that could compensate for the drag and maintain a high lift-to-drag (L/D) ratios.

Plate 25 (see the color section) displays the same results where the graphic postprocessing capability of displaying local axial force is used instead of pressure. The high-drag regions on the sides of the fuselage nose are again shown by the red regions. Note also the high-drag region on the aft fuselage which was not evident in the previous plate.

The goal in creating the blended-body configurations was to reduce the drag by both reducing the surface pressures and by inclining the surfaces to enable the high pressures on the lower half of the fuselage to contribute to the configuration lift more effectively. It was determined that the fuselage nose angles, and thus drag, could be further reduced by reducing the fuselage volume and increasing the wing volume so that a significant portion of the fuel may be contained in the wing. Also, to meet the constraints of the propulsion system, the modified configurations maintained the flattened lower surface required to provide inlet compression to the engine and expansion for the nozzle.

An intermediate solution for a blended-body configuration developed early in this study is shown in Plate 26 (see the color section). The fuselage nose drag has been decreased by the blended upper surface, but most of the volume is still located in the fuselage. Guided by the computed and displayed flow properties, further design changes were made, resulting in the configuration shown in Plate 27 (see the color section). Here the high-drag regions are further minimized, and the volume is more evenly distributed between the wing and fuselage. The similarity of the HABP pressure results to SCRAM are shown in Plates 28 and 29 (see the color section). The success of the process is shown in Table 1 by the 34% increased L/D of the final configuration.

Prior to the CFD and postprocessing development, the capability to develop configurations similar to this did not exist and the aerodynamic improvements were unavailable.

Case 2: Sonic Boom

Basic sonic boom prediction theory consists of two components: 1) the near-field pressure or source function determined by the geometry of the configuration and 2) the subsequent propagation of this pressure field to the ground. Once the source function has been determined, by whatever means, the theory for the propagation of this pressure field through the atmosphere to the ground is well established. The atmospheric propagation component of sonic boom theory can be considered a mature technology, even though certain aspects, such as turbulence, are not included in most models.

The source function calculation, unlike atmospheric propagation, has a strong dependency on aircraft shape and speed. Before the development of nonlinear CFD, aircraft configuration pressure fields were defined using a linear theory developed by Whitham.[3] The near-field pressure signature was defined by calculating a term defined as the Whitham F function. The F function uses the axial distributions of volume and lift to determine the normalized pressure field away from the aircraft body. For aerodynamically smooth aircraft flown at Mach numbers less than approximately 3,

Table 1 L/D improvement summary

Baseline wing-body	4.83	⌐
Interim blended-body	6.23	+34%
Final blended-body	6.50	⌐

the F function method has proven to be very accurate. However, when strong nonlinearities occur in the flowfield because of hypersonic speeds or nonslender configurations, the F functions method is not accurate. The introduction of fully nonlinear advanced CFD and numerical methods provides the opportunity for accurate sonic boom estimates for these cases.

An early application of CFD to sonic boom nonlifting predictions was first successfully accomplished by Ashby[4] for a nonlifting case. Here a finite-difference model was used to predict the aerodynamic near-field pressure signature for an axisymmetric blunt-body wind-tunnel model at speeds up to Mach 6. The calculations showed good agreement to measured wind-tunnel data. The shortcoming of Ashby's finite-difference model was that it only applied to axisymmetric bodies. Complete aircraft configurations with wings and lift were beyond the scope of his model.

The advanced CFD codes now available allow for calculation of the aerodynamic near-field pressures of a complete lifting aircraft. A typical example of the undertrack pressure field for a Mach 5.0 concept is shown in Plate 30 (see the color section). This type of CFD near-field solution has been combined with acoustic propagation codes and compared to wind-tunnel measurements. Plate 31 (see the color section) shows a CFD solution for a simplified test case of the undertrack flowfield for a cone.[5] This data was read at one body length from the cone and extrapolated out to 16 body lengths using acoustic propagation. Figure 2 shows the comparison of the resultant pressure field to wind-tunnel data collected at 16 body lengths. The positive overpressure region is very well predicted with this method, whereas the rear position is not. The deficiency in the negative overpressure region is explained by the difficulty in modeling the sharp edge of the wind tunnel model in the CFD code.

Another advantage of CFD methods for sonic boom prediction is that complex geometries can be analyzed that are outside of the scope of current linear codes, such as the example shown previously in Plate 30. With these CFD methods the complexity of the geometry that can be analyzed is only limited by the grid-generation and flow-solver capability. Furthermore, features such as blunt-leading-edge airfoils can be assessed with nonlinear, numerical codes, but not by their simple, linear counterparts. These advantages make CFD a more attractive method for sonic boom prediction than the basic linear methods.

The primary difficulties encountered using CFD solutions for sonic boom predictions are the cost of the computer time and the grid cell size. The solutions cells must be small enough to accurately capture the formation of shocks without smearing them, but the solution must extend to the far field so that all asymetric effects due to lift have been resolved. This requires a fine grid over a large area, and hence is costly. There is no doubt that the future of sonic boom prediction lies with numerical methods. Current problems with accuracy and cost should be solved through improvements in supercomputer performance and advanced CFD codes. This will open up the hypersonic flight regime for accurate sonic boom prediction. Further, advanced CFD methods will become applied with increasing regularity in the design of supersonic/hypersonic aircraft, and the sonic boom may be calculated along with the other aerodynamic characteristics.

Case 3: Supersonic Laminar Flow Control and Drag Due to Lift

The process of designing a wing for minimum supersonic cruise drag using laminar flow control (LFC) technology for a given airplane size and flight condition requires the accurate computation of both the outer inviscid flowfield and the boundary layer. Of the various methods proposed for boundary-layer stabilization, wall suction is the most versatile and practical in most instances. However, the size, weight, and power requirements of the suction system depend on the total suction airflow. The suction airflow required for laminarization is therefore a fundamental consideration in the design of optimal LFC wings.

Conditions in the leading-edge region are critical to wing laminarization. On swept wings, in the absence of supercritical roughness elements such as rivet heads, insect debris, steps and gaps, etc., there are three mechanisms acting to limit the extent of laminar flow in the leading-edge region: 1) pre-existing turbulence propagating along the attachment line from locations upstream, 2) two-dimensional instabilities in the spanwise boundary layer along the attachment line, and 3) crossflow instabilities associated with the swept acceleration region. All of these mechanisms are strengthened by increased sweep, bluntness, and unit Reynolds number, but can be countered using wall suction.

The amount of suction required in the leading-edge region depends strongly on the geometry and pressure distribution and accounts typically for more than 50% of the total suction flow. It is therefore critical to have analysis and design codes that can accurately model the leading-edge region.

Until the development of CFD the only methods available for designing supersonic cruise wings were based on linearized small-disturbance theory. This includes the Woodward[6] and PAN AIR[7] codes. The small-perturbation assumption of these methods precluded proper modeling of the flow in the leading-edge (attachment line) region of the subsonic leading-edge wing. In addition, for supersonic leading edges the assumption is made that shocks are isentropic. Therefore, successful supersonic/hypersonic LFC designs are dependent on the development of advanced nonlinear CFD codes. Euler codes such as SCRAM[2] and FLO67[8] are free of the small-disturbance approximation in the leading-edge region and are inherently more accurate. In principle, they also properly account for changes occurring across a shock wave and allow for the vorticity found downstream if the shock is curved.

An example showing the improvement of the Euler codes over the small-disturbance potential codes will be illustrated using a Mach 2.2 cruise advanced supersonic transport (AST) configuration, shown in Fig. 3. This vehicle was designed using linear theories before CFD codes were available. Computed pressure distributions for this initial configuration using a newly developed Euler code are shown in Figs. 4–6 for selected stations (wind-tunnel data are compared to the solution to illustrate accuracy). This Euler solution provided a much more detailed look at the inviscid flow in the leading-edge region than had been previously attainable. These more detailed Euler solutions allowed more accurate input for the calculating of the wing boundary-layer development and a determination of the minimum suction levels required to maintain it in a laminar state.

Inspection of the chordwise pressure distributions revealed areas for improvement for laminarization and for drag due to lift. For LFC the minimum required wall suction corresponds to a pressure distribution that accelerates the flow from attachment line conditions to nearly constant velocity in the shortest possible surface distance. This minimizes the growth of boundary-layer crossflow waves and hence the suction required to subdue them.

From a standpoint of drag due to lift the ideal supersonic wing would have a high aspect ratio, little taper, and be swept a few degrees beyond the Mach line. This allows leading-edge suction and trailing-edge recovery for the normal component of flow. Of course, such a wing could not hold sufficient fuel, nor would it be structurally efficient. The low-aspect delta or arrow wing can approach the high-aspect wing aerodynamically if a large part of the upper-surface contribution to lift can be carried in a narrow band along the leading edge. With the use of previously available linearized methods, the leading edge of the AST wing did not carry upper-surface lift as far forward as was possible as shown previously in Figs. 4–6; thus potential reductions in some of the drag due to lift were not being exploited. This also is consistent with LFC requirements as discussed earlier. The disadvantage of the low-aspect wing with its high surface area and skin-friction drag can be reduced by using the benefits of LFC, thereby synergistically approaching the ideal.

After several design iterations using the Euler code to satisfy the low suction LFC and drag-due-to-lift requirements, a modification to the original wing was selected. Color-graphic displays of the pressures for the original wing and for the modified wing are shown in Plates 32 and 33 (see the color section). Plate 32 indicates that the region of high pressure at the leading edge of the original wing was eliminated in the new design which favored both LFC and drag due to lift. Plate 33 shows the before and after pressures for the lower surface. A comparison of the chordwise pressure distributions at the 60% semi-span location is shown in Fig. 7. Here again it can be seen that the pressure distributions have been redesigned to favor LFC and drag due to lift. The drag characteristics of the modified wing with LFC and the original turbulent wing are shown in Fig. 8. It is unlikely that the drag improvement of 26% would have been achieved without the CFD capability.

Case 4: High Mach Engine Inlet Design

Viscous boundary-layer effects within high-speed engine inlet systems must be controlled in order to produce high inlet performance over a wide range of Mach numbers. Aircraft configuration design principles typically minimize the occurrence of shock waves, but for inlets, strong shocks are a design objective for airflow velocity reduction prior to entering the engine, and their interaction with the wall boundary layers must be carefully controlled. This is an ideal application of advanced Navier-Stokes

CFD. Control of the viscous effects in the inlet have been achieved historically through the use of boundary-layer bleed, that is, the removal of a portion of the boundary-layer flow in regions of high-pressure rise that could lead to boundary-layer separations. Boundary-layer separations are to be avoided in high-speed inlets. These separations result in energy or total pressure losses and the potential for the introduction of additional compressions and expansions brought about by the displacement effects of the boundary layers occurring on the internal surfaces. Also, the separated flow will contribute to distorted and unsteady flows entering the engine. The higher the Mach number, the more dominant the viscous effects.

Historically, designs have relied on a combination of inviscid method of characteristics (MOC) and viscous boundary-layer (BL) codes. The design methodology was based on the concept of adjusting the compression contours for the viscous displacement effects that were expected to occur over the range of Mach numbers. The design objective was to create contours that achieve a relatively undistorted flowfield in the throat of the inlet over a wide range of Mach numbers. The contour finally selected is typically arrived at on the basis of a tradeoff between a mild compression rate to minimize the boundary-layer removal to reduce bleed airflow momentum drag and a relatively rapid compression of the inviscid flow to shorten the inlet and reduce weight. The design of a Mach 5 inlet[9] shown in Fig. 9 will be used to illustrate how the design process can be enhanced by the use of CFD methods that were not available using these previous methods. The knowledge of the three-dimensional viscous flows provided by the CFD codes can be used to design the surface contours and select the type and location of appropriate boundary-layer control concepts.

The Mach 5 inlet is two dimensional; that is, the compression surfaces consisting of the ramp and cowl are the only ones that produce a geometric contraction. The ramp is a multiple-element system, whereas the cowl is a single contoured surface. The resulting aerodynamic cross-sectional design using the MOC-BL theories for the Mach 5 inlet is shown in Fig. 10. The X and Y dimensions are nondimensionalized to the cowl lip height. Mach numbers in the various flow regions are shown for the cruise (Mach 5) condition. At cruise conditions the freestream airflow is at an angle of 9 deg relative to the first ramp surface. The resulting first compression wedge of 9 deg and Mach 5 freestream conditions gives a local Mach number of 4.1 on the first inlet wedge. The Mach conditions are successively reduced by additional wedges to obtain Mach 3.1 on the final external ramp surface. A cowl shock and additional distributed compression and a terminal shock are employed for internal supersonic compression. Since one of the main driving factors in the design was length, the design features a cowl shock to be canceled at the inlet shoulder followed by a strong cowl-generated compression fan. The design compression split is about 85% external (with four ramps) and 15% internal. The ramp is articulated according to a schedule of Mach numbers that both decreases the ramp angles and enlarges the throat area as the Mach number is decreased.

The design process interactively accounted for inviscid flowfield, the displacement effects due to the boundary layer, and the boundary-layer bleed system. The boundary-layer code used in the original design of the bleed system could not be used to calculate through boundary-layer separations and reattachments, but it did account for boundary-layer bleed. Thus, boundary-layer separation, its potential upstream influence, and unwanted displacement effects were eliminated in the design process. Boundary-layer bleed was used in regions of pressure rises associated with the shock-wave/boundary-layer interactions to control boundary-layer separation. Although the inlet was designed as a two-dimensional compression system, it was realized that parallel sidewalls would be added to the inlet and that the boundary layer on these sidewalls would have to be accounted for by estimating the boundary-layer displacement effects. Accounting for the ramp, cowl, and sidewall viscous displacement effects with bleed, the inlet was designed to have a throat Mach number of 1.2.

After the original design was completed, a newly developed three-dimensional parabalized Navier-Stokes code (PNS) called PEPSIS[10] was used to analyze the inlet flowfield. This code was a significantly more powerful analysis tool. The results from the three-dimensional calculation are presented in Figs. 11–13 corresponding to the stations indicated on Fig. 10. Because of flow symmetry, different solution parameters are shown for each side of the inlet centerline. The left side of the figure shows Mach number contours, and the right side shows the secondary velocity vectors within the survey plane. Plate 34 (see the color section) shows a color representation of Fig. 13 and Plate 35 (see the color section) shows the Mach contours for several locations in the inlet (the higher total pressures and Mach numbers are shown in red). The survey planes are perpendicular to the ramp and cowl surfaces.

On the solid surface ramp, cowl, and sideplate, the development and growth of the boundary layer is indicated by a concentration of Mach contours near these surfaces. Shock waves are noted by a concentration of Mach contours away from the solid surfaces. They can also be detected by an abrupt change in the velocity vectors.

The secondary velocity vectors indicate that a crossflow is induced along the sideplate and feeding forward of the inviscid shock location. Each of the succeeding compression ramps increases the secondary flow. In Fig. 12 a circulation region has been generated near the sideplate corner pulling in flow along the ramp surface. The last calculated cross section shown in Fig. 13 indicates that, as the flow enters the cowl, a shock wave is generated by the cowl lip, as indicated by the horizontal lines in the Mach contours. The strong secondary flow moving up the sideplate encounters the internal cowl surface, and the velocity vectors indicate that this flow turns through the corner formed by the cowl and sideplate. Figure 13 shows that the secondary flow rolls up into a vortical flow region, and the low-energy flow is concentrated in the corner. Since the internal surface of the cowl had been shaped to further compress the flow, the low-energy flow in the corner is subjected to an adverse pressure gradient created by this turning, and a large separation occurs. The midstream ramp boundary layer, which started at zero thickness at the leading edge of the ramp, is also very thick at this station. If this boundary layer would be allowed to proceed down the inlet to where the cowl shock, or later the normal shock, would interact with it, another separation would likely occur.

The bleed amounts and locations that were originally determined for the inlet based on the MOC analysis were then modified based on the new flow information. The bleed regions near the throat were expanded as shown in Fig. 14. Bleed regions were added to the sidewalls and cowl corners in the areas shown by the shaded bands in Fig. 15. The sidewall bleed in Fig. 15 is intended to remove the low-energy sidewall boundary layer before it is captured by the cowl, and the corner bleed is intended to remove the low-energy flow once it has been captured by the cowl.

It can be seen that flow within a relatively simple-looking two-dimensional inlet is highly three dimensional, and the three-dimensional CFD codes gave a much greater understanding of the three-dimensional viscous flow details. This additional knowledge guided the selection and location of cowl and sidewall bleed regions in an attempt to control the three-dimensional viscous flow requirements.

After the PEPSIS analysis was completed, a new, more powerful three-dimensional explicit, time-accurate full Navier-Stokes code[11] became available and was used to study the inlet flowfield with and without bleed in the regions previously described. The new code also allowed bleed flow to be simulated with the solution so that bleed regions and amounts could be accurately determined. Figure 16 shows the Mach number contours using this code without bleed on a cross plane at about the same station as the last station shown in the PEPSIS solution (Fig. 13). The ramp boundary-layer separation, cowl boundary layer, and cowl shock wave can be seen. The vortical flow (as evidenced by the disturbance to the nominally two-dimensional flowfield) is rather extensive at this station without any boundary-layer control and duplicates the PEPSIS solution very well.

In order to prevent the ramp boundary-layer separation, bleed was applied on the ramp just upstream of the shoulder in the same region as predicted by the MOC-BL code. Mass flow rates through the ramp surface used corresponded to about 4% of the inlet's theoretical capture mass flow. Also, attempts were made to control the three-dimensional effects of the vortical flow in the cowl sidewall corner through classical boundary-layer bleed techniques by applying 3% bleed on the cowl surface in the corners at the sidewall/cowl intersection. Figure 17 shows the calculated Mach number contours for this case. The ramp separation is not present, and it can also be seen that the bleed imposed on the cowl surface appears to have some effect on the extent of the vortical flow. Although about 3% of the inlet's capture mass flow has been removed from the solution at the cowl bleed location, and the extent of the vortical effects are reduced, much of the vortical flow remains. Since this bleed configuration showed the most promising results of those tried to date in this study, it appears that the development of a novel method to control this vortical flow is required.

An inlet test program was conducted to validate the analysis procedure (test setup shown previously in Fig. 9). Figure 18 compares the results of a symmetry plane Navier-Stokes solution at the end of the third ramp with test data. The analysis closely predicts the boundary-layer thickness. Figure

19 shows the nondimensionalized pitot pressure profiles for a cowl corner rake located just downstream of the cowl lip in the area where analyses predicted vortical flow and possible flow separation. The circles are data points at design airflow conditions with the ramp shoulder bleed open but with both sidewall and corner regions closed. The squares represent data points at the same conditions with all of the bleed regions fully open. For the analysis there is no sidewall bleed simulated. Because of the different bleed configurations, exact comparisons cannot be made, but it is encouraging that the shapes of the analytical and experimental curves for the bleed open configurations are very similar. This would tend to indicate that the experimental results are demonstrating similar flowfields as those being predicted by analysis.

As we have seen, inlets designed previously using the traditional MOC code with boundary-layer corrections can have inadequate operational capabilities because the details of the viscous three-dimensional flows could not be computed and visualized. The three-dimensional viscous analytical codes have overcome these shortcomings and have increasingly become a part of the inlet design process.

Conclusions, Implications, Perspective on CFD Applications in the Future

The impact of CFD on supersonic/hypersonic aircraft design has been significant, as the four examples have illustrated. Hypersonic vehicle configuration design lift-to-drag ratios have increased 34%, sonic boom calculations for complex configurations and hypersonic speeds are now possible, wing designs at Mach 2.2 when coupled with LFC have 26% less drag, and the performance of high Mach inlet designs is enhanced by virtue of the ability to understand the complex viscous three-dimensional flows. Important progress has been made, but additional capabilities are still required.

One area for improvement is for additional components of the configurations to be included in the Euler solutions on a "user-friendly" basis to allow numerous/efficient design options to be evaluated.

The integration of the propulsion system holds significant promise for performance improvement if the propulsion nacelle can be accurately modeled with the vehicle including the presence of the multienergy flows. Jet noise is a major design barrier for these vehicles, and the ability to model and better understand the internal and external mixing of jet exhausts holds the promise of a breakthrough in this area.

Low-speed aerodynamic performance of the highly swept planforms typical of supersonic/hypersonic designs is dependent on viscous induced separated vortical flows. The ability to calculate and understand these three-dimensional viscous flow phenomena (as was the case with the inlet) with deflected vehicle leading and trailing edges will lead to performance improvements that will increase the economic and community noise aspects of supersonic/hypersonic concepts.

Flight structural design criteria are typically set by aerodynamic loads at the edges of the flight envelope. At these conditions the vehicle is maneuvering so that large normal forces, or accelerations, are present. Viscous, separated vortical supersonic/hypersonic flows are present at these conditions. The capability to compute the fluid flow with deflected vehicle surfaces to minimize the aerodynamic and hence structural load will lead to reductions in the vehicle structural weights.

Throughout the process, the increased computer graphics capability as illustrated in the hypersonic vehicle design example is needed to "communicate" the large amount of data to the designer in a clear and understandable manner so that configuration design changes using fundamental aerodynamic theories can be made. Of course, inverse design capability would prove invaluable.

The impact of CFD in the past has been significant and will continue to be. The increased code and computer capabilities will present the aerodynamicist with more and more detail and understanding of the fluid flows, which in turn can only lead to further vehicle performance improvements.

Acknowledgments

The author wishes to acknowledge and express his appreciation to those who contributed to this chapter. First, we thank Elaine Anderson, whose patience permitted several drafts. Second, we thank the following researchers who conducted the analysis and contributed these sections: G. S. Page and J. A. Page of McDonnell Douglas (Hypersonic Vehicle Design and Integration); R. A. Sohn and K. B. Dailey of McDonnell Douglas (Sonic Boom); A. G. Powell and A. A. Killeen of McDonnell Douglas (Laminar Flow Control and Drag due to Lift); and R. E. Coltrin and T. J. Benson of NASA Lewis and W. E. Rose of Rose Engineering Research (High Mach Engine Inlet Design).

References

[1]Gentry, A. E., Smyth, D. N., and Oliver, W. R., "The Mark IV Supersonic-Hypersonic Arbitrary-Body Program," Air Force Flight Dynamics Lab. TR-73-159 Vol. 1–3, Nov. 1973.

[2]Verhoff, A., and O'Neill, P. J., "Accurate, Efficient Prediction Method for Supersonic/Hypersonic Inviscid Flow," AIAA Paper 87-1165, 1987.

[3]Whitham, G. B., "The Flow Pattern of a Supersonic Projectile," *Communications on Pure and Applied Mathematics*, Vol. 5, 1952, pp. 301–348.

[4]Ashby, G. C., "A Study of the Sonic Boom Characteristics of a Blunt Body at a Mach Number of 6," NASA TP-1737, Dec. 1980.

[5]Carlson, H. W., "An Investigation of the Influence of Lift on Sonic Boom Intensity by Means of Wind Tunnel Measurement of the Pressure Fields of Several Wing Body Combinations at A Mach Number of 2.01," NASA TN-D-881, July 1961.

[6]Woodward, F. A., "An Improved Method for the Aerodynamic Analysis of Wing-Body-Tail Configurations in Subsonic and Supersonic Flow, Part I—Theory and Applications," NASA CR-2228, May 1973.

[7]Carmichael, R. L., and Erikson, L. L., "Pan Air—A Higher Order Panel Method for Predicting Subsonic and Supersonic Linear Potential Flows about Arbitrary Configurations," AIAA Paper 81-1255, June 1981.

[8]Jameson, A., "A Vertex Based Multigrid Algorithm for Three-Dimensional Compressible Flow Calculation," ASME Symposium on Numerical Methods for Compressible Flow, Anaheim, CA, Dec. 1986.

[9]Perkins, E. W., Rose, W. C., and Harie, G., "Design of a Mach 5 Inlet System Model," NASA CR-3830, Aug. 1984.

[10]Buggeln, R. C., McDonald, H., Levy, R., and Kreskowsky, J. P. "Development of a Three-Dimensional Supersonic Inlet Flow Analysis," NASA CR-3218, Jan. 1980.

[11]Kumar, A., "Numerical Simulation of Flow Through Scramjet Inlets Using a Three-Dimensional Navier-Stokes Code," AIAA Paper 85-1664, July 1985.

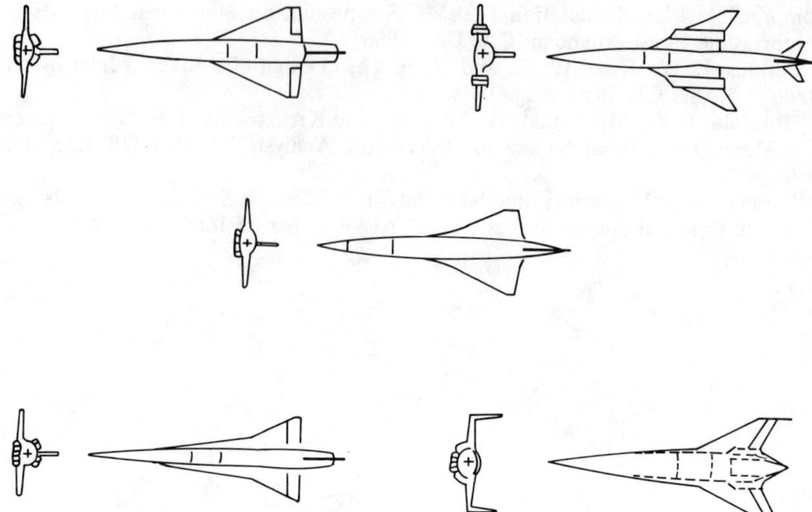

Fig. 1   Previous wing-body designs, c. 1979 (Morris and Brewer, NASA CR-15826).

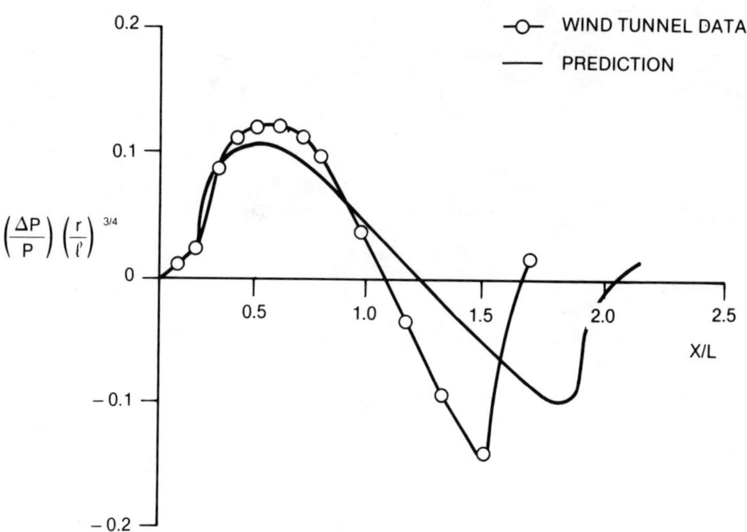

$$\left(\frac{\Delta P}{P}\right)\left(\frac{r}{\ell}\right)^{3/4}$$

Fig. 2   Comparison of predicted and measured pressure fields 16 body lengths from the cone body.

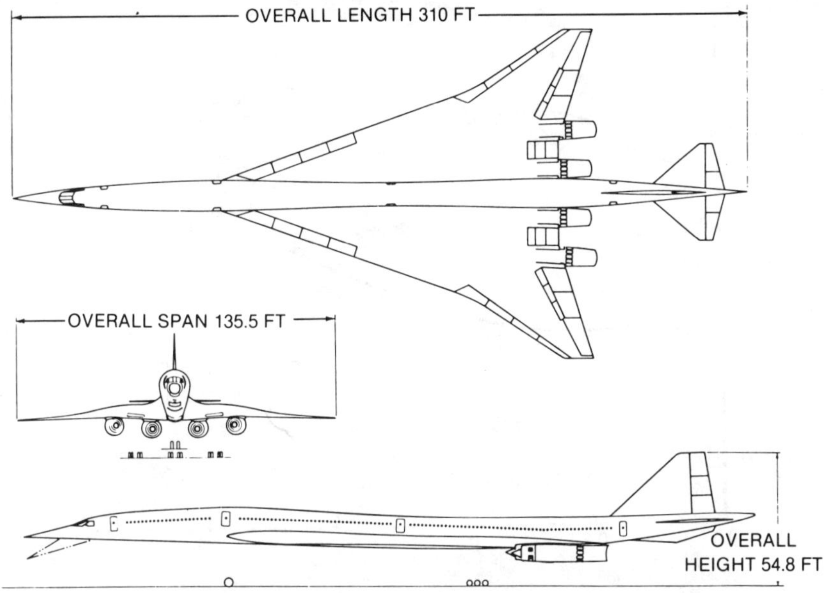

Fig. 3   Advanced supersonic transport baseline configuration.

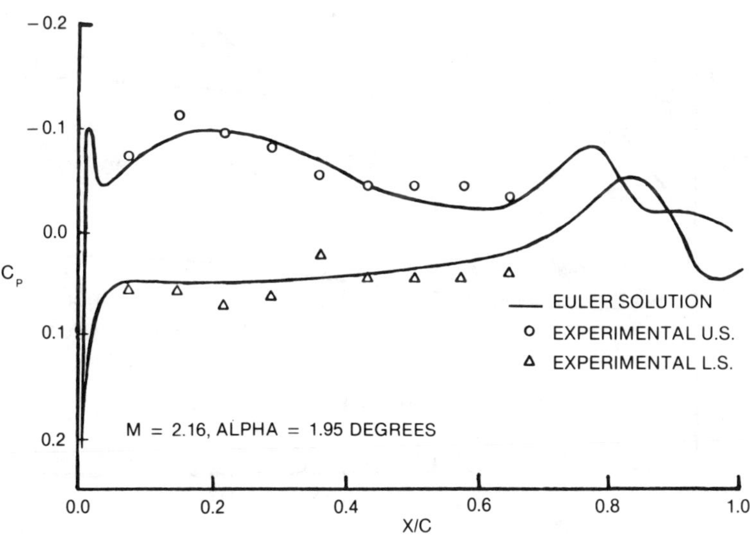

Fig. 4   AST Euler pressure distribution for 12.5% semispan.

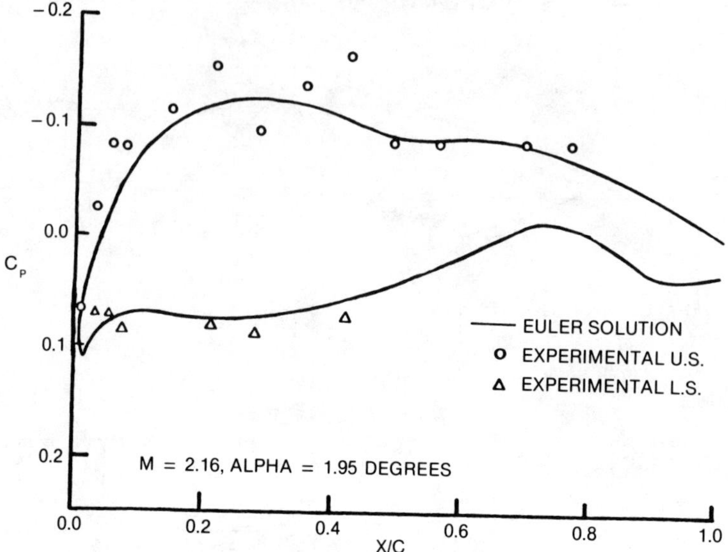

Fig. 5   AST Euler pressure distribution for 43.3% semispan.

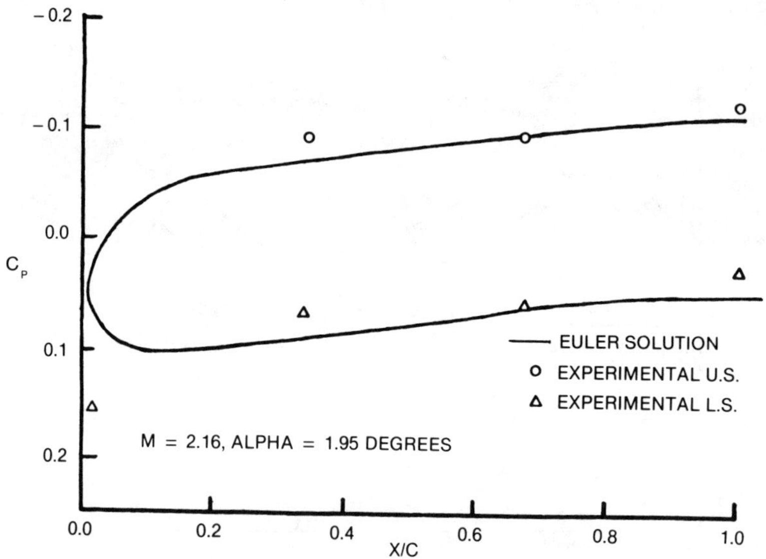

Fig. 6   AST Euler pressure distribution for 94.4% semispan.

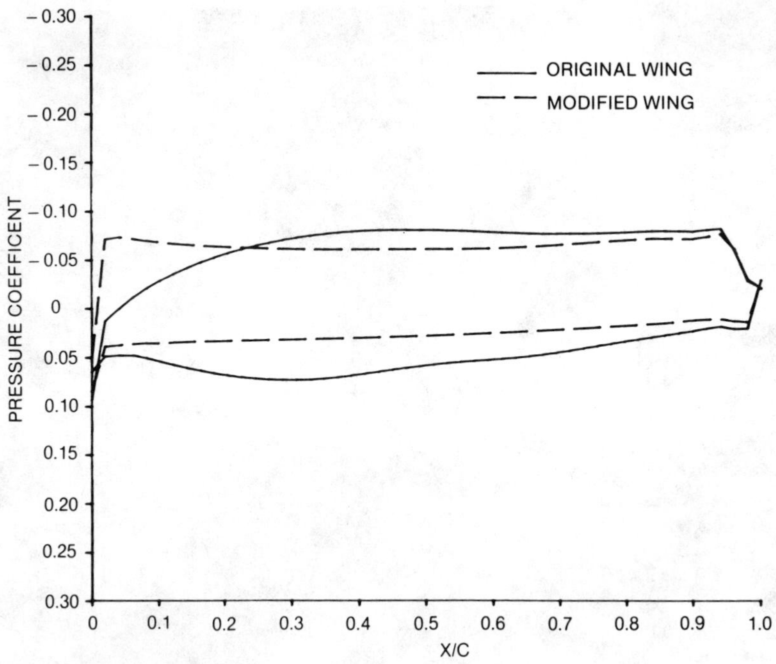

Fig. 7 Modified chordwise pressure distribution at 60% semispan station ($M = 2.2$, $\alpha = 2$ deg).

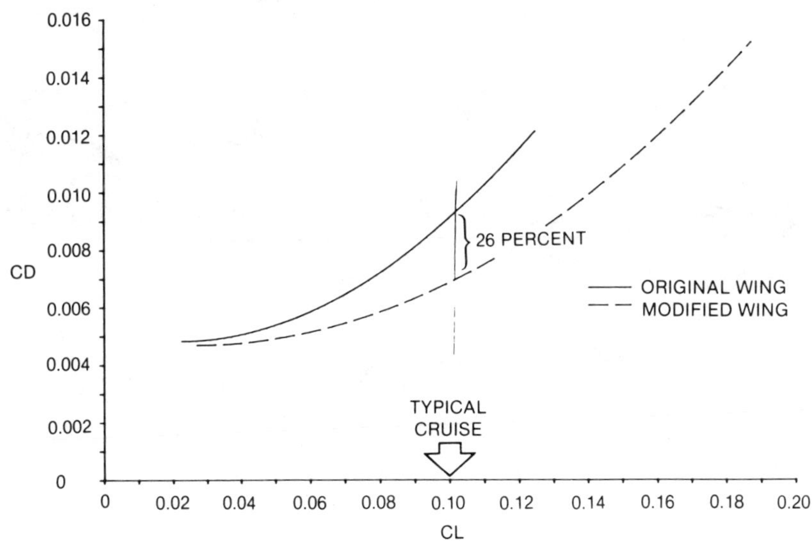

Fig. 8 Modified wing drag improvement ($M = 2.2$).

Fig. 9   Mach 5 inlet and test installation.

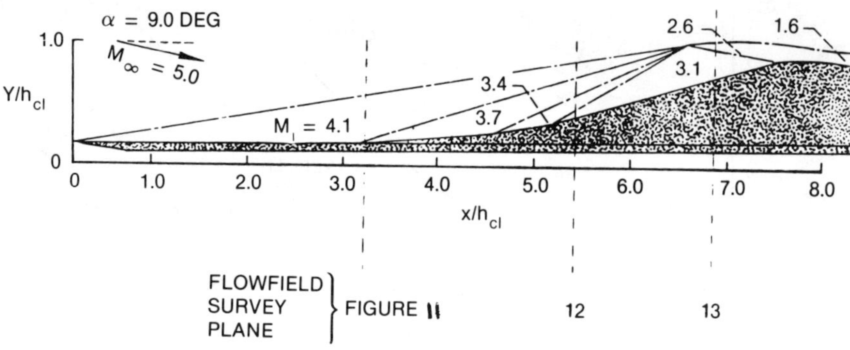

Fig. 10   Mach 5 inlet aerodynamic design.

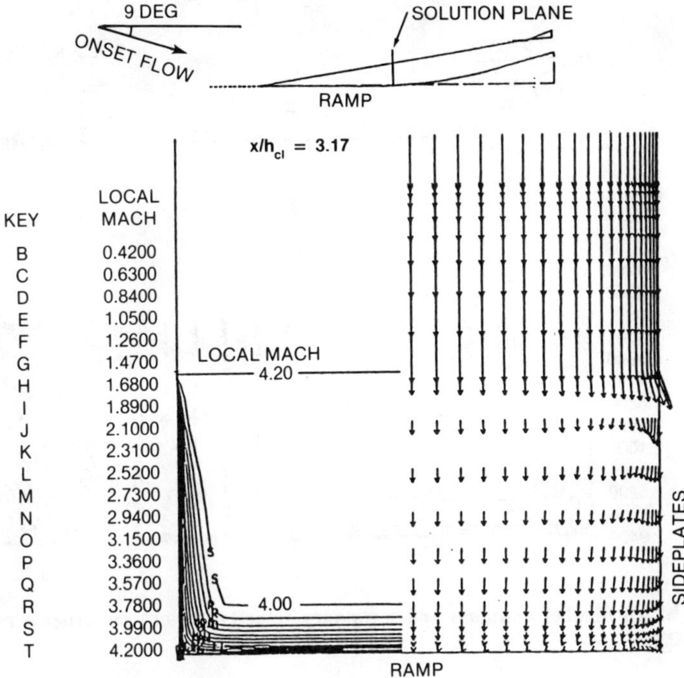

Fig. 11 **Mach number contours and secondary velocity vectors before second ramp using PEPSIS code.**

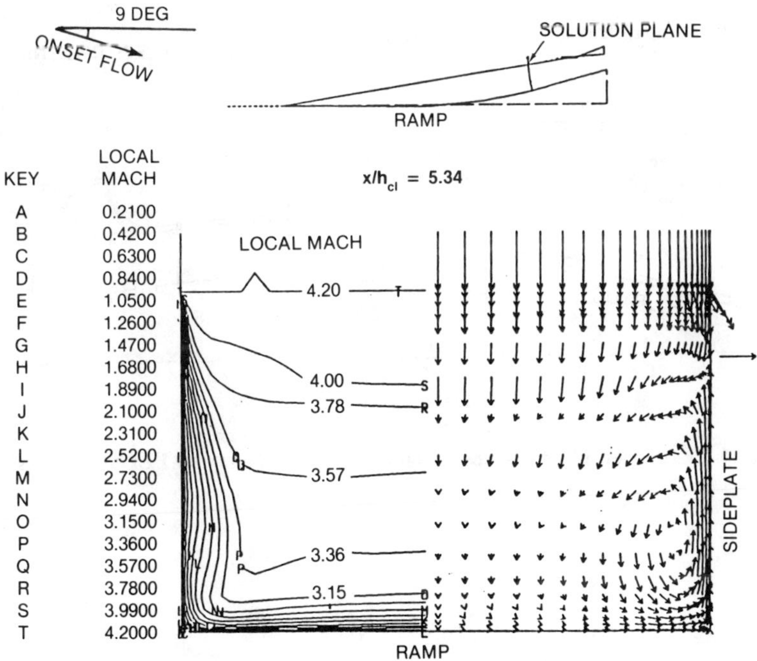

Fig. 12 **Mach number contours and secondary velocity vectors aft of fourth ramp using PEPSIS code.**

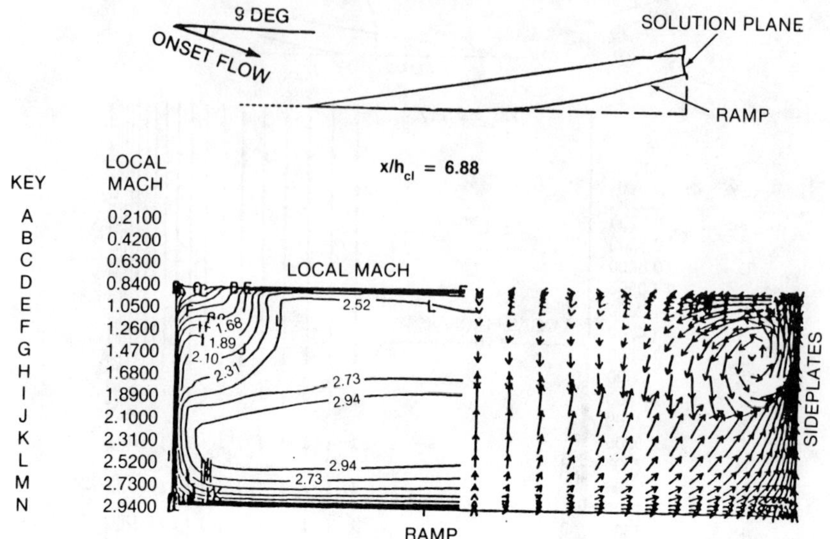

Fig. 13   Mach number contours and secondary velocity vectors at corner separation using PEPSIS code.

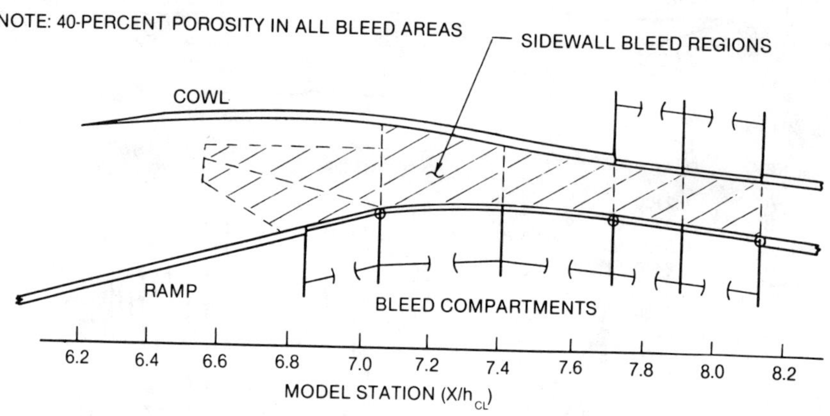

Fig. 14   Expanded bleed regions designed into inlet based on PEPSIS analysis.

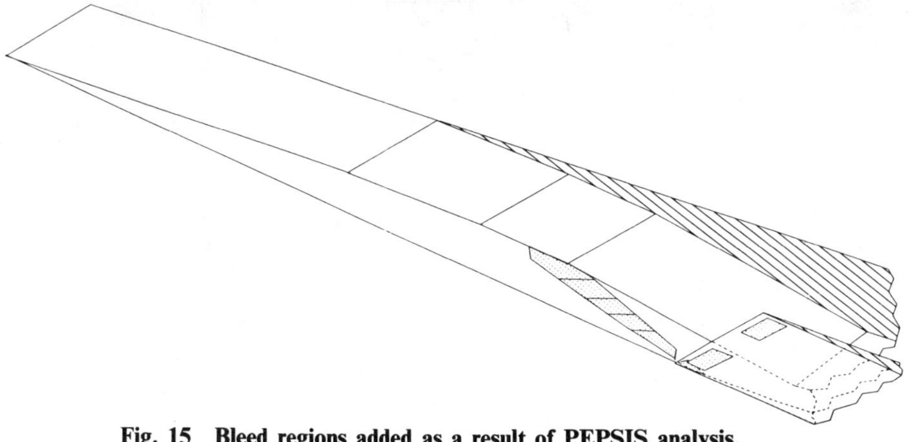

Fig. 15   Bleed regions added as a result of PEPSIS analysis.

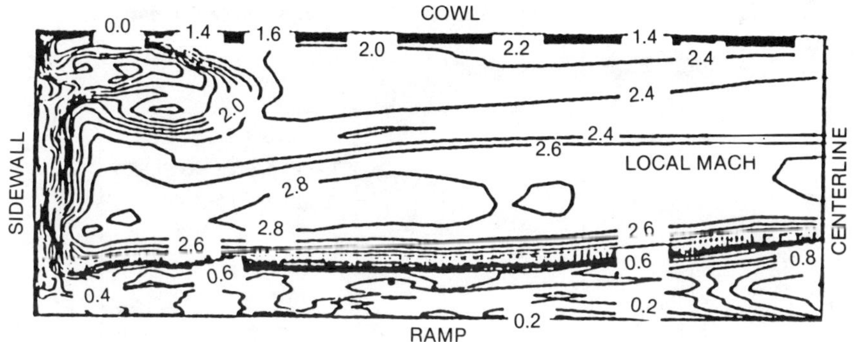

Fig. 16   Full Navier-Stokes solution for inlet location $x/h_{cl} \sim 7.0$ without any bleed (corresponds to PEPSIS solution in Fig. 13).

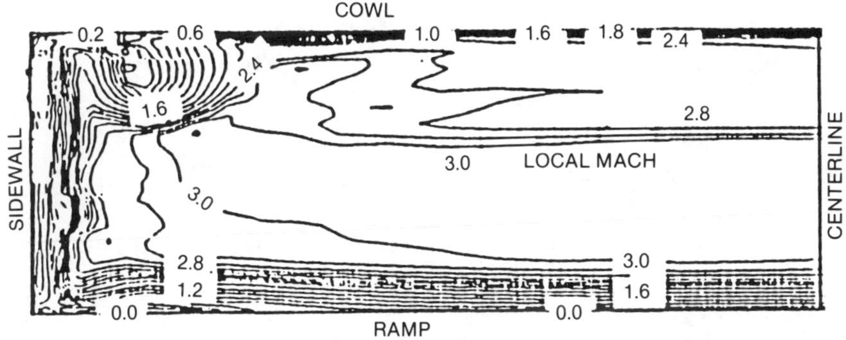

Fig. 17   Full Navier-Stokes solution for inlet location $x/h_{cl} \sim 7.0$ with centerbody and cowl (but no sidewall) bleed.

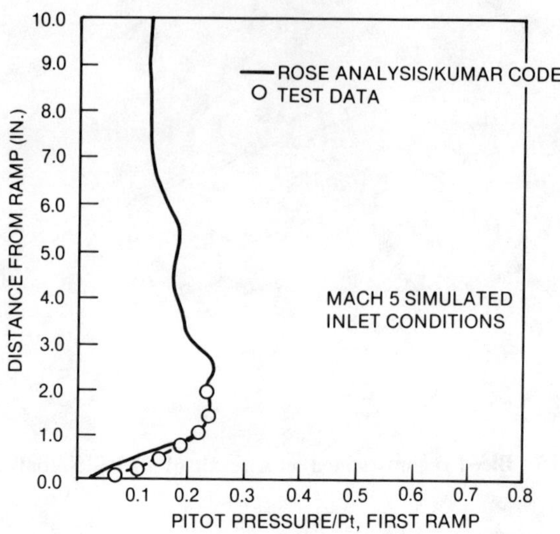

Fig. 18   Third ramp centerline rake: code analysis/data comparison.

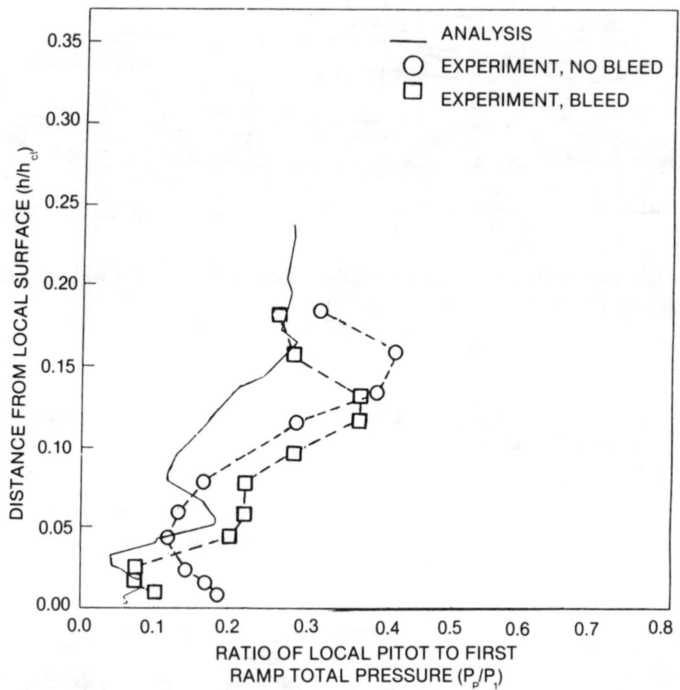

Fig. 19   Cowl corner rake: CFD analysis/data comparison.

Chapter 23

Supersonic/Hypersonic Euler Flowfield Prediction Method for Aircraft Configurations

A. Verhoff,* D. Stookesberry,† B. Hopping,† and T. Michal‡
McDonnell Douglas Corporation, St. Louis, Missouri

Introduction

A METHOD for accurate, efficient prediction of supersonic and hypersonic inviscid flowfields is presented that uses a spatial marching procedure to numerically approximate the solution of the QAZ1D form of the Euler equations presented in Ref. 1. This formulation is unique in that the equations are written in terms of Riemann-type variables using a natural streamline coordinate system. Because of this, the equations model wave propagation in a very physical manner with no Mach number limitations. Equally important is the ease with which the associated numerical solution procedure can be coded and the high efficiency that can be achieved on a vector processor such as the Cray XMP or Convex C1/C2.

Because the QAZ1D Euler formulation describes the physical process of wave propagation and has no inherent Mach number restrictions, it lends itself readily to supersonic and hypersonic flowfield analysis. A computer program called Streamline Coordinate Riemann Axial Marching (SCRAM) has been developed[2,3] based on the QAZ1D formulation. For fully supersonic inviscid flow, three-dimensional flowfield analysis can be reduced to a sequence of two-dimensional problems by using a spatial marching solution procedure.

Many existing marching procedures use the longitudinal body axis as the primary direction of information propagation. This requires the velocity in the body axis direction to be supersonic. The QAZ1D formulation uses a local streamline-oriented coordinate system at each grid point in the flowfield to propagate information downstream in a much more physically accurate manner. As a result, flowfields about configurations having wings, tails, fins, etc., can be accurately predicted using relatively coarse computational grids. Supersonic flow is required, but the velocity in the marching direction is only required to be positive.

Copyright © 1989 by the American Institute of Aeronautics and Astronautics, Inc. All rights reserved.

*Principal Technical Specialist-Aerodynamics.
†Senior Engineer-Aerodynamics.
‡Engineer-Aerodynamics.

Results from the SCRAM code are presented for several realistic aircraft configurations, including cases with a nonzero yaw angle. Predicted force and moment coefficients and surface pressures are compared with available test data. Predictions of control surface effectiveness are also presented. Real-gas SCRAM results may be found in Refs. 2 and 3.

Governing Equations

The fluid dynamic equations on which the SCRAM code is based are the three-dimensional Euler equations that describe inviscid flow. In terms of (characteristic-type) extended Riemann variables in a natural streamline coordinate system (s,n,m), these equations are[1]

$$\frac{\partial Q}{\partial t} + (q+a)\frac{\partial Q}{\partial s} = -\frac{\gamma-1}{2}a\left(S - \frac{2}{\gamma-1}\right)\left[\frac{\partial}{\partial s}\left(q - \frac{2}{\gamma-1}a\right)\right]$$
$$-\frac{\gamma-1}{2}qaS\left[\frac{\partial\theta}{\partial n} + \cos\theta\frac{\partial\phi}{\partial m}\right] \tag{1a}$$

$$\frac{\partial R}{\partial t} + (q-a)\frac{\partial R}{\partial s} = +\frac{\gamma-1}{2}a\left(S - \frac{2}{\gamma-1}\right)\left[\frac{\partial}{\partial s}\left(q + \frac{2}{\gamma-1}a\right)\right]$$
$$+\frac{\gamma-1}{2}qaS\left[\frac{\partial\theta}{\partial n} + \cos\theta\frac{\partial\phi}{\partial m}\right] \tag{1b}$$

$$\frac{\partial S}{\partial t} + q\frac{\partial S}{\partial s} = 0 \tag{1c}$$

$$\frac{\partial\theta}{\partial t} + q\frac{\partial\theta}{\partial s} = -\frac{a^2}{\gamma q}\frac{\partial \ell np}{\partial n} \tag{1d}$$

$$\frac{\partial\phi}{\partial t} + q\frac{\partial\phi}{\partial s} = -\frac{a^2}{\gamma q\cos\theta}\frac{\partial \ell np}{\partial m} \tag{1e}$$

The extended Riemann variables are defined as

$$Q \equiv q + aS \tag{2a}$$

$$R \equiv q - aS \tag{2b}$$

where q is the velocity magnitude, and a is the speed of sound. Time is denoted by t. The modified entropy S is defined in terms of pressure p, density ρ, and the ratio of specific heats γ by the relation

$$S \equiv \frac{1}{\gamma(\gamma-1)}\left[2\gamma - \ell n\left(\frac{p}{\rho^\gamma}\right)\right] \tag{3}$$

The preceding relationships assume that the gas is perfect.

The flow angles θ and ϕ are defined as shown in Fig. 1. Also shown is the natural streamline coordinate system (s,n,m) with respect to a fixed rectangular Cartesian system (x,y,z). The natural system is orthogonal in that s is measured in the direction of flow, n lies in the plane defined by the y axis and the s-coordinate direction, and m is normal to the (s,n) plane. At each point in the computational grid, the system of equations is expressed in the streamline coordinate system local to that point.

Solution Methodology

The various elements of the solution methodology used in the SCRAM code are described in the following subsections.

Numerical Solution Algorithm

Within each successive cross-sectional plane, the QAZ1D Euler equations [Eqs. (1)] are relaxed to steady state using explicit time integration. Streamwise spatial derivatives are approximated by three-point one-sided (upstream) finite differences to properly model the wave propagation. Derivatives normal to the streamline direction are decomposed and approximated with three-point one-sided differences determined by the local direction of wave propagation. The characteristic analysis for the decomposition can be outlined simply for the case of isentropic two-dimensional flow. In matrix notation the QAZ1D equations under these assumptions have the form

$$\{F\}_t + [A]\{F\}_s + [B]\{F\}_n = 0 \tag{4}$$

where $\{F\}$ is the column vector of dependent variables, and $[A]$ is the diagonal matrix of characteristic velocities. Subscripts s and n refer to derivatives along and normal to the local streamline direction. In order to distinguish directions of propagation normal to the local streamline direction, the B matrix is transformed according to

$$[B] = [X][\lambda][X]^{-1} \tag{5}$$

This provides the rule for splitting the B matrix into

$$[B] = [B^+] + [B^-] \tag{6}$$

where $[B^+]$ and $[B^-]$ each have a unique direction of propagation based on the eigenvalue sign. Splitting the B matrix allows numerical differencing normal to the streamline direction that is consistent with the physical direction of wave propagation. It acts to inhibit any odd-even decoupling of the solution (sawtooth pattern) sometimes observed with central-difference approximations. Although it produces a dissipative truncation error similar to "artificial viscosity," it is not artificial because the coefficients of the dissipative terms are not arbitrary. A more detailed description of the characteristic analysis, can be found in Ref. 4.

To accelerate the convergence rate in each cross plane, local maximum time steps are used along with a checkerboard (odd-even) scheme. The resulting formulation is almost completely vectorizable on Cray XMP and CONVEX computers. Nonphysical input parameters are not used in the solution algorithm.

Starting Solution

A supersonic starting plane solution is required to initiate the downstream marching procedure. For sharp-nosed configurations it is assumed that a small region about the nose may be modeled by conical flow. The local conical flowfield solution is obtained by guessing a starting plane solution and shock geometry. An iterative "step back" procedure is then used. A downstream step is taken assuming the geometry is conical, and the solution is converged in this plane. This solution is transferred back to the original starting plane location under the assumption of conical flow. This process is repeated until the solution and shock geometry being transferred back to the starting plane are unchanged from the previous cycle. This starting plane solution is then used to continue the solution downstream for the actual geometry.

Surface Boundary Conditions

During the transient solution process in each cross plane, a local velocity vector tangent to a solid boundary surface will respond to the local pressure gradient on the surface. Since the streamlines are constrained to the body surface, the velocity vector can only rotate in the local tangency plane. Because of this, the flow angles θ and ϕ are algebraically related. With the QAZ1D formulation, the surface boundary calculation is reduced locally to a two-dimensional problem on the tangent plane at a given surface grid point. The pressure gradient normal to the streamline in this plane dictates the deflection required of the local velocity vector. Once the new velocity vector orientation is known, the corresponding changes to θ and ϕ are easily determined.

Shock Wave Treatment

Bow shock waves can be fitted within the SCRAM formulation using the Rankine-Hugoniot relations to obtain flow conditions behind the shock. As the solution is converged in each cross plane, the grid outer boundary is adjusted along radial grid lines to conform to the shock shape. The grid is then rescaled within SCRAM to the new outer-boundary shape. Secondary shock waves are captured isentropically. These waves tend to be weaker than the bow wave and can usually be approximated as isentropic. An option also exists to capture the bow shock isentropically. An evaluation of isentropic shock capturing error using this option is given in Ref. 2. Within the typical supersonic operating envelope of fighter aircraft, a bow shock wave captured isentropically results in local pressure errors of well below 1%. Although more calculations are required for bow shock fitting, the overall number of grid points is reduced because the grid need not extend beyond the shock wave. Because the shock fitting calculations vectorize, there is only a slight increase in computational cost when the shock fitting option is used.

Grid Generation

Accurate calculations of flowfield quantities is highly dependent on the quality of the computational grid regardless of the flowfield solution procedure. The grid must accurately resolve the body shape in addition to being nearly orthogonal and smoothly varying, particularly near the body surface.

The surface-conforming grid-generation method that was developed for the SCRAM code evolved from the method presented in Ref. 5. The grid is generated by solving an elliptic system of partial differential equations. Forcing functions of the type described in Ref. 6 allow strong localized control of node spacing and grid orthogonality on and near grid boundaries. Moreover, this grid control near boundaries allows multiple grid blocks to be generated independently and matched smoothly at their common boundaries. This feature is particularly useful for the geometries characteristic of hypersonic vehicle concepts.

For the H-O topology used by the SCRAM code, the entire three-dimensional grid is constructed by successive generation of two-dimensional grids at each longitudinal station of interest. The two-dimensional grid code automatically sections a complex domain into a number of simpler domains on the basis of singular geometry points such as sharp internal or external corners. A high-quality grid is then generated for each of the simpler sections. The sections are matched smoothly at their common boundaries due to the nearly complete control of the grid in the vicinity of boundaries. The procedure is automatic, not user interactive.

Capabilities and Applications

Numerical solutions have been computed for many different geometries in order to demonstrate the ability of SCRAM to predict flowfields about realistic aircraft configurations. Comparisons with test data have been made whenever possible. Representative examples are presented in the following subsections.

Blended-Body Shape

The Advanced Manned Interceptor (AMI-X) described in Ref. 7 is useful for validating supersonic flowfield codes because of its associated wind-tunnel data base. A sketch is shown in Fig. 2. The model was tested at Mach numbers of 4 and 6 with and without the propulsion module. Calculations were carried out for the engine-absent/tail-off geometry. The computational grid is shown in Fig. 3; it has half-plane dimensions 51 (circumferential) × 25 (radial) with 24 axial stations. Force and moment predictions are shown in Fig. 4. The agreement with test data is good, especially for inviscid supersonic drag. An estimate of the skin-friction drag has been subtracted from the drag data. Each case required about 3 min execution time on a Cray XMP-18 computer, including input and output processing.

Predicted surface pressure distributions at four axial stations are shown in Fig. 5. The corresponding off-body pressure contours for both the bow shock capturing and bow shock fitting options are also shown to illustrate solution quality. The bow shock is cleanly resolved when shock capturing is used. For both options the bow shock is much stronger on the upper side due to the relatively steep body surface slope. Stations 3 and 4 indicate that the wings are near the bow shock, which causes local regions of high pressure at the leading edge. This interaction is responsible for the "spikes" seen in the surface pressure distribution.

Wing/Body Shapes

A high-speed wind-tunnel test of a wing/body configuration was conducted in 1986 by McDonnell Aircraft Company.[8] The primary objective of the test was to provide an extensive supersonic data base for use in computational fluid dynamics (CFD) code development and validation. The fully configured model is shown in Fig. 6. The model had two removable forebodies with and without a canopy. The wing was removable for fuselage-alone testing so that component interactions could be quantified in detail. The forebodies were pressure-instrumented. The SCRAM code was applied to the four possible configurations, and the results were compared to test data at Mach numbers of 1.8, 2.5, and 5.0. Surface definition provided by the computational grid for the fully configured model is shown in Fig. 7. The grid had half-plane dimensions of 61×31 with 35 axial stations. Note that the fuselage extends beyond the wing trailing edge, which is also swept.

Force and moment predictions for the fully configured model (canopy on) are compared with test data in Fig. 8. An estimate of the skin-friction drag was again subtracted from the drag data. SCRAM accuracy is very good for all Mach numbers. A typical calculation for the fully configured model required 3 min of execution time on a Cray XMP-18 computer.

Accurate force and moment predictions require accurate prediction of surface pressures. SCRAM pressure predictions at several stations along the forebody with and without the canopy are compared with test data in Figs. 9 and 10. Agreement with test data is good. The SCRAM results also provide flowfield detail near the canopy/forebody intersection that would be difficult to measure experimentally. Note that this accuracy is achieved using a relatively coarse grid (61×31) in the crossflow half-plane. For good accuracy, SCRAM generally requires only enough points to resolve the geometry adequately.

Cambered Shapes

Flowfield predictions were made for the cambered fighter forebody of Ref. 9. This geometry has a highly cambered strake and a sharp wing leading edge. The surface grid is shown in Fig. 11. The grid is composed of three sections with half-plane dimensions of $61 \times 19 \times 6$, $85 \times 35 \times 11$, and $85 \times 41 \times 21$, respectively. Longitudinal patching allowed a smooth transfer of flowfield data between grids while maintaining adequate grid definition of the surface. Force and moment predictions are shown in Fig. 12 for a freestream Mach number of 1.8. An estimate of the skin-friction drag has been added to the SCRAM results. The agreement with test data is good, especially for inviscid supersonic drag. Predicted spanwise surface pressure distributions compared with test data at three axial stations are shown in Fig. 13 along with the corresponding off-body pressure contours. The SCRAM predictions agree well with test data, although the crossflow shock location is slightly in error. Predicted centerline pressures are also compared with test data.

Sideslip Capability

Most SCRAM computations can be carried out using a half-plane grid with symmetry boundary conditions to reduce computer time. Because of its low CPU and computer memory requirements, SCRAM can efficiently run full-plane grids for nonsymmetric vehicles or sideslip conditions. The hypersonic research vehicle of Ref. 10 provides a good wind tunnel data base for validating the sideslip capability of SCRAM. Figure 14 shows the surface grid that had full-plane dimensions of $121 \times 15 \times 25$. Flowfield predictions were made at a freestream Mach number of 3.46 and sideslip angle β of -4.0 to 11.0 deg. The solutions were obtained using the shock-fitting option in SCRAM. Lateral force and moment predictions are shown in Fig. 15 for 0.0- and 4.0-deg angle of attack. The comparison with test data is very good for both angles of attack throughout the range of sideslip angle. The SCRAM side force prediction begins to depart slightly from test data at the largest sideslip angle. This is probably due to the onset of separation on the tail surface. Computer run times for the different cases were between 12 and 15 min on a Cray XMP-18. Compared to the AMI-X case described previously, fewer radial grid points were used, but the shock fitting obtains greater accuracy for about the same cost. Aside from the difference in circumferential grid size (121 vs 51), the difference in run times was primarily due to the slower convergence associated with the more complex geometry.

Arbitrary Nose Shapes

For blunt-nosed configurations a subsonic flow region exists between the bow shock and the body. A starting plane solution for SCRAM must be provided at a location where the flow has become fully supersonic. An option in SCRAM computes a three-dimensional blunt-nosed flow solution with a fully iterative procedure. Solution of the Euler equations [Eqs. (1)] is carried out simultaneously at a number of cross-sectional planes encompassing the subsonic region. This procedure allows information from all directions to be utilized in the calculation. The space-marching procedure is then used downstream from where the flow becomes supersonic. By using the QAZ1D formulation [Eqs. (1)] and the shock-fitting option, consistency is provided between the nose tip solution and the marching solution.

The starting procedure has been validated for various blunt-nosed geometries. Figure 16 shows the predicted surface pressure distribution for a blunt cone/cylinder geometry at a freestream Mach number of 4.95. The predictions agree well with flight-test data[11] up to the sharp expansion corner. A small region of flow separation exists downstream of the corner which an inviscid code such as SCRAM will not predict. The predicted shock-wave shape for a cylinder with a spherical nose cap at a freestream Mach number of 3.0 is shown in Fig. 17. The comparison with test data[12] is very good. For both of these computed results, a smooth transition is obtained between the fully iterative starting solution and the marching solution.

As a further demonstration of SCRAM capability, the flowfield about the lifting-body geometry of Ref. 13 was computed at a freestream Mach number of 3.0. Figure 18 shows the surface portion of the computational grid that had half-plane dimensions of $61 \times 13 \times 47$. Pressure contours on the surface and centerplane are also shown. Two shock waves are visible: the primary bow shock and a secondary shock produced by the canopy. The primary shock is fitted, whereas the secondary shock is isentropically captured. The starting solution extends to the middle of the canopy in order to encompass a region of subsonic flow produced by the canopy shock. The transition between the starting solution and the marching solution is smooth.

An option is also included in SCRAM to generate a starting plane solution for wedge-type forebodies. The approach assumes that the forebody leading-edge region is a two-dimensional wedge and the bow shock angle and flow quantities on the upper and lower leading-edge surfaces are obtained using two-dimensional oblique shock and expansion wave theory. The downstream marching procedure is then initiated from this "slit" of double-valued initial conditions. Force and moment predictions for a wedge-type forebody are presented in Fig. 19. A half-plane grid of $49 \times 15 \times 9$ was used. The SCRAM predictions compare well with the test data of Ref. 14. An adjustment for skin-friction drag was not made to the SCRAM predictions.

Inlet/Nozzle Capability

Figure 20 presents results for the supersonic persistent fighter (SSPF) model of Ref. 15. This model has engine inlets and was used to validate the SCRAM flowing inlet capability. At the inlet face, the mass flow is removed from the flowfield assuming no spillage. This mass flow may be added back at a nozzle face to simulate an exhaust plume. Surface pressure comparisons with test data at a freestream Mach number of 1.8 are shown at two stations downstream of the inlet. The results compare well with the test data.

At the nozzle exit face, arbitrary flow conditions can be imposed. To demonstrate the nozzle capability, a generic shape with an elliptic inlet and nozzle was constructed. Figure 21 shows the predicted Mach number contours within the centerplane for a freestream Mach number of 4.0. Constant pressure, Mach number, and temperature conditions were prescribed at the nozzle. Uniform growth of the plume and expansion along the reflexed underside of the aft end are evident. The slip surface between the plume and external flowfield was captured naturally by the QAZ1D formulation.

Flow-Through Boundary

For vehicle geometries that have aft-swept trailing edges or overhanging empennages, isolated solid surfaces appear in the cross-plane flowfield. A wake surface is added between the solid surfaces, and a flow-through boundary condition is applied. Force and moment predictions are presented in Fig. 22 for an arrow wing/body configuration at a freestream Mach number of 3.0. The predictions compare well with the test data of Ref. 16. An estimate for skin-friction drag was not added to the SCRAM predictions. Figure 23 shows a cross-sectional view of the flow-through boundary. Pressure contours are shown along with the shock-fitted grid. Note that the contour lines are relatively smooth at the flow-through boundary.

Control Surface Deflections

The effect of small deflections of control surfaces can be computed using SCRAM. The control surface can be any trailing-edge-type surface (i.e., aileron, elevon, rudder). To account for the surface deflection, SCRAM adjusts the flow angle on the vehicle surface similar to a surface blowing or suction boundary condition. The grid is not altered from the zero-deflection case and cannot account for the gap between the control surface and the adjacent geometry. Grid points on a control surface are flagged, and the boundary condition is adjusted based on the grid point location, an input hinge line, and an input deflection angle. The control surface can be full span or partial span, and the hinge line may be swept. By restarting the solution upstream of the control surface and changing the deflection angle in the input file, a series of deflection angles can be run efficiently.

Elevon control power predictions were made for the SSPF model of Ref. 17. Calculations were made using a faired inlet forebody/tails-off geometry. Figure 24 shows the surface grid and the actual model configuration. Predicted elevon control power increments are compared with test data in Fig. 25 for a freestream Mach number of 2.0. Control surfaces were deflected 5 and 10 deg. The agreement with test data is good. The baseline case required approximately 12 min of execution time on a Cray XMP-18 computer. Subsequent deflected surface cases required about 2 min each.

Subsonic Regions

Occasionally at a canopy- or wing-fuselage juncture a small region of subsonic flow may occur that will cause a breakdown in the marching (supersonic) solver. An example is the lifting-body geometry described earlier (see Fig. 18). When a subsonic point is detected, SCRAM will write a restart file from the last supersonic station and stop. The solution procedure may be restarted using a fully iterative routine (similar to the blunt-nosed solver) over a specified number of stations. When a longitudinal station is reached where the flowfield again becomes fully supersonic, SCRAM will switch back to the marching solver to save execution time and computer memory. For the case of the lifting body, the blunt-nosed routine would be used to start the solution. The marching solver would then be used up to the canopy. At the front of the canopy the subsonic routine would be used, and then SCRAM would switch back to the marching solver for the rest of the body.

Conclusions

Numerical solutions of the Euler equations for supersonic flowfields about various aircraft configurations have been obtained using the SCRAM code. SCRAM is based on the QAZ1D form of the Euler equations and has been shown to be an accurate, efficient prediction method for supersonic/hypersonic inviscid flow about complex configurations. The code has been coupled with a versatile grid generation procedure for construction of high-quality computational grids about such shapes. Pressure distributions, forces, and moments compare well with test data for configurations having aft-swept wing trailing edges, vertical tails, engine inlets and nozzles, and control surface deflections. A fully iterative procedure is used to obtain starting solutions for blunt-nosed geometries.

References

[1]Verhoff, A., and O'Neil, P. J., "A Natural Formulation for Numerical Solution of the Euler Equations," AIAA Paper 84-0163, Jan. 1984.

[2]Verhoff, A., and O'Neil, P. J., "Accurate, Efficient Prediction Method for Supersonic/Hypersonic Inviscid Flow," AIAA Paper 87-1165, June 1987.

[3]Verhoff, A., Stookesberry, D. C., Hopping, B. M., and Michal, T. R., "Supersonic/Hypersonic Euler Flowfield Prediction Method for Aircraft Configurations," Fourth Symposium on Numerical & Physical Aspects of Aerodynamic Flows, Long Beach, CA, Jan. 16–19 1989.

[4]Agrawal, S., Vermeland, R., Verhoff, A., and Lowrie, R. B., "Euler Transonic Solutions over Finite Wings," AIAA Paper 88-0009, Jan. 1988.

[5]Thompson, J. F., Thames, F. C., and Mastin, C. W., "Automatic Numerical Generation of Body-Fitted Curvilinear Coordinate System for Field Containing Any Number of Arbitrary Two-Dimensional Bodies," *Journal of Computational Physics*, Vol. 15, No. 3, July 1974, pp. 299–319.

[6]Sorenson, R. L., "A Computer Program To Generate Two-Dimensional Grids About Airfoils and Other Shapes by the Use of Poisson's Equation," NASA TM-81198, May 1980.

[7]Woods, J., Scheller, D. M., and Bear, R. L., "Experimental Study of Advanced Manned Interceptor Configurations," McDonnell Douglas MDC Rept. A0397, Vol. I, Sept. 1971.

[8]Schwimley, S. L., and Roesch, M. T., "Aerodynamic Effects of Forebody Camber and Canopy," McDonnell Douglas MDC Rept. B0552, Aug. 1987.

[9]Pittman, J. L., and Bonhaus, D. L., "A Strake Design Method for Supersonic Speeds and Low Lift," AIAA Paper 87-2638, Aug. 1987.

[10]Hart, N. E., Dickson, C. R., Boyer, K. V., and Pieper, E. M., "Wind Tunnel Tests on a Hypersonic Research Model in the McDonnell Polysonic Wind Tunnel, Series I," McDonnell Aircraft Co. MCAIR Rept. No. 9981, Dec. 1963.

[11]McConnell, D. G., "Free-Flight Observation of a Separated Turbulent Flow Including Heat Transfer Up to Mach 8.5," NASA TN D-278, Oct. 1961.

[12]Baer, A. L., "Pressure Distributions on a Hemisphere Cylinder at Supersonic and Hypersonic Mach Numbers," Arnold Engineering Development Center TN-61-96, 1961.

[13]Rakich, J. V., "Aerodynamic Performance and Static-Stability Characteristics of a Blunt-Nosed, Boattailed 13_0 Half-Cone at Mach Numbers from 0.60 to 5.0," NASA TM-X-570, July 1961.

[14]Krieger, R. J., "A Technique for Developing Low Drag Nose Shapes for Advanced Supersonic Missile Concepts," AIAA Paper 80-0361, Jan. 1980.

[15]Haynes, D. A., Miller, D. S., Klein, J. R., and Louie, C. M., "Design and Experimental Verification of an Equivalent Forebody to Produce Disturbances Equivalent to Those of a Forebody with Flowing Inlets," AIAA Paper 88-0195, Jan. 1988.

[16]Holdaway, G. H., and Mellenthin, J. A., "Investigation at Mach Numbers of 0.20 to 3.50 of Blended Wing-Body Combinations of Sonic Design with Diamond, Delta, and Arrow Plan Forms," NASA TM-X-372, Aug. 1960.

[17]Wind Tunnel Test Data, "Aerodynamic Characteristics of Four Supersonic Cruise Configurations for Mach 1.6 thru Mach 2.16," NASA Langley Research Center Unitary Plan Wind Tunnel Test 1424, Sept. 1982.

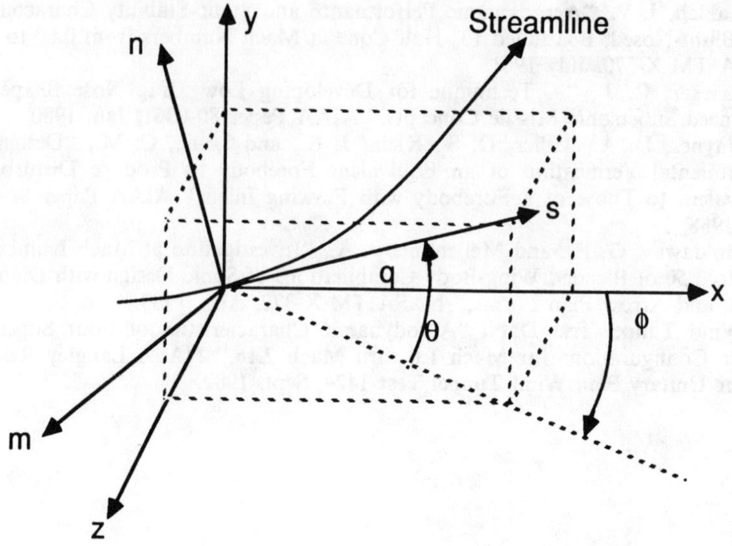

Fig. 1   Definition of flow angles and streamline coordinates.

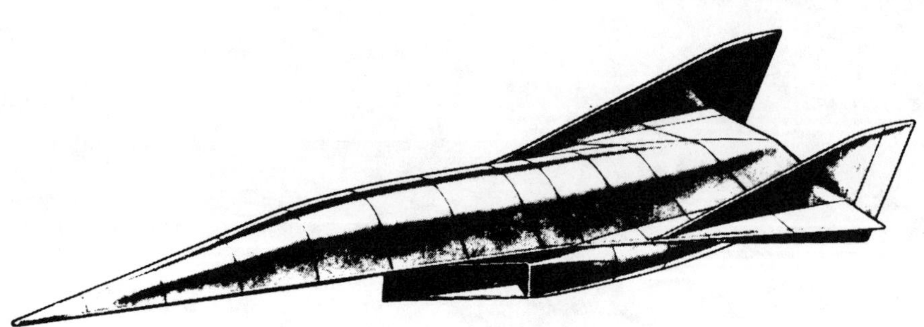

Fig. 2   AMI-X geometry.

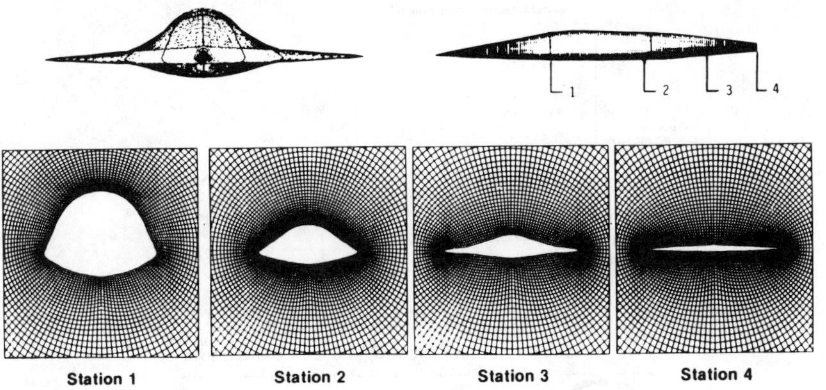

Fig. 3 Computational grid for AMI-X blended-body (vertical tails and propulsion module off).

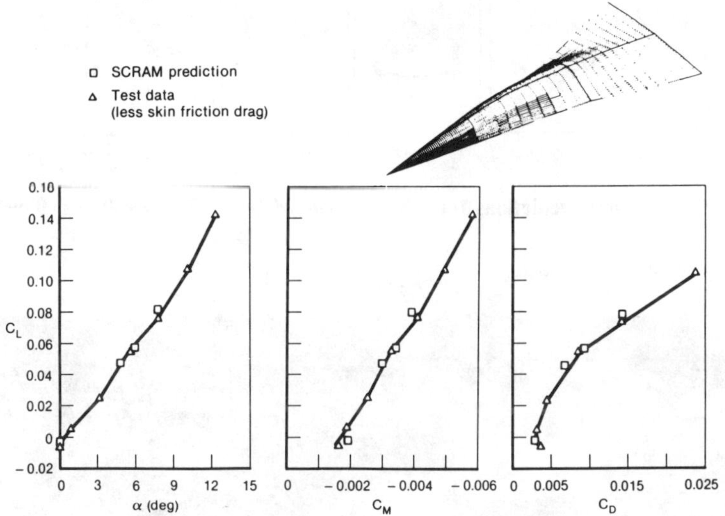

Fig. 4 Force and moment predictions for AMI-X blended body ($M_\infty = 6.0$).

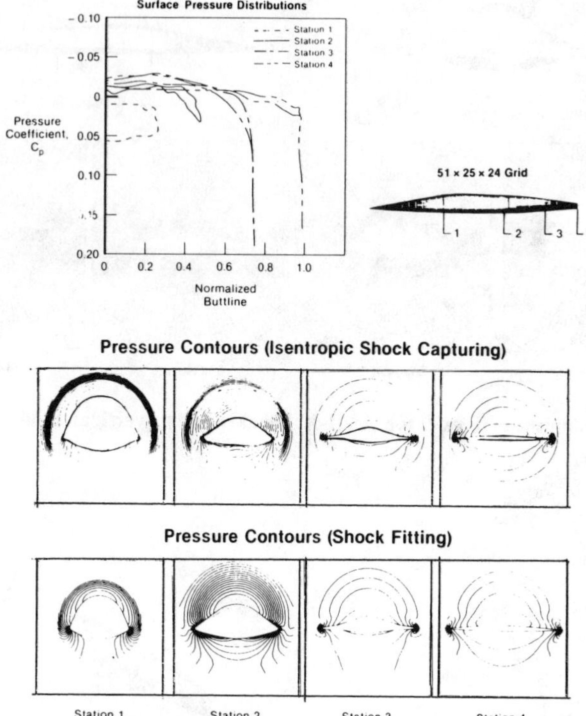

Fig. 5 Pressure predictions for AMI-X blended body ($M_\infty = 6.0$, $\alpha = 0$ deg).

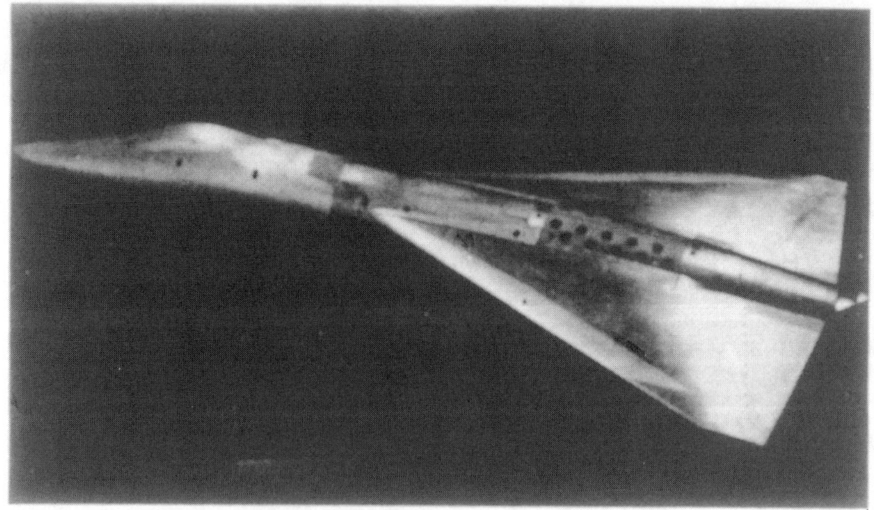

Fig. 6 High-speed research model.

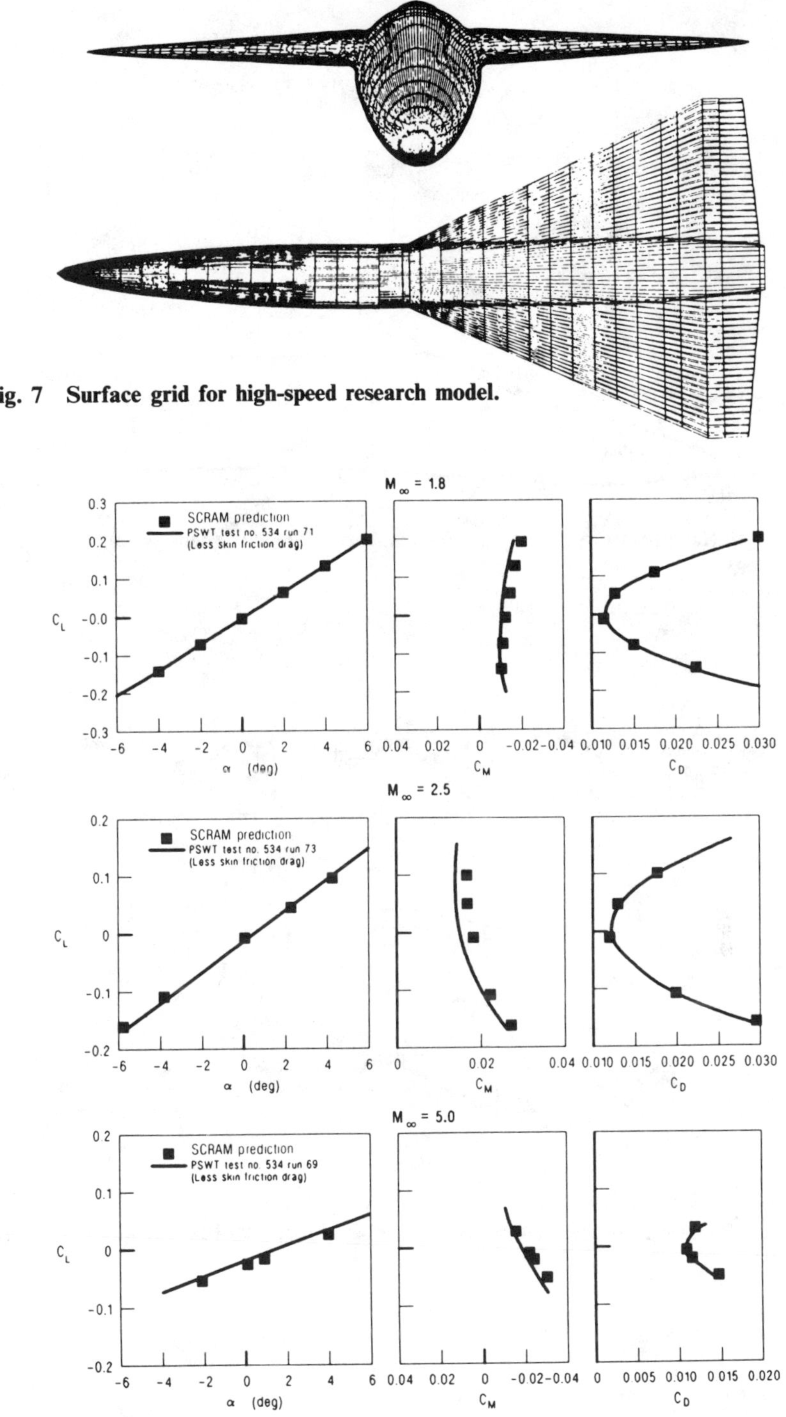

Fig. 7   Surface grid for high-speed research model.

Fig. 8   Force and moment predictions for high-speed research model.

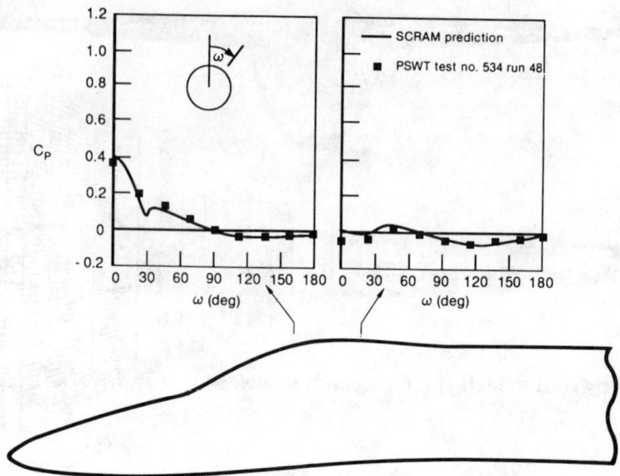

Fig. 9 **Surface pressure predictions for cambered forebody (canopy on, $M_\infty = 2.5$, $\alpha = 0$ deg).**

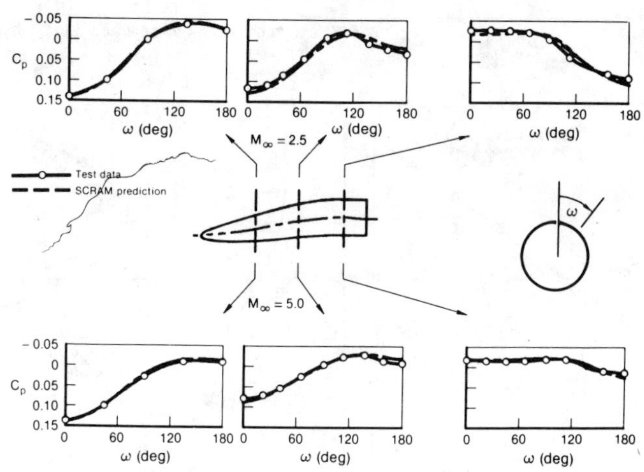

Fig. 10 **Surface pressure predictions for cambered forebody (canopy off, $\alpha = 0$ deg).**

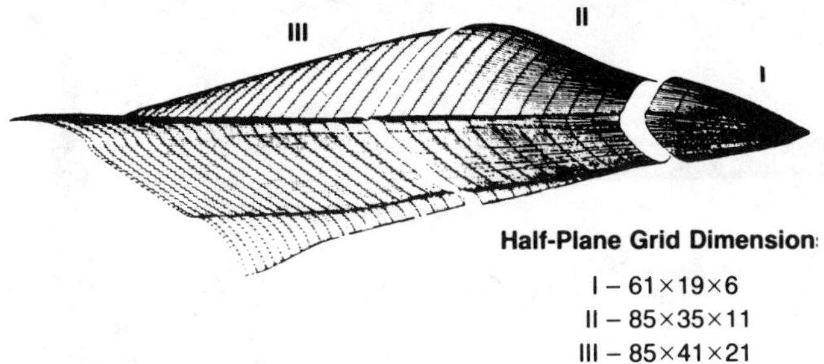

Fig. 11 Surface grid for cambered fighter forebody.

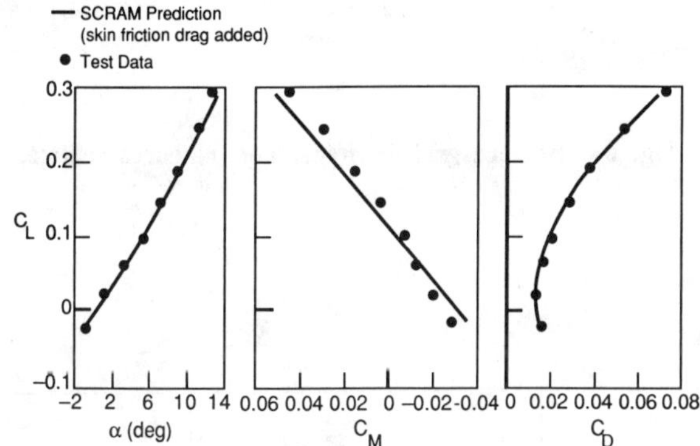

Fig. 12 Force and moment predictions for cambered fighter forebody ($M_\infty = 1.8$).

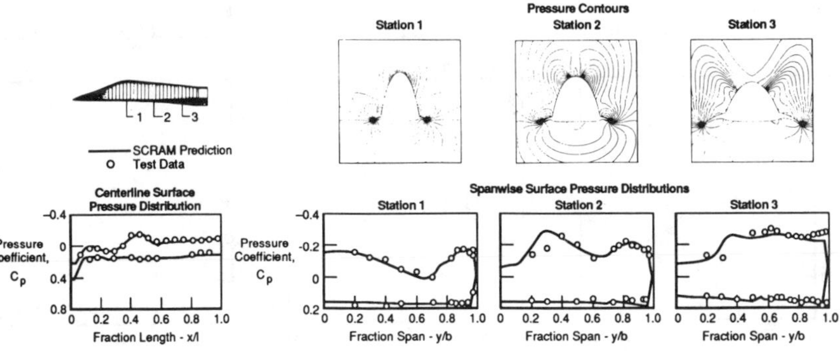

Fig. 13 SCRAM pressure predictions for cambered fighter forebody ($M_\infty = 1.8$, $\alpha = 13$ deg).

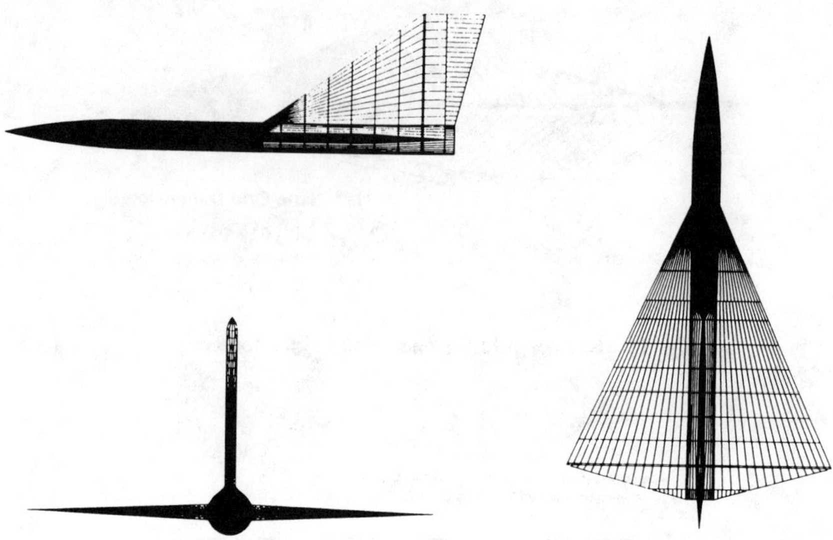

Fig. 14 Surface grid for hypersonic research vehicle.

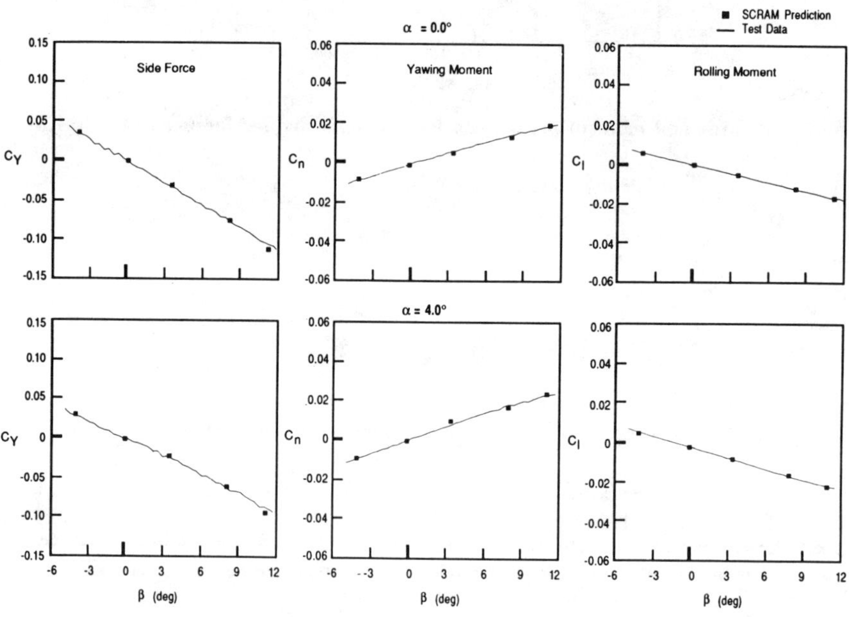

Fig. 15 Lateral force and moment predictions for hypersonic research vehicle ($M_\infty = 3.46$).

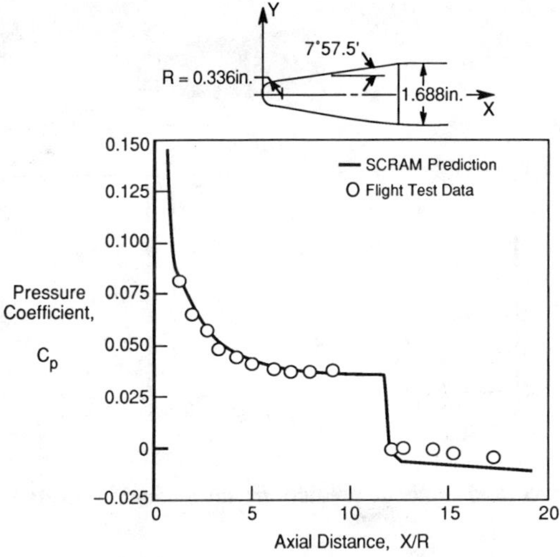

Fig. 16 Surface pressure distribution for blunt cone/cylinder ($M_\infty = 4.95$, $\alpha = 0$ deg).

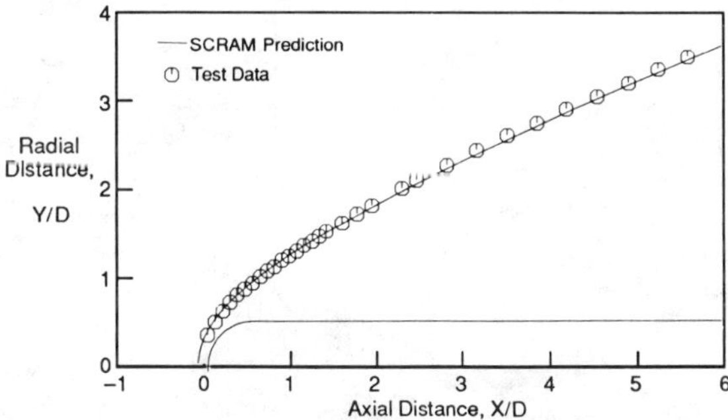

Fig. 17 Shock location for hemisphere/cylinder ($M_\infty = 3.0$, $\alpha = 0$ deg).

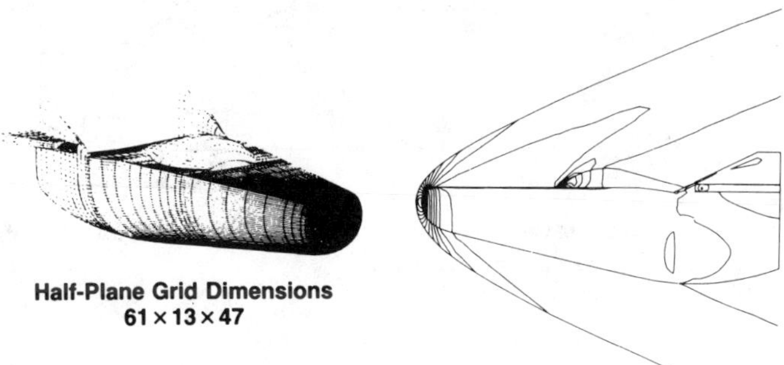

Fig. 18 Pressure contours for lifting-body geometry ($M_\infty = 3.0$, $\alpha = 0$ deg).

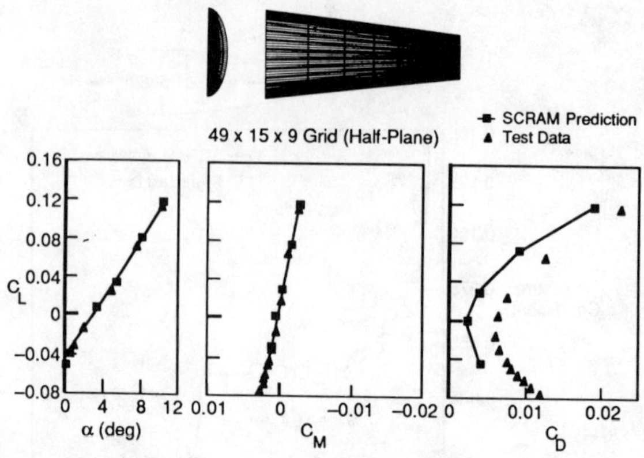

Fig. 19   Force and moment predictions for wedge forebody ($M_\infty = 5.0$).

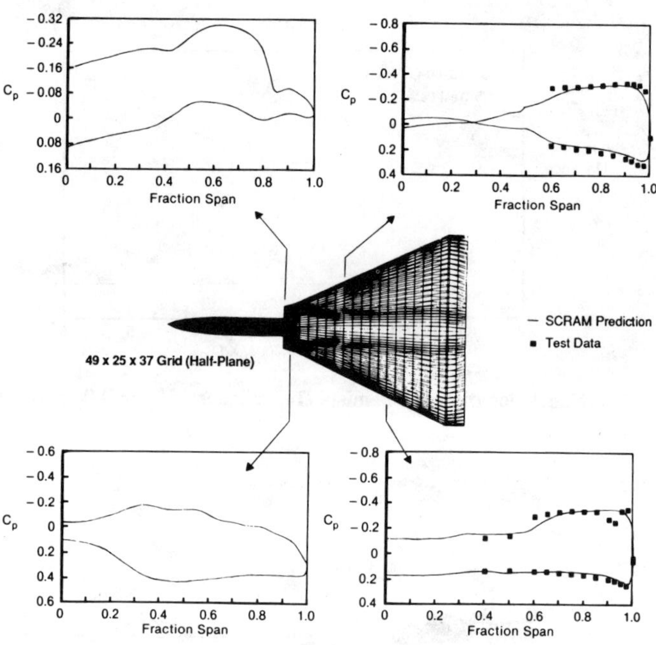

Fig. 20   Surface pressure predictions for supersonic persistent fighter (SSPF) model with engine inlets ($M_\infty = 1.8$, $\alpha = 10.2$ deg).

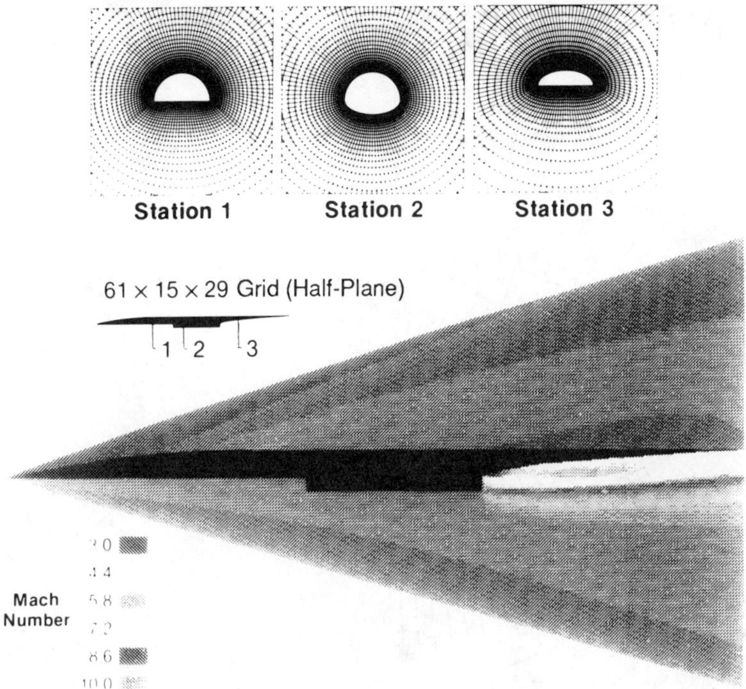

Fig. 21 Mach number predictions for generic shape with inlet and nozzle ($M_\infty = 4.0$, $\alpha = 0$ deg, $P_E/P_\infty = 4.0$, $M_E/M_\infty = 1.5$).

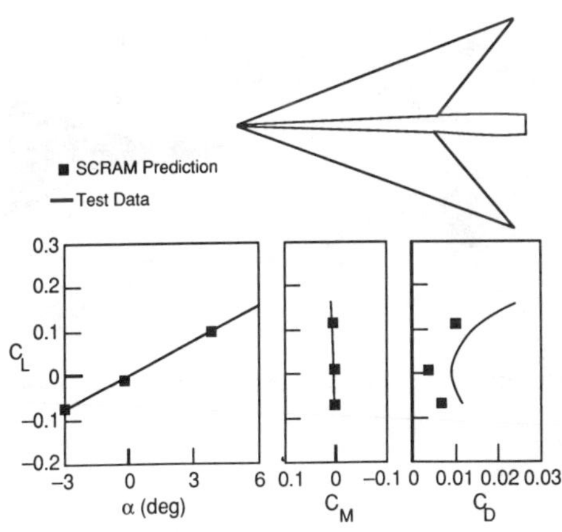

Fig. 22 Force and moment predictions for arrow wing/body ($M_\infty = 3.0$).

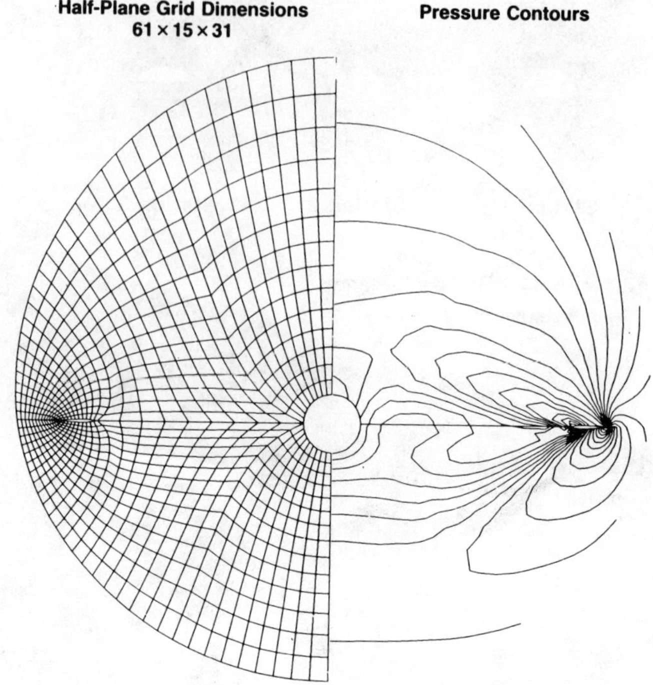

Fig. 23 Arrow wing/body flow-through boundary illustration ($M_\infty = 3.0$, $\alpha = 4$ deg).

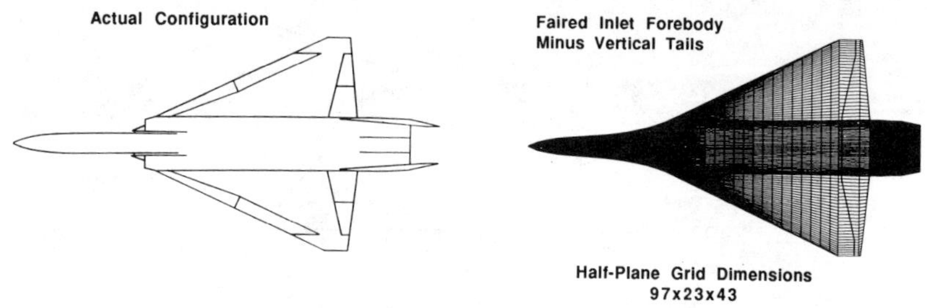

Fig. 24 Surface grid for supersonic persistent fighter (SSPF) model.

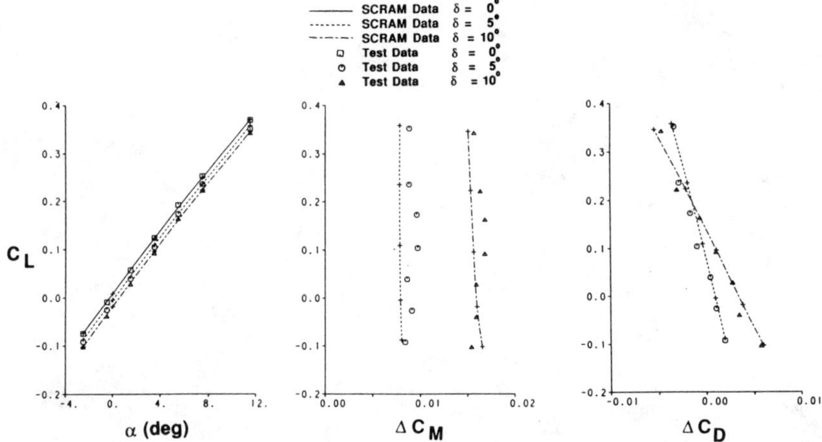

Fig. 25 SSPF elevon control power ($M_\infty = 2.0$).

Chapter 24

Flow Computations for the Space Shuttle in Ascent Mode Using Thin-Layer Navier-Stokes Equations

F. W. Martin Jr.*
NASA Lyndon B. Johnson Space Center, Houston, Texas
and
J. P. Slotnick†
Lockheed Engineering and Sciences Company, Houston, Texas

Introduction

ACCURATELY predicting the Space Shuttle ascent aerodynamic environment has proven to be a very elusive goal. Only recently, through the application of state-of-the-art numerical techniques on modern, large-memory supercomputers, does this goal appear obtainable. In contrast to the numerical nature of the current effort, an elaborate wind-tunnel test program was conducted during the 1970s in order to define the Space Shuttle's force and moment characteristics and corresponding pressure distribution. Over 27,000 h were expended in developing and defining the current ascent configuration. In spite of this effort, anomalies were apparent during the first mission when the vehicle lofted during first stage, causing solid rocket booster (SRB) separation to occur 3000 m higher than predicted. This effect was traced to increases in both vehicle normal force and a nose-down pitching moment. The centroid of the anomaly was found to be located at the Orbiter's lower surface as shown in Fig. 1. Also, the base pressures for the Orbiter and external tank (ET) were more positive than expected.

The force and moment coefficients, which are required for proper ascent guidance, were corrected by adding a flight-extracted increment, thereby accounting for the anomaly. An equivalent, straightforward approach to correcting the pressure distribution was not available. The pressure measurements that were obtained during the orbital flight test (OFT) program were intended to confirm the preflight estimates and thus were far too

---

Copyright © 1989 by the American Institute of Aeronautics and Astronautics, Inc. No copyright is asserted in the United States under Title 17, U. S. Code. The U. S. Government has a royalty-free license to exercise all rights under the copyright claimed herein for Governmental purposes. All other rights are reserved by the copyright owner.

* Senior Engineer, Aerosciences Branch, Advanced Programs Office.

† Engineer, CFD Group, Aerosciences Department.

sparse to resolve the discrepancies that were encountered. Following STS-5, a considerable amount of engineering judgement was used to create a third version of the integrated vehicle baseline configuration (IVBC-3) aerodynamic loads data base by combining wind-tunnel data and flight-measured strains and surface pressures. Structural analysis, based on the IVBC-3 pressure loading, has resulted in *predictions* of Orbiter payload bay structural deformations that greatly exceed those encountered in flight. Sufficient information did not exist to isolate the discipline that was contributing large uncertainties to the structural analysis.

The application of computational fluid dynamics (CFD) techniques to the Shuttle ascent environment was begun, after the STS 51-L accident, in order to provide a new source of aerodynamic information for both nominal and abort conditions. Realistic simulations of the integrated Space Shuttle vehicle (Orbiter, SRBs, and ET) are feasible due to the availability of the NASA Numerical Aerodynamic Simulation Program (NAS) Cray 2 and Cray YMP supercomputers, provided reasonable engineering approximations are made. Because of the complexity of the integrated vehicle, the chimera composite grid approach[1-4] was chosen for the discretization process. In this approach, an overset body-conforming grid is used to represent each geometric component (Orbiter, external tank, SRBs, and attach hardware) as well as special flow regions (Orbiter base and SRB plumes). An implicit approximately factored finite-difference procedure[5,6] has been used to solve the three-dimensional thin-layer Navier-Stokes equations, and all calculations have been carried out on the NASA NAS computers.

To reduce the complexity of the numerical simulations, various geometric simplifications are used such as the elimination or idealization of the attach hardware between the ET and the Orbiter and the ET and SRBs. The main fuel and oxidizer feedlines and various other small protuberances are also not represented. In addition, plumes are either represented with stings or treated as hot air jets.

The computational procedures and issues are briefly discussed in the section on computational procedure. Computational results and comparison with experimental data, wind tunnel and flight are described in the results section, along with preliminary unsteady flow results for SRB separation. As noted in the concluding remarks, further refinements and evaluation of numerical errors is needed, although relatively good agreement is obtained with the experimental data.

Computational Procedure

Geometry Modeling

A view of the integrated Space Shuttle Orbiter with the hardware that attaches the Orbiter and SRBs with the ET is given by Fig. 2. Details of the external fuel lines, Space Shuttle main engine (SSME) nozzles, and various other protuberances are also evident. Because simulation of a configuration with this complexity is considerably beyond our current capability of modeling and computational resource, various simplifications have been made. An effort has been made to make approximations so that useful engineering data can still be extracted and so that refinements can be included as experience and capability is gained as demonstrated in Fig. 3. The initial grid system, which was intended for supersonic simulations, treated the base regions as solid stings and excluded all attach hardware and protuberances. The grid system has undergone several levels of enhancement, culminating in a system that includes a simplified representation of the Orbiter/ET attach hardware, a smooth Orbiter base, an ET base, SRB base/plume, and the SRB/ET aft attach ring with the integrated electronics assembly (IEA) protuberance. Several key features that remain to be included are the external 17-in. propellant feedlines, the Orbiter vertical tail, and the SSME nozzle/plume grids. The geometry definition for the integrated vehicle was supplied by Shmuel Ben-Shmuel of Rockwell International.

Chimera Approach

The chimera overset grid scheme[1-4] was used to allow treatment of the multiple-body integrated Space Shuttle configuration. This composite grid approach allows an existing ordered grid code for the thin-layer Navier-Stokes equations and an existing single-mesh grid-generation procedure to be used to obtain solutions about a multibody configuration. Consequently, highly vectorized computer codes can be employed.

The chimera overset grid approach essentially allows each component of the configuration to be gridded separately and overset onto a main grid to form the complete model. Usually there is a major grid that is stretched over the entire field, and often it is generated about a dominant boundary or body surface (here the ET). Minor grids are generated about remaining portions of the body configuration or any other special feature such as an intense vortical region. The minor grids are used to resolve features of the geometry or flow that are not adequately resolved by the major grid. They are generated somewhat independently of the major grid and are overset on the major grid without requiring mesh boundaries to join in any special way.

A major piece of software that is needed with an overset grid approach is a routine that interconnects the grids. Essentially it needs to identify when grid points of one grid fall within a body boundary of another (grid hole points) and supply interconnection data so one grid can provide boundary data to another. Various algorithms can be devised for performing these tasks automatically.[1,7,8] In the current work the code Pegasus (versions 2.0 and 3.0) as provided by CALSPAN of Arnold Engineering Development Center (AEDC) has been used. This code has limited documentation; thus, a collaboration with J. Benek, B. Dietz, and N. Suhs of CALSPAN was established to help implement the code.

Grid Generation

Because of the use of the overset grid approach as well as the use of simplified geometry, the task of grid generation has been reduced in complexity. For the integrated Space Shuttle vehicle, each grid component was generated independently using a hyperbolic grid-generation procedure.[9-11] Details of the grid-generation procedure are given in Ref. 11, and specific details for generating the grid for the Orbiter have been reported in Refs. 12 and 13. Figure 4 shows various computational planes of the Orbiter grid.

Composite Grids

The body-conforming grids used for each component are shown in Fig. 5 at their respective planes of symmetry. The Orbiter and SRB grids are overset onto the ET grid, which extends to the far field. Approximately 1 million grid points distributed among seven grids have been used. Figure 6 shows a $\zeta = $ const plane for the Orbiter and for the ET projected onto an $x = $ const plane. Whenever points of a grid, say, grid 1, fall within a circumscribing body boundary of another grid, say, grid 2, the points of grid 1 are cut out, forming a hole in grid 1. The hole boundary data of grid 1 is then supplied from grid 2. Hole grid points have been removed from view in this figure. All hole cutting and information on how to supply boundary data are provided by the AEDC Pegasus code.

Governing Equations and Numerical Algorithm

For body-conforming coordinates and high-Raynolds-number flow, the thin-layer approximation can be made and the conservation equations of mass momentum and energy can be represented as in Ref. 14:

$$\partial_\tau \hat{Q} + \partial_\xi \hat{F} + \partial_\eta \hat{G} + \partial_\zeta \hat{H} = Re^{-1} \partial_\zeta \hat{S} \tag{1}$$

where the viscous terms in ζ have been collected into the vector $\hat{S}$, and the equations are in nondimensional form. Here the independent variables τ, ξ, η, and ζ are chosen to map a curvilinear body conforming discretization into a uniform computational space with ζ the coordinate away from the surface. The conservative form of the equations is maintained chiefly to capture the Rankine-Hugoniot shock jump relations as accurately as possible. The Baldwin-Lomax algebraic turbulence model[15] has been used. Although not shown, an option to also include thin-layer viscous terms in η is available.

An implicit approximately factored scheme for the thin-layer Navier-Stokes equations that uses central differencing in the η and ζ directions and upwinding in $\xi^{5,6}$ is written in the form

$$[I + i_b h \delta^b_\xi (\hat{A}^+)^n + i_b h \delta_\zeta \hat{C}^n - i_b h \, Re^{-1} \bar{\delta}_\zeta Ja^{-1} \bar{M}^n Ja - i_b D_i|_\zeta]$$

$$[I + i_b h \delta^f_\xi (\hat{A}^-)^n + i_b h \delta_\eta \hat{B}^n - i_b D_i|_\eta] \Delta \hat{Q}^n$$

$$= -i_b \Delta t \{ \delta^b_\xi (\hat{F}^+)^n + \delta^f_\xi (\hat{F}^-)^n + \delta_\eta \hat{G}^n + \delta_\zeta \hat{H}^n - Re^{-1} \bar{\delta}_\zeta \hat{S}^n \}$$

$$- i_b (D_e|_\eta + D_e|_\zeta) \hat{Q}^n \tag{2}$$

where $h = \Delta t$ or $(\Delta t)/2$ for first- or second-order time accuracy. Here δ is typically a three-point second-order-accurate central-difference operator, and $\bar{\delta}$ is a midpoint operator used with the viscous terms. The flux F has been split based on the eigenvalues of the Jacobian matrix A, allowing the use of backward- and forward-difference operators δ_ξ^b and δ_ξ^f. The flux differences themselves are midpoint differenced, and backward or forward weights of the split fluxes are used in the manner of Thomas et al.[16] This differencing maintains the freestream divergence in general curvilinear coordinates when coupled with consistently differenced centered metrics[14] or finite-volume metrics.[17] The matrices $\hat{A}$, $\hat{B}$, $\hat{C}$, and $\hat{M}$ result from local linearization of the fluxes about the previous time level. Here Ja denotes the Jacobian of the coordinate transformation. Dissipation operators D_e and D_i are used in the central space differencing directions (the $D_e|_\zeta$ term is described later in connection with differencing near embedded holes in the grid). For the differencing scheme 2, the factored left-hand-side operators can be readily inverted by sweeping in ζ with simultaneous inversion of tridiagonal matrices with 5×5 blocks. The two-factor implicit scheme is readily vectorized or multitasked in planes of $\xi = \text{const}$.

In the chimera approach there is the possibility of having arbitrary holes in the grid, that is, points that lie within a body (or user-specified boundary zone) from another grid. At such points the partial difference equations are shut off to leave the points unchanged. (The hole includes fringe points that are later updated by interpolating the solution from the overset grid about the body that created the hole.) To shut off the differencing scheme at hole points, an array of values i_b has been introduced into the scheme such that $i_b = 1$ at normal grid points and $i_b = 0$ at hole points. In the scheme Eq. (2), i_b multiplies the dissipation terms and the time step h. Thus, when $i_b = 1$ the normal scheme is maintained, but when $i_b = 0$ the scheme reduces to

$$\Delta \hat{Q}^n = 0$$

or $\hat{Q}^{n+1} = \hat{Q}^n$, and thus $\hat{Q}$ is not changed at a hole point.

If only three-point central spatial differencing is used, for example, the algorithm requires no other modifications. However, when differencing points adjacent to a hole boundary, operators that require more information than immediate neighbors of the differenced point will use information inside the hole. Since only the hole boundary points are updated via interpolation from other grids, interior hole points do not contain meaningful data. Consequently, it is necessary to limit the difference operators adjacent to holes so that points inside the hole are not used. This is simply accomplished using i_b, but does require some additional coding modifications.

Boundary Conditions

Boundary conditions for the current integrated vehicle are imposed with calls to modular routines that include viscous wall conditions, outflow conditions, axis conditions, wing cut conditions, and symmetry plane conditions. Although the code would benefit by use of far-field approximations, none has been coded as yet for this project. Presently, uniform conditions are set in the far field (except for outflow), and the grid outer boundary is stretched far away from the Shuttle configuration. Explicit boundary condition routines have been used. The overset grid interface hole and outer boundary conditions are also treated as boundary condition routines and are updated using trilinear interpolation.

Solution Procedure

The chimera approach has the advantage that a flow simulation code developed for a single general curvilinear grid can be readily adapted for composite overset grids. One simply writes an outer loop control program that calls in grids, supplies boundary condition routines to handle grid interface data, and inserts logic to blank out hole points as described earlier. Boundary interconnection information consists of several arrays: the Q_{BC} array, which holds boundary values for each grid that are supplied from other grids, arrays with corresponding indexes into the grids (where the Q_{BC} data come from and where they go), and pointers into the Q_{BC} and index arrays for the start of each grid. All arrays except Q_{BC} are produced by the Pegasus code.

As implemented here, at each time step a grid and its flow quantities are fetched from secondary memory [high-speed disk or solid-state memory device (SSD)] into the primary memory. Boundary interface arrays are also brought in. The solution on the current grid is then updated or advanced in time using the flow algorithm, the Q_{BC} boundary values, and the usual boundary condition routines (tangency, outflow, etc.) that are applicable. Overset boundary data that the current grid sends to other grids are then copied into Q_{BC}, and the updated solution is sent back to secondary memory. The next grid is fetched, and so on.

Computational Issues

Most of the computations have been carried out on the NAS Cray 2, but a few preliminary results have been obtained on the NAS Cray YMP. The NAS Cray 2 is a four-processor machine with 256 million words of memory. The machine is run in an interactive mode, and a large number of users are logged in during the day. Jobs with smaller time and memory requirements are given higher priority.

The flow solver used for these computations is named F3D. As coded, it requires 35 words of storage per grid point plus a variety of temporary arrays that are used to enhance vectorization. At least 38 million words (megawords) of memory would be needed to carry out a simulation on a one-million-point grid, if all data were memory-resident. However, the memory required for a calculation on only one grid of the composite grid is a fraction of this. The F3D program is coded to use secondary memory as described earlier: for the Shuttle flow simulations, the largest grid has about 350,000 points; 16 megawords of primary memory are needed. Although this process is less efficient than keeping all data in primary memory (which is quite feasible on the Cray 2), it considerably enhances computer access, and on the NAS system it costs no more than about 5%. Thus, the composite grid approach allows the use of hierarchical memory, permitting the code to run at a high priority. On a machine with limited primary memory, it makes it possible for the code to run at all.

The NAS YMP as presently configured has 8 processors, 32 megawords of memory, and a 256-megaword SSD. Batch jobs are limited to 8 megawords; the interactive limit is 2 megawords. By partitioning a single grid into a set of smaller grids, the code has been made to fit within 4 million words without extensive recoding and data management. However, each grid interface boundary is updated explicitly; the overall efficiency of the code is degraded with increased grid partitioning. The vector temporary arrays and boundary interface arrays also become a larger percentage of the overall memory as grid partitioning is increased. On the current F3D code a single YMP processor appears to be about twice as fast as a single Cray 2 processor.

Results

A plane of symmetry has been used to save computer time, although the code has the capability for sideslip. Grid sizes ranging from 250,000 to one million points have been used. The emphasis of the calculations is to predict pressure loading, and viscous effects are modeled with this end in mind.

In flight the maximum dynamic pressure is encountered between Mach 1.0 and 1.6, and SRB separation occurs above $M_\infty = 4$. Computations have been carried out at freestream Mach numbers of 0.6, 0.9, 1.05, 1.4, 1.55, 2.0, 2.5, and 4.5 at flight angles of attack but at wind-tunnel Reynolds numbers. Only the $M_\infty = 1.05$ and 2.5 cases have been computed on the large grid and with the SRB plume approximation. All other computations either did not include the SRB plumes or they used less accurate geometry modeling as described in Ref. 18. Consequently, only data from the $M_\infty = 1.05$ and 2.5 cases will be described here.

Simulation Enhancement

Consistent progress in enhancing the numerical simulation by improving the grid system and increasing the complexity of modeled features is illustrated in Figs. 7a–7e and Plates 36–42 (see the color section). Orbiter surface pressure contours, showing the difference between the solution obtained on the previous grid system and that obtained on the following grid system, are interspersed with images of the appropriate grid system, for reference. The computations were performed for $M_\infty = 1.05$ and angle of attack $\alpha = -3$ deg, which is a nominal ascent flight condition of considerable interest. Figure 7a shows the initial grid system, which was employed for $M_\infty = 1.05$. Note that the Orbiter and SRB cross sections are extended aft as a solid sting. However, the ET grid follows the base profile until it turns parallel to the ET centerline, forming a small cylinder, for which a flow-through boundary condition is applied. Further simplifications include the elevons being set to 0 deg instead of their flight positions of 10 deg inboard and 9 deg outboard (positive down) and the lack of any attach hardware, protuberances, or vertical tail.

The next grid system, shown in Fig. 7b, includes an approximation of the Orbiter/ET forward bipod and aft thrust diagonal/cross beam. The elevons are set to their nominal flight attitude of 10 deg/9 deg; however, the gaps between the fuselage and the elevons are not modeled. Also, the Orbiter grid resolution has been doubled in the streamwise direction. Plate 36 shows the difference between the solutions obtained from these two grid systems. On the upper surface the largest difference is due to the deflected elevons. On the lower surface there are large differences, just behind the nose, associated with the forward bipod, and the aft end of the vehicle shows large areas influenced by both the aft attach structure and the elevons.

The next improvement was to resolve the Orbiter's blunt base. This was accomplished by placing a grid inside of the Orbiter "sting" such that the forward face of the new grid formed the Orbiter's base, as shown in Fig. 7c. This enhancement resulted in a 0.1–0.2 decrease in pressure coefficient behind and outboard of the Orbiter/ET aft attach area as shown in Plate 37. Other isolated differences are due to a change in streamwise resolution, with the $98 \times 77 \times 48$ Orbiter grid being the baseline grid system.

In preparation for including the SRB plumes, an additional grid, shown in Fig. 7d, was inserted inside the SRB sting such that its forward face formed the nozzle exit plane. This change influenced the Orbiter in approximately the same pattern as the Orbiter base, but the effect was an order of magnitude smaller as shown in Plate 38 in the color section. The SRB plume flowfield was computed with an axisymmetric solver,[19] which modeled the plume as hot air exhausting into a uniform freestream. As a first approximation of the plume's influence on the launch vehicle, the axisymmetric solution was rotated about the axis of symmetry and used as the initial conditions for the SRB "plume" grid, whereas the solution at the nozzle exit plane (which also came from the axisymmetric solution) was fixed, creating the plume boundary condition. The resulting effect on the Orbiter is to raise the surface pressure by 0.1–0.2 C_p in the same areas that were affected by the inclusion of the Orbiter base and SRB plume grids, as can be seen in Plate 39. Work to include more of the attach hardware is underway. The SRB/ET aft attach ring and the IEA protuberance have been incorporated by using a geometry definition and surface grid provided by S. Johnson of the NASA Johnson Space Center. In addition, the ET base was refined in order to eliminate the previous small sting, as can be seen in Fig. 7e. The aft attach ring has a profound influence on the Orbiter wing loads, as shown in Plate 40, which illustrates the plume-off differences due to the attach ring. When integrated over the wing and evaluated at flight dynamic pressure, the differences produce an 18,150-kg (40,000 lb) increase in wing root shear.

Comparison with Experimental Data

Plate 41 (see the color section) shows a comparisons between the computation and wind-tunnel test IA-105[20] for $M_\infty = 1.05$ at an angle of attack $\alpha = -3$ deg and using the wind-tunnel Reynolds number of $Re = 4.0 \times 10^6/\text{ft}$ for the computations. This test used a 3% scale model of the integrated vehicle which was instrumented with 1538 pressure taps—a number sufficient for extracting surface pressure contours from the experimental data. However, neither the SRB or the SSME plume was simulated, thus allowing the model to be supported from downstream by dual, small-diameter stings that entered the SRB nozzles. Comparisons of surface pressure coefficient between wind-tunnel data and the computation are shown in Plate 41 (top and side views) to illustrate three-dimensional effects and the extent of expansion and compression regions.

The overall variational agreement between the computation and experiment appears to be reasonably good. The magnitude of extreme expansion regions, such as that behind the Orbiter canopy and the SRB nose cone, is not accurately captured, however. Mach contours on a constant z cut through the ET and SRB centerlines are also shown in Plate 41 and are used to highlight the SRB plume, which was modeled as a hot air jet.

A limited amount of flight-test data[21] are also available for comparison. During the first five Shuttle flights, pressure data were taken at various locations on the Orbiter Columbia. Much of this data were taken along the side of the fuselage around a constant-angle station, and Fig. 8 shows pressure comparisons between flight, wind tunnel, and computation at $\phi = 70$ deg (see figure insert). At this particular Mach number, 1.05, the computation is in better agreement with the flight data than the wind-tunnel data, apparently because of wind-tunnel wall interference. This computation required about 15 h of computer time using a single processor of the Cray 2.

Comparison with IVBC-3PL

The IVBC-3PL (payload bay) data base is a derivative of IVBC-3, which was created to rectify deficiencies that were apparent at supersonic Mach numbers. Structural analysis, using the aero loads from IVBC-3 at $M_\infty = 1.55$, predicted unrealistic deformations in the payload bay doors. The data base was altered by replacing the pressure distribution on the doors with the data obtained in the IA-105 wind-tunnel test. This corrected the supersonic situation, but led to equally unrealistic predictions for the payload bay truss elements in the transonic range, where it is recognized that IA-105 suffers from wall interference and/or tunnel blockage effects. A more in-depth discussion of IVBC-3 is outside the scope of this work.

Considerable insight into the transonic aerodynamic loads can be gained by comparing the transonic computational results that were previously described, IVBC-3PL, and the available experimental data as shown in Fig. 9 for the Orbiter upper pitch plane. The computation compares well with the only two flight measurements that were on the vehicle upper centerline, and good agreement is shown with the IA-308 wind-tunnel test, which used a 2.25% model that included hot gas simulations of the SRB and SSME plumes. This comparison is somewhat qualitative, since the flight data, IVBC-3PL, and computation (which neglected the vertical tail) are all at $M_\infty = 1.05$, whereas the IA-308 test was conducted at $M_\infty = 1.1$. A similar comparison is shown on the fuselage side in Fig. 10. The computational results follow the solid line that varies in ϕ, whereas the dashed line shows the IVBC-3PL trace.

Predictions around the fuselage at the forward end of the payload bay are shown in Fig. 11. Negative pressure coefficients are plotted outside the cross section, with length proportional to magnitude, whereas positive pressures are plotted inside the cross section. In general, good agreement is shown between the aero data base and the numerical simulation. The apparent excellent agreement with the flight measurement, V07P9860, is slightly misleading, as the instrument was located 1 m forward of this cross section. As one moves aft toward the center of the payload bay, the agreement deteriorates, as can be seen in Fig. 12. Whereas the computation shows a smooth, slightly negative distribution around the payload bay doors and fuselage side wall, the IVBC-3PL data base indicates significant suction on the payload bay doors, which abruptly changes to near zero on the side wall. The flight measurements, which straddle this location, both indicate a small positive pressure in this area. Unfortunately, they were both immediately upstream of an Orbiter vent door and may have been biased (in a positive direction) by the venting. The last plot in this sequence, shown in Fig. 13, indicates very large differences between the computational results and the IVBC-3PL data base on the fuselage side wall, while showing good agreement between the flight data and the computation. The quality of the numerical simulations on the upper fuselage has been very encouraging, though this is somewhat to be expected as this is the windward surface during ascent. Unfortunately, the agreement on the Orbiter wing is not as good, primarily due to the simplified attach hardware representation that was used in the preceding computations. The various components of attach hardware shown in Fig. 1 create considerable blockage, which strongly influences the aerodynamic loads on the Orbiter wing, especially in the transonic flow environment. The sensitivity of the wing pressure distribution to the relatively small aft attach hardware components and protuberences can be seen in Figs. 14 and 15. The SRB/ET aft attach ring/IEA has made a significant influence on the wing lower surface and a subtle effect on the upper surface. Although the improvement is very encouraging, the simulation still underpredicts the pressure on the wing.

Supersonic Simulations

After debris damage to the STS-27 Atlantis flight, a numerical simulation at $M_\infty = 2.5$ and $\alpha = 3$ deg was performed to provide a supersonic flowfield in support of a debris trajectory analysis. At this flight condition, Orbiter inboard elevons are set to 0 deg and outboard elevons to -5 deg (positive down). In the computation, elevon settings were left at the Mach 1.05 flight settings of 10 and 9 deg. Computed limiting streamlines (simulated oil flow) for this higher Mach number condition (Fig. 16) indicate the presence of streamwise flow separation before the SRB attach ring. Quantitative comparisons between the computations and flight data from STS-5 are shown in Figs. 17 and 18. Figure 16 shows a comparison between measured flight pressure data and the computation at various points along the side of the Orbiter (see Orbiter insert for locations), and Fig. 18 shows a similar comparison for a wing station directly over the centerline of the SRB. Wing pressures from a coarse grid calculation, without any attach hardware and with elevons set to 0 deg inboard and outboard, are also shown. This geometry and grid, illustrated in Fig. 19, was set up for a computation of SRB separation. Although agreement on the upper surface is very good, lower-surface pressures are not well represented. Further investigation of these differences is needed, and efforts to correct geometric discrepancies are underway.

Solid Rocket Booster Separation

Solid rocket booster separation occurs just after 2 min into the flight, at an altitude near 50,000 m and a Mach number above 4. Use of the composite overset grid approach makes it is possible to simulate SRB separation from the Orbiter and ET without regridding at each time advance of the flowfield. Preliminary results illustrating this capability are shown in Plate 42 for $M_\infty = 4.5$. Here a time-accurate calculation was carried out for one-third of a second of real flight time while the SRBs are moved along a prescribed path. To reduce computer time, a more simplified geometry model (Fig. 19) was used in which the Orbiter, ET, and SRBs are represented without attach hardware. Stings behind the Orbiter and ET are used for the same reason. The end of the SRB body is closed off with a spherical cap. The composite grid contains approximately 350,000 points. The initial solution at time $t = 0$ and solutions corresponding to $t = 0.204$ and $t = 0.367$ s are shown in Plate 42. The sequence of solutions illustrates the pressure contours on the ET, SRB, and Orbiter at these times. Six hours of (single processor) Cray 2 time was used over 250 time steps. Additional details on this approach and code validation are given in Ref. 22.

Concluding Remarks

Overall, the calculations show reasonably good agreement in both flow structure and surface pressure with the available wind-tunnel and flight-test results. The results have been used to provide a new source of information in the transonic and supersonic regime, which enabled inconsistencies in the aerodynamic loads data base to be resolved before the STS-26 return to flight mission. Strain-gauge data, obtained during the mission on the orbiter payload bay truss members, increased the confidence in the numerical simulations. The computational results appear to be more accurate and less conservative than the IVBC-3PL data base. Similar insight into the wing aerodynamic loads is being obtained as the fidelity of the geometry model improves. It appears that accurate modeling of the relatively small aft attach hardware components and protuberances will be required. Also, the grid system needs to be enhanced at the wall so that calculations at flight Reynolds numbers, which are two orders of magnitude greater than typical wind-tunnel values, can be carried out. These issues are being pursued, as is the modeling of unsteady flow effects.

Acknowledgments

The authors are indebted to Joe Steger and the Ames Space Shuttle Flow Field Simulation Group; Shmuel Ben-Shmuel of Rockwell International; and Jack Benek, Bill Dietz, and Norm Suhs of CALSPAN. Financial support was given by the National Space Transportation System Program Office, NASA Johnson Space Center, and computer resources were provided by the Numerical Aerodynamic Simulation facility, NASA Ames Research Center. This study was performed in collaboration with the Ames Spaces Shuttle Flow Field Simulation Group (S. Obayashi, Y. M. Rizk, J. L. Steger and M. Yarrow).

References

[1]Steger, J. L., Dougherty, F. C., and Benek, J. A., "A Chimera Grid Scheme," *Advances in Grid Generation*, edited by K. N. Ghia and U. Chia, American Society of Mechanical Engineers, New York, FED-5, 59–69, 1983.

[2]Benek, J. A., Buning, P. G., and Steger, J. L., "A 3-D Grid Embedding Technique," AIAA Paper 85-1523, July 1985.

[3]Dougherty, F. C., Benek, J. A., and Steger, J. L., "On Application of Chimera Grid Scheme to Store Separation," NASA TM-88193, Oct. 1985.

[4]Benek, J. A., Donegan, T. L., and Suhs, N. E., "Extended Chimera Grid Embedding Scheme with Application to Viscous Flows," AIAA Paper 87-1126-CP, June 1987.

[5]Ying, S. X., Steger, J. L., Schiff, L. B., and Baganoff, D., "Numerical Simulation of Unsteady, Viscous, High Angle of Attack Flows Using a Partially Flux-Split Algorithm," AIAA Paper 86-2179, Aug. 1986.

[6]Steger, J. L., Ying, S. X., and Schiff, L. B., "A Partially Flux-Split Algorithm for Numerical Simulation of Compressible Inviscid and Viscous Flow," *Proceedings of the Workshop on Computational Fluid Dynamics*, Institute of Nonlinear Sciences, Univ. of California, Davis, CA, 1986.

[7]Benek, J. A., Steger, J. L., Dougherty, F. C., and Buning P. G., "Chimera: A Grid Embedding Technique," Arnold Engineering Development Center, Tullahoma, TN, TR-85-64, 1986.

[8]Steger, J. L., and Benek, J. A., "On the Use of Composite Grid Schemes in Computational Aerodynamics," *Computer Methods in Applied Mechanics and Engineering*, Vol. 64, Nos. 1–3, Oct. 1987, pp. 301–320.

[9]Steger, J. L., and Chaussee, D. S., "Generation of Body Fitted Coordinates Using Hyperbolic Partial Differential Equations," *SIAM Journal on Scientific and Statistical Computing*, Vol. 1, No. 4, Dec. 1980, pp. 431–437.

[10]Kinsey, D. W., and Barth, T. J., "Description of a Hyperbolic Grid Generation Procedure for Arbitrary Two-Dimensional Bodies," AFWAL TM-84-191-FIMM, July 1984.

[11]Steger, J. L., and Rizk, Y. M., "Generation of Three Dimensional Body Fitted Coordinates Using Hyperbolic Partial Differential Equations," NASA TM-86753, June 1985.

[12]Rizk, Y. M., and Ben-Schmuel, S., "Computation of the Viscous Flow Around the Shuttle Orbiter at Low Supersonic Speeds," AIAA Paper 85-0168, Jan. 1985.

[13]Rizk, Y. M., Steger, J. L., and Chaussee, D., "Use of a Hyperbolic Grid Generation Scheme in Simulating Supersonic Viscous Flow About Three-Dimensional Winged Configurations," NASA TM-86776, July 1985.

[14]Pulliam, T. H., and Steger, J. L., "On Implicit Finite-Difference Simulations of Three Dimensional Flow," AIAA Paper 78-10, Jan. 1978.

[15]Baldwin, B. S., and Lomax, H., "Thin Layer Approximation and Algebraic Model for Separated Turbulent Flow," AIAA Paper 78-257, Jan. 1978.

[16]Thomas, J. L., Taylor, S. L., and Anderson, W. K., "Navier-Stokes Computations of Vortical Flows over Low Aspect Ratio Wings," AIAA Paper 87-0207, Jan. 1987.

[17]Vinokur, M., "An Analysis of Finite-Difference and Finite-Volume Formulations of Conservation Laws," NASA Contractor Rept. 177416, June 1986.

[18]Buning, P. G., Chiu, I. T., Obayashi, S., Rizk, Y. M., and Steger, J. L., "Numerical Simulation of the Integrated Space Shuttle Vehicle in Ascent," AIAA Paper 88-4359-CP, Aug. 1988.

[19]Obayashi, S., "Numerical Simulation of Underexpanded Plumes Using Upwind Algorithms," AIAA Paper 88-4360-CP, Aug. 1988.

[20]Spangler, R. H., "Results of Tests Using a 0.03 Scale Model (47-OTS) of the Space Shuttle Integrated Vehicle in the AEDC 16 Foot Transonic Propulsion Wind Tunnel (IA105A)," Vol. 2, NASA Contractor Rept. 160851, Oct. 1981.

[21]Laspasa, F. S., "STS-5 Postflight Report on Ascent External Aerodynamic Loads," Rockwell International Internal Letter SAS/AERO/83-094, March 1983.

[22]Meakin, R. L., and Suhs, N., "Unsteady Aerodynamic Simulation of Multiple Bodies in Relative Motion," AIAA Paper 89-1996, June 1989.

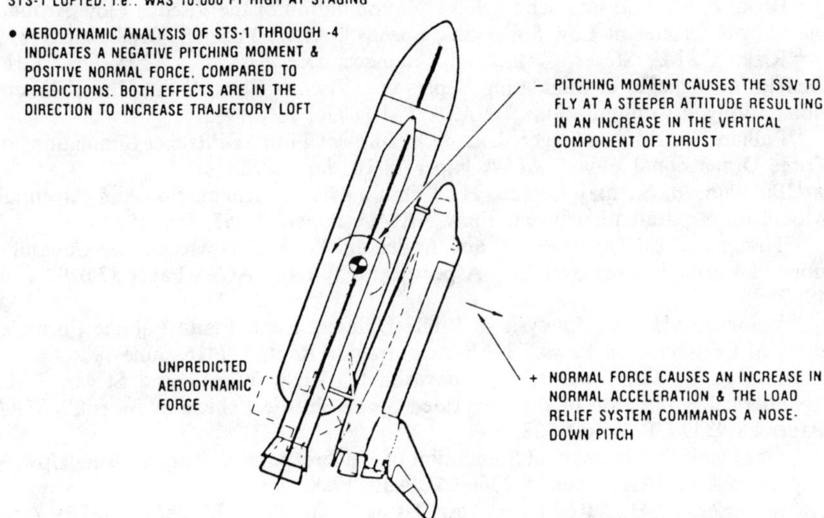

Fig. 1   Aerodynamic anomaly during the first Space Shuttle mission.

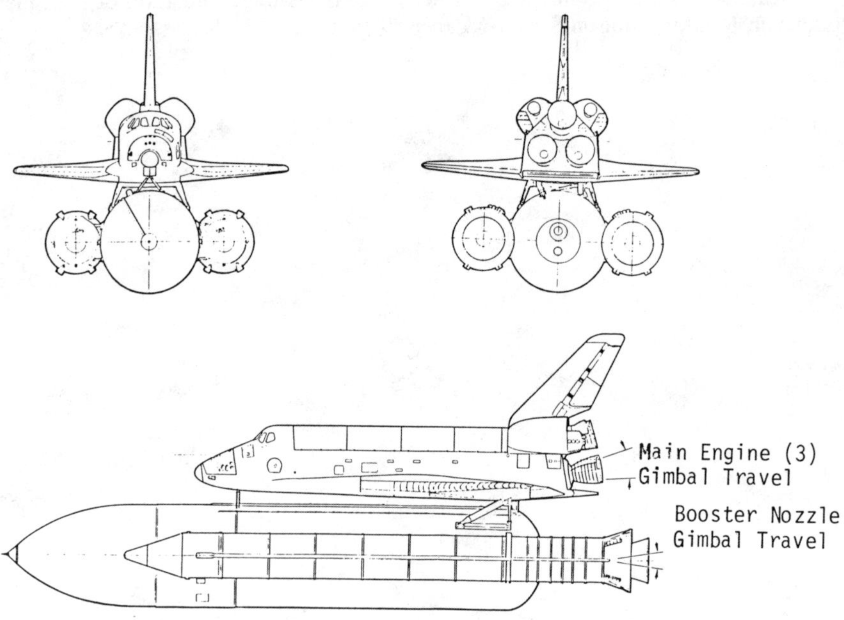

Fig. 2   Detail views of the integrated Space Shuttle.

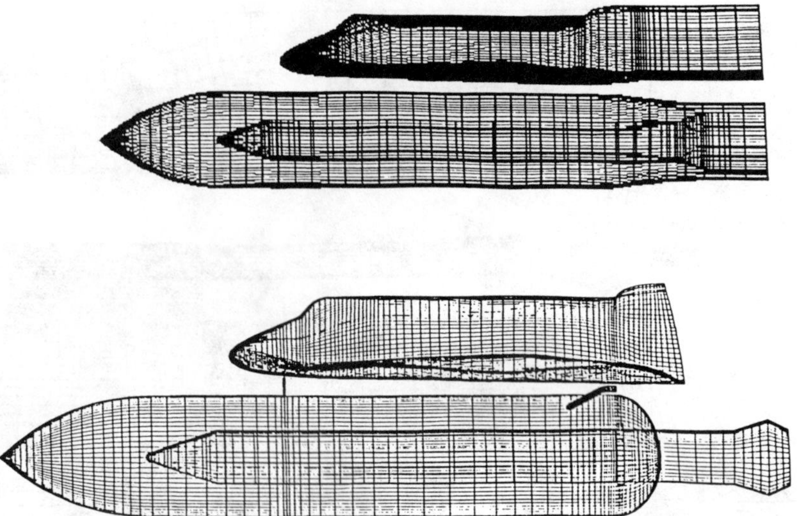

Fig. 3 Comparison of initial configuration and surface grid point distributions, with the current grid system.

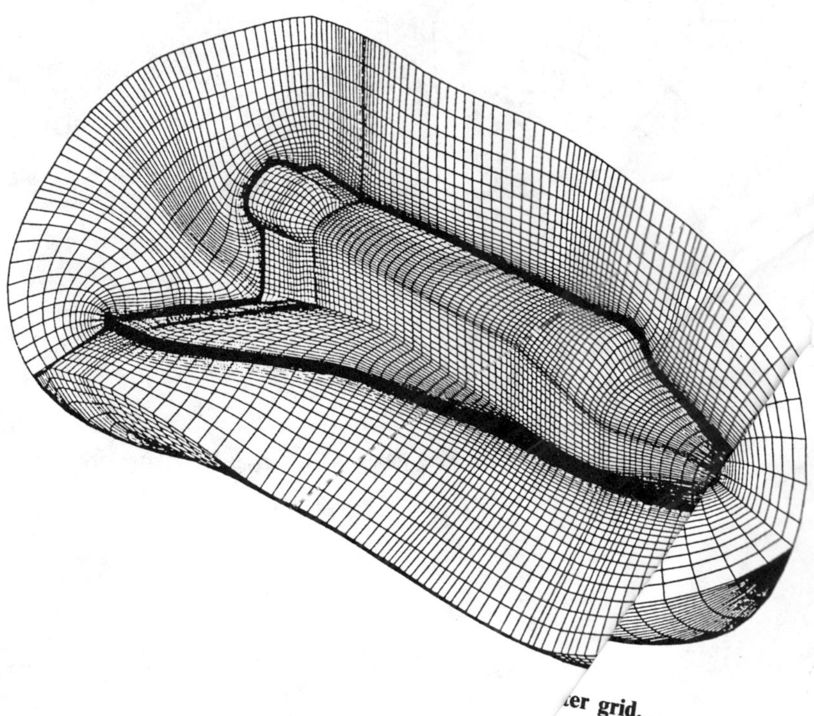

Fig. 4 Various computational planes of the outer grid.

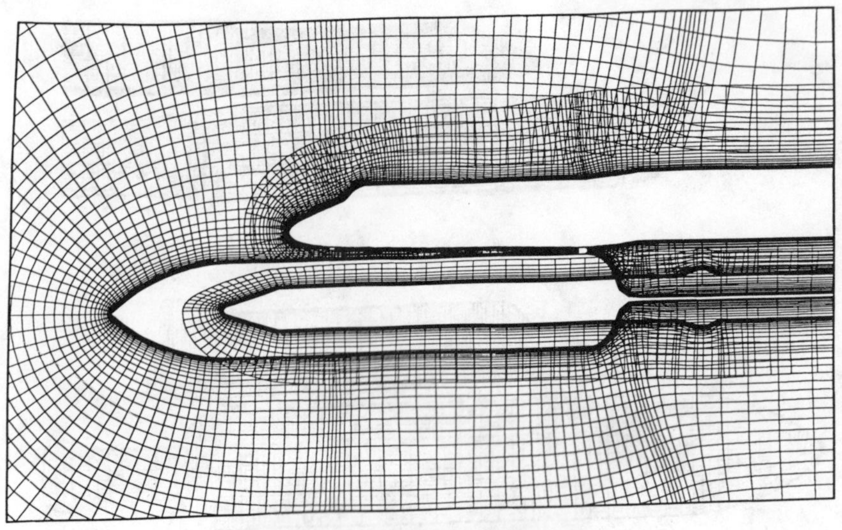

Fig. 5　Symmetry planes of all grids.

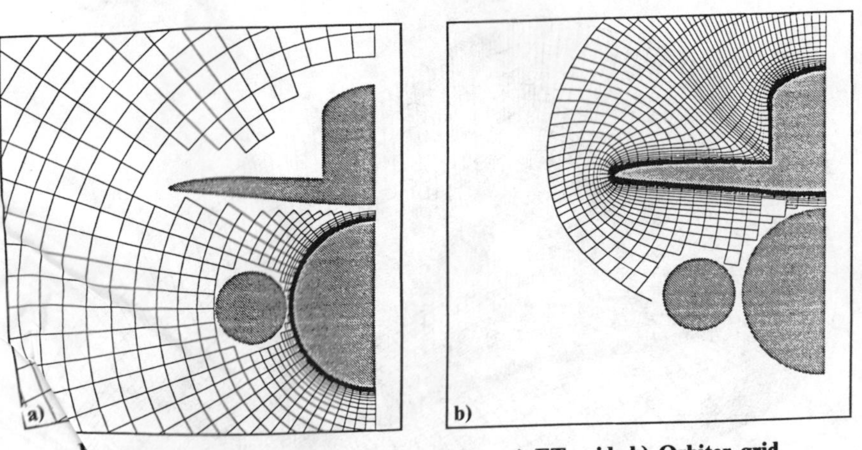

Fig. 6　Cross section of grids showing holes: a) ET grid; b) Orbiter grid.

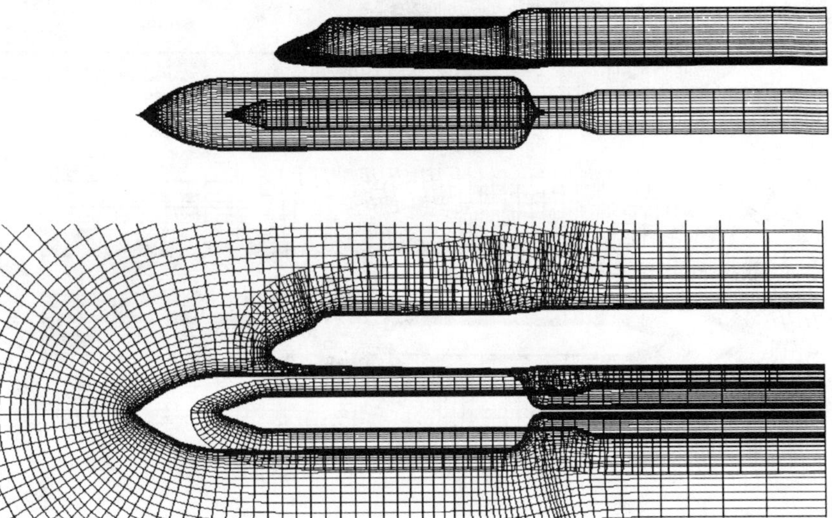

Fig. 7a Initial grid system for the $M_\infty = 1.05$ computation.

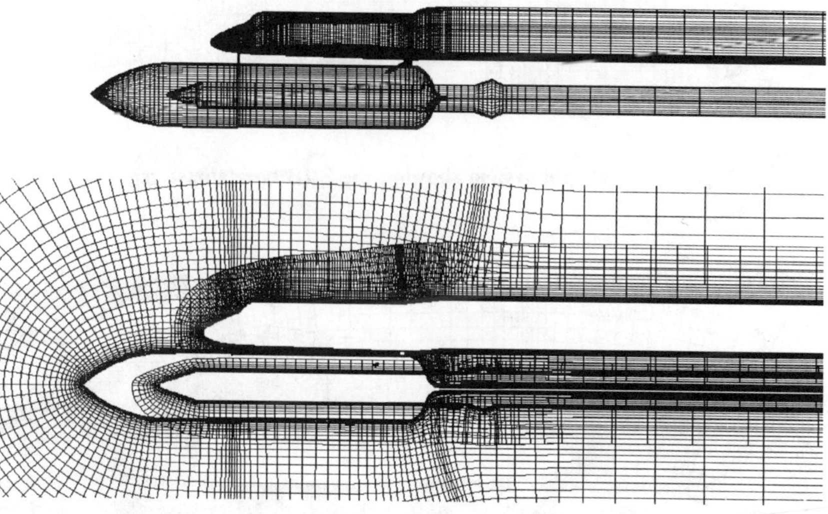

Fig. 7b Enhanced grid system showing the Orbiter/ET attach hardware approximation, deflected elevons, and reduced radius of the SRB sting.

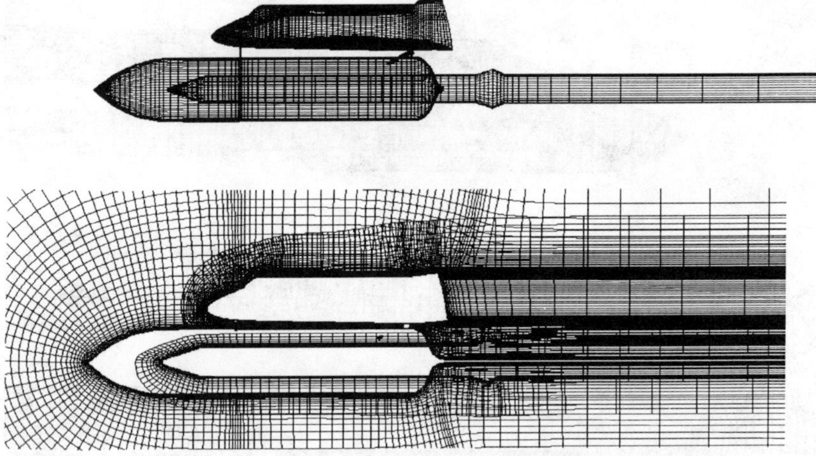

Fig. 7c   Grid system showing the Orbiter base.

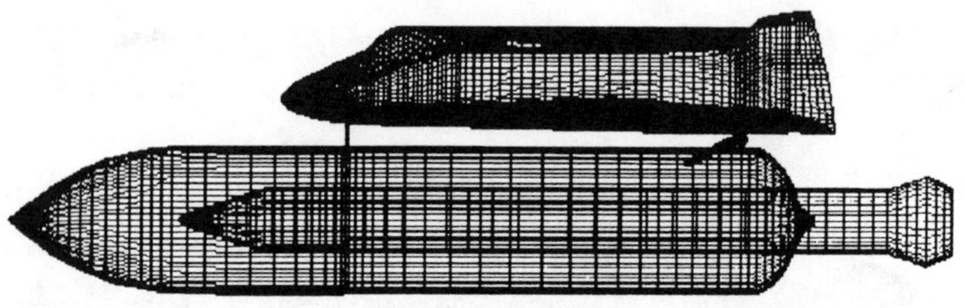

Fig. 7d   Final grid system showing the SRB nozzle/base region.

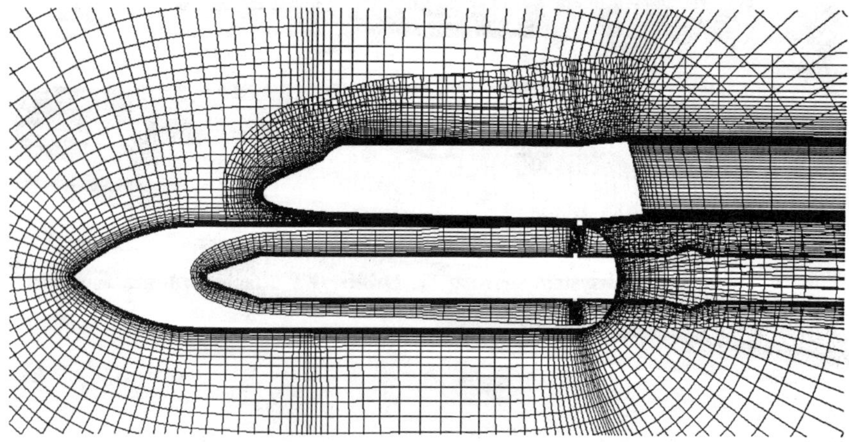

Fig. 7e   Grid system showing SRB/ET aft attach ring and IEA.

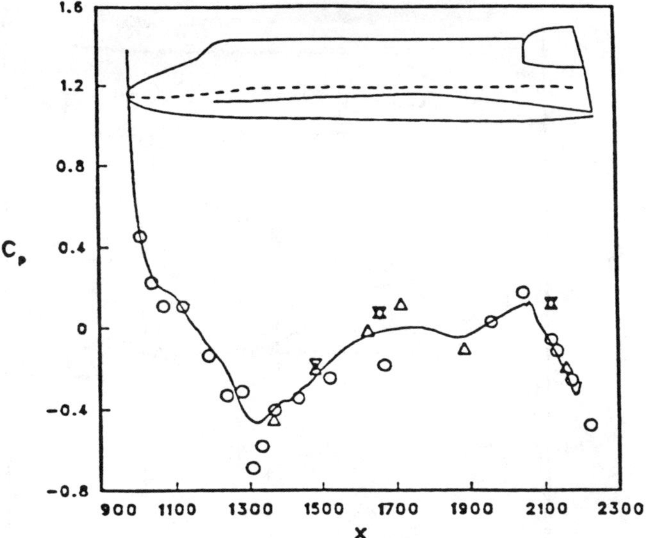

Fig. 8 Comparison of C_p showing computation (–), wind tunnel ($\bigcirc$), and flight test (∇, right side; $\triangle$, left side) along the $\phi = 70$ deg line of the Orbiter fuselage ($M_\infty = 1.05$ and $\alpha = -3$ deg, $Re = 4.0 \times 10^6$/ft, and 10 deg/9 deg elevon deflection).

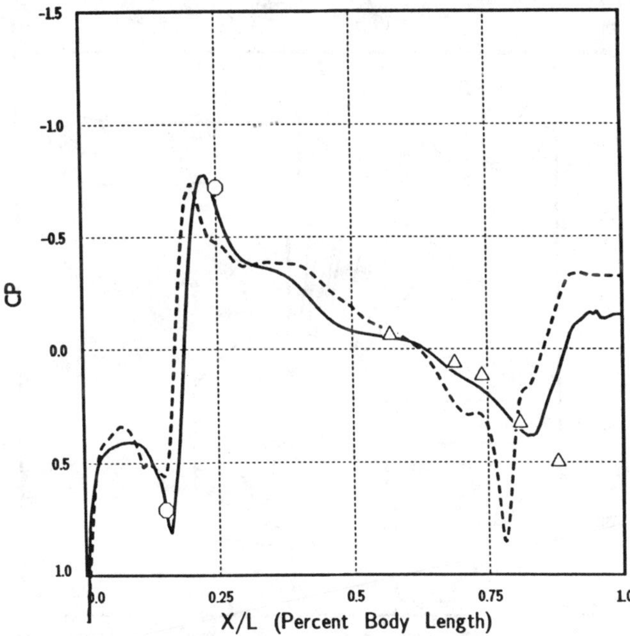

Fig. 9 Comparison of pressure coefficient on the Orbiter upper pitch plane between the computation (–), IVBC-3PL (––), STS-5 flight data ($\bigcirc$), IA-308 wind-tunnel data (at $M_\infty = 1.1$) ($\triangle$), for $M_\infty = 1.05$, $\alpha = -3$ deg.

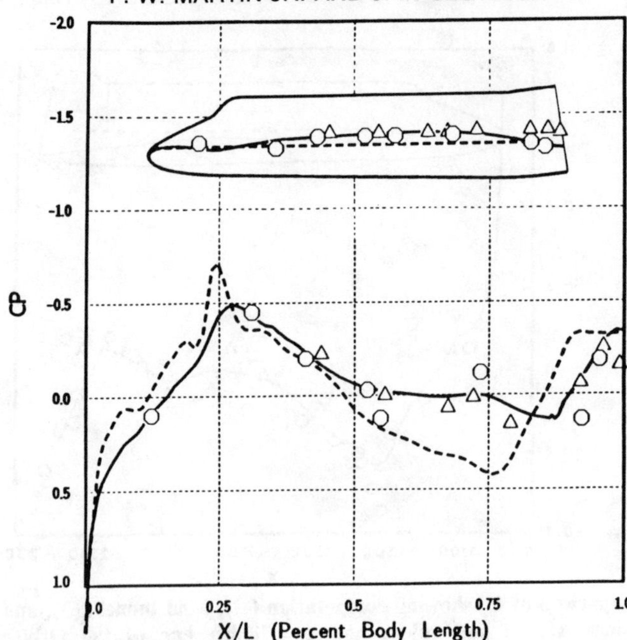

Fig. 10 Comparison of pressure coefficient along the Orbiter fuselage showing computation (–), IVBC-3PL (– –), STS-5 flight data ($\bigcirc$), IA-308 wind-tunnel data (at $M_\infty = 1.1$) ($\triangle$) for $M_\infty = 1.05$, $\alpha = -3$ deg.

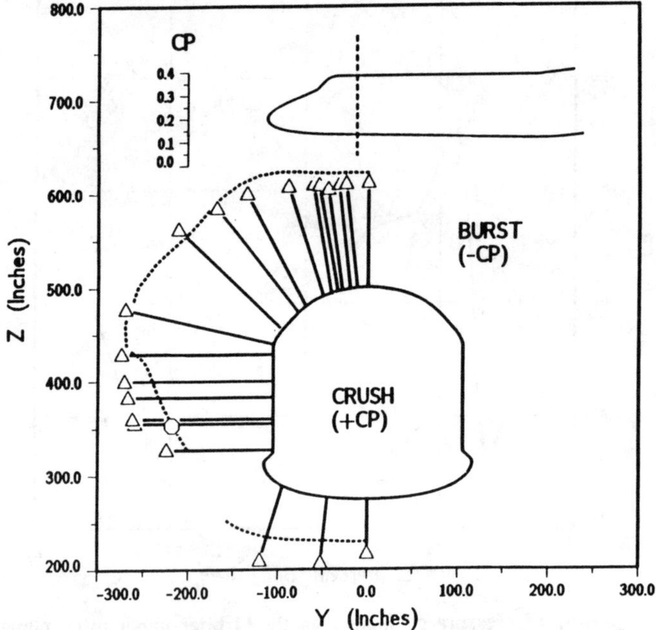

Fig. 11 Comparison of pressure coefficient between the computational results and IVBC-3PL at Orbiter station 590 for $M_\infty = 1.05$, $\alpha = -3$ deg: computation (– –), IVBC-3PL, ($\triangle$), STS-5 flight data ($\bigcirc$).

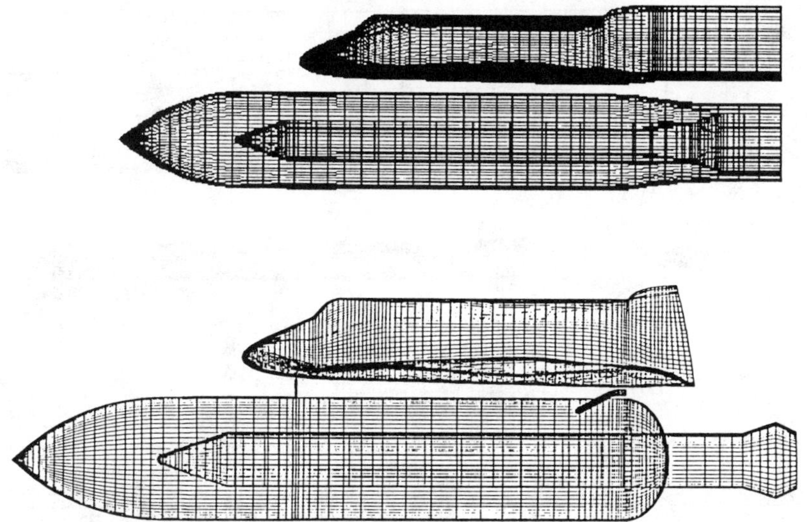

Fig. 3　Comparison of initial configuration and surface grid point distributions, with the current grid system.

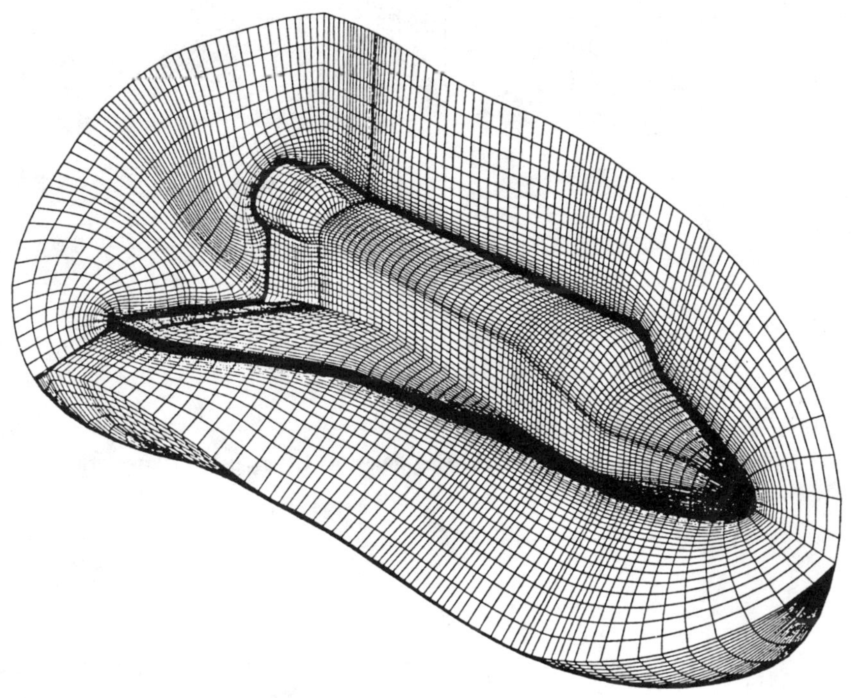

Fig. 4　Various computational planes of the Orbiter grid.

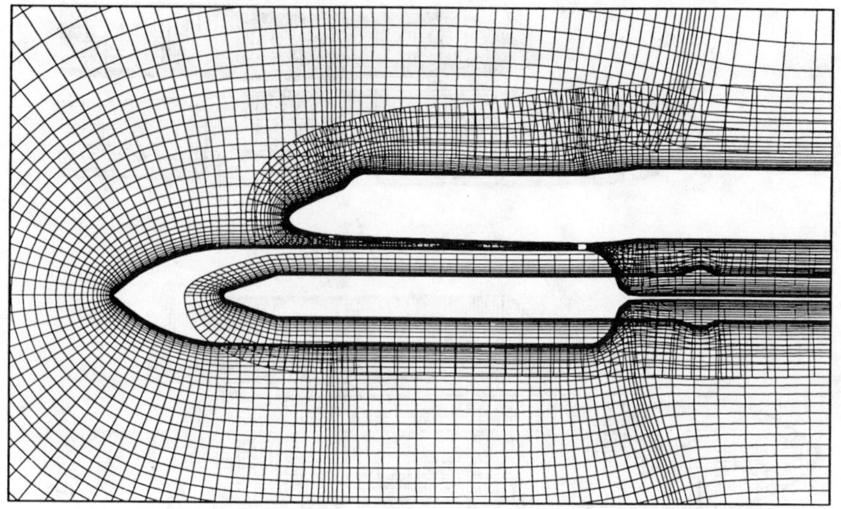

Fig. 5   Symmetry planes of all grids.

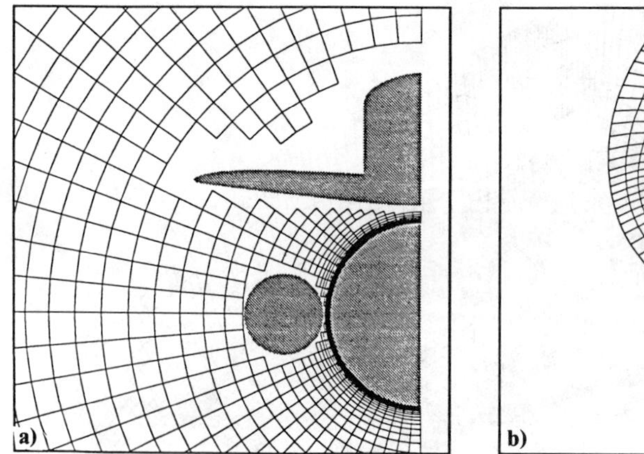

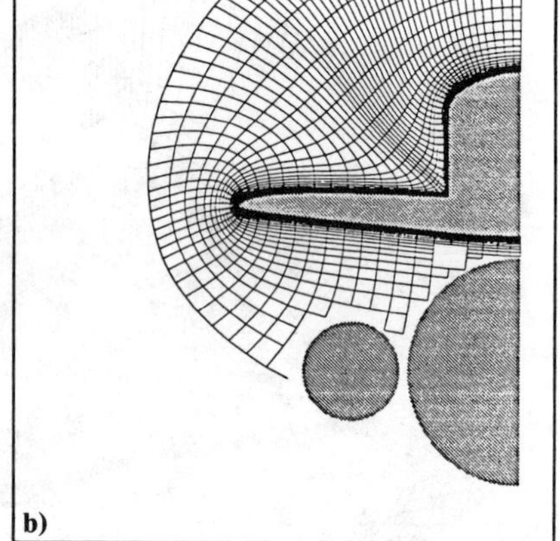

Fig. 6   Cross section of grids showing holes: a) ET grid; b) Orbiter grid.

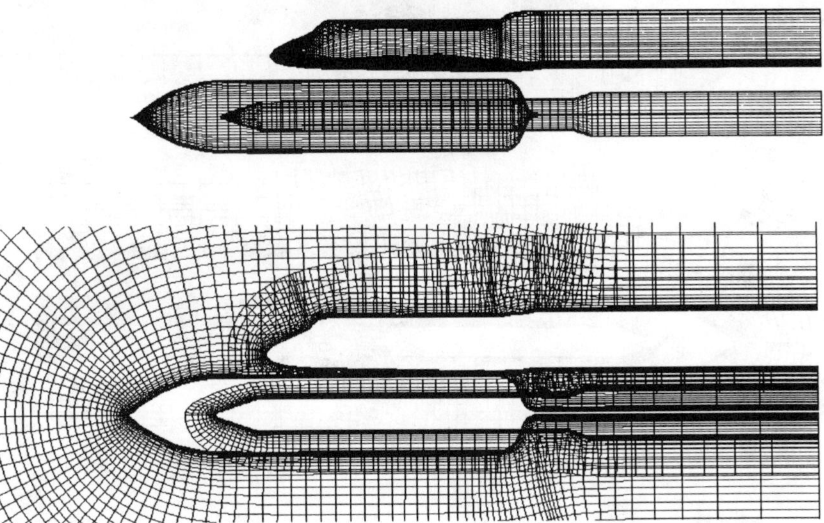

Fig. 7a Initial grid system for the $M_\infty = 1.05$ computation.

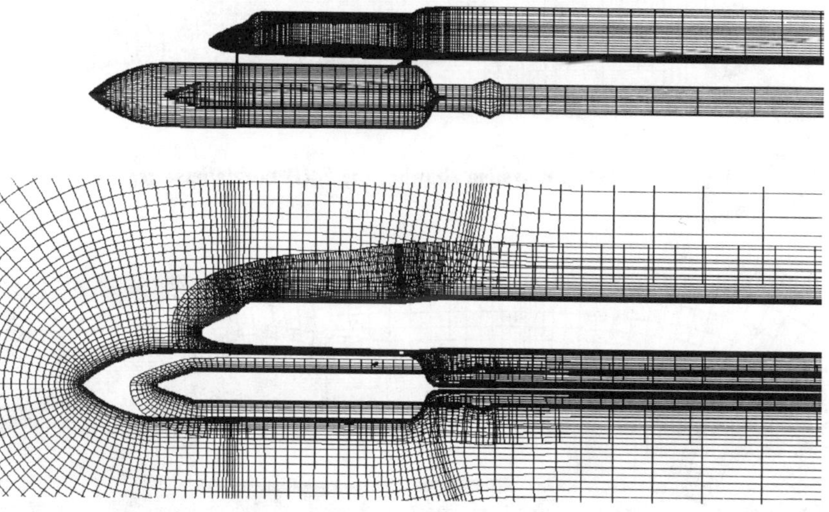

Fig. 7b Enhanced grid system showing the Orbiter/ET attach hardware approximation, deflected elevons, and reduced radius of the SRB sting.

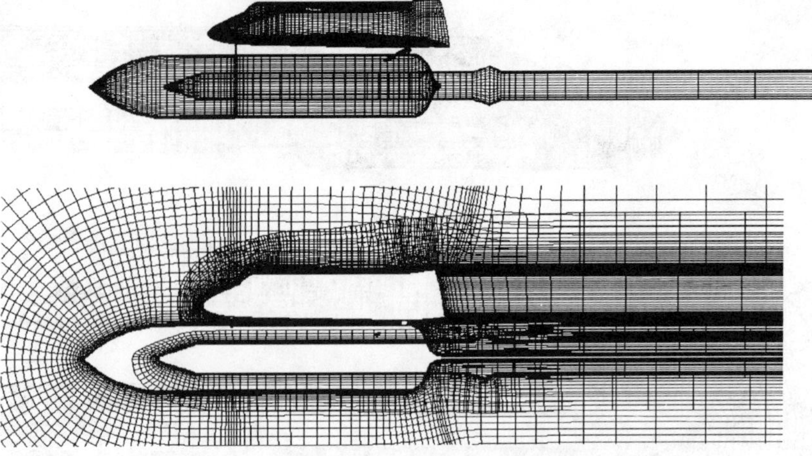

Fig. 7c   Grid system showing the Orbiter base.

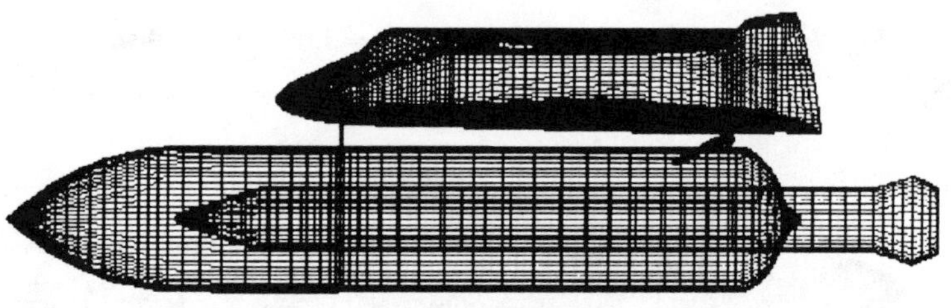

Fig. 7d   Final grid system showing the SRB nozzle/base region.

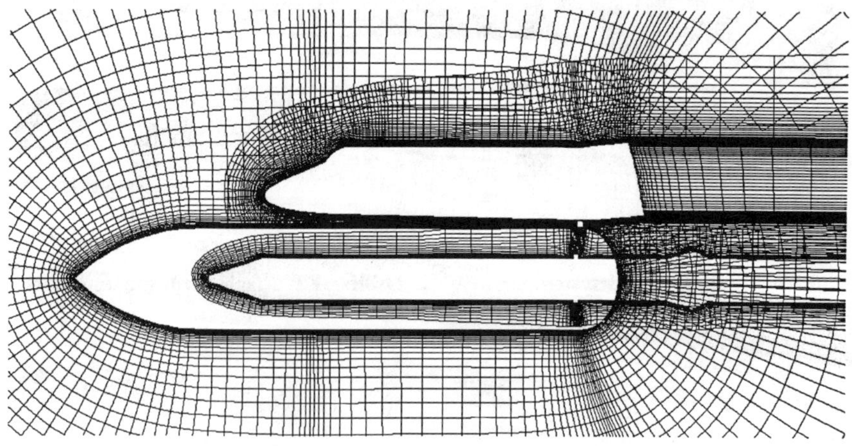

Fig. 7e   Grid system showing SRB/ET aft attach ring and IEA.

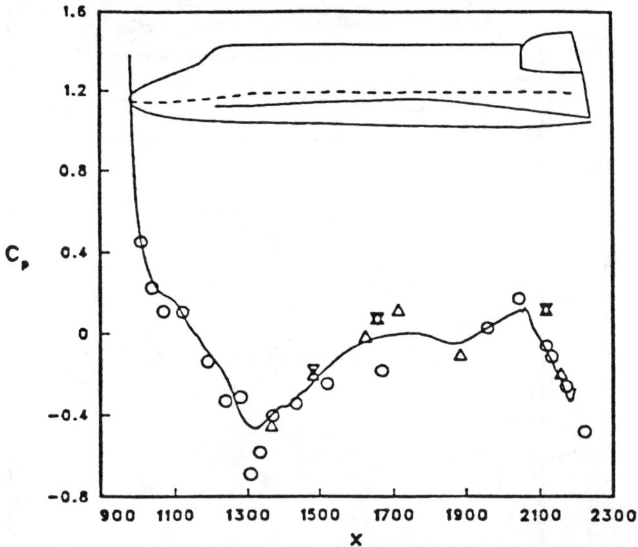

Fig. 8 Comparison of C_p showing computation ($-$), wind tunnel ($\bigcirc$), and flight test (∇, right side; $\triangle$, left side) along the $\phi = 70$ deg line of the Orbiter fuselage ($M_\infty = 1.05$ and $\alpha = -3$ deg, $Re = 4.0 \times 10^6$/ft, and 10 deg/9 deg elevon deflection).

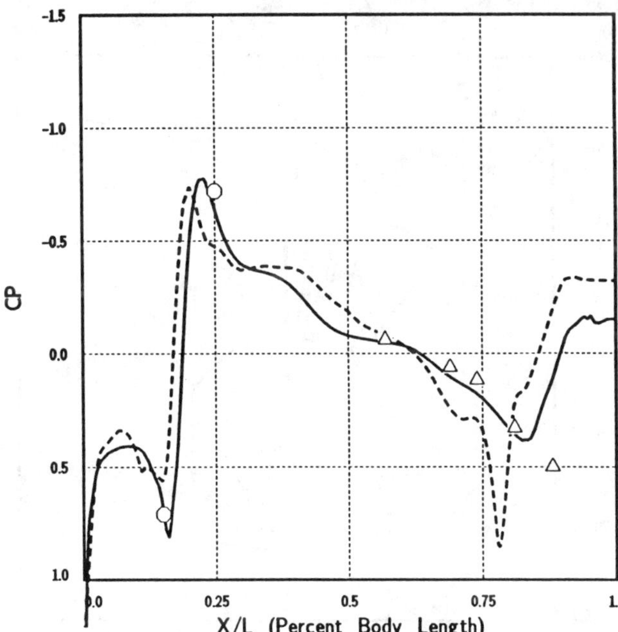

Fig. 9 Comparison of pressure coefficient on the Orbiter upper pitch plane between the computation ($-$), IVBC-3PL ($--$), STS-5 flight data ($\bigcirc$), IA-308 wind-tunnel data (at $M_\infty = 1.1$) ($\triangle$), for $M_\infty = 1.05$, $\alpha = -3$ deg.

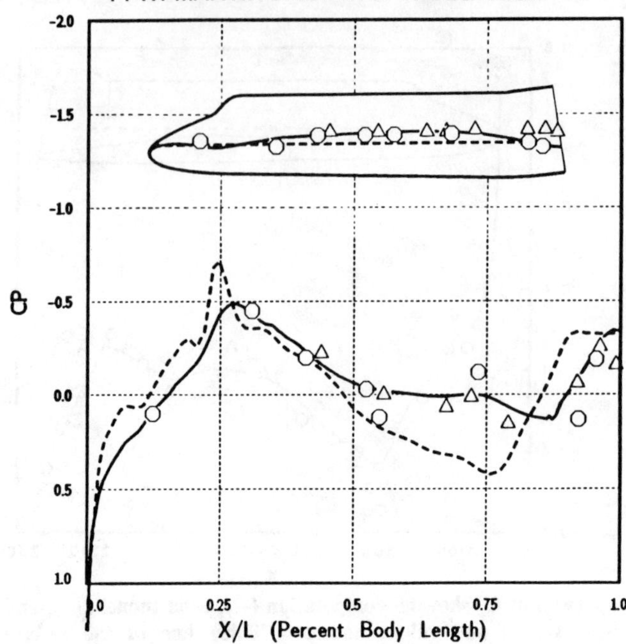

Fig. 10 Comparison of pressure coefficient along the Orbiter fuselage showing computation (−), IVBC-3PL (−−), STS-5 flight data (○), IA-308 wind-tunnel data (at $M_\infty = 1.1$) (Δ) for $M_\infty = 1.05$, $\alpha = -3$ deg.

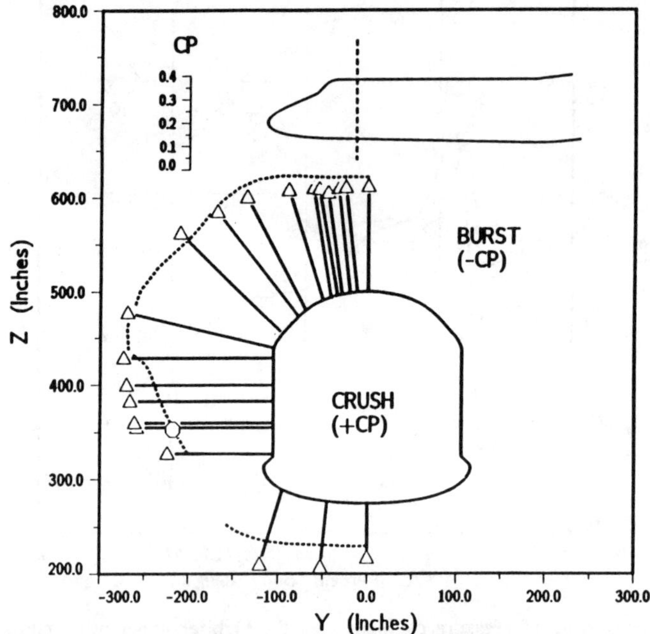

Fig. 11 Comparison of pressure coefficient between the computational results and IVBC-3PL at Orbiter station 590 for $M_\infty = 1.05$, $\alpha = -3$ deg: computation (−−), IVBC-3PL, (Δ), STS-5 flight data (○).

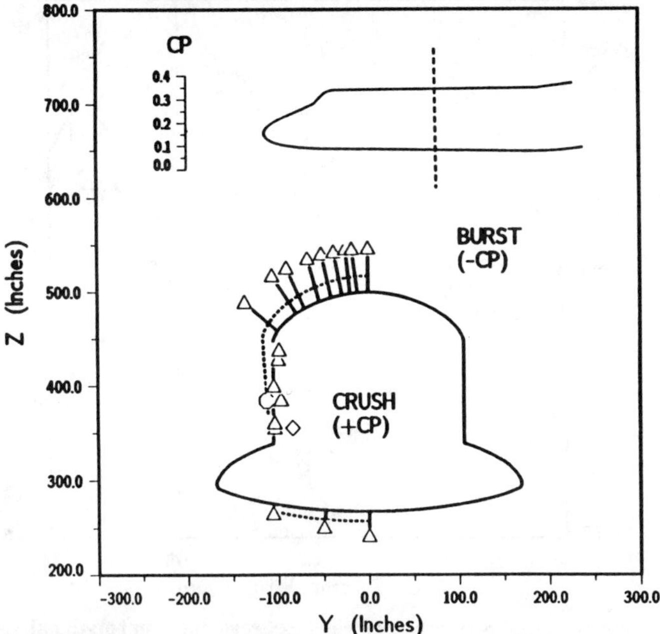

Fig. 12 Comparison of pressure coefficient between the computational results and IVBC-3PL at Orbiter station 905 for $M_\infty = 1.05$, $\alpha = -3$ deg: computation (--), IVBC-3PL (Δ), STS-5 flight data ($\bigcirc$, $\diamond$).

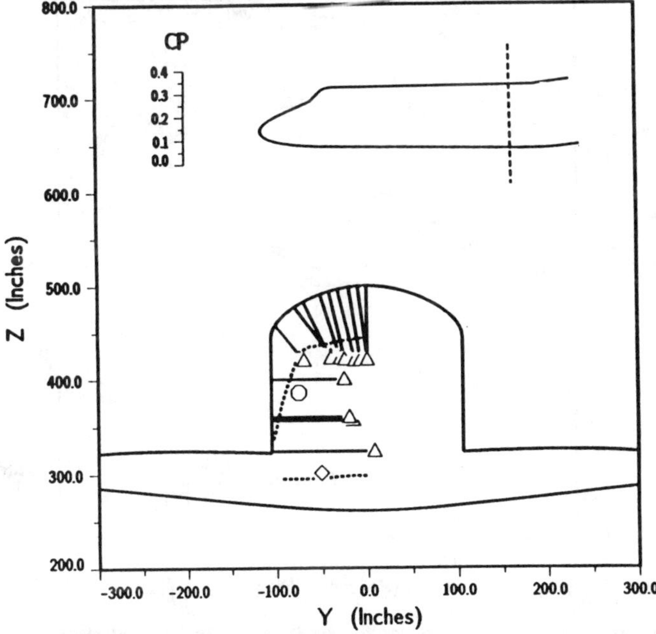

Fig. 13 Comparison of pressure coefficient between the computational results and IVBC-3PL at Orbiter station 1214 for $M_\infty = 1.05$, $\alpha - 3$ deg: computation (--), IVBC-3PL (Δ), STS-5 flight data ($\bigcirc$, $\diamond$).

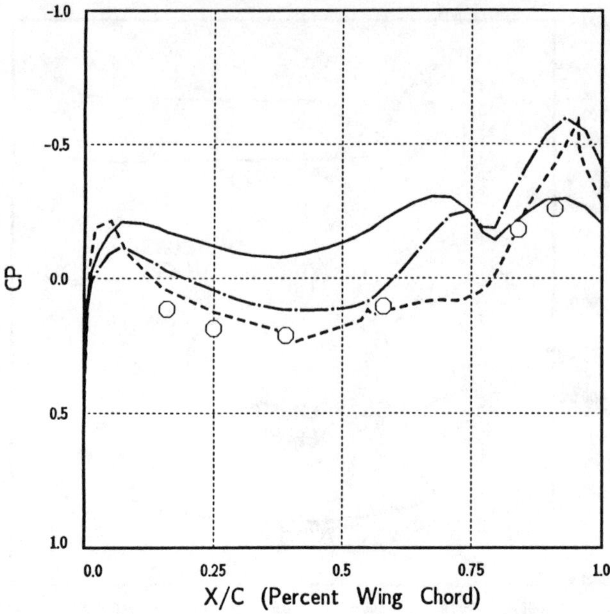

Fig. 14 Comparison of pressure coefficient between the computational results and IVBC-3 along the lower wing for $M_\infty = 1.05$, $\alpha = -3$ deg: computation without SRB/ET aft attach ring/IEA (−), with SRB/ET aft attach ring/IEA (− · −), IVBC-3PL (−−), STS-5 flight data (○).

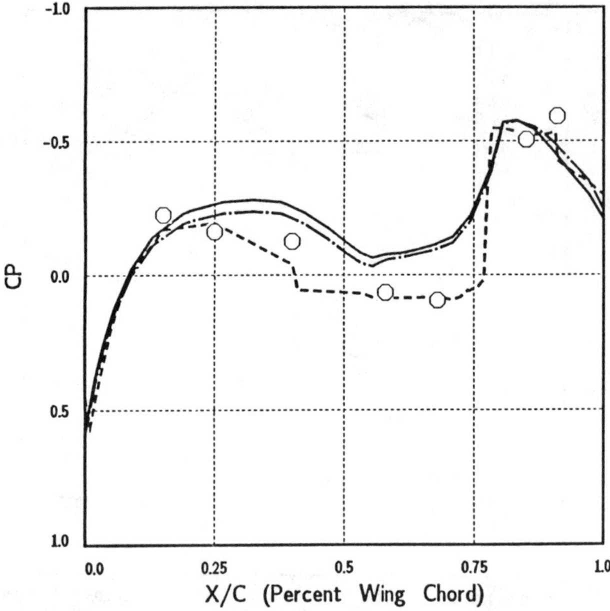

Fig. 15 Comparison of pressure coefficient between the computational results and IVBC-3 along the upper wing for $M_\infty = 1.05$, $\alpha = -3$ deg: computation without SRB/ET aft attach ring/IEA (−), with SRB/ET aft attach ring/IEA (− · −), IVBC-3PL, (−−), STS-5 flight data (○).

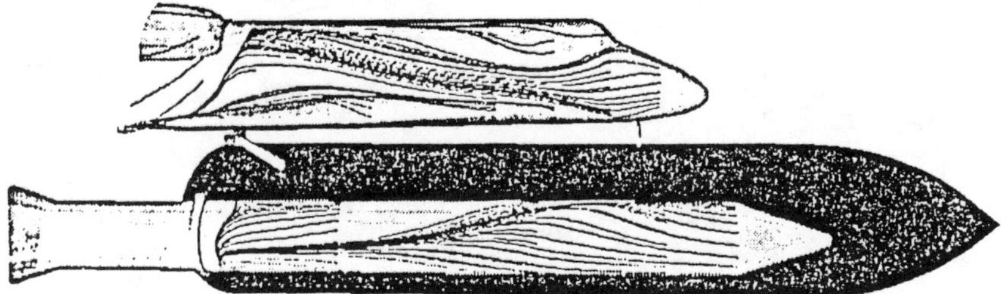

Fig. 16 Computed limiting streamlines for $M_\infty = 2.5$, $\alpha = 3$ deg, and $Re = 3.0 \times 10^6$/ft.

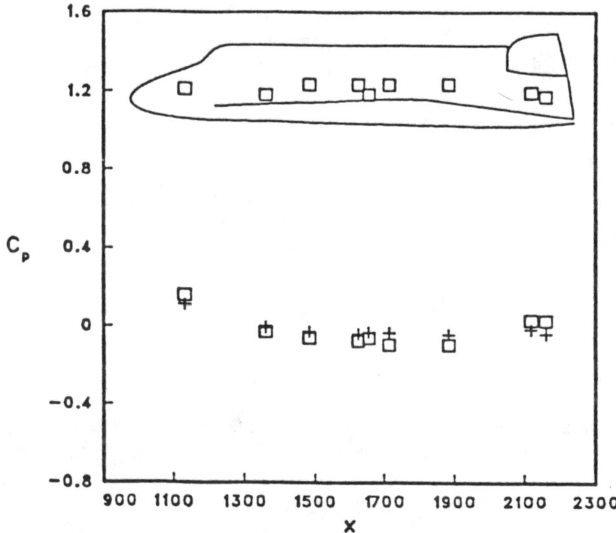

Fig. 17 Comparison of pressure coefficient between computation $(+)$ and flight test $(\Box)$ at stations on the side of the Orbiter fuselage $(M_\infty = 2.5$, $\alpha = 3$ deg, and $Re = 3.0 \times 10^6$/ft).

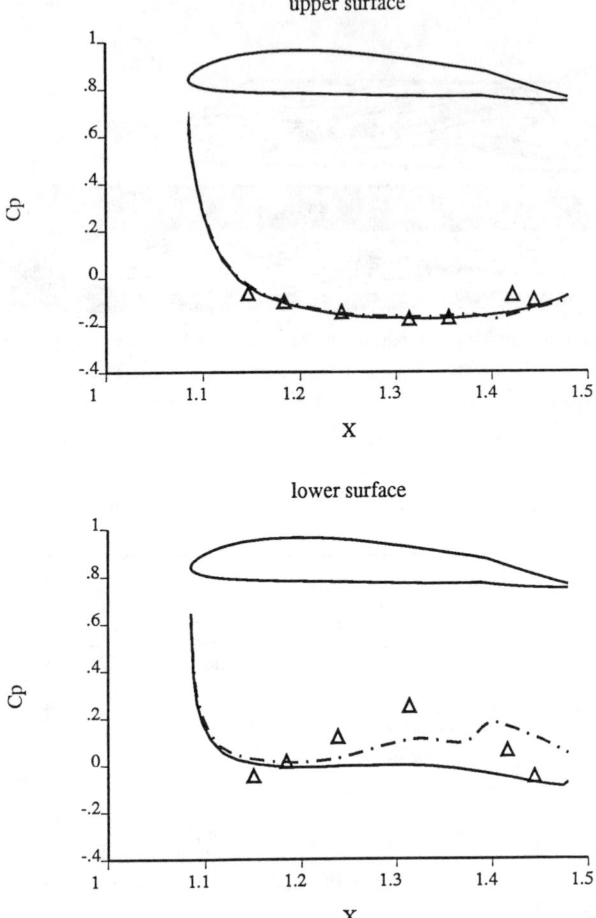

Fig. 18   Comparison of pressure coefficient between computation and flight test at a wing station directly over the centerline of the SRB ($M_\infty = 2.5$, $\alpha = 3$ deg, and $Re = 3.0 \times 10^6$/ft): fine grid with 10 deg/9 deg elevon deflections ($-\cdot-$), coarse grid with 0°/0° elevon deflections ($-$), STS-5 flight data (Δ).

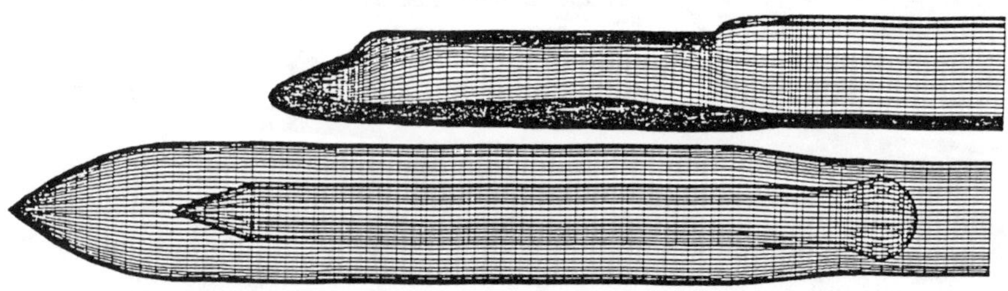

Fig. 19   Simplified configuration used for SRB separation calculations.

Part 8.   Forecast

Chapter 25

Future Directions for Applied Computational Fluid Dynamics

Richard G. Bradley*
General Dynamics, Fort Worth, Texas

Introduction

COMPUTATIONAL fluid dynamics (CFD) is becoming an increasingly powerful tool in the aerodynamic design of aerospace systems because of improvements in numerical algorithms, geometric modeling, grid generation, and physical parameter modeling, as well as dramatic improvements in supercomputer processing speed and memory. With the realization of the potential of forthcoming supercomputers, much of the current CFD work will be expanded to address more complex configurations, geometries, flight regimes, and applications. Thus, CFD is emerging as an important aerospace design and development tool.

In order for CFD to fulfill its apparent destiny as the premier aerodynamics design tool and to be fully accepted by the designer, a number of practical considerations are important. This chapter is intended to examine these considerations in light of the designer's requirements and, in turn, to focus on some directions for the future.

Some of the concepts and definitions presented in the text are taken from two studies: one sponsored by the National Research Council to examine the status of CFD in the United States and another by NASA's Aeronautics Advisory Committee as an ad hoc panel study on CFD validation.

Design Applications

The development of aerospace vehicles over the years has been an evolutionary process in which engineering progress in the aerospace community was generally based on prior experience and databases obtained through wind-tunnel and flight testing. Advances in the fundamental understanding of flow physics and wind-tunnel and flight-test capability and new mathematical insights into the governing flow equations have been translated into

---

Copyright © 1989 by the General Dynamics Corporation. All rights reserved. Published by the American Institute of Aeronautics and Astronautics, Inc., with permission.
*Director, Aerospace Technology, Fort Worth Division.

improved air vehicle design. Two notable examples of this evolutionary process that resulted in significant improvements to air vehicles are the area rule and supercritical wing technology. These improvements evolved from a combination of wind-tunnel tests and analytical advances. The analytical advances resulted in the ability to obtain solutions to simplified forms of the appropriate supersonic and transonic flow equations. The modern-day field of CFD is seen as a continuation of this growth in analytical capability and the digital mathematics needed to solve more rigorous forms of the fluid flow equations.

From a design manager's point of view, one obtains a perspective completely different from that of the CFD code developer. The aerospace designer is responsible for defining the best configuration to meet performance specifications in the shortest time and at the lowest cost. He has traditionally relied on extensive test results to guide his design decisions and to supply the flight envelope database needed for comprehensive vehicle performance calculations. Admittedly, wind-tunnel test data have many limitations, but designers have years of experience to help in understanding those limitations. The designer of high-performance aircraft must decide when and how to introduce CFD into the design process.

The potential of CFD is well known. For the first time in the history of aeronautics, the designer has the opportunity to generate solutions for complex flowfields and to examine the detailed features of the flow that influence a design. Today's challenge is to provide usable, believable, cost- and schedule-effective codes for design application, and to integrate these codes into the designer's toolset. This is both a technical and a management challenge. However, a responsible designer raises legitimate concerns about the benefits of CFD in helping him to meet his requirements, as illustrated in Fig. 1, taken from Ref. 1.

Some useful definitions are in order:

Capability, in the context of this discussion, pertains to the usefulness of a code in modeling flowfields about geometries of interest over a wide range of flow conditions producing results in a form that are meaningful to the designer.

Turnaround represents the span time required to set up the geometry and the grid meshes and to obtain converged solutions relative to the schedule-driven design exercise.

Availability (or usability) of the selected CFD code concerns the level of expertise required to generate the required flowfield solutions and the availability of the needed skill level for a particular project.

Cost represents the cost effectiveness of a CFD approach relative to other analytical and/or experimental approaches.

Confidence (or believability) relates to the dependability of the code to give accurate solutions over the range of design variables.

These concerns should be focal points for future development of CFD codes that are intended to be integrated into the design process. Although all are vitally important, perhaps the most critical at this time is confidence. The confidence level is especially important to the tactical aircraft designer since experience in the use of CFD is low and the flowfields of tactical aircraft are very complicated.

Mature CFD Capability

In light of the designer's concerns, it is clear that a number of important steps need to be taken to continue CFD development in the future. The objective is a fully mature CFD capability. It is understood that a mature CFD capability results in codes that can be used routinely in design applications by engineers.

A study conducted by the National Research Council and completed in 1986 examined the status of CFD in the United States and suggested future directions for development work.[2] An extensive survey of codes, either under development or in use, formed a basis for evaluation of the state of affairs in the field. A five-phase development cycle for CFD was presented as a guide for assessing CFD capabilities. This development cycle, shown in Fig. 2, illustrates the steps generally followed to achieve a mature CFD capability.

The predominant effort over the last two decades has been focused on phases I, II, and III. Code developers have made outstanding progress in developing algorithms and physical models that allow solution of inviscid and viscous flow equations for a large number of flows of interest. In recent years industry has placed significant emphasis on learning how to use the evolving capability, phase IV. Although the success in algorithm development for the potential equations and Euler equations has been noteworthy, most codes are equipped with "knobs" that control convergence and smooth the solutions. Unfortunately, these artifices may introduce unacceptable errors. On the other hand, Navier-Stokes solvers rely heavily on physical models, e.g., for turbulence, that are not well understood and require additional research.

The mature CFD capability, phase V, involves increasing the understanding and verification of the code's sensitivities to grids, convergence characteristics, numerical accuracy, reliability, robustness, ease of use, and cost effectiveness. The physical models employed for turbulence, chemical kinetics, transition, and heat-transfer mechanisms are critical elements for some flows and require considerable attention. In a pragmatic sense, one key feature of maturity in CFD capability involves careful comparison with experimental measurements.

It is, of course, important to note that CFD codes can be used in analysis and design applications long before the codes are considered to be mature. Engineers have always been able to use less than perfect tools to provide design guidance by using them in conjunction with experience and calibration to known physical quantities. Calibration provides an error band or correction factor that enhances the capability of a particular code or analytical method to predict specific parameters that are important to the design objectives for a particular design without verifying that all other features of the flow are modeled accurately. For example, one might calibrate a code to predict shock location, lift, and moment on a wing without any assurance that the flowfield off the surface and the wake behind the wing are properly modeled. Or one may calibrate a code to compute gross pressure loss through a supersonic inlet-duct combination without concern for the accuracy of the distribution distribution at the compressor face. Although the use of calibrated CFD solutions is risky because of the subtle interactions that are sensitive to geometry and flowfield, skilled engineers can obtain useful design information and guidance from relatively immature codes.

CFD Validation

CFD has opened the door for precise solutions to governing fluid flow equations without many of the simplifying assumptions required previously for analytical treatment. As a result, complex flows can be calculated to a level of detail not possible in the past. On the one hand, this unique capability gives the designer the ability to "step inside" a flowfield in order to understand its impact on the design. On the other hand, the integrity of the computed flowfield is a key factor in the usefulness of the CFD solution for design decisions.

A concept for gaining full confidence in CFD has been suggested by an ad hoc committee formed by NASA's Aeronautics Advisory Committee. The concept, called "CFD validation," has been reported in Ref. 3. CFD validation involves detailed comparison of computed flowfields with experimental data to verify the code's capability to accurately model the critical physics of the flow. These comparisons require close coordination between the code developer and the experimentalist to ensure that the accuracy and limitations of the experiments, as well as those of the numerical algorithm and grid densities, are understood and are taken into account. Through validation, one assures that the numerical physics of the code truly represents the flow physics being modeled.

The following definitions have been developed to help clarify the validation concept and to identify the associated experimental requirements[3]:

CFD Code Validation: Detailed surface and flowfield comparisons with experimental data to verify the code's ability to *accurately model* the *critical physics* of the flow. Validation can occur only when the accuracy and limitations of the experimental data are known and thoroughly understood and when the accuracy and limitations of the code's numerical algorithms, grid-density effects, and physical basis are equally known and understood over a *range* of specified parameters.

CFD Validation Experiment: An experiment that is designed to provide detailed building block data for developing CFD codes. This objective requires that the data be taken in the form and detail consistent with CFD modeling requirements and that the accuracy and limitations of the experimental data be thoroughly understood and documented.

Validated CFD Code: A code whose *accuracy* and *range* of validity has been determined by detailed comparison with CFD validation experiments so that it can be applied, without calibration, directly to a geometry and flow condition of engineering interest with a high degree of confidence.

CFD Code Calibration: The comparison of CFD code results with experimental data for realistic geometries that are similar to the ones of design interest, made in order to provide a measure of the code's capability to predict specific parameters that are of importance to the design objectives without necessarily verifying that *all* of the features of the flow are correctly modeled.

Often the terms "validation" and "calibration" have been loosely used to describe any comparison with experiment. The preceding definitions are given to clearly distinguish between the two. Although it is desirable to have mature, validated codes for design applications, it is recognized that code calibration remains an important part of the design process, especially since most of the codes being applied today are in the development stage.

One observes that full validation places stringent requirements on both computational and experimental accuracy. Accuracy for a CFD code is defined pragmatically as its capability to predict the physical features of a given flow in much the same way experimental accuracy is reflected in the ability to measure the flow physics.

In this definition the accuracy of a CFD solution has four limitations: 1) The appropriateness of the *governing equations* to represent the dominant features of the flow is, of course, a key consideration that can limit a solution. 2) The *numerical algorithm*, including smoothing routines and convergence acceleration schemes, can be a source of numerical error. 3) *Grid resolution* can lead to error in that generated grids may not capture the scale of the physical phenomena throughout the solution domain. 4) Many solutions require *physical models* to represent phenomena that are not well understood (e.g., turbulence), which can lead to significant sources of error. The net effect of these limiting factors is of primary concern for the designer who must gain confidence in a code's accuracy for specific applications.

Numerical error in CFD solutions is a strong function of the geometry, the algorithms selected, and the grid density. The truncation error decreases with an increasing number of mesh points; however, the practical limitations of today's computers do not permit total elimination of this error source. Furthermore, introduction of numerical "viscosity" terms to speed convergence and provide smoothing to the solution introduces a source of error that must be understood and evaluated. Grid size distributions must be matched to the physical scales of the flow to help ensure accuracy in the solution. Often such optimum distribution is not known a priori and can be a source of error. Confidence in the solution stems from an understanding of these limitations and their effect on the solution.

Much work remains to be done in physical modeling of turbulence, transition, combustion kinetics, and heat-transfer mechanisms, including wall catalycity. Today's CFD must rely on various levels of numerical approximation to represent these basic physical phenomena, which are not completely understood. In the case of transition and turbulence, full Navier-Stokes equation solutions are providing insight, but even this application is limited by practical computer limitations today. Much work remains to be done before one can have implicit confidence in the physical models required for a mature CFD capability.

In light of known limitations inherent in CFD codes, validation must entail a thorough understanding of how these limitations affect the ultimate accuracy of a solution in a given application. Thus, validation is seen to include the *bounds* of accuracy in terms of a *range* of numerical and physical parameters for which an acceptable error band may be achieved.

This level of confidence, then, must be related to experimental evidence that is, of course, subject to its own band of uncertainties.

Equally critical to the concept of CFD validation is the reliability of experimental data. To provide the basis for CFD validation, the experiments must provide measurements, with adequate accuracy and resolution, under flow conditions that are representative of those for which CFD will be used. The available capabilities of various facilities to simulate realistic flow conditions of Mach number, Reynolds number, turbulence, gas composition, and other flow parameters is an important consideration. An understanding of the effect of flow gradients in the test section, sting and wall interference effects, blockage, and noise is required to ensure the integrity of the flow measurements and to provide the proper flow and boundary conditions for the accompanying CFD calculations.

It is also essential that critical details of the flow are measured to a level of accuracy comparable to the accuracy expected from the calculations and in a form that is compatible with the CFD solutions. An understanding of the *bounds* of accuracy associated with the various measurements is essential. Instrumentation to provide redundant measurements for both the surface and the flowfield is highly desirable for validation experiments.

In summary, CFD validation requires a disciplined form of experimentation that is closely coupled to the requirements of the codes being validated. This experimentation requires close cooperation between the code developer and the experimentalist. Each must be thoroughly familiar with the requirements and limitations of the other. Future maturation of CFD will depend on greater emphasis on validation. NASA has proceeded to place added emphasis on CFD validation activities, as noted in Ref. 4, for example.

Challenges for the Future

The future for design applications of CFD is bright. Continued development of algorithms and computer technology will certainly provide the designer with an unparalleled capability for understanding flow details and for translating this understanding into more efficient aerodynamic vehicles and fluid systems. However, the path to a mature CFD capability does lead through a number of obstacles that provide a significant challenge to the code developer and aerodynamics researcher. These challenges result not only from the numerical mathematics of CFD itself, but also, interestingly enough, from some of the tough problems of fluid dynamics that have been a nemesis for the aerodynamicist over the years. In the following paragraphs some of these challenges are highlighted.

Computer Development

An in depth discussion of expectations in computer hardware development is certainly beyond the scope of this review. However, the future for applications of CFD to real design problems is very much dependent on the continued rapid development of computers. Of course, the issues of algorithm development, grid generation, and computer hardware improvements are interrelated. For example, the numerical algorithm and the discretization of the physical space where solutions of the Navier-Stokes equations are desired will, to a large extent, be controlled and limited by the architecture of available computers.

Many revolutionary changes have occurred in computers during the past 20 years. In memory development alone, major improvements in capacity have occurred. In 1975 the 4-million-word memory of the Cray X-MP/24 would have been considered huge. However, in solving problems in the present environment, this memory size presents a serious limitation. Fortunately, improvements in memory technology provide today's computers with large central memories, and even greater capacities are expected in the future as a result of continuing advances in silicon-based random access memory technology. The average increase in density has been an order of magnitude every five years, and memories of billions of words should be a reality within the next decade. Thus, memory will in all probability not be a limiting factor in attacking many of the computational fluid dynamic problems of interest.

Improvements in CPU capacity have been less rapid due to the limitations in miniaturization and cycle times due to finite signal propagation speeds. This has resulted in the introduction of parallel machines using multiple vector pipeline architectures. The future of the development of such parallel devices is likely to follow two paths. One path is to employ larger numbers of vector pipeline units similar to those of present supercomputers. The limitation for such units is likely to be economic in the sense that each processor in such devices is envisioned to be extremely high performance and, accordingly, very expensive. Consequently, the number of processors in such an advanced supercomputer may be in the 32/64 range. Another path is to continue the development of massively parallel machines that can take advantage of expected microprocessor advances. Such massively parallel devices typically employ of the order of 64,000 processors to achieve performance in the Gflop range. The continued improvement in microprocessor power will likely provide large improvements in performance, particularly if better ways of internodal communication are developed.

Future systems may be constructed for solving restricted classes of problems by using custom-designed VLSI chips. Such an approach implements the numerical algorithms in silicon and promises tremendous increase in speed. One can thus envision massively parallel systems with each processor being a hard-wired solver for the Navier-Stokes equations. In such a machine, solutions that in today's environment require tens of hours of CPU time can require minutes. Work on the technology has already started, and rapid progress is expected. However, there are also some serious pitfalls that must be recognized. The major disadvantage in implementing a Navier-Stokes solver in silicon is the lack of robustness. In order to be cost effective, a dedicated solver must be capable of reliably producing solutions for a wide class of problems. At this time, numerical stability and convergence characteristics of existing algorithms are not robust enough. However, the technology should be developed, and improvements in algorithm design will make dedicated solver technology practical for the future.

Grid Generation

Grid generation is very tightly interwoven with computer architecture and algorithm development. Future grid structures will most likely continue to be created using four different philosophies: block structured, unstructured, a combination of the two, and adaptive.

Block-Structured

The present state of the art of grid generation is probably best represented by block grid techniques used to discretize domains surrounding complex geometries. Since this approach relies on structured grids where connectivity of adjacent points is easily provided through standard array indexing, either explicit or implicit numerical schemes can be used efficiently. However, there are several areas associated with block interfaces where future work must be directed. For example, most block schemes have been designed to provide at least grid point continuity across block boundaries, but this may not always be the optimal type of boundary interface. If point continuity is not enforced, interpolation routines typically are employed to enforce conservation through the common boundary. Better ways are needed to efficiently provide data exchange through zonal boundaries.

Also, block-structured schemes limit the efficiency of implicit algorithms. Block-structured approaches cannot, in general, be viewed as globally implicit but, rather, are implicit by blocks, thus limiting the span of each implicit sweep. This is a difficult limitation to remove, and innovative ideas will be required to produce globally implicit methods on block grids.

Unstructured

With the increasing popularity of finite-volume methods, future grid research will also include heavy activity in the unstructured grid-generation area. Unstructured grids have the advantage that any domain can be discretized without regard to computational domain considerations. However, before unstructured grids can become practical, a number of areas of research are of major importance. First, the totally unstructured grid presents difficulties in easily identifying neighboring points or cells. Thus, a connectivity matrix must be created as an identification device. Research on optimal cell identification schemes minimizing matrix bandwidth will continue. An associated problem is that of vectorizing numerical algorithms on unstructured grids. Progress has been made in vectorizing calculations by cycling through computations on each cell face. However, this process may create a connectivity identification scheme that reduces the efficiency of other parts of the numerical calculation. Second, a difficult problem with unstructured meshes is that they are generally not suited for use with implicit algorithms. Standard matrix solvers cannot be used in the same way as with unstructured grids. Explicit methods are easily and naturally implemented on such meshes with much less difficulty. When applied with a massively parallel computer, unstructured grids are a much more attractive alternative.

Combined

Grid-generation ideas based on combined structured and unstructured grids will be a focus of the future, and two schools of thought exist. On one hand, an unstructured grid may be used to fill in the region between the body and a smooth structured grid near the body. In this way, the difficulty of creating body-conforming structured grids is avoided, but the efficiency of the structured scheme may be used over the majority of the domain. The disadvantage is that very high mesh point densities parallel and normal to the body surface will result from the unstructured mesh, and the structured mesh will most likely dilate excessively at large distances from the body. On the other hand, a structured grid may be used to discretize the domain near the body with an unstructured grid outside of this grid. In this way the excessive resolution parallel to the body required in an unstructured mesh and the lack of resolution far away from the structured mesh can be avoided. Research of these concepts will undoubtedly continue into the future.

Adaptive

Finally, research will continue to uncover ways of providing error reduction in numerical solutions by using adaptive grids. Two techniques exist that need continued exploration. The first tries to determine how a fixed number of grid points should be distributed in a domain to provide the test solution for a given problem. The other approach is to refine the mesh locally in regions where error is large. This creates a data management problem and presents some difficult issues in maintaining global connectivity for the mesh. Both of these ideas deserve continued study. The ultimate solution will no doubt involve both grid point movement ideas and mesh refinement.

Algorithms

Improvements in areas specific to numerical algorithms provide a continuing challenge. The recent trend in constructing upwind schemes to more accurately treat the convection terms of the Navier-Stokes equations has improved the robustness of modern methods. However, significant improvement is still needed since user skill in applying codes remains a major factor in attaining usable results. Future progress in algorithm construction should emphasize calculation of accurate, reliable solutions with a minimum of user intervention.

Another area of concern that must be addressed is that of convergence or, in particular, the time necessary to obtain time-asymptotic solutions. Although modern upwind techniques appear to improve reliability, they are generally slow to converge when applied to complex problems. They also require much more work per iteration than the older technology (central difference) codes; hence, the overall cost of a computation can be greatly increased. Upwind techniques require less user intervention than codes based on the older algorithms and produce solutions with reduced effects of artificial dissipation. Methods of accelerating convergence are needed, such as multigrid, local iteration, and other new approaches.

The use of Riemann and approximate Riemann solvers have provided improved simulation of fluid physics, particularly in convection-dominated flows. Accurate, reliable solutions to problems where shock capturing is employed are now a reality. These solvers treat a sequence of one-dimensional problems rather than a single multidimensional calculation. Although the approach seems to provide acceptable results, work on such solvers in a true three-dimensional setting is necessary. Associated with some of these techniques, continued effort must also be expended in studying entropy conditions, dissipation models, and flux limiters.

Implementation of boundary conditions continues to be an area that deserves attention. Recent progress in formulating consistent inflow/outflow boundary treatment has improved the quality of numerical solutions. In many cases, difficulties associated with the application of appropriate boundary conditions are still encountered. Inflow/outflow conditions are usually based on one-dimensional characteristic information. Future work needs to include formulation of conditions more appropriate to a fully three-dimensional flow. Pole boundaries (i.e., $r = 0$ in cylindrical geometries) and grid singularities prove, at best, to be inconvenient to treat and always require special attention. Even accepted solid boundary surface conditions are difficult to apply accurately in regions of high surface curvature when mesh densities are less than optimum. New and improved techniques for applying boundary conditions are necessary for the construction of improved algorithms. This area will require a significant effort for the foreseeable future.

Transition and Turbulence

The aerodynamic phenomena of transition from laminar to turbulent flow and the subsequent development of turbulent flow remains a major limitation to the application of CFD. In many practical cases the vehicle is engulfed in a fully turbulent flow due to the large Reynolds numbers of flight. Thus, numerical simulation, as in wind-tunnel testing, may be based on the assumption of fully developed turbulent flow from a point very near the leading edge or nose of the configuration. In other cases, such as analysis of a laminar flow control design or high-speed/altitude applications where transition may be delayed significantly, the flow assumptions for CFD must include regions of laminar, transitional, and turbulent flow. Applications of CFD to the fully turbulent case is certainly difficult today. However, far more difficult is the application of CFD methods to mixed laminar transitional turbulent flow conditions.

The full, time-dependent Navier-Stokes equations are considered adequate for modeling the complete viscous flowfield. However, practical solutions of the full Navier-Stokes equations are in an infantile state of development at the current time, and several orders of magnitude of improvement in computational hardware capability will be required before they become useful methods for applications. However, incompressible solutions to the Navier-Stokes equations that are achievable and large-eddy simulation approaches are useful for fundamental studies that can lead to improved physical modeling for use in solving the turbulence-averaged Navier-Stokes equations.

In the case of turbulence-averaged Navier-Stokes solutions, the calculation of transition for practical application is left to correlations from experiment and/or applications of (usually linear) stability theory. Either of these approaches leaves a great deal to be desired at the present time and presents a major challenge for the future in CFD applications.

Engineering calculations for turbulent flow are almost always based on solutions of the turbulence-averaged Navier-Stokes equations. These equations are derived by assuming some form of turbulence averaging, and their solution requires physical models for the Reynolds stresses and turbulent heat fluxes to replace the turbulent transport terms in the equations. The formulation of these physical models involves statistical correlations between fluctuating components measured at the same point in space time. A number of approaches to turbulent modeling have been attempted with varying degrees of success depending on the type of flow being studied and whether it is attached or separated. A good review of a number of these models may be found in Ref. 2.

Not surprisingly, no single model for turbulence has been found that is satisfactory for the many varied types of flows associated with aircraft configurations. One model may work well for attached boundary layers and not be satisfactory at all for separated flows, corner flows, or vortex flow, and so forth. Thus, the application to design requires a significant calibration of the CFD solution with experimental data and a lot of experience on the part of the aerodynamic designer.

The challenge for the future, then, continues to be the development of usable, trustworthy, and validated physical models for transition and turbulence. Such modeling is expected to become the pacing item in the development of mature CFD codes for many engineering design applications both for external and internal flow application.

Aerodynamic Drag

The ultimate objective for design of aerospace vehicles is to maximize the performance. The ultimate measure of merit is often aerodynamic drag. Thus, the accurate prediction of drag is the appropriate goal for a mature CFD capability. Unfortunately, accurate and consistent drag prediction from CFD solutions for practical configurations is not possible at the present time and thus presents a major challenge for the future.

In 1988 the Fluid Dynamics Panel of AGARD organized a technical status review on drag prediction and analysis from computational fluid dynamics.[5] The intent of this forum was to obtain a survey of the state of the art from the NATO countries. The main conclusions reached by this body confirmed that "accurate and consistent computation through CFD of absolute drag levels for complex configurations is, not surprisingly, beyond reach for a considerable time to come." Fundamentally, the reason for inadequate drag prediction results from the inability to accurately predict all of the pressure and friction forces acting on the configuration and then to integrate these forces over the vehicle. The pacing items for drag prediction are those factors that have already been noted as necessary to achieve a mature CFD capability. However, for drag prediction, the importance of such things as turbulence modeling, grid generation, and grid resolution to accurately capture boundary-layer and separated-flow phenomena are amplified considerably.

The prediction of drag for simple configurations with attached boundary-layer flow have met with some limited success. Such flows as two-dimensional airfoils, simple wings, or wing bodies in some cases may be analyzed to a degree of accuracy using potential or Euler codes for the inviscid flow and adding attached boundary-layer calculations to provide the skin friction. Prediction of drag due to lift for highly swept wings with strong leading-edge vortices has also met with some success using Euler codes. In general, pressure drag predicted from solutions of the current Euler codes is hampered significantly by oversensitivity to grid density and grid resolution as well as by the artificial damping or dissipation terms required to achieve a solution in many cases.

Drag prediction from the solutions of the Navier-Stokes codes have not been successful to date except for some simple two-dimensional codes. Grid density and resolution play an important part in attached Navier-Stokes solutions, and at the current state of the art they cannot do a better job than zonal methods involving potential flow coupled with boundary layers. Separated flows appear to require turbulence modeling that is more reliable than currently available. In addition, appropriate grid clustering and refinements present a problem area even for two-dimensional airfoil flow.

It should be noted that, in spite of the uncertainty and limitations associated with CFD drag prediction today, useful design data can be obtained from incremental changes in drag from an established drag level that result from small geometry changes. However, in the long run, accurate calculation of absolute levels of aerodynamic drag must be the criteria that measures the maturity of CFD as a fully acceptable design tool.

Acknowledgments

The author gratefully acknowledges the contributions of the researchers and scientists who served on committees and performed the studies from which several concepts and definitions in this text were taken. The support of the General Dynamics Computational Fluid Dynamics Group is also acknowledged.

References

[1]Bradley, R. G., Bhateley, I. C., and Howell, G. A., "Computational Fluid Dynamics—Transition to Design Applications," NASA Conference on Supercomputing in Aerospace, NASA Ames Research Center, Palo Alto, CA, March 1987.

[2]"Current Capabilities and Future Directions in Computational Fluid Dynamics," National Research Council, National Academy Press, Washington, DC, 1986.

[3]Bradley, R. G., "CFD Validation Philosophy," AGARD Symposium on Validation of Computational Fluid Dynamics, Lisbon, Portugal, May 1988.
[4]NASA CFD Validation Workshop, NASA Ames Research Center, Moffett Field, CA.
[5]"Drag Prediction and Analysis From Computational Fluid Dynamics," AGARD, Fluid Dynamics Panel, Technical Status Review 256, 1989.

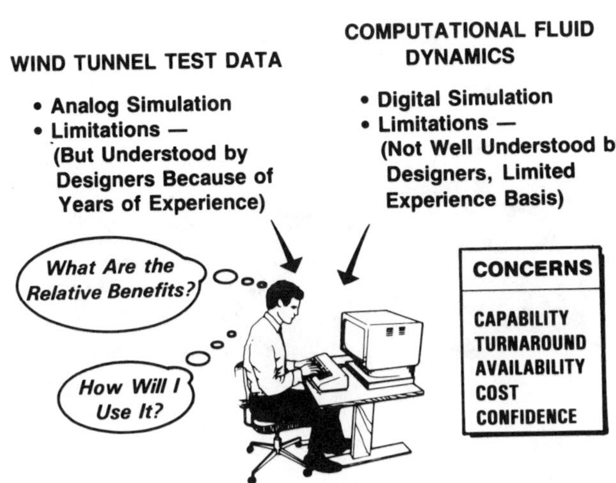

Fig. 1   Designer's dilemma.

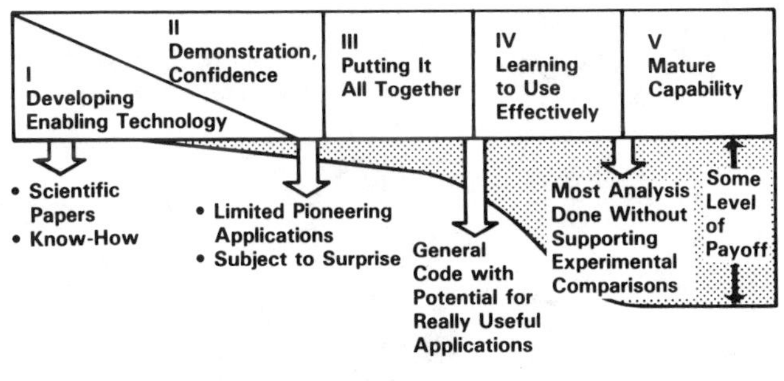

Fig. 2   CFD development cycle.

Epilogue

The works presented in this volume represent, for the most part, a cross section of computational aerodynamics applications. The applications range from classical airfoil studies to the aerodynamic evaluation of complete aircraft. These applications indicate the tremendous progress that has been made in the field of computational aerodynamics in the last 30 years. The ability to calculate nonlinear flows about complete aircraft configurations is evident in a number of chapters.

However it is easy to confuse first applications with proven, user-friendly, well-documented, well-understood capabilities. Clearly, a certain expertise or knowledge base is still required to apply most computational procedures in a design application environment. The treatment of computational aerodynamics methods as black boxes is risky at best. Computational methods can easily be applied improperly or applied beyond the region of basic assumption validity. Computational results must still pass judgment tests imposed by experienced aerodynamicists. Aerodynamic design experience is just as necessary today as it was 30 years ago. The solution of aerodynamic design problems is facilitated by computational methods. Hence, such methods represent an additional tool, similar to wind-tunnel testing and flight testing, to be utilized by the aerodynamicist. Proper application of such a tool is not necessarily easy even though the tool may be very sophisticated and user friendly.

Examples of computational methods application have been presented in which insights and developments were made that could not be done or would be prohibitively expensive with a wind-tunnel test or flight-test approach. Hence, computational approaches are being used to promote innovation. New ideas can be explored in a much more timely manner. High-risk or expensive ideas, which might not otherwise be pursued, have a better opportunity if they are first investigated computationally. Innovation is one aspect of computational aerodynamics that will probably become increasingly important. Although such studies are typically highly proprietary or classified, an increasing number of such developments is becoming evident.

Several of the chapters reviewed inverse or design as well as direct or analysis capabilities. The ability to specify aerodynamic characteristics numerically and calculate the appropriate aerodynamic surface provides exceptional advantages in a design application environment. Inverse capabilities allow a designer to quickly establish a suitable configuration without an extended cut-and-try process. Desirable aerodynamic characteristics can be imposed as an initial boundary condition rather than an ultimate goal. Just as computational methods complement experimental methods, inverse methods complement direct methods. An effective design operation makes use of both in resolving design problems.

The volume and quality of investigative design information that can be generated computationally is unprecedented. Flow quantities evaluated at tens of thousands of points on the aerodynamic surface are orders of magnitude beyond technical and economic feasibility experimentally. On-body and off-body computational flow visualization is adding remarkable understanding to complex flows. The pre- and postprocessing graphics environment utilizing color, shaded surfaces, and transparencies of complex configurations and flowfields is a necessity in effectively handling such volumes of data. Several chapters included examples in which these pre- and postprocessing environments are clearly the keys to making the flow computation meaningful.

In closing, it should be noted that computational aerodynamics methods have achieved broad acceptance in the aerodynamics community. Applications are abundant in every phase of vehicle development and use. Computational schemes have been applied in conceptual and preliminary design, development and tailoring, loads definition, certification and acceptance analyses, and in-service modification. The applications presented in this Progress Series volume represent but a few of these examples.

A special note of appreciation goes out to each of the contributors to the volume. Each chapter represents a sizeable effort for the authors and each of their efforts is recognized as a unique and key part of the volume.

Author Index

PROGRESS IN ASTRONAUTICS AND AERONAUTICS
SERIES VOLUMES

***1. Solid Propellant Rocket Research** (1960)
Martin Summerfield
Princeton University

***2. Liquid Rockets and Propellants** (1960)
Loren E. Bollinger
Ohio State University
Martin Goldsmith
The Rand Corp.
Alexis W. Lemmon Jr.
Battelle Memorial Institute

***3. Energy Conversion for Space Power** (1961)
Nathan W. Snyder
Institute for Defense Analyses

***4. Space Power Systems** (1961)
Nathan W. Snyder
Institute for Defense Analyses

***5. Electrostatic Propulsion** (1961)
David B. Langmuir
Space Technology Laboratories, Inc.
Ernst Stuhlinger
NASA George C. Marshall Space Flight Center
J.M. Sellen Jr.
Space Technology Laboratories, Inc.

***6. Detonation and Two-Phase Flow** (1962)
S.S. Penner
California Institute of Technology
F.A. Williams
Harvard University

***7. Hypersonic Flow Research** (1962)
Frederick R. Riddell
AVCO Corp.

***8. Guidance and Control** (1962)
Robert E. Roberson,
Consultant
James S. Farrior
Lockheed Missiles and Space Co.

***9. Electric Propulsion Development** (1963)
Ernst Stuhlinger
NASA George C. Marshall Space Flight Center

***10. Technology of Lunar Exploration** (1963)
Clifford I. Cummings
Harold R. Lawrence
Jet Propulsion Laboratory

***11. Power Systems for Space Flight** (1963)
Morris A. Zipkin
Russell N. Edwards
General Electric Co.

***12. Ionization in High-Temperature Gases** (1963)
Kurt E. Shuler, Editor
National Bureau of Standards
John B. Fenn,
Associate Editor
Princeton University

***13. Guidance and Control – II** (1964)
Robert C. Langford
General Precision Inc.
Charles J. Mundo
Institute of Naval Studies

***14. Celestial Mechanics and Astrodynamics** (1964)
Victor G. Szebehely
Yale University Observatory

***15. Heterogeneous Combustion** (1964)
Hans G. Wolfhard
Institute for Defense Analyses
Irvin Glassman
Princeton University
Leon Green Jr.
Air Force Systems Command

16. Space Power Systems Engineering (1966)
George C. Szego
Institute for Defense Analyses
J. Edward Taylor
TRW Inc.

17. Methods in Astrodynamics and Celestial Mechanics (1966)
Raynor L. Duncombe
U.S. Naval Observatory
Victor G. Szebehely
Yale University Observatory

18. Thermophysics and Temperature Control of Spacecraft and Entry Vehicles (1966)
Gerhard B. Heller
NASA George C. Marshall Space Flight Center

---

*Out of print.

***19.   Communication Satellite Systems Technology** (1966)
Richard B. Marsten
Radio Corporation of America

***20.   Thermophysics of Spacecraft and Planetary Bodies: Radiation Properties of Solids and the Electromagnetic Radiation Environment in Space** (1967)
Gerhard B. Heller
NASA George C. Marshall Space Flight Center

21.   Thermal Design Principles of Spacecraft and Entry Bodies (1969)
Jerry T. Bevans
TRW Systems

22.   Stratospheric Circulation (1969)
Willis L. Webb
Atmospheric Sciences Laboratory, White Sands, and University of Texas at El Paso

23.   Thermophysics: Applications to Thermal Design of Spacecraft (1970)
Jerry T. Bevans
TRW Systems

24.   Heat Transfer and Spacecraft Thermal Control (1971)
John W. Lucas
Jet Propulsion Laboratory

25.   Communication Satellites for the 70's: Technology (1971)
Nathaniel E. Feldman
The Rand Corp.
Charles M. Kelly
The Aerospace Corp.

26.   Communication Satellites for the 70's: Systems (1971)
Nathaniel E. Feldman
The Rand Corp.
Charles M. Kelly
The Aerospace Corp.

27.   Thermospheric Circulation (1972)
Willis L. Webb
Atmospheric Sciences Laboratory, White Sands, and University of Texas at El Paso

28.   Thermal Characteristics of the Moon (1972)
John W. Lucas
Jet Propulsion Laboratory

29.   Fundamentals of Spacecraft Thermal Design (1972)
John W. Lucas
Jet Propulsion Laboratory

30.   Solar Activity Observations and Predictions (1972)
Patrick S. McIntosh
Murray Dryer
Environmental Research Laboratories, National Oceanic and Atmospheric Administration

31.   Thermal Control and Radiation (1973)
Chang-Lin Tien
University of California at Berkeley

32.   Communications Satellite Systems (1974)
P.L. Bargellini
COMSAT Laboratories

33.   Communications Satellite Technology (1974)
P.L. Bargellini
COMSAT Laboratories

34.   Instrumentation for Airbreathing Propulsion (1974)
Allen E. Fuhs
Naval Postgraduate School
Marshall Kingery
Arnold Engineering Development Center

35.   Thermophysics and Spacecraft Thermal Control (1974)
Robert G. Hering
University of Iowa

36.   Thermal Pollution Analysis (1975)
Joseph A. Schetz
Virginia Polytechnic Institute
ISBN 0-915928-00-0

37.   Aeroacoustics: Jet and Combustion Noise; Duct Acoustics (1975)
Henry T. Nagamatsu, Editor
General Electric Research and Development Center
Jack V. O'Keefe, Associate Editor
The Boeing Co.
Ira R. Schwartz, Associate Editor
NASA Ames Research Center
ISBN 0-915928-01-9

38.   Aeroacoustics: Fan, STOL, and Boundary Layer Noise; Sonic Boom; Aeroacoustics Instrumentation (1975)
Henry T. Nagamatsu, Editor
General Electric Research and Development Center
Jack V. O'Keefe, Associate Editor
The Boeing Co.
Ira R. Schwartz, Associate Editor
NASA Ames Research Center
ISBN 0-915928-02-7

**39. Heat Transfer
with Thermal Control
Applications** (1975)
M. Michael Yovanovich
University of Waterloo
ISBN 0-915928-03-5

**40. Aerodynamics of
Base Combustion** (1976)
S.N.B. Murthy, Editor
J.R. Osborn,
Associate Editor
Purdue University
A.W. Barrows
J.R. Ward,
Associate Editors
*Ballistics Research
Laboratories*
ISBN 0-915928-04-3

**41. Communications
Satellite Developments:
Systems** (1976)
Gilbert E. LaVean
*Defense Communications
Agency*
William G. Schmidt
CML Satellite Corp.
ISBN 0-915928-05-1

**42. Communications
Satellite Developments:
Technology** (1976)
William G. Schmidt
CML Satellite Corp.
Gilbert E. LaVean
*Defense Communications
Agency*
ISBN 0-915928-06-X

**43. Aeroacoustics: Jet
Noise, Combustion and
Core Engine Noise** (1976)
Ira R. Schwartz, Editor
*NASA Ames
Research Center*
Henry T. Nagamatsu,
Associate Editor
*General Electric Research
and Development Center*
Warren C. Strahle,
Associate Editor
*Georgia Institute
of Technology*
ISBN 0-915928-07-8

**44. Aeroacoustics: Fan
Noise and Control; Duct
Acoustics; Rotor Noise**
(1976)
Ira R. Schwartz, Editor
*NASA Ames
Research Center*
Henry T. Nagamatsu,
Associate Editor
*General Electric Research
and Development Center*
Warren C. Strahle,
Associate Editor
*Georgia Institute
of Technology*
ISBN 0-915928-08-6

**45. Aeroacoustics:
STOL Noise; Airframe
and Airfoil Noise** (1976)
Ira R. Schwartz, Editor
*NASA Ames
Research Center*
Henry T. Nagamatsu,
Associate Editor
*General Electric Research
and Development Center*
Warren C. Strahle,
Associate Editor
*Georgia Institute
of Technology*
ISBN 0 915928-09-4

**46. Aeroacoustics:
Acoustic Wave
Propagation; Aircraft
Noise Prediction;
Aeroacoustic
Instrumentation** (1976)
Ira R. Schwartz, Editor
*NASA Ames
Research Center*
Henry T. Nagamatsu,
Associate Editor
*General Electric Research
and Development Center*
Warren C. Strahle,
Associate Editor
*Georgia Institute
of Technology*
ISBN 0-915928-10-8

**47. Spacecraft Charging
by Magnetospheric
Plasmas** (1976)
Alan Rosen
TRW Inc.
ISBN 0-915928-11-6

**48. Scientific
Investigations on the
Skylab Satellite** (1976)
Marion I. Kent
Ernst Stuhlinger
*NASA George C. Marshall
Space Flight Center*
Shi-Tsan Wu
University of Alabama
ISBN 0-915928-12-4

**49. Radiative Transfer
and Thermal Control**
(1976)
Allie M. Smith
ARO Inc.
ISBN 0-915928-13-2

**50. Exploration of the
Outer Solar System** (1976)
Eugene W. Greenstadt
TRW Inc.
Murray Dryer
*National Oceanic and
Atmospheric Administration*
Devrie S. Intriligator
*University of Southern
California*
ISBN 0-915928-14-0

**51. Rarefied Gas
Dynamics, Parts I and II**
(two volumes) (1977)
J. Leith Potter
ARO Inc.
ISBN 0-915928-15-9

**52. Materials Sciences
in Space with Application
to Space Processing** (1977)
Leo Steg
General Electric Co.
ISBN 0-915928-16-7

73. Combustion Experiments in a Zero-Gravity Laboratory (1981)
Thomas H. Cochran
NASA Lewis Research Center
ISBN 0-915928-48-5

74. Rarefied Gas Dynamics, Parts I and II (two volumes) (1981)
Sam S. Fisher
University of Virginia
ISBN 0-915928-51-5

75. Gasdynamics of Detonations and Explosions (1981)
J.R. Bowen
University of Wisconsin at Madison
N. Manson
Université de Poitiers
A.K. Oppenheim
University of California at Berkeley
R.I. Soloukhin
Institute of Heat and Mass Transfer, BSSR Academy of Sciences
ISBN 0-915928-46-9

76. Combustion in Reactive Systems (1981)
J.R. Bowen
University of Wisconsin at Madison
N. Manson
Université de Poitiers
A.K. Oppenheim
University of California at Berkeley
R.I. Soloukhin
Institute of Heat and Mass Transfer, BSSR Academy of Sciences
ISBN 0-915928-47-7

77. Aerothermodynamics and Planetary Entry (1981)
A.L. Crosbie
University of Missouri-Rolla
ISBN 0-915928-52-3

78. Heat Transfer and Thermal Control (1981)
A.L. Crosbie
University of Missouri-Rolla
ISBN 0-915928-53-1

79. Electric Propulsion and Its Applications to Space Missions (1981)
Robert C. Finke
NASA Lewis Research Center
ISBN 0-915928-55-8

80. Aero-Optical Phenomena (1982)
Keith G. Gilbert
Leonard J. Otten
Air Force Weapons Laboratory
ISBN 0-915928-60-4

81. Transonic Aerodynamics (1982)
David Nixon
Nielsen Engineering & Research, Inc.
ISBN 0-915928-65-5

82. Thermophysics of Atmospheric Entry (1982)
T.E. Horton
University of Mississippi
ISBN 0-915928-66-3

83. Spacecraft Radiative Transfer and Temperature Control (1982)
T.E. Horton
University of Mississippi
ISBN 0-915928-67-1

84. Liquid-Metal Flows and Magnetohydrodynamics (1983)
H. Branover
Ben-Gurion University of the Negev
P.S. Lykoudis
Purdue University
A. Yakhot
Ben-Gurion University of the Negev
ISBN 0-915928-70-1

85. Entry Vehicle Heating and Thermal Protection Systems: Space Shuttle, Solar Starprobe, Jupiter Galileo Probe (1983)
Paul E. Bauer
McDonnell Douglas Astronautics Co.
Howard E. Collicott
The Boeing Co.
ISBN 0-915928-74-4

86. Spacecraft Thermal Control, Design, and Operation (1983)
Howard E. Collicott
The Boeing Co.
Paul E. Bauer
McDonnell Douglas Astronautics Co.
ISBN 0-915928-75-2

87. Shock Waves, Explosions, and Detonations (1983)
J.R. Bowen
University of Washington
N. Manson
Université de Poitiers
A.K. Oppenheim
University of California at Berkeley
R.I. Soloukhin
Institute of Heat and Mass Transfer, BSSR Academy of Sciences
ISBN 0-915928-76-0

88. Flames, Lasers, and Reactive Systems (1983)
J.R. Bowen
University of Washington
N. Manson
Université de Poitiers
A.K. Oppenheim
University of California at Berkeley
R.I. Soloukhin
Institute of Heat and Mass Transfer, BSSR Academy of Sciences
ISBN 0-915928-77-9

102. Numerical Methods for Engine-Airframe Integration (1986)
S.N.B. Murthy
Purdue University
Gerald C. Paynter
Boeing Airplane Co.
ISBN 0-930403-09-6

103. Thermophysical Aspects of Re-Entry Flows (1986)
James N. Moss
NASA Langley Research Center
Carl D. Scott
NASA Johnson Space Center
ISBN 0-930403-10-X

104. Tactical Missile Aerodynamics (1986)
M.J. Hemsch
PRC Kentron, Inc.
J.N. Nielsen
NASA Ames Research Center
ISBN 0-930403-13-4

105. Dynamics of Reactive Systems Part I: Flames and Configurations; Part II: Modeling and Heterogeneous Combustion (1986)
J.R. Bowen
University of Washington
J.-C. Leyer
Université de Poitiers
R.I. Soloukhin
Institute of Heat and Mass Transfer, BSSR Academy of Sciences
ISBN 0-930403-14-2

106. Dynamics of Explosions (1986)
J.R. Bowen
University of Washington
J.-C. Leyer
Université de Poitiers
R.I. Soloukhin
Institute of Heat and Mass Transfer, BSSR Academy of Sciences
ISBN 0-930403-15-0

107. Spacecraft Dielectric Material Properties and Spacecraft Charging (1986)
A.R. Frederickson
U.S. Air Force Rome Air Development Center
D.B. Cotts
SRI International
J.A. Wall
U.S. Air Force Rome Air Development Center
F.L. Bouquet
Jet Propulsion Laboratory, California Institute of Technology
ISBN 0-930403-17-7

108. Opportunities for Academic Research in a Low-Gravity Environment (1986)
George A. Hazelrigg
National Science Foundation
Joseph M. Reynolds
Louisiana State University
ISBN 0-930403-18-5

109. Gun Propulsion Technology (1988)
Ludwig Stiefel
U.S. Army Armament Research, Development and Engineering Center
ISBN 0-930403-20-7

110. Commercial Opportunities in Space (1988)
F. Shahrokhi
K.E. Harwell
University of Tennessee Space Institute
C.C. Chao
National Cheng Kung University
ISBN 0-930403-39-8

111. Liquid-Metal Flows: Magnetohydrodynamics and Applications (1988)
Herman Branover,
Michael Mond, and
Yeshajahu Unger
Ben-Gurion University of the Negev
ISBN 0-930403-43-6

112. Current Trends in Turbulence Research (1988)
Herman Branover,
Michael Mond, and
Yeshajahu Unger
Ben-Gurion University of the Negev
ISBN 0-930403-44-4

113. Dynamics of Reactive Systems Part I: Flames; Part II: Heterogeneous Combustion and Applications (1988)
A.L. Kuhl
R & D Associates
J.R. Bowen
University of Washington
J.-C. Leyer
Université de Poitiers
A. Borisov
USSR Academy of Sciences
ISBN 0-930403-46-0

114. Dynamics of Explosions (1988)
A.L. Kuhl
R & D Associates
J.R. Bowen
University of Washington
J.-C. Leyer
Université de Poitiers
A. Borisov
USSR Academy of Sciences
ISBN 0-930403-47-9

115. Machine Intelligence and Autonomy for Aerospace (1988)
E. Heer
Heer Associates, Inc.
H. Lum
NASA Ames Research Center
ISBN 0-930403-48-7

116. Rarefied Gas Dynamics: Space-Related Studies (1989)
E.P. Muntz
University of Southern California
D.P. Weaver
U.S. Air Force Astronautics Laboratory (AFSC)
D.H. Campbell
University of Dayton Research Institute
ISBN 0-930403-53-3

117. Rarefied Gas Dynamics: Physical Phenomena (1989)
E.P. Muntz
University of Southern California
D.P. Weaver
U.S. Air Force Astronautics Laboratory (AFSC)
D. Campbell
University of Dayton Research Institute
ISBN 0-930403-54-1

118. Rarefied Gas Dynamics: Theoretical and Computational Techniques (1989)
E.P. Muntz
University of Southern California
D.P. Weaver
U.S. Air Force Astronautics Laboratory (AFSC)
D.H. Campbell
University of Dayton Research Institute
ISBN 0-930403-55-X

119. Test and Evaluation of the Tactical Missile (1989)
Emil J. Eichblatt Jr.
Pacific Missile Test Center
ISBN 0-930403-56-8

120. Unsteady Transonic Aerodynamics (1989)
David Nixon
Nielsen Engineering & Research, Inc.
ISBN 0-930403-52-5

121. Orbital Debris from Upper-Stage Breakup (1989)
Joseph P. Loftus Jr.
NASA Johnson Space Center
ISBN 0-930403-58-4

122. Thermal-Hydraulics for Space Power, Propulsion and Thermal Management System Design (1989)
William J. Krotiuk
General Electric Co.
ISBN 0-930403-64-9

123. Viscous Drag Reduction in Boundary Layers (1990)
Dennis M. Bushnell
Jerry N. Hefner
NASA Langley Research Center
ISBN 0-930403-66-5

124. Tactical and Strategic Missile Guidance (1990)
Paul Zarchan
Charles Stark Draper Laboratory, Inc.
ISBN 0-930403-68-1

125. Applied Computational Aerodynamics (1990)
P.A. Henne
Douglas Aircraft Company
ISBN 0-930403-69-X

(Other Volumes are planned.)

Table of Contents for Colored Plates

Plate 1   Euler solution for McDonnell Douglas MD-11 transport.

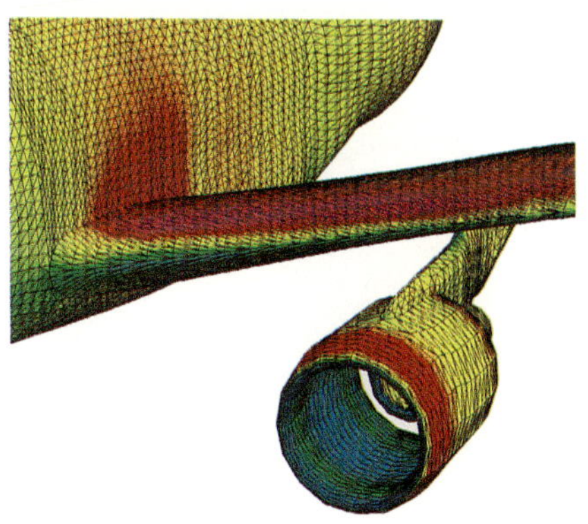

Plate 2   Detailed solution for McDonnell Douglas MD-11 transport fuselage/wing/pylon/nacelle.

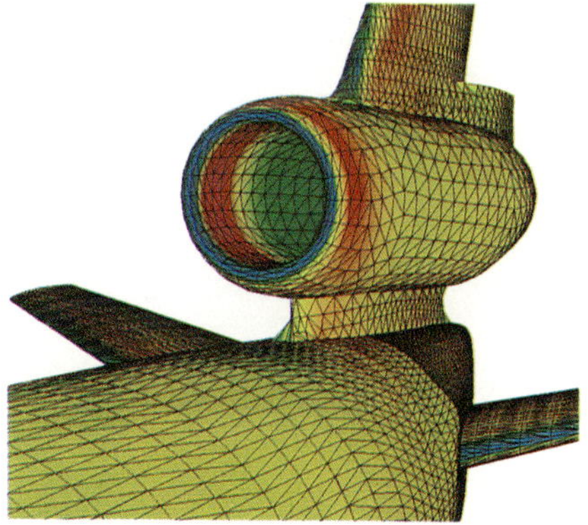

Plate 3   Detailed solution for McDonnell Douglas MD-11 transport empennage/aft nacelle.

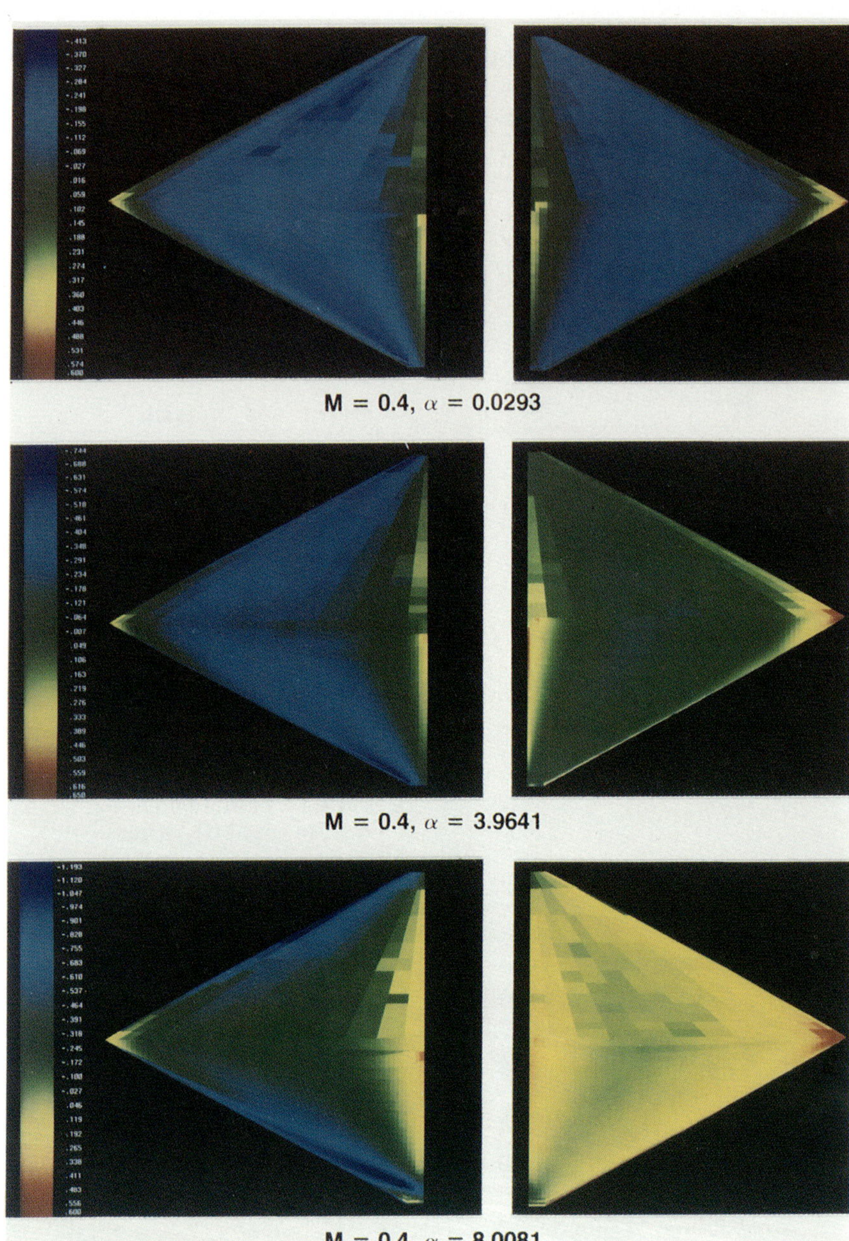

Plate 4 Surface pressure distributions for delta wing, $M = 0.40$. Upper half of each surface shows experimental pressures.

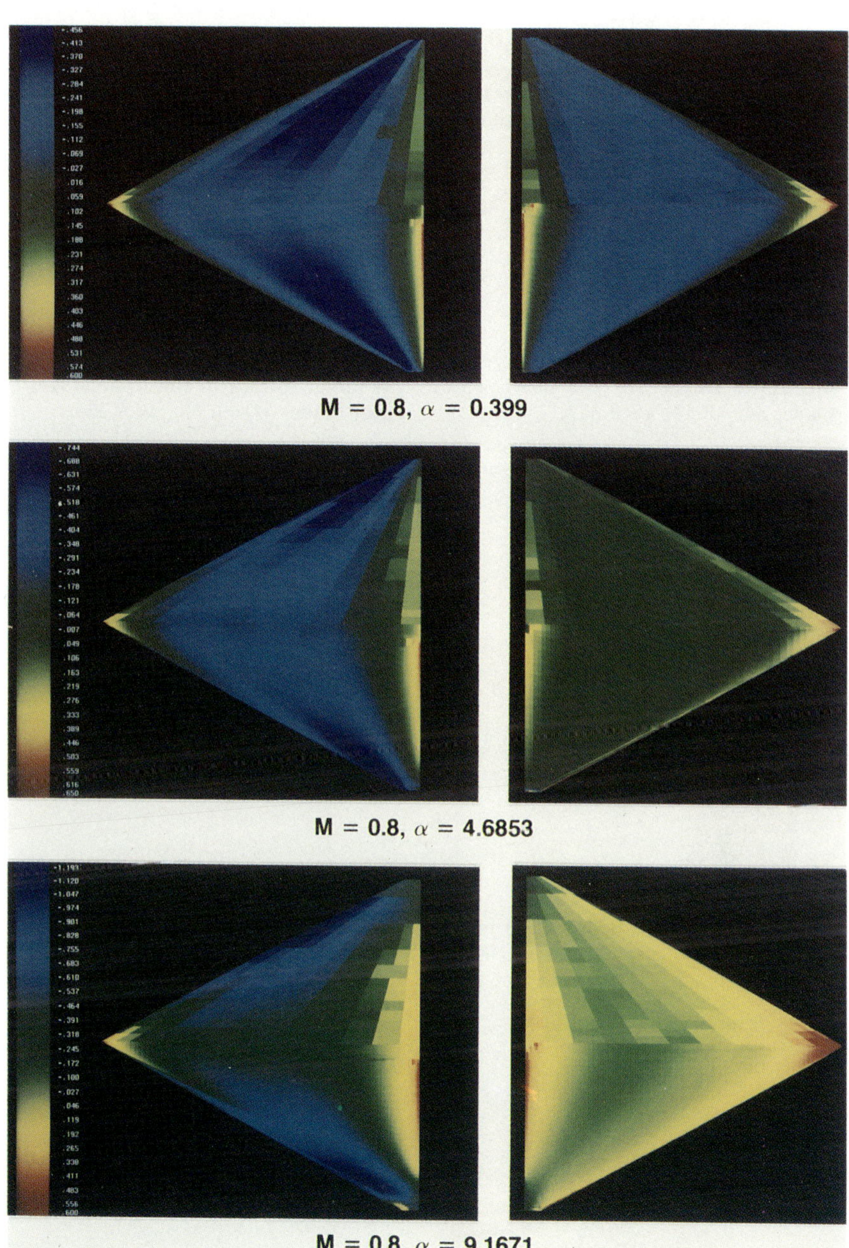

M = 0.8, α = 0.399

M = 0.8, α = 4.6853

M = 0.8, α = 9.1671

Plate 5   Surface pressure distributions for delta wing, $M = 0.80$. Upper half of each surface shows experimental pressures.

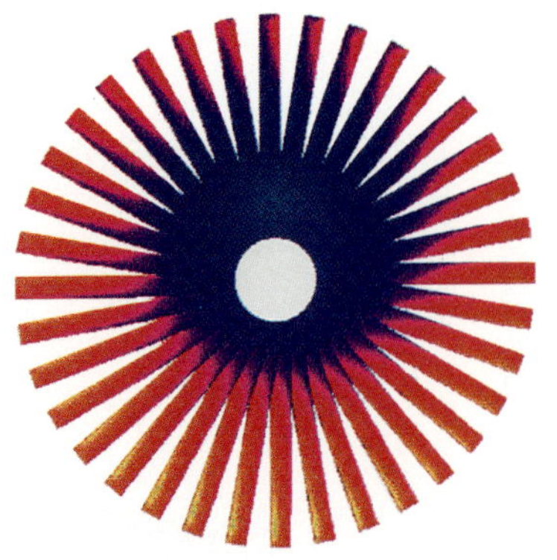

Plate 6   Mach number distribution on the upper surface of the 500-E rotor blade; $M_t = 0.58$, $\mu = 0.312$.

Plate 7   Mach number distribution on the upper surface of the Apache AH-64 rotor blade: $M_t = 0.65$, $\mu = 0.326$.

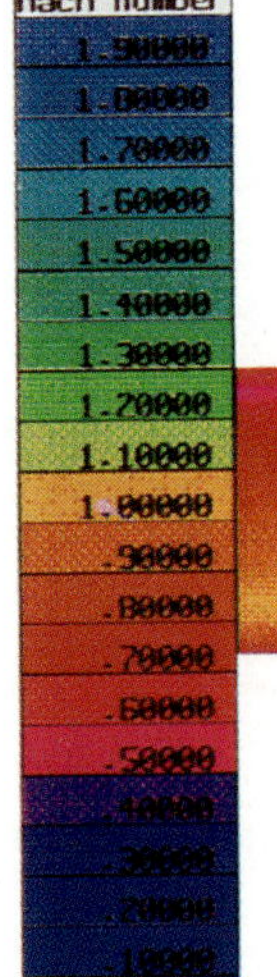

Plate 8   Mach number distribution on the upper surface of the Apache AH-64 blade at $\psi = 90$ deg, $M_t = 0.65$, $\mu = 0.326$.

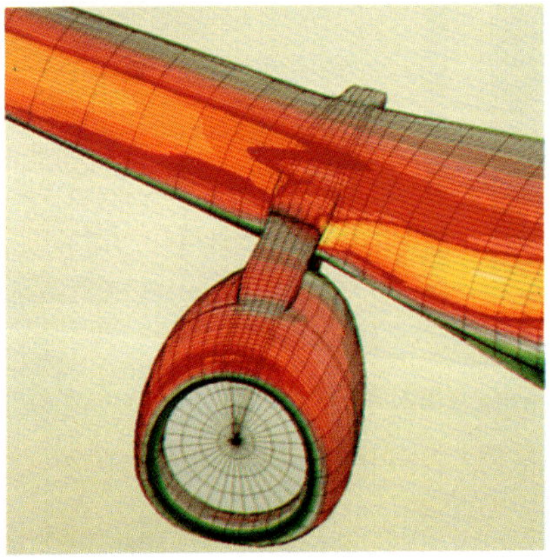

Plate 9   Wing pressure distribution adjacent to engine installation. 737-400—Tranair solution; $M = 0.74$.

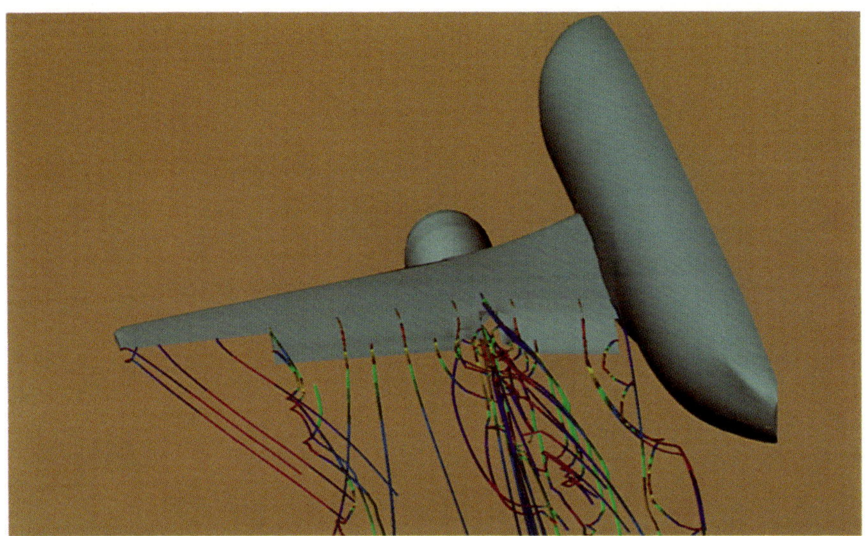

Plate 10 Computational analysis used to support 737-300 precertification flight test program.

Plate 11 Euler solution for advanced propfan transport.

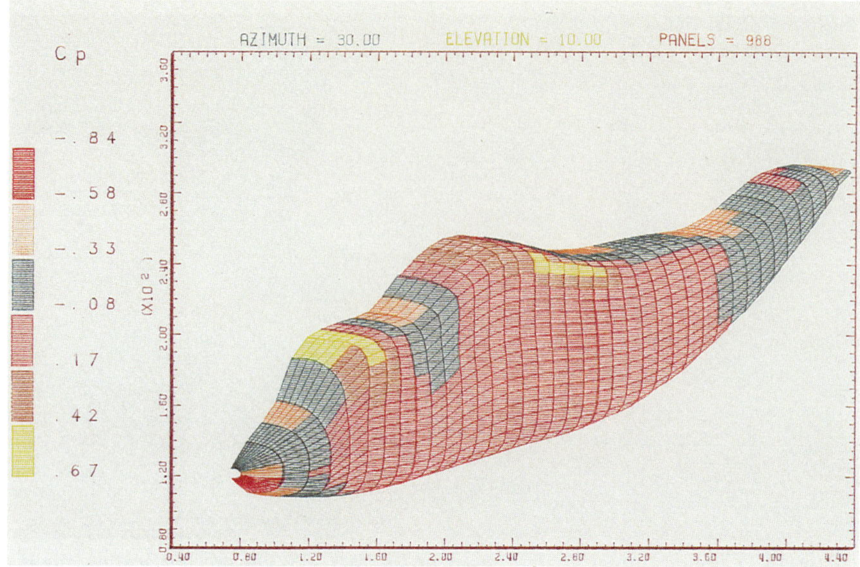

Plate 12   Panel model of the modified fairing produces a smooth pressure distribution.

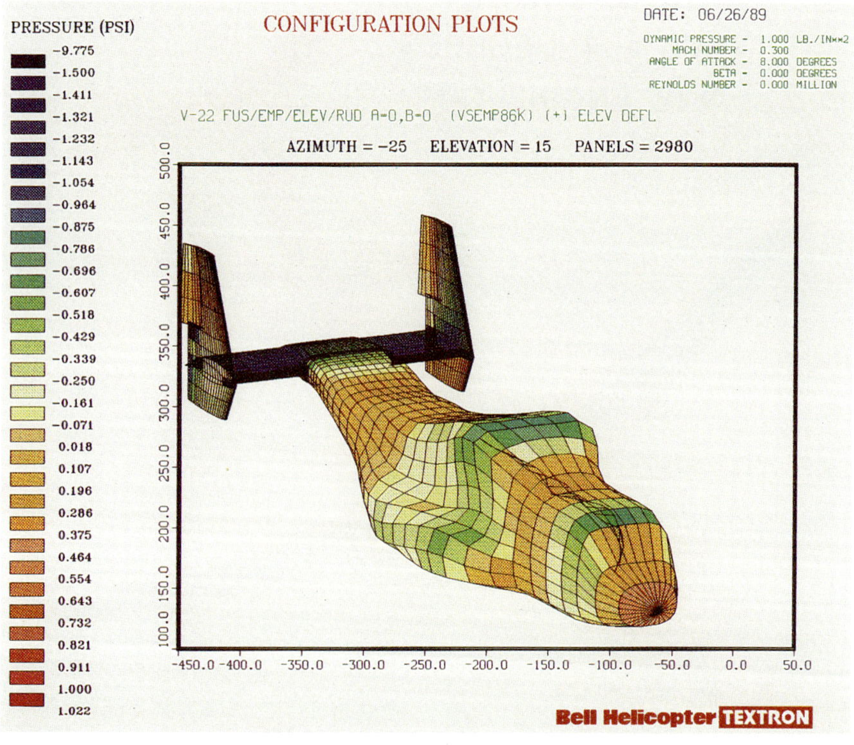

Plate 13   Special models were developed to generate airload distributions on V-22 components.

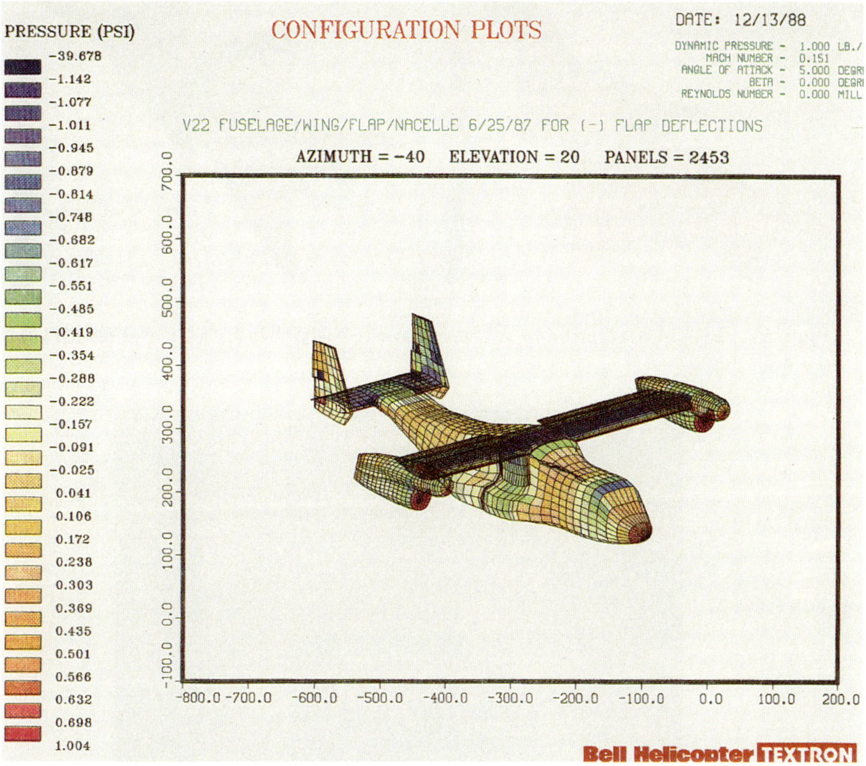

CONFIGURATION PLOTS

Plate 14 More complete panel method models of the V-22 generate less detailed information.

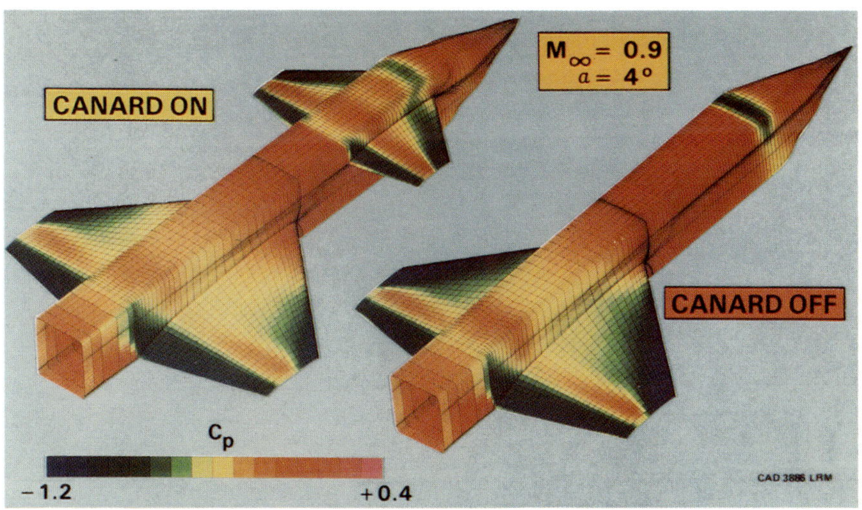

Plate 15 Computed surface pressure distributions for canard-on and canard-off configuration ($M = 1.2$, $\alpha = 4$ deg).

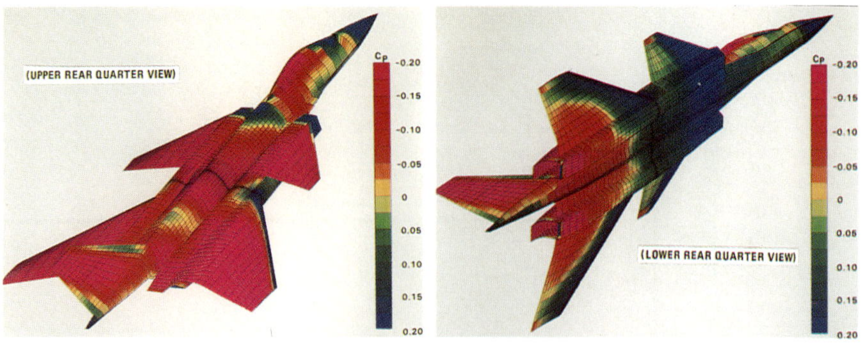

Plate 16   TEAM prediction of surface pressure distribution; ANC fighter ($M = 1.2$, $\alpha = 4.85$ deg).

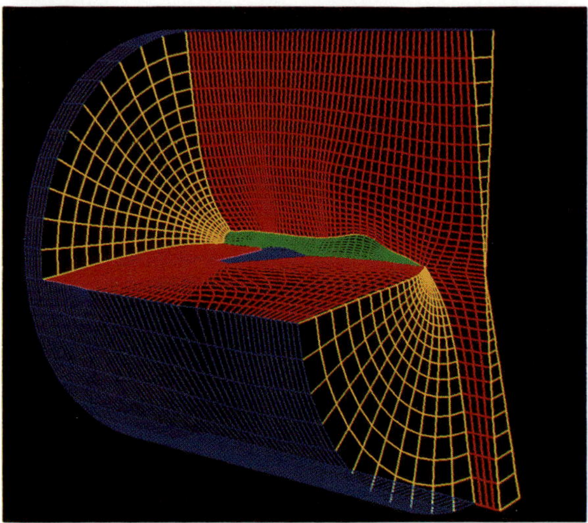

Plate 17   Coarse, base grid about the F-16A geometry showing the general grid topology (taken from Flores and Chaderjian[40]).

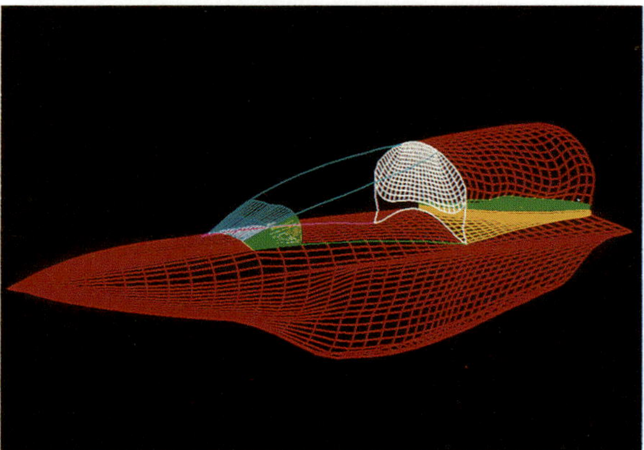

Plate 18   F-16A fuselage-inlet surface grid and inlet zonal topology (taken from Flores and Chaderjian[40]).

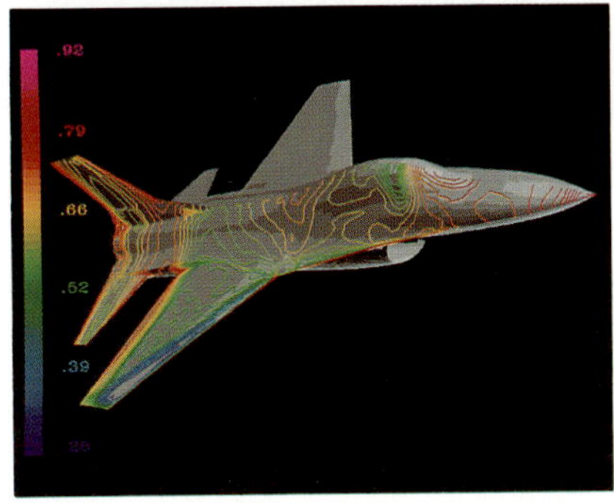

Plate 19   Computed pressure contours on the upper surface of the F-16A geometry, $M_\infty = 0.90$, $\alpha = 6.0$ deg, $Re_c = 4.5 \times 10^6$ (taken from Flores and Chaderjian[40]).

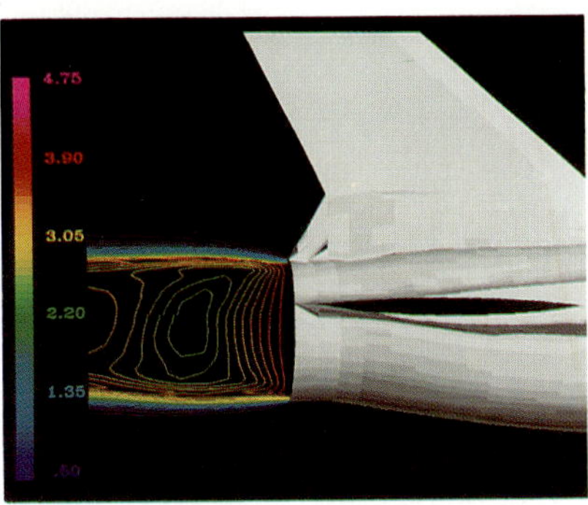

Plate 20   Computed temperature contours on the exhaust plume symmetry plane, F-16A geometry, $M_\infty = 0.90$, $\alpha = 6.0$ deg, $Re_c = 4.5 \times 10^6$ (taken from Flores and Chaderjian[40]).

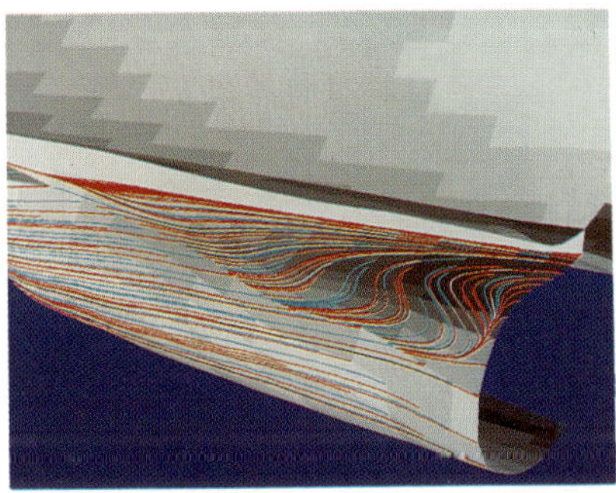

Plate 21   Simulated oil flow on the upper-inlet surface, F-16A geometry, $M_\infty = 0.90$, $\alpha = 6.0$ deg, $Re_c = 4.5 \times 10^6$ (taken from Flores and Chaderjian[40]).

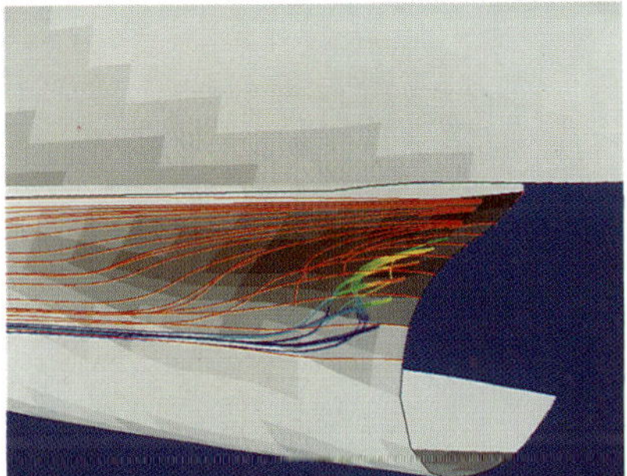

Plate 22   Three-dimensional particle traces around the upper-inlet region, F-16A geometry, $M_\infty = 0.90$, $\alpha = 6.0$ deg, $Re_c = 4.5 \times 10^6$ (taken from Flores and Chaderjian[40]).

Plate 23   Three-dimensional particle traces around the F-16A aircraft in sideslip, $\beta = 5.0$ deg, $M_\infty = 0.90$, $\alpha = 6.0$ deg, $Re_c = 4.5 \times 10^6$ (taken from Flores and Chaderjian[40]).

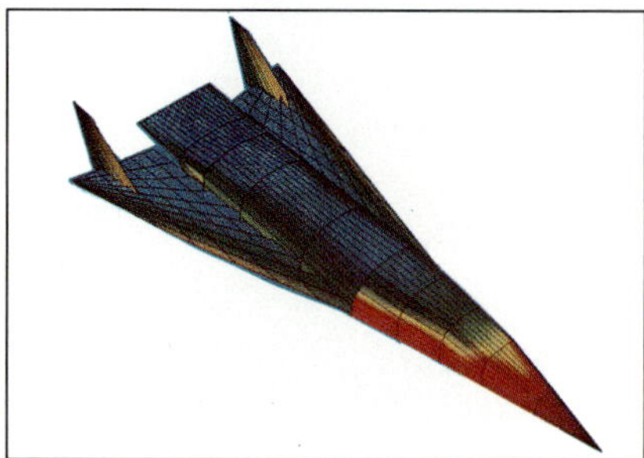

Plate 24 Surface pressure definition for baseline wing-body using HABP.

Plate 25 Axial force definition for baseline wing-body using HABP.

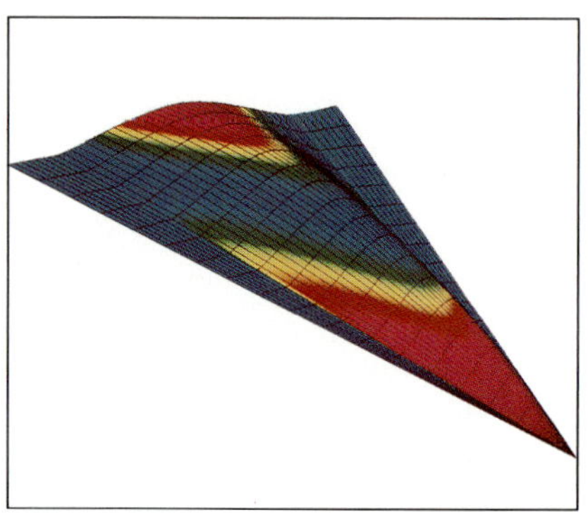

Plate 26 Axial force distribution for intermediate definition of blended body: HABP solution.

Plate 27 Axial force distribution for final blended body: HABP solution.

Plate 28 Surface pressures for final blended body: HABP solution.

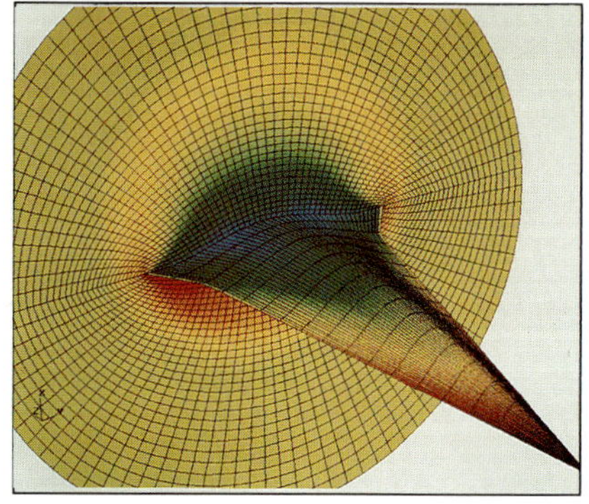

Plate 29 Pressure solution for final blended body using SCRAM: Euler solution.

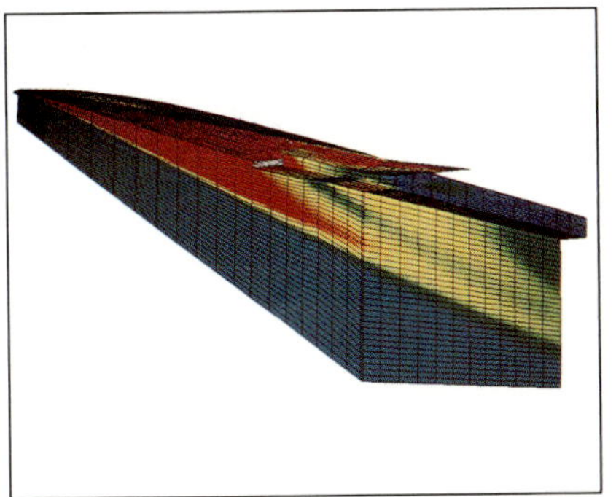

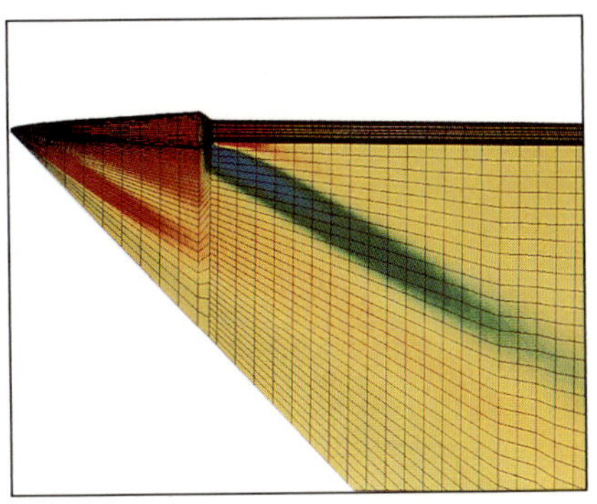

Plate 30   Undertrack flowfield for Mach 5 vehicle for sonic boom analysis.

Plate 31   Sonic boom methods verification flowfield for a cone test case.

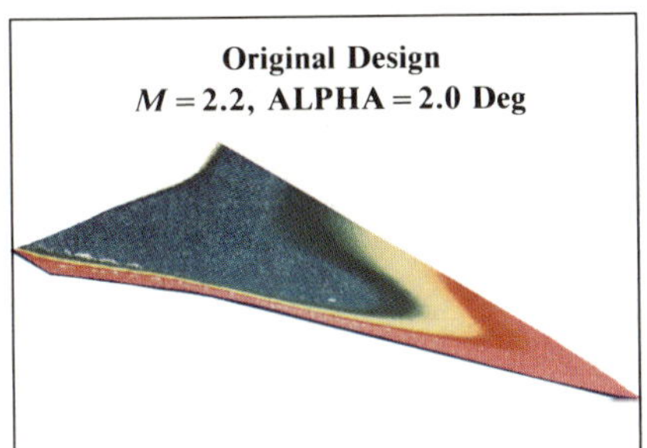

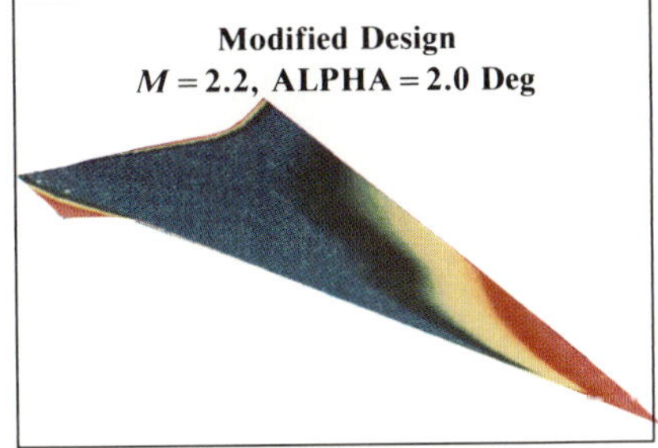

Plate 32   Euler surface pressures for upper-surface $M = 2.2$ AST.

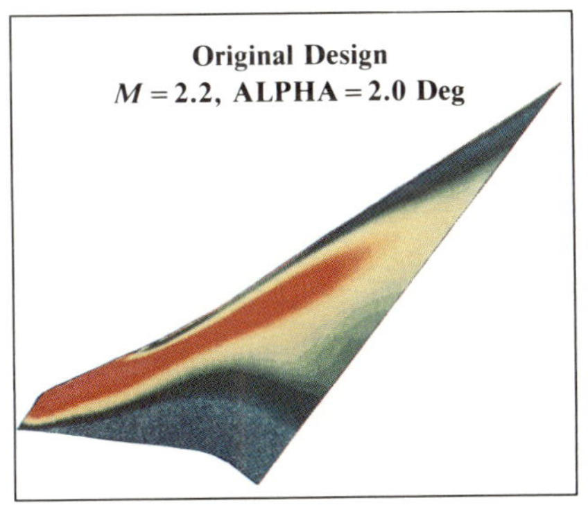

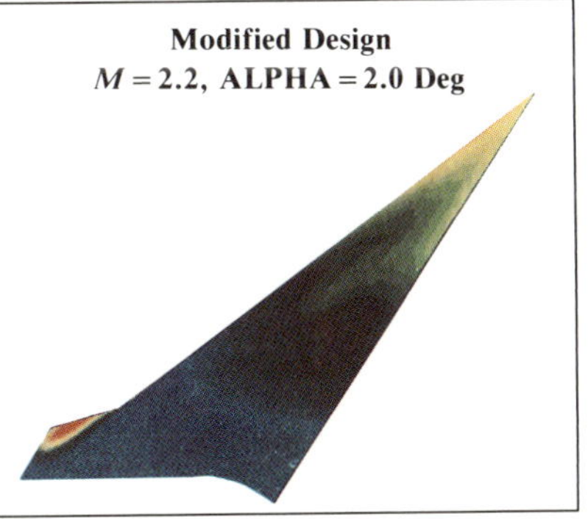

Plate 33   Euler surface pressures for lower-surface $M = 2.2$ AST.

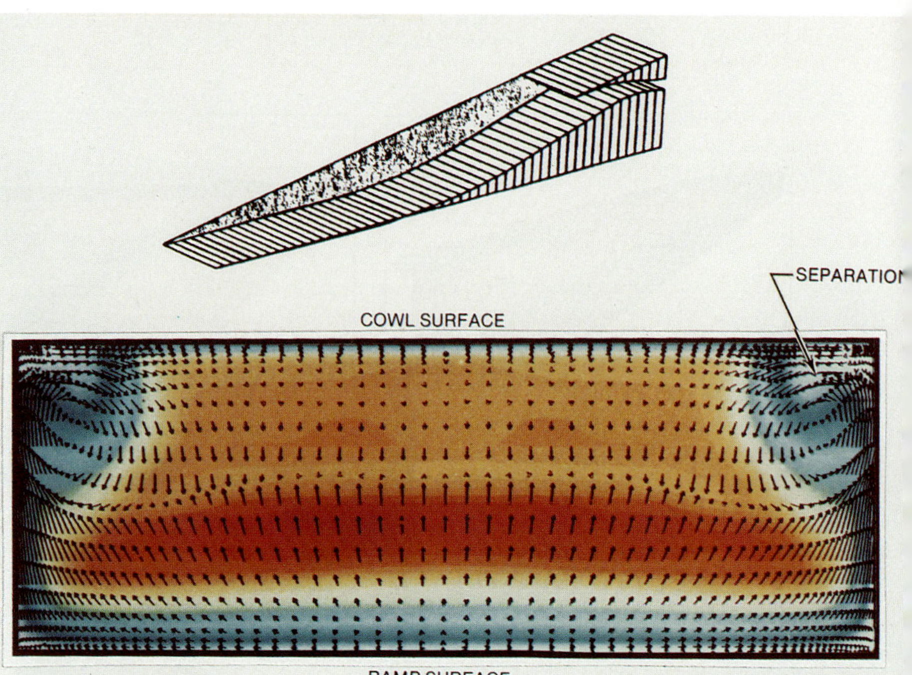

Plate 34   Pepsis total pressure and secondary velocity vectors for survey plane shown in Fig. 13 of Chapter 22 by Welge.

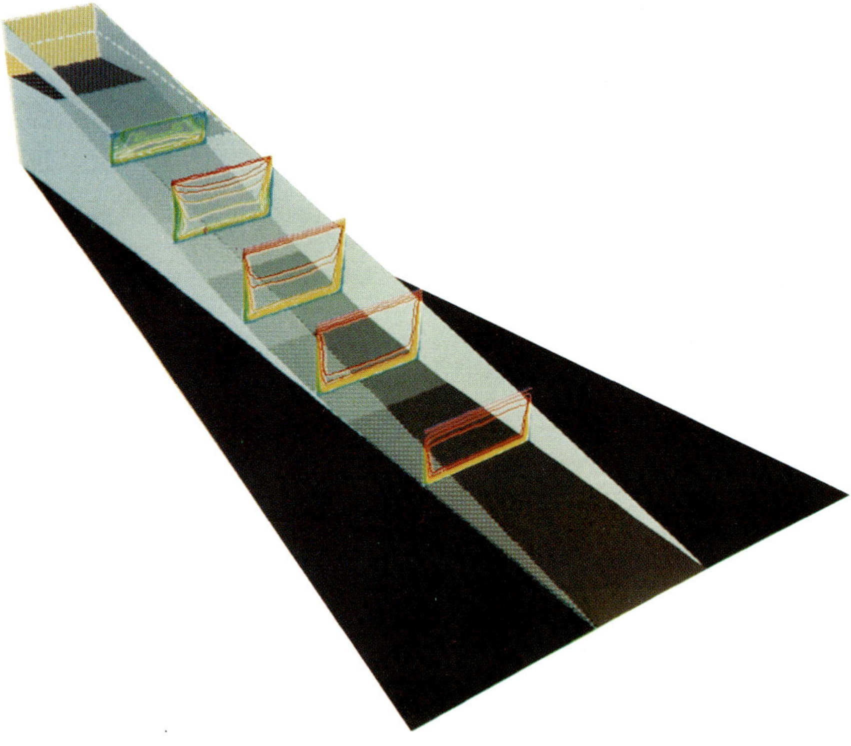

Plate 35   Mach contours at several locations within the inlet.

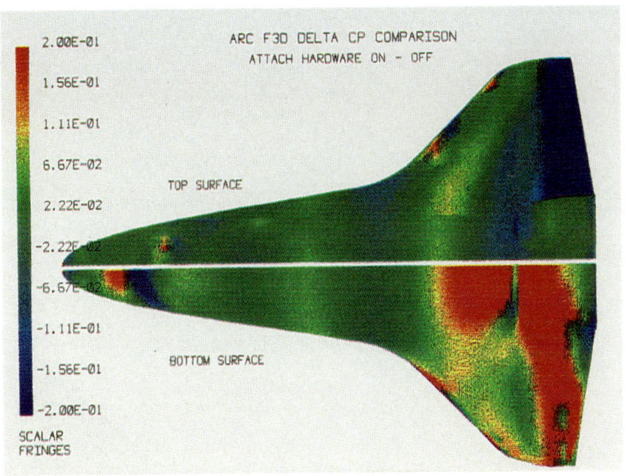

Plate 36   Differences in surface pressure coefficient between the solution obtained with the grid in Fig. 7a and that obtained with the grid in Fig. 7b.

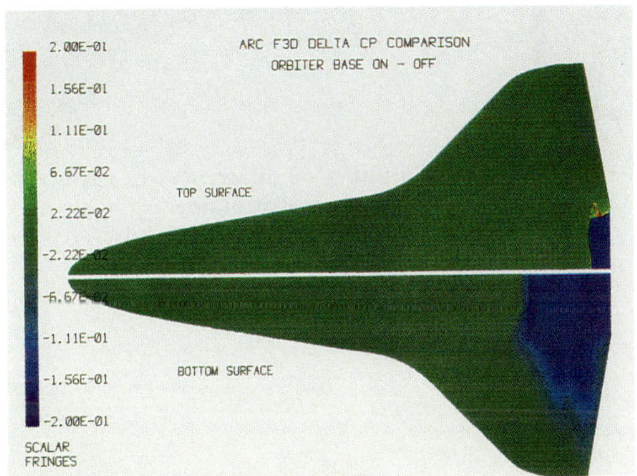

Plate 37   Contours of differences in surface pressure due to including Orbiter base.

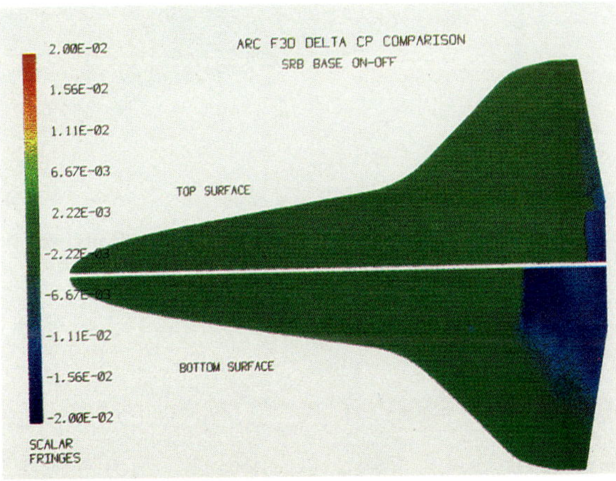

Plate 38   Contours of differences in surface pressure due to including the SRB nozzle/base grid.

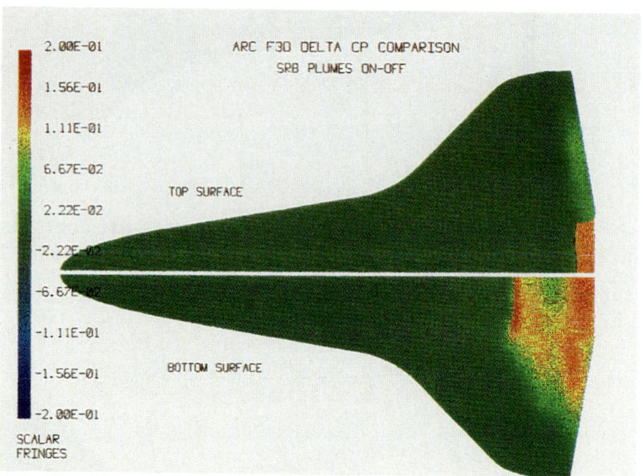

Plate 39   Contours of differences in surface pressure due to simulating the SRB plume.

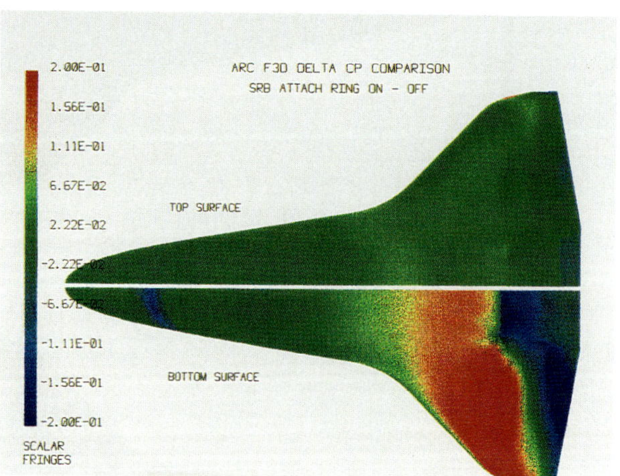

Plate 40   Contours of differences in surface pressure due to adding SRB/ET aft attach ring/IEA.

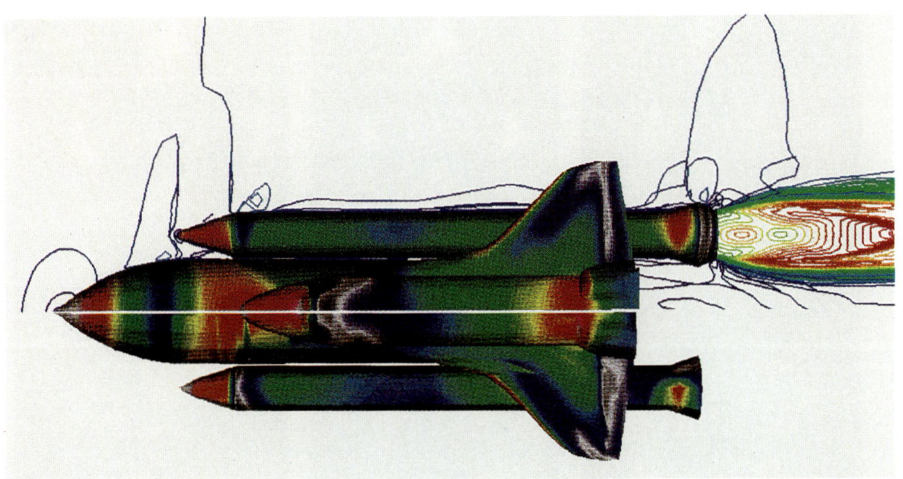

a) Top view

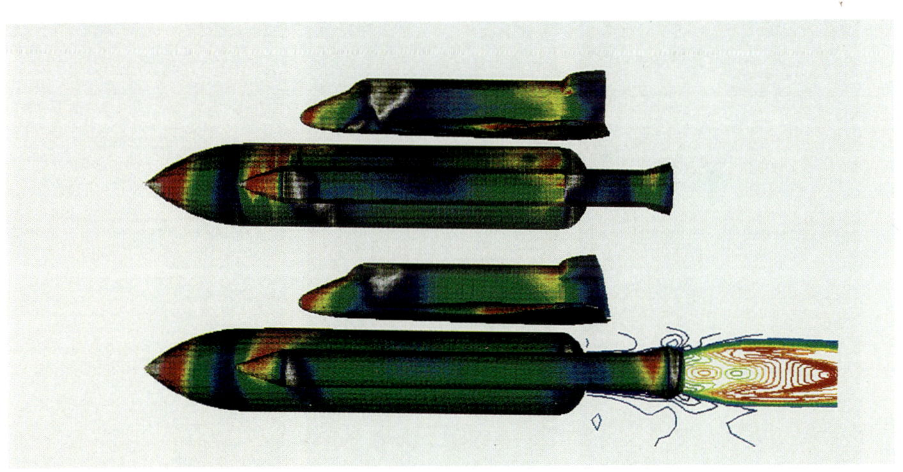

b) Side view

Plate 41   Comparison of pressure coefficient between computation and wind tunnel, 3% model ($M_\infty = 1.05$, $\alpha = 3$ deg, and $Re = 4.0 \times 10^6$/ft).

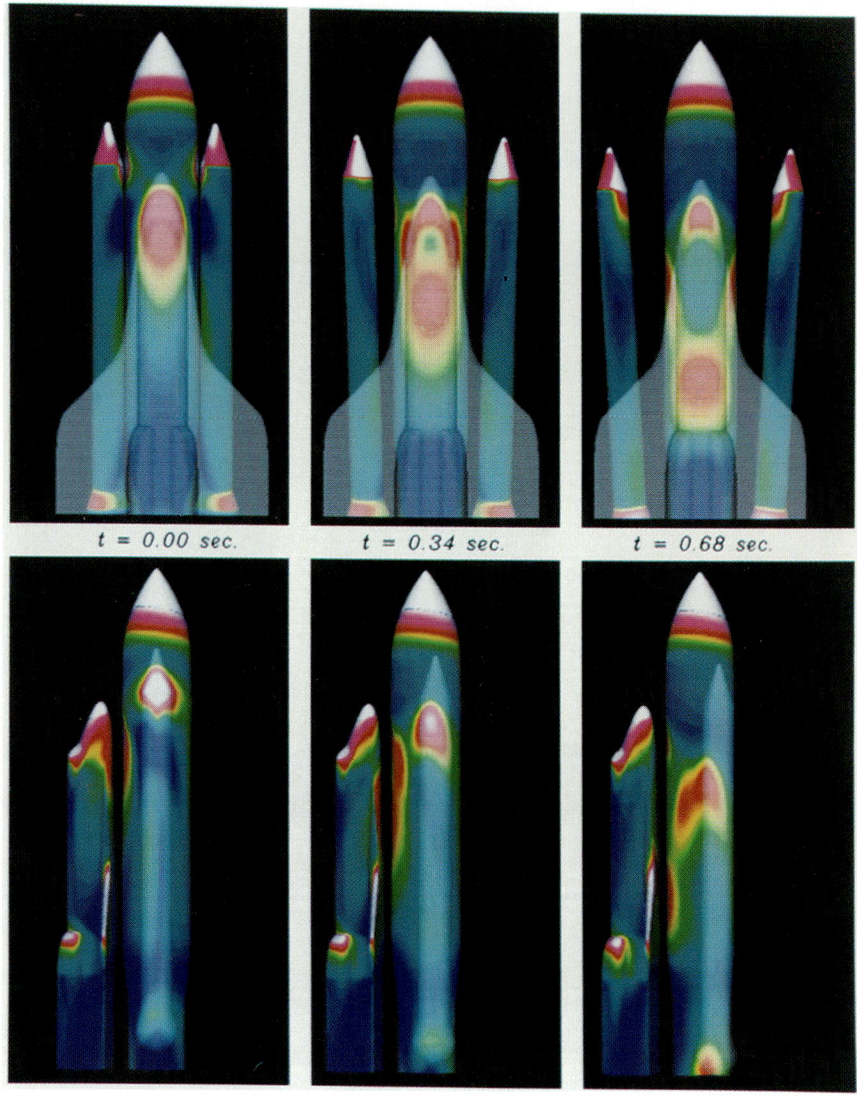

Plate 42 Surface pressure distributions computed at $t = 0$, 0.204, and 0.367-during initial SRB separation at $M_\infty = 4.5$, $\alpha = 3$ deg, using a prescribed flight path. In the top views, the Orbiter is transparent, whereas, in the side views, the SRB is transparent.